General and Inorganic CHEMISTRY

PART I

General and Inorganic CHEMISTRY

PART I

Third Edition

An Advanced Text for the Undergraduates across the globe with exposure to Postgraduate Level and for students of Materials Science and Environmental Studies

Ramaprasad Sarkar
Reader in Chemistry [Retd.], New Alipore College
University of Calcutta, India

New Central Book Agency (P) Ltd

Co-Published by

Bijay Publications

E-mail : bijaypublications@gmail.com

Exclusively Distributed by

Bijay Pustak Bhandar (P) Ltd

90/6A M. G. Road, Bhawani Dutta Lane, Kolkata-700007

✆ : (033) 4605-7856, 8777898576, 8274916104

E-mail : bijaypustakbhandar@gmail.com

NCBA

REGD OFFICE
8/1 Chintamoni Das Lane, Kolkata 700 009, India
email: ncbakolkata@gmail.com

OVERSEAS
NCBA (UK) Ltd, 149 Park Avenue North, Northampton, NN3 2HY, UK
email: ncbauk@yahoo.com

CORPORATE OFFICE
212 Shahpur Jat, New Delhi 110 049
email: ncbaexports@ncbapvtltd.com

BIJAY PUBLICATIONS
90/6A M. G. Road, Bhawani Dutta Lane, Kolkata-7
E-mail : bijaypublications@gmail.com

BIJAY PUSTAK BHANDAR (P) LTD
90/6A M. G. Road, Bhawani Dutta Lane, Kolkata-7
E-mail : bijaypustakbhandar@gmail.com

GENERAL AND INORGANIC CHEMISTRY [Part I] • Ramaprasad Sarkar

First Published: 1984
Second Edition: 1992
Reprinted: 1995, 1997, 1999, 2001
Thoroughly Revised and Enlarged Third Edition: July 2011
Reprinted: 2012, 2014, 2016, 2018, 2020, 2022, **2024**

PUBLISHER • TYPESETTER • PRINTER
New Central Book Agency (P) Ltd
8/1 Chintamoni Das Lane, Kolkata 700 009

CO-PUBLISHED BY
Bijay Publications
90/6A M G Road, Bhawani Dutta Lane, Kolkata 700007

PROJECT TEAM
Suman Dutta and Prabhat Jas

ISBN: 978 81 7381 680 2

Price ₹ 985.00

To the memory of
my parents and elder brother
and
to my teachers and students

CONTENTS

Preface to the third edition

The thoroughly revised and rewritten third edition of General and Inorganic Chemistry Part I, is a consequence of repeated demand from students and teachers at different levels. The book is supposed to be useful to students taking chemistry major or honours in the undergraduate level and beyond; students of other disciplines like biochemistry, agricultural chemistry, materials science and environmental science will also find the book helpful.

With minimum disturbance in the original format, the following major changes in this edition are noteworthy:

1. Portions now covered in the +2 level have been largely omitted except when recapitulation or reinforcement was felt absolutely necessary.
2. A brief history of the development of chemistry as a discipline of study has been included to generate interest in the subject.
3. Genesis of the elements and their distribution in nature have been included to provide an overview of the vast domain of inorganic chemistry.
4. IUPAC recommendations for nomenclature of inorganic chemistry, 2005, has been introduced in the beginning as there are certain notable changes from the previous nomenclature system. However, it may take time for the students and teachers for complete adaptation to the new system (for example, the use of hydron for proton). According, a dual mention will be found in many cases.
5. Elementary ideas on symmetry and point groups have been included to provide a better understanding of bonding. The treatment has been restricted to a level where beginners will get just an essence to the approach without bothering about details of group theory.
6. Chemical bonding and structure of crystalline solids have been given wider coverage than in the earlier editions.
7. The relevance of inorganic chemistry to our everyday life in the form of new materials developed has been emphasised; for example, LCD, semiconductors, superconductors, solid electrolytes, room temperature ionic liquids, and so on.
8. The problem of radiation hazard and potential risks of nuclear power plants have been emphasized specially in view of the recent nuclear disaster in Japan. An up-to-date nuclear power prospect of India has also been included.
9. In-chapter problems and end-chapter exercises have been thoroughly revised and extended to meet students' requirements. The hints provided with these problems (both in-chapter and end-chapter) will be illuminating.

While carrying out these changes, the author has always faced the dilemma of deciding about the extent of the coverage. There are a large number of excellent books with in-depth review of many aspects of each and every part of a theory or a problem. But the author has whole-heartedly resisted the temptation to drag the students at this stage deep into any particular topic. This is mainly in view of the fact that the main purpose of this book is to give a modest acquaintance with the major avenues through the vast domain of the subject—at least as large as a safari park if not too small as a zoo or too big as a jungle! The students have to study many other subjects concurrently and in the present education system, in most places, the intricacy and fineness of a treatment often fail to illuminate them; rather, the meaningless hugeness of the subject dismays them—as may be observed from their ultimate performance.

As before, I have borrowed ideas freely from different established authors to whom I am indebted to an endless extent. I convey my grateful acknowledgements and heartiest thanks to all of them. A list of the important books consulted appears at the end.

Most sincerely and gratefully I also acknowledge everyone—teachers, friends and students—for their inspiration, suggestions and criticisms at all stages. In particular, I wish to record my sincere gratitude to the following persons:

Prof. Rupendranath Bandyopadhyay who carefully edited the first few chapters and many helpful suggestions. Prof. Sanjib Bagchi and Dr. Arindam Rana also read a few chapters and suggested some changes.

Dr. Ashoke Bandyopadhyay, Dr. Arunabha Sen, Dr. Gautam Mukherjee, Dr. Subhendu Gupta, Dr. Rana Sen, and Dr. Soma Sen who inspired me all through my work.

Last but not the least, I express my heartfelt thanks to Mr Amitabha Sen of New Central Book Agency (P) Ltd whose endless patience and continuous urge have been a driving force for me in completing this work. I also pay my homage to the memory of Late Jogendranath Sen, the former managing director of this company, who was a source of constant inspiration to me.

I also extend my thanks to the project team lead by Mr Prabhat Jash and Suman Dutta who untiringly worked through the jungle of my manuscript to give the final shape—and that too with a smiling face!

Finally, I accept all responsibilities for errors and misconceptions of all sorts which are likely to have crept in this present edition. All constructive criticisms will be highly appreciated.

Ramaprasad Sarkar

June 12, 2011
Kolkata 700114

Extract from preface to the first edition

A jungle or a zoo

A jungle or a zoo, by whatever name it is called, chemistry has been expanding its horizon by leaps and bounds over the last few decades. A radical change has been marked in the very approach to the subject from a largely descriptive one to a probe essentially interpretative in nature. Excellent books have been produced by eminent chemists and famous writers to cope with this changed outlook. Unfortunately, however, the mode of teaching of this subject in our country has hardly been able to keep pace with this situation. The modern trend of the subject, the syllabii prescribed by the universities and the aptitudes of most of the students entering the general stream of education—all form nearly skew lines; it is very difficult to converge them into a single focal point. Consequently, chemistry remains to be a jungle of facts to the average learners, save when the magic wand of an efficient teacher does the miracle turning it into a zoo. A visit to this zoo becomes pleasant and meaningful only when a balanced textbook is available as a guide. The present book has been written with this intention.

RELATIVE ATOMIC MASSES (^{12}C = 12.000)

Element	Z		Element	Z		Element	Z	
Actinium, **Ac**	89	(227)	Hassium, **Hs**	108	(269)	Promethium, **Pm**	61	(145)
Aluminium, **Al**	13	26.98	Helium, **H**	2	4.00	Protactinium, **Pa**	91	231.04
Americium, **Am**	95	(243)	Holmium, **Ho**	67	164.93	Radium, **Ra**	88	(226)
Antimony, **Sb**	51	121.76	Hydrogen, **H**	1	1.0079	Radon, **Rn**	86	(222)
Argon, **Ar**	18	39.95	Indium, **In**	49	114.82	Rhenium, **Re**	75	186.21
Arsenic, **As**	33	74.92	Iodine, **I**	53	126.90	Rhodium, **Rh**	45	102.91
Astatine, **At**	85	(210)	Iridium, **Ir**	77	192.22	Rubidium, **Rb**	37	85.47
Barium, **Ba**	56	137.33	Iron, **Fe**	26	55.85	Ruthenium, **Ru**	44	101.07
Berkelium, **Bk**	97	(247)	Krypton, **Kr**	36	83.80	Rutherfordium, **Rf**	104	(261)
Beryllium, **Be**	4	9.01	Lanthanum, **La**	57	138.91	Samarium, **Sm**	62	150.4
Bismuth, **Bi**	83	208.98	Lawrencium, **Lr**	103	(262)	Scandium, **Sc**	21	44.96
Boron, **B**	5	10.81	Lead, **Pb**	82	207.19	Seaborgium, **Sg**	106	(263)
Bromine, **Br**	35	79.90	Lithium, **Li**	3	6.94	Selenium, **Se**	34	78.96
Cadmium, **Cd**	48	112.41	Lutetium, **Lu**	71	174.97	Silicon, **Sl**	14	28.09
Caesium, **Cs**	55	132.91	Magnesium, **Mg**	12	24.31	Silver, **Ag**	47	107.87
Calcium, **Ca**	20	40.08	Manganese, **Mn**	25	54.94	Sodium, **Na**	11	22.9898
Californium, **Cf**	98	(251)	Meitnerium, **Mt**	109	(268)	Strontium, **Sr**	38	87.62
Carbon, **C**	6	12.011	Mendelevium, **Md**	101	(258)	Sulfur, **S**	16	32.066
Cerium, **Ce**	58	140.12	Mercury, **Hg**	80	200.59	Tantalum, **Ta**	73	180.95
Chlorine, **Cl**	17	35.45	Molybdenum, **Mo**	42	95.94	Technetium, **Tc**	43	98.91
Chromium, **Cr**	44	51.996	Neodymium, **Nd**	60	144.24	Tellurium, **Te**	52	127.60
Cobalt, **Co**	27	58.93	Neon, **Ne**	10	20.18	Terbium, **Tb**	65	158.92
Copper, **Cu**	29	63.55	Neptunuim, **Np**	93	(237)	Thalium, **Tl**	81	204.38
Curium, **Cm**	96	(247)	Nickel, **Ni**	28	58.69	Thorium, **Th**	90	232.04
Dysprosium, **Dy**	66	162.50	Niobium, **Nb**	41	92.91	Thulium, **Tm**	69	168.93
Einsteinium, **Es**	99	(252)	Nitrogen, **N**	7	14.0067	Tin, **Sn**	50	118.71
Erbium, **Er**	68	167.26	Nobelium, **No**	102	(259)	Titanium, **Ti**	22	47.87
Europium, **Eu**	63	151.96	Osmium, **Os**	76	190.2	Tungsten, **W**	74	183.84
Fermium, **Fm**	100	(257)	Oxygen, **O**	8	15.9994	Uranium, **U**	92	238.03
Fluorine, **F**	9	18.998	Palladium, **Pd**	46	106.42	Vanadium, **V**	23	50.94
Francium, **Fr**	87	(223)	Phosphorus, **P**	15	30.9738	Xenon, **Xe**	54	131.29
Gadolinium, **Gd**	64	157.25	Platinum, **Pt**	78	195.07	Ytterbium, **Yb**	70	173.04
Gallium, **Ga**	31	69.72	Plutonium, **Pu**	94	(244)	Yttrium, **Y**	39	88.91
Germanium, **Ge**	32	72.61	Polonium, **Po**	84	(209)	Zinc, **Z**	36	65.39
Gold, **Au**	79	196.97	Potassium, **K**	19	39.098	Zirconium, **Zr**	40	91.22
Hafnium, **Hf**	72	178.49	Praseodymium, **Pr**	59	140.91			

Values in parentheses are mass numbers of most stable isotopes

Some Basic and Derived Units

Basic SI Units

Quantity	Unit
Length	meter, m
Mass	kilogram, Kg
Time	second, s
Electric current	ampere, A
Thermodynamic temperature	kelvin, K
Luminous intensity	candela, cd

Some Derived SI Units and their Conversion Factors

Quantity	Unit	Expression	Conversion
Force	newton, N	kg m s^{-2}	1 dyne = 10^{-5} N
Energy	joule, J	NM (= kg m^2s^{-2})	1 erg = 10^{-7} J
			1 cal = 4.184 J
			1 cm^{-1} (kayser, K) = 1.986 × 10^{-19} J $\Rightarrow$ 11.96 J mol^{-1}
kk			1 kK = 11.96 kJ mol^{-1}
			1 eV = 1.602 × 10^{-19} J $\Rightarrow$ 96.485 kJ mol^{-1}
Pressure	pascal, Pa	N m^{-2}	Hartree energy (E_h) = 4.3597 × 10^{-18} J
			1 atm = 101.325 kPa = 760 Torr
			1 bar = 10^5 Pa
Dipole moment	coulomb meter	C m	1 Debye (D) (= 10^{-18} esu cm) = 3.3356 × 10^{-30} Cm
Magnetic flux	weber, Wb	V s	
Magnetic flux density	tesla, T	Wb m^{-2}	1 gauss = 10^{-4} T
Volume	meger³/ m³		1 L = 1 dm³
Electric capacitance	farad, F	CV^{-1}	
Electrical conductance	siemens, S	ohm^{-1}	
Frequency	hertz, Hz	s^{-1}	
Radioactivity	becquerel, Bq	s^{-1}	
Absorbed does of radiation	gray, Gy	J kg^{-1}	
Power	watt, W	J s^{-1}	

SOME FUNDAMENTAL PHYSICAL CONSTANTS

Constant	Symbol	Value
Avogadro's constant	N_A	$6.022\ 14 \times 10^{23}$ mol^{-1}
Bohr magneton	$\mu_B = e/2m_e$	9.274×10^{-24} J T^{-1}
Bohr radius	a_0	5.292×10^{-11} m
Boltzmann's constant	k	1.381×10^{-23} J K^{-1}
		8.617×10^{-5} eV K^{-1}
Compton wavelength of electron	$\lambda_C = h/m_e c$	2.426×10^{-12} m
Electron charge	e	1.602×10^{-19} C
Electron rest mass	m_e	9.109×10^{-31} kg
		0.0005486 u
		0.5110 MeV/c^2
Electronvolt	eV	1.602×10^{-19} J
Hydrogen atom rest mass	m_H	1.673×10^{-27} kg
		1.007825 u
Mass unit	u	1.6606×10^{-27} kg
		931.5 MeV/c^2
Nuclear magneton	μ_N	5.04×10^{-27} J T^{-1}
Neutron rest mass	m_n	1.675×10^{-27} kg
		1.0086654 u
		939.6 MeV/c^2
Permeability of free space	μ_0	1.257×10^{-6} T.m/A
	$\mu_0/4\pi$	10^{-7} T.m/A
Permittivity of free space	ε_0	8.854×10^{-12} C^2 $N^{-1}.m^{-2}$
	$1/4\pi\varepsilon_0$	8.988×10^{9} $N.m^2C^{-2}$
Planck's constant	h	6.626×10^{-34} J.s
		4.136×10^{-15} eV.s
	$= h/2\pi$	1.055×10^{-34} J.s
		6.582×10^{-16} eV.s
Proton rest mass	m_p	1.672×10^{-27} kg
		1.0072766 u
		938.3 MeV/c^2
Rydberg constant	R	1.097×10^{7} m^{-1}
Speed of light in free space	c	2.998×10^{8} $m\,s^{-1}$

Extended Form of the Periodic Table

s-block | *d*-block | *p*-block

Groups[1]	IA	IIA	IIIB / IIIA	IVB / IVA	VB / VA	VIB / VIA	VIIB / VIIA	VIIIB / VIII			IB	IIB	IIIA / IIIB	IVA / IVB	VA / VB	VIA / VIB	VIIA / VIIB	VIIIA / 0
IUPAC	1	2	3	4	5	6	7	8	9	10	11	12	13	14	15	16	17	18
Periods 1	^{1}H 1.008																	^{2}He 4.00
2	^{3}Li 6.94	^{4}Be 9.01											^{5}B 10.81	^{6}C 12.01	^{7}N 14.01	^{8}O 16.00	^{9}F 19.00	^{10}Ne 20.18
3	^{11}Na 22.99	^{12}Mg 24.31											^{13}Al 26.98	^{14}Si 28.09	^{15}P 30.97	^{16}S 32.06	^{17}Cl 35.45	^{18}Ar 39.95
4	^{19}K 39.10	^{20}Ca 40.08	^{21}Sc 44.96	^{22}Ti 47.87	^{23}V 50.94	^{24}Cr 52.00	^{25}Mn 54.94	^{26}Fe 55.84	^{27}Co 58.93	^{28}Ni 58.69	^{29}Cu 63.55	^{30}Zn 65.41	^{31}Ga 69.72	^{32}Ge 72.64	^{33}As 74.92	^{34}Se 78.96	^{35}Br 79.90	^{36}Kr 83.80
5	^{37}Rb 85.47	^{38}Sr 87.62	^{39}Y 88.91	^{40}Zr 91.22	^{41}Nb 92.91	^{42}Mo 95.94	^{43}Tc 98.00	^{44}Ru 101.07	^{45}Rh 102.90	^{46}Pd 106.42	^{47}Ag 107.87	^{48}Cd 112.41	^{49}In 114.82	^{50}Sn 118.71	^{51}Sb 121.76	^{52}Te 127.60	^{53}I 126.90	^{54}Xe 131.29
6	^{55}Cs 132.91	^{56}Ba 137.33	^{57}La* 138.91	^{72}Hf 178.49	^{73}Ta 180.95	^{74}W 183.84	^{75}Re 186.21	^{76}Os 190.23	^{77}Ir 192.22	^{78}Pt 195.08	^{79}Au 196.97	^{80}Hg 200.59	^{81}Tl 204.38	^{82}Pb 207.2	^{83}Bi 208.98	^{84}Po 209	^{85}At 210	^{86}Rn 222
7	^{87}Fr (223)	^{88}Ra (226)	^{89}Ac† (227)	^{104}Rf (261)	^{105}Db (262)	^{106}Sg (263)	^{107}Bh (262)	^{108}Hs (265)	^{109}Mt (266)	^{110}Ds (271)	^{111}Rg (272)	^{112}Uub	^{113}Uut	^{114}Uuq	^{115}Uup	^{116}Uuh	^{117}Uus	^{118}Uuo

Atomic number → 1
Symbol → H
Atomic weight → 1·008
(^{12}C = 12·000)

f-block

Lanthanoids* (Lanthanides)	^{58}Ce 140.12	^{59}Pr 140.91	^{60}Nd 144.24	^{61}Pm 145	^{62}Sm 150.36	^{63}Eu 151.96	^{64}Gd 157.25	^{65}Tb 158.93	^{66}Dy 162.50	^{67}Ho 164.93	^{68}Er 167.26	^{69}Tm 168.93	^{70}Yb 173.04	^{71}Lu 174.97
Actinoids† (Actinides)	^{90}Th 232.04	^{91}Pa 231.04	^{92}U 238.03	^{93}Np (237)	^{94}Pu (244)	^{95}Am (243)	^{96}Cm (247)	^{97}Bk (247)	^{98}Cf (251)	^{99}Es (252)	^{100}Fm (257)	^{101}Md (258)	^{102}No (259)	^{103}Lr (262)

1. The IUPAC recommendation abolished A and B subgroups and numbered the groups from 1–18

1

INTRODUCTION

1.1 GENESIS OF CHEMISTRY

The early days

Chemistry gained its own position in human civilization much before it became recognized as an area of systematic scientific knowledge. Early man became acquainted with fire, air, water and earth. Metals like copper and gold were used as early as 4500-4000 B.C. Subsequently many chemical changes were carried out to obtain various useful materials, but mostly without any precise knowledge of the reactions involved. Glass, including coloured glass, ceramics, pigments and dyes were developed; distillation and other techniques were used to prepare wines, herbal extracts and a variety of "healing substances" or medicines. Ancient civilizations of Babylon, Egypt, India, China and Greece bear ample evidences to this effect. The practice of such chemical processes was often associated with some religious rites and beliefs. Attempts to understand the physical background of the universe and its constituents were also made, notably among the ancient Hindu, Jain, Greek and Chinese philosophers.

The major developments of chemistry seem to have emerged from its association with "good health" of people in general. The very term "*Rasayana*" probably evolved from the word "*Ayashyani*" (meaning long life and good health), a word found in the Atharvaveda (~2000 B.C.). In ancient India, remarkable progress was achieved in the field of medicine and surgery from 6th century B.C. onward by Susruta, Jeevaka, Charaka and many others. The methods of preparation of medicines developed during this time gave a collateral development of chemistry. The word chemistry is said to have had its origin from the word "*chemeia*"—probably the Greek transliteration of an Egyptian word; a similar word in ancient hieroglyphic means "*Egyptian knowledge*"—knowledge of the techniques acquired by the Egyptians (~100 A.D. in Alexandria) regarding metals, minerals and other compounds known at the time. Such techniques were often mystified by vague attempts and claims to transform inferior metals to gold and perform other magics. Such practices were also known in China around 300-400 B.C. The Arabs called these techniques "The Chemie" or '*Al-kimiya*' (*Al* ≡ the; *Kimiya* ≡ alchemy). The term alchemy now refers to early chemistry in general.

Among other things, the principal aim of early alchemy, be it in India, China or Alexandria, appears to have been the preparation of a substance that will impart immortality. Mercury, the "living metal", gold, cinnabar (HgS) and a number of chemicals were familiar in those days. Side by side, the alchemists believed in transmutation of elements, particularly in the conversion of inferior metals to gold.

The earliest types of reactions studied by the alchemists were those of metals embedded in powdered chemicals, for example, the formation of brass from copper heated in powdered zinc ore; or the carbonization of wrought iron by charcoal. Mention is also found of an "Egyptian method" of exposing a metal or compound to the vapours of certain substances. A treatise by Bolos, a Graeco-Egyptian, describes the 'making' of gold, silver, gems and purple with data compiled from Egyptian, Jewish, Babylonian and Persian sources.

Theoretical ideas, too, were forwarded for the various chemical processes. The four-elements theory of the Greeks (fire, air, water and earth) stressed on the "qualities" of these elements rather than the elements themselves. Plato and other philosophers believed in a neutral primary matter which turns into a specific element by absorbing the "quality" of the latter. In the hands of the practical alchemists, these 'qualities' became more and more materialistic characteristics which may be added to or subtracted from the neutral sub-stratum. By chemical reactions they attempted to separate the 'essence' or 'spiritus' from an element or substance and transfer it to another, with an ultimate goal for 'transmutation' of such substances to gems or metals. Their futile attempts, however, led to the development of many vessels and apparatus for various chemical processes like distillation, boiling, digestion, etc. which were proved to be very useful to later chemists.

As the practice of alchemy slowly percolated into the society, there appeared a group of artisans or practical alchemists and also several recipe books for useful chemical products. Around 300 A.D., Zosimos compiled the whole of alchemical doctrines and literature in 28 volumes. Over hundreds of years, some alchemists even tried to associate religion with alchemy. For example, Steppanos and his school (8th century A.D.) believed that the transmutation of metals stood for the regenerating force of religion in transforming the human soul. Similar involvement of chemistry in religion was also observed around the same time among the followers of Tantric school in India.

In the meantime the study of alchemy had been taken up by the Arabs (~8th century A.D.) who made significant developments in the subject. Jabir ibn Hayyān (~720–813 A.D.) was a famous Arab alchemist. Many elements and their compounds were prepared and studied, though the central focus still remained on the "touch stone" and an "elixir of life". The state of affairs continued even when the study of alchemy re-entered Europe in the twelfth century. Even Roger Bacon (1214–1292), recognized as one of the first advocates of the importance of experiment in science, wrote extensively about the magical powers of alchemy. Even kings patronized the practice of alchemy in their craze for gold and perpetual youth. The fraudulent practice of alchemy reached such a state that Pope John XII prohibited such practice in 1317. Secret practice, however, continued up to the sixteenth century.

In the fifteenth century, a new school of chemistry developed in Europe with the aim of healing and was called *Iatrochemistry* (iatro ≡ healing) or medicinal chemistry. Paracelsus (1493–1541) was a pioneer in this field. The iatrochemists developed several medicinal formulations using mainly mercury, sulfur and various salts but they were also somewhat motivated to find an elixir of life.

The door to the modern systematic chemistry was opened only in the post-renaissance period, sixteenth century onward. Francis Bacon (1561–1625) laid stress on the inductive method of studying science. A major breakthrough was made by Robert Boyle (1627–91) who overthrew the idea of four elements of Aristotle which had dominated for the past two thousand years. Boyle described elements "as things which cannot be decomposed further". However, the real momentum towards modern chemistry was gained only in the late eighteenth century—nearly 100 years later. Scheele and Priestley independently prepared oxygen (1774); Cavendish showed that water is a compound of hydrogen and oxygen (1781) and not an element as thought earlier. Cavendish also determined the composition of water though its complete interpretation was due to Lavoisier (1785). In fact Lavoisier (1743–1794) is called the founder of modern chemistry. He introduced the use of balance in chemical studies and established the law of indestructibility of matter. The role of oxygen in combustion was first recognized by him and proved beyond doubt from several experimental results. This came as a blow on the widely believed theory of a hypothetical substance called *phlogiston* (*phlogistos* = inflammable). According to this phlogiston theory (Bacher and Stahl), phlogiston is released when a substance, for example a metal, is heated. The residue obtained (*Calx*) may be reconverted to the metal by heating it with carbon, a substance which is rich in phlogiston. Lavoisier showed that a metal actually gains weight on heating in air rather than losing the weight of phlogiston. Oxygen was actually involved in this chemical change—he concluded. It is a pity that Lavoisier met a tragic end to his life at a premature age of 51 under the guillotine (1794). It took a long time to completely eradicate the phlogiston theory from the mind of contemporary scientists.

A rapid and widespread development of chemistry marks the nineteenth century—both in theory and practice. The century began with the landmark theory of John Dalton (1803) on atoms. There were enough data to frame the laws of chemical combination: the law of definite proportion was framed in 1799 (Proust); Dalton himself framed the law of multiple proportion (1802). Dalton's logical thought to explain these laws led to the atomic theory of matter—a practical theory that puts atoms to direct experiment, and does not just leave them in the misty concepts of early philosophers. The rest of the story is well-known and we save space by mentioning only a few landmark contribututions that followed the atomic theory.

1811	Amedeo Avogadro put the idea of molecules. The idea, however, remained unappreciated till 1856 when Cannizzaro successfully established its importance.
1860-65	Stas determined the atomic weights of several elements with a high degree of accuracy. These were refined by Richards in 1904.
1869	Mendeleef and Lothar Meyer independently arranged the elements in a periodic arrangement.
1898	Madame Curie made extensive studies on radioactivity—a phenomenon observed by Becquerel in 1895.

Besides these landmark developments which guided the course of chemistry, we are also overwhelmed by the widespread development in both practical and theoretical fields which occurred throughout the nineteenth century. Only a few examplary mentions will be made:

1828	Wohler prepared urea from ammonium cyanate and disproved the 'theory of vital forces'.
1832	Faraday published the laws of electrolysis.
1865	Kekule proposed the ring structure of benzene.
1886	van't Hoff studied osmotic pressure of solutions.
1887	Raoult studied other colligative properties.
1893	Werner forwarded his theory of complex compounds.

A tragic end

Like any other process of development, the progress of science was not smooth along the entire nineteenth century. The atomic theory was well accepted by the chemists, specially after Cannizzaro reaffirmed Avagadro's idea of molecules and thus cleared some of the inconsistencies in relative atomic masses obtained by various methods. Determination of the value of Avogadro number (Loschmidt, 1865) enabled one to estimate the mass of a single atom from bulk measurement.

But still a group of physicists were not convinced of the atomic theory in absence of any direct proof. The kinetic theory of gases, developed between 1848 to 1898 by Maxwell, Boltzmann and others, presupposed the existence of atoms and molecules: Ludwig Boltzmann was the leader of these "atomists". On the other hand, there were Mach, Ostwald and others who did not believe in the reality of atoms and molecules and considered them simply as useful models for mathematical application in a world based entirely on transformations of energy. This latter group of "energetists" practically cornered Boltzmann as evident from his writing in 1898: "*I am conscious of being an individual struggling weakly against the stream of time*" (Lectures on Gas Theory). Even in 1904, nearly hundred years after Dalton framed his atomic theory, a group of physicists rejected the idea of atoms at a conference in St. Louis. Boltzmann was not even invited to it. Some people think that the resulting depression might be the chief reason for his suicide (September 5, 1906). However, Boltzmann attained immortality through his pioneering works in statistical thermodynamics.

The final establishment of the atomic theory, however, followed soon. Einstein applied (1905) the kinetic theory to the random motion of small particles suspended in a stationary fluid (Brownian motion). His equations were supported by the experimental results of Perrin in 1908. This confirmed the reality of atoms and molecules.

The twentieth century

The dawn of the twentieth century was marked by a large number of developments in quick succession. The structure of the atom was explored, dual nature of radiation and matter was established, quantum mechanics was developed. Widespread application of quantum mechanics to atomic structure and bonding theories provided excellent interpretations of many chemical facts. Side by side, advancement in instrumentation and other technologies enabled chemists to prepare and study a wide variety of compounds with novel structural features.

In the later half of the twentieth century, the activity of chemists intensified to develop new drugs, new materials and new catalysts. Investigations on the chemistry of living systems opened a new horizon of biochemistry. The role of metals and other elements in biological systems opened a further new area of study: bioinorganic chemistry. The last lap of the past century and the current first decade of twentifirst century also witnessed some emerging fields. Growing concern about the environment and its protection cumulated to environmental chemistry; closely related to it, our realization of the harmful effects of several age-old chemical reactions in both laboratory and industry is seeking safe alternatives in green chemistry. The same period is also marked by very rapid growth of nanotechnology and its widespread applications in chemistry. We eagerly wait to see the new horizons unveiled by the application of nanotechnology to all fields of chemistry.

Nearly one hundred and eighty years ago (1828), Wöhler had shown that urea, a so-called "organic" compound, can be prepared from an "inorganic" compound. Subsequent developments have definitely established the interplay of same fundamental principles of bonding in all chemical compounds. Nevertheless, academicians found it convenient to compartmentalize chemistry as organic, inorganic, physical and so on. This demarcation has some advantages as well as disadvantages in the graduate level of study after which the lines of demarcation gradually begin to fade out. As a matter of convenience and convention, we accept the demarcation. Our discussion will be limited to the field of traditional inorganic chemistry but we shall have glimpses of bioinorganic chemistry and nanochemistry.

Inorganic chemistry is the chemistry of *all* known elements, including carbon. Only excluded are the carbon compounds in which emphasis on functional groups provides a convenient generalized treatment—the 'organic compounds'. To begin with, we shall have a brief survey of the elements, their origin and occurrence in nature.

The roll strength of the elements in the periodic table now extends up to 118, with scopes for further addition. However, only 75 elements occur in the earth's crust in concentrations greater than 10^{-4} ppm. Their abundances show wide variations in the earth as well as in the universe as a whole. A study of this abundance of the elements in various isotopic compositions present an interesting starting point for the study of their origin and chemical behaviour.

1.2 GENESIS OF THE ELEMENTS

The origin of the elements must be traced back to the origin of the universe itself. The Big Bang theory seems to be the most dominant one in this regard. According to this theory, the entire matter and energy of the universe was cumulated in the form of a cosmic egg of very high density ($\sim 10^{96}$ g cm^{-3}) and temperature (10^{32} K). It is not clear how this nucleus was formed. Perhaps the "matter" present in it was actually in the form of radiations. Now, a tremendous and sudden explosion (big bang) occurred approximately 1.8×10^{10} years ago*. The universe started to expand after this explosion and the temperature began to fall very rapidly. For example, after one second of the explosion, the temperature is supposed to have fallen to $\sim 10^{10}$ K. From this stage onward, the universe was densely populated by elementary particles like neutrons, protons and electrons. During the next 10–500 seconds, these particles started to condense, like nuclear fusion reactions, into the nuclei of different elements.

The hot Big Bang theory is supported by the existence of 2.7 K radiation, experimentally verified in 1965 (Penzias and Wilson; Nobel Prize 1978 for this discovery). The universe must have been filled initially with radiation corresponding to its very high temperature, $\sim 10^{10}$ K. As the universe expanded, its temperature came down. It has been estimated that after 1.8×10^{10} years, the temperature of the universe should be around 3 K. In 1965, Penzias and Wilson built a highly sensitive horn-shaped radio-receiver for use in a space program. While testing it, they discovered a faint whisper of radio noise coming from all directions in outer space. This was actually a microwave radiation of wavelength 1.285 cm. This wavelength corresponds to isotropic thermal black body radiation at a temperature of 2.7 K. Hence, the cosmic microwave background radiation can be considered as a remnant of the Big Bang.

Whatever be the origin of the universe in its earliest form, our knowledge about the formation of the elements is more compact in terms of series of several thermonuclear reactions. The distribution of the various elements throughout the entire universe and their isotopic compositions estimated by wide range spectroscopic studies on the solar system, as well as on stars, galaxies, nebulae and interstellar space can be understood through such schemes of nuclear reactions.

In the process of formation of elements, the initial event appears to have been the formation of neutrons which decayed quickly (half-life = 11.3 min) into electrons, protons and antineutrinos:

$$ {}^{1}_{0}n \longrightarrow {}^{1}_{1}\mathrm{H} + {}^{0}_{-1}e + \overline{\nu} $$

After about 1 second, the universe was filled with neutrons, protons, electrons, antineutrino and of course, photons. The temperature was too high to allow the particles to combine—they were present in the plasma state.

As the temperature dropped, the protons could capture electrons to form H atoms. The H atoms could now condense to form mainly helium nuclei. It is estimated that within the first few seconds, ${}^{4}_{2}\mathrm{He}$ constituted nearly 25 per cent of the mass of the universe. The atoms gathered together to form galactic clusters and then more dense stars. The process of combi-

* Estimated from the observed present rate of expansion of the universe, 18 km s^{-1} per 10^6 light years; the rate is supposed to have remained constant from the beginning.

nation of hydrogen nuclei (fusion) continued (*hydrogen burning*) and huge amount of energy was released in the form of radiation. Initially the outward thermal pressure counteracted the gravitational force on the gaseous mass, but when approximately 10 per cent of the hydrogen in a star of about the same size as the sun has been used up, the thermal pressure of radiation became insufficient to counteract the gravitational pull. Consequently, the star contracted and the temperature rose higher. When the temperature reached about 10^8 K (100 MK), fusion occurred between helium nuclei forming nuclei of heavier elements (*helium burning*). Similar processes in large stars gave rise to the nuclei of still heavier elements. A few representative schemes of such nuclear reactions are summarized below:

Process	Typical Reactions
H-burning	${}^{1}_{1}\mathrm{H} + {}^{1}_{1}\mathrm{H} \longrightarrow {}^{2}_{1}\mathrm{H} + e^{+} + \nu$
$\sim 10^7$ K	${}^{2}_{1}\mathrm{H} + {}^{1}_{1}\mathrm{H} \longrightarrow {}^{3}_{2}\mathrm{He} + \gamma$
	${}^{3}_{2}\mathrm{He} + {}^{3}_{2}\mathrm{He} \longrightarrow {}^{4}_{2}\mathrm{He} + 2{}^{1}_{1}\mathrm{H}$
He-burning	${}^{4}_{2}\mathrm{He} + {}^{4}_{2}\mathrm{He} \longrightarrow {}^{8}_{4}\mathrm{Be}$
$\sim 10^8$ K	${}^{8}_{4}\mathrm{Be} + {}^{4}_{2}\mathrm{He} \longrightarrow {}^{12}_{6}\mathrm{C} + \gamma$
	${}^{12}_{6}\mathrm{C} + {}^{4}_{2}\mathrm{He} \longrightarrow {}^{16}_{8}\mathrm{O} + \gamma$
	${}^{16}_{8}\mathrm{O} + {}^{4}_{2}\mathrm{He} \longrightarrow {}^{20}_{10}\mathrm{Ne} + \gamma$
	${}^{20}_{10}\mathrm{Ne} + {}^{4}_{2}\mathrm{He} \longrightarrow {}^{24}_{12}\mathrm{Mg} + \gamma$
C-burning	${}^{12}_{6}\mathrm{C} + {}^{12}_{6}\mathrm{C} \longrightarrow {}^{24}_{12}\mathrm{Mg} + \gamma$
$\sim 10^9$ K	${}^{12}_{6}\mathrm{C} + {}^{12}_{6}\mathrm{C} \longrightarrow {}^{23}_{11}\mathrm{Na} + {}^{1}_{1}\mathrm{H}$
	${}^{12}_{6}\mathrm{C} + {}^{12}_{6}\mathrm{C} \longrightarrow {}^{20}_{10}\mathrm{Ne} + {}^{4}_{2}\mathrm{He}$
O-burning	${}^{16}_{8}\mathrm{O} + {}^{16}_{8}\mathrm{O} \longrightarrow {}^{28}_{14}\mathrm{Si} + {}^{4}_{2}\mathrm{He}$
$\sim 10^9$ K	${}^{16}_{8}\mathrm{O} + {}^{16}_{8}\mathrm{O} \longrightarrow {}^{31}_{15}\mathrm{P} + {}^{1}_{1}\mathrm{H}$
	${}^{16}_{8}\mathrm{O} + {}^{16}_{8}\mathrm{O} \longrightarrow {}^{31}_{16}\mathrm{S} + {}^{1}n$
α-process	${}^{20}_{10}\mathrm{Ne} + \gamma \longrightarrow {}^{16}_{8}\mathrm{O} + {}^{4}_{2}\mathrm{He}$
$> 10^9$ K	${}^{20}_{10}\mathrm{Ne} + {}^{4}_{2}\mathrm{He} \longrightarrow {}^{24}_{12}\mathrm{Mg} + \gamma$
	${}^{24}_{12}\mathrm{Mg} + {}^{4}_{2}\mathrm{He} \longrightarrow {}^{28}_{14}\mathrm{Si} + \gamma$
Neutron absorption	slow(*s*–): ${}^{56}_{26}\mathrm{Fe} + {}^{1}n \longrightarrow {}^{57}_{27}\mathrm{Co} + \beta^{-}$
$\sim 10^9$ K	rapid(*r*–): ${}^{57}_{27}\mathrm{Co} + {}^{1}n \longrightarrow {}^{58}_{28}\mathrm{Ni} + \beta^{-}$
	(see note 4)

Note 1. ^4He may also result as the net product from a cycle consisting of C, N and O (Bethe and Weizsacker):

$$^{12}C + {}^1H \longrightarrow {}^{13}N + \gamma \qquad {}^{13}N \longrightarrow {}^{13}C + e^+ + \nu$$

$$^{13}C + {}^1H \longrightarrow {}^{14}N + \gamma \qquad {}^{14}N + {}^1H \longrightarrow {}^{15}O + \gamma$$

$$^{15}O \longrightarrow {}^{15}N + e^+ + \nu \qquad {}^{15}N + {}^1H \longrightarrow {}^{12}C + {}^4He$$

It has been estimated that about 10% of the energy of the sun comes from this process. Most of the rest comes from the straightforward H-burning.

2. Besides the processes mentioned above, rapid proton capture by heavier nuclides (*p*-process) may lead to proton-rich nuclei.

3. Spallation reactions (chapter 14) also give rise to some light elements. Cosmic rays consist of a wide variety of atomic particles, from hydrogen to uranium. While travelling great distances in the galaxies, the heavier particles occasionally collide with atoms of the interstellar gas—largely ^1H and ^4He. As a result, fragmentation occurs and the lighter elements are formed. High speed α-particles may also collide with interstallar iron group elememts, inducing spallation. Such processes, together with ^{13}C (p, α) ^{10}B and ^{14}N (p, α) ^{11}C reactions, followed by β-decay of ^{11}C to ^{11}B, account for the abundance of the lighter isotopes of Li, Be and B.

4. Neutron absorption β emission processes are most significant for the elements beyond iron. The reactions are primarily (n, γ) type, the unstable nuclide formed subsequently undergoing β-decay. In the *s*-process, neutron capture is slow compared with β-emission while in the *r*-process, neutron capture is rapid. The nuclides formed by the *s*-process are controlled by the neutron capture cross section of the precursor nuclide. The stable nuclei corresponding to neutron magic numbers 50, 82 and 126 have very low neutron capture cross sections. This explains the relatively high abundances of $^{89}_{39}Y$, $^{90}_{40}Zr$, $^{138}_{56}Ba$, $^{140}_{58}Ce$, $^{208}_{82}Pb$ and $^{209}_{83}Bi$. In the *r*-process, a large number of neutrons are added successively into a single nucleus in a very short time; for example, some 200 neutrons might be added to an iron nucleus in 10–100 s. Eventually the product becomes excessively neutron-rich and a cascade of 8–10 β-emissions result in the formation of a stable nuclide. The abundances of nuclides of mass numbers 80, 130, 194 as well as ^{36}S, ^{46}Ca, ^{48}Ca are explained in this manner.

Such schemes of proposed nuclear reactions have been considerably developed to account for the observed abundances of different nuclides throughout the universe. Let us now have a glance at the abundances of the elements.

1.3 ABUNDANCE OF THE ELEMENTS

The average relative content of an element in any natural system is called its abundance. Spectroscopic analysis of the celestial bodies and analysis of meteorites provide knowledge about the abundance of the elements over the whole universe, or the cosmos; this is often referred to as the cosmic abundance of the elements (sec. 1.3.1). Abundances of the elements on the earth are somewhat different and are referred to as the terrestrial abundance of the elements (sec. 1.3.2).

1.3.1 Cosmic Abundance

A plot of the abundances of elements (in terms of number of atoms per 10^6 atoms of silicon) against their atomic numbers is shown in Fig. 1.1. Certain selected values (A. Cameron) are also given in Table 1.1. The values are approximate and disputed. Yet the general features presented by them are interesting:

(i) The abundance data may be clearly divided into two separate curves, one lower and one upper, as shown. The upper curve (solid line, blue) connects the nuclides of even atomic number which are clearly more abundant than the nuclides of neighbouring elements with odd Z (broken line, red).

(ii) Abundances show a rapid exponential decrease from elements of small atomic number to molybdenum ($Z = 42$). After this, abundances remain more or less constant.

(iii) Hydrogen and helium are most abundant of all elements. Carbon and oxygen come next in the sequence.

(iv) The abundances of lithium, boron (both with odd Z) and also beryllium are unusually low. These nuclides are easily transmuted by nuclear bombardment. Thus $^{8}_{4}Be$ formed by fusion of two $^{4}_{2}He$ are readily converted to more stable $^{12}_{6}C$ (see helium burning reactions).

(v) Nuclides whose mass numbers are multiples of 4 (e.g., ^{16}O, ^{20}Ne, ^{24}Mg, ^{28}Si, ^{32}S, ^{36}Ar, ^{40}Ca, ^{48}Ti) are more abundant than their immediate neighbours (rule of Oddo). This shows the stability of the 2-proton-2-neutron combination ("*alpha-particle nuclides*").

(vi) Iron (atomic number 26) is marked by a peak in the curve, ^{56}Fe being remarkably abundant in comparison to its immediate neighbours ($\sim 10^4$ times). We shall see later that this nuclide has a very high nuclear binding energy (Chapter 14).

(vii) Most of the peaks in the curve correspond to magic numbers of protons and neutrons: $^{4}_{2}He$, $^{16}_{8}O$, $^{40}_{20}Ca$, $^{90}_{40}Zr$, $^{119}_{50}Sn$, $^{138}_{56}Ba$, $^{208}_{82}Pb$ (Chapter 14).

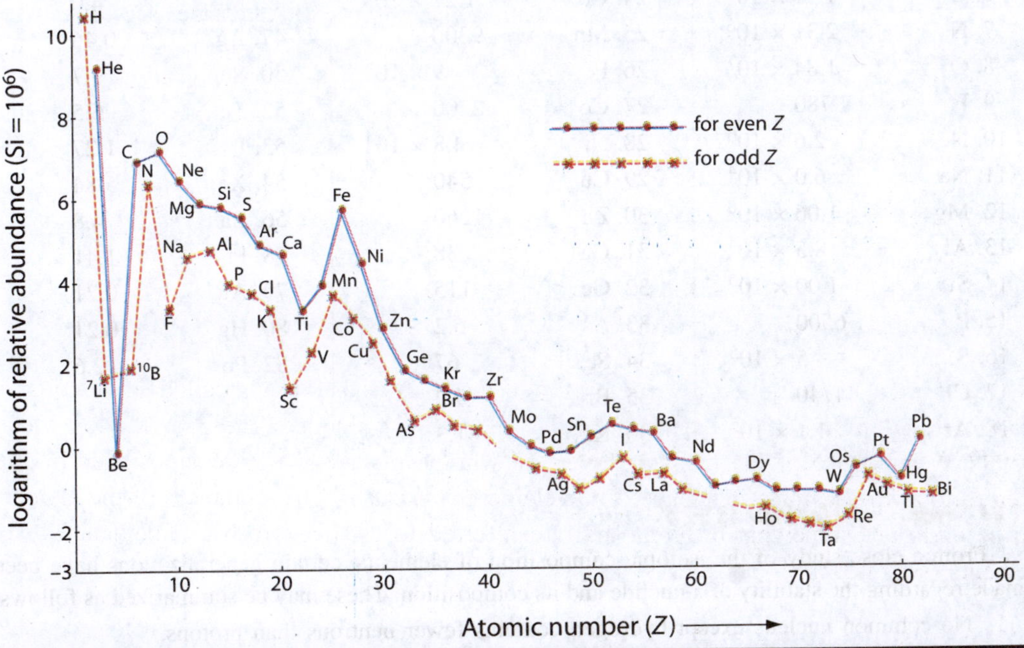

Fig.1.1: Nature of variation of cosmic abundance of elements relative to Si = 10^6. Adapted from Chemistry of the Elements, by N.N. Greenwood and A. Earnshaw, 2nd edition; Butterworth-Heinemann, Oxford, 1997.

(viii) The most common isotope of a given element is that with an even number of neutrons. Isotopes containing protons and neutrons in odd numbers are less abundant, e.g., ${}^{2}_{1}H$, ${}^{40}_{19}K$.

Plot of abundances against mass numbers (not shown here) further reveals that: (a) atoms of heavy elements tend to be rich in neutrons; heavy proton-rich nuclides are rare. This lends support to the hypothesis of primordial elemental synthesis by neutron absorption. (b) Among the heavier elements, abundance maxima occur in pairs for mass numbers 80, 90; 130, 138; 196, 208. This shows the stability of a magic number of nucleons (Chapter 14).

Study of the abundances of elements throws light on the mode of formation of the elements (nucleogenesis) as outlined in the last section. The ultimate abundance of an element depends on several factors, particularly the probability of the nuclear process involved and the stability of the various isotopic species.

Table 1.1

Relative abundances of some elements (approx.)

(Atoms per 10^6 atoms Si)

Element	Abundance	Element	Abundance	Element	Abundance
1. H	2.7×10^{10}	19. K	3500	37. Rb	6.1
2. He	1.8×10^{9}	20. Ca	6.2×10^{4}	38. Sr	22.9
3. Li	60	21. Sc	31	39. Y	4.8
4. Be	1.2	22. Ti	2400	40. Zr	12
5. B	9	23. V	254	41. Nb	0.9
6. C	1.11×10^{7}	24. Cr	1.27×10^{4}	42. Mo	4.0
7. N	2.31×10^{6}	25. Mn	9300	47. Ag	0.46
8. O	1.84×10^{7}	26. Fe	9×10^{5}	50. Sn	3.7
9. F	780	27. Co	2200	52. Te	6.5
10. Ne	2.6×10^{6}	28. Ni	4.8×10^{4}	53. I	1.27
11. Na	6.0×10^{4}	29. Cu	540	54. Xe	5.84
12. Mg	1.06×10^{6}	30. Zn	1260	56. Ba	4.8
13. Al	8.5×10^{4}	31. Ga	38	78. Pt	1.41
14. Si	1.00×10^{6}	32. Ge	115	79. Au	0.21
15. P	6500	33. As	6.2	80. Hg	0.21
16. S	5×10^{5}	34. Se	67	82. Pb	2.6
17. Cl	4740	35. Br	9.2		
18. Ar	1.1×10^{5}	36. Kr	41.3		

The "rules" of Harkins (1928)

From a close study of the isotopic composition of elements, certain generalizations have been made regarding the stability of a nuclide and its composition. These may be summarized as follows:

1. No common nucleus except hydrogen contains fewer neutrons than protons.
2. Elements with even number of nuclear charge (or protons) are more abundant and more stable than those with odd charges. They are also richer in isotopes.

3. Nuclei with even numbers of neutrons are more abundant and more stable than nuclei having odd numbers of neutrons.
4. Nuclei with even mass numbers are more abundant than nuclei having odd mass numbers.

The applicability of these generalizations may be readily appreciated with reference to any table of isotopic abundance. The numbers of stable nuclides known for different odd and even combinations of atomic number (*Z*) and neutron number (*N*) are also noteworthy (also see section 14.1):

Combination	**Number** (approx.)
Z even, *N* even	163
Z even, *N* odd	56
Z odd, *N* even	50
Z odd, *N* odd	5 ($^{2}_{1}H$, $^{6}_{3}Li$, $^{10}_{5}B$, $^{14}_{7}N$, $^{50}_{23}V$)

1.3.2 Terrestrial Abundance

A very small fraction of the mass of the earth is available for direct analysis and study—the crust (see box), the hydrosphere and the atmosphere. They make up less than 1 per cent of the earth by mass. Obviously, indirect methods have to be used to estimate the composition of the earth. One may reasonably assume that the sun and the planets and meteorites of the solar system were derived via the same set of nucleosynthetic events. Hence the abundance of the non-volatile elements are in the same proportion in the earth as they are in the sun or meteorites. Related arguments and studies have led to several estimates of the abundances of the elements in the crust of the earth. The estimates differ considerably from one another, but their overall trend is not very different from the cosmic abundance of the elements. Some approximate figures are given in Table 1.2.

Seismic data and other studies primarily suggest that the earth consists of several zones based on density and perhaps chemical affinity: (a) the crust (~36 km. in depth), (b) the mantle (36–2900 km), and (c) the core (2900 km. to centre). Primarily, the crust consists of a heterogeneous layer of silicates and oxides of different elements; the mantle has a fairly homogeneous combination of different silicates, mainly of magnesium and iron. Geophysical evidence indicates that the core should contain elements with the properties of molten iron (and some nickel); a mean atomic number of 22 is explained by the supposed presence of some lighter elements like sulfur; carbon, silicon, and oxygen may also be present.

Table 1.2

Abundance of some elements (approximate) in crustal rocks

(Grams per metric ton or ppm.)

Element	Crustal average	Element	Crustal average	Element	Crustal average
1. H	1500	11. Na	28,000	19. K	25,900
2. He	0.003	12. Mg	20,900	20. Ca	36,500
3. Li	20	13. Al	81,300	22. Ti	4300
6. C	200	14. Si	277,000	23. V	135
7. N	20	15. P	1100	24. Cr	120
8. O	466,000	16. S	340	25. Mn	1000

Element	Crustal average	Element	Crustal average	Element	Crustal average
26. Fe	52,000	30. Zn	75	79. Au	0.004
27. Co	29	47. Ag	0.08	80. Hg	0.08
28. Ni	100	78. Pt	0.01	82. Pb	13
29. Cu	68				

Certain features of the table are interesting:

(i) Eight elements constitute the major part of the crust—O, Si, Al, Na, Fe, Ca, Mg and K. They make nearly 98.5% of the total.

(ii) Of these eight elements, oxygen is the most predominant. The crust consists almost entirely of oxygen compounds—mainly silicates of the six metals out of these eight elements. In terms of volumes of the atoms, oxygen alone occupies more than 90% of the total volume occupied by the elements. The crust may thus be considered as a packing of oxygen anions—bonded by silicon and the common metals.

(iii) The abundance of certain common elements are much lower than or comparable to those of many less familiar elements; for example (crustal abundances in ppm in parenthesis):

Cu (68) < **Zr** (162) ; **Pb** (13) < **Ga** (19)

Hg (0.08) << **La** (35) or **Tb** (1.2) ; **Zn** (75) ≃ **Rb** (78)
or any "rare earth"

Sn (2) << **V** (135) ; **As** (1.8) < **Sc** (25)

I (0.5) < **Hf** (2.8)

The abundance of an element in terms of its average percentage in the crust of the earth is expressed in **clarke** (symbol C). The clarke of oxygen is ~46, that of silicon ~28.

1.4 OCCURRENCE OF THE ELEMENTS

The abundance of an element is different from its mode of occurrence, or availability. Some elements present in the crust in considerable amounts never occur in any concentrated form; they are dispersed in common minerals and are called *dispersed elements*. Thus, copper, much less abundant than rubidium, forms rich ore deposits. On the other hand, rubidium is found dispersed in potassium minerals. Similarly, gallium is dispersed in aluminium minerals; but chromium and lead, both comparatively less abundant metals, are known to form large scale ores. Indium, hafnium, and rhenium are similarly dispersed elements.

The availability of an element depends largely on its ability to form concentrated mineral sources. Even then, elements with rich "ore deposits" also have a larger fraction unavailable as dispersed amount. Extraction of the element from such dispersed states is impractical for technical and economic reasons. Thus magnesium in large quantities (~30%) is present in vast deposits of olivine. But it is extracted from sea-water, where it is present only to the extent of 0.03%. Sometimes, a relatively rare element gains importance due to its usefulness, e.g., Be (see text). Sometimes, a rare element may be obtained as an industrial by-product. Electrolytic refining of copper can provide large quantities of tellurium; about 1000 tons of gallium and germanium may be obtained annually from the ashes of certain coals. It is time that we should explore possibilities of utilizing these elements on a commercial scale.

The main forms of existence of the common elements on the earth is shown in Table 1.3. It appears that oxygen compounds, mainly in the form of silicates, predominate on the earth's crust. Carbonates, sulfides and sulfates occur in gradually decreasing quantities.

Table 1.3
Common forms of occurrence of the elements

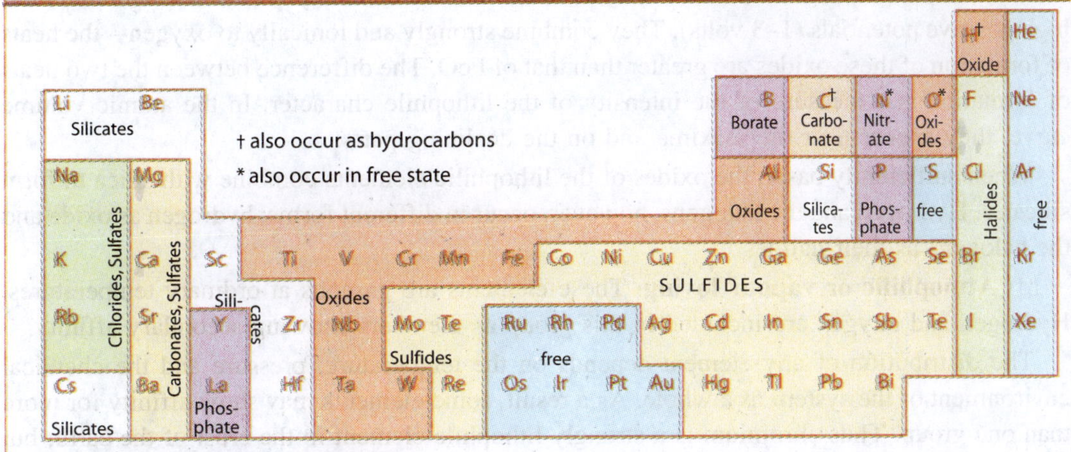

Note. Colours used are fictitious

1.4.1 Geochemical Classification of the Elements

According to Goldschmidt, distribution of the elements in the crust, mantle and the core of the earth was initiated by formation of different layers of varying densities during the solidification of the earth. Different elements were then distributed among these layers according to their affinity. On this basis he classified the elements into four groups according to their relative affinities for metallic, sulfidic, silicate and gaseous phases. These are:

(a) **Siderophilic or iron-loving (Greek *sideros* ≡ iron)**: These elements are believed to occur free in the core of the earth together with iron. However, many of them occur in the crust in other combined forms. Except Ru, Rh, Pd, Os, Ir, Pt, Au and C, the remaining elements occur in combination with sulfur in the crust of the earth (chalcophilic). Ru, Rh, Pd and Os also show some affinity for sulfur. Carbon occurs as carbonates in rocks (lithophilic). Some elements, viz., Fe, Co, Ni, Cu, Ga, Ge and Sn are also found as their oxides in rocks (lithophilic).

This group includes elements which have incompletely filled outermost shells of electrons. They are relatively inert chemically—the triads of group VIII (Mendeleef classification) are most important in this category (Table 1.3).

The siderophilic elements have high negative electrochemical potentials and they occur near the minima in the atomic volume (vs. atomic no.) curve. The heats of formation of oxides of these elements are lower than that of FeO.

(b) **Chalcophilic or copper loving (Greek *chalcos* ≡ copper)**: This group generally contains elements whose ions contain 18 electrons in the outer shell and whose ionization energies are greater than about 600 kJ mol^{-1} (6–7 eV) and which form covalently bonded sulfides. Notable among them are Mo, Fe, Cu, Ag, Zn, Cd, Hg, Ga, In, Tl, Pb, As, Sb, Bi, S, Se, Te. Since copper is chiefly found as its sulfide mineral, these elements are also found principally as sulfides (Table 1.3). These elements have intermediate potentials in the electrochemical series. In the atomic volume curve, they occupy regions in which atomic volume increases with atomic number.

(c) **Lithophilic or stone-loving (Greek *lithos*, stone)**: These elements readily form ions with an octet in the outermost shell. They have low ionization energies (< ~ 600 kJ mol^{-1}) and high positive potentials (1–3 volts). They combine strongly and ionically to oxygen—the heats of formation of these oxides are greater than that of FeO. The difference between the two heats of formation is a measure of the intensity of the lithophile character. In the atomic volume curve, they appear near the maxima and on the declining parts.

When sufficiently basic, the oxides of the lithophilic elements combine with silica to form silicates. Hydrogen and the halogens, however, occur in different forms: hydrogen as oxide and the halogens as their halides.

(d) **Atmophilic or vapour loving**: These elements are gaseous at ordinary temperatures. Hydrogen and oxygen are included in this group as members showing secondary affinity.

The distribution of any element depends on the temperature, pressure and the chemical environment of the system as a whole. As a result, some elements may show affinity for more than one group. Thus chromium is a strongly lithophile element in the crust of the earth; but it becomes chalcophilic in an oxygen-deficient atmosphere, as is found in some meteorites. Carbon and phosphorus become siderophilic under strongly reducing conditions.

Table 1.4

Geochemical classification of elements according to Goldshmidt

Siderophilic	Chalcophilic	Lithophilic	Atmophilic
$Fe^{c,l}$, $Co^{c,l}$, $Ni^{c,l}$	Cu, Ag	Li, Na, K, Rb, Cs	H, N, O
Ru, Rh, Pd	Zn, Cd, Hg	Be, Mg, Ca, Sr, Ba	He, Ne, Ar
Os, Ir, Pt	Ga, In, Tl	B, Al, Sc, Y, La	Kr, Xe
$Cu^{c,l}$, Au	Ge, Sn, Pb	Si, Ti, Zr, Hf, Th	
C^{l}, $Ga^{c,l}$, $Ge^{c,l}$	As, Sb, Bi	P, V, Nb, Ta	
As^{c}, Sb^{c}, Mo^{c}, W^{c}	S, Se, Te	O, Cr, U	
	Fe, Mo, Os	H, F, Cl, Br, I	
	Ru, Rh, Pd, Pt	Fe, Mn, Zn, Ga	

c → chalcophilic in the crust of the earth. *l* → lithophilic in the crust of the earth.

In the last three groups, the affinity is much less pronounced for the elements in normal (*not bold*) types.

The above geochemical classification of an element refers to its behaviour in liquid phase equilibria in melts. The actual mineralogy of an element is somewhat different because there are many secondary factors responsible for mineral formation. Thus, thallium, a chalocophile element, forms sulphide minerals. But the greater part of thallium in the earth's crust is found in potassium minerals where Tl^+ ions replace K^+ ions. In the following section, we shall briefly survey the principal factors responsible for mineral formation.

1.4.2 The Formation of Minerals

Minerals are naturally occurring substances, generally solid and inorganic, whose atoms and ions are arranged in regular, three-dimensional patterns that give each mineral a characteristic set of physical properties. A mineral may be composed of either a single element (graphite) or of several elements. Minerals may have widely different chemical compositions and structures. They present a great diversity of forms, structures and physical and chemical properties. We identify a mineral by its form, colour, transparency, hardness, lustre, crystal structure etc.

About 3300 different kinds of minerals have been known to be present in nature—of which only a few hundred are common.

A mineral rarely exists alone in the crust. Usually two or three are combined in regular pattern and definite proportion to form different rocks. According to their origin of formation we group various rocks into igneous, sedimentary and metamorphic rocks. The mother of these rocks may be identified as the magma, or molten rock material, mainly silicate. Temperature in the crust increases at the rate of about 1°C for each 33 m depth and it does not become steady until the depth is very great. From 60 km onward, the temperature ranges from 500–1200 °C, or even 2000 °C. This temperature is sufficient to melt any rock. But at the very high pressure of the upper crust (~ 500,000 atmosphere), rocks do not melt entirely but turn into a thick fluid, the magma.

Igneous rocks. Igneous rocks result from freezing of magma. They may be of two types—intrusive rocks and extrusive rocks. The magma intrudes upward through cracks into surrounding rocks—with a sharp fall in temperature and pressure it differentiates and crystallizes to form minerals. The "frozen" magma now becomes a body of intrusive rocks. The lava poured out from a volcano solidifies into extrusive rocks.

Almost all igneous rocks are crystalline. Intrusive rocks are formed by slow cooling and are usually coarse-grained. But extrusive or volcanic rocks are formed by rapid cooling, and so large crystals cannot be formed; they are either fine-grained or glassy.

Granites and basalts are common examples of igneous rocks. Chemically, silicates are the main constituent of these rocks. Pegmatites are another kind of igneous rock which crystallize during the last stages of cooling and solidification of magma. They consist of coarser grains than granites and contain many rare elements (see Li). Igneous rocks are the major kinds of rocks in the crust, making up about 65 per cent of the total by weight.

Sedimentary rocks. Sedimentary rocks are formed by the erosion of igneous rocks by exogenic forces. Rocks exposed on the surface are acted upon by sun rays, ice freezing, wind, rains, plant roots, etc. With this sort of continuous weathering they get transformed into debris, sand and soil. These are then carried out by water current, wind, glacier, etc. and get deposited in rivers, lakes, sea-beds and the like. The sediments grow in thickness over years. As the pressure and temperature increase, water is squeezed out of the pores in the sediments and the pores are filled by silica, iron oxide or carbonates from solution. Their cementing action converts the loose grainy sediment to sedimentary rocks. Mudstone, sandstone, limestone, dolomite, etc. are examples of sedimentary rocks. Such rocks can be readily identified by their layer structure or stratification.

Sedimentary rocks are nearly 8% of the total weight of the crust. But they are widely distributed over continental surfaces and sea-beds. Almost 75% of the continental surfaces are covered by them.

Metamorphic rocks. These rocks are formed by the transformation of igneous and sedimentary rocks. The latter, buried at great depths in the crust, get heated by intruding hot magma. At the high temperature and pressure, the rocks undergo a change of structure, texture and ingredients, too; ultimately they are converted to metamorphic rocks. Slate and marble are common examples of such rocks. They make up about 27% of the whole crust.

Factors controlling the formation of minerals

The actual processes involved in the formation of rocks are a series of complicated chemical reactions taking place over a very long period of time under varying external conditions. As stated earlier, the elements may be broadly classified according to their geochemical affinity into lithophilic, siderophilic, etc. groups. This classification may be used as a rough starting point for the formation of minerals.

As the earth gradually attained its present form, a phase differentiation was brought about by gravity. The compositions of these phases were determined by the most abundant elements—Fe, Mg, Si, O, S and Al. The amount of oxygen was evidently more than that of sulfur; again the total oxygen plus sulfur was insufficient to combine with the entire amount of electropositive elements. Among them, iron was more abundant than magnesium and silicon, and possessed the greatest affinity for sulfur. At the same time, it was most easily reduced to the free metal. As a consequence, three distinct phases were formed—free iron (left uncombined), iron sulfide and iron-magnesium silicate. Elements more electropositive than iron displaced iron from the silicates. The less electropositive elements were displaced *by* iron from the silicate phase. Elements which form essentially homopolar bonds with sulfur—the sulfide group metals in analytical chemistry and the metalloids—were accommodated in the sulfide phase.

This interpretation is also in keeping with the classification of elements as class-a and class-b (see hard and soft acids and bases). The class-a metals are usually found associated with oxygen and oxoanions such as phosphates, silicates and carbonates. The class-b metals are usually found as their sulfides, e.g., zinc, cadmium and mercury. It is likely that in the reducing atmosphere prevailing when the earth's crust solidified, these and other chalcophilic metals separated out in the sulfide phase. Subsequently, as rocks were weathered, some of the sulfides were leached out and eventually precipitated as carbonate, silicate or phosphate. This idea appears to be very much applicable to the minerals of zinc.

Further insight into the process of mineral formation involves detailed consideration of the various factors determining the formation and stability of compounds in general. The atoms and ions are striving toward stability through stages of redistribution and recombination under different prevailing conditions. The principal controlling factors for such changes may be broadly generalized as—(i) structural control (ii) thermodynamic control and (iii) kinetic control.

Ionic size, ease of packing and electrical charge balance are the most important structural factors in the formation of minerals. Thus ions of similar radii but *different chemical character* often occur together in silicate minerals. We have already mentioned the existence of thallium in potassium minerals which may be attributed to similar ionic radii (K^+ 0.14 nm; Tl^+ 0.15 nm). Some other examples are shown in Table 1.5.

Table 1.5

Ions occurring together on the basis of similarity in ionic radius

Ionic radius (nm)	Ions
0.05–0.07	Li^+, Mg^{2+}, Al^{3+}, Fe^{3+}, Ti^{4+}, Mo^{6+}, W^{6+}
0.07—0.09	Ni^{2+}, Co^{2+}, Fe^{2+}, Zn^{2+}, Zr^{4+}, Sn^{4+}, Nb^{5+}
0.09—0.11	Na^+, Ca^{2+}, Cd^{2+}, Ce^{4+}, Th^{4+}, U^{4+}
0.11—0.14	K^+, Sr^{2+}, La^{3+} —Eu^{3+}

Examples of structural control in mineral formation

The ease of packing of cations and anions in a particular mineral is largely determined by radius-ratio effects (chapter 8). When a cation has a counter-ion with radius-ratio lying near the boundary between two types of coordination, it may occur in both coordination numbers. The temperature and pressure at which crystallization took place gains importance in such cases. High temperature and low pressure favour the lower coordination number. Thus aluminium has a radius-ratio value 0.38 with oxide ion. It assumes C.N. 4 in typically high-temperature minerals and substitutes silicon (e.g., feldspars). In minerals formed at lower temperatures, it tends to attain C.N. 6.

The importance of electrical stability in crystals is nicely summarized in the following rule of Pauling: "*In a stable structure the total strength of the valency bonds which reach an anion from all the neighbouring cations is equal to the charge on the anion*". Thus feldspars (see silicates) contain 4-coordinated Si or Al with oxide ions. Only one-fourth of the valency of an oxygen atom is left unsatisfied by the Si or Al. Mg or Fe in sixfold coordination would contribute at least 2/6 or one-third unit of charge. This would be greater than the "charge-requirement" of the oxygen atom and hence Mg or Fe cannot enter such structures. But Na, K, etc. univalent ions and large bivalent cations like Ca^{2+} have radius-ratios (with the oxide ion) greater than 0.71, nearly that required for C.N. 8. They can meet the residual charge requirement of the oxide ion. So we find feldspars with Ca, Na and K but not with Mg or Fe in their structures.

Formation of isomorphous crystals or solid solutions and atomic substitution are other examples of structural control on the formation of minerals. Examples of isomorphism are already wellknown. Solid solution formation is independent of isomorphism, e.g., between FeS and ZnS. In atomic substitution, atoms or ions replace others of similar size and charge. Sometimes the charge may be balanced by the inclusion of an additional ion, or by way of coupled substitution: $Mg^{2+} - Si^{4+}$ ion-pairs in diopside ($CaMgSi_2O_6$) are replaced in part by $Al^{3+} - Al^{3+}$. Substitution does not usually occur when the ionic charges differ by more than unity: Y^{3+} not by Na^+; nor Zr^{4+} by Mn^{2+}. The presence of iron and manganese in dolomites (calcium magnesium carbonate) is now interpreted in terms of atomic substitution of magnesium. This is often referred to as *diadochy*, the ability of different elements to occupy the same lattice position.

Crystal field stabilization energy provides another interesting illustration of various stabilizing factors involved in mineral formation. Pending our introduction to the nature of this energy in the chapter on coordination chemistry, we simply accept that certain ions are preferentially stabilized in octahedral sites, e.g., the divalent ions of Ti, V, Cr, Fe, Co, Ni and Cu. The non-transition elements and Mn(II), Zn (II) ions lack this stabilization. The order of decreasing octahedral site preference energy for some of the +2 and +3 transition metals is

+2 : Ni > Co > Mn ;	**+3** : Cr > V > Sc

This order suggests that Ni^{2+} would be more readily depleted from the magma during crystallization of olivine and pyroxene than would Co^{2+}. The structures of inverse spinels also point out the importance of crystal field stabilization (see chapter 27).

ESSENTIALS

***Genesis of Chemistry*:** Chemistry developed as an applied knowledge of preparing materials useful to early civilization—e.g., metals, glass, pigments and several other substances capable of healing ailments. The foundation of modern chemistry was laid in late eighteenth century by Lavoisier.

***Genesis of the elements*:** Elements were formed through a series of nuclear reactions at the very high temperature prevailing after the Big Bang. The elements combined between themselves to form various minerals.

Nuclides with even numbers of protons and even numbers of neutrons are more stable and more abundant then nuclides containing any of these particles in odd number.

Hydrogen and helium are the most abundant elements in the universe.

EXERCISE

1. What is the likely root of the word '*Rasayana*'?
2. What is Iatrochemistry? In which period of time did it gain prominence?
3. What is phlogiston theory? Who contradicted it from experimental results?
4. Write the main thermonuclear reactions leading to the formation of $^{4}_{2}He$ and $^{24}_{12}Mg$. Which require a higher temperature?
5. $^{16}_{8}O$, $^{24}_{12}Mg$, $^{28}_{14}Si$, $^{32}_{16}S$, $^{40}_{20}Ca$ are more abundant compared to other isotopes of respective elements. Comment.
6. The abundances of Li, Be and B are very low. Explain.
7. Feldspar, a silicate mineral, contains Na, K and Ca but not Mg or Fe in their structures—explain.

2

NOMENCLATURE OF INORGANIC CHEMISTRY

*IUPAC RECOMMENDATIONS, 2005**

2.1 INTRODUCTION

By the time chemistry emerged as a systematic science in the late eighteenth century, it had accumulated a rich vocabulary during its long journey over nearly 2000 years through different countries. In 1782, Guyton de Morveau rightly expressed the need for a "method of denomination which helps the intelligence and relieves the memory". He received quick response from eminent scientists like Lavoisier, Berthollet and later, Berzelius. But their venture could not cope with the huge number of compounds added to chemistry during the nineteenth century when several *ad hoc* systems of naming had to be adopted.

* Adapted from Nomenclature of Inorganic Chemistry IUPAC RECOMMENDATIONS 2005 edited by Neil G. Connelly, Ture Damhus, (with R M. Hartshorn and A T. Hutton); published by The Royal Society of Chemistry, Cambridge, UK for the IUPAC; read with corrections up to 12-10-2008.

The first systematic effort to an international nomenclature in ionrganic chemistry was seen in 1913 when a commission was set up by the Council of International Association of Chemical Societies, though the working of the commission ended abruptly owing to World War I. In 1921, IUPAC again set up a commission on nomenclature and the first comprehensive report of the Inorganic Commission was published in 1940. Since then, recommendations from IUPAC on inorganic nomenclature appeared in 1959, 1971, 1990, 2000 and the latest one in 2005 (with amendments in October, 2008).

The aim of all nomenclature systems is to provide an unambiguous, easily recognizable and yet uncomplicated representation of the chemical elements and compounds, in formula as well as in naming. However, it has not always been possible to provide a single unique name to a compound, since the purpose of naming varies with the context to which it refers. Accordingly, more than one acceptable names may be assigned to a particular compound, but such names should adhere to a general system of nomenclature and be readily recognizable. As new discoveries make fresh demands for revision and extension of the nomenclature systems in a continuous process, the system of nomenclature is dynamic and ever-expanding. The Division of Chemical Nomenclature and Structure Representation of IUPAC, formed in 2001, is working with this mission.

A chemical nomenclature may serve multiple functions. At the *first level*, the nomenclature gives some information about the substance named but does not reveal its composition. The common names of many compounds are of this type which are significant only to the chemists or their users. At the *second level*, a name may be used to express the stoichiometric formula of a compound. This provides a systematic prediction about the nature of the compound. At *the third* and more sophisticated level, a name may be used to convey some information regarding the three-dimensional structure of a compound but this is meaningful only to the specialists willing to know the structure at the cost of sophistication. At a *fourth level*, a nomenclature may have to be derived in a unique manner to facilitate compilation and extensive indexing.

Three systems of nomenclature are important in inorganic chemistry. These are (a) compositional nomenclature; (b) substitutive nomenclature and (c) additive nomenclature.

(a) ***Compositional nomenclature*** is based on the stoichiometry or composition of a chemical species (*stoichiometric name*). The constituents are formally divided into electropositive and electronegative categories and arranged with multiplicative prefixes. The electropositive constituents are usually named first. Alphabetical order is maintained within each category (see ordering in Grammar, section 2.2).

O_3 : trioxygen — $MgCl(OH)$: magnesium chloride hydroxide

PCl_3 : phosphorus trichloride — Na_3Bi_5 : trisodium pentabismuthide

More examples will be found later.

(b) ***Substitutive nomenclature*** is based on the concept of a parent hydride in which hydrogen atoms have been substituted by other atom(s) or groups. Thus PCl_3 may be looked upon as derived from the hydride PH_3 (phosphane); hence the substitutive name of PCl_3 will be trichlorophosphane.

There are some grammatical rules for naming the parent compounds and substituents and providing proper order of the substituents (section 2.2).

(c) ***Additive nomenclature*** is constructed by selecting a central atom or atoms and associating ligands with this central atom. The coordination compounds are best examples of this type of nomenclature though a wide variety of compounds may be covered by it, making it the most generally applicable system of nomenclature in inorganic chemistry. Rules regarding the central atom and the ligands are mentioned later (section 2.2).

$CoCl_3(NH_3)_3$: triamminetrichloridocobalt(III)

PCl_3 : trichloridophosphorus

$HNO_3 = [NO_2(OH)]$: hydroxidodioxidonitrogen

$[NO_2]^-$: dioxidonitrate(1−)

$H_3SO_4^+$ ($\equiv [SO(OH)_3]^+$: trihydroxidooxidosulfur(1+).

The choice between the three systems of nomenclature actually depends upon the type of information one desires to convey and also on the class of compound. Acceptable non-systematic names for many compounds are also in wide usage.

Whatever be the function of nomenclature, it has to be constructed from names of the elements or their modifications in accordance with certain uniform rules. These rules will determine the sequence of the elements in the name together with prefixes, affixes, etc. and provide a uniform code of punctuation using hyphen, italics, enclosing marks and so on.

In the limited scope of this treatment, we shall mention only some of the basic rules essential to the beginners with particular emphasis on suggested changes in the new IUPAC recommendations, 2005.

2.2 THE GRAMMAR OF NOMENCLATURE

Just as we write a sentence using different words arranged according to an accepted norm (the syntax), we write the name of a compound using the names of atoms according to some accepted procedure. In writing a sentence, we use prepositions, and various punctuation marks to establish the inter-relation between the words. In chemical nomenclature too, specific prefixes or suffixes and some punctuation marks like hyphens, enclosing marks, etc. are used to describe a compound.

2.2.1 Suffixes, Prefixes and Infixes

In writing the name of a compound from the names of elements or roots derived therefrom, certain affixes are added to these names to convey specific informations. Any modifying affix before the name is called a prefix while suffixes are placed at the end of root names or systematic names to convey some information on the nature of combination involved.

The term 'ending' is generally used in a broader sense for systematic names of different classes of compounds. Some common suffixes and endings are very familiar to us, for example the endings *–ane, –ene and –yne* in organic compounds, the ending *–ide* in chloride, bromide, iodide, etc. and so on. Infixes are similar syllables placed *within* the name. Some common suffixes and endings with their major areas of use are mentioned below:

Table 2.1
Common suffixes used in nomenclature

a Added at the end of an atom (its name) when the atom substitutes other atoms in a skeleton, e.g., carbon atoms replaced by oxygen (oxa), nitrogen (aza), boron atoms in boron hydride based names like carba, thia, etc.

For use in substitutive nomenclature, names of all elements are modified by the suffix 'a' according to specific conventions like

aluminium	alumina	antimony	stiba	barium	bara
beryllium	berylla	bismuth	bisma	boron	bora
cadmium	cadma	calcium	calca	chlorine	chlora
deuterium	deutera	iron	ferra	nitrogen	aza**
oxygen	oxa	sulfur	thia*	tin	stanna
tritium	trita	uranium	urana	xenon	xenona
yttrium	yttra				

* from theion; ** from azote

ane Ending used in names of neutral saturated parent hydrides of elements of Groups 13–17 e.g., borane (BH_3), methane (CH_4), azane (NH_3), oxidane (H_2O), fluorane (HF). See substitutive nomenclature.

ate Ending in names of anions and salts/esters of oxoacids with "–ic" in their names, for example nitrate, carbonate, acetate, etc.

In additive names of anions, specially coordination compounds; often followed by charge number in parentheses;

$[Fe(CN)_6]^{4-}$: hexacyanidoferrate (4–); $[AlH_4]^-$: tetrahydridoaluminate (1–);

ato Ending used when an anion having suffix –ate acts as a ligand, e.g., sulfato, acetato etc.

ic Ending in names of many acids. Suffix to names of elements in compounds with a higher oxidation state (no longer acceptable), e.g., ferric chloride.

ide Ending of names of homoatomic anions (e.g. chloride, triiodide) and some acceptable nonsystematic heteropoly anions (e.g., cyanide, hydroxide). Suffix to names of anions formed by removal of one or more hydrons (protons) from a hydride, e.g., methanide (CH_3^-), hydrazinide (H_2NNH^-), hydrazine-1, 2-diide ($^-HNNH^-$).

ido Ending used when anions with ide suffix act as ligands, e.g., chlorido (formerly chloro), hydroxido (formerly hydroxo).

Also ending of some prefixes for anionic substituent groups, e.g., oxido ($-O^-$).

inous Suffix to parent names of acids having the skeleton $H_2X(OH)$, X = N, P, As, Sb. $H_2Sb(OH)$: Stibinous acid (dihydridohydroxidoantimony).

onate Ending in names of anions and esters of oxoacids with 'onic' modifiers : tetrathionate, phosphonate $[PHO_3]^{2-}$, (hydridotrioxidophosphate (2–)).

onato Modification of 'onate' anions when they act as ligands.

Ending of prefixes for anionic substituent groups, e.g., phosphonato, $-P(=O)(O^-)_2$; sulfonato, $-S(=O)_2(O^-)$.

Contd. to next page

one	Suffix specifying the substituent =O for two hydrogen atoms on the same atom, e.g., phosphanone, HP=O.
	Suffixes for =S, thione ; =Se, selenone ; =Te, tellurone.
onic	Ending in parent names of acids of type $HXO(OH)_2$, X = N, P, As, Sb; type $HXO_2(OH)$, X = S, Se, Te; e.g., stibonic acid, sulfonic acid etc. Also, boronic acid $HB(OH)_2$.
ono	Ending of prefixes when 'onic'–acids act as substituent groups by removal of an atom of hydrogen e.g. $-P(=O)(OH)_2$: phosphono.
	But $-S(=O)_2OH$ is sulfo, not sulfono.
onous	Ending in parent names of acids of the type $HX(OH)_2$, X = N, P, As, Sb; e.g., $HSb(OH)_2$, stibonous acid.
	Ending in parent names of acids like dithionous, trithionous, etc.
onoyl	Ending of prefix for substituent groups obtained by removal of all hydroxy groups from 'onic' acids e.g. phosphonoyl, selenoyl.
onyl	Ending of prefixes for substituent groups > $X(=O)_2$, X = S, Se, Te; e.g., sulfonyl, selenonyl, etc.
oryl	Ending of prefixes for substituent groups formed when all hydroxy groups are removed from 'oric' acids, e.g., phosphoryl, ⩾P(O).
ous	Ending in parent names of some acids.
	Ending of some element names to indicate lower oxidation states like ferrous, cuprous, etc. (no longer acceptable).
y	End modifiers for prefixes of some substituent groups, e.g., carboxy (–COOH), hydroxy (–OH), oxy (–O–).
	Terminal vowel in prefixes to specify atoms in chains and rings for inorganic chain and ring compounds.
yl	Suffix used to indicate removal of one or more hydrogen atoms from a parent hydride resulting in radicals or substituent groups, e.g., $H_2NNH^{\bullet}$ or H_2NNH-, hydrazinyl; $^{\bullet}HNNH^{\bullet}$ or –HNNH–, hydrazine –1, 2– diyl.

Prefixes are placed in the beginning of systematic names to provide a closer description of the compound. Thus, prefixes may be used to simply state the substituents in substitutive nomenclature, e.g., chlorosilane, or with multiplicative prefixes (e.g. dichlorosilane) or with numerals to specify points of substitution, e.g., 1,2,3-trichloropropane.

In additive nomenclature, prefixes may simply indicate the ligands e.g., aquacobalt or with multiplicative prefixes, indicate the number of ligands or constituents e.g., hexaamminecobalt.

Prefixes are also used to convey structural information on a compound. We shall encounter many such prefixes (like *nido*, *closo*, etc.) in boron compounds. *cis* and *trans* are also very familiar structural prefixes. Greek alphabets kappa(κ), lambda (λ, Λ), mu (μ), etc. are also used frequently.

Multiplicative prefixes like mono, bi, di, tri, etc. are used to indicate the number of identical chemical entities. With composite ligand names or when necessary to avoid ambiguity, the alternative prefixes bis, tris, etc. are used (Table 2.2).

Table 2.2

Multiplicative prefixes recommended by IUPAC

1. mono[1]	8. octa, octakis	21. henicosa	50. pentaconta
2. di[2], bis[3]	9. nona, nonakis	22. docosa	100. hecta
3. tri, tris	10. deca, decakis	23. tricosa	200. dicta
4. tetra, tetrakis	11. undeca	30. triaconta	500. pentacta
5. penta, pentakis	12. dodeca	31. hentriaconta	1000. kilia
6. hexa, hexakis	13. trideca	32. dotriaconta	2000. dikilia
7. hepta, heptakis	20. icosa	40. tetraconta	

Note:

1. Monoxide is an acceptable alternative for monooxide. In other cases, no elisions are made: e.g., tetraaqua (*not tetraqua*), hexaoxide (*not* hexoxide).
2. For ligands using two donor atoms, the term 'bidentate' (vs. 'didentate') is recommended as per prevailing practice.
3. Alternative prefixes bis, tris, etc. are shown for 2–10 but they continue throughout. These are used with composite ligand names or in order to avoid ambiguity:

   Na_2S_3, disodium (trisulfide); the anion is the trisulfide ion S_3^{2-}.

   Fe_2S_3, di-iron tris(sulfide) or iron(III) sulfide; the anion is S^{2-}.

   $Ba_3(PO_4)_2$, tribarium bis(phosphate); $Ca_2P_2O_7$, calcium diphosphate.

   $[Co(en)_3]Cl_3$, tris(ethane-1,2-diamine)cobalt(III) chloride
4. In building composite multiplicative prefixes, units are cited first, followed by tens, hundreds and so on; e.g., 135 is written pentatriacontahecta (ending with kis, when appropriate).

2.2.2 Enclosing Marks

Enclosing marks are used extensively in formulae and names to isolate an entity from the rest. Square brackets ([ ]) and parentheses (( )) are the most frequently used enclosing marks. We shall come across numerous usages of these marks throughout this chapter and the entire book. Only a few typical examples of common use of these enclosing marks are given below.

Parentheses are extensively used in formulae to enclose formulae for groups of atoms to avoid ambiguity or when the group is multiplied. With common ions like nitrate, sulfate, etc., parentheses are recommended though not mandatory. Abbreviation of ligand names in formulae are also enclosed by parenthesis.

In formulae, square brackets are used mainly to enclose an entire formula or a whole coordination entity, neutral or charged. Parentheses and braces ({ }) are then used alternately.

Examples

1. $Fe(PO_4)$ or $FePO_4$; $Ca_3(PO_4)_2$
2. $[PH_2(O)(OH)]$(i.e., H_3PO_2)
3. $[Co(NH_3)_6]Cl_3$; $[Co(en)_3]^{3+}$
4. $[\{Cr(NH_3)_5\}_2(\mu\text{-}OH)]^{5+}$

In solid-state chemistry, square brackets are used to indicate an atom or a group of atoms in an *octahedral* site while parentheses are used for *tetrahedral* site:

$$(Mg)[Cr_2]O_4.$$

In writing names, parentheses are used to enclose substituent groups or ligand names to eliminate ambiguity or when used with prefixes bis, tris, etc.

$[Hg(CHCl_2)Ph]$	(dichloromethyl)(phenyl)mercury
$[CuCl_2(NH_2Me)_2]$	dichloridobis(methylamine)copper(II)
$[Co(en)_3]Cl_3$	tris(ethane-1,2-diamine)cobalt(III) trichloride

The oxidation state of the metal or the charge number of an ionic coordination species is also enclosed in parentheses.

2.2.3 Numerals

Both Arabic and Roman numerals are extensively used in nomenclature—in formulae as well as names.

Arabic numerals

(a) Arabic numerals are principally used in formulae as right subscripts to indicate the number of each kind of atom or ion. Unity is not indicated, for example, H_2O ; $(NH_4)_2SO_4$.

(b) As right superscript in formulae, the Arabic numerals indicate the charge number of a species; unity is excluded : Cl^-, S^{2-}, Fe^{3+}

(c) Composition of addition compounds and non-stoichiometric compounds are also expressed by Arabic numerals written on the same line as the symbols. Unity is again omitted and each component is separated by a dot (·), for example,

$Na_2CO_3 \cdot 10H_2O$; $8H_2S \cdot 46H_2O$; $AlCl_3 \cdot 4EtOH$.

(d) We are also familiar with the use of Arabic numerals in several other cases, for example, to indicate the mass number and atomic number of nuclides (e.g., $^{16}_{8}O$), hapticity of ligands (e.g., η^5-C_5H_5) or bridging multiplicity of ligands (e.g., μ_3-CO).

(e) In names also, Arabic numerals are extensively used for various descriptive purposes.

1. As locants for numbering skeletal atoms in parent hydrides to indicate substitution, unsaturation, etc. or in replacement/additive nomenclature.
2. To indicate the number of hydrogen atoms in the parent boron compound:

   $B_{10}H_{14}$: decaborane(14) ; B_2H_6 : diborane(6).
3. To indicate the charge number of a species; $[AlCl_4]^-$ tetrachloridoaluminate (1–). The number '1' is included to avoid ambiguity with sign of optical rotation.
4. As subscript to μ and superscript to η, κ, etc. to indicate the bridging mode or hapticity of a ligand (μ, η) or to indicate the number of donor atoms of a particular type bound to a central atom. Examples to such applications will appear in due course.
5. As stoichiometric descriptor in names of formal addition compounds:

$3CdSO_4 \cdot 8H_2O$	cadmium sulfate-water (3/8)
$Al_2(SO_4)_3 \cdot K_2SO_4 \cdot 24H_2O$	aluminium sulfate-potassium sulfate-water (1/1/24)

Roman numerals

Roman numerals are used mainly to indicate the formal oxidation states of elements—as a right superscript in formulae and beside the name of the qualifying elements in parentheses.

$Fe^{II}Fe^{III}_2O_4$ iron(II)diiron(III) oxide

$[Cr^{VI}O_4]^{2-}$ tetraoxidochromate(VI)

2.2.4 Italic Letters

Italic letters are used in names to highlight certain features—some major uses are mentioned below.

(a) Geometrical and structural prefixes like *cis, trans, nido*, etc. are written in italics.

(b) In names of polynuclear compounds, the symbols of central atoms in the bonding descriptor are written in italics within parentheses:

$[Mn_2(CO)_{10}]$ bis(pentacarbonylmangenese)(*Mn—Mn*)

(c) In additive names a bond between two central atoms is indicated by including the atoms in parentheses and joined by an em-dash (at the end of the name). Thus, dithionite (IUPAC approved non-systematic name), $S_2O_4^{2-} = O_2SSO_2^{2-}$: bis(dioxidosulfate)(*S—S*)(2–).

(d) The structural type of mixed oxides and hydroxides is often mentioned in italics within parentheses after the formula

$MgTiO_3$ (*ilmenite type*)

(e) Italic letters are also used to designate the symbols of the coordinating atoms of a ligand, with κ (kappa).

For example, N-bonded NCS: thiocyanato-κ*N* ; S-bonded NCS: thiocyanato-κ*S*.

(f) Italic lower case letters are used to represent an undefined number, e.g., $(NaPO_3)_n$, $-\!\!(S)\!\!-_n$.

(g) Italic letters are also used in crystal system symbols and polyhedral symbols.

2.2.5 Order of Citation

In writing the name or formula of a compound, the successive order of placing the symbols or names of elements is maintained according to some general principles. These principles also apply to groups of atoms and ligands.

In general, cations precede anions in both formulae and names. Within each category, alphabetical order is usually followed. But in compositional names, priority is determined from an accepted norm based mainly on electropositive character. In coordination compounds, names of ligands are cited alphabetically before the name of the metal.

Ordering based on the periodic table

In compositional names and formulae of binary compounds, elements are arranged according to the priority sequence shown in Table 2.3. Actually, the table shows the *reverse order of priority* and starts with the most electronegative element fluorine and moves vertically downward along a group and then again begins at the top of the group immediately left. Hydrogen occupies a special position and the noble gases are placed to the left of alkali metals. The sequence thus moves from the electronegative to the electropositive elements. For purposes of compositional names and formulae, *the element appearing last in the sequence is placed first*.

Table 2.3

IUPAC recommended sequence of elements

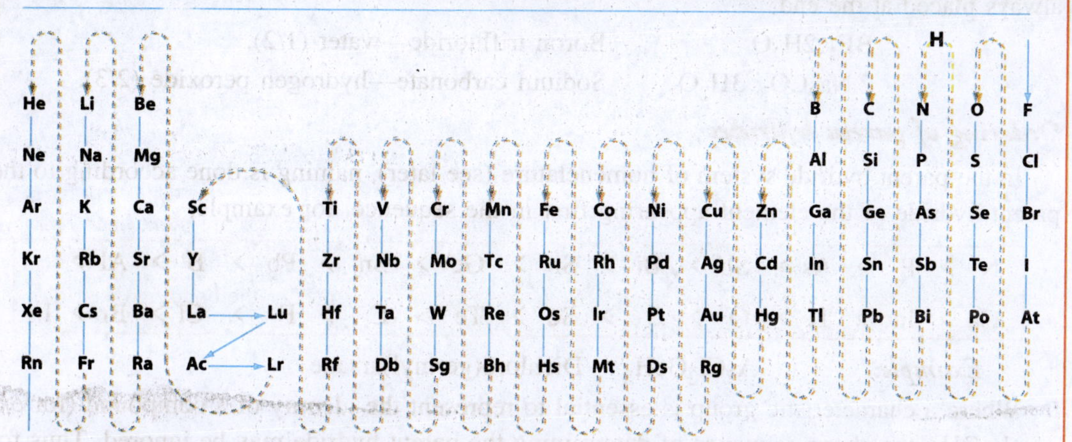

S_2Cl_2	disulfur dichloride
O_2Cl	dioxygen chloride

The above sequence of elements is also followed in a few other cases, e.g., additive names of polynuclear compounds, chains, rings etc.

Alphabetical order

In formulae of salts and double salts, cations are placed before anions. Within each category, element symbols are cited alphabetically. Exceptions arise when specific structural informations are desired to be conveyed, for example,

$CaTiO_3$ and $SrFeO_3$, both *perovskite* type.

In coordination compounds also, ligands are placed in alphabetical order of their formulae or abbreviations.

More examples of the application of these rules are given later.

In names, alphabetical order of the names of elements is followed within each category of electropositive elements and electronegative elements. In compositional names, elements of the first category (electropositive) are placed first. *The names (in alphabetical order) may now differ from the order of appearance of the elements in the formula* (where element sequence is based on the periodic table):

$KMgF_3$	Magnesium potassium fluoride
ZnI(OH)	Zinc hydroxide iodide
$SrFeO_3$	Iron strontium oxide

In additive names, ligands are mentioned in alphabetical order regardless of the number of each ligand. Substituent groups in substitutive nomenclature are also named in alphabetical order.

$K[AuS(S_2)]$	Potassium (disulfido)sulfidoaurate(1–)
$[CoCl_2(NH_3)_4]^+$	Tetraamminedichloridocobalt(1+)

In case of formal addition compounds, the names of individual components are arranged first according to increasing number of the components and then alphabetically. '*Water*' is always placed at the end.

$BF_3.2H_2O$	Boron trifluoride—water (1/2).
$2\ Na_2CO_3.3H_2O_2$	Sodium carbonate—hydrogen peroxide (2/3).

Ordering of parent hydrides

In the parent hydride system of nomenclature (see later), naming is done according to the parent hydride of the element appearing first in the sequence, for example,

N > P > As > Sb > Bi > Si > Ge > Sn > Pb > B > Al > Ga > In > Tl > O > S > Se > Te > C > F > Cl > Br > I.

Example: $AsCl_2GeH_3$ Dichloro(germyl)arsane

Where a characteristic group is essential to represent the identity of a compound (for example OH), the above sequence of determining the parent hydride may be ignored. Thus for HTeOH, the OH group is maintained as a suffix and the name is based on tellane (TeH_2): tellanol.

More ordering principles and details of grammar will be appended in due course.

2.3 NAMES AND SYMBOLS OF ELEMENTS

The names of elements with atomic numbers 1–111 appear in the periodic table (chapter 5) with respective symbols. Element 112 and any further newly discovered element(s) may be represented by names derived from their atomic number and corresponding three-letter symbols according to following conventions.

0 = nil	1 = un	2 = bi	3 = tri	4 = quad
5 = pent	6 = hex	7 = sept	8 = oct	9 = enn

Element 112 = Ununbium, symbol Uub.

The element caesium is alternatively spelt as cesium.

In chemical formulae, each element is represented by its symbol appearing in the periodic table. Isotopes are represented by their mass numbers (and atomic numbers) with the same symbol, for example, $^{18}_{8}O$ or ^{18}O for oxygen-18 isotope. For hydrogen isotopes of mass number two and three, the alternative symbols D and T respectively are also used, but ^2H and ^3H, are preferred because D and T may disturb the alphabetical ordering of elements in formula. The combination of a muon and an electron is sometimes treated as a light isotope of hydrogen in the name muonium, symbol Mu.

Ions derived from an element A are represented as A^{n+} or A^{n-} (not A^{+n} or A^{-n}).

Proton vs. Hydron

The cation derived from hydrogen is commonly known as proton which is used in many senses, for example, for the isotopically pure $^1H^+$ ion derived from ^1_1H and also for the naturally occurring isotopic composition (containing some $^2H^+$ and $^3H^+$). To avoid confusion, it is now recommended that the undifferentiated isotope mixture be designated by the general name hydron derived from hydrogen. The name proton is reserved for $^1H^+$. The cations $^2H^+$, $^3H^+$ and Mu^+ are also known as deuteron, triton and muon, respectively.

Atoms, molecules and allotropic modifications of an element are named by the name of the element with appropriate prefixes where necessary. When an element *does not normally occur* in the monoatomic state, the prefix 'mono' is used:

		Systematic name	Alternative name
1.	Ar	argon	
2.	H	monohydrogen	
3.	H_2	dihydrogen	hydrogen
4.	O_2	dioxygen	oxygen
5.	O_3	trioxygen	ozone
6.	P_4	tetraphosphorus	white phosphorus
7.	S_6	hexasulfur	ε-sulfur
8.	S_8	*cyclo*-octasulfur	α-sulfur, β-sulfur, γ-sulfur
9.	S_n	polysulfur	μ-sulfur, plastic sulfur
10.	C_{60}	hexacontacarbon	[60] fullerene

Crystalline allotropic modifications of an element may be assigned more sophisticated systematic names with structural parameters. For example, diamond is represented as carbon (*cF*8) as it has a cubic (*c*) all-face-centered (*F*) lattice containing eight (8) atoms of carbon per unit cell. *cF* is the *Pearson symbol* of the crystalline variety.

In the periodic classification of elements, certain collective names are IUPAC approved: alkali metals, alkaline earth metals, pnictogens (or pnicogens; N, P, As, ···), chalcogens (O, S, ···), halogens, noble gases, *lanthanoids* and *actinoids*. *The last two names are preferred over lanthanides and actinides since the suffix -ide refers to a negative ion.*

Roots derived from the names of elements are frequently used in nomenclature. While most of the roots are derived from the common accepted name of the element, some roots are derived from ancient names of the elements. For example, the root '*az*' for nitrogen derives from the name '*azote*' (Lavoisier; *a* = no; *zoe* = life). Similarly, the Greek name *theion* provides the root '*thi*' for sulfur.

2.4 FORMULAE

A formula is used for simple, brief and quick representation of a compound. For general acceptability and unambiguity, certain norms have been recommended by the IUPAC.

The empirical formula of a compound is primarily meant to express the simplest mole ratio of different elements present in alphabetical order with appropriate numerical subscripts. ('1' is not mentioned). Carbon-containing compounds are exceptions to this norm where C and H are usually cited first and second, respectively; for example, C_2H_5Br, $C_{10}H_{10}ClFe$.

The molecular formula represents the actual composition of a compound consisting of discrete molecules and is more informative than empirical formula. The order of citation of the elements in such formula is mostly based on electronegativity criterion explained earlier. This is particularly useful for binary compounds.

Empirical formula	Molecular formula	Empirical formula	Molecular formula
S	S_n, S_8	H_3N	NH_3
ClS	S_2Cl_2	ClO_2	O_2Cl
ClHg	Hg_2Cl_2	HO^-	HO^-
F_6S	SF_6		(conventionally OH^-)

However, for most species containing more than two atoms, formal treatment as coordination compound is found convenient. The order of citation of central atoms is again based upon electronegativity. Ligands or their abbreviations are now arranged alphabetically. Where possible the ligand formula is so written that the donor atom is closest to the symbol of the central atom.

$PBrCl_2$ $[BH_4]^-$ $[Al(OH)(OH_2)_5]^{2+}$

A moiety of atoms functioning as an entity is not split and written as such, e.g.,

$POBr_3$ (not PBr_3O) UO_2Br_2 (not UBr_2O_2)

In the parent hydride system, the remaining unsubstituted hydrogen atoms of the parent hydride are cited first, e.g.,

SiH_2BrCl GeH_2F_2 B_2H_5Cl

For carbaboranes, boron atoms are cited first; carbon atoms replacing skeletal boron atoms are placed next regardless of what other elements are present, e.g.,

$B_3C_2H_4Br$

Formulae of inorganic oxoacids are conventionally written according to the traditional order of citation : replaceable hydrogen atoms (i.e., those bound to O), central atom, non-replaceable hydrogen atoms, finally oxygen. These acids may also be represented in the format of coordination compounds.

Traditional formulae	'Coordination' formulae
HNO_3	$[NO_2(OH)]$
H_2PHO_3 (H_3PO_3)	$[PHO(OH)_2]$
$(HBO_2)_n$	$\text{—}(B(OH)O)\text{—}_n$

For chain compounds containing three or more different elements, the atom to atom connectivity of the elements is generally represented in the formula, though it may violate other ordering principles (alphabetical order or order based on electronegativity).

NCS^- or SCN^- ; BrSCN ; HOCN ; HNCO

A compound may be treated as a generalized salt when its formula is not naturally assigned in the categories just described. As mentioned earlier (sec. 2.2: Grammar), all electropositive constituents precede all electronegative constituents (Table 2.3). Within each category, alphabetical order is followed:

$$MgCl(OH) \quad Na[HPHO_3] \quad FeO(OH) \quad Na(UO_2)_3[Zn(H_2O)_6](O_2CH_3)_9$$

Deviation from alphabetical order within the same category is allowed if it is intended to emphasize similarities between compounds:

$$CaTiO_3 \quad \text{and} \quad ZnTiO_3.$$

Formulae of addition compounds and compounds which can be regarded as such (including clathrates and multiple salts), are written *in order of increasing mole ratio* of the components with the mole ratio expressed by arabic numerals (except 1) before each component. The components are separated by a dot (·). ***Components with equal numbers are arranged alphabetically***. Water, when present, is conventionally shown at last.

$Al_2(SO_4)_3.K_2SO_4.24H_2O$ **(not $K_2SO_4.Al_2(SO_4)_3.24H_2O$).** $C_6H_6.NH_3.Ni(CN)_2$

For isotopically modified compounds, the modified nuclide(s) is represented in their formula. Such compounds may be of two types—*isotopically substituted compounds* and *isotopically labelled compounds.*

An isotopically substitued compound has all the molecules substitued for the indicated nuclide(s) at each designated position. The mass numbers of substituted nuclides are shown by left superscript.

$$H^3HO \quad H^{36}Cl \quad {}^{235}UF_6$$

An isotopically labelled compound is essentially a mixture of one or more analogous isotopically substituted compound with an isotopically unmodified, that is, "normal" compound. When a unique isotopically substituted compound is added formally to the analogous 'normal' compound (i.e., isotopically unmodified), one gets a *specifically labelled compound.* This is indicated by enclosing the relevant nuclidic symbol(s) (with multiplicative subscripts, if any) in square brackets.

$$H[^{36}Cl] \quad [^{15}N]H_2[^2H] \quad [^{32}P]O[^{18}F_3]$$

A mixture of specifically labelled compounds are classed as *selectively labelled compounds.* This is shown in formula by placing the symbol of the relevant nuclide(s) as a prefix in square bracket. No multiplicative subscript is used.

$$[^2H]PH_3 \quad [^{36}Cl]SOCl_2$$

Besides empirical and molecular formulae, a structural formula gives partial or complete information about the connectivity of atoms and their spatial arrangement. A line formula showing the sequence of atomic symbols is the simplest representation, meaningful only to one who knows, for example,

HOCN HNCO HOOH

A more complex structure may be represented in line formulae using appropriate enclosing marks:

$$SiH_3[SiH_2]_8SiH_3 \quad K[Os(N)_3] \quad -\!\!(S)\!\!-_n$$

A formula or part of a formula which represents a molecular entity may be placed in enclosing marks. Except in case of repeating polymeric units, square brackets are used to enclose the entire formula. Thus one may represent sodium chloride as

NaCl [NaCl]

NaCl stands for the solid sodium chloride with composition NaCl. [NaCl] represents the molecular compound consisting of one atom of sodium and one atom of chlorine; the additive name 'chloridosodium' may be used specifically for the molecular compound when it is looked upon as a coordination entity.

Several modifiers to all types of formulae are used to provide specific information like oxidation state (Roman numerals I, II, etc.), formal ionic charge, free radicals, optical rotations, structural descriptors (e.g., *cis/trans*) and excited states (* in right superscript).

2.5 COMPOSITIONAL NOMENCLATURE

In its simplest form, a compositional name is a stoichiometric name showing proportions of constituent elements by multiplicative prefixes. Oxidation numbers or charge numbers may be indicated.

(i) ***An element*** is named as such with multiplicative prefix (e.g., S_8 : octasulfur ; O_3 : trioxygen).

(ii) ***Cations*** derived from an element, monoatomic or polyatomic, are also named by the name of the element unchanged. Only hydrogen forms an exception.

Multiplicative prefixes (except mono) and charge numbers are used for complete description.

H^+	hydrogen(1+), hydron	O_2^+	dioxygen(1+)
$^1H^+$	protium(1+), proton	$^3H^+$	tritium(1+), triton
$^2H^+$	deuterium(1+), deuteron	Hg_2^{2+}	dimercury(2+)

(iii) ***Anions*** derived from an element (monoatomic or homopolyatomic) end their names with 'ide', followed by charge number in parentheses. When there is no ambiguity, the latter may be dropped. In the following examples, acceptable alternative names are shown in green.

Cl^-	chloride	O_2^-	dioxide(1–) **(superoxide)**
O_2^{2-}	dioxide(2–) **(peroxide)**	N_3^-	trinitride(1–) **(azide)**
S_2^{2-}	disulfide(2–)	Pb_9^{4-}	nonaplumbide(4–)

*** Note:** To avoid ambiguity, oxidation numbers are no longer recommended in names of homopolyatomic ions. So we write dimercury(2+) and not dimercury(I); dioxide(2–) instead of dioxide (–I).

(iv) ***In naming a binary compound***, one element is taken as the electropositive constituent according to the priority criterion explained before (sec. 2.2.5 Grammar). The name starts with the unmodified name of the element. The name of the other element is modified by the suffix '–ide' and placed separately from the name of the first element. Multiplicative prefixes apply as usual. When there is no ambiguity regarding the stoichiometry of the compound, prefixes may be dropped.

HCl	hydrogen chloride	N_2O_4	dinitrogen tetraoxide
OCl_2	oxygen dichloride	O_2Cl	dioxygen chloride
Ca_3P_2	tricalcium diphosphide or, calcium phosphide		

(v) For compounds or species containing more than two elements, *generalized stoichiometric names* are assembled from the names of the constituents. These constituents may be described by any system of nomenclature like stoichiometric, substitutive or additive. Since such constituents may be monoatomic or polyatomic (homo- or hetero-), we first discuss the naming of heteropolyatomic ions. Naming of homopolyatomic ions has been mentioned in (ii) and (iii).

Heteropolyatomic cations are commonly named substitutively or additively. In substitutive names the suffix itself indicates the charge and no separate charge number is required. In the following examples, acceptable non-systematic names are shown in green.

BH_2^+	:	boranylium (subst)	H_3O^+	:	oxidanium (subst) oxonium (not hydronium)
NH_4^+	:	azanium (subst) ammonium	PH_4^+	:	phosphanium (subst) (not phosphonium)
NMe_4^+	:	tetramethylazanium (subst) tetramethylammonium	PH_2^+	:	phosphanylium
$N_2H_5^+$	:	diazanium hydrazinium	SbF_4^+	:	tetrafluorostibanium (subst) or tetrafluoridoantimony(1+) or tetrafluoridoantimony(V) (both additive)
$N_2H_6^{2+}$	:	diazanediium hydrazinediium			

Heteropolyatomic anions are also named both substitutively and additively. Some examples are given with acceptable non-systematic names in green.

NH_2^-	:	azanide (subst); dihydridonitrate(1–) (additive); amide
$HO^-(OH^-)$	:	oxidanide (subst); hydridooxygenate(1–) (additive); hydroxide
HS^-	:	sulfanide (subst); hydrogen(sulfide)(1–) (additive)
SO_3^{2-}	:	trioxidosulfate(2–) (additive); sulfite
OCl^-	:	chloridooxygenate(1–) (additive); hypochlorite
ClO_3^-	:	trioxidochlorate(1–) (additive); chlorate
HCO_3^-	:	hydrogen(trioxidocarbonate)(1–) (additive); hydrogencarbonate
$H_2PO_4^-$	:	dihydrogen(tetraoxidophosphate)(1–) (additive); dihydrogenphosphate

Generalized stoichiometric names are assembled from the names of the constituents with the electropositive ones (including cations) preceding the electronegative ones (and anions). *Alphabetical order is maintained within each category except hydrogen.* If classified as an electropositive constituent, it is always cited after all other electropositive constituents. The name of each constituent is separated by space. Charge number and oxidation number may be indicated in deserving cases.

PBrClI	phosphorus bromide chloride iodide
PCl_3O	phosphorus trichloride oxide
$NaNH_4[HPO_4]$	ammonium sodium hydrogenphosphate
$KMgCl_3$	magnesium potassium trichloride

$(UO_2)_2SO_4$	bis(dioxidouranium) tetraoxidosulfate
or	dioxidouranium(V) sulfate
$[Co(NH_3)_6]Cl(SO_4)$	hexaamminecobalt(3+) chloride sulfate
or	hexaamminecobalt(III) chloride sulfate
$Fe_4[Fe(CN)_6]_3$	iron(III) hexacyanidoferrate(II)
or	iron(3+) hexacyanidoferrate(4–)

(vi) ***In naming addition compounds*** including clathrates and multiple salts, the names of individual components are first ascertained by any suitable system. The names are then arranged according to increasing number of moles of the components and then alphabetically. The names of the components are separated by em-dash (—). Water is named at the last as an exception. The mole proportions of various components are shown in parentheses by arabic numerals.

$Al_2(SO_4)_3.K_2SO_4.24H_2O$ aluminium sulfate—potassium sulfate—water (1/1/24)

The class name 'hydrate' is acceptable for addition compounds containing water as a component but 'deuterate' (or 'tritiate') is not acceptable for addition compound of ^2H_2O (or ^3H_2O).

2.6 SUBSTITUTIVE NOMENCLATURE

This system is based on the names of parent hydrides defined for a particular valency of each element in periodic groups 13–17 (Table 2.4). Replacement of one or more hydrogen from these hydrides is indicated by suffixes or prefixes (see examples).

Table 2.4

Parent hydrides in substitutive nomenclature

BH_3 borane	CH_4 methane[1]	NH_3 azane[2]	H_2O oxidane[3, 5]	HF fluorane[4]
AlH_3 alumane[5]	SiH_4 silane	PH_3 phosphane[6]	H_2S sulfane[5]	HCl chlorane[4]
GaH_3 gallane	GeH_4 germane	AsH_3 arsane[6]	H_2Se selane[5]	HBr bromane
InH_3 indigane[7]	SnH_4 stannane	SbH_3 stibane[6]	H_2Te tellane[5]	HI iodane[4]
TlH_3 thallane	PbH_4 plumbane	BiH_3 bismuthane[5]	H_2Po polane[5]	HAt astatane[4]

Note:

1. Because of the universal popularity of the name 'methane', the systematic analogue 'carbane' is not recommended.
2. 'Azane' is derived from the name 'azote' for nitrogen. 'Ammonia' is the acceptable common name used widely. 'Azane' is used mainly in substitutive nomenclature.

3. Intended for use in naming derivatives of water. The name 'oxane' presents other problems (see note 5).
4. The compositional names hydrogen fluoride, hydrogen chloride, etc. cannot be used as parent names in nomenclature. 'Fluorane', 'chlorane', etc. are the basis of forming substitutive names.
5. The names aluminane, bismane, oxane, thiane, selenane, etc. are the names of saturated six-membered heteromonocycles (Hantzsch-Widman system); hence, these cannot be given to the hydrides. 'Alane' is also not acceptable because $-AlH_2$ would then be 'alanyl'; the same term is associated with the acyl group derived from the amino acid alanine.
6. The names 'phosphine', 'arsine' and 'stibine' are no longer acceptable.
7. The name 'indane' is the name for 2, 3-dihydroindene, and hence unacceptable. 'Indiane' also presents problems in naming unsaturated derivatives. The accepted name 'indigane' is derived from indigo—the flame colour of indium.

Homopolynuclear acyclic parent hydrides in which all atoms have their standard bonding number are named using multiplicative prefixes:

HOOH	dioxidane (*hydrogen peroxide*);	H_2PPH_2	diphosphane
H_2NNH_2	diazane (*hydrazine*);	HSeSeSeH	triselane

If, for the purpose of substitutional nomenclature, an atom has to be assigned a non-standard bonding number (other than that shown in Table 2.4) this is shown by λ and a superscript:

PH_5 λ^5-phosphane $\qquad$ SH_6 λ^6-sulfane

Atoms in a homonuclear acyclic hydride showing non-standard valence state are shown by locants as prefixes. Lower value of locant(s) and lower bonding number(s) are preferred:

$$\overset{1}{H_5S}\overset{2}{S}\overset{3}{S}H_4\overset{4}{S}H$$

$1\lambda^6, 3\lambda^6$-tetrasulfane

S-atoms in 1 and 3 show bonding number 6. Others have standard bonding number 2. Numbering from other end would give locant numbers 2, 4; hence unacceptable.

There are more rules but all of them cannot be discussed. Boron hydrides form a separate class and their nomenclature has been discussed in their chemistry.

We have already found several examples of substitutive nomenclature in the previous section. A brief overview of the main principles are given here with some more examples.

(i) *Substitutive names of derivatives of parent hydrides*

Replacement of hydrogen atoms in parent hydrides is indicated by appropriate prefixes like 'hydroxy', 'nitro', etc. Halogen atoms and some other substituents are always cited as prefix. Multiplicative prefixes are inserted as necessary. Except for 'hydro', prefixes are arranged in alphabetical order.

When there are more than one substituent, the highest ranking substituent is shown in the suffix and the remaining substituents as prefixes.

If there is more than one choice of parent hydride, priority is determined in the sequence:

N > P > As > Sb > Bi > Si > Ge > Sn > Pb > B > Al > Ga > In > Tl > O > S > Se > Te > C > F > Cl > Br > I.

***Example*:** (Some additive names are given for comparison)

SiH_3OH	silanol
$Si(OH)_4$	silanetetrol; tetrahydroxidosilicon (*additive*)
SF_6	hexafluoro-λ^6-sulfane; hexafluoridosulfur (*additive*)
SiH_3NH_2	silanamine; amidotrihydridosilicon (*additive*)
$PbEt_4$	tetraethylplumbane; tetraethyllead (*additive*)
$MePHSiH_3$	methyl(silyl)phosphane *not* (methylphosphanyl)silane *or otherwise* since P > Si > C.

Substitutive names of cations and anions derived from parent hydrides have been mentioned previously (compositional nomenclature). Some more examples are given:

PH_2^+	phosphanylium
SiH_3^+	silylium
BH_2^+	boranylium
$Si_2H_5^+$	disilanylium
$[NF_4]^+$	tetrafluoroazanium, tetrafluoroammonium
$[PCl_4]^+$	tetrachlorophosphanium
$[SEtMePh]^+$	ethyl(methyl)phenylsulfanium

$[BH_4]^-$	boranuide (from borane), tetrahydridoborate(1-) (*additive*)
NH^{2-}	azanediide or, imide
SiH_3^-	silanide
$^-HNNH^-$	diazane -1,2- diide or, hydrazine -1,2- diide
H_2NN^{2-}	diazane -1,2- diide or, hydrazine -1,1- diide
$SnCl_3^-$	trichlorostannanide (*subst*) or, trichloridostannate(1-) (*additive*)
$[PF_6]^-$	hexafluoro –λ^5– phosphanuide (*subst*) or, hexafluoridophosphate(1–) (*additive*)

2.7 ADDITIVE NOMENCLATURE

In additive names, a compound is regarded as being composed of a central atom (or atoms) surrounded by other groups of atoms or ligands. Examples of additive names for a large number of compounds appear throughout previous sections. Only some major principles are summarized here.

Any metal atoms(s) present in the compound is chosen as the central atom. In general, an element which is most electropositive in the sequence shown in Table 2.3 is chosen as the central atom. With more than one central atom, a compound is considered dinuclear or polynuclear. The name is constructed by placing the ligand name(s) in alphabetical order before the name of the central atom.

Names of neutral and cationic ligands are used unchanged except in a few special cases like water ⟶ aqua ; ammonia ⟶ ammine ; carbon monoxide bound through carbon ⟶ carbonyl ; nitrogen monoxide bound through nitrogen ⟶ nitrosyl.

Some ligands are conventionally regarded as neutral, e.g., amines, phosphanes and ligands derived from hydrocarbons after removal of a hydrogen atom (methyl, benzyl etc.).

In mentioning anionic ligands in prefixes, their endings are modified as

ide ⟶ ido	(hydroxido, chlorido, fluorido etc. Formerly these were placed as hydroxo, chloro etc.);
ate ⟶ ato	(sulfato, carbonato etc.);
ite ⟶ ito	(e.g., nitrito);

For cationic or neutral species, the overall name ends with the name of the metal unchanged. In cations, the charge number is shown in parentheses. Oxidation numbers may also be indicated in Greek numeral.

In anionic species, the name of the central element at the end is modified by 'ate'. In the overall name, cations precede anions as usual.

Several examples already appear in previous sections. A few more examples are given here.

Table 2.5

Names of some common compounds

[**Note**: Names, where given, are in the order (i) stoichiometric name (in black) (ii) substitutive name (in blue) (iii) additive name (in orange) (iv) acceptable nonsystematic name (in green).

1. $AlCl_3$, $[AlCl_3]$	(i) aluminium trichloride (ii) trichloroalumane (iii) trichloridoaluminium
2. $[AlCl_4]^-$	(ii) tetrachloroalumanuide (iii) tetrachloridoaluminate(1–)
3. Al_2Cl_6 $[Cl_2Al(\mu\text{-}Cl)_2AlCl_2]$	(iii) di-μ-chlorido-bis(dichloridoaluminium)
4. $AsH_3 \cdot [AsH_3]$	(i) arsenic trihydride (ii) arsane (iii) trihydridoarsenic
5. AsH_4^+	(ii) arsanium (iii) tetrahydridoarsenic(1+)
6. AsO_4^{3-}	(iii) tetraoxidoarsenate(3–) (iv) arsenate
7. BH_4^-	(ii) boranuide (iii) tetrahydridoborate(1–)
8. $H_3BO_3 = [B(OH)_3]$	(iii) trihydroxidoboron (iv) boric acid
9. $H_2BO_3^- = [BO(OH)_2]^-$	(iii) dihydroxidooxidoborate(1–) (iv) dihydrogenborate
10. $[BO_3]^{3-}$	(iii) trioxidoborate(3–) (iv) borate
11. $(BO_2^-)_n$ = $\text{-}(OBO)\text{-}_n^{n-}$	(iii) catena-poly[(oxidoborate-μ-oxido)(1–) (iv) metaborate
12. $H_2CO_3 = [CO(OH)_2]$	(iii) dihydroxidooxidocarbon (iv) carbonic acid
13. $HCO_3^- = [CO_2(OH)]^-$	(iii) hydroxidodioxidocarbonate(1–) (iv) hydrogencarbonate
14. $[CO_3]^{2-}$	(iii) trioxidocarbonate(2–) (iv) carbonate
15. $HOCN = [C(N)OH]$	(iii) hydroxidonitridocarbon (iv) cyanic acid
16. $HNCO = [C(NH)O]$	(iii) azanediidooxidocarbon, (hydridonitrato)oxidocarbon (iv) isocyanic acid
17. $Si(OH)_4$	(ii) silanetetrol (iii) tetrahydroxidosilicon (iv) silicic acid
18. $HNO_3 = [NO_2(OH)]$	(iii) hydroxidodioxidonitrogen (iv) nitric acid

Contd. to next page.

19. $[NO_3]^-$ (iii) trioxidonitrate(1–) (iv) nitrate
20. $HNO_2 = [NO(OH)]$ (iii) hydroxidooxidonitrogen (iv) nitrous acid
21. $[NO_2]^-$ (iii) dioxidonitrate(1–) (iv) nitrite
22. $H_2NHO = [NH_2OH]$ (iii) dihydridohydroxidonitrogen (iv) hydroxylamine
23. $H_3PO_4 = [PO(OH)_3]$ (iii) trihydroxidooxidophosphorus (iv) phosphoric acid
24. $H_2PO_4^- = [PO_2(OH)_2]^-$ (iii) dihydrooxidodioxidophosphate(1–) (iv) dihydrogenphosphate
25. $H_3PO_3 = [P(OH)_3]$ (iii) trihydroxidophosphorus (iv) phosphorous acid
$= [PHO(OH)_2]$ (iii) hydridodihydroxidooxidophosphorous (iv) phosphonic acid
26. $H_3PO_2 = [PH_2O(OH)]$ (iii) dihydridohydroxidooxidophosphorous (iv) phosphinic acid
27. $H_2SO_4 = [SO_2(OH)_2]$ (iii) dihydroxidodioxidosulfur (iv) sulfuric acid
28. SO_4^{2-} (iii) tetraoxidosyulfate(2–) (iv) sulfate
29. $HSO_4^- = [SO_3(OH)]^-$ (iii) hydroxidotrioxidosulfate(1–) (iv) hydrogensulfate
30. $H_2SO_3 = [SO(OH)_2]$ (iii) dihydroxidooxidosulfur (iv) sulfurous acid
31. $[SO_3]^{2-}$ (iii) trioxidosulfate(2–) (iv) sulfite
32. $H_2S_2O_7 = [(HO)(O)_2SOS(O)_2(OH)]$ (iii) μ-oxido-bis(hydroxidodioxidosulfur) (iv) disulfuric acid
33. $H_2S_2O_6 = [(HO)(O)_2SS(O)_2(OH)]$ (iii) bis(hydroxidodioxidosulfur)(*S–S*) (iv) dithionic acid
34. $H_2S_2O_4 = [(HO)(O)SS(O)(OH)]$ (iii) bis(hydroxidooxidosulfur)(*S–S*) (iv) dithionous acid
35. $H_6TeO_6 = [Te(OH)_6]$ (iii) hexahydroxidotellurium (iv) orthotelluric acid
36. $HClO_4 = [ClO_3(OH)]$ (iii) hydroxidotrioxidochlorine* (iv) perchloric acid
37. ClO_4^- (iii) tetraoxidochlorate(1–)* (iv) perchlorate
38. $HClO_3 = [ClO_2(OH)]$ (iii) hydroxidodioxidochlorine (iv) chloric acid
39. ClO_3^- (iii) trioxidochlorate(1–)* (iv) chlorate
40. $HClO_2 = [ClO(OH)]$ (iii) hydroxidooxidochlorine (iv) chlorous acid
41. $[ClO_2]^-$ (iii) dioxidochlorate(1–)* (iv) chlorite

Contd. to next page.

42. $HClO = [O(H)Cl]$ (iii) chloridohydridooxygen (iv) hypochlorous acid
43. $[OCl]^-$ (iii) chloridooxygenate(1–)* (iv) hypochlorite
44. $H_5IO_6 = [IO(OH)_5]$ (ii) pentahydroxy-λ^7-iodanone (iii) pentahydroxidooxidoiodine (iv) orthoperiodic acid
45. IF_4^+ (ii) tetrafluoro-λ^3-iodanium (iii) tetrafluoridoiodine(1+)
46. IF_4^- (ii) tetrafluoro-λ^3-iodanuide (iii) tetrafluoridoiodate(1–)
47. SCN^- (iii) nitridosulfidocarbonate(1–) (iv) thiocyanate
48. CrO_4^{2-} (iii) tetraoxidochromate(2–) (iv) chromate
49. $Cr_2O_7^{2-} = [O_3CrOCrO_3]^{2-}$ (ii) heptaoxidodichromate(2–) (iii) μ-oxidobis(trioxidochromate)(2–) (iv) dichromate
50. $CrO_5 = [CrO(O_2)_2]$ (iii) oxidodiperoxidochromium
51. $FHO = [HOF]$ (ii) fluoranol (iii) fluoridohydridooxygen
52. $HMnO_4 = [MnO_3(OH)]$ (iii) hydroxidotrioxidomanganese
53. MnO_4^- (iii) tetraoxidomanganate(1–) (iv) permanganate
54. $H_2O = [OH_2]$ (i) dihydrogen oxide (ii) oxidane
55. ^1H_2O (i) diprotium oxide (iv) ($^1H_2$) water
56. H_2O_2 (i) dihydrogen peroxide (ii) dioxidane (iii) bis(hydridooxygen)(*O–O*) (iv) hydrogen peroxide
57. KO_2 (ii) potassium dioxide(1–) (iv) potassium superoxide
58. KO_3 (ii) potassium trioxide(1–) (iv) potassium ozonide
59. K_2O (ii) dipotassium oxide
60. K_2O_2 (ii) dipotassium dioxide(2–) (iv) potassium peroxide

***Note:**

1. The element sequence in Table 2.3 places oxygen *after* all halogens in the direction of arrow. So systematic additive names should be based on oxygen as central atom. But the chlorine atom occupies the central position in the structure of the ion and, therefore, is accepted as the central atom from symmetry consideration. Hence the oxidohalate names. In OCl^- (hypochlorite), oxygen is the central atom.
2. Similar namings apply to corresponding acids of Br and I.

Naming of coordination compounds follows additive system, as shown earlier. Rules for construction of such names have also been stated. These are also summarized in the chapter on coordination compounds. The only important difference from previous system of nomenclature lies in the prefixes for anionic –ide ligands which are modified as –ido, e.g., oxido, chlorido, bromido, cyanido, etc.

$[CoCl(NH_3)_5]Cl_2$ pentaammine*chlorido*cobalt(2+) chloride (previously *chloro*)

$[AuXe_4]^{2+}$ tetra*xenonido*gold(2+)

We have also seen use of κ (kappa) to indicate the number of ligating points and of μ for bridging ligands. More examples will be found in appropriate context.

Nomenclature of organometallic compounds and boron cage compounds will be discussed in relevant sections.

ESSENTIALS

The aim of all nomenclature systems is to provide an unambiguous and easily recognizable representation of the elements and compounds.

Compositional nomenclature is based on the stoichiometry or composition of a species. The constituents are formally divided into electropositive and electronegative categories and arranged with the electropositive constituents first. Alphabetical order is maintained within each category.

MgCl(OH) : magnesium chloride hydroxide; Na_3Bi_5 : trisodium pentabismuthide

Substitutuve nomenclature is based on the concept of a parent hydride in which H atoms have been substituted by other atoms.

$SiH_3SiH_2SiHF_2$: 1,1-diflurotrisilane

Additive nomenclature is constructed by selecting a central atom and associating ligands with this central atom. The coordination compounds are best examples of this type though other compounds may also be covered by it

$CoCl_3(NH_3)_3$: triamminetrichloridocobalt(III)

$HNO_3 \equiv [NO_2(OH)]$: hydroxidodioxidonitrogen

Significant changes in the IUPAC recommendations, 2005

1. *Names of cations*

In naming cations derived from parent hydrides according to substitutive nomenclature, charge numbers in parentheses appear redundant when appropriate endings are used. For example, $N_2H_6^{2+}$ was previously named hydrazinium (2+). But the ending 'ium' itself denotes addition of hydrogen (1+) and hence implies the charge. So the cation is named hydrazinediium or diazanediium.

$N_2H_6^{2+}$ hydrazinium(2+) ⟶ *hydrazinediium, diazanediium*

2. *Names of anions*

Names of anions are given consistently following the rules stated below:

(i) Compositional names of homopolyatomic anions end in 'ide':

I_3^- triiodide(1–) O_2^{2-} dioxide(2–)

(ii) Anions formed by formal removal of hydrogen (1+) from the parent hydride also end their names in 'ide'- (parent hydride nomenclature).

$^-HNNH^-$: hydrazine –1,2–diide previously hydrazide (2–)

$MeNH^-$: methanaminide

(iii) Additive names of anions end in 'ate' : PS_4^{3-}, tetrasulfidophosphate (3–).

Additive names of radical ions are also named accordingly, e.g.,

$HSSH^{\bullet-}$ bis(hydridosulfate)(*S*–*S*) (•1–) | previously bis(hydridosulfide) (*S*–*S*) (•1–) |

3. ***Sequence of elements***

The exceptional position assigned to oxygen in the previous system has been abolished. The element sequence in Table 2.3 is strictly adhered to and *oxygen is treated as more electropositive relative to any halogen* in the construction of compositional names and corresponding formulae of binary compounds.

O_2Cl dioxygen chloride | previously, ClO_2 chlorine dioxide |

The element sequence in Table 2.3 is also adhered to in the construction of additive names of polynuclear compounds.

4. ***Names of anionic ligands***

In modifying ligand names in additive nomenclature of formal coordination entities, anion names ending in 'ide', 'ite' and 'ate' are changed respectively to 'ido', 'ito' and 'ato'. Exceptions allowed previously for historically known ligands are now withdrawn. Some of these changes are noteworthy; for example,

~~fluoro~~ →	*fluorido*	~~hydroxo~~ →	*hydroxido*
~~chloro~~ →	*chlorido*	~~hydro~~ →	*hydrido*
~~bromo~~ →	*bromido*	~~cyano~~ →	*cyanido*
~~iodo~~ →	*iodido*	~~oxo~~ →	*oxido*

The ligand S^{2-} is now named **sulfido**; 'thio' is now reserved for functional replacement nomenclature.

Consequent to changes in names of anions, the names of corresponding ligands have also changed. For example,

$^{-}HNNH^{-}$ hydrazine –1,2–diido; $MeNH^{-}$ methanaminido

5. ***Formulae of coordination entities***

In the formulae of formal coordination entities, ligands are now arranged in alphabetical order of their abbreviation or formula, *irrespective of charge*. For example, Zeise's salt is now represented

$$[Pt(\eta^2\text{-}C_2H_4)Cl_3]^- \text{ rather than } [PtCl_3(\eta^2\text{-}C_2H_4)]^-, \text{ since } C > Cl$$

6. ***Additive names of polynuclear entities***

In additive names of dinuclear and polynuclear entities, the order of citation of central atoms in names should follow the order in table 2.3, the element occurring later being cited first. Kappa convention is used in the same order specifying which ligator atom coordinates to which central atom with locants. The atom symbols at the end to indicate metal-metal bonding are similarly ordered. For example,

$[(CO)_5ReCo(CO)_4]$ is nonacarbonyl –1κ^5*C*, 2κ^4*C*–rheniumcobalt (*Re–Co*) *rather than* nonacarbonyl –1κ^5*C*, 2κ^4*C*–cobaltrhenium (*Co–Re*).

7. ***Names of inorganic acids***

The main principle is to use additive nomenclature for deriving systematic names for inorganic acids. Thus, the systematic name for the dihydrogenphosphate ion ($H_2PO_4^-$) is dihydroxidodioxidophosphate (1–). Names of common acids and some of their representative anions have been given earlier. Names

like tetraoxosulfuric acid, trioxochloric(V) acid are now abandoned. Names using **'hydrogen nomenclature'** have also been modified; for example,

$HCrO_4^-$ hydrogen(tetraoxidochromate)(1–)

$H_2NO_3^+$ dihydrogen(trioxidonitrate)(1+)

Parentheses and charge number may be omitted in a few cases like hydrogencarbonate, dihydrogenphosphate, etc.

8. *Addition compounds*

Names are constructed by ordering the *components alphabetically*. The formulae are also arranged in *alphabetical order of the symbols*. Water is always placed at the end.

$BF_3.2H_2O$ boron trifluoride—water (1/2)

$KCl.MgCl_2.6H_2O$ magnesium chloride—potassium chloride—water (1/1/6).

EXERCISE

1. Write the formulae

   (i) magnesium potassium fluoride (ii) zinc hydroxide iodide (iii) disodium (trisulfide) (iv) hydroxidooxidonitrogen (v) arsanium ion (vi) silanetetrol (vii) oxidane (viii) dioxidane (ix) iron(II) diiron(III) oxide (x) hexaamminecobalt(III) chloride sulfate (xi) phosphane

2. Write IUPAC recommended names (in any system)

   (i) $PbEt_4$ (ii) $Cr_2O_7{}^{2-}$ (iii) $MnO_4{}^-$ (iv) CrO_5 (v) $BF_3.2H_2O$ (vi) O_2Cl (vii) $ClO_2{}^-$ (viii) $Cr_{23}C_6$ (ix) UO_2SO_4 (x) $(UO_2)_2SO_4$ (xi) $Na(NH_4)[HPO_4]$ (xii) Ca_3P_2

3. What are the IUPAC accepted common names of

   (i) H_3O^+ (ii) oxidane (iii) azane (iv) diazanium (v) chloridooxygenate (1–) (vi) diazane (vii) trioxidosulfate (2–) (viii) trihydroxidoboron?

4. What information is conveyed (or not conveyed) by the two representations in each group?

   (i) NaCl, [NaCl] (ii) TlI_3, $Tl(I_3)$ (iii) K_5CuSb_2, $K_5[CuSb_2]$ (iv) $MgCr_2O_4$, $(Mg)[Cr_2O_4$ (v) $[Pt_2(\eta^3\text{-}C_2H_4)_2Cl_4]$, $[\{Pt(\eta^2\text{-}C_2H_4)Cl(\mu\text{-}Cl)\}_2]$.

5. Write the following formulae in IUPAC recommended form

   (a) $Zn(H_2O)_6(UO_2)_3Na(OAc)_9$ (Ac = CH_3CO) (b) $K_2SO_4.Al_2(SO_4)_3$, $24H_2O$

   (c) $NH_4Na[HPO_4]$ (d) UCl_2O_2

   (e) Mg(OH)Cl (f) $[CoN_3(NH_3)_5]^{2-}$

   (g) $[Pt(NH_3)(py)Cl_2]$ (h) $[Pt(NH_3)BrCl(NO_2)]^-$

   (i) CNS^- (j) BrCNS

   Ans. (a) $Na(UO_2)_3[Zn(H_2O)_6](OAc)_9$ (b) $Al_2(SO_4)_3.K_2SO_4.24H_2O$

   (c) $NaNH_4[HPO_4]$ (d) UO_2Cl_2

   (e) MgCl(OH) (f) $[Co(NH_3)_5(N_3)]^{2-}$

   (g) $[PtCl_2(NH_3)(py)]$ (h) $[PtBrCl(NH_3)(NO_2)]^-$

   (i) NCS^- or SCN^- (i) BrSCN

6. What are the names of $[CoCl_3(NH_3)_3]$ and $[CoCl_3(NH_2Me)_3]$?

   Ans. Triamminetrichloridocobalt(III) and trichloridotris(methylamine)cobalt(III).

3

EXTRANUCLEAR STRUCTURE OF THE ATOM

PART I: *MODELS FOR THE ATOM*

3.1 INTRODUCTION : THE BACKGROUND

Philosophical ideas about an atomic constituent of matter have been forwarded from ancient times in different parts of the world, notably by the Hindu, Jain and Greek philosophers. But they failed to establish the identity of an atom on a general experimental framework. Around 1803, John Dalton conceived of atoms as the smallest indestructible part of an element retaining the characteristic properties of each; these atoms are able to combine with other atoms in a chemical manner. Dalton assigned a characteristic mass to each kind of atom. This satisfactorily explained most experimental facts about chemical combination.

Dalton's atomic theory is still regarded as the gateway to understand the material world; his concept of atoms as the smallest part of an element maintaining its individual characteristics still holds good for chemical changes. However, Dalton's idea of an atom as the smallest, indivisible stable particle of matter has been revised to account for many later scientific observations. Several experiments decidedly established the identity of a number of subatomic particles and their role as common constituents of all types of atom. To save space and avoid repetition, we mention only a few of them in Table 3.1. Some of these will be examined thoroughly in subsequent sections.

Table 3.1

Dalton to Rutherford: *Towards a model for the atom*

1803	***John Dalton*** (England) established the atomic theory. Laws of chemical combination were explained.
1833	***Michael Faraday*** (England) established the laws of electrolysis. The idea of an electron was latent in these laws (Stoney, 1891).
1859	***H. Geissler and J. Plücker*** (Germany) observed "cathode rays" in gas discharge tubes.
1876	***E. Goldstein*** (Germany) gave the name "cathode rays" and made further studies. Also discovered positive rays.
1879	***W. Crookes*** (England) made many remarkable observations on the discharge of electricity through gases.
1891	***J. Stoney*** (England) suggested the name *electron* for a unit negative charge in electrolysis.
1896	***Henri Becquerel*** (France) discovered radioactivity.
1897	***J. J. Thomson*** (England) determined *e/m* for cathode ray particles. The value was same irrespective of the nature of the cathode.
1904	***J. J. Thomson*** forwarded a model of the atom.
1909-1911	***R. A. Millikan*** (USA) determined the charge of electrons.
1911	***Ernest Rutherford*** (England) forwarded his nuclear model of the atom.

In the beginning of the twentieth century, only electrons were characterized and established as a common constituent of matter. The characterization of protons came later (1920) though positive rays or canal rays had been observed much earlier. The neutron was discovered still later in 1932 (Chadwick; see Chapter 14). Subsequently, many other subatomic particles have been characterized, some of them having been predicted theoretically before they were discovered. The electrons, protons and neutrons, however, appear to be the principal constituents of atoms.

Table 3.2

Some properties of electrons, protons and neutrons

Particle	Symbol	Rest mass u^1	Rest mass kg	Charge coulomb (C)
Electron	${}_{-1}^{0}e$	$5{\cdot}4858 \times 10^{-4}$	$9{\cdot}10938 \times 10^{-31}$	$1{\cdot}602176 \times 10^{-19}$*
Proton	${}_{1}^{1}p$	1·0073	$1{\cdot}67262 \times 10^{-27}$	,,
Neutron	${}_{0}^{1}n$	1·0087	$1{\cdot}67493 \times 10^{-27}$	0

1. 1 atomic mass unit (u) = $1{\cdot}66054 \times 10^{-27}$ kg.

* This is equal to the elementary charge e.

Studies on the interaction of matter (atoms in bulk) with energy, particularly radiant energy like heat and light provided immense information regarding the structure of atoms (and molecules). In the next section, we shall review the developments in physics regarding the nature of radiation and its interaction with matter.

3.2 RADIATION: THE ELECTROMAGNETIC THEORY

Light is the most common form of radiation. There have been many speculations about its nature from ancient times. In the seventeenth century Newton forwarded his corpuscular (particle nature) theory of light. In the beginning of the nineteenth century, diffraction and interference studies (pioneered by Thomas Young in 1801) established that light propagates in the form of waves. A wave means vibration of some kind in space and in time. What is then actually vibrating in light waves? At first they were supposed to be waves in an imaginary elastic medium, the ''ether'', extending throughout space. But the mechanical properties required for such a medium were unacceptable. The final answer was found in the 1860s from Maxwell's electromagnetic theory of radiation.

Faraday had shown that a changing magnetic field can induce electric current in a wire loop. In other words, a changing magnetic field gives rise to an electric field. Maxwell proposed that a changing electric field should also be able to produce a magnetic field. If an oscillating electric field is coupled with an oscillating magnetic field, then, according to Maxwell's theory, the periodically changing electric and magnetic fields will interact with one another by both

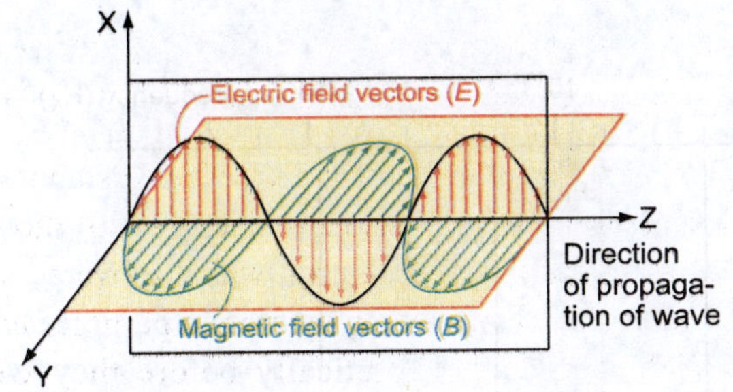

Fig.3.1: An electromagnetic wave: The electric field vectors (E), the magnetic field vectors (B) and the direction of propagation of the wave are mutually perpendicular. The wave-pattern shown is for 1½ cycles of a *plane-polarized wave* meaning that the E vectors all lie in one plane (and so also the B vectors). In a collection of heated atoms or molecules, the particles act independent of one another and the radiation emitted by them contains the E vectors in random orientation, producing *unpolarized* waves (for example, sunlight). A plane polarized wave may be generated by the to and fro oscillation of electrons in a straight wire.

electromagnetic induction and its converse mechanism that he proposed. This will result in transfer of energy between the electric and magnetic fields. Maxwell deduced mathematically that such energy would be transmitted from one place to another in free space (i.e., without any medium) with a constant speed c given by

$$c = \frac{1}{\sqrt{\mu_0 \varepsilon_0}},$$

where μ_0 and ε_0 are respectively the permeability and permittivity of free space. We know,

$$\mu_0 = 4\pi \times 10^{-7}\,\mathrm{N\,C^{-2}s^2}; \qquad \varepsilon_0 = 8\cdot 854 \times 10^{-12}\ \mathrm{C^2N^{-1}m^{-2}}\ \ (\mathrm{F\,m^{-1}})\ \ (\mathrm{F} = \text{Farad}).$$

Hence, $c = (4\pi \times 10^{-7} \times 8\cdot 854 \times 10^{-12})^{-1/2}\ \mathrm{m\,s^{-1}} = 2\cdot 9979 \times 10^8\,\mathrm{m\,s^{-1}}$ (conveniently taken as $3{\cdot}0 \times 10^8$ m s^{-1}). This is same as the speed of light measured experimentally. Maxwell thus concluded that light waves are electromagnetic waves produced by oscillating electric and magnetic fields.

Unfortunately, Maxwell died in 1879 at the age of forty-eight and could not rejoice the experimental verification of his theory in 1888 by Heinrich Hertz (Germany).

Electromagnetic radiations of all frequencies travel with the same speed c in vacuum. The speed varies when the radiation passes through different media depending on the nature of the medium and the frequency. The speed c' in a medium (for light of a given frequency) is related to the *refractive index* (n) of the substance for that frequency (ν):

$$n = c / c'$$

As c changes, λ also changes but ν remains the same. Some values of refractive index for light of sodium D line (λ = 589·3 nm in vacuum) at 25 °C and 1 atm. are given:

Air	Water	Benzene	CS_2	C_2H_5OH	Glass
1·0003	1·33	1·50	1·63	1·36	1·5–1·9

The electromagnetic waves cover a wide range of wavelength and frequency; in fact, there is no upper and lower limit to the values of λ and ν of electromagnetic waves. Depending on their applications, electromagnetic waves are conveniently classified over various regions of wavelength (or frequency). This spectrum of electromagnetic waves is commonly referred to as the electromagnetic spectrum (Fig. 3.2). The visible region covers only a small portion of the entire spectrum.

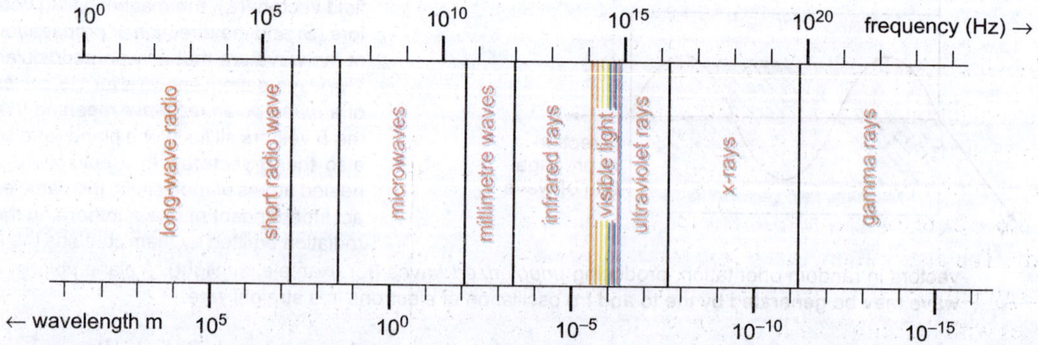

Fig.3.2: The electromagnetic spectrum (on a logarithmic scale). Boundaries between adjacent regions are not sharp as shown.

3.3. THE QUANTUM THEORY OF RADIATION

The interpretation of radiation as electromagnetic waves got a setback towards the end of the nineteenth century with some new experimental observations. The emission of radiant energy by matter, for example, by the action of heat or light, required a new look at the nature of radiation.

3.3.1 Blackbody Radiation

It is our common experience that a heated solid emits "light" or radiant energy. If we take, for example, a piece of coal or iron sufficiently heated to emit visible radiation, we observe that the substance changes colour as the temperature is raised (from red to yellow to white).

This shows that radiations of different wavelength are emitted to different extent at different temperatures. It was observed that for certain "ideal" emitters the intensity and distribution of radiant energy of different wavelength depend on the temperature alone and not on the nature of the source. Such an ideal emitter was called a "blackbody"; the radiation emitted by it was called blackbody radiation.

An ideal blackbody is one which absorbs all of the radiant energy falling upon it—its absorptivity (fraction of radiant energy incident on its surface which gets absorbed) is unity and hence maximum. This is also related to the emissive power of the body—so a blackbody is also the best emitter of thermal radiation. Experimentally, a uniformly heated hollow enclosure with a small opening may function as a blackbody. Any radiation entering through the hole will be repeatedly reflected inside the cavity and will be nearly completely absorbed. When heated to a constant temperature, the walls of the cavity will emit radiation. A portion of this will escape through the hole and can be studied for blackbody radiations.

The radiant energy emitted per unit area of emitter per unit time in the wavelength range λ and $\lambda + d\lambda$ is known as the monochromatic emissive power or energy density (E_λ). The total emissive power E at any temperature is the sum over all wavelength:

$$E = \int_0^\infty E_\lambda d\lambda$$

Characteristics of blackbody radiation

The radiant energy from a blackbody at a constant temperature is distributed over different wavelengths (or frequencies). This is independent of the nature of the substance making the black body. The distribution, that is, the fraction of radiant energy emitted in different ranges of wavelength, varies with the temperature (Fig. 3.3). Some features of this distribution attracted the interest of scientists:

(i) The distribution is Gaussian in nature, similar to the distribution of molecular velocities. As shown in Fig 3.3(a), the plot of energy emitted against λ at any temperature passes through a maximum, showing that the radiations emitted at that wavelength are most intense at that temperature. As the temperature is raised, the position of the maximum shifts towards shorter wavelength (higher frequency).

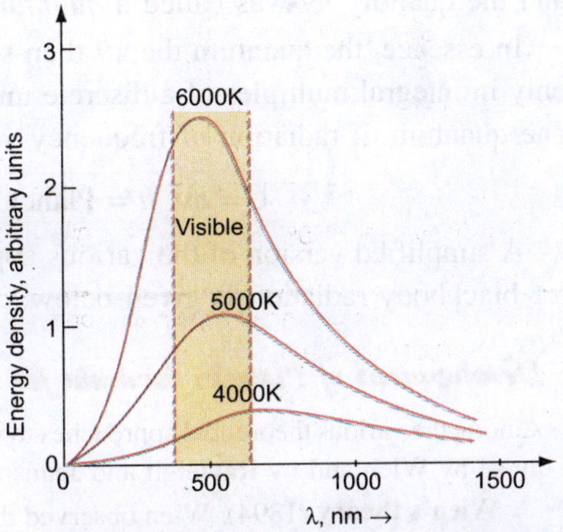

Fig.3.3(a): Nature of spectral distribution of energy of a blackbody. The maximum at each temperature shifts to lower wavelength as the temperature is raised.

(ii) The emissive power E of a blackbody at any temperature is proportional to fourth power of the absolute temperature of the blackbody (**Stefan-Boltzmann law**):

$$E = \sigma T^4 \quad (\sigma = \text{Stefan-Boltzmann constant} = 5{\cdot}669 \times 10^{-18}\ \text{W m}^{-2}\ \text{K}^{-4})$$

(iii) The product of the wavelength corresponding to maximum intensity (λ_{max}) is inversely proportional to the absolute temperature of the blackbody (**Wien's law;** *also*, **Wien's displacement law**).

$$\lambda_{max} T = \text{constant} = 2 \cdot 8976 \times 10^{-3}\,\text{m K}$$

These observations could not be explained by the classical electromagnetic theory of radiation. According to classical physics, an oscillating charge produces energy in the form of electromagnetic radiation. Such energy takes on a continuous range of values and a system is capable of losing or gaining any amount of energy (energy is a continuous variable). At first, scientists assumed the material particles on the wall of the blackbody as electrically charged; these particles oscillated to produce electromagnetic waves which ultimately gave rise to black body radiation. But their conclusions about the nature of spectral distribution of black body radiation did not match the type of experimental curve shown in Fig. 3.3(a).

In an attempt to derive a distribution formula which fits the experimental results, Planck arrived at a situation in which the idea of continuous absorption and emission of energy by the charged oscillators of a blackbody had to be rejected. He considered that the walls of the blackbody contained electric charges oscillating at various frequencies and producing electromagnetic waves. A charge oscillating at frequency ν will emit radiation of that frequency. Planck's derivation of a suitable distribution function now required the assumption that the energy of each oscillating charge could only take on the possible values 0, $h\nu$, $2h\nu$, $3h\nu$, ··· where h is a constant having the dimension of energy × time. Thus, the energy of each oscillator was restricted to whole number multiples of $h\nu$. The energy of the system was thus *quantized**, and the quantity $h\nu$ was called a *quantum of energy* (Latin neuter of *quantus* = how much).

In essence, the quantum theory then stated that a blackbody can absorb or emit radiations only in integral multiples of a discrete unit of quantum (ε, say). The energy corresponding to one quantum of radiation of frequency ν is given by the relation

$$\varepsilon = h\nu,\ h = \text{Planck's constant} = 6{\cdot}626 \times 10^{-34}\ \text{J s.}$$

A simplified version of the various approaches towards an acceptable distribution function for blackbody radiation is given below.

Development of Planck's formula for energy distribution in blackbody radiation

Among the various theoretical approaches to interpret the experimental results on black body radiation, those by Wien and by Rayleigh and Jeans are noteworthy.

Wien's theory (1894). Wien observed the similarity between the graphs of black body radiation (Fig. 3.3) and the graphs of Maxwellian distribution of speed among gas molecules. He argued that if the molecules of a black body are thermally agitated, they may be expected to have a distribution of speed similar to the Maxwellian distribution. If these molecules are charged, then their acceleration

*Quantization is now referred to restriction of any property to discrete values.

will give rise to radiation according to the electromagnetic theory. Starting from these ideas, he deduced an expression

$$E_\lambda d\lambda = \frac{a.\lambda^{-5}}{e^{b/\lambda T}} d\lambda \qquad a, b = \text{arbitrary constants.}$$

This was the mathematical expression of Wien's displacement law. With suitable choice of a and b, this relation matched the experimental curves of Fig. 3.3(a) at shorter wavelengths. But it did not work well at longer wavelengths (Fig. 3.3(b)). Moreover, the formula was essentially an empirical one and did not clearly explain the nature of radiation.

The Theory of Rayleigh and Jeans (1900)

Lord Rayleigh and Jeans assumed that the electromagnetic radiations inside an isothermal black body undergo repeated reflection from the walls to form a system of standing electromagnetic waves. Each standing wave of radiation may be supposed to be generated by an electric dipole acting as a linear harmonic oscillator with frequency ν. The electromagnetic waves emitted by a large number of such atomic oscillators constitute the entire black body radiation. The number of modes of vibrations of these oscillators (dn_λ), per unit volume of space in the wavelength range λ and λ + $d\lambda$ was shown to be:

$$dn_\lambda = 8\pi \frac{d\lambda}{\lambda^4}.$$

Now each mode of vibration introduces two degrees of freedom (one for its potential energy and one for its kinetic energy), and is therefore associated with a total energy kT. So the monochromatic emissivity (E_λ) will be simply the product of the number of modes and the energy associated with each mode:

$$E_\lambda d\lambda = 8\pi kT \frac{d\lambda}{\lambda^4}.$$

Though theoretically very elegant, the expression so obtained fitted the experimental curves only at very long wavelengths. For shorter wavelengths (i.e., in the UV region), the mismatch was so prominent that this was referred to as the *ultraviolet catastrophe* (Fig. 3.3(b)).

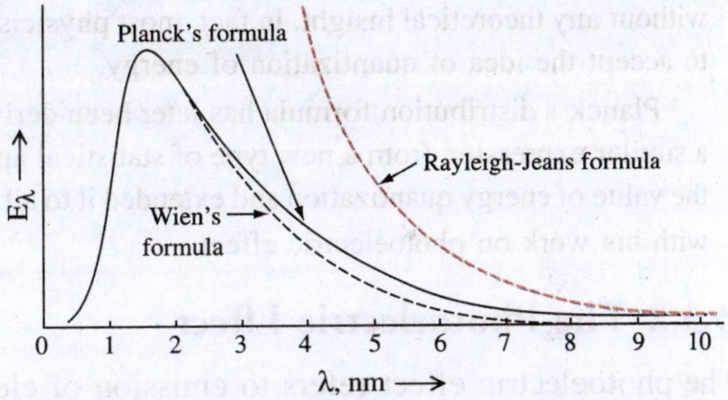

Fig.3.3(b): Graphs of different radiation laws

The Theory of Max Planck(1901)

Planck first tried to fit both Wien's and Rayleigh-Jeans formula into a single expression which will satisfactorily explain the energy-distribution curves of black body radiation in both long and short wavelength regions. He slightly changed Wien's formula by putting a (–1) in the denominator:

$$E_\lambda d\lambda = \frac{a.\lambda^{-5}}{e^{b/\lambda T} - 1} d\lambda.$$

This relation reduces to the Rayleigh-Jeans formula at long wavelengths and can be fitted to the experimental curves with greater success. Since $e^x - 1 \simeq x$ when x is small, $e^{b/\lambda T} - 1 \simeq b/\lambda T$ for high values of λ. This gives $E_\lambda d\lambda = (a/b)\, T\, \lambda^{-4} d\lambda$. Next Planck tried to derive this relation. This could be done only with a new postulate.

In the Rayleigh-Jeans treatment it was assumed that each harmonic mode of vibration contributed an average energy kT. Planck assumed that each mode can have energy in discrete steps of a fundamental amount, say ε. So the energies may have values 0, ε, 2ε, 3ε, ··· in general $n\varepsilon$, where n is an integer. The average energy now became

$$\bar{\varepsilon} = \frac{\varepsilon}{e^{\varepsilon/kT} - 1}.$$

When this is multiplied by dn_λ (= $8\pi d\lambda/\lambda^4$; see above), one gets,

$$E_\lambda d\lambda = \frac{8\pi}{\lambda^4} \cdot \frac{\varepsilon}{e^{\varepsilon/kT} - 1} d\lambda,$$

This relation is very similar to the Wien formula or its ad hoc modification. If a value of ε is chosen to make the powers of the exponential terms equal, we can write

$$b/\lambda T = \varepsilon/kT$$

$$\text{or, } \varepsilon = \frac{bk}{\lambda} = \frac{bk}{c}\nu.$$

c is the speed of electromagnetic waves and ν is the frequency of the oscillator and is equal to the frequency of the emitted radiation. The constant (bk/c) may be replaced by another constant h to obtain $\varepsilon = h\nu$.

Thus Planck introduced the new idea that the energy states of an oscillator should be an integral multiple of the fundamental unit $h\nu$, which is called a bundle or packet of energy, a *quantum*. h is known as **Planck's constant**. It may be calculated from Wien's constant b, but it is also determined from the photoelectric effect. $h = 6{\cdot}6256 \times 10^{-34}$ J s.

It is amusing to note that the idea of quantization was arrived at from purely calculational device, without any theoretical insight. In fact, most physicists, including Planck himself (!), were reluctant to accept the idea of quantization of energy.

Planck's distribution formula has later been derived in several other ways. S. N. Bose arrived at a similar expression from a new type of statistical approach, the *Bose statistics*. Einstein recognized the value of energy quantization and extended it to all types of electromagnetic radiation in connection with his work on photoelectric effect.

3.3.2 The Photoelectric Effect

The photoelectric effect refers to emission of electrons from metal surfaces by the action of electromagnetic radiation. As early as 1887, Hertz had observed during his experiments to produce electromagnetic waves that when one of the metal balls in his transmitter equipment was exposed to UV rays, spark occurred more readily in the air gap. Subsequently it was established that electrons are actually emitted by the action of UV radiation on the metal surface. Thus, the very apparatus which demonstrated that light is an electromagnetic wave, also provided a clue to the particle nature of light.

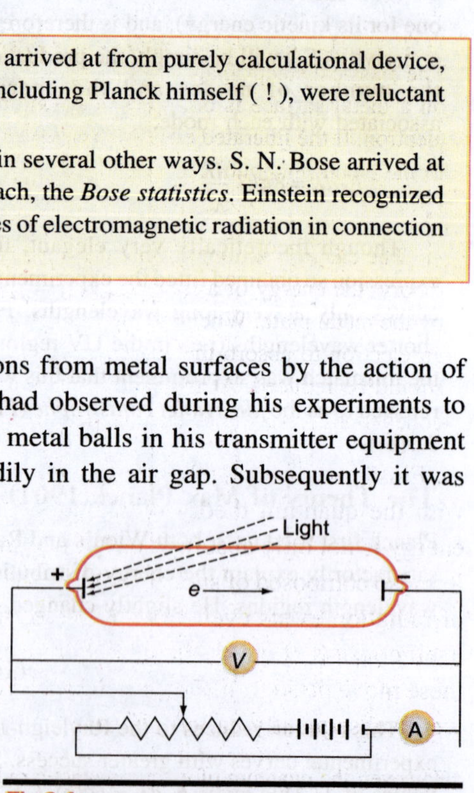

Fig.3.4: Scheme of an arrangement for the study of photoelectric effect.

The photoelectric effect may be readily demonstrated by an arrangement schematically shown in Fig. 3.4. The metal anode in an evacuated glass tube is irradiated by light. The potential difference (V) between the two electrodes may be varied and the current flowing across the electrodes measured by the ammeter A. Some of the electrons emitted by the anode have sufficient energy to reach the cathode inspite of its negative polarity. The current drops as the retarding potential V is increased; eventually a point comes when no electrons reach the cathode. (This potential is called the *stopping potential*; see later).

Some important observations in connection with the photoelectric effect may be summarized as follows:

(i) For a given metal surface there exists a *minimum value of the frequency* of the incident light which can liberate electrons from the surface. With light of smaller frequency, electrons are not emitted, no matter how great the intensity is or how long does light fall on the surface. This minimum frequency is referred to as the *photoelectric threshold frequency*.

(ii) With light of frequency greater than the threshold frequency, the total photoelectric current is proportional to the intensity of the incident light.

(iii) The *maximum kinetic energy* of the emitted electrons is *independent of the intensity* of the incident radiation.

(iv) The *maximum kinetic energy* of the ejected electrons *depends linearly* on the *frequency* of the incident radiation.

(v) The effect is almost instantaneous; with light of frequency greater than the threshold frequency, the time interval between irradiation and electron-emission is of the order of 10^{-9} s only.

The classical electromagnetic theory of light was in no way able to explain these facts. If an electron on a metal surface is bound by a potential energy W, this energy must be supplied to release the electron. If the liberated electron has kinetic energy equal to $mv^2/2$, the total energy E_T to be supplied to the electron should be

$$E_T = W + \frac{1}{2}mv^2$$

The electron should acquire this energy from the incident beam of light. According to the wave theory, the energy of a light beam depends on its intensity and it should be absorbed continuously by the metal plate. When the intensity is weak, a relatively long time interval should be required for an electron to absorb the energy needed for its ejection. If the intensity is increased, the time gap should be reduced. At the same time, there should not be any threshold frequency. Clearly, these conclusions from the electromagnetic theory do not support the experimental facts.

Einstein explained the facts about the photoelectric effect (1905) with the quantum theory of radiation. He proposed that radiant energy is not only absorbed or emitted in discrete units of quanta; it is also composed of such discrete units. This "quantum structure" of radiation exists even when it passes through space. Thus light itself consists of quanta or small corpuscles of energy equal to $h\nu$; these move through space with the speed of light. These corpuscles were subsequently called photons*.

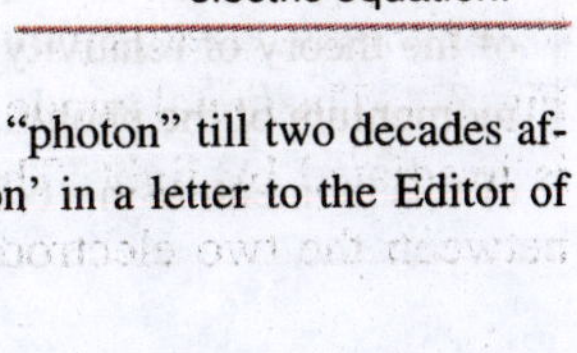

Fig.3.5: Illustration of photoelectric equation.

*However, the quantum of electromagnetic radiation was not referred to as the "photon" till two decades after Einstein's seminal paper! In fact, G. N. Lewis first coined the word 'photon' in a letter to the Editor of "Nature" (1926 vol. 118. p.874-875).

According to Einstein, when a light beam strikes a metal surface, the photons, each carrying energy $h\nu$, strike the electrons. Now an electron will absorb the entire quantum of energy. If this energy of the quantum is greater than the binding energy (W) of the electron to the metal surface, the electron will be ejected instantaneously; the excess of the energy received from the photon will appear as the maximum kinetic energy of the electron. This is expressed in the **photoelectric equation**

$$h\nu = \tfrac{1}{2}mv^2 + W.$$

Clearly, no electron will be ejected when $h\nu < W$. The threshold frequency ν_0 is then given by

$$h\nu_0 = W.$$

W is known as the **photo-electric workfunction** of the metal.

Greater intensity means larger number of incident quanta per unit time. This will increase the number of electrons emitted and hence the current will be proportional to the intensity. But so long as the energy of a quantum $h\nu$ is fixed, the kinetic energy of the ejected electrons will remain the same.

Work function (W) of some metals

	Na	K	Mg	Cu	Ag
W(eV)	2.30	2.25	3.7	4.8	4.3

The photoelectric work function represents the energy needed to remove a loosely bound surface electron of the metal and is nearly half the energy required to remove an electron from a free gaseous atom of the metal (ionization energy).

The kinetic energies of the emitted electrons were measured by applying a variable positive potential to the metallic surface. The potential V *just strong enough* to prevent the emission of electrons was noted. At this point the energy of the electron is just equal to the electrical energy applied on the electron of charge e:

$$\tfrac{1}{2}mv^2 = Ve.$$

V is known as the **stopping potential**. From the photoelectric equation, one may write

$$Ve = h\nu - W; \quad \text{or,} \quad V = (h/e)\nu - (W/e)$$

If the potential V is plotted as a function of ν, the frequency of the incident light, one should get a straight line with a slope h/e. This was experimentally verified by Millikan (1916). Since the value of e was known with precision, the value of h could be determined from such studies. The values were in close agreement with those obtained by Planck for black body radiation.

The idea of photons points to another interesting aspect of light. The mass energy equivalence of the theory of relativity gives $\varepsilon = mc^2$. Combining this with $\varepsilon = h\nu$, one gets $mc = h\nu/c$; mc is the momentum of the photon. Since a photon possesses a momentum, it should be able to transfer a part of this momentum in a collision with another particle. This was demonstrated in the *Compton Effect*.

The electronvolt

We have seen how the maximum kinetic energy of photoelectrons may be measured by noting the stopping potential V. This leads us to a new unit of energy called *electronvolt* (eV); this is the energy gained by one electronic charge (e) when it moves through a potential difference of one volt. 1 eV = $1{\cdot}602 \times 10^{-19}$ C × 1 V = $1{\cdot}602 \times 10^{-19}$ J.

3.3.3 The Compton Effect

The particle nature of electromagnetic radiation received another direct confirmation from the **Compton effect** (A.H. Compton, 1923). Compton measured the angle of scattering and the change in wavelength of high frequency x-rays by a carbon block. The results could be explained by assuming billiard-ball like collisions between an x-ray photon and loosely bound electrons in the carbon atoms. If radiation consists of photons, the collision of a photon with an electron would result in exchange of momentum between the two particles. As a result, the photon will be scattered with a lower momentum or smaller frequency. The situation is schematically shown in Fig. 3.6.

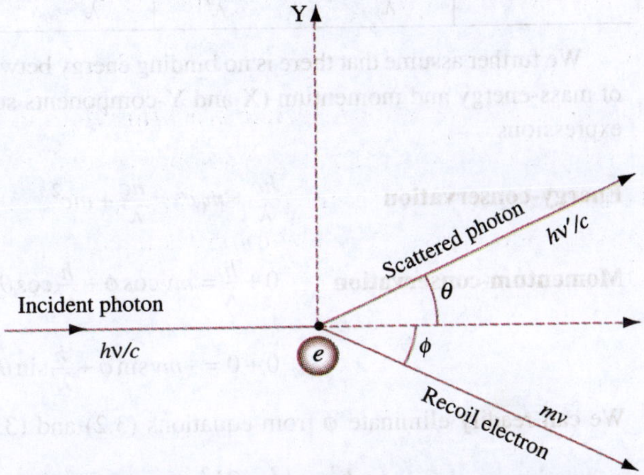

Fig.3.6: Schematic representation of the Compton effect. Conservation of energy and momentum gives $\Delta\lambda = \lambda' - \lambda = h(1 - \cos\theta)/m_0c$; m_0 = rest mass of the electron. h/m_0c is called the *Compton wavelength* of the electron.

In this collision, the photon transfers a part of its momentum to the electron, and hence its initial energy ($h\nu$) is lowered (to $h\nu'$, say). Consequently, the wavelength of the radiation increases after scattering. On the other hand, the electron gains momentum and recoils in a different direction. Compton measured the angle of scattering and the increase of wavelength of x-rays scattered. Recoil electrons were also detected. The results were in agreement with the laws of conservation of energy and momentum (see box).

The Compton effect

The increase in wavelength of scattered x-rays after undergoing Compton scattering may be expressed theoretically from the principle of conservation of energy and momentum of two colliding particles, in this case an incident x-ray photon and an electron in the scattering material. For the sake of simplicity, we assume that the electron is initially at rest at the origin of the coordinate system (Fig. 3.6). Let the photon, with an associated wavelength λ (frequency ν) strike it along the X-axis in a manner such that after collision, the photon is scattered at an angle θ with the X-axis; its wavelength changes to λ' (frequency ν'). The electron, after collision, is scattered in a direction (opposite to that of the photon) which makes an angle ϕ with the X-axis, with a velocity v. The energy of the electron before collision is only its rest-mass energy, m_0c^2; after collision it changes to mc^2 (m = mass at velocity v). The energy of the photon before collision is $h\nu = hc/\lambda$, and after collision $h\nu' = hc/\lambda'$. Similarly, we can compute the momenta of the particles before and after collision, separately for the X- and Y-components. These are summarized in the following tabular form:

	Energy		Momentum			
			X-component		Y-component	
	Before	*After*	*Before*	*After*	*Before*	*After*
Electron	m_0c^2	mc^2	0	$mv\cos\phi$	0	$-mv\sin\phi$
Photon	$\frac{hc}{\lambda}$	$\frac{hc}{\lambda'}$	$\frac{hc}{\lambda}$	$\frac{hc}{\lambda'}\cos\theta$	0	$\frac{hc}{\lambda'}\sin\theta$

We further assume that there is no binding energy between the particles on impact. Conservation of mass-energy and momentum (X and Y-components separately) now leads us to the following expressions:

Energy-conservation $\quad \frac{hc}{\lambda} + m_0c^2 = \frac{hc}{\lambda'} + mc^2$. .. .. [3.1]

Momentum-conservation $\quad 0 + \frac{h}{\lambda} = mv\cos\phi + \frac{h}{\lambda'}\cos\theta$ .. .. [3.2]

$$0 + 0 = -mv\sin\phi + \frac{h}{\lambda'}\sin\theta \quad .. \quad .. \quad [3.3]$$

We can readily eliminate ϕ from equations (3.2) and (3.3) by squaring and adding:

$$m^2v^2 = \frac{h^2}{\lambda'^2} + \frac{h^2}{\lambda^2} - \frac{2h^2}{\lambda\lambda'}\cos\theta. \quad .. \quad .. \quad [3.4]$$

From the relativistic interdependence of m_0 and m by the relation

$$m = m_0 / \sqrt{1 - v^2/c^2}\text{, we get, } m^2v^2 = m^2c^2 - m_0^2c^2 \quad .. \quad .. \quad [3.5]$$

Now we eliminate v^2 from Eqns. (3.4) and (3.5)

$$m^2c^2 - m_0^2c^2 = \frac{h^2}{\lambda^2} + \frac{h^2}{\lambda'^2} - \frac{2h^2}{\lambda\lambda'}\cos\theta. \quad .. \quad [3.6]$$

Next we evaluate m from Eqn. (3.1) and calculate m^2c^2. We put this value in Eqn. [3.6]. Rearrangement now gives

$$\Delta\lambda = \lambda' - \lambda = \frac{h(1-\cos\theta)}{m_0c} \quad .. \quad .. \quad [3.7]$$

h/m_0c (= $2{\cdot}426 \times 10^{-12}$m) is called the **Compton wavelength** of the electron. It appears that the greatest value of $\Delta\lambda$ will occur when $\theta = 180°$: $\Delta\lambda = 2h/m_0c = 4{\cdot}852$ pm. For heavier particles, $\Delta\lambda$ will be still smaller. Such small changes are observable only with x-rays.

In actual experiment, a substantial proportion of x-rays do not undergo any change in wavelength. Only free or loosely bound electrons in the scattering material can modify the wavelength of scattered x-rays significantly. If a photon strikes a tightly bound electron, the atom recoils as a whole (instead of the electron). The value of m_0 in the expression of $\Delta\lambda$ (Eqn. 3.7) should now be the rest mass of the atom. As this is many times larger than the rest mass of the electron, the shift in wavelength will be negligible.

For very low *energy* photons, e.g., with visible light, almost all the electrons of an atom will appear bound and hence no Compton scattering is expected. Instead, the photoelectric effect becomes important. As the photon energy increases, more and more electrons appear free and an increasing fraction of the incident photons undergo Compton scattering. However, $\Delta\lambda$ *is always independent of the wavelength of the incident radiation.* With x-rays of energy less than 0.1 MeV, photoelectric effect is stronger than the Compton effect. With x-rays of higher energy, the Compton scattering becomes more important. *But with very high energy radiation,* say gamma rays, the gamma ray photon converts itself into a pair of material particles—a positron and an electron (negatron). This is known as *pair formation.* According to the mass energy relation $E = mc^2$, the energy equivalent of an electron mass is 0·511 MeV. The gamma ray photon must be equivalent to twice the electron mass, so its energy should be at least 1.022 MeV. In the reverse process of pair formation, **positron annihilation**, the charge of a positron is neutralized by that of an electron; the rest energy of the two particles is converted into the energy of a *photon pair.* Each photon flies in opposite directions with the speed of light.

The essential differences between the photoelectric effect and the Compton effect may now be summarized:

(i) Photoelectric effect occurs with radiations of relatively longer wavelengths, e.g., visible light or UV radiation or x-rays of energy less than 0.1 MeV. Compton effect occurs with x-rays of higher energy (0.1 MeV – 1 MeV).

(ii) In photoelectric effect, the entire energy of a photon is transferred to an electron in an atomic system. A photon cannot give up all its energy to a free electron. So, the photoelectric effect occurs only when photons strike bound electrons. The reverse is true for Compton effect.

Summary of the properties of photon

1. Photons are created when an excited atom or molecule loses energy and changes its energy state or when an energetic electron strikes the target of an x-ray tube.
2. Photons may be destroyed (annihilated) on being absorbed by matter; its energy is converted into vibrational energy of the constituent particles of the absorbing matter.
3. The photon is a particle carrying one quantum of energy of a beam of electromagnetic radiation. For radiation of frequency ν and wavelength λ, the energy (ε) of the photon is $h\nu$ and its momentum is $h\nu/c = h/\lambda$ (λ = Planck's constant).
4. Photons travel with velocity c, the speed of light in vacuum. Its rest mass (m_0) is thus zero because otherwise its mass would have been infinite. (From the relation $m_v = m_0\big/\sqrt{1-v^2/c^2}$, m_v becomes infinite when $v = c$).
5. Photons can be created and annihilated *singly*. Other particles, like positrons and electrons, can only be created or destroyed in *pairs*.
6. Photons possess intrinsic angular momentum, or spin, of magnitude $h/2\pi$.

3.3.4 The Dual Nature of Radiation

The electromagnetic theory views radiation in terms of electromagnetic waves which may be absorbed or emitted by matter continuously. The theory can explain the phenomena of diffraction and interference. Real existence of such electromagnetic waves is confirmed beyond doubt by their formation and transmission in several experimental set-ups.

But the theory cannot explain the facts of black body radiation and photoelectric effect. Here we are led to view radiation as consisting of discrete particles (photons as they are called) bearing a finite momentum which they can share with other material particles, as demonstrated in the Compton effect.

Thus we have two versions of light or radiation—one as waves which are not localized in space, and the other as particles which are localized in space. The photon picture leads one to the idea of quantization of light energy. However, the wave and particle models are mutually exclusive. Thus, Einstein's expression for the energy (ε) of a photon

$$\varepsilon = h\nu$$

contains the energy of a quantum of radiation (ε) on the left and the frequency of the radiation on the right. Apparently this appears as a self-contradiction.

However, the special theory of relativity expresses the energy of the photon in terms of its mass:

$$\varepsilon = mc^2,$$

where m is the relativistic mass of the photon and c is the speed of light. The rest mass of the photon is zero but at speed c its mass is non-zero. Its momentum is given by h/mc which we used to interprete the results of Compton scattering. As we shall presently realize, such wave-particle duality is also manifest in matter and the same basic equations may be adapted to obtain expressions for matter waves.

Problems

Problem 3.1

Calculate the energy of one quantum of radiation of (*a*) frequency 5×10^{14} Hz (*b*) wavelength 400 nm.

Solution: (*a*) Energy (E) of a photon $= 6{\cdot}626 \times 10^{-34}\ \text{J s} \times 5 \times 10^{14}\ \text{s}^{-1} = 3{\cdot}313 \times 10^{-19}$ J.

$$(b)\ E = h\nu = \frac{hc}{\lambda} = \frac{6{\cdot}626\times10^{-34}\ \text{Js} \times 3\times10^{8}\ \text{m s}^{-1}}{400\times10^{-9}\ \text{m}} = 4{\cdot}969 \times 10^{-19}\ \text{J}.$$

Problem 3.2

(*a*) A lamp emits red light of wavelength 760 nm and another emits yellow light of wavelength 560 nm. Which lamp emits more photons per second if both lamps are 100 watt?

(*b*) Calculate the number of photons emitted per second by the yellow lamp. What assumption is implicit in your calculation?

Solution: (*a*) Energy E of a photon is expressed as $E = hc/\lambda$.

Since $\lambda_{yellow} < \lambda_{red}$, energy of a photon of yellow light is *greater* than that of red light. Hence the *number of photons* per joule of energy will be *smaller for yellow light.*

(*b*) Power of the lamp = 100 W = 100 J s^{-1}

$$\text{Energy of a photon of yellow light} = \left(\frac{hc}{\lambda}\right) = \frac{6{\cdot}626\times10^{-34}\ \text{Js}\times3\times10^{8}\ \text{m s}^{-1}}{560\times10^{-9}\ \text{m}} = 3{\cdot}55\times10^{-19}\ \text{J}.$$

$$\text{No. of photons emitted (per second)} = \frac{100\ \text{J.s}^{-1}}{3{\cdot}55\times10^{-19}\ \text{J}} = 2{\cdot}8\times10^{20}\ \text{s}^{-1}.$$

It is assumed that the lamps emit radiation of a particular wavelength (monochromatic) and their efficiency is 100%.

Problem 3.3

Light of wavelength 5000 Å falls on a metal having photoelectric work function 2 eV. Find (*a*) the energy of the photon in eV, (*b*) the K.E. of the most energetic photoelectron in eV and in joules, and (*c*) the stopping potential.

Solution: (*a*) Energy of photon = $hc/\lambda \Rightarrow 3.98 \times 10^{-19}$ J ÷ 1.60×10^{-19} eV J^{-1} = 2.49 eV.

(*b*) From the law of conservation of energy,

Energy of the photon = K.E. of the electron (max) + work function.

$\therefore E_k(\text{max}) = 2.49 \text{ eV} - 2 \text{ eV} = 0.49 \text{ eV} \equiv 7.84 \times 10^{-20}$ J.

(*c*) If the stopping potential is V volt, for stopping of electron emission

$V \times e = E_k$ (max), where e = charge of an electron.

$\therefore V = 0.49 \text{ eV} \div e = 0.49$ V.

Problem 3.4

The photoelectric work function of a metal is 2.0 eV. (*a*) Calculate the thereshold wavelength. (*b*) What will be the velocity of the most energetic electrons when the metal is irradiated with light of wavelength 4000 Å?

Solution: (*a*) At threshold wavelength, energy of the photon = work function (W).

$\therefore hc/\lambda = W$ or, $\lambda = hc/W \Rightarrow 6.216 \times 10^{-7}$ m = 6216 Å.

(*b*) Maximum K.E. of electrons $\left(\frac{1}{2}mv^2\right)$ = Energy of a photon − work function.

$$\therefore \frac{1}{2}mv^2 = \frac{hc}{\lambda} - W = 1.77 \times 10^{-19} \text{ J.}$$

Putting $m = 9.11 \times 10^{-31}$ kg. one gets, $v = 6.23 \times 10^5$ m s^{-1}.

3.4 ATOMIC SPECTRA

A spectrum is an ordered array of the components of any composite electromagnetic radiation. The composite radiation may be split or dispersed by a prism or fine grating and the components may appear as bands of different colours (as in the spectrum of white light) or spaced lines corresponding to definite frequencies or wavelengths. The radiations studied may belong to any region in the electromagnetic spectrum. The term has been extended to other similar resolved patterns; in mass spectrum, for example, isotopes are resolved according to their mass in electric and magnetic fields.

The study of spectrum provides essential information regarding absorption or emission of radiant energy by matter and thus the spectrum acts as a fingerprint of the structure of matter. From the outset, atomic spectra provided valuable guidance to develop our knowledge on the structure of atom.

Different types of spectra

The study of spectrum may be carried out in either of two ways. (1) The radiation emitted by a material to be studied under excited conditions may be examined directly, when we get the *emission spectra*. (2) The light coming out from a source and showing a continuous emission spectrum may be sent through an absorbing material and then analysed. The emerging radiation will show the *absence* of certain wavelengths absorbed by the absorbing material; the spectra obtained is called *absorption spectra*.

Both emission or absorption spectra may be of three types: continuous, band or line. The nature of spectra depends on the nature of the substance emitting (or absorbing) the radiation under study. On heating to incandescence, a solid body changes its colour continuously, as may be observed with a piece of iron. The radiations emitted are spread continuously over a wide range of wavelength; but radiations of different wavelengths are emitted to varying extent at different temperatures. At each temperature, maximum intensity occurs at a particular wavelength; the colour at that temperature corresponds to this wavelength. As the wavelengths of the radiations change gradually and continuously, a continuous spectrum is obtained in such cases. Solids and liquids give rise to continuous emission and absorption spectra covering a wide range of wavelengths without any sharp discontinuity.

In contrast, substances in gaseous or vapour state, when sufficiently excited, emit radiations of discontinuous nature. The spectra obtained consist of discrete lines or bands, corresponding to emission of radiations of certain definite wavelengths. Atoms produce series of sharp lines, each line corresponding to a definite wavelength. Molecules, on the other hand, give spectra containing a series of bands which really consist of a large number of closely spaced lines.

In order to get the atomic spectrum of a gaseous element, a discharge tube containing the gas is usually employed. In the high voltage applied, a discharge is started and the molecules split into atoms; some of these atoms acquire excess internal energy which they radiate as visible, ultraviolet and infrared radiation. This radiation passes through a slit and after collimation falls on a prism or grating where it is dispersed into its component wavelengths. *Images of the slit* are formed by each resolved component of the radiation which appear as discrete illuminated lines on a dark background when viewed through a telescope. A photographic plate may also be placed at the back focal plane of the telescope objective to record the spectrum.

An atom absorbs the same radiations as are emitted by it in the excited state. However, the emission spectra are often more convenient to study. Excitation necessary for obtaining the emission spectra of atoms may also be caused by other means. A suitable compound may be heated strongly to dissociate it into atoms, e.g., heating sodium chloride in a flame produces the characteristic spectra of sodium atom. Alternatively, an arc may be used, or the electrodes forming the arc may be coated with suitable compounds containing the elements. Each element gives rise to a unique line emission spectrum (Kirchhoff, 1859). These lines may be used to identify the presence of the element in an unknown sample. Helium was discovered in the sun from its spectrum. However, the number, intensity and exact position of the lines in the spectrum of an element depend on several factors like motion of the source, presence or absence of electric or magnetic fields, temperature and pressure.

Sufficient advancement in spectroscopy took place in the early part of the nineteenth century; very fine diffraction gratings were devised and wavelengths of several lines in many atomic spectra were measured accurately. Among these, the spectrum of atomic hydrogen was found to be the simplest in appearance. A closer study of this spectrum provided important clues to the structure of atom.

The spectrum of atomic hydrogen

The spectrum of atomic hydrogen was known to consist of four main lines in the visible region, ranging from about 650 nm (red) to 410 nm (violet). Photography revealed many more lines coming closer together in the ultraviolet end; they converged to a limit at 364.6 nm. It was soon found that these wavelengths fall into a definite set; this was called a spectral series.

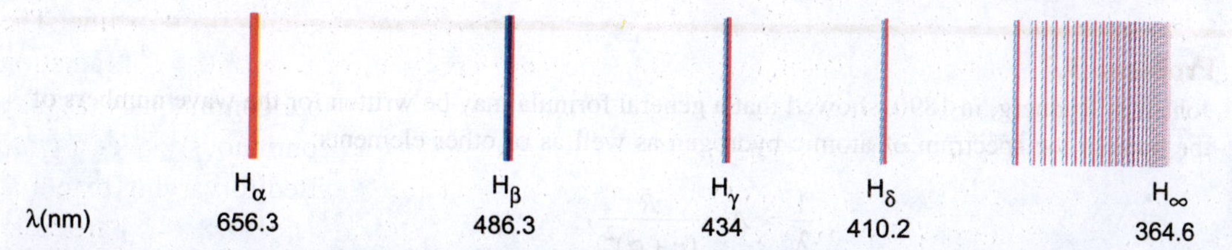

Fig.3.7: The Balmer series in the spectrum of atomic hydrogen (schematic).

Balmer, in 1885, showed that the wavelengths in the visible spectrum of hydrogen could be expressed by a common formula

$$\frac{1}{\lambda} = R_H\left(\frac{1}{2^2} - \frac{1}{n^2}\right) \qquad n = 3, 4, 5, \cdots$$

where R_H was a constant and n an integer greater than 2. Putting $n = 3$ in this relation, one gets λ for the first line (656.3 nm), also called the H_α line. The value of R_H fitting this relation is 1.09677×10^7 m^{-1} and is known as the **Rydberg constant**. The relation reproduces the values of wavelengths of all the lines in the series, now named the **Balmer series**, within 1 part in 5000. From the relation $c = \lambda\nu$, $1/\lambda = \nu/c = \bar{\nu}$. $1/\lambda$ (or $\bar{\nu}$) is called the **wave number** and is equal to the number of waves per unit length.

The energy of a quantum of radiation of 1 cm^{-1} wave number is 1.987×10^{-23}J. It is also used as a unit of energy called a **Kayser, K**.

Problem 3.5

Balmer first proposed his formula in the form : $\lambda = K \cdot \frac{n^2}{n^2 - 4}$.

Change this to the common form involving wave number ($\bar{\nu} = 1/\lambda$) and the Rydberg constant for hydrogen R_H. Find a relation between K and R_H.

Hint: $$\bar{\nu} = \frac{1}{\lambda} = \frac{(n^2 - 4)}{Kn^2} = \frac{1}{K} - \frac{4}{Kn^2} = \frac{4}{K}\cdot\frac{1}{4} - \frac{4}{K}\cdot\frac{1}{n^2} = \frac{4}{K}\left(\frac{1}{2^2} - \frac{1}{n^2}\right).$$

By comparison, $R_H = \frac{4}{K}$, $K = \frac{4}{R_H}$.

Problem 3.6

Calculate the wavelength for the series limit in the Balmer series. $R_H = 1.097 \times 10^7$ m^{-1}

Solution: The series limit occurs at $n = \infty$ in the general expression

$$\frac{1}{\lambda} = R_H\left(\frac{1}{2^2} - \frac{1}{n^2}\right).$$

For $n = \infty$, $\frac{1}{\lambda} = \frac{R_H}{4}$, when $\lambda = 364.7$ nm.

Problem 3.7

Johannes Rydberg, in 1890, showed that a general formula may be written for the wave numbers of the lines in the spectrum of atomic hydrogen as well as of other elements:

$$\frac{1}{\lambda} = A - \frac{R}{(n+\alpha)^2},$$

where R, A and α are constants for a particular series and the elements. For hydrogen, $R = R_H$ as before. Show that this relation can be arrived at from Balmer's original formula:

$$\lambda = Kn^2/(n^2 - 4).$$

Solution: From Problem 3.5, $\frac{1}{\lambda} = \frac{1}{K} - \frac{4}{Kn^2} = A - \frac{R_H}{(n+\alpha)^2}$,

where $A = \frac{1}{K}$ and $R_H = \frac{4}{K}$; $\alpha = 0$.

Besides the Balmer series in the visible region, the spectrum of atomic hydrogen also consists of the Lymann series in the ultraviolet region and three others in the infrared region – Paschen, Brackett and Pfund. Another series (Humphreys) lies in the far-infrared region.

Soon it was shown that the wavelengths of the lines in all the series in the spectrum of atomic hydrogen may be described by the simple relation:

$$\frac{1}{\lambda} = R_H\left(\frac{1}{n_1^2} - \frac{1}{n_2^2}\right) \quad . . . (n_2 > n_1),$$

where R_H is Rydberg constant for hydrogen and n_1, n_2 are integers. For a particular series, n_1 remains constant; successive variation in n_2 ($n_2 > n_1$) gives the wave numbers of the lines in that series.

Lyman series	$\frac{1}{\lambda} = R_H\left(\frac{1}{1^2} - \frac{1}{n^2}\right)$	$n = 2, 3, 4, \cdots$
Balmer series	$\frac{1}{\lambda} = R_H\left(\frac{1}{2^2} - \frac{1}{n^2}\right)$	$n = 3, 4, 5, \cdots$
Paschen series	$\frac{1}{\lambda} = R_H\left(\frac{1}{3^2} - \frac{1}{n^2}\right)$	$n = 4, 5, 6, \cdots$
Brackett series	$\frac{1}{\lambda} = R_H\left(\frac{1}{4^2} - \frac{1}{n^2}\right)$	$n = 5, 6, 7, \cdots$
Pfund series	$\frac{1}{\lambda} = R_H\left(\frac{1}{5^2} - \frac{1}{n^2}\right)$	$n = 6, 7, 8, \cdots$

The longest wavelength (or highest frequency) in a series is obtained when the smallest value of n (applicable to that series) is used, for example, $n = 3$ in Balmer series, $n = 4$ in Paschen series and so on. The series limit in all cases is obtained by the first term alone, because this occurs at $n = \infty$. We shall understand the reason behind this common formula in connection with Bohr's theory.

Ritz combination principle **(1908)**

According to this principle, the wave number of any spectral line for a particular atom may be represented by the combination of two terms, one of which is constant for a particular series while the other varies regularly through the series. This may be expressed as

$$\bar{\nu} = \frac{R}{(m+\beta)^2} - \frac{R}{(n+\alpha)^2},$$

where R, α and β are constants for a particular element; m and n are integers. m remains constant for a particular series when variation of n produces the wave numbers of the lines in that series. For hydrogen, α and β are equal to zero; $R = R_H$. Hence the relation reduces to

$$\bar{\nu} = R_H\left(\frac{1}{m^2} - \frac{1}{n^2}\right)$$

For other elements, the relation may be applied with proper values of α and β. R is obtained by the relation $R = R_H \cdot Z^2$, where Z = atomic number of the element (see Bohr's theory).

Problem 3.8

The Ritz combination principle was applied to predict the Paschen series in the spectrum of atomic hydrogen. Explain.

Solution: Consider the first two lines H_α and H_β in the Balmer series:

$$\bar{\nu}_\alpha = R_H\left[\frac{1}{2^2} - \frac{1}{3^2}\right]; \quad \bar{\nu}_\beta = R_H\left[\frac{1}{2^2} - \frac{1}{4^2}\right]$$

The difference: $\bar{\nu}_\beta - \bar{\nu}_\alpha = R_H\left[\frac{1}{3^2} - \frac{1}{4^2}\right]$ is also an expression of the type $R_H\left(\frac{1}{m^2} - \frac{1}{n^2}\right)$. Hence it should represent a series in the spectrum where successive lines will be given by varying the second term.

3.5 THE NUCLEAR ATOM MODEL

Up to about the first decade of twentieth century (1910) the following major characteristics of atoms were recorded:

(i) Atoms are stable.
(ii) They are electrically neutral but they contain oppositely charged constituents.
(iii) They are exceedingly small. (The approximate radius of an atom was calculated to be in the range of 10^{-10} m from the kinetic theory of gases).
(iv) Atoms on excitation give rise to discontinuous line spectrum.

An early atom model suggested by J. J. Thomson (1904) considered the electrons as embedded in a sphere of uniform positive charge. This was unsatisfactory for many reasons, specially due to its failure to explain the atomic spectrum and was ultimately discarded.

3.5.1 The Atom Model of Rutherford

The first correct approach to atomic structure was made only when it was established that atoms posses a small nucleus. The corresponding atom model, first forwarded by Lord Rutherford

(1911), is called the nuclear model of the atom. In spite of several extensions and modifications, the basic idea of a nuclear atom is still maintained.

Experimental background

In Rutherford's laboratory, experiments were being carried out on the scattering of alpha-particles by thin metal foils (Geiger and Marsden, 1909). Alpha-particles have a mass four times that of a hydrogen atom and carry two units of positive charge. The massive energetic α-particles from a radioactive source were allowed to strike a very thin gold foil, 0.0004 mm in thickness. The position of the particles after passing through the foil was ascertained by the flash produced on a zinc sulfide screen.

Marsden, a young scholar of Rutherford, was asked to see if any alpha-particle could be scattered through a large angle. Actually most of the alpha particles passed without any appreciable deviation. But all on a sudden, one or two particles out of thousands (approximately 1 in 20,000) returned after hitting the gold foil. To quote Rutherford:

"It was quite the most incredible event that has ever happened to me in my life. It was almost as incredible as if you fired a 15 inch shell at a piece of tissue paper and it came back and hit you."

Analysis of the results

According to Rutherford, the wide angle scattering of the alpha-particle could not occur by the additive effect of a number of small scatterings. It was due to a *single* collision. From the angle of scattering, Rutherford made an idea about the charge of the target which scattered the alpha particle. He concluded that there exists a central position in the atom; the entire mass and positive charge of the atom is concentrated therein. This was termed the *nucleus* of the atom.

As already stated, the chance of such a collision was extremely rare. But the alpha particles passed through a layer of about 1000 gold atoms. This can be easily estimated : atoms were known to have diameters in the range of 10^{-10} m; nearly 4000 such atoms would therefore, produce a thickness of the gold foil used, i.e., 0.0004 mm or 4×10^{-7} m. The rare chance of large angle scattering could, therefore, be explained by assuming the nucleus to be very small in comparison to the total size of the atom.

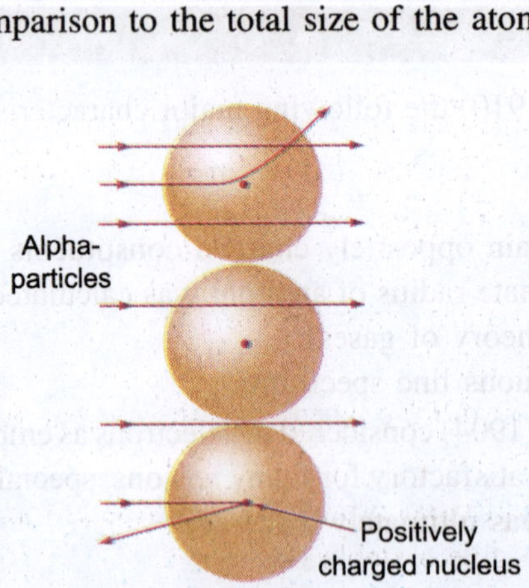

Fig.3.8: Scattering of α-particles as interpreted by Rutherford.

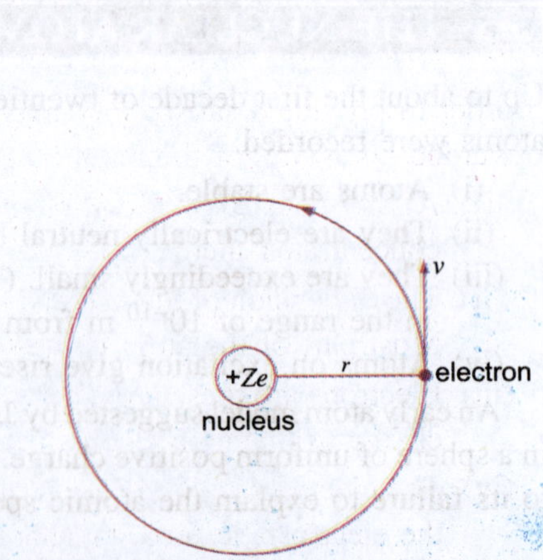

Fig.3.9: The nuclear atom of Rutherford.

Size and charge of the nucleus

Rutherford estimated the charge (Z) in the nucleus from scattering data. This came out nearly half the atomic weight. Accurate determination of nuclear charge was carried out later by Moseley (1913).

The dimension of the nucleus was also calculated. For this purpose it was supposed that the striking α-particle commences to reverse its direction when the kinetic energy of the particle $\left(\frac{1}{2}mv^2\right)$ just balances the potential energy due to repulsion.

The charge of an α-particle is $2e$; let that of the nucleus be Ze. (Here e represents the unit positive charge, equal to the charge of the electron, but opposite in sign.)

The repulsive force at a distance r from the nucleus is: $\dfrac{2e.Ze}{4\pi\varepsilon_0 r^2} = \dfrac{Ze^2}{2\pi\varepsilon_0 r^2}$,

where e and r are expressed in SI units (coulomb and metre respectively).

The corresponding potential energy is obtained by calculating the work done in bringing the α-particle from infinity to distance r from the nucleus. The work has to be done against the repulsive force. Hence the potential energy of the α-particle is equal to

$$-\int_{r=\infty}^{r=r} \frac{Ze^2}{2\pi\varepsilon_0 r^2}.dr = \frac{Ze^2}{2\pi\varepsilon_0 r}$$

When the α-particle approaches close to the nucleus, its direction is reversed at a stage when its kinetic energy is just balanced by this potential energy, i.e.,

$$\frac{1}{2}mv^2 = \frac{Ze^2}{2\pi\varepsilon_0 r}$$

The value of r satisfying this equation may be approximately taken as the radius of the nucleus. With appropriate values of v (velocity of the α-particle), m (mass), Z and e, and putting ε_0 equal to 8.854×10^{-12} kg^{-1} m^{-3} s^4 A^2, r comes out to be around 10^{-15} m.

Rutherford's atom model

In order to provide a satisfactory explanation to the scattering of alpha-particles, Rutherford forwarded his model of the atom in 1911. The basic features of this model are as under:

(i) Atoms consist of a central nucleus, the entire positive charge of the atom being concentrated there.

(ii) The dimension of the nucleus is extremely small in comparison to the size of the atom. Most part of the atom is thus vacant.

(iii) Electrons are present outside the nucleus; their number is equal to the nuclear positive charge to maintain electrical neutrality of the atom.

(iv) The electrons revolve around the nucleus in circular orbits. The centripetal force and the electrostatic force balance each other to give a stable orbit:

$$\frac{mv^2}{r} = \frac{Ze^2}{4\pi\varepsilon_0 r^2}$$

Some later extensions to the model were:

(v) The atomic mass (*A*) of an element was found to be nearly double its nuclear positive charge (*Z*). It was assumed that its nucleus contained *Z* protons and (*A* – *Z*) electrons.

(iv) Later in 1920 Rutherford assumed the existence of neutrons to account for the mass of an atom. They were, however, confirmed only in 1932 by Chadwick (Ch-14).

Merits and demerits

The Rutherford model satisfactorily explained the results of the scattering experiment. It also explained the electrical neutrality of atoms.

But the model suffered from a serious drawback in view of the electromagnetic theory of radiation dominating at that time. According to this theory, accelerated charged particles should continuously emit radiations. An electron revolving around the nucleus is also an accelerated charged particle; so it should also emit radiation continuously and the energy of radiation will be derived from the energy of its motion.

> Both velocity and acceleration are vector quantities, having specific directions. When an electron moves in a circular orbit, its direction changes continuously even when the electron moves with a constant speed. As the direction changes, so also does the velocity. Since the rate of change of velocity is termed acceleration, the electron is accelerated.

As a consequence of this emission of radiant energy, the electron will gradually lose energy. Its distance from the nucleus will decrease continously and ultimately it will fall into the nucleus along a spiral path. The atomic system conceived by Rutherford would thus get destroyed.

Another discrepancy of the model is worth noticing. According to the electromagnetic theory the frequency of radiation from a charged body will be equal to its frequency of revolution. In that case the frequency of the radiation emitted by the electron should change continuously, because it approaches the nucleus along a spiral path of continuously diminishing radius. The observed spectra of such an atom should then be continuous. But the spectra of elements already obtained indicated that atoms emit discontinuous radiations of isolated wavelengths. Thus the Rutherford atom model could not explain the observed line spectra of atoms.

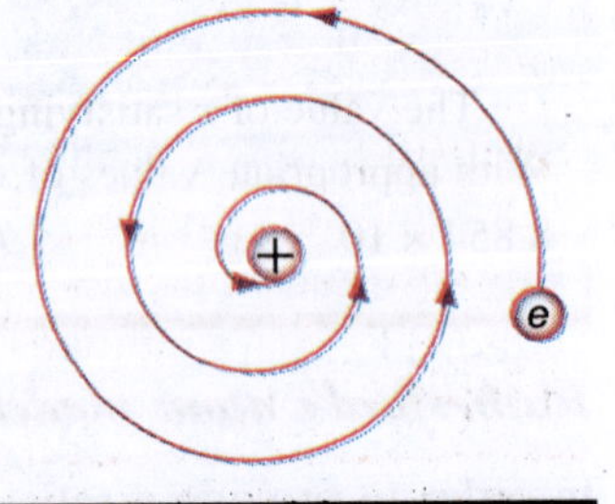

Fig.3.10: The fate of an electron according to the electromagnetic theory (schematic).

One was, therefore, forced to conclude that either (i) the Rutherford atom model was wrong or (ii) the electromagnetic theory was inadequate in this case. But Rutherford's nuclear model was based on definite experimental background. So the position of the electromagnetic theory was reconsidered by the scientists.

3.5.2 The Bohr Atom Model

The problem of stability of the atom in the Rutherford model was ultimately solved by Niels Bohr (1913). He set aside the eletromagnetic theory and applied the quantum theory of radiation successfully to interpret the line spectrum of hydrogen.

Basic postulates of Bohr's theory

(i) In a Rutherford type atom model, the electrons can revolve around the nucleus only along certain *selected* orbits. An electron does neither absorb nor emit any radiation as long as it moves in such a selected orbit. As the energy of an electron in such orbits ramains ulaltered, such orbits are also called *stationary orbits.*

The criterion of selecting such an orbit follows from the principle: *angular momentum of the electron should be an integral multiple of h/2π* (h = Planck's constant). If the mass and the velocity of the electron be respectively m and v at an orbit of radius r, its angular momentum will be mvr. Accordingly r may have only those values for which

$$mvr = n.\frac{h}{2\pi}, \text{ where } n \text{ is a non-zero positive integer, i.e., } 1, 2, 3, \cdots\cdots \text{ etc.}$$

The angular momentum of the electron is thus said to be "quantized", i.e., subdivided into small units of $h/2\pi$. Now the angular momentum can increase only by steps of $h/2\pi$.

The integer n used in the quantization of angular momentum was termed the *quantum number* of the particular orbit. Subsequently this has been termed the principal quantum number. Its significance will be discussed later.

> However, imposition of the condition $mvr = nh/2\pi$ for selecting the orbits was rather ad hoc; Bohr did not provide any satisfactory explanation for this except that this explains the spectral lines so nicely. We shall find (Chapter 4) that the condition follows automatically from the wave nature of electron.

(ii) When an electron jumps from one such selected orbit to another, the difference of energy between the two states is emitted or absorbed in accordance with the quantum theory of radiation. Thus, when an electron jumps from an orbit with energy E_2 to an orbit of energy E_1 ($E_1 < E_2$), the difference of energy $E_2 - E_1$ is emitted in the form of quantized radiation. The frequency ν of this emitted radiation will be given by the relation:

$$E_2 - E_1 = h\nu.$$

Thus the appearance of discrete lines in the spectrum of atomic hydrogen can be explained in terms of jumps of electrons between different selected orbits; radiations of various definite frequencies are emitted thereby (see later).

(iii) The atomic system obeys classical mechanics (Newtonian) in all other aspects.

Simple mathematical treatment

The treatment here is shown for the hydrogen atom as well as *hydrogen-like species* which contain a single orbital electron outside a nuclear charge of $+Z_1e$. He^+ ($Z = 2$), Li^{2+} ($Z = 3$), Be^{3+} ($Z = 4$), etc. are hydrogen-like or hydrogenic.

The values of e, m, ε_0 and h (in their usual meaning) are:

$e = 1.602 \times 10^{-19}$ C ; $m = 9.109 \times 10^{-31}$ kg ; $\varepsilon_0 = 8.854 \times 10^{-12}$ $C^2N^{-1}m^{-2}$;

$4\pi\varepsilon_0 = 1.11265 \times 10^{-10}$ $C^2N^{-1}m^{-2}$;

$1/4\pi\varepsilon_0$ (**Coulomb's law constant**) = 8.9875×10^9 N $m^2C^{-2} \simeq 9 \times 10^9$ N m^2C^{-2}.

Radii of the Bohr orbits

In Bohr's theory, the basic Rutherford model is maintained. Hence the condition of equibrium between centripetal force and coulombic force still applies to these orbits. For an electron moving in an orbit of radius r around a nucleus with Z units of positive charge, we have

$$\frac{mv^2}{r} = \frac{Ze^2}{4\pi\varepsilon_0 r^2} \tag{3.8}$$

At the same time, the Bohr condition of angular momentum requires

$$mvr = \frac{nh}{2\pi}. \tag{3.9}$$

Eliminating v, we get $$r = \frac{\varepsilon_0 n^2 h^2}{\pi m Z e^2} \tag{3.10}$$

This is the desired expression of the radii of permitted Bohr orbits.

For a hydrogen atom, $Z = 1$. The radius of the smallest possible orbit in this atom, r_0, may be calculated by putting $n = 1$ and other known values in the expression of r:

$$r_0 = \frac{\varepsilon_0 h^2}{\pi m e^2} = \frac{8.854\times10^{-12}\,\text{C}^2\text{N}^{-1}\text{m}^{-2}\times(6.626\times10^{-34})^2\,\text{J}^2\text{s}^2}{3.141\times9.109\times10^{-31}\,\text{kg}\times(1.602\times10^{-19})^2\,\text{C}^2}$$

$$= 5.293\times10^{-11}\,\text{m}\ (= 0.05293\ \text{nm})$$

This is the radius of the **first Bohr orbit** (denoted by a_0). Clearly, $r_n = r_0 \times n^2$

The value of the first Bohr radius of the hydrogen atom, i.e., 0.05293 nm is by definition the **atomic unit of length**, sometimes called the **Bohr**.

Velocity and orbital frequency of an electron

The speed of an electron in a Bohr orbit (v) is simply

$$v = \frac{nh}{2\pi m r} \quad \text{where } r = \frac{\varepsilon_0 n^2 h^2}{\pi m e^2 Z}$$

So, $$v = \frac{Ze^2}{2\varepsilon_0 nh}; \quad \text{for hydrogen, } v = \frac{e^2}{2\varepsilon_0 nh}$$

The orbital frequency f of an electron is $= \dfrac{\text{electron speed}}{\text{circumference of the orbit}} = \dfrac{v}{2\pi r}$

or, $f = \dfrac{Z^2 m e^4}{4\varepsilon_0^2 n^3 h^3}$ (substituting values of v and r; for hydrogen, $Z = 1$).

In c.g.s. system of unit, $r = \dfrac{n^2h^2}{4\pi^2 m Z e^2}$; $v = \dfrac{nh}{2\pi m r}$; $f = \dfrac{4\pi^2 m e^4 Z^2}{n^3 h^3}$

(See Problem 3.15)

Energy of an electron in a Bohr orbit

An electron in the atom possesses (i) kinetic energy by virtue of its motion and (ii) potential energy due to its attachment to the nucleus. The total energy is the sum of these two energy terms. The two energy terms will be calculated separately.

Kinetic energy: $E_k = \frac{1}{2}mv^2 = \frac{Ze^2}{8\pi\varepsilon_0 r}$. [3.11]

Potential energy: The potential energy of an electron in an atom is calculated with reference to its position at infinity from the nucleus; at this point the potential energy is assigned zero value. The potential energy is then given by the total work *done* in bringing the electron from infinity to its position in the orbit of radius *r*. The force involved is the coulombic force of attraction by the nucleus, $Ze^2/4\pi\varepsilon_0 r^2$.

The work done (W) and hence the potential energy (E_p) is then given as

$$E_p = W = \int_{r=\infty}^{r=r} \frac{Ze^2}{4\pi\varepsilon_0 r^2} = -\frac{Ze^2}{4\pi\varepsilon_0 r}. \qquad [3.12]$$

The negative sign implies that work has to be done to remove the electron away from the nucleus.

Total energy of the electron: Since the total energy (E) of the electron is the sum of its kinetic energy (E_k) and potential energy (E_p), we may write

$$E = \frac{Ze^2}{8\pi\varepsilon_0 r} - \frac{Ze^2}{4\pi\varepsilon_0 r} = -\frac{Ze^2}{8\pi\varepsilon_0 r} = -\frac{Z^2 me^4}{8\varepsilon_0^2 n^2 h^2}. \qquad [3.13]$$

In c.g.s system of unit, $E = -\frac{2\pi^2 Z^2 me^4}{n^2 h^2}$

For precise calculations, m, the mass of the electron, should be replaced by μ, the reduced mass where $\mu = m/(1 + M)$, M = mass of the nucleus (see motion of the nucleus).

The negative sign of the energy expression is very significant. This means that the electron is actually bound to the atomic system by this amount of energy. If the electron is to be freed from the influence of the nucleus, work equal to the magnitude of E has to be done *on* it (see ionization energy).

The presence of n^2 in the denominator suggests that in an atom the *magnitude* of E decreases with increasing value of n. Thus the magnitudes of E for the first, second and third orbits in an atom will be in the ratio of

$$\frac{1}{1^2} : \frac{1}{2^2} : \frac{1}{3^2} \text{ or } 1 : \frac{1}{4} : \frac{1}{9}.$$

It is clear that as n becomes gradually larger, the separation between the energy levels becomes smaller; the energy levels thus crowd together and tend to merge as n becomes very large (also see Fig.3.12).

The actual energy of the electron, however, increases with increasing n due to the negative sign of E. Thus an orbit with a higher n value corresponds to higher energy of the electron.

The energy of an electron in the first Bohr orbit of hydrogen ($n = 1$, $Z = 1$) is

$$E_1 = -\frac{me^4}{8\varepsilon_0^2 h^2} = -2.18 \times 10^{-18}\ \text{J} = -13.6\ \text{eV}.$$

This is often called the **ground state energy**. The energy in any other orbit of principal quantum number n is

$$E_n = -\frac{me^4}{8\varepsilon_0^2 n^2 h^2} = -\frac{E_1}{n^2} = -2.18 \times 10^{-18} \times \frac{1}{n^2}\ \text{J} \equiv -\frac{13.6}{n^2}\ \text{eV}.$$

For other hydrogenic species,

$$E_n = -2.18 \times 10^{-18} \times \frac{Z^2}{n^2}\ \text{J} \equiv -13 \cdot 6 \frac{Z^2}{n^2}\ \text{eV}.$$

Problem 3.9

Calculate the size of the He^+ ion from Bohr's theory.

Solution: The size of the He^+ ion is given by the distance at which the outermost electron revolves around the nucleus. This is equal to the first Bohr radius calculated for He^+.

We know, $r = \dfrac{\varepsilon_0 n^2 h^2}{\pi m Z e^2}$. Here $Z = 2$ (for helium); $n = 1$ (first orbit).

$$\therefore \quad r = \frac{8.854 \times 10^{-12}\ \text{C}^2\text{N}^{-1}\text{m}^{-2} \times (6.626 \times 10^{-34})^2\ \text{J}^2\text{s}^2}{3.141 \times 9.109 \times 10^{-31}\ \text{kg} \times 2 \times (1.602 \times 10^{-19})^2\ \text{C}^2} = 2.648 \times 10^{-11}\ \text{m}.$$

[This is half the first Bohr radius for hydrogen, as in this case only Z has been doubled.]

Problem 3.10

Show that for any hydrogen-like ion, $r_n = \dfrac{52.9 n^2}{Z}$ pm $= \dfrac{n^2 a_0}{Z}$ pm,

where terms have their usual meaning.

Try yourself.

Problem 3.11

Calculate the velocity of an electron excited to the third Bohr orbit in a hydrogen atom.

Solution: From the relations $mvr = \dfrac{nh}{2\pi}$ and $r = \dfrac{\varepsilon_0 n^2 h^2}{\pi m e^2 Z}$

$$v = \frac{e^2 Z}{2\varepsilon_0 nh}; \quad \text{here, } Z = 1, n = 3.$$

Putting usual values of other terms,

$$v = 7.29 \times 10^5\ \text{m s}^{-1}.$$

Note: (1) The velocity of the electron in the first Bohr orbit of hydrogen will be thrice this value ($n = 1$).

(2) The velocity of an electron in the He^+ ion will be twice its velocity in an orbit of same n in the hydrogen atom as Z for He^+ is 2.

Problem 3.12

Using the Bohr model, calculate the ratio of the cross-sectional area of a hydrogen atom in the excited $n = 10$ state to that of one in the n = 1 state.

Solution: From Bohr model (Eqn. 3.10), $r = \varepsilon_0 n^2 h^2/\pi m Z e^2$, where the different notations have their usual significance. When other quantities are constant, i.e., for the same atom, $r \propto n^2$. Again, cross-sectional area (A, say) = πr^2 for a circle. $\therefore A \propto n^4$.

$$\text{Hence } \frac{A(n=10)}{A(n=1)} = \frac{10^4}{1} = 10000:1.$$

Problem 3.13

What would be the size of a Ne^{9+} ion according to the Bohr model?

Solution: The Ne^{9+} ion consists of a single electron moving outside a nucleus containing 10 units of positive charge ($Z = 10$). So we may apply the expression of r from Bohr model (Eqn. 3.10) and substitute $Z = 10$ and the value of other quantities. The value of n will be 1 since electrons are ionized from the outermost inward; the remaining electron in Ne^{9+} occupies the lowest energy state with $n = 1$.

$$\therefore r\left(Ne^{9+}\right) = \frac{\varepsilon_0 . 1^2 . h^2}{\pi . m . e^2 . 10} = 5.293 \times 10^{-12} \text{ m.}$$

Interpretation of the spectrum of atomic hydrogen

Let us consider two orbits with quantum numbers n_1 and n_2 in a Bohr type atom, such that $n_2 > n_1$; let the corresponding energy values be E_{n_1} and E_{n_2} respectively; $E_{n_2} > E_{n_1}$.

If an electron initially excited to the higher orbit (n_2) returns to the lower orbit (n_1), the difference of energy ($E_{n_2} - E_{n_1}$) will be emitted in the form of radiation and give rise to a spectral line of frequency ν :

$$E_{n_2} - E_{n_1} = h\nu.$$

Since $Z = 1$ for hydrogen, $E_{n_1} = -\dfrac{me^4}{8\varepsilon_0^2 h^2 n_1^2}$; $\quad E_{n_2} = -\dfrac{me^4}{8\varepsilon_0^2 h^2 n_2^2}$.

$$\therefore E_{n_2} - E_{n_1} = h\nu = \frac{me^4}{8\varepsilon_0^2 h^2}\left[\frac{1}{n_1^2} - \frac{1}{n_2^2}\right]$$

or, $\quad \nu = \dfrac{me^4}{8\varepsilon_0^2 h^3}\left[\dfrac{1}{n_1^2} - \dfrac{1}{n_2^2}\right]$ $\quad$ and $\quad \bar{\nu} = \dfrac{\nu}{c} = \dfrac{me^4}{8\varepsilon_0^2 c h^3}\left[\dfrac{1}{n_1^2} - \dfrac{1}{n_2^2}\right]$.

Substituting values for m, e, ε_0 and h, $\bar{\nu} = 1.096775 \times 10^7 \left[\dfrac{1}{n_1^2} - \dfrac{1}{n_2^2}\right] \text{m}^{-1}$.

This expression agrees excellently with the experimentally established relation of Rydberg for the spectrum of atomic hydrogen:

$$\bar{\nu} = R\left\{\frac{1}{n_1^2} - \frac{1}{n_2^2}\right\}; \quad R = 1.096776 \times 10^7 \text{ m}^{-1}.$$

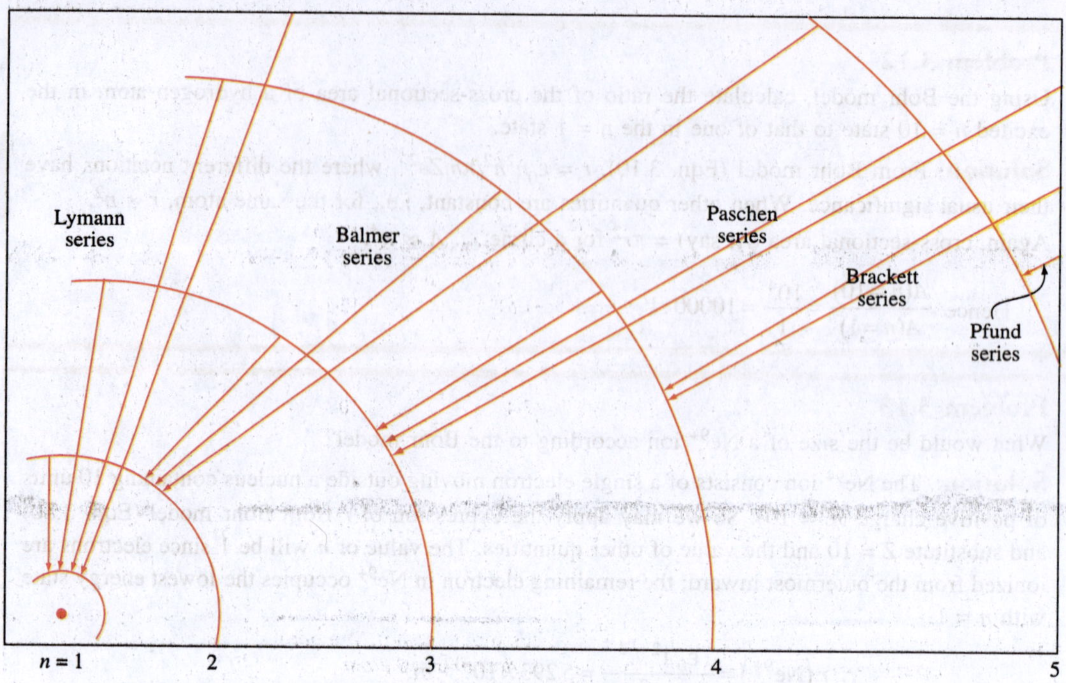

Fig.3.11: Transitions which give rise to different series in the spectrum of atomic hydrogen (schematic). One electron can undergo only one transition at any instant of time giving rise to a line of one particular frequency or wavelength. In an aggregate of atoms, different transitions occur in different atoms.

Thus the value of R, the Rydberg constant, has been arrived at theoretically from Bohr's postulates. Different values of $\bar{\nu}$ may now be obtained by using separate pairs of values for n_1 and n_2 as illustrated in Fig. 3.11. Bohr's theory can thus reproduce the lines in the spectrum of atomic hydrogen with appreciable accuracy. A diagrammatic representation of the energy-levels corresponding to various Bohr orbits is shown in Fig. 3.12.

It may be noted in Fig. 3.12 that the energy levels become gradually closer as n increases and finally converges to a limit. Correspondingly, the lines in the spectrum approach each other more closely until they merge into a continuum. This corresponds to a region of continuous absorption or emission.

In fact, the nuclear field practically disappears when r becomes very large. The series limit approximately represents the range of effective nuclear field. Beyond this an electron is no more guided by the quantum restrictions and it can emit or absorb radiation *continuously*.

The height of the series limit above the ground state represents the ionization energy of an atom in the ground state.

Ground and excited states; excitation and ionization energy. We have seen that the energy of an electron in the ground state of the H-atom is

$$E = -me^4 / 8\varepsilon_0^2 h^2 = -2.18 \times 10^{-18} \text{ J } (-13.60 \text{ eV})$$

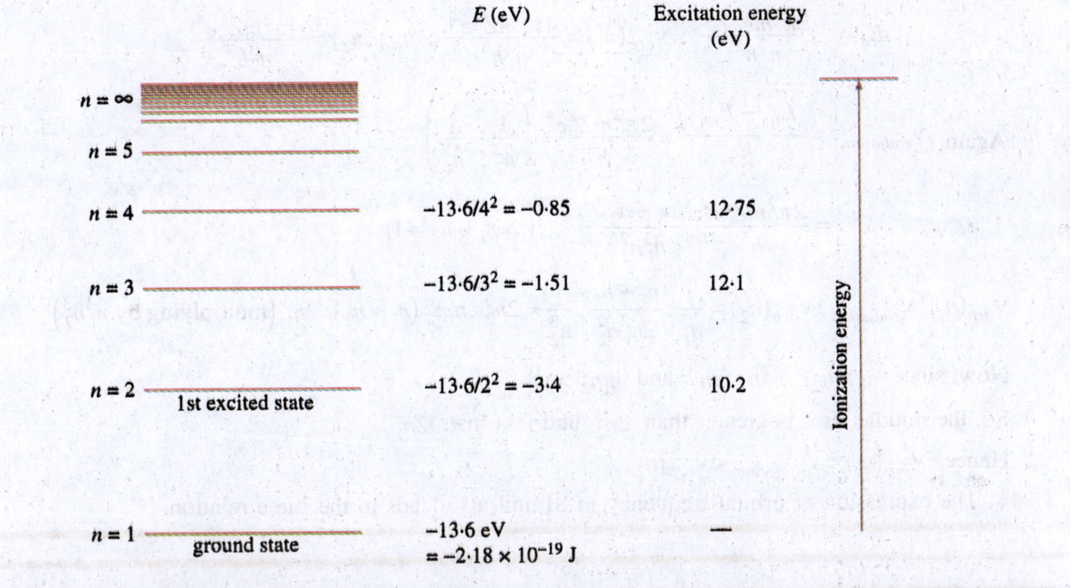

Fig.3.12: Energy-levels corresponding to the Bohr orbits (not to scale). The ground state corresponds to $n = 1$. The state corresponding to $n = 2$ is called the ***first excited state.***
The energy required to promote an electron from the ground state to an excited state is called the *excitation energy*.
The energy difference between the levels $n = 1$ and $n = \infty$ is the *ionization energy.*

In order to separate this electron from the atom (that is, to ionize it), this 13.60 eV of energy is to be supplied; this is known as the **ionization energy**. Converted to a mole, the total energy of all the electrons in 1 mole of H-atom in ground state becomes $21.79 \times 10^{-19} \times 6.02 \times 10^{23}$ J mol^{-1} = −1312 kJ mol^{-1}. So the ionization energy would be 1312 kJ mol^{-1}.

Problem 3.14

Show that the ground state energy E_1 of an electron in the hydrogen atom is $-R_H \cdot ch$, where R_H is the Rydberg constant for hydrogen.

Solution: $R_H = \dfrac{me^4}{8\varepsilon_0^2 ch^3}$ $\qquad E_1 = -\dfrac{me^4}{8\varepsilon_0^2 h^2} = R_H . ch$

Problem 3.15

Show that the frequency of revolution of an electron in a Bohr orbit (quantum number n) is given by the expression

$$\nu_{orb} = \frac{4\pi^2 mZ^2 e^4}{n^3 h^3} \quad \text{(terms have their usual meaning.)}$$

Hence show that the frequency ν of the emitted radiation for transition from n_1 to n_2 ($n_1 - n_2 = 1$) is intermediate between the frequencies of orbital revolution in these two orbits.

Solution: The expression for orbital frequency has been deduced earlier. Absence of ε_0 in the expression given here suggests non-SI approach. So we use the expressions:

$$E_n = -\frac{2\pi^2 me^4 Z^2}{n^2h^2} \; ; \; \nu_{orb}(n_1) = \frac{4\pi^2 mZ^2e^4}{n_1^3h^3} \; ; \; \nu_{orb}(n_2) = \frac{4\pi^2 mZ^2e^4}{n_2^3h^3}$$

Again, $$\nu_{radiation} = \frac{E_{n_1} - E_{n_2}}{h} = \frac{2\pi^2 mZ^2e^4}{h^3}\left(\frac{1}{n_2^2} - \frac{1}{n_1^2}\right)$$

$$= \frac{2\pi^2 mZ^2e^4}{h^3} \cdot \frac{n_1 + n_2}{n_1^2 n_2^2} \qquad (\because n_1 - n_2 = 1)$$

$$\nu_{orb}(n_1) : \nu_{radiation} : \nu_{orb}(n_2) = \frac{1}{n_1^3} : \frac{n_1 + n_2}{2n_1^2 n_2^2} : \frac{1}{n_2^3} = 2n_2^3 : n_1 n_2 (n_1 + n_2) : 2n_1^3 \; \left(\text{multiplying by } n_1^3 n_2^3\right)$$

Now, since $n_1 > n_2$, $n_1^2 n_2 > n_2^3$ and $n_1 n_2^2 > n_1^3$.

So, the middle term is greater than $2n_2^3$ and less than $2n_1^3$.

Hence, $\nu_{orb}(n_1) < \nu_{radiation} < \nu_{orb}(n_2)$.

Note: The expression of orbital frequency in SI unit also leads to the same relation.

Bohr's correspondence principle

Though the wave numbers of spectral lines were given by the Bohr's theory, no inference regarding the nature and intensity of the lines could be made from it. Bohr, therefore, established a principle linking the classical theory and the quantum theory. This enabled him to make use of the results of the classical theory. The principle, known as Bohr's correspondence principle, states that for large quantum numbers the behaviour of the atom tends to that which would be expected from the classical theory. Accordingly, for very large values of *n*, the frequency of emitted radiations are nearly the same as the orbital frequency.

Motion of the nucleus

According to a principle of dynamics, whenever a particle moves in a circle about another of finite mass, the central part must also move in a circle about the common centre of gravity. When this principle of dynamics is applied to the motion of the electron in the Bohr model, the situation gets complicated by the motion of the nucleus. A solution to this problem is to use the *reduced mass* μ of the electron in place of m used above; μ is expressed as

$$\mu = mM / (m + M) = m / (1 + m/M),$$

where M = mass of the nucleus. This gives better agreement with experimental data. It has been of particular importance in explaining many hydrogen-like spectra, e.g., the spectra of He^+.

It should be understood, however, that the mass of the electron does not really decrease owing to motion of the nucleus. The reduced mass is only a parameter having the unit of mass and a value slightly less than the rest mass of the electron. It is used in the Bohr equation to adjust the latter for the effect of nuclear motion.

Consideration of the motion of the nucleus modifies the expressions of energy, the Rydberg constant and other related terms:

$$E_n = -\frac{\mu e^4 Z^2}{8\varepsilon_0^2 n^2 h^2} \; ; \quad R_H = \frac{\mu e^4}{8\varepsilon_0^2 ch^3} = \frac{me^4}{8\varepsilon_0^2 ch^3} \cdot \frac{1}{1 + m/M} = \frac{R_\infty}{1 + m/M}.$$

R_∞ is the Rydberg constant calculated assuming a nucleus of infinite mass. In general, the value of the Rydberg constant will slightly increase from R_H for heavier elements/isotopes and in the limit converge to the value of R_∞ (1.0973731×10^7 m^{-1}). The difference in the value of R for two isotopes gives rise to a splitting of the spectral lines (see Problem 3.19). The isotope deuterium was actually discovered by this method.

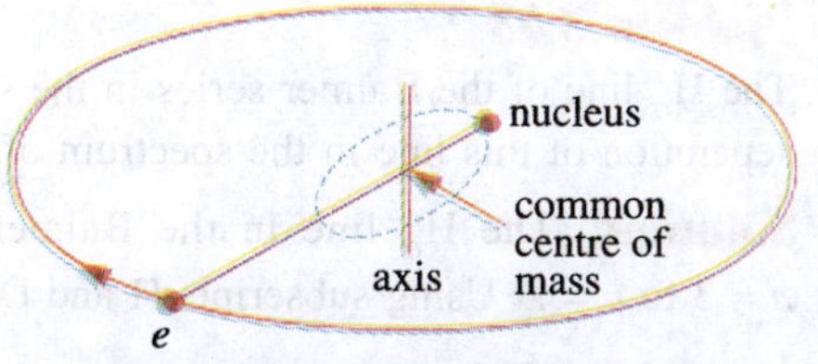

Fig.3.13: Motion of the electron and the nucleus around their common centre of mass.

Problem 3.16

What will be the wavelength of the second line in the visible spectrum of atomic hydrogen?

Hint: Here the series is not named. But we know that in the spectrum of atomic hydrogen, only the Balmer series lies in the visible range. So we are to calculate the wavelength of the second line in the Balmer series.

For the Balmer series, $n_1 = 2, n_2 = 3, 4, \ldots$, etc. For the second line, $n_2 = 4$, $R = 1.097 \times 10^7$ m^{-1}. This gives $\bar{\nu} = 1/\lambda = 2.06 \times 10^6$ m^{-1}. or, $\lambda = 4.86 \times 10^{-7}$ m.

Problem 3.17

Calculate the potential difference (p.d.) through which an electron must be accelerated in order to (*a*) raise the energy of a hydrogen atom from the ground state to the second excited state; (*b*) ionize the atom.

Hint: (*a*) From equation (3.13) and (3.14), the energy of the electron in the ground (n = 1) and second excited state (n = 3) will be

$$E_1 = -13.6 \text{ eV} \qquad E_3 = -13.6/3^2 \text{ eV} = -1.51 \text{ eV}.$$

So the excitation energy will be $E_3 - E_1 = (-1.51 + 13.6)$ eV = 12.09 eV.

An electron acquires this energy when it is accelerated through a p.d. of 12.09 V.

(*b*) The ionization energy corresponds to complete removal of the electron from the ground state.

Problem 3.18

The Rydberg constants for hydrogen and helium are 1.096776×10^7 m^{-1} and 1.097224×10^7 m^{-1} respectively. Assuming that a proton is 1836 times heavier than an electron, find the relative mass of the helium nucleus.

Solution: Putting $m = M_H/1836$, $M_{He}/M_H = 3.95$.

$$\frac{R_H}{R_{He}} = \frac{1096776}{1097224} = \frac{1 + m/M_{He}}{1 + m/M_H} \Rightarrow m/M_{He} = 0.00013.$$

Problem 3.19

The H_α line of the Balmer series in the spectrum of hydrogen occurs at 656.28 nm. Calculate the separation of this line in the spectrum of deuterium.

Solution: The H_α line in the Balmer series arises due to transition of the electron from $n = 3$ to $n = 2$. Using subscripts H and D for hydrogen and deuterium respectively, we may write,

$$\bar{\nu}_H = R_H\left(\frac{1}{2^2}-\frac{1}{3^2}\right) \text{ and } \bar{\nu}_D = R_D\left(\frac{1}{2^2}-\frac{1}{3^2}\right).$$

$$\therefore \quad \frac{\bar{\nu}_H}{\bar{\nu}_D} = \frac{R_H}{R_D} = \frac{1+m/M_D}{1+m/M_H} = \frac{1+m/2M_H}{1+m/M_H} = 0.9997277$$

$$\text{or, } \frac{\bar{\nu}_D - \bar{\nu}_H}{\bar{\nu}_D} = 0.0002723 \qquad \text{or, } \frac{1/\lambda_D - 1/\lambda_H}{1/\lambda_D} = \frac{\Delta\lambda}{\lambda_H} = 0.0002723.$$

$$\therefore \quad \Delta\lambda = 0.0002723 \times 656.28 \text{ nm.} = 0.179 \text{ nm.}$$

Problem 3.20

The Balmer series in the spectrum of atomic hydrogen arises from transitions between $n = 2$ state and higher states. Compare the wavelengths for the first three lines in the Balmer series with those expected for similar transitions in Be^{3+} ion ($R_H = 10967758 \text{ m}^{-1}$).

Solution: Using subscripts H and Be for the two species, we can write, for transitions giving rise to the Balmer series

$$\bar{\nu}_H = R_H\left(\frac{1}{2^2}-\frac{1}{n^2}\right) \text{ and } \bar{\nu}_{Be} = R_{Be} \times 4^2 \times \left(\frac{1}{2^2}-\frac{1}{n^2}\right); \ (Z = 4 \text{ for Be}; n > 2).$$

$$\therefore \quad \frac{\bar{\nu}_{Be}}{\bar{\nu}_H} = \frac{\lambda_H}{\lambda_{Be}} = \frac{R_{Be}}{R_H} \times 16.$$

If we ignore the slight variation in the value of the Rydberg constant, we may conclude that $\lambda_H = 16 \times \lambda_{Be}$, that is the transitions in Be^{3+} will occur at wavelengths 1/16 times the hydrogen wavelengths.

A more refined estimate may be obtained by calculating R_{Be} from the relation

$$R_{Be} = R_\infty / (1 + m / M_{Be}).$$

$$\text{For } {}^9_4\text{Be}, M_{Be} = 9M_H = 9 \times 1836\, m; \ R_\infty = 10973731 \text{ m}^{-1} \Rightarrow R_{Be} = 10973065 \text{ m}^{-1}.$$

$$\therefore \quad R_{Be} / R_H = 10973065 \div 10967758 = 1.00004836 \text{ or, } \lambda_H / \lambda_{Be} = 16.00774.$$

Now we may readily calculate the wavelengths for the first three transitions in hydrogen by putting $n = 3$, 4 and 5 respectively in the relation

$$\frac{1}{\lambda_H} = R_H\left(\frac{1}{2^2}-\frac{1}{n^2}\right).$$

Merits and demerits of the Bohr atom model

Bohr's model of the atom successfully explained (i) the stability of atoms and (ii) the origin of discontinuous line spectrum by them. It also helped in understanding some related phenomena like the production of characteristic x-rays by elements. Satisfactory agreement with experimental results was observed for the Rydberg constant, excitation potential of atoms, characteristic x-rays, etc.

But the model had its own drawbacks also. Using instruments of high resolving power, several of the apparent "single" lines in the spectrum were found to consist of a number of closely spaced lines. This fine structure of the spectral lines has no explanation in Bohr's theory. Also, its agreement with experiment in the case of higher elements was far from satisfactory.

Bohr's introduction of the concept of integers (n) in selecting the stable orbits for electrons was also arbitrary, merely to fit experimental results. There was no theoretical justification for it.

Bohr discarded the classical behaviour of the electron as regards the emission of radiation. But the motion of the electron was still considered by the classical Newtonian mechanics. The position and the momentum of the electron in a Bohr orbit could be evaluated accurately at the same time. Later developments in physics showed that this is not possible for a tiny particle like the electron; there always exists some uncertainty in these quantities. Bohr's theory was thus inadequate to cope up with later developments of science.

3.5.3 Sommerfeld's Extension

The observed fine structure of the spectrum of atomic hydrogen stood against Bohr's concept of fixed energy levels for the electrons. If, according to Bohr, an electron undergoes transition from one definite energy level to another, the difference of energy should come out as radiation of a particular frequency—giving rise to a single line in the spectrum. The fine structure, therefore, suggests the presence of more energy levels for any given value of n. This indicates the presence of energy sublevels corresponding to each Bohr orbit.

Principal aspects of Sommerfeld's treatment

In 1915 Sommerfeld proposed that the general shape of an electron's orbit was elliptical; corresponding to each principal quantum number (n), several orbits of varying ellipticity were possible. The nucleus was now situated at one of the focii of the ellipse describing the electron's orbit. When the two axes of an ellipse became equal, a circular orbit was obtained.

The position of an electron moving in an elliptical orbit is determined by two quantities (Fig. 3.14): (i) the radius-vector (OA) and (ii) the angle subtended by the radius-vector with the semi-

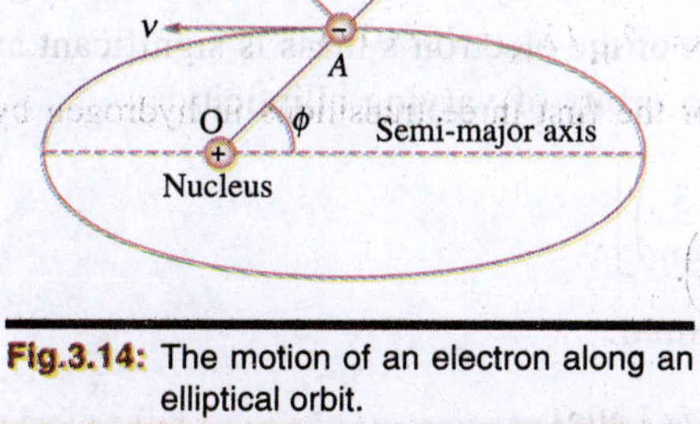

Fig.3.14: The motion of an electron along an elliptical orbit.

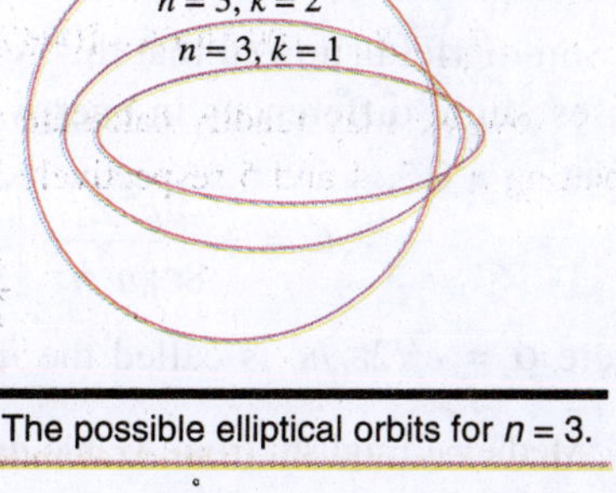

Fig.3.15: The possible elliptical orbits for $n = 3$.

major axis (ϕ). ϕ is called the *azimuthal* angle*. The velocity (v) of the electron is directed tangentially to the curve; accordingly, its momentum can be resolved into two components: (i) one component along the radius vector (*AB*) (ii) and the other perpendicular to it (*AC*). The second part now represents the angular momentum. Two quantum numbers, n_r and k were now necessary to separately quantize each resolved part of the momentum. These are (i) radial quantum number, n_r and (ii) azimuthal quantum number, k. These were related to the principal quantum number (n) and the ellipticity of the orbit by the following relations:

$$\text{(i) } n = n_r + k, \qquad \text{(ii) } \frac{n}{k} = \frac{\text{length of semi-major axis } (a)}{\text{length of semi-minor axis } (b)}.$$

The ellipticity of the orbits was thus determined by the relative values of n and k. The eccentricity of the ellipse ε is expressed as

$$\frac{b}{a} = \frac{k}{n} = \sqrt{1-\varepsilon^2}$$

For any given n, permissible values of n_r and k are:

$$n_r = n, (n-1), (n-2), \cdots, 1, 0; \; k = n, (n-1), (n-2), \cdots, 1.$$

Calculation shows that the total energy of an electron in an orbit of principal quantum number n is given by

$$E_n = -\frac{me^4Z^2}{8\varepsilon_0^2 n^2} = -\frac{me^4Z^2}{8\varepsilon_0^2 (n_r + k)^2}$$

This implies that the energy of an electron would be same in any orbit of principal quantum number n, whatever be the ellipticity. At this point Sommerfeld considered relativistic variation of the mass of the electron during its motion in an elliptical orbit and showed that the energy of the electron will be different for different values of k.

To maintain balance with the varying electrostatic force with varying distance from the nucleus, an electron moving in an elliptical orbit has to change its velocity continuously. The velocity is greatest when the electron is closest to the nucleus; it decreases gradually as the electron moves farther. According to the theory of relativity, the mass of the electron also changes according to the relation

$$m = \frac{m_0}{\sqrt{1 - v^2 / c^2}}$$

where m_0 is the rest mass of the electron, m its mass at velocity v and c is the speed of light.

Sommerfeld showed that this relativistic variation of the electron's mass is significant and causes slight differences in energy of the electron in orbits of varying ellipticity:

$$E = -\frac{\mu e^4 Z^2}{8\varepsilon_0^2 n^2 h^2}\left[1 + \frac{\alpha^2 Z^2}{n}\left(\frac{1}{k} - \frac{3}{4n}\right)\right],$$

where $\alpha = e^2/2\varepsilon_0 hc$ is called the *fine structure constant*.

*Medieval English; from Arabic *as-sumüt*, the way (*via* Latin).

Relativistic variation of the mass of the electron has a further consequence: each time the electron passes near the nucleus, the equilibrium force between the electron and the nucleus is slightly perturbed and the major axis of the ellipse shifts, i.e., undergoes *precession*. The resulting path of the electron appears like a *rosette* (Fig. 3.16).

The fine structure of the spectra was now understood by assuming transitions involving orbits of different n and k. However, experimentally it was found that only those fine-structure lines are actually observed which correspond to $\Delta k = \pm 1$. This has been used as a ***selection rule*** for transitions which are allowed: $\Delta k = \pm 1$. This puts a limit to the number of fine structure lines (Fig. 3.17).

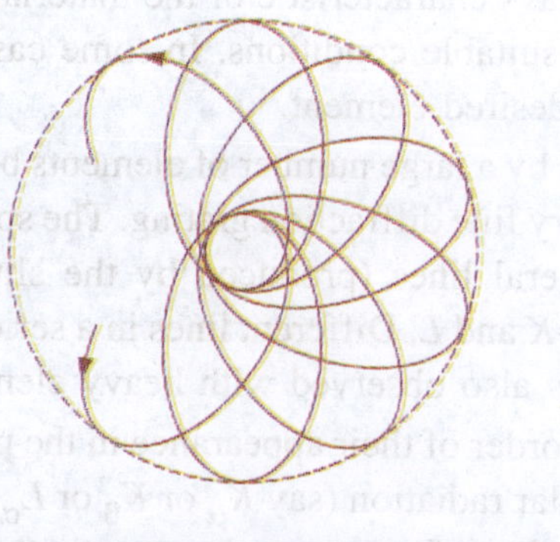

Fig.3.16: The electron's path represented by a precessional ellipse.

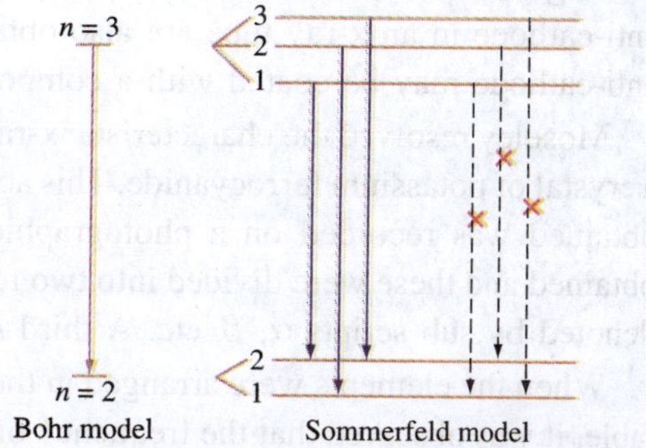

Fig.3.17: Transitions responsible for fine structure of H_α line in the Balmer series of hydrogen. Dashed lines are not allowed by the selection rules.

The azimuthal quantum number k has subsequently been replaced by the symbol l for certain mathematical advantages. The values of l are fixed in relation to k as $l = k - 1$. Thus for any given value of n, l may have values 0, 1, $\cdots$ to $(n - 1)$.

Sommerfeld's theory was supported by experimental data; the agreement, however, was not perfect in all cases. At the same time, the objections raised against the Bohr idea of discrete orbits by later developments in physics are equally applicable to Sommerfeld's theory. Also, the correct number of fine structure lines was not available from this theory and further quantum numbers were required for this purpose.

The existence of definite energy levels within the atom has been further confirmed from successive excitation potentials of electrons and corresponding spectral observations. The x-ray spectra of elements were also explained satisfactorily from the Bohr-Sommerfeld atom model (Section 3.5.4).

Further interpretation of atomic spectra introduced two more quantum numbers. The magnetic quantum number was introduced to explain the observed splitting of spectral lines when the source of spectra was placed in a magnetic field (*Zeeman effect*). Further details of the spectra led to the spin quantum number (Chapter 4). It was soon realized that an atomic system has its own peculiarities originating from the very microscopic nature of the system. These could be more rationally interpreted only with a new approach based on wave properties of matter. This will be discussed in the next chapter.

3.5.4 Characteristic X-ray Spectra of Elements

Experiments on the scattering of α-particles revealed that for most elements (specially the light ones), the nuclear charge was almost half their atomic weights. A. van den Broek (1913) suggested that the nuclear charge is equal to the serial number of appearance of an element in the periodic table, allowance being made for vacant spaces. This idea was later confirmed by the works of H.G.J. Moseley (1913) on x-rays produced by different elements.

The x-rays produced in an x-ray tube generally consist of a wide range of frequencies. When these x-rays strike various materials, secondary x-rays are produced; their frequencies depend upon the nature of atoms present in the target material. These x-rays, characteristic of the target material, are called *characteristic x-rays*. x-rays characteristic of the material of the anti-cathode in an x-ray tube are also obtained under suitable conditions. In some cases, the anti-cathode may be coated with a compound of the desired element.

Moseley resolved the characteristic x-rays produced by a large number of elements by using a crystal of potassium ferrocyanide. This acted like a very fine diffraction grating. The spectrum obtained was recorded on a photographic plate. Several lines (produced by the slit) were obtained and these were divided into two main series—*K* and *L*. Different lines in a series were denoted by sub-scripts α, β, etc. A third *M* series was also observed with heavy elements.

When the elements were arranged in the sequential order of their appearance in the periodic table, it was observed that the frequency of any particular radiation (say K_α or K_β or L_α or M_α) in the x-ray spectrum shifts gradually to higher values from one element to the other (Fig. 3.18). The serial number of appearance of an element in the periodic table was taken as its atomic number *Z*. Moseley showed that for all elements the frequency (ν) of any particular line in the spectrum followed the relation

$$\sqrt{\nu} = a(Z - \sigma),$$

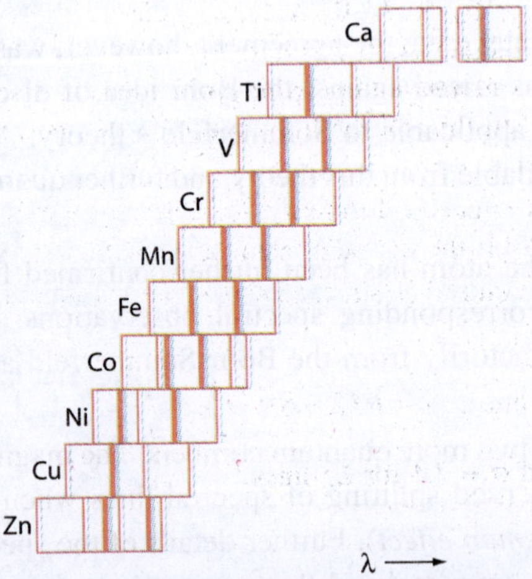

Fig.3.18: Type of characteristic X-ray spectra of a few elements. Note how the position of a particular line, say K_α, shifts gradually to lower wavelength.

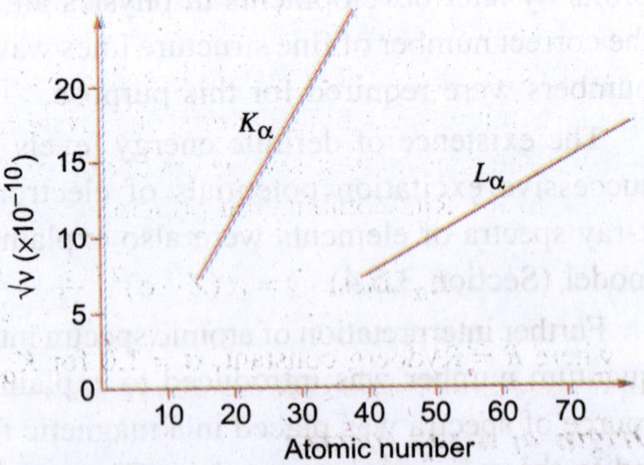

Fig.3.19: Nature of variation of the frequencies of characteristic x-rays with atomic number.

where a is a proportionality constant and σ is a constant characteristic of the series (for K-series, $\sigma = 1.0$; for L, $\sigma = 7.4$. Plot of $\sqrt{\nu}$ against Z produced straight lines (Fig. 3.19). But no such regular relation could be established with the atomic weights of the elements.

This showed that the atomic number Z of an element was more significant than its atomic weight. Moseley concluded that this was a fundamental property of the elements. His conclusion was confirmed when nuclear charges of elements were directly determined by Chadwick (1920) and found to be equal to their respective atomic number.

In Mendeleev's periodic table, the priority of an element with lower atomic weight was violated in four cases. Chemical analogy justified this violation, but no general reason could be placed in support. Determination of atomic number showed that the so-called reversed orders were actually in accordance with increasing Z. The atomic number of potassium (19) was one more than that of argon (18); and so for all other "misfit" pairs. The fundamental importance of atomic number in explaining the periodicity of properties of the elements was thus realized and the periodic law was restated as:

Physical and chemical properties of elements are periodic functions of their atomic number.

Besides establishing the importance of atomic number (Z) of an element, the study on characteristic x-rays contributed a lot more towards systematization of the periodic table:

(i) New elements were discovered and characterized by their atomic numbers: technetium ($_{43}$Tc), rhenium ($_{75}$Re), promethium ($_{61}$Pm) and hafnium ($_{72}$Hf).

(ii) The fourteen rare-earth elements from cerium (58) to lutetium (71) resemble each other very closely and often occur together. It was a very difficult task to isolate each of them and establish their individuality by chemical methods. The x-ray spectrum provided a quick and sanguine method to identify them.

(iii) The actual number of missing elements in the periodic table could be ascertained from the determination of atomic numbers and any claim towards the discovery of a new element could be readily verified.

X-ray spectra vs. atomic spectra

(i) X-ray spectra consist of very few lines and hence are much simpler than atomic spectra.

(ii) In atomic spectra, successive lines can be obtained by gradually increasing the excitation energy, i.e., promoting the electrons to different higher energy levels. But in the case of x-ray spectrum, the energy of excitation must exceed a *minimum value corresponding to a series limit.* Below this, no emission occurs; beyond this, all the lines in that series appear simultaneously.

(iii) X-ray spectra of different elements are similar in appearance—differing only in the frequencies. Atomic spectra differ from one group of element (in the periodic table) to another.

(iv) The lines in the x-ray spectra of different elements follow certain general expressions similar to the Ritz combination principle:

$$K_\alpha \text{ lines;}\quad \bar{\nu} = R(Z-\sigma)^2\left(\frac{1}{1^2}-\frac{1}{2^2}\right);\quad L_\alpha \text{ lines:}\quad \bar{\nu} = R(Z-\sigma)^2\left(\frac{1}{2^2}-\frac{1}{3^2}\right)$$

where R = Rydberg constant; $\sigma = 1.0$ for K_α lines and $\sigma = 7.4$ for L_α lines.

Origin of x-ray spectra

The origin of characteristic x-rays may be understood from the Bohr atom model. For example, the emission of K-series may be explained as follows:

When an atom in the target material absorbs sufficient energy, one of its K-shell electrons (n = 1) may be ejected out. Since the next higher shells (n = 2, 3, ..., etc.) may be completely filled, the ejected electron must have sufficient energy to move to a very high level or to escape from the atomic system. The vacancy created in the *K*-shell may now be filled by an electron *"dropping"* from the higher levels, the difference of energy being emitted in the form of secondary x-rays. An electron moving from the L-shell ($n = 2$) gives rise to the K_α line; transition from the M-shell ($n = 3$) produces the K_β line; and so on.

Ejection of electrons from *L*- or *M*-shells similarly produces higher series of x-rays. The above mechanism is justified by the validity of Ritz combination type equations for the wave numbers of radiation emitted. The term σ accounts for the screening of the nuclear charge by other electrons (see later). $(Z - \sigma)$ then represents the effective nuclear charge.

The K_α x-ray line corresponds to transition of an electron from the $n = 2$ to $n = 1$ state. The charge *Ze* of the nucleus is screened by the electron still remaining in the $n = 1$ state; the effective nuclear charge on the electrons in the $n = 2$ shell is approximately $(Z - 1)\,e$. This gives

$$\frac{1}{\lambda} = \bar{\nu} = R\,(Z-1)^2\left(\frac{1}{1^2}-\frac{1}{2^2}\right) \qquad (R = \text{Rydberg constant})$$

It can also be shown that the energy of a K_α x-ray photon is

$$E(K_\alpha) = 10.2\ (Z - 1)^2\ \text{eV}.$$

Problem 3.21

The limit of the Balmer series for hydrogen lies at 364.7 nm. Calculate the wavelength of the shortest x-ray that may be obtained from Tc($Z = 43$).

Solution: Limit of the Balmer series corresponds to transition between $n = 2$ and $n = \infty$. Hence

$$\bar{\nu} = \frac{1}{\lambda} = R\left(\frac{1}{2^2}-\frac{1}{\infty^2}\right);\ \text{or, } R = \frac{4}{\lambda} = \frac{4}{364.7\times10^{-9}} = 1.097\times10^{7}\ \text{m}^{-1}.$$

The shortest x-ray corresponds to the *K*-series for which $\sigma = 1$ and $n = \infty$.

$$\text{From } \frac{1}{\lambda} = R\,(Z-1)^2\left(\frac{1}{1^2}-\frac{1}{n^2}\right) \Rightarrow \lambda = 0.517\ \text{Å}.$$

ESSENTIALS

Electromagnetic radiation has dual character. Sometimes it behaves as waves—the frequency ν, wavelength λ and speed *c* are related as

$$c = \lambda\nu\ ;\ \text{wave number } 1/\lambda = \nu/c.$$

In explaining some experimental facts one has to assume that the radiation acts as dicrete packets of energy, called quanta (quantum theory of radiation). The energy ε of a quantum of radiation is

$$\varepsilon = h\nu \qquad h = \text{Planck's constant} = 6.626\times10^{-34}\ \text{J s}.$$

The relation is a bridge between the corpuscular (particle) nature and wave nature of radiation.

Einstein generalized the quantum theory of radiation for all types of radiation and explained the photoelectric effect.

$$h\nu = \tfrac{1}{2}mv^2 + W\ \ (\text{work function})$$

$$W = h\nu_0\ ;\ \nu_0 = \text{threshold frequency}.$$

Rutherford put forward his nuclear atom model based on experiments on the scattering of α-particles by thin gold foils. But the model could not explain the observed discrete spectra of atoms from the electromagnetic theory of radiation.

Niels Bohr adopted the quantum theory of radiation and introduced the idea of quantized orbits and interpreted the spectra of hydrogen like atoms.

$mvr = nh/2\pi$, n is an integer (1, 2, 3, ⋯,)

$r = \varepsilon_0 n^2 h^2/\pi m Z e^2$; $v = Ze^2/2\varepsilon_0 nh$.

$E = -Z^2 me^4/8\varepsilon_0{}^2 n^2 h^2 = -13.6\ Z^2/n^2$ eV.

$R_\infty = Z^2 me^4/8\varepsilon^2 ch^3 = 10973731\ \text{m}^{-1}$. $R = R_\infty/(1+m/M)$.

EXERCISE

1. What are Bohr's postulates regarding the extranuclear structure of an atom? Show that the product of the radius of a Bohr orbit and its energy is a constant. Find the value of this constant for hydrogen. [**Ans:** $E.r = -Ze^2/8\pi\varepsilon_0$. For H, $E.r = -1.15 \times 10^{-28}$J]

2. You have been asked to prepare a large exhibit to demonstrate the Bohr atom model. You are to place it on a table of size 1 m × 1 m. What will be the radius of the smallest and the largest orbit if you are to place the first six orbits in your model?
[**Hint:** The sixth orbit will be of 1 m diameter or 0.5 m radius. $r_n = r_0 \times n^2 \Rightarrow r_0 = 1.4$ cm]

3. Estimate the potential energy of an electron in a hydrogen atom in terms of r and e.

4. Obtain an expression for the kinetic energy of an electron in the He^+ ion.

5. Calculate the wave number of the emitted spectral line when an electron undergoes transition from (i) $n = 3$ to $n = 1$ and (ii) $n = 2$ to $n = 1$ in a hydrogen atom.

6. Calculate the wavelengths of the first two lines in the Paschen series of the spectrum of atomic hydrogen ($R = 1.097 \times 10^7\ \text{m}^{-1}$). [**Ans:** 1st line 1.876×10^{-6} m]

7. What new line results from the combination of the first two lines in the Paschen series? What will be its wavelength? [**Ans:** 1st line in Brackett series 4.054×10^{-6} m]

8. How can one estimate the ionization energy of hydrogen from the study of its atomic spectrum?
[**Hint:** The ionization energy corresponds to the energy difference between $n = 1$ and $n = \infty$ levels. This may be obtained from the series limit in the Lyman series.]

9. Which labelling of the following sketch by two students is correct regarding line spectrum of hydrogen (not to scale)?

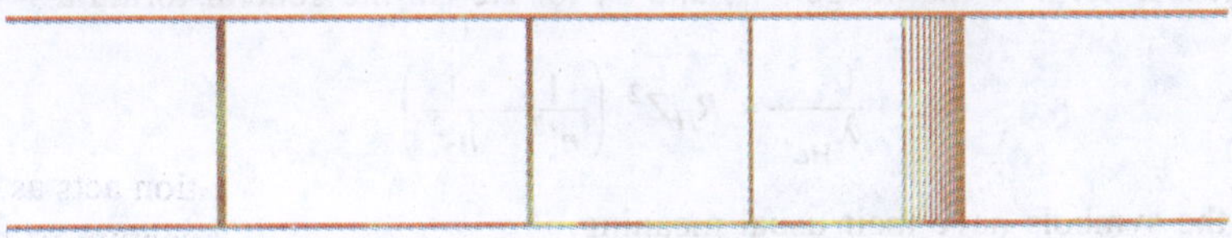

Student *A*: λ ⟶ ; Student *B*: ν ⟶

[**Hint:** The series limit occurs in the direction of increasing energy. *B* is correct.]

10. (a) What is the significance of the negative sign in the expression of the energy of an electron in a hydrogen-like atom or ion?

(b) How did Rutherford conclude that the nucleus is (i) massive, (ii) positively charged?

(c) How many total orbits are possible, according to Sommerfeld, for $n = 4$? How many of them will be elliptical? What will be their k values and l values?

(d) What is a Bohr? What is its magnitude in (i) angstrom, (ii) metre?

(e) What is wave number? How is it related (i) to the wavelength? (ii) to the energy of a quantum?

11. What should be the separation between two energy-levels in a hydrogen-like system if an electron undergoing a transition between these levels emits a radiation of 1 m wavelength?

[**Ans:** 1.987×10^{-25} J.]

12. The Rydberg constants for hydrogen and He^+ are 10967758 m^{-1} and 10972226 m^{-1} respectively. Given $M_{He} = 3.9726\ M_H$, calculate M_H/m (m = mass of the electron. M = mass of respective nuclei).

13. Calculate the radius of the Li^{2+} ion in the first excited state.

14. Calculate the first three energy levels for the electron in Be^{3+}. What will be the ionization potential of Be^{3+}? What will be the ionization energy in joules per mole?

15. Calculate the wavelength of all spectral lines that could be observed from hydrogen atoms excited by a potential of 11V.

[**Hint:** See Fig. 3.12. Only $n = 2$ state will be populated.]

16. The hydrogen atom consists of a single electron. How can you explain the appearance of so many lines in the spectrum of atomic hydrogen?

[**Hint:** A collection of H-atoms are studied.]

17. Suppose the nucleus of a hydrogen atom is replaced by a positron, a particle with the same mass as that of the electron; it has a charge equal in magnitude to that of the electron but opposite in sign. Show that the energy levels of the new system (positronium) will be of the same type as those for hydrogen but their magnitudes will be smaller by a certain factor. Find the value of this factor.

[**Hint:** Find the reduced mass of the electron in the hydrogen atom and in positronium.]

[**Ans:** $\frac{1}{2}(1 + m/M_H)$]

18. The series limit of the Lyman series for hydrogen is at 91.2 nm. What will be the approximate wavelength of the K_α line in the x-ray spectrum of zinc, $Z = 30$?

[**Ans:** 0.145 nm.]

19. The effective charge on an M-shell ($n = 3$) electron in an atom of an element (atomic no = Z) is $(Z - 7.4e)$ approx. Show that the frequency of the L_α x-rays of this element is given by

$$\nu = 5cR(Z - 7.4)^2/36.$$

20. Some lines in the spectrum of He^+ match almost exactly with the lines in the Balmer series of hydrogen. Determine the integers n_1' and n_2' for He^+ in the general formula

$$\frac{1}{\lambda_{He^+}} = R_H Z^2 \left(\frac{1}{n_1'^2} - \frac{1}{n_2'^2}\right)$$

where the symbols have their usual meaning.

Hint: For Balmer lines of hydrogen: $\frac{1}{\lambda_H} = R_H\left(\frac{1}{2^2} - \frac{1}{n^2}\right)$, $n = 3, 4, \ldots$

For $\lambda_{He^+} = \lambda_H$, $\left(\frac{1}{2^2} - \frac{1}{n^2}\right) = 4\left(\frac{1}{n_1'^2} - \frac{1}{n_2'^2}\right)$ $\quad (Z = 2)$.

4

EXTRANUCLEAR STRUCTURE OF THE ATOM

PART II: *WAVE MECHANICAL INTERPRETATION*

4.1 WAVE NATURE OF THE ELECTRON

Recognition of the dual nature of radiation had a far-reaching consequence. This guided Louis de Broglie to conceive of a similar dual character for matter (1923). He observed that integers like 1, 2, 3, ... occur naturally in physics in connection with waves. Again, these integers were used by Bohr in connection with the electron orbits. So, the electrons might also possess wave like properties. In his Nobel Prize address de Broglie said:

> "*... determination of the stable motion of electrons in the atom introduces integers; and upto this point the only phenomenon involving integers in Physics were those of interference and of normal modes of vibration. This fact suggested to me the idea that electrons, too, could not be regarded as corpuscles, but that periodicity must be assigned to them also.*"

The hypothesis (theory) of de Broglie

De Broglie suggested that the dual nature of radiation should have a counterpart in matter. According to him, the relation between the wavelength and momentum of a photon is completely general and applies equally to both photons and material particles. He assumed that a moving body has its associated matter wave. For a material particle, the wavelength of such waves is given by

$$\lambda = h/mv$$

where m and v are respectively the mass and velocity of the particle. Subsequently, λ has been called the *de-Broglie wavelength* of the particle.

The argument

The fundamental law of relativity connecting energy E, rest mass m_0 and momentum p of a particle is

$$E = \sqrt{p^2c^2 + m_0^2c^4}$$

The rest mass of a photon is zero. Putting $m_0 = 0$, we get the momentum of a photon as

$$p = E/c = h\nu/c = h/\lambda.$$

If this expression is adopted for a material particle of mass m and velocity v, its momentum p ($= mv$) will be related to its wavelength λ as

$$mv = h/\lambda \text{ or } \lambda = h/mv.$$

Alternatively, we may combine the relations for a photon (terms having their usual significance):

$$E = h\nu \quad \text{and} \quad E = mc^2$$

whence, momentum of the photon, $mc = \dfrac{h\nu}{c} = \dfrac{1}{\lambda}$.

Application to a particle of velocity v now gives $\lambda = h/mv$.

The theory received experimental support (1927) when Davisson and Germer obtained diffraction patterns of electrons with a single crystal of nickel. Diffraction can be explained only by wave properties as a consequence of interference of waves by an object in their path.

Nature of de Broglie waves. The physical significance of de Broglie waves is not straightforward. We try to explore it in the following manner.

Let us first consider the velocity of the de Broglie waves associated with a particle (mass m, velocity v). Combining the relations $E = h\nu$ and $E = mc^2$, we get $\nu = mc^2/h$. The velocity w of the de Broglie wave, which is different from the particle velocity v, is:

$$w = \lambda\nu = mc^2/h \times h/mv = c^2/v.$$

Since the velocity (v) of the particle is always less than c, w must be greater than c! This result, however, *does not violate* the principles of relativity, since the speed of light is the limiting speed for mass-energy. w is *not* the speed of the mass-energy of the particle. It is called the *phase velocity* of the associated waves.

Detailed consideration of wave properties has led to another more useful characteristic of a group of large number of waves. When many waves travel through the same medium, the resultant wave form is determined by the superposition of the various contributing waves. We know that when two tuning forks of nearly the same frequency are sounded together, "beats" occur at regular intervals due to superposition of the two wavetrains. These beats correspond to successive wave groups characterized by interference of the two wave trains (a succession of similar waves at equal intervals. See Fig 4.1). The velocity of propagation of this modulation is called the *group velocity*. Extending to the superposition of a large number of matter waves, it can be shown that this group velocity is identical with the velocity of the material particle concerned. Thus a moving particle may also be treated in terms of its associated wave groups.

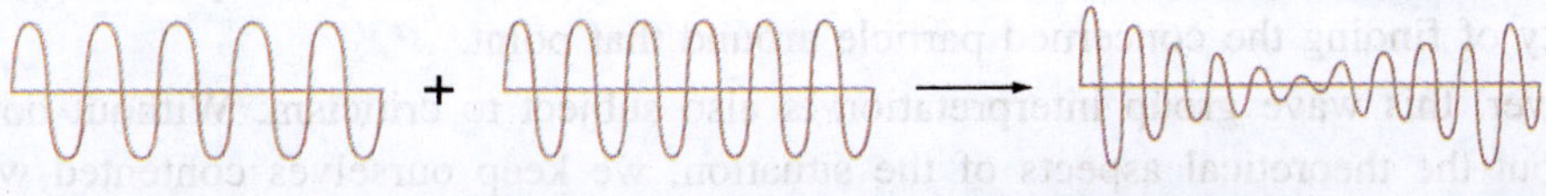

Fig.4.1: Superposition of two waves with nearly equal frequencies produces beats.

Equations for wave motion

Let us consider a wave travelling in the x-direction along a stretched string whose vibrations are along the y-axis (Fig 4.2).

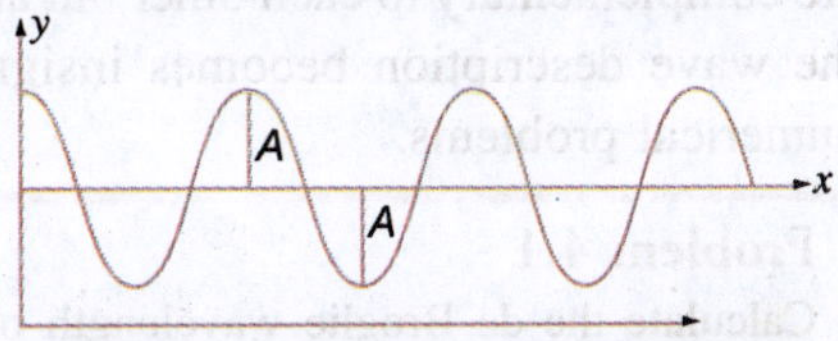

Fig.4.2: A wave propagating along the x-axis.

A = maximum amplitude, ν = frequency,

w = speed of the wave $= \lambda\nu$.

The displacement y at any point x at time t is given by the equation

$$y = A \cos 2\pi\nu \left(t - \frac{x}{w}\right).$$

$$= A \cos 2\pi\left(\nu t - \frac{\nu x}{w}\right) = A \cos 2\pi\left(\nu t - \frac{x}{\lambda}\right) \quad [\because w = \lambda\nu]$$

Waves are often described in terms of *angular frequency* ω in radian per second. Motion along a complete circle sweeps 360° or 2π radian. So a particle moving ν times around the circle in one second will sweep $2\pi\nu$ radians. Hence

$$\text{Angular frequency } \omega = 2\pi\nu \text{ radian s}^{-1}.$$

Since there are 2π radian in one complete wave, the number of radians corresponding to a wave train of length 1m is the *wave number*, k.

$$\text{Wave number } k = \frac{2\pi}{\lambda} \text{ radian m}^{-1}$$

$$= \frac{\omega}{w} \text{ (from } \omega = 2\pi\nu \text{ and } \nu = w/\lambda).$$

The wave equation in terms of ω and k is then

$$y = A \cos(\omega t - kx).$$

During the formation of a wave group by amplitude modulation, as in de Broglie waves for a body moving with a velocity v, it can be shown that

Phase velocity $w = \dfrac{\omega}{k} = \dfrac{c^2}{v}$

Group velocity $u = \dfrac{d\omega}{dk} = v.$

A wave motion implies a periodic variation of some parameter. In water waves, the height of the water surface (amplitude) varies periodically; in electromagnetic waves, the electric and magnetic fields vary. In case of matter waves, we cannot directly get an analogy from our physical world. But such waves do really exist and become significant at times, as shown by diffraction experiments with electrons. So we conclude that an amplitude-analog in matter waves undergoes such periodic variation. We shall call it the *wave function*, a function of the space coordinates of a point and of time. We shall see that this function may be related to the probability of finding the concerned particle around that point.

However, this wave group interpretation is also subject to criticism. Without bothering much about the theoretical aspects of the situation, we keep ourselves contented with the information that certain properties of microscopic particles can be explained by associating wave properties with these particles. But a material particle cannot *simultaneously* manifest both particle and wave properties in one experiment. The wave and particle descriptions are our own mental device to look at things and describe their behaviour—both these "model"s are complementary to each other but always partial. For macroscopic bodies, i.e., large objects, the wave description becomes insignificant. This may be appreciated from the following numerical problems.

Problem 4.1

Calculate the de Broglie wavelength of a bullet ($m = 2 \times 10^{-3}$ kg) moving with a speed of 300 m s^{-1}.

Solution: $\lambda = \dfrac{h}{mv} = \dfrac{6.626 \times 10^{-34}\ \text{J s}}{2 \times 10^{-3}\ \text{kg} \times 300\ \text{m s}^{-1}}$

$= 1.1 \times 10^{-33}$ m (1.1×10^{-23} A°)

Problem 4.2

Calculate the de Broglie wavelength of an electron moving with a speed nearly 1/10th that of light ($v = 3 \times 10^7$ m s^{-1}).

Solution: $\lambda = \dfrac{h}{mv} = \dfrac{6.626 \times 10^{-34}\ \text{J s}}{9.109 \times 10^{-31}\ \text{kg} \times 3 \times 10^{7}\ \text{m s}^{-1}}$

$= 2.425 \times 10^{-11}$ m (= 0.2425 Å).

The results of the above two problems are very significant. We find that the wavelength of a rifle bullet is negligibly small, particularly in comparison to the target it strikes. Similar conclusions may be drawn for any object of the "macro world"— be it a cricket ball or a racing car. In every case, the wave nature would come out to be very much insignificant in comparison to the particle nature of the object. But for an electron, and so for any other object of the "micro world", the wavelength falls within the dimensions of the atomic system to which the electron belongs. Consequently, the wave nature of the electron (or any micro object) can no longer be overlooked.

Matter waves are inherently different from electromagnetic waves. Electromagnetic waves are essentially energy waves — they provide satisfactory explanation of many observations regarding emission, propagation and absorption of radiant energy. The waves travel with the same speed in a medium. Matter waves or de Broglie waves, on the contrary, have varying speed and are always associated with their source. The idea of matter waves offers a convenient interpretation of certain behaviour of microparticles, as we shall presently find in connection with atomic structure.

Electron waves and Bohr orbits

The Bohr condition for stable orbits of the electron in an atom was set by the relation $mvr = nh/2\pi$. The integral values of n were *assumed arbitrarily* merely to fit spectral data. But the same condition for stable orbits is obtained as a natural consequence of the wave nature of the electron.

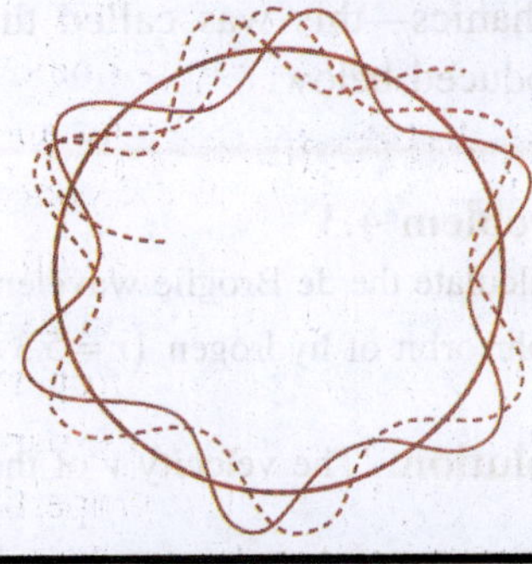

Fig.4.3: A Bohr orbit in terms of electron-waves.
As the waves become smaller, the circular Bohr orbit (bold line) is obtained as a limiting case.

Suppose an electron wave is constrained to move around a nucleus. The waves will complete a full cycle around the nucleus and the next course of waves will exactly superimpose on the track of the initial waves only when the circumference of the electron orbit contains an *integral* number of waves (Fig. 4.3: solid line). In this case the waves constitute a stable orbit around the nucleus. If this condition is not maintained, successive wave trains will not superimpose exactly on one another (Fig. 4.3: broken lines); the waves will cancel each other by interference.

For a stable orbit of radius r, the circumference is $2\pi r$. If λ is the de Broglie wavelength of the electron, the above condition for stable orbit requires

$$2\pi r = n\lambda, \text{ where } n \text{ is an integer.}$$

Substituting $\lambda = h/mv$ (terms having their usual significance), one readily gets

$$mvr = nh/2\pi.$$

Bohr orbits in hydrogen atom: Let us find an expression for the Bohr orbits in a hydrogen atom which conform to de Broglie wavelengths of electrons. As shown above, stable Bohr orbits should follow the relation

$$2\pi r_n = n\lambda \qquad n = 1, 2, 3, \ldots$$

where r_n stands for radii of cooresponding orbits. λ is the de Broglie wavelength of the electron:

$$\lambda = \frac{h}{mv},$$

where v is the velocity of the electron and m is the electron mass.

For a circular orbit of radius r_n in an H-atom (Z = 1), we have (sec. 3.5.2),

$$\frac{mv^2}{r_n} = \frac{e^2}{4\pi\varepsilon_0 r_n^2} \quad \text{or,} \quad v = \frac{e}{\sqrt{4\pi\varepsilon_0 m r_n}}$$

Using $2\pi r_n = n\lambda$ and $\lambda = h/mv$, we get $r_n = \dfrac{\varepsilon_0 n^2 h^2}{\pi m e^2} \quad (n = 1, 2, 3, \ldots)$

The expression is same as the one derived earlier from Bohr's postulates (sec. 3.5.2).

The condition of stable electron orbits in terms of the integer n thus appears from a more general theory. The failure of Bohr's theory to explain the details of spectroscopic observations initiated scientists to investigate this new theory in more details. An important principle known as the uncertainty principle shortly emerged. This ultimately led to the development of a new mechanics—this was called the quantum mechanics. Both of these topics will be briefly introduced below.

Problem 4.3

Calculate the de Broglie wavelength of an electron in the first Bohr orbit of hydrogen $\left(r = 5.3 \times 10^{-11}\,\text{m}\right)$.

Solution: The velocity v of the electron: $v = \dfrac{e}{\sqrt{4\pi\varepsilon_0 mr}}$

Now, $\lambda = h/mv = \dfrac{h}{e}\sqrt{\dfrac{4\pi\varepsilon_0 r}{m}} = 3.3 \times 10^{-10}\,\text{m}$ or, 3.3Å.

Comment: The circumference of the orbit is equal to one complete de Broglie wave of the electron ($\lambda = 2\pi r$) joined head to tail (Fig 4.4).

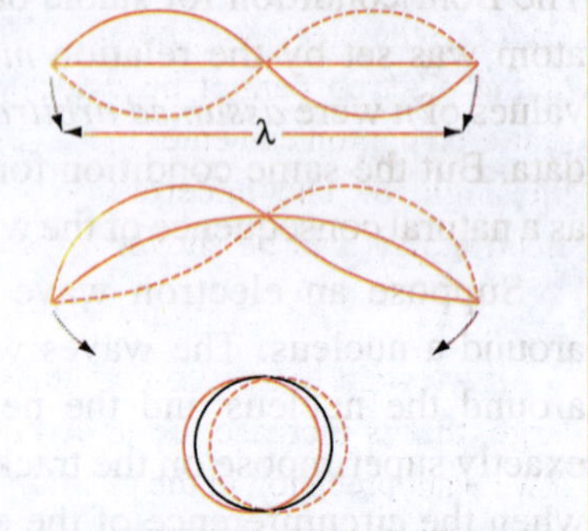

Fig. 4.4: The first Bohr orbit of hydrogen in terms of de Broglie wavelength λ.

Problem 4.4

What will be the wavelength of electrons accelerated through a potential difference of 20 kV?

Hint: Energy E of the electron = 20,000 electron volt

$$= 20{,}000 \times 1.602 \times 10^{-19}\ \text{J}.$$

Now, $E = p^2/2m$ gives $p = \sqrt{2mE}$; $\lambda = h/p = \dfrac{h}{\sqrt{2mE}}$ [Ans: 8.65×10^{-12} m]

4.2 THE UNCERTAINTY PRINCIPLE

For large or macroscopic objects in motion, the exact position and velocity at any instant of time can be expressed by the laws of Newtonian mechanics. But for a very small (submicroscopic) object like the electron, involvement of the wave character does no more permit the application of Newtonian mechanics. We shall first try to appreciate this point by considering a hypothetical situation.

A "thought" experiment. Suppose an imaginary microscope of high resolving power is designed to measure the position and momentum of an electron simultaneously. The resolving power of a microscope may be expressed in terms of the distance Δx between two points which can just be resolved by the microscope. Δx is approximately given by

$$\Delta x \simeq \lambda/2 \sin\theta$$

where λ is the wavelength of the light used to illuminate the object and θ is the half angle of the lens used (Fig. 4.5.) Clearly, the position of the electron cannot be measured more precisely than Δx : so the minimum uncertainty in the position of an electron in this experiment will be Δx. If we want to keep Δx as small as possible, the wavelength λ of the light used should be as small as practicable—e.g., x-rays or γ-rays.

Now the process requires that an electron is hit by a photon and the photon is scattered into the field of the microscope. In Compton scattering we learned that in such an interaction the photon transfers a part of its momentum ($h\nu/c$) to the electron. The exact change in the momentum of the electron cannot be determined precisely since the scattered photon can enter the microscope anywhere within the angle θ. The uncertainty in the momentum of the electron, Δp, can be shown to be of the order

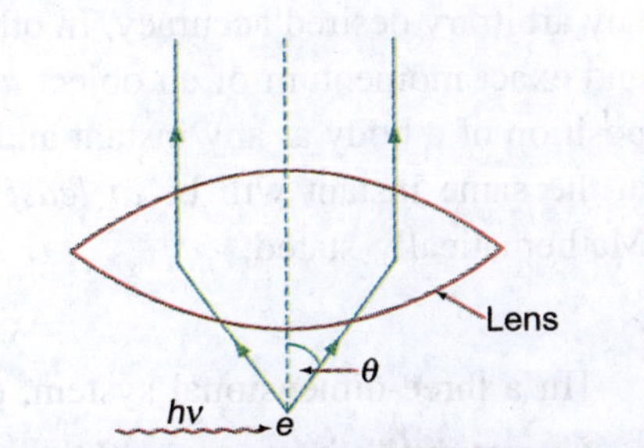

Fig.4.5: Compton-scattering of a photon by an electron into the field of a microscope (schematic).

$$\Delta p \simeq h\nu \sin\theta/c = h \sin\theta/\lambda.$$

If we want to keep Δp small, λ should be very large. As we have seen above, this will make Δx large, that is increase the uncertainty in the measured position of the electron. On the other hand, if we want precision in the position measurement, the small value of λ will result in greater uncertainty in the momentum determination. Combining the expressions of Δx and Δp,

$$\Delta x \, \Delta p \simeq h.$$

Δx and Δp are the *minimum uncertainties* in respective cases; the actual uncertainty may be greater. So we write,

$$\Delta x \, \Delta p \geq h.$$

A more rigorous argument gives

$$\Delta x \, \Delta p \geq h/4\pi \quad \text{or} \ \frac{\hbar}{2} \qquad \left(\text{where } \hbar = \frac{h}{2\pi}\right)$$

In a three-dimensional system, similar relations hold for each coordinate axis. In cartesian coordinates, the relations may be stated as

$$\Delta p_x \, \Delta x \geq \frac{h}{4\pi} \qquad \Delta p_y \, \Delta y \geq \frac{h}{4\pi} \qquad \Delta p_z \, \Delta z \geq \frac{h}{4\pi}$$

Δp_x, Δp_y and Δp_z are uncertainties in the components of momentum along respective axes and Δx etc. are corresponding uncertainties in position.

This uncertainty is a direct consequence of wave-particle duality. A moving particle may be considered in terms of a wave group resulting from superposition of a large number of waves. At a given time, the particle may be located anywhere within the group. The narrower the wave group, the more precisely can we locate the position of the particle. But in a narrow wave group we do not have enough waves to measure the wavelength precisely. Hence the momentum ($= h/\lambda$) becomes uncertain. Similarly, in a wide wave group, we can determine the wavelength (hence the momentum) with greater precision but in that case the position becomes more uncertain. Thus, the uncertainty arises as an essential consequence of the wave-particle dualism and not due to any limitation in our measuring device (W. Heisenberg, 1927).

The uncertainty principle

From an analysis of the wave-particle dualism of matter, W. Heisenberg (1927) stated: it is not possible to determine *simultaneously* both the position and the momentum of a body with any arbitrary desired accuracy. In other words, it is impossible to know both the exact position and exact momentum of an object *at the same time*. The product of the uncertainty Δx in the position of a body at any instant and the uncertainty Δp in the x-component of its momentum at the same instant will be *at least* of the order of h, the Planck constant (precisely $h/4\pi$). Mathematically stated,

$$\Delta x \, \Delta p \geq h/4\pi.$$

[In a three-dimensional system, equivalent expressions apply to other coordinate axis.]

Uncertainties in energy and time. The uncertainty relation holds between any two conjugate properties like energy and time. An alternative expression for the uncertainty principle may be arrived at from the consideration of energy measurement of a particle. We might wish to measure the energy E emitted sometime during the time interval Δt in an atomic process. Here the uncertainty in the time is Δt; this will introduce an uncertainty ΔE in the energy measurement according to the relation

$$\Delta E \, \Delta t \geq h/4\pi$$

This uncertainty relation is also applicable to any time-dependent system. Suppose we wish to measure the energy E of a free particle moving with a velocity v in the x-direction. The energy E is equal to the kinetic energy of the particle:

$$E = \frac{1}{2}mv_x^2 = \frac{p_x^2}{2m} \qquad (m = \text{mass}, \quad p_x = \text{momentum})$$

If the momentum has an uncertainty Δp_x and the corresponding uncertainty in energy be ΔE, then

$$\Delta E = \left(\frac{\partial E}{\partial p_x}\right) . \Delta p_x = \frac{p_x}{m} . \Delta p_x = v_x . \Delta p_x .$$

But velocity v_x = displacement / time $= \dfrac{\Delta x}{\Delta t}$.

$$\therefore \ \Delta E = \frac{\Delta x}{\Delta t} . \Delta p_x \quad \text{or,} \quad \Delta E \, \Delta t = \Delta x \, \Delta p_x$$

Since the right hand side is $\geq \dfrac{h}{4\pi}$, $\Delta E \Delta t \geq \dfrac{h}{4\pi}$.

Thus, the product of the uncertainty in energy and the uncertainty in time at which a particle passes over any point is equal to or greater than $h/4\pi$.

Position and momentum or energy and time are two pairs of *canonically conjugate* variables. In classical mechanics, the dynamical situation of a particle is completely defined by any pair of these variables. The physical quantities in each pair are complementary to one another; they are precisely known at the same time in classical mechanics. The prominent involvement of wave nature for microscopic particles now puts a limit to this precision in the form of the uncertainty principle. We are thus led to the development of a new mechanics, called the quantum mechanics.

Problem 4.5

Calculate the minimum uncertainty in the location of (a) a 1-g mass moving with a speed of 1.5 m s^{-1} and (b) an electron moving with a speed of 2.2×10^6 m s^{-1}. In each case the uncertainty in the momentum measurement is 0.1 per cent, i.e., $\Delta p = 10^{-3} p$.

Solution: (a) Momentum $(p) = 1 \times 10^{-3}$ kg $\times$ 1.5 m s^{-1} = 1.5×10^{-3} kg m s^{-1}.

$\therefore\ \Delta p = 1.5 \times 10^{-6}$ kg m s^{-1}.

$\therefore\ \Delta x = h/(4\pi\Delta p) = 3.5 \times 10^{-29}$ m = 3.5×10^{-19} A°.

(b) Using the rest mass of the electron = 9.11×10^{-31} kg,

$p = 9.11 \times 10^{-31} \times 2.2 \times 10^6$ kg m s^{-1}

$\Delta p = 9.11 \times 2.2 \times 10^{-28}$ kg m s^{-1}

$\therefore\ \Delta x = h/(4\pi\Delta p) = 2.63 \times 10^{-8}$ m = 263 A°

[**Note**: The uncertainty in (a) is really regligible. But in (b) it is serious. Note that the velocity of the electron is equal to that in the first Bohr orbit of the hydrogen atom. The radius of this orbit is 0.53 A° only. If we reverse the problem and attempt to locate the electron within this distance, the corresponding uncertainty in momentum and hence, in velocity, would be very large. See problem 4.6.]

Problem 4.6

What is the minimum uncertainty in the velocity determination of an electron if we want to locate it within 0.01 A° of the first Bohr radius in a hydrogen atom?

Solution: Here $\Delta x = 0.01$ A° = 10^{-12} m

$\therefore\ \Delta p = m\Delta v = h/4\pi\Delta x = 5.27 \times 10^{-23}$ kg m s^{-1}.

Using the rest mass of the electron as an approximation

$\Delta v \simeq 5.8 \times 10^7$ m s^{-1}.

[This is greater than the velocity of the electron which is of the order of 10^6 m s^{-1} !]

Problem 4.7

What is the uncertainty in the location of a photon of wavelength 5000 A° if the wavelength is known to an accuracy of 1 part in a million?

Solution: The momentum of a photon, p, is equal to $h\nu/c = h/\lambda$.

Let an uncertainty $\Delta\lambda$ in wavelength (λ) give rise to an uncertainty Δp in the momentum. Since p decreases as λ increases, we can write

$$p - \Delta p = \frac{h}{\lambda + \Delta\lambda} \quad \text{or,} \quad p + \Delta p = \frac{h}{\lambda - \Delta\lambda}$$

Substituting $p = h/\lambda$ in either expression, $\Delta p = \dfrac{h\Delta\lambda}{\lambda^2 \pm \lambda\Delta\lambda} \simeq \dfrac{h\Delta\lambda}{\lambda^2} \quad (\Delta\lambda << \lambda)$

$\therefore$ Uncertainty in position, $\Delta x = \dfrac{h}{4\pi\Delta p} = \dfrac{\lambda^2}{4\pi\Delta\lambda}$

Here $\Delta\lambda$ is 1 in 10^6. In 5000 Å, this amounts to $\dfrac{5000 \times 10^{-10}\,\text{m}}{10^6} = 5 \times 10^{-13}\,\text{m}$.

Substitution now gives uncertainty in location, $\Delta x = 0.04$ m.

Problem 4.8

The average time gap between the excitation of an atom and consequent emission of a photon is 10^{-8} s. Calculate the uncertainty in the (i) energy measurement (ii) frequency determination.

Solution: $\Delta t = 10^{-8}$ s.

$\therefore\ \Delta E \geq h/4\pi\Delta t = 5.3 \times 10^{-27}$ J.

Since $E = h\nu$ or $\Delta E = h\Delta\nu$, the uncertainty in frequency $\Delta\nu \geq \Delta E/h = 8 \times 10^6$ Hz.

[**Note**: This uncertainty in the determination of frequency of radiation from an excited atom is partly responsible for the *broadening of spectral lines* from $\nu - \Delta\nu$ to $\nu + \Delta\nu$. However, other factors may also contribute, like: (i) *pressure broadening* caused by interaction with the electric fields of neighbouring atoms or molecules (ii) *Doppler broadening* arising from motion of the radiating atom or molecule relative to the observer.]

Problem 4.9

The position of an electron is determined at some instant of time with an accuracy of ± 1 Å. Find the uncertainty in the electron's position 1 s later.

Solution: The uncertainty in momentum $\Delta p = h/4\pi\Delta x$; $\Delta x = 10^{-10}$ m (given).

The uncertainty in the velocity of the electron (Δv) is

$$\Delta v = \Delta p/m = h/4\pi m\Delta x.$$

Putting $m = 9.1 \times 10^{-31}$ kg, $\Delta v = 3.15$ km s^{-1}.

The uncertainty in position after time t is $\Delta v \times t$; here t = 1s. $\therefore\ \Delta v \times t = 3.15$ km.

From the problems given above, we conclude that we cannot express simultaneously the exact position and momentum (or energy and time) of an electron in the atom with the help of classical Newtonian mechanics. In fact, scientists felt the need to develop a mechanics which would be able to describe microscopic systems in a befitting manner even before the uncertainty principle was stated as such in 1927.

4.3 DEVELOPMENT OF QUANTUM MECHANICS

The limitation of the Newtonian mechanics for atomic systems led to the development of a new approach suitable for systems with prominent wave character. Two independent developments appear significant in this connection: Heisenberg, Born and Jordan approached the problem with the properties of matrix (1925) while Schrödinger tried to develop a mechanics based on the properties of waves (1926). Both approaches are called *quantum mechanics* because they developed around the central idea of quantization of energy (energy was supposed to be continuous in classical mechanics). Here we shall have a brief introduction to Schrödinger's approach to quantum mechanics, popularly known as *wave mechanics*.

4.3.1 The Schrödinger Equation

Classical mechanics

In classical mechanics, the state of a system is defined by specifying the forces acting on it and all the position coordinates and velocities (or momenta) of the constituent particles. The position and momentum of any particle at any instant of time may be stated precisely by equations derived from Newton's laws of motion. But we cannot derive these laws of motion. We simply accept them and then apply them to a real observation. We recall the second law of Newton which summarily leads to

$$\text{Force, } F = \text{mass} \times \text{acceleration} = m\, d^2x/dt^2$$

for a particle of mass m moving in the x-direction. This is a second order differential equation involving x, the position of the particle with respect to time t. If we know the force F and the mass m, we can find x by integrating the above equation twice. But this will introduce two integration constants. These constants may be evaluated by two pieces of information like the position and velocity of the particle at some given time, say at the beginning of the observation ($t = 0$). Thus we have our familiar relations like (in usual symbols)

$$v = u + ft \quad \text{or} \quad s = ut + \tfrac{1}{2}ft^2 \quad \text{and so on.}$$

These relations enable us to predict the precise position and momentum of the particle at any future time.

As already explained, such prediction of position and momentum with absolute precision is not possible for microscopic particles because of the inherent uncertainty of the system itself. We try to describe such systems through their associated matter waves. Before that, let us briefly summarize our understanding of wave motion.

Wave motion

Any common example of wave motion involves periodic variation of some property of the system; the property is thus a function of time. When we throw a small piece of stone in a tranquil pond, waves of water propagate in the form of concentric rings. The height of water (compared to the undisturbed level) at any point varies periodically with time. This displacement of any point in the wave (ϕ) may be expressed as a function of the space coordinates of the point and time t:

$$\phi \equiv f(x, y, z, t)$$

This description holds for any other wave motion. In electromagnetic waves, for example, the electric and magnetic fields vary periodically. The dependence of ϕ on the space coordinates and time is expressed by the relation

$$\frac{\partial^2\phi}{\partial x^2} + \frac{\partial^2\phi}{\partial y^2} + \frac{\partial^2\phi}{\partial z^2} = \frac{1}{v^2}\cdot\frac{\partial^2\phi}{\partial t^2}$$

where v is the velocity of the wave. If we restrict our attention to a wave propagating along the x-direction, we get the relation

$$\frac{\partial^2\phi}{\partial x^2} = \frac{1}{v^2}\cdot\frac{\partial^2\phi}{\partial t^2}$$

We may readily visualize such waves in a string attached at one end when a wave motion propagates along the x-direction. Such wave patterns (Fig. 4.6) are typical for simple harmonic motion and are expressed by sine or cosine functions,

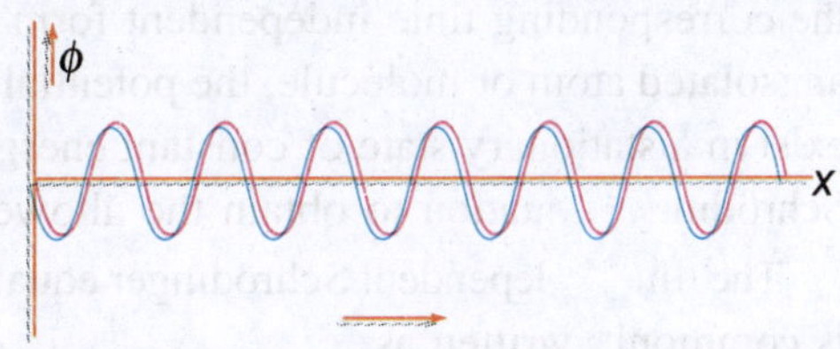

Fig.4.6: A wave propagating along the x-axis.

$$\phi = \sin(kx - \omega t)$$

where, k = wave number = $2\pi/\lambda$ and ω = angular frequency = $2\pi\nu$.

(λ and ν are wavelength and frequency respectively.)

Matter wave

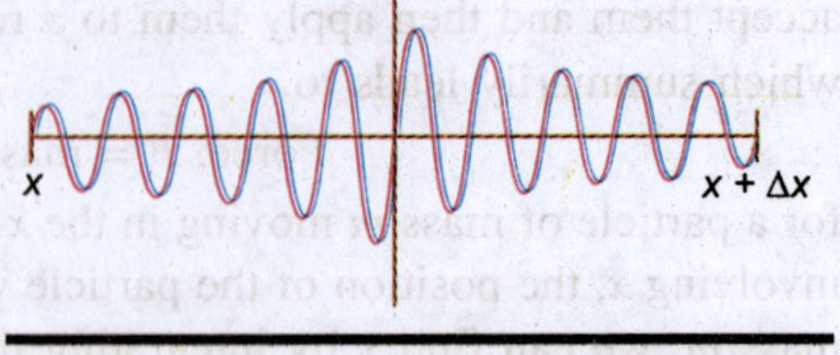

Fig.4.7: A wave group confined in the region x and $x + \Delta x$ may be taken to represent a matter wave.

Now we want to describe a microscopic particle by its associated de Broglie waves. Clearly, such waves cannot be extended over a large span as shown in Fig. 4.7. We want that the wave will remain as close to the particle as possible. In other words, the wave function ϕ should be significant only around the actual position of the particle, say between x and $x + \Delta x$. Such description in terms of waves involves wave groups formed by superposition of waves and represented by linear combination of several sine or cosine waves (Fig 4.7). The wave number k of the contributing simple harmonic waves will vary slightly from one another and so also will λ ($= 2\pi/k$).

An equation suitable for describing matter waves was arrived at by Schrödinger from the classical wave equation and de Broglie's idea of matter waves but with the introduction of some new basic postulates. The equation cannot be derived from any more fundamental principles.

In this new approach, the *state* of a system is defined by a mathematical function Ψ (capital psi) which is a function of the space coordinates of the particles in the system and of time. For a single particle,

$$\Psi = \Psi\ (x, y, z, t)$$

where x, y and z are the space coordinates of the particle. The function Ψ may be real or imaginary and hence is abstract itself; but it can be related to measurable properties of the system and hence carries a meaningful significance to be discussed shortly.

If the total energy of a system remains unchanged with time, we may express Ψ as the product of *two* functions, one of the space coordinates (ψ) and the other a function of time t.

$$\Psi\ (x, y, z, t) = \psi\ (x, y, z)\ .\ f(t)$$

The function $\psi\ (x, y, z)$ conforms to a wave which depends only on the space coordinates like a stationary wave created in a string bound at the two ends. The displacement of the string at any point may be obtained from solution of the general wave equation putting the appropriate boundary conditions. We shall restrict ourselves to such time-independent wave functions and the corresponding time-independent form of Schrödinger equation. This is not unjustified. In an isolated atom or molecule, the potential energy V is independent of time and the system can exist in a stationary state of constant energy. In such cases we can apply the time-independent Schrödinger equation to obtain the allowed energy levels (see later).

The time independent Schrödinger equation, also called the steady state Schrödinger equation, is commonly written as

$$\nabla^2\psi + \frac{8\pi^2 m}{h^2}(E - V)\psi = 0 \qquad [4.1]$$

Where ∇^2 (del squared) $\equiv \dfrac{\partial^2}{\partial x^2} + \dfrac{\partial^2}{\partial y^2} + \dfrac{\partial^2}{\partial z^2}$ in cartesian coordinates.

The equation is thus a second order differential equation involving a function ψ. ψ is called the wave function which is a function of the space coordinates of a point around which we intend to locate a particle of mass m, total energy E and potential energy V.

For a single particle (mass m) moving in the x-direction, the Schrödinger equation [4.1] gets the form

$$\frac{\partial^2 \psi}{\partial x^2} + \frac{8\pi^2 m}{h^2}(E - V)\psi = 0 \qquad [4.2]$$

Substituting $h/2\pi = \hbar$ (h bar or h cross) equation 4.2 may be rearranged to get

$$-\frac{\hbar^2}{2m}\frac{\partial^2 \psi}{\partial x^2} + V\psi = E\psi \qquad [4.3]$$

or,
$$\left[-\frac{\hbar^2}{2m}\frac{\partial^2}{\partial x^2} + V\right]\psi = E\psi \qquad [4.4]$$

The corresponding equation in three dimensions would be

$$\left[-\frac{\hbar^2}{2m}\nabla^2 + V\right]\psi = E\psi \qquad [4.5]$$

If we replace the part in the square brackett in equations 4.4 or 4.5 by H, we get

$$H\psi = E\psi \qquad [4.6]$$

H is an operator called the energy operator or *Hamiltonian operator* which, when operated on Ψ, gives the energy E of the system (see later).

All the equations from 4.1 to 4.6 are expressions of the Schrödinger equation (time-independent).

For collision between two molecules or for a molecule subjected to varying electric and magnetic fields of electromagnetic radiation, the condition of stationary state is no more applicable and one must deal with the time-dependent Schrödinger equation. In such cases, the wave function is obtained as

$$\Psi = \psi . e^{-iEt/\hbar}$$

As already mentioned, the equation of Schrödinger based on matter-wave cannot be derived from any more fundamental principle. It has to be taken as the basic equation or postulate of quantum mechanics. The final form of the equation has to be appreciated from its agreement with experimental facts and not by its genesis. A simplified approach to the construction of the equation is given in the box.

The Schrödinger equation

1. Wave Motion. A simple example of wave motion in one dimension can be provided by a displacement moving along a string fixed at one end only. Suppose the string vibrates with a frequency ν, producing waves of length λ which travel in the positive x-direction with a velocity u. The amplitude y, i.e., the height of the wave at any point x and time t is a function of both x and t, that is, $y = f(x, t)$.

Now, such a wave motion in one dimension can be represented mathematically by the general partial differential equation

$$\frac{d^2y}{dx^2} = \frac{1}{u^2}\frac{d^2y}{dt^2} \qquad [4.7]$$

It can be shown by direct substitution that this equation is satisfied by any function of the form

$$y = f(x - ut)$$

For a wave travelling in the negative x-direction, the solution would be

$$y = g(x + ut)$$

Both f and g are arbitrary functions.

2. Standing waves. Vibration of a string fixed at its two ends produce standing waves or stationary waves. The length (l) of the string is now fixed, say between $x = 0$ and $x = l$. Any wave formed by the string will be guided by the condition that its amplitude must be zero at the two ends. This leads to the conclusion that all standing waves must satisfy the condition

$$n \times \lambda/2 = l, \text{ where } n \text{ is an integer.}$$

For any standing wave formed within the string, the function of x always has the same form irrespective of the value of t. The situation is analogous to a sine curve : at certain values of x, there is a node in the wave at all times and at certain other values of x, the wave always has its maximum amplitude. (The magnitude of this maximum amplitude, however, oscillates with time). In this sense, a standing wave is an oscillatory motion; it is not strictly a wave motion as there is no propagation of energy along the string.

By actual application, it has been found that the equation representing a standing wave is most suitable to describe an electron wave in the atom. We may proceed to obtain the desired equation as follows:

With separated variables, the amplitude y of the wave may be expressed as

$$y = f_1(x) f_2(t), \text{ where } f_1(x) \text{ is a function only of } x$$
$$\text{and } f_2(t) \text{ can be written as } f_2(t) = A \sin 2\pi\nu t \qquad [4.8]$$

Here A is a constant equal to the maximum amplitude of the wave.

Therefore, $\quad y = f_1(x)\, A \sin 2\pi\nu t$

Differentiating y twice with respect to t at constant x, we get

$$dy/dt = f_1(x)\, 2\pi\nu\, A \cos 2\pi\nu t$$
$$d^2y/dt^2 = -4\pi^2\nu^2 f_1(x)\, A \sin 2\pi\nu t$$

From 4.8, $\dfrac{d^2y}{dt^2} = -4\pi^2\nu^2 f_1(x) f_2(t)$ [4.9]

Therefore, from 4.7, $$\frac{d^2y}{dx^2} = -\frac{4\pi^2\nu^2}{u^2} f_1(x) f_2(t) \qquad [4.10]$$

Again, we differentiate y ($= f_1(x)f_2(t)$) twice with respect to x at constant t:

$$\frac{d^2y}{dx^2} = \frac{d^2 f_1(x)}{dx^2} \cdot f_2(t).$$

Comparison with 4.10 now gives, $$\frac{d^2 f_1(x)}{dx^2} = -\frac{4\pi^2 v^2}{u^2} f_1(x) \quad [4.11]$$

The velocity of the wave, u is related to the frequency, ν, by the relation $u = \lambda.\nu$, where λ is the corresponding wavelength. Hence, from 4.11,

$$\frac{d^2 f_1(x)}{dx^2} = -\frac{4\pi^2}{\lambda^2} f_1(x) \quad [4.12]$$

This equation is applicable to a wave moving in one direction only. Extending this to a wave moving in three dimensions, $f_1(x)$ is replaced by a new function depending on the three cartesian coordinates : $\psi(x, y, z)$. Writing simply ψ for $\psi(x, y, z)$, equation 4.12 is transformed to

$$\frac{\partial^2 \psi}{\partial x^2} + \frac{\partial^2 \psi}{\partial y^2} + \frac{\partial^2 \psi}{\partial z^2} = -\frac{4\pi^2}{\lambda^2}\psi \quad \text{or,} \quad \nabla^2 \psi = -\frac{4\pi^2}{\lambda^2}\psi \quad [4.13]$$

The symbol ∇^2 (del squared) is often used to represent the differential operator (Laplacian operator); in cartesian coordinates,

$$\nabla^2 \equiv \frac{\partial^2}{\partial x^2} + \frac{\partial^2}{\partial y^2} + \frac{\partial^2}{\partial z^2}$$

3. Matter waves: As already stated, the fundamental *postulate* of Schrödinger was that an equation of type 4.12 or 4.13 may be applied to all particles. The equations may be further adjusted by inserting the mass (m) and velocity (v) of the particle concerned from the de Broglie relationship $\lambda = h/mv$. From 4.13,

$$\nabla^2 \psi = -\frac{4\pi^2 m^2 v^2}{h^2}\psi \quad [4.14]$$

The total energy E of the particle must be equal to the sum of its potential energy V and kinetic energy $\frac{1}{2}mv^2$:

$$E = V + \tfrac{1}{2}mv^2; \qquad \text{or,} \quad mv^2 = 2(E - V)$$

Substituting in equation 4.14, one gets, $$\nabla^2 \psi + \frac{8\pi^2 m(E - V)}{h^2}\psi = 0$$

This is the wave equation of Schrödinger in the time-independent form (Eqn. 4.1).

We may rearrange the above equation to write

$$-\nabla^2 \psi + \frac{8\pi^2 mV}{h^2}\psi = \frac{8\pi^2 mE}{h^2}\psi$$

or, $$-\frac{h^2}{8\pi^2 m}\nabla^2 \psi + V\psi = E\psi$$

or, $$\left[-\frac{\hbar^2}{2m}\nabla^2 + V\right]\psi = E\psi. \qquad (\hbar = h/2\pi)$$

The part within square brackett may be represented by H to obtain Eqn. 4.6:

$$H\psi = E\psi$$

where H is called the energy operator or Hamiltonian operator.

Interpretation of ψ

How do we interpret the wave function ψ? By this time we have seen that ψ is a function of the space coordinates of the position of a particle of the system and is essentially the amplitude analogue of the associated matter wave. We may take the analogy of electromagnetic waves to get further insight into the nature of ψ.

For electromagnetic waves, the intensity of radiation at any point is proportional to the square of the amplitude of the wave at that point. Again, the greater the intensity of radiation at a point, greater is the number of photons striking that point. Combining these two relations, greater amplitude of electromagnetic waves at a point suggests a greater number of photons at that point. An alternative expression for this would be : *the greater the amplitude of a light wave at a point, greater is the probability of a photon to be found there.* The probability is thus related to the square of the amplitude.

Parallel arguments may be applied to a matter wave, e.g., the wave of an electron. Let the amplitude of this wave at a point (x, y, z) be ψ. The probability of finding the electron near that point should be proportional to ψ^2, the square of the amplitude function. As ψ may also be complex, the probability is expressed by including its complex conjugate ψ^*.

$\psi^*\psi$ is real and represents the probability density, that is, probability per unit volume. The probability that an electron may be found in a small volume element dv will be given by $\psi^*\psi dv$. If a particle can move only along the x-axis and is characterized by the wave function $\psi(x)$, the probability density is $\psi^*(x)\psi(x)$. The probability that the particle may be located between x_1 and x_2 will be

$$\int_{x_1}^{x_2} \psi^*(x)\psi(x)\,dx$$

The total probability of locating the particle between $x = -\infty$ to $x = +\infty$ must be unity, i.e.,

$$\int_{-\infty}^{\infty} \psi^*(x)\psi(x)\,dx = 1.$$

Since the probability $\psi^*(x)\psi(x)dx$ is dimensionless, $\psi(x)$ should have the dimension of length$^{-1/2}$ or the unit m$^{-1/2}$. Wave functions in three dimensions will have the unit m$^{-3/2}$.

Max Born postulated (1926) that $|\Psi|^2$ gives the probability density of finding any particle at given locations in space at time t where Ψ is, as before, a function of the space coordinates of the particle and time t. Here $|\Psi|$ is the absolute value of Ψ. If $\Psi = a + ib$, the absolute value of Ψ is defined as $(a^2 + b^2)^{1/2}$. Also,

$$\Psi^* = a - ib \text{ and } \Psi^*\Psi = (a - ib)(a + ib) = a^2 + b^2 = |\Psi|^2$$

For a one-particle, one-dimensional system, the probability of finding the particle in the region d_1 to d_2 is given by

$$\int_{x=d_1}^{x=d_2} |\Psi|^2\,dx$$

For stationary states, we have similar expressions using the space-only function ψ.

In fact, the implications of the wave function are not limited to position only. The function provides information on the outcome of a measurement of any property of the system, position, momentum, energy, angular momentum etc.

Finally, the wave function is not to be associated with any physical wave. It is an abstract mathematical entity though it provides information on physical quantities which may be verified experimentally.

The Schrödinger equation thus specifies the laws of motion relevant to particles of microscopic systems. The theory includes Newtonian mechanics as a special case. In the next section we shall learn some applications of the equation to a few simple systems. Prior to that, we shall discuss a few more properties of the wave function.

Eigenfunctions and Eigenvalues

We recall the form of the time-independent Schrödinger equations 4.5 or 4.6:

$$\left[-\frac{\hbar^2}{2m}\nabla^2 + V\right]\psi = E\psi \qquad \text{or,} \quad H\psi = E\psi$$

We mentioned the part in the square bracket as the Hamiltonian operator H (also written $\hat{H}$) or the energy operator, since it yields the energy E of the system when applied on ψ. Here H operates on ψ to give E times ψ. In general when an operator $\hat{A}$ (a hat, to indicate that A is an operator) operates on a function ψ_i to give back the function multiplied by some constant a_i, that is,

$$\hat{A}\psi = a_i\psi_i \qquad \text{operator (function)} = \text{constant} \times \text{same function}$$

we say that ψ_i is an *eigenfunction* of the operator $\hat{A}$ and a_i is its *eigenvalue* (eigen ≡ appropriate). In case of the Hamiltonian operator $\hat{H}$ mentioned above, $\psi(x)$ is an eigenfunction and the several values of E that are obtained for various wavefunctions are eigenvalues.

(d/dx) of e^{kx} is $k.e^{kx}$. Hence e^{kx} is an eigenfunction of (d/dx) with an eigenvalue k.

Well-behaved wave functions

The Schrödinger equation in its steady-state or time-independent form may have more than one solution for ψ. But all of them may not correspond to physical reality. We want only those solutions of the Schrödinger equation which will give us information on physically measurable properties of the system (though ψ itself is an abstract entity). Thus an acceptable wave function or well-behaved wave function should not only satisfy the Schrödinger equation, but also obey certain conditions pertinent to the system. Such acceptable solutions for the wave function (eigenfunctions) are selected according to the following conditions:

***Condition* 1:** *ψ must be single-valued.* This means ψ may have one and only one value at any point in space. This is necessary to ensure that the corresponding probability at any point may have only one value.

***Condition* 2:** *ψ must be finite.* The probability density at any point in space must be finite— an infinite probability density at any point signifies that the particle is present precisely at that point only. Accordingly ψ should vanish at infinity.

***Condition* 3:** *ψ must be continuous*. Certain properties like the momentum are calculated by taking the derivative of ψ. If ψ is discontinuous at some point, its first derivative at that point will be infinite. But values of physical properties cannot be infinite.

***Condition* 4:** *ψ must be square integrable*. This allows multiplication of ψ by a constant to make it *normalized*; this means the total probability of finding a particle described by ψ over entire space must equal unity:

$$\int_{-\infty}^{\infty} |\psi|^2 \, dx\,dy\,dz = 1$$

If ψ is a solution of the Schrödinger equation, so also is $k\psi$, where k = constant. This is called the *normalization constant*.

We shall learn the implications of these conditions further in connection with the particle in a box problem.

4.3.2 Particle in a Box

The simplest illustration of the application of Schrödinger equation is provided by the particle in a one-dimensional box problem. Let us consider a particle trapped in a region of space in one-dimension (along the x-axis, say); it is restricted to move between $x = 0$ and $x = L$ by infinitely high impenetrable walls. The situation is analogous to a particle confined in a box or a square well (one-dimensional) whose walls are infinitely high [Fig. 4.8]. The particle does not lose energy when it collides with such walls; hence its total energy remains constant. The potential energy V of the particle within the box is also constant—for convenience we *assume* it to be zero. On both sides of the box, the potential is infinite. *We assume that the behavior of the particle in the box can be described by the Schrödinger wave equation* (time-independent form). Since the energy of the particle is finite, it cannot exist outside the box, that is in regions with $x \leq 0$ and $x \geq L$. *So the wave function of the particle,* ψ, is zero in these regions.

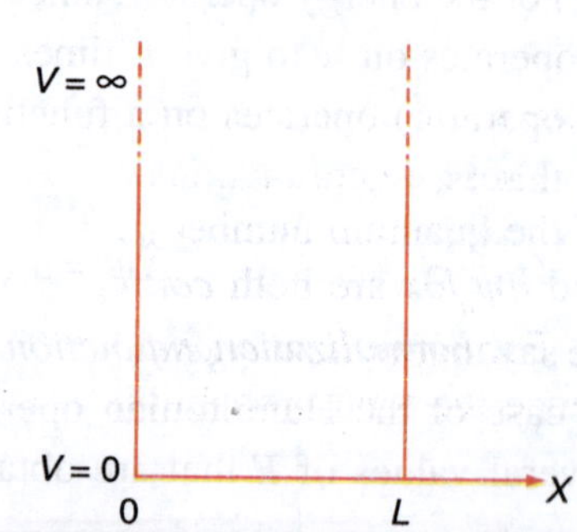

Fig.4.8: A one-dimensional "box" with infinitely high rigid boundaries.

Inside the box, $V = 0$; the Schrödinger equation (Equation 4.2) becomes

$$\frac{d^2\psi}{dx^2} + \frac{8\pi^2 m}{h^2} E\psi = 0.$$

The differential equation may be written as

$$\frac{d^2\psi}{dx^2} = -k^2\psi \text{ where } k^2 = \frac{8\pi^2 mE}{h^2}.$$

The general solution of this differential equation is

$$\psi = A \sin kx + B \cos kx.$$

Boundary conditions: At $x = 0$, and at $x = L$, ψ must be zero.

$$0 = A\sin(0) + B\cos(0) \text{ or } B = 0.$$

The equation thus reduces to $\psi = A \sin kx$.

At the other wall, i.e., $x = L$, $\psi = 0$; Hence, $0 = A \sin kL$.

A cannot be zero since then ψ itself vanishes (trivial solution); so we have $\sin kL = 0$. This occurs when,

$$kL = n\pi, \quad \text{or, } k = \frac{n\pi}{L} \qquad n = 1, 2, 3, \ldots$$

$$\Rightarrow \psi = A\sin\frac{n\pi x}{L}$$

From the value of k^2, we get

$$E_n = \frac{n^2h^2}{8mL^2}\ ; \quad n = 1, 2, 3, \ldots \qquad [4.15]$$

We see at once that the energy *has been quantized* in terms of the integral values of n. The particle can now assume only discrete energy values E_n corresponding to each integral value of n. These are the energy eigenvalues corresponding to the eigenfunctions

$$\psi_n = A\sin(n\pi x/L) = A\sin\frac{\sqrt{8\pi^2 mE_n}}{h}x.$$

Here n quantizes the energy; so it becomes a quantum number.

These eigenfunctions satisfy the conditions for an well-behaved function. For each value of the quantum number n, ψ_n has a definite value at any point x. : it is thus *single-valued*. ψ_n and $\partial\psi_n/\partial x$ are both *continuous*.

For *normalization*, we note that the particle is confined within the region $x = 0$ to $x = L$ and hence the total probability of finding the particle in this region must equal unity:

$$\int_0^L A^2\sin^2\left(\frac{n\pi x}{L}\right)dx = 1 \qquad \text{or,} \quad A^2\int_0^L \tfrac{1}{2}\left(1-\cos\frac{2n\pi x}{L}\right)dx = 1.$$

Simplifying, we get $A = \sqrt{2/L}$. The normalized wave functions of the particle are therefore

$$\psi_n = \sqrt{\frac{2}{L}}\sin\frac{n\pi x}{L} \qquad \left(\sqrt{\frac{2}{L}} \text{ is the normalization constant}\right)$$

The nature of plot of the wave functions for n = 1, 2 and 3 and the corresponding probability densities, $|\psi_1|^2$, $|\psi_2|^2$, are shown in Fig. 4.9. We observe that ψ_n may be both positive or negative, but $|\psi_n|^2$ is always positive.

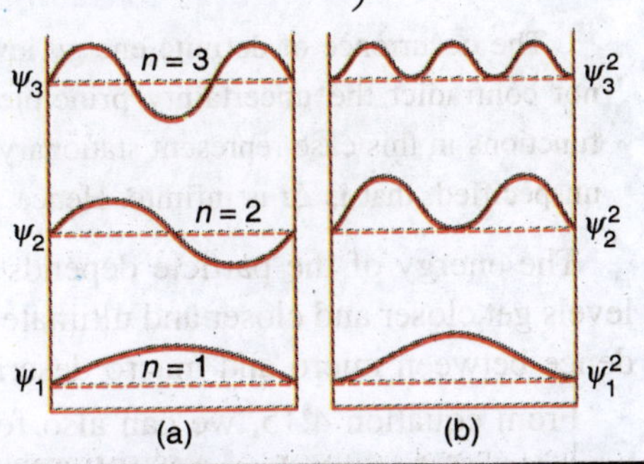

Fig.4.9: (a) Wave functions and (b) probability densities of a particle confined in a one-dimensional box with infinitely high rigid walls.

The classical momentum p_x of the particle corresponding to the energy value $E_n (= p_x^2/2m)$ gives

$$p_x = \pm\frac{nh}{2L}; \quad n = 1, 2, 3, \ldots$$

From the de Broglie relation, $\lambda = h/p_x$,

$$\lambda = \frac{h}{nh/2L} = \frac{2L}{n} \quad \text{i.e., } L = \frac{n\lambda}{2}.$$

Thus for each ψ_n, the length of the box should contain an integral number of half-wavelengths, as seen in Fig 4.9(a). The waves thus resemble the possible modes of vibration of a stretched string.

The plots of the probability densities are also interesting (Fig 4.9(b)). For $n = 1$, the most probable position of the particle is at $L/2$; but for $x = 2$, this point corresponds to a node, i.e., the probability of finding the particle at the point becomes zero. For every ψ_n, there are $n-1$ number of nodes (in addition to those at the boundaries). The particle has high probability of being found on either side of these inner nodes. This necessitates a change in our classical outlook for the "particle" when it is subjected to wave mechanical treatment : the ''particle'' cannot simply "roll" along the x-axis showing maxima at some point and nodes at some other points.

Classically, the particle can roll back and forth within the walls—the probability of finding it anywhere between $x = 0$ and $x = L$ is constant. The classical probability function has a value of $1/L$. The quantum mechanical treatment shows n number of peaks of probability where n is the quantum number. As n becomes very large, the number of peaks increase and the separation between them decreases. Ultimately they will merge and the situation will become identical to the classical description. There is thus *correspondence between the classical and the wave mechanical descriptions.*

The same correspondence is revealed when we consider the energy of the particle (Fig. 4.10). Classically a particle inside the one-dimensional box may have any positive energy. But our wave-mechanical treatment spontaneously leads to discrete quantized energy levels ($E_n = n^2h^2/8mL^2$; $n = 1, 2, 3, \ldots$). Since n cannot be zero (in that case the whole wavefunction will disappear), the particle in the box has a minimum energy of $h^2/8mL^2$ ($= E_0$) even at absolute zero—its residual energy or *zero-point energy.*

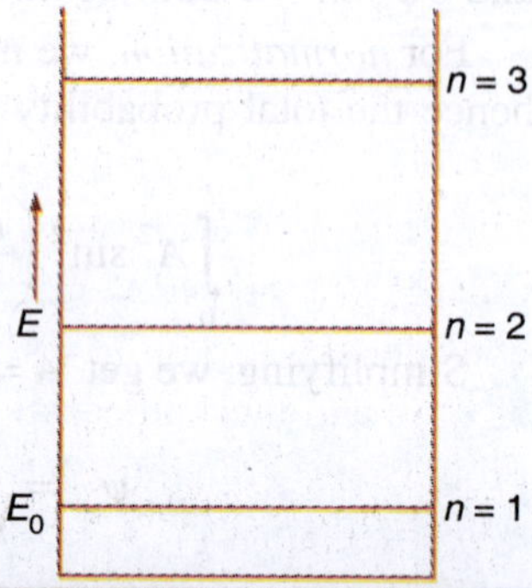

Fig.4.10: The nature of discrete energy levels of a particle in a one-dimensional box (not to scale).

The occurrence of definite energy levels for the particle does not contradict the uncertainty principle: $\Delta E \Delta t \geq h$. The wave functions in this case represent stationary states for which time is unspecified, that is Δt is infinite. Hence ΔE will be zero.

The energy of the particle depends directly on n^2. As n increases, the successive energy levels get closer and closer and ultimately they merge into a continuum. This shows correspondence between micro and macro descriptions of a system.

From equation 4.15, we can also follow the dependence of energy of the particle on the mass (m) of the particle and length (L) of the box. The energy of the particle depends inversely on m and L^2. Hence a particle with the mass of an electron in a linear region of the order of a molecule will have much wider separation between the energy levels than a molecule in a one-dimensional box of few centimeter length. The same reasoning will apply to a cubic box. In fact the energy levels of thermal motions of molecules (in three dimensions) are so close that their kinetic energy appears to vary continuously.

Let us express the energy of the particle for $n = 1$ state by E_0.

$$\text{Then,} \quad E_0 = h^2/8mL^2; \quad E_n = n^2 E_0.$$

$$\therefore \quad E_{n+1} - E_n = \left[(n+1)^2 - n^2\right] E_n = (2n+1)\, E_0$$

For a particle of $m = 1$ g(0.001 kg) in a box of length 10 cm (0.1 m), with a velocity 10 cm s^{-1},

$$E_0 = \left(h^2/8mL^2\right) = 5 \cdot 5 \times 10^{-63} \text{ J}; \text{ kinetic energy } \left(\tfrac{1}{2}mv^2\right) = 5 \times 10^{-6} \text{ J}.$$

The quantum number n corresponding to this energy would be given by:

$$n^2 = E_n / E_0 = 5 \times 10^{-6} \div 5.5 \times 10^{-63} \simeq 10^{57} \Rightarrow n \simeq 3 \times 10^{28}.$$

The spacing between the energy levels in this high quantum number range will be so small (~10^{-35} J) that they will appear continuous. Here again we see the correspondence between macro and micro systems.

For a two and a three-dimensional box, the wave function would separate into two and three equations respectively, each having the same type as the one for one-dimensional box. The allowed energies may then be expressed as (L = side of the box):

$$\text{two-dimensional box: } E = \frac{h^2}{8mL^2}\left(n_x^2 + n_y^2\right)$$

$$\text{three-dimensional box: } E = \frac{h^2}{8mL^2}\left(n_x^2 + n_y^2 + n_z^2\right).$$

4.3.3 The Hydrogen Atom

The hydrogen atom consists of a central proton (charge $+e$) and an extranuclear electron (charge $-e$). Since the mass of a proton is nearly 1836 times the mass of an electron, one may consider the proton to be stationary while the electron moves around it under a balance of electric and mechanical forces. As shown in connection with Bohr's theory, the motion of the proton may be adjusted by replacing the mass m of the electron by its reduced mass μ (= 0.999456 m).

To begin with, we have the time-independent Schrödinger equation (for stationary states of the atom) with the wave function of the electron ψ which is a function of the space coordinates x, y, z and the potential energy V:

$$\psi \equiv \psi\,(x, y, z, V)$$

$$\frac{\partial^2 \psi}{\partial x^2} + \frac{\partial^2 \psi}{\partial y^2} + \frac{\partial^2 \psi}{\partial z^2} + \frac{2m}{\hbar^2}(E - V)\,\psi = 0.$$

E is the total energy of the electron; the potential energy V of the electron at a distance r from the nucleus is:

$$V = -\frac{e^2}{4\pi\varepsilon_0 r}.$$

Since V is a function of r, we cannot substitute it directly in the Schrödinger equation written above using cartesian coordinates. The situation may be tackled by writing

$$V = -\frac{e^2}{4\pi\varepsilon_0 \sqrt{x^2+y^2+z^2}}$$

An alternative, and also a more realistic approach in view of the spherical symmetry of the system, would be to express the Schrödinger equation using spherical polar coordinates. For this, we have to understand the interrelation between the cartesian and polar systems of coordinates and the corresponding modification of the Laplacian operator ∇^2 (see Box).

The Schrödinger equation in polar coordinates

The interrelation between the Cartesian and spherical polar coordinates of a point (P) may be understood with reference to Fig. 4.11.

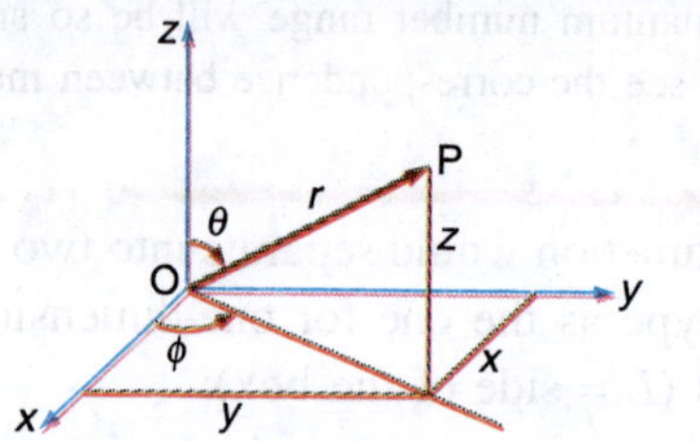

Fig.4.11: Cartesian and polar coordinates.

r = length of radius vector from origin O to point P

$= \sqrt{x^2+y^2+z^2}$;

θ = zenith angle (angle between radius vector and +z axis).

$= \cos^{-1}\dfrac{z}{\sqrt{x^2+y^2+z^2}}$

ϕ = azimuth angle = angle between the projection of the radius vector on the xy plane and the $+x$ axis = $\tan^{-1}(y/x)$.

The coordinates x, y, z and r, θ, ϕ for the point P in the two systems (origin at the nucleus O of the atom) are related as

$$x = r\sin\theta\cos\phi; \quad y = r\sin\theta\sin\phi; \quad z = r\cos\theta \quad \text{and} \quad r^2 = x^2+y^2+z^2$$

Accordingly, the Laplacian operator ∇^2 becomes

$$\nabla^2 = \frac{1}{r^2}\frac{\partial}{\partial r}\left(r^2\frac{\partial}{\partial r}\right) + \frac{1}{r^2\sin\theta}\frac{\partial}{\partial\theta}\left(\sin\theta\frac{\partial}{\partial\theta}\right) + \frac{1}{r^2\sin^2\theta}\frac{\partial^2}{\partial\phi^2}.$$

Using this expression for ∇^2 and putting $V = -e^2/4\pi\varepsilon_0 r$, the Schrödinger equation for a one electron system like the hydrogen atom has the form

$$\frac{1}{r^2}\frac{\partial}{\partial r}\left(r^2\frac{\partial\psi}{\partial r}\right) + \frac{1}{r^2\sin\theta}\frac{\partial}{\partial\theta}\left(\sin\theta\frac{\partial\psi}{\partial\theta}\right) + \frac{1}{r^2\sin^2\theta}\frac{\partial^2\psi}{\partial\phi^2} + \frac{8\pi^2 m}{h^2}\left(E + \frac{e^2}{4\pi\varepsilon_0 r}\right)\psi = 0. \quad [4.16]$$

E is the total energy of the electron; its mass m should be actually replaced by the reduced mass μ as explained earlier. For other hydrogen-like species (see Bohr's theory) the potential energy will be $-\dfrac{Ze^2}{4\pi\varepsilon_0 r}$. The equation may also be expressed in the following form by multiplying throughout by $r^2\sin^2\theta$

$$\sin^2\theta\frac{\partial}{\partial r}\left(r^2\frac{\partial\psi}{\partial r}\right) + \sin\theta\frac{\partial}{\partial\theta}\left(\sin\theta\frac{\partial\psi}{\partial\theta}\right) + \frac{\partial^2\psi}{\partial\phi^2} + \frac{8\pi^2 m r^2\sin^2\theta}{h^2}\left(E + \frac{e^2}{4\pi\varepsilon_0 r}\psi\right) = 0. \quad [4.17]$$

The Schrödinger equation in the polar form has an advantage : one can express ψ as a product of three functions ψ_r (or $R(r)$), ψ_θ (or $\Theta(\theta)$) and ψ_ϕ (or $\Phi(\phi)$) which depend only on r, θ and ϕ respectively:

$$\psi(r, \theta, \phi) = \psi(r)\,\psi(\theta)\,\psi(\phi) = R(r)\,\Theta(\theta)\,\Phi(\phi) \qquad [4.18]$$

$R(r)$ or ψ_r is a function of r, the radial distance from the nucleus. This is therefore called the *radial part* of the wave function, or simply the radial function and abbreviated by R. $\psi(\theta)$ or $\Theta(\theta)$ is a function of θ and $\psi(\phi)$ or $\Theta(\phi)$ is a function of ϕ. The last two functions represent the *angular part* of the wave function. They may also be combined into a single function $\psi(\theta, \phi)$:

$$\psi(r, \theta, \phi) = \psi(r)\,\psi(\theta, \phi). \qquad [4.19]$$

Solution of equation 4.18 leads to three quantum numbers to describe the electron in a hydrogen atom. In Bohr's treatment only one such quantum number (n) was introduced. Similarly, the state of a particle in a one-dimensional box also introduced one quantum number. The motion of the electron considered in a three-dimensional space now introduces three quantum numbers for meaningful and well-behaved solution of the Schrödinger equation. They have the same set of values as assigned before from spectroscopic studies. These are

(i) The Principal quantum number (n) = 1, 2, 3 ...

(ii) The Orbital quantum number (l) = 0, 1, 2 ... (n − 1)

(iii) The Magnetic quantum number (m_l or m) = 0, ± 1, ± 2 ... ± l.

A fourth quantum number (spin) was introduced later. We shall discuss all four quantum numbers in a later section. For a given energy state, the radial function R depends on the values of n and l. This is often expressed by writing n and l as subscripts, i.e., as R_{nl}. Similarly, the Θ-function depends on l and m while Φ depends on m only. Summarily, the hydrogen-like wave functions are written as

$$\psi = R_{nl}(r)\,\Theta_{lm}(\theta)\,\Phi_m(\phi) \qquad [4.20]$$

These well-behaved one-electron spatial wave functions (obtained as solutions of the Schrödinger equation with one-electron Hamiltonian) for hydrogen and hydrogen-like species are called ***orbitals*** (Mulliken). The state of an electron in the atom is described by a characteristic wave function or orbital, designated by appropriate n and l values when we say that the electron "occupies" the orbital. These orbitals are conveniently expressed by the numerical values of n and certain letters (s, p ...) for the l values:

l	=	0	1	2	3	4	...
Symbol	=	s	p	d	f	g	...

Thus an 1s orbital means a state represented by a wave function having n = 1 and l = 0.

Since a hydrogen-like atom has only one electron, all the hydrogen-like wave functions are orbitals. For many electron atoms we may find an approximate one-electron wave function by subjecting an electron to a combined field of the nucleus and all other remaining electrons.

Energy of the electron

Solutions of the $R(r)$ or radial part of the wave equation for the hydrogen (and hydrogen-like) atom must satisfy the physical condition that $R(r)$ should vanish as $r \to \infty$. This leads to discrete energy values of the system depending upon n, the principal quantum number:

$$E_n = -\frac{\mu e^4 Z^2}{2(4\pi\varepsilon_0)^2 \hbar^2 n^2}$$

The energy is negative as explained earlier (see Bohr's theory) and has the same expression as derived before. It may also be expressed as

$$E_n = -\frac{Z^2}{n^2} \cdot \frac{e^2}{(4\pi\varepsilon_0)\,2a} \qquad \text{where, } a = \frac{\hbar^2 (4\pi\varepsilon_0)}{\mu e^2}$$

Substitution gives $a = 0.5295$Å (against 0.5292Å for a_0, the first Bohr radius; see box). Clearly, $a \simeq a_0$ when $\mu = m$ (for a supposedly infinite mass of the nucleus).

The ground state energy for hydrogen may be calculated by putting $Z = 1$ and $n = 1$ in the above expression of E_n when we get,

$$E = -2.179 \times 10^{-18}\text{ J} = -2.179\text{ aJ (a for } atto,\ 10^{-18})$$

$$= -13.60\text{ eV}$$

The *ionization energy* of a hydrogen atom would be accordingly 13.60 eV.

The energy for other values of n and Z in different hydrogen-like species (He^+, Li^{2+} etc.) may be found by the generalized formula

$$E_n = -\frac{Z^2}{n^2} \times 13.60\text{ eV}.$$

Here we have ignored the variation in μ owing to change in mass of the nucleus.

Atomic units

In quantum chemistry, some conventional non-SI units are found for length and energy. The first Bohr radius of hydrogen (a_0) is taken as the **atomic unit of length** and called 1 bohr.

$$\textbf{1 bohr} = a_0 = \hbar^2(4\pi\varepsilon_0)/me^2 = 0\cdot5292\text{ Å}$$

The potential energy of two electrons separated by the Bohr radius is called the Hartree energy (E_h) which is the **atomic unit of energy**.

$$\textbf{1 Hartree}\ (E_h) = e^2/(4\pi\varepsilon_0)a_0 = 27\cdot211\text{ eV} \quad \text{or,} \quad 4\cdot3598 \times 10^{-18}\text{ J.}$$

If we approximate a by a_0, the energy of the hydrogen atom (ground state) is $-\frac{1}{2}$ hartree. This value is also used as a unit in many cases, called a rydberg (Ry).

$$\textbf{1 rydberg} = E_h/2 = 2\cdot18 \times 10^{-18}\text{ J.}$$

Radial wave function, R_{nl}

The radial part of the wave function (R_{ml} or simply R) depends only on n and l and is independent of θ and ϕ. Some typical examples of hydrogenic radial wave functions are

n	l	orbital	
1	0	$1s$	$R(1s) = 2\left(\frac{Z}{a}\right)^{3/2} e^{-Zr/a}$
2	0	$2s$	$R(2s) = \frac{1}{2\sqrt{2}}\cdot\left(\frac{Z}{a}\right)^{3/2}\left(2-\frac{Zr}{a}\right)e^{-Zr/2a}$
2	1	$2p$	$R(2p) = \frac{1}{2\sqrt{6}}\left(\frac{Z}{a}\right)^{3/2}\cdot\frac{Zr}{a}\,e^{-Zr/2a}$

Radial functions are also expressed conveniently in terms of the dimensionless quantity ρ:

$$\rho = \frac{2Zr}{na}, \qquad a = \frac{\hbar^2(4\pi\varepsilon_0)}{\mu e^2}.$$

$$R(1s) = 2\left(\frac{Z}{a}\right)^{3/2} e^{-\rho/2} \quad ; R(2s) = \frac{1}{2\sqrt{2}}\left(\frac{Z}{a}\right)^{3/2}(2-\rho)\,e^{-\rho/2}$$

$$R(2p) = \frac{1}{2\sqrt{6}}\left(\frac{Z}{a}\right)^{3/2}\rho\,e^{-\rho/2} \quad ; \quad R(3s) = \frac{1}{\sqrt{243}}\left(\frac{Z}{a}\right)^{3/2}(6-6\rho+\rho^2)\,e^{-\rho/2}$$

$$R(n, l) = r^l\,.\,e^{-Zr/a}\left(b_0 + b_1 r + b_2 r^2 + \ldots + b_{n-l-1}\,r^{n-l-1}\right),$$

where b_0, b_1 ... are constants whose values depend on n and l.

The important general features of these functions are understood better from their plot against the radial distance r. For convenience, the functions may be multiplied by some suitable factor in the plot. Fig. 4.12*a–c* shows some typical plots of $R(r)\ (a/Z)^{3/2}$ vs (Zr/a) for some hydrogenic orbitals.

Some important general features of radial functions

(i) All the functions contain an exponential term which decreases with r (''decays'') as exp $(-Zr/na)$. So $R(2s)$ has a slower rate of decay with r than $R(1s)$. *Thus the wave function of an orbital with higher n extends to a larger distance from the nucleus.*

(ii) The R-function for the $2s$ orbital passes through zero when $(2-Zr/a) = 0$, i.e., at $r = 2a/Z$, $R(2s) = 0$. The point at $r = 2a/Z$ thus represents a ***radial node***. The function changes its sign beyond this point. It can be shown that an orbital designated by n and l has $n-l-1$ number of radial nodes (the node at the origin is not counted). *A node is a surface where the wave function passes through zero as it changes sign*, as at $r = 2a$ for the $2s$ orbital in hydrogen atom. This is an example of a *radial node* or *spherical node* as the radial function passes

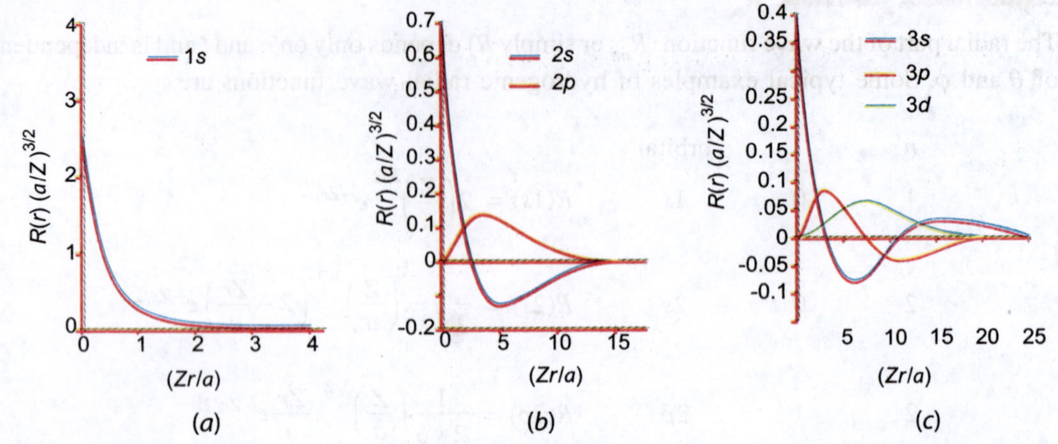

Fig.4.12: Nature of variation of the R-functions with r for various hydrogenic orbitals (approximate). (a) 1s (b) 2s and 2p (c) 3s, 3p and 3d.

through zero over a spherical surface at this distance from the nucleus. The number of such spherical nodal surfaces arising from a radial function are (n–l–1), as shown below:

1s	0	2p	0	3d	0
2s	1	3p	1	4d	1
3s	2	4p	2	5d	2

Besides such radial nodes, there are also angular nodes (discussed later).

Radial distribution function

In wave mechanical interpretation the probability of finding a particle at some point is proportional to the square of its wave function ψ for that point. So the dependence of probability on the radial distance from the nucleus will be obtained from the square of the radial part of the wave function, $\psi(r)^2$ or R^2. We may consider the atom to be composed of concentric layers of thin shells of thickness dr and calculate the volume dv of such a shell between r and $r + dr$:

$$V = 4\pi r^3/3; \; dV = 4\pi r^2 dr.$$

The probability that an electron will be found in this volume is R^2dV or $4\pi r^2R^2dr$. The function $4\pi R^2r^2$ or more generally R^2r^2 (for non-spherical orbitals) is called the **radial distribution function,** $P(r)$. When multiplied by dr, the function gives the probability of finding the electron anywhere in the shell of thickness dr at a distance r from the nucleus. In this sense $P(r)$ may be considered the **radial probability density**.

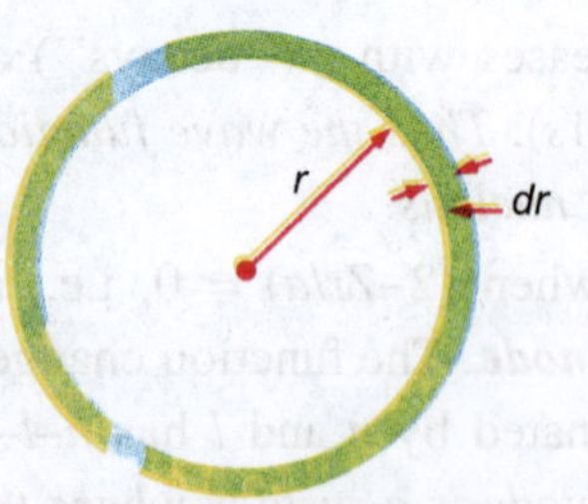

Fig.4.13: A spherical shell of thickness dr.

For the 1s orbital of hydrogen, (Z = 1), we may write, from

$$R = 2\left(\frac{Z}{a}\right)^{3/2} e^{-Zr/a} \quad , \quad \Rightarrow R_{1s}(\mathrm{H}) = 2.\left(\frac{1}{a}\right)^{3/2} e^{-r/a}$$

So the radial distribution function $P(r)$ will be

$$R_{1s}^2 . r^2 = \frac{4}{a^3} e^{-2r/a} . r^2 . \quad \text{... [4.20]}$$

Fig. 4.14 shows the nature of plot of $P(r)$ vs r. We observe that

(i) The radial distribution function R^2r^2 is zero at the nucleus because of the r^2 term, since, $r = 0$ at this point.

In contrast, the R-function had a very high value at $r = 0$.

(ii) Whereas the R-function decreases exponentially but r^2 increases rapidly with r, the resultant plot of R^2r^2 passes through a maximum at an intermediate value of r. This is called the *most probable radius*, r_{mp}. For hydrogen, this occurs at 53 pm (0.53 Å) which is the radius of the first Bohr orbit. This also represents the most probable distance (r_{max}) which decreases with nuclear charge in hydrogen-like ions: $r_{max} = a/Z$. The exponential term in the R function, exp $(-Zr/na)$, also suggests that R^2r^2 will decrease less slowly as n increases. Accordingly, r_{max} increases as n increases.

(iii) Owing to the presence of the exponential term, $P(r) \rightarrow 0$ as $r \rightarrow \infty$.

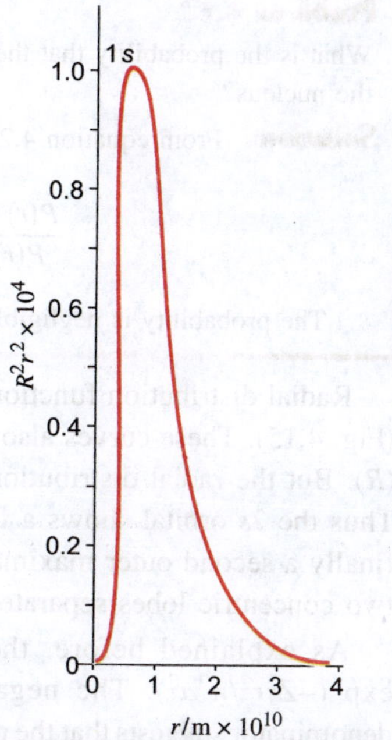

Fig.4.14: Plot of radial distribution function for the 1*s* orbital of hydrogen

Problem 4.10

R_{1s} for hydrogen is $\frac{2}{a^{3/2}} e^{-r/a}$.

Show that the most probable radius (r_{mp}) occurs at a.

Hint: The radial distribution function $P = 4\pi R^2r^2 = \frac{16\pi}{a^3} . e^{-2r/a} . r^2$

We find the expression of r_{mp} for maximum value of P from $dP/dr = 0$.

For $dp/dr = 0$, $2r_{mp} - \frac{2r_{mp}^2}{a} = 0 \Rightarrow r_{mp} = a$

Problem 4.11

Calculate the probability of finding an electron in a shell of thickness 0.001Å at $r = 0.1a_0$ ($a_0 = 0.529$ Å) in an atom of hydrogen. $R_{1s} = \frac{2}{a^{3/2}} e^{-r/a}$

Hint: The probability is $4\pi r^2R^2dr$ where $dr = 0.001$ A°, $r = 0.0529$ A° and $R = R_{1s}$.

We can put $a = a_0$ without any compromise in precision.

Problem 4.12

What is the probability that the electron in a hydrogen atom will be found at a distance $10a$ from the nucleus?

Solution: From equation 4.20

$$\frac{P(r) \text{ for } r = 10a}{P(r) \text{ for } r = a} = \frac{(10a)^2}{a^2} \cdot \frac{e^{-20}}{e^{-2}} = 100\, e^{-18} = 1\cdot 5 \times 10^{-6}$$

The probability is negligibly small but nevertheless finite.

Radial distribution functions for orbitals higher than 1*s* may be similarly plotted against *r* (Fig. 4.15). These curves also show nodes corresponding to the nodes in their radial function (*R*). But the radial distribution function is always positive and shows one or more maxima. Thus the 2*s* orbital shows a first maxima at $0.8a$ (0.04 nm) followed by a node (at $2a$) and finally a second outer maxima at $5.2a$ (0.28 nm). This shows that the 2*s* orbital has actually two concentric lobes separated by a nodal surface.

As explained before, the distribution functions decay exponentially according to exp $(-Z^2r^2/n^2a^2)$. The negative sign with n^2 in the denominator suggests that the probability maxima will shift to right as *n* increases. Also, the rate of decrease will be slower as *n* increases. One can also calculate the *mean radius* of an orbital which is the expectation value of *r* calculated as

$$\langle r \rangle = \int \psi^* r \psi \, d\tau = \int r |\psi|^2 \, d\tau$$

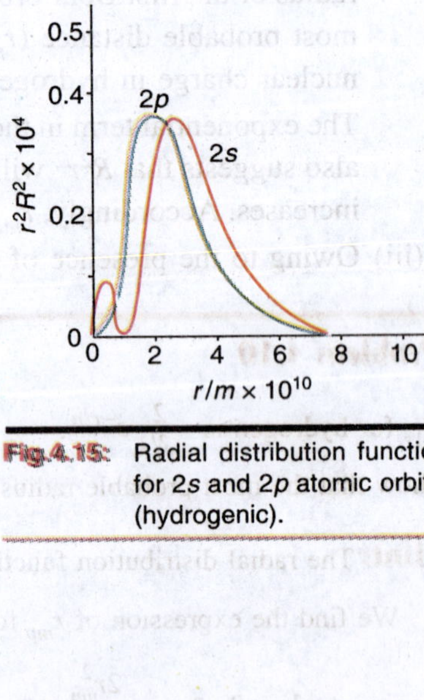

Fig.4.15: Radial distribution functions for 2*s* and 2*p* atomic orbitals (hydrogenic).

The 1*s* atomic orbital in a hydrogen atom has a mean radius equal to $3a/2$ or 0.79 A°. For other hydrogen-like species in their ground-state, $\langle r \rangle = 3a/2Z$.

For an orbital with quantum numbers *n* and *l*, the mean radius is given by the general expression

$$r_{av} = n^2 \left\{ 1 + \tfrac{1}{2}\left(1 - \frac{l(l+1)}{n^2} \right) \right\} {}^{a}\!/_{Z}$$

For a given value of *n*, the mean radius decreases as *l* increases. Thus, the 2*p* orbital ($l = 1$) has a smaller mean radius than the 2*s* orbital ($l = 0$).

Penetration. The radial distribution functions of the 2*s*, 2*p*, 3*s* etc. orbitals show small but finite presence within the inner orbitals. Comparison of Figures 4.14–4.16 reveals that the radial density of the 2*s* orbital spreads into the curve for the 1*s* orbital; the 3*s* orbital

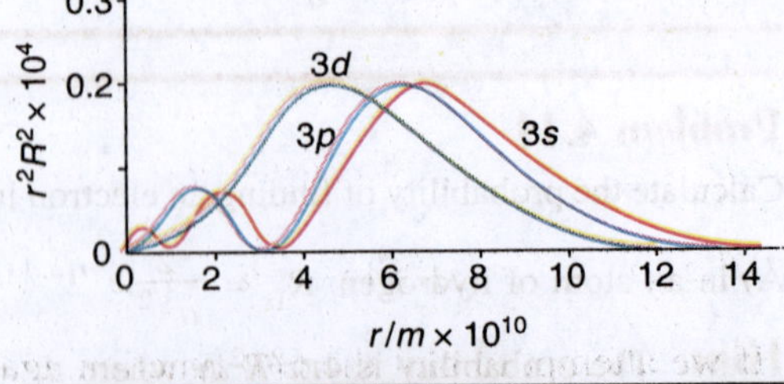

Fig.4.16: Radial distribution functions for 3*s*, 3*d* and 3*p* atomic orbitals (hydrogenic).

spreads into the 2*s* and 1*s* orbitals and so on. This is termed "*penetration*" of orbitals—that is, their distribution into inner electron cores. As a consequence of penetration, an electron in an outer orbital does not get the full screening by the inner electrons from the nuclear charge. If the radial distribution of the 2*s* orbital had been distributed completely "outside" the 1*s* orbital, the 2*s* electron(s) would be completely screened by the 1*s* electrons. Since the 2*s* orbital penetrates somewhat into the core of 1*s* electrons, the screening will be less. The extent of penetration decreases from *s* to *f* orbitals:

$$s > p > d > f.$$

This variation in penetration influences mutual screening of the electrons and hence controls the effective nuclear charge as well as the relative energy of the orbitals (see later). The consequences are far-reaching, as we shall encounter in connection with ionization potential, electronegativity, size, oxidation state etc.

The radial distribution functions shown above are for $Z = 1$. Enhanced nuclear charge does not change the gross shapes of the curves, but their extension into space is reduced. This is called *contraction of the orbitals*. Such contraction is determined by several other factors, particularly the presence of other electrons in the system. We shall consider them separately.

Angular parts of the wave function

The angular parts of hydrogenic wave functions (equation 4.18) are of the type shown in Table 4.1. As we shall soon appreciate, it is useful to consider them together as the product function $\Theta(\theta)\Phi(\phi)$ (abbreviated $\Theta\Phi$).

Table 4.1

Angular parts of some hydrogenic wave functions

Orbital	$\Theta(\theta)$	$\Phi(\phi)$	$\Theta\Phi$
s	$1\sqrt{2}$	$1/\sqrt{2\pi}$	$1/\sqrt{4\pi}$
p_x	$\sqrt{3/4}\sin\theta$	$(1/\sqrt{\pi})\cos\phi$	$\sqrt{3/4\pi}\sin\theta\cos\phi$
p_y	$\sqrt{3/4}\sin\theta$	$(1/\sqrt{\pi})\sin\phi$	$\sqrt{3/4\pi}\sin\theta\sin\phi$
p_z	$\sqrt{3/2}\cos\theta$	$1/\sqrt{2\pi}$	$\sqrt{3/4\pi}\cos\theta$
d_{z^2}	$\sqrt{5/8}(3\cos^2\theta-1)$	$1/\sqrt{2\pi}$	$\sqrt{5/16\pi}(3\cos^2\theta-1)$
d_{xy}	$\sqrt{15/16}\sin^2\theta$	$(1/\sqrt{\pi})\sin 2\phi$	$\sqrt{15/16\pi}\sin^2\theta\sin 2\phi$
d_{yz}	$\sqrt{15/4}\sin\theta\cos\theta$	$(1/\sqrt{\pi})\sin\phi$	$\sqrt{15/4\pi}\sin\theta\cos\theta\sin\phi$
d_{zx}	$\sqrt{15/4}\sin\theta\cos\theta$	$(1/\sqrt{\pi})\cos\phi$	$\sqrt{15/4\pi}\sin\theta\cos\theta\cos\phi$
$d_{x^2-y^2}$	$\sqrt{15/16}\sin^2\theta$	$(1/\sqrt{\pi})\cos 2\phi$	$\sqrt{15/16\pi}\sin^2\theta\cos 2\phi$

If we attempt to plot the $\Theta\Phi$ function against the independent variables θ and ϕ, we shall require a three-dimensional coordinate system: one for θ, one for ϕ and the other for the dependent variable $\Theta\Phi$. As noted earlier (Fig.4.1a), θ ranges from 0 to π and ϕ ranges from

0 to 2π. We may even use a cartesian coordinate system and choose to plot θ along the x-axis, ϕ along the y-axis and $\Theta\Phi$ along the z-direction. The $\Theta\Phi$ values will now describe a surface. At any point ($\theta\,\phi$), the height of the surface from the xy plane will give the corresponding value of $\Theta\Phi$. The angular part of the probability density in the direction $\theta\,\phi$ will now be given by $\Theta^2\Phi^2$.

Spherical polar coordinates are more suitable for understanding the variation of $\Theta\Phi$(or $\Theta^2\Phi^2$) with θ and ϕ. Conventionally, the function is plotted along the r-coordinate, measured from the origin (supposed to be at the nucleus of the atom). θ and ϕ are plotted along their respective coordinates. The plot, the locus of the points corresponding to $\Theta\Phi$ (or their square) now describe a characteristic surface for each type of orbital. The shape of the surface tells us about the nature of variation of the angular function (or the angular probability density) with direction.

As an example, we first consider the 1s orbital (hydrogenic). From Table 4.1, $\Theta\Phi$ for this wave function is independent of θ or ϕ and is equal to $1/\sqrt{4\pi}\,(=0.282)$. So the function $\Theta\Phi$ will have the same value for the full range of θ (0–π) and ϕ (0–2π). Before we appreciate the nature of the three dimensional plot, we begin with a two-dimensional plot of $\Theta\Phi$ on the xz-plane. What we simply do is that we draw radial lines from the z-axis ($\theta = 0°$) at regular small intervals, say 15° upto $\theta = 180°$. We mark the distance 0.282 from the origin on these lines (any unit). We shall find that the dots (or their locus) form a semicircle (Fig. 4.17a) covering $\theta = 0$–$180°$ and $\phi = 0°$. If we continue plotting similar dots on the $\phi = 180°$ half-plane, we get a circle. Since the value of $\Theta\Phi$ is independent of θ and ϕ, the plot will look the same on any vertical plane containing the z-axis. So we can mentally get the 3-dimensional plot of

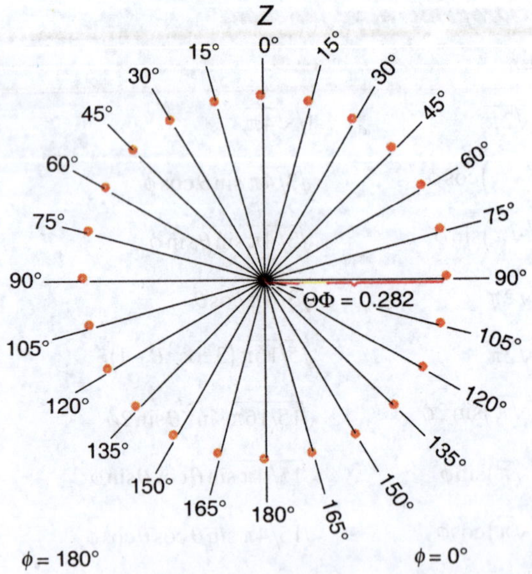

(a)

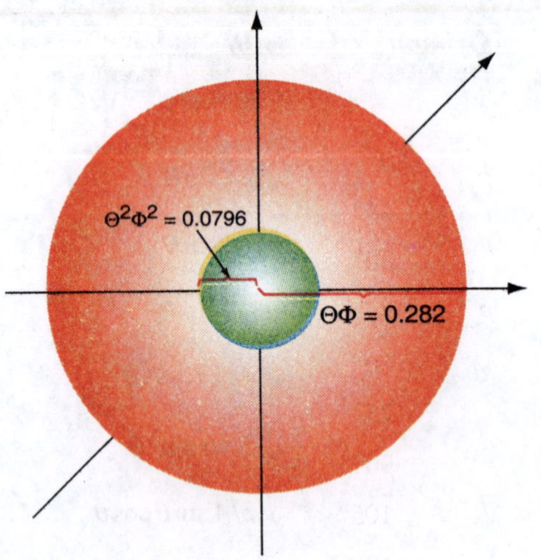

(b)

Fig.4.17: (a) Plot of angular part of wave function for 1s orbital in xy-plane
(b) Three-dimensional representation of $\Theta\Phi$ (outer sphere) and $\Theta^2\Phi^2$ (inner sphere) for 1s.

the ΘΦ function by rotating the circle about the z-axis. We actually get the cross-section of a sphere of radius 0.282 (Fig. 4.17b). Similarly we can get the plot of $\Theta^2\Phi^2$ as a sphere of radius $(0.282)^2$ (= 0.0796) (Fig.4.17b; green core). Since the values of ΘΦ and hence $\Theta^2\Phi^2$ do not change with direction, the angular probability density will be same in all directions. We say that the 1s orbital is *spherically symmetrical.*

Similar plots may be done for the angular part ΘΦ or its square for other orbitals. The procedures for p_z and p_x orbitals are illustrated in the box.

Angular functions of 2p-orbitals

2p_z-orbital: We begin with the 2p_z-orbital as its angular function depends only on θ ($\sqrt{3/4\pi}\cos\theta$; Table 4.1). So for a given value of θ, the function ΘΦ will have the same value on any plane containing the z-axis. We choose to plot the function on the xz-plane and draw radial lines from the origin at intervals of 15° for θ (Fig. 4.18). Now we mark the values of ΘΦ for different θ on these lines (Table 4.2). We find that the points for ΘΦ fall on a semicircle between θ = 0°–90°, the function becoming zero at θ = 90°. The points between θ = 90° to 180° repeat a similar semicircle; here the values of the function are all negative. As the function is independent of ϕ, we expect a similar semicircle on the left half-plane. Thus we have two circles touching one another at θ = 90°. The circumference of one circle describes positive values of the angular function while in the other, the circumference covers negative values of the function. Since the angular function is independent of ϕ, it will have the same value on all vertical planes passing through the z-axis. So, we can get the three-dimensional description of the angular function by rotating these circles along the z-axis. We get two hollow spheres touching one another at the origin. The surfaces of these spheres convey the angular variation of the ΘΦ function. Any straight line drawn through the origin at any direction will

Table 4.2
Values of ΘΦ and $\Theta^2\Phi^2$ for 2p_z at different θ

θ	cos θ	ΘΦ	$\Theta^2\Phi^2$
0°	1	0.489	0.239
15°	0.966	0.472	0.223
30°	0.866	0.423	0.179
45°	0.707	0.346	0.119
60°	0.500	0.244	0.060
75°	0.259	0.127	0.016
90°	0	0	0
105°	– 0.259	– 0.127	0.016
120°	– 0.500	– 0.244	0.060
135°	– 0.707	– 0.346	0.119
150°	– 0.866	– 0.423	0.179
165°	– 0.966	– 0.472	0.223
180°	– 1	– 0.489	0.239

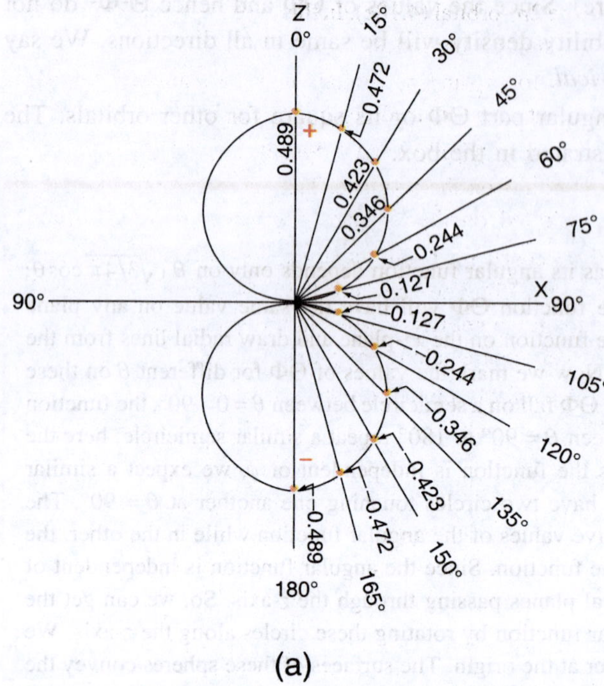

(a)

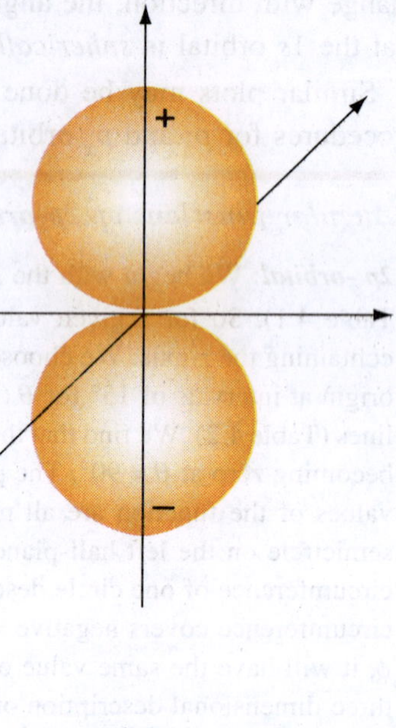

(b)

Fig.4.18: (a) Plot of ΘΦ for $2p_z$ orbital on *xz* plane.

(b) Angular part of the $2p_z$ orbital in three dimensions.

(c) Plot of $\Theta^2\Phi^2$ for $2p_z$ *on* xz-plane (not in the same scale as (a).

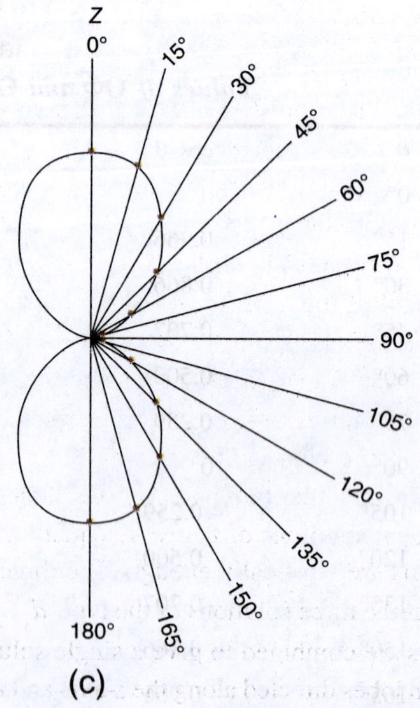

(c)

touch the surfaces of these two spheres at two opposite points which give us the values of the angular function in that direction. There is no information inside the spheres as such.

Instead of ΘΦ, one may also plot $\Theta^2\Phi^2$ for the $2p_z$-orbital (4.18c). In this case the values (squares) are all positive and it is redundant to write '±' sign on the lobes.

The 2p$_x$ (and 2p$_y$) orbitals: Angular functions for p_x and p_y orbitals may be similarly plotted. However, the two angular functions are different and we need separate treatment for each. The function for p_x orbital ($\sqrt{(3/4\pi)}$ $\sin\theta\cos\phi$) depends on both θ and ϕ. Since $\cos\phi = 0$ at $\phi = \pi/2$ or $3\pi/2$, the angular function becomes zero at all points on the *yz*-plane. This provides a nodal plane. The p_x angular function is advantageously plotted on the *xy*-plane for which $\theta = \pi/2$. Since $\sin\pi/2 = 1$, the angular function becomes $\sqrt{3/4\pi}\cos\phi$. We recall that the range of ϕ is 0°–360° and draw radial lines along directions of ϕ varying by 15°. The values of the ΘΦ function (calculated for each ϕ) are now marked on these lines. Again we get two spheres in three dimension as in the case of the p_z orbital but these spheres have their centres along the *x*-axis (Fig. 4.19).

Angular function for the p_y orbital may be similarly plotted on the *xy* plane.

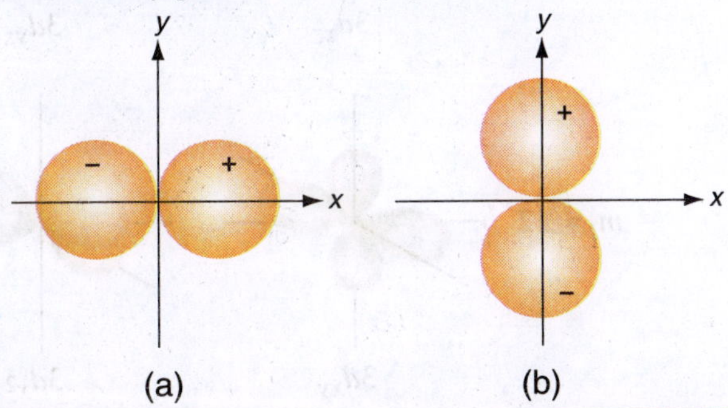

Fig.4.19: Angular parts of (a) $2p_x$ and (b) $2p_y$ on *xy*-plane.

Angular nodes result when the angular function (ΘΦ) becomes equal to zero and are planar or conical. There are *l* angular nodes in any orbital. The d_{z^2} orbital has two conical surfaces meeting at the nucleus and passing between the two sets of lobes; this is counted as 2. The number of angular nodes for *s*- *p*- and *d*- orbitals are as follows:

s-orbital: 0 ; *p*-orbitals: 1 plane for each

d-orbitals: 2 planes for each, the conical surface in d_{z^2} being counted as 2.

An orbital has a total of (*n*–1) nodes: (*n*–*l*–1) radial nodes and *l* angular nodes.

***d*-orbitals.** There are a set of five *d* orbitals (Fig. 4.20) labeled according to their orientation in a cartesian frame. The d_{xy}, d_{yz} and d_{zx} have their symmetry axes lying on the respective planes but pointing between the concerned axes. The symmetry axes of the $d_{x^2-y^2}$ orbital lie in the *xy* plane with its lobes along the *x* and *y*-axis. Two analogous wave functions can be written for the other two pairs of axes, lebeled as $d_{y^2-z^2}$ and $d_{z^2-x^2}$. But there are only five independent solutions of the wave equation corresponding to the *d* type ($l = 2$), i.e., only five *nd* orbitals are physically enough. The three solutions of the type d_{xy} are independent of each other, but the three solutions of the type $d_{x^2-y^2}$ are not. By convention, the $d_{y^2-z^2}$ and $d_{z^2-x^2}$ functions are combined to give a single solution labeled as d_{z^2}. This function corresponds to two main lobes directed along the *z*-axis and a collar perpendicular to the *z*-axis and symmetrical about the *xy* plane. Now d_{xy}, d_{zx}, d_{yz}, $d_{x^2-y^2}$ and d_{z^2} describe five independent *d*-orbitals. Each has two nodal planes passing through the nucleus.

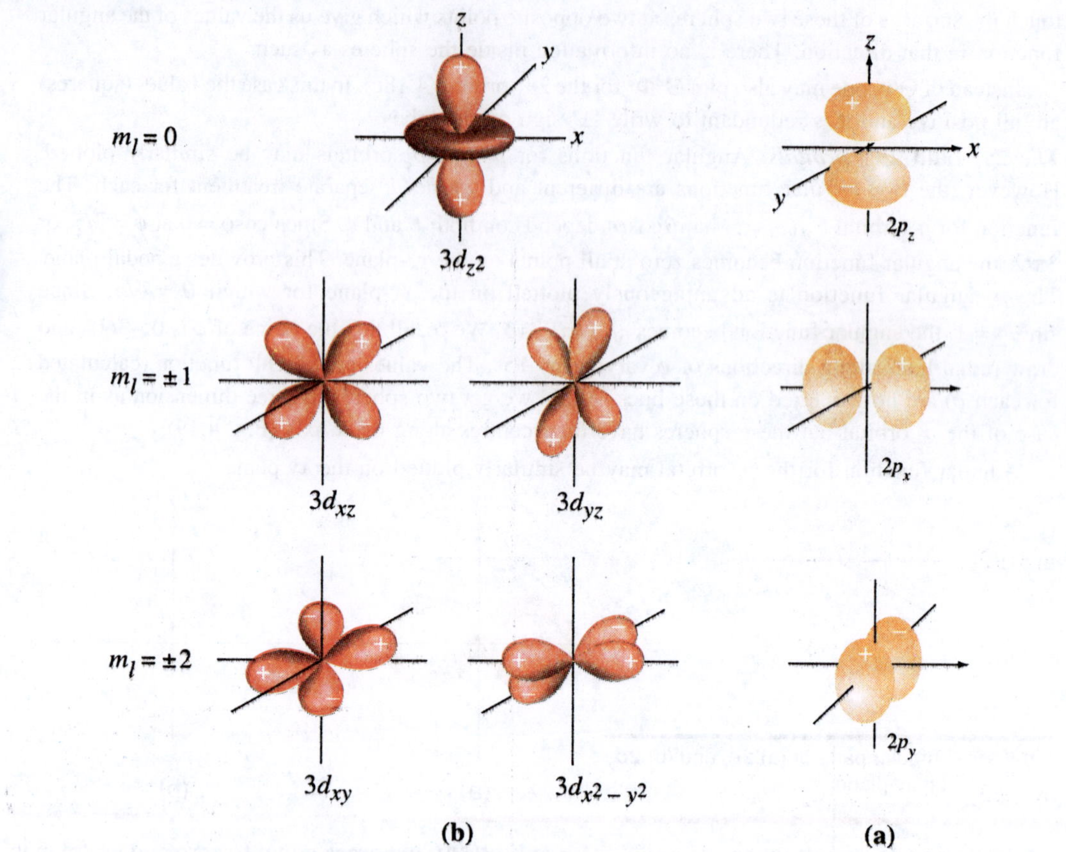

Fig.4.20: Nature of angular dependence of wave functions for (a) $2p$ and (b) $3d$ orbitals.

The *f*-orbitals have angular functions corresponding to six lobes and three nodes passing through the nucleus. Seven such orbitals are physically meaningful (Fig. 4.21).

Angular probability function

The probability of finding an electron in a given direction, summed over all distances from the nucleus, is proportional to the square of the angular functions. $\Theta^2\Phi^2$ is thus called the angular probability function. For an *s*-orbital, the spherical symmetry remains unchanged on squaring. But the sign of the wave function is obviously dropped since the probability is always positive (Fig. 4.22). For *p* orbitals, the envelope becomes more elongated (Fig. 4.22b); the '+' and '–' signs necessarily disappear. Similar changes occurs for the *d* and *f* orbitals.

Total wave function (space-part)

The total wave function (time-independent) for any hydrogenic orbital is obtained by combining the radial and angular functions as

$$\psi(r, \theta, \phi) = R(r)\,\Theta(\theta)\,\Phi(\phi)$$

Some examples for the hydrogen atom ($Z = 1$) are given in Table 4.3.

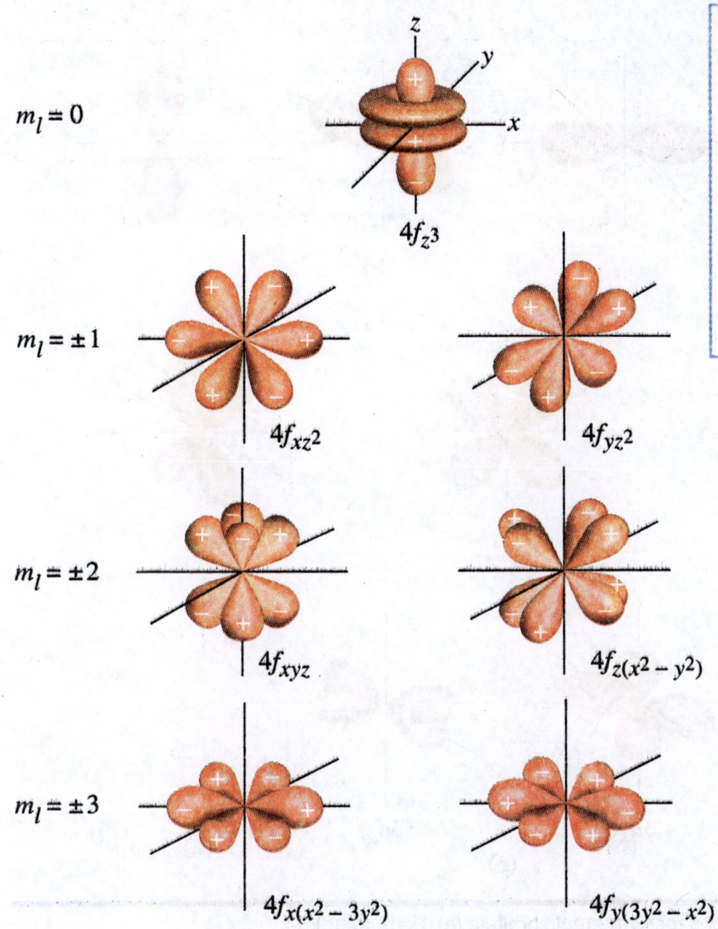

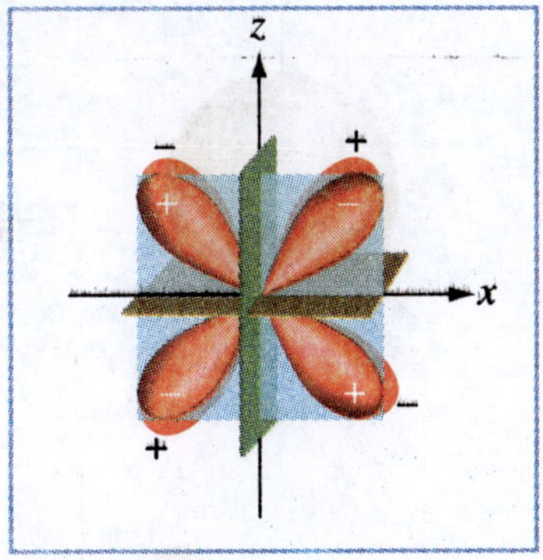

Fig.4.21: Angular dependence of the 4*f* orbitals (hydrogenic). Inset: Nodal planes of $4f_{xyz}$ orbital (viewed along the *y*-axis).

Table 4.3
$\psi(r, \theta, \phi)$ *for hydrogenic orbitals*

orbital	Θ	Φ	$\psi(r, \theta, \phi)$
1*s*	$\frac{1}{\sqrt{2}}$	$\frac{1}{\sqrt{2\pi}}$	$\frac{1}{\sqrt{\pi}a^{3/2}}e^{-r/a}$
2*s*	$\frac{1}{\sqrt{2}}$	$\frac{1}{\sqrt{2\pi}}$	$\frac{1}{4\sqrt{2\pi}a^{3/2}}\left(2-\frac{r}{a}\right)e^{-r/2a}$
$2p_z$ ($m_l = 0$)	$\frac{\sqrt{6}}{2}\cos\theta$	$\frac{1}{\sqrt{2\pi}}$	$\frac{1}{4\sqrt{2\pi}a^{3/2}}\left(\frac{r}{a}\right)e^{-r/2a}\cos\theta$
$2p_x$ } $m_l = \pm 1$	$\frac{\sqrt{3}}{2}\sin\theta$	$\frac{1}{\sqrt{\pi}}\cos\phi$	$\frac{1}{4\sqrt{2\pi}\,a^{3/2}}\left(\frac{r}{a}\right)e^{-r/2a}\sin\theta\cos\phi$
$2p_y$ } $m_l = \pm 1$	$\frac{\sqrt{3}}{2}\sin\theta$	$\frac{1}{\sqrt{\pi}}\sin\phi$	$\frac{1}{4\sqrt{2\pi}\,a^{3/2}}\left(\frac{r}{a}\right)e^{-r/2a}\sin\theta\sin\phi$

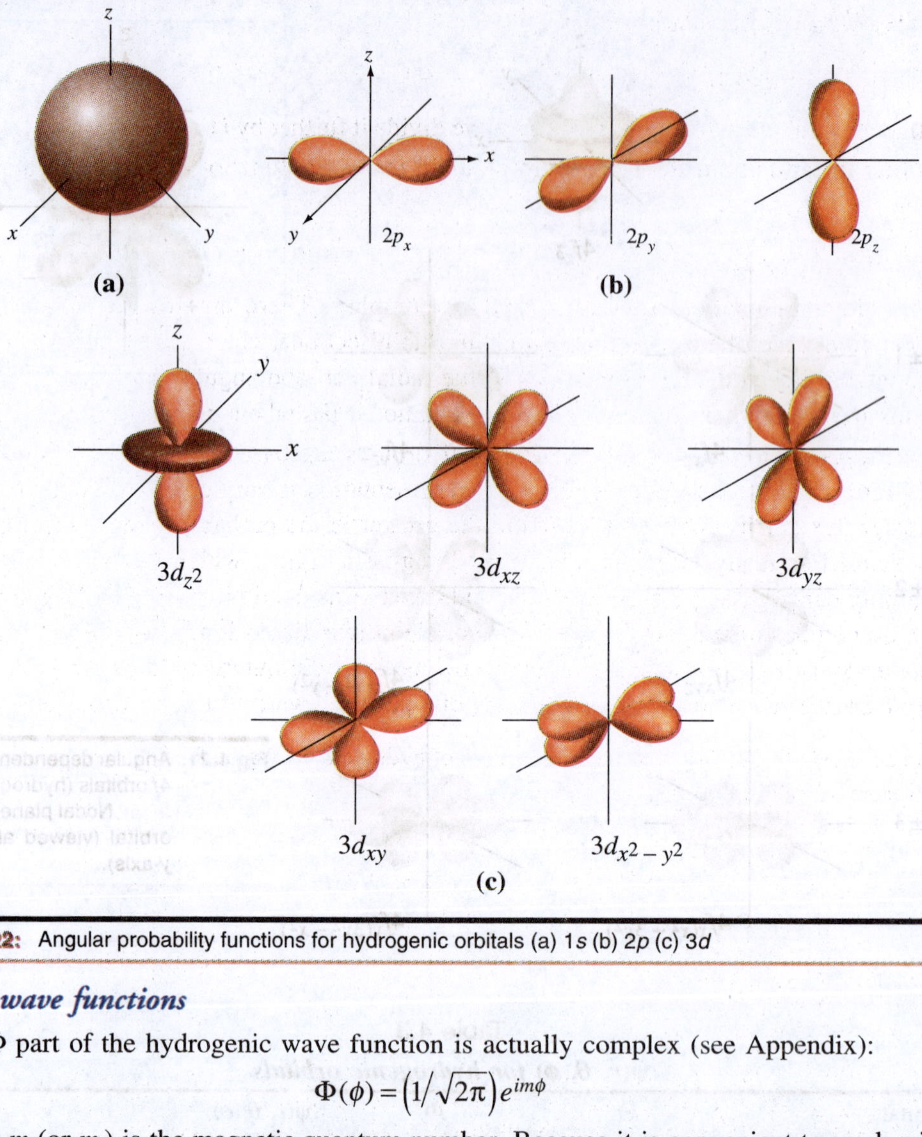

Fig.4.22: Angular probability functions for hydrogenic orbitals (a) 1*s* (b) 2*p* (c) 3*d*

Real wave functions

The Φ part of the hydrogenic wave function is actually complex (see Appendix):

$$\Phi(\phi) = \left(1/\sqrt{2\pi}\right)e^{im\phi}$$

where m (or m_l) is the magnetic quantum number. Because it is convenient to work with real functions, we can convert them to real functions with the help of another theorem in quantum mechanics:

Any linear combination of two or more stationary-state wave functions belonging to the same degenerate energy level will be an eigenfunction of the Hamiltonian operator with the same energy value. For example, the $2p$ orbitals have $m_l = 0, \pm 1$. Conventionally $m_l = 0$ refers to the z-axis taken as reference; hence $\Phi\ (2p_z)$ is simply $1/\sqrt{2\pi}$. For p-orbitals with $m_l = \pm 1$, we take the sum and difference of the two Φ functions:

$$\Phi(2p_x) = \left(1/\sqrt{2}\right)\left(\Phi_{+1} + \Phi_{-1}\right) = \left(1/2\sqrt{\pi}\right)\left(e^{i\phi} + e^{-i\phi}\right)$$

$$\Phi(2p_y) = \left(1/\sqrt{2}\right)\left(\Phi_{+1} - \Phi_{-1}\right) = \left(1/2\sqrt{\pi}\right)\left(e^{i\phi} - e^{-i\phi}\right)$$

Since $e^{\pm i\phi} = \cos\phi \pm i\sin\phi$, we get

$$\Phi(2p_x) = \left(1/\sqrt{\pi}\right)\cos\phi \;; \qquad \Phi(2p_y) = \left(i/\sqrt{\pi}\right)\sin\phi$$

To get the real wave function for $\Phi(2p_y)$, we divide it further by i to get $\Phi(2p_y) = (1/\sqrt{\pi})\sin\phi$.

Similar linear combinations are used to obtain real wave functions for higher states of the H-atom.

Representation of orbitals

Orbitals are mathematical functions, real or complex. There are several ways for their representation which help us in understanding the directional character of bonds formed by atoms. We have seen the representations of the radial part and angular part of different wave functions and also of their squares. Each representation is partial but useful in different contexts.

The most popular way of representation of orbitals are based on the idea of probability density through plot of ψ^2. The electron dot representation is very common in this respect—the greater the density of dots at a region, the greater is the probability of an electron to be found there. For ready reference, we often imagine a sphere which contains most of the cprobability density, say 90 or 99 per cent, and take this as the limiting size of the ortibal (Fig. 4.23a). Sometimes only the boundary line (a circle in two dimension) is retained (Fig. 4.23b). For higher s orbitals, this representation corresponds to the outermost maxima in the radial density curve. There will be radial nodes as one passes inward to the nucleus. (Fig. 4.23c).

It can be shown that for the $1s$ orbital in an atom of hydrogen, the region for 90 per cent probability is a sphere of radius 1.40 A°.

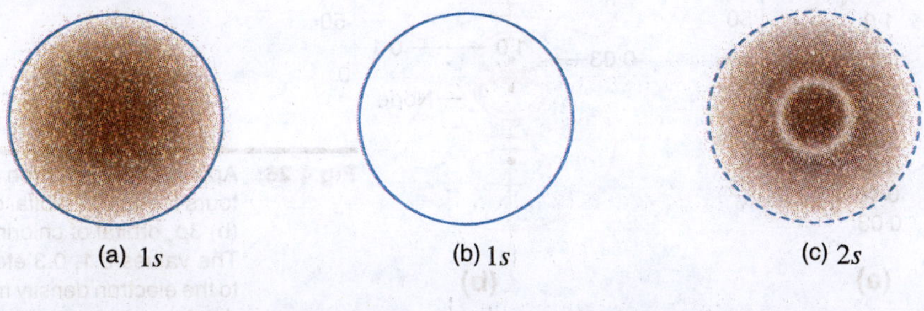

Fig. 4.23: Various representations of the hydrogen $1s$ orbitals.

Problem 4.14

Fig 4.24 shows a cross-sectional sketch of an ns orbital (hydrogenic) in the xy plane. Identify the orbital.

Fig.4.24

Hint: Count the number of radial nodes.

Ans: $4s$.

Contour diagrams are obtained by joining points at which ψ or ψ^2 has the same value. The construction of the contour diagram for ψ_{1s} is illustrated in Fig. 4.25. In three dimensions, these will correspond to concentric spheres. The electron density contours for 2*p* and 3*p* orbitals are shown in Fig. 4.26. In three dimensions these lines will correspond to cross-sectional view of curved surfaces. Computerised contour diagrams are now commonly used.

(a) Plot of ψ_{1s} against *r*.

(b) Contours of constant ψ value.

(c) Boundary surface within which ψ is greater than an arbitrarily chosen value.

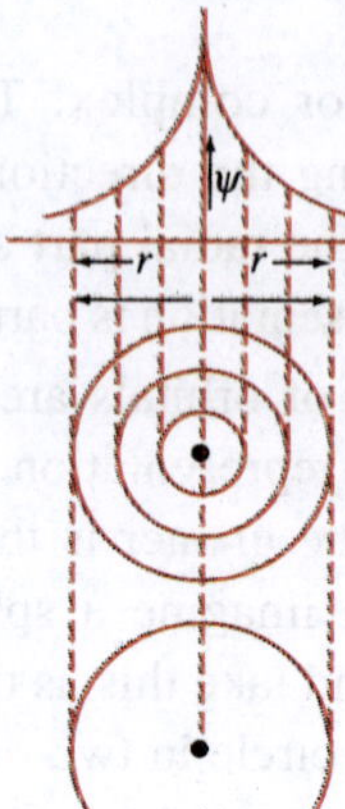

Fig.4.25: Contour diagram for H_{1s} orbital

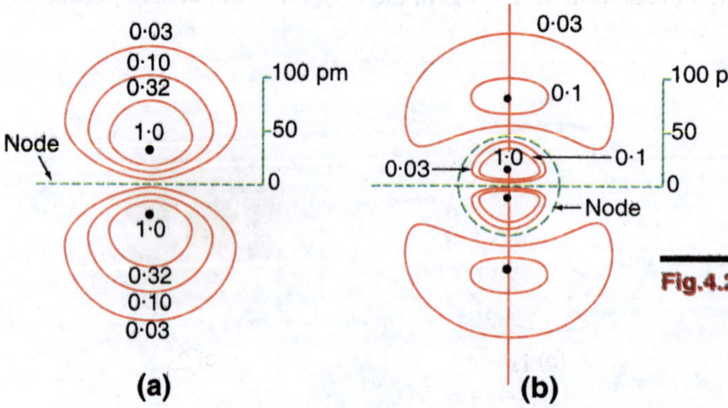

Fig.4.26: Approximate electron density contours for (a) $2p_z$ orbital of carbon and (b) $3p_z$ orbital of chlorine. The values 0.1, 0.3 etc. are relative to the electron density maxima (black dots).

The symmetry properties of the orbitals are interesting. In case of *s*- and *d*-orbitals, the wave functions have the same sign at the same distance in opposite directions from the centre of symmetry (see later). These orbitals are thus *centrosymmetric* and are said to be *gerade* (German, even); they are labeled by the letter *g*. The wave functions for *p*-orbitals have opposite signs on either side at the same distance from the centre—they change sign on inversion about this centre; they are therefore said to be *non-centrosymmetric* or *ungerade* (German, uneven) and labeled by *u*. Symmetric and antisymmetric wave functions are discussed later (section. 4.5).

Orbits versus Orbitals. In the Rutherford-Bohr atom model, orbits represent discrete trajectories of electrons around the nucleus. The behaviour of an electron in such quantized orbits can be described by the laws of Newtonian mechanics applied to an electron as a particle.

The position, momentum and hence the energy of an electron in such orbits may be expressed simultaneously with any degree of precision. Although the idea of orbits could explain the gross features observed in spectra of atoms and related behaviour like ionization energy etc., the concept was found inadequate when wave-particle dualism was established in the behaviour of the electron and the uncertainty principle followed as a necessary consequence.

The idea of orbitals gets introduced in the wave mechanical atom model. The extranuclear part of the atom is supposed to be subject to the fundamental wave equation of Schrödinger framed to suit material particles. The equation correlates the energy of a system with a function (ψ) of the space coordinates of a point at some instant of time. ψ is the amplitude analogue of general wave motion in physics; $\psi^2 dv$ gives the probability of finding an electron in the volume element dv at some time. Each physically meaningful solution of the equation is now termed an orbital; they are specified in terms of four quantum numbers—three from the time-independent Schrödinger equation. The spin quantum number came later (see next section).

The orbitals thus *correspond* to a description of the probability of finding an electron in various regions in space in an atom. Probably this is the source of the popular misconception often accumulated from the school days: orbits are two-dimensional while orbitals are three-dimensional. In fact, orbitals are abstract mathematical functions having little physical significance except the energy level or state specified by each. Each orbital has a definite energy value expressed in terms of certain quantum numbers. The energy expressions happen to be the same as those arrived at by Bohr for hydrogen like species. The maximum in the radial density function of the 1*s* orbital of a hydrogen atom also coincides with the radius of the first Bohr orbit.

But there always exists a basic difference in approach. If we recall the example of a particle in a box treated by the wave equation, we find that there may be nodes where the probability of finding the particle is zero. Yet there may be regions on either sides of these nodes where such probability is high. Similar description for electron does not have any parallel interpretation in terms of orbits where the electron is treated exclusively as a particle. The question: "how can an electron simultaneously exist on both sides of a point through which it never passes" has no sense when we deal with its orbital description. This is merely one of many examples of novel situations arising from the application of quantum mechanics. A similar situation will be found in the idea of "tunnelling" (see later).

4.3.4 The Quantum Numbers

In the Bohr-Sommerfeld atom model and its later extensions, four quantum numbers (n, l, m_l and s) were introduced from time to time to explain the spectral features of atoms. The acceptable values of these quantum numbers were selected rather arbitrarily just to fit experimental data.

The quantum numbers appear logically in the quantum mechanical atom model. We adopt the general principle that the Schrödinger wave equation provides the correct behaviour of the atomic system. The acceptable solutions of the wave equation then gives the different ψ-functions or orbitals; they represent the various allowed energy states of an electron in the atom. It is found that the eigenfunctions $\psi(n, l, m_l)$ for a single electron in the field of a nucleus are specified by three quantum numbers. Another quantum number, the fourth one, arises from the quantum mechanical treatment of Dirac including relativistic effects. The acceptable values of these quantum numbers then appear as a natural consequence of the mathematical requirements for a well-behaved wave function pertinent to the atomic system (see Appendix).

1. *The principal quantum number : n*

This quantum number determines the successive major energy levels of electrons in an atom. The energy E is expressed in terms of n by the relation

$$E = -\frac{Z^2 me^4}{8\varepsilon_0^2 n^2 h^2} \quad \text{(more precisely } \mu \text{ in place of } m\text{)}$$

where m, e, ε_0 and h have their usual significance. Successive values of E are obtained by putting n = 1, 2, 3, ..., etc. integers in the above expression. This expression of E and the integral nature of n follow from the solution of the Schrödinger equation; however, they also correspond to the conclusions arrived at in the Bohr's theory.

As n increases, the magnitude of E decreases; but due to the negative sign, the actual energy of the electron increases. Increasing values of n thus indicate successive higher values of the energy of the electron. The energy of an electron in an orbital is frequently referred to as the energy of the orbital also.

The implication of n in the radial function (and radial distribution) has been mentioned earlier.

2. *The azimuthal or angular momentum quantum number : l*

This quantum number determines the magnitude L of the orbital angular momentum of the electron:

$$L = \sqrt{l(l+1)}\, h/2\pi$$

The corresponding kinetic energy of angular motion of the electron is of course limited by the total energy of the electron. The acceptable values of l are related to n. Theory and experiment indicate that l may assume all integral values from 0 to $(n - 1)$. Thus, for $n = 1$, $l = 0$; for $n = 2$, $l = 0$ or 1; when $n = 3$, $l = 0$, 1 or 2; and so on. $h/2\pi$ is the natural unit of angular momentum and is equal to 1.054×10^{-34} J.s.

In the designation of orbitals, it is customary to indicate l values 0, 1, 2, 3, 4, ... etc. by the letters *s*, *p*, *d*, *f*, *g*, ... respectively. Orbitals are represented by their n and l values as 1*s*, 2*s*, 3*d* etc.

3. *The magnetic quantum number : m_l*

This quantum number determines the orientation of the orbital angular momentum vector of the electron with respect to an externally applied magnetic field. The allowed orientations are those for which the components of the angular momentum vector along the direction of the magnetic field (conventionally the z-axis) have values equal to $m_l\, h/2\pi$ where m_l may have all integral values in the range $+l, +(l-1), \cdots, -l$. m_l is called the magnetic quantum number.

The magnetic quantum number thus introduces the concept of *space quantization*, that is, it specifies certain selected directions in space for orientation of the orbital angular momentum vector of the electron.

Classically, the direction of the angular momentum vector is normal to the plane of the electron's orbit. Accordingly, the magnetic quantum number also restricts the orientation of the plane of the electron's orbit in space (Fig. 4.27a). As the electron is negatively charged, its

> In quantum mechanics, orbital angular momentum is treated in terms of its corresponding operators. Such treatment is beyond the scope of the present approach and we shall merely state the results. It appears that the magnitude | L | of the angular momentum vector of the electron and its z-component L_z in the H-atom (described by the state ψ_{nlm}) are
>
> $$|L| = \sqrt{l(l+1)}\frac{h}{2\pi} \quad \text{and} \quad L_z = m_l \frac{h}{2\pi}$$
>
> where l and m_l have their usual significance. For s-states, $l = 0$ and the orbital angular momentum of the electron is zero. This is difficult to understand from the classical idea of orbital angular momentum arising from orbital motion of the electron. In other cases, analogy with the classical idea of an electron moving in orbits appears useful in understanding the significance of the magnetic quantum number. Such description must not be taken literally but analogically.

motion gives rise to a magnetic moment at the same time (see below). The vector of this magnetic moment is also normal to the orbital plane, but acts in a direction opposite to that of the angular momentum vector, that is, they are antiparallel due to the charge of the electron. When the atom is placed in a magnetic field, the resultant interaction tends to align the angular momentum vector parallel to the magnetic field. As a consequence, the angular momentum vector undergoes precession around the direction of the magnetic field and so also does the electron's orbit (*Larmor precession*, Fig. 4.27). It is found that the angular momentum vector can orient itself only in certain specified angles with respect to the magnetic field (conventionally along the z-axis) so that its components along the magnetic field are quantized; they may have values $m_l\ h/2\pi$ where m_l is the quantum number for magnetic components of l; m_l (or m) is called the *magnetic quantum number*. As mentioned before, m_l may have only integral values between $-l$ ... 0 ... $+l$, total $2l + 1$ in number.

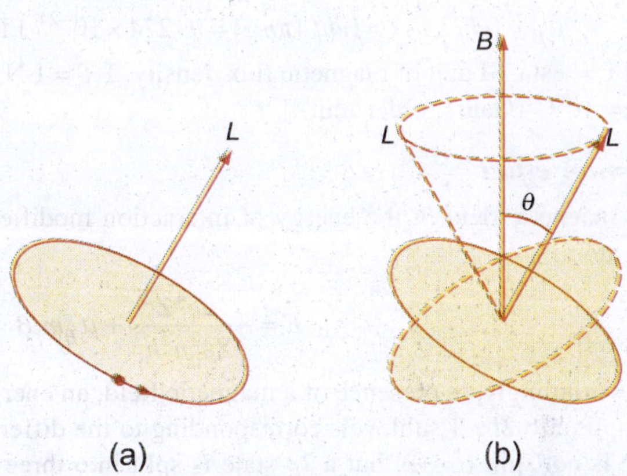

Fig.4.27: (a) Vector representation of orbital angular momentum (b) Larmor precession of an electron's orbit in presence of an external magnetic field B.

The magnetic quantum number was first introduced to explain the Zeeman effect (1896)—splitting of spectral lines when the source is placed in a magnetic field (see box). Some elements like sodium produced additional lines than predicted by this theory (anomalous Zeeman effect) and this led to the idea of spin.

Magnetic moment due to orbital motion of the electron

The magnetic dipole moment μ_L arising from the orbital motion of the electron is given by

$$\mu_L = \gamma_e L = -(e/2m_e)\,L = -\frac{e\hbar}{2m_e}\sqrt{l(l+1)} \qquad ...\left(L = \sqrt{l(l+1)}\,\hbar\right)$$

The mass of the electron is expressed by m_e to avoid confusion with the magnetic quantum number m_l, which is sometimes also written as m. γ_e ($= -e/2m_e$) is called the *magnetogyric ratio* of the electron. The negative sign arises from the charge of the electron; it shows that the orbital angular momentum and the magnetic moment are antiparallel (Fig. 4.28).

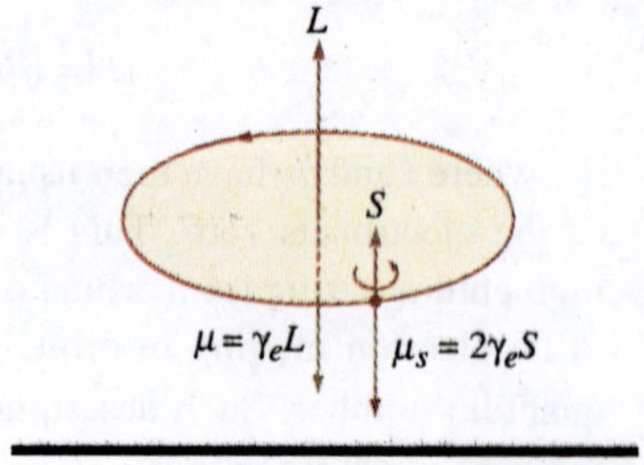

Fig.4.28: Antiparallel arrangement of orbital angular momentum and the magnetic moment arising from it.

For the z-components of μ and L, we have

$$\mu_{L(Z)} = \gamma_e L_z = -(e/2m_e)\,L_z$$

Since $L_z = m_l\hbar$, $\mu_{L(Z)} = -(e\hbar/2m_e)\,m_l$

Putting $e\hbar/2m_e = \mu_e$, $\mu_{L(Z)} = -\mu_B m_l$.

μ_B is called the *Bohr magneton* which is a natural unit of magnetic moment. It may be considered as the quantum of magnetic moment.

$$\mu_B = (e\hbar/2m_e) = (eh/4\pi m_e) = 9\cdot274\times10^{-24}\ \mathrm{JT^{-1}}.$$

[T = tesla, SI unit of magnetic flux density. 1 T = 1 N $A^{-1}m^{-1}$. So $\mu_B = 9.274\times10^{-24}$ A m^2. Also, 1 T = 10^4G (Gauss, older unit).]

Zeeman effect

In a magnetic field B, the energy of interaction modifies the energy of the electron in a hydrogen like atom:

$$E = -\frac{\mu e^4 Z^2}{8\varepsilon_0^2 n^2 h^2} + \mu_B m_l B$$

Accordingly, in presence of a magnetic field, an energy level specified by a given value of n and l is split into $2l + 1$ sublevels corresponding to the different m_l values (see Exercise, Q. 21). The 1s state is not split ($l = 0$), but a 2p state is split into three sublevels corresponding to $m_l = -1$, 0 and +1. For the transition $2p \to 1s$, we now expect *three* lines instead of one line seen in absence of any magnetic field. This is in accordance with the observations made by Zeeman (Zeeman effect; 1896).

Precession of the angular momentum vector

The angular momentum vector L cannot be fixed in space with definite values for all three components L_x, L_y and L_z. In that case we would confine the electron's orbit (in the vector model) in a definite plane perpendicular to L. If, for example, L points in the z-direction, the orbit of the electron will be confined to the xy plane and the uncertainty in the z-component of the momentum of the electron will become infinite. This will violate the uncertainty principle. Simultaneous specification of L_x, L_y and L_z is thus contrary to the uncertainty principle. Once we fix L_z, L_x and L_y become indeterminate. The L vector may align it in any plane containing the z-axis maintaining an angle θ for which L_z will be equal to $m_l\hbar$. In the vector model, this is equivalent to precession of the L-vector about the z-axis sweeping the surface a cone. (Fig. 4.29b).

In quantum mechanical terms, the very nature of the operators for the three components of L do not commute with one another. So we can specify only any one of them at a time (except when $l = 0$). L_x, L_y and L_z are thus said to be complementary observables.

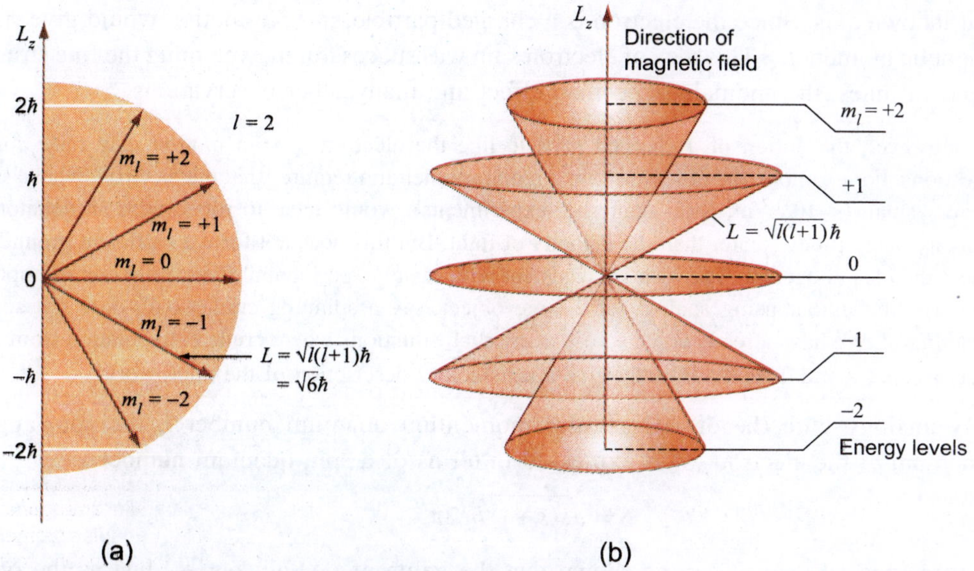

Fig.4.29: (a) Space quantization of orbital angular momentum. (b) Orientation of the orbital angular momentum vector for $l = 2$ with respect to an external magnetic field.

The selection of the magnetic field as the frame of reference is one of convenience. The space quantization of the orbital angular momentum vector L may be expressed with reference to any arbitrary direction—in all cases the component of L along the direction would equal $m_l h/2\pi$. The magnetic field actually provides an experimentally meaningful reference direction. The internuclear axis in a hydrogen molecule is an example of an alternative reference direction. The components of the angular momenta of the H-atoms along this line will also be given by the m_l values.

Problem 4.13

The angular momentum vector L cannot point in the direction of an applied magnetic field—explain.

Hint: $L_z = m_l h/2\pi$ where $m_l = -l, ...0... +l$. So the maximum value of L_z is $lh/2\pi$.

But $L = \sqrt{l(l+1)}\, h/2\pi$ which is always greater than $lh/2\pi$.

Note: This is also in agreement with the uncertainty principle. If the angular momentum vector gets pointed at any unique axis, the orbital electron is confined to a plane perpendicular to this direction.

4. *The spin quantum numbers (m_s and s)*

Arthur Compton had the first idea (1921) that an electron might possess an intrinsic angular momentum or spin and therefore possess magnetic properties. In 1925, W. Pauli showed that the observed fine structure (doublet) in the spectra of alkali metals could be explained if an electron could exist in two distinct states. Uhlenbeck and Goudsmit showed (1925) that the angular momentum of the electron in these two states were different. They proposed that the electron possesses an *intrinsic angular momentum* independent of any orbital angular momentum. This angular momentum may be interpreted in terms of an assumed spin motion of the electron about its own axis. Since the electron is a charged particle, its spin motion would give rise to a magnetic moment μ_S. This idea of electron spin was successful in explaining the fine structure of spectral lines, the anomalous Zeeman effect and many other observations.

However, the notion of a charged particle like the electron in spin motion may raise some questions. For example, the observed spin angular momentum requires that the electron, whose size is very small ($< 10^{-16}$ m; from scattering experiments), would have to rotate with an equatorial velocity many times greater than the velocity of light. But this does not form a sufficient ground to reject the idea altogether. The idea is consistent with many experimental facts; only our attempt to model its behaviour using analogy of a large object, say a spinning cricket-ball, does not seem meaningful. We have already come across many such situations where crude extrapolation from the macroscopic world does not directly give a satisfactory description of the microworld.

By analogy with the orbital angular momentum quantum number l, the spin angular momentum of the electron (S) is expressed in terms of a spin quantum number s:

$$S = \sqrt{s(s+1)}\, h/2\pi.$$

From spectral data, it can be shown that the value of s is always $\frac{1}{2}$. Just as the orbital angular momentum vector can have $2l+1$ orientations in a magnetic field from $+l$ to $-l$, the spin angular momentum vector can have $2s + 1 = 2$ orientations with respect to the magnetic field, taken conventionally along the z-axis. This space-quantization of the electron spin is described by a *spin magnetic quantum number* m_s: the component of the spin angular momentum of the electron along the z-direction (S_z) is given by

$$S_z = m_s . h/2\pi. \quad \text{where } m_s = \pm\frac{1}{2}$$

$$S_z = m_s \frac{h}{2\pi} = \pm\frac{h}{4\pi} \quad \text{or } \pm\frac{\hbar}{2} \qquad \left(\text{since } m_s = \pm\frac{1}{2}\right)$$

Complete specification of the state of a hydrogen atom now involves the four quantum numbers n, l, m_l and m_s.

$$S = \tfrac{1}{2}, m_s = +\tfrac{1}{2}, -\tfrac{1}{2} \text{ for electrons, protons and neutrons.}$$

Magnetic moment associated with electron spin

The spin magnetic moment μ_S of the electron is proportional to the spin angular momentum S. But unlike the orbital magnetic moment μ_L, μ_S involves an additional factor of about 2:

$$\mu_S = -g_e \gamma_e S = -g_e \frac{e}{2m_e} S \qquad g_e = 2.002\,319$$

g_e is called the ***g-value*** of the electron and is calculated spectroscopically.

Since $S = \sqrt{s(s+1)}\,\hbar$ and $s = \frac{1}{2}$, $\mu_S = -g_e . e\hbar\sqrt{3}/4m_e = \left(-g_e\sqrt{3}/2\right)\mu_B$.

μ_B is the Bohr magneton introduced earlier. Theoretically, the g-value may be interpreted from (i) relativistic effects and (ii) the interaction of the electron with electromagnetic fluctuations in the surrounding free space.

The z-component of S, $S_z = m_s\hbar, \quad m_s = \pm\frac{1}{2}$

So, the z-component of μ_S, $\mu_{S(z)}$ is: $\mu_{S(z)} = -g_e\gamma_e S_z = -g_e \frac{e\hbar}{2m_e} m_s = -g_e\mu_B m_s$

If we take $g_e \simeq 2, \quad \mu_{S(z)} = \pm\mu_B$.

In a magnetic field B in the z-direction, the contribution of spin to the energy of the electron is

$$E_{\text{spin}} = g_e\mu_B m_s B \simeq \pm\mu_B B$$

Thus a magnetic field B splits an otherwise degenerate energy state into two energy levels separated by $2\mu_B B$ (actually $g_e\mu_B B$). Transition between these two levels is studied in electron paramagnetic resonance (EPR) or electron spin resonance (ESR) spectroscopy.

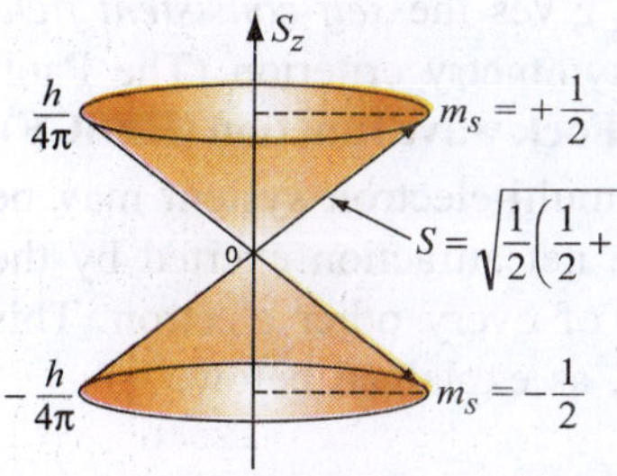

Fig.4.30: Space quantization of electron spin vector. The S-vector also precesses about the z-axis and describes the surface of a cone (compare Fig. 4.29). Chemists often use the symbol ↑ for $m_s = +\frac{1}{2}$ and ↓ for $m_s = -\frac{1}{2}$.

Experimental verification. The idea of electron spin could successfully explain the observations of Stern and Gerlach (1921) in connection with the effect of an inhomogeneous magnetic field on a beam of silver atoms. The field divided the beam into two distinct parts—as shown by their action on a photographic plate. The silver atom has only one unpaired s-electron ($l = 0$) in its outermost shell. So the entire magnetic moment of a silver atom is due to the spin of this electron. The spinning electron acts like a little magnet with two poles. In a uniform magnetic field, such a dipole would experience a torque that would tend to align it with the field. But when the magnetic field is inhomogeneous, the two different poles of the tiny magnet experiences forces of different magnitude.

The resultant of these forces depends on the orientation of the spin-magnetic dipole with respect to the external magnetic field. The splitting of the beam of silver atoms into two distinct parts corresponds to the two permitted alignments of the spin angular momentum vector as given by the two values of m_S, $+\frac{1}{2}$ and $-\frac{1}{2}$. The experiment may also be performed with atomic beams of alkali metals, copper or gold—all with one unpaired electron. Zinc, cadmium, mercury etc. have no effect on similar treatment (no unpaired electron). As we shall see later, for a single s-electron, $l = 0$, $S = \frac{1}{2}$ and $J = \frac{1}{2}$. $2J + 1$ or 2 alignments are thus observed.

However, the spin quantum number does not follow directly from the solution of the time-independent Schrödinger wave equation for the hydrogen atom. In 1929, Paul Dirac developed relativistic quantum mechanics. In the nonrelativistic treatment the energy E of the electron was expressed as $E = p^2/2m + V$. Dirac started with the relativistic equation $E = \sqrt{m_0^2c^4 + p^2c^2} + V$ and showed that a particle with the mass and charge of the electron should possess an intrinsic angular momentum and magnetic moment of the same magnitude as established by Uhlenbeck and Goudsmit.

4.4 MANY ELECTRON ATOMS

The Schrödinger wave equation becomes increasingly complex as the number of electrons in an atom increases. Several approximations have been proposed to overcome the situation.

An early attempt to such approximation by Hartree (1928) and later by Fock (and Slater, 1930) starts with one-electron hydrogenic wave functions for each electron ignoring inter-electron repulsion. The potential energy function for any one electron is obtained by assuming a spherical distribution of the charges of *all other* electrons. One thus gets a first approximation function for that particular electron. This function is now used to improve the potential energy function for the second electron, resulting in an improved wave function for this one. The process is continued for all *n* electrons in the system and then again recycled from electron 1 until results of two successive stages do not show any gross variation. The *n* one-electron wave functions are now said to be self-consistent and their product gives the *self-consistent field* (**SCF**) wave function for the system. Inclusion of spin and symmetry criterion (The Pauli exclusion principle, sec. 4.5) gives the further improved **Hartree-Fock wave function (HFSCF)**.

The effect of inter-electron repulsion on the energy of a multi-electron system may be conveniently tackled in terms of a fractional reduction of the net attraction exerted by the nuclear positive charge on an electron in the neighbourhood of every other electron. This modified nuclear charge is called the effective nuclear charge, as explained below.

4.4.1 Effective Nuclear Charge

It has been observed that the radial part of the wave function changes appreciably with atomic number *Z* but the angular parts of the wave functions remain nearly the same as with the hydrogenic functions. So the basic shapes of *s*, *p*, ... etc. orbitals may be roughly taken to be the same for all atoms. The radial functions are expressed by modifying the actual nuclear charge (*Z*; in terms of *e*) by an effective nuclear charge Z^* or Z_{eff} which is equal to the nuclear charge *Z minus* a shielding (or screening) factor σ arising from the presence of other electrons in the atom.

$$Z_{eff} = Z - \sigma \qquad \text{(in units of } e\text{)}$$

where σ = sum of the shielding contributions from all *other* electrons.

The shielding factor represents the sum of all the reductions of the nuclear charge caused by electrons closer than or at about the same distance from the nucleus relative to the electron under consideration.

Let us consider the lithium atom as an example. The 2*s* electron will be shielded by the inner 1*s* electrons and hence experience a lower effective nuclear charge than the 1*s* electrons. Again as the radial distribution curve of the 2*s* orbital has some finite distribution inside the 1*s* shell (penetration), the 2*s* electron does not get complete shielding by the 1*s* electrons. Slater proposed (1930) a set of empirical rules for estimating the shielding contributions by electrons in different orbitals. These are known as Slater's rules.

Slater's rules for estimation of shielding contributions

(i) The electrons of the atom are divided into groups like

(1*s*), (2*s* 2*p*), (3*s* 3*p*), (3*d*), (4*s* 4*p*), (4*d*), (4*f*), (5*s* 5*p*), (5*d*), ... etc.

The fraction (of *e*) contributed to shielding on any electron in a particular electron group by other electrons is now estimated as

(ii) electrons *in groups on the right* of the one under consideration: **zero.**

(iii) each *other* electron *in the same group*:

0.30 if the group is 1*s*; otherwise **0.35.**

(iv) if the electron under consideration belongs to an (*ns*, *np*) group, *all electrons with a principal quantum number less by one* : **0.85 each.**

All electrons in groups further left : **1.0 each**

(v) if the electron under consideration belongs to a *d* or *f* group, *all electrons in groups lying left of this group* : **1.0 each.**

Electron group	Contribution to shielding by each electron in			
	all higher groups	same group	groups $(n-1)$	groups $\leq (n-2)$
1*s*	0	0.30	–	–
(*ns*, *np*)	0	0.35	0.85	1.00
(*nd*) or (*nf*)	0	0.35	1.00*	1.00

**for all groups* on left, including *s* or *p* electrons of the same *n*. See illustration.

***Illustration*:** Let us consider an atom of iron in the ground state configuration $1s^2\ 2s^22p^6\ 3s^23p^63d^64s^2$ ($Z = 26$). The electrons are first separated into orbital groups:

$$(1s)^2\ (2s2p)^8\ \ (3s3p)^8\ \ (3d)^6\ \ (4s)^2.$$

The shielding contribution and Z_{eff} for an electron (in *e*) may now be calculated:

(i) **1*s* electron:** This is shielded by only one *other* electron in 1*s*.

Electrons in groups to the right of 1*s*	= 0
The remaining electron in 1*s*	= 0.30
Total (σ)	= 0.30

Hence, $Z_{eff} = 26 - 0.30 = 25.70.$

(ii) **2*s* or 2*p* electron:**

Electrons in groups to the right	= 0
7 *other* electrons in the 2*s*2*p* group; 0.35 *each* : (7 × 0.35)	= 2.45
2 electrons in the $(n-1)$ group i.e., 1*s*; 0.85 *each* : (2 × 0.85)	= 1.70
Total (σ)	= 4.15

$Z_{eff} = 26 - 4.15 = 21.85.$

(iii) **3*s* or 3*p* electron**

Electrons in 3*d* and 4*s*	= 0
7 *other* electrons in 3*s*3*p* group, 0.35 *each*: 7 × 0.35	= 2.45
8 electrons in $(n-1)$ group i.e., (2*s*2*p*), 0.85 *each* : 8 × 0.85	= 6.80
2 electrons in group further left (1*s*), 1.0 *each* : 2 × 1.0	= 2.00
Total (σ)	= 11.25

$Z_{eff} = 26 - 11.25 = 14.75.$

(iv) **3d electron**

Electrons in 4s	= 0
5 *other* electrons in the 3d group, 0.35 *each* : 5 × 0.35	= 1.75
18 electrons in all *the groups to the left* of the 3d group, 1.0 *each* : 18 × 1.0	= 18.0
Total (σ)	= 19.75

$$Z_{eff} = 26 - 19.75 = 6.25.$$

(v) **4s electron.**

One *other* electron in the same group	= 0.35
14 electrons in (n – 1) groups, (3s 3p) and 3d : 0.85 *each* : 14 × 0.85	= 11.90
10 electrons in groups further left i.e., ≤ (n – 2), 1.0 *each* : 10 × 1.0	= 10.00
Total (σ)	= 22.25

$$Z_{eff} = 26 - 22.25 = 3.75$$

Slater applied these values of Z_{eff} to obtain a set of *Slater orbitals* corresponding to each hydrogenlike orbital. The angular dependence of these orbitals is identical to that of the hydrogenlike orbital but the radial function is of the type

$$R(r) = N.r^{n^*-1}.e^{-Z_{eff}r/n^*} \quad (N = \text{normalization constant})$$

n^* is called the "*effective principal quantum number*" (n^*) having the following values:

n	=	1	2	3	4	5	6
n^*	=	1.0	2.0	3.0	3.7	4.0	4.2

The difference between n and n^* is called the **quantum defect**.

The Slater type orbitals have no radial nodes (except at $r = 0$ and $r = \infty$) and also do not depend explicitly on the value of l. So ns and np orbitals will have the same radial functions. However, the functions are found to be useful in the range of covalent bond distances in the calculation of maximum radial charge density, ionization energy etc. of an atom or ion.

Calculation of ionization energy and electron affinity

According to Slater's treatment, the energy of an electron in an atom is given by the empirical equation

$$E = -\frac{me^4}{8\varepsilon_0^2 h^2}\left(\frac{Z_{eff}}{n^*}\right)^2 = -13.6\left(\frac{Z_{eff}}{n^*}\right)^2 \text{eV.}$$

where Z_{eff} is the effective nuclear charge and n^* is the effective principal quantum number.

The energy of all the electrons in an atom or ion will be obtained by summation of the $(Z_{eff}/n^*)^2$ terms for each electron. This amount of energy will be required to remove all the electrons from the atom/ion to infinity. The ionization energy of an atom or ion will be equal to the difference of energy between the energies before and after ionization. For lithium, for example, the first ionization energy I will be

$$I = E_{Li^+} - E_{Li}$$

Similarly, the electron affinity of an element X will be given by

$$E.A. = E_{X^-} - E_X.$$

where the E terms represent the sum of energies of all the electrons in the species.

$$E_{Li} = -13.6\left(\frac{Z_{eff}}{n^*}\right)^2_{2s^1} - 13.6\left(\frac{Z_{eff}}{n^*}\right)^2_{1s^2} \; ; \; E_{Li^+} = -13.6\left(\frac{Z_{eff}}{n^*}\right)^2_{1s^2}$$

The value of Z_{eff} has to be calculated *separately* for each type of electron. For Li, Z_{eff} for the 1s electrons is same in Li and Li^+ since the 2s electron in Li does not screen the inner 1s electron. Hence

$$I = E_{Li^+} - E_{Li} = 13.6\left(\frac{Z_{eff}}{n^*}\right)^2_{2s^1}.$$

For the 2s electron, $n = 2$, $n^* = 2$. $Z_{eff} = 3 - \sigma$, where σ is the total screening by the two 1s electrons. So $\sigma = 2 \times 0.85$ and $Z_{eff} = 1.30$. I comes out to be 5.75 eV (experimental value 5.4 eV). The agreement is good.

Problem 4.14

Calculate the first I.E. of oxygen using Slater's rules.

Hint: Ionization energy $I = E_{o^+} - E_o$.

E terms are calculated for all the electrons in the species by Slater's relation:

$$E = -13.6\,(Z_{eff}/n^*)^2 \text{ eV}.$$

The electron group will be O : $(1s)^2(2s2p)^6$; O^+ : $(1s)^2\,(2s2p)^5$.

The 1s electrons are not screened by the electrons in 2s2p group. So the effective nuclear charge on 1s electrons remains the same in O and O^+; the contribution of the 1s electrons to the energies of the atom and ion are equal. So we consider only the energy of the electrons in 2s2p groups in the atom and ion. Z_{eff} will be different for the two species.

In O^+, each electron in the 2s2p group is screened by four *other* electrons in the same group and two 1s electrons.

$\therefore \sigma = 4 \times 0.35 + 2 \times 0.85$; $Z_{eff} = 8 - 3.10 = 4.90$.

In O, similarly we have $\sigma = 5 \times 0.35 + 2 \times 0.85$; $Z_{eff} = 8 - 3.45 = 4.55$.

$n^* = n = 2$ for all electrons. The E-terms are now calculated separately.

Finally, $I = E_{o^+} - E_o \Rightarrow 14.16$ eV.

[**Note**: Experimental value = 13.6 eV; Slater's rules do not consider spin correlations; only electrostatic repulsion is considered.]

Problem 4.15

Calculate the electron affinity of fluorine applying Slater's rules.

Hint: *E.A.* of F = $E_{F^-} - E_F$.

F : $1s^2\ (2s,2p)^7$; F^- : $1s^2\ (2s2p)^8$.

$Z^* = 9 - (6 \times 0.35 + 2 \times 0.85)$ $Z^* = 9 - (7 \times 0.35 + 2 \times 0.85)$

$n^* = n = 2$ gives *E.A.* = + 3.80 eV.

Shielding contributions by different electron shells have been more accurately estimated using SCF functions (Clementi and Raimondi, 1963), though involving complex computation with a larger number of parameters, including penetration of outer orbitals into inner cores. Some values of Z_{eff} calculated in this manner are given in Table 4.4.

Table 4.4

Effective nuclear charge: Z_{eff} (in units of e)

		H							**He**
Z		1							2
Z_{eff}	1s	1.00							1.69
		Li	**Be**	**B**	**C**	**N**	**O**	**F**	**Ne**
Z		3	4	5	6	7	8	9	10
Z_{eff}	1s	2.69	3.68	4.68	5.67	6.66	7.66	8.65	9.64
	2s	1.28	1.91	2.58	3.22	3.85	4.49	5.13	5.76
	2p			2.42	3.14	3.83	4.45	5.10	5.76
		Na	**Mg**	**Al**	**Si**	**P**	**S**	**Cl**	**Ar**
Z		11	12	13	14	15	16	17	18
Z_{eff}	1s	10.63	11.61	12.59	13.57	14.56	15.54	16.52	17.51
	2s	6.57	7.39	8.21	9.02	9.82	10.63	11.43	12.23
	2p	6.80	7.83	8.96	9.94	10.96	11.98	12.99	14.01
	3s	2.51	3.31	4.12	4.90	5.64	6.37	7.07	7.76
	3p			4.07	4.29	4.89	5.48	6.12	6.76

4.4.2 Sequence of Energy Levels

The sequence of energy levels or the order of increasing energy of the orbitals in a many electron atom is primarily determined by the effective nuclear charge and spin-correlations between the electrons in an orbital. We have gained some idea about the first factor while spin-correlation will be treated later. Summarily, the energy of the hydrogenic orbitals is found to depend on the principal and azimuthal quantum numbers *n* and *l* as follows:

(i) orbital energies increase as $(n + l)$ increases;

(ii) if there are two orbitals with the same value of $(n + l)$, the one with the smaller *n* has the lower energy.

Accordingly the increasing order of energy among the hydrogenic orbitals is

$$1s < 2s < 2p < 3s < 3p < 4s \simeq 3d < 4p < 5s \simeq 4d < 5p < 6s \simeq 5d \simeq 4f < 6p < 7s \simeq 6d \simeq 5f.$$

This may be reproduced with the diagram in Fig 4.31.

However, this generalization is *valid only for the one electron wave functions for each separate orbital*. When these orbitals get successively occupied by electrons in a multielectron atom, the effective nuclear charge gets gradually modified by varying penetration and shielding effects at each and every addition of a proton to the nucleus and an electron to the available orbital. The situation is further complicated by contraction of the inner electron core by the enhanced nuclear charge and consequent change in penetration by outer orbitals into the inner core. Summarily, in such cases, the criterion of lowest available energy of an orbital to an incoming electron is determined by the *total energy of the atomic system before and after addition of the electron and requisite nucleons*. An approximate nature of variation of the energy of different orbitals with increasing atomic number is shown in Fig. 4.32.

We note that the energies of different orbitals drop to different extent with increase in atomic number. This is consistent with the effect of increasing nuclear charge on the radial distribution functions as shown in Fig. 4.15 and 4.16. The energy of d and f orbitals are particularly interesting; at some Z or other they overlap in energy with s and p orbitals of the

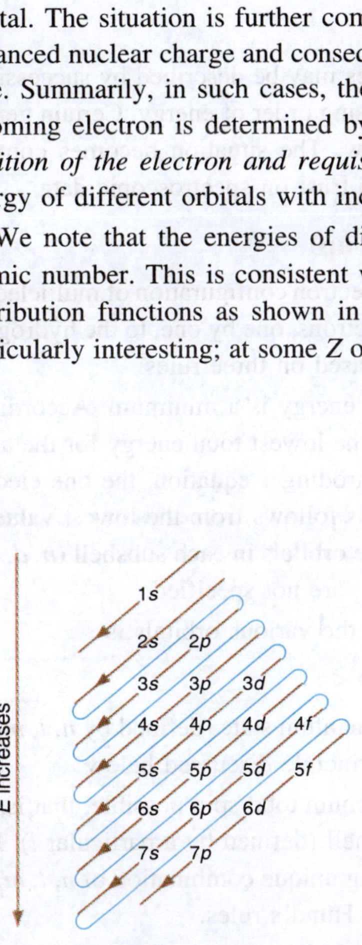

Fig.4.31: Sequence of energy levels of the individual orbitals in a many electron atom.
The sequence is obtained by starting with the uppermost arrow and successively placing the orbitals connected by each arrow.

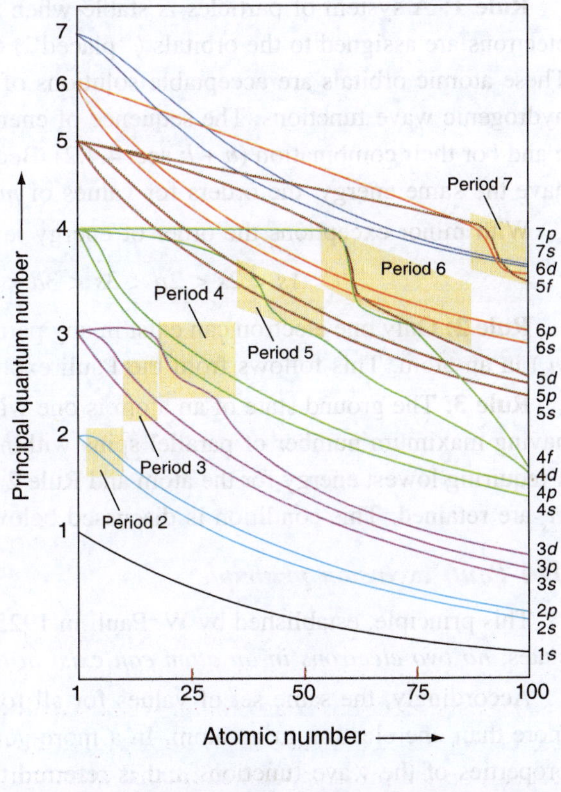

Fig.4.32: Nature of variation of orbital energy with atomic number.

next higher principal quantum number. Over certain range of atomic number, the energies of *ns* orbitals become nearly equal to those of $(n-1)d$ and $(n-2)f$ orbitals.

The energy of an orbital in a free atom does not depend on the value of *m*. The three *p* orbitals (or five *d* orbitals or seven *f* orbitals) of the same principal quantum number are of equal energy.

In the next section we shall see how these energy levels are successively filled in the ground state of different elements.

4.4.3 Electron Configuration of Atoms

The electron configurations of atoms in their ground states may be described by successively assigning the electrons to the atomic orbitals in the increasing order of energy. Certain general principles established for the purpose are outlined below. The situation becomes complex when orbitals with $n > 3$ are involved. There we have to trust on spectroscopic data.

The Aufbau principle (German *aufbau* = building up)

The aufbau principle concerns gradual building up of the electron configuration of multielectron atoms starting from hydrogen and successively adding electrons, one by one, to the hydrogenic orbitals in increasing order of energy. The principle is based on three rules:

Rule 1: A system of particles is stable when its total energy is a minimum. Accordingly electrons are assigned to the orbitals (''placed'') to give the lowest total energy for the atom. These atomic orbitals are acceptable solutions of the Schrödinger equation, the one electron hydrogenic wave functions. The sequence of energy levels follows from the lowest values of *n* and *l* or their combination ($n + l$; sec. 4.4.2). Because the orbitals in each subshell (*p*, *d*, etc.) have the same energy, the orders for values of m_l and m_s are not specified.

With minor exceptions the order of energy levels for the various orbitals is

$$1s < 2s < 2p < 3s < 3d < 4s < 4p < \ldots$$

Rule 2: Only one electron can exist in any particular quantum state (defined by n, l, m_l and m_s) in an atom. This follows from the Pauli exclusion principle discussed below.

Rule 3: The ground state of an atom is one with maximum total spin possible, that is, one having maximum number of parallel spins within a subshell (defined by a particular *l*). Rule 1 requiring lowest energy for the atom and Rule 2 requiring unique combination of n, l, m_l and m_s are retained. This condition is discussed below under Hund's rules.

The Pauli exclusion principle

This principle, established by W. Pauli in 1925 from studies on atomic spectra, essentially states: *no two electrons in an atom can exist in the same quantum state.*

Accordingly, the same set of values for all four quantum numbers cannot be assigned to more than one electron in an atom. In a more general form the principle involves symmetry properties of the wave functions and is referred to as the **Pauli principle** (see section 4.5).

The principle restricts the maximum number of electrons which may be assigned to a given orbital. Thus, for an 1*s* orbital, the possible values of the four quantum numbers are $n = 1$,

$l = 0$, $m_l = 0$ and $m_s = \pm\frac{1}{2}$. The exclusion principle at once tells us that there may be only two *unique* sets of these quantum numbers:

$$1, 0, 0, +\tfrac{1}{2} \text{ and } 1, 0, 0, -\tfrac{1}{2}.$$

According to the principle we can assign only one electron to each set. Hence, the maximum number of electrons which may be ''placed'' in the 1*s* orbital is two. Any third electron would mean a repetition of either of these two sets, which violates the principle. The same conclusion applies to any *s* orbital. With two electrons of opposite spin, the *s*-orbital is said to be ''closed''.

For the *p*-subshell, similar application of the exclusion principle shows that a maximum of six electrons can be accommodated in it. The possible combination of quantum numbers and the total number of electrons which may be accommodated in each principal energy level are worked out in Table 4.5 for $n = 1$ to 4.

Table 4.5

Arrangement of electrons in quantum states up to n = 4

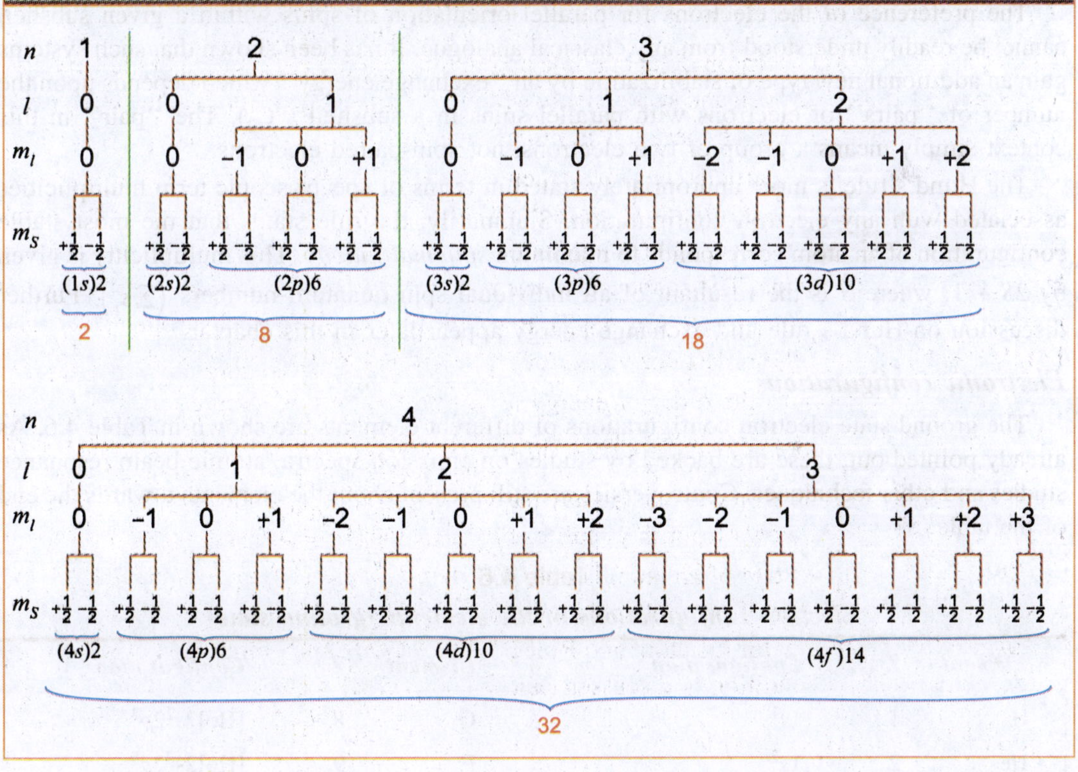

It is observed from the Table that a maximum number of 2, 8, 18 and 32 electrons can be accommodated in shells with $n = 1, 2, 3$ and 4 respectively.

The exclusion principle, therefore, helps one to write down the electronic configuration of an atom. As soon as the maximum capacity of an orbital is reached, the next electron has to be placed in the next higher orbital. Successive orbitals are thus filled in their gradual sequence giving rise to a regular pattern in the electronic configuration of the elements. Though empirical in nature, there are no exceptions known to the exclusion principle. Every particle with half-integral spin (*fermion*) like the electron obeys this principle, whether nuclear or extranuclear, excited or normal.

While the aufbau principle and the exclusion principle guide us to obtain the electron distribution of an atom in its various shells (designated by n) and subshells (designated by l), another rule determines the distribution of electrons *within* a subshell in the ground state of an atom. This is known as the Hund's rule.

Hund's rules (1925)

The rule may be stated as follows:

The ground state of an atom should contain the maximum number of unpaired electrons (within the same subshell) with their spins parallel.

As an illustration, a system with two or three *np*-electrons will have one electron in different *p*-orbitals with spins parallel. The same principle applies to *d* and *f* orbitals.

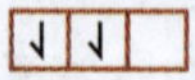 and not 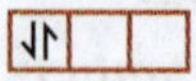 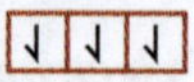 and not 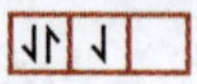

The preference of the electrons for parallel orientation of spins within a given subshell cannot be readily understood from any classical analogue. It has been shown that such systems gain an additional new type of stabilization by an ''exchange energy'' which depends upon the number of ''pairs'' of electrons with parallel spins in a subshell ($^nC_2$). The ''pair'' in this context simply means a group of two electrons, not spin-paired electrons.

The Hund's rule is more appropriately stated in terms of spectroscopic term multiplicities associated with any electron configuration. Summarily, the rule states that the most stable configuration of an atom corresponds to maximum *spin multiplicity*. This multiplicitiy is given by $2S + 1$, where S is the resultant of all individual spin quantum numbers $\left(\sum s_i\right)$. Further discussion on Hund's rule and exchange energy appear later in this chapter.

Electronic configurations

The ground state electron configurations of different elements are shown in Table 4.6. As already pointed out, these are backed by studies on emission spectra, atomic beam resonance studies and other techniques. Controversies are still present about the elements towards the end of the table.

Table 4.6

Electron configurations of the elements (ground state)

Element	*Z*	*Configuration*	*Element*	*Z*	*Configuration*
H	1	$1s^1$	O	8	$[He]2s^22p^4$
He	2	$1s^2$	F	9	$[He]2s^22p^5$
Li	3	$[He]2s^1$	Ne	10	$[He]2s^22p^6$
Be	4	$[He]2s^2$	Na	11	$[Ne]3s^1$
B	5	$[He]2s^22p^1$	Mg	12	$[Ne]3s^2$
C	6	$[He]2s^22p^2$	Al	13	$[Ne]3s^23p^1$
N	7	$[He]2s^22p^3$	Si	14	$[Ne]3s^23p^2$

continued ...

Element	Z	Configuration	Element	Z	Configuration
P	15	$[\text{Ne}]3s^{2}3p^{3}$	In	49	$[\text{Kr}]4d^{10}5s^{2}5p^{1}$
S	16	$[\text{Ne}]3s^{2}3p^{4}$	Sn	50	$[\text{Kr}]4d^{10}5s^{2}5p^{2}$
Cl	17	$[\text{Ne}]3s^{2}3p^{5}$	Sb	51	$[\text{Kr}]4d^{10}5s^{2}5p^{3}$
Ar	18	$[\text{Ne}]3s^{2}3p^{6}$	Te	52	$[\text{Kr}]4d^{10}5s^{2}5p^{4}$
K	19	$[\text{Ar}]4s^{1}$	I	53	$[\text{Kr}]4d^{10}5s^{2}5p^{5}$
Ca	20	$[\text{Ar}]4s^{2}$	Xe	54	$[\text{Kr}]4d^{10}5s^{2}5p^{6}$
Sc	21	$[\text{Ar}]3d^{1}4s^{2}$	Cs	55	$[\text{Xe}]6s^{1}$
Ti	22	$[\text{Ar}]3d^{2}4s^{2}$	Ba	56	$[\text{Xe}]6s^{2}$
V	23	$[\text{Ar}]3d^{3}4s^{2}$	La	57	*$[\text{Xe}]5d^{1}6s^{2}$
Cr	4	*$[\text{Ar}]3d^{5}4s^{2}$	Ce	58	*$[\text{Xe}]4f^{1}5d^{1}6s^{2}$
Mn	25	$[\text{Ar}]3d^{5}4s^{2}$	Pr	59	$[\text{Xe}]4f^{3}6s^{2}$
Fe	26	$[\text{Ar}]3d^{6}4s^{2}$	Nd	60	$[\text{Xe}]4f^{4}6s^{2}$
Co	27	$[\text{Ar}]3d^{7}4s^{2}$	Pm	61	$[\text{Xe}]4f^{5}6s^{2}$
Ni	28	$[\text{Ar}]3d^{8}4s^{2}$	Sm	62	$[\text{Xe}]4f^{6}6s^{2}$
Cu	29	*$[\text{Ar}]3d^{10}4s^{1}$	Eu	63	$[\text{Xe}]4f^{7}6s^{2}$
Zn	30	$[\text{Ar}]3d^{10}4s^{2}$	Gd	64	$[\text{Xe}]4f^{7}5d^{1}6s^{2}$
Ga	31	$[\text{Ar}]3d^{10}4s^{2}4p^{1}$	Tb	65	$[\text{Xe}]4f^{9}6s^{2}$
Ge	32	$[\text{Ar}]3d^{10}4s^{2}4p^{2}$	Y	66	$[\text{Xe}]4f^{10}6s^{2}$
As	33	$[\text{Ar}]3d^{10}4s^{2}4p^{3}$	Ho	67	$[\text{Xe}]4f^{11}6s^{2}$
Se	34	$[\text{Ar}]3d^{10}4s^{2}4p^{4}$	Er	68	$[\text{Xe}]4f^{12}6s^{2}$
Br	35	$[\text{Ar}]3d^{10}4s^{2}4p^{5}$	Tm	69	$[\text{Xe}]4f^{13}6s^{2}$
Kr	36	$[\text{Ar}]3d^{10}4s^{2}4p^{6}$	Yb	70	$[\text{Xe}]4f^{14}6s^{2}$
Rb	37	$[\text{Kr}]5s^{1}$	Lu	71	$[\text{Xe}]4f^{14}5d^{1}6s^{2}$
Sr	38	$[\text{Kr}]5s^{2}$	Hf	72	$[\text{Xe}]4f^{14}5d^{2}6s^{2}$
Y	39	$[\text{Kr}]4d^{1}5s^{2}$	Ta	73	$[\text{Xe}]4f^{14}5d^{3}6s^{2}$
Zr	40	$[\text{Kr}]4d^{2}5s^{2}$	W	74	$[\text{Xe}]4f^{14}5d^{4}6s^{2}$
Nb	41	*$[\text{Kr}]4d^{4}5s^{1}$	Re	75	$[\text{Xe}]4f^{14}5d^{5}6s^{2}$
Mo	42	*$[\text{Kr}]4d^{5}5s^{1}$	Os	76	$[\text{Xe}]4f^{14}5d^{6}6s^{2}$
Tc	43	$[\text{Kr}]4d^{5}5s^{2}$	Ir	77	$[\text{Xe}]4f^{14}5d^{7}6s^{2}$
Ru	44	*$[\text{Kr}]4d^{7}5s^{1}$	Pt	78	*$[\text{Xe}]4f^{14}5d^{9}6s^{1}$
Rh	45	*$[\text{Kr}]4d^{8}5s^{1}$	Au	79	*$[\text{Xe}]4f^{14}5d^{10}6s^{1}$
Pd	46	*$[\text{Kr}]4d^{10}$	Hg	80	$[\text{Xe}]4f^{14}5d^{10}6s^{2}$
Ag	47	*$[\text{Kr}]4d^{10}5s^{1}$	Tl	81	$[\text{Xe}]4f^{14}5d^{10}6s^{2}6p^{1}$
Cd	48	$[\text{Kr}]4d^{10}5s^{2}$	Pb	82	$[\text{Xe}]4f^{14}5d^{10}6s^{2}6p^{2}$

continued ...

Element	Z	Configuration	Element	Z	Configuration
Bi	83	$[Xe]4f^{14}5d^{10}6s^26p^3$	Es	99	$[Rn]5f^{11}7s^2$
Po	84	$[Xe]4f^{14}5d^{10}6s^26p^4$	Fm	100	$[Rn]5f^{12}7s^2$
At	85	$[Xe]4f^{14}5d^{10}6s^26p^5$	Md	101	$[Rn]5f^{13}7s^2$
Rn	86	$[Xe]4f^{14}5d^{10}6s^26p^6$	No	102	$[Rn]5f^{14}7s^2$
Fr	87	$[Rn]7s^1$	Lr	103	$[Rn]5f^{14}6d^17s^2$
Ra	88	$[Rn]7s^2$		or	$[Rn]5f^{14}7p^17s^2$
Ac	89	* $[Rn]6d^17s^2$	Rf	104	$[Rn]5f^{14}6d^27s^2$
Th	90	* $[Rn]6d^27s^2$	Db	105	$[Rn]5f^{14}6d^37s^2$
Pa	91	* $[Rn]5f^26d^17s^2$	Sg	106	$[Rn]5f^{14}6d^47s^2$
U	92	* $[Rn]5f^36d^17s^2$	Bh	107	$[Rn]5f^{14}6d^57s^2$
Np	93	* $[Rn]5f^46d^17s^2$	Hs	108	$[Rn]5f^{14}6d^67s^2$
Pu	94	$[Rn]5f^67s^2$	Mt	109	$[Rn]5f^{14}6d^77s^2$
Am	95	$[Rn]5f^77s^2$	Ds	110	* $[Rn]5f^76d^17s^1$
Cm	96	* $[Rn]5f^76d^17s^2$	Rg	111	$[Rn]5f^{14}6d^97s^1$
Bk	97	$[Rn]5f^97s^2$	Uub	112	$[Rn]5f^{14}6d^{10}7s^2$
Cf	98	* $[Rn]5f^96d^17s^2$	Uuo	118	$[Rn]5f^{14}6d^{10}7s^27p^6$

* Elements with configurations that do not follow the simple order of orbital filling.

The general trend in filling of successive orbitals may be largely understood from the discussion on the variation of orbital energies with effective nuclear charge (Fig. 4.32). To illustrate, the 3*d* orbital has the same energy as the 3*s* and 3*p* orbitals in the hydrogen atom. Since it penetrates little into the inner core of electrons in $n = 1$ and $n = 2$, it is almost completely shielded from the increasing nuclear charge as these shells are being filled and its energy remains unchanged (see Fig. 4.32) upto $Z = 10$. On the other hand, the 4*s* and 4*p* orbitals, which are much higher in energy than the 3*d* orbitals in the lighter elements, do penetrate appreciably into the inner electron core and hence their energy drops quickly as *Z* increases. At about $Z = 15$, the energy of the 4*s* orbital falls below that of the 3*d*. The last electrons in potassium and calcium thus enter the 4*s* orbital outside the argon core. The added 4*s* electrons do not effectively shield the 3*d* orbital. The 3*d* orbital now drops in energy in comparison to the 4*p* level. The next ten electrons now enter the 3*d* level which lies below the 4*s* level. However, these two levels are very close in energy and so frequent exchange of electrons occur between them, as shown for chromium ($3d^54s^1$) and copper ($3d^{10}4s^1$). Obviously these configurations are more stable than the alternatives ($3d^44s^2$ and $3d^94s^2$ respectively).

Similar effects, to varying extent, determine the electron population in the higher orbitals. The *f* orbitals are even less penetrating than the *d* orbitals and are not filled before $Z = 58$. At this stage the energy of the 5*d* level also lies very close to the 4*f* level and is involved in some electron configurations. The stability of the half-filled and full-filled *d* and *f* shells is not always the determining factor in the later elements, as may be seen for Sm, W, Pt etc. The overall situation involves an intricate energy balance for the atom as a whole.

Spin correlation

Exchange energy: The concept of spin leads to a new conclusion in quantum mechanics having no classical counterpart : two particles with the same spin tend to avoid each other in space more than two particles with opposite spins. This spin-quantum effect is observed for any particle that possesses spin angular momentum—even for the chargeless particle neutron which does not experience any electrostatic repulsion.

Spin-correlation of electron motion leads to a new type of quantum mechanical interaction known as the *exchange energy*, E_{ex}. This is proportional to the number of pairs of electrons of the same spin that can be arranged from n electrons with parallel spins. The pair in this context means simply a group of two electrons with same spin quantum number whose number is equal to ${}^nC_2$.

$$E_{ex} = {}^nC_2 = \frac{n(n-1)}{2}K \qquad (K = \text{a constant})$$

$n =$	1	2	3	4	5	6
$E_{ex}/K = n(n-1)/2 =$	0	1	2	6	10	15

This exchange energy provides the basis of Hund's rule that electrons in the same orbital have their spins parallel as far as possible.

Illustration: Let us calculate the exchange energy (in terms of K) for three p electrons when (i) Hund's rule is obeyed and (ii) maximum pairing occurs.

***Distribution* 1** (following Hund's rule): ↿ ↿ ↿

For electrons with spin ↿, $n = 3$; $E_{ex} = \frac{3.2}{2}K = 3K$.

There are no electrons with spin ⇂. Hence total $E_{ex} = 3K$.

***Distribution* 2** (with maximum pairing): ↿⇂ ↿ _

For electrons with spin ↿, $n = 2$ $E_{ex} = \frac{2.1}{2}K = K$.

There is only one electron with spin ⇂. $E_{ex} = 0$. Total $E_{ex} = K$

Thus distribution 1 has greater stabilization from exchange energy.

Problem 4.15

Calculate the exchange energy of four p-electrons with maximum pairing and maximum unpairing.

Hint: Distribution 1 . (*maximum unpairing*): ↿⇂ ↿ ↿

For spin ↿, $E_{ex} = 3K$. For the single electron with spin ⇂. $E_{ex} = 0$.

Distribution 2. (*maximum pairing*): ↿⇂ ↿⇂ _

For spin ↿, $E_{ex} = \frac{2.1}{2}K = K$. For spin ⇂, $E_{ex} = \frac{2.1}{2}K = K$.

Total E_{ex} for distribution 1 is $3K$ vs. $2K$ for distribution 2.

The magnitude of the exchange energy is relatively small, yet it may become a determining factor for electron configuration when similar species are compared. Let us take the example of chromium. If there are four electrons in the 3d shell obeying Hund's rule, the exchange energy would be $6K$. For d^5 configuration, this would be $10K$. This difference of $4K$ in exchange energy is sufficient to compensate for the destabilization in orbital energy in moving an electron from the $4s$ orbital to a $3d$ orbital. Hence the ground state configuration of chromium is written (outside the argon core) as $3d^5 4s^1$ and not $3d^4 4s^2$. This provides a basis for the phrase "*stability of the half-filled shell*".

Similar exchange energy considerations explain the "stability of the full-filled shell" as observed in Cu ($3d^{10}4s^1$), Ag ($4d^{10}5s^1$) and Au ($5d^{10}6s^1$). These configurations give rise to an additional exchange energy equal to $4K$ ($20K - 16K$; try and calculate) in each.

Similar preference for half-filled and full-filled shells or subshells is also exhibited by elements with f-electrons. However, there appear several anomalies, for example:

W $5d^4 6s^2$ **Ni** $3d^8 4s^2$ **Pd** $4d^{10} 5s^0$ **Pt** $5d^9 6s^1$.

In such cases the balance of orbital energy favours the configurations adopted.

The extra stability of half-filled and full-filled shells and subshells is also reflected in the properties of the atoms (ionization energy, for example) and ions (e.g., Mn^{II} is particularly stable with a d^5 configuration). Many examples of this kind will be found in subsequent chapters.

4.5 SYMMETRIC AND ANTISYMMETRIC WAVE FUNCTIONS

We have seen that the space part of the wave function ψ of the electron in an atom of hydrogen may be expressed as a product of three functions:

$$\psi(r, \theta, \phi) = \psi_{I}(r)\, \psi_{II}(\theta)\, \psi_{III}(\phi).$$

Similarly, the wave function of a system of n *non-interacting particles* may be expressed as the product of the wave functions for each individual particle. Representing these by $\psi(1)$, $\psi(2)$ ... etc, one can write,

$$\psi(1, 2, 3, ..., n) = \psi(1)\, \psi(2)\, \psi(3) \, ... \, \psi(n).$$

Let us apply this relation to a system of two identical particles. We suppose that one particle is in quantum state a and the other in state b. The wave function of the system would be

$$\psi_I = \psi_a(1)\, \psi_b(2).$$

If we now exchange the particles between the two states, the system will be described by another wave function:

$$\psi_{II} = \psi_a(2)\, \psi_b(1).$$

Since the particles are identical, exchange of particles between the two states would make no difference in the probability density $|\psi|^2$ of the system. So we require that $\psi_I^2 = \psi_{II}^2$ or $\psi_I = \pm\, \psi_{II}$.

Therefore, either, $\psi_I = \psi_{II}$ i.e., $\psi_a(1)\, \psi_b(2) = \psi_a(2)\, \psi_b(1)$

or, $\psi_I = -\psi_{II}$ i.e., $\psi_a(1)\, \psi_b(2) = -\,\psi_a(2)\, \psi_b(1)$

In the first case, the wave function of the system is unaffected by an exchange of particles between the two states and is said to be *symmetric*. In the second case, exchange of particles causes a reversal of sign in the wave function of the system; this is called an *antisymmetric* wave function. This generalization holds for any number of identical particles which are non-interacting (i.e., repulsion between the electrons is ignored).

Since the particles 1 and 2 are identical, it is not possible to distinguish between them, that is, they are indistinguishable. So we do not know which particle is in which state at any moment. The likelihood that particle 1 is in state a and particle 2 is in state b is equal to the other alternative. So both ψ_I and ψ_{II} have equal probability of describing the system. Actually a linear combination of these two functions is found to represent the system satisfactorily. Two such combinations are possible:

$$\textit{Symmetric}: \psi_S = \frac{1}{\sqrt{2}}[\psi_a(1)\psi_b(2) + \psi_a(2)\psi_b(1)]$$

$$\textit{Antisymmetric}: \psi_A = \frac{1}{\sqrt{2}}[\psi_a(1)\psi_b(2) - \psi_a(2)\psi_b(1)]$$

The factor $1/\sqrt{2}$ appears from normalization of the two functions. The symmetric function does not change if the particles 1 and 2 are interchanged. The antisymmetric combination reverses its sign under such exchange.

Does it make any difference if we describe the system by the symmetric or the antisymmetric wave function? If the system is described by the symmetric wave function, we can put $a = b$, when ψ_S becomes $\sqrt{2}\,[\psi_a(1)\,\psi_b(2)]$. Both the particles can now simultaneously exist in the same quantum state. But if the system is described by the antisymmetric wave function, setting $a = b$ makes $\psi_A = 0$ or the system becomes non-existent. In this case the two particles cannot be simultaneously in the same quantum state.

Pauli deduced a general form of this principle from relativistic quantum field theory (1940) though its essential aspects were realized by Dirac and Heisenberg much earlier (1926). The principle, known as **the Pauli Principle**, essentially states:

For a system of identical fermions (particles with half-integral spin), the total wave function including space and spin parts must be antisymmetric with respect to interchange of all the coordinates (space and spin) of any two particles.

For a system of identical bosons (particles with integral spin), the total wave function is required to be symmetric under such exchange.

Validity of the principle has been experimentally established.

> Protons, neutrons, mesons and ^{13}C nuclei are examples of fermions with spin (s) 1/2; the ^{35}Cl nucleid is a fermion with spin 3/2. Photons (s = 1), α-particles and ^{12}C nuclei (S = 0) are examples of bosons.

We now recall the exclusion principle of Pauli: *no two electrons in an atom can exist in the same quantum state.* This may be called empirical because Pauli arrived at this conclusion from the analysis of atomic spectra of elements other than hydrogen. Lines corresponding to transitions to and from states with certain combinations of quantum numbers were never observed. We have seen that a *system containing two electrons is appropriately described by antisymmetric wave functions.* In general it can be said that any atomic system containing several optically active electrons must be described by antisymmetric wave functions. The complete quantum state of an electron in an atom is defined by the four quantum numbers, n, l, m_l and m_s. The corresponding wave function may be resolved into two parts:

$$\psi = \psi_{orbital}\ \psi_{spin}, \quad \text{or,} \quad \psi_{space}\ \psi_{spin}$$

The orbital part is determined by the quantum numbers, n, l and m_l while the spin part is determined by the spin quantum number m_s. If any one of these parts be antisymmetric, the total wave function will be antisymmetric. When two electrons in an atom have the same values for n, l, and m_l, they have the same ψ_{space}; so the spin parts of the wave function must be different, or simply they must have opposite spins. This is our familiar form of the exclusion principle.

4.6 ATOMIC ENERGY STATES AND TERM SYMBOLS

Atomic spectra of many-electron atoms are very complex in nature—they contain multiple lines indicative of several transitions. The energy states of the atom corresponding to these transitions cannot be calculated as easily as in the case of hydrogen-like species with only one electron. The reason is obvious—one has to consider the mutual interaction between the electrons. The wave mechanical procedure to tackle this situation is extremely lengthy and complicated. However, experimental spectroscopists have developed a procedure to assign the various energy states of an atom corresponding to observed spectral transitions without any reference to any theoretical description of electron motion. There is good correspondence between the procedure with experiment.

The vector atom model

Although the wave formalism abandons the picture of specific orbits for the electrons, the idea of angular momentum of an electron in an orbital is retained in this model. Next the orbital angular momentum and the spin angular momentum of the individual electrons in a multi-electron system may be combined according to the principle of vector addition to obtain the resultant spin and orbital angular momenta of the atomic system; these may be utilized to obtain different energy states of the atom. This is the basis of the *vector model of the atom.* In this model, the different energy states of an atom are designated by particular term symbols derived from the resultant orbital and spin angular momenta of the extranuclear electrons. (A further interaction with the spin angular momentum of the nucleus will give rise to still finer splitting of the energy levels, but this is neglected in the present context.)

The interactions between the spin and orbital angular momenta of a many electron system may be considered in terms of the one-electron quantum numbers (n, l, m_l and m_s) of the individual electrons. If we consider a two-electron system, for example, there may be three types of interaction: (a) *orbit-orbit coupling*, in which the orbital angular momenta of the two electrons couple; (ii) *spin-spin coupling*, in which the spin angular momenta of the two electrons couple; and (iii) *spin-orbit coupling*, in which the spin and orbital angular momenta of the *same* electron interact with one another. The coupling of the spin of one electron with the orbital angular momentum of a different electron is small and can be ignored. There may be two ways to consider such interactions.

(i) Russel-Saunders scheme (*L–S* scheme). In this scheme it is assumed that the strength of interaction varies as

spin-spin coupling > orbit-orbit coupling > spin-orbit coupling.

The individual orbital angular momenta of the electrons are first combined to give a total orbital angular momentum of the system. By analogy with the orbital angular momentum of a single electron ($=\sqrt{l(l+1)}\,h/2\pi$), this is expressed by means of a total orbital angular momentum quantum number L; the total orbital angular momentum is equal to $\sqrt{L(L+1)}\,h/2\pi$.

Similarly, the resultant spin angular momentum of the atom is expressed by a corresponding quantum number S as $\sqrt{S(S+1)}\,h/2\pi$.

The total angular momentum (spin + orbital) is now represented by a corresponding quantum number J and is equal to $\sqrt{J(J+1)}\,h/2\pi$. The procedures for obtaining L, S and J will be described below.

This coupling scheme works fairly well for the lighter elements (say upto atomic number 30). As the radial function in heavier atoms decays more *slowly to a larger distance in space*, the interaction between the spin and orbital angular momenta of two electrons decreases and the *L–S* scheme becomes less appropriate.

(ii) *j-j* coupling scheme. In this scheme, the spin and orbital angular momenta of individual electrons are first coupled to obtain a resultant ''spin + orbital'' angular momentum for each; the corresponding quantum number is denoted by j. A weaker coupling of the individual electron angular momenta now gives the total angular momentum for the atom. The relative importance of spin-orbit coupling increases in a gradual manner with atomic number and there is a smooth transition from the *L–S* to the *j–j* scheme.

Term Symbols (*L–S* scheme): First of all we recall the salient features of the single electron quantum numbers l, m_l and m_s.

(i) The orbital quantum number l has integral values from 0 to $(n-1)$ for any principal quantum shell n and the orbital angular momentum is given by $\sqrt{l(l+1)}\,h/2\pi$.

(ii) The orbital angular momentum has $2l+1$ possible directions in space (with reference to the direction of an external magnetic field) specified by m_l, the magnetic quantum number: $m_l = l, (l-1), \ldots 0 \ldots -(l-1), -l$.

(iii) The spin angular momentum is given by $\sqrt{s(s+1)}\,h/2\pi$ where s is the spin quantum number $\left(\frac{1}{2}\right)$. The z-component of this vector (chosen arbitrarily) is given by m_s $(h/2\pi)$ where m_s is the spin magnetic quantum number; this can be $+s$ or $-s$, i.e., $+\frac{1}{2}$ or $-\frac{1}{2}$.

Now we return to our case of a many electron atom and apply an analogous treatment.

(i) The resultant orbital quantum number, *L*. This gives the resultant orbital angular momentum of the system and is obtained by vectorial addition of individual l values. The different L values, 0, 1, 2, 3, 4, etc. are represented by a letter symbol (like $s, p, d, \ldots$ for l values):

L =	0	1	2	3	4	5	6	7
Symbol =	S	P	D	F	G	H	I	K

But how should one combine the individual l-values? The angular momentum vectors related to l are, in general, not parallel and so direct addition of such vectors is very difficult. But the individual m_l values represent the component of the orbital angular momentum vectors ($m_l h/2\pi$) oriented in the same direction (conventionally the z-axis) and so their vector addition can be done simply by algebraic summation. Such algebraic summation for all the electrons in the system gives the resultant orbital angular momentum along the z-axis, which may be represented by $M_L(h/2\pi)$. $M_L = (\Sigma m_l)$ may be called the resultant orbital magnetic quantum number.

Just like m_l, there are $(2L + 1)$ values for M_L: $L, (L-1), \cdots 0 \cdots -(L-1), -L$.

Different possible distribution of electrons among different m_l values gives different M_L values (Σm_l) and hence L-values. These correspond to different terms or energy states for the system (see examples).

(ii) The resultant spin quantum number, *S*. Similar to L, we may introduce a resultant spin quantum number S which gives the resultant spin angular momentum as $\sqrt{S(S+1)}\, h/2\pi$. Algebraic summation of m_s values of individual electrons gives a resultant spin magnetic quantum number $M_S = (\Sigma m_s)$.

$M_S = (\Sigma m_s)$ may have $2S + 1$ values: $S, (S-1), \cdots 0 \cdots -(S-1), -S$. The quantity $(2S + 1)$ is known as the *spin multiplicity* of a state.

Note that S is used for the resultant spin quantum number as well as the symbol for a state with $L = 0$.

(iii) The inner quantum number, *J*. This expresses the total angular momentum as $\sqrt{J(J+1)}\, h/2\pi$. J can have the values $L + S, L + S - 1, \ldots |L - S|$. The modulus notation indicates that only the magnitude of $(L - S)$ is involved. Like j for a single electron, J can be either positive or zero. So two cases may arise:

$L > S$: there are $2S + 1$ possible values of J.

$L < S$: there are $2L + 1$ possible values of J.

The complete term symbol (or state symbol) for an energy state is written using $(2S + 1)$, L and J as

$$^{2S+1}L_J$$

Instead of the numerical value of L, we use the corresponding symbols : $S(L = 0)$, $P(L = 1)$ etc. Thus, $^3P_0$ (triplet-P-zero) stands for a system in which $L = 1$, $2S + 1 = 3$ and $J = 0$. Terms which only differ in J values differ slightly in energy. The energy state is therefore often expressed by ^{2S+1}L with the J-values omitted. The different J values now give different components of that state.

Now we shall work out the ground-state terms for a few species:

Hydrogen atom ($1s^1$). $l = 0$; hence $L = 0$. This corresponds to an S state. Now for the single electron, $S = \frac{1}{2}, 2S + 1 = 2.$ and $J = \frac{1}{2}$. So the ground state term symbol is $^2S_{1/2}$.

Helium atom ($1s^2$). For both electrons $l = 0$. So $L = 0$. Since the electrons are spin-paired, $m_s = +\frac{1}{2}$ and $-\frac{1}{2}$ and $S = 0$. $2S + 1 = 1$, $J = 0$. So the state symbol is $^1S_0$.

Note: For any ns^2 or np^2 configuration or the like, i.e., a *closed* subshell (or a closed shell), M_S and M_L will be zero. The spins will cancel one another, so also will the '+' and '–' values of m_l. So for lithium and beryllium, the 1*s* shell does not contribute anything to the term symbol. Li ($1s^22s^1$) has the same term symbol as hydrogen ($^2S_{1/2}$) and Be($1s^22s^2$) has the same term symbol as helium ($^1S_0$).

Boron ($1s^22s^22p^1$): Here also, only the *p* electron need be considered. $l = 1$, so $L = 1$, i.e., a *P* state is indicated. $S = \frac{1}{2}$ gives $2S + 1 = 2$, a doublet state. *J* may be $L + S$ or $L - S$, i.e., $\frac{3}{2}$ or $\frac{1}{2}$. So the state symbols are $^2P_{3/2}$ or $^2P_{1/2}$. The ground state term will be $^2P_{1/2}$ (see later).

Carbon ($1s^22s^22p^2$): As explained above, we consider only the two *p*-electrons. For each, $l = 1$ *and* $m_l = +1, 0$ or -1. Let us, for convenience, label the electrons as 1 and 2. Then the possible values of m_l for electron(1) and electron(2) are

$$m_l(1) \quad : \quad +1, \qquad 0, \qquad -1$$
$$m_l(2) \quad : \quad +1, \qquad 0, \qquad -1$$

The possible values of M_L ($= m_l(1) + m_l(2)$) are then 2, 1, 0, 0, –1, 0, –1, –2. These can be arranged and the different *L* values (hence the term symbols) derived therefrom as follows:

$M_L = 2, 1, 0, -1, -2$	$L = 2$	*D*-term
$M_L = 1, 0, -1$	$L = 1$	*P*-term
$M_L = 0$	$L = 0$	*S*-term

The *D*-term: For $M_L = +2$ or -2, $m_l(1) = m_l(2) = 1$ or $m_l(1) = m_l(2) = -1$. Thus, *n*, *l*, and m_l for the two electrons become identical; hence the Pauli principle requires that the spins of the two electrons must be opposite, that is $M_S = +\frac{1}{2} + \left(-\frac{1}{2}\right) = 0$. So $S = 0$ and $2S + 1 = 1$; it is a singlet state. $L + S = 2$ and $L - S = 2$, so that $J = 2$. The complete term symbol is thus $^1D_2$.

The *P*-terms: For each value of M_L, the two electrons may have spins $\left(\frac{1}{2}, \frac{1}{2}\right)$, $\left(\frac{1}{2}, -\frac{1}{2}\right)$, $\left(-\frac{1}{2}, -\frac{1}{2}\right)$, giving $M_L = +1, 0$, or -1. We have 9 possible combinations of M_L and M_S.

M_L	M_S		
+1	+1	0	–1
0	+1	0	–1
–1	+1	0	–1

When $M_S = +1, 0$ and -1, $S = 1$ and $2S + 1 = 3$. The term symbol is thus ^3P. Next we find that $L + S = 1 + 1 = 2$ and $L - S = 1 - 1 = 0$. So *J* may have values 2, 1, 0. The complete term symbols will be $^3P_2$, $^3P_1$ and $^3P_0$.

The *S*-term: This corresponds to $M_L = 0$ and $M_S = 0$, and hence $S = 0$. The complete symbol will be $^1S_0$.

Microstates. In working out the term symbols for the carbon atom, we have ultimately considered the various possible distribution of the *p*-electrons among the m_l and m_s values obeying the exclusion principle. Each such combination of m_l and m_s for the two electrons is called a microstate. In Table 4.7, these permitted combinations (15 all total) are shown with the resultant M_L and M_S value for each:

Table 4.7

Allowed m_l and m_s values for np^2 configuration.
Electron spins are represented as ↑ for $+\frac{1}{2}$ and ↓ for $-\frac{1}{2}$.

m_l =	+1	0	−1	M_L	M_S
1.	↑↓			2	0
2.		↑↓		0	0
3.			↑↓	−2	0
4.	↑	↑		1	1
5.	↑		↑	0	1
6.		↑	↑	−1	1
7.	↓	↓		1	−1
8.	↓		↓	0	−1
9.		↓	↓	−1	−1
10.	↑	↓		1	0
11.	↓	↑		1	0
12.	↑		↓	0	0
13.	↓		↑	0	0
14.		↑	↓	−1	0
15.		↓	↑	−1	0

These 15 microstates can be grouped into three sets: (i) a set of five with $M_L = 2, 1, 0, -1, -2$, all having $M_S = 0$; these correspond to the $^1D_2$ term. $M_J = 2, 1, 0, -1, -2$, i.e., $J = 2$. (ii) a set of nine microstates corresponding to the ^3P terms, that is. $M_L = 1, 0$ or -1 and $M_S = 1, 0$ or -1. These can be subdivided into five with $J = 2$ ($^3P_2$), three with $J = 1$($^3P_1$) and one with $J = 0$ ($^3P_0$). (iii) a single combination with $M_L = 0$ and $M_S = 0$, corresponding to the term $^1S_0$. However, there are three combinations (serial no 2, 12 and 13) for $M_L = 0$, $M_S = 0$ and there is no way to decide which $M_L - M_S$ combination should be assigned to which state. It is not meaningful either.

The total number of mocristates can be ascertained very easily. The first electron can be placed in any of the three *p*-orbitals with either of two spin quantum numbers; this can be done in six different ways (these may be called six *spin orbitals*). The second electron can be placed in the particular *p*-orbital already occupied by the first electron with an opposite spin only; in the other two *p*-orbitals it may have either spin. So this electron may be placed in the system in five different

continued ...

manners. Combining, there are 6 × 5 = 30 different ways. But since the electrons are indistinguishable, the actual number will be 30 ÷ 2 = 15. The number of microstates N arising from 'e' number of electrons filling a set of orbitals of same n and l values are

$$N = \frac{N_l!}{e!(N_l - e)!}$$

where N_l is the number of $m_l - m_s$ combinations for a single electron in the orbital set [= 2(2l+1)]. N_l = 6 for np orbitals, 10 for nd orbitals etc.

Also, each term with quantum numbers L and S consists of $(2L + 1) \cdot (2S + 1)$ numbers of $M_L - M_S$ pairs or microstates.

Nitrogen ($1s^22s^22p^3$). We consider only the $2p^3$ configuration and label the electrons as 1, 2 and 3. The possible values of m_l are 1, 0, −1 for each. This gives the possible L-values as 3, 2, 1, 0. But L = 3 means M_L should also have the value 3 (or −3). This will occur only if all three electrons belong together to m_l = +1 (or −1). This is not permitted since m_s may have only two values $+\frac{1}{2}$ or $-\frac{1}{2}$ and there may be at best two electrons for each m_l. So L cannot be 3 and there cannot be any F term.

L = 2 indicates a D-term. M_L = 2, 1, 0, −1, −2. M_L can be equal to 2 only when two of the electrons are spin-paired (check yourself). So M_S can be only $+\frac{1}{2}$ or $-\frac{1}{2}$. This gives $S=\frac{1}{2}$ and $2S+1=2$. $L+S=2+\frac{1}{2}$; $L-S=2-\frac{1}{2}$; so the allowed values of J are $\frac{5}{2}$ and $\frac{3}{2}$. So the complete term symbols for the D-term will be $^2D_{5/2}$ and $^2D_{3/2}$. From L = 2 and $S=\frac{1}{2}$, there will be $(2S + 1)(2L + 1)$ i.e., 10 microstates for the D-terms. Each J-value has $2J + 1$ microstates. (6 and 4 respectively).

L = 1 corresponds to a P-term. Here M_L = 1, 0, −1. For a combination of three electrons this again gives $M_S=\frac{1}{2}$ or $-\frac{1}{2}$ and $S=\frac{1}{2}$. $2S + 1 = 2$ and J values are $\frac{3}{2}$ and $\frac{1}{2}$. The term symbols are $^3P_{3/2}$ and $^3P_{1/2}$ (six microstates).

L = 0 means M_L = 0. But all three electrons cannot be together in m_l = 0 as that would violate the Pauli principle. So they must arise from combinations like $m_l(1)$ = 1, $m_l(2)$ = −1 and $m_l(3)$ = 0. Since the m_l values are different, m_s may be same. So the possible values of M_S are $\frac{3}{2}, \frac{1}{2}, -\frac{1}{2}, -\frac{3}{2}$. This gives $S=\frac{3}{2}$ and $J=\frac{3}{2}$. The complete term symbol is $^4S_{3/2}$ (four microstates).

Oxygen and Fluorine. Discussion of the $2p^4$ and $2p^5$ configurations becomes easier when we introduce a useful simplification known as the ***hole formalism***. According to this, the $2p^4$ configuration may be looked upon as a $2p^6$ configuration *plus* two positrons to annihilate two electrons. We know that positrons differ from electrons only with respect to their charge. So two positrons will give rise to the same states as given by two electrons. The filled p^6 configuration does not contribute anything to the terms and so we find that the p^4 configuration will have the same terms as the p^2 configuration. Similarly, the p^5 configuration will have the same terms as the p^1 configuration. In general, for any subshell which is more than half-filled, p^{6-n} or d^{10-n} configurations will give rise to the same terms as p^n or d^n configurations respectively.

Ground state term : Hund's rules. Once the energies of various states for an electron configuration are labeled by respective term symbols, it becomes necessary to know the relative energies of these states. A number of rules have been framed for the purpose from analysis of spectroscopic data. These are collectively known as Hund's rules which were summarily stated earlier. Where Russel-Saunders coupling holds ($Z \leq 30$), the rules essentially state that

1. *The energy of the states decreases as the spin multiplicity (and hence S) increases*. The ground state of an atom should therefore show *highest spin multiplicity* and hence possess the maximum number of unpaired spins in parallel orientation.
2. When the value of S is same for more than one state, the state with *the highest value of L has the minimum energy*.
3. For all states having same S and L, the following order of energy will hold:
   (i) *The subshell is less than half-filled*: the component with the smallest value of J has the least energy.
   (ii) *The subshell is more than half-filled or half-filled*: the component with the highest value of J has the least energy.

For boron, we obtained two terms—$^2P_{3/2}$ and $^2P_{1/2}$. The ground state term will be $^2P_{1/2}$ (less than half-filled, least value of J).

For carbon, the triplet state ^3P will be most stable as it has the highest spin multiplicity. Since the electron configuration is less than half-filled, the $^3P_0$ state, with lowest value of J, will be the ground state.

One can derive the ground state term for a given configuration using Hund's rules, without deriving all the possible terms. Thus for carbon ($2s^22p^2$), we consider only the $2p$ electrons as the closed shell $2s^2$ contributes nothing (see before). We have

m_l =	1	0	−1
	1	1	_

Maximum $M_L = 1$; $L = 1$, $S = 1$. Spin multiplicity = $2S + 1 = 3$. $L = 1$ denotes a P state. Since we are dealing with a less than half-filled configuration, the lowest value of J will give the ground state, which is $^3P_0$.

Problem 4.17

Determine the ground-state term for a d^8 configuration.

Hint:

m_l =	2	1	0	−1	−2
	⥮	⥮	⥮	1	1

It will be convenient if we use the hole formalism discussed before (see oxygen and fluorine). d^8 is equivalent to d^{10-8} or d^2.

Maximum $M_L = 3$, so $L = 3$ and $S = 1$. Spin multiplicity = $2 \times 1 + 1 = 3$.

$L = 3$ denotes an F-state. Since we have a configuration which is more than half-filled, the ground state term will have the highest value of J or $L + S$, in this case, 4. So the ground state is $^3F_4$.

Problem 4.18

Derive the term for a hydrogen atom in the excited states (i) $2s^1$, (ii) $2p^1$.

Solution: (i) For an s-orbital, $L = 0$;

Here $S = \frac{1}{2}$, $2S + 1 = 2$. $J = \frac{1}{2}$. The term will be $^2S_{1/2}$.

(ii) $L = 1$, $S = \frac{1}{2}$. So, $J = \frac{1}{2}, \frac{3}{2}$

The states will be $^2P_{3/2}$ and $^2P_{1/2}$.

The ground state term symbols of the first few elements are given below:

Table 4.8
Ground state terms for some elements

H	$1s^1$	$^2S_{1/2}$			
He	$1s^2$	$^1S_0$			
Li	$[He]\ 2s^1$	$^2S_{1/2}$	Na	$[Ne]\ 3s^1$	$^2S_{1/2}$
Be	$[He]\ 2s^2$	$^1S_0$	Mg	$[Ne]\ 3s^2$	$^1S_0$
B	$[He]\ 2s^22p^1$	$^2P_{1/2}$	Al	$[Ne]\ 3s^23p^1$	$^2P_{1/2}$
C	$[He]\ 2s^22p^2$	$^3P_0$	Si	$[Ne]\ 3s^23p^2$	$^3P_0$
N	$[He]\ 2s^22p^3$	$^4S_{3/2}$	P	$[Ne]\ 3s^23p^3$	$^4S_{3/2}$
O	$[He]\ 2s^22p^4$	$^3P_2$	S	$[Ne]\ 3s^23p^4$	$^3P_2$
F	$[He]\ 2s^22p^5$	$^2P_{3/2}$	Cl	$[Ne]\ 3s^23p^5$	$^2P_{3/2}$
Ne	$[He]\ 2s^22p^6$	$^1S_0$	Ar	$[Ne]\ 3s^23p^6$	$^1S_0$

Problem 4.19

Examine whether the ground-state term $^7S_3$ for Cr corresponds to the $3d^44s^2$ configuration.

Solution: The closed-shell s^2 configuration does not contribute to the term. For $3d^4$ configuration we get

$$L = 2 + 1 + 0 + (-1) = 2 \quad S = 4 \times \tfrac{1}{2} = 2.$$

$$2S + 1 = 5; \quad J_{max} = 4.$$

Hence the term would be $^5D_4$ which does not match the term $^7S_3$.

(For $3d^54s^1$ configuration $L = 0$, $S = 6 \times \frac{1}{2} = 3$ which gives the $^7S_3$ term).

Terms, Levels and States

A term symbol summarily gives the symbolic representation of the state of an atom. However, the various words used in this context are used to imply different meanings.

An atomic **term** conveys definite values of the total orbital angular momentum quantum number L and the total spin angular momentum quantum number S. The L value is represented by a code letter (S for $L = 0$, P for $L = 1$ etc.). S is shown by $2S + 1$ (spin multiplicity) as a left superscript. J is not included in the term but the quantum numbers n and l may precede the term. A term is thus designated by the quantum numbers n, l, L and S.

Levels arise from splitting of terms according to different values of J (fine structure). A given value of L gives the levels of its term according to the values of J. A level is thus designated by the quantum numbers n, l, L, S and J. When $L \geq S$, the multiplicity $(2S + 1)$ is the number of levels of the term. As we shall find below, a level consists of a group of states with a common value of J.

In magnetic fields, levels can further split into **states.**

Illustration: Show the terms and levels for (i) $[Ne]3s^1$ (ii) $[Ne]3p^1$ (iii) the ^3D term of $[Ne]2p^13p^1$.

(i) For $3s^1$, $L = l = 0$; $\quad S = s = \frac{1}{2}$; $\quad J = \frac{1}{2}$.

Term = $2S$. There is only one level for $J = \frac{1}{2}$ which we denote $^2S_{1/2}$.

(ii) For $3p^1$, $L = l = 1$; $S = s = \frac{1}{2}$; $J = \frac{3}{2}, \frac{1}{2}$

Term = $2P$. There are two levels, $^2P_{3/2}$ and $^2P_{1/2}$.

(iii) For ^3D term, $L = 2$, $S = 1$. So, $J = 3, 2, 1$. The levels are $^3D_3$, $^3D_2$, $^3D_1$.

Parity

A spectroscopic term symbol sometimes contains an additional information about *parity*. The term parity in general refers to how a wave function changes when all the spatial coordinates undergo an inversion in space (for example we replace x, y, z by $-x$, $-y$ and $-z$ respectively).

$$\psi(x, y, z) = \psi(-x, -y, -z): \text{ even parity}$$

$$\psi(x, y, z) = -\psi(-x, -y, -z): \text{ odd parity}$$

We may introduce a quantum number P which may have values +1 for even parity and –1 for odd parity:

$$\psi(x, y, z) = P\psi(-x, -y, -z).$$

For $y = \cos x$, $f(x) = +f(-x)$ $P = +1$; for $y = \sin x$, $f(x) = -f(-x)$ $P = -1$

Since atoms are symmetrical, its associated wave functions must either remain unchanged or change signs under inversion. Orbitals of odd l value are said to have odd parity (p, f etc). An odd number of p or f electrons thus leads to an odd parity. An even number of p and f electrons leads to an even parity of the term. Thus we simply add the l values of the electrons to see whether a function is odd or even.

For even parity (P = +1), no indication appears on the term. Odd parity (P = –1) is sometimes indicated by writing '0' as a right superscript in the term. Thus, the term symbol for Cl (ground state) is written $^2P^o_{3/2}$.

Selection rules for spectral transitions

Not all possible spectral transitions between different states of an atom are actually observed. At first some empirical selection rules were proposed to account for such "unseen" transitions which were declared forbidden. Subsequently, these and other related selection rules have been largely rationalized from various theoretical approaches.

A photon has a spin of unity. For left and right circularly polarized light, m_s = +1 and –1 respectively. When an atom absorbs a photon, the angular momentum of the photon is transferred to the electrons of the atom; when a photon is emitted in a transition, its angular momentum must come from a change in angular momentum of the electron undergoing transition. Summarily, interaction of the electric vector of the radiation with instantaneous electric dipoles in the atom (or molecule) results in oscillating electric field component of the light, giving rise to *electric dipole transitions*. Oscillating magnetic field components of light will similarly give rise to magnetic dipole transitions which are usually much weaker. Since angular momentum has to be conserved in emission or absorption, Δl should be equal to ±1 for the atom.

The selection rules set the criterion for allowed transitions. The criterion referred to in the selection rules may vary in different cases. For the simplest one electron species like hydrogenlike atoms, the selection rules summarily state that for allowed transitions

$$\Delta l = \pm 1 \text{ and } \Delta m_l = 0, \pm 1.$$

There is no restriction on Δn (the principal quantum number).

Thus a d electron ($l = 2$) can make radiative transition to an orbital with $l = 1$ (p-orbital or $l = 3$ (f-orbital). It cannot undergo a transition to an s-orbital ($l = 0$) or another d-orbital ($l = 2$). Similarly, a $4s$ electron can make radiative transition to np orbitals only.

We note that an allowed transition is accompanied by a change in parity. The parity of a polyelectronic atom is even if Σl_i is even over all electrons; parity is odd if Σl_i is odd. When only one electron is promoted from the ground configurations, $\Delta l = \pm 1$ results in a change in parity.

For many electron atoms, the states of the atom, and spectral transitions are stated using term symbols. However, the selection rules are applicable when Russel-Saunders coupling is valid. The following criterion are used to identify allowed transitions:

i. $\Delta S = 0,$

ii. $\Delta L = 0, \pm 1$ but $L = 0 \not\leftrightarrow L = 0$ (*not allowed*)

iii. $\Delta J = 0, \pm 1$ but $J = 0 \not\leftrightarrow J = 0$ (*not allowed*)

iv. $\Delta l = \pm 1.$

Illustration: Let us consider the *principal series* of sodium corresponding to electronic transitions from various p-excited states to the $3s$ ground electronic state.

Ground state ($3s^1$) : $l = 0$, $s = \frac{1}{2}$ give $L = 0$, $S = \frac{1}{2}$ and $J = \frac{1}{2}$. Term symbol : $^2S_{1/2}$

Excited states (np^1) : $l = +1$, $s = \frac{1}{2}$ give $L = 1$, $S = \frac{1}{2}$ and $J = \left(1+\frac{1}{2}\right), \left(1+\frac{1}{2}-1\right) = \frac{3}{2}, \frac{1}{2}$

Term symbols : $^2P_{3/2}$, $^2P_{1/2}$ ($^2P_{1/2}$ will be lower in energy).

The electronic spectra should consist of emission lines for transitions between the $2P$ states and the ^2S state. We consider the selection rules one by one: $\Delta S = 0$; $\Delta L = \pm 1$ ($P \longleftrightarrow S$) ; $\Delta J = \pm 1$ ($\frac{3}{2} \longleftrightarrow \frac{1}{2}$) ; 0 ($\frac{1}{2} \longleftrightarrow \frac{1}{2}$); $\Delta l = \pm 1$. Thus all the selection rules are satisfied.

The golden-yellow "D-lines" of sodium around 589 nm (~17000 cm^{-1}) are actually doublets at 589.16 nm and 589.76 nm. The difference of 0.6 nm (17 cm^{-1}) corresponds to splitting between the $^2P_{3/2}$ and $^2P_{1/2}$ states.

When the sodium atom (or any atom with $S = 0$) is placed in a magnetic field parallel to the z-axis assigned to the atom, the multiplets will be split into $2L + 1$ components according to the M_L values: $M_L = \Sigma m_i = L, L - 1, - L$ (in this case +1, 0, −1). The allowed transition will now obey the additional rule

$$\Delta M_L = 0, +1.$$

This is observed in normal Zeeman effect (section 4.3.4).

The Zeemann effect becomes complicated for atoms with $S \neq 0$.

Problem 4.20

Do you expect any transition between the $3s$ and $4s$ atomic orbitals of sodium?

Hint: The transition in question is $^2S_{1/2} \longleftrightarrow {}^2S_{1/2}$. As per selection rule (ii), $L = 0 \leftrightarrow L = 0$ transitions are not allowed.

The applicability of the selection rules depends largely on the coupling of spin and orbital angular momenta of the electrons. The L–S coupling scheme or the Russel-Saunders term symbols are useful guides for lighter atoms. For heavier atoms, spin-orbit interaction of individual electrons become more dominating (j–j coupling) and we find breakdown of the above selection rules. For example, transitions between singlet and triplet states ($\Delta S = \pm 1$) gain prominence in heavier atoms. Such "forbidden transitions" usually show weaker intensity.

The relative intensity of allowed transitions depend upon ΔL and ΔJ:

Most intense :	$\Delta L = -1, \Delta J = -1$	*Very weak* :	$\Delta L = +1, \Delta J = 0$
Less intense :	$\Delta L = -1, \Delta J = 0$	(*Forbidden* :	$\Delta L = +1, \Delta J = -1$
Weak :	$\Delta L = +1, \Delta J = +1$		$\Delta L = 1, \Delta J = +1$)

ESSENTIALS

Wave nature of the electron (Louis de Broglie; France; 1924). Moving material particles are also associated with characteristic waves of wavelength given by $\lambda = h/mv$. The wavelength is insignificant for large objects but is important for micro objects like the electron. Diffraction of a beam of electrons (Davisson and Germer; USA; 1927) confirmed the wave nature of electrons.

Quantum mechanics (Born, Heisenberg, Jordan; Germany; 1925) was developed as an alternative to the classical Newtonian mechanics and was based on the idea of matter waves.

The uncertainty principle (W. Heisenberg; Germany; 1927). There is a limit to the precision upto which the position and momentum of a particle can be determined simultaneously : $\Delta x \, \Delta p \geq h$ (approx.) where Δx and Δp represent the uncertainties in position and momentum respectively in x-direction. In terms of energy and time, this becomes, $\Delta E \Delta t \geq h$.

Wave mechanics (Schrödinger; Austria; 1926): The behaviour of the electron in an atom of hydrogen was described by comparison with standing waves through a fundamental wave equation named after Schrödinger. In this equation, the amplitude ψ of the wave associated with the electron is related to its kinetic and potential energy. ψ is a function of the space coordinates of a point around which the probability of finding the electron in a volume element dv is given by $\psi^2 dv$. The physically meaningful solutions for ψ represent the orbitals of an electron which correspond to the permissible energy states for it in the atom.

For mathematical convenience the Schrödinger equation for H atom is written with the space coordinates expressed in the polar system using r, θ and ϕ. The total wave function can now be resolved into three separate functions depending exclusively on r, θ and ϕ. The wave function is written $\psi(r, \theta, \phi) = R(r)\, Y(\theta, \phi)$.

The quantum numbers n, l and m_l appear now as a logical consequence from the solution of the wave equation. Their role in specifying the state of electron remain the same as used earlier in spectral interpretation.

The idea of spin motion of the electron (Uhlenbeck and Goudsmit; Netherlands; 1925) and an intrinsic magnetic moment was forwarded to explain certain fine structure of atomic spectra. This led to a new quantum number for spin (s and m_s). This could be harmonized with wave mechanics through relativistic quantum mechanics.

For many-electron atoms, the Schrödinger wave equation cannot be solved exactly and approximations have to be done. Slater's rules provide one such approximation for modifying the nuclear charge (Ze) of an atom by other electrons present.

The Aufbau principle consists of gradual building up of the electron shells of heavier atoms from hydrogen by successive addition of electrons to the lowest available orbital. Two major guidelines in this respect are the Pauli exclusion principle (W. Pauli; 1925) and Hund's rule of maximum spin multiplicity.

The Pauli exclusion principle states that no two electrons in any atom may have the same values for all four quantum numbers n, l, m_l and m_s.

The Pauli principle **states that when the labels of any two identical fermions (half-integral spin) are exchanged, the total wave function changes sign. (For any two identical bosons (integral spin), the sign remains same under exchange).**

The energy state of an atom is represented by the term symbol ^{2S+1}L ; $2S + 1$ is called the spin multiplicity of the state ($S = \Sigma m_s$). L ($= \Sigma m_l$) is represented by letter codes: S for $L = 0$; P for $L = 1$ etc.

EXERCISE

1. (a) What will be the de Broglie wavelength of a body of mass 1 mg whose speed is 100 m s^{-1}?

(b) How many waves are completed by an electron in an orbit of $n = 3$ in the hydrogen atom?

2. What will be the total energy of a free electron whose de Broglie wavelength is 10^{-10} m?

3. In an He$^+$ ion, an electron moves in an orbit having circumference of four times the de Broglie wavelength of the electron. Find the energy of the electron in the atomic unit.

[**Hint:** $E = -Z^2/2n^2$ a.u. Here $n = 4$.]

4. (a) An electron is brought closer to the nucleus in an atom. Will its de Broglie wavelength increase, decrease or remain the same?

(b) Show that the de Broglie wavelength of an electron in the first Bohr orbit of hydrogen is $2\pi a_0$.

5. What is the speed of a particle whose de Broglie wavelength is the same as its Compton wavelength?

[**Hint:** $h/m_0c = h/mv$; $m = m_0 / \sqrt{1 - v^2/c^2}$.] [**Ans.** $c/\sqrt{2}$.]

6. Compare the wavelengths of an electron and a neutron each with kinetic energy of 1 MeV.

7. Compare the uncertainties in the velocities of an electron and a proton in an 1 Å box.

8. Suppose the uncertainty in the location of a particle is equal to its de Broglie wavelength. Show that the minimum uncertainty in its velocity is nearly 8% of its velocity.

9. Express the uncertainty principle in terms of energy and time. Show that there is no uncertainty in the energy of an atom in the stationary state.

10. Calculate, using Slater's rules, the effective nuclear charge on (i) a 4s electron and (ii) a 3d electron in the bromine atom.

11. State the uncertainty principle and explain situations in which it becomes insignificant.

12. Write Schrödinger wave equation in Cartesian and polar coordinates and explain the terms. What are eigenfunctions? What are their characteristics?

13. What do you mean by radial and angular parts of the wavefunction for the electron in the hydrogen atom?

Show the rough shape of the radial functions for 1s, 2s, and 3s atomic orbitals in hydrogen.

14. How many radial and angular nodes are present in the 3p orbital (hydrogenic)?

15. What is radial density as applied to an atomic system? Explain, with rough sketches, the plots of radial density versus radial distance for the 3s, 3p and 3d orbitals in a hydrogen atom.

16. $\psi(r)$ for the 1s orbital in a hydrogen atom has a very high value at $r = 0$. But the plot of radial density *vs* r passes through zero at $r = 0$ and has a maximum at 0.53 Å. Explain.

17. The energylevels of the *pi*-electrons in a conjugated hydrocarbon chain may be roughly obtained by analogy with those of an electron in a one dimensional box. Calculate the energy required for excitation of butadiene to the first excited state in terms of the chain length l_b (in Å). How will this energy change when butadiene is dimerized to octatetraene?

[**Hint:** In a box of length l(Å), the energy levels are given by $E_n = n^2h^2/8ml^2 = 37.7n^2/l^2$ eV. Butadiene has four pi-electrons, so the highest occupied energy level has $n = 2$. First excitation energy corresponds to the separation between $n = 3$ and $n = 2$: ΔE(eV) $= 37.6\ (3^2 - 2^2)/l^2_b$].
In case of octatetraene, $l = 2l_b$ and highest occupied level has $n = 4$ (eight pi-electrons). Find the corresponding frequencies.]

18. What are quantum numbers?
Which quantum number is used to express the orbital angular momentum of the electron?
How do we represent space quantization of the orbital angular momentum vector of an electron?

19. "The orbital angular momentum of an electron does not give its total angular momentum"—explain. Which quantum number is used to supplement the orbital angular momentum?

20. (a) What is the value of the orbital angular momentum (L) of a $3p$ electron in a hydrogenlike atom?
(b) What will be the values of L_z?

(a) $1{\cdot}49 \times 10^{-34}$ J s (b) $\pm 1{\cdot}054 \times 10^{-34}$ J s.

21. (a) What is the value(s) of L and L_Z of a $3d$ electron in a hydrogenlike atom?
(b) What will be the relative values of the possible energy levels when the atom is placed in a magnetic field of 1T (along the Z-axis)?

Hint: $L = \sqrt{6}\hbar$; $L_Z = -2\hbar, -\hbar, 0, \hbar, 2\hbar$
$E = -\mu_B m_l B$ ($m_l = 0, \pm 1, \pm 2$; B = 1T).
[See section 4.3.4: magnetic moment due to orbital motion of the electron.]

22. The $3s$ and $3p$ orbitals have identical energies in the hydrogen atom, but in the chlorine atom their energies are much different. Explain with the idea of penetration and shielding. Use rough skteches.

23. Comment on the following ground state electron configurations

K	:	[Ar] $4s^1$	Cr	:	[Ar] $3d^5 4s^1$
Ni	:	[Ar] $3d^8 4s^2$	Cu	:	[Ar] $3d^{10} 4s^1$

24. Which of the following electronic transitions are allowed for a hydrogenlike atom?
(a) $1s \longleftrightarrow 2s$ (b) $1s \longleftrightarrow 2p$ (c) $2s \longleftrightarrow 3d$ (d) $3p \longleftrightarrow 3d$.

5

PERIODIC PROPERTIES OF ATOMS

5.1. INTRODUCTION

Chemistry involves study of the elements—now more than a hundred in number, and their compounds which are beyond counting. As in all other fields of science, in this case also, generalization of facts stands essential for a systematic study of the subject. Fortunately, the observed chemical analogy of the elements initiated a sort of classification for them; it was observed that the properties of elements repeat themselves after a certain interval (though variable) as one proceeds from the simplest, lightest element hydrogen to the last and heaviest element known. This periodicity in properties, expressed in different ways, gave rise to a number of approaches to classify the elements—each with an attempt to study the essentials of chemistry in a very compact and readily accessible form. The underlying reason for this periodicity of properties has also been explored; it is now known that elements having similar patterns of electron configuration in their outermost shell resemble each other closely. In this chapter, we shall survey the main systems of periodic classification, the basis for such classification, as well as the principal properties of atoms which exhibit periodicity.

Among the many approaches toward a convenient periodic classification of the elements, the independent contributions by Lothar Meyer and Dimitri Mendeleev deserve special mention. The table presented by Mendeleev (1869) has proved to be the most useful one. It was a very long and thorough search in the jungle of chemical facts together with an almost prophetic insight into the properties of elements that enabled Mendeleev to successfully convert the jungle into a zoo. At about the same time, Lothar Meyer independently made the remarkable observation that "atomic volumes" (= gram atomic weight/density) of different elements presented a periodic pattern (section 5.3.3).

Mendeleev based his periodic table (Table 5.1) on the periodic law: *Physical and chemical properties of elements are periodic functions of their atomic weights.* He arranged the elements into eight main groups according to their chemical similarities; each group was again subdivided into two subgroups. However, it was later observed that the arrangement of his table is best understood in the sequence of atomic numbers (Z) of the elements.

The periodic table initiated by Mendeleev made the first successful and comprehensive classification of the elements and their compounds. In spite of a few minor drawbacks, the table has served the purpose of the only road map in the vast land of chemistry for nearly a century. At present it has been more rationally extended in terms of electron configurations of the atoms. But its simple and direct correlation of basic chemical behaviour and main valence states of an element have found no parallel.

Some trivial names for different groups of elements in the periodic table

Alkali metals	Li, Na, K, Rb, Cs, Fr
Alkaline earth metals	Ca, Sr, Ba, Ra
Pnictogens (*pnictides*)	N, P, As, Sb, Bi
Chalcogens (pronounce *kalkojens*)	O, S, Se, Te, Po
Halogens	F, Cl, Br, I, At
Noble gases (*inert gases or rare gases*)	He, Ne, Ar, Kr, Xe, Rn
Rare earth elements	Sc, Y, La,—Lu
Lanthanoids (*Lanthanides*)	Ce—Lu
Actinium series	Ac—Lw
Actinoids (*actinides*)	Th—Lw
Transuranium elements	Elements beyond U
Coinage metals	Cu, Ag, Au
Platinum metals	Ru, Rh, Pd, Os, Ir, Pt
Noble metals	Platinum metals and Au (sometimes also Ag, Re or Hg).

Table 5.1

Modern version of Mendeleev's periodic table

(only atomic numbers of elements shown)

Group / Period	0	I A	I B	II A	II B	III A	III B	IV A	IV B	V A	V B	VI A	VI B	VII A	VII B	VIII
1		H 1														
2	He 2	Li 3		Be 4			B 5		C 6		N 7		O 8		F 9	
3	Ne 10	Na 11		Mg 12			Al 13		Si 14		P 15		S 16		Cl 17	
4	Ar 18	K 19	Cu 29	Ca 29	Zn 30	Sc 21	Ga 31	Ti 22	Ge 32	V 23	As 33	Cr 24	Se 34	Mn 25	Br 35	Fe 26, Co 27, Ni 28
5	Kr 36	Rb 37	Ag 47	Sr 38	Cd 48	Y 39	In 49	Zr 40	Sn 50	Nb 41	Sb 51	Mo 42	Te 52	Tc 43	I 53	Ru 44, Rh 45, Pd 46
6	Xe 54	Cs 55	Au 79	Ba 56	Hg 50	La* 57–71	Tl 81	Hf 72	Pb 82	Ta 73	Bi 83	W 74	Po 84	Re 75	At 85	Os 76, Ir 77, Pt 78
7	Rn 86	Fr 87		Ra 88			Ac† 89–103	Rf 104		Db 105		Sg 106		Bh 107		Hs 108, Mt 109

THE RARE EARTHS and ACTINOIDS

*Lanthanoid series	Ce 58	Pr 59	Nd 60	Pm 61	Sm 62	Eu 63	Gd 64	Tb 65	Dy 66	Ho 67	Er 68	Tm 69	Yb 70	Lu 71
†Actinoid series	Th 90	Pa 91	U 92	Np 93	Pu 94	Am 95	Cm 96	Bk 97	Cf 98	Es 99	Fm 100	Mv 101	No 102	Lr 103

5.2 CLASSIFICATION BASED ON ELECTRON CONFIGURATION

Many attempts have been made to provide a better and systematic classification of the elements. But they mostly lack the simplicity of the Mendeleev classification (modified version). When the extranuclear structure of the atom was established, it was natural to attempt at classification of the elements on the basis of electron configuration of their atoms. One such arrangement (after Longuet-Higgins) is shown in Table 5.2.

Table 5.2

A periodic table based on electron configuration

PERIODS	Subshell	Elements	Configuration
7	6d	Rf Db Sg Bh Hs Mt Ds Rg	$6d^2$ to $6d^9$
	5f	Th Pa U Np Pu Am Cm Bk Cf Es Fm Md No Lr	$5f^1$ to $5f^{14}$
	6d	Ac	$6d^1$
	7s	Fr Ra	$7s^1$ $7s^2$
6	6p	Tl Pb Bi Po At Rn	$6p^1$ to $6p^6$
	5d	Hf Ta W Re Os Ir Pt Au Hg	$5d^2$ to $5d^{10}$
	4f	Ce Pr Nd Pm Sm Eu Gd Tb Dy Ho Er Tm Yb Lu	$4f^1$ to $4f^{14}$
	5d	La	$5d^1$
	6s	Cs Ba	$6s^1$ $6s^2$
5	5p	In Sn Sb Te I Xe	$5p^1$ … $5p^6$
	4d	Y Zr Nb Mo Tc Ru Rh Pd Ag Cd	$4d^1$ … $4d^{10}$
	5s	Rb Sr	$5s^1$ $5s^2$
4	4p	Ga Ge As Se Br Kr	$4p^1$ … $4p^6$
	3d	Sc Ti V Cr Mn Fe Co Ni Cu Zn	$3d^1$ … $3d^{10}$
	4s	K Ca	$4s^1$ $4s^2$
3	3p	Al Si P S Cl Ar	$3p^1$ … $3p^6$
	3s	Na Mg	$3s^1$ $3s^2$
2	2p	B C N O F Ne	$2p^1$ … $2p^6$
	2s	Li Be	$2s^1$ $2s^2$
1	1s	H He	$1s^1$ $1s^2$

Table 5.3
Extended form of the periodic table

s-block | *d*-block | *p*-block

Groups[1]	IA	IIA	IIIB / IIIA	IVB / IVA	VB / VA	VIB / VIA	VIIB / VIIA	VIIIB / VIII			IB	IIB	IIIA / IIIB	IVA / IVB	VA / VB	VIA / VIB	VIIA / VIIB	VIIIA / 0
IUPAC	**1**	**2**	**3**	**4**	**5**	**6**	**7**	**8**	**9**	**10**	**11**	**12**	**13**	**14**	**15**	**16**	**17**	**18**
Periods 1	^{1}H 1.008																	^{2}He 4.00
2	^{3}Li 6.94	^{4}Be 9.01											^{5}B 10.81	^{6}C 12.01	^{7}N 14.01	^{8}O 16.00	^{9}F 19.00	^{10}Ne 20.18
3	^{11}Na 22.99	^{12}Mg 24.31											^{13}Al 26.98	^{14}Si 28.09	^{15}P 30.97	^{16}S 32.06	^{17}Cl 35.45	^{18}Ar 39.95
4	^{19}K 39.10	^{20}Ca 40.08	^{21}Sc 44.96	^{22}Ti 47.87	^{23}V 50.94	^{24}Cr 52.00	^{25}Mn 54.94	^{26}Fe 55.84	^{27}Co 58.93	^{28}Ni 58.69	^{29}Cu 63.55	^{30}Zn 65.41	^{31}Ga 69.72	^{32}Ge 72.64	^{33}As 74.92	^{34}Se 78.96	^{35}Br 79.90	^{36}Kr 83.80
5	^{37}Rb 85.47	^{38}Sr 87.62	^{39}Y 88.91	^{40}Zr 91.22	^{41}Nb 92.91	^{42}Mo 95.94	^{43}Tc 98.00	^{44}Ru 101.07	^{45}Rh 102.90	^{46}Pd 106.42	^{47}Ag 107.87	^{48}Cd 112.41	^{49}In 114.82	^{50}Sn 118.71	^{51}Sb 121.76	^{52}Te 127.60	^{53}I 126.90	^{54}Xe 131.29
6	^{55}Cs 132.91	^{56}Ba 137.33	^{57}La* 138.91	^{72}Hf 178.49	^{73}Ta 180.95	^{74}W 183.84	^{75}Re 186.21	^{76}Os 190.23	^{77}Ir 192.22	^{78}Pt 195.08	^{79}Au 196.97	^{80}Hg 200.59	^{81}Tl 204.38	^{82}Pb 207.2	^{83}Bi 208.98	^{84}Po 209	^{85}At 210	^{86}Rn 222
7	^{87}Fr (223)	^{88}Ra (226)	^{89}Ac† (227)	^{104}Rf (261)	^{105}Db (262)	^{106}Sg (263)	^{107}Bh (262)	^{108}Hs (265)	^{109}Mt (266)	^{110}Ds (271)	^{111}Rg (272)	^{112}Uub	^{113}Uut	^{114}Uuq	^{115}Uup	^{116}Uuh	^{117}Uus	^{118}Uuo

Atomic number → 1
Symbol → H
Atomic weight → 1·008
(^{12}C = 12·000)

f-block

Lanthanoids* (Lanthanides)	^{58}Ce 140.12	^{59}Pr 140.91	^{60}Nd 144.24	^{61}Pm 145	^{62}Sm 150.36	^{63}Eu 151.96	^{64}Gd 157.25	^{65}Tb 158.93	^{66}Dy 162.50	^{67}Ho 164.93	^{68}Er 167.26	^{69}Tm 168.93	^{70}Yb 173.04	^{71}Lu 174.97
Actinoids† (Actinides)	^{90}Th 232.04	^{91}Pa 231.04	^{92}U 238.03	^{93}Np (237)	^{94}Pu (244)	^{95}Am (243)	^{96}Cm (247)	^{97}Bk (247)	^{98}Cf (251)	^{99}Es (252)	^{100}Fm (257)	^{101}Md (258)	^{102}No (259)	^{103}Lr (262)

1. The IUPAC recommendation abolished A and B subgroups and numbered the groups from 1—18

A modern version of the long periodic table consists of eighteen groups or columns (Table 5.3). These are obtained by separating the subgroups A and B of Mendeleev's table; the group VIII consists of three columns. The speciality of the table lies in the placement of the transition elements: each of them has been assigned a separate column in the middle of the table; Groups 1 and 2 are placed to the left of these transition elements. The remaining groups lie on the right hand side of the table. The inert gases, previously appearing in Group 0, are placed at the extreme right under group 18.

The lanthanoid and actinoids, are, however, placed separately at the bottom of the table, as was done in the Mendeleev's table. These elements could also be included in the main table but that would make it excessively long.

The arrangement of the long periodic table may be very easily understood from the aufbau principle. Each period starts with an element having its last electron in an *ns* orbital; n = 1 for H, n = 2 for Li and so on. The period ends with an element whose last electron completes ns^2np^6 or ns^2 $(n-1)$ $d^{1-10}np^6$ configuration.

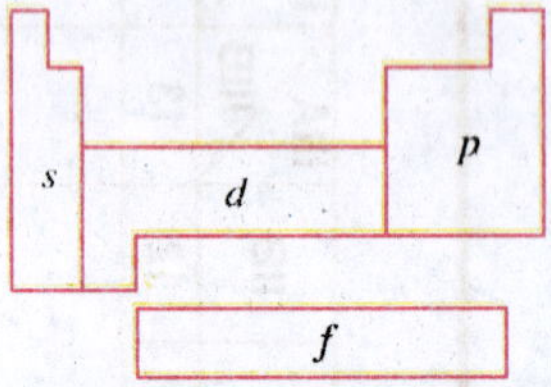

Fig.5.1: Four regions in the periodic table.

With this background, the long form of the periodic table may be separated into four distinct regions. Each region corresponds to filling of the *s*-, *p*-, *d*- or *f*- sublevels, Accordingly the regions may be termed *s*- *p*-, *d*- or *f*- blocks (Fig. 5.1).

The eight groups of elements belonging to the *s*- and *p*- blocks are often called the *representative elements*; those in the *d*-block are called the *transition elements*, and the members of the *f*-block are termed the *inner-transition elements*. However, Zn, Cd, Hg and some other *f*-block elements do not show the characteristics of transition elements.

A and B subgroups. This classification continued from the Mendeleev table. In groups I and II, the main group elements (Li and Be) were classified into A-subgroup; the transition elements (Cu and Zn groups) were classified into B-subgroup. But the classification is not clear. In the more common usage, the main groups are termed B, but the reverse system is also followed. Thus Group VB might mean the nitrogen group in some books and the vanadium group of transition elements in some.

An attempt was made to replace the notations 'A' or 'B' by M (main) or T (transition). But this use was limited to English language only and has not been accepted universally.

The 18-group system has now been recommended by the IUPAC. The groups are numbered serially from left to right (in the extended form) by numerals 1, 2, 3, ... 18. Though unambiguous, the system lacks the simple relation between the maximum valency of an element with its group number. One easy way to correlate the new group numbers in the *p*-block with earlier group numbers is to drop the '1' in 13, 14, 15, ... etc. Thus group 13 is previous group III.

Prediction of missing elements

Chemists have not satisfied themselves merely with the study of the available elements and their compounds. Their attempts to systematize the chemistry have always inspired them to predict missing elements which would fill certain gaps in their generalization. The monumental work of Mendeleev in this connection is already known to us. The first credit of correctly predicting a missing element, however, seems to go to Newlands, who predicted (1864) a missing element of atomic weight 73 between silicon and tin (germanium). In the following eighty years, a large number of "missing" elements were predicted and discovered. The discovery of the noble gases (1894-98) is remarkable as their existence was not "predicted" from the periodic table of Mendeleev. Moseley's work on the characteristic x-rays of elements and the determination of their atomic number (1913) provided a direct method of ascertaining the actual number of missing elements between hydrogen and uranium. A significantly new step in the prediction of missing elements was taken by Seaborg in 1944, (eighty years after Newlands and seventyfive years after Mendeleev) when he forwarded his actinide hypothesis and predicted the existence of 14 elements upto $Z = 103$.

Some important achievements concerning the prediction and discovery of elements upto this point have been summarized in Table 5.4.

Table 5.4

***Major events in the prediction, discovery and synthesis of elements* (upto $Z = 103$)**

1864	Prediction of a missing element between silicon and tin. Estimated atomic weight = 73	Newlands
1869–71	Prediction of Ge, Ga, Sc, Tc, Re, Po, Fr, Ra, Ac, Pa. Surprisingly correct predictions for Ge, Ga, Sc and their compounds.	Mendeleev
1894–98	Noble gases: (He), Ne, Ar, Kr and Xe discovered.	Lord Rayleigh, W. Ramsay, M. W. Travers.
1898	Discovery of Po, then Ra.	Marie Curie Pierre Curie
1899	Actinium discovered	Debierne
1900	Radon discovered	Dorn
1913	Explanation of the periodic table on the basis of of atomic structure	Bohr
	Work on characteristic x-rays and atomic number.	Moseley
1917–45	Discovery of Pa (1917), Hf (1923), Re (1925), Tc (1937), Fr (1939), and Pm (1945).	Hahn, Meitner, Coster, Hevesy, Noddack, Segre and others
1940	Synthesis of the first trans-uranium element ^{93}Np.	McMillan and Abelson
1944	The actinide hypothesis–prediction of 14 elements upto $Z = 103$.	G. T. Seaborg

The trans-actinoid elements

The last actinoid element with $Z = 103$ (lawrencium, Lr*) was synthesized in 1961. Attempts to synthesize elements beyond lawrencium continued and elements 104 and 105 were reported by two groups of workers—A. Ghiorso (and others) in Berkeley, California and G. N. Flerov (and others) in Dubna, near Moscow (1970). Priority in these discoveries was disputed, so also were the names of these elements. Elements 106–109 were synthesized between 1972-1984. Element 106 was detected by both the American and Russian groups of scientists (1972), while element 107 (1981) was claimed singly by the Russians. Synthesis of element 108 (1982) was claimed on the basis of three decay events from long-term bombardment of Pb with ^{58}Fe: an isotope of mass number 265 has a half-life of 3 minutes. Element 109 was similarly claimed (1984) from a single decay event after bombarding a bismuth target with ^{58}Fe for three weeks. An isotope of mass number 226 has a half-life of about 5 minutes.

Controversies on naming a newly synthesized element began since element 102. An international team of scientists working at the Nobel Institute for Physics in Stockholm claimed to have synthesized the element in 1957. But their results were not reproducible. The same synthesis was reported to have been carried out in 1958 by Ghiorso and others. The earlier name Nobelium was however retained. But Soviet workers later claimed that this element was rightly characterized by them (1963-66). They suggested the parallel name joliotium (in honour of F. Joliot-Curie). The wave of priority controversy and disputed naming continued to the elements discovered later.

The IUPAC (1977) suggested a new method of naming these elements from atomic number. The digits appearing in the atomic number of an element are serially expressed using the following numerical roots:

0	1	2	3	4	5	6	7	8	9
nil	un	bi	tri	quad	pent	hex	sept	oct	enn

The roots are put together and the name is completed by adding *ium* (irrespective of whether the element is expected to be a metal or a nonmetal). To avoid repetition, the final '*i*' of *bi* and *tri* are dropped when they appear before *ium*; similarly, the final '*n*' of *enn* is dropped when it occurs before *nil*.

The symbol of an element is derived from its name using the initial letters of the numerical roots in the proper sequence, as shown in Table 5.5 for elements 111–118. The trans-actinoid elements ($Z = 104$-118) are shown in Table 5.5. Names of elements up to $Z = 110$ have been finalized by the IUPAC. Elements up to $Z = 118$ have been synthesized till January 2010, mostly using accelerated ^{48}Ca as projectile against the appropriate actinoid nuclide. Element 114 was first synthesized in 1998 followed by 116 and 118 (2000–2002). Elements 113 and 115 were synthesized between 2003–2004. Element 117 is the last one to have been synthesized in January 2010. These elements formally complete the seventh period ending at element 118 (*eka*-radon).

*Formerly Lw; Lr was adopted by the IUPAC in 1965.

Table 5.5
Elements with Z = 104 to 118

Z (Year)	Name, Symbol (previous names)	Suggested electron config.
104 (1969)	Rutherfordium, **Rf** (*Kurchatovium, Dubnium*)	[Rn] $5f^{14}6d^27s^2$
105 (1970)	Dubnium, **Db** (*Nielsbohrium, Hahnium, Joliotium*)	[Rn] $5f^{14}6d^37s^2$
106 (1974)	Seaborgium, **Sg** (*Rutherfordium*)	[Rn] $5f^{14}6d^47s^2$
107 (1981)	Bohrium, **Bh** (*Nielsbohrium*)	[Rn] $5f^{14}6d^57s^2$
108 (1984)	Hassium, **Hs** (*Hahnium*)	[Rn] $5f^{14}6d^67s^2$
109 (1982)	Meitnerium, **Mt**	[Rn] $5f^{14}6d^77s^2$
110 (1994)	Darmstadtium, **Ds**	[Rn] $5f^{14}6d^87s^2$
111 (1994)	Roentgenium, **Rg**	[Rn] $5f^{14}6d^97s^2$
112 (1996)	Ununbium, **Uub**	[Rn] $5f^{14}6d^{10}7s^2$
113 (2003)	Ununtrium, **Uut**	[Rn] $5f^{14}6d^{10}7s^27p^1$
114 (1998)	Ununquadium, **Uuq**	[Rn] $5f^{14}6d^{10}7s^27p^2$
115 (2004)	Ununpentium, **Uup**	[Rn] $5f^{14}6d^{10}7s^27p^3$
116 (2000)	Ununhexium, **Uuh**	[Rn] $5f^{14}6d^{10}7s^27p^4$
117 (2010)	Ununseptium, **Uus**	[Rn] $5f^{14}6d^{10}7s^27p^5$
118 (2002)	Ununoctium, **Uuo**	[Rn] $5f^{14}6d^{10}7s^27p^6$

5.3 PERIODIC PROPERTIES OF THE ELEMENTS

Many chemical and physical properties of the elements vary periodically with their atomic number. In most cases, the general nature of this variation can be followed from the electronic configuration of atoms. Appreciation of this trend helps us to a great extent to correlate, remember and predict the chemistry of individual elements. A few selected properties of atoms like melting and boiling points, size, ionization energy etc. are discussed here.

5.3.1 Melting and Boiling Points

A solid melts when the force holding its constituent units in position is overcome by thermal energy gained from an increase in temperature. The melting temperature is determined by a number of different factors: nature of packing of the atoms, ions, or molecules, the lattice energy, association through hydrogen bond, etc. Boiling involves complete separation of the constituents of a liquid. Melting and boiling points of an element thus indicate the nature and extent of binding between the atoms of that element. A better understanding is obtained from the standard enthalpy of atomization values. These also show a periodic trend, as revealed in Table 5.6. In spite of the contribution of a large number of variable factors in controlling melting and boiling points, a striking periodicity is observed in their values. Fig. 5.2 represents a plot of melting points of some elements against their respective atomic numbers. Melting points of the individual elements are given in Table 5.6. The general shape of the plot in Fig. 5.2 indicates periodic variation. It may be observed that—

(i) noble gases have very low melting points;

(ii) the melting point of elements in any period increases gradually to Group IV(14) where it becomes maximum; it then decreases to the next noble gas;

(iii) the melting points of elements occurring in a group in the *s*- and *p*- block usually decrease downwards, as one moves to higher atomic numbers.

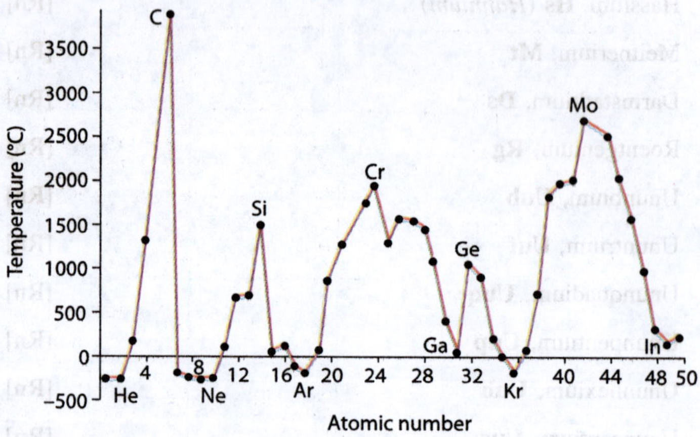

Fig.5.2: Periodic variation of melting point of elements.

Explanation

We try to explain these observations in a simple manner with all the limitations of our complete knowledge regarding the various contributing factors. (i) The noble gas atoms contain a completed electron shell. So there exists little tendency for these atoms to combine between themselves. This is why they occupy the lowest range of melting points. (ii) The alkali metals have only one electron in their outermost shells. Their contribution to metallic bonding is, therefore, weak (see metallic bond), resulting in low melting points. (iii) As the number of electrons participating in metallic bonding increases, the melting point also increases: Be > Li; Mg > Na. (iv) Boron, carbon and silicon form giant three-dimensional structures; breaking of these arrangements require large amount of energy. Hence they have very high melting points. (v) Nitrogen, oxygen, fluorine—all form stable diatomic molecules containing triple, double and single bonds respectively. They do not form giant three-dimensional lattices—may be due to greater repulsion between the non-bonding electrons and the greater stabilities of these multiple bonds. At the same time, very poor binding exists among the molecules themselves. Consequently they have very low melting points. (vi) As we proceed downward along a group, the elements involve increasingly higher *s*-orbital for binding their atoms. Poorer bonding,

Table 5.6

Density, enthalpy of atomization, MP and BP of some elements

Line 1: Density of the element, g cm^{-3}. ($\times 10^3$ kg m^{-3})

Line 2: Enthalpy of atomization, ΔH°_{atom}, kJ mol^{-1}

Line 3: Melting point °C

Line 4: Boiling point °C

Element	ρ	ΔH°_{atom}	MP	BP
Li	0.54	160	180.6	1326
Be	1.85	324	1289	2472
B	2.34	563	2300	3658
C	2.27[1]	717	3547(d)	4827 sub
N	—	473	−210	−195.8
O	—	247	−229	−183
F	—	79	−219.6	−188.1
Na	0.97	108	97.8	883
Mg	1.74	146	651	1090
Al	2.70	326	660.5	2327
Si	2.33	456	1410	~2355
P	1.82[6]	315	44.1[α]	280.5
S[2]	2.1	278	112.8	444
Cl	—	121.7	−101	−34
K	0.86	90	63.7	756
Ca	1.55	178	843	1484
Sc	3.0	390	1541	2831
Ti	4.5	469	1660	3287
V	6.11	514	1887	3377
Cr	7.14	397	1857	2672
Mn	7.43	281	1244	1962
Fe	7.87	416	1535	2750
Co	8.90	425	1495	2870
Ni	8.90	4297	1456	2732
Cu	8.96	338	1083	2567
Zn	7.14	131	419.5	907
Ga	5.91	277	29.8	~2250
Ge	5.32	377	937	2830
As	5.78	302	816[3]	615 (sub)
Se	4.79(grey)	207	217	685
Br	3.12	112	−7.3	59.5
Rb	1.53	81	39.1	688
Sr	2.54	164	769	1384
Y	4.5	425	1522	3338
Zr	6.51	609	1852	4377
Nb	8.57	726	2468	4742
Mo	10.22	658	2617	4612
Tc	11.5	—	2200	4567
Ru	12.41	643	2310	3900
Rh	12.4	557	1966	3727
Pd	12.0	378	1552	3140
Ag	10.49	284	962	2212
Cd	8.65	112	320.8	765
In	7.31	244	156.6	2070
Sn	7.27[β]	302	232	2260
Sb	6.70	262	630.7	1635
Te	6.25	192	450	990
I	4.94	—	113.6	185.2
Cs	1.87	76	28.4	690
Ba	3.59	178	729	1637
La	6.14	431	921	3457
Hf	13.28	619	2230	5197
Ta	16.65	782	2996	5534
W	19.3	849	3422	5657
Re	21.0	770	3180	5627
Os	22.6	791	3054	~5027
Ir	22.4	665	2410	4130
Pt	21.4	565	1772	3827
Au	19.32	336	1064	2808
Hg	13.53	61	−38.9	357
Tl	11.85	182	303.5	1553
Pb	11.35	195	327	1744
Bi	9.75	207	271.4	1560
Po	9.32	145	~250	962
At	—	—	302[4]	337[4]

Lanthanoids

Element	ρ	ΔH°_{atom}	MP	BP
Ce	8.24	423	799	3426
Pr	6.77	356	931	3512
Nd	7.01	328	1021	3074
Pm	7.22	301	1168	2727[4]
Sm	7.52	207	1077	1791
Eu	5.23	175	822	1597
Gd	7.90	398	1313	3266
Tb	8.23	389	1356	3123
Dy	8.55	290	1412	2562
Ho	8.8	301	1474	2700
Er	9.07	317	1529	2868
Tm	9.32	232	1545	1950
Yb	6.97	152	819	1196
Lu	9.84	—	1663	3402

Wide variation is observed among different books. Values quoted here are mostly compiled from 1. The Elements: Emsley, J; ELBS-OUP edition; 2. Advanced Inorganic Chemistry; Cotton, F.A.; Wilkinson, G; Murillo, C; Bochmann, M. Sixth edition, J. Wiley & Sons.

1. For graphite; 2. For α (see text); 3. At 39 atm pr. Sublimes at b.p. ; 4. Estimated ; 5. For β, γ, δ; 8.24 for α ; 6. For P_4; red P 2.2; black 2.69

therefore, results in lowering of melting points (Li–Na–K–Rb–Cs). Exceptions to this take place in Groups 15 to 17 (e.g., P > N; S > O; I > Br > Cl > F). Phosphorus and sulfur have a diminished tendency to bond multiplicity (larger size, poor lateral overlap). They form single bonded polyatomic structures having higher melting points. Among the halogens, polarizability of the halogen atoms increases with their size; stronger van der Waals interaction is established, resulting in gradually higher melting point. Thus bromine is a liquid and iodine is a solid at room temperature.

The variation of melting point among the *d*- and *f*-block elements is not regular, as expected from the complex nature of filling of the valence orbitals and related effects. Nevertheless, their plots show a periodic variation as seen in Fig. 5.3. The plots of standard enthalpy of atomization (ΔH°_{atom}) also show a similar (again irregular) periodic variation.In each transition series, both melting point and ΔH°_{atom} first rise with the number of *d*-electrons up to group 5 (V, Nb, Ta). The values decrease at groups 6(Cr ...) and 7(Mn ...) but again tend to rise at group 8 (Fe ...). Thereafter the values decrease markedly at groups 11(Cu ...) and 12(Zn ...).

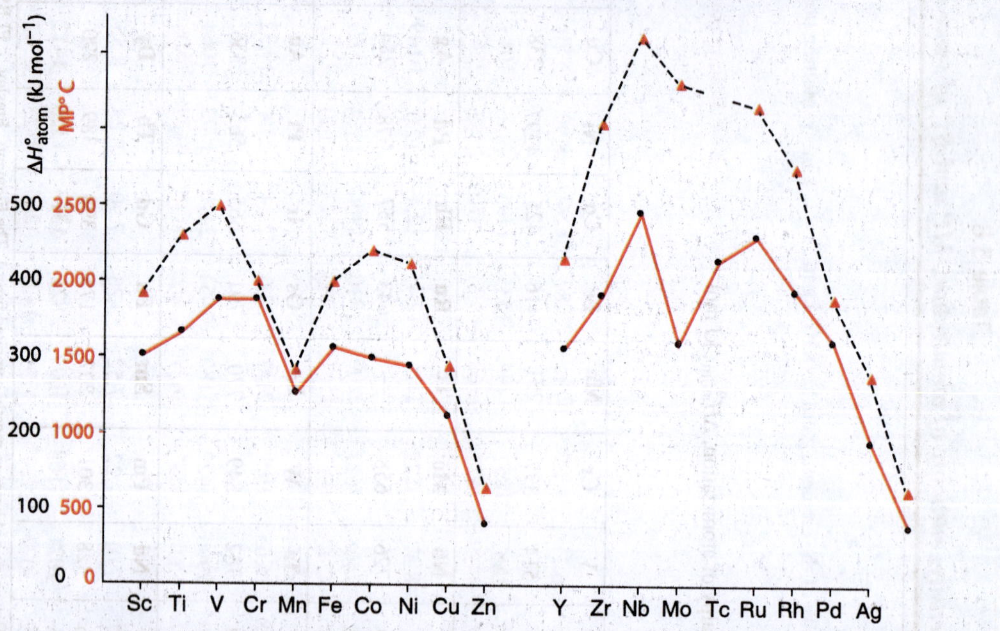

Fig.5.3: Plot of enthalpy of atomization (ΔH°_{atom}; - - -) and melting point °C; —) of first and second row elements of the *d*-block.

Explanation

The binding force between the atoms in the solid depends upon the size and nature of packing of the atoms as well as on the contribution of valence electrons. The role of valence electrons in the bonding in metals is described by the band theory (Chapter 8). In a simplified approach, we may understand the initial rise in melting point and ΔH°_{atom} as more and more electrons occupy the lower energy bonding levels in the metallic band produced primarily by the *d*-orbitals. The bonding energy levels should be completely occupied in metals providing five electrons per atom. After this, the electrons will enter energy levels in the band which are nonbonding or antibonding in nature. So this will weaken the net bonding interaction among metal atoms. However, the simple trend expected in this manner is subject to many alterations from a number of different factors. Some of these factors are common for most atomic properties. Proceeding horizontally along any transition series, effective nuclear charge increases steadily from one element to the next owing to poor screening by the *d*-electrons. Consequently, the $(n-1)d$ orbitals are gradually lowered in energy—the orbitals "contract" and the electrons occupying them ultimately become a part of the core electrons as one reaches group 12 (Zn, Cd ...). The participation of $(n-1)d$ and ns valence electrons in bonding the metal atoms is thus greatly influenced. The melting point and $\Delta H^{\circ}_{\text{atom}}$ both show a maxima at group 5. Vanadium ($3d^3 4s^2$) effectively uses all five electrons per atom to fill the lower half (bonding) of the band of energy levels. The situation gets complicated in later elements due to several factors: (i) the separation between $(n-1)d$ and ns orbitals gradually increases; (ii) exchange energy increases for the d^5 configuration; (iii) electron-electron repulsion also increases to force some electrons to occupy the upper antibonding region in the energy band ("non-aufbau occupancy"). This weakens the net bonding interaction as we observe in the very low melting point of Mn (and to a smaller extent in Tc). In Tc, and also in Re, inter-electron repulsion decreases owing to larger size of the orbitals. So the depression is not great. (iv) Along the third transition series, the increase in effective nuclear charge is further enhanced by the very poorly screening 4*f* electrons.

Moving vertically down any group, we observe that both melting point and $\Delta H^{\circ}_{\text{atom}}$ remarkably rise from 3*d* to 5*d*, indicating stronger metal-metal bonding. It appears that the extent of overlap between valence-shell *d*-orbitals increases as $3d$–$3d < 4d$–$4d < 5d$–$5d$. This trend is reverse of the trend observed in groups 1 and 2. It is likely that the *d*-orbitals can provide better overlap through their longer spatial distribution and lower inter-electron repulsion.

Elements of the *f*-block also display periodic variation of melting point and $\Delta H^{\circ}_{\text{atom}}$ (discussed in Chapter 33). The variations are understandable on the basis of the previous discussion on *d*-block, with due attention to the very poor screening power of the 4*f* electrons. As expected, minima are observed in f^7 and f^{14} configurations which emphasize the importance of exchange energy for half-filled and full-filled configurations.

5.3.2 Density

The densities of the elements in the solid state also exhibit periodic variation when plotted against their atomic numbers. It may be observed from Fig. 5.4. that the densities of the alkali metals occupy the minima in the curve; the density first rises and then decreases along a period, the elements in the middle having the highest density values. Densities of the solid and liquid elements are given in Table 5.6.

The reason underlying the periodic variation of density is largely complicated by the nature of packing of the atoms. A simple interpretation may be provided by remembering that stronger and compact binding leads to high density. The binding force is again dependent on the number of valence electrons provided by each atom. This is why the densities of the central elements in any period are among the highest.

The increase of density while moving down any group in the periodic table is similarly explained from the increasing number of neutrons and protons in the nuclei.

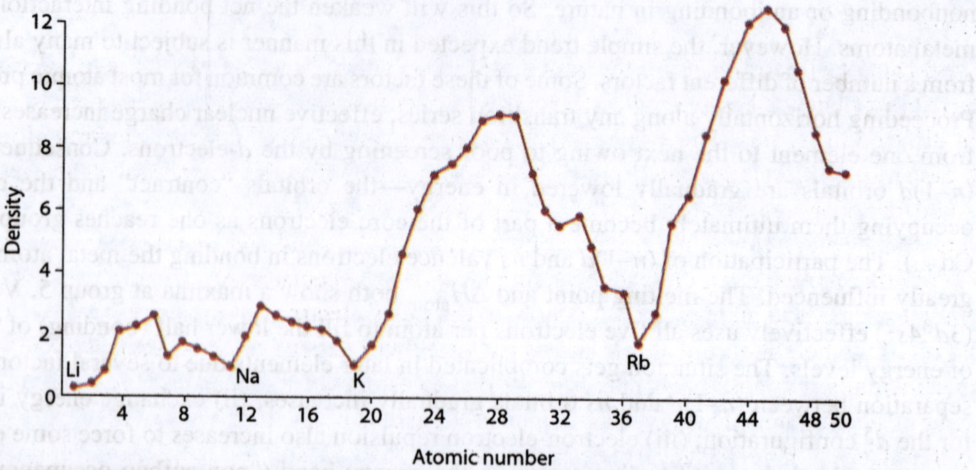

Fig.5.4: Periodic variation of density with atomic number.

5.3.3. Sizes of Atoms

Sizes of atoms may be compared in a number of ways. An early attempt based on atomic volume by Lothar Meyer (Germany) in 1870 is historically important in this connection.

Atomic volume

Lothar Meyer independently introduced the idea of periodicity in properties of the elements from study of their atomic volumes. The atomic volume is the volume of one mole of atoms of an element; it was obtained as: atomic volume (c.c. mole^{-1}) = atomic weight/density. This is only a rough indication of atomic size since the density of an element is determined by several factors like physical state, allotropic form, number of atoms per molecule, nature of packing and so on. Yet a periodic variation is observed when the atomic volumes of elements are plotted against their atomic numbers (Fig. 5.5).

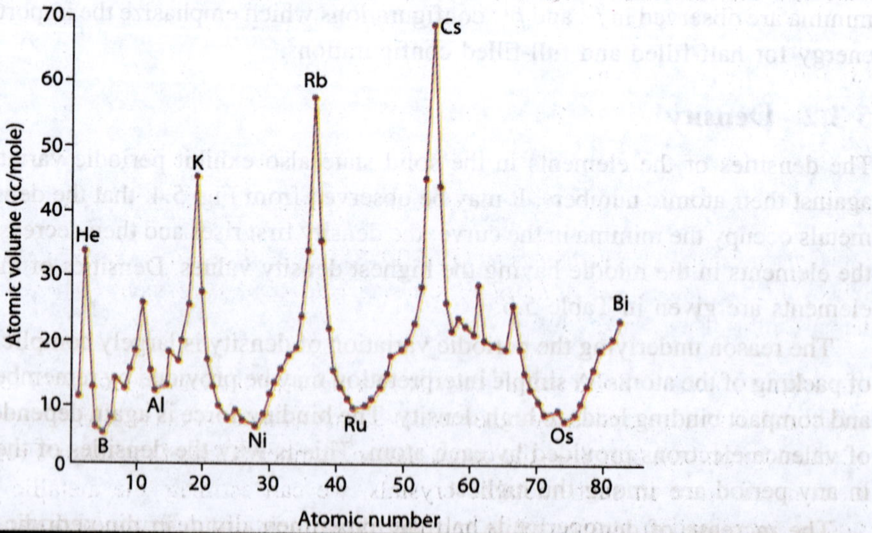

Fig.5.5: Periodic variation of atomic volume.

Atomic radius

We cannot directly measure the actual size of an atom. What we can do is to measure the internuclear distance between two atoms; this is obtained from spectroscopic or x-ray or electron diffraction studies. The outer part of the atom is made up of diffuse "electron clouds"—the density of this cloud falls exponentially with the radius. So there cannot be any well-defined boundary of an atom and the radius of an isolated atom has no meaning. We can calculate the effective volume occupied by an atom in the solid and liquid state in an environment of other atoms. Even then, the situation is not simple because the nature of packing differs from one element to another. In a compound, the outer electrons of an atom are subject to the field of all the neighbouring atoms and so its "size" will depend on the chemical environment. Experimentally it is possible to measure the distances between atoms in a close-packed aggregate or in a compound. This gives us an idea of the radius of an atom which will of course differ in different environments. Nevertheless, such atomic radii data reproduce observed bond lengths in molecules to a fair degree of accuracy. Any gross discrepancy between the two figures suggests that some change in bond type or some other factor must have occurred.

Different sets of radii of atoms have been worked out depending on their environment: van der Waals radii, metallic radii, covalent radii and ionic radii.

Metallic radii (for metals) and covalent radii (mainly for nonmetals) are often referred to as atomic radii.

(i) ***Van der Waals radius.*** This is obtained from the shortest distance to which two nonbonded atoms can approach before repulsion between their electron clouds comes into action. In solid xenon, for example, the measured interatomic distance is 436 pm. The van der Waals radius of xenon is thus estimated to be 218 pm. Steric effects between different parts of a molecule are governed mainly by this radius.

Table 5.7
Some van der Waals radii, pm

						H 120	**He** 180
Li 180*	**Be** —	**B** 208	**C** 185	**N** 154	**O** 140	**F** 135	**Ne** 160
Na 231	**Mg** 170*	**Al** 205	**Si** 200	**P** 190	**S** 185	**Cl** 181	**Ar** 191
K 231	**Ca** —	**Ga** 190*	**Ge** —	**As** 200	**Se** 200	**Br** 195	**Kr** 198
Rb 244	**Sr** —	**In** 190*	**Sn** 220*	**Sb** 220	**Te** 220	**I** 215	**Xe** 220*

Values from Emsley, J. *The Elements*; OUP (ELBS), 1989 except those marked * which are from Huheey, J.E., Keiter and Keiter; *Inorganic Chemistry*: *Principles of Structure and Reactivity*, 4th ed., Harper-Collins, New York, 1993.

(ii) ***Metallic radius.*** Solid metals consist of a close-packed arrangement of the atoms—each metal atom being surrounded by other metal atoms as nearest neighbours whose number may be 12(most common), 6 or 8 — the coordination number (CN) of the atom. From the observed interatomic distance in such metallic crystals, we can estimate the metallic radius for that particular coordination number: it is half the experimentally determined distance between the centres of two nearest neighbour atoms in the solid. This distance varies slightly with the

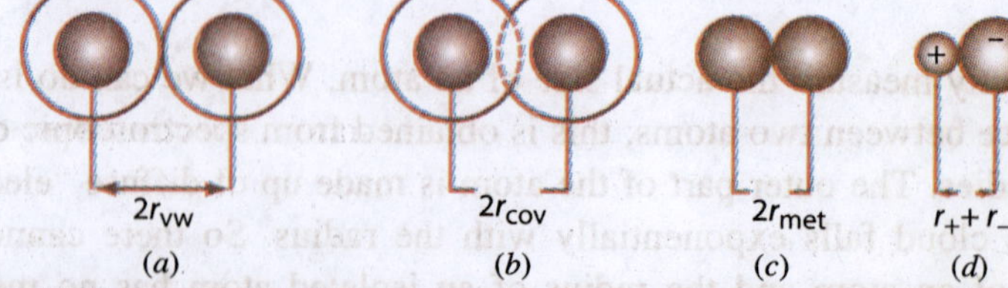

Fig.5.6: Schematic representation of (a) van der Waals radius (r_{vw}) (b) Covalent radius (r_{cov}) (c) metallic radius r_{met} and (d) ionic radius ($r_+ + r_-$).

coordination number: as the coordination number increases, crowding of the metal atoms introduce more repulsive interaction and the atoms are slightly "pushed back", resulting in a larger value of the metallic radius. To provide a uniform common reference for comparison, Goldschmidt suggested that all radii may be put on a relative scale based on the 12-coordinated radius as unity (Table 5.7).

Table 5.7

Relative values of metallic radius for various coordination numbers (C.N.)

C.N.	Relative radius
12	1
8	0.97
6	0.96
4	0.88

The values so obtained from other coordination numbers are the radii the metals would have in coordination number 12. 12-coordinate metallic radii of *s*, *p* and *d*-block elements are given in Table 5.8. Radii of *f*-block elements are given in Table 5.12.

Problem 5.1:

The interatomic distance between two sodium atoms in C.N. 8. is 370 pm. Calculate the metallic radius of Na in C.N. 12.

Hint: r_8 = 370/2 pm = 185 pm; r_{12} = 185/0.97 pm = 191 pm

Table 5.8

12-coordinate metallic radii, pm

Li 157	**Be** 112												
Na 191	**Mg** 160											**Al** 143	
K 235	**Ca** 197	**Sc** 164	**Ti** 147	**V** 135	**Cr** 129	**Mn** 137	**Fe** 126	**Co** 125	**Ni** 125	**Cu** 128	**Zn** 137	**Ga** 153	**Ge** 139
Rb 250	**Sr** 215	**Y** 182	**Zr** 160	**Nb** 147	**Mo** 140	**Tc** 135	**Ru** 134	**Rh** 134	**Pd** 137	**Ag** 144	**Cd** 152	**In** 167	**Sn** 158
Cs 272	**Ba** 224	**La** 188	**Hf** 159	**Ta** 147	**W** 141	**Re** 137	**Os** 135	**Ir** 136	**Pt** 139	**Au** 144	**Hg** 155	**Tl** 171	**Pb** 175

Values from Porterfield, W.W., *Inorganic Chemistry—A Unified Approach*, 2nd ed, Academic Press (*An Imprint of Elsevier*), 1993.

Periodic trend

In the *s*- and *p*-block, metallic radii increase vertically downward as electrons occupy new shells. The increase from period 4 to 5 (e.g., K–Rb) is much less compared to corresponding figures from period 3 to 4 (e.g., Na–K). This is due to interposition of the poorly shielding 3*d*-electrons in period 4. Moving horizontally to right along any period, say, Na–Mg–Al, we observe a steady decrease in radius resulting from increasing nuclear charge.

Among the *d*-block elements, the horizontal trend of decreasing size is very prominent upto Group 10; this is expected from poor shielding by the *d*-electrons and consequent rapid rise in Z_{eff} along the series. Mn is again an exception owing to its unique structure. After group 10, the size slightly increases, from Ni to Cu for example, presumably due to increase in inter electron repulsion in the crowded valence shell. The trend continues to Gr 13 (e.g., Zn–Ga) where electrons enter the outer *np* orbital. The effect of increasing nuclear charge again appears to dominate in decreasing the size (Ga–Ge, In–Sn).

Moving vertically down any group in the *d*-block, we observe that the increase in size is much less from 3*d* to 4*d* series (for example, Ti–Zr, V–Nb and so on), again a consequence of the poor screening provided by the *d*-electrons. The effect is more prominent as we move from 4*d* to 5*d* series; the size remains practically unchanged (for example Nb–Ta; Mo–W). The effect of increase in atomic number is largely cancelled by a rapid increase in Z_{eff} consequent to very poor screening provided by the 4*f* electrons in the interposed lanthanoid elements.

Among the *f*-block elements (Table 5.11), metallic radius decreases from left to right in the series, again emphasizing the effect of rapidly increasing effective nuclear charge consequent to poor screening ability of 4*f* electrons. This effect is more distinct in the radii of the M^{3+} ions (conventionally represented as Ln^{3+}) and is referred to as the *lanthanoid contraction* (see Table 5.12; ionic radii). Another effect, called the *relativistic effect* becomes significant for these heavier elements (see later). Europium ($4f^7$) and ytterbium ($4f^{14}$) are somewhat exceptional to the general trend.

(iii) ***Covalent radius***. The covalent radius of an element may be taken as half the internuclear distance between two atoms of the same element joined by a single bond. For elements forming single bonded units, the experimental internuclear distance may be halved to obtain the covalent radius. In other cases, the covalent radius may be calculated from the internuclear distance in single bond to another element of known radius.

Some covalent radii of common nonmetals are given in Table 5.9. The values are subject to change depending on bonding environment. Again, some of the values have been empirically adjusted to fit experimental bond lengths in common compounds. The values for hydrogen and fluorine are noteworthy in this connection:

Molecule	Bond distance (pm)	r_{cov} (pm) (*estimated*)	r_{cov} (pm) (*accepted*)
H–H	74	37	29
F–F	142	71	64

Table 5.9

Covalent radii, pm

H																
29																
Li	**Be**											**B**	**C**	**N**	**O**	**F**
123	89											82	77	70	66	64
Na	**Mg**											**Al**	**Si**	**P**	**S**	**Cl**
154	136											125	117	110	104	99
K	**Ca**	**Sc**	**Ti**	**V**	**Cr**	**Mn**	**Fe**	**Co**	**Ni**	**Cu**	**Zn**	**Ga**	**Ge**	**As**	**Se**	**Br**
203	174	144	132	122	118	117	116.5	116	115	117	125	125	122	121	117	114
Rb	**Sr**	**Y**	**Zr**	**Nb**	**Mo**	**Tc**	**Ru**	**Rh**	**Pd**	**Ag**	**Cd**	**In**	**Sn**	**Sb**	**Te**	**I**
216	192	162	145	134	129	127	125	125	128	134	141	150	140	141	136	133
Cs	**Ba**	**La**	**Hf**	**Ta**	**W**	**Re**	**Os**	**Ir**	**Pt**	**Au**	**Hg**	**Tl**	**Pb**	**Bi**	**Po**	**At**
235	198	169	144	134	130	128	126	126	129	134	144	155	147	146	(146)	(145)

Values from Emsley, J. *The Elements*, OUP (ELBS), 1989

The observed bond length between two different atoms in a covalent single bond is often shorter than the sum of the covalent radii of the atoms. This shortening depends on the difference of electronegativity of the elements concerned and may be expressed by the empirical **Schomaker-Stevenson relation:**

$$r_{A-B} = r_A + r_B - 9(x_A - x_B)$$

r_A and r_B are the covalent radii (pm) of A and B respectively; r_{A-B} is the estimated single bond length (pm) between A and B; x_A and x_B are electronegativities of the respective elements.

A modified version of this relation has been proposed to fit many bond lengths:

$$r_{A-B} = r_A + r_B - 7(x_A - x_B)^2$$

According to these relations, greater the polarity of a bond (i.e., higher electronegativity difference), the shorter will be the bond length. One important contribution to this effect is likely to come from the attraction between the opposite partial charges built up on the atoms. This is reflected in the following series of compounds where the C–F bond distance decreases gradually as substitution by fluorine increases:

	CH_3F	CH_2F_2	CHF_3	CF_4
C–F distance (pm)	139.1	135.8	133.2	132

When two atoms are joined by multiple bonds, the bond length is shorter than the sum of their single bond covalent radii. From the observed bond-lengths, it is possible to assign "double bond radii" or "triple bond radii" to the atoms. Thus the carbon-carbon bond length of 135 pm in ethylene leads to a double bond radius of carbon equal to 67 pm. Some typical values are given below.

	B	**C**	**N**	**O**
Double bond radius	71	67	60	56
Triple bond radius	64	60	55	

Periodic trend: Along a period, covalent atomic radii gradually decrease from left to right, as expected from the influence of gradually increasing effective nuclear charge. The effect is most prominent in all *s*-block elements and in the *p*-block elements of periods 2 and 3 (B–C–N··· and Al–Si–P···). In periods 4 and 5, some *p*-block elements with $(n–1)\ d^{10}$ core show slight anomaly (Ga–Ge–As; In–Sn–Sb).

Top to bottom in any group, atoms generally become larger with the addition of new shells. But the trend is somewhat checked in the *p*-block elements after the *d*-block: the size increases from B to Al but then actually decreases to Ga. There is little increase in covalent radius from Si to Ge and similar elements in groups 15–18. The trend in size, related mainly to effective nuclear charge, is actually a continuation of the effect that started from scandium with the filling of the 3*d* level and continued along the transition series owing to poor shielding provided by the *d*-electrons. Sometimes this has been called *scandide contraction* of the elements Ga to Kr, though this is rather a situation of "non-expansion" or less than expected expansion. The increase in atomic radius between Ga–In, Ge–Sn and so on for period 4 and period 5 elements in other groups is distinct, though small, because the elements in both periods have $3d^{10}$ or $4d^{10}$ electron population and the increase in Z_{eff} is comparable. Between the elements in period 5 and period 6 (for example In–Tl or Sn–Pb etc.), the increase in size is again very small, as expected from the introduction of an additional $4f^{14}$ level which gives even less shielding than the d^{10} level. We observed a similar situation with metallic radii among the *d*-block elements of period 5 and 6. These may be ascribed to an aftereffect of *lanthanoid contraction*, since the effect started after electrons started to enter the 4*f* level after La.

(iv) *Ionic radius.* X-ray diffraction patterns from ionic crystals provide us the interionic distance in crystals. But there is no obvious way to apportion this distance into cationic and anionic radii. Several indirect methods were proposed (see below) to obtain sets of ionic radii consistent with experiments. However these radii are also susceptible to change with the actual environment in which an ion occurs. More recently, it has been possible to draw detailed electron density contours from x-ray data on a number of crystals. The minima in the electron density now correspond to the effective boundary of the domain of each ion and hence, their radii. However, these values are also uncertain to some extent since the plot of electron density against distance often gives a curve with a flat bottom.

Classically, two sets of ionic radii were proposed (1926–29) by Goldschmidt and by Pauling. Goldschmidt assumed that in crystals formed by large anions with small cations, e.g., LiI, the anions would touch each other (Fig. 5.7). One half the distance between the anions should correspond to the radius of the anion. The interionic distance in crystals of this anion with larger cations (e.g., NaI, KI etc.) will now provide us with the radii of the cations, Na^+, K^+ etc. These figures may in turn be used to obtain radii of other anions like Cl^-. Now we can return to the radius of Li^+ from measurements on LiCl. Goldschmidt started with the assumption that the molar refractivities of the ions in alkali metal halides MX and alkaline earth metal oxides

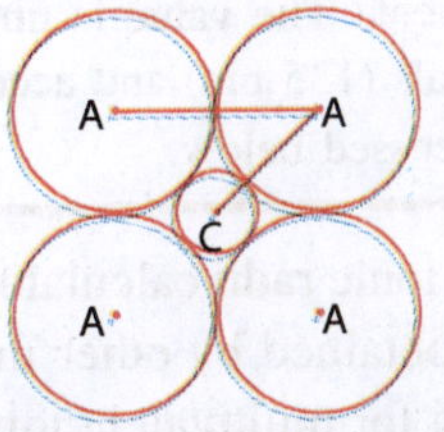

Fig.5.7: A small cation (C) surrounded by large anions (A). r_A = AA/2. From measurement on silicate minerals, the radius of the oxide ion was thus estimated to be 135 pm.

could be apportioned on the basis of ionic volume which led to the radii of F^- and O^{2-} ions as 133 pm and 132 pm respectively. From these, he calculated a number of ionic radii.

Pauling's approach. Pauling based his calculation on the assumption that in crystals with similar structure (same r_+/r_- ratio) the radii of a series of isoelectronic ions should vary inversely with their effective nuclear charge:

$$r = \frac{C}{Z_{eff}} = \frac{C}{Z - s}$$

where C is a proportionality constant, Z = atomic number and s = screening contribution by inner electron core.

He then started with the observed interionic distances in the crystals of NaF, KCl, RbBr, CsI and Li_2O. All of them have r_+/r_- value around 0.75 and adopt the NaCl structure (Chapter 8).

In NaF, for example, both Na^+ and F^- ions have the neon configuration. The screening contribution s for these ions will be

[Ne]: $1s^2(2s2p)^8$: $s = 8 \times 0.35 + 2 \times 0.85 = 4.5$ (other calculations give $s = 4.48$.)

We can now write, $r_{Na^+} = \dfrac{C}{11-4.5}$; $r_{F^-} = \dfrac{C}{9-4.5}$

From the observed interionic distance 231 pm, we get

$$\frac{C}{6.5} + \frac{C}{4.5} = 231 \text{ or, } C = 614 \text{ pm and } r_{Na^+} = 95 \text{ pm}; r_{F^-} = 136 \text{ pm}.$$

Proceeding in the same manner, Pauling calculated the radii of other ions (pm):

K^+ : 133; Rb^+ : 148; Cs^+ : 169; Cl^- : 181; Br^- : 195; I^- : 216.

Problem 5.2:

Calculate the radius of the oxide ion, O^{2-}, according to Pauling's formula.

Solution: The O^{2-} ion has the neon configuration; hence C = 614 pm (see above).

$$\therefore r_{O^{2-}} = \frac{614}{8-4.5} \text{pm} = 175 \text{ pm}.$$

Comment: The value is unusally large, compared to that estimated from measurements on some minerals (135 pm) and accepted by Goldschmidt (132 pm). The reason for this apparent anomaly is discussed below.

The ionic radii calculated by Pauling's procedure for univalent ions agree satisfactorily with values obtained by other methods and also with observed interionic distances in crystals. But the radii for multivalent ions obtained in this manner are unusually large, as pointed out above for the O^{2-} ion. The interionic distance in compounds like CaS and MgO are much shorter than the sum of ionic radii calculated according to this procedure.

Pauling remarked that the radii so calculated for multivalent ions are actually their "*univalent radii*", the radii these ions would possess if they were to retain their electron distribution but enter into coulomb interaction *as if they were univalent*. If the multivalent ions, with charges

Z_+e and Z_-e, say, are present in a hypothetical crystal lattice with coulomb interactions equal to that between charges $+e$ and $-e$, their interionic separation will be given by the sum of their univalent radii.

The crystal radii (r_c) which conform to actual distances between ions in crystals, may be obtained from their univalent radii (r_1) using the relation

$$r_c = r_1(Z)^{-2/(n-1)}$$

where Z is the charge number of the ion (e.g., 2 for O^{2-} and 3 for Al^{3+}) and n is the Born exponent which has a particular integral value for each electron configuration (see chapter 8). For neon configuration (Na^+, F^-, Mg^{2+}, O^{2-}), n is 7; for argon configuration (K^+, Cl^-, Ca^{2+}, S^{2-}), n is 9.

Problem 5.3

The univalent radius of Ca^{2+} is 117 pm; that for S^{2-} is 219 pm. Calculate the effective crystal radii of the ions.

The observed interionic distance in CaS is 284 pm. Comment.

Hint: For both ions, $Z = 2$ and $n = 9$. Hence, the crystal radii (r_c) are found as

$$r_c\ (Ca^{2+}) = 117 \times 2^{-(2/8)} = 98.4 \text{ pm};\ r_c\ (S^{2-}) = 219 \times 2^{-(2/8)} = 184 \text{ pm}.$$

The estimated interionic distance would be (184 + 98.4)pm = 282.4 pm. This agrees satisfactorily with the observed value of 284 pm.

Problem–5.4

The univalent radius of the oxide ion is 175 pm (see Problem 5.2). Calculate the crystal radius of the O^{2-} ion.

Hint: For the oxide ion (neon configuration), $n = 7$ and $Z = 2$. $r_c = 175 \times 2^{-(2/6)}$ pm = 139 pm.

Pauling calculated the radii of most common ions taking the radius of the oxide ion as 140 pm (Table 5.10). The oxide ion offers a good standard for correlating ionic radii as it occurs in combination with a wide range of ions and is reasonably unpolarizable. Its size dose not change appreciably when the accompanying cation is changed. However, there are also other accepted values for the radius of the oxide ion. Goldschmidt, for example, compiled his set of ionic radii taking the radius of the oxide ion as 132 pm. Shannon and Prewitt compiled a set of "crystal radii" (Table 5.11) based on the radius of oxide ion as 126 pm (6 coordination).

Table 5.10

Radii of some common ions (pm) as estimated by Pauling (6:6 coordination)

Li^+	60	Mg^{2+}	65	Cr^{3+}	55	H^-	208
Na^+	95	Ca^{2+}	99	Fe^{3+}	53	F^-	136
K^+	133	Sr^{2+}	113	Mn^{2+}	80	Cl^-	181
Rb^+	148	Ba^{2+}	135	Fe^{2+}	75	Br^-	195
Cs^+	169	Al^{3+}	50	Hg^{2+}	110	I^-	216

continued...

Cu^+	96	Ga^{3+}	62	Co^{2+}	72	O^{2-}	140
Ag^+	126	In^{3+}	81	Ni^{2+}	69	S^{2}	184
Au^+	137	Tl^{3+}	95	Zn^{2+}	74	Se^{2-}	198
Be^{2+}	31	Ti^{3+}	76	Cd^{2+}	97	Te^{2-}	221

Table 5.11

Crystal radii (Shannon and Prewitt), pm

Ion	C.N. 4	C.N. 6	C.N. 8
Li^+	73	90	106
Na^+	113	116	132
K^+	151	152	165
Cu^+	74	91	–
Ag^+	114	129	142
	(C.N. 2 81)		
Au^+	–	151	–
Be^{2+}	41	59	–
Mg^{2+}	71	86	103
Ca^{2+}	114	126	148
Sr^{2+}	–	132	140
Ba^{2+}	–	149	156
Al^{3+}	53	68	–
Cr^{2+}	–	87/94*	–
Cr^{3+}	–	75	–
Mn^{2+}	80	81/97*	–
Fe^{2+}	77	75/92*	–
Fe^{3+}	63	69/78*	–
Co^{2+}	72	79/89*	–
Co^{3+}	–	68/75*	–
Ni^{2+}	69	83	–
Cu^{2+}	–	87	–
Au^{3+}	82 (sq)	99	–
Zn^{2+}	74	88	104
Cd^{2+}	92	109	124
Hg^{2+}	110	116	128
F^-	117	119	–
Cl^-	–	167	–
Br^-	–	182	–
I^-	–	206	–
OH^-	–	123	–
O^{2-}	124	126	128
S^{2-}	–	170	–

* Smaller value for low-spin state (Chapter 27).

As expected, the radius of an ion increases with coordination number—greater repulsion among the surrounding ions effectively increases the size of the central ion.

The radii of transition metal ions depend, in addition to coordination number, on the spin state of the ion (chapter 27). In the high spin state, the size of the ion is effectively greater.

General trend

Ionic radii show the same general trend as observed earlier with atomic radii. To summarize:

a. Ionic radii generally increase down a group. In the *d*-block, the increase between 4*d* and 5*d*-series metal ions is arrested by the introduction of poorly screening 4*f* level.

b. Radii of ions decrease from left to right along a period owing to higher effective nuclear charge as well as higher charge of the cation in its common oxidation state. Along any *d*- or *f*-series, the radii decrease steadily from left to right for ions in same oxidation state—a consequence of poor screening by *d* and *f* electrons. The effect is most prominent in the *f*-block and deserve special mention as lanthanoid (lanthanide) and actinoid (actinide) contraction.

c. As expected, anions will be larger than cations owing to additional extranuclear electrons. In the same line, radii of anions will increase with increasing anion charge. Obviously, the reverse trend applies to cations (e.g., $Fe^{2+} > Fe^{3+}$).

Thermochemical radii

For polyatomic ions, the variation of size with environment is very uncertain and assignment of a particular ionic radius is often very difficult. In such cases, the radius may be "backcalculated" from experimental enthalpy of formation of an ionic crystal together with its estimated lattice energy (see Ch. 8). Such radii have been called thermochemical radii (Yatsimiriskii; Table 5.12). The idea of a radius for such ions is however an approximation since many of these ions are not spherical, e.g., the CH_3COO^- ion.

Table 5.12
Radii of some polyatomic ions, pm.

Ion	*Radius*	*Ion*	*Radius*	*Ion*	*Radius*
NH_4^+	151	IO_3^-	108	PO_4^{3-}	238
CH_3COO^-	148	NO_2^-	178	SO_4^{2-}	244
ClO_3^-	157	NO_3^-	165	SeO_4^{2-}	235
ClO_4^-	226	CO_3^{2-}	164	CrO_4^{2-}	242
HCO_3^-	142	OH^-	119	MnO_4^-	215

Lanthanoid (lanthanide) contraction

Among the post-lanthanum elements of the 4*f*-series, the metallic radii as well as the radii of tripositive ions (Ln^{3+}, Ln = a lanthanoid element) show a prominent steady decrease (Table 5.13 and Fig. 5.8)—the phenomenon is popularly referred to as lanthanoid contraction. The reason for this steady decrease in size lies primarily on the filling of the 4*f* subshell. The 4*f* orbital penetrates least into the inner electron core and successive electrons added to this subshell do not efficiently screen the next incoming electrons from the progressively increasing nuclear charge. The consequent rise in effective nuclear charge essentially contracts the spatial distribution of the orbitals and hence the size of the atoms or ions. The effect of lanthanoid contraction is far reaching—it extends beyond the 4*f* series: we remember that the sizes of the metals and their ions in 4*d* and 5*d*-series show little variation. The relativistic effect also contributes to the decrease in size of the heavier lanthanoids.

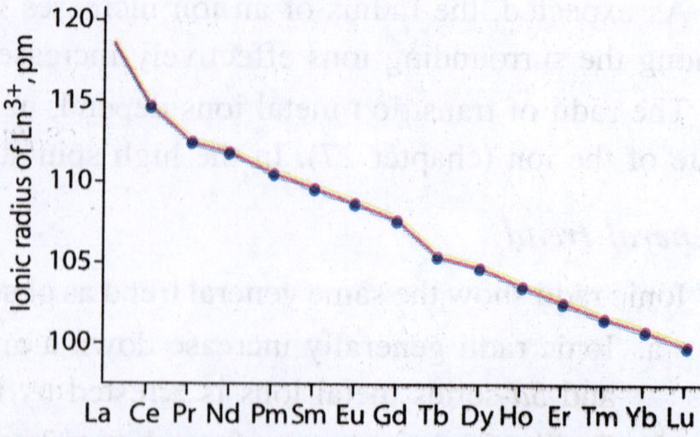

Fig.5.8: Variation of ionic radii of Ln^{3+} ions.

The second row of *f*-block elements, the actinoids, practically repeat the same trend which is referred to as ***actinoid contraction***. Relativistic effects have a greater influence along this series.

Table 5.13

(a) Metallic radii and (b) Ionic radii of M^{3+} ions **(C.N. 6 *for both*)** ***of the f-block elements.***

4*f*	Ce	Pr	Nd	Pm	Sm	Eu	Gd	Tb	Dy	Ho	Er	Tm	Yb	Lu
a) r_{met}, pm	182	182	181	181	180	208	180	177	178	176	176	175	193	174
b) $r_{M^{3+}}$, pm	115	113	112	111	110	109	108	106	105	104	103	102	101	100
5*f*	**Th**	**Pa**	**U**	**Np**	**Pu**	**Am**	**Cm**	**Bk**	**Cf**	**Es**	**Fm**	**Md**	**No**	**Lr**
a) r_{met}, pm	180	164	154	155	159	173	174	170	186	186	194	194	194	171
b) $r_{M^{3+}}$, pm	108	118	117	114	115	112	111	110	109					

The Relativistic effect

In heavier elements ($Z > 80$), relativistic variation of the mass of the electron becomes prominent so as to cause an effective contraction in the spatial distribution of the orbitals. This phenomenon is called the relativistic effect.

The Bohr model of the atom helps us to understand the relativistic effect. In this model, the radial velocity v of an electron in an orbit of principal quantum number n is given by

$$v = Ze^2/2\varepsilon_0 nh$$

From the known values of ε_0, h and e, the velocity v of an 1*s* electron in hydrogen ($Z = 1$, $n = 1$) comes out as 2.19×10^6 m s^{-1} = 0.0073 c.

For an element with $Z = 80$ (Hg), similar calculation for the 1*s* electron ($Z = 80$, $n = 1$) gives $v/c = 0.58$. According to the relativistic relation

$$m = \frac{m_0}{\sqrt{1 - v^2/c^2}}.$$

The electron mass m now becomes nearly 1.2 times m_0, that is the mass of the 1s electron in Hg increases by about 20 per cent relative to its rest mass! The expression of radius,

$$r = Ze^2/4\pi\varepsilon_0 mv^2$$

now tells us that the radius of the orbit will also "shrink", that is, become smaller by the same factor.

Since the velocity of an electron is inversely proportional to n, the relativistic effect will be most pronounced for electrons closer to the nucleus (low value of n). Transferred to the wave mechanical atom model, this implies that the relativistic effect will be most pronounced for s-orbitals which penetrate most into the inner electron core. The p-orbitals, which penetrate less into the inner core, will undergo relativistic contraction to a smaller extent. The d- and f-orbitals, having radial distribution pattern with insignificant penetration, are thus expected to be almost immune to such effect. However, these orbitals are now subjected to more effective screening by the contracted inner core of s- and p-orbitals. The effective nuclear charge is thus diminished and the orbitals undergo *relativistic expansion*.

The relativistic effect becomes prominent in the sixth period, particularly on the 6s electrons. The contracted 6s orbital is lowered in energy and gets largely separated from the 6p orbital. The 5d orbital, on the other hand, suffers relativistic expansion and its effective size becomes comparable to that of the 6s orbital. The 5d electrons in the second transition series elements can be removed easily and we find an extensive range of compounds in higher oxidation states. If we compare the d-block elements in the 4d-series with corresponding elements in the 5d-series, we find that elements in the two series closely resemble one another in the early part (e.g., Zr–Hf; Nb–Ta. See text). But the similarity gradually disappears as we proceed to right along the d-block : increasing disparity is observed instead (Pd–Pt; Ag–Au, for example). The 5d-electrons in the latter elements are shielded better by the contracted inner core and can participate more in bonding compared to corresponding elements in the 4d-series. Thus Pd(IV) and Ag(III) are much less stable compared to their 5d counterparts—Pt(IV) and Au(III) respectively. Similarly, Pt(VI) and Au(V) have no analogue among Pd and Ag.

Properties of gold and mercury are largely influenced by the relativistic effect. Electronegativity of these elements are much higher than the corresponding elements above, silver and cadmium. Lanthanide contraction alone is not enough to justify this large enhancement. As mentioned before, the 6s orbitals in these elements are contracted significantly and get largely separated from the 6p orbitals in energy. The higher energy 6p orbitals in these elements are thus reluctant to take part in valence. Thus mercury has its valence electron configuration confined to $5d^{10}6s^2$—virtually a "closed shell" configuration! The very low melting point of mercury, related to poor interatomic forces, may also be related to this electron configuration. In this consideration mercury is comparable to the noble gases characterized by a closed shell configuration. Gold is one electron short of the $5d^{10}6s^2$ configuration, thus comparable to the halogen atoms. Indeed, gold forms a number of compounds containing the auride ion, Au^-, having a size comparable to that of the Br^- ion.

The consequences of relativistic effect are also prominent among the f-block elements. The 5f orbitals undergo large relativistic expansion and are thus lowered in energy. Uranium is known in several higher oxidation states compared to its 4f counterpart, Nd. Effects comparable to that on Au seem to recur at Fm.

5.3.4 Ionization Energy

The ionization energy (I) of an element is the minimum energy required to remove an electron from a free gaseous atom of the element in its ground state. The molar internal energy change for the process

$$A(g) \xrightarrow{I_1} A^+(g) + e(g)$$

at absolute zero is called the first ionization energy (I_1) of element A. Subsequent removal of electrons, one at a time, involves second, third etc. ionization energies (I_2, I_3 and so on).

$$A^+(g) \xrightarrow{I_2} A^{2+}(g) + e(g)$$

An atom has as many ionization energies as it has electrons.

We have seen earlier that the ground state energy of a hydrogen atom ($n = 1$) is $E_1 = -R_H.ch$ where R_H is the Rydberg constant for hydrogen. Accordingly, the energy required to remove the electron to $n = \infty$ requires energy $-E_1 = R_H.ch$ which is the ionization energy of hydrogen per atom.

From the values of R_H, c and h, we have,

$$-E_1 = 2.18 \times 10^{-18} \text{ J } (2.18 \text{ attojoules, } atto = 10^{-18})$$

The ionization energy I of hydrogen is, therefore,

$$I = 2.18 \times 10^{-18} \times 6.022 \times 10^{23} \text{ J mol}^{-1} = 1312 \text{ kJ mol}^{-1}.$$

Thus ionization energies may be determined spectroscopically from the series limit in the emission spectrum of an element. We know (Bohr's theory, chapter 3) that the energy of the electron in an atom of H for different values of n is given by

$$E_n = -\frac{R_H.hc}{n^2} \quad (n = \text{principal quantum number})$$

The wave number corresponding to transitions to the $n = 1$ level from $n = n$ level is

$$\bar{\nu} = R_H\left(\frac{1}{1^2} - \frac{1}{n^2}\right) = \frac{I}{hc} - \frac{R_H}{n^2} \quad (\text{since } I = R_H ch \text{ per atom})$$

So, a plot of $\bar{\nu}$ against $1/n^2$ will give a straight line of slope $-R_H$ and intercept I/hc. Values of $\bar{\nu}$ may be fitted to this relation to evaluate the ionization energy. Ionization energy is also expressed (traditionally) in electron volt (eV). This value is numerically equal to the *ionization potential* (volt) which is the voltage at which one or more electrons may be just detached from a gaseous atom (that is, with no kinetic energy for the resulting ion or the electron). We have seen (Bohr atom model; Fig. 3.12) that the ionization energy of hydrogen is 13.6 eV while the ionization potential is 13.6 V. The values in eV may be converted to kilojoules per mole using the equivalence 1 eV ≡ 96.485 kJ mol^{-1}.

Ionization energies may also be obtained by x-ray photoelectron spectroscopy. In the technique, a sample of the element is irradiated with x-rays and the kinetic energies of ejected electrons are analyzed. For example, suppose a sample of neon gas is irradiated with x-rays of frequency high enough to knock out an 1s electron from the neon atom. The energy required for ejecting an electron, that is, the ionization

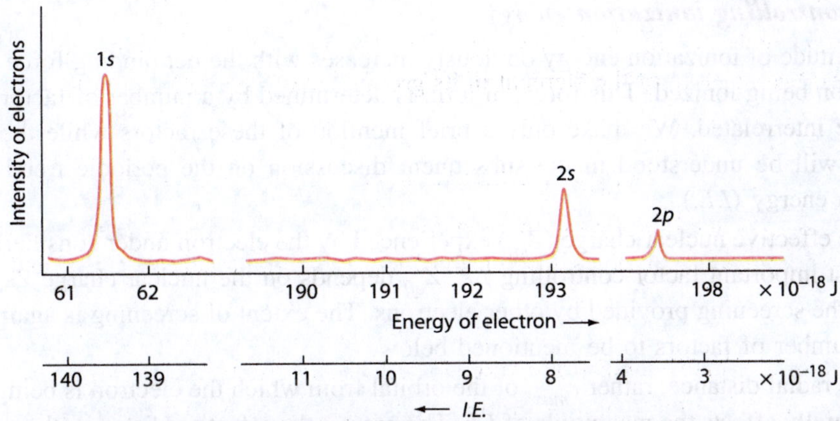

Fig.5.9: Nature of x-ray photoelectron spectrum of neon (approx). Wavelength of x-ray used = 9.894×10^{-10} m. Energy of each x-ray photon = 200.9×10^{-18} J.

energy (*I.E.*) for that electron, will be derived from the energy ($h\nu$) of the x-ray photon. The kinetic energy (E_k) of the electron will now be given by

$$E_k = h\nu - I.E.$$

For a particular frequency of the x-ray, E_k depends linearly on the ionization energy.

In an actual experiment, the intensity of electrons is plotted against their kinetic energy (Fig. 5.9). The photoelectric spectrum shows peaks corresponding to different occupied orbitals in the neon atom. From the known frequency of the x-ray, one can calculate the ionization energies for an 1*s*, 2*s* or 2*p* electron in neon.

In thermodynamic calculations we often need the ***standard enthalpy of ionization*** (ΔH°_{ion}), conventionally referred to as enthalpy of the ionization process at 298 K. This is slightly greater than the ionization energy *I* defined earlier for the same ionization process at 0 K.

ΔH at 298 K is related to ΔH at 0 K by Kirchoff's equation

$$\Delta H_{298} = \Delta H_0 + \int_0^T \Delta C_P dT$$

where C_P is molar heat capacity at constant pressure. For the process $A(g) = A^+(g) + e(g)$

$$\Delta C_P = C_P(A^+) + C_P(e) - C_P(A)$$

Assuming, A, A^+ and *e* to be ideal monatomic gases, C_P for each will be $\frac{5}{2}R$. Hence ΔC_P is $\frac{5}{2}R$. Since ΔH at 0 K is equal to the ionization energy *I*, we have

$$\Delta H_{298} = I + \frac{5}{2}RT$$

Thus, the standard enthalpy of ionization is greater than *I*, the first ionization energy, by an amount $\frac{5}{2}RT$ or $\frac{5}{2} \times 8.314 \times 298\,\text{J} \equiv 6.19\,\text{kJ}$.

Since first ionization energies usually lie in the range of ~400 (alkali metals) to ~2000 kJ (neon), the difference is often negligible. Successive ionization energies are still higher.

Note: The difference $\frac{5}{2}RT$ appears for each mole of gaseous electron appearing on the right hand side. For $A(g) \rightarrow A^{2+}(g) + 2e(g)$, the difference between ΔH° and $(I_1 + I_2)$ will be $5RT$.

Factors controlling ionization energy

The magnitude of ionization energy obviously increases with the net binding force exerted on the electron being ionized. This force, in turn, is determined by a number of factors, many of which are interrelated. We make only a brief mention of these factors while their intricate interplay will be understood in the subsequent discussion on the periodic trend shown by ionization energy (*I.E.*).

i. The effective nuclear charge (Z_{eff}) experienced by the electron under consideration is the most important factor controlling *I.E.* Z_{eff} depends on the nuclear charge Ze as well as on the screening provided by other electrons. The extent of screening is again related to a number of factors to be mentioned below.

ii. The radial distance, rather r_{max}, of the orbital from which the electron is being removed, naturally affects the magnitude of *I.E.* The greater the effective distance, the easier it will be to remove the electron. We remember that r_{max} is primarily determined by the principal quantum number *n*.

iii. The extent of penetration of outer orbitals into the inner electron core affects screening. As seen earlier, penetration depends on, among other things, on *n* and *l* of an orbital.

iv. Relative stabilities of the electron configurations before and after ionization also contribute significantly to the magnitude of *I.E.* In fact, several factors are responsible for such stabilities, for example, simple electrostatic repulsion, spin correlation, exchange energy and, of course, overall energy for the relevant electron configuration.

Periodic trends in ionization energy

Ionization energies (I_1, I_2 and I_3) of elements hydrogen to radon are shown in tables 5.14 and 5.15. A plot of first ionization energy of these elements against atomic number Z shows a clear periodic variation (Fig. 5.10).

Table 5.14

Ionization energy of some s- and p- block elements (kJ mol^{-1})

	H							He
I_1	1312							2373
I_2								5250
	Li	Be	B	C	N	O	F	Ne
I_1	513	899	801	1086	1402	1314	1681	2081
I_2	7298	1757	2427	2352	2856	3388	3374	3952
I_3	11815	14848	3660	4620	4578	5300	6050	6122
	Na	Mg	Al	Si	P	S	Cl	Ar
I_1	496	738	577	787	1012	1000	1251	1520
I_2	4562	1451	1817	1577	1903	2251	2297	2665
I_3	6912	7733	2745	3231	2912	3361	3826	3928

	K	Ca	Ga	Ge	As	Se	Br	Kr
I_1	419	590	579	762	947	941	1140	1351
I_2	3051	1145	1979	1537	1798	2044	2104	2350
I_3	4411	4910	2963	3302	2735	2974	3500	3565
	Rb	**Sr**	**In**	**Sn**	**Sb**	**Te**	**I**	**Xe**
I_1	403	550	558	709	834	869	1008	1170
I_2	2632	1064	1821	1412	1794	1795	1846	2046
I_3	3900	4210	2704	2943	2443	2698	3200	3097
	Cs	**Ba**	**Tl**	**Pb**	**Bi**	**Po**	**At**	**Rn**
I_1	376	503	589	716	703	812	930	1037
I_2	2420	965	1971	1450	1610	–	1600	–
I_3	3400	3619	2878	3082	2466	2700	2900	–

Table 5.15

***Ionization energies of* d*-block elements and the lanthanoids* (kJ mol^{-1})**

	Sc	Ti	V	Cr	Mn	Fe	Co	Ni	Cu	Zn
I_1	631	658	650	653	717	759	760	737	745	906
I_2	1235	1310	1414	1592	1509	1561	1646	1753	1958	1733
I_3	2389	2652	2828	2987	3248	2957	3232	3393	3554	3833
	Y	**Zr**	**Nb**	**Mo**	**Tc**	**Ru**	**Rh**	**Pd**	**Ag**	**Cd**
I_1	616	660	664	685	702	711	720	805	731	868
I_2	1181	1267	1382	1558	1471	1617	1743	1875	2072	1631
I_3	1980	2218	2416	2621	2850	2747	2997	3176	3361	3616
	La	**Hf**	**Ta**	**W**	**Re**	**Os**	**Ir**	**Pt**	**Au**	**Hg**
I_1	538	642	761	770	760	840	880	870	890	1007
I_2	1067	1440	1500	1700	1260	1600	1680	1791	1980	1810
I_3	1850	2250	2100	2300	2510	2400	2600	2800	2899	3300

	Ce	Pr	Nd	Pm	Sm	Eu	Gd	Tb	Dy	Ho	Er	Tm	Yb	Lu
I_1	527	523	530	536	543	547	593	565	572	581	589	597	603	524
I_2	1047	1018	1035	1052	1068	1085	1166	1112	1126	1139	1151	1163	1176	1341
I_3	1949	2086	2130	2150	2260	2404	1990	2114	2213	2204	2194	2285	2415	2022

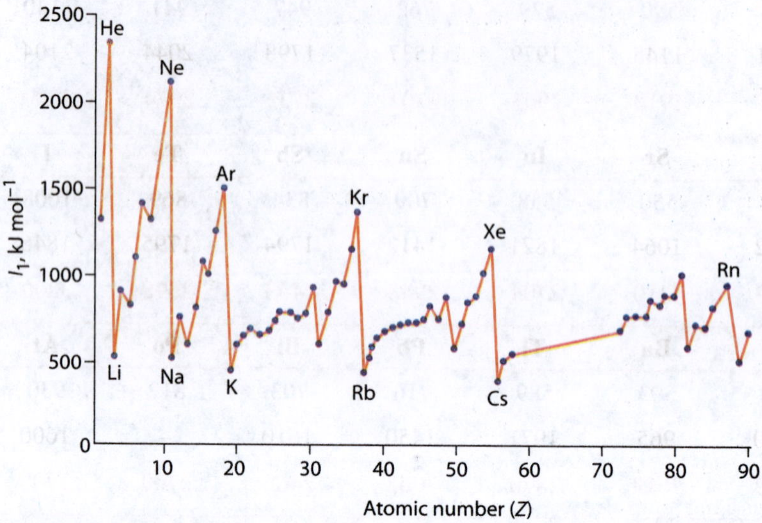

Fig.5.10: Periodic variation of first ionization energy.

The successive ionization energies of an element, I_1, I_2, I_3 ... always increase since removal of an electron from a positively charged ion always requires greater energy than from a neutral atom or a cation with lower positive charge.

Side by side, removal of an electron also results in an increase in Z_{eff} and a consequent extra contraction in the size of the cation. Hence removal of the next electron from this more compact and smaller cation becomes increasingly difficult. However, the extent of variation from I_1 to I_2 to I_3 and so on depends upon the very nature of the shell from which the electron is expelled. A plot of I_1, I_2 etc. (I_n) shows distinct jumps when a stable configuration is broken (Fig. 5.11 and 5.12). In Fig. 5.11, the plots of I_n for Mg shows a sharp jump at I_3 when an electron ionizes from the Mg^{2+} ion ($2s^22p^6$ configuration). The plots in Fig. 5.12 show a break at I_2 ($3s^2$ configuration) and a sharp jump at I_4 ($2s^22p^6$ configuration) for aluminium.

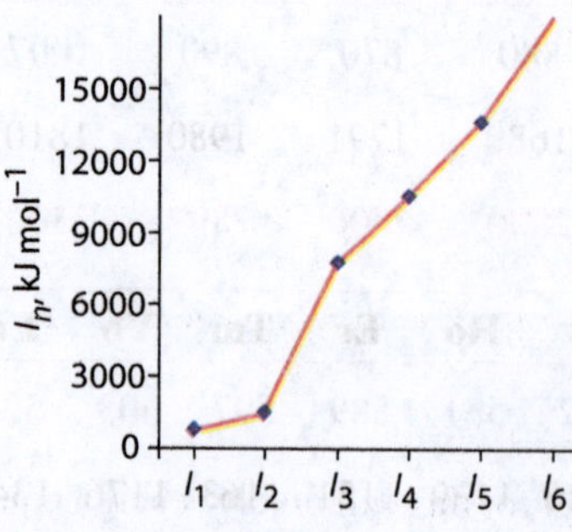

Fig.5.11: Plot of successive ionization energies of Mg.

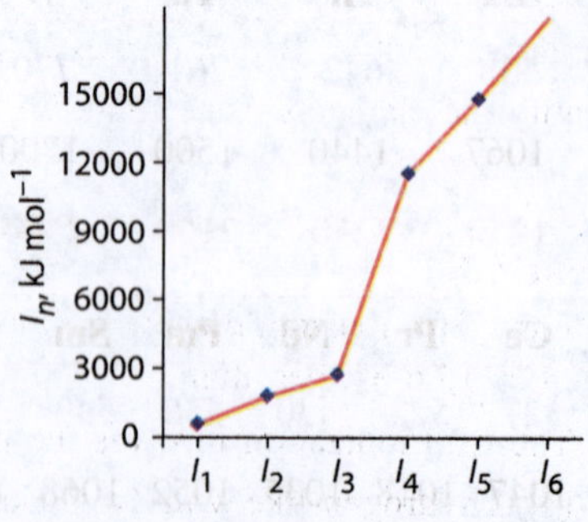

Fig.5.12: Plot of successive ionization energies of Al.

Problem 5.5

Look at table 5.14. I_2 of Li is more than ten times its I_1; for Be I_2 is nearly twice its I_1. Comment.

Hint: I_2 of Li involves removal of an electron from the $1s^2$ core.

Note

1. Similar large jump of I_2 from I_1 is observed in all alkali metals. But the ratio I_2/I_1 gradually falls in the case of heavier members.
2. A similar large jump in the magnitude of I_2 from I_1 is also prominent among the elements of Gr 13(III): B, Al etc. The main reason is obvious.

Problem 5.6

Look at the following I_n values (approx) for elements X and Y.

	I_1	I_2	I_3	I_4	I_5	I_6	(kJ mol^{-1})
X	590	1150	4900	6500	8100	10,500	
Y	1010	1900	2900	5000	6300	21,300	

Plot the I_n values against the number of electrons ionized. Comment.

Hint: X shows a big jump after I_2, suggesting disruption of an inner core. It is likely to be an element with ns^2 configuration, belonging to Group 2(II).

Y shows a sudden big jump after I_5. It probably belongs to Group 15(V).

Look at Table 5.14 for Ca and P.

Problem 5.7

Successive ionization energies (I_n in kJ mol^{-1}) for three main group elements X, Y and Z are given below. Comment on the likely electron configuration and the periodic groups to which the elements belong.

	I_1	I_2	I_3	I_4	I_5	I_6	I_7 (kJ mol^{-1})
X	800	2427	3660	25033	32824	–	–
Y	1086	2352	4620	6229	37838	47277	–
Z	1012	1903	2912	4957	6274	21269	25397

Comment: X shows a sudden jump at I_4 and I_5. It is likely that its valence shell consists of only three electrons, the first one of which is removed relatively easily. X may have an ns^2np^1 electron configuration in the ground state and is likely to belong to group 13(III).

Similarly: Y: ns^2np^2; Group 14(IV); Z: ns^2np^3; Group 15(V)

Identify the elements from Table 5.14.

Problem 5.8

Estimate the fifth ionization energy of boron applying Bohr's theory. Use the ionization energy of hydrogen as 13.6 eV per atom.

Hint: The fifth ionization involves the process $B^{4+}(g) \rightarrow B^{5+}(g) + e(g)$

$B^{4+}(g)$ is hydrogen-like with $Z = 5$. Hence the ionization energy for this will involve energy equal to $RchZ^2$ where $Z = 5$ and Rch amounts to the ionization energy of H, i.e., 13.6 eV.

$\therefore I = 13.6 \times 5^2$ eV $= 340$ eV $\Rightarrow 340 \times 96.485$ kJ mol^{-1} $= 32.8$ MJ mol^{-1}.

Vertical trends. As one proceeds down a given group of elements in the *s*- and *p*-block, the first ionization energies show a gradual decrease except in a few cases (see below). The effective nuclear charge increases down a given group of elements; at the same time, the maxima in the radial distribution function for the valence electron also shifts outward from the nucleus. This second factor appears to dominate and results in an overall decrease in ionization energy down a group. However, differences in penetration and consequent shielding effects of various intermediate core electrons influence the magnitude of ionization energy in the same manner as encountered before with atomic size. Thus, in group 13(III), Ga shows a slightly higher ionization energy than Al owing to "*transition metal contraction*" along the 3*d* series. Similarly, Tl in the same group shows a higher value of I_1 consequent to "*lanthanoid contraction*" noted earlier. Similar (apparent) anomalies are observed in the ionization energies of Group 14 elements but the effect appears to ease out to Groups 15 onward. As the number of *p*-electrons in the valence shell increases, their screening effect makes a more direct influence on the increasing nuclear charge and the ease with which the first electron gets ionized. At Group 17(VII), the ionization energies decrease in a more regular fashion.

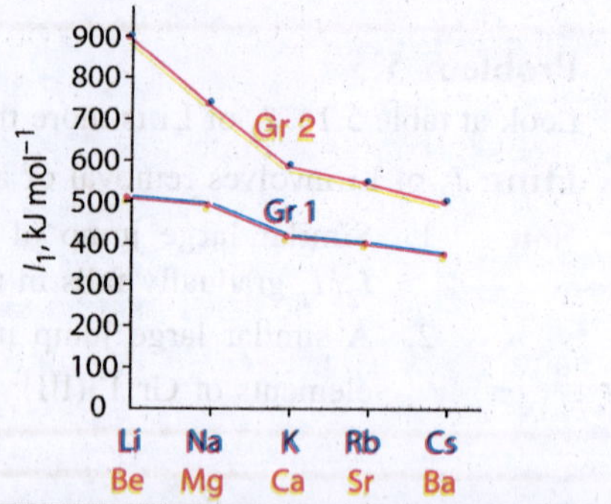

Fig.5.13: Variation of I_1 for Groups 1 and 2.

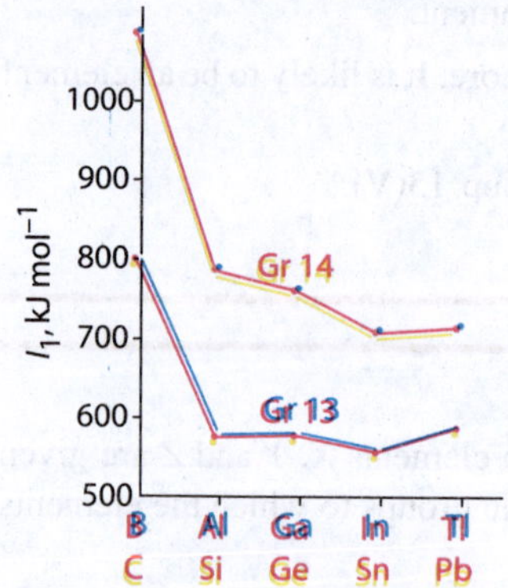

Fig.5.14: Variation of I_1 for Groups 13, and 14.

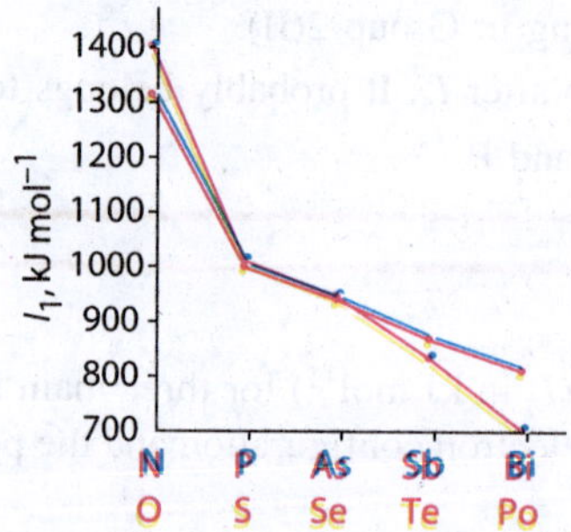

Fig.5.15: Variation of I_1 for Groups 15 and 16.

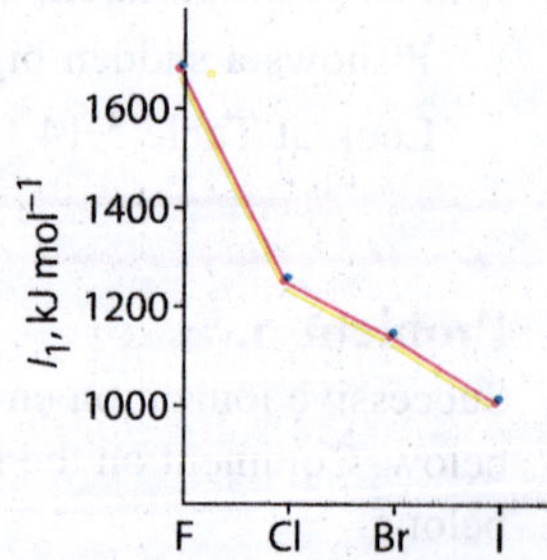

Fig.5.16: Variation of I_1 for Group 17.

Among the *d*-block elements, there is no consistent vertical trend in the first ionization energies (Fig. 5.17). I_1 increases downward in groups, 5(V) and 6(Cr) but in groups 7(Mn), 8(Fe) and 9(Co), I_1 first decreases to the second (4*d*-) transition series and

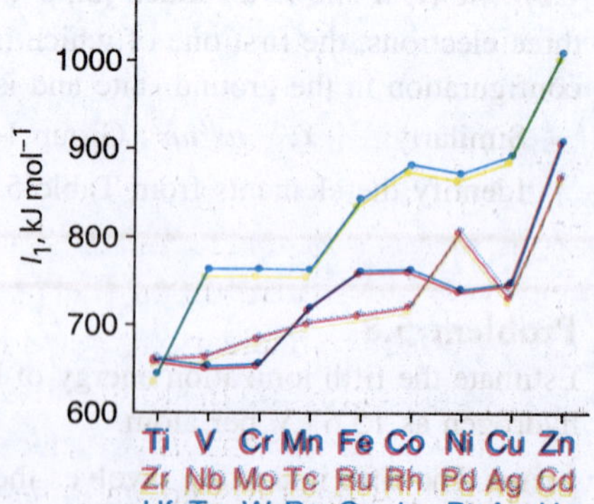

Fig.5.17: Plot of first ionization energies of the *d*-block elements.

then again rises to the third transition series (5*d*-). The same trend is observed in groups 11(Cu) and 12(Zn) but in group 10(Ni), I_1 rises steadily along Ni—Pd—Pt.

Several factors appear responsible in determining the ionization energies among the *d*-block elements. The extent to which the (*n*-1) *d*-orbitals penetrate the inner core compared to *ns* and *np* orbitals varies from $n = 3$ to $n = 5$. The consequent variation in shielding and effective nuclear charge plays a major role in determining the ionization energy. We have also mentioned earlier the effect of lanthanoid contraction and relativistic effect among the heavier *d*-block elements. From an overall consideration we expect the ionization energy to increase down any vertical group. But there are also other factors to add to the complicacy of the situation; we mention only two points—(i) inter-electron repulsion, which decreases in general with increasing principal quantum number and (ii) the exchange energy, which stabilizes a configuration with parallel spins. When an electron ionizes from a neutral transition metal atom, the *nd* orbital is stabilized to a greater extent than the $(n + 1)s$-orbital. The resulting $M^+(g)$ ions now all have d^n electron configurations:

M(g)	$\longrightarrow$	**M⁺(g)**	
$3d^n4s^2$	$\longrightarrow$	$3d^{n+1}$	M = Ti, V, Mn, Co, Ni
$3d^n4s^1$	$\longrightarrow$	$3d^n$	M = Cr, Cu

The stabilization gained in the $M^+(g)$ ion is different for the different transition series and also in different parts of the same series. This may, at least partly, account for the irregular trend observed in the variation of I_1 among the three series. We shall discuss the influence of exchange energy later in connection with the horizontal trend.

Horizontal trend: The first ionization energies of the *s*- and *p*-block elements show a more or less similar increasing trend along a horizontal row in the periodic table (Fig. 5.18). The overall increase in ionization energy along a period is a result of the increasing effective nuclear charge and consequent contraction in atomic size. Elements with ns^2np^1 and ns^2np^4 electron configurations (ground state) deviate from the general trend, their ionization energies being lower than expected from simple extrapolation. The

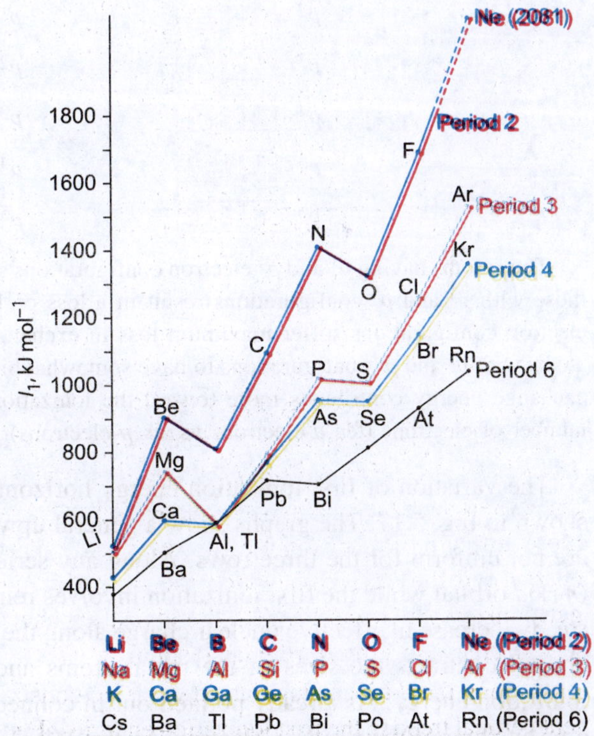

Fig.5.18: Plot of first ionization energies (I_1) of the second, third, fourth and sixth row of elements (*s*- and *p*-block) in the periodic table.

discontinuity at ns^2np^1 configurations for B, Al and Ga may be attributed to the lower binding energy of the *np* orbitals relative to the *ns* orbitals—we recall the fact that the *np* orbitals, being less penetrating, get better shielding from the enhanced nuclear charge. The discontinuity gradually decreases to heavier elements in the group—at row 6 we find thallium to maintain the usual upward trend. This may be correlated to lanthanoid contraction and relativistic contraction mentioned earlier.

In the ns^2np^4 ground state electron configuration, the last electron enters an orbital which already contains one electron and thus experiences an additional electrostatic repulsion. Since the spin of this electron is opposite to that of the pre-existing electron, there is no gain from exchange energy either. Hence there is a drop in I_1 for this configuration. The inter-electron repulsion factor diminishes as one moves to larger valence orbitals with higher '*n*' down the group: Te and Po do not show any downward trend in their I_1 values.

The variation in exchange energy stabilization consequent to ionization has a small but significant contribution to ionization energy. The change (in terms of *K*; section 4.4.3) may be readily estimated from the number of electrons with parallel spin before and after ionization:

Electron configuration of *M*(*g*)		Electron configuration of $M^+(g)$	Loss in exchange energy
p^1	[↑][ ][ ]	p^0	0
p^2	[↑][↑][ ]	p^1	*K*
p^3	[↑][↑][↑]	p^2	2*K*
p^4	[↑↓][↑][↑]	p^3	0
p^5	[↑↓][↑↓][↑]	p^4	*K*
p^6	[↑↓][↑↓][↑↓]	p^5	2*K*

Thus atoms having p^1 and p^4 electron configurations suffer no loss in exchange energy on ionization; those with p^2 and p^5 configurations result in a loss of 1*K* in exchange energy; atoms with p^3 and p^6 electron configurations suffer maximum loss in exchange energy (2*K*) on ionization. Thus we expect atoms with p^3 and p^6 configurations to have somewhat higher ionization energy. Soon we shall find that exchange energy contributes more toward the ionization of transition elements which have a higher number of electrons (ten *d*-electrons vs six *p*-electrons).

The variation of first ionization energy horizontally across the three rows in *d*-block was shown in Fig. 5.17. The graphs show a general upward trend but with many deviations which are not uniform for the three rows. Along any series, electrons are successively added to the (*n*–1)*d* orbital while the first ionization involves removal of an electron from the *ns* level. The steady increase in effective nuclear charge along the series (the *d*-electrons have poor shielding ability) contracts the sizes of the metal atoms and leads to an overall increase in the first ionization energy. As already pointed out in connection with the vertical trend, there is often a reshuffling of the residual electrons in the $M^+(g)$ ion to attain best available orbital binding energy for the electrons. The corresponding change in exchange energy influences the ionization energy to various extent depending on the configuration attained after ionization. We recapitulate the exchange energies (in terms of *K*) for various d^n configurations:

d^1	0	d^6	$10K$
d^2	K	d^7	$11K$
d^3	$3K$	d^8	$13K$
d^4	$6K$	d^9	$16K$
d^5	$10K$	d^{10}	$20K$

The second ionization energies of the *d*-block elements present a comparatively regular trend since most of them involve removal of an electron from the d^n configuration: $M^+(g;\ d^n) \rightarrow M^{2+}(g;\ d^{n-1})$. The corresponding loss in exchange energy is easily calculated:

Electron configuration and exchange energy		Loss in exchange energy (K)	Electron configuration and exchange energy		Loss in exchange energy (K)
$M^+(g)$	$M^{2+}(g)$		$M^+(g)$	$M^{2+}(g)$	
d^1 (0)	d^0 (0)	0	d^6 ($10K$)	d^5 ($10K$)	0
d^2 (K)	d^1 (0)	K	d^7 ($11K$)	d^6 ($10K$)	K
d^3 ($3K$)	d^2 (K)	$2K$	d^8 ($13K$)	d^7 ($11K$)	$2K$
d^4 ($6K$)	d^3 ($3K$)	$3K$	d^9 ($16K$)	d^8 ($13K$)	$3K$
d^5 ($10K$)	d^4 ($6K$)	$4K$	d^{10} ($20K$)	d^9 ($16K$)	$4K$

We observe that $M^+(g)$ ions having d^5 and d^{10} configurations undergo maximum loss in exchange energy on losing one more electron. Accordingly we expect that the second ionization energies of corresponding metals will be high. The values of I_2 in Table 5.15 support this conclusion for all three rows in the *d*-block. Indeed, plot of I_2 values of the metals in the first and second row (Fig. 5.19) shows distinct peaks at Cr/Mo (d^5 configuration of M^+) and at Cu/Ag (d^{10} configuration of M^+).

As we shall encounter in subsequent chapters, ionization energies are extremely helpful in understanding the chemistry of an element. The likely oxidation states of an element may be largely anticipated from the successive ionization energies of an element (see Problems 5.6–5.8). We can also understand the possibility of formation of an ionic compound through a relevant Born-Haber cycle (see later). However, in solution and in coordination compounds, additional factors like hydration energy, crystal field stabilization energy etc. introduce additional parameters for the ultimate chemical behaviour of an element.

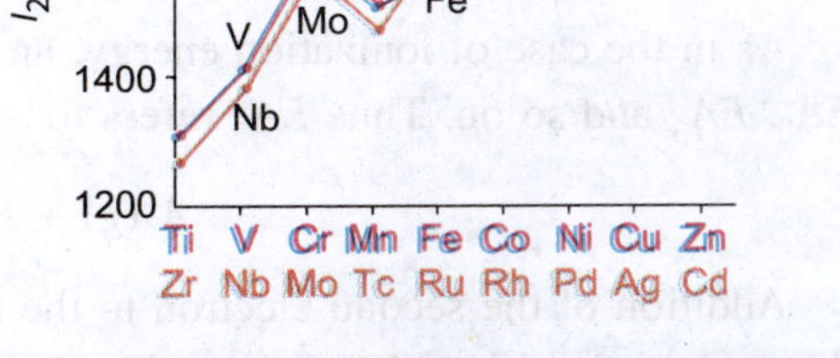

Fig.5.19: Plot of second ionization energy (I_2) of the 1st and 2nd row elements in *d*-block.

Valence Orbital Ionization Potential (VOIP)

Valence orbital ionization potentials are estimated from the energy difference between the various states of an atom as obtained from its atomic spectra. The energies of different orbitals are calculated quantum mechanically and the energy differences are correlated to appropriate electron configurations. The energy corresponding to photoionization of the atom is then estimated for various valence shell configurations and expressed as the VOIP (in eV; see Table 7.4). Trends in the VOIP values reflect the influence of Z_{eff}/r on the various orbitals (r is the calculated radial maxima for the orbital) and are essentially parallel to the trends of ionization energies.

5.3.5 Electron Affinity

The electron affinity of an element is traditionally defined as the energy change accompanying the addition of an electron to a free gaseous atom of the element in its ground state:

$$A(g) + e(g) \rightarrow A^{-}(g) \qquad \text{E.A. (electron affinity)} = -\Delta U$$

Note the sign convention; a positive electron affinity implies energy is released when the atom A gains an electron, that is, the ion $A^{-}(g)$ has a lower internal energy than the neutral atom, $A(g)$. However, this is *opposite* to the standard thermodynamic convention where energy released by the system is assigned a negative sign. An alternative definition of electron affinity is the energy of ionization of the $A^{-}(g)$ ion, that is, ΔU for the process

$$A^{-}(g) \rightarrow A(g) + e(g)$$

This definition does not require a sign change. Because this process is similar to the ionization process, in fact one step backward from the first ionization process of $A(g)$, electron affinity is sometimes described as the ***"zeroth ionization energy"***.

For thermodynamic calculations we often need the corresponding standard enthalpy of electron gain or electron attachment, $\Delta H^{\circ}_{\text{E.A.}}$ which obviously has the opposite sign of conventional electron affinity. Since the thermodynamic standard state is set at 298K, here also we have a small difference from $-\Delta U_{\text{E.A.}}$ which is defined at 0K. Applying the same argument as in the case of ionization energy (section 5.3.4), we may arrive at the relation

$$\Delta H^{\circ}_{\text{E.A.}} = -U - \tfrac{5}{2}RT$$

In a thermodynamic cycle, however, use of both ionization enthalpy and electron attachment enthalpy will result in cancellation of the $\frac{5}{2}RT$ terms in the two quantities; we are thus left with the relevant ΔU terms—ionization energy and electron affinity (with proper sign).

As in the case of ionization energy, an atom may have successive electron affinity values EA_2, EA_3 and so on. Thus EA_2 refers to $-\Delta U$ for the process

$$A^{-}(g) + e(g) \rightarrow A^{2-}(g).$$

Addition of the second electron to the negative A^{-} ion involves repulsion that outweighs the nuclear attraction; EA_2 is therefore negative, or ΔU is positive, implying that $A^{2-}(g)$ has a higher energy compared to free $A(g)$ or $A^{-}(g)$. Obviously, the same conclusion applies to all successive EA_n values.

The magnitude of electron affinity is largely determined by the nature of the lowest *unfilled* atomic orbital which will accomodate the added electron. (During ionization, an electron is lost from the highest occupied atomic orbital. Both types of orbitals are called *frontier orbitals*). An element will normally show a high electron affinity when the added electron enters a shell where it experiences a strong effective nuclear charge and low repulsion from the existing electrons. This generalization is reflected in the electron affinity values of different elements (Table 5.16 and Figures 5.20, 5.21).

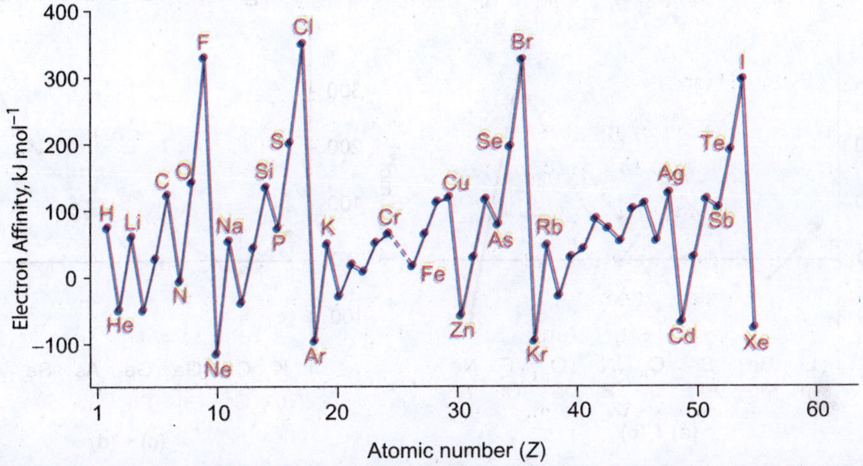

Fig.5.20: Variation of electron affinity along the periodic table.

Table 5.16

Electron affinities of elements. ($-\Delta U$ for $A(g) + e \rightarrow A^-(g)$* in kJ mol^{-1}).

Groups	1	2	3	4	5	6	7	8	9	10	11	12	13	14	15	16	17	18
PERIODS	H																	He
	72.8																	−50
	Li	Be											B	C	N	O	F	Ne
	59.6	−50											26.7	121.9	−7	141.0	328.0	−116
																−780*		
	Na	Mg											Al	Si	P	S	Cl	Ar
	52.9	−39											42.6	133.6	72.0	200.4	349.0	−97
																	−492*	
	K	Ca	Sc	Ti	V	Cr	Mn	Fe	Co	Ni	Cu	Zn	Ga	Ge	As	Se	Br	Kr
	48.4	−29	18.1	7.6	50.7	64.3	<0.0	15.7	63.8	111.5	118.5	−58	29	115.8	78	195	325	−97
	Rb	Sr	Y	Zr	Nb	Mo	Tc	Ru	Rh	Pd	Ag	Cd	In	Sn	Sb	Te	I	Xe
	46.9	−29	29.5	41.1	86.2	72.1	53.1	101.2	109.7	53.7	125.6	−68	29	116	103	190	295	−77
	Cs	Ba	La	Hf	Ta	W	Re	Os	Ir	Pt	Au	Hg	Tl	Pb	Bi	Po	At	Rn
	45.5	−29	48	0	31	78.6	14.5	106	151	205.3	222.8	−48	19	35.1	91.3	183	270	−68

* Values in second line for $A^-(g) + e \rightarrow A^{2-}(g)$.

Values from Miessler, G.L. and Tarr, Donald A, *Inorganic Chemistry*, 3rd ed, 2004, Pearson Education, Inc.

We observe that the electron affinity values vary in a periodic manner but the variation is not very regular. The noble gas elements, with their closed shell configurations, have the lowest electron affinities. Elements of Group 2 (ns^2 configuration) and Group 12 ($(n-1)d^{10}ns^2$) are also marked by negative electron affinities. Electron affinity tends to increase among the *p*-block elements from left to right along a row, reaching maxima at the halogens. Nitrogen, with its half filled $2p^3$ configuration has a negative electron affinity presumably due to pairing of the added electron with an electron in the same subshell and the consequent repulsion. The electron affinity decreases from the first row to second row in group 1 (Li–Na–K ...) as effective

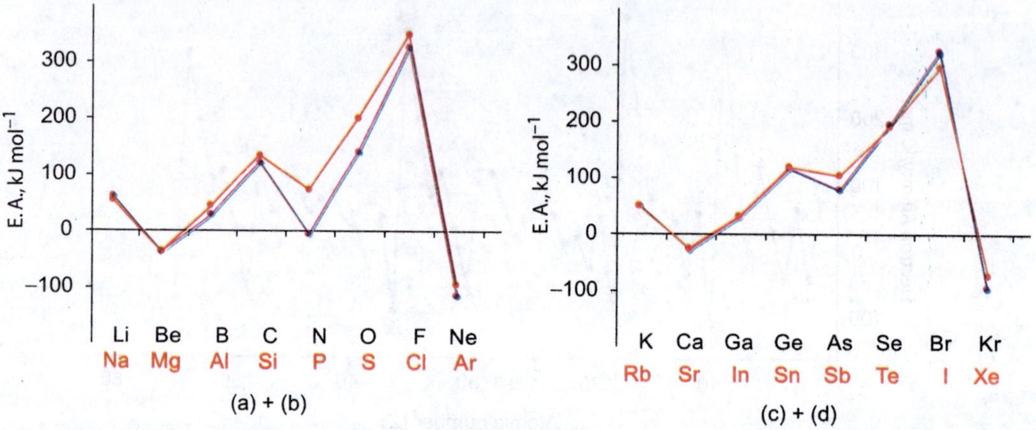

Fig.5.21: Variation of electron affinity horizontally along (a) second (b) third (c) fourth and (d) fifth row of main group elements.

nuclear charge decreases; repulsion from the single ns^1 electron is insignificant in these atoms. But in other groups, electron affinity values first show an increasing trend from the first to the second row of elements owing to reduced repulsion. Thus there is a distinct increase in electron affinity from the first row in the *p*-block (B–F) to the corresponding second row element (Al–Cl) in groups 13–17; as the new electron enters a less compact electron cloud in larger orbitals, repulsion becomes less. This is prominently illustrated by the pairs N–P, O–S and F–Cl. (Interestingly, the second element in each pair is marked by a lower electronegativity). The variation of electron affinity in lower rows of the periodic table, including the *d*-block, shows a complex interplay of effective nuclear charge as a consequence of varying extent of screening and the nature and population of the acceptor orbital.

Problem 5.7

Comment on the electron affinity values (kJ mol^{-1}): Na 53; Mg −39.

Solution: In sodium ([Ne]$3s^1$) the added electron enters the 3*s* orbital which gets fulfilled. The positive electron affinity value of 53 kJ mol^{-1} implies that the $Na^-(g)$ ion is more stable than the free Na(*g*) atom by this amount of energy.

In case of magnesium ([Ne]$3s^2$), the incoming electron must be placed in the 3*p* orbital which gets screened by the $3s^2$ core. Hence the last electron is much less tightly bound, making $Mg^-(g)$ unstable by 39 kJ mol^{-1} with respect to the free atom.

5.3.6 Electronegativity

The concept of electronegativity was first introduced by Linus Pauling (1930s) to describe the ability of an atom to attract electrons to itself in a chemical bond with another element. Relative numerical values of electronegativity (also called *electronegativity coefficients*) of different elements were compiled by Pauling from available bond energies (see later). Subsequently, a number of other approaches have been made to describe the same property of an element, giving rise to different "*electronegativity scales*".

The electronegativity values of elements in the same scale show periodic variation. However, electronegativity differs from other periodic properties described before (e.g., ionization energy, electron affinity etc.) as it is not an inherent property of an isolated atom. Hence the value of electronegativity assigned to an element depends on the theoretical basis of its assignment and also on the bonding environment of the atom. The value of electronegativity in the same scale may also vary depending upon the physical data used like bond energy, electron affinity etc. Hence electronegativity values are relatively less precise. Nevertheless, electronegativity values within the same scale provide satisfactory qualitative guideline to understand the chemical behaviour of an element or its compounds. Quantitative correlations are also quite satisfactory in many cases, for example the Schoemaker-Stevenson rule for the estimation of bond length (sec. 5.3.3).

Let us first summarize the main electronegativity scales:

(i) ***The Pauling scale*** (1932): This is the first and by far the most widely accepted scale, based on bond energies.

If two elements A and B have electronegativities x_A and x_B respectively, then

$$|x_A - x_B| = 0{\cdot}102\sqrt{\Delta}$$

where Δ is obtained from the bond energies (E, in kJ mol^{-1}; actually enthalpies) as:

$$\Delta = E_{A-B} - (E_{A-A} \cdot E_{B-B})^{1/2}$$

(ii) ***The Mulliken scale*** (1934): Average of ionization potential and electron affinity, both in electronvolts (eV) per atom, is taken as a measure of electronegativity.

$$x_M = \frac{1}{2}\,(\text{I.E.+E.A.})$$

where x_M is the electronegativity of an element in the Mulliken scale.

The ionization potential and electron affinity refer to the *valence state* of the element in a compound and not to the ground state of the atom (see later).

Mulliken electronegativity values (x_M) may be related to the Pauling electronegativity (x_P) through the relation

$$x_P = 1.35\,(x_M)^{1/2} - 1.37$$

(iii) ***The Allred and Rochow scale*** (1958): It relates electronegativity to the electric field at the surface of an atom (covalent radius = r).

$$x_{AR} = 0.744 + \frac{3590 Z_{eff}}{(r/pm)^2}$$

(iv) ***The Allen scale*** (1989): Electronegativity is expressed in terms of weighted average of the ionization energies (eV) of all the valence shell electrons of the free atom (ground state). As the ionization energies are obtained spectroscopically, the electronegativity calculated in this manner is called **spectroscopic electronegativity**. When adjusted to the range of Pauling scale, Allen's values agree well with Pauling values for *s*- and *p*-block elements. Difficulties arise with *d*-block elements as precise determination of number of *p*- and *d*-electrons in valence shell becomes a problem.

$$x_{\text{Allen}} = 0.169 < I.E._{val} > /\text{eV}.$$

(v) ***The Nagle scale*** (1990): Electronegativity is described in terms of polarizability (α) of an atom. For an element with *n* number of valence shell electrons

$$x_{\text{N}} = 1 \cdot 66 \left[n \frac{4\pi\varepsilon_0}{\alpha} \text{Å}^3 \right]^{1/3} + 0 \cdot 37$$

Several other electronegativity scales have also been proposed, for example those by *Sanderson* (1952, electron densities of atoms); *Pearson* (1990, average of electron affinity and ionization energy); *Jaffe* (1962, idea of orbital electronegativity from valence shell ionization energy and electron affinity) and others. Space limitation restricts us only to a brief discussion on the first three scales mentioned above.

Electronegativity values of the elements in different scales are given in Tables 5.17–5.18. The Pauling values given are those most commonly quoted, but some later modifications based on recent bond energy data have also been suggested. The values in *Mulliken* and *Allred-Rochow* scales have been suitably adjusted to conform to the values in Pauling scale.

The Pauling scale

Let us consider two covalent molecules, A—A and B—B. It is generally observed that reactions of the type

$$\text{A—A + B—B} \rightarrow \text{A—B + A—B}$$

are almost always exothermic; the bond A—B is thus more stable and hence stronger than either of the bonds A—A or B—B. If the bond A—B were purely covalent (like the bonds in A—A or B—B), that is, the electron pair forming the bond would have been equally shared by A and B, the bond A—B would not get this extra stability. The extra stability, Pauling suggested, was derived from unequal sharing of the bonding electrons between A and B, giving rise to some ionic character. The extent of this ionic character, as measured by the extra stability of the bond, is naturally dependent on the inequality in sharing the bonding electrons. The inequality arises from the relative electron-attracting powers of the atoms A and B in the bond. Pauling termed this property as electronegativity (x) and correlated the extra stability of the bond A—B to the difference of electronegativity between A and B. In terms of bond energies (enthalpies), the extra stability may be written as the difference between (i) the actual (thermochemical) bond energy of the bond A—B, $E_{\text{A—B}}$ and (ii) the hypothetical energy of the bond A—B if it were fully covalent, E_{cov}:

$$E_{\text{A—B}} - E_{\text{cov}} = \Delta.$$

Table 5.17
Electronegativity of s- and p-block elements

	H							He
P	2.20	P: Pauling scale						
M	3.06	M: Mulliken scale						
A.R.	2.20	A.R.: Allred-Rochow scale						
	Li	**Be**	**B**	**C**	**N**	**O**	**F**	**Ne**
P	0.98	1.57	2.04	2.55	3.04	3.44	3.98	—
M	1.28	1.99	1.83	2.67	3.08	3.21	4.42	4.60
A.R.	0.97	1.47	2.01	2.50	3.07	3.50	4.10	5.10
	Na	**Mg**	**Al**	**Si**	**P**	**S**	**Cl**	**Ar**
P	0.93	1.31	1.61	1.90	2.19	2.58	3.16	—
M	1.21	1.63	1.37	2.03	2.39	2.65	3.54	3.36
A.R.	1.01	1.23	1.47	1.74	2.06	2.44	2.83	3.30
	K	**Ca**	**Ga**	**Ge**	**As**	**Se**	**Br**	**Kr**
P	0.82	1.00	1.81	2.01	2.18	2.55	2.96	—
M	1.03	1.30	1.34	1.95	2.26	2.51	3.24	2.98
A.R.	0.91	1.04	1.82	2.02	2.20	2.48	2.74	3.10
	Rb	**Sr**	**In**	**Sn**	**Sb**	**Te**	**I**	**Xe**
P	0.82	0.95	1.78	1.96	2.05	2.10	2.66	2.6
M	0.99	1.21	1.30	1.83	2.06	2.34	2.88	2.59
A.R.	0.89	0.99	1.49	1.72	1.82	2.01	2.21	2.40
	Cs	**Ba**	**Tl**	**Pb**	**Bi**			
P	0.79	0.89	2.04	2.33	2.02			
M	0.70	0.90	1.80	1.90	1.90			
A.R.	0.86	0.97	1.44	1.55	1.67			

Table 5.18
Electronegativity of d- and f-block elements

	Sc	Ti	V	Cr	Mn	Fe	Co	Ni	Cu	Zn
P	1.3	1.5	1.6	1.7	1.5	1.8	1.9	1.9	1.9	1.6
A.R.	1.20	1.32	1.45	1.56	1.60	1.64	1.70	1.75	1.75	1.66
	Y	**Zr**	**Nb**	**Mo**	**Tc**	**Ru**	**Rh**	**Pd**	**Ag**	**Cd**
P	1.2	1.4	1.6	1.8	1.9	2.2	2.9	2.2	1.9	1.7
A.R.	1.11	1.22	1.23	1.30	1.36	1.42	1.45	1.35	1.42	1.46
	La	**Hf**	**Ta**	**W**	**Re**	**Os**	**Ir**	**Pt**	**Au**	**Hg**
P	1.1	1.3	1.5	1.7	1.9	2.2	2.2	2.2	2.4	1.9
A.R.	1.08	1.23	1.33	1.40	1.46	1.52	1.55	1.44	1.42	1.44

	Ce	Pr	Nd	Pm	Sm	Eu	Gd	Tb	Dy	Ho	Er	Tm	Yb
A.R.	1.08	1.07	1.07	1.07	1.07	1.01	1.11	1.10	1.10	1.10	1.11	1.11	1.06
	Lu	**Th**	**Pa**	**U**	**Np**	**Pu**							
A.R.	1.14	1.11	1.14	1.22	1.22	1.22							

Δ is also called the ionic-covalent resonance energy. In valence bond description, the energy of a bond is calculated theoretically by assuming resonance among several ionic and covalent structures (see later).

Pauling suggested that Δ is related to the difference of electronegativity of A and B, $|x_A - x_B|$:

$$|x_A - x_B| = 0 \cdot 208\sqrt{\Delta} \text{ where bond energies are in kcal mol}^{-1}.$$

or, $|x_A - x_B| = 0 \cdot 102\sqrt{\Delta}$ for bond energies in kJ mol^{-1}.

The factor 0.208 appears from conversion of bond energy data in electron-volt *per molecule* to *kilocalories per mole* (the electron volt was the polpular unit earlier).

$$1 \text{ eV molecule}^{-1} \equiv 23.06 \text{ kcal mol}^{-1}; \quad 1/\sqrt{23 \cdot 06} = 0 \cdot 208$$

Similarly, when bond energies are expressed in kJ mol^{-1}, we have,

$$1 \text{ eV molecule}^{-1} \equiv 96.485 \text{ kJ mol}^{-1}; \quad 1/\sqrt{96 \cdot 485} = 0 \cdot 102$$

E_{cov}, the energy of the hypothetical 100% covalent bond was first taken as the arithmetic mean of the covalent bond energies in A_2 and B_2 molecules, $E_{A—A}$ and $E_{B—B}$, respectively:

$$E_{cov} = (E_{A—A} + E_{B—B})/2$$

But this led to some difficulties which could be avoided by taking the geometric mean:

$$E_{cov} = (E_{A—A} . E_{B—B})^{1/2}, \text{ giving } \Delta = E_{A—B} - (E_{A—A} . E_{B—B})^{1/2}$$

The final expression for the electronegativity difference thus becomes

$$x_A - x_B = 0.102\sqrt{E_{A—B} - (E_{A—A} . E_{B—B})^{1/2}} \qquad \text{(see background)}$$

The bond energies (kJ mol^{-1}) represented by the respective *E*-terms are obtained thermochemically. Pauling next assigned (arbitrarily) a value of 4.0 for the electronegativity of fluorine and prepared a scale of electronegativity for other elements. The present values of electronegativity in the Pauling scale are somewhat different from Pauling's original values; these have been calculated by Pauling's procedure using later thermochemical data. Some further revisions are also available.

Background

Pauling observed some interesting trends in bond dissociation energies of gaseous elements (where bonds involve two like atoms) and of compounds, mostly gaseous halides. He noted that

(i) polar bonds formed by various elements (A) with halogens (X) are almost always stronger than the nonpolar (single) bonds between the elements themselves (A—A) or between the halogens themselves (X–X).

(ii) The gaseous halides of alkali metals, actually ion-pairs, have the highest bond energies for any given halogen.

(iii) The decreasing order of bond energy appears as

ion-pair > polar covalent bond > non-polar covalent bonds.

Pauling assumed that the bond energy appears as a sum of covalent and ionic contributions. The covalent contribution may be taken as the average of the element-element and halogen-halogen covalent bond energies. The ionic contribution may be taken as proportional to the partial charges δq_+ and δq_- developed on the atoms from unequal sharing of the bonding pairs; these charges operate at the bond distance r;

$$E_{\text{ionic}} = k.\frac{\delta q_+ \times \delta q_-}{r} \qquad (k = \text{constant})$$

In general, the bond energy (actually enthalpy) E_{A-B} between two unlike atoms A and B may now be expressed as: $E_{A—B} = E_{\text{cov}} + E_{\text{ionic}}$.

If we neglect the variability of the energy on the bond distance, r may be included in the constant k. So one gets

$$E_{A—B} = \frac{1}{2}[E_{A—A} + E_{B—B}] + k.\delta q_+ \delta q_-.$$

According to the definition of electronegativity accepted, each of the partial positive and negative charges are proportional to the difference of electronegativity between A and B, $(x_A - x_B)$:

$$\delta q_+ = k'(x_A - x_B); \qquad \delta q_- = k'(x_A - x_B).$$

Hence, $$E_{A-B} = \frac{1}{2}[E_{A-A} + E_{B-B}] + k.k'^2(x_A - x_B)^2$$

Setting kk'^2 equal to 96.5 kJ eV^{-1} and rearranging, one gets

$$96.5\,(x_A - x_B)^2 = E_{A-B} - \frac{1}{2}[E_{AA} + E_{BB}]$$

or, $$|x_A - x_B| = \frac{1}{\sqrt{96.5}}\left[E_{A-B} - \frac{E_{AA} + E_{BB}}{2}\right]^{1/2} = 0.102\left[E_{A-B} - \frac{E_{AA} + E_{BB}}{2}\right]^{1/2}$$

The arithmetic mean was subsequently replaced by the geometric mean:

$$|x_A - x_B| = 0.102\left[E_{A-B} - \sqrt{E_{AA}.E_{BB}}\right]^{1/2}$$

Discussion: The Pauling scale of electronegativity has several merits as well as demerits, as is to be expected from any pioneering work. The approach is straightforward and involves simple calculations based on available data. The scale offers a good first hand knowledge of bond polarity and hence helps to predict the course of many reactions with reasonable correctness (see Applications). The basis of 'extra bond energy' fits such speculations because most reactions are enthalpy driven and exothermic at room temperature. Though imprecise to some extent, the Pauling values may even be used to predict bond distances within reasonable limits as shown in Schoemaker Stevenson rule (sec.5.3.3). We shall come across many applications of this scale in this and subsequent chapters.

The main demerit of Pauling's approach seems to be the idea of a single invariant electronegativity value for an element. It is easy to understand that the attraction exerted by any atom on bonding electrons should depend upon the effective charge acquired by the atom in its chemical environment in the molecule. Oxidation state, state of hybridization and the nature of orbitals are equally involved in determining the electronegativity of an element. Another drawback of the scale is the uncertainty in bond enthalpy values. To obtain reliable values, one has to average over a large number of compounds to cancel out experimental errors and other variables.

The Mulliken scale

In this scale, the electronegativity of an element (x_M) is taken as the average of its ionization energy and electron affinity, both expressed in electronvolts.

$$x_M = \frac{I.E. + E.A.}{2} \quad \text{(both in eV)}$$

x_M may be related to Pauling electronegativity values (x_P) through a relation like

$$x_P = 1.35\sqrt{x_M} - 1.37$$

Other linear relations have also been suggested.

Background

In a molecule AB, the tendency of any atom to attract electron to itself and become more negative may be measured by the relative stabilities of structures A^+B^- and A^-B^+. Using the free atom ionization energy (*I*) and electron affinity (*EA*), the energy difference between the structures A^+B^- and A^-B^+ may be expressed as

$$E_{A^+B^-} - E_{A^-B^+} = (I_A - EA_B) - (I_B - EA_A) = (I_A + EA_A) - (I_B + EA_B)$$

Mulliken defined the electronegativity difference between atoms A and B as $\left(E_{A^+B^-} - E_{A^-B^+}\right)/2$. Accordingly the electronegativity of an element in the Mulliken scale (x_M) is given by

$$x_M = \frac{I + EA}{2}$$

The implications of Mulliken's theory may be far-reaching, as shown by Jaffé (1962). In the original version of Mulliken, the ionization energy and electron affinity of an element refer to the ground state neutral atom. According to Mulliken-Jaffé interpretation, the electron-attracting ability of an atom in a compound should be calculated from the *valence state ionization energy* and *valence state electron affinity* of the element in the valence state actually present in a molecule. Naturally, this involves the nature of the particular orbital to which an atom pulls the electron charge. Electronegativity is thus more a property of the orbital than an invariant property of the neutral atom.

Though theoretically elegant, the Mulliken-Jaffé scale involves complicated calculations. Availability of reliable electron affinity data also stands against wide applicability of the method. Assignment of hybridization of the element is also not unambiguous in every case.

The Mulliken electronegativity of an element has been termed the *absolute electronegativity* by Pearson who related it to the absolute hardness of a chemical species in terms of the energies of highest occupied molecular orbital (HOMO) and lowest unoccupied molecular orbitals (LUMO) (Chapter 10).

The Allred and Rochow scale

In this scale, the electronegativity of an atom is measured in terms of the electrostatic force F exerted by the nucleus of the atom on an electron at a distance of the covalent radius (r_{cov}) from the nucleus. Z_{eff}, the effective nuclear charge of the atom (in units of e), may be calculated using Slater's rule for *all* the electrons in the atom. The force F may be expressed as

$$F \propto \frac{Z_{eff}.e^2}{r_{cov}^2}$$

Some empirical adjustments are then made to bring the calculated electronegativity coefficients in the same range of values as in the Pauling scale. The Allred-Rochow electronegativity values (x_{AR}) are then expressed by the relation

$$x_{AR} = \frac{0.3590\ Z_{eff}}{(r_{cov}\ /\ \text{Å})^2} + 0.744 = \frac{3590\ Z_{eff}}{(r_{cov}\ /\ \text{pm})^2} + 0.744.$$

The electronegativity calculated using available values of r_{cov} are in fair agreement with Pauling values in most cases (Tables 5.18 and 5.19).

Problem 5.9

Calculate the electronegativity of bromine in the Allred-Rochow scale. r_{cov} = 114 pm.

Solution: With r_{cov} in pm, the A—R electronegativity is given by

$$x_{AR} = \frac{3590 \times Z_{eff}}{(r_{cov}\ /\ \text{pm})^2} + 0.744$$

Z_{eff} is calculated by taking into account the screening by all electrons:

$[Br = (1s)^2\ (2s2p)^8\ (3s3p3d)^{18}\ (4s4p)^7]$

$Z_{eff} = 35 - [0.35 \times 7 + 0.85 \times 18 + 1.00 \times 10] = 7.25$

$$\therefore\ x_{AR}(\text{Br}) = \frac{3590 \times 7.25}{114^2} + 0.744 = 2.75$$

Problem 5.10

Calculate A—R electronegativity of Au(Z = 79). r_{cov} = 134 pm

Hint: $Z_{eff} = 79 - [1 \times 0.35 + 18 \times 0.85 + 60 \times 1] = 3.35$; $x_{AR} = 1.41$

The Allred-Rochow interpretation of electronegativity rests directly on two fundamental properties of the atom — the effective nuclear charge and the covalent radius. As such it appears theoretically more elegant than the Pauling approach based solely on thermochemical data. The Allred-Rochow approach enables calculation of electronegativity values of atoms for which no thermochemical data are available, for example, the noble gases. The A—R values are also consistent with the electron configurations of the atoms. The horizontal and vertical trends in electronegativity are at par with similar other periodic properties like size and ionization energy which are also influenced strongly by effective nuclear charge. Some typical examples will be discussed shortly.

However, the Allred-Rochow approach also suffers from the difficulty in obtaining precise value for the covalent radius of an atom. Also, the covalent radius of an atom is susceptible to change with oxidation state and coordination number.

Periodic trend

Though electronegativity is not a precisely measurable quantity and the same element is assigned various values in different scales, the overall trend of variation of electronegativity in any one scale across the periodic table reflects the same general trend as observed for enthalpies of ionization and electron attachment (Fig. 5.22 and 5.23).

1. *Moving horizontally* along any period, electronegativity generally increases from left to right. This is very prominent among the *s*- and *p*-block elements. Among the *d*-block elements,

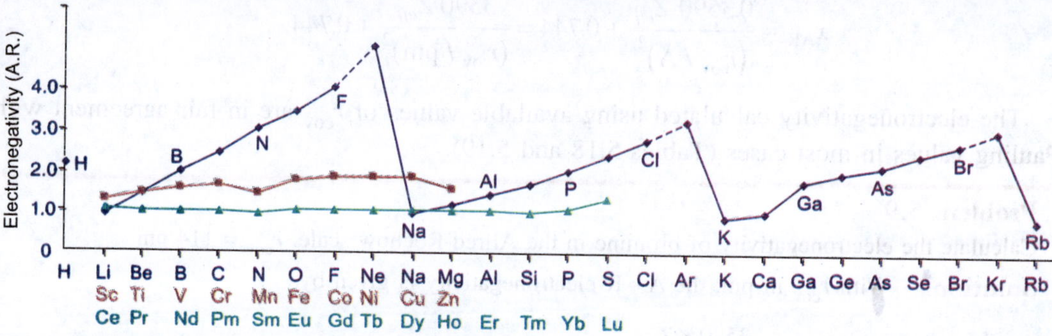

Fig.5.22: Variation of electronegativity (A.R.) horizontally along the periods

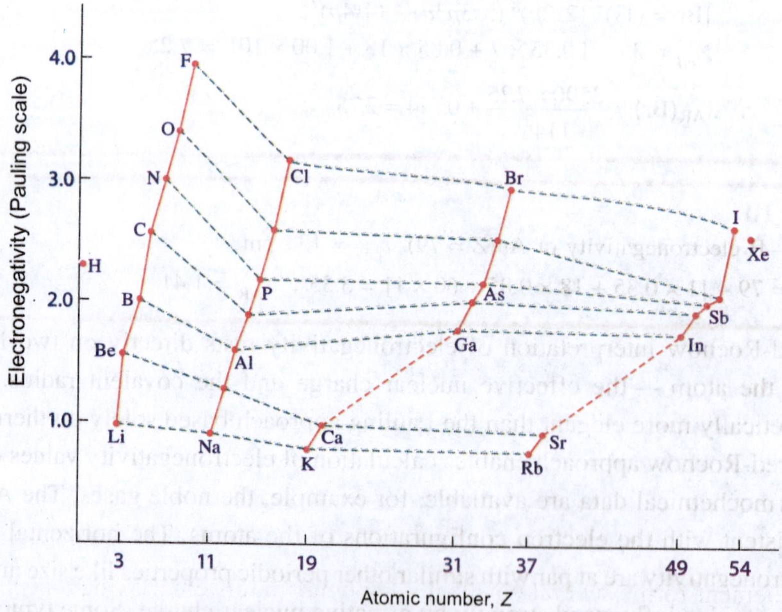

Fig.5.23: Variation of electronegativity (Pauling) along periods (———) and groups (- - -)

the increase is relatively slower. Among the *f*-block elements, electronegativity remains more or less constant in a series.

The observed trend is by and large consistent with increasing effective nuclear charge and decreasing radius of the atoms.

2. *Down a column of the periodic table*, electronegativities of the elements generally decrease. Discontinuities are observed due to rapid variations of size and effective nuclear charge along the *d*- and *f*-block elements. Consequently, elements following these blocks have electronegativity values which are higher than expected: Ga, Ge, As, Se and Br have high electronegativity because of *d*-level contraction or scandide contraction. Tl, Pb and Bi behave similarly because of lanthanoid contraction.

In the d-block, the variation of electronegativity among the three transition series is interesting. For corresponding elements in the three transition series, electronegativity varies as

1st transition series > 3rd transition series > 2nd transition series.

Elements of the third series, which have nearly the same size as the elements of the second series, have higher effective nuclear charge owing to the intermediate 4*f* subshell. Relativistic effect also contributes to the increase in electronegativity.

3. Electronegativity falls sharply from the second row elements (C, N, O, F) to the corresponding third row element (Si, P, S, Cl). The variation is less marked on passing to subsequent rows. Rather, the *p*-block elements of the fourth row show slightly higher electronegativity as explained above. This has strong influence on the chemistry of the elements in the fourth period (Ge—Br).

4. Hydrogen is marked by a high electronegativity compared to the alkali metals with which it is conventionally placed in the periodic table. We remember that hydrogen is actually unique in many of its properties—its single electron just outside a bare proton results in a high effective nuclear charge operative at a small distance, resulting in its high electronegativity.

Similarly, calculations suggest that the electronegativities of the noble gases may be larger than those of the preceding halogen. Greater effective nuclear charge and smaller size (compared to the preceding halogen atom) do really lead to a high estimate of their electronegativity. However, the estimated values vary significantly according to the method adopted. For example, a suggested alternative for the electronegativity of neon is 4.79 (vs. 5.10).

Variation of electronegativity of an element

The electronegativity of an element varies with its bonding situation —mainly oxidation state and the nature of hybrid orbitals involved.

A higher positive oxidation state results in a higher electronegativity, as illustrated by the following electronegativity values (A–R):

Fe(II)	1.83	Tl(I)	1.62	Sn(II)	1.80	Pb(II)	1.87
Fe(III)	1.96	Tl(III)	2.04	Sn(IV)	1.96	Pb(IV)	2.33

When an element forms several covalent compounds in different oxidation states with the same element, the bond enthalpy varies. One can use such data to estimate the variation of electronegativity. The following problem will be illustrative.

Problem 5.11

Calculate the Pauling electronegativity (x) of iodine in IF and IF_7 from the following bond enthalpy data (kJ mol^{-1}). Take $x_F = 4.0$.

IF: I—F = 278 IF_7: I—F = 231 I_2 = I—I = 149 F_2 = F—F = 155

Solution: The Pauling electronegativity of I (x_I) and of F(x_F) are related as

$$x_F - x_I = 0.102\left(E_{I-F} - \sqrt{E_{I-I}.E_{F-F}}\right)^{1/2}$$

In IF, $x_I = x_F - 0.102\left(E_{I-F} - \sqrt{E_{I-I}.E_{F-F}}\right)^{1/2} = 4.0 - 0.102\,(278-152)^{1/2} = 2.86$

In IF_7, $x_I = 4.0 - 0.102\,(231-152)^{1/2} = 3.09$

Tabulated value of electronegativity of I in Pauling scale = 2·66

When an element forms compounds with different types of hybrid orbitals, its electronegativity varies according to the composition of the hybrid orbitals. We know that s-orbitals penetrate most into the inner electron core and experience higher effective nuclear charge compared to other orbitals (penetration decreases as $s > p > d > f$). Accordingly, a hybrid orbital with greater s-character should be able to exert greater effective nuclear charge on the bonding electrons and thus show higher electronegativity. This is readily exemplified by the estimated values for electronegativities of carbon and nitrogen in compounds formed with different hybridizations (Pauling scale).

	Tabulated value	sp^3	sp^2	sp
C	2.55	2.5	2.75	3.3
N	3.04	3.7	3.9	4.7

The effect of increased electronegativity is seen among the acidic characters of hydrocarbons and basic characters of amines:

	sp^3	sp^2	sp	
	CH_4	C_2H_4	C_2H_2	*(acidity*
pK_a	48	44	25	*increases)*
	Me_3N:	C_5H_5N:	$Me-C\equiv N$:	*(basic*
pK_b	4.2	8.8	—	*character decreases)*

The partial charge induced by neighbour atoms may also significantly alter the electronegativity of an atom. Consider the reactions

$$CH_3I + OH^- \longrightarrow CH_3OH + I^-, \quad \text{but} \quad CF_3I + OH^- \longrightarrow CF_3H + IO^-$$

In the second reaction, the presence of fluorine atoms induces a partial positive charge on the carbon and effectively increases its electronegativity. Ultimately carbon becomes more electronegative than iodine, inducing a positive charge on the iodine atom. Nucleophilic attack by the OH^- ion now takes place at the positive iodine, giving OI^-.

A similar situation arises in the following reactions

$$CH_3I + Na^+[Mn(CO)_5]^- \longrightarrow NaI + CH_3Mn(CO_5)$$

$$2CF_3I + Na^+[Mn(CO)_5]^- \longrightarrow NaI + Mn(CO)_5I + C_2F_6$$

Group electronegativity

The overall effect of charge, substituents, hybridization of the bonding orbitals etc. of a group on electronegativity may be summarily expressed in terms of its group electronegativity. We have already seen how CH_3I and CF_3I differ in their reactions with alkali. Here the CH_3 and CF_3 groups may be assigned separate overall electronegativity values characterizing each group—this will represent the electronegativity of carbon as modified by the attached hydrogen or fluorine atoms. Such estimates may be made from the atomic electronegativities or from other physical measurements like x-ray photoelectron spectroscopy, infrared spectroscopy, dipole moment measurement, kinetic studies etc. Some representative values in Pauling scale are:

—CH_3 2.34;	—CF_3 3.35;	—CCl_3 3.03;	—COOH 2.8-3.5;	—NO_2 3.4;	C_6H_5—3.0

The value of electronegativity for the methyl group suggests that it should be slightly more electronegative than hydrogen. Yet the methyl group (and other alkyl groups) behaves as a better donor of electron density presumably because of its larger size which makes it more easily polarizable.

Optical electronegativities have been assigned to metals and ligands in connection with ligand to metal charge transfer bands (sec. 27.2).

Applications

Electronegativity has found widespread applications in all spheres of chemistry in spite of its imprecise nature. It provides a rough but useful guide to the electron distribution in a bond and hence the bond type. An ionic compound is usually formed by two elements if they differ in electronegativity by more than two units. In covalent bonds, electronegativity helps us to understand the bond polarity. For example, electronegativity of carbon and silicon (Pauling scale) indicates the following charge distribution:

δ^+		δ^-		δ^-		δ^+
Si	—	H		C	—	H
1.9		2.2		2.55		2.2

The hydrogen atoms in silicon hydrides are thus negatively charged; they are readily attacked by protons (hydrons) of water forming H_2 molecules. Hydrocarbons, on the other hand, are inert toward such hydrolytic reaction.

For enthalpy controlled and exothermic reactions, the direction of reaction between molecular species may be expected to be the one in which bonds are formed between the most electronegative and the least electronegative element, particularly when the total number of bonds on the two sides of the reaction are equal and similar:

$$BBr_3 + PCl_3 \rightarrow BCl_3 + PBr_3 \qquad 2HI + Cl_2 \rightarrow 2HCl + I_2$$

$$SiCl_4 + LiAlH_4 \rightarrow SiH_4 + LiCl + AlCl_3 \qquad TiCl_4 + 4C_2H_5OH \rightarrow Ti(OC_2H_5)_4 + 4HCl.$$

While examples of this type are numerous, exceptions are also frequent owing to interplay of several factors other than electronegativity—for example, entropy effect, packing efficiency, shift of equilibrium and kinetic effects. One very commonly quoted example is the reaction

$$CsF + LiI \rightarrow CsI + LiF$$

which proceeds to right in spite of cesium being the *least* electronegative element and fluorine being the *most* electronegative element. This apparent paradox (*Pearson Pauling paradox*, see Hard and Soft Acids and Bases) may be rationalized by the higher lattice energy of LiF.

Since the concept of electronegativity was developed by Pauling on the basis of bond enthalpies, one may expect that a small difference of electronegativity between two elements will give rise to a weak bond between these elements and hence an unfavourable enthalpy of formation for their binary compound. This is consistent with the endothermic nature of many compounds, as for example:

Compound	ΔH_f° (kJ mol^{-1})	Δx_P	*Compound*	ΔH_f° (kJ mol^{-1})	Δx_P
PH_3	+5.4	0.01	CS_2	+89.7	0.03
AsH_3	+66.44	0.02	NCl_3	+232	0.12

But electronegativity difference alone cannot be taken as an exclusive guide to enthalpy of formation of a compound. The latter quantity (ΔH_f°) is determined by the difference in bond enthalpies in the elements and in the compound. The bond enthalpies are again determined by the efficiency of overlap between the valence orbitals of the combining atoms. Thus, the large increase in enthalpy of formation from PH_3 to AsH_3 actually reflects poorer overlap due to mismatch in size and energy between the combining orbitals. The very high positive enthalpy of NCl_3 results from the high bond dissociation enthalpy of $N \equiv N$. The importance of overlap is also helpful in understanding the stability of covalent bonds between two atoms of the same element where the difference of electronegativity is zero.

A knowledge of electronegativity often helps us understand the properties of a compound in a qualitative manner, but with due caution to possible involvement of other factors. Thus we can readily understand the progressively increasing acidic character along the following series

$$CH_4 < NH_3 < OH_2 < HF \quad \text{or,} \quad HOCl < HClO_2 < HClO_3 < HClO_4.$$

But the reverse trend cannot be understood from electronegativity alone:

$$HF < HCl < HBr < HI \quad \text{or,} \quad BF_3 < BCl_3 < BBr_3$$

In the first series, efficiency of overlap and hence bond energy plays the key role; in the second series also, efficiency of π-overlap from the halogen to boron decreases to increase the Lewis acidity (see later).

Apart from these and many other qualitative predictions on property from bond polarity, electronegativity difference may also be used quantitatively to estimate the ionic character of a covalent bond. In a heteronuclear covalent bond A—B, the ionic character may be related to the difference of electronegativities x_A and x_B (Pauling scale):

$$\text{fractional ionic character} = 0.16(x_A - x_B) + 0.035\,(x_A - x_B)^2$$

In HCl $x_H = 2.20$ and $x_{Cl} = 3.16$ gives

$$\text{fractional ionic character} = 0.16\,(3.16 - 2.20) + 0.035\,(3.16 - 2.20)^2 = 0.186$$

or, percent ionic character of the H–Cl bond = 18.6%

The agreement with calculation based on dipole moment (16.6%; see later) is fair, specially when we remember the imprecise nature of electronegativity coefficients.

A plot of percent ionic character against electronegativity difference shows that a bond corresponding to an electronegativity difference of 2.1 is nearly 50% ionic (Fig. 5.24; solid line). The original relationship used by Pauling produced a slightly different curve (Fig. 5.24; broken line); this corresponds to an electronegativity difference of 1.7 for 50% ionic character.

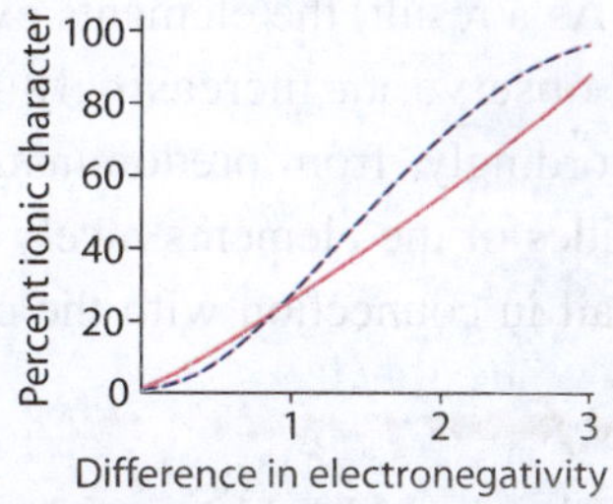

Fig.5.24: Percent ionic character of a covalent bond as a function of electronegativity difference between the bonded atoms

The Schomaker-Stevenson relationship (sec. 5.3.3) is another example where electronegativity values have been quantitatively utilized to predict bond lengths from covalent radii of bonded atoms. Bond lengths may be predicted within about 0.03Å in many cases using this empirical relationship.

5.4 RESUME; THE FUTURE PERIODIC TABLE

We have studied how different properties of atoms vary periodically with atomic number. The properties we studied mainly belong to three categories: (i) size, ionization energy and electron affinity which are measurable properties of atoms; (ii) density, melting and boiling points etc. which are also measurable properties but are actually properties of aggregates of atoms; (iii) electronegativity which is an assigned property of atoms, not measurable directly but can be related to other measurable properties like ionization energy. The overall periodic variation of these different types of properties is really interesting. As we have seen in the previous sections, the gross features of such trend may be correlated to two main factors; the nuclear charge and screening of outer electrons by inner electrons.

Down a group, shielding by added electrons overrides the influence of increasing nuclear charge. As a consequence, it becomes progressively easier to remove the outer electrons. This is reflected in increasing metallic nature of the elements, ionic character of their compounds and basic nature of the oxides. Beyond these gross features, one may also observe some intricate interplay of other factors like interposition of an inner *d*- or *f*- subshell, relativistic effect etc. We have discussed some of these aspects in connection with the increase of ionization energy and electronegativity of elements like gallium, germanium or arsenic. Toward the end of a group, such effects combine with relativistic effects to dominate the overall chemistry of an element, as for example, in thallium, gold, mercury and the *f*-block elements.

The term *secondary periodicity* has been proposed for the observed fluctuation of different properties of the elements in the fourth and sixth period in a group due to interposition of *d*- and *f*- subshells (e.g., Ga–Ge–As and Tl–Pb–Bi).

Across a period, protons (hydrons) are successively added to the nucleus but the corresponding electrons are added to an incomplete shell until we reach a noble gas configuration at the end of each period. The screening provided by the electrons in a shell for the next incoming electron from the enhanced nuclear charge is small—the screening constant s in Slater's treatment is only 0.30 (or 30%) for electrons in the same shell.

As a result, the elements exhibit successively higher ionization energy across a period and we observe an increasingly nonmetallic character. The nature of bonding also changes accordingly, from predominantly ionic to predominantly covalent. The hydrides, oxides and halides of the elements nicely illustrate this variation. We shall discuss these aspects in more detail in connection with the chemistry of the elements.

The future periodic table

Elements 110–112 were synthesized in quick succession in November 1994, December 1994 and February 1996 respectively. The half-lives of these synthetic elements are exceedingly small—just one atom of 112 was first produced which lasted for 0.00028 s. In fact, the elements were identified through their decay products. As such, studies on the chemical properties of these elements appear far from reality. Chemical properties appear to be known only up to element 105 (Dubnium, Db). The nature of the last few known elements up to $Z = 118$ (Table 5.5) are only extrapolated.

The d-block in the seventh period is completed in element 112 ($7s^2$). With pronounced relativistic effect (greater than that among the elements in the sixth period), it is presumed that 112, placed below Hg, will also be a liquid. The next six elements in this period (113–118) belong to the p-block, again expected to be liquids. At 118, we expect to find a liquid "noble gas" held by large van der Waals forces.

Elements of the eighth period are still under speculation. If we look at Table 5.2, we find that in periods 1, 2, 4 and 6 a new type of orbital begins to be occupied: s in period 1, p in period 2, d in period 4 and f in period 6. In the eighth period, the 'g'-orbitals are expected to be occupied for the first time. The $5g$ orbitals, deeply buried inside the $8s$ orbitals, will present a new block of 18 elements with nearly equal radii and identical properties. Without relativistic considerations, the order of filling would be $8s < 5g < 6f < 7d < 8p$. At 168, one might expect a liquid with noble gas configuration. But the picture may be quite different when relativistic effects are considered. Strong spin-orbit interactions will widely split the $8p$ orbitals and the $8p_{1/2}$ level may become low enough to get filled after $8s$. The $8p_{3/2}$ level, on the other hand, will be so much high that it may not be filled at all. The period may then end with the filling of $7d$ orbitals in 164 (two places below Hg).

Even though synthesis of superheavy elements is extremely difficult and their half-lives are very small, study on "magic numbers" of neutrons and protons suggests one or more "islands of stability" for the superheavy elements. Calculations suggest that after lead (82), 114 and 164 protons might fill the next two proton energy levels. Similarly after 126 neutrons in ^{208}Pb, 184 and 196 neutrons might fill the next two neutron levels. Combining the proton magic number 114 with the neutron magic number 184, it is hoped that $^{298}_{114}Uuq$ will have a stable isotope with long estimated life (with large uncertainty, though). Calculations suggest the possibility of existence of long-lived isotopes of other superheavy elements, too.

ESSENTIALS

A periodic table is an arrangement of the elements according to their atomic number that is designed to indicate chemical similarities of the elements. The most useful table is the long periodic table which is consistent with the aufbau principle and consists of eighteen vertical columns or groups. These are distributed in three regions (*s*-, *p*- and *d*-) according to the subshell of the last added electron. The *f*-block elements are placed separately below the main region.

Periodic properties of the elements include a number of physical and chemical properties that vary periodically. The general nature of this variation can be followed from the electron configurations of the atoms. Sizes of atoms and ions, ionization energy, electron affinity and electronegativity are some of the important periodic properties of atoms.

Down a group, shielding of outer electrons by inner electrons dominates over the influence of increasing nuclear charge. Decreasing ionization energy results in increasing metallic character of the elements, ionic character of their compounds and basic nature of the oxides. Interposition of an inner *d*- or *f*- subshell and relativistic effect (for the heavy elements) interferes with the monotonic trend in properties.

Across a period, the screening provided by the electrons is small; the increasing nuclear charge causes a gradual increase in ionization energy and the elements become increasingly nonmetallic. The elements gradually involve more covalent type of bonding.

The first ionization energy (I_1) is the minimum energy required to remove an electron from an atom in the gas phase. In many-electron atoms, successive ionization energies (I_n) are possible.

The ionization potential is the minimum potential gradient necessary for ionization to occur from a gaseous atom or ion. The corresponding ionization energy is expressed in electronvolt (eV). 1 eV ≡ 96.485 kJ mol^{-1}.

Ionization energies vary periodically among the elements, determined mainly by effective nuclear charge, electron configuration before and after ionization and relativistic effects (for heavy atoms).

The electron affinity (*E.A.*) is the energy released when an electron is added to a gas-phase atom, conventionally expressed as a positive quantity (negative of the energy change). Thermodynamically, we use the electron attachment energy with its usual sign. Electron affinity values also vary periodically.

Electronegativity (x) is an assigned property of an element related to the power of an atom to attract electrons in a molecule. Values of electronegativity may be related to other measurable properties of the atom like ionization energy, electron affinity and bond energy. Electronegativity of the elements vary periodically, depending largely on effective nuclear charge.

EXERCISE

1. 5-I shows an approximate plot of first *I.E.* of a few consecutive elements A—J.

(a) Which element has the largest atomic radius?

(b) Which element is most metallic?

(c) Which two elements are expected to have similar chemical properties?

(d) Which element has an ns^2np^4 electron configuration in the outermost shell of its atom?

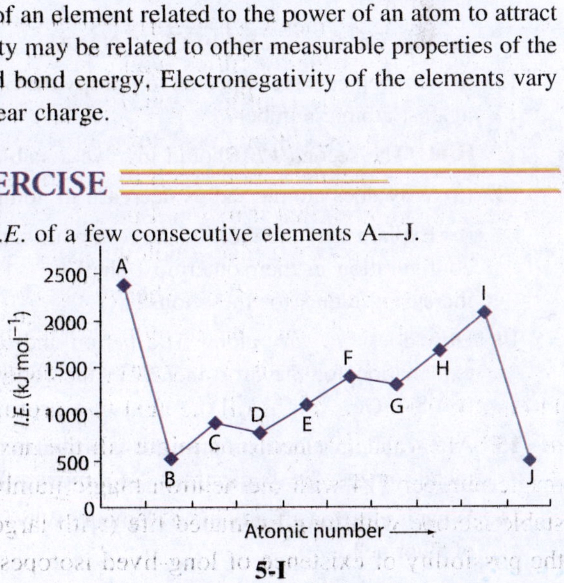

5-I

2. Compare the ionization energies (kJ mol^{-1}) of and comment on any irregularity, if present.
(a) I_1: Ca (590); Sr (550); Ba (503); Ra (509).
(b) I_2: Ti (1310); V (1365); Cr (1592); Mn (1509).
3. Explain the variation of the following properties across Period 3 (long form):
(a) ionization energy (b) electronegativity.
4. What is meant by ionic radius? Discuss Pauling's method for the determination of univalent radii.
5. (a) What do you mean by univalent radius of the oxide ion (O^{2-})? How would you modify this to obtain a radius consistent with observed values?
(b) The atomic radii of Zr and Hf are almost identical—explain.
6. Using Pauling's method, calculate the radii of K^+ and Cl^- ions. The observed K^+—Cl^- distance in KCl crystal is 314 pm.
7. Compare (qualitatively) the first ionization energies of Ca and Zn, both having $4s^2$ configuration of their valence shell.
8. Arrange the different types of radii in order of increasing size for a given element: van der Waals radius, cation radius, anion radius, covalent radius, metallic radius.
9. According to a rough estimate, the radii of different categories for an element increase by about 40–60 pm for each category. Given the covalent radius of sodium as 154 pm, estimate its (a) metallic radius (b) cation radius (c) anion radius and (d) v-W radius. The increment between each category is about 40 pm.
Hint: (a) r_{met} = 154 + 40 = 194 pm (b) r_{Na+} = 154 – 40 = 114 pm (c) $r_{Na^-} = r_{met}$ + 40 = 234 pm (d) $r_{vw} = r_{met}$ + 40 = 234 pm.
10. Calculate the A–R electronegativity of Zn taking its covalent radius as 125 pm.
Hint: Zn: $(1s2s2p)^{10}$ $(3s3p3d)^{18}$ $(4s)^2$; Z_{eff} = 30 – (2×0.35 + 18 × 0.85 + 10 × 1.0) = 4.00
$x_{A.R.}$ = 0.744 + (3590 × 4.00/125^2) = 1.66
11. Without the help of any table, arrange each of the following sets in order of increasing electronegativity.
(a) Mg, Ca, Ba, Be, Sr (b) Cl, P, Na, Al.
12. Which elements are expected to occupy the peaks when the *second I.E.* of elements are plotted against atomic number?
Hint: The second *I.E.* should involve a stable closed-shell configuration.
13. (a) Why does atomic radius decrease in going from left to right along a period?
(b) Explain why the radii of the following ions will be different though each possesses $2s^22p^6$ configuration in their outermost shell O^{2-}, F^-, Na^+, Mg^{2+}, Al^{3+}? What will be the order of increasing radius for these ions?
14. How does *I.E.* vary along (i) a period and (ii) a group in the periodic table? Why? Suggest explanation for the drop in *I.E.* in the following cases: (i) Be—B (ii) N—O, (iii) Ne—Na (iv) Si—Ge.
15. After calcium, electrons enter the 4*s* orbital before going to the 3*d* orbitals. But when a transition metal ionizes, the 4*s* electrons are removed first. Why?
16. What is electron affinity of an element? Why is the electron affinity of chlorine greater than that of fluorine?

17. (a) Define electronegativity. Is it an inherent property of the free atom? Briefly explain how the Pauling scale of electronegativity was framed.

(b) What are the limitations of the concept of electronegativity?

18. Which of the following reactions should proceed spontaneously in the gas phase?

(i) $Xe + He^+ \rightarrow Xe^+ + He$

(ii) $Si + Cl^+ \rightarrow Si^+ + Cl$

(iii) $F^- + I \rightarrow I^- + F$

Hint: Consider relative magnitudes of I.E./E.A.

Ans. (i) and (ii). Reaction (iii) should proceed in the reverse direction.

19. Comment on the following melting point (°C) sequences:

(a)	Cu	1083	Ag	961	Au	1064						
(b)	Zn	419.5	Cd	320.8	Hg	–38.9						
(c)	Ga	29.8	In	156.6	Tl	303.5						
(d)	Cr	1900	Mn	1244	Fe	1535						
(e)	Sm	1072	Eu	826	Gd	1312	Tm	1546	Yb	826	Th	1750

Hint: Melting point is a reflection of the attractive force holding the constituent particles of the solid element. In metals, this force depends primarily on the size and nature of packing of the atoms in the crystal and on the contribution of valence shell electrons in bonding. The lower melting point of Ag suggests poor participation of the *d*-electrons in metallic bonding. This is also consistent with its common oxidation state of +1 (vs +2 for Cu and +3 for Au).

The lower melting points of Zn-Cd-Hg in comparison to Cu-Ag-Au are also indicative of *d*-orbital contraction from increasing effective nuclear charge; in Hg this is reinforced by still poorer shielding by the *f*-electrons.

The unusually low melting point of Ga has no simple explanation, though the influence of *d*-orbital contraction is apparent. The single *p*-electron outside the [Ar] $3d^{10}4s^2$ core does not seem to provide strong bonding. The increasing melting points of In and Tl are consistent with their enhanced metallic character.

The drop in melting point at Mn, Eu and Yb may also be correlated to the stability of d^5, f^7 and f^{14} electron configuration.

6

COMBINATION BETWEEN ATOMS

PART I: *FOUNDATIONS*

6.1. THE ELECTRONIC THEORY OF VALENCY

Attempts to explain the nature of forces governing chemical combination of atoms have been made since the beginning of the nineteenth century. Decomposition of substances by electric current was known at that time. Davy, and later Berzelius (1812), suggested that the chemical forces might be electrical in origin. The idea faced some opposition after the recognition of organic compounds as a separate class, but its essence continued to induce later scientific thoughts.

Studies were also made on different aspects of the combining capacity of an element, without any direct aim at the origin of such capacity. The combining capacity of an element was measured by the number of hydrogen atoms with which an atom of the element could combine and was expressed as its valency (Frankland, 1852; from Latin *valentia* = capacity). Several aspects of this *valency* were studied in the next fifty years: positive and negative valency; ionizable or primary valency and nonionizable or secondary valency; spatial distribution of valency etc. But the key to all these observations was to be found in the electron—in the twentieth century.

After the discovery of the electron, J. J. Thomson realized that the chemical properties of an element were in some manner dependent on the electrons present in its atom. But all modern refinements in the theories of valency may be traced back to the two independent publications of Lewis and Kossel in 1916. Kossel attempted to explain the nature of bonds in electrolytes while Lewis centered his ideas among the nonpolar molecules. Both ideas were based on the distribution of electrons in the Bohr atom model and were collectively called *the electronic theory of valency*. The theory had the octet rule as its starting point.

Inspired by the stable outer electron configuration of the atoms of noble gases (except He), it was assumed that other atoms would also tend to gain stability by attaining an effective octet of electrons in their outermost shell (two for hydrogen), giving rise to various modes of combination (valency): electrovalency, covalency and coordinate covalency.

The octet rule worked well to explain the formation as well as some properties of a large number of compounds. But many exceptions were also observed, for example:

$$Cu^+ \ (2.8.18),\ Zn^{2+} \ (2.8.18);\ BF_3,\ BeCl_2;\ PCl_5,\ SF_6.$$

The electronic theory does not mention or predict any shape or geometry of a molecule. We do not know whether the molecule H_2O is linear or bent; whether the molecules CH_4 and NH_3 are planar or three-dimensional. The theory also fails to consider the energy factors involved in attaining a "*stable*" noble gas configuration.

In spite of all these limitations, the Lewis octet theory is extremely useful in making a two-dimensional description of the bonding in a large number of compounds, particularly those of the non-transition elements. Such description involves a suitable scheme of sharing electrons in a reasonable atomic skeleton with the valence electrons represented by dots—hence the name electron dot formula or "*structure*". The procedure for writing such structures is illustrated in the next section.

6.1.1 Lewis Electron Dot Formula

The Lewis electron-dot formula of a compound is similar to its structural formula showing the atom-to-atom connectivity and also the location of non-bonded pairs of electrons (lone pairs). Bonding electron pairs are represented by dots or a dash, while non-bonding electron pairs are shown by dots. The main steps for arriving at such electron-dot formula are outlined below for molecules (or ions) of main group elements (*s*- and *p*-block).

1. The skeleton structure showing atom to atom connectivity is ascertained from experiment, but in simple cases may be reasonably predicted.
   (a) If one atom is different from the others, it is positioned in the centre with other atoms arranged around it. Usually the more electropositive atom occupies the centre; hydrogen, however, cannot be placed at the centre since it cannot form more than one electron-pair bond.

   CH_4 NH_3 SO_3 PCl_5 NO_3^-

Fluorine also never occurs at the centre (most electronegative). Other halogens also do not appear at the centre unless linked to O or more electronegative halogens. Oxygen also is not commonly placed at the centre unless it is connected to two other atoms as in Cl_2O, H_2O or H_2O_2.

(b) If there is one atom for each of two elements, the one with the higher atomic number will be placed at the centre: $POCl_3$, $SOCl_2$

(c) In oxoacids, O atoms (and other electronegative atoms) are linked to the central atom with one or more H atom bonded to the O atom. O-atoms without any attached H-atom are mostly linked by a double bond, as shown later (see expansion of octet).

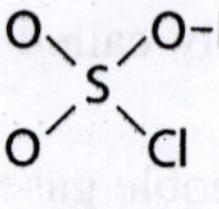
HSO_3Cl
(i)

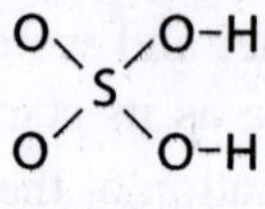
H_2SO_4
(ii)

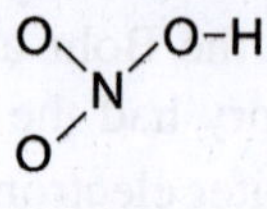
HNO_3
(iii)

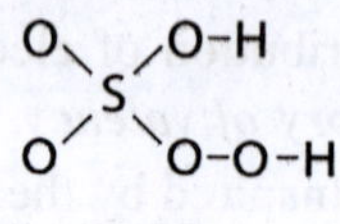
H_2SO_5
(iv)

Peroxoacids, however, contain the –O–O– group linked to the central atom (example iv).

(d) The most likely skeleton structure may also be judged by inspection of the formal charges on the atoms (see below). Same kind of charge on adjacent atoms make a structure less favourable. The most likely structure is one that has uniform and low formal charges spread over all atoms.

2. Once the skeleton structure is known, the Lewis electron-dot structure can be established as follows:

(a) Find the total number of valence electrons for all atoms.

For *s*-block elements (Groups 1 and 2), the number of valence electrons is equal to the number of *s*-electrons; for *p*-block elements (Gr 13–18), the number of valence electrons is the sum of *s*- and *p*- electrons. For *d*-block elements, the sum of *ns*- and $(n–1)d$- electrons is usually taken.

For a negavitely charged ion, add one electron for each negative charge to the sum obtained above. For a cation, similarly subtract one electron for each positive charge.

BF_3		NH_4^+		ClO_4^-
1 B = 3		1 N = 5		1 Cl = 7
3 F = 21		4 H = 4		4 O = 24
Total = 24		9		31
	Less for '+' charge	1	Add for '–' charge	1
		Total = 8		Total = 32

(a) In the examples taken above, we write the skeleton structures as

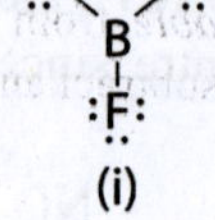
(i)

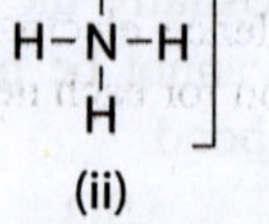
(ii)

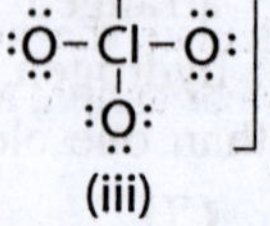
(iii)

(b) In the skeleton structure, assign electrons to the surrounding atoms to satisfy octet (including the bond pair to the central atom).

(c) Now put the remaining electrons in pairs on the central atom. If the central atom does not meet the octet (including the bond pairs), a multiple bond is most likely present. Shift one or more pairs from a surrounding atom (except H) to make a double bond or triple bond. Some molecules like BeH_2, $BeCl_2$, BF_3, BCl_3, $AlCl_3$ etc. are formally electron-deficient.

In BF_3, we require $3 \times 6 = 18$ electrons to satisfy the octet of three F atoms. Three bond pairs count for 6 electrons. The total $18 + 6 = 24$ leaves no more electrons to be placed on the B-atom. In NH_4^+, four bonds require eight electrons, with none left in hand.

In ClO_4^- also, the peripheral O-atoms require $4 \times 6 = 24$ electrons and four bonds require eight electrons, making a total of $24 + 8 = 32$, with none left.

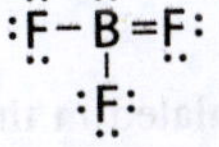

Except in BF_3, the central atoms N or Cl have their octet satisfied in the structures so arrived. To satisfy the octet of boron, we shift one pair of electrons from a terminal F atom to form a double bond to get the structure shown.

Problem 6.1

Write Lewis structures for NF_3 and $COCl_2$.

Solution: Step 1. Skeleton structure (more electropositive atom at centre):

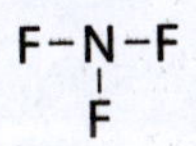

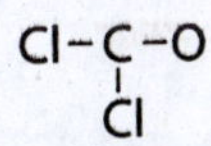

Step 2. (a) Electron count: $5 + 3 \times 7 = 26$ $\quad$ $4 + 6 + 2 \times 7 = 24$

(b) Distribution of electrons to surrounding atoms:

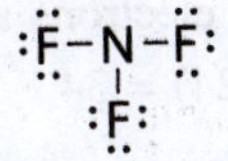

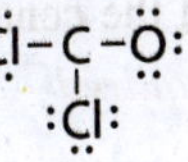

(c) In NF_3, we have (26–24) = 2 electrons in hand. So we place one pair of electrons on the central N atom. The resulting structure satisfies octet for all atoms.

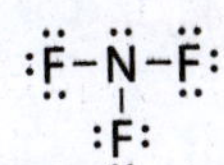

In $COCl_2$, all electrons (24) have been assigned to the surrounding O- and Cl- atoms. But the central carbon is still short of 2 electrons from an octet. So we move a pair of electrons from the O-atom to make a double bond between C and O. This gives us the final Lewis structure.

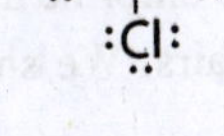

Both O and C use 2*p* atomic orbitals to give better π-overlap.
Also, this structure is favoured from formal charge distribution (see below).

An alternative procedure

In this approach, we first calculate the number of electrons normally *required* for satisfying the valence shell of each atom—2 for hydrogen and 8 for other atoms.

Next we calculate the number of valence electrons actually *available* from the atoms in the molecule. In case of an ion, add an electron for each negative charge or subtract an electron for each positive charge.

The *difference* between the numbers obtained above will give the number of shared electrons. The number of bonds will be *half* this number, to start with.

Let us take the examples of H_2O and NO_3^-.

Required	**H_2O:** 2 H (4) + 1 O (8) = 12;	**NO_3^-:** 1 N (8) + 3 O (24) = 32
Available	2 H (2) + 1 O (6) = 8;	1 N (5) + 3 O (18) + 1 '–ve' charge(1) = 24
Difference	12 – 8 = 4 ≡ 2 bonds	32 – 24 = 8 ≡ 4 bonds

We sketch the molecule/ion with the number of bonds so obtained, maintaining the general procedure outlined earlier (for central atom etc.). In case of H_2O, we have two bonds to connect two hydrogens to the oxygen. In the NO_3^- ion, we have four bonds to connect three oxygen atoms; so we use a double bond to connect one oxygen. Now we complete octet around each atom (except H) and get the Lewis structures:

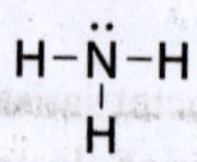

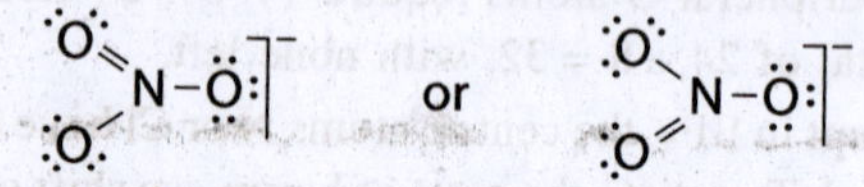 etc.

Expansion of octet becomes necessary when the number of shared electrons calculated in the above manner is not enough for bonds required to connect all atoms. This is applicable when the central atoms in the molecule is silicon onward ($Z \geq 14$). (Though the popular explanation of involvement of *d*-orbitals in bonding is now under dispute, the method works in most cases). Let us take two examples:

Required	**ClF_3:** 3 F (24) + 1 Cl (8) = 32;	**SF_6:** 6 F (48) + 1 S (8) = 56
Available	3 F (21) + 1 Cl (7) = 28;	6 F (42) + 1 S (6) = 48
Difference	32 – 28 = 4 ≡ 2 bonds. Insufficient to form three bonds to 3 Cl atoms.	56 – 48 = 8 ≡ 4 bonds. Insufficient to form six bonds to 6 F atoms.

To provide the minimum number of bonding electrons, we increase the number of electrons around the central Cl or S atom in steps of 2 electrons at a time:

Required	Cl (expanded; 8+2) + 3 F (24) = 34	S (expanded; 8 + 4) + 6 F (48) = 60
Available	28	48
Difference	34 – 28 = 6 ≡ 3 bonds	60 – 48 = 12 ≡ 6 bonds

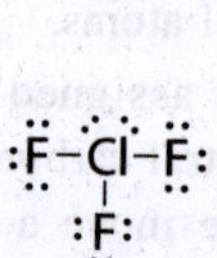

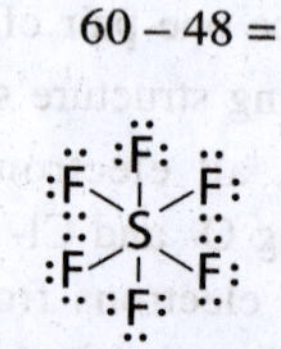

The chlorine atom has two lone pairs plus three bond pairs of electrons around it. Sulfur has six bond pairs. We shall discuss the shapes of these molecules in the following section.

Formal charge

The *formal charge* on an atom in a Lewis formula is the hypothetical charge acquired by the atom assuming equal sharing of bonding electrons and complete possession of nonbonding electrons, if any. The formal charge is calculated as follows:

(a) Assign half of the electrons in all bonds to the atom under consideration ($n_b/2$) to the atom.

(b) Assign both electrons of lone pair(s) (n_l) to the atom.

(c) Ascertain the number of valence electrons of the free atom (n_v).

The formal charge on the atom is now calculated as:

Formal charge = $n_v - (n_l + n_b/2)$

Let us calculate the formal charges on all atoms of NF_3, $COCl_2$ and CO_3^{2-} (Problems 6.1 and 6.2).

Atom	n_v	n_b	n_l	Formal charge	Atom	n_v	n_b	n_l	Formal charge
N	5	6	2	$5-(2+\frac{1}{2}\times 6)=0$	O*	6	2	6	$6-(6+\frac{1}{2}\times 2)=-1$
F	7	2	6	$7-(6+\frac{1}{2}\times 2)=0$	O**	6	4	4	$6\text{-}(4+\frac{1}{2}\times 4)=0$
C	4	8	0	$4-(0+\frac{1}{2}\times 8)=0$	Cl	7	2	6	$7-(6+\frac{1}{2}\times 2)=0$

* two singly bonded O in CO_3^{2-} ; ** doubly bonded O in $COCl_2$ and (one) in CO_3^{2-}.

Question 6.2

Draw Lewis structures for NO_3^- and CO_3^{2-} ions.

Hint: Both have 24 valence electrons (isoelectronic). N and C are central atoms. One O atom in each ion forms a double bond. This gives three resonating structures for each. One is shown.

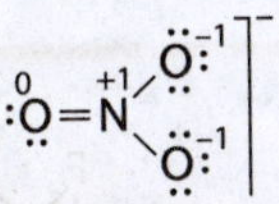

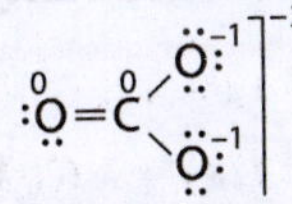

Problem 6.3

Write reasonable Lewis structures for BF_4^-, FCO_2^-, and ICl_2^+ showing formal charges.

Solution: (i) BF_4^- : The valence-electrons are counted as: 1 B (3) + 4 F (28) + 1 ('–' charge) = 32

B will be the central atom. The 32 electrons may be placed in 16 pairs around the four atoms as shown.

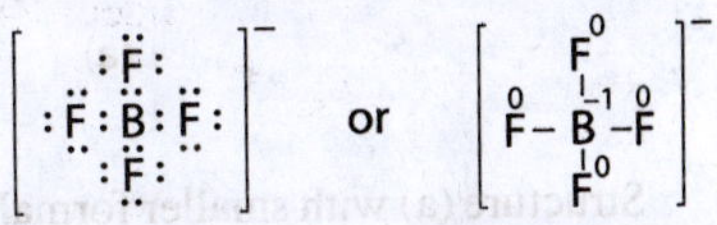

(ii) Total number of valence electrons = 24. The central atom will be carbon. The electrons may now be distributed as

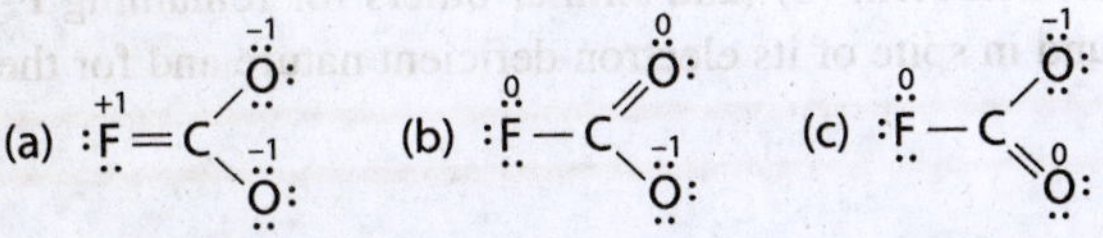

[In the next section, we shall find that structures like (a) are unlikely; structures (b) and (c) are known as canonical forms or resonance structures.]

(iii) Total number of electrons = 20 (3 × 7 – 1; remember the '+' charge). These are simply distributed as shown.

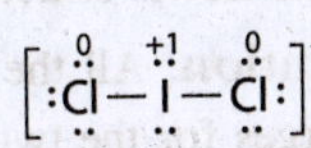

The electroneutrality principle

Generally, the structure with smallest formal charges on the atoms is the most stable one. This is consistent with Pauling's *electroneutrality principle*: in a molecule, electrons are so distributed that the residual charge on each atom becomes zero or nearly zero; hydrogen and the most electropositive metals may acquire partial positive charges and the most electronegative atoms can acquire partial negative charges. This principle may be widely used in deciding the relative importance of different structures for a species. To summarize,

(i) structures with small formal charges are more stable than those with larger formal charges;

(ii) nonzero formal charges of same sign on adjacent atoms make an unfavourable structure;

(iii) for better stability, more electronegative atoms should bear negative formal charge;

(iv) The most favourable structure should have the maximum number of bonds and highest extent of π-bonding, where possible.

(v) The most stable structure is also that for which the sum of the electronegativity differences of adjacent atoms is maximum. Thus, for the arrangements HOCl and HClO, the electronegativities of the atoms yield:

$$\overset{2.20}{\text{H}} - \overset{3.44}{\text{O}} - \overset{3.16}{\text{Cl}} : \quad \Sigma = (3.44 - 2.20) + (3.44 - 3.16) = 1.52$$

$$\overset{2.20}{\text{H}} - \overset{3.16}{\text{Cl}} - \overset{3.44}{\text{O}} : \quad \Sigma = (3.16 - 2.20) + (3.44 - 3.16) = 1.24$$

H–O–Cl is thus the most stable structure. This is also consistent with formal charge distribution and electroneutrality principle.

Problem 6.4

Which structure (a or b) is more likely?

(a) 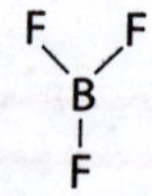 (b) 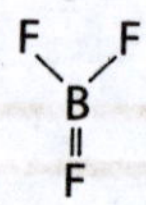

Solution: Placing lone pairs, the formal charges on the atoms in the two structures are calculated as shown:

(a) 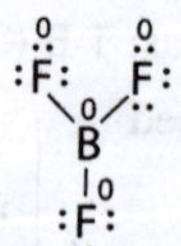 (b) 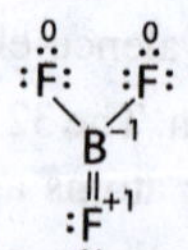

Structure (a) with smaller formal charge on the atoms should have lower energy and be dominating. However, small contribution form (b) (and similar others for remaining F-atoms) accounts for the stability of the compound in spite of its electron deficient nature and for the shortness of the bonds.

Problem 6.5

Write Lewis structures for CNO^- and OCN^- ions and comment on each.

Solution: All the elements are limited by the octet rule. The electron dot structures and formal charges for the two ions are as follows.

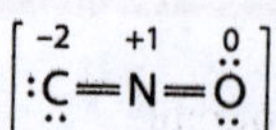
(a)

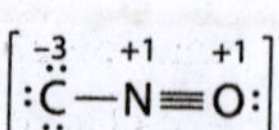
(b)

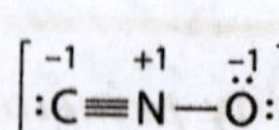
(c)

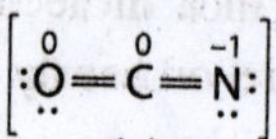
(a)

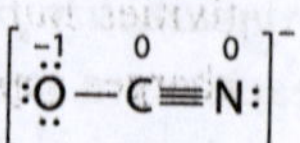
(b)

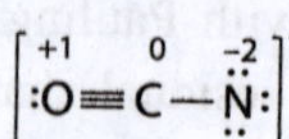
(c)

For CNO^-, structure (c) has the most favourable charge distribution, followed by (a); structure (b) is worst.

For OCN^-, both structures (a) and (b) have favourable charge distribution, but structure (b) should be the best—the negative charge resides on the more electronegative oxygen atom.

From the structures written, we also conclude that CNO^- (fulminate ion) should be less stable than OCN^- (cyanate ion). This is really the case.

Problem 6.6

(a) Why N–O–N is not a likely skeleton for nitrous oxide?

(b) Compare the different electron dot structures for the skeleton N–N–O.

Solution: (a) We write electron dot structures for N–O–N obeying octet with formal charges:

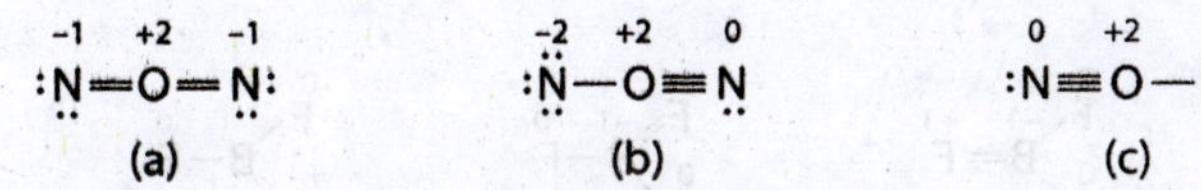

The charge distribution is unfavourable in every case with positive charge on the more electronegative oxygen atom.

(b)

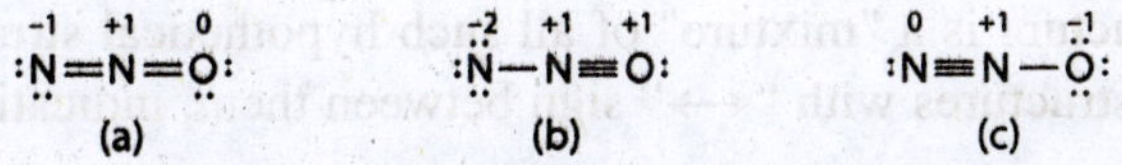

From charge distribution, structure (c) is most favourable (–ve charge on O); structure (a) is next; structure (b) is worst (+ve charge on O; like charge on adjacent atoms).

Problem 6.7

Comment on the following electron dot structures for SCN^- ion:

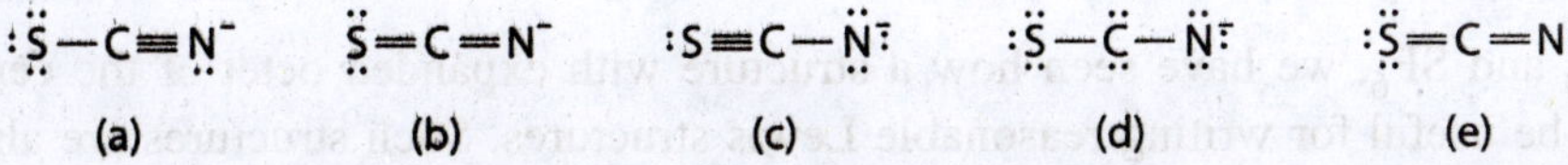

Solution: Only structures *a* – *c* fit the electron-dot formalities. Structure *d* exceeds the total number of valence electrons; structure *e* does not satisfy the octet of N.

Next we count the formal charges on different atoms in structures *a* – *c*.

	S	C	N	*Comment*
(*a*)	–1	0	0	–ve charge not on more electronegative atom (N); unfavourable.
(*b*)	0	0	–1	–ve charge on N; favourable.
(*c*)	+1	0	–2	Formal charges consistent with relative electronegativities but the charge on N is rather high. Also, charges separated widely. Less favourable than '*a*'.

Problem 6.8

Which of the following skeleton structures are favoured for $SOCl_2$ from formal charge consideration? (Complete the Lewis structures).

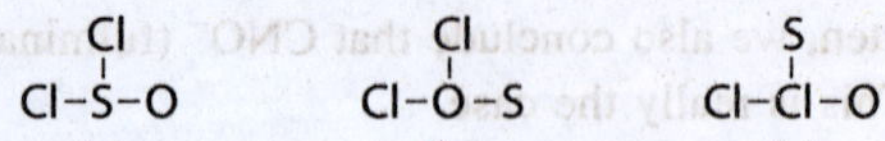

(a) (b) (c)

Ans. Skeleton (a) with S = O.

Question. Draw Lewis structures and assign formal charges on the atoms in HN_3 and N_3^-.

For some chemical species, it is possible to write several Lewis structures. In case of BF_3, for example, these structures have the same distribution of formal charge and all are equally favourable:

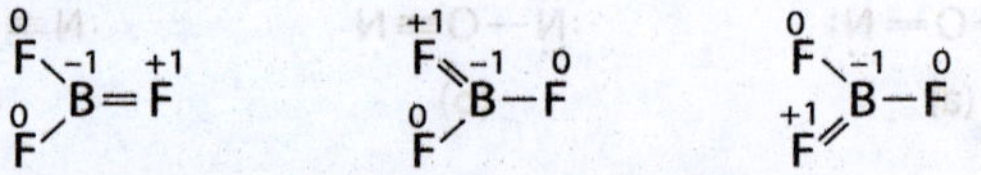

In some other cases, for example, N_2O and SCN^-, different structures have different charge distribution and all of them are not equally favourable. In such cases of more than one possible Lewis structure, the actual structure is a "mixture" of all such hypothetical structures. This is represented by writing all the structures with "⟷" sign between them, indicating *resonance*,

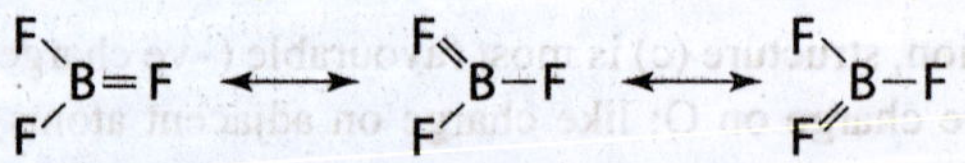

a term used to represent virtual movement or delocalization of electrons over the same atomic skeleton. The idea of resonance has its origin in the valence bond theory which attempts to explain covalence on the basis of localized bonds appearing in various Lewis structures.

Expanded octet

In ClF_3 and SF_6, we have seen how a structure with expanded octet of the central atom appears to be useful for writing reasonable Lewis structures. Such structures are also favourable from formal charge distribution. This is applicable to large and heavy central atoms for which expansion of the valence shell permits structures with multiple bonds, often stabilized by resonance.

Let us consider, for example, the sulfate $\left(SO_4^{2-}\right)$ and sulfite $\left(SO_3^{2-}\right)$ ions. We first write Lewis structures satisfying octet:

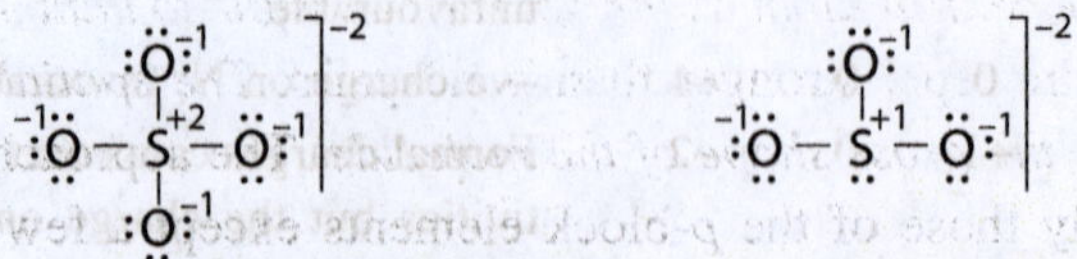

SO_4^{2-} SO_3^{2-}

With expanded shell for the sulfur atom, the structures shown minimize formal charges through S–O multiple bonds. Shifting of double bond(s) to other oxygen atoms gives other resonating structures. The observed structures are consistent with such descriptions.

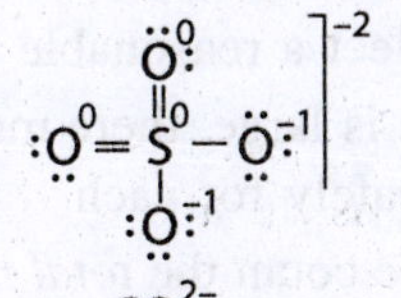

SO_4^{2-}

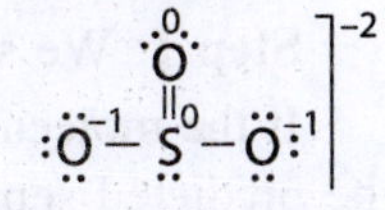

SO_3^{2-}

Problem 6.9

Draw Lewis structures for XeO_3 having (a) an octet around xenon; (b) an expandend valence shell around xenon. Assign formal charges. Which structure is more acceptable?

Hint: (a) 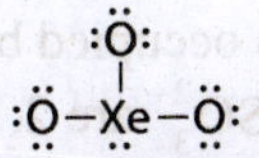 (b) 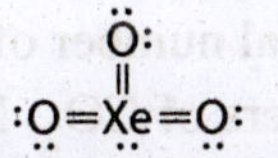 (zero formal charges on all atoms.)

Problem 6.10

Draw reasonable Lewis structures for (a) SNF_3 (b) SO_2Cl_2.

Hint: Expanded shell of sulfur gives zero formal charge on all atoms. Use S≡N and S=O.

Problem 6.11

Draw a reasonable Lewis structure for $IO_2F_2^-$.

Hint: I is the central atom. It is not subject to octet limitation. We may involve pi bonds for best utilization of valence electrons and a favourable distribution of formal charge. The positions of π-bonds and the lone pair have been discussed later.

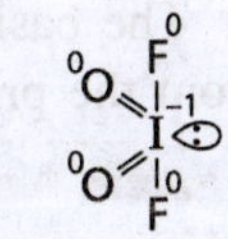

6.2 SHAPES OF MOLECULES

Although the Lewis electron dot structure does not directly convey the shape of a molecule, one can start to predict the shape of a molecule from such structures. Among several approaches to achieve this aim, the VSEPR theory is most well practised.

The VSEPR theory (*Valence Shell Electron Pair Repulsion* theory, sometimes called VESPAR theory) predicts the shape of a molecule or ion on the basis of minimum electrostatic repulsion between the negatively charged electrons in the valence shell of the central atom, present either as covalent bond pairs or as unshared pairs of electrons. First mentioned by Sidgwick and Powell (1940) and later developed by Gillespie and Nyholm (1957), *the VSEPR theory assumes that all the valence shell electron pairs* (bonding and non-bonding) *are oriented in space around the central atom as far apart as possible to minimize repulsion*. It is further assumed that lone pairs repel stronger than bond pairs. *The spatial distribution of the bond pairs now determines the gross shape of the molecule*. The approach works well for covalent molecules, particularly those of the *p*-block elements except a few having large atoms (see later).

The theory may be applied to predict the geometry of a molecule by the following steps.

Step 1: We select a reasonable Lewis structure for the species.

If the molecule is large, there may be more than one central atom and the geometry has to be predicted separately for each.

Step 2: Next we count the *total* number of lone pairs (nonbonded pairs/unshared pairs) and number of atoms linked to the central atom, irrespective of single or multiple bonds involved in the link.

In CO_2 (O = C = O), the central atom has no lone pairs; only two O-atoms are linked to it. NH_3 and SO_3^{2-} have one lone pair and three atoms linked to the central atom in each.

The total number of lone pairs and atoms bonded to the central atom is called the **steric number (SN)** of the central atom. It is the total number of positions occupied by atoms or lone pairs around a central atom. The steric numbers of CO_2, NH_3 and SO_3^{2-} are 2, 4 and 4 respectively.

For a molecule represented as AX_mE_n where A = central atom; X = any other atom/group of atom and E = a lone pair on A, the steric number (SN) is $m + n$.

The sum $m + n$ is also called the *total coordination number*. The coordination number is simply m.

One may also proceed to determine the geometry from the σ valence pairs which is equal to the steric number. The σ valence pairs are obtained by counting the total valence pairs (bond pairs and lone pairs) and then subtracting the π valence pairs: $VP_\sigma = VP_{tot} - VP_\pi$.

The basic distribution of the electron pairs in space around the central atom now follows from the principle of minimum repulsion (maximum angular separation) as follows:

Steric Number	Arrangement	Steric Number	Arrangement
2	Linear	6	Octahedral
3	Trigonal planar	7	Pentagonal bipyramid
4	Tetrahedral	8	Square antiprism
5	Trigonal bipyramidal		

The arrangement suggested by the steric number gives the "electron geometry" of the species, not its actual structure which is determined by the position of nuclei only. When a species does not possess any lone pair of electrons, the geometry suggested above represents the gross structure of the species. When lone pairs are present, we have to follow step 3 before deciding the structure.

Step 3: In molecules containing lone pairs, the actual structure is determined by the positions of the atoms only, lone pairs are *not* included in describing the shape. But the position of the lone pair(s) is often important in deciding the relative positions of other atoms.

When the gross structure suggested from steric number has all its positions equivalent, the lone pair(s) may be placed in any position and the shape determined from the position of the atoms. The following examples will be illustrative:

S.N. = 3

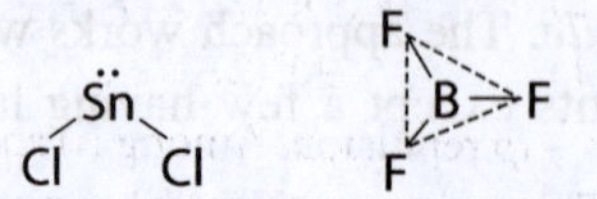

Bent (V-shaped) Trigonal planar

S.N. = 4

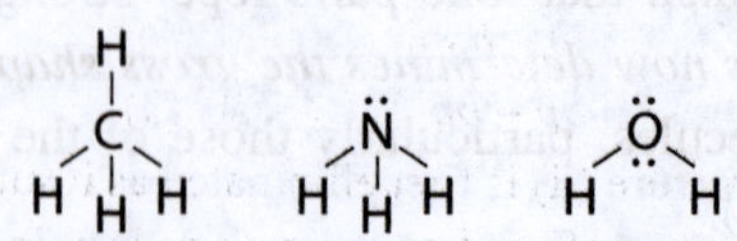

Tetrahedral Trigonal pyramid Bent (V-shaped)

In structures where all positions are not same, one has to distinguish lone pairs from bond pairs.

Since a lone pair of electrons belongs to a single nucleus, it represents a region of high electrostatic charge density and therefore repels other electron pairs strongly. A bond pair of electrons, on the other hand, is spread between two nuclei and hence exerts less electrostatic repulsion on other electron pairs. Hence the decreasing order of repulsion may be written

lone pair-lone pair > lone pair-bond pair > bond pair-bond pair
(lp – lp) > (lp – bp) > (bp – bp)

The effect of greater repulsion by the lone pair is demonstrated by the molecules CH_4—NH_3—H_2O, all with total of four electron pairs around the central atom (Steric Number 4). CH_4, with no lone pair, has its bonds at the equilibrium position of 109.5°. In NH_3, one lone pair exerts stronger *lp – bp* repulsion than the *bp – bp* repulsions and hence the bond angle is slightly squeezed from 109.5°. The two lone pairs in H_2O repel the bond pairs strongly and reduce the bond angle to 104.5°.

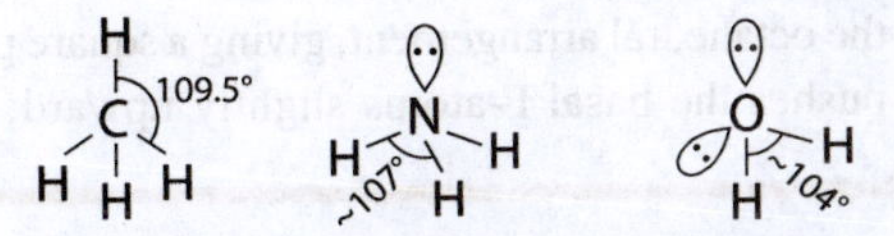

4 bond pair | 3 bond pairs, 1 lone pair | 2 bond pairs, 2 lone pairs

The trigonal bipyramid arrangement (S.N. = 5) has two possible locations of lone pairs, axial and equatorial. In an equatorial position, a lone pair has two other electron pairs in the equatorial plane at 120° angular separation and two more at axial positions at 90°. A lone pair at an axial position, on the other hand, has three electron pairs at 90° and one at 180°. Hence a lone pair at an equational position will have minimum interaction with other lone/bond pairs compared to a lone pair at an axial position. This is illustrated by the molecule SF_4; the lone pair at an equatorial position squeezes the ∠FSF (eq) to ~101.6° (from 120°). The bond pairs at the axial positions are repelled away from the lone pair, the angle ∠FSF (axial) being compressed to 173° (from 180°). The shape of the molecule is that of a sawhorse (or a seasaw).

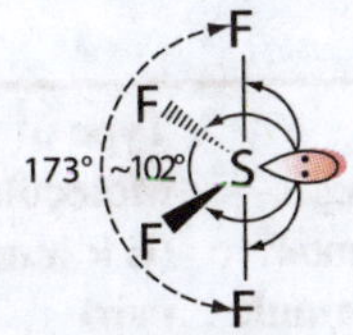

Problem 6.12

Predict the structure of ClF_3 according to the VSEPR theory.

Hint: Steric number = 5; three bond pairs and two lone pairs; trigonal bipyramidal.

Next we consider the interactions between various electron pairs in structures i, ii and iii.

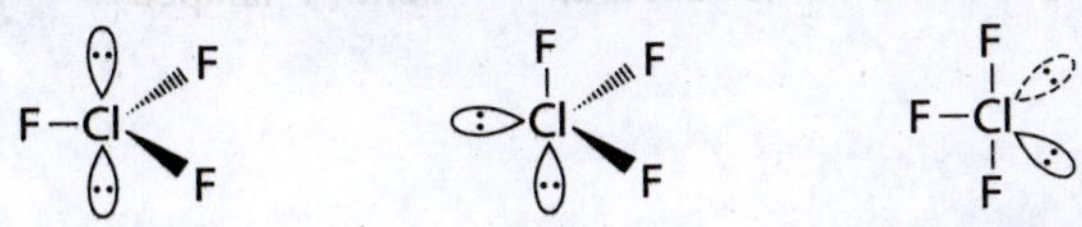

(i) (ii) (iii)

	(i)	(ii)	(iii)
lp – lp	180°	90°	120°
lp – bp	90°(6)	90°(3), 120°(2)	90°(4), 120°(2)
bp – bp	120°(3)	90°(2), 120°(1)	90°(2)

Structure (ii) is first eliminated as it suffers maximum *lp – lp* repulsion. Among (i) and (iii), structure (iii) is preferred as it involves less *lp – bp* repulsion. This agrees with the observed angles:

∠FClF(ax) = 175°.

Problem 6.13

XeF_2 is linear. Explain from the VSEPR theory.

Hint: SN = 5, two bond pairs and three lone pairs. Lone pairs occupy equatorial position.

Problem 6.14

In square pyramidal IF_5, the iodine atom is slightly below the basal plane of F-atoms. Explain.

Hint: SN = 6, 5 bond pairs and one lone pair. We may place the lone pair at any one vertex of the octahedral arrangement, giving a square pyramidal structure (AX_5E; Table 6.1). L.P.–B.P. repulsion pushes the basal F-atoms slightly upward; the I-atom now lies below the plane of F-atoms.

The geometry of some typical molecules and ions is shown in Table 6.1.

Table 6.1

Geometry of some typical molecules (VSEPR theory)

Representative formula	Type of Molecule (E = lone pair)	No. of lone pairs	S.N.	Distribution of electron-pairs	Shape of Molecule	Example(s)
AX_2	AX_2	0	2	linear	linear	O=C=O ($BeCl_2$, HgF_2, ZnI_2)
	AX_2E	1	3	trigonal planar	bent (V-shaped)	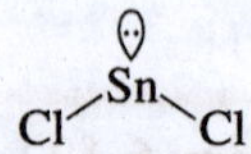 (O_3, NO_2^-)
	AX_2E_2	2	4	tetrahedral	bent (V-shaped)	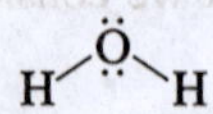 (SCl_2, ClO_2^-)
	AX_2E_3	3	5	trigonal bipyramid	linear	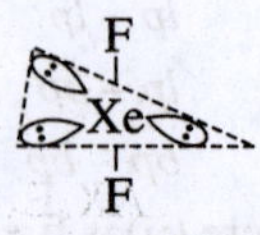 (ICl_2^-, I_3^-)

Continued...

Table 6.1 (*contd.*)

AX_3	AX_3	0	3	trigonal planar	triangular	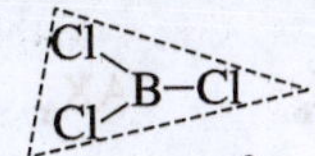 (NO_3^-, CO_3^{2-})
	AX_3E	1	4	tetrahedral	trigonal pyramid	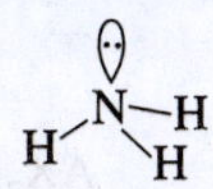 (NF_3, H_3O^+, ClO_3^-)
	AX_3E_2	2	5	trigonal bipyramid	T-shape	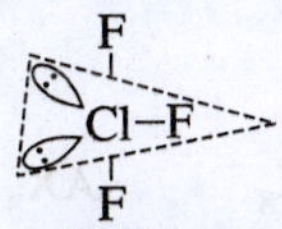 (BrF_3)
AX_4	AX_4	0	4	tetrahedral	tetrahedral	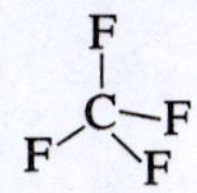 (ClO_4^-, SO_4^{2-})
	AX_4E	1	5	trigonal bipyramid	tetrahedron (irregular; sawhorse)	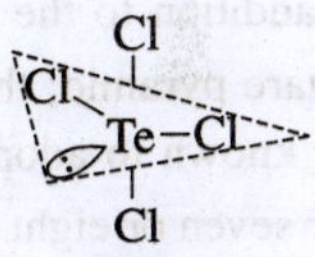 (SF_4)
	AX_4E_2	2	6	octahedral	square planar	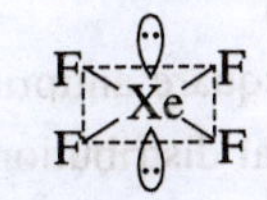 (ICl_4^-)
AX_5	AX_5	0	5	trigonal bipyramid	trigonal bipyramid	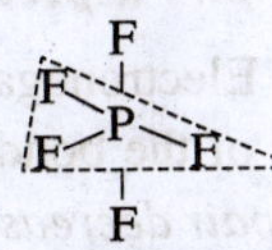 (PCl_5 (*g*))
				square[1] pyramid	square pyramid	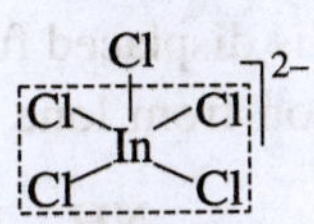 (Ph_5Sb)
	AX_5E	1	6	octahedral	square pyramid	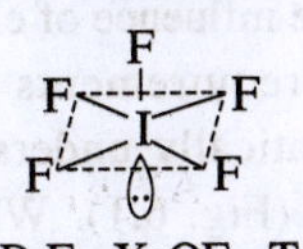 (BrF_5, $XeOF_4$, TeF_5^-)
	AX_5E_2	2	7	pentagonal planar	pentagonal planar	

Continued...

Table 6.1 *(contd.)*

AX_6	AX_6	0	6	octahedral	octahedral	 (PCl_6^-)
AX_7	AX_7	0	7	pentagonal bipyramid[2]		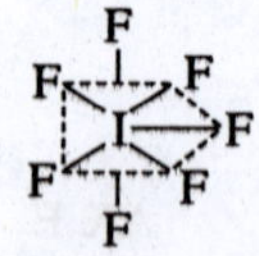
AX_8	AX_8	0	8	square antiprism[3]		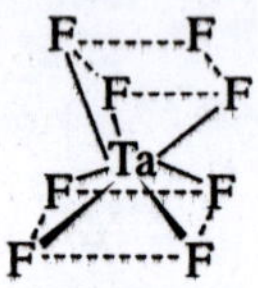 (IF_8^-)

Note:

1. In addition to the TBP structure, five electron pairs may also be arranged at the four corners of a square pyramid; though this arrangement has a slightly higher energy, some species (e.g., $InCl_5^{2-}$) are known to adopt this structure.
2. For seven or eight electron pairs, other types of distribution are also possible: the capped octahedron and the capped trigonal prism have one additional atom along one face-centre of the octahedron or prism.
3. The square antiprism structure results on twisting the top face of a cube by 45° along a diagonal.
4. Spatial distribution of upto 14 electron pairs have been predicted but their examples are very rare.

Some more aspects of electron pair repulsion

(i) Electronegavity of different atoms bonded to the central atom largely controls the domain of the bond pair and hence the extent of bond pair-bond pair repulsion. *Repulsion by a bond pair decreases as the electronegativity of the atom bonded to the central atom increases.*

As the bonded atom becomes more electronegative, the bonding electron pair (or its charge cloud) is displaced further away from the central atom. Hence *bp – bp* repulsion decreases and repulsion from lone pair(s) dominates to squeeze the bond angle:

NH_3 107.3 NF_3 102° OH_2 104.5° OF_2 103.2°

The influence of electronegativity on relative sizes or space requirements of bond pair electrons may be schematically understood in terms of the solid cone angles (Fig. 6.1). When shared with a more electronegative atom X, the charge cloud of the bond pair electrons is dragged towards X, making a smaller cone angle (*b* vs *c*). The largest cone angle is made by the unshared pair of electrons (lone pairs), which is also included (a) as reference.

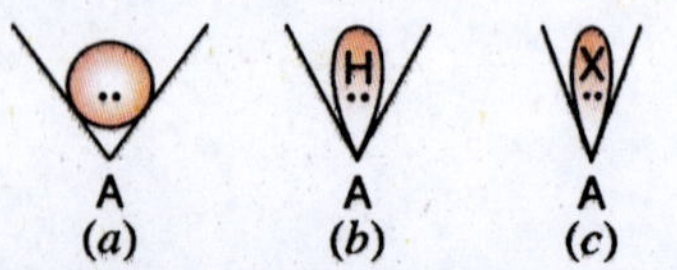

Fig.6.1: Schematic drawing to illustrate cone angles made by (a) lone pair (b) bond pair to H and (c) bond pair to a more electronegative element X (not to scale).

This influence of electronegative substituents is also observed in trigonal bipyramid molecules containing different peripheral atoms, like PCl_3F_2. The bond pairs to more electronegative F atoms require less space compared to the bond pairs to Cl atoms. As a consequence, the Cl atoms occupy equatonal positions in the TBP geometry. We shall find later that the same conclusion is arrived at from the empirical Bent's rule (Ch. 7).

(ii) Another aspect of electron pair repulsion is: *electron pairs in filled shells repel stronger than electron pairs in incomplete shells*. Elements of third period onward have their valence shell only partly filled—the 3*d*, 4*d* etc. orbitals do not fully participate in valence. Accordingly, these atoms (e.g., S, P) offer lower *bp* – *bp* repulsion compared to second period elements like N, O etc. The stronger *lp* – *bp* repulsion dominates again to squeeze the bond angle to near 90°.

NH_3	107.3	PH_3	93.3°	H_2O	104.5°	H_2S	92.2°

(iii) *Lone pairs may sometimes be transferred from a filled shell of one atom to an unfilled shell of another bonded atom*. For example, in PF_3, the valence shell of the phosphorus atom (including 3*d*) is incompletely filled. The high charge-density of electrons on F-atom causes a partial shift of its lone pairs to the phosphorus atom *via* pi-bond formation. The P–F bonds attain some double bond character; the bond length is shortened (as in BF_3, SiF_4 etc. also). The shift of electrons in the P–F bonds is now contrary to our expectation from the electronegativity of fluorine; increased b.p.–b.p. repulsion now increases the FPF bond angle. Similar is the case with AsF_3.

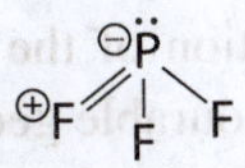

PF_3	97.8°	AsF_3	96.2°
PH_3	93.3°	AsH_3	91.8°

In Cl_2O, similar delocalization of the lone pair electrons of oxygen to chlorine *decreases* the repulsion by lone pairs and *increases* the repulsion between bond pairs. The bond angle thus becomes very large.

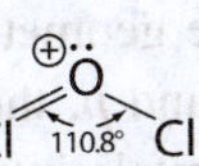

	F_2O	103.2°
vs.	H_2O	104.5°

In SCl_2, the lone pairs on S need not be delocalized; the angle is only 102°.

(iv) *Multiple bond orbitals repel other orbitals more strongly than single bond orbitals*.

The multiple bond orbitals possess extra electron density and naturally exert stronger repulsion on a neighbouring electron pair than a single bond does. Thus a double bond containing four electrons exerts a greater repulsive effect on other sigma bonds present in the molecule and result in a smaller bond angle as shown below.

$O{=}PF_3$	∠FPF	102.5°	$O{=}CF_2$	∠FCF	112.5°
$O{=}PCl_3$	∠ClPCl	103.5°	$O{=}CCl_2$	∠ClCCl	111.3°
$O{=}PBr_3$	∠BrPBr	108°	$O{=}CMe_2$	∠MeCMe	119.6°

On the left, the bond angles deviate from the tetrahedral angle of 109°29′; for compounds on the right, deviations occur from 120°, the angle expected for a trigonal structure. In both the series, maximum deviation occurs when fluorine is present. The bond pairs are removed farthest from the phosphorus or carbon atoms, b.p. – b.p. repulsion becomes minimum and the repulsion from the double bond makes the bond angle least. The multiple bonds thus tend to take the same position as lone pairs to minimize repulsion; they preferentially occupy equatorial position and squeeze the other bond angles (see SOF_4). When a species contains both lone pairs and multiple bonds, the lone pairs prefer equatorial position (e.g., $IF_2O_2^-$ and IF_3O).

Problem 6.15

Suggest the structure of $H_2C = SF_4$.

Hint: There are five sigma bond pairs around the sulfur atom—a trigonal bipyramid distribution. The bulky π-bond should be in an equatorial position, suppressing ∠FSF (eq) as well as ∠FSF (axial). (Actually these angles are 97° and 170° respectively).

The π-lobes of C = S bond are parallel to the equatorial plane (nodal plane) to avoid congestion. The H-atoms joined to the sp^2-hybridized carbon should be in the axial plane.

VSEPR theory and hybridization

The VSEPR theory does not presuppose any hybridization of the valence orbitals of the central atom in a molecule. But once we decide the most favourable geometry of the molecule, we may associate relevant hybrid orbitals with this geometry to describe its localized sigma bonding skeleton. A detailed description of the involvement of various atomic orbitals in bonding can be obtained only by thorough quantum mechanical calculations though such procedure does not provide an easy view of the structure. An MO treatment assuming one of the familiar set of hybrid orbitals consistent with the geometry is also very useful. When the starting atomic orbitals (basis sets) are limited to *s* and *p*, the hybrid orbitals may be readily identified from *sp*, sp^2 or sp^3. Involvement of *d*-orbitals introduces a fairly extensive set of hybrid orbitals (Ch 7; tables 7.1 and 7.2).

Merits and limitations of the VSEPR approach

The VSEPR approach provides a simple as well as quick means to arrive at molecular geometries consistent with experimental observations in a large number of cases. The starting principles are very easy to remember and no detailed consideration of hybrids, molecular orbitals etc. is required. However, difficulties arise when more than one distribution of similar energy is possible for a steric number. Thus five electron pairs may have both trigonal bipyramid or square pyramid arrangement around a central atom. Many conformations of similar energy are also possible for seven electron pairs (S.N. = 7). Some examples of such diverse structures are given in Table 6.1.

Another major limitation of the VSEPR theory is observed with some compounds of heavy *p*-block elements. The lone pair of electrons on the central atom of these molecules (or ions) does not seem to take part in deciding the geometry; the lone pairs thus appear *stereochemically inert*. A number of species with six bond pairs and one lone pair (type AB_6E) have octahedral structure, suggesting that the lone pair of electrons is stereochemically inactive; for example, $[SbBr_6]^{3-}$, $[SeCl_6]^{2-}$, $[SeBr_6]^{2-}$, $[TeCl_6]^{2-}$, $[TeBr_6]^{2-}$, $SeCl_4py_2$ are all octahedral.

It has been suggested that the observed departure from predictions may be a consequence of packing of the closely spaced peripheral atoms. In effect, this is equivalent to dominance of bond pair-bond pair repulsion. The inactivation of the lone pairs has also been ascribed to the presence of soft donor atoms (Ch. 10). Hard donor atoms (O, F) usually result in stereochemically active nonbonding electron pairs. On the other hand, soft donor atoms possess orbitals in which the nonbonded electron pairs can be delocalized. At the same time, longer bond length and more diffuse charge density reduces lone pair–bond pair repulsion.

XeF_6 also possesses six bond pairs and a lone pair (AB_6E type) and do not follow VSEPR prediction. The gaseous molecule has a distorted octahedral structure in which the lone pair "*shuttles*" between several face centers, making XeF_6 a fluxional molecule (Ch. 25).

The theory does not apply to many compounds of the transition metals and systems having extensively delocalized π-electrons.

Stereochemical nonrigidity: No molecule is rigid in the strictest sense since all of them execute a set of vibrations even at absolute zero; the atoms oscillate about their average positions with amplitudes of a few tenths of an angstrom. Again a molecule like ammonia can undergo a simple inversion in which the nitrogen atom quickly oscillates through the plane of the three hydrogen atoms just as an umbrella can turn inside out. The potential energy of the molecule is higher when the N-atom is in-plane with the H-atoms during transition. However, the potential energy barrier to this inversion is very low (~24 kJ mol^{-1}) and the process is very rapid (frequency ~2.39×10^{10}Hz) at ordinary temperatures. Such inversion does not allow optical isomers for chiral molecules like $NR_1R_2R_3$ to be isolated. The potential energy barrier increases with decreasing bond angle in case of heavier elements or with bulky substituents.

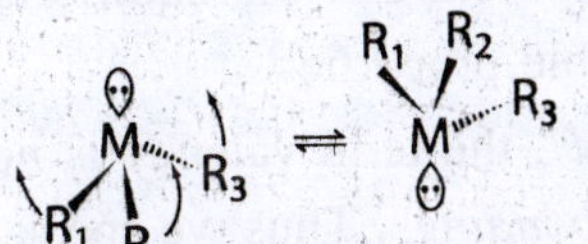

For M $(CH_3)_3$ compounds, the energies of inversion (kJ mol^{-1}) are

N $(CH_3)_3$	34
P $(CH_3)_3$	133
As $(CH_3)_3$	122
Sb $(CH_3)_3$	112

It is possible to resolve a pyramidal $MR_1R_2R_3$ species into optical isomers if inversion takes place slowly enough (inversion leads to racemization). Calculations show that pure optical isomers can be stable under normal conditions when the energy barrier is at least of the order of 100 kJ mol^{-1}. Hence asymmetric tertiary phosphanes (phosphines) and arsanes (arsines) may be resolved. Thus AsEtMePh (ethylmethylphenylarsane) has an inversion energy 177 kJ mol^{-1} and can be resolved into the optical isomers.

In addition, there are many molecules which undergo rapid structural rearrangement in which atoms interchange places. In PF_5, for example, the axial and equatorial fluorine atoms rapidly interchange position (See Ch. 7; Berry pseudorotation for mechanism). The process involves no net chemical change—only interchange of indistinguishable nuclei takes place. These molecules are the most important class of stereochemically nonrigid molecules and are called fluxional molecules. XeF_6, metal carbonyls, many organometallic and organic compounds are known to exhibit stereochemical nonrigidity.

6.3 THE CHEMICAL BOND

While considering the limitations of the electronic theory of valency, we certainly feel the necessity of a realistic interpretation of the combining capacity of an element in terms of associated energy changes and consequent disposal of the "*valencies*" in space. The idea of valency then seems to be somewhat abstract in nature. The energy changes accompanying a chemical reaction may be interpreted mostly in terms of the linkages broken and formed in the reaction. The geometry of a molecule may be predicted if we could guess how the different linkages between the constituent atoms in the molecule are oriented in space. With these points in mind, one would appreciate that the idea of a chemical bond is more realistic than the idea of valency. A chemical bond is responsible for holding the constituent atoms in a chemically distinct species; so a chemical change is always accompanied by breaking and formation of such bonds. Sometimes, it is possible to conceive of individualized bonds existing between the atoms in a molecule; however, the scope of the term is also extended to cover substances where the nature of this force is collective, and no individual bonds may be assigned (the ionic "bond" and the metallic "bond", for example).

The physical significance of a chemical bond may be appreciated more directly when we speak in terms of the energy required to rupture a chemical bond. The same amount of energy is released during the formation of this bond. This is in keeping with the universal tendency of all systems to attain the state of lowest energy. Modern theories of bonding start from this general principle. Atoms combine to attain a state of lower energy (potential) than that in the isolated ones. When the energy of an aggregate of atoms is lower than that of the isolated atoms by about 42 kJ mol^{-1} (10 kcal), the aggregated species bears distinct and characteristic chemical properties. The constituent atoms in the species are then said to be bound to one another by chemical bonds. The bond energy, the force operative in the bond and the equilibrium internuclear distance of the atoms held by the bond are measurable properties.

In view of the wider significance of the term "*chemical bond*", the term valency is now largely replaced by *valence*, meaning "*associated with chemical bonding*". Thus we speak of valence electrons and valence theories. It is also used as a noun having more or less the same meaning as "*valency*", but in a wider sense covering all aspects of bonding.

6.3.1 Bond Length

The equilibrium distance between the centres of two bonded atoms may be obtained from x-ray diffraction, infrared or microwave spectroscopy and electron diffraction. All such measurements indicate that for the same type of bonding, the distance between a pair of atoms in different molecules remain very nearly the same. This gives us the idea of a definite distance or length for each bond. The **bond length** represents the equilibrium distance between the centres of two bonded atoms. In fact this is the average distance between the nuclei as they continuously vibrate with respect to each other.

Some typical values of bond distances are given in Table 6.2.

Table 6.2

Some bond lengths, pm

H—H	(H_2^+)	106	C—C	(diamond)	154	N—F	(NF_3)	137
	(H_2)	74		(C_2H_6)	153	N—Cl	(NCl_3)	175
H—C	(CH_4)	110	C=O	(CO_2)	116	O—O	(O_2)	121
H—N	(NH_3)	101	C—N	(CH_3NH_2)	147		(O_2^+)	112
H—O	(H_2O)	97	C—O	(CO)	113		(H_2O_2)	148
H—F	(HF)	92		(CO^+)	112		(F_2O_2)	122
H—Si	(SiH_4)	148	C=C	(C_2H_4)	133	O—F	(OF_2)	142
H—P	(PH_3)	142	C≡C	(C_2H_2)	120		(O_2F_2)	158
H—S	(H_2S)	135	C—Cl	(CCl_4)	177	F—F	(F_2)	141
H—Cl	(HCl)	127	N—N	(N_2)	110	Si—Si	(Si_2)	225
H—Br	(HBr)	141		(N_2^+)	112	S—S	(S_8)	205
H—I	(HI)	161		(N_2H_4)	145	S—F	(SF_6)	156
B—F	(BF_3)	129	N—O	(NO)	115	S—O	(SO_2)	143
B—Cl	(BCl_3)	175		(NO^+)	106	Cl—Cl	(Cl_2)	199
B—Br	(BBr_3)	187		(NH_2OH)	146	I—I	(I_2)	267
B—I	(BI_3)	210		($NaNO_3$)	124		(I_3^-)	291

Bond lengths range from 74 pm in H_2 to 465 pm in $CS_2(g)$. Bonds between elements in the first, second and third periods are usually in the range 100—200 pm. As expected, bond lengths between the same pair of atoms vary with the bond order or bond multiplicity and also with the nature of hybrid orbitals involved in forming the bond. For example, the C—C single bond lengths (pm) in the following bond types are all different.

$$\geqslant C \overset{154}{—} C \leqslant \qquad \geqslant C \overset{151}{—} C \lessgtr \qquad \geqslant C \overset{146}{—} C \equiv \qquad \equiv C \overset{137}{—} C \equiv$$

Here bond distance decreases as the fractional *s*-character of the hybrid orbital used in bonding increases on going from sp^3 to sp.

As mentioned earlier, there are many instances where the actual bond distance between two atoms differs significantly from the sum of the covalent radii (e.g., The Schomaker-Stevenson rule, section 5.3.3).

Problem 6.16

The covalent radius of fluorine is 60 pm while the F—F bond distance in F_2 is 140 pm. Comment.

Hint: Strong lone pair–lone pair repulsion.

Remark: O—O and N—N bonds are similarly longer than twice their respective covalent radius.

Problem 6.17

Comment on the N—N and N—O bond lengths:

N_2H_4 : N—N = 145 pm	NH_2OH : N—O = 146 pm
$N_2H_6^{2+}$: N—N = 140 pm	Me_3N—O : N—O = 136 pm

Hint: L.P.—L.P repulsion is largely reduced in the second species in each column:

$$\overset{+}{H_3N} - \overset{+}{NH_3}\ (\text{vs. } H_2\ddot{N} - \ddot{N}H_2) \quad Me_3\overset{+}{N} - O^- \ (\text{vs. } H_2\ddot{N} - \ddot{\underset{..}{O}} - H)$$

6.3.2 Bond Strength

Energy is required to separate the units linked by a chemical bond. Consequently, a parameter called bond energy or bond dissociation energy may be assigned to each chemical bond. The two terms, applied in a slightly different sense as we shall appreciate soon, are actually expressed in terms of the enthalpy change for breaking one mole of bonds of any particular type between two atoms in the gaseous state.

$$H_2(g) = H(g) + H(g): \Delta H^\circ = 432 \text{ kJ mol}^{-1}\ ;\ O_2(g) = O(g) + O(g): \ \Delta H^\circ = 498 \text{ kJ mol}^{-1}$$

It is assumed that the main contribution to ΔH° for the process in gas phase comes from the change in electronic energy associated with bond breaking. Changes in translational, rotational and vibrational energies are neglected as their contributions are small.

The enthalpy change for breaking the bond in a diatomic molecule gives the bond dissociation energy or the bond energy for that bond, both terms having the same connotation.

$$HCl(g) = H(g) + Cl(g) \qquad \Delta H^\circ = 431.0 \text{ kJ mol}^{-1}$$

For polyatomic molecules, the stepwise bond dissociation enthalpies are not equal:

Step 1: $H_2O(g) = H(g) + OH(g) \qquad \Delta H^\circ = 502 \text{ kJ mol}^{-1}$

Step 2: $OH(g) = H(g) + O(g) \qquad \Delta H^\circ = 427 \text{ kJ mol}^{-1}$

Both enthalpy terms involve dissociation of an O—H bond, but the bond dissociation energies differ due to change in chemical environment. The bond energy (D) of the O—H bond is then expressed by the average of the bond dissociation enthalpies.

$$D_{O—H} = (502 + 427)/2 = 464.5 \text{ kJ mol}^{-1}$$

Problem 6.18

Calculate the average C—H bond energy from the following bond dissociation energies (ΔH° in kJ mol^{-1}):

$$CH_4(g) \rightarrow CH_3(g) + H(g) \qquad \Delta H^\circ = 421 \text{ kJ mol}^{-1}$$
$$CH_3(g) \rightarrow CH_2(g) + H(g) \qquad \Delta H^\circ = 470 \text{ kJ mol}^{-1}$$
$$CH_2(g) \rightarrow CH(g) + H(g) \qquad \Delta H^\circ = 415 \text{ kJ mol}^{-1}$$
$$CH(g) \rightarrow C(g) + H(g) \qquad \Delta H^\circ = 335 \text{ kJ mol}^{-1}$$

[**Ans:** 410 kJ mol^{-1}.]

The bond energy thus represents the approximate energy required to split a chemical bond in any combination. It is also the approximate energy released when the bond is formed from the neutral gaseous atoms. Bond dissociation enthalpies may be obtained thermochemically from the enthalpy of reactions and enthalpy of atomization of the concerned elements. The following examples are illustrative.

Problem 6.19

Calculate the C—C bond energy from the following standard enthalpy values (kJ mol^{-1}; in parenthesis)

C(*s*; graphite) → C(*g*) (712.9); $CH_4(g) \rightarrow C(g) + 4H(g)$ (1660); $H_2(g) \rightarrow 2H(g)$ (435.9);

2C (*s*; graphite) + $3H_2(g) \rightarrow C_2H_6(g)$ (–79.9).

Solution: The data on methane gives us the C—H bond energy: (1660 ÷ 4 = 415 kJ mol^{-1}). Now, ethane has six C—H and one C—C bonds. If we can calculate the enthalpy change for breaking all these bonds, subtraction of the 6 C—H bond energies will give us the C—C bond energy. We require $\Delta H°$ for the process:

$$C_2H_6(g) \rightarrow 2C(g) + 6H(g).$$

We may arrive at this process by manipulating the supplied enthalpy data.

	$2\,C(s) \rightarrow 2C(g)$	2 × 712.9 kJ mol^{-1}
	$3\,H_2(g) \rightarrow 6H(g)$	3 × 435.9 kJ mol^{-1}
Add,	$2\,C(s) + 3H_2(g) \rightarrow 2\,C(g) + 6\,H(g)$	2733.5 kJ mol^{-1}
Again,	$2\,C(s) + 3H_2(g) \rightarrow C_2H_6(g)$	–79.9 kJ mol^{-1}
Subtract, rearrange,	$C_2H_6(g) \rightarrow 2\,C(g) + 6\,H(g)$	2813.4 kJ mol^{-1}

From this, we subtract 6 × 415 kJ for the breaking of 6 C—H bonds.

Hence C—C bond energy = 2813 – (6 × 415) kJ mol^{-1} = 323.4 kJ mol^{-1}.

Problem 6.20

Calculate the O—O bond energy in H_2O_2 from relevant $\Delta H°$ values taken from the following:

Enthalpy of formation ($\Delta H_f°$, kJ mol^{-1}): $H_2O(g)$: –242; $H_2O(l)$: –286; $H_2O_2(g)$: –136; $H_2O_2(l)$: –188.

Enthalpy of bond breaking (kJ mol^{-1}): $H_2(g)$ 432, $O_2(g)$ 494.

Solution: H_2O_2, H—O—O—H, contains two O—H bonds and one O—O bond. To estimate the O—O bond energy we should know the O—H bond energy and the enthalpy change for the process

$$H_2O_2(g) = 2\,H(g) + 2\,O(g).$$

Estimation of O—H *bond energy*: The *average* O—H bond energy is 921 ÷ 2 = 460.5 kJ mol^{-1}. This can be done with the data given for $H_2O(g)$, $H_2(g)$ and $O_2(g)$ as follows:

			$\Delta H°$
	$H_2O(g)$	$= H_2(g) + \frac{1}{2}O_2(g)$	242
	$H_2(g)$	$= 2\,H(g)$	432
	$\frac{1}{2}\,O_2(g)$	$= O(g) + \frac{1}{2} \times 494$	247
Add,	$H_2O(g)$	$= 2\,H(g) + O(g)$	921 kJ

Estimation of O—O bond energy:

			ΔH°	
	$H_2O_2(g)$	$= H_2(g) + O_2(g)$	136	
	$H_2(g)$	$= 2H(g)$	432	
	$O_2(g)$	$= 2\,O(g)$	494	
Add,	$H_2O_2(g)$	$= 2\,H(g) + 2\,O(g)$	1062	kJ

We get the O—O bond energy by subtracting the energy required for the two O—H bonds:

$$E_{O-O} = 1062 - 921 \text{ kJ mol}^{-1} = 141 \text{ kJ mol}^{-1}.$$

In these examples, we have tacitly assumed that the C—H and O—H bond energies are same in the compounds chosen: CH_4 and C_2H_6 for C—H bond; H_2O and H_2O_2 for O—H bond. This is only an approximation since the presence of one kind of bond may strengthen or weaken the other kind of bond in a molecule. For the same reason, direct thermochemical data for any compound is always preferable over indirect calculation.

Problem 6.21

Calculate separately the N—N single bond energy from the following two sets of data (all in gaseous state):

(i) $N_2F_4 \rightarrow 2N + 4F$ $\Delta H^\circ = 1305$ kJ mol^{-1}; $NF_3 \rightarrow N + 3F$ $\Delta H^\circ = 850$ kJ mol^{-1}.

(ii) $N_2H_4 \rightarrow 2N + 4H$ $\Delta H^\circ = 1703$ kJ mol^{-1}; $NH_3 \rightarrow N + 3H$ $\Delta H^\circ = 1158$ kJ mol^{-1}.

Ans. (i) 172 kJ mol^{-1} (ii) 159 kJ mol^{-1}.

Some mean bond enthalpies are given in Table 6.3.

Table 6.3

***Some bond enthalpies* (kJ mol^{-1})**

H—H	432	C=C	614	N≡N	946	S—H	347
H—F	566	C≡C	839	N—F	283	S—S (S_8)	226
H—Cl	431	C—F	485	N—Cl	193	S—Cl (S_2Cl_2)	255
H—Br	366	C—Cl	327	N—O	200	Si—H	326
H—I	299	C—O	358	N=O	678	Si—Si	226
B—H	381	C—N	305	O—H	459	Si—F	582
B—F	644	C=N	615	O—O	146	Si—Cl	391
B—Cl	444	C≡N	891	O=O	498	Si—O	452
B—Br	377	C≡O	1074	P—F (PF_3)	490	F—F	159
B—B	175	N—H	390	P—Cl (PCl_3)	326	Cl—Cl	242
C—H	411	N—N	160	P—H (PH_3)	322	Br—Br	193
C—C	348	N=N	415	P—O	335	I—I	151

Some distinct trends may be observed among the bond energies of *p*-block elements. Bond energy usually decreases with increase in size of the combined atoms, but exceptions may arise in certain cases. The following examples are illustrative (ΔH° in kJ mol^{-1}):

C—H	411	N—H	390	O—H	459	C—Cl	327
Si—H	326	P—H	328	S—H	347	Si—Cl	391
Ge—H	288	As—H	247	Se—H	305	Ge—Cl	340

The first three columns illustrate the general expectation that bond strength should decrease as sizes of the atoms increase and result in poor overlap. In the last column, this trend is reversed. Large difference of electronegativity of the atoms and possible delocalization of the lone pair electrons of chlorine to the silicon may be important in this case.

The influence of lone pair-lone pair repulsion on bond enthalpies ($\Delta H°$, kJ mol^{-1}) is also noteworthy.

C—C	348	N—N	160	F—F	159	N—Cl	193
Si—Si	226	P—P	201	Cl—Cl	242	P—Cl	326
Ge—Ge	188	As—As	146	Br—Br	193	As—Cl	293

For the bonds in the first column, the bonded elements (C, Si, Ge) have no unshared pair of electrons on the atoms. The bond enthalpy decreases down the group as per our expectation from poorer overlap. In other columns, the bonds at the top show remarkably low bond enthalpy owing to repulsion between lone pair electrons on the atoms.

Though the values of bond enthalpies quoted in Table 6.3 are approximate, a study of bond enthalpy may be often useful to interpret the course of a reaction which are usually accompanied by release of energy (exothermic). Thus favourable C—C and C—H bond enthalpy explains why carbon shows such exceptionally high catenation property. Silicon, on the other hand, has a high Si—O bond enthalpy over Si—Si and Si—H bond enthalpies; as such, catenated silicon compounds are fewer and unstable in air (see Group 14).

The influence of lone pairs of electrons is again observed in periodic groups 15 (N vs P) and 16 (O vs S). Here the first element in each group forms stable diatomic molecules N_2 and O_2; the second element (P or S) cannot form stable multiple bond owing to poor π-overlap and forms species linked by single bonds (e.g., P_4 and S_8). For the same reason, sulfur forms a number of catenated species having no oxygen analog. We shall come across many other applications of bond enthalpy data in subsequent chapters.

However, one must be cautious in drawing conclusion from bond enthalpy data alone since the ultimate feasibility or potential of a chemical reaction is determined by the Gibbs potential change, ΔG. The magnitude of ΔG depends on the enthalpy term ΔH as well as an entropy term ΔS. For a spontaneous process, ΔG (= $\Delta H - T\Delta S$) should be negative (sec. 6.3.3).

Vibrations of chemical bonds

A chemical bond is not like a solid stick connecting two atoms. Sometimes it becomes useful to compare a bond with a stiff spring connecting the two nuclei (Fig. 6.2). Any attempt to stretch the spring through a distance x produces a restoring force F which is proportional to the displacement x when x is small:

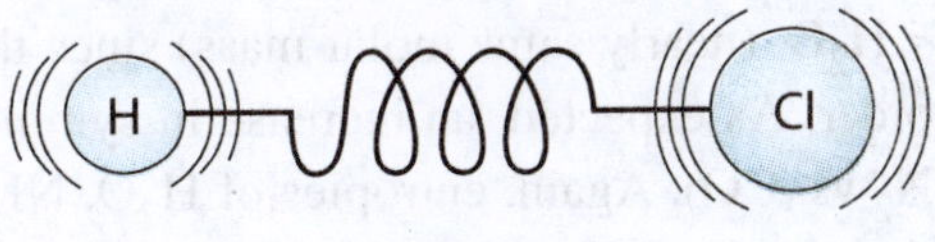

Fig.6.2: A "spring-model" of the H–Cl bond. The bond vibrates at a frequency of 8.652×10^{13} vibrations per second.

$$F = -kx$$

(The minus sign appears because the restoring force and displacement are in opposite directions.)

k is called the *force constant* of the bond. The stiffer the bond, the greater will be the value of k. The force constant thus provides a measure of the strength of a chemical bond. A higher force constant implies a stronger bond. In vibrational spectroscopy, the system is treated quantum mechanically as a harmonic oscillator whose energies are obtained by solving the appropriate Schrödinger equation (see later).

The energies of such vibrations fall in the infrared (IR) region. When a molecule is exposed to IR radiation, it shows strong absorptions at frequencies which match the energies for different modes of vibrations of the bonds present in the molecule. A study of such absorption pattern in the IR spectrum of the compound enables us to characterize the bonds and calculate their stretching frequencies (conventionally expressed as stretching wave numbers, ν/c). These stretching frequencies provide a measure of the strength of a bond. Any change in the bonding environment is reflected in a change in the characteristic stretching frequency of the bond, as we shall see in subsequent discussions.

6.3.3 Enthalpy, Entropy and Gibbs Energy

In this section we discuss the major thermodynamic aspects of a chemical reaction—bond breaking and bond making. The kinetic aspects of a chemical reaction are also very important in understanding the course of a chemical reaction. However, we defer our discussion on this aspect to a later chapter (Chapter 15).

The **enthalpy** H of a system is an extensive state property defined as

$$H = U + PV$$

(U = internal energy; P, V are pressure and volume of the system).

For any process at constant pressure, the enthalpy change ΔH is the heat change Q_P:

$$\Delta H = \Delta U + P\Delta V = Q_P$$

The **entropy** S of a system is also a state function, that is, its change ΔS in a process depends only on the initial and final states of the system:

$$\Delta S = \frac{Q_{rev}}{T} \qquad (Q_{rev} = \text{reversible heat change})$$

Entropy is related to the chaotic dispersal of energy and hence the randomness of any system and is expressed in terms of the thermodynamic probability w:

$$S = k \ln w \qquad (k = \text{Boltzmann constant})$$

Entropy increases with increasing complexity of molecules, e.g., along the series Ne < HF < H_2O (nearly same molar mass) since the degrees of rotational freedom increase in the same order. As expected, an increase in symmetry (see later) results in a decrease in entropy (e.g., N_2 vs. CO). Again, entropies of H_2O, NH_3 and CH_4 are nearly same in spite of an increase in the number of atoms in the molecules (with nearly equal M). Increasing symmetry of the molecules appear to counteract the rise in entropy.

Solids with strong binding energies have low entropy, e.g., diamond and silicon. Graphite has a relatively higher entropy due to its lower cohesive energy. Similarly, entropies of $KClO_3$, NH_4Cl etc. are higher than those of MgO, NaCl etc. due to greater complexity of the units involved.

ΔS°_{298} for some common reactions are given below (J mol^{-1} K^{-1}):

(i) C (graphite) $+ O_2(g)$	$= CO_2(g)$		2.86
(ii) $\frac{1}{2}$C (graphite) $+ \frac{1}{2}CO_2$	$= CO(g)$		87.93
(iii) $CaCO_3(s)$	$= CaO(s) + CO_2(g)$		160.5
(iv) $\frac{3}{2}H_2(g) + \frac{1}{2}N_2(g)$	$= NH_3(g)$		−103.1

We observe that (i) the change in entropy is very small when the number of moles of gaseous substances remains the same. (ii) The entropy increases when the number of moles of gaseous substances in the system increases, and hence, the volume increases at constant pressure. (iii) Entropy decreases when the number of moles of gaseous products is reduced, i.e., $\Delta V_p < 0$.

The Gibbs energy* (G) of a system is a property of the state of the system defined as

$$G = H - TS$$

For a finite change at constant T,

$$\Delta G = \Delta H - T\Delta S.$$

It can be easily shown that the decrease in Gibbs energy of a system for any transformation at constant temperature and pressure ($-\Delta G_{T,\,P}$) is equal to the maximum available work from the process, i.e., the net available work other than expansion work.

For any spontaneous process, the standard Gibbs energy change ΔG° (conventionally at 1 atm pressure and 298K) should be negative. At equilibrium, $\Delta G^{\circ} = 0$. However, a process with a negative value of ΔG° has the thermodynamic potentiality or feasibility to take place, though it may not occur owing to kinetic reasons, as in the reactions of carbon or hydrogen with oxygen at 298 K:

$$C(s) + O_2(g) = CO_2(g) \qquad \Delta G^{\circ} = -394.36 \text{ kJ mol}^{-1}$$

$$H_2(g) + O_2(g) = H_2O(g) \qquad = -228.57 \text{ kJ mol}^{-1}$$

We may proceed a bit further to examine the contributions of the ΔH and ΔS terms to make the overall ΔG term negative for a chemical process. We recall the relation $\Delta G = \Delta H - T\Delta S$ and consider three ways to attain a negative ΔG.

Case I. Both ΔH and ΔS make ΔG negative. This occurs when $\Delta H < 0$ and $\Delta S > 0$.

A large number of combustion reactions illustrate this case, e.g.,

	$C(s)$	+	$\frac{1}{2}O_2(g)$	=	$CO(g)$	
ΔH°:	0		0		−110.5	kJ mol^{-1}
S°:	5.7		205		197.67	J mol^{-1} K^{-1}

$$\Delta G^{\circ} = -110.5 - 298 \times [197.67 - (5.7 + ½ \times 205)] \times 10^{-8} \text{ kJ mol}^{-1}$$

$$= -137.16 \text{ kJ mol}^{-1}.$$

* Proviously known as Gibbs function, Gibbs free energy or simply free energy; also represented by F. The IUPAC agreement is to use the term **Gibbs energy** and the symbol G.

Explosives and propellants involve reactions favoured by both ΔH and ΔS (see box).

EXPLOSIVES AND PROPELLANTS

Reactions favoured by high negative ΔH and positive ΔS have been used in making explosives and propellants. A large negative enthalpy change with several moles of gaseous products are main features of most explosives; the common products are stable gases like N_2, CO, CO_2 and also O_2. A few common reactions of this type are:

(i) $2\,N_2H_4\,(l) + N_2O_4(l) \rightarrow 4\,H_2O\,(g) + 3\,N_2(g)$

$\Delta H° = -1040$ kJ; $\Delta S° = +912\ JK^{-1}$ per mol reaction.

The reaction is self-igniting (*hypergolic*) and is used in rocket propulsion.

(ii) $18\,C + 5S + 14\,KNO_3 \rightarrow 5\,K_2CO_3 + K_2SO_4 + K_2S + 3\,S + 10\,CO_2 + 3CO + 7N_2$

$\Delta H° = -4948$ kJ; $\Delta S° = +3260\ JK^{-1}$ per mol reaction.

The reaction forms the basis of the classic "*black powder*", used in guns and canons since 13th century and even in popular fireworks known as *toobri*.

The energy liberated per kg of reactant is nearly one-third of that for reaction (i). Moreover, the reaction produces a large amount of solid reaction products which produce dense smoke and obscure visibility.

(iii) The detonation of nitroglycerine may be expressed by the approximate reaction

$$4\,C_3H_5N_3O_9 \rightarrow 6\,N_2 + 10\,H_2O + 12\,CO_2 + O_2$$

$\Delta H° = -1809$ kJ and $\Delta S° \simeq +920\ JK^{-1}$, both per mole of nitroglycerine. Nitroglycerine contains more oxygen than required to convert all its carbon and hydrogen to CO_2 and H_2O. It is said to be *overoxidized*. Carbon-rich explosives like nitrocellulose (gun-cotton) are *underoxidized* as they do not contain enough oxygen. The deficiency may be overcome by mixing it with nitroglycerin—a soft form is used as *blasting gelatin* and a harder form is used as a solid rocket propellant.

Ammonium nitrate yields N_2, O_2 and H_2O on rapid thermal decomposition and is often used to meet the oxygen balance of underoxidized explosives. A mixture of amonium nitrate and fuel oil, ANFO, is used in mining.

(iv) Ammonium nitrate or amonium perchlorate is also used with aluminium powder in rubber-bound solid propellants.

$$6\,NH_4ClO_4 + 10\,Al \rightarrow 3\,N_2 + 9\,H_2O + 5\,Al_2O_3 + 6\,HCl.$$

$\Delta H° = -9337$ kJ; $\Delta S° = +2246\ JK^{-1}$ per mol reaction. The enthalpy-yield of the reaction per kg is higher than that of reaction (i).

Case II. ΔS is negative, so that the $T\Delta S$ term opposes ΔH. But ΔH is more negative to give an overall negative ΔG. For example,

	$SO_2(g)$ +	½$O_2(g)$ =	$SO_3(g)$.	
$\Delta H°$	–296.8	0	–395.7	kJ mol^{-1}.
$S°$	248.2	205	256.76	J mol^{-1} K^{-1}.

$$\Delta G° = [-395.7 - (-296.8 + 0)] - 298\,[256.76 - (248.2 + ½ \times 205)] \times 10^{-3}$$
$$= -70.9\ \text{kJ mol}^{-1}$$

Case III. The reaction is endothermic i.e., $\Delta H > 0$, but ΔS is positive so that the $T\Delta S$ term makes ΔG negative. Examples of such processes are numerous. To begin with, we may take note of the dissolution of many crystalline salts (e.g., NaCl, KNO_3 etc.) in water. As explained later with ionic compounds (Chapter 8), these salts have comparable values of lattice enthalpies and solvation enthalpies. For NaCl, an experimental heat of solution (ΔH) is +4.9 kJ mol^{-1}. The dissolution is thus mainly entropy-driven. Similarly, extraction of a metal salt like uranyl nitrate from acidic aqueous solution with tributyl phosphate is an entropy-driven process. Such processes are also applied in the separation of metals like Nb from Ta, lanthanoids and actinoids.

The following endothermic reaction reportedly occurs at ordinary temperature due to the large positive ΔS term accompanying the formation of a large number of gaseous products:

$$CoCl_2, 6\,H_2O(s) + 6\,SOCl_2(l) \rightarrow CoCl_2(s) + 6\,SO_2(g) + 12\,HCl(g)$$

But we note with interest that many endothermic processes are favoured at higher temperatures when the $T\Delta S$ term outweighs the ΔH term. We consider the reaction

$$CaCO_3(s) = CaO(s) + CO_2(g) \quad \Delta H^\circ = +178 \text{ kJ mol}^{-1};\ \Delta S^\circ = 160.48 \text{ J mol}^{-1}\text{ K}^{-1}.$$

At 1500 K, $\Delta G = 178 - 1500 \times 160.48 \times 10^{-3} \text{ kJ mol}^{-1} = -62.6 \text{ kJ mol}^{-1}$.

So, one should expect solid calcium carbonate to decompose at high temperatures. At 25°C, however, the *reverse process* has a large negative ΔG°.

$$CaO(s) + CO_2(g) = CaCO_3(s).$$

$$\Delta G^\circ = -178 - (298 \times -160.48) \times 10^{-3} \text{ kJ mol}^{-1} = -130.18 \text{ kJ mol}^{-1}.$$

We shall find a very useful role of the $T\Delta S$ term in the extraction of metals by carbon reduction (Ellingham diagram, Chapter 15).

Many interesting synthesis have been carried out by condensation of atoms or molecular fragments generated through high temperature entropy-driven processes. BF, SiF_2, C_2, C_3 etc species provide numerous examples of this type in the gaseous phase.

$$BF_3(g) + 2\,B(s) \rightarrow 3\,BF\,(g).$$

$$BF + BF_3 \rightarrow B_2F_4\ ;\ B_2F_4 + BF \rightarrow B_3F_5$$

$Si + SiF_4 \rightarrow 2SiF_2\ ;\ SiF_2 + C_2H_4 \rightarrow$ 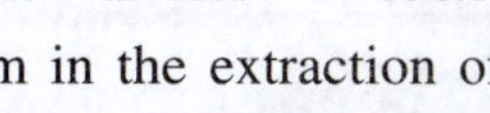 (also others)

Two familiar endothermic and entropy-driven reactions occur at moderate temperatures and provide the starting materials for many boron-nitrogen and sulfur-nitrogen compounds.

$$6\,S_2Cl_2 + 4\,NH_4Cl \xrightarrow{160^\circ C} S_4N_4 + S_8 + 16\,HCl$$

$$\Delta H^\circ = +428 \text{ kJ} \quad \Delta S^\circ \simeq 1080 \text{ JK}^{-1} \text{ per mol } S_4N_4.$$

$3\,BCl_3 + 3\,NH_4Cl \xrightarrow{175^\circ C}$ 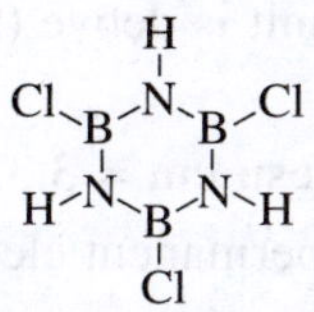 $+ 9\,HCl$

$$\Delta H^\circ = +322 \text{ kJ} \quad \Delta S^\circ \simeq 1000 \text{ JK}^{-1} \text{ per mol reaction.}$$

We shall discuss some of these reactions again in the chemistry of individual elements.

Problem 6.22

Try to guess whether the following reactions are enthalpy-driven or entropy-driven. Don't search bond enthalpy data.

(a) $BCl_3(g) + BBr_3(g) \rightarrow BBr_2Cl(g) + BCl_2Br(g)$

(b) $2d\text{-}[Co(en)_3]^{3+} \rightarrow d\text{-}[Co(en)_3]^{3+} + l\text{-}[Co(en)_3]^{3+}$ (en = ethylenediamine)

Hint: Both should be entropy-driven since there are only rearrangement of bonds with ΔH nearly zero. A lower symmetry in the products results in a positive ΔS in each case.

Problem 6.23

Is it thermodynamically meaningful to attempt at the preparation of hydrazine by oxidizing aqueous NH_3 by NO? $\Delta G°$ (kJ mol^{-1}): $H_2O(l)$: – 237.13; $N_2H_4(aq)$: 128.03; NO(g): 86.57; $NH_3(aq)$: –26.50.

Solution: The reaction involved is

$$8\,NH_3(aq) + 2\,NO(g) \rightarrow 5\,N_2H_4(aq) + 2\,H_2O(l)$$

$$\Delta G° = 5 \times 128.03 + 2 \times (-237.13) - 8 \times (-26.50) - 2 \times 86.57 \text{ kJ per reaction.}$$

$$= +\,204.75 \text{ kJ per reaction.}$$

It is suggested that the reaction will not be favoured thermodynamically.

The compound may be prepared in an alternative route.

6.3.4 Polarity of Bonds

In a covalent bond between two atoms of the same element, the bonding electrons are shared equally by the two nuclei concerned. But for bonds between atoms of two different elements, the distribution of the bonding electrons is unequal between the two nuclei. The atom with higher electronegativity has a greater share of the bond pair. The bond thus becomes *polar*.

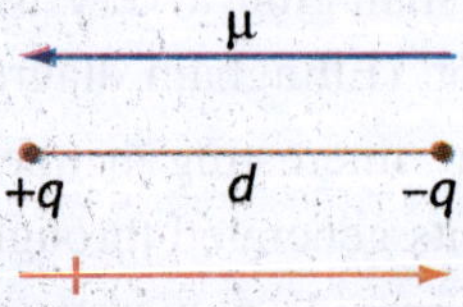

Fig. 6.3: An electric dipole. The direction of the vector μ is from the negative to the positive charge centre. Sometimes an arrow with a 'plus'-tail, ⟴ is used in the Lewis structures to indicate the direction of electron shift. The direction of ⟴ is *opposite* to that of the vector μ.

Polarity of a bond increases with the separation of the positive and negative charge centres. This is expressed in terms of an *electric dipole moment* vector μ. For an electric dipole consisting of two charges $+q$ and $-q$ separated by a distance d, the magnitude of the electric dipole moment μ is defined as

$$\mu = qd$$

directed from the *negative to the positive charge*. The SI unit of dipole moment is coulomb-meter (C m). A commonly used unit is debye (D; in honour of Peter Debye, a pioneer in this field).

$$1\text{ D} = 10^{-18} \text{ esu cm} = 3.33564 \times 10^{-30} \text{ C m}$$

When a molecule possesses a permanent electric dipole moment, it is said to be polar. The polarity arises from unbalanced arrangement of polar bonds, together with other bonding features like the polarity of an unshared pair of electrons. Thus dipole moment is related to the symmetry properties of molecules (section 6.4). In a symmetrical molecule, the constituent bond moments cancel one another and the molecule becomes nonpolar, as in linear CO_2 (O=C=O) and planar C_6H_6. All homonuclear diatomic molecules have zero dipole moment; the triatomic homonuclear molecule of ozone, (O_3, angular) is polar.

As a first approximation, the dipole moment of a molecule may be roughly estimated from vector sum of the contributing electric dipole moments of the various bonds. These bond moments are, in turn, calculated from experimental dipole moments of molecules with known geometry. Some commonly accepted values of bond moments are given in Table 6.3.

Table 6.3

***Some individual bond dipole moments*: (a) in debye, D (b) in C m × 10^{30}**

	H—O	H—N	H—Cl	H—C*	C—Cl	C—O	C═O	C—N	C≡N
(a)	1.5	1.3	1.03	0.4	1.5	0.8	2.5	0.5	3.5
(b)	5.00	4.33	3.44	1.33	5.00	2.67	8.34	1.67	11.67

*Based on traditionally assumed polarity H^+—C^-.

Problem 6.24

The dipole moment of chlorobenzene is 1.57 D. Estimate the ratio of the dipole moments of 1,2- (*ortho*) and 1,3- (*meta*) dichlorobenzenes.

Hint:

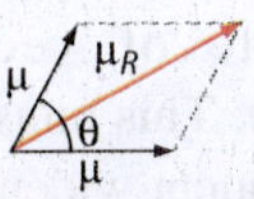

$$\mu_R^2 = 2\mu^2 (1 + \cos\theta)$$

For 1,2-dichlorobenzene, θ = 60°; for 1,3-dichlorobenzene, θ = 120°.

$$\mu_R(1,2-)/\mu_R(1,3-) = 3^{1/2} = 1.73$$

Note: The ratio holds for any two identical substituent groups.

The resultant of two nonequal vectors μ_1 and μ_2 is given by $\mu_R = \left(\mu_1^2 + \mu_2^2 + 2\mu_1\mu_2 \cos\theta\right)^{1/2}$

Calculation and measurement of dipole moment

The extent of charge separation in a polar molecule may be measured by placing it in an electric field between two charged plates. The molecules, being electrically neutral as a whole, will not move toward either plate. But they will try to orient themselves in the field so that the positive ends are close to the negative plate of the applied field and the negative ends of the molecules are closer to the positive plate. Such orientation will change the potential energy of the system. This small energy change can be measured in a number of ways and from the results one may calculate the electric dipole moment of the molecules.

For an ionic compound like Na^+Cl^-, the dipole moment may be calculated by assuming the charges $+e$ and $-e$ situated at the centres of the spherical ions. If the equilibrium distance between the centres be d, the dipole moment will be equal to ed. For covalent molecules, calculation of dipole moment requires knowledge of partial charge on all atoms and their locations. Such calculation may be found in advanced Physical Chemistry texts.

Molecular dipole moments can be obtained from microwave spectra. In presence of an electric field, the rotational energy levels of a gaseous polar molecule are split (*The Stark effect*) according to the square of the permanent electric dipole moment. But the technique is applicable to molecules which may be studied in the gaseous phase by rotational spectroscopy.

Measurement of relative permeability, ε_r, (dielectric constant) at a series of temperatures provides a good method for the determination of dipole moment. The molar polarization P_m is calculated by the *Debye equation*

$$P_m = \frac{\varepsilon_r - 1}{\varepsilon_r + 2} \cdot \frac{M}{\rho}$$

where M = molar mass of the substance of density ρ. The molar polarization is also related to the polarizability α of the molecule and the dipole moment μ:

$$P_m = \frac{N_0 \alpha}{3\,\varepsilon_0} + \frac{N_0 \mu^2}{9\,\varepsilon_0 kT} \qquad [N_0 = \text{Avogadro number}]$$

[α, the *polarizability* of the molecule, is the ratio of the induced dipole moment μ* to the applied electric field *E* which induces the moment: $\alpha = \mu^*/E$. A greater value of α means the applied electric field can distort the molecule and align its permanent electric dipole moment.]

A plot of P_m vs 1/T gives a straight line of slope $N_0\mu^2/9\,\varepsilon_0$ from which μ can be calculated. The intercept at 1/T = 0 gives $N_0\alpha/3\,\varepsilon_0$ and hence α.

Calculation of dipole moment is more complex than suggested by mere vector addition of the bond moments given in Table 6.3. For example, vector addition leads one to expect that the dipole moments of $CHCl_3$ and CH_3Cl should be equal. However, the actual dipole moment of methyl chloride (1.87 D) alone agrees with this principle of vector sum (1.5 D + 0.4 D); chloroform has a much smaller dipole moment (1.01 D). This unusually low value might result from repulsion between the chlorine atoms and consequent widening of the Cl—C—Cl bond angle. But electron diffraction studies indicate that the actual bond angle (~111°) is not large enough to explain such low dipole moment. It is now believed that each C—Cl bond dipole induces an opposite dipole on the other two C—Cl bonds. Each C—Cl bond moment is thereby reduced, ultimately lowering the overall dipole moment of the molecule.

The effect of lone pairs of electrons on dipole moment of a molecule is revealed in the dipole moments of NH_3, NF_3 and H_2O. In all of these, lone pair(s) of electrons on the central atom occupy sp^3 hybrid orbitals which are strongly directional. In NH_3 and H_2O, higher electronegativity of the central atom (N or O) causes the bond moments to reinforce the moment of the lone pair. In NF_3, the resultant of the N—F bond moments (fluorine is now the negative end) act in opposition to the lone pair moment ; the dipole moment of the whole molecule is thereby lowered.

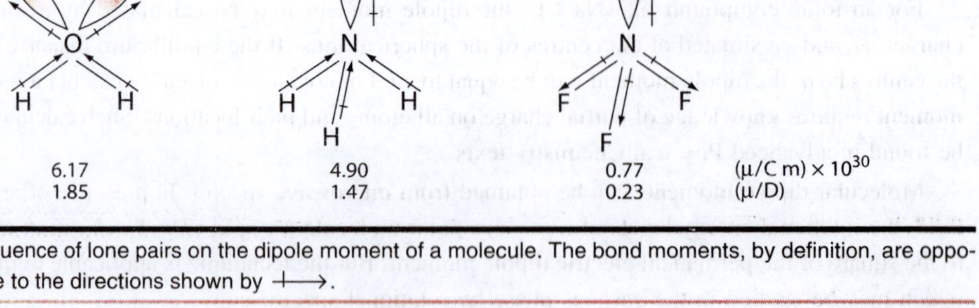

Fig.6.4: Influence of lone pairs on the dipole moment of a molecule. The bond moments, by definition, are opposite to the directions shown by ⟶.

Dipole moments of a few common compounds are given in Table 6.4.

Table 6.4

Dipole moments of some gaseous molecules
($\mu \times 10^{30}$ C m; *values in debye in parenthesis*)

H_2O	6.17	(1.85 D)	C_2H_5OH	5.64	(1.69 D)	HF	6.37	(1.91 D)
NH_3	4.90	(1.47 D)	CH_2Cl_2	5.24	(1.57 D)	HCl	3.60	(1.08 D)
CO	0.390	(0.117 D)	CH_3Cl	6.24	(1.87 D)	HBr	2.67	(0.80 D)
CH_3OH	5.70	(1.71 D)	$CHCl_3$	3.37	(1.01 D)	HI	1.40	(0.42 D)

Polarity of molecules is responsible for a large number of bulk properties of matter like intermolecular association, H-bonding, mixing with another substance or a solvent, surface phenomenon etc. and of course, their chemical behaviour. The shapes and interactions of biomolecules and macromolecules are also regulated by the polarity of bonds in different parts of the molecule—in some cases so crucial for their typical function.

In spite of its simplicity, the point-charge approximation for estimation of dipole moment has its own drawbacks. The major difficulty appears to be proper evaluation of the role of lone pairs of electrons. A lone pair may sometimes make a large contribution to the dipole moment of a molecule and even change the direction of the moment. An estimated charge distribution as $C^{+0.057}\ O^{-0.057}$ suggests that the CO molecule should have a dipole moment of more than 0.3 D. The experimental dipole moment is only 0.11 D acting in the opposite direction—the carbon end of the molecule is negative. This can be rationalized only when one considers the carbon lone pair to have a greater role in the overall charge distribution compared to the oxygen lone pair. We shall discuss this point further in connection with the MO description of CO (sec. 7.3.3).

6.4 ELEMENTARY IDEA OF SYMMETRY

Consciously or unconsciously, we encounter numerous examples of symmetry in nature and in our daily life—small plants, leaves, flowers and fruits, all possess some sort of symmetry in their appearance. Our appreciation of a thing of beauty is often associated with an inherent sense of symmetry in our subconscious mind: just think about the spider's web! In chemistry also, we have encountered different types of symmetry in the angular distributions of atomic orbital wave functions and in the shapes of various molecules. A systematic study of such symmetry properties lead to certain important generalizations and hence, useful simplifications. Therefore, such generalizations have become indispensable in the study of the ever-expanding field of chemistry. We shall develop here the preliminary line of approach for the application of the ideas of symmetry in chemistry. To appreciate the utility of taking this additional "*trouble*", one may have a look at the following list (incomplete) of applications of the principles of symmetry and its direct outcome, the group theory.

1. Classification of molecular orbitals and molecular structure
2. Classification of electronic states of molecules.
3. Construction of hybrid orbitals.
4. Prediction of splitting of electronic levels in electric fields of various symmetry.

5. Classification of the normal modes of vibration of molecules.
6. Prediction of allowed transitions in spectra.

However, within the limited scope of this book we shall restrict ourselves to a mere familiarity with the symmetry operations and symmetry elements which is likely to provide a better starting point for the study of group theory in the next higher course of study.

6.4.1 Symmetry Operations and Symmetry Elements

A molecule (or any chemical species) is said to possess symmetry if it has two or more orientations in space which cannot be distinguished from one another. We may go from one orientation to the other by different processes like reflection on a mirror plane, rotation about an axis, or a combination of these. Each such process is called a symmetry operation. *A symmetry operation is a process (real or imaginary) which when carried out on an object leaves that object in an indistinguishable spatial arrangement of the atoms (not necessarily identical) with respect to its initial arrangement.* In crystallography, the additional operation of translation is permitted.

The symmetry present in molecules and crystals is described with the help of several types of symmetry elements. Each symmetry operation is associated with a corresponding *symmetry element*, about which the symmetry operation is carried out. These may be a point, line or plane containing the molecule.

Two systems of notations are used to represent symmetry elements. (i) *The Schönflies system* is commonly used for molecular geometry and group theory. (ii) *The Hermann-Mauguin system* or *International system* is used in crystallography. The method of representation in each system will be found in the following discussion though we shall mainly follow the Schönflies system.

Five symmetry elements and operations are important for single molecules:

1. Identity (E). This symmetry element is included for theoretical interest only. The identity operation corresponds to doing *nothing* to a molecule so as to leave it unchanged. At best this may be described as a rotation by 360° about some symmetry axis. All molecules possess this symmetry element. Though meaningless at first sight, the operator is important as a limit for the repetitions of a given operation. The symbols E and $\hat{E}$ (E with a caret overhead) are used to represent the element and the operation respectively (*Schönflies*). The Herrmann-Mauguin symbol for the identity element is 1.

2. Centre of symmetry (i). If one starts from any atom of a chemical species having a centre of symmetry, moves in a straight line to this centre, and then proceeds an equal distance

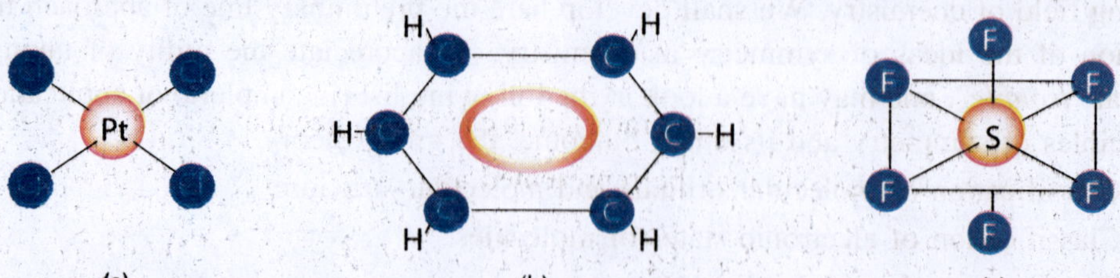

(a) (b) (c)

Fig.6.5: Species containing centre of symmetry.

on the other side of the centre, one encounters a similar atom. For example, the square planar $PtCl_4^{2-}$ ion, benzene and SF_6 (octahedral)—possess a centre of symmetry. *The inversion operator is represented by i* (Schönflies) or $\bar{1}$ (Hermann-Mauguin).

The inversion operation transforms the coordinates *x, y, z* of each point into their respective negatives: −*x*, −*y* and −*z*. This may be represented by

$$i\begin{bmatrix} x \\ y \\ z \end{bmatrix} = \begin{bmatrix} -x \\ -y \\ -z \end{bmatrix}$$

Two successive applications of *i* (or any even number of times) reproduce the original configuration:

$$i\,.\,i\begin{bmatrix} x \\ y \\ z \end{bmatrix} = i\begin{bmatrix} -x \\ -y \\ -z \end{bmatrix} = \begin{bmatrix} x \\ y \\ z \end{bmatrix}$$

This is equivalent to the *identity operation, E*, which leaves the orientation unchanged. We may write $i^k = E$ when *k* is even. For *k* odd, $i^k = i$.

3. Proper axis of rotation, C_n. If a molecule assumes an indistinguishable appearance by rotation about an axis through an angle of $2\pi/n$, (counter-clockwise by convention), the molecule is said to possess an *n*-fold rotational axis of symmetry. *The rotation operator is represented by C_n* (*n* in Hermann-Mauguin). The water molecule possesses one two-fold axis of symmetry, The highest-fold rotation axis of a molecule is conventionally taken as the vertical *z*-axis. A C_2 operation about the *z*-axis transforms the coordinate *x, y, z* of a point into −*x*, −*y*, *z*, i.e.

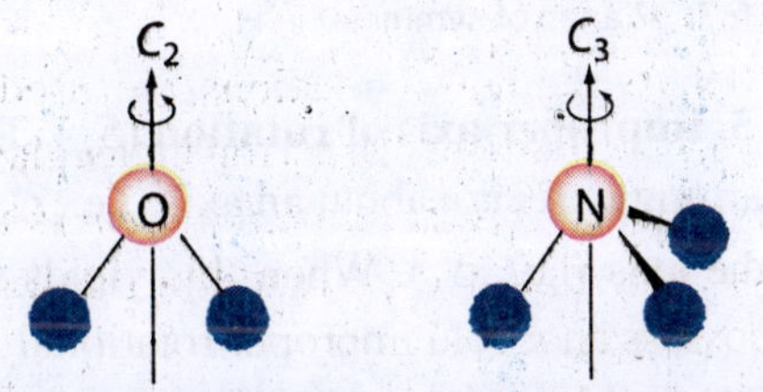

Fig.6.6: Rotational axis of symmetry in H_2O and NH_3.

$$C_2\begin{bmatrix} x \\ y \\ z \end{bmatrix} = \begin{bmatrix} -x \\ -y \\ -z \end{bmatrix}$$

Two successive applications of the C_2 operator or three successive applications of the C_3 operator will bring a molecule back to its original position (equivalent to the identity operation, *E*). Symbolically,

$$C_2^2 = C_3^3 = E$$

For a fourfold symmetry axis (C_4 operator), it is easy to realize that $C_4^2 = C_2^1$ and $C_4^4 = E$. The operators of a fourfold axis may be represented as

$$\begin{matrix} C_4^1 & C_4^2 & C_4^3 & C_4^4 \\ & C_2^1 & & E \end{matrix}$$

4. Plane of symmetry (mirror plane, σ). If reflection of all atoms of a molecule through a plane in it gives an indistinguishable configuration, the molecule is said to possess a symmetry plane. *Both the symmetry plane and the reflection operator are represented by* σ (*m* in Hermann-Mauguin). Two successive applications of σ (or any even number of times) give back the original configuration. Hence, $\sigma^k = E$, k even and $\sigma^k = \sigma$, k odd. The water molecule possesses two planes of symmetry. One plane bisects the HOH angle and another plane contains the three atoms.

> If there are several rotational axes of the same highest symmetry, e.g., three two fold axes, the z-axis is so chosen that it passes through the greatest number of atoms.

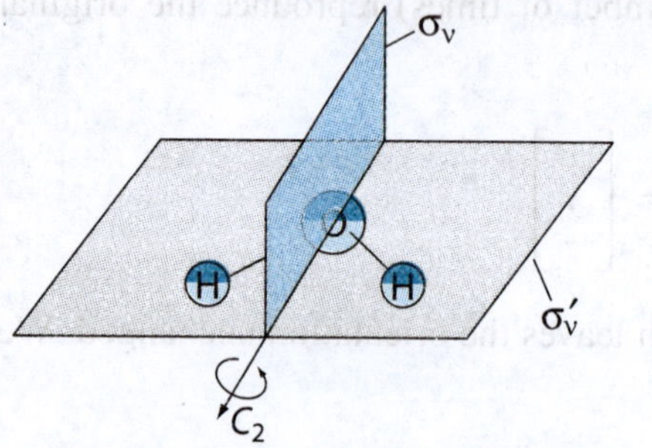

Fig.6.7: Planes of symmetry in H_2O.

Symmetry planes containing the principal C_n axis are called vertical symmetry planes, usually symbolized as σ_v. The particular symmetry planes which bisect the angles between successive twofold axis are usually called diagonal planes σ_d. A mirror plane perpendicular to the vertical axis is the horizontal plane, σ_h. (e.g., BCl_3, Fig. 6.11). The symmetry planes of the water molecule contain the C_2 symmetry axis and may be represented by σ_v and σ_v' (Fig. 6.7). The ammonia molecule has three σ_v containing the C_3 axis.

5. Improper axis of rotation (S_n). The operation of improper rotation consists of a rotation by an angle of $2\pi/n$ about an axis (i.e., $\hat{C}_n$), followed by reflection through a plane perpendicular to the axis (i.e., $\hat{\sigma}_h$). When this yields an indistinguishable configuration, the species is said to possess an n-fold improper rotational axis or rotation-reflection (*rotoreflection*) axis, S_n. The Hermann-Mauguin symbol for this operation is $\tilde{n}$ (tilday en'). The staggered configuration of ethane contains an S_6 axis coincident with the C_3 axis. The methane molecule has three S_4 axis coincident with the C_2 axis (Fig. 6.8).

The operation of improper rotation consists of two successive operations, rotation (C_n) followed by reflection (σ). Hence, $S_n = \sigma C_n$.

But we shall see that to have the S_n symmetry element, a molecule need not possess the symmetry elements σ and C_n. It is further interesting to note:

An S_1 element is equivalent to a plane of symmetry (σ).

An S_2 element is equivalent to a centre of symmetry (*i*).

When n is odd, $S_n^{2n} = E$; for n even, $S_n^n = E$.

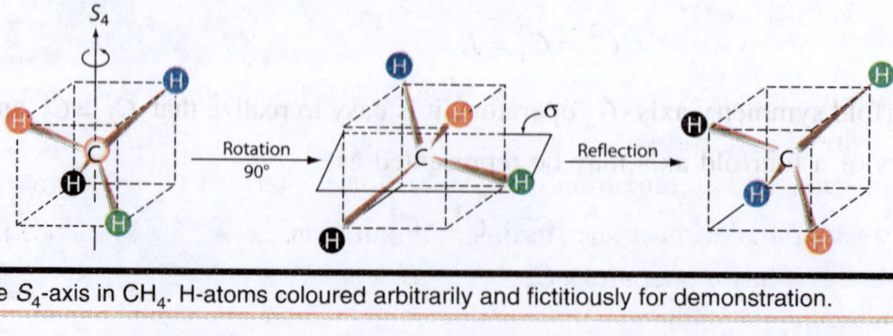

Fig.6.8: One S_4-axis in CH_4. H-atoms coloured arbitrarily and fictitiously for demonstration.

6.4.2 Point Groups

The complete set of symmetry operations that can be carried out on a molecule constitutes a mathematical group (see box). These are called ***symmetry groups*** or sometimes ***point groups***. All the operations leave the molecule fixed at a certain point in space—hence the name point group.

Thus, no translational motions are considered in point groups. There may be other groups of symmetry operations, such as those applied to crystal structures. There the individual units/molecules move from one location to another (translate) to generate a space lattice.

A set of operations belongs to a mathematical group only when they are interrelated by the following conditions:

(i) If A and B are two symmetry operations of the same group, their product AB (equal to F, say) will also be an operation of the group. Operation AB means operation B followed by A. This leads to the idea of multiplication of operations. A multiplication table may be prepared for each symmetry group, the one for the C_{2v} group being given in Table 6.5. In this particular group, the product AB is always equivalent to BA, i.e., when the order of application of the operators is reversed. Here the operators are said to commute, i.e., they are *commutative*: $AB = BA$. In other cases, $AB \neq BA$.

(ii) Each group will have an identity operator E. For any other operation A,

$$AE = EA = A.$$

(iii) For each operation A, there will be an inverse operation A^{-1} which will return the object to its original arrangement: $A^{-1}A = AA^{-1} = E$. The inverse of a C_2, σ or i operation is the operator itself. The inverse of C_3^1 is C_3^2; the inverse of S_4^1 is S_4^3.

(iv) The associative law of multiplication will hold.

$$A\,(BC) = AB\,(C).$$

Table 6.5

Multiplication table for the group C_{2v}

		Operation B			
		E	C_2^1	σ_v	σ_v'
	E_1	E	C_2^1	σ_v	σ_v'
Operation A:	C_2^1	C_2^1	E	σ_v'	σ_v
	σ_v	σ_v	σ_v'	E	C_2^1
	σ_v'	σ_v'	σ_v	C_2^1	E

The major point group systems and their symbols (Schönflies) are given below:

1. C_n, C_{nh} and C_{nv} groups

The groups generated by repetition of C_n operation are termed C_n point groups; n repetitions of this operation gives back the starting configuration, i.e., $C_n^n = E$. Hence the order of the group is n and it has n operations.

If a horizontal mirror plane σ_h is added to the C_n operation, we get C_{nh} groups. The C_{1h} group is called C_s.

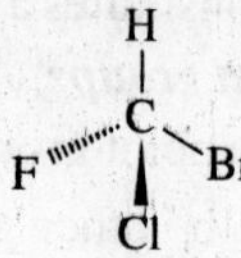
C_1

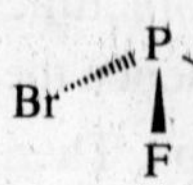
C_1

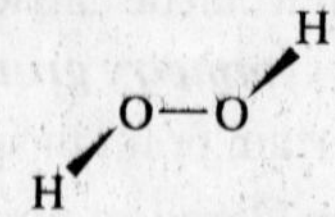
C_2

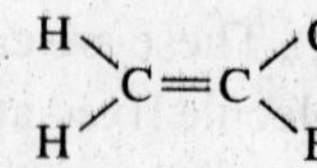
C_s

C_s

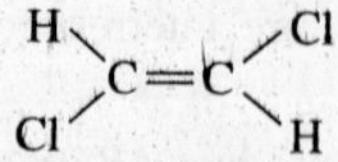
C_{2h}

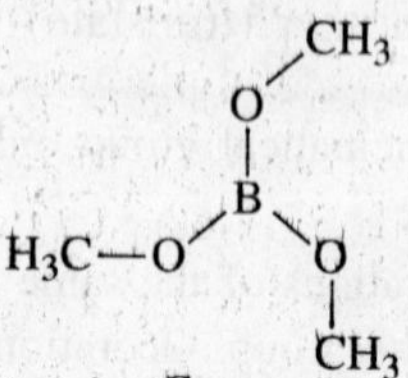
C_{3h}

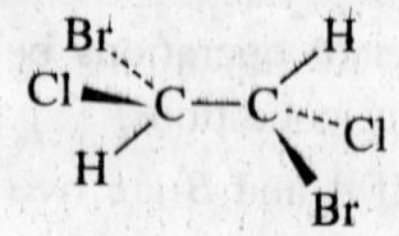
C_i

Addition of a vertical mirror plane (σ_v) to the C_n group gives a C_{nv} group. Both C_{nv} and C_{nh} groups have $2n$ elements or operations.

C_{2v}

C_{3v}

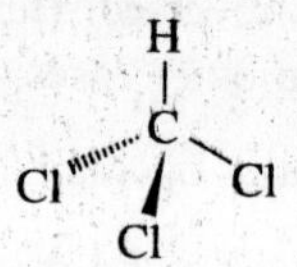
C_{3v}

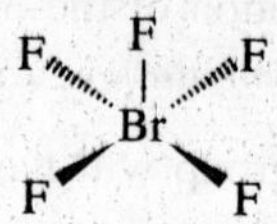
C_{4v}

2. D_n, D_{nh} and D_{nd} groups

When a vertical C_n axis is accompained by a set of n C_2 axes perpendicular to it, we get a *dihedral group, D_n*. There are $2n$ symmetry elements in this group. Molecules in this group are optically active and have a zero dipole moment. $\left[Co(en)_3\right]^{3+}$ (en = $NH_2CH_2CH_2NH_2$, shown as N⌒N, assumed to form a planar ring with Co) is an example of D_3 group.

Addition of a σ_h operation (reflection in a horizontal plane of symmetry) to a D_n group generates a D_{nh} group, with $4n$ symmetry elements.

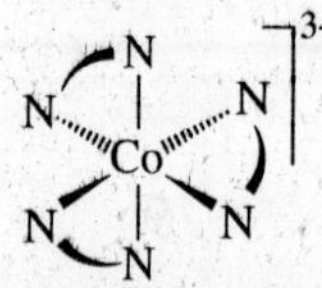
D_3

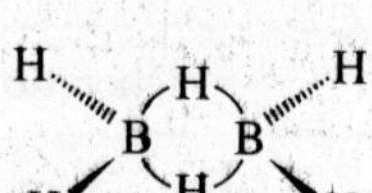
D_{2h}

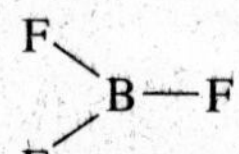
D_{3h}

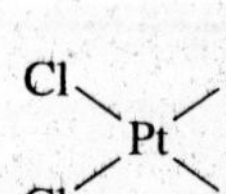
D_{4h}

Addition of vertical mirror planes to a D_n group which bisects the angles between pairs of C_2 axes generates a D_{nd} group. Such groups have no σ_h. The mirror planes are referred to as σ_d, dihedral planes, which are n in number. There are $4n$ symmetry elements in a D_{nd} group.

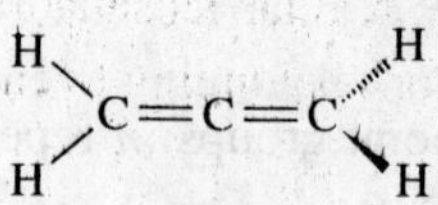
Allene
D_{2d}

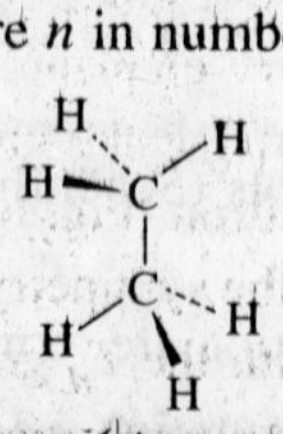
Ehane (staggered)
D_{3d}

Ferrocene (staggered)
D_{5d}

Osmocene (eclipsed)
D_{5h}

3. The S_n groups

The S_n groups are generated by the S_n operation. For odd values of n, $S_n^{2n} = E$; in such cases the S_n group is the same as the C_{nh} group. When n is even, new groups may arise, though they may be written in another way, e.g., $S_4^2 \equiv C_2, S_2 = i$. Molecules of S_n symmetry are not very common.

4. The $C_{\infty v}$ and $D_{\infty h}$ groups

Linear molecules possess an ∞-fold axis of rotation passing through the nuclei. Simultaneously there are an infinite number of reflection planes (σ_v) passing through the atoms. If no horizontal mirror plane (σ_h) is present (*alternatively*, no C_2 axis perpendicular to the C_∞ axis), the molecule belongs to the $C_{\infty v}$ group: HCl, HCN, CO etc. When a C_2 axis is present perpendicular to the C_∞ axis, the species belongs to the $D_{\infty h}$ group: N_2, O_2, CO_2 etc.

Assigning molecules to symmetry groups

Several schemes have been suggested for assigning molecules to their respective symmetry groups. The following outline may be helpful:

(i) Linear molecules (C_∞) may be easily classified as $C_{\infty v}$ (no i) or $D_{\infty h}$ (i present).

(ii) For others, try to locate a C_n axis. *If there is none*, the molecule belongs to one of the *low symmetry groups*:

C_s, if a σ element is present.

C_i, if no σ, but an i is present.

C_1, if neither σ nor i is present.

For example, CHFClBr has no symmetry other than E and therefore belongs to the C_1 group. $H_2C = CClBr$, with only one mirror plane, has C_s symmetry. HClBrC—CHClBr in the staggered conformation has only a centre of inversion; it belongs to the C_i symmetry group.

(iii) If a C_n axis is present, find the highest order. Then look for other noncoincident C_n axes with $n > 2$. If they are present, refer to Table 6.7 for point groups of higher symmetry. Otherwise proceed as follows:

(iv) If there is a C_2 axis perpendicular to the C_n axis, the molecule goes to the D_n group. (There will be n C_2 axis). Proceed as shown in Table 6.6.

(v) If there is no other noncoincident axis of rotation, the molecule belongs to the C_n group. Then proceed as shown in Table 6.6.

5. The Cubic and Icosahedral groups (*T*, *O*, *I* and their derivatives)

Many common chemical species possess more than one principal axis of symmetry. For example, the ammonium ion possesses four C_2 axes, one along each N—H bond (Fig. 6.9). The methane molecule also possesses the same symmetry elements. Such molecules (or species) belong to groups of very high symmetry. The group T_d is the group of the regular tetrahedron and the group O_h is the group of the regular octahedron (Fig. 6.10). The regular icosahedron belongs to the I_h group. The actual evaluation of all the symmetry operations belonging to each such group is a bit complicated. We illustrate it with the T_d group.

[The T_d group has three S_4 axes, four C_3 axes and six reflection planes. These symmetry elements give rise to the following symmetry operations:

Table 6.6

A scheme to select point-group symbol

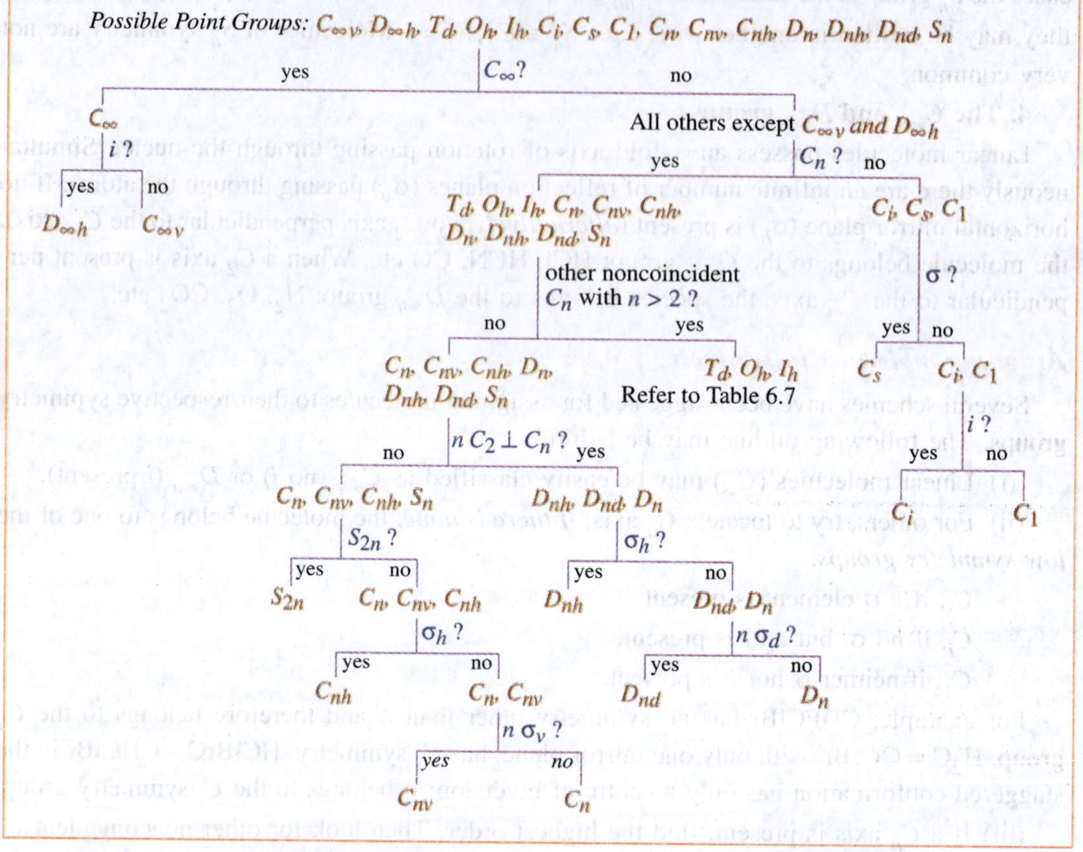

Each S_4 axis gives rise to the operations S_4, $S_4^2(=C_2)$, S_4^3 and $S_4^4(=E)$. Neglecting S_4^4 , we have 3 operations for each of the three S_4 axis, making a total of $3 \times 3 = 9$ operations.

Each C_3 axis gives rise to the operations C_3, C_3^2 and $C_3^3(=E)$. Neglecting the last, we have, for the four C_3 axes, $4 \times 2 = 8$ operations.

Each reflection plane gives rise to a σ_d operation, contributing $6 \times 1 = 6$ operations.

Thus we have a total of $9 + 8 + 6 +$ one identity operation $= 24$ operations.

Similarly the O_h group has 48 operations. The I_h group (icosahedron and pentagonal bipyramid) has 120 operations].

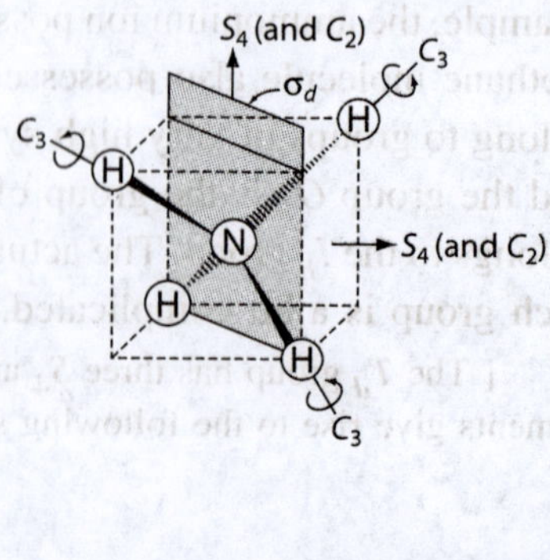

Fig.6.9: Some of the symmetry elements of the NH_4^+ ion (T_d point group). There are three S_4 axes at right angles to one another, but the third one (projected to the reader) has not been shown. Similarly, there are four C_3 axes (at angles of 109°28′) along each N—H bond, but only three are shown. Also there are six vertical σ_d planes, each of which passes through the nitrogen and two hydrogens and relates the other two hydrogens to each other. Only one is shown.

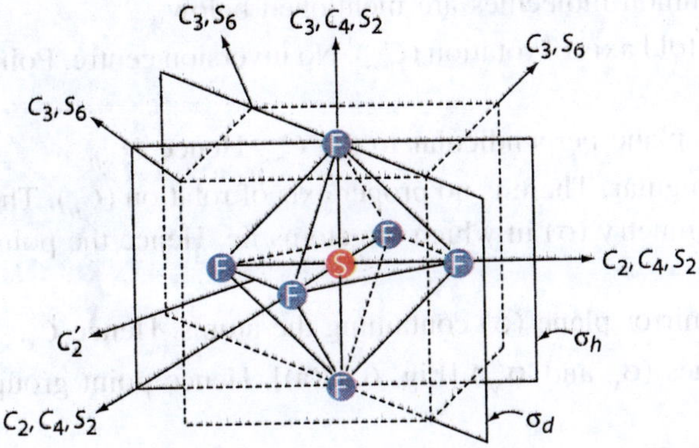

Fig.6.10: The symmetry elements of SF_6 (O_h group). The octaherdron is inscribed in a cube to show that both have the same elements of symmetry. There are three C_4 axes at right angles to one another; three S_4 and three C_2 axes are coincident with the C_4 axes. Four C_3-axes (and four coincident S_6 axes) are situated along the body-diagonals. There are six C_2' axes bisecting opposite edges, three planes of σ_h type and six of the type σ_d.

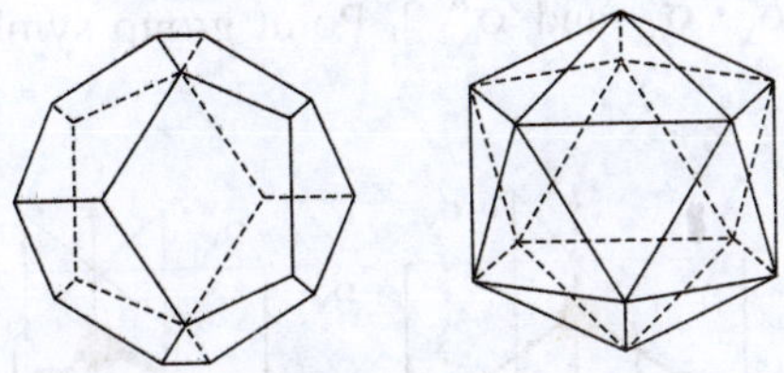

Fig.6.11: (a) Pentagonal bipyramid and (b) icosahedron.

However, it is possible to classify a molecule (or any species) into its point group formalism from the following scheme in Table 6.7.

(i) Find the highest C_n axis.

(ii) See if there are other noncoincident C_n with $n > 2$. n may be equal to 3, 4 or 5.

(iii) $n = 3$: It belongs to the tetrahedral (T) group. If there is an i or σ_h symmetry element, the point group is T_h.

If there is no i or σ_h but 6 σ_d elements, the group is T_d.

If there is none of the i or σ elements, the point group is simply T.

(iv) $n = 4$. The point group belongs to the O category. If there is a σ or i element, the point group will be O_h, otherwise simply O.

(v) $n = 5$. The point group belongs to the icosahedron, I. Again, if there is a σ or i element, the point group will be I_h, otherwise simply I.

Table 6.7

Classification of T, O and I point group

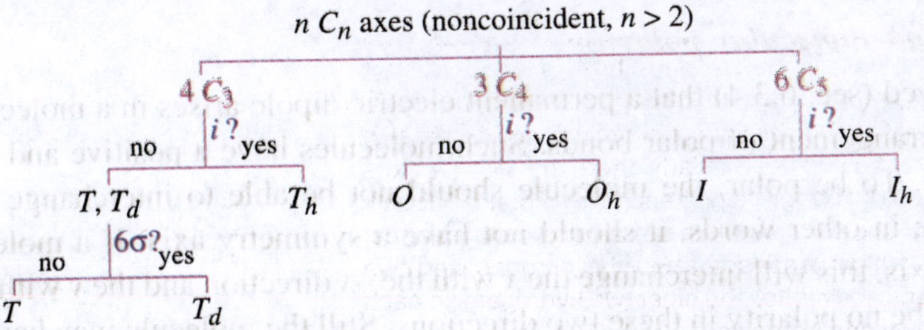

The point group symbols of some common molecules are mentioned below.

1. HCl. A linear molecule, hence an ∞-fold axis of rotation (C_∞). No inversion centre. Point group $C_{\infty v}$.

2. H_2. Linear molecule, C_∞. Inversion plane perpendicular to the C_∞. Hence $D_{\infty h}$.

3. BFClBr. The molecule is planar triangular. There is no proper axis of rotation (C_n). The only symmetry element is the plane of symmetry (σ) in which the atoms lie. Hence the point group is C_s.

4. HOCl. Here also, no C_n, only the mirror plane (σ) containing the atoms. Hence C_s.

5. H_2O. One C_2 axis, two mirror planes (σ_v and σ'_v) [Fig. 6.12(a)]. Hence point group symbol C_{2v}.

6. NH_3. There is only one C_3 axis and three mirror planes containing this axis [Fig. 6.12(b), σ_v, σ'_v and σ''_v]. Point group symbol C_{3v}.

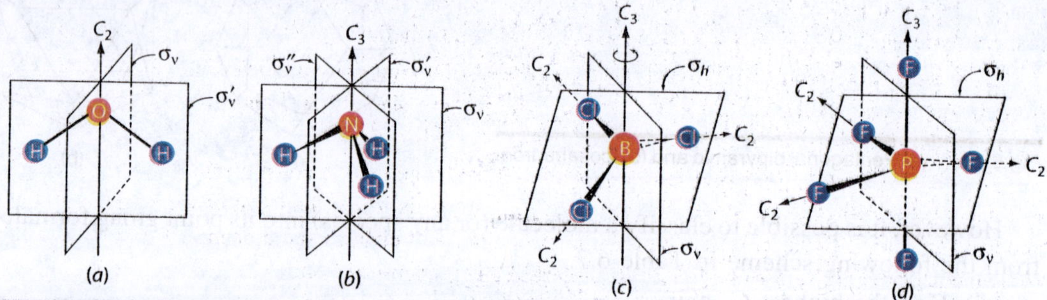

Fig.6.12: Symmetry elements of some common molecules.
(a) H_2O (C_{2v}) (b) NH_3 (C_{3v}) (c) BCl_3 (D_{3h}) (d) PF_5 (D_{3h}).

7. BCl_3. There is a C_3 axis. Three C_2 axes lie perpendicular to this C_3 axis, one along each B—Cl bond. There is also a mirror plane containing the atoms (σ_h), and perpendicular to the C_3 axis. In addition there will be three mirror planes perpendicular to the σ_h plane, one along each B—Cl bond. Only one such σ_v plane is shown in figure 6.12c, others being excluded for clarity. The total symmetry elements are then C_3, $3C_2 \perp C_3$, $\sigma_h \perp C_3$, and $3\sigma_v \perp \sigma_h$. Point group symbol D_{3h}.

8. PF_5. The molecule is trigonal bipyramidal (Fig. 6.12d). There is a C_3 axis and perpendicular to this are three C_2 axes. There is a mirror plane in the equatorial plane containing three P—F bonds, and perpendicular to the C_3 axis (σ_h). Three vertical mirror planes are present perpendicular to this σ_h ($3\sigma_v$), one along each equatorial P—F bond. Only one such plane is shown. The point group symbol will be D_{3h}.

Symmetry and molecular polarity

We observed (sec. 6.3.4) that a permanent electric dipole arises in a molecule mainly from unbalanced arrangement of polar bonds. Such molecules have a positive and negative end, at least partially. To be polar, the molecule should not be able to interchange its positive and negative ends; in other words, it should not have a symmetry axis. If a molecule posesses a principal C_n axis, this will interchange the x with the $-x$ direction and the y with the $-y$ direction and there can be no polarity in these two directions. Still the molecule may have polarity along the z-direction. But if the molecule has an additional C_2 axis perpendicular to the principal axis, or has a σ_h plane, these will interchange the z with $-z$ direction. Hence molecules belonging to a C_{nh} or any D or any normal higher order point groups cannot develop polarity. Only molecules belonging to the groups C_n, C_{nv} and C_s will have a permanent electric dipole moment.

For a molecule belonging to the C_n group with $n > 1$, there cannot be any dipole moment perpendicular to the symmetry axis. In the H_2O molecule, for example, it will be clear from Fig. 6.6 that the component of one O—H bond moment in a direction perpendicular to the C_2 axis will be cancelled by an equal but opposite component from the other O—H bond moment. So the dipole moment of the molecule must lie along its twofold symmetry axis. The same conclusion applies to any C_n and C_{nv} group.

ESSENTIALS

The electronic theory of valency is based on the octet rule which states that atoms of elements other than the noble gases tend to achieve a stable configuration of eight electrons in their outermost shell (except hydrogen which tends to attain the configuration of helium).

There are a large number of ions and molecules which do not obey the octet rule: e.g., Cu^+ (18 electrons in outermost shell), $BeCl_2$ (4 electrons in the valence shell of Be) and SF_6 (12 electrons in the valence shell of S).

The chemical bond: It is said to be responsible to hold together the constituent atoms in a chemically distinct species. When the energy of an aggregate of atoms is lower than that of the isolated atoms by about 42 kJ mol^{-1}, the aggregated species bears distinct and characteristic chemical properties and its constituent atoms are said to be bound to one another by chemical bonds.

Bond energy is the average energy required (kJ mol^{-1}) to split the atoms linked by a chemical bond.

Lewis electron-dot formulae represent the atom-to-atom connectivity and the unshared pairs of electrons. The total number of valence electrons provided by all the atoms in a molecule (with necessary adjustment for ions) is distributed judiciously to satisfy octet to each element (except H) and a reasonable charge distribution. The more electropositive atom is usually placed at the centre. Expansion of octet may be necessary for elements of period 3 onward.

The formal charge on an atom in a Lewis formula is the hypothetical charge acquired by it assuming equal sharing of bonding electrons and complete possession of nonbonding electrons. If an atom with n_v valence electrons has n_b electrons in bonds around it and n_l number of electrons as unshared pairs, the formal charge is calculated as $n_v - (n_l + n_b/2)$.

The electroneutrality principle sets the criterion for most favourable Lewis structure:

(i) low formal charges on atoms;
(ii) absence of same sign of nonzero charges on adjacent atoms;
(iii) negative charges on more electronegative atoms;
(iv) maximum number of bonds and highest extent of π-bonding.

The VSEPR (valence shell electron pair repulsion) theory predicts the shape of a molecule (or ion) on the basis of minimum electrostatic repulsion between the electrons in bond pairs and unshared (lone) pairs of electrons in the valence shell of the central atom.

The steric number (SN) of an atom in a molecule is the sum of the number of bonded atoms and the number of lone pairs. The most favourable electron distribution in a species with different steric numbers are

SN =	2	3	4	5	6	7
	linear	trigonal planar	tetrahedral	trigonal bipyramid	octahedral	pentagonal bipyramid

The *actual geometry of the molecule* is then determined by the position of bonded atoms.

It is assumed that lone pairs (*lp*) require more angular space compared to bond pairs (*bp*); hence the order of repulsion is: *lp—lp* > *lp—bp* > *bp—bp*.

For higher steric numbers, specially 5 and 7, more than one geometry are observed.

Bond length, Bond dissociation energy (ethalpy) and stretching force constants are measurable parameters of a chemical bond.

Changes in Enthalpy (H), Entropy (S) and Gibbs energy (G) provide useful thermodynamic guideline for the feasibility of a chemical reaction. A reaction is considered thermodynamically favourable when $\Delta G = \Delta H - T\Delta S < 0$. Both negative ΔH and positive ΔS make ΔG negative but in certain cases a positive ΔH may be overpowered by a negative $T\Delta S$ term.

Polarity of a bond results from unequal sharing of the bonding electrons between two atoms having different electronegativity. The extent of polarity is expressed by an electric dipole moment vector μ directed from the *negative to the positive charge*. For two charges $+q$ and $-q$ separated by a distance d, $\mu = qd$.

The SI unit of dipole moment is coulomb meter (C m) but the traditional unit debye (D) is also used. 1 D = 10^{-18} esu cm = 3.3356×10^{-30} C m.

Polar molecules possess a permanent electric dipole moment due to unbalanced arrangement of polar bonds, together with other contributions like those from unshared electron pairs.

A symmetry operation is a process which when carried out on an object leaves that object in an indistinguishable spatial arrangement of the atoms with respect to its initial arrangement.

Each symmetry operation is associated with a corresponding *symmetry element* about which the symmetry operation is carried out. For a discrete object like a molecule, five symmetry elements are found to be significant:

(i) the axis of symmetry, *C*, (ii) the plane of symmetry, σ, (iii) the centre of inversion, *i*, (iv) the improper axis of rotation, *S and* (v) the identity *E*.

The corresponding symmetry operations are

(i) C_n, rotation of the molecule through an angle 360°/*n* about a symmetry axis. *n* is called the *order* of rotation. The highest order rotation axis is called the principal axis of rotation; it is usually placed vertically along an imaginary *z*-axis.

(ii) σ, reflection of all atoms through a mirror plane.

(iii) *i*, inversion of all atoms through a point of the molecule.

(iv) S_n, rotation of the molecule by 360°/*n*, followed by reflection in a plane perpendicular to the axis.

(v) *E*, the identity operation which leaves the molecule unchanged.

Planes of reflection which are perpendicular to the principal axis are called horizontal planes (*h*). Similarly, planes of reflection containing the principal axis are called vertical planes (*v*) or dihedral planes (*d*) when they bisect two 2-fold axes.

The complete set of symmetry operations that can be performed on a molecule is called the ***symmetry group* or *point group*** of the molecule.

EXERCISE

1. Consider the following structures for the ion $S_3N_2Cl^+$ found in the salt $(S_3N_2Cl)Cl$.

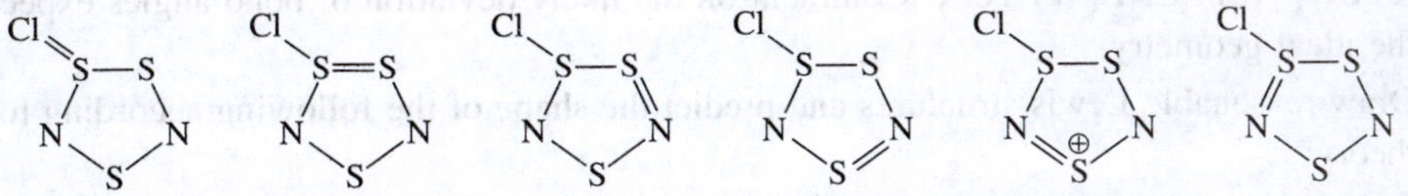

(a) (b) (c) (d) (e) (f)

(i) Place nonbonding electrons as necessary and show formal charges (nonzero) on the atoms.
(ii) Which structure should be most favoured?
(iii) Which structure(s) would be unimportant?

Ans. (i) (a) 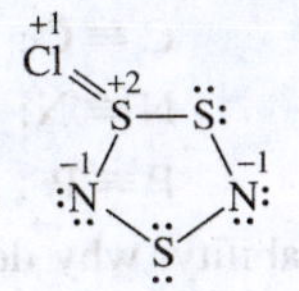 (b) 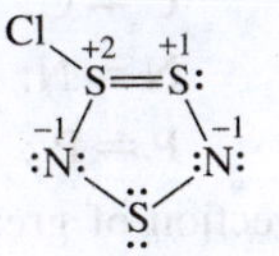 (c) 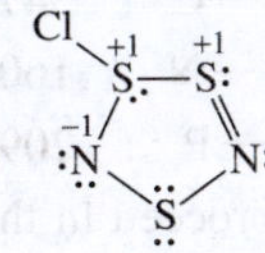 . etc. (ii) *d* (iii) *a, b, c*

2. (a) Why do you consider the structures 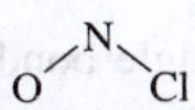 and 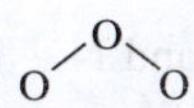 as wrong? Write the correct structures for each and assign formal charges. For O_3, the charges on the terminal atoms will be given by the average values of charges in the two resonating structures.

Ans. $-\frac{1}{2}$ for each terminal O in O_3.

3. The I_3^- ion is said to be hypervalent and the structure is written with an expanded octet for the central I: $[:\ddot{I}-\ddot{I}-\ddot{I}:]^-$. Show that the ion may also be represented by two equivalent resonance structures obeying octet.

Ans. I—I $I^- \leftrightarrow \cdots$. This is an example of no-bond resonance (see chapter 7).

4. NO_2^+ is isoelectronic with CO_2 as both have same number of valence electrons (and also same number of atoms). Give examples of two other neutral molecules and two anions which are isoelectronic with them.

Ans. HNCO, N_2O, NCO^-, N_3^-.

5. Draw reasonable Lewis octet structures and show formal charges (non-zero only):
(a) Al_2Cl_6, (b) $SnCl_3^-$, (c) NS^+, (d) SO_3F^-, (e) $S_4N_3^+$ (cyclic with one S—S bond), (f) BrF_4^+.

Ans. (a) 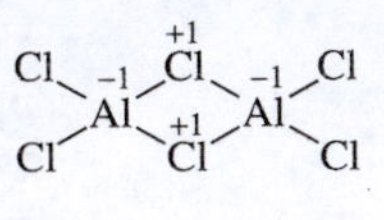 (e) 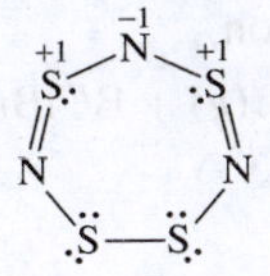 (f) 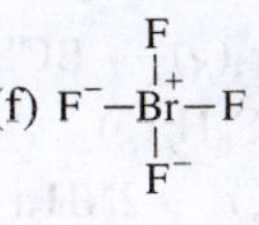 etc. (no bond resonance).

6. Select the most likely skeleton structure for $SOCl_2$ from (i)–(iii) with reason. Write the most appropriate Lewis formula with the selected skeleton. The central atom may have an expanded octet.

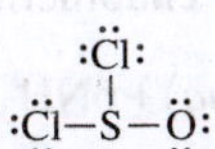 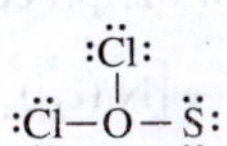 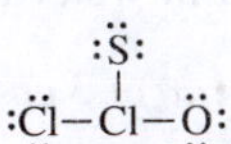

(i) (ii) (iii)

7. Write Lewis formula for $HClO_4$, H_2SO_4, H_3PO_4 and H_2SO_5. The last one contains an –O–O– bond.
8. Apply VSEPR theory to predict the geometry of the following species: (a) NO_2^- (b) ClO_2^- (c) SO_3^{2-} (d) PCl_2F_3 (e) TeF_7^-. Comment on the likely deviation of bond angles expected from the ideal geometry,
9. Draw reasonable Lewis structures and predict the shape of the following according to VSEPR theory:

$$SOF_4,\ ClO_2F_3,\ IOF_4^-,\ IOF_3,\ XeO_3F_2 \text{ and } XeOF_4.$$

Comment on the following bond angles;

$F_{eq}SF_{eq}$ in SOF_4 = 110°; OIF in $IOF_4^- \simeq 89°$; $F_{eq}IO$ = 98° and $F_{ax}IF_{ax}$ = 168° in IOF_3,
10. Use the following bond energies (kJ mol^{-1}) to predict whether C_2H_2, C_2H_4, N_2, NH_3, PH_3 and P_2 should form polymers (containing single bonds).

C – H :	414;	C – C :	347;	C = C:	611;	C ≡ C:	837.
N – H:	391;	N – N :	160;	N = N:	418;	N ≡ N:	946,
P – H :	322;	P – P :	209;	P = P :	335;	P ≡ P :	485.

11. If chemical changes always proceed in the direction of greater stability, why do endothermic reactions occur at all?
12. Calculate the N – H and N – N single bond energies from the following enthalpy data: Enthalpy of formation (kJ mol^{-1}): $NH_3(g)$ –46; $N_2H_4(g)$ +95.4; Enthalpy of bond breaking (kJ mol^{-1}) H_2 : 436; N_2 : 945.6.

Ans. E_{N-H}: 391 kJ mol^{-1}, E_{N-N}: 158.2 kJ mol^{-1}.
13. State, with reasons, whether you expect $\Delta S°$ for the following processes to be +ve or –ve,

(i) ice → water.

(ii) $P_4(g) + 10F_2(g) = 4PF_5(g)$.

(iii) $I_2(s) + Cl_2(g) = 2ICl(g)$.

(iv) $BF_3(g) + NH_3(g) = H_3NBF_3(g)$,

(v) $Zn(s) + 2H^+(aq) = Zn^{2+}(aq) + H_2(g)$

Comment: In the last reaction, one might expect $\Delta S° > 0$ as one mole of gaseous product is formed. But the hydration of Zn^{2+} ion orients dipoles of H_2O around Zn^{2+} and lowers the entropy. The overall $\Delta S°$ is < 0.
14. What should be the main thermodynamic criterion for the following processes to proceed from left to right?

(i) $NH_4NO_3(s)$ + water → solution.

(ii) $BCl_3(g) + BBr_3(g) \rightarrow BCl_2Br(g) + BClBr_2(g)$.

(iii) $3C_2H_2(g) \rightarrow C_6H_6(g)$

(iv) $N_2(g) + 3H_2(g) \rightarrow 2NH_3(g)$

(v) $SnCl_2 + Cl_2 \rightarrow SnCl_4$.

(vi) LiI + CsF → LiF + CsF,
15. Explain the role of entropy in the following spontaneous processes:

(i) KI(*s*) dissolves in water though the process is endothermic;

(ii) $[Ni(NH_3)_6]^{2+}(aq) + 3\,en\,(aq) \rightarrow [Ni(en)_3]^{2+}(aq) + 6NH_3(aq)$

en = 1,2-diaminoethane, $H_2NCH_2CH_2NH_2$

(iii) $HCO_3^-(aq) + H^+(aq) \rightarrow H_2O + CO_2(g)$

Hint: (i) The process is endothermic owing to poor hydration enthalpy of the large I^- ions. The process is entropy driven since breaking of the crystal lattice makes ΔS largely positive.

(ii) ΔH° value of the reaction is indeed very small (-12 kJ mol^{-1}) since both sides involve six Ni^{2+}–N bond. A net increase in the number of molecules in solution makes ΔS° large and positive (84 J mol^{-1}). The $T\Delta S^\circ$ term contributes about 25 kJ mol^{-1} at 298 K (see chelate effect, Ch-27).

(iii) The reaction is accompanied by a drop in temperature and hence endothermic. The entropy gain from production of 1 mole of gaseous CO_2 seems to drive the reaction.

16. It is proposed to synthesize compounds of the type N_2X_4 (X = halogen). The structures are similar to that of N_2H_4. With which halogen would you suggest starting? On a purely thermochemical basis, would you expect the synthesis to be feasible?

17. Do you consider it worthwhile to attempt production of ethene from carbon dust and methane (in natural gas) through the following reaction?

$$C(s) + CH_4(g) \rightarrow C_2H_4(g) : \Delta G^\circ \text{ (kJ mol}^{-1}) : C_2H_4 : 68.1; CH_4 : -50.8.$$

[**Hint:** ΔG° for the reaction = +118.9 kJ mol^{-1}]

18. In diamond, each carbon atom is bonded to four other carbon atoms. But the enthalpy of sublimation of diamond (~715 kJ mol^{-1}) is nearly double the C—C bond enthalpy (348 kJ mol^{-1}). Comment.

[**Hint:** $2N$ bonds are broken during evaporation of 1 mol of diamond.]

19. Which of the following molecules are expected to have permanent electric dipole moment?

SO_2, SF_2, SF_4, S_2Cl_2, C_2H_2, SiF_4, N_2F_4, PF_5, BrF_5, XeF_4.

20. The dipole moment of water is 6.17×10^{-30} C m. The HOH angle is 104° and the O—H distance is 96 pm. Calculate the percent ionic character of the O—H bond.

[**Hint:** The moment of each O—H bond, [p(O—H)], may be calculated from the principle of vector addition:

$$2p_{(O—H)} \times \cos 52^\circ = 6.17 \times 10^{-30} \text{ C m}.$$

The dipole moment of 100% ionic bond would be 1.602×10^{-19} C $\times$ 96 $\times 10^{-12}$ m i.e., 15.4 $\times 10^{-30}$ C m.

[**Ans.** 32.5 per cent]

21. Assign point groups to the following:

(i) CFBrClI; (ii) $SOCl_2$; (iii) N_2 (iv) HCl (v) $SiHDBr_2$ (vi) SiFClBrI (vii) SiH_2Br_2 (viii) letter T (ix) $OPCl_3$ (x) $B_3N_3H_3Cl_3$ (B, B, B-trichloroborazine).

Ans. (i) only E, C_1 (ii) $E + \sigma$, C_s (iii) $C_2 \perp C_\infty$, $D_{\infty h}$ (iv) no $C_2 \perp C_\infty$ or no i, $C_{\infty v}$ (v) C_s (vi) C_1 (vii, viii) C_{2v} (ix) C_{3v} (x) D_{3h}.

22. Using symmetry arguments discuss whether the molecules CH_3Cl, CH_2Cl_2 and $CHCl_3$ will have dipole moments.

Ans: CH_3Cl, $CHCl_3$, C_{3v}; have dipole moment along the C_3 axis (also line of intersection of the $3\sigma_v$); CH_2Cl_2; C_{2v}, has dipole moment along the C_2 axis (also line of intersection of σ_v planes).

7

COMBINATION BETWEEN ATOMS

PART II: *Wave Mechanical Approach*

7.1 INTRODUCTION

We have discussed the application of quantum mechanics to atomic systems. The first successful application of quantum mechanics to explain the bonding in molecules, particularly that in dihydrogen (H_2), was made in 1927 by Heitler and London (the valence bond method). Since then quantum mechanical methods have developed tremendously to enable calculation and prediction of electronic structures of molecules and other associated properties to a fair degree of accuracy, thanks to the availability of advanced computer technology.

In applying the Schrödinger equation to a molecular system, we must consider the various interactions inherent to the system. The Hamiltonian or energy operator, constructed accordingly, may then be used to obtain the energy E of the system by solving the Schrödinger equation

$$\hat{H}\psi = E\psi$$

provided we have a knowledge of the relevant wave function ψ.

The actual situation is more difficult than stated above. First, the Hamiltonian operator becomes very complicated when all electron-nucleus, nucleus-nucleus and electron-electron interactions are taken into account and the Schrödinger equation becomes mathematically unmanageable for solution. Second, we do not know the wave function ψ for the molecular system.

Two basic simplifying procedures have been adopted to overcome these difficulties:

(i) The Born-Oppenheimer approximation, and,

(ii) The Variation principle.

The Born-Oppenheimer approximation

According to this approximation, the motion of the heavy nuclei in ordinary molecular vibrations are considered to be very slow compared to the motion of electrons; hence the nuclei are assumed to be stationary or fixed while calculating the electronic energy states of a molecule. Thus, to a good approximation, the motion of the nuclei and vibration/rotations of the molecule can be treated as separate problems. The internuclear distances and the relative position of the nuclei may be set as a parameter (quantities which are constant for one set of circumstances but may vary when circumstances are changed) and the Hamiltonian operator may be written for the electronic part of the molecule. Solution of the corresponding Schrödinger equation using suitable electronic wave function ψ will give the electronic energy for the molecular system. The nuclear motion of the molecule becomes significant in the study of vibration-rotation spectra.

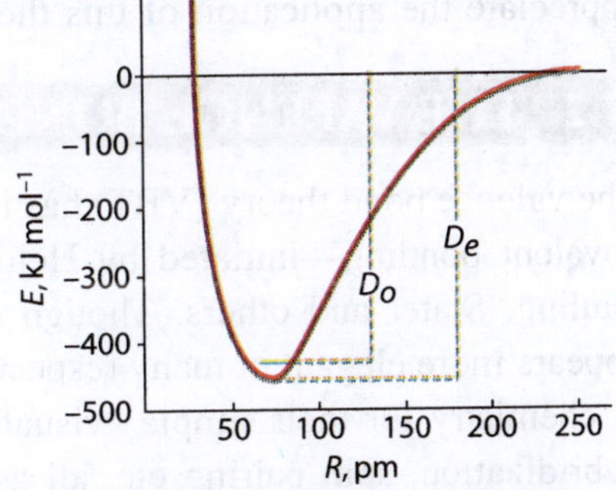

Fig.7.1: Nature of the molecular potential energy curve for the dihydrogen molecule (H_2). The curve shows a minima at –458 kJ mol^{-1} for an internuclear separation of 74 pm. The equilibrium bond dissociation energy (D_e) is 458 kJ mol^{-1}. The spectroscopic bond dissociation energy is 432 kJ mol^{-1}. The zero point vibrational energy is 26 kJ mol^{-1}.

A molecule may have many possible different electronic states with different electronic wave functions and energy which vary with the nuclear arrangement. One may plot the variation of electronic energy (E) for different internuclear separations (R) to get a molecular potential energy curve. The curve for the dihydrogen molecule (H_2) is shown in Fig. 7.1. The minima in the curve corresponds to the equilibrium internuclear separation (R_e) or the bond length and the minimum value of the potential energy D_e. The spectroscopic bond dissociation energy D_o is slightly above the depth of the well; the separation ($D_e - D_o$) is the zero point vibrational energy of the molecule.

The variation theorem

The energy (electronic) of a molecule may be obtained from the expression

$$E = \int \psi^* \hat{H} \psi \, d\tau$$

where ψ is the wave function (electronic) suitable for all the electrons in the molecule and $\hat{H}$ is the electronic Hamiltonian operator or energy operator based on the total kinetic and potential energy of the electrons in the given nuclear framework. For molecules with many electrons, exact solution of the Schrödinger equation becomes almost impossible. It is useful to intuitively select several trial wave functions and calculate the energy of the molecule. In this connection comes the variation theorem which may be stated as: *no trial wave function (normalized and well-behaved) of the coordinates of the particles of a quantum-mechanical system will yield an energy which is lower than the true ground-state energy of the system.* If E_o be the ground state energy of the molecule and ϕ is any trial wave function, then,

$$\int \phi^* H \phi \, d\tau > E_o$$

This principle, which may be proved quantum mechanically, allows us to construct a suitable wave function for a molecule from the participating atomic orbital wave functions and improve it to fit experimental findings without any risk of "overcorrection". We shall soon appreciate the application of this theorem in connection with the valence bond theory.

7.2 THE VALENCE BOND THEORY (VBT)

The valence bond theory (VBT) has its origin in the quantum mechanical procedure to explain covalent bonding—initiated by Heitler and London (1927) and subsequently developed by Pauling, Slater and others. Though a later procedure, namely the molecular orbital theory, appears more elegant in many respect, certain concepts conveyed by the VBT are widely used in chemistry for their simple "visual" appeal—for example, orbital overlap, σ and π bonds, hybridization, spin pairing etc, all associated with the very idea of localized bonds holding atoms in space. Here we shall discuss only the essential concepts of VBT; for the actual quantum mechanical procedures one may consult any textbook of quantum chemistry.

7.2.1 The Basic Approach in VBT

The valence bond theory leads to the idea of bond formation through spin-pairing of electrons between the participating atoms. To appreciate this, we begin with the simplest possible electron pair bond, the one that is present in the dihydrogen molecule, H_2.

We want to apply the Schrödinger equation ($\hat{H}\psi = E\psi$) to calculate the energy E of a system of two hydrogen atoms, say H_A and H_B, for various fixed internuclear separations (R_{AB}, say). Next we compare the minimum value of E and the corresponding internuclear separation with those in the molecular potential energy curve shown in Fig. 7.1.

We now need two things: the appropriate Hamiltonian operator $\hat{H}$ and a befitting well-behaved wave function ψ. The Hamiltonian operator consists of terms involving attraction of each electron for each of the two nuclei, repulsion between the nuclei and repulsion between the electrons. We skip the manipulation of the Hamiltonian operator which will be dealt with in physical chemistry units. We shall briefly describe the steps to develop a suitable wave function for the system as this will provide us a better understanding of various terminologies like spin-pairing, overlap, resonance, hybridization etc.

In its search for a suitable wave function for the molecular system, the VB theory uses two principles:

(i) If ψ_A and ψ_B are the wave functions for two independent systems A and B respectively, then the total wave function ψ for their combination may be expressed as a simple product

$$\psi = \psi_A \psi_B$$

and the total energy (E) of the system will be the sum of the individual energies:

$$E = E_A + E_B.$$

(ii) If ψ_1, ψ_2, ψ_3 etc. are different acceptable wave functions for the same system, then the true wave function ψ may be obtained from a linear combination of these wave functions, i.e.,

$$\psi = c_1\psi_1 + c_2\psi_2 + c_3\psi_3 + \cdots$$

c_1, c_2 etc. are coefficients which may be suitably adjusted to find a ψ corresponding to lowest energy. The relative contribution of each wave function to the true wave function may be qualitatively compared from the squares of the coefficients of respective wave functions.

Let us consider two widely separated hydrogen atoms and mark them as H_A and H_B. We assume that the electron in H_A is labelled '1', and described by the atomic orbital (1*s*) wavefunction $\psi_A(1)$ for the isolated atom in its ground state. Similarly the electron in H_B is marked '2' and described by the wavefunction $\psi_B(2)$.

Step I. ***Electrons confined to respective nuclei***

The simplest trial wave function for the combined system of two electrons and two nuclei may be written as

$$\psi = \psi_A(1)\ \psi_B(2) \qquad \text{(arrangement } H_A^1\ H_B^2) \qquad \text{... [7.1]}$$

Using this trial wave function, the energy of the system may be calculated as a function of internuclear separation (R). The plot (Fig 7.2, graph *a*) shows a slight depression at about –24 kJ mol^{-1} at R = 90 pm. This is in poor agreement with the experimental values (–458 kJ at R = 74 pm). Nevertheless, the lowering in energy obtained theoretically is significant.

Step II. ***Exchange of electrons between the nuclei***

Now we may try to improve the wave function by one practical consideration. When the hydrogen atoms are sufficiently close for bond formation, there is really no justification for assigning electron '1' to atom A and electron '2' to atom B. In fact electrons are indistinguishable and the system of two H-atoms may be represented equally in any of the two following descriptions.

$$\psi_I = \psi_A(1)\ \psi_B(2) \quad \text{for arrangement } H_A^1\ H_B^2 \qquad \text{... [7.2]}$$

$$\psi_{II} = \psi_A(2)\ \psi_B(1) \quad \text{for arrangement } H_A^2\ H_B^1. \qquad \text{... [7.3]}$$

ψ_I and ψ_{II} differ only by *exchange* of the electrons arbitrarily marked '1' and '2' between the atoms A and B. According to quantum mechanics, the true wave function ψ is better represented by a combination of ψ_I and ψ_{II}

$$\psi = c_1\psi_I \pm c_2\psi_{II}$$

It is found that the combination $c_1\psi_I + c_2\psi_{II}$ corresponds to a lowering of energy for the system and hence bonding between the two H atoms (see symmetric and antisymmetric wave functions). The valence bond wave function for the H_2 molecule (unnormalized) thus takes the form

$$\psi = \psi_I + \psi_{II} = \psi_A(1)\,\psi_B(2) + \psi_A(2)\,\psi_B(1) \quad \text{... [7.4]}$$

The plot of energy calculated using this ψ against internuclear distance is shown in curve (*b*) in Fig. 7.2. We find a dramatic improvement in our result, the minimum energy is now found at –303 kJ mol^{-1} for a separation of 87 pm.

From comparison with plot (*a*), the extra stabilization gained by mere spatial exchange of the electrons is equal to the difference in the two minima: (303 – 24) = 279 kJ mol^{-1}. This is often called the exchange energy.

(*a*) Step I	$H_A^1\, H_B^2$	–24 kJ, 90 pm
(*b*) Step II	$H_A^1\, H_B^2 \;\; H_A^2\, H_B^1$	–303 kJ, 87 pm
(*c*) Step III	Screening	–365 kJ, 74.3 pm
(*d*) Step IV	Ionic contribution	–388 kJ, 74.9 pm
(*e*) Experimental		–458 kJ, 74 pm

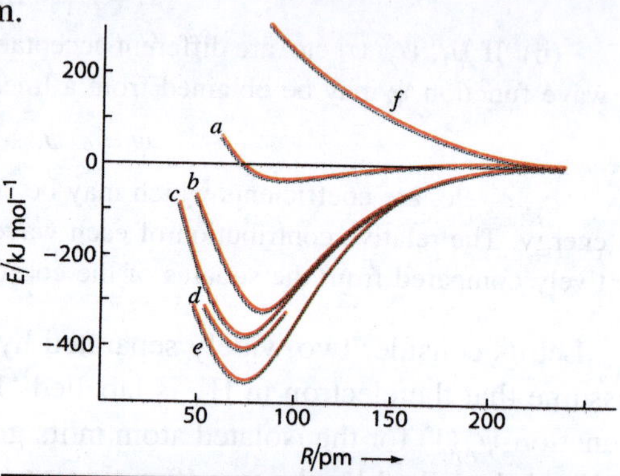

Fig. 7.2: Energy of the H_2 molecule calculated using various trial wave functions in VBT. Energy minima and corresponding internuclear separation for each are shown on the left.

Step III. ***Inclusion of effective nuclear charge***

In an isolated H atom, the electron experiences the full nuclear charge of the proton. In the H_2 molecule, one electron screens the other electron. The effect may be included in the wave function by using effective nuclear charge (Z_{eff}). The improvement in ψ is evident from the occurrence of energy minima at –365 kJ mol^{-1} at R = 74.3 pm. (graph *c*, Fig. 7.2).

Step IV. ***Contribution of ionic character***

When the two H atoms are close to one another, it is quite likely that both electrons 1 and 2 may be momentarily associated with a particular nucleus, giving rise to ionic structures like

$$H_A^{1,2}\, H_B \;\text{ or }\; H_A^-\, H_B^+ \quad\text{and}\quad H_A\, H_B^{1,2} \;\text{ or }\; H_A^+\, H_B^-$$

with equal probability. In the total wave function, these are included by two ionic terms (ψ_{ion}) as

$$\psi_{ion} = \psi_A(1)\,\psi_A(2)\left(\text{for } H_A^{1,2}\, H_B\right) \;+\; \psi_B(1)\,\psi_B(2)\left(\text{for } H_A H_B^{1,2}\right).$$

The total wave function (unnormalized) for the H_2 molecule now becomes

$$\psi = (1-\lambda)\,\psi_{cov} + \lambda\,\psi_{ion} \quad \text{... [7.5]}$$

where λ (<1) is called the mixing coefficient. The energy minimum is now calculated at –388 kJ mol^{-1} at R = 74·9 pm (curve *d*, Fig. 7.2).

The theoretically predicted energy of the H_2 molecule is however still higher compared to the experimental value of – 458 kJ mol^{-1}. Various refinements on the wave function ψ yield values of energy further close to the experimental value.

Symmetric and antisymmetric wave functions

As mentioned earlier, two wave functions ψ_I and ψ_{II} may be combined to give the true wave function ψ as $\psi = c_1\psi_I + c_2\psi_{II}$, where c_1 and c_2 are adjustable coefficients.

Substitution of ψ_I and ψ_{II} in equation 7.4 for our supposed H_2 molecule now gives

$$\psi = c_1\psi_A(1)\,\psi_B(2) + c_2\psi_A(2)\,\psi_B(1).$$

In the symmetrical H_2 molecule, the component wave functions ψ_I and ψ_{II} must contribute with equal weight. As the weight is proportional to the square of the coefficient, we have,

$$c_1^2 = c_2^2 \quad \text{or} \quad c_1 = \pm c_2$$

Since the coefficients are relative quantities, we can put c_1 = 1. Hence, c_2 = ±1. So we have two representations for the true wave function ψ (spatial) of the H_2 molecule.

Symmetric: $$\psi_S = \psi_A(1)\psi_B(2) + \psi_A(2)\,\psi_B(1) \quad \text{... [7.6]}$$

Antisymmetric: $$\psi_A = \psi_A(1)\,\psi_B(2) - \psi_A(2)\,\psi_B(1). \quad \text{... [7.7]}$$

The representation ψ_S is called symmetric, since the wave function remains unchanged if the electrons 1 and 2 are interchanged. The representation ψ_A changes sign if the electrons are interchanged and is called antisymmetric (verify). ψ_S leads to a minima in the potential energy diagram (Fig. 7.2, curve *b*). The potential energy diagram for ψ_A is shown in curve *f* which gradually rises as the atoms come closer.

To develop a net attraction between the two H atoms, it is necessary that the electron density between the two nuclei gets increased. This means the two electrons should spend most of the time between the two nuclei, that is, the space region they occupy will be same or nearly the same. The space coordinates of the two electrons will then be very close to one another. Accordingly, we may assume $\psi_A(1) \simeq \psi_A(2)$ and $\psi_B(1) \simeq \psi_B(2)$ under such conditions. When this substitution is made in the combination wave functions (unnormalized, for the sake of simplicity), we get

$$\psi_S = 2\psi_A(1)\,\psi_B(2) \quad \text{and} \quad \psi_A = 0.$$

Hence the antisymmetric combination (ψ_A) cannot represent a bonding electron pair as it effectively reduces the electron population between the nuclei and implies stronger repulsion between them. This explains the nature of curve *f*. Only the symmetric combination ψ_S (Eqn. 7.6) fits a bonding situation.

One may also look at the situation like this: in ψ_S, the wave pattern represented by ψ_I (ψ_A (1) ψ_B (2)) interferes constructively with the wave pattern represented by ψ_{II} ($\psi_A(2)$ $\psi_B(1)$) and there is an increase in the value of ψ in the region between the two nuclei. Since the probability of finding the electrons is proportional to ψ^2, we conclude that the electrons will have a higher probability of being found between the nuclei, binding them together. The interaction is very similar to our classical idea of covalent bonding. Actually ψ_S is often referred to as ψ_{cov} (unnormalized). Soon we shall find how ψ_S leads to the idea of spin pairing.

Spin pairing

The Pauli principle (Chapter 4) requires that the total wave function for the two electrons in the H_2 molecule should be antisymmetric, that is, the wave function should change sign when the labels of the two electrons are interchanged. As we have seen, the orbital part of the wave function should be symmetric to ensure bonding. Hence the spin part of the wave function must be antisymmetric to make the overall wave function (space plus spin) antisymmetric. This implies that the two bonding electrons described by the symmetric valence bond wave function (eqn. 7.6) must have opposite spins. Thus the idea of spin pairing appears necessary for explaining bond formation in VBT. Conversely, when the electrons in the H_2 molecule are described by the antisymmetric orbital wave function (eqn. 7.7), the spin part must be symmetric; the electrons will then have parallel spins. As mentioned earlier, such electrons increase the energy of the molecule owing to repulsion, as shown in curve *f* in Fig. 7.2.

For the two electrons marked 1 and 2 (for arguments sake)

$$\psi_{\text{Total}}(1,2) = \psi_{\text{orbital}}(1,2) \times \psi_{\text{spin}}(1,2) = \{\, \psi_A(1)\, \psi_B(2) + \psi_A(2)\, \psi_B(1)\,\}\, \psi_{\text{spin}}(1,2)$$

When we interchange the labels 1 and 2, we get,

$$\psi_{\text{Total}}(2,1) = \{\, \psi_A(2)\, \psi_B(1) + \psi_A(1)\, \psi_B(2)\}\, \psi_{\text{spin}}(2,1)$$

$$= \{\, \psi_A(1)\, \psi_B(2) + \psi_A(2)\, \psi_B(1)\,\}\, \psi_{\text{spin}}(2,1)$$

Since Pauli principle requires

$\psi_{\text{Total}}(1,2) = -\psi_{\text{Total}}(2,1)$, we conclude, $\psi_{\text{spin}}(1,2) = -\psi_{\text{spin}}(2,1)$

which corresponds to paired electron spins (see box).

Let the spin wave functions for the two possible spins (m_s) of the individual electrons be α and β. For the two electrons in the H_2 molecule, several combinations are possible:

(i) Both electron 1 and electron 2 have same spins, both α or both β; the corresponding wave functions may be written as:

$$\alpha(1)\,\alpha(2) \quad \text{or} \quad \beta(1)\,\beta(2)$$

Both functions are symmetric as they do not change when the electrons 1 and 2 are interchanged.

(ii) Electrons 1 and 2 have different spins, one α and the other β; the corresponding spin wave functions will be

$$\alpha(1)\,\beta(2) \quad \text{or} \quad \beta(1)\,\alpha(2)$$

Since both situations are equally probable, linear combinations of these two functions are used to write the combined spin wave functions (unnormalized):

$$\alpha(1)\,\beta(2) + \beta(1)\,\alpha(2) \quad \text{or} \quad \alpha(1)\,\beta(2) - \beta(1)\,\alpha(2).$$

The first function (left) is symmetric as it does not change when the electrons 1 and 2 are interchanged. The other function (right) is antisymmetric from the same standpoint:

$$\psi_{\text{spin}}(1,2) = \alpha(1)\,\beta(2) - \beta(1)\,\alpha(2)$$

$$\psi_{\text{spin}}(2,1) = \alpha(2)\,\beta(1) - \beta(2)\,\alpha(1) = -\,\psi_{\text{spin}}(1,2).$$

Hence only this antisymmetric spin wavefunction fits the symmetric orbital wave function of the electrons in the bond in H_2 to give an overall antisymmetric wave function as required by the Pauli principle. Clearly, the spins of the two electrons have to be different when they form a bond.

7.2.2 Some Useful Concepts of VBT

(i) *Sigma (σ) and pi (π) bonds*

In the H_2 molecule, the bonding electron pair described by the valence bond wave function has cylindrically symmetrical distribution around the internuclear axis (Fig. 7.3a). The bond in H_2 is called a σ (sigma) bond. As we shall find soon, σ bonds may also be formed by overlap of other orbitals in proper orientation (Fig. 7.3b). A σ bond always possesses an axis of symmetry along the internuclear axis; the overall electron probability density responsible for the bond has cylindrical symmetry around this axis.

We may apply the VB theory to other molecules also. In N_2, for example, we start with two nitrogen atoms, each having the valence electron configuration $2s^2 2p_x^1 2p_y^1 2p_z^1$.

The electrons in *p*-orbitals of one atom may be paired with corresponding orbitals of the other atom to produce three electron pair bonds. But the electron distribution of these three bonds will not be identical. If we imagine two nitrogen atoms approaching one another along the *z*-axis (conventionally chosen), we realize that only the $2p_z$ -orbitals on each will overlap to form a σ-bond. The $2p_y$ orbital of one atom with its axis perpendicular to the internuclear axis, may overlap with $2p_y$ orbital of the other atom sidewise or laterally (Fig. 7.3c). The resulting electron distribution has a nodal plane containing the internuclear axis and is symmetrically distributed on the two sides of this plane. The $2p_x$ orbitals on each N-atom are also suitably oriented to form such a bond by lateral overlap. Such bonds are called π-bonds. A pi-bond possesses a nodal plane containing the internuclear axis. The N_2 molecule has one σ bond and two π-bonds. We shall discuss more on bond types in section 7.3.1.

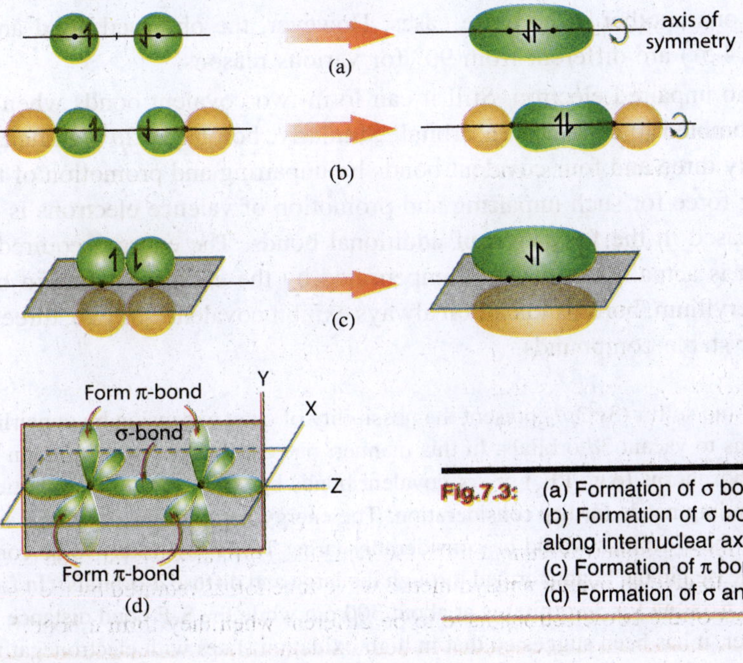

Fig.7.3: (a) Formation of σ bond in H_2;
(b) Formation of σ bond by overlap of two *p*-orbitals along internuclear axis;
(c) Formation of π bond by overlap of two p_y orbitals;
(d) Formation of σ and π bonds in N ≡ N.

The valence bond theory predicts a double bond (one σ and one π) between two oxygen atoms in the O_2 molecule. This description does not agree with the presence of two unpaired electrons in O_2 which is paramagnetic in nature. As we shall find soon, the molecular orbital theory provides a satisfactory explanation for this.

(ii) *Shapes of polyatomic molecules*

The basic idea of orbital overlap and spin pairing of electrons between two atoms leading to a chemical bond may be extended to heteronuclear diatomic molecules and polyatomic molecules as well. The formation of σ and π bonds is visualized in the same manner as with homonuclear diatomics.

A direct consequence of orbital overlap is the directional nature of covalent bonds . We recall that the atomic orbitals are functions of the space coordinates. Only the *s* atomic orbitals are spherically symmetrical while others have their characteristic angular dependence. Overlap by such orbitals would take place most efficiently along the directions in which the electron probability densities are high. Accordingly, the resulting covalent bonds will also link the atoms in a definite geometrical arrangement with respect to one another. The shape of a covalent molecule will then be determined by the spatial distribution of the constituent atoms in space. Thus, the molecule NH_3 is expected to be pyramidal and the molecule H_2O bent, the bonds being at right angles to one another in both the cases. However, the observed bond angles (NHN: 107·3°; HOH: 104·5°) are different from 90° for various reasons.

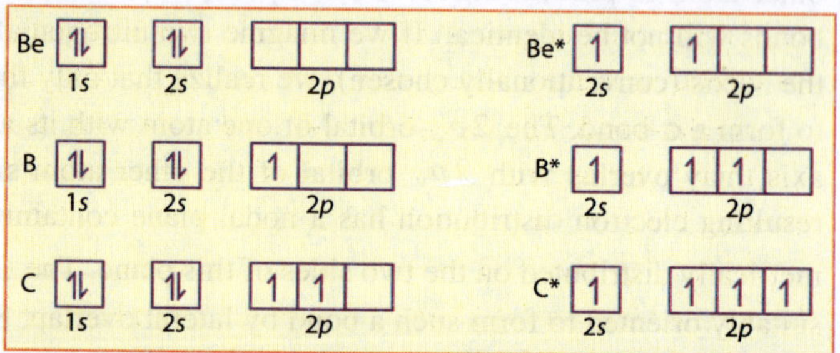

Fig.7.4: Ground state (left) and excited state (right) electron configuration of Be, B and C. Only outermost shells shown on right.

Beryllium ($2s^2$) has no unpaired electron. Still it can form two covalent bonds when one of the 2*s* electrons is promoted to a vacant 2*p* orbital. Similarly, boron ($2s^22p^1$) and carbon ($2s^22p^2$) form respectively three and four covalent bonds by unpairing and promotion of their 2*s* electrons. The driving force for such unpairing and promotion of valence electrons is provided by the energy released in the formation of additional bonds. The energy required for unpairing and promotion is actually more than compensated by the energy set free in bond formation. This is why beryllium, boron and carbon always exhibit covalence of two, three and four respectively in their stable compounds.

Phosphorus ($3s^23p^3$) and sulfur ($3s^23p^4$) present the possibility of octet expansion by unpairing and promotion of electrons to vacant 3*d* orbitals. In this manner, one can explain the formation of five covalent bonds by phosphorus (e.g., PF_5) or six covalent bonds by sulfur (SF_6). Such participation of *d*-orbitals in bonding needs further consideration. The energy required for unpairing and promotion of electrons to the *d*-orbitals would be considerably large. This is not likely to be compensated by the energy released in the bonds formed through the large and diffuse *d*-orbitals. (In free S-atom, the radial density maxima for 3*d* orbital is at about 300 pm while the S-F bond distance in SF_6 is ~156 pm). However, it has been suggested that in high oxidation states with electronegative substituents, the partial charge induced on the central P and S atom can substantially lower the energy of the *d*-orbitals and contract them to smaller compact orbitals suitable for hybridization and bonding. Though theoretically still an open controversy, this interpretation works well to explain the observed geometry of the molecules and the nonexistence of compounds like PH_5 or SH_6 (low electronegativity of hydrogen and small bond energy). Large size of chlorine aı d lower S—Cl bond energy similarly excludes SCl_6. Nitrogen, with no low energy vacant orbital, has little possibility to expand its octet and NCl_5 is not formed. In the molecular orbital theory, formation of molecules like PCl_5 and SF_6 is explained through three-centre m.o.s without the necessity of *d*-orbital participation (see later).

(iii) *Overlap and bond strength*

The strength of a covalent bond is approximately proportional to the extent of overlap between the combining orbitals. Greater overlap thus makes a bond stronger. So far we have encountered two main types of overlap—(i) direct overlap giving rise to σ-bonds and (ii) lateral overlap giving rise to π-bonds. Obviously, the lateral overlap in a π-bond cannot be as effective as the overlap in a σ–bond. Hence, a σ-bond is always stronger than a π-bond.

Again, a σ-bond may arise from various possible combinations among *s*-, *p*- and *d*-orbitals (or their hybrids, as discussed shortly). So far we have encountered three main types of overlap that lead to a σ-bond: (i) *s*—*s* overlap, as in H_2; (ii) *s*—*p* overlap, as in H_2O or NH_3. (iii) *p*—*p* overlap, as in N_2. These σ-bonds will not be of equal strength. The extent of overlap, and hence the bond strength, in each case will depend upon the distribution of charge cloud of the participating orbitals. Two *s*-orbitals cannot overlap very strongly with their spherical charge clouds. On the other hand, the charge clouds of the *p*-orbitals are concentrated along either of the *x*, *y* or *z*-axis. Their projected lobes can overlap more effectively with other *s* or *p*-orbitals. Consequently, an *s*—*p* bond is stronger than an *s*—*s* bond; the *p*—*p* bonds are still stronger. Relative bond strengths of the *s*—*s*, *s*—*p* and *p*—*p* bonds are

s—*s*	*s*—*p*	*p*—*p*
1	1.73	3.

The extent of overlap of two atomic orbitals ψ_A and ψ_B on two different atoms A and B is expressed in terms of the *overlap integral S*:

$$S = \int \psi_A \psi_B \, d\tau$$

The dependence of bond strength on the extent of overlap is revealed in the decreasing stability of the hydrides down any group in the periodic table. As the principal quantum number of the valence shell electrons increases, the orbitals (*p* or *sp* hybrid) become larger in size; consequently, their overlap with the small 1*s* orbital of hydrogen becomes gradually poorer. Thus stability decreases along either of the following series: H_2O, H_2S, H_2Se, H_2Te; HF, HCl, HBr, HI; NH_3, PH_3, AsH_3, SbH_3, BiH_3.

(iv) *Hybridization*

We know that the four C—H bonds in methane are all alike and they are arranged symmetrically around the central carbon atom directed along the four corners of a tetrahedron. This necessitates a concept of mixing the *s*-orbital of carbon with its three *p*-orbitals before overlap. The resulting equivalent orbitals, each having one-fourth *s*-character and three-fourth *p*-character, may now undergo overlap with four hydrogen 1*s* atomic orbitals to form four equivalent C—H bonds. We may reasonably extend this concept to interpret the equivalence of the two bonds in BeF_2 or the three bonds in BCl_3. This procedure of prior mixing or hybridization of the orbitals has been given rigorous mathematical formulation.

Hybridization is a concept of mixing different atomic orbitals of comparable energy resulting in an equal number of orbitals with mixed character. The resulting hybrid orbitals undergo better overlap and form stronger bonds than the pure orbitals in conformity with the most stable geometry for a molecule. The hybrid orbitals formed by a given set of "*pure*" orbitals are mostly equivalent but they may not be so; (see later). The formation of four equivalent C—H bonds by carbon in forming methane may then be conceived of in terms of the following successive steps:

At first, a 2*s* electron of the carbon atom gets unpaired and promoted to a 2*p* orbital. The 2*s* and the three 2*p* orbitals are now hybridized to give four equivalent orbitals, each possessing one part *s* character to three parts *p* character in their wave function, directed to the corners

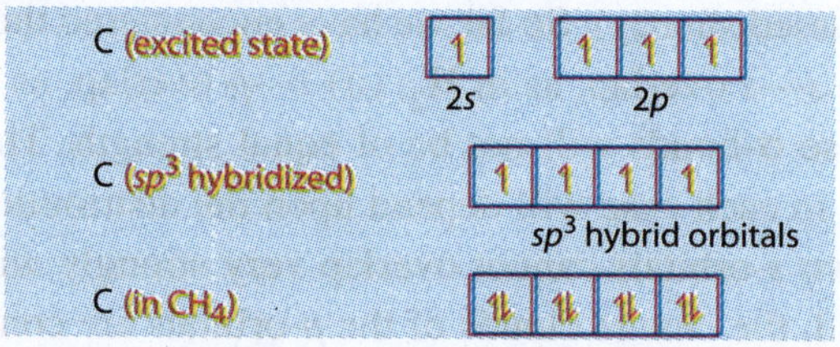

of a regular tetrahedron. This process also requires energy. These sp^3 hybrid orbitals now form four equivalent C—H bonds distributed tetrahedrally around the carbon atom. The energy released by the formation of bonds through the hybrid orbitals more than compensates the energy required for promotion and hybridization.

But one must be cautious to avoid misconceptions like "*methane is tetrahedral because the carbon atom is sp^3 hybridized*". Methane is tetrahedral simply because the energy of the molecule is lowest, and hence the molecule is most stable in that configuration. This is principally because of increased bond energies and decreased repulsion energies which may be explained (according to the Valence Bond Theory) by introducing the concept of hybridization.

The wave functions of the hybrid orbitals are obtained by suitable combination of the contributing wave functions that corresponds to the most favourable energy of formation. Qualitatively this can be explained by stating that the hybrid orbitals permit a wider distribution of the electronic charge cloud and more effective overlap. Overlap integrals calculated for C—C and C—H bonds as a function of the percent *s*-character of the carbon orbitals show maximum values at hybridizations intermediate between pure *s* and pure *p* orbitals. As one moves from pure *p* orbital to *sp* hybrid orbitals (the common range found in molecules), the overlap increases with *s*-character. Since bond strength is proportional to overlap integral, we expect that bond strength should increase as the *s*-character of a hybrid orbital increases. This gain in bond energy works behind the use of hybrid orbitals.

We know that compared to a *p*-orbital, an *s* orbital is lower in energy and has a higher electronegativity. Therefore we expect the effective electronegativity of an *s—p* hybrid orbital to increase as the percent *s*-character increases. In NH_3, if we assume that the *s*-character is distributed equally among the lone pair and σ-bonding orbitals, the lone pair of electrons would occupy an sp^3 hybrid orbital. Similar logic would suggest an sp^2 hybrid orbital for the lone pair electrons of the nitrogen in pyridine, C_5H_5N: and an *sp* hybrid orbital for the lone pairs in dinitrogen, :N ≡ N: . The relation of electronegativity with *s* content of a hybrid now suggests that ammonia should be a much stronger base than pyridine, which, in turn, would be a much stronger base than molecular dinitrogen.

The angles in the molecules CH_4, NH_3 and H_2O are close to tetrahedral angles but they fall sharply to nearly 90° in the hydrides of heavier elements in all groups. We discuss the situation with reference to a particular pair, say H_2O and H_2S. If the O—H bonds in water involved pure *p*-orbitals, the HOH angle would have been 90°; the oxygen would now contain one lone pair of electrons in the 2*s* orbital and one lone pair in one 2*p* orbital. This is precisely equivalent to place the two lone pairs in two *sp* hybrid orbitals directed at right angles to the HOH plane. The strong repulsion between the lone pairs and the bond pairs may be somewhat relaxed if the bond angles are widened. In that case the bonding orbitals will have to acquire some *s*-character and lose some *p*-character. Correspondingly, there will be some decrease (from the 50% in *sp* hybrids) in the *s*-character of the lone pair orbitals and gain in *p*-character. This means that we are virtually promoting a fraction of an electron from the 2*s* to the 2*p* orbital. Thus energy will be required to bring the oxygen atom from its ground state to the hypothetical state of hybridization. This energy would be compensated by (i) reduced inter-electron repulsion and (ii) increase in bond energy due to an increase in *s*-character of the bonding orbitals. Such balance of hybridization energy, electron repulsion energy and bond energy determines the ultimate geometry of the molecule. Now we come to H_2S. Calculation shows that the energy required to raise the sulfur atom to the tetrahedrally hybridized state is still appreciable but the inter-electron repulsion has decreased considerably (large size). So the system involves bonding through *p*-rich orbitals having low hybridization energy. The same argument applies to the hydrides of P, As, Sb, etc. However, the suppression of some bond angles to slightly below 90° cannot be explained in this manner.

The observed continuous variation of bond angle from 180° for *sp* hybrid orbitals (50% *s*) through 120° for sp^2 (33.3% *s*) to 109.5° for sp^3 (25% *s*) suggests that bond angle should decrease gradually with decrease in *s*-character and increase in *p*-character. A simple relationship connecting the bond angle θ in equivalent *s–p* hybrid orbitals and their composition is

$$\cos\theta = \frac{S}{S-1} = \frac{P-1}{P}$$

where *S* or *P* respectively represent the fractional *s*- or *p*-character in the bond. This formula is well supported by some common bond angles:

$$CH_4 : \theta = 109.5°; \quad \cos\theta = -0.333 \Rightarrow S = 0.25, \quad P = 0.75.$$

$$H_2O : \theta = 104.5°; \quad \cos\theta = -0.250 \Rightarrow S = 0.20, \quad P = 0.80.$$

It is now possible to calculate the *s*-characters of the lone pairs in H_2O and the angle between their axes:

The *total s*- and *p*-character of the four hybrid orbitals of oxygen must be in the ratio 1:3; the total of *s*-characters (in fraction) over the four hybrids must also equal 1. The two bonding orbitals use 2 × 0.20 or 0.40 fraction of *s*-character. So each lone pair orbital has ½(1 – 0.40) or 0.30 fraction of *s*-character. The angle between the axes of these orbitals (say α) will be given by

$$\cos\alpha = \frac{0.30}{0.30-1.00} = -0.43 \quad \text{or,} \quad \alpha = 115.5°.$$

Calculation of *s*- or *p*-character in hybrids involving *d*-orbitals is complicated.

The common hybridization schemes and the resulting spatial distribution of the hybrid orbitals are shown in Table 7.1.

As per convention, the *z*-axis is taken as the internuclear axis or vertical axis (also see Table 7.2 for alternative hybridization schemes).

Table 7.1
Common hybridization schemes

Orbitals involved	Type	Resulting Geometry
s, p_z	sp	linear
s, p_x, p_y	sp^2	trigonal planar (*xy* plane)
s, p_x, p_y, p_z	sp^3	tetrahedral (Td)
$(n-1)\,d_{x^2-y^2}, ns, np_x, np_y$	dsp^2 (also sp^2d)	square planar (*xy* plane)
$ns, np_x, np_y, np_z, nd_{z^2}$	sp^3d	trigonal bipyramid (TBP)
$ns, np_x, np_y, np_z, nd_{x^2-y^2}$	sp^3d (also dsp^3)	square pyramid
$ns, np_x, np_y, np_z, nd_{x^2-y^2}, nd_{z^2}$	sp^3d^2 (also d^2sp^3)	octahedral (Oh)
$d_{x^2-y^2}, d_{xy}, d_{z^2}, s, p_x, p_y, p_z$	d^3sp^3	pentagonal bipyramid

Schematic representation of the overlap of atomic orbitals for some hybridization schemes involving *s* and *p* orbitals and their corresponding wave functions are given below.

sp hybridization

One *s* and one *p* orbital combine to give two *sp* hybrid orbitals. Their formation is illustrated in Fig. 7.5 using the boundary lines of the constituent orbitals.

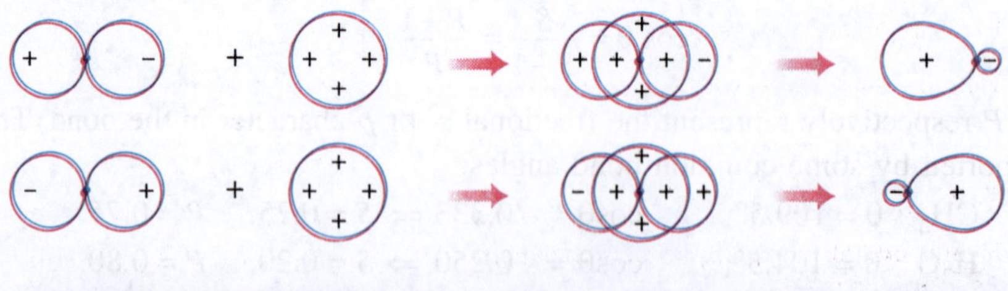

Fig.7.5: Combination of *s* and *p* atomic orbital forming two *sp* hybrid orbitals. The wave functions are $\psi_{sp}(1) = (1/\sqrt{2})\,(\psi_{2s} + \psi_{2p})$; $\psi_{sp}(2) = (1/\sqrt{2})\,(\psi_{2s} - \psi_{2p})$

As usual, the '+' and '–' signs represent the sign of the wave function. We find that the lobes of the *p*-orbital may combine with the *s*-orbital in two ways. In each case, portions with same sign of the two wave-functions "*expand*", implying greater probability of finding the electron over that place. Conversely, portions with opposite sign of the two wave functions get largely reduced. The two *sp* hybrid orbitals are distributed symmetrically around the nucleus. The enlarged lobes in both protrude further along the axis than either of the "*pure*" *s* or *p* orbitals; as such, these hybrid orbitals form stronger bonds. Beryllium chloride, $BeCl_2$; mercuric Chloride, $HgCl_2$ and acetylene (ethyne), C_2H_2 provide examples of *sp* hybridization.

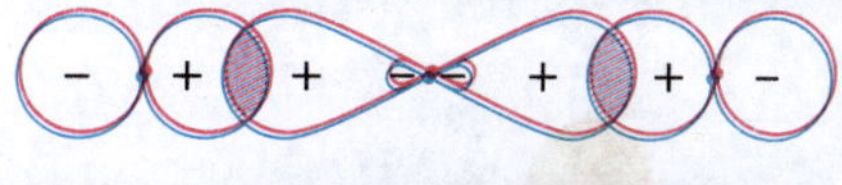

Fig.7.6: Overlap of two *sp* hybrid orbitals of beryllium with *p* orbitals of chlorine

$BeCl_2$ is polymeric in the solid state. But individual molecules may be present in the vapour state. In such molecules, two *sp* hybrid orbitals of beryllium overlap with the *p*-orbitals of two chlorine atoms (Fig. 7.6).

sp hybridization is also helpful in explaining the observed dipole moment and bond strength of the hydrogen halides. In HCl, for example, one may expect the 1*s* atomic orbital of hydrogen to overlap with a pure 3*p* orbital (say, the $3p_z$, *z*-axis being taken as the molecular axis) of the chlorine atom. But there is reason to believe that the chlorine atom uses *sp* hybrid orbitals, formed by its 3*s* and $3p_z$ atomic orbitals. In that case, the hydrogen 1*s* orbital overlaps with one *sp* hybrid orbital, forming a stronger bond. At the same time, the other *sp* hybrid orbital contains the lone pair originally assigned to the pure 3*s* orbital of chlorine. The lone pair is thus further removed from the hydrogen atom, causing greater charge separation in the HCl molecule. This gives rise to a larger dipole moment.

sp² hybridization

One *s* and two *p* orbitals combine to give three equivalent sp^2 hybrid orbitals. These are coplanar and directed at 120° angle to one another. Their formation is illustrated in Fig. 7.7 from the 2*s*, $2p_x$ and $2p_y$ orbitals (taken arbitrarily).

Fig.7.7: Formation of sp^2 hybrid orbitals.

(i) Hybrid orbital formed by the combination $2s + 2p_x$. The $2p_y$ orbital does not contribute as its two lobes make "equal" but "opposite" overlap with the *s* orbital (shown in dotted line).

$$\psi_{sp^2}(1) = \frac{1}{\sqrt{3}}\psi_{2s} + \sqrt{\frac{2}{3}}\psi_{2p_x}$$

(ii) Hybrid orbital formed by the combination $2s - 2p_x + 2p_y$

$$\psi_{sp^2}(2) = \frac{1}{\sqrt{3}}\psi_{2s} - \frac{1}{\sqrt{6}}\psi_{2p_x} + \frac{1}{\sqrt{2}}\psi_{2p_y}$$

(iii) Hybrid orbital formed by the combination $2s - 2p_x - 2p_y$

$$\psi_{sp^2}(3) = \frac{1}{\sqrt{3}}\psi_{2s} - \frac{1}{\sqrt{6}}\psi_{2p_x} - \frac{1}{\sqrt{2}}\psi_{2p_y}$$

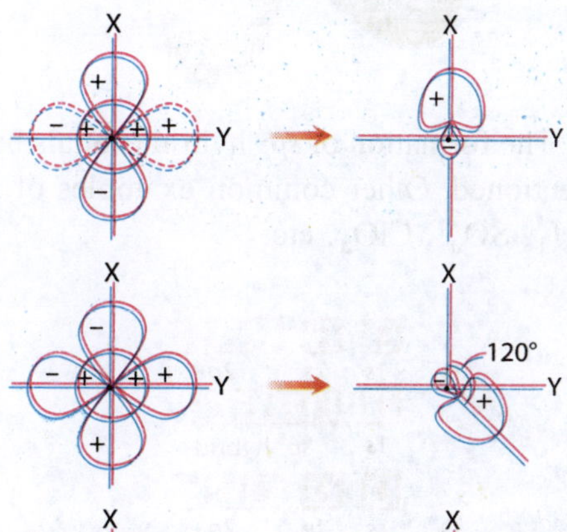

Boron trifluoride, BF_3, contains three coplanar B—F bonds formed by the overlap of three sp^2 hybrid orbitals of boron with the $2p$ orbitals from three fluorine atoms (Fig. 7.8 and Fig. 7.9). Other boron trihalides also involve sp^2 hybridization.

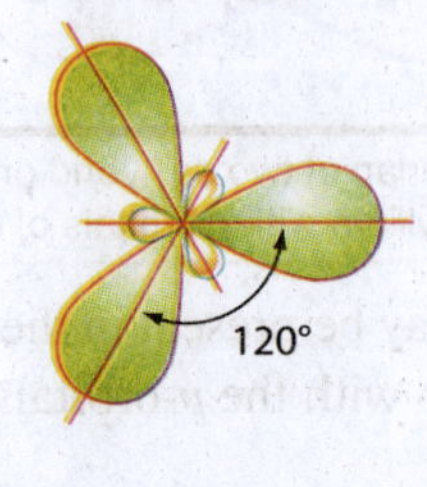

Fig.7.8: Three sp^2 hybrid orbitals of boron

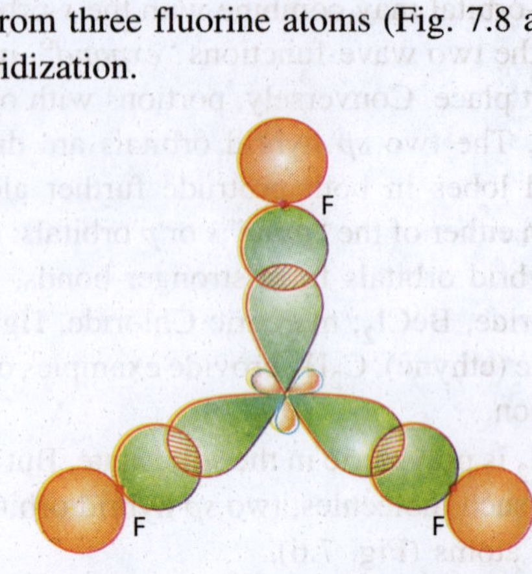

Fig.7.9: Overlap of the sp^2 hybrid orbitals of boron with p-orbitals of fluorine atom

sp^3 hybridization

Combination of one s and three p orbitals gives rise to four equivalent hybrid sp^3 orbitals. These are directed toward the corners of a regular tetrahedron (Fig. 7.10). The angles between the orbitals are all equal to 109.28′.

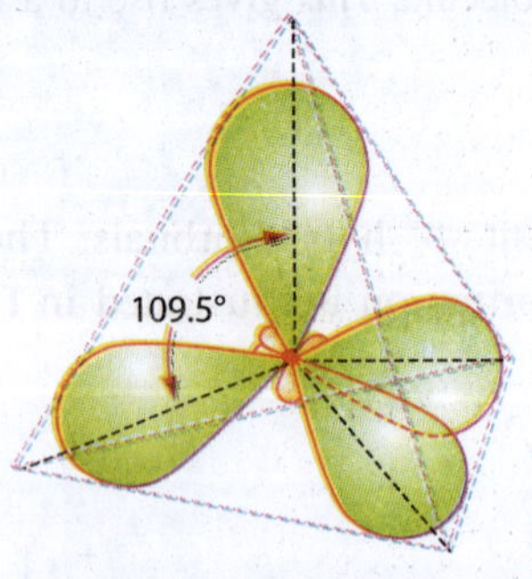

Fig.7.10: Distribution of four sp^3 hybrid orbitals in space.

$$\psi_{sp^3}(1) = \tfrac{1}{2}\left(\psi_{2s} + \psi_{2p_x} + \psi_{2p_y} + \psi_{2p_z}\right)$$
$$\psi_{sp^3}(2) = \tfrac{1}{2}\left(\psi_{2s} + \psi_{2p_x} - \psi_{2p_y} - \psi_{2p_z}\right)$$
$$\psi_{sp^3}(3) = \tfrac{1}{2}\left(\psi_{2s} - \psi_{2p_x} + \psi_{2p_y} - \psi_{2p_z}\right)$$
$$\psi_{sp^3}(4) = \tfrac{1}{2}\left(\psi_{2s} - \psi_{2p_x} - \psi_{2p_y} + \psi_{2p_z}\right)$$

The formation of sp^3 hybrid orbitals by a carbon atom in forming methane has already been mentioned. Other common examples of sp^3 hybridization are NH_3, H_2O, SiH_4, $SnCl_4$, BF_4^-, NH_4^+, SO_4^{2-}, ClO_4^-, etc.

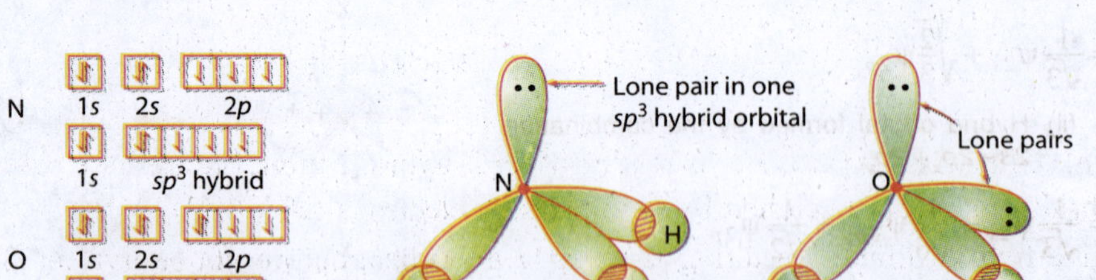

Fig.7.11: (a) Hybrid orbitals of nitrogen and oxygen. (b) Geometry of ammonia and water molecule.

In NH_3 and H_2O, the central nitrogen or oxygen atom involves all the 2*s* and 2*p* atomic orbitals in bonding rather than the singly occupied 2*p* orbitals only. The non-bonding electron pairs now occupy sp^3 hybrid orbitals—one in NH_3 and two in H_2O. The remaining hybrid orbitals form bonds with hydrogen. The observed bond angles and the polarity of the molecules is in good agreement with this scheme. However, the bond angles deviate somewhat from that observed in CH_4—the ideal tetrahedral bond angle of 109·28′. The values are ∠HNH:107°45′; ∠HOH: 104°27′. The deviations result from the repulsion of the non-bonding pair(s) with the bond pairs of electrons.

The energies of the hybrid orbitals are given by the weighted average of the energies of the combining orbitals. For carbon, energies (*E*) of the 2*s* and 2*p* orbitals are $E_{2s} = -1878$ kJ mol^{-1} and $E_{2p} = -1028$ kJ mol^{-1}. The energy of an sp^3 hybrid orbital is then calculated as

$$E_{sp^3} = (E_{2s} + 3E_{2p}) \div 4 = -1240 \cdot 5 \text{ kJ mol}^{-1}$$

The *hybridization energy* is the energy difference of the valence electrons between the hybridized state and the unhybridized state. In this case

$$E_{hy}(sp^3) = 4E_{sp^3} - (2E_{2s} + 2E_{2p}) = \left[4 \times (-1240 \cdot 5) - (2 \times -1878 + 2 \times -1028) \text{ kJ mol}^{-1}\right]$$

$$= [-4962 - (-5812)] \text{ kJ mol}^{-1} = 850 \text{ kJ mol}^{-1}.$$

sp^3d hybridization

One *s*, three *p* and one *d* orbital $\left(d_{z^2}\right)$ may be hybridized to give five hybrid orbitals directed towards the corners of a trigonal bipyramid (Fig. 7.12).

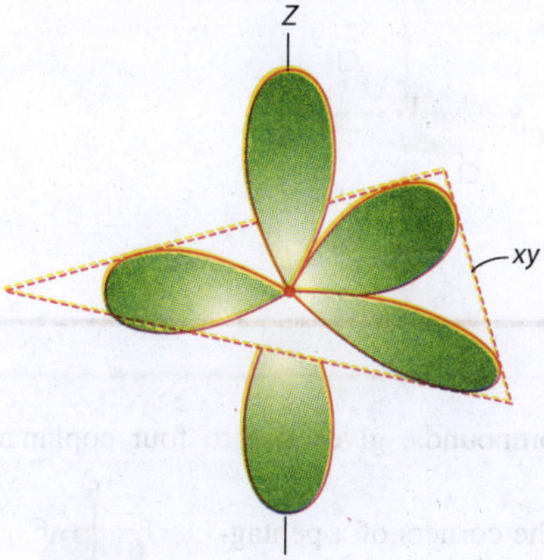

Fig. 7.12: Spatial distribution of five sp^3d hybrid orbitals in trigonal bipyramid arrangement (*s*, p_x, p_y, p_z, and d_{z^2})

Equatorial : $\psi_1 = \frac{1}{\sqrt{3}}\left(\psi_s + \sqrt{2}\psi_{p_x}\right)$

$$\psi_2 = \frac{1}{\sqrt{3}}\left(\psi_s - \frac{1}{\sqrt{2}}\psi_{p_x} + \sqrt{\frac{3}{2}}\,\psi_{p_y}\right)$$

$$\psi_3 = \frac{1}{\sqrt{3}}\left(\psi_s - \frac{1}{\sqrt{2}}\psi_{p_x} - \sqrt{\frac{3}{2}}\,\psi_{p_y}\right)$$

Axial : $\psi_4, \psi_5 = \frac{1}{\sqrt{2}}\left(\psi_{p_z} \pm \psi_{d_{z^2}}\right)$

$PCl_5(g)$, ClF_3, SF_4, I_3^- provide examples of sp^3d hybridization; but in these cases, one or more hybrid orbitals are occupied by lone pairs of electrons (Fig. 7.13).

The five hybrid orbitals resulting from sp^3d hybridization are not equivalent. The three orbitals directed trigonally, known as the *equatorial* orbitals are slightly different from the two orbitals directed linearly perpendicular to the plane of these three equatorial orbitals; these two orbitals are known as *axial* orbitals. The equatorial orbitals may be looked upon as sp^2 hybrids $(sp_x p_y)$ with $33\frac{1}{3}\%$ *s* character; the axial ones are *dp* hybrids $\left(p_z d_{z^2}\right)$ with 50% *p*-character.

The bonds formed by one set of orbitals are different from the bonds formed by the other set, even when the bonded atoms are all identical. Thus in PCl_5, the equatorial P—Cl bonds are stronger than the axial P—Cl bonds. The axial bonds are also longer by about 15 pm.

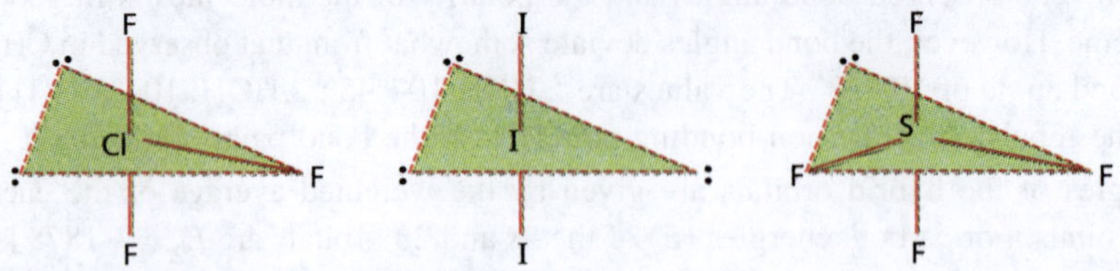

Fig.7.13: Distribution of bonds in ClF_3, I_3^- and SF_4. The position of the lone pairs may be explained on the basis of minimum inter-electron repulsion (VSEPR theory).

In mixed chlorofluorides, e.g., PCl_3F_2, the fluorine atoms occupy the axial positions. This is in accordance with the empirical **Bent's rule** (1960): more electronegative substituents "*prefer*" hybrid orbitals having less *s* character (see later).

sp^3d^2 (or d^2sp^3) hybridization

One *s*, three *p* and two *d* orbitals may hybridize (Table 7.1) to give six hybrid orbitals directed at the corners of a regular octahedron. The hybrid orbitals are all equivalent with $16\frac{2}{3}\%$ *s*, 50% *p* and $33\frac{1}{3}\%$ *d* character. SF_6, IF_5 and ICl_4^- involve this type of hybrid orbitals. In the last two one or two positions respectively are occupied by lone pairs.

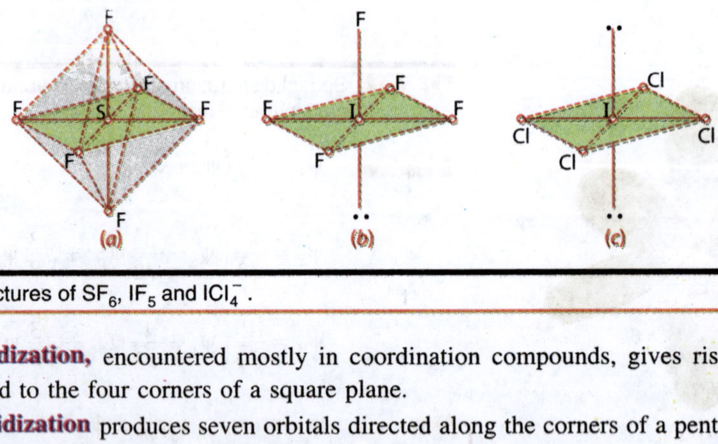

Fig.7.14: Structures of SF_6, IF_5 and ICl_4^-.

***dsp*² hybridization,** encountered mostly in coordination compounds, gives rise to four coplanar orbitals directed to the four corners of a square plane.

***sp*³*d*³ hybridization** produces seven orbitals directed along the corners of a pentagonal bipyramid, IF_7 providing an example.

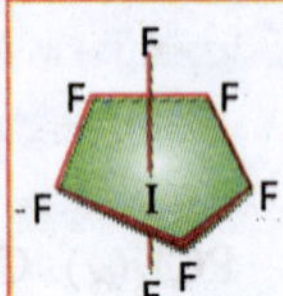

7-I

The sets of hybrid orbitals discussed above are commonly used for qualitative sigma-bonding descriptions. Orbitals of the central atom remaining outside hybridization may form pi bonds with orbitals of matching symmetry on the atoms joined by sigma bonds. We have used such descriptions for the bonding in species like C_2H_4, C_2H_2, N_2 etc. In these species, all *p* orbitals outside hybridization get engaged in pi bonding. In some cases, the central atom using hybrid orbitals of *s* and *p* orbitals may enter π-bonding through their vacant *d*-orbitals with filled donor orbitals on the sigma bonded atoms. The Si—X bonds in SiX_4 are not sigma only in this sense (Chapter 21, Group 14).

The hybridization schemes illustrated above are not unique for the associated structures. There may be other less common hybridization schemes resulting in the same spatial distribution of the hybrid orbitals. Some of these are shown below:

Table 7.2

Some alternative and less common hybridization schemes

Coordination number	Arrangement	Scheme
2	linear	*pd, sd*
3	trigonal planar	p^2d
	unsymmetrical	*spd*
	trigonal pyramidal	pd^2
4	tetrahedral	sd^3
	irregular tetrahedral	spd^2, p^3d, pd^3
	square planar	p^2d^2
5	trigonal bipyramidal	spd^3
	tetragonal pyramidal	sp^2d^2, sd^4, pd^4, p^3d^2
	pentagonal planar	p^2d^3
6	trigonal prism	spd^4, pd^5
	trigonal antiprism	p^3d^3
7	capped trigonal prism	d^4sp^2
8	dodecahedral	d^4sp^3
	square antiprismatic	d^5p^3

Equivalent and nonequivalent hybrid orbitals

In the beginning, hybridization was introduced as a hypothetical concept evoked to explain the equivalence of the four bonds in CH_4. Each of the four sp^3 hybrid orbitals of carbon contain 25 percent *s* and 75 percent *p* character and are distributed along the four corners of a regular tetrahedron. This is an idealized situation where all bonds hold identical atoms and there are no lone pairs of electrons. We have already seen that the $s : p$ ratio of an sp^3 hybrid orbital changes from the ideal integral ratio of 1:3 when lone pairs of electrons are present—one in NH_3 and two in H_2O. These hybrid orbitals become non-equivalent from the hybrid orbitals containing bond pairs of electrons, though all of them are still labelled sp^3. Among the bonding orbitals also, non-equivalence may arise when the atoms or groups bonded vary widely in electronegativity. This is observed in the variations in bond angles and bond lengths on substitution by electronegative atoms. Thus in CH_2F_2, the $\widehat{HCH}$ angle is 111·9° while the $\widehat{FCF}$ angle is 108·3°, suggesting less than 25% *s*-character in the C–F bonds. The variation in bond lengths in the following examples is noteworthy:

Bond		***Molecule* (Bond length in pm)**
C—F	:	CH_3F (139.1) CF_4 (132.3)
C—Cl	:	CH_3Cl (178.3) CH_2Cl_2 (177.2)
		CF_3Cl (175.1) $CHCl_3$ (176.4) CCl_4 (176.6)

C—C : C_2H_6 (153.6) C_2F_6 (151)
C = O : H_2CO (122.5) HFCO (119) F_2CO (117)

These data indicate that replacement of H by F or Cl results in a decrease in adjacent bond lengths. We recall that bond length decreases and bond strength increases with increasing participation of *s* orbitals in a bond. In other words, *s*-character increases in the orbitals forming bonds to less electronegative substituents and *p* character concentrates in the orbitals directed toward the electronegative substituents. This is because an electron in a *p* orbital is on the average farther from the nucleus than an electron in an *s* orbital and is more readily shared by an orbital centered on the more electronegative atom. Hence replacement of a substituent by a more electronegative one causes a fine change in the hybridization—*p* character shifts to the bonds connecting the electronegative atoms and correspondingly there is a shift in *s* character in the adjacent bonds, which become stronger and shorter. This generalization forms the basis of the ad hoc ***Bent's rules***, to be discussed shortly.

Certain hybridization schemes like sp^3d give rise to inherently nonequivalent hybrid orbitals. The sp^3d hybrid orbitals may be looked upon as a combination of two sets: (i) a set of three sp_xp_y hybrid orbitals distributed in the equatorial *xy* plane of a trigonal bipyramid (*z*-axis taken as the 3-fold rotation axis) and (ii) a set of two $p_z - d_{z^2}$ hybrid orbitals directed axially. Substituents at the equatorial plane are bonded through the *s*-rich (sp^2) hybrid orbitals; the bond lengths in this plane are thus expected to be shorter. If there are any lone pair(s) of electrons, they will also preferentially occupy the axial *s*-rich orbitals. (We arrived at similar conclusions in the VSEPR theory from the principle of minimization of repulsion.) The following structures (Fig. 7.15) support this conclusion:

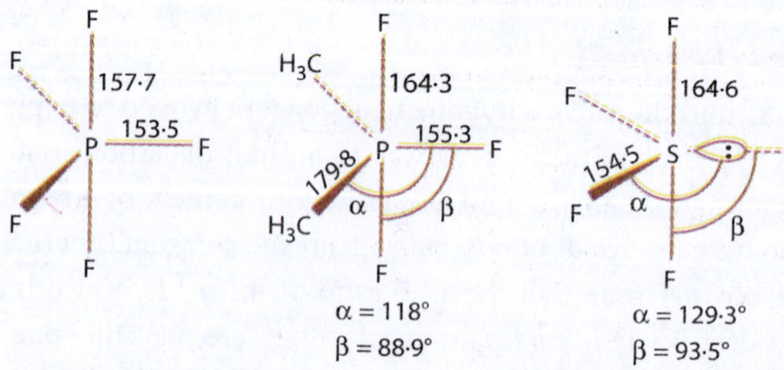

Fig.7.15: Skeleton structures of PF_5, $PF_3(CH_3)_2$ and SF_4. Distances in pm. The distortion from trigonal bipyramid structure is seen in the values of α and β.

However, the difference in bond length and environment of the axial and equatorial substituents can be understood distinctly when very quick observation techniques are used, for example, electron diffraction studies in which the time span is of the order of 10^{-20}s. Observed under a longer time span, for example using nmr spectroscopic studies (time scale 10^{-1}–10^{-9} s), the axial and equatorial groups may appear equivalent in some cases. For example, ^{19}F nmr spectrum of PF_5 and of $PF_4(CH_3)$ shows signal of only one kind of fluorine atom. This is because the axial and apical fluorine atoms may undergo rapid interchange or scrambling (more than 10,000 times per second) via a square pyramid intermediate (Berry pseudorotation; see box). The rate of scrambling is greater than the difference in the nmr frequencies of the two types of fluorine atoms; a weighted average of these two types of atoms is actually observed.

Berry Pseudorotation

The rapid interchange of axial and equatorial groups in a trigonal bipyramid (TBP) structure may be explained by a mechanism suggested by R.S. Berry (1960). It has been proposed that molecules of the type AX_5 with no lone pairs of electrons on the central atom(A) usually adopt the TBP structure corresponding to sp^3d hybridization of the central atom. However, such structures are not stereochemically very stable with respect to other configurations of comparable energy. Accordingly, a trigonal bipyramid structure may readily convert to a square pyramid transition state and then back to a new TBP structure. During this process, one equatorial group (marked with *) acts as a *pivot* and remains unchanged while the other two equatorial groups interchange their positions with the axial groups. The whole molecule may be said to have undergone a rotation about an axis connecting the pivot group and the central atom (Fig. 7.16).

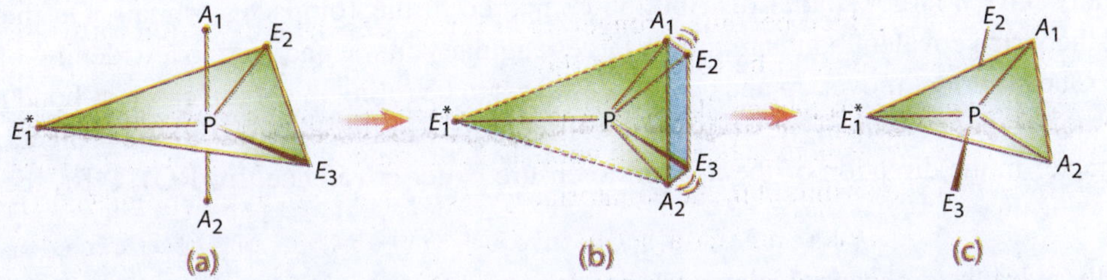

Fig.7.16: Berry pseudorotation in a pentacovalent phosphorus compound. (a) E_1, E_2, E_3 form the equatorial plane in the initial TBP structure. A_1, A_2 are axial groups. (b) The square pyramid intermediate. E_2, E_3 and A_1, A_2 form the square base. E_1^* acts as the pivot. (c) The new TBP structure after interchange of axial groups with two equatorial groups. E_2 and E_3 now occupy axial positions; E_1^*, A_1 and A_2 form the equatorial plane which is now on the plane of the paper. Note that the initial equatorial plane in (a), perpendicular to the plane of the paper, has undergone a rotation (90°) about the E_1^*—P axis.

CH_3PF_4 can undergo scrambling or interchange of axial and equatorial fluorine atoms by rotation about the CH_3—P axis. The CH_3 group acts as the pivot group (position E_1^* in Fig. 7.16(a)).

Pseudorotation of $(CH_3)_2PF_3$ would force one electropositive methyl group (initially in an equatorial position, E_2 or E_3) into an apical position in the new TBP structure (Fig. 7.16(c)). Presumably this involves a high energy configuration; the molecule *does not undergo pseudorotation* and retains both methyl groups in equatorial position.

Bent's Rules

We have seen how bonding of an electronegative atom or group is favoured through hybrid orbitals rich in *p*-character (or having less *s*-character). *s*-orbitals, with greater penetration into inner electron core and influenced by a higher effective nuclear charge of the central atom, are less likely to be distorted or drawn closer to the electronegative atoms forming bonds. Since an electronegative atom or group introduces polarity in bonds, we may also associate higher ionic character or poor covalence of a bond with greater *p*-character in the hybrid orbital of the central atom. At the same time poor covalence may also result from poor overlap consequent to mismatch in energy and many other factors. These observations are summarily expressed in Bent's rules:

(i) more electronegative substituents prefer hybrid orbitals of the central atom with greater *p*-character and less *s*-character.

(ii) the central atom involves hybrid orbitals with higher *s*-character to develop higher covalence and less *s*-character in bonds with greater ionic character.

Examples for applications of Bent's rules have already been given. We have just seen that in compounds like PCl_3F_2, $PF_4(CH_3)$ or $PF_3(CH_3)_2$, the fluorine atoms occur at axial positions. The aptitude of *s*-rich orbitals for greater covalence thus gives rise to a structural motivation in a molecule with highly electronegative atoms.

Though proposed as an empirical rule, Bent's rule has been substantiated by molecular orbital calculations. The rule also agrees with the conclusions drawn from the VSEPR theory (see for example, the bond angles of CH_2F_2). An electronegative substituent will displace the bond pairs of electrons away from the central atom to itself, thereby reducing bond pair-bond pair repulsion. This is most favoured by orbitals with more *p*-character, since the *p*-orbitals generally have a larger spatial distribution compared to the compact *s*-orbitals. On the other hand, the *s*-rich covalent bonds require a larger angular volume and lead to a widening of bond angle. The criterion of overlap and covalence seems to be ruling in certain cases. π-bond results in a shorter bond with higher covalence. In such cases, the central atom projects more *s*-character in the direction of the π-bond to ensure better covalence. In $(RO)_2\overset{\displaystyle O}{\overset{\|}{P}}Br$, the tetrahedrally hybridized phosphorus atom does not extend higher *s*-character in its bond with bromine, the least electronegative element in the molecule. Rather it exerts more *s*-character in the P=O link for the sake of better covalence and efficient overlap with the shortened double bond.

The isolobal relationship

The concept of hybridization has been successfully applied to recognize structurally analogous fragments of molecules which have been termed *isolobal* (from the lobe-like shape of a hybrid orbital in the molecular fragment). Elaborated by R. Hoffmann in his Nobel lecture in 1982, two molecular fragments are said to be isolobal "if the number, symmetry properties, approximate energy and shape of the frontier orbitals and the number of electrons in them are similar—not identical, but similar". Isolobality is expressed by the symbol $\leftarrow\!\!\circ\!\!\rightarrow$. As we shall soon find, isolobal fragments may originate from apparently unrelated molecules but they suggest the same pattern of bonding.

Imagine we simply remove one H atom from the molecule CH_4 *retaining* the hybridization and spatial distribution of the sp^3 orbitals of carbon. In the resulting 7-electron fragment CH_3, three of the sp^3 orbitals of carbon are involved in σ bonding with the hydrogens; the fourth hybrid, singly occupied, is at a higher energy than the σ-bonding orbitals. This fourth hybrid orbital has the same σ symmetry as a hydrogen 1*s* orbital and both have the same electron occupancy, one in each. Hence H is considered to be isolobal with CH_3.

Let us next consider the species $Mn(CO)_5$, a species apparently different from H or CH_3. It is a 17 electron species. The σ-interactions between the ligands and Mn may be considered to involve five of the d^2sp^3 hybrid orbitals centered on Mn; the singly occupied sixth hybrid orbital is at higher energy than the five σ-bonding orbitals. This sixth hybrid orbital thus resembles the singly occupied fourth hybrid orbital of CH_3. We conclude that $Mn(CO)_5$ is isolobal with CH_3. Both are one electron short of a filled shell octet or 18*e* configuration. Similarly we can identify families of other isolobal fragments as illustrated below.

In all these examples, the characterizing hybrid lobe contains a single electron. We can readily rationalize and sometimes successfully guess the formation of molecules like H—H, H_3C—Br,

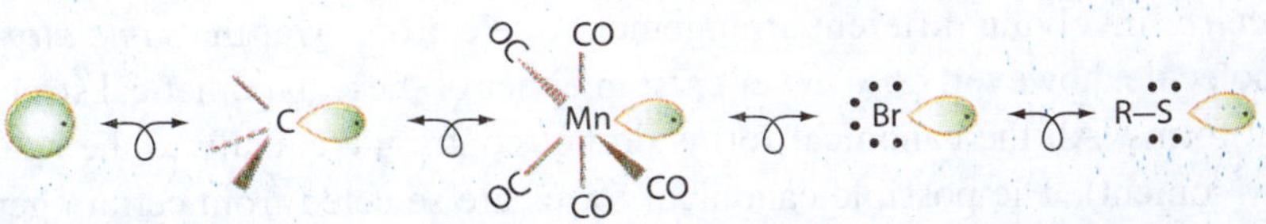

$H_3C—CH_3$, $(CO)_5Mn—CH_3$ and $(CO)_5Mn—Mn(CO)_5$. We shall find many applications of this concept in connection with organometallic compounds and metal clusters.

Removal of two hydrogens from CH_4 (retaining the sp^3 hybrid orbitals) generates the 6-electron fragment CH_2, two electrons short of a completed octet and having two single electrons in two sp^3 hybrid orbitals. Similar removal of two ligands from the octahedral ML_6 complex gives rise to the fragment ML_4, again two electrons short of the 18-*e* configuration and having two single electrons in two vacant octahedrally hybridized orbitals. Hence we conclude that CH_2 is isolobal with ML_4. Similarly we may have isolobal CH and ML_3, each having three singly occupied lobes.

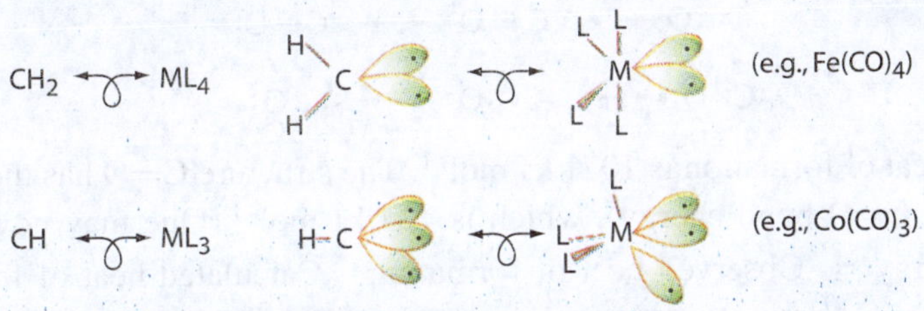

Isolobal fragments provide useful synthetic possibilities but they must be supplemented by experiments. Also, one must be careful while establishing isolobal analogies and drawing inferences therefrom since several other factors may be involved at each step.

(v) *Resonance*

The concept of resonance developed as an essential part of the valence bond theory. We have seen how the observed lowering of energy in the formation of a hydrogen molecule was arrived at theoretically by considering a number of *hypothetical structures* for the hydrogen molecule. The "*structures*" were obtained by assuming possible interchange in the positions of the electrons between the two nuclei.

$$H_A^1\ H_B^2 \qquad H_A^2\ H_B^1 \qquad \overset{1,2}{H_A^-}\ H_B^+ \qquad H_A^+\ \overset{1,2}{H_B^-}$$
$$\text{(i)} \qquad\quad \text{(ii)} \qquad\quad \text{(iii)} \qquad\quad \text{(iv)}$$

The wave functions corresponding to each of the hypothetical structures (i)—(iv) could be suitably combined to obtain the energy of the H_2 molecule which is lower than the energy of any of the individual structures. This calculated value was found to be in good agreement with experimental results. The hydrogen molecule was then said to involve resonance among the structures (i)—(iv), which were called its canonical forms. The difference of energy between the lowest energy canonical form and the actual molecule is the stabilization gained through resonance—known as the *resonance energy* or sometimes as the *exchange energy*.

Resonance is a mechanism called for theoretically to explain the stability and other properties of a (multinuclear) chemical species which cannot be assigned any single unique electron dot structure. The best description of the actual structure is obtained from a number of *hypothetical structures* involving different arrangement of electrons over the *same atomic skeleton.* The actual molecule, however, does never exist in either of these hypothetical structures, called the canonical forms. All the canonical forms do not contribute equally to the actual structure (or its energy content). The possible canonical forms are selected from certain general principles. They follow the same principles as mentioned earlier with Lewis structures (Ch. 6).

The resonance hybrid should *not* be considered to be *intermediate* between the resonating structures. The energy of the resonance hybrid is lower than the energy of any of the contributing structures. The actual structure is thus more stable than any of the canonical forms. The stability of the resonance hybrid is expressed in terms of *resonance energy*. This is the difference between (i) the energy of the actual species (obtained from observed heat of formation) and (ii) the energy of the most stable canonical form (calculated from the bond energy data relevant to the written structure). Thus for carbon monoxide one can write the canonical forms

$$\ddot{C}=\ddot{O} \longleftrightarrow {}^{-}C\equiv O^{+} \longleftrightarrow {}^{+}C-O^{-}$$

$$:C::\ddot{O}: \longleftrightarrow \bar{:}C:::\overset{+}{O}: \longleftrightarrow \overset{+}{:}C:\ddot{\underset{\cdot\cdot}{O}}\bar{:}$$

The observed heat of formation is 1071 kJ mol^{-1}. The structure C=O has the lowest energy (obtained from the C=O bond energy), which is 724 kJ mol^{-1}. One may now write

$$\text{Resonance Energy} = \text{Observed heat of formation} - \text{Calculated heat of formation}$$
$$= (1971 - 724)\ \text{kJ mol}^{-1} = 347\ \text{kJ mol}^{-1}$$

Several examples of resonance have been presented earlier in connection with Lewis structures (Chapter 6). More examples will be found later.

The several primary structures involved in resonance cannot be written at random. The different canonical forms should fulfil few necessary conditions. These are:

(i) They must have the same atomic skeleton, i.e., the same relative position of the nuclei. Thus N–N–O and N–O–N cannot contribute to the same resonance hybrid.

(ii) The number of unpaired electrons must be the same in all the contributing structures.

(iii) Only those structures which possess similar energies (hence stabilities) can contribute appreciably to a given resonance hybrid. Thus ${}^{2-}\!:\ddot{N}:\overset{+}{N}:::\overset{+}{O}:$ is not likely to contribute to the resonance hybrid of nitrous oxide. Considerable electrostatic energy would presumably be required to have two adjacent atoms with like charges and a doubly negatively charged atom on the other end; the energy of this structure would be consequently very high.

A special kind of resonance, called "*no bond resonance*" or "hyperconjugation" is sometimes invoked to rationalize extraordinary bonding situations. In such canonical forms there is an increase in the order of one bond at the cost of the order of another bond:

$$A-X-B \longleftrightarrow {}^{+}A=X\ \ B^{-}$$

The unusually long N–F bond in ONF_3 may be explained through no bond resonance involving canonical forms of the type

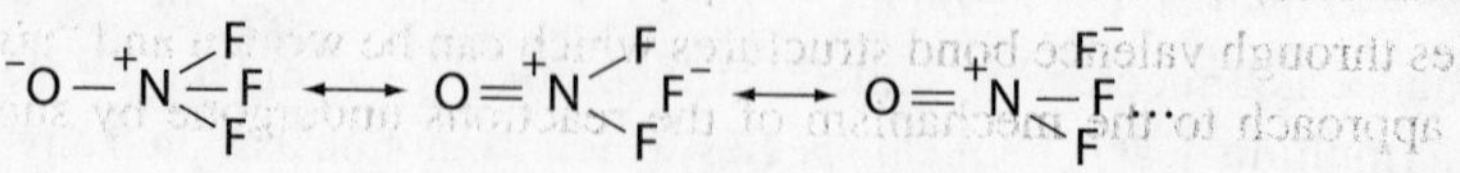

Though these canonical forms are ad hoc, they still bear formal charge distribution consistent with electronegativity.

Another example of no bond resonance or hyperconjugation appears in explaining the very strong acid-base interaction between BH_3 and CO :

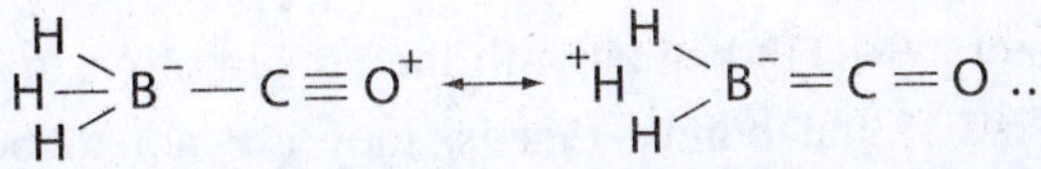

Resonance differs from tautomerism in the hypothetical nature of the canonical forms. In *tautomerism*, a compound *actually exists* as an equilibrium mixture of different isomeric molecules, each having *different atomic skeletons*. The different forms of a tautomeric mixture are capable of easy interconversion. Under appropriate conditions, one or other forms may be made to dominate in a chemical reaction and may also be isolated. Sulfurous acid, H_2SO_3, exists as a tautomeric mixture of the molecules HO.SO.OH and $HO.SO_2H$; diethyl esters of both the forms have been prepared.

Uses and misuses of the concept of resonance

The idea of resonance has been widely used in describing the properties, and hence the structures of numerous organic compounds and simple inorganic compounds. Though it was introduced basically as a part of the valence bond theory in its quantum mechanical approach to interpret covalent bond, its idea may be used largely qualitatively. The idea helps us to "*explain*" the observed properties of many compounds in terms of physically realizable "*pictures*" of the structure, though the individual "*pictures*" are purely imaginary in nature.

Actually this emphasis on the pictorial nature of resonance has led to many misuses of the concept. Isolated from these misconceptions for which resonance was not actually introduced, the hypothetical as well as arbitrary nature of the canonical forms constitute the main weak point of this popular concept.

7.2.3 Summary and Discussion on VBT

The valence bond theory (VBT) begins with the atomic orbital wave functions of the combining atoms arranged according to definite valence bond structures. The theory can successfully interpret the bonding and properties of a large number of covalent molecules. The geometry of the molecules is also explained quite satisfactorily. The idea of resonance also appears useful in the interpretation of many reactions and their mechanisms.

However, actual calculations in VBT for larger molecules become extremely complicated—the alternative molecular orbital theory offers relatively simple mathematical handling. At the same time, VBT is solely dependent on spin pairing for bond formation. This cannot explain the existence of the dioxygen molecule with two unpaired electrons. Similarly, odd electron species like H_2^+, He_2^+ etc. and the bonding in metals cannot be covered by VBT.

VBT emphasizes only the positive overlap between atomic orbitals but as we shall soon realize, negative overlap between atomic orbitals giving rise to antibonding molecular orbitals also plays an important role in explaining the properties of molecules. This is particularly true in the study of electronic spectroscopy where VBT fails to offer any ready explanation of the various higher orbitals to which the electrons are promoted on excitation.

Nevertheless, the valence bond theory provides an easily realizable bonding description in molecules through valence bond structures which can be written and "*visualized*". This permits an easy approach to the mechanism of the reactions undergone by such molecules.

7.3 THE MOLECULAR ORBITAL THEORY (MOT)

7.3.1 Introduction to MOT

The molecular orbital theory (MOT) begins with the presumption that the electrons in a molecule do not belong to individual bonds, that is, they are not to be regarded as localized. Instead, the electrons are treated as spreading throughout the molecule. Just as each electron in a single atom can be represented by a wave function corresponding to a particular atomic orbital, so each electron in a molecule can be described by wave functions determined jointly by the set of nuclei present in a system. These wave functions are then said to represent *molecular orbitals*, which bear more or less the same significance as the atomic orbitals in an atom. They describe the energy states and spatial distribution of the electrons in the combined field of the nuclei. The Pauli principle and the Hund's rule of maximum multiplicity applies to these molecular orbitals. Thus, each molecular orbital can accommodate a maximum number of two electrons with opposite spins. Following the aufbau principle, electrons may be assigned one by one to these molecular orbitals from the lowest available energy state. Each type of molecular orbital gives rise to its characteristic spatial distribution.

We recall that in connection with atomic orbitals, the nature of the wave functions was first ascertained from an exact solution of the Schrödinger wave equation for the hydrogen atom which contained only one electron. For atoms with many electrons, it was next assumed that the orbitals have the same form, their relative energies being only affected by the shielding exerted on one electron by every others. Each electron was thus treated as if it moved in the effective field produced by the nucleus and all the other electrons. In the molecular orbital theory also, the orbital of an electron is first calculated by assuming that only *one electron* moves in the combined field of the different nuclei present. If we now have a system involving two electrons whose one-electron molecular orbital wave functions are ψ_1 and ψ_2 respectively, the molecular orbital wave function for the total system will be

$$\Psi_{MO} = \psi_1 \psi_2$$

But how to obtain ψ_1 and ψ_2, the one-electron molecular orbital wave functions? The one electron atomic orbital wave functions were obtained by exact solution of the wave equation for the hydrogen atom. The case of an electron moving in the field of several nuclei cannot be solved exactly. First of all, we can separate the electronic part by applying the Born-Oppenheimer approximation and work with a fixed nuclear framework. Even then one has to adopt several approximations for the purpose. A very convenient method of approximation is provided by the **Linear Combination of Atomic Orbitals (LCAO)** procedure.

The LCAO method works with the idea that when an electron moves around in a nuclear framework, at any instant of time it will be close to one nucleus and relatively far from others. During this segment of time, the electron may be supposed to be "controlled" mostly by the nucleus close to it and its wave function will be given by the wave function of the atomic orbital for that nucleus alone. The resultant wave function of the electron in the nuclear framework may then be represented by an algebraic sum of the various atomic orbital wave functions for different nuclei in the framework. This leads us to the linear combination of atomic orbitals. If an electron is described by the one-electron wavefunction ψ_A for nucleus A and ψ_B for nucleus B, then the wave function of the electron in a system containing the two nuclei A and B may be represented by a superposition of ψ_A and ψ_B:

$$\Psi_{\pm} = N(\psi_A \pm \psi_B) \qquad (N \text{ is a normalizing constant})$$

The hydrogen molecule-ion

We assume that the hydrogen molecule-ion consists of a single electron in the field of two nuclei marked A and B, as depicted in 7-II. The internuclear separation R is assumed fixed according to the Born-Oppenheimer approximation.

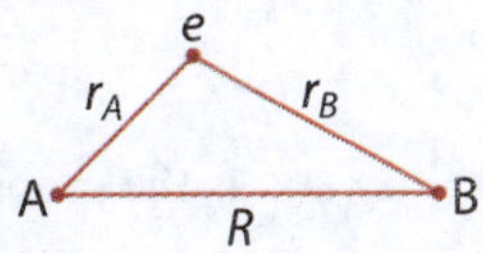

7-II

When the electron is close to either A or B, it may be approximately described by the ground state atomic orbital wave function ψ_A of atom A or ψ_B or atom B, as the case may be:

$$\psi_A = \psi_A(1s) = (1/a_0)^{3/2}\pi^{-1/2}e^{-r_A/a_0} \qquad \psi_B = \psi_B(1s) = (1/a_0)^{3/2}\pi^{-1/2}e^{-r_B/a_0}$$

The approximate one electron wave function for the H_2^+ ion (ground state) may now be obtained by a linear combination of the atomic orbital wave functions ψ_A and ψ_B:

$$\Psi = c_A\Psi_A + c_B\Psi_B$$

The c-coefficients (c_A and c_B) indicate the extent to which each atomic orbital contributes to the molecular orbital. The greater the value of c^2, the greater will be the contribution of that atomic orbital to the molecular orbital.

The energy of the system is obtained by solving the Schrödinger equation $H\Psi = E\Psi$ where H is the relevant Hamiltonian (electronic part):

$$H = \frac{\hbar^2}{2m_e}\nabla^2 + V$$

The potential energy term V involves attraction on the electron by each nucleus and repulsion between the nuclei

$$V = -\frac{e^2}{4\pi\varepsilon_0}\left(\frac{1}{r_A} + \frac{1}{r_B} - \frac{1}{R}\right)$$

The energy of the system calculated in this manner for varying internuclear separation is shown in Fig. 7.18. The results agree only approximately with actual values.

The one-electron wave functions obtained by solving the equation $H\Psi = E\Psi$ are called molecular orbitals (MO or LCAO-MO). The distribution of the electron in the molecule is obtained from $|\Psi^2|$. To get the complete wavefunction (electronic) one should multiply this Ψ (space function) with a spin function (α or β).

The coefficients c_A and c_B may be determined by the variation procedure. In this case, we have two identical nuclei. So the electron probability density should be equal on each side of the molecule which requires $c_A^2 = c_B^2$ or $c_A = \pm c_B$. This gives us two possible combinations of ψ_A and ψ_B:

$$\Psi_+ = c_A\psi_A + c_A\psi_B \qquad \Psi_- = c_A\psi_A - c_A\psi_B.$$

c_A can be evaluated from normalization conditions.

The probability density of the electron in the system is now given by

$$\Psi_+^2 = c_A^2\left(\psi_A^2 + \psi_B^2 + 2\psi_A\psi_B\right) = c_A^2\left(\psi_A^2 + \psi_B^2\right) + 2c_A^2\psi_A\psi_B$$

$$\Psi_-^2 = c_A^2\left(\psi_A^2 + \psi_B^2 - 2\psi_A\psi_B\right) = c_A^2\left(\psi_A^2 + \psi_B^2\right) - 2c_A^2\psi_A\psi_B$$

$c_A^2\psi_A^2$ is the probability density of the electron with nucleus A alone; similarly $c_B^2\psi_B^2$ is the probability density with nucleus B. $c_A^2\left(\psi_A^2 + \psi_B^2\right)$ is thus equivalent to the electron probability density of two separate (noninteracting) hydrogen 1*s* atomic orbitals. In Ψ_+^2, we have an additional $2c_A^2\psi_A\psi_B$ term. The magnitude of this term will be large where both ψ_A and ψ_B are large. So this additional term increases the probability density of the electron in the region between the two nuclei. This results in greater electrostatic attraction of the electron from the two nuclei and lowers the potential energy of the system. The integral $\int\psi_A\psi_B d\tau$ is defined as the overlap integral (*S*). The combination c_A $(\psi_A + \psi_B)$ represents a *bonding molecular orbital* whose energy is lower than the energy of either ψ_A or ψ_B.

An analogous logic helps us to realize that the $-2c_A^2\psi_A\psi_B$ term in Ψ_-^2 will reduce the electron probability density from the region between the nuclei. This will cause greater repulsion between the two nuclei and raise the potential energy of the system. As such, the function $\Psi_- = c_A(\psi_A - \psi_B)$ represents a higher energy state for H_2^+ and is called an *antibonding molecular orbital*. If we now consider a plane perpendicular to the internuclear axis and placed exactly midway between the two nuclei, we have $r_A = r_B$ for any point on this plane. This gives $\psi_A = \psi_B$, so that Ψ_-^2 becomes zero anywhere on this plane. This is called a *nodal plane*.

However, the stabilization in energy gained (relative to the isolated atomic orbitals) in the bonding molecular orbital is *less* than the destabilization or increase in energy of the antibonding molecular orbital. *Thus the antibonding m.o. is more antibonding than the bonding effect provided by the bonding m.o.* This may be associated, at least in part, to the difference in internuclear repulsion in the two situations. Though this repulsion raises the energy of both the bonding and antibonding m.o.s, the latter seems to be more sensitive to such repulsion as there is less electron probability density in the internuclear region.

The nature of variation of the LCAO-MO wave functions Ψ_+ and Ψ_- along the internuclear axis is shown in Fig. 7.17. Since r_A and r_B vary with internuclear separation *R* (7-II), Ψ_+ and Ψ_- also vary with *R*. Plots of corresponding squares show the build up of electron probability density between the nuclei (for Ψ_+) and the reduced electron probability density for Ψ_-. Note the occurrence of nodes in the plots of Ψ_- and Ψ_-^2.

The variation of energy (electronic) of the system calculated from Ψ_+ and Ψ_- is approximately shown, in Fig. 7.18.

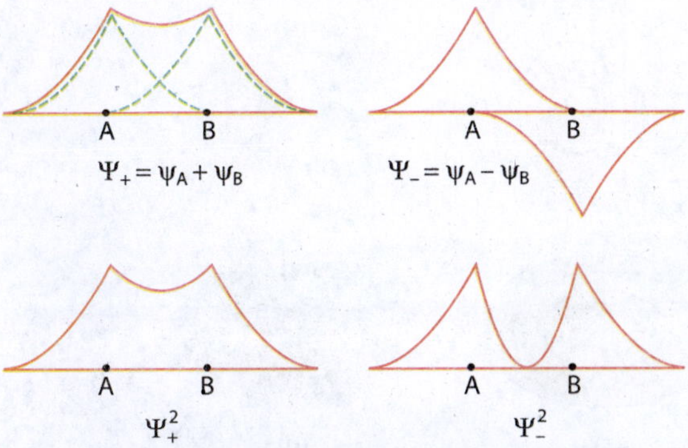

Fig.7.17: Nature of variation of Ψ_+ and Ψ_- and their squares along the internuclear axis in H_2^+.

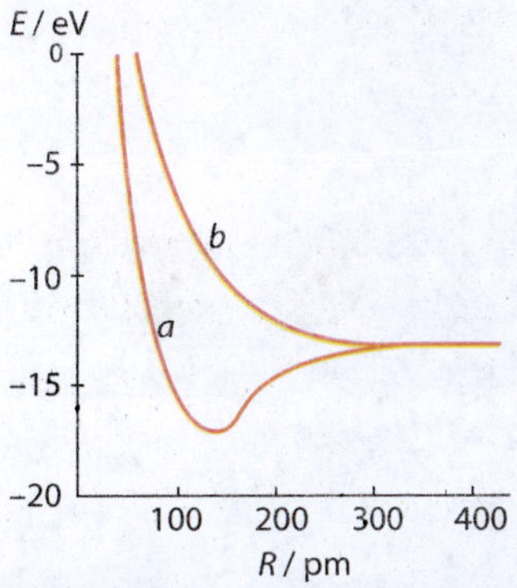

Fig.7.18: Nature of variation of electronic energy with internuclear distance in H_2^+ (approximate).

(*a*) Calculated from Ψ_+. A minimum occurs at R_e = 132 pm with D_e (bond dissociation energy) 1.8 eV. (Actual values 106 pm and 2.8 eV).

(*b*) Calculated from Ψ_-. The curve has no minimum.

Conditions for linear combination of atomic orbitals

In order that atomic orbitals may interact and form molecular orbitals, three conditions must be fulfilled. These are:

(i) *The atomic orbitals must have comparable energies.*

(ii) *They must overlap appreciably.*

(iii) *They must have the same symmetry with respect to the bonding molecular axis.*

(i) Atomic orbitals having large difference of energy will not combine to form molecular orbitals. Thus, in elements of the second row of the periodic table (Li to Ne), the 1*s* atomic orbital is far below the 2*s* and 2*p* atomic orbitals and do not form molecular orbitals with them. Similarly the *s* and *p* atomic orbitals do not combine to form molecular orbitals in homonuclear diatomic molecules when their energy difference is large.

(ii) The combining atomic orbitals must overlap to a considerable extent. This is also contained in the principle of maximum overlap explained earlier. The greater the overlap of the atomic orbitals involved, the greater will be the electron probability density between the two nuclei.

(iii) Atomic orbitals of comparable energy would combine only when they have the same symmetry about the internuclear axis. We shall soon appreciate that an *s* orbital can not effectively combine with a p_x or p_y atomic orbital if the molecular axis lies along the *z*-axis.

Different types of molecular orbitals

As we have done with atomic orbitals, molecular orbitals may also be represented by contours of constant probability density in a plane containing the nuclei (Fig. 7.19). The three-dimensional nature of the orbitals may be imagined by rotating the contours about the internuclear axis.

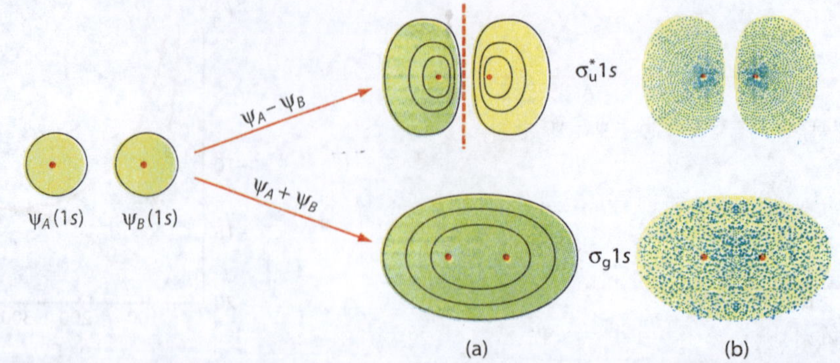

Fig.7.19: Formation of molecular orbitals from two hydrogen 1*s* atomic orbitals. (a) Contours of constant probability density. The dashed line indicates the nodal plane. (b) Electron probability density according to charge cloud interpretation. The labels $\sigma_g 1s$ and $\sigma_u^* 1s$ are explained in the next section.

In the hydrogen molecule ion (H_2^+) in its ground state, the single electron occupies the lowest energy bonding m.o. which is $\sigma_g 1s$ or simply $1\sigma_g$; the numerical 1 now denotes the position of the m.o. from lower to higher energy. The electron configuration is written as $1\sigma_g^1$.

Bonding, antibonding and nonbonding m.o.s.

A bonding molecular orbital is represented by a wave function that corresponds to (through its square) an increased electron probability density between two nuclei, resulting from constructive interference between the participating atomic orbitals. An electron (or electrons) occupying this molecular orbital helps to bind the two atoms together. The energy of the bonding m.o. is lower than the energies of the contributing atomic orbitals. For two hydrogen 1*s* atomic orbitals, the combination ($\psi_A + \psi_B$) represents a bonding molecular orbital (Fig. 7.19).

An antibonding molecular orbital is represented by a wave function that stands for a decrease in electron probability density between two nuclei arising from destructive interference between the participating atomic orbitals. When occupied by an electron (or electrons), antibonding orbitals cause reduction in the net binding between the nuclei. The energy of an antibonding orbital is higher than the energies of combining orbitals. The combination ($\psi_A - \psi_B$) for two hydrogen 1*s* atomic orbitals represents an antibonding m.o. (Fig. 7.19). Such m.o-s are often marked by an asterisk (*).

Nonbonding molecular orbitals result when two atomic orbitals combine to zero net overlap (7-III). The energy of a nonbonding m.o. is essentially that of the atomic orbitals, either because the atomic orbitals do not match in symmetry or the energy of the molecular orbital matches that of the atomic orbital by coincidence. Thus a nonbonding m.o. is a molecular orbital that essentially consists of a single atomic orbital from one participating atom. Occupation of this orbital does not stabilize or destabilize a molecule but such orbitals may become important in certain cases. We shall get several examples of nonbonding orbitals later.

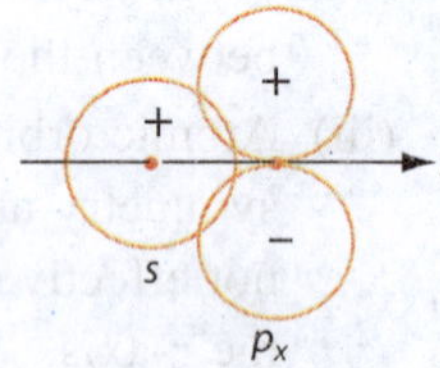

7-III

Bonding, antibonding and nonbonding orbitals are important in the M.O. theory for conservation of orbitals. In assembling *N* atomic orbitals to form molecular orbitals by linear combination, we obtain N molecular orbitals; no atomic orbitals are "*lost*". But the existence of such molecular orbitals, however, does not mean that all of them are occupied in normal states of molecules.

The overlap integral, S

The extent of overlap between two atomic orbitals ψ_A and ψ_B is expressed in terms of the overlap integral S:

$$S = \int \psi_A \psi_B \, d\tau$$

Clearly, the value of S will be large in the region of space where the values of both ψ_A and ψ_B are significant. In the case of two hydrogen 1s orbitals, the region between the nuclei (say A and B) will provide maximum overlap. The magnitude of S will vary with the internuclear distance R. At R = 0, $\psi_A(1s) = \psi_B(1s)$ and S = 1. At $R = \infty$, ψ_A and ψ_B do not overlap and S = 0. The nature of variation of S in the intermediate region is shown is Fig. 7.20.

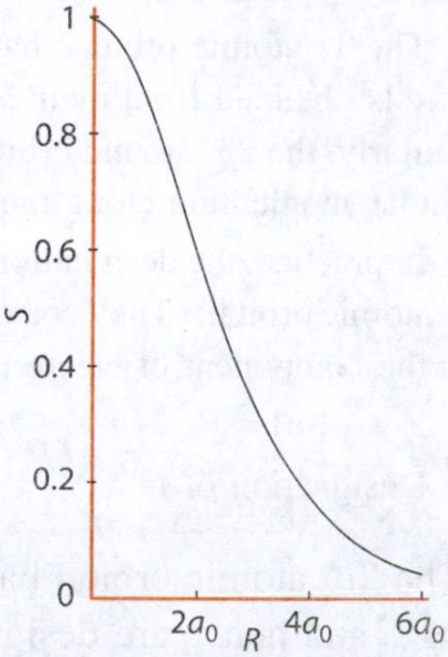

Fig.7.20: Nature of variation of S with internuclear distance between two hydrogen nuclei.

σ, π, δ (and φ) m.o.s

Molecular orbitals are classified as σ, π, δ and φ m.o.s. according to their rotational symmetry about the internuclear axis. The molecular orbitals resulting from the 1s atomic orbitals in Fig. 7.19 have cylindrical symmetry about the internuclear axis (conventionally z) and are called σ m.o.s. σ molecular orbitals are formed from overlap between atomic orbitals that have cylindrical symmetry about the internuclear axis. Thus an s-orbital may combine with a p_z orbital to from σ molecular orbitals. Some other combinations leading to σ m.o.s. are $p_z + p_z$, $s + d_{z^2}$, $p_z + d_{z^2}$, $d_{z^2} + d_{z^2}$ etc. (z-axis taken as internuclear axis in all cases). The combination of two $2p_z$ atomic orbitals is represented in Fig. 7.21.

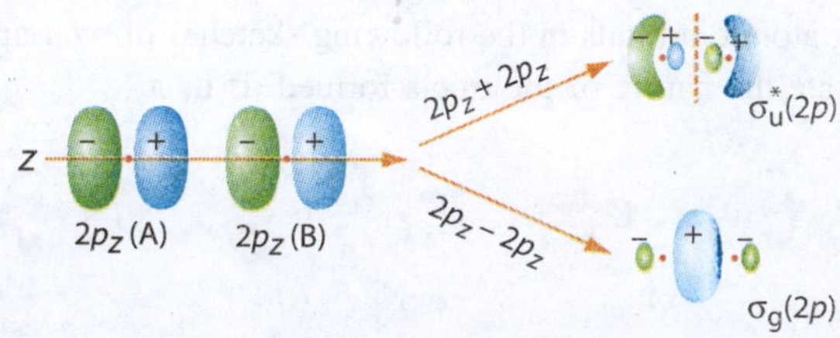

Fig.7.21: Formation of molecular orbitals from two $2p_z$ atomic orbitals.
Note the opposite signs (of wave functions) on the right lobe of $2p_z$(A) and the left lobe of $2p_z$(B). The linear combination $2p_z$(A) + $2p_z$(B) thus represents an antibonding m.o. with a nodal plane midway between the nuclei (dashed line). The linear combination $2p_z$(A) – $2p_z$(B) results in a bonding m.o. Antibonding m.o.-s are marked by *. The meaning of subscripts 'g' and 'u' are explained later.

Both s- and σ-orbitals have no nodal planes passing through the nucleus or internuclear axis; p and π orbitals have one such nodal plane while the d and δ orbitals have two such nodal planes. Both f and φ orbitals have three such nodal planes.

In the simplest case, a π m.o. results from the combination of two $2p_x$ atomic orbitals on two separate atoms (Fig. 7.22), the z-axis being again taken as the internuclear axis. It is easy to realize that two $2p_y$ atomic orbitals may also combine to form two π m.o.s. The π m.o.-s are characterized by a nodal plane containing the internuclear axis.

The $1s$ atomic orbitals have zero orbital angular momentum (electronic) and so the molecular orbitals obtained from them also have zero orbital angular momentum along the internuclear axis. Similarly, the $2p_z$ atomic orbitals have $m = 0$ and $L_z = 0$ (chapter 4) and so the component of orbital angular momentum along the internuclear axis (z) is also 0.

In practice, the designations σ, π, δ ··· for m.o.-s are Greek equivalents of the letters s, p, d, f used for atomic orbitals. They correspond to the values of $|m|$ where $m = 0, \pm 1, \pm 2$ ···. (The possible values for the component of electronic orbital angular momentum along the z-axis are $L_z = m\hbar$).

$|m| =$	0	1	2	3
Designation of m.o.	σ	π	δ	φ

The $2p_x$ atomic orbital has $|m| = 1$. The m.o.-s derived from $2p_x$ atomic orbitals also have $|m| = 1$ and hence are designated as π m.o.-s.

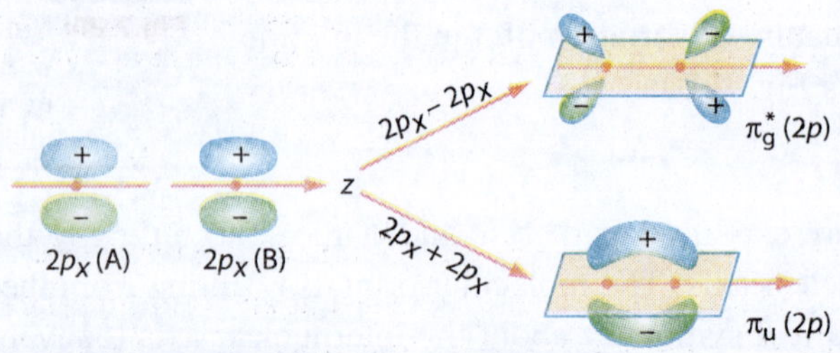

Fig. 7.22: Combination of $2p_x$ atomic orbitals to form π m.o.s. Combination of $2p_y$ atomic orbitals will produce similar m.o.s. but rotated by 90° about the internuclear axis.
The antibonding π m.o. has an additional nodal plane perpendicular to the nodal plane containing the internuclear axis.

Problem 7.1

Identify the participating atomic orbitals in the following sketches of overlap (z-axis taken as internuclear axis) and also state the nature of the m.o.s formed (σ or π),

(a) (b) (c) (d)

Ans. (a) $s + p_z$ (σ) (b) $p_z + d_{z^2}$(σ)
(c) $p_x + d_{zx}$ or $p_y + d_{yz}$ (π) (d) $d_{xz} + d_{xz}$ or $d_{yz} + d_{yz}$ (π)

Problem 7.2

Sketch the overlap of the following combinations of orbitals (z-axis as internuclear axis) and comment on the nature of m.o. formed.

(a) $p_z + d_{xz}$ (b) $s + d_{z^2}$ (c) $s + d_{yz}$

Ans. (a), (c): no net overlap; (b) σ

The molecular orbital for a delta bond (δ-bond) has two mutually perpendicular nodal planes containing the internuclear axis and has $|m| = 2$ as mentioned before. $d_{x^2-y^2}$ orbitals or d_{xz} orbitals on two atoms in parallel planes overlap to give rise to δ-bond, (Fig. 7.23).

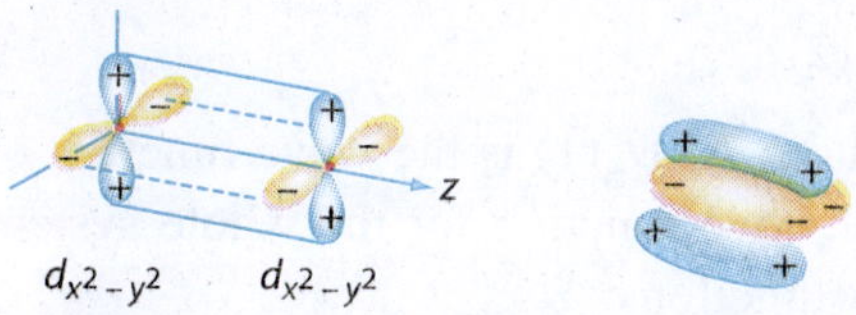

Fig.7.23: Formation of δ-bond by combination of two $d_{x^2-y^2}$ a.o.-s.

ϕ type molecular orbitals are comparatively rare and may be formed from two *f* orbitals as sketched in Fig. 7.24.

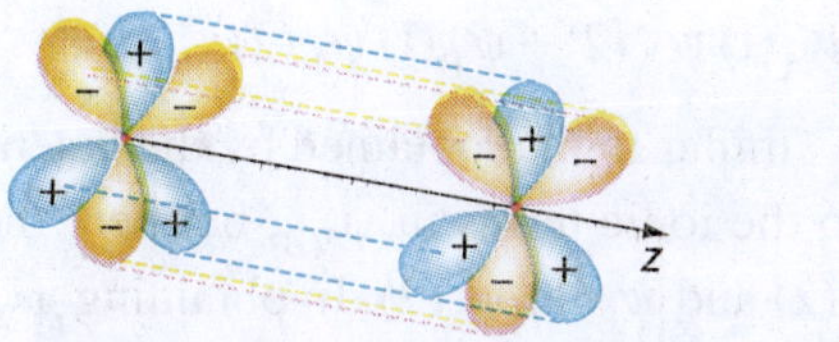

Fig.7.24: Formation of ϕ_u m.o. by overlap between $f_x(x^2-3y^2)$ and $f_y(y^2-3x^2)$ atomic orbitals. As mentioned before, they possess three nodal planes containing the internuclear axis and have $|m| = 3$.

g and u notations

The subscripts *g* and *u* (from German *gerade* = even; *ungerade* = odd) are used to label a molecular orbital by its inversion symmetry. If we start at any point on the bonding σ orbitals in Fig. 7.19 and 7.20, proceed to the centre of the molecule (centre of inversion) along a straight line and project it out on the other side by an equal distance, we arrive at an identical value of the wave function. The same is true for the antibonding π molecular orbital in Fig. 7.21. All these wave functions are thus even or '*g*'-type orbitals. The point diagonally opposite *x, y, z* are $-x, -y, -z$. For the '*g*' functions, $\psi(x, y, z) = \psi(-x, -y, -z)$.

In the antibonding σ* orbitals in Fig. 7.19 and 7.20 or in the bonding π orbital in Fig. 7.21, two points diagonally opposite and equidistant from the inversion centre have values of the orbital differing by a factor of −1; such functions are called odd and labelled by '*u*' as a subscript. For an odd function, $\psi(-x, -y, -z) = -\psi(x, y, z)$.

The *g-u* classification of molecular orbitals on the basis of orbital symmetry does not apply to diatomic molecules containing atoms of two different elements (heteronuclear) since such molecules do not have a centre of inversion.

In labelling molecular orbitals with *g* and *u* subscripts, each set of orbitals having the same symmetry (either *g* or *u*) is labelled separately. In Fig. 7.19, the two m.o.-s in increasing order of energy are labelled as $1\sigma_g$ and $1\sigma_u^*$ respectively. The asterisk may be dropped. This sequence becomes important for diatomic molecules of heavier elements.

7.3.2 Homonuclear Diatomic Molecules

The hydrogen molecule : H_2

The principles of molecular orbital theory as laid down in the preceding section may now be applied to other systems. To begin with, we first consider the hydrogen molecule, H_2. We have seen how we can handle a system of two hydrogen nuclei and one electron. For a hydrogen molecule, there will be two electrons. As in the case of H_2^+, suppose the atoms are marked A and B and the electrons are marked by numbers 1 and 2. The one electron wave functions for them will be of the type (omitting normalization constants)

$$\psi_1 = \psi_A(1) + \psi_B(1); \; \psi_2 = \psi_A(2) + \psi_B(2)$$

where $\psi_A(1)$ is the wave function of electron 1 on atom A, $\psi_B(1)$ is the wave function of electron 1 on atom B, and so on. The molecular orbital wave function for the whole system is then obtained as the product of the two one electron functions:

$$\psi_{MO} = \psi_1\psi_2 = [\psi_A(1) + \psi_B(1)]\,[\psi_A(2) + \psi_B(2)]$$

Multiplying out and rearranging, we get,

$$\psi_{MO} = \psi_A(1)\,\psi_B(2) + \psi_B(1)\,\psi_A(2) + \psi_A(1)\,\psi_A(2) + \psi_B(1)\,\psi_B(2).$$

It is interesting to find that the total wave function is similar to that obtained by the valence bond method. The last two terms in ψ_{MO} correspond to the ionic terms in ψ_{VB}, but here they are weighted as heavily as the covalent terms $\psi_A(1)\,\psi_B(2)$ and $\psi_B(1)\,\psi_A(2)$. In obtaining ψ_{VB}, the contribution of the ionic terms was considered less important owing to interelectron repulsion. But in obtaining ψ_{MO}, we did not take into account such repulsion. Thus, the molecular orbital theory places too much emphasis on the ionic terms. However, the presentation made here is too elementary and the wave function may be considerably improved by inserting suitable parameters to give proper weight to the various terms.

At this point, it will be interesting to compare the results of calculation of bond dissociation energy and equilibrium internuclear distance in the VBT and MOT.

Table 7.3

Calculated values of bond dissociation energy (D_e) and equilibrium internuclear distance (R_e) for H_2

VBT	D_e (kJ mol^{-1})	R_e (pm)	MOT	D_e (kJ mol^{-1})	R_e (pm)
(i) $\psi = \psi_A\psi_B$ (uncorrected)	24	90	(i) $\psi = \psi_A + \psi_B$ (uncorrected)	260	85
(ii) Heitler London	303	87	(ii) Addition of shielding	337	73
(iii) Addition of shielding	365	74.3	(iii) Hartree-Fock wave function	351	73
(iv) Ionic contribution added	388	74.9	(iv) Observed	458	74.1
(v) Observed	458	74.1			

We observe that the MOT, in this form, is less satisfactory in reproducing the bond dissociation energy. However, with further improvement in the one-electron functions it has been possible to obtain excellent agreement with experimental values. In many other cases, the MOT has provided better interpretation of experimental facts.

The molecular orbital energy level diagram for H_2 is the same as that of H_2^+ (Fig. 7.25). Since the system contains two electrons, these may be placed in the lowest available orbital $1\sigma_g$ with spins paired (see below). The ground state m.o. description of H_2 is thus $1\sigma_g^2$.

Spin pairing in MOT

To satisfy the Pauli principle (see Ch. 4 and the VBT), the overall molecular orbital wave functions must be antisymmetric. The space part of the wave functions for two electrons in the bonding molecular orbital is of type $\psi(1)\psi(2)$ where $\psi(1)$ and $\psi(2)$ are the one electron wave functions. The function $\psi(1)\psi(2)$ is symmetric with respect to interchange of the labels 1 and 2. So this function must be multiplied by an antisymmetric spin function.

As shown in connection with VBT, an antisymmetric spin state may be written as $\alpha(1)\,\beta(2) - \beta(1)\,\alpha(2)$, so that

$$\psi_{MO}\,(\text{Total}) = \psi(1)\,\psi(2)\,\{\alpha(1)\,\beta(2) - \beta(1)\,\alpha(2)\}$$

becomes unsymmetrical. The spin function $\alpha(1)\,\beta(2) - \beta(1)\,\alpha(2)$ corresponds to opposite spins of the two electrons in the $1\sigma_g$ m.o. of H_2.

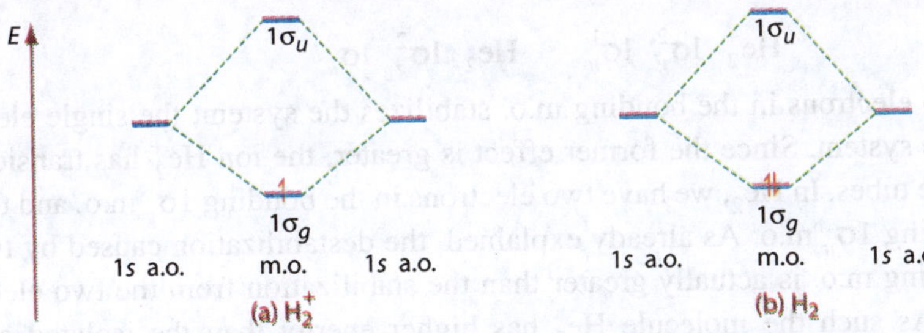

Fig.7.25: Molecular orbital energy level diagrams for (a) H_2^+ and (b) H_2 in ground state electron occupancy. The energy scale is arbitrary.

Before we proceed to other diatomic species, we recapitulate the aufbau or building up principle as applicable for obtaining ground state electron distribution in molecules.

1. Only two electrons of opposite spin can be assigned to a molecular orbital.
2. Electrons are assigned one by one to the available molecular orbital of lowest energy.
3. In case of degenerate orbitals (more than one orbital of same energy), each orbital is first occupied singly by electrons with parallel spins.

The molecular orbitals or their energy levels may be designated by the letters σ, π, δ, etc. with the mention of contributing atomic orbitals, with or without parentheses, e.g., σ_{1s} or $\sigma(1s)$; antibonding molecular orbitals are shown with an asterisk. e.g., σ_{1s}^*. In modern usage, the asterisk may be dropped and the m.o.-s labelled with the notations g or u as subscripts. As explained earlier, these refer to inversion symmetry. σ_g and π_u represent bonding m.o.-s while

σ_u and π_g refer to antibonding m.o.s In each category (σ, π, ···), the m.o.s are serially numbered, as we have done in H_2. The lowest energy σ-m.o. is $1\sigma_g$ and the next higher m.o., also σ, but antibonding, is $1\sigma_u$.

He_2 and He_2^+:

Since the helium atom consists of two 1*s* electrons, the molecular orbitals for He_2 and He_2^+ will resemble the m.o.-s discussed for H_2^+ and H_2 from the combination of two 1*s* atomic orbitals (Fig. 7.26). Since He_2^+ contains a total of three electrons and the $1\sigma_g$ level can contain

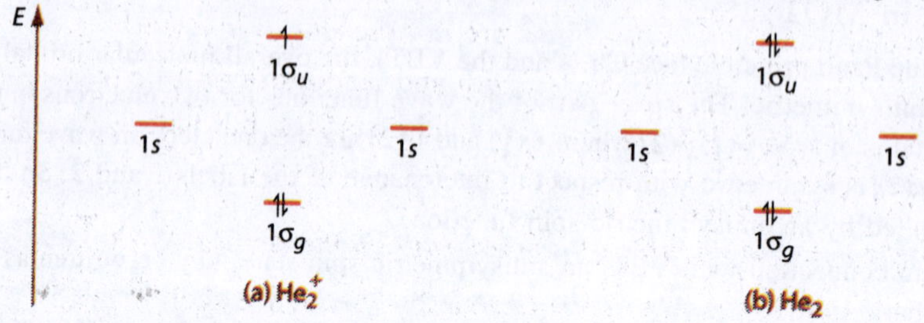

Fig.7.26: Electron occupancy (ground state) of (a) He_2^+ and (b) He_2. The energy scale is arbitrary.

at best two electrons of opposite spins, the third electron must be assigned to the next higher energy level $1\sigma_u$ (antibonding). For He_2, there will be two electrons in $1\sigma_g$ and two electrons in $1\sigma_u$.

$$He_2^+ \ 1\sigma_g^2 \ 1\sigma_u^1 \qquad He_2 \ 1\sigma_g^2 \ 1\sigma_u^2$$

In He_2^+, two electrons in the bonding m.o. stabilizes the system; the single electron in $1\sigma_u$ destabilizes the system. Since the former effect is greater, the ion He_2^+ has transient existence in gas discharge tubes. In He_2, we have two electrons in the bonding $1\sigma_g$ m.o. and two electrons in the antibonding $1\sigma_u$ m.o. As already explained, the destabilization caused by two electrons in the antibonding m.o. is actually greater than the stabilization from the two electrons in the bonding m.o. As such the molecule He_2 has higher energy than the isolated atoms and is unlikely to exist. However, He_2 has been detected only at very low pressure and low temperature molecular beams. The very low binding energy reported (~ 0.01 J mol^{-1}) suggests that the two atoms may be actually held by weak van der Waals type forces. It is also possible to have a He_2 molecule in the excited state in which an electron from the $1\sigma_u$ m.o. is excited to the bonding σ_g m.o. formed by the 2*s* a.o.s.

Bond order

In the molecular orbital theory, bond order is defined as the number of electron pairs in bonding m.o.'s *minus* the number of electron pairs in antibonding m.o.'s. That is,

Bond order = No. of electron *pairs* in bonding m.o.'s – No. of electron pairs in antibonding m.o.'s.

= $\frac{1}{2}$[No. of electrons in *bonding* m.o.'s (n)
– No. of electrons in *anti-bonding* m.o.'s (n^*)]

= $\frac{1}{2}(n - n^*)$

Thus, in H_2 the number of electrons in the bonding m.o. is two and there is no electron in the antibonding m.o. Hence the bond order in H_2 is ½ (2–0) = 1. This is equivalent to a single bond according to electron pair bond formalism (VBT). Bond order helps us in book-keeping the electrons contributing to bonding, but is different from the electron pair bond in VBT. Unlike VBT, bond order may be fractional. In H_2^+ for example, the bond order will be ½(1–0) = ½. He_2^+ also falls in this category. The bond order of He_2 is naturally zero.

Again, in MOT, an integral value of bond order does not necessarily mean that there are as many electron pair bonds. We shall see later that the bond order in the dioxygen molecule (O_2) is two, though two electron pair bonds are not present in it.

Bond order may be correlated to the strength and length of bonds, e.g., bond dissociation enthalpies or bond stretching frequencies and bond distances, but with due consideration for other aspects (see later).

Homonuclear diatomic molecules of second row elements (Li – Ne).

Atoms of the elements lithium to neon have a 2*s*–2*p* valence shell outside the $1s^2$ core. In a homonuclear diatomic molecule of any of these elements, it is possible to have overlap between (i) the 2*s* a.o.-s on both A and B, 2*s*(A) ± 2*s*(B) and (ii) the 1*s* a.o. on A with 2*s* a.o. on B or vice-versa: 1*s*(A) ± 2*s*(B) or 2*s*(A) ± 1*s*(B). The combinations under possibility (ii) may be ruled out since the 1*s* atomic orbitals are much lower in energy compared to the 2*s* atomic orbitals and the two are unlikely to make any effective overlap. (Such overlap is also ruled out by the requirement of symmetry of the point group). Similarly we can also neglect any overlap between the atomic 1*s* orbitals of each atom from consideration of poor overlap. The MO-s constructed from these two core atomic orbitals will have no effect on the total bonding situation. In fact, these MOs will be virtually indistinguishable from the 1*s* AOs on each atom.

The 2*s* atomic orbital on one atom will combine with the 2*s* atomic orbital on another atom (both *g*-type) to give rise to two σ-molecular orbitals, one bonding ($\sigma_g 2s$) and another antibonding ($\sigma_u 2s$). These will be similar to the sketches for $1\sigma_g$ and $1\sigma_u$ m.o.s shown in Fig 7.19 but the signs of wave functions will now refer only to the outer region. (The 2*s* atomic orbital has an inner part of the wave function with opposite sign separated by a spherical nodal surface). In any case, we get two energy levels for these two combinations (Fig. 7.27).

Next we consider the three 2*p* atomic orbitals (*u*-type) on each of the two atoms. We have seen (Fig. 7.21 and 7.22) that these may be combined to give two σ m.o.s and two pairs of π m.o.s. For the time being we assume that the 2*s* and 2*p* atomic orbitals on the same atom are separated by large energy difference and they do not interact.

We have thus obtained a total of eight molecular orbitals for the diatomic system from the eight contributing atomic orbitals (one *s* + three *p* on each).

Next we have to consider the order of these energy levels. We know that the 2*p* atomic orbitals of an atom have identical energy which is higher than that of the 2*s* atomic orbital. Since we have further assumed that the 2*s* and 2*p* atomic orbitals do not interact with one another, the order of the MO energy levels is determined by the energies of parent AO-s and the extent of overlap. Again, σ-overlap being more effective than π-overlap, bonding σ orbitals

(σ_g) will be more stabilized compared to bonding π-orbitals (π_u). Similarly, the antibonding σ_u orbitals will be more destabilized than the antibonding π_g orbitals. The order of increasing energy among the molecular orbitals is thus

(no s-p mixing assumed) $\sigma_g 2s \langle \sigma_u 2s \langle \sigma_g 2p_z \langle \pi_u (2p_x, 2p_y) \langle \pi_g (2p_x, 2p_y) \langle \sigma_u 2p_z$

The order is shown in Fig. 7.27(a) where they are numbered from $1\sigma_g$. The core MOs from $1s$ AOs are not taken into consideration.

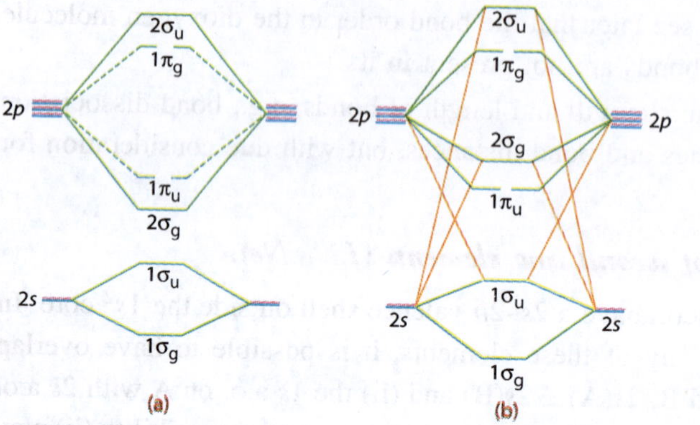

Fig.7.27: (a) Energy level diagram (arbitrary scale) for homonuclear diatomic molecules (or ions derived from them) with valence shell 2*s* and *p* atomic orbitals of participating atoms. The separation between 2*s* and 2*p* atomic orbitals is assumed large enough to eliminate 2*s*–2*p* mixing. The core MOs are not counted.
(b) Same diagram with small separation between 2*s* and 2*p* atomic orbitals, allowing *s*–*p* mixing (see below).

However, this order of increasing energy among the molecular orbitals is slightly altered when the 2*s* and 2*p* atomic orbitals are close in energy. Actually, molecular orbitals of same symmetry type and similar energy (that is when they are derived from atomic orbitals with energies close to each other) will also interact between themselves. This may be looked upon as a sort of '*orbital mixing*' or '*hybridization*' (σ–σ interaction). It is schematically represented in Fig. 7.28.

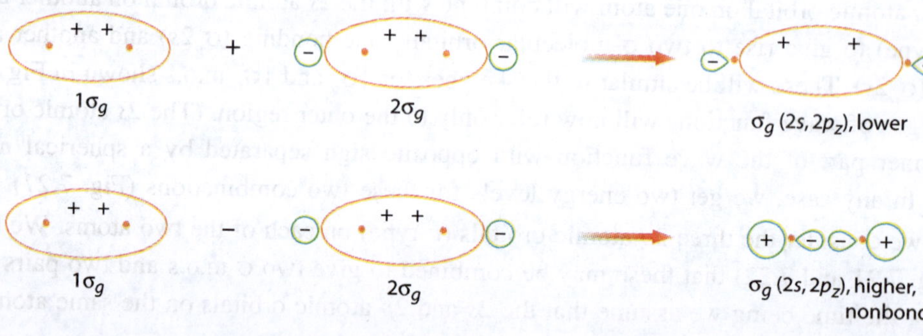

Fig.7.28: (Schematic representation of the effect of σ–σ interaction on the energies of $1\sigma_g$ and $2\sigma_g$ molecular orbitals. The energy of the lower lying $1\sigma_g$ MO (derived initially from 2*s* AO-s on the two atoms) is further lowered while the energy of the higher lying $2\sigma_g$ MO (initially derived from $2p_z$ AOs) is further raised.

In fact, as a consequence of this $1\sigma_g$–$2\sigma_g$ interaction (hybridization) the energy of the initial $2\sigma_g$ MO is now so raised that it becomes essentially a nonbonding MO $\left(2\sigma_g^{nb}\right)$ and is placed above the $1\pi_u$ bonding MOs. This is shown in Fig. 7.27(b).

A similar interaction among the $1\sigma_u$ antibonding MO (derived from 2*s* AOs) and the $2\sigma_u$ antibonding MO (derived from $2p_z$ AOs) somewhat stabilizes the $1\sigma_u$ MO and destabilizes the $2\sigma_u$ MO.

To summarize, two molecular orbitals of the same symmetry type and close in energy interact with one another in a manner in which the lower energy MO is further lowered in energy and the higher energy MO is further raised in energy.

As we proceed along the second row elements in the periodic table from Li to Ne, the effective nuclear charge (Z_{eff}) increases remarkably owing to poor screening effect of the *p*-electrons. As a result, the separation between the 2*s* and 2*p* atomic orbitals increases from Li to F. Some representative values for this separation in kJ mol^{-1} are

B : 550 C : 849 N : 1197 O : 1595 F : 2077

Hence the extent of this *s–p* mixing or σ–σ interaction decreases among the later elements in the row. By means of photoelectron spectroscopy it has been estabilished that the order of energy level shown in Fig.7.27(b) persists upto N_2 (see box). In F_2 and O_2, the order shown in Fig 7.27(a) ignoring *s-p* mixing becomes operative.

Photoelectron spectroscopy

The principle of photoelectron spectroscopy was mentioned in connection with measurement of ionization energy (sec. 5.3.4). The photoelectron spectrum of a molecule consists of a plot of the rate of electron emission versus their kinetic energy (or *I.E.*). In ultraviolet photoelectron spectroscopy (UPS), the ionizing radiation is usually the 58·4 nm radiation from a helium discharge tube. The radiation arises from the transition $1s^12s^1 \rightarrow 1s^2$ giving photons with kinetic energy equal to 21·2 eV.

The different ionization energies of a molecule in its photoelectron spectrum may be related to the energy of the electron in different molecular orbitals. Here we accept **Koopman's theorem**: the observed ionization energy of an electron is equal to the negative of its orbital energy in the molecule, *assuming that* the remaining part of the molecule does not reorganize after ejecting the electron. In fact, the ions left after photoionization may slightly readjust the energy levels in the changed circumstances of interelectron and internuclear repulsion. However, the theorem holds pretty well in most cases and provides an experimental method for determining the energies of different molecular orbitals.

As an example, the ionization energy of the H_2 molecule is found to be 15.45 eV. We may assign this energy to the HOMO of the molecule, that is, the $1\sigma_g$ m.o. We also observe that this M.O. is stabilized with respect to the 1*s* A.O. on either H atom (ionization energy 13.60 eV).

The actual photoelectron spectrum of H_2 does not show a single absorption precisely at 15.45 eV. Instead the spectrum shows many evenly spaced lines at higher energies. This is because photoionization of one σ_g electron results in a vibrationally excited H_2^+ ion which has a longer bond length at equilibrium. The additional absorption lines at higher energies correspond to extra energies absorbed to excite vibrations of the molecule to attain the equilibrium bond length. The spacing between the lines correspond to the energy equivalent to H—H stretching frequency in H_2^+. We expect similar multiplets when a molecule ejects an electron from an antibonding m.o; the resulting ion now vibrates to the equilibrium internuclear distance which is shorter than that in the unionized molecule. On the other hand, if an electron is removed from a nonbonding M.O., or an M.O. which is less involved in bonding, we do not expect any major change in bond order or bond length in the resulting ion. Hence there will be little or no excited vibrations to reach the equilibrium bond length and no fine structure. The presence or absence of multiplet fine structure in the photoelectron spectrum thus provides a clue regarding the nature of M.O. involved.

For molecules having several occupied MOs, the photoelectron spectra consist of various groups of lines or *bands*. Each band corresponds to removal of electrons from a different valence shell M.O. The lowest energy ionization energy corresponds to removal of electrons from the highest occupied M.O. The nature of the photoelectron spectra of N_2 and O_2 are shown in Fig. 7.29[(a) and (b)].

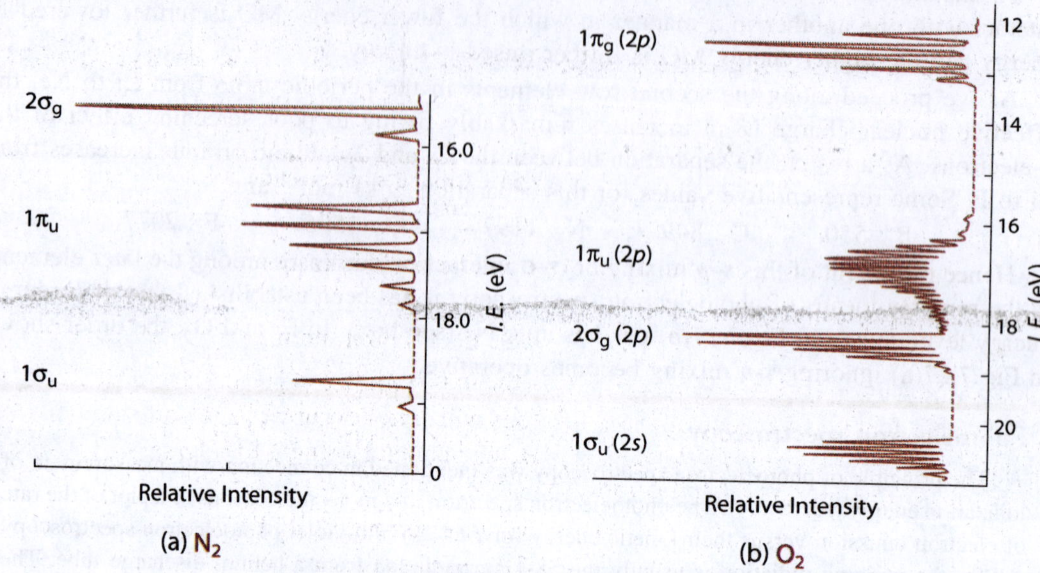

Fig.7.29: Sketch of photoelectron spectrum of (a) N_2 and (b) O_2 showing the gross nature and approximate position of bands. The relavant MOs of origin of the bands are also shown.

The lowest ionization energy band for N_2 occurs at 15.6 eV which corresponds to removal of an electron from the HOMO $2\sigma_g$ in the valence shell (Fig. 7.31c). The absence of significant fine structure shows that the level is less involved in bonding. We reached the same conclusion from consideration of σ-σ interaction (Fig. 7.27b). The spectrum shows a second band with several fine structure at 17 eV which results from the loss of an electron from the $1\pi_u$ M.O. A third band at 19 eV corresponds to loss of an electron from the $1\sigma_u$ antibonding M.O. The presence of fine structure in these two bands indicate that the N_2^+ ion is produced in a vibrationally excited state. Each line in a band relates to a different vibrational level of the appropriate electronic state of N_2^+. The separation between the lines in a band is approximately equal to $h\nu$ where ν is the vibrational frequency of N_2^+ in the corresponding electronic state.

The photoelectron spectrum of O_2 shows four bands at a wider separation compared to the bands in N_2. In O_2, the difference in *I.E.* between the first and the third band is about 6 eV (vs.3 eV with N_2). All the bands show multiplet structures, indicating strong involvement of the originating M.O. in bonding. The first two bands correspond to removal of an electron from the $1\pi_g$ M.O. and the $1\pi_u$ M.O. respectively (Fig. 7.32a).

The electron occupancy of the molecular orbitals of the homonuclear diatomic species Li_2 to Ne_2 are discussed below.

Li_2. Each lithium atom contributes one 2*s* electron, making a total number of two electrons. These can be accommodated in the 1σ m.o. (Fig. 7.30). The bond order will be 1. Such Li_2 molecules exist in the vapour state. Though the bonding is analogous to that in H_2, the bond in Li_2 is longer (270 pm) and weaker (105 kJ mol^{-1}) than in H_2 (74 pm; 432 kJ). This is because the 2*s* and 2*p* orbitals of lithium are larger in size and more diffuse than the 1*s* orbital of hydrogen. Their overlap is therefore poor and takes place at a greater internuclear distance than with hydrogen. Repulsion between the filled 1*s* shells is also likely to prevent closer approach of the two lithium nuclei.

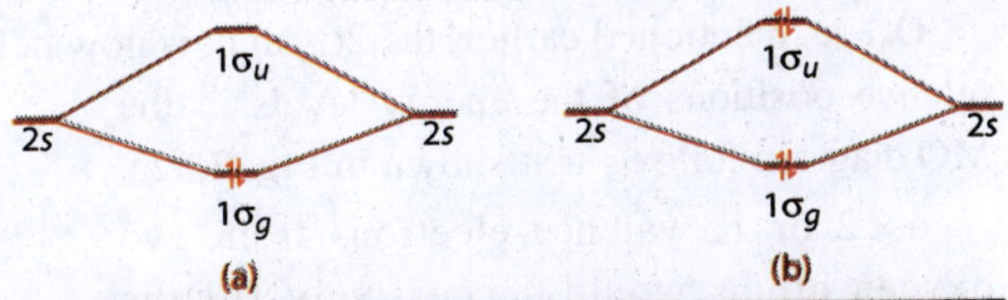

Fig.7.30: Ground state electron occupancy of the MOs in (a) Li_2 (b) Be_2 (only valence shell).

Be_2. Each beryllium atom would provide two 2*s* electrons. The total four electrons would then occupy the $1\sigma_g$ and $1\sigma_u$ m.o.'s (Fig. 7.30). The bond order would be ½ (2 – 2) = 0. So there will be no stabilization and the molecule is not likely to exist.

B_2. B_2 molecules exist in the vapour state. Each boron atom contributes three valence electrons; the total six electrons are placed in the m.o.-s as $1\sigma_g^2\, 1\sigma_u^2\, 1\pi_u^2$ (Fig. 7.31). Since the $1\pi_u$ level is doubly degenerate, the two electrons will *singly* occupy the two orbitals. The bond order will be ½ (4–2) = 1. So the molecule B_2 will contain a single bond essentially made up of two half π-bonds and be stable (289 kJ mol^{-1}).

Experimentally, the presence of two unpaired electrons in B_2 has been confirmed from magnetic studies. This shows that the order of the m.o.-s suggested here ($1\pi_u < 2\sigma_g$) is correct. If the $2\sigma_g$ m.o. were lower than the $1\sigma_u$ m.o., the two electrons would have been paired.

C_2. A total of eight valence electrons for the species C_2 may be arranged in its valence shell molecular orbitals $1\sigma_g^2\, 1\sigma_u^2\, 1\pi_u^4$. The bond order is ½ (6 – 2) = 2, equivalent to a double bond. This is consistent with the bond energy (630 kJ mol^{-1}) of C_2 occurring in flames and electric discharges through carbon-containing gases.

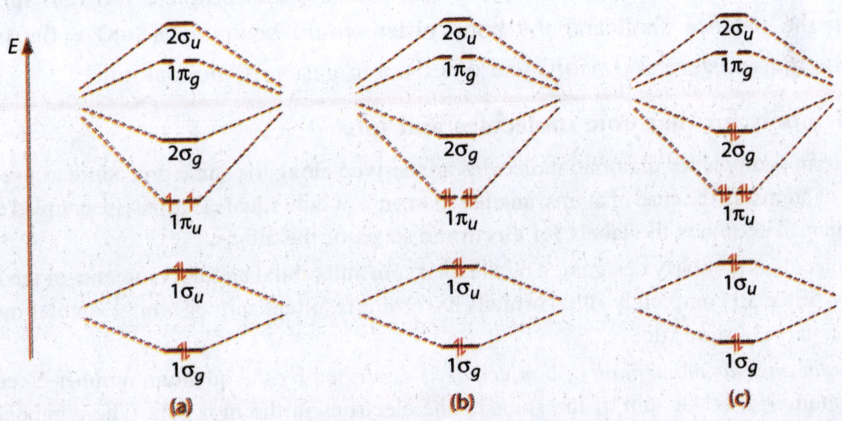

Fig.7.31: MO occupancy for (a) B_2, (b) C_2 and (c) N_2 in ground state (only valence shell). Not to scale.

N_2. Two nitrogen atoms contribute 5+5 = 10 valence electrons; these may be accomodated in the valence shell molecular orbitals of N_2 as $1\sigma_g^2\, 1\sigma_u^2\, 1\pi_u^4\, 2\sigma_g^2$.

The nature of the m.o.-s in N_2 have been discussed under photoelectron spectroscopy. The bond order is now [½ (8–2)] or 3, which indicates a strong stable bonding between the atoms (946 kJ mol^{-1}).

O_2. As mentioned earlier, the $2\sigma_g$ m.o. is now below the $1\pi_u$ m.o.'s (after nitrogen) and the relative positions of the energy levels in the MO diagram follows that shown in Fig. 7.32a.

6 × 2 or 12 valence electrons from two oxygen atoms would successively fill the molecular orbitals as shown in Fig. 7.32(a). The doubly degenerate $1\pi_g$ contains only two electrons; following Hund's rule, each orbital in the π^* m.o. should be singly occupied. The presence of two unpaired electrons in O_2 has been confirmed by its paramagnetic moment which supports the MO occupancy. The bond order becomes ½ (8–4) = 2, which is consistent with its bond energy (493 kJ mol^{-1}).

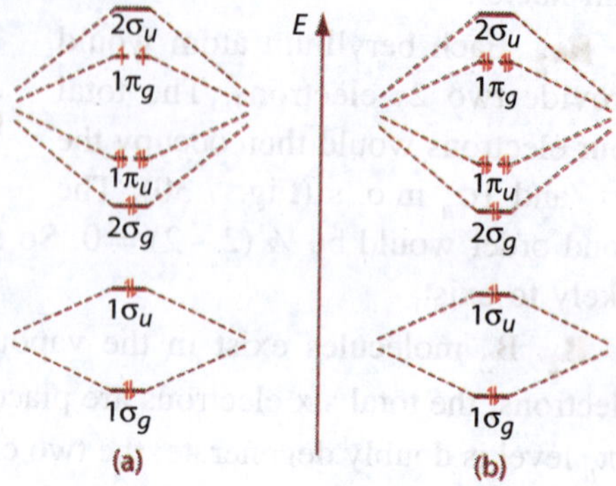

Fig.7.32: Ground state electron occupancy in the MO-s of (a) O_2 and (b) F_2 (only valence shell). Not to scale.

This was one of the early triumphs of the MO theory over the VB theory. A bond order of 2 is formally equivalent to a double bond but one need not depend solely on spin-pairing to explain bond formation.

F_2. The 14 valence electrons completely fill the molecular orbitals upto the $1\pi_g$ level and the bond order becomes ½(8–6) = 1. The bond energy of 158 kJ mol^{-1} is consistent with this bond order.

Ne_2. A total number of sixteen valence electrons would completely fill all the molecular orbitals in the valence shell and the bond order would become ½(8–8) = 0. As such, the molecule is non-existent and neon, like other noble gases, is monoatomic.

Term symbols for diatomic molecules and ions

The term symbols for diatomic molecules are derived along the same procedure as used for term symbols of atoms. In the case of atoms, angular momenta of individual electrons are coupled together—the resultant determines the labels for electronic states of the atoms.

A molecule is similarly designated by the total spin and orbital angular momenta of the individual electrons. Since all completely filled orbitals have zero resultant spin or orbital angular momentum, the task is largely simplified.

The spin angular momentum of a molecule is described by the quantum number Σ (counterpart of S for atoms); it is the sum of the spins of the electrons in the molecule. The component of the combined spin angular momenta along the direction of the internuclear axis is $\Sigma\hbar$. The spin multiplicity of a state is expressed by $2\Sigma + 1$. For a closed shell, $\Sigma = 0$, multiplicity is 1 and we call it a *singlet state*. Similarly, a single unpaired electron gives $\Sigma = ½$ and multiplicity of 2 (*a doublet*).

The component of the combined orbital angular momenta of all electrons along the direction of the internuclear axis is represented by $\Lambda\hbar$. The orbital angular momenta of all electrons couple to give the resultant angular momentum L. The component of L along the internuclear axis is the sum of the contributions by all electrons (in terms of $\lambda\hbar$).

$$M_L = \lambda_1 + \lambda_2 + \lambda_3 \cdots$$

$|M_L|$, the absolute value of M_L, is given the symbol Λ.

A cylindrically symmetrical σ orbital has no angular node when viewed along the internuclear axis. Hence σ electrons do not contribute anything to Λ ($\lambda = 0$). A π electron contributes one unit of angular momentum ($1\hbar$). When we have one electron in each of two degenerate π orbitals, one contributes +1 unit and the other −1 unit to the orbital angular momentum. If the π orbitals are taken as π_x and π_y, these contributions actually refer to the linear combinations ($\pi_x + i\,\pi_y$) (simply π^+) and ($\pi_x - i\,\pi_y$) (or π^-) respectively. In this case, $\Lambda = 0$. If both the electrons occupy the same π orbital (both has $\lambda = +1$), $\Lambda = 2$.

Finally, we write the term symbols of molecules in the same manner in which we wrote term symbols for atoms. The multiplicity is written as a left superscript and the value of Λ is transformed into symbols (as we did for L):

$\|\Lambda\|$	0	1	2	3
	Σ	Π	Δ	Φ

(The code Σ (upright) is used in the term symbol for $\Lambda = 0$. Σ (italics) is used to represent the component of total spin angular momentum along the direction of the internuclear axis.)

The term symbol for a molecule is then primarily represented as

$$^{2\Sigma+1}\Lambda_{\Omega}$$

where Ω (= $|\Lambda + \Sigma|$) is the quantum number for the component of *total* angular momentum (orbital plus spin) around the internuclear axis.

For homonuclear diatomic molecules or ions, right hand subscripts g or u are added in the term to indicate the symmetry of the wave function with respect to inversion at the centre (parity). A state with odd number of u orbitals is indicated by the subscript u while states with even number of u orbitals are designated by g. Representing g by +1 and u by −1, we may write

$$g \times g = g \qquad u \times u = g \qquad u \times g = u$$

For Σ terms, the symmetry of the wave function with respect to reflection in a plane passing through the nuclei is shown by a + or − sign as right superscript. As we shall see just now, the ground state of H_2 is $^1\Sigma_g^+$; the same term applies to N_2 also.

But the ground state of O_2 is $^3\Sigma_g^-$. We assume that one electron in the HOMO ($1\pi_g$) of O_2 is in $1\pi_g(x)$ and the other electron in $1\pi_g(y)$. Now the $1\pi_g(x)$ orbital changes sign under reflection from the yz plane while the electron in $1\pi_g(y)$ does not change sign under reflection from the same plane. The overall reflection symmetry of the O_2 molecule is thus

$$\text{(closed shell)} \times (+) \times (-) = (+) \times (+) \times (-) = (-)$$

Problem 7.3

Find the term symbols for the ground and the first excited state of H_2.

Ans.: The ground state electron configuration of H_2 is $1\sigma_g^2$. σ electrons do not contribute any orbital angular momentum. Hence $\Lambda = 0$. Since the electrons must be paired, Σ is also zero. The multiplicity is $2 \times 0 + 1 = 1$. We thus arrive at the symbol $^1\Sigma$. Since the $1\sigma_g$ orbital remains unchanged on inversion at the centre of the molecule, we shall add g as a right subscript. Similarly, the $1\sigma_g$

orbital does not change on reflection in a plane containing the molecular axis, and we add '+' sign as a right superscript. The complete term symbol is thus $^1\Sigma_g^+$.

Promotion of one electron to $1\sigma_u$ generates the configuration $1\sigma_g^1, 1\sigma_u^1$. Λ is still zero. Since the electrons are in different orbitals, their spins may be parallel or antiparallel. For spins parallel, $\Sigma = 1$. For antiparallel spins, $\Sigma = 0$.

Since one orbital is g and the other u, we have $g \times u = u$. Both the orbitals are unchanged on reflection in a plane containing the internuclear axis, hence the term will be '+'. Thus we get two terms: $^1\Sigma_u^+$ and $^3\Sigma_u^+$.

Problem 7.4

Derive the terms for the ground and excited states of O_2.

Hint.: The valence shell M.O. has the electron distribution $1\sigma_g^2\, 1\sigma_u^2\, 2\sigma_g^2\, 1\pi_u^4\, 1\pi_g^2$. We need consider only the highest occupied $1\pi_g$ M.O. (I); the lower filled M.O.-s contribute zero to either Λ or Σ.

The $1\pi_g$ M.O. contains two electrons in two degenerate orbitals. So, $\Sigma = 1$, multiplicity = 3. As the contributions to orbital angular momentum by the π_x and π_y orbitals (say) are +1 and –1, $\Lambda = 0$, corresponding to a Σ term. There are no u electrons. So the total wave function will be g. As explained above, the wave functions are antisymmetric with respect to reflection. We show this by including a '–' sign. The complete term symbol for O_2 is thus $^3\Sigma_g^-$.

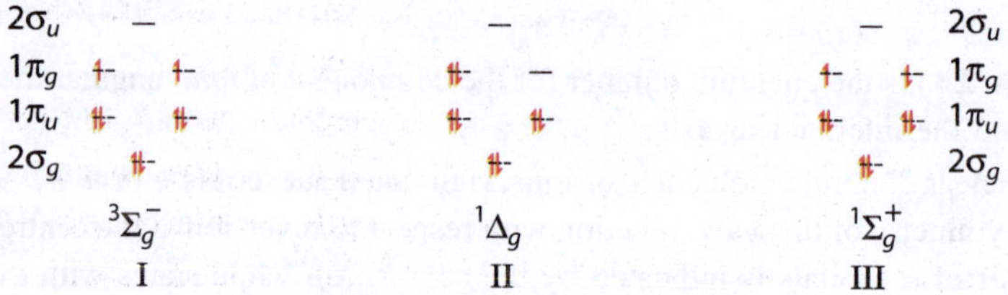

The[illegible] tions for the excited states as shown (II and III). In II, the electrons [illegible] their spins paired, hen[illegible] 0 (a singlet) The [illegible] nta of the electrons now add giving [illegible] m. The term [illegible] a Σ term; hence we need [illegible] ury. The s[illegible] dure gives $\Sigma = 0$ and $\Lambda = 0$ for the excited state shown in III. The term symbol is $^1\Sigma_g^+$.

The two excited states of O_2 discussed above are summarily called *singlet* O_2 (see Group 16). The $^1\Delta_g$ state is 92 kJ mol^{-1} above the ground state while the $^1\Sigma_g^+$ state is 155 kJ mol^{-1} above the ground state. Singlet O_2 has many interesting applications in chemistry, as we shall see later.

Frontier orbitals

The valence shell electron occupancy of a molecule provides useful information regarding the spectral and chemical behaviour of the species. Thus, the response of a molecule to electron loss in Lewis base and redox behaviour is concerned with the energy and population of the ***highest occupied molecular orbital*** (abbreviated ***HOMO***) and also that of the resulting ion. Similarly, Lewis acid and redox behaviour related to electron gain is associated with the nature of the ***lowest unoccupied molecular orbital*** (***LUMO***). Again, redox behaviour of the free radical type is likely to involve a ***singly occupied molecular orbital*** (***SOMO***), as is found in the ground state of dioxygen molecule (two unpaired electrons in $1\pi_g$). All these orbitals, HOMO, LUMO and SOMO are collectively known as ***frontier orbitals*** which play significant role in the chemistry of a species. We shall get several examples to this effect in subsequent sections.

The ultraviolet and visible spectra of a molecule or ion are similarly dependent on the energies of the HOMO and LUMO. When a molecule absorbs a photon, the lowest energy absorption occurs for the promotion of an electron from the HOMO to the LUMO. Such transitions may occur in the visible or UV region of the spectrum as we shall observe in the colour of diatomic halogen molecules. The intensity of the transition (and hence of the colour) depends on the symmetry properties of the MO from which an electron is being excited (in the ground state) to that of the MO being occupied in the excited state. The colours of the halogen molecules will again serve as an illustration of this (Group 17).

Ions derived from homonuclear diatomic molecules

The electron distribution in the valence shell of an ion derived from the diatomic molecules by addition or removal of electron(s) may be obtained from the electron occupancy of the parent molecule. We assume that there is no major reorganization of the energy levels in the resulting ion, which holds in most cases. Loss of an electron occurs from the most unstable energy level of the molecule or HOMO. Similarly an electron is added to the lowest available energy level or LUMO. The corresponding changes in bond order, bond length and bond dissociation energy are in fair agreement with this description.

As an example, we consider the introduction of an electron into O_2 to form O_2^-. The electron would enter one of the singly occupied $1\pi_g$ M.O.s. Since this level is antibonding we expect a decrease in bond order and an increase in bond length. Similarly, when an electron is removed from O_2 to form O_2^+, an antibonding MO ($1\pi_g$) is vacated, resulting in an increase in bond order and decrease in bond length. The following data are illustrative:

	O_2	O_2^+	O_2^-	N_2	N_2^+	N_2^-
Bond energy (kJ mol^{-1})	498	643	395	946	841	765
Bond length (pm)	121	113	128	110	112	119

Problem 7.5

Calculate the bond orders of the ionic species mentioned above.

However, one should remember that the observed changes in bond length and bond energy on ionization or electron attachment may be well influenced by other factors like change in inter-electron repulsion, environment of the ionic species and so on.

7.3.3 Heteronuclear Diatomic Molecules

The procedure for constructing M.O.s of homonuclear diatomic molecules may be extended to heteronuclear diatomic molecules with due consideration for the conditions necessary for linear combination of atomic orbitals like the criterion of matching symmetry, comparable energy and appreciable overlap. The molecular orbitals of heteronuclear diatomics will differ from those of homonuclear species because of unequal contributions from the participating atomic orbitals. In the linear combination of two unlike atomic orbital wave functions, say ψ_A and ψ_B (of elements A and B respectively), summarily represented as $\Psi_{MO} = c_1 \psi_A + c_2 \psi_B$, the coefficients c_1 and c_2 will be unequal. In general, the atomic orbital with lower energy, i.e., one contributed to by the more electronegative element, will have the larger coefficient in the bonding M.O. and a smaller coefficient in the antibonding M.O. This will place the bonding electrons close to the more electronegative atom. The antibonding M.O. will be naturally close to the A.O. from the less electronegative atom.

Before we proceed to construct the MO energylevel diagram for a heteronuclear species, it will be advantageous to look at some of the valence orbital energies of the first few elements in the periodic table. The valence orbital ionization potentials (VOIP; sec. 5.3.4) are obtained from spectroscopic states. The orbital energy is the negative of VOIP in eV (Table 7.4).

Table 7.4

Potential energies of valence orbitals of the elements H to Ar (eV)

	H							**He**
1*s*	–13.6							–24.59
	Li	**Be**	**B**	**C**	**N**	**O**	**F**	**Ne**
2*s*	–5.39	–9.32	–14.05	–19.43	–25.56	–32.38	–40.17	–48.47
2*p*			–8.30	–10.66	–13.18	–15.85	–18.65	–21.59
	Na	**Mg**	**Al**	**Si**	**P**	**S**	**Cl**	**Ar**
3*s*	–5.14	–7.65	–11.32	–15.89	–18.84	–22.71	–25.23	–29.24
3*p*			–5.98	–7.78	–9.65	–11.62	–13.67	–15.82

The molecular orbital occupancy of a few heteronuclear diatomic molecules and ions is discussed below.

HF. By convention we choose the molecular axis along the *z*-axis. We consider the following valence shell atomic orbitals with their orbital energies in eV:

$$\textbf{H}\ 1s(\text{H}); -13.6 \quad ; \quad \textbf{F}\ 2s(\text{F}); -40.2\ \ 2p_x(\text{F}), 2p_y(\text{F})\ 2p_z(\text{F}); -18.6$$

Clearly, the core 1*s* A.O. on fluorine (<< –40.2 eV) is excluded on the ground of extreme mismatch in energy with the 1*s*(H) A.O. The 1*s*(F) A.O. will appear as a nonbonding M.O. with unchanged energy and is omitted from our diagram. One may include this as a nonbonding 1σ M.O. when the serial numbers of the successive MOs will change accordingly.

The 2*s* AO of fluorine is much lower in energy (–40.2 eV) compared to the 1*s* AO of hydrogen. So this is also expected to be mainly nonbonding in character localized near fluorine. However, it may enter some small degree of hybridization with other MOs formed.

The three 2*p* AOs on fluorine are energetically comparable with the 1*s*(H) AO. When we consider the signs of the wave functions in the lobes of *p*-A.O.-s and their orientation with respect to the molecular axis, we readily conclude that the $2p_x$(F) and $2p_y$(F) A.O.s will have no net overlap with the 1*s*(H) AO. These fluorine AO-s have $|m| = 1$ and one nodal plane containing the internuclear axis. Hydrogen has no valence shell AO with | m | = 1. Hence $2p_x$(F) and $2p_y$(F) AOs will give rise to nonbonding π MOs (π^{nb}) with unchanged energy.

The $2p_z$ AO on fluorine has the right symmetry to combine with the 1s(H) AO. Both have $m = 0$ and no nodal plane containing the internuclear axis. So we expect two σ MOs from linear combination of these AOs:

$$\sigma = c_1 1s(\text{H}) + c_2 2p_z(\text{F}) \qquad \sigma^* = c_1{}'1s(\text{H}) - c_2{}'2p_z(\text{F})$$

The σ MO results in buildup of electron probability density ("charge") between the two nuclei and is bonding. c_2 is substantially greater than c_1, placing the bonding MO closer in energy to the fluorine AO. The σ^* MO causes depletion of charge between the nuclei and is antibonding. Here c_1' is greater than c_2' and the antibonding MO is closer to 1s(H) AO.

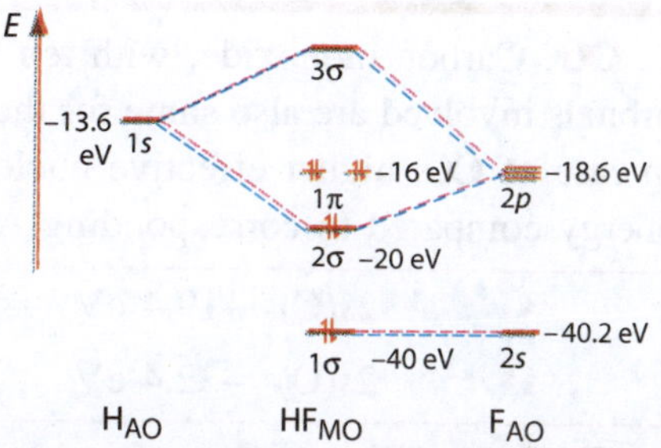

Fig.7.33: MO energy level diagram of HF (not to scale). The energy of the antibonding 3σ MO (LUMO) is uncertain.

The resulting energylevel diagram is shown in Fig. 7.33. We assign the eight available valence electrons from the lowest energy MO and obtain the configuration $1\sigma^2\, 2\sigma^2 1\pi^4$. The HOMO is the 1$\pi$ level, the LUMO is the antibonding 3σ MO. Of the three occupied MOs, only the 2σ MO is bonding and we have a bond order 1, equivalent to a single bond in the electron dot representation or valence bond description. The other six electrons remain as three nonbonding pairs on the fluorine atom. Summarily, all the occupied energy levels remain predominantly on the fluorine atom and we expect the HF molecule to be polar with a partial negative charge on the fluorine atom.

In the notation for heteronuclear diatomic molecules, the g, u designations do not apply. The lowest σ MO is called the 1σ MO, the next lowest σ MO is the 2σ MO and so on. Similarly the lowest π energy level is called the 1π MO.

The MO electron distribution in HF agrees well with its photoelectron spectrum which shows a sharp band at –16.0 eV without any vibrational fine structure. This we can readily assign to the nonbonding 1π MO (HOMO). The next band appears at –20 eV which is broadened due to fine structure and corresponds to the bonding 2σ MO. A band at – 40 eV may be assigned to the nonbonding 1σ MO which is very close to the 2s(F) AO at – 40.2 eV.

The energy of the LUMO (3σ) is not available from photoelectron spectra. Computational estimates vary widely with values as high as +6.0 eV.

Problem 7.6

The MO energy level diagram for HCl can be arrived at by procedure similar to that applied for HF. Construct the diagram and assign the valence electrons to the MOs.

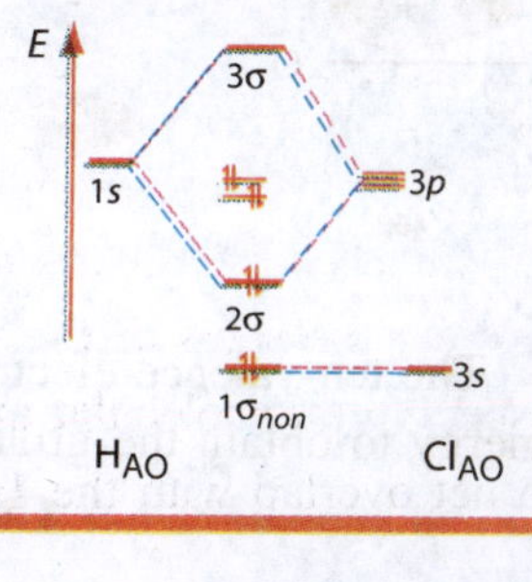

Hint: Consider the orbital energies of Cl (eV), 3s : –25.2 ; 3p : (–13.7). For the sake of simplicity one may consider chlorine 3s as practically nonbonding. Hydrogen 1s and chlorine 3p_z will produce one σ and one σ^* MO (z-axis as molecular axis). 3p_x and 3p_y remains nonbonding.

Problem 7.7

The photoelectron spectrum of HCl shows a sharp peak at approximately 12.8 eV followed by a broad band at 16.5 eV (approx). Comment.

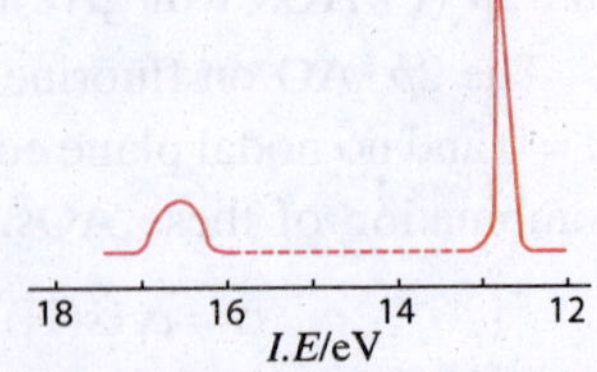

Hint: The HOMO is a nonbonding π MO (1π). The next higher in energy is the bonding 2σ M.O.

CO. Carbon monoxide, with ten valence electrons, is isoelectronic with N_2. The atomic orbitals involved are also same for the two molecules—2*s*, $2p_x$, $2p_y$ and $2p_z$ on each atom. But in case of CO, higher effective nuclear charge of oxygen places its atomic orbitals at lower energy compared to corresponding AO-s on carbon.

C :	2*s*(C) −19.4 eV	$2p_x$, $2p_y$, $2p_z$ (C) −10.7 eV
O :	2*s*(O) −32.4 eV	$2p_x$, $2p_y$, $2p_z$ (O) −15.9 eV

The large difference in energy between 2*s*(C) and 2*s*(O) is noteworthy. Taking *z*-axis as the internuclear axis, we may construct the MOs as in the case of N_2. The same sequence of MOs do occur but the *g* and *u* labels do not apply to the heteronuclear molecule. In this case also, possibility exists for σ–σ interaction or *s*–*p* hybridization of the four σ MOs resulting from the four atomic orbitals 2*s*(C), 2*s*(O), $2p_z$(C) and $2p_z$(O). The lowest energy σ MO, 1σ (we consider only the valence shell), is closer to the lowest energy 2*s*(O) AO on oxygen which it closely resembles (Fig. 7.35). Similarly, the highest energy antibonding σ MO, 4σ, is located near the highest energy parent AO, $2p_z$(C), and is largely influenced by this AO. Between these MOs lie (i) the antibonding 2σ MO, formally from 2*s*(C) and 2*s*(O) with some *sp* mixing; and (ii) the "bonding" 3σ MO, formally from the $2p_z$(C) and $2p_z$(O), with *sp* mixing. As is to be expected from the energy of the contributing AOs, these MOs have reduced overlap between the carbon and oxygen atoms and practically tend to be nonbonding. A similar situation was encountered with N_2.

Besides these σ MOs, there is a pair of bonding π MO (1π) and a pair of antibonding π MO (2π) formally arising from the $2p_x$ and $2p_y$ AOs on the two atoms. These π MOs do not undergo any hybridization. The bonding π MOs are closer in energy to the $2p_x$ and $2p_y$ AOs on oxygen and have larger contribution from these AOs. Similarly the antibonding π MOs are closer in energy to the carbon $2p_x$ and $2p_y$ AOs which make greater contribution to these MOs. (Fig. 7.34).

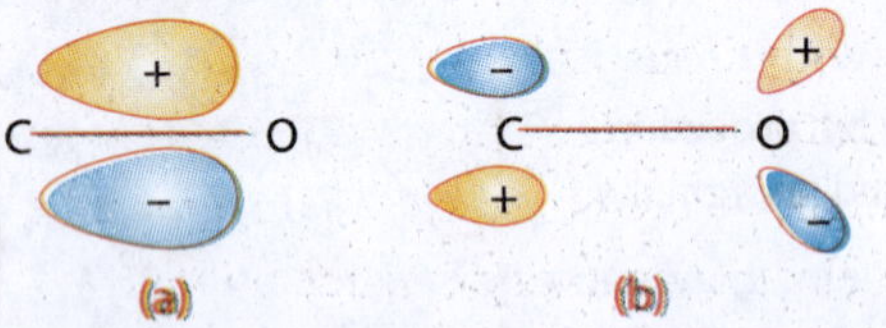

Fig.7.34: Diagrammatic sketches of (a) one bonding π MO (1π) and (b) one antibonding π MO (2π) in CO. In (a), the wave function has higher amplitude around O while in (b), the wave function has higher amplitude around C. The two sets of π MOs merge to a cylindrical distribution around the C—O axis.

The ten valence electrons may be assigned to these eight MOs in the increasing order of energy to obtain the ground state valence shell MO configuration (Fig. 7.35):

CO: $1\sigma^2 2\sigma^2 1\pi^4 3\sigma^2$; Bond order = ½ (8–2) = 3.

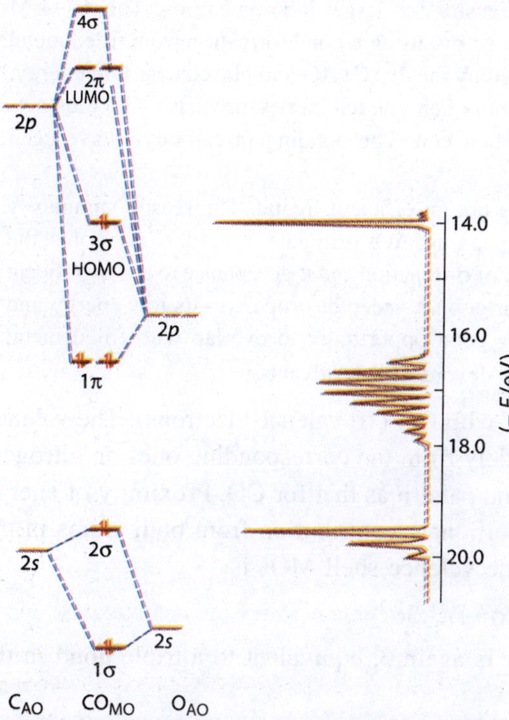

Fig.7.35: Valence shell MO energy levels in CO with sketch of photoelectron spectrum.

In the formal counting of bond order we have taken the 2σ MO as antibonding and the 3σ MO as bonding. This leaves two σ electrons and four π electrons for bonding—equivalent to one σ bond and two π bonds in the valence bond description.

However, we have noted earlier that the 2σ and 3σ MOs are largely nonbonding in character. The 3σ MO forms the HOMO in the molecule; its nonbonding character is revealed in the sharp peak in photoelectron emission spectrum around 14.0 eV. There is another sharp band around 19.5 eV corresponding to the 2σ MO. The lack of fine structure in both cases supports the nonbonding role of the electrons in these MOs. Emission of electrons from the intermediate 1π MO (bonding) is marked by a band with multiple fine structure centered at 17.0 eV.

The LUMO in CO is the antibonding 2π MO with greater participation from the carbon 2*p* AOs. We shall soon find that this plays an important role in the chemistry of CO, particularly in the metal carbonyls.

Dipole moment of CO

The anomalous dipole moment of CO was mentioned in chapter 6. The experimental electric dipole moment of CO (0.1 D) is apparently quite small in view of the large electronegativity difference between C and O. At the same time, the negative end of the dipole lies on the less electronegative carbon atom. The composition and occupancy of the MOs are mainly responsible for this apparent anomaly. The HOMO 3σ is actually largely nonbonding; it is derived mainly from the $2p_z$(O) and $2p_z$(C) AOs. The low-lying $2p_z$(O) atomic orbital also contributes to the antibonding 2σ and 4σ MOs. The higher lying $2p_z$(C) AO, on the other hand, contributes mainly to the 3σ and 4σ MOs. Consequently, the contribution of $2p_z$(O) is distributed among three MOs; as such, the 3σ and 4σ MOs get smaller contributions from this AO. The $2p_z$(C) AO contributes only to the 3σ (HOMO) and the antibonding 4σ MOs. So its contribution to these MOs is stronger. In fact, the HOMO in CO has a higher electron density and larger lobe on carbon and not on oxygen as expected from simple electronegativity considerations.

The lone pair electrons on carbon thus appear to dominate over the bond pair electrons (localized on O) and give rise to the apparent anomaly in dipole moment.

The ligand characteristics of CO toward transition metal atoms may be understood from the MO description. While coordinating to a vacant acceptor orbital on the metal, CO donates an electron pair from its HOMO which has a higher electron density and larger lobe on carbon. This C → M σ donation is not very strong as such but it is reinforced by strong π bond formation from filled metal orbitals to the vacant LUMO on CO, again dominated by the $2p_z$(C) AO and placed near it in energy. The σ donor and π acceptor behaviour mutually strengthen one another (synergism). CO thus acts as a good σ donor and π acceptor through its carbon end. The bonding in carbonyls have been discussed later under organometallic compounds.

Dinitrogen (N_2) is isoelectronic with CO but is not an efficient ligand. The reason summarily rests on the nonpolar character of N_2. Since the participating AOs originate from identical atoms, the resulting MO energy levels do not offer the advantage of σ-donation and π-acceptance to any significant extent. The LUMO in CO is largely responsible for good π acceptor property—its low energy and high amplitude on its carbon donor atom provides good opportunity to overlap with filled metal *d*-orbitals. The symmetrical N_2 molecule is clearly devoid of this advantage.

CN^-. The cyanide ion (CN^-) is isoelectronic with CO (10 valence electrons). The valence shell atomic orbitals on carbon do not differ widely from the corresponding ones on nitrogen. The MO energy level diagram will have the same pattern as that for CO. Proximity of energy between participating AOs suggests more proportionate contribution from both atoms in the resulting MOs. The electron configuration of the valence shell MOs is

$$CN^- : \; 1\sigma^2\, 2\sigma^2\, 1\pi^4\, 3\sigma^2$$

The 2σ MO is antibonding. The bond order is again 3, equivalent to a triple bond in the valence bond description.

As in CO, CN^- is also expected to be a good ligand donating primarily through its HOMO localized near the carbon atom. The negative charge on the ion will favour σ donation but at the same time it will reduce the efficiency as a π acceptor. The lower electronegativity of nitrogen compared to oxygen will also make CN^- a poorer π acid. The coordination chemistry of CN^- has been discussed later.

NO. Nitric oxide has eleven valence electrons, one electron more than CO, or one electron less than O_2. The pattern of overlap and the nature of molecular orbitals are largely similar to those in CO. The separation between the 2*s* and 2*p* AOs in oxygen is large enough (~ 16 eV) to reduce σ–σ interaction. So the order of MOs follows the same sequence as in the case of O_2 (Fig. 7.36). We get one bonding σ MO (1σ) with pronounced oxygen character and one σ antibonding MO (2σ) with pronounced nitrogen character. This is followed by the bonding 3σ MO resulting from the $2p_z$ AOs on both atoms (*z*-axis taken as internuclear axis). This MO has predominantly oxygen character. A pair of bonding π MOs (1π) and a pair of antibonding π MOs (2π) are placed in succession. The latter forms the HOMO with one unpaired electron. As expected, the bonding 1π MOs have greater oxygen participation while the antibon-

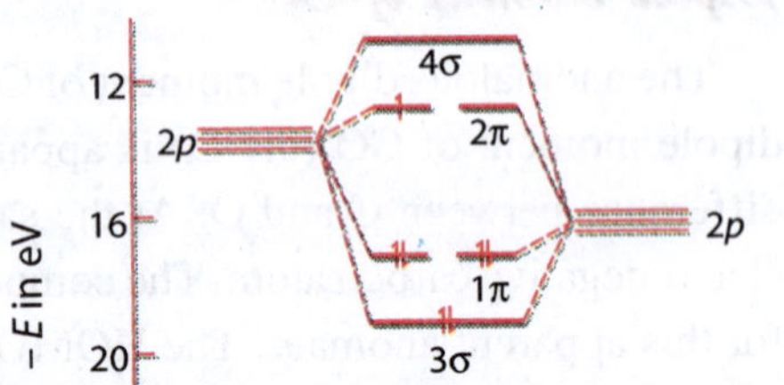

Fig.7.36: Approximate MO energy level diagram for NO.

ding 2π MO is localized near the nitrogen $2p$ AOs and gets greater participation from nitrogen. On the top lies the antibonding 4σ MO, again with pronounced participation from nitrogen. With eleven valence electrons, we have the ground state configuration for the valence shell MOs as

NO : $1\sigma^2\ 2\sigma^2\ 3\sigma^2\ 1\pi^4\ 2\pi^1$. Bond order = ½(8–3) = 2.5

The 2σ and 2π MOs are antibonding. The antibonding 2π MO is the HOMO as well as the LUMO.

The experimental bond dissociation energy of NO is 626 kJ mol^{-1} with a bond length of 115 pm. Removal of the odd electron results in NO^+ which has a bond order of 3. This is consistent with the increase in bond dissociation energy to 1047 kJ mol^{-1} and a smaller bond length (106 pm).

The presence of the unpaired electron in the antibonding 2π MO makes NO a reactive radical. It has several important biological functions.

7.3.4 Polyatomic Molecules

The molecular orbital wave functions for polyatomic molecules can be constructed from the participating atomic orbitals with due recognition for symmetry and energy. The AOs which actually contribute to a given MO is determined by the symmetry properties of the involved AOs and the resulting MO. We remember that the MOs of a diatomic molecule were classified as σ, π, δ, ... etc. types according as they possessed respectively 0, 1, 2, ... nodal planes containing the internuclear axis. Atomic orbitals having no such nodal planes contributed to a σ MO; AOs with one nodal plane combined to give a π MO and so on.

We may apply the LCAO approach to polyatomic molecules to obtain the overall Ψ_{MO} from the participating atomic orbitals ψ_i as

$$\Psi_{MO} = \sum_i c_i \psi_i$$

For diatomic molecules, things were easier since they are always linear. With polyatomic molecules, we should also consider the most favourable geometry for combination of the AOs. One may start with an approximate geometry consistent with VSEPR theory and common hybridization schemes closest to observed bond angles. The favourable geometry of a small molecule may also be approached with the *Walsh diagrams* which show how the energies of molecular orbitals vary with geometry. Though the total energy of the molecule is different from the energy of the MOs, the prediction of shapes of small molecules is often satisfactory (see later). Various software packages are now available for deciding on the geometry for most favourable energy. We begin our discussion with two typical triatomic molecules—one linear, BeH_2 and one bent, H_2O. The presence of hydrogen as peripheral atoms makes our task further easier.

BeH_2. The valence orbitals of Be are $2s$, $2p_x$, $2p_y$ and $2p_z$.

The valence orbitals of the two H atoms are $1s_A$ and $1s_B$ (say).

Actually, the *minimal basis set of AOs*, or the AOs to be considered for any calculation of MO, should include all AOs in the inner shell and the valence shell. Accordingly, the $1s$ AO of Be also needs inclusion. However, our purpose is limited to qualitative understanding of the nature of MOs and their relative energy sequence. For this purpose, we may, as in the case of diatomic molecules, neglect the inner core AO. In fact, it will appear with almost unchanged energy from the parent AO. But its inclusion will modify the serial number of the σ MOs.

We may reasonably assume the molecule BeH_2 in the linear and symmetrical arrangement HBeH which will give minimum repulsion between the bond pairs of electrons. Accurate theoretical calculations support this arrangement.

Qualitative prediction of molecular geometries of AH_2 type molecules is also provided by *Walsh's rules*: an AH_2 molecule (or ion) will be linear if it contains three or four valence electrons (BeH_2 has four); the molecule will be bent if it contains one, two or five to eight valence electrons. These arrangements conform to minimum total electronic energy of the molecule (or ion) as the valence electrons occupy the MOs in the increasing order of energy. Thus we predict that LiH_2^+ (2 valence electrons) will be bent; BH_2 (5 valence electrons) will be bent but BH_2^+ will be linear (see Walsh diagram).

In the symmetrical HBeH arrangement, the BeH_2 molecule has a centre of symmetry and the labels *g* or *u* will be applicable to the MOs. We assume *z*-axis to be the internuclear axis.

The 2*s* AO on Be has no nodal plane containing the internuclear axis and has the same sign of wave function at diagonally opposite points of the Be nucleus (Fig. 7.37). So it has σ_g symmetry. The $2p_z$(Be) AO has also σ symmetry but it has opposite signs of the function on two sides. Hence it has σ_u symmetry. The $2p_x$(Be) and $2p_y$(Be) AOs have π_u symmetry.

Both 2*s* and 2*p* AOs on Be and the 1*s* AOs on the two H atoms, say $1s(H_A)$ and $1s(H_B)$ have similar energies. The combinations $1s(H_A) \pm 1s(H_B)$, have no nodal planes containing the internuclear axis; they can participate in forming σ MOs. The combination $1s(H_A) + 1s(H_B)$ is *g*-type and the combination $1s(H_A) - 1s(H_B)$ is *u* type. (Such combinations of atomic orbitals are called *symmetry-adapted linear combination* SALC. SALCs are very useful in constructing MO wave functions in polyatomic molecules).

Thus the basis set functions for BeH_2 (in addition to 1*s*(Be)) are

2*s*(Be)	$2p_z$(Be)	$2p_x$, $2p_y$(Be)	$1s(H_A) + 1s(H_B)$	$1s(H_A) - 1s(H_B)$
σ_g	σ_u	π_u	σ_g	σ_u

The lowest energy σ_g MO, $1\sigma_g$, is obtained by combining the lowest energy 2*s*(Be) AO with $1s(H_A) + 1s(H_B)$.

$$1\sigma_g = c_1\,2s(\text{Be}) + c_2\,[(1s(H_A) + 1s(H_B)].$$

The corresponding antibonding MO is obtained as

$$2\sigma_g = c_3 2s(\text{Be}) - c_4[(1s(H_A) + 1s(H_B)].$$

The $2p_z$(Be) AO combines with $1s[(H_A) - 1s(H_B)]$ to form a bonding and an antibonding MO. These are *u* type as seen from the signs on the lobes on two sides of Be:

$$1\sigma_u = c_5 2p_z(\text{Be}) + c_6[(1s(H_A) - 1s(H_B)]$$

$$2\sigma_u = c_7 2p_z(\text{Be}) - c_8[(1s(H_A) - 1s(H_B)]$$

The $2p_x$ and $2p_y$ AOs on Be (both π_u) form two nonbonding π_u MOs which have the same energy as the original AOs and lie above the bonding σ MOs ($1\sigma_g$ and $1\sigma_u$; the antibonding $2\sigma_g$ and $2\sigma_u$ MOs will be of higher energy).

Sketches for the above combinations and the corresponding MO energy level diagram are shown in Fig. 7.37. The four valence electrons may be assigned to the $1\sigma_g$ and $1\sigma_u$ bonding MOs, implying two single bonds to two H atoms. The ground state MO configuration is thus $1\sigma_g^2\ 1\sigma_u^2$. Though it is convenient to "*visualize*" the BeH_2 molecule as having two localized Be—H bonds, the sketches for overlap suggest that the functions are actually delocalized over the entire molecule. In each bonding MO, the electron probability density (ψ^2) is large and continuous between adjacent atoms but the wave functions spread over the entire molecule and is shared by all three atoms, not merely by any adjacent pair. The $1\sigma_u$ bonding MO has a node passing through the centre of the molecule. The antibonding MOs contain nodes *between* adjacent pairs of nuclei. In both bonding and antibonding MOs, number of nodes increases as we move to higher energy.

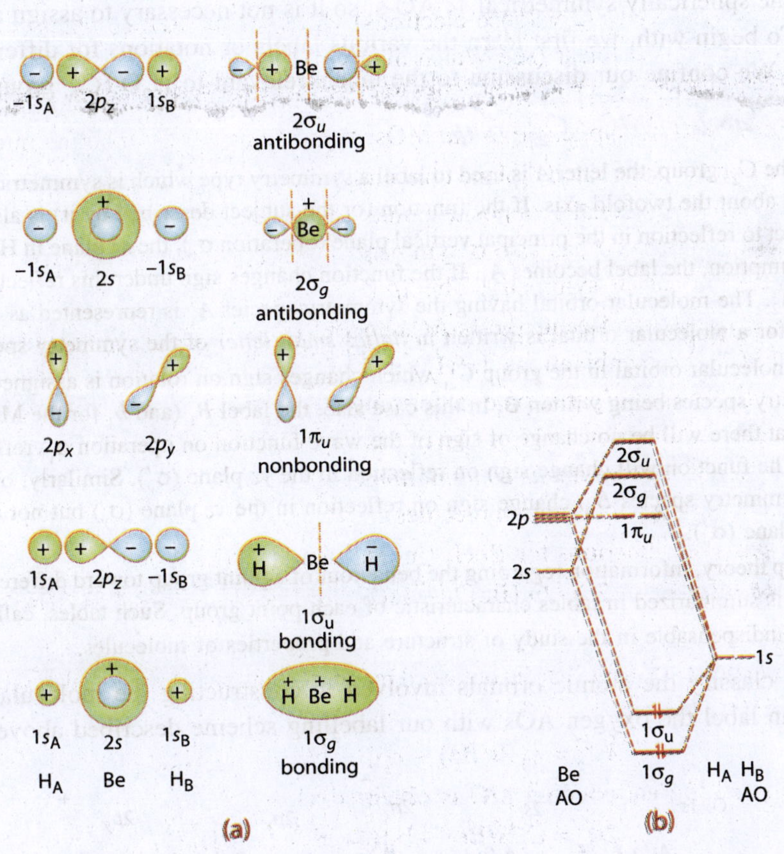

Fig.7.37: (a) Sketch for formation of MOs in BeH_2 (b) MO energy level diagram (not to scale). As explained before, σ_g and σ_u MOs are assigned separate serial numbers. Inclusion of the σ_g MO from Be 1*s* A.O. (not shown) will modify the serial numbers of σ_g MOs.

H_2O. The molecule H_2O is a bent triatomic molecule with oxygen at the centre. The atomic orbitals involved are 1*s* AOs of two hydrogens and the 2*s* and 2*p* AOs of oxygen. The 1*s* AO of oxygen is also a constituent of the minimal basis set but for our purpose of drawing the qualitative energy level diagram, it may be omitted.

The orbital energies of the valence electrons are as follows.

$1s(\text{H}) : -13.6$ eV ; $2s(\text{O}) : -32.4$ eV ; $2p_x, 2p_y, 2p_z(\text{O}) : -15.9$ eV.

The large difference between $1s$(H) and $2s$(O) is noteworthy. The $2p$ orbitals of oxygen are quite close to the $1s$ orbital of hydrogen.

Next we have to consider the symmetry criterion of the orbitals for the formation of MOs. H_2O belongs to the C_{2v} point group (Ch. 6) having a two fold rotational axis of symmetry and two perpendicular mirror planes. *Conventionally, the highest order rotation axis of a molecule is chosen as the z-axis of the molecule.* Accordingly, we choose the C_2 axis as the z-axis and xz plane as the plane of the H atoms (the molecular plane). One may also consider the yz plane as the molecular plane. In that case the b_1 and b_2 labels of the molecular orbitals will be interchanged. The meaning of these labels is discussed below. The hydrogen atoms have only the spherically symmetrical $1s$ AO-s; so it is not necessary to assign any axis to the H atoms. To begin with, we first learn the various labels or notations for different symmetry operations. We confine our discussion to the labels relevant to H_2O (C_{2v} group).

Symmetry labels of MOs: C_{2v} group

(i) In the C_{2v} group, the letter A is used to label a symmetry type which is symmetric with respect to rotation about the twofold axis. If the function (or the subject described by it) is also symmetric with respect to reflection in the principal vertical plane (operation σ_v), the xz plane in H_2O according to our assumption, the label becomes A_1. If the function changes sign under this reflection, one uses the label A_2. The molecular orbital having the symmetry species A_1 is represented as an a_1 orbital. The label for a molecular orbital is written in *italics small letter* of the symmetry species.

(ii) A molecular orbital in the group C_{2v} which changes sign on rotation is assigned the label b, the symmetry species being written B. In this case also, the label B_1 (and b_1 for the MO) is used to indicate that there will be no change of sign of the wave function on operation σ_v, reflection in the xz plane. The function will change sign on reflection in the yz plane (σ_v'). Similarly, orbitals of the type b_2 (symmetry species B_2) change sign on reflection in the xz plane (σ_v) but not on reflection in the yz plane (σ_v').

In group theory, information regarding the behaviour of a point group toward different symmetry operations is summarized in tables characteristic of each point group. Such tables, called character tables, are indispensable in the study of structure and properties of molecules.

We now classify the atomic orbitals involved in constructing the molecular orbitals for H_2O. We can label the oxygen AOs with our labelling scheme described above:

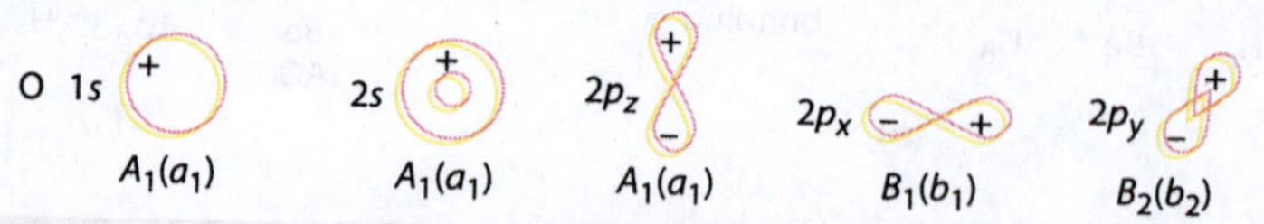

The two hydrogen $1s$ AOs (marked $1s_A$ and $1s_B$, arbitrarily though), may be combined to form the symmetry adapted linear combinations (SALC): $\phi_1 = 1s_A + 1s_B$ and $\phi_2 = 1s_A - 1s_B$.

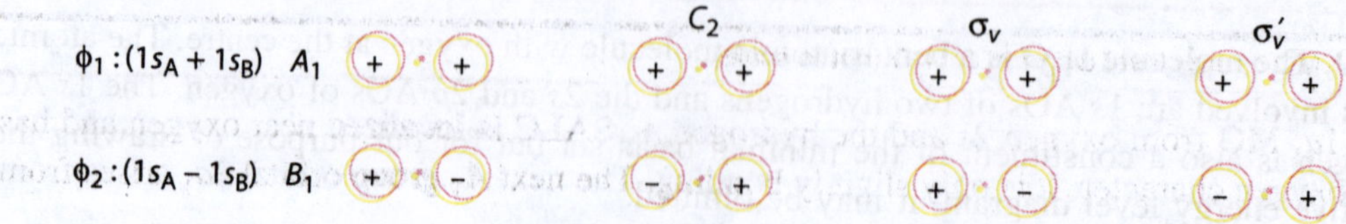

ϕ_1 does not change on rotation about the z-axis (C_2) or on any of the reflections in any of the planes xz or yz (σ_v and σ_v'). Hence ϕ_1 has the symmetry type A_1 and is labeled a_1. ϕ_2 changes sign on the operations C_2 and σ_v' and hence is labeled b_2.

We are now in a position to combine the oxygen AOs and hydrogen group orbitals (the SALC) to form MOs according to matching symmetry and of course, energy.

The oxygen $1s$, as mentioned earlier, is only of theoretical importance and omitted for our qualitative purpose. The MOs formed by the remaining functions are shown in Fig. 7.38.

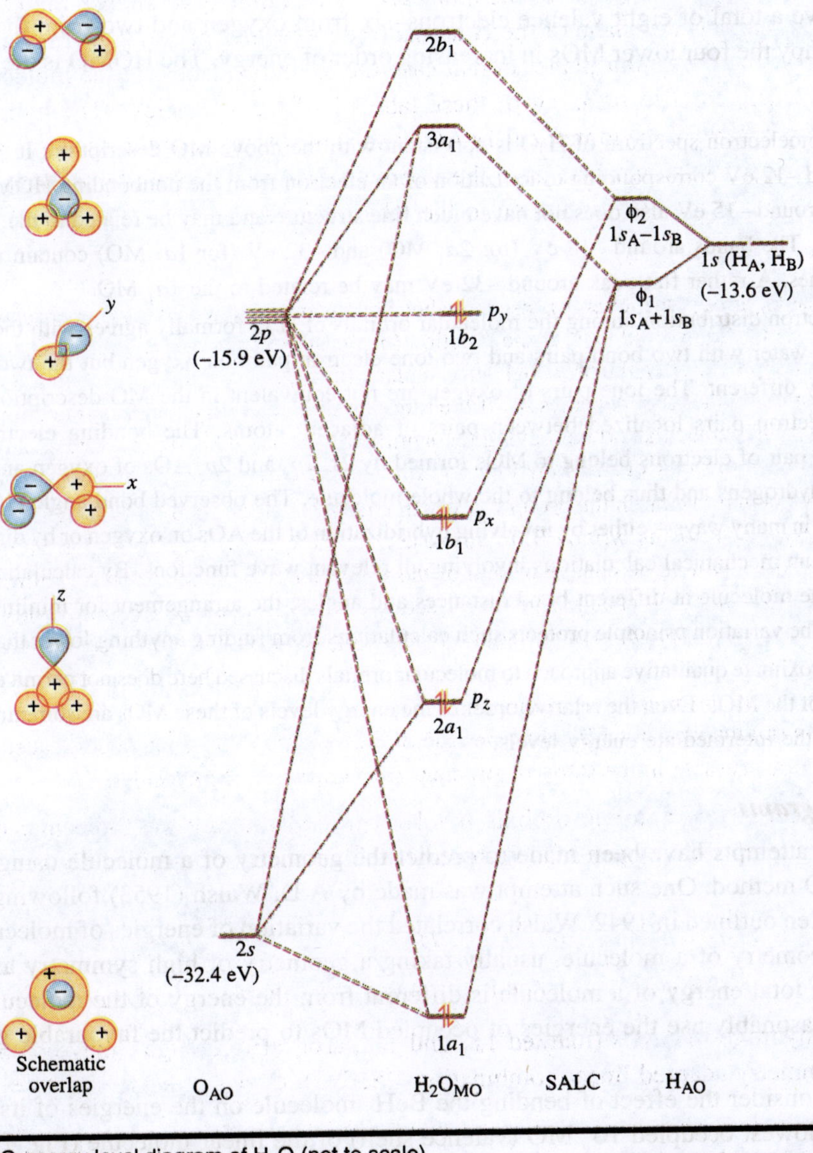

Fig.7.38: MO energy level diagram of H_2O (not to scale).

The $1a_1$ MO from oxygen $2s$ and the hydrogen A_1 SALC is localized near oxygen and has mainly oxygen character. It is only slightly bonding. The next A_1 group orbital $2a_1$ arises from oxygen $2p_z$ and the A_1 SALC; it is also bonding to some extent. The third A_1 group orbital, $3a_1$ is antibonding.

There are two B_1 type MOs formed by the $2p_x$ AO of oxygen and B_1 type SALC of hydrogen (ϕ_2): one is bonding ($1b_1$) and one is antibonding ($2b_1$).

The $2p_y$ AO of oxygen (symmetry B_2) forms a nonbonding MO ($1b_2$) as there is no matching SALC of the H atoms.

Three combining orbitals with A_1 symmetry ($2s$(O), $2p_z$(O) and ϕ_1 SALC) give rise to three molecular orbitals of type a_1.

We have a total of eight valence electrons–six from oxygen and two from the hydrogens. These occupy the four lower MOs in increasing order of energy. The HOMO is the nonbonding $1b_2$ M.O.

The photoelectron spectrum of H_2O is consistent with the above MO description. It shows a sharp band around –12 eV corresponding to ionization of an electron from the nonbonding HOMO ($1b_2$). The next band around –15 eV also does not have much fine structure and may be related to the $1b_1$ MO from oxygen $2p_x$. The bands around –18 eV (for $2a_1$ MO) and –32 eV (for $1a_1$ MO) contain multiplet fine structure lines. A rather flat peak around –32 eV may be related to the $1a_1$ MO.

The electron distribution among the molecular orbitals of H_2O formally agrees with the electron dot structure of water with two bond pairs and two lone electron pairs on oxygen but the two descriptions are basically different. The lone pairs of oxygen are not equivalent in the MO description nor are the bonding electron pairs localized between pairs of adjacent atoms. The bonding electrons and one nonbonding pair of electrons belong to MOs formed by $2s$, $2p_x$ and $2p_z$ AOs of oxygen and the $1s$ AOs of the two hydrogens and thus belong to the whole molecule. The observed bond angle (104.5°) can be rationalized in many ways—either by involving hybridization of the AOs on oxygen or by direct computer aided quantum mechanical calculations involving all relevant wave functions. By calculating the overall energy of the molecule at different bond distances and angles, the arrangement for minimum energy is found out. The variation principle protects such calculations from finding anything lower than the lowest.

The approximate qualitative approach to molecular orbitals discussed here does not permit determination of energies of the MOs. Even the relative order of the energy levels of these MOs are sometimes disputed, particularly the intermediate energy levels.

Walsh diagrams

Several attempts have been made to predict the geometry of a molecule using the simple LCAO–MO method. One such attempt was made by A.D. Walsh (1953) following an idea of R.S. Mulliken outlined in 1942. Walsh correlated the variation of energies of molecular orbitals with the geometry of a molecule, usually taking a geometry of high symmetry as reference. Though the total energy of a molecule is different from the energy of the molecular orbitals, one may reasonably use the energies of occupied MOs to predict the favourable geometry of a molecule.

Let us consider the effect of bending the BeH_2 molecule on the energies of its MOs (Fig. 7.39). The lowest occupied $1\sigma_g$ MO (valence shell) of the linear molecule (Fig. 7.37) is now relabeled as $1a_1$ in the bent geometry ($\angle$HBeH = 90°) as explained under H_2O. As the $1\sigma_g$ MO is formed from overlap of AOs which are everywhere positive (symmetric combination), bending will result in better overlap between the hydrogen AOs (they become closer; Fig. 7.39). The energy of this MO, now relabeled $1a_1$ is thus expected to decrease slightly relative to $1\sigma_g$. The hydrogen 1*s* AOs will also overlap better with the *p* orbital of Be on the twofold molecular axis, (may be p_x as we had earlier chosen the *z*-axis as the molecular axis in linear geometry). As *p*-orbitals have higher energy than *s*-orbitals, mixing in of *p*-orbitals will actually raise the energy of the $1a_1$ MO. However, this will be a minor component of $1a_1$ as we have assumed, for simplicity, no mixing between 2*s* and 2*p* AOs on Be.

The next higher MO in linear BeH_2, $1\sigma_u$, is formed from p_z-orbital of Be (molecular axis *z*) with asymmetric combination of the hydrogen 1*s* AOs (Fig. 7.37). In the bent geometry, the two 1*s* orbitals, with opposite signs, give a negative overlap. At the same time, their overlap with the Be–p_z AO also decreases. The energy of the $1\sigma_u$ MO thus increases in the bent structure; the MO is now relabeled $1b_2$.

The increase in energy of $1\sigma_u$ is greater than the decrease in energy of $1\sigma_g$. Hence the bent geometry of BeH_2 would have a higher energy for the MO occupancy $1a_1^2 1b_2^2$ compared to the MO occupancy in linear BeH_2 ($1\sigma_g^2 1\sigma_u^2$). Hence BeH_2 is expected to be linear.

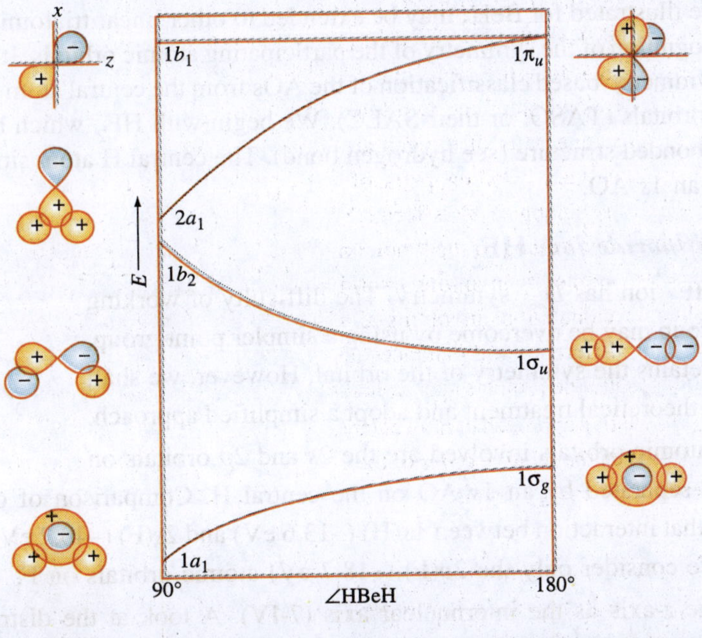

Fig.7.39: Walsh diagram for BeH_2 (not to scale). Only valence MOs considered.

The MOs in any XH_2 species may be treated similarly to predict the favourable geometry. The MOs above the occupied ones in linear BeH_2 are the doubly degenerate $1\pi_u$ orbitals (nonbonding) resulting from the two *p*-orbitals on X perpendicular to the molecular axis. In the bent geometry, the *p*-orbital on X which lies along the two-fold molecular axis (assumed *x* in BeH_2 bent) will overlap with a symmetric combination of hydrogen 1*s* orbitals: the resulting a_1 MO ($2a_1$ according to our numbering which counts only valence MOs) will get some bonding character and fall below the $1\pi_u$ (linear) MO. The other *p*-orbital on X (p_y in our case) perpendicular to the molecular plane gives no net overlap with hydrogen 1*s* AOs and remain unchanged in energy ($1b_1$ MO). The decrease in energy from $1\pi_u$ to $2a_1$ is actually steeper than the rise in energy of $1b_2$ above $1\pi_u$ because of greater overlap in $2a_1$.

In the first excited state of BeH_2, one electron from $1\sigma_u$ MO will be promoted to the next higher MO. Since the distribution $1a_1^2 1b_2^1 2a_1^1$ in angular geometry has a lower energy than the distribution $1\sigma_g^2\, 1\sigma_u^1\, 1\pi_u^1$ (linear), a bent structure is inferred.

H_2O, with a total of eight valence electrons, is expected to be bent. Occupancy of the $2a_1$ and $1b_1$ MOs offers greater stabilization relative to the linear structure where these electrons occupy the nonbonding $1\pi_u$ MOs.

The Walsh diagrams thus provide the basis for Walsh's rules mentioned earlier. However, the diagrams are based only on approximate calculations and it is assumed that the *s* and *p* orbitals on the central atom X are widely separated. Internucleus and interelectron repulsions are also not taken into account. Hence such correlations lead to approximate geometry only; the bond angles can be worked out only through more accurate calculations.

More linear polyatomic species

The procedure illustrated for BeH_2 may be extended to other linear triatomic molecules and ions with due recognition of the symmetry of the participating atomic orbitals. It is advantageous to begin with a symmetry-based classification of the AOs from the central atom and the terminal atom symmetry orbitals (TASO; or their SALC). We begin with HF_2^- which has a symmetric linear hydrogen-bonded structure (see hydrogen bond). The central H atom simplifies our task as it offers only an 1*s* AO.

The hydrogendifluoride ion: HF_2^-

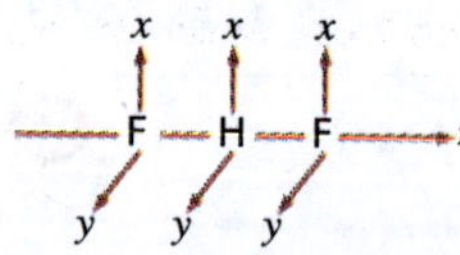

7-IV

The linear FHF^- ion has $D_{\infty h}$ symmetry. The difficulty of working with this point group may be overcome by using a simpler point group, D_{2h}, which still retains the symmetry of the orbital. However, we shall bypass the group theoretical treatment and adopt a simplified approach.

The valence atomic orbitals involved are the 2*s* and 2*p* orbitals on the two F atoms separated by an 1*s* AO on the central H. Comparison of orbital potential energies suggest that interaction between 1*s*(H) (–13.6 eV) and 2*s*(F) (–40.2 eV) will be indeed very small. So we consider only the 2*p*(F) (–18.7 eV) atomic orbitals on F.

We choose the *z*-axis as the internuclear axis (7-IV). A look at the distribution of axes further simplfies our job since the p_x and p_y atomic orbitals on fluorine will produce no net overlap with the hydrogen 1*s* AO. Only one SALC of two

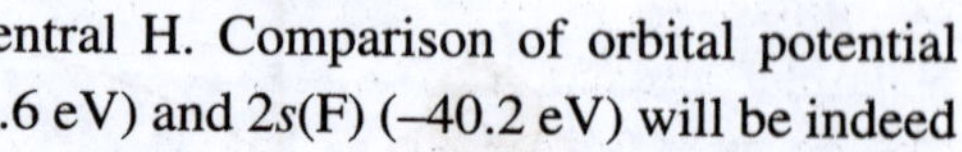

A_g B_{1u}

fluorine $2p_z$ orbitals (symmetry A_g) has the right symmetry to combine with the 1*s*(H) AO, producing one bonding and one antibonding MO. The other SALC (B_{1u}) will not give any net overlap with hydrogen 1*s* and will remain nonbonding. The overall MO diagram (Fig. 7.40) essentially contains 2 electrons (bond pair) from the F—H bond in a bonding MO and 2 electrons (lone pair) from F^- in a nonbonding MO. The remaining electrons remain essentially nonbonding with their energies unchanged from the original AOs. The scheme is often referred to as a 3-centre-4-electron (3*c*-4*e*) MO.

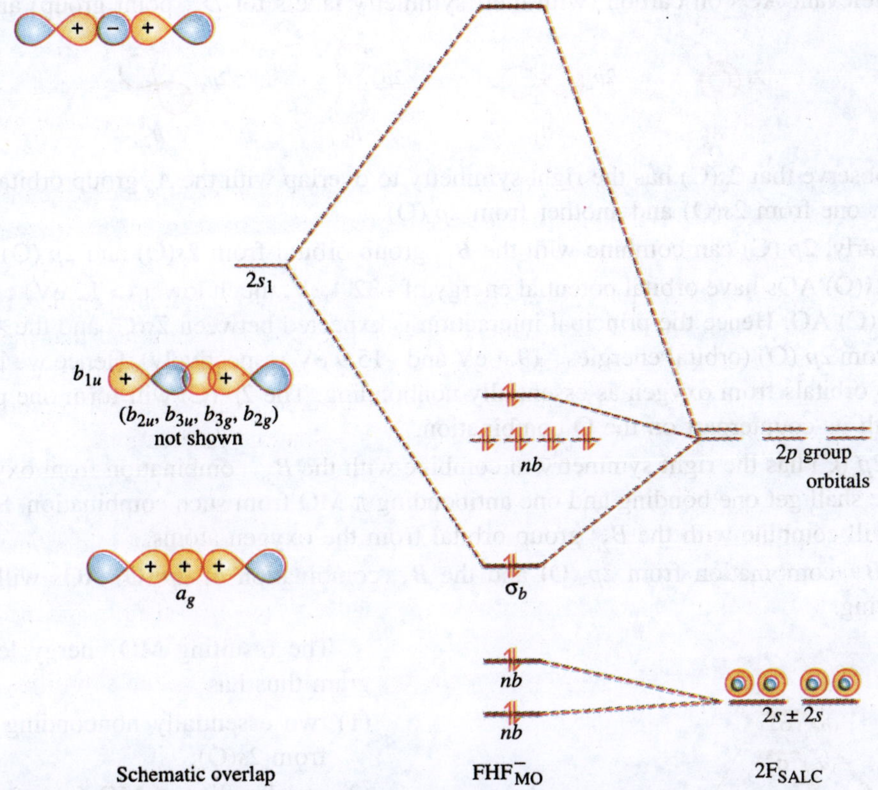

Fig.7.40: Molecular orbital energylevel diagram for FHF^- (not to scale).

Carbon dioxide, CO_2

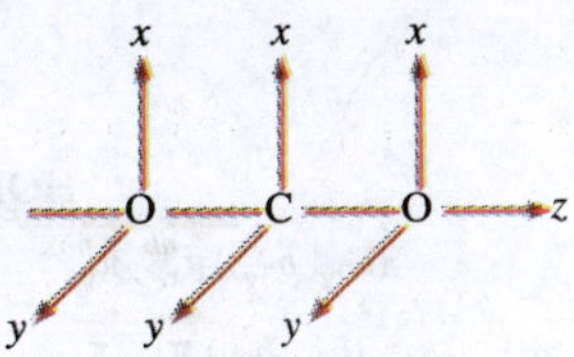

7–V

CO_2 is a linear molecule with C at the centre and belongs to the $D_{\infty h}$ group. As in FHF^-, the D_{2h} symmetry group may be used as a simpler alternative. The choice of axes and the group orbitals formed by various oxygen orbitals are shown:

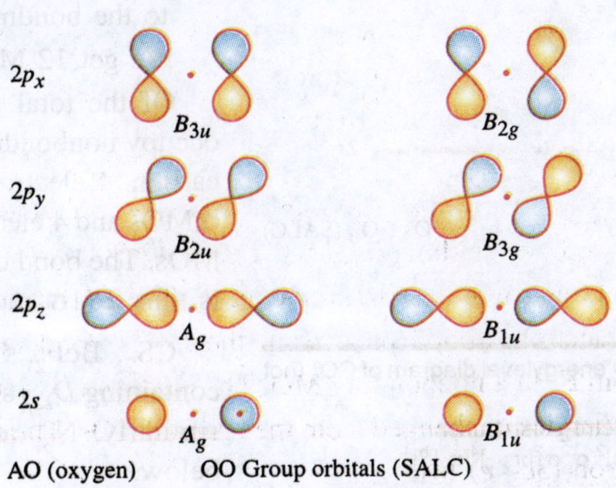

The relevant AOs on carbon (with their symmetry labels for D_{2h} point group) are

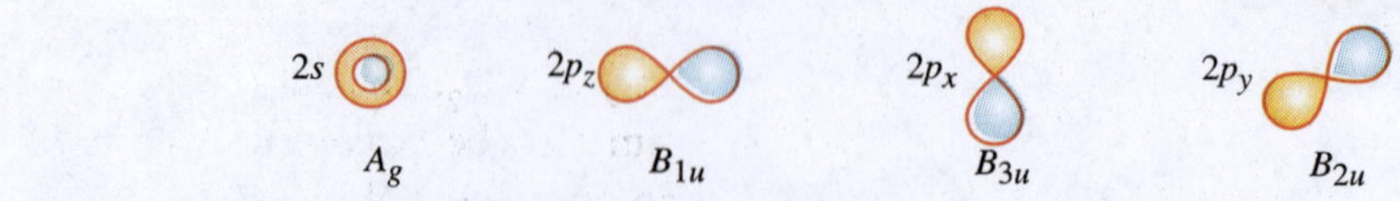

We observe that 2*s*(C) has the right symmetry to overlap with the A_g group orbitals of the O-atoms, one from 2*s*(O) and another from $2p_z$(O).

Similarly, $2p_z$(C) can combine with the B_{1u} group orbital from 2*s*(O) and $2p_z$(O).

But 2*s*(O) AOs have orbital potential energy of –32.4 eV, much lower (> 12 eV) compared to the 2*s*(C) AO. Hence the principal interaction is expected between 2*s*(C) and the A_g group orbital from $2p_z$(O) (orbital energies –19.4 eV and –15.9 eV respectively). Hence we leave the 2*s* group orbitals from oxygen as essentially nonbonding. The $2p_z$(C) will form one pair of σ MOs with its counterpart on the O combination.

The $2p_y$(C) has the right symmetry to combine with the B_{2u} combination from oxygen $2p_y$ AOs. We shall get one bonding and one antibonding π MO from such combination. Similarly $2p_x$(C) will combine with the B_{2u} group orbital from the oxygen atoms.

The B_{3g} combination from $2p_y$(O) and the B_{2g} combination of $2p_x$(O) AOs will remain nonbonding.

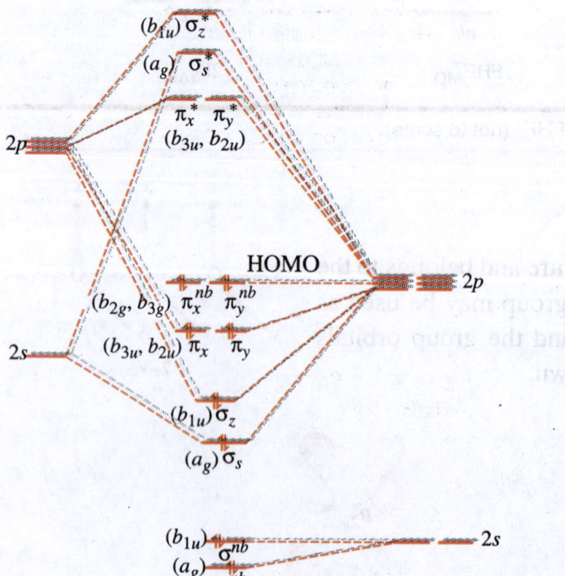

Fig.7.41: Approximate MO energylevel diagram of CO_2 (not to scale).
Ground state electron distribution
$[2s(O_A)]^2\ [2s(O_B)]^2\ \sigma_s^2\sigma_z^2[\pi_{x,y}]^4[\pi_{x,y}^{nb}]^4$

The resulting MO energy level diagram thus has

(1) two essentially nonbonding σ MOs from 2*s*(O);

(2) one bonding σ MO from 2*s*(C) and $2p_z$ SALC of O;

(3) one bonding σ MO from $2p_z$(C) and $2p_z$ group orbital from $2p_z$(O);

(4) two bonding π-orbitals from $2p_x$(C) and $2p_y$(C);

(5) two nonbonding orbital from B_{2g} and B_{3g} combinations of oxygens;

(6) four antibonding MOs corresponding to the bonding MOs.

We get 12 MOs from 12 AOs.

Of the total 16 valence electrons, 8 occupy nonbonding MOs on oxygen and carbon. 4 electrons occupy two bonding σ MOs and 4 electrons occupy bonding π MOs. The bond order for each C–O bond is thus 2 (1σ and 1π).

CS_2, BeF_2, OCN^-, N_3^- etc. species containing $D_{\infty h}$ symmetry may be treated similarly. N_3^- and BeF_2 are discussed below.

N_3^- *(the azide ion)*

The linear azide ion has the same symmetry as CO_2. The MO energylevel diagram may be arrived at from combination of the central nitrogen orbitals with group orbitals (SALC) from terminal N atoms. Since all atoms have equal orbital energies, participation of the 2*s* AOs in bonding becomes important. The overall MO energylevel diagram has the same symmetry labels as those in CO_2.

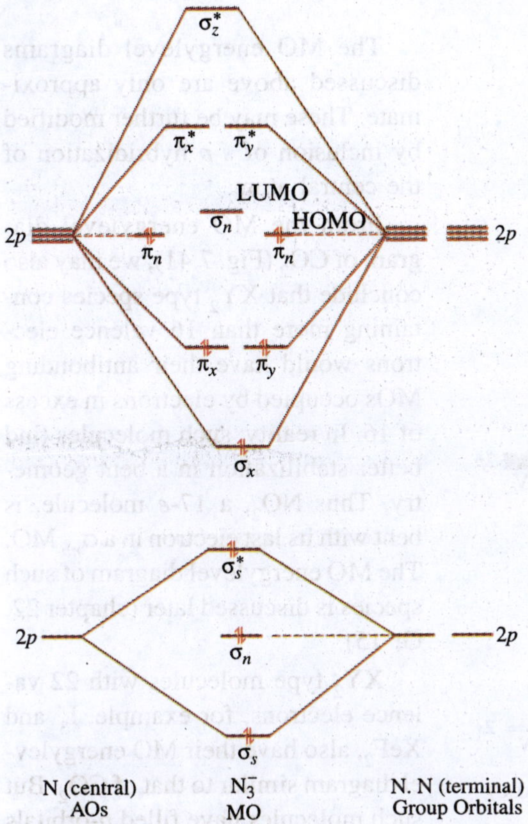

Fig.7.42: Approximate MO energylevel diagram of N_3^- (azide ion) (not to scale). The nonbonding σ MO (LUMO) is slightly higher than the π_n MOs owing to antibonding interaction with 2*s* AO of central N.

BeF_2

BeF_2 has 16 valence electrons like CO_2 and the same symmetry of the participating AOs as shown:

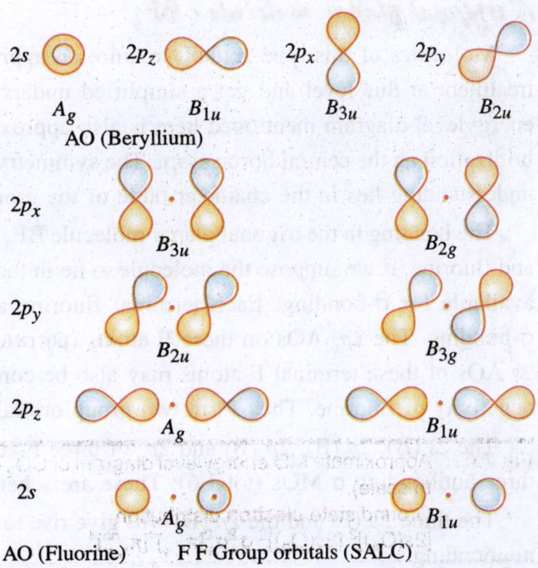

Proceeding in the same manner as in CO_2, we would obtain an energylevel diagram similar to Fig. 7.41. Assignment of 16 valence electrons as in CO_2 would suggest a bond order of 2 (1σ and 1π) for each Be–F bond. In reality, however, the 2*p* AOs on Be and F differ so widely in energy that π-bonding becomes insignificant. Hence the SALC of $2p_x$(F) and $2p_y$(F) AOs give rise to essentially nonbonding MOs localized near F. The four electron pairs assigned to these MOs are comparable to the lone pairs of electrons on F atoms in the Lewis dot structure. The bonds in BeF_2 molecule are now equivalent to two σ bonds, one for each Be–F bond.

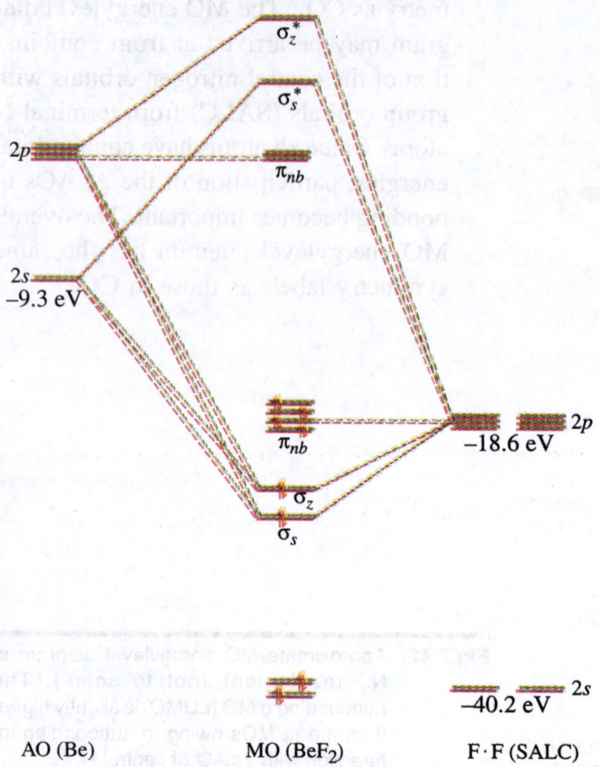

Fig.7.43: Approximate MO energylevels in BeF_2 (not to scale).

The MO energylevel diagrams discussed above are only approximate. These may be further modified by inclusion of *s-p* hybridization of the central atom.

From the MO energylevel diagram of CO_2 (Fig. 7.41), we may also conclude that XY_2 type species containing more than 16 valence electrons would have their antibonding MOs occupied by electrons in excess of 16. In reality, such molecules find better stabilization in a bent geometry. Thus NO_2, a 17-*e* molecule, is bent with its last electron in a σ_{nb} MO. The MO energylevel diagram of such species is discussed later (chapter 22, Gr 15).

XY_2 type molecules with 22 valence electrons, for example, I_3^- and XeF_2, also have their MO energylevel diagram similar to that of CO_2. But such molecules have filled *p*-orbitals which lead to equal population of the bonding and antibonding π-MOs.

A trigonal planar molecule : BF_3

Molecules of this type actually require an approach based on group theory. We shall bypass such treatment at this level and get a simplified understanding of the molecular orbitals formed. The MO energylevel diagram mentioned here is also approximate and may be modified by inclusion of *s-p* hybridization on the central boron atom. The symmetry labels mentioned earlier are also omitted since their understanding lies in the character table of the group.

The bonding in the trigonal planar molecule BF_3 involves the 2*s* and 2*p* valence orbitals of both boron and fluorine. If we suppose the molecule to lie in the *xy* plane, we have the 2*s*, $2p_x$ and $2p_y$ AOs of boron available for σ-bonding. Each terminal fluorine atom may be supposed to provide its $2p_x$ AO for σ bonding. The $2p_y$ AOs on these F atoms (perpendicular to the B–F bonds) remain nonbonding. The 2*s* AOs of these terminal F atoms may also be considered as nonbonding in view of the high electronegativity of fluorine. They form two group orbitals as shown in Fig. 7.44.

The 2*s* (B), $2p_x$(B), $2p_y$(B) and $2p_x$ of three F-atoms (say F_A, F_B, F_C) give rise to three bonding and three antibonding σ MOs (total 6). These are schematically shown in Fig. 7.44.

The three $2p_z$(F) and the $2p_z$(B) will give rise to four π MOs–one bonding, one antibonding and two nonbonding.

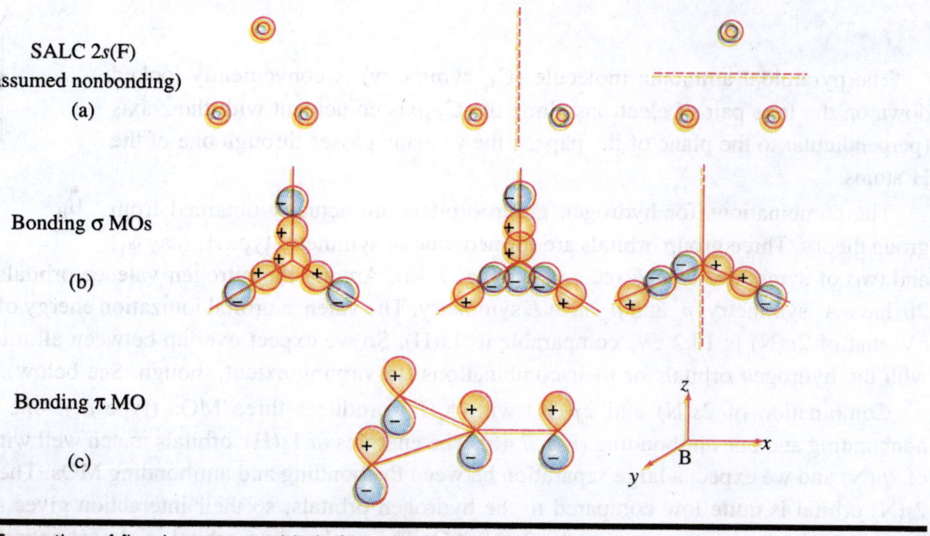

Fig.7.44: (a) Formation of fluorine group orbitals from 2*s*(F). These are considered nonbonding; (b) Schematic overlap for 3 B–F σ MOs. (c) Schematic overlap for B–F π MO.

The approximate MO energylevel diagram of BF_3 is given in Fig. 7.45. The LUMO is a π* MO which gets involved when BF_3 acts as a Lewis acid.

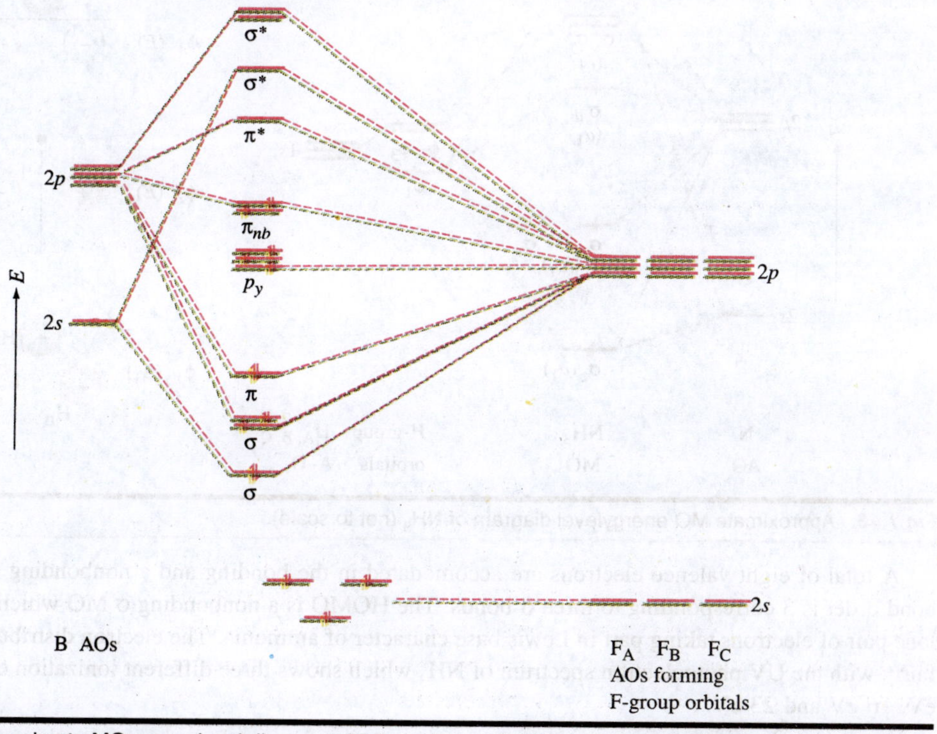

Fig.7.45: Approximate MO energylevel diagram of BF_3 (not to scale). The electron distribution (ground state) is: $[2s(F)_{A,B,C}]^6\,\sigma_s^2\,\sigma_{p_xp_y}^{2\,2} \times \pi_{p_z}^2\,[p_y(F)_{A,B,C}]^6\,\pi_{nb}^4$

Molecular orbitals for other trigonal planar molecules or ions like SO_3, NO_3^- and CO_3^{2-} may be described in an analogous manner.

NH_3

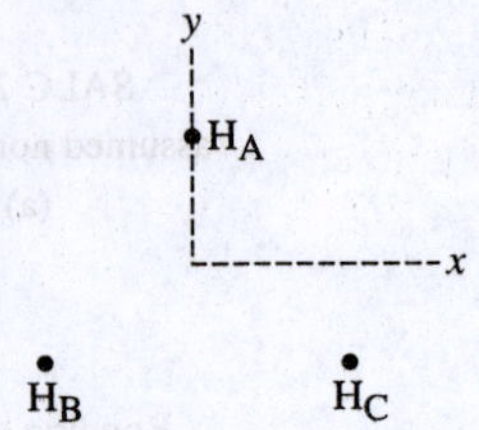

The pyramidal ammonia molecule (C_{3v} symmetry) is conveniently looked down on the lone pair of electrons along the C_3 axis coincident with the *z*-axis (perpendicular to the plane of the paper); the *yz* plane passes through one of the H atoms.

The combinations for hydrogen group orbitals are actually obtained from group theory. Three group orbitals are formed: one of symmetry type A_1 (say ϕ_1) and two of symmetry type *E* (ϕ_2 and ϕ_3; Fig. 7.46). Among the nitrogen valence orbitals, both 2*s* and $2p_z$ have A_1 symmetry; p_x and p_y have *E* symmetry. The valence orbital ionization energy of 2*s*(N) is 25.6 eV; that of 2*p*(N) is 13.2 eV, comparable to 1*s*(H). So we expect overlap between all nitrogen orbitals with the hydrogen orbitals or their combinations (to varying extent, though. See below).

Combination of 2*s*(N) and $2p_z$(N) with ϕ_1(H) produces three MOs (type a_1): one bonding, one nonbonding and one antibonding (Fig. 7.46). The energies of 1*s*(H) orbitals match well with the energies of 2*p*(N) and we expect a large separation between the bonding and antibonding MOs. The energy of the 2*s*(N) orbital is quite low compared to the hydrogen orbitals; so their interaction gives rise to an MO having nearly the same energy as the 2*s*(N) AO. The nonbonding orbital is almost entirely nitrogen p_z.

The nitrogen p_x and p_y orbitals combine with the *E* group orbitals of hydrogen (ϕ_2 and ϕ_3) giving two bonding and two antibonding MOs.

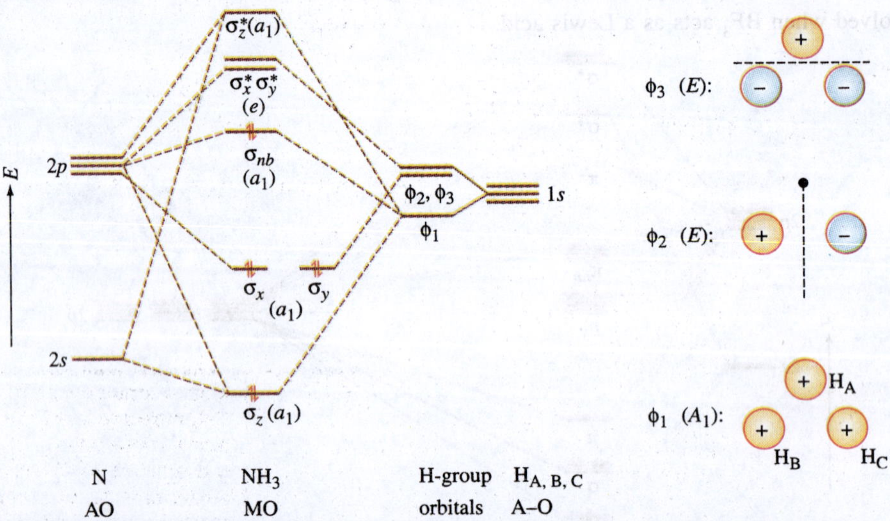

Fig.7.46: Approximate MO energylevel diagram of NH_3 (not to scale).

A total of eight valence electrons are accomodated in the bonding and a nonbonding σ MO. Total bond order is 3 corresponding to three σ bonds. The HOMO is a nonbonding σ MO which contains the lone pair of electrons taking part in Lewis base character of ammonia. The electron distributation agrees fairly with the UV photoelectron spectrum of NH_3 which shows three different ionization energies at 11 eV, 16 eV and 23 eV.

CH_4

The tetrahedral methane molecule can be related to the Cartesian axes by joining diagonally opposite corners of a regular cube (Fig. 7.47a). The hydrogen atoms occupy these corners with carbon atom at the centre of the cube.

The $2s$, $2p_x$, $2p_y$, and $2p_z$ orbitals of carbon and $1s$ orbitals of the four hydrogen atoms (A, B, C, D) take part in forming the MOs of CH_4. The four $1s$ AOs form four hydrogen group orbitals–one of A_1 symmetry (ϕ_1) and three of T_2 symmetry (ϕ_2, ϕ_3, ϕ_4):

$$\phi_1(A_1) : 1s_A + 1s_B + 1s_C + 1s_D$$
$$\phi_2(A_1) : 1s_A + 1s_B - 1s_C - 1s_D$$
$$\phi_3(T_2) : 1s_A - 1s_B + 1s_C - 1s_D$$
$$\phi_4(T_2) : 1s_A - 1s_B - 1s_C + 1s_D$$

The valence orbital ionization energy of a $2s$ electron in carbon is 19.4 eV and that of a $2p$ electron is 10.7 eV. So we expect that both AOs will combine with the hydrogen group orbitals. The $2s$(C) AO has the right symmetry to combine with the ϕ_1 group of orbitals from hydrogen. Their overlap is shown in Fig. 7.47(b). The nonsymmetric combinations ϕ_2, ϕ_3 and ϕ_4 separately match the p_x, p_y or p_z AO of carbon. One such combination with p_x orbital is shown in Fig. 7.47(c). Similar combinations can be drawn for other p orbitals. The resulting MO energylevel diagram is shown below.

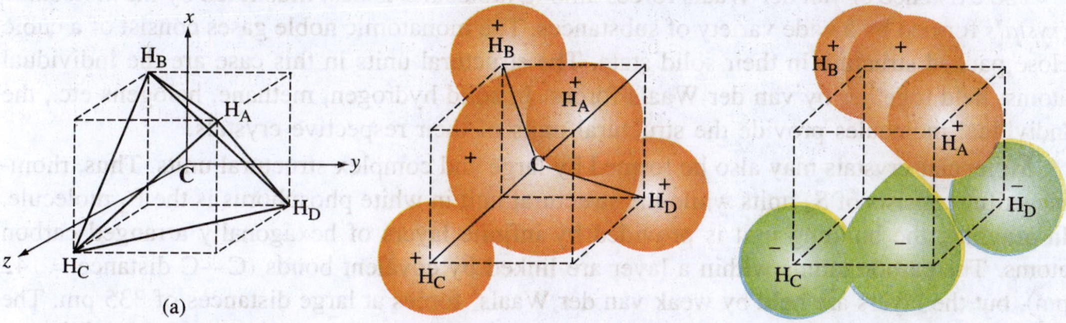

$\sigma_s = 2s(C) + 1s_A + 1s_B + 1s_C + 1s_D$

(b)

$\sigma_{px} = 2p_x(C) + 1s_A + 1s_B - 1s_C - 1s_D$

(c)

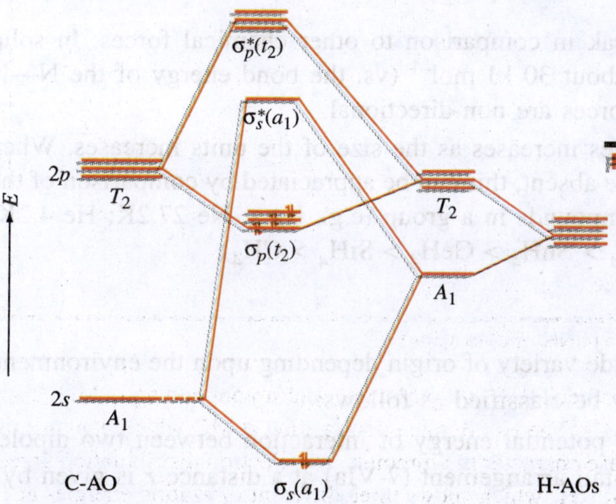

Fig.7.47: (a) Mutual arrangement of atoms of CH_4 molecule in a Cartesian system of coordinates.
(b) σ overlap of $2s$(C) with A_1 group orbital of H atoms. There will be corresponding antibonding overlap.
(c) σ overlap of $2p_x$(C) with one T_2 group orbital of H atoms. There will be a corresponding antibonding overlap.
The other $2p$ orbitals of carbon similarly produce two MOs each.

The electron distribution among the MOs is consistent with the UV photoelectron spectrum of CH_4 which shows two absorption bands at ~23 eV and ~14 eV.

The MO energylevel diagram for CH_4 may be applied to BH_4^-, NH_4^+, ClO_4^- or PCl_4^+ etc. tetrahedral species (without π-bonding) with minor modifications for the orbitals used by the terminal atoms and their relative energies.

7.4 INTERMOLECULAR FORCES

7.4.1 Introduction

We have enough reasons to believe that a net attractive force operates between the molecules of a gas. Though very weak in nature, this force is ultimately responsible for liquefaction and solidification of gases. But we cannot explain the nature of this force from the simple ideas of ionic or covalent bond, particularly when we think of nonpolar molecules like H_2, CH_4, He, Ne, Ar etc. The existence of intermolecular attraction in gases was first recognized by van der Waals (1813) and accordingly intermolecular forces have been termed *van der Waals forces*. It has been established that such forces are also present in the solid and liquid states of many substances. Collectively they have also been termed *London forces* since their nature was first explained by Fritz London using wave mechanics (1937).

The existence of van der Waals forces among molecules is best illustrated by the *molecular crystals* formed by a wide variety of substances. The monatomic noble gases consist of a cubic close packed structure in their solid state. The structural units in this case are the individual atoms, held together by van der Waals forces. In solid hydrogen, methane, halogens etc., the individual molecules provide the structural units of their respective crystals.

Molecular crystals may also be formed by large and complex structural units. Thus, rhombic sulfur consists of S_8 units while the structural unit in white phosphorus is the P_4 molecule. In graphite, the building unit is provided by infinite layers of hexagonally arranged carbon atoms. The carbon atoms within a layer are linked by covalent bonds (C—C distance = 142 pm), but the layers are held by weak van der Waals' forces at large distances of 335 pm. The layers easily slide over one another; hence the use of graphite as a solid lubricant. Molybdenum sulfide, MoS_2 forms similar layer structure and is also used as a lubricant at high temperatures under the name ***Molyslip***.

Van der Waals' forces are very weak in comparison to other chemical forces. In solid ammonia, for example, it amounts to about 30 kJ mol^{-1} (vs. the bond energy of the N—N covalent bonds = 389 kJ mol^{-1}). The forces are non-directional.

The strength of van der Waals' forces increases as the size of the units increases. When other factors (like hydrogen bonding) are absent, this can be appreciated by comparison of the melting or boiling points of similar compounds in a group (e.g., b.p.s: Ne 27.2K; He 4.2K; H_2Se 231K; H_2S 213K; similarly, $PbH_4 > SnH_4 > GeH_4 > SiH_4 > CH_4$).

7.4.2 Origin

Intermolecular forces may have a wide variety of origin depending upon the environment. The different interactions involved may be classified as follows:

1. Dipole-dipole interactions. The potential energy of interaction between two dipoles having moments μ_A and μ_B in head-to-tail arrangement (7-VIa) at a distance r is given by

$$E = -(\mu_A \mu_B / 2\pi\varepsilon_0 r^3)$$

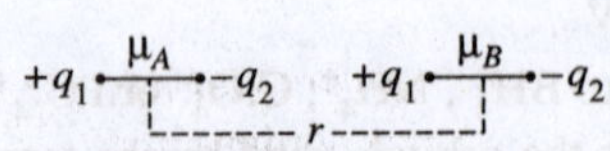

7-VI(a)

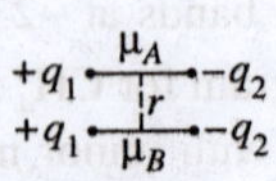

7-VI(b)

Clearly, the extent of interaction depends upon the orientation of the dipoles. When the dipoles are in the arrangement shown in 7-VIb, the potential energy is $\mu_A\mu_B/4\pi\varepsilon_0 r^3$. Such forces are responsible for association of polar liquids like HF, NF_3 etc. Hydrogen bonding may be considered a special type of such interaction.

2. Ion-dipole interactions. The potential energy of interaction between an ion of charge number Z with a dipole of moment μ at a distance r is given as

$$E = -Z\mu e / 4\pi r^2 \varepsilon_0.$$

Such interactions are operative in solvation and dissolution of ionic compounds in polar solvents.

3. Induced dipole interactions. Charged ions can polarize neutral molecules and distort their electronic atmosphere to induce a dipole moment. The energy of such an interaction is

$$E = -Z^2 \alpha e^2 / 2r^4$$

where Z = numerical charge of the ion and r is the distance between the two; α is the polarizability or "*softness*" of the molecule. A strong dipole of moment μ may also take the role of an ion ($E = -\mu^2\alpha/r^6$). Such interactions fall rapidly with distance and are involved in the dissolution of ionic or polar compounds in non-polar solvents.

4. Instantaneous dipole-induced dipole interactions. In molecules having no permanent dipole moment the electronic atmosphere is supposed to be symmetrical. But this is true only on a time-average consideration. Rapid and continuous changes in the intensities of charge concentration in the electron atmosphere may give rise to instantaneous and fluctuating dipoles. These dipoles may induce further dipoles on adjacent molecules. In terms of polarizability α and mean instantaneous dipole moment $\bar{\mu}$, the potential energy varies inversely with sixth power of distance:

$$E \simeq -2\bar{\mu}\alpha / r^6 .$$

The resulting weak, short-range forces are called ***London forces*** or ***dispersion forces***. The forces increase with increasing polarizability (and hence molar volume) among comparable species (e.g., CH_4 and SiH_4). Usually these are related to molecular weight but the polarizability is always important. Thus CD_4 has a slightly lower boiling point than CH_4; the fluorocarbons also have lower or equal boiling points compared to corresponding hydrocarbons with much lower molecular weight.

7.5 THE HYDROGEN BOND (H-BOND)

7.5.1 Introduction to H-bond

An atom of hydrogen linked covalently to a strongly electronegative atom can establish an extra weak attachment to another electronegative atom (in the same or in a different molecule). This secondary attachment is called a hydrogen bond (Latimer and Rodebush, 1920). To distinguish from a normal covalent bond, a hydrogen bond is represented by a broken line, e.g., $\overset{\delta-}{X}—\overset{\delta+}{H}\cdots Y$ where X and Y are two electronegative atoms, the hydrogen atom being covalently linked to X. One may also consider XH as a donor of hydrogen (hydron, proton) and Y as an acceptor of the hydron (proton); usually Y has one or more lone pairs of electrons or a polarizable electron cloud (Fig. 7-VIIa, b). Such interaction is strongest when both X and Y are first members in their respective periodic groups. F, N and O form prominent H-bonds as both X and Y; weaker H-bonds are formed when X is carbon or a second row element (P, S, Cl, or even Br, I). Carbon never acts as Y (other than in certain π-electron system). Cl, Br or I are less efficient in the role of Y unless negatively charged. S and P also participate weakely in the role of Y. Enhancement of the electronegativity of atom X by other substituents will also give stronger H-bonds. Thus strength of H-bond increases from aliphatic alcohols to phenols, from NH_4^+ to substituted ammonium ions (not quaternery) and from methyl chloride to methylene chloride to chloroform. Transition metal hydrides MH_n may also function as X. The O atoms in metal alkoxides, $ROML_n$, may act as base Y towards alcohols (ROH) as hydrogen donors.

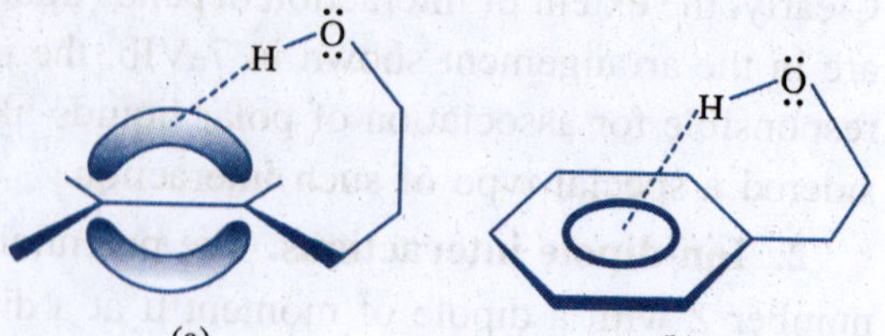

(a) 7-VII (b)

The H-bond in X—H···Y may be linear or bent. Bifurcated (X–H<) and trifurcated (X–H<−) H bonds are also known (7-VIII, 7-IX).

Now it has been established that hydrogen may enter into several other modes of bonding interactions outside normal covalent bond formation. In fact, the hydrogen bond described above is now considered as the "*classical hydrogen bond*". In metal-hydrogen agostic interactions (Ch 17), C—H...M, the C—H bond actually supplies electrons to the metal M.

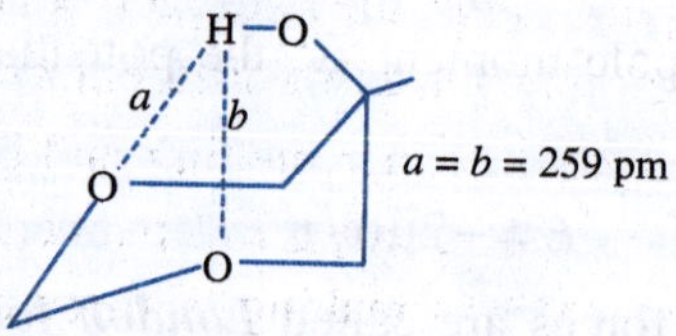

1, 3-dioxanol-5

7-VIII

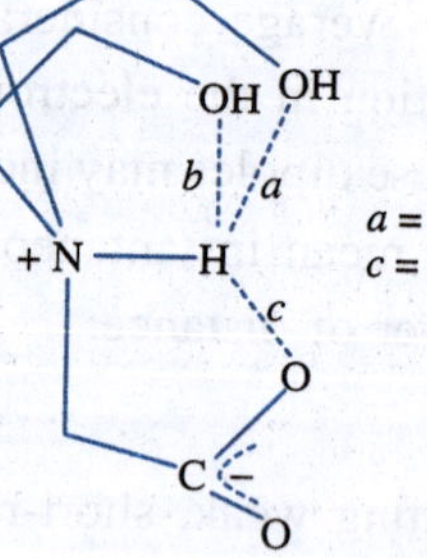

N, N-bis(2-hydroxyethyl)glycine

7-IX

The strength of a hydrogen bond may vary widely which may be measured by the enthalpy change for the process:

$$X{-}H{\cdots}Y \longrightarrow X{-}H + Y$$

The enthalpy change is small for weak interactions, 10–50 kJ mol^{-1}; strong interactions are marked by enthalpy changes around 50–100 kJ mol^{-1}. Very strong H-bonding has ΔH values above 100 kJ mol^{-1}, e.g., in FHF^- ΔH is as high as 165 kJ mol^{-1}.

Enthalpy changes for dissociation of H bond may be obtained from measurement of properties of hydrogen bonded systems over a range of temperature. Ion cyclotron resonance studies also provide reliable values for energies of H bonds. Quantum mechanical calculations are now applied to supplement experimental data. Some representative values of enthalpies for dissociation of H bonds are given in Table 7.6.

Table 7.6

A few hydrogen bond dissociation enthalpies* (kJ mol^{-1}) *in the gaseous phase at 298 K

FH·····F^-	165	FH·····OH_2	38	H_2NH·····NH_3	17
H_2OH^+·····OH_2	151	ClH·····OMe_2	30	NCH·····NCH	16
HOH·····F^-	98	FH·····FH	29	HSH·····SH_2	7
HCO_2H·····F^-	~200 (calc.)	HOH·····OH_2	22		
HOH·····Cl^-	55	MeOH·····OHMe	19		

7.5.2 Effect of Hydrogen Bonding

Effects of hydrogen bonding are amazingly numerous. Through weak but extensive interactions, H bonding influences several physical properties like melting and boiling temperatures, diectric constant, viscosity, solubility, miscibility with other solvents and even sensitivity to chromatographic separation. The structural influence of H bonding has far reaching consequences. Only a few of these effects can be mentioned within this limited scope.

Molecules of water, ammonia, hydrogen fluoride, ethyl alcohol etc. are all associated through hydrogen bonding in the liquid state. The bond also persists in many solid crystals, e.g., in ice, boric acid and sodium bicarbonate. Apart from these *intermolecular hydrogen bonding*, there may also be *intramolecular hydrogen bonds*; the hydrogen bond now takes place between two electronegative atoms in the same molecule as in different *ortho* aromatic compounds.

Intermolecular association leading to higher melting and boiling temperatures is the first direct effect of hydrogen bonding. In water, only a few per cent of the water molecules appear not to be hydrogen bonded even at 90°C. Breaking of these hydrogen bonds throughout the entire liquid requires appreciable heat energy. This is indicated in the relatively higher boiling points of hydrogen bonded compounds compared to similar compounds with little H bonding (Table 7.7).

Table 7.7

***Boiling points* (°C) *for some hydrogen bonded compounds* (1st row)**

H_2O	100	HF	19	C_2H_5OH	64.5
H_2S	– 60	HCl	– 85	C_2H_5SH	5.8

The same conclusion may be drawn from the general trend in melting and boiling points of the hydrides along a group in the periodic table (Fig. 7.48). For covalent compounds, melting and boiling points should rise gradually with increase in molecular weight owing to increasing London force. This is true for the hydrides of Gr. 14 elements. But in Groups 15, 16 and 17, the hydrides of the first element in each group (which is the most electronegative) have unusually higher boiling points than the extrapolated value from other group members.

It appears surprising that NH_3 (liq.) is associated through extensive hydrogen bonding whereas this effect is negligible, if at all, in liquid HCl, although the electronegativities of nitrogen and chlorine are same (3.0 in the Pauling scale). One probable explanation lies in the larger size of chlorine atom, and its consequent low charge density.

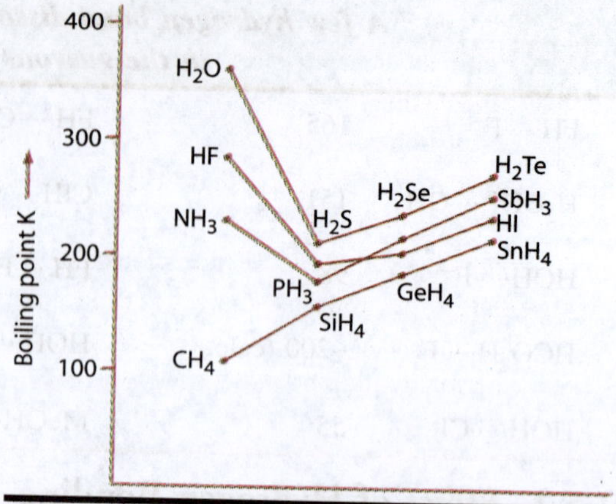

Fig.7.48: Boiling points of some hydrides.

Crystalline hydrogen fluoride consists of the polymer $(HF)_n$. This has a zigzag chain structure (Fig. 7-X) involving H-bonds. The H—F—H bond angle is nearly 120°. The H bonding persists in liquid as well as in the vapour —even at its boiling point HF consists of species upto $(HF)_6$ (see exercise).

Ice contains extensive hydrogen bonding. The structure of ice is complex and has been discussed in Chapter 17. In the ordinary form of ice, each oxygen atom is surrounded tetrahedrally by four hydrogen atoms—two H-atoms by usual covalent bonds, the other two by longer hydrogen bonds (Fig. 7-XI). When ice melts,

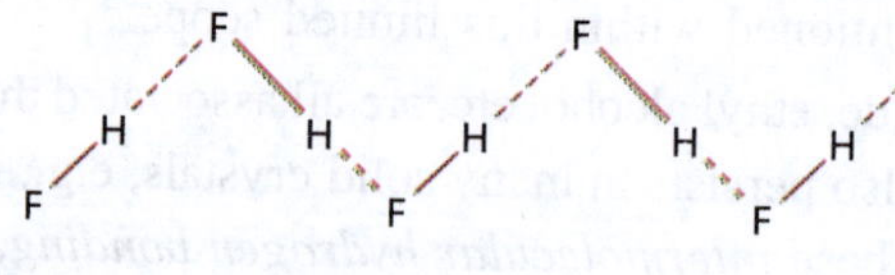

7-X

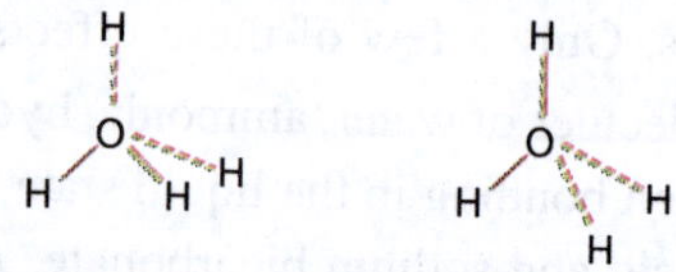

7-XI

some of these hydrogen bonds break,; the water molecules can now come closer (than in the rather stretched arrangement in ice). Hence the volume decreases on melting. When water at 0°C is heated, breaking of such ordered H-bonds still continues up to about 4°C and the contraction caused is greater than the thermal expansion (molecular agitation). So the volume goes on decreasing up to this point. Normal expansion of volume on heating starts only after this temperature. Consequently water has its highest density at 4°C.

The influence of H-bonding on the electrical properties of some solids and liquids is noteworthy. The high electrical conductance of H^+ and OH^- ions in aqueous solutions is associated with rapid formation and rupture of H-bonds between neighbouring water molecules. The dielectric constants of H-bonded liquids are also unusually large owing to the orienting effect of the H-bonds—the molecular dipoles reinforce one another rather than cancelling each other due to random thermal motion. A permanent electric polarization arising from ordered H-bonds is found to operate in some ferroelectric crystals (see box).

> A ferroelectric crystal has an electric dipole moment even in the absence of an external electric field. The centers of positive and negative charges in such crystals do not coincide and the dipole moment can be very high due to presence of a stable permanent electric polarization. Such polarization may arise from (a) ordering by H-bonding, as in KH_2PO_4 and KH_2AsO_4 or (b) a displacement of one sublattice with respect to another, as in barium titanate, $BaTiO_3$ and potassium niobate, $KNbO_3$.

Ferroelectrics have many practical applications in electronic industry, e.g., as miniature ceramic capacitors and in modulating and deflecting laser beams. Ferroelectrics are also piezoelectrics (acquire or alter an electric polarization under external mechanical stress and vice versa) and hence can convert mechanical pulses into electrical ones and vice versa. Hence their use in ultrasonic generators, microphones, frequency controllers and in computer technology.

In crystalline KH_2PO_4, each tetrahedral $[O_2P(OH)_2]^-$ unit is joined to four other $[O_2P(OH)_2]^-$ units by hydrogen bonds. Two short O—H bonds and two long O···H bonds are present around each PO_4 group. Below 121K, all the short O—H bonds are on the same side of the PO_4 groups. By appropriate application of an electric field, the polarization of the hydrogen bonds can be reversed. This spontaneous and reversible electric polarization gives rise to the ferroelectric behaviour. Above 121K, thermal agitation breaks down the ordering due to H-bonds and the crystal is said to become para-electric.

The crystal structures of ice, KH_2PO_4 and HF(*s*) are only a few of the many examples of widespread ordering influence of H bonding on the structures of solids. We shall find more examples of this kind in course of our discussion of different elements and their compounds, as for example, in the structures of sodium bicarbonate, boric acid, $Te(OH)_6$, NH_4F etc. We shall also realize that H bonding plays a vital regulatory influence on the conformation, structure and reactivity of several biomolecules (section 7.5.4 and chapter 33).

Intramolecular hydrogen bonding

Intramolecular hydrogen bonding also gives rise to certain distinctive properties in many compounds. The case of *ortho*-nitrophenol is well known. It forms intramolecular H bonds while the *para-* and *meta* isomers cannot do so. Consequently, *o*-nitrophenol has a lower boiling point (214°C) than the others (*m*-nitrophenol 290°C; *p*-nitrophenol 279°C). These latter compounds are able to form *intermolecular* hydrogen bonds while *o*-nitrophenol is not. For the same reason, *o*-nitrophenol does not form hydrogen bonds with water; it is therefore less soluble in water and is *steamvolatile*. The *m-* and *p*-nitrophenols, on the other hand, establish hydrogen bonds with water and are more soluble; they are *not* steamvolatile. *o*-hydroxybenzaldehyde, *o*-chlorophenol and *o*-hydroxybenzoic acid also behave similarly.

Impact on structural parameters

When an X—H bond gets engaged in hydrogen bonding, there are prominent changes in the bond parameters which may be studied by vibration spectroscopy (infrared and Raman), proton nmr spectroscopy and x-ray or neutron diffraction studies. Such studies provide valuable information regarding the hydrogen bond.

On forming a hydrogen bond, the X—H stretching frequency shifts to lower wave numbers as the bond is partially weakened. For example, the O—H frequency in ice is about 3250 cm^{-1} as against 3750 cm^{-1} in $H_2O(g)$. The breadth and intensity of the band also show a marked increase, often more than tenfold. H bonding also results in an increase in the bending mode of the X—H bond. Sometimes new stretching and bending modes of the hydrogen bond itself may be observed at very low wave numbers (20–200 cm^{-1}).

In proton nmr spectroscopic studies, an X—H bond involved in H bonding is marked by a chemical shift to low field. In the formation of an X—H···Y bond, the proton slightly shifts to Y while the bonding electrons are repelled by Y towards X. The electron probability density around H is thus lowered and the H atom gets less shielding. The shielding is further reduced by other factors initiated by Y.

Hydrogen bonding also influences UV and visible spectra but their correlation with the H bond is often difficult. In general, the n—π^* transition of the donor Y shifts to higher frequency on forming H bond, $\Delta\nu$ being in the range ~300–4000 cm^{-1} for bands in the range 1500–3000 cm^{-1}. π–π^* transitions on the base Y usually shifts to lower frequencies by about 500–2300 cm^{-1} in the band 3000–47000 cm^{-1}.

The interionic separation in crystals involving H bond is also modified by hydrogen bonding. There are two types of H bonds—symmetrical and unsymmetrical. In symmetrical H bond, which is strong and less common, the H atom is midway, as in FHF^- ion (F—H = 213 pm). In contrast, in the unsymmetrical $ClHCl^-$ ion, Cl—H = 137 pm and 185 pm.

7.5.3 Theory of Formation of Hydrogen Bond

Since hydrogen bonding occurs only with electronegative atoms, it was first assumed that it involves mainly electrostatic interactions. An electronegative element X highly polarizes the X—H bond as $X^{\delta-}$–$H^{\delta+}$. This partially positively charged hydrogen may now give rise to ion-dipole or dipole-dipole interactions forming a hydrogen bond. This simple electrostatic model is in fair agreement with some experimental data on the energy of the hydrogen bond and dipole moment. Also, the hydrogen bonds are mostly linear as required for maximum electrostatic interaction.

Deviations from linearity may, however, result when constraints of crystal packing or other geometrical factors are involved. Side by side, it has been suggested that an X—H ···Y hydrogen bond is formed preferentially in the direction in which the charge cloud of Y is mostly concentrated.

If unshared pairs of electrons in Y are concentrated along the direction of *hybrid orbitals*, the partially positively charged hydrogen will approach the atom Y preferentially along this direction. In other words, the donor atom Y will now appear as an "*atomic dipole*" rather than a spherical accumulation of negative charge. This hypothesis of preferred direction can explain the zigzag structure of $(HF)_n$ polymer in the crystalline state.

But the simple electrostatic theory cannot explain the marked increase of intensity of the stretching vibration $\nu_{X—H}$ after H bond formation nor the lower intensity of the bending mode. Electronic transitions are also largely unexplained.

Some covalent character may be assigned to the hydrogen bond in terms of resonance hybrids of the type:

$$X—H : Y \leftrightarrow \overset{-}{X} : H—\overset{+}{Y}$$

But detailed calculations show that contribution of such covalent character to the hydrogen bond is really very small. This contribution becomes somewhat significant only for the strongest, shortest bonds like those in the bifluoride ion, HF_2^- (symmetrical). In VB approach, one can invoke resonance among the structures

$$F—H\cdots\bar{F} \leftrightarrow \bar{F}\cdots H—F$$

In MO treatment, H bonding in HF_2^- may be approached from a 3-centre 4-electron description as illustrated earlier.

Simplified MO *description of the bonding in* HF_2^-

The essentials of the MO energy level diagram of FHF^- (Fig. 7.40) may also be arrived at from the MO diagram of HF (Fig. 7.34) and a matching orbital of F^-. Neglecting nonbonding MOs and the electrons occupying them, we simply combine the σ and σ* MOs of HF (a_1 type) with the p_z orbital of F^- (also having σ symmetry with respect to the molecular axis, *z*). We get three MOs: one bonding, one antibonding and one nonbonding (Fig. 7.49). The additional electron pair from F^- now occupies this essentially nonbonding MO. So this MO becomes the HOMO. The antibonding MO becomes the LUMO. We observe that the resulting FHF^- has a total energy lower than the sum of the energies of HF and F^-.

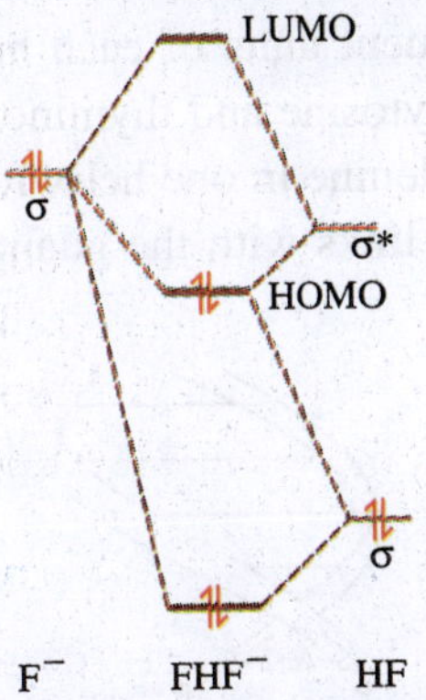

Fig.7.49: Simplified MO diagram for FHF^-.

This observation may be generalized for any hydrogen bonding of the type X—H···Y where XH replaces HF and Y replaces F^-. If the HOMO of Y and the LUMO of HX have comparable energies, both occupied orbitals in X—H···Y will be lower than either of the interacting orbitals. Consequently the hydrogen bonded unit X—H···Y will have a lower total energy compared to XH and Y in a 3-centre-4-electron MO system.

If X is a highly electronegative element, its highest occupied orbital will lie much lower in energy than the 1*s* orbital of H and the H—X bond will be polar with major electron density near X. This will favour H bonding with Y whereby the energy of the bonding MO of HX will be further lowered.

If the energy of the HOMO of Y lies far below that of the LUMO of XH, the HOMO in XH···Y will be above the Y-orbital and will not serve to stabilize the system. A still greater mismatch in the energies of HOMO and LUMO will stabilize the system in favour of YH^+ and X^- by transfer of a hydron (proton). We shall find more applications of such HOMO-LUMO interaction in connection with acids and bases.

The hydrogen bond differs from the "*hydrogen bridge bond*" involved in explaining the bonding in B_2H_6 and other boranes (see text). The B—H—B hydrogen bridge bond involves a 3-centre-2-electron (3*e*-2*e*) system.

It is not difficult to understand why other atoms cannot form bonds similar to hydrogen bonds. For greatest interaction, either electrostatic or covalent, the participating atoms should approach as close as possible. Hydrogen has two favourable points in this respect: (i) its small size and (ii) absence of inner electron shells.

This is why in a hydrogen bond X—H···Y, X and Y can approach very close to each other without involving large repulsion. Replacement of H by a larger atom, say Na($3s^1$), would cause enough repulsion from the inner shell of electrons to offset the net binding.

7.5.4 Hydrogen Bond in Biological Systems

Hydrogen bonding plays a vital role in biological systems. Proteins contain chains of amino acids. The amino acid units are arranged in a spiral form, somewhat like a stretched coil spring (forming a helix). The N—H group of each amino acid unit and the fourth C=O group following it along the chain establishes N—H...O hydrogen bonds. These bonds are partly responsible for the stability of the spiral structure.

Nucleic acids also contain extensive hydrogen bonding. DNA (deoxyribonucleic acid) consists of two intertwinning helices (a double helix) held together by hydrogen bonding. The constituent units of each helix, called nucleotides, contain different bases like adenine, guanine, cytosine and thymine. These basic parts interlink the two helices by hydrogen bonding, e.g., adenine in one helix forms hydrogen bonds to the thymine of another helix; the cytosine of one links with the guanine of another (Fig. 7.50).

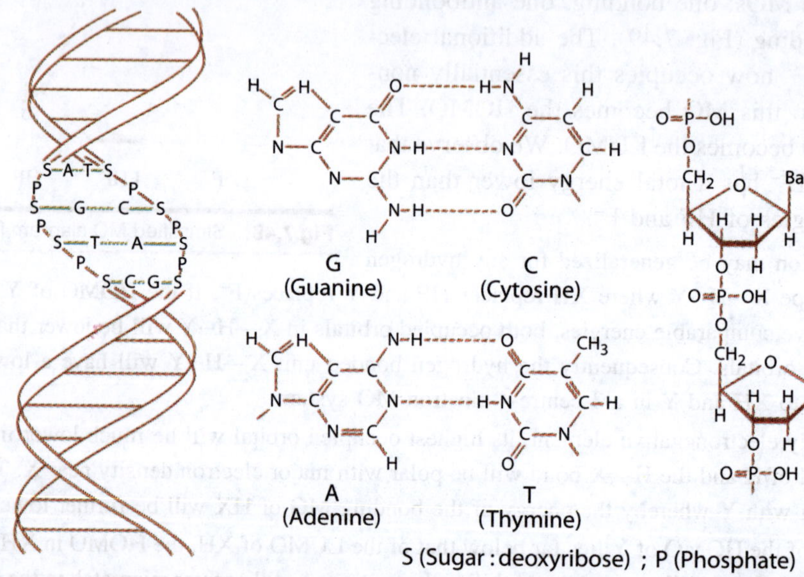

Fig.7.50: A double helix of DNA molecule with structural units shown enlarged on the right.

The functioning of many essential metalloenzymes (e.g., carbonic anhydrase and carboxypeptidase, Chapter 33) is regulated by H bonds. Proper orientation of the aminoacid residues around the metallobiosite, which appears crucial for initiating the enzyme activity, is maintained by H bonds.

ESSENTIALS

The variation principle states that the energy of a molecular system calculated by any arbitrary wave function can never be less than the actual energy.

The Born-Oppenheimer approximation asumes the nuclear framework as stationary in calculating the potential energy of molecular electrons.

In the valence bond theory (VBT) a localized bond is formed between two atoms when an electron in an atomic orbital of one atom pairs its spin with another electron in an atomic orbital of the other atom. Directional character of atomic orbitals gives rise to directional character of the bonds.

The VBT involves the idea of *hybridization* or mixing of atomic orbitals to explain the geometry of molecules.

In the molecular orbital theory (MOT), electrons are assigned to definite energy levels or molecular orbitals (MO) formed by combination of different contributing atomic orbitals.

The combining atomic orbitals should possess similar energies and matching symmetry around the internuclear axis. The total number of MOs formed is always equal to the total number of participating AOs.

A bonding MO, when occupied, lowers the energy of the system relative to the participating atomic orbitals. They are said to contribute to form a bond between the atoms concerned. *An antibonding MO* has higher energy relative to the contributing atomic orbitals; when occupied, they increase the energy of the molecule and decrease the strength of bonding between the atoms. *A nonbonding MO* has nearly the same energy as the combining atomic orbitals and does not contribute to bonding.

A σ (sigma) molecular orbital possesses an axis of symmetry coincident with the internuclear axis. Its orbital angular momentum about the internuclear axis is zero.

A π (pi) molecular orbital possesses a nodal plane passing through the internuclear axis. It has one unit of orbital angular momentum around the internuclear axis.

Molecular orbitals containing two and three mutually perpendicular nodal planes intersecting on the internuclear axis are designated as δ (delta) and ϕ (phi) MO-s respectively.

In homonuclear diatomic molecules from Li_2 to N_2, the $2s$ and $2p$ AOs on the same atom are sufficiently close to interact with one another. This results in higher energy of the *σ ($2p_z$) MO relative to the π ($2p_x$* and $2p_y$) MOs. This explains the paramagnetic nature of B_2.

In a bond between two atoms with different electronegativity, the lowest energy MO gets greater contribution from the AO of the more electronegative atom. The highest energy MO gets greater contribution from the AO on the less electronegative atom.

Molecular orbitals for polyatomic molecules are constructed from symmetry adapted combination of the concerned atomic orbitals.

Van der Waals interaction between closed-shell molecules involve weak interactions involving ion-dipole, dipole-dipole and instantaneous dipole-induced dipole type interactions (London dispersion interaction).

A hydrogen bond is an interaction of the type X—H···Y where X and Y are mostly electronegative atoms. Though considerably weaker than usual covalent bonds, extensive H-bonding has profound influence on structure and orientation of molecules.

EXERCISE

1. Starting with the $1s$ atomic orbital wave functions of two separate H atoms, how can you proceed to develop a wave function for the space part of the electrons in the H_2 molecule? Do all such wave functions imply an H—H bond?

2. How does the idea of spin pairing appear essential in the valence bond approach to the H_2 molecule?

3. A and B are any two atoms. Illustrate, with rough sketches, how the following MOs will be formed (A and B may be different pairs of atoms in a, b, and c):

(a) σ and σ* from $2s$(A) and $3p$(B); (b) σ and σ* from $3d$(A) and $3d$(B); (c) δ and δ* from $3d$(A) and $3d$(B). Take z-axis as internuclear axis.

Hint: (a) σ: 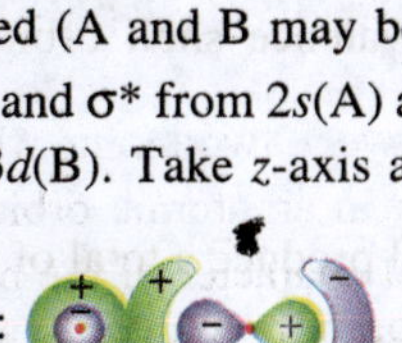

($2s + 3p_z$) (b) $\hat{\sigma}$* : 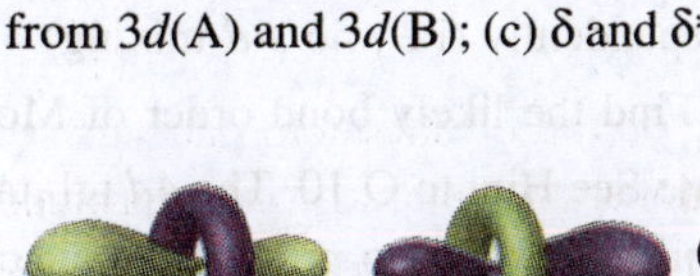
($3d_{z^2} - 3d_{z^2}$)

(c) δ* : 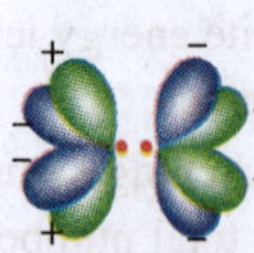 $(3d_{x^2-y^2} + 3d_{x^2-y^2}$ or $3d_{xz} + 3d_{xz})$

4. Guess, with rough sketches, if you expect any net overlap in the following combinations of AOs. In case of any fruitful overlap, also guess the type of MO formed (*z*-axis as internuclear axis). (a) $(3d_{z^2} + 3d_{yz})$ (b) $(3d_{yz} + 3d_{xz})$ (c) $2s + 3d_{x^2}$.

5. With rough MO energy level diagram, show the ground-state electron configurations of (a) H_2^- (b) N_2 (c) B_2^-. Find the bond order in each case.

6. Give the ground state electron distribution in MO-s of ClF, CO and CS.

7. Which of the molecules are expected to be stabilized by (a) addition of an electron (b) by removal of an electron? (a) NO (b) C_2 (c) O_2 (d) CN. Answer by qualitative MO description.

Ans. (a) C_2, CN (b) NO, O_2.

8. A radiation of 100 nm wavelength is used to eject an electron from an orbital of ionization energy 10.9 eV. What will be the energy of the electron?

Ans. 2.4×10^{-19} J.

9. (a) Construct a qualitative MO energy level diagram for linear H_3. (b) Calculate the total bond order for each of the species H_3^+, H_3 and H_3^-. (c) Why is H_3 unstable?

[**Hint**: (a) The MO energy level diagram is formed as shown. (b) Total bond order for each = 1. Bond order per link = 0.5. (c) $2H_3 \longrightarrow 3H_2$ is exothermic (consider change in total bond order)].

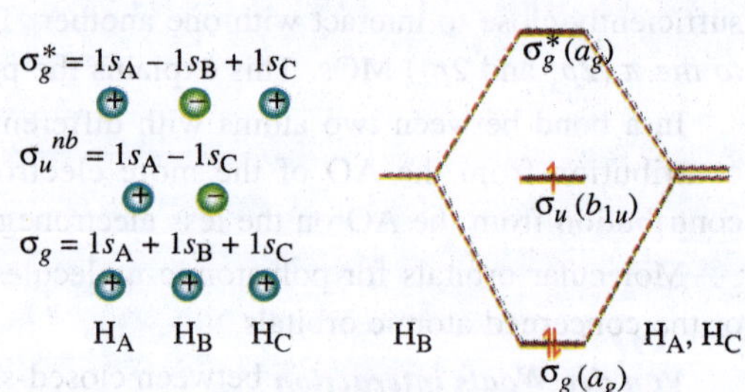

7-XII

10. (a) With rough sketches, show the formation of σ, π and δ bonds by overlap of appropriate *d*-orbitals of two identical *d*-block atoms (*z*-axis as internuclear axis).

(b) Do you expect the effectiveness of such *d-d* overlap to increase from period 4 (3*d* series) to period 5(4*d*)? Explain.

Hint: (a) σ, σ* (d_{z^2}); π, π* (d_{xz}, d_{yz}); δ, δ* $(d_{xy}, d_{x^2-y^2})$.

(b) For elements in the 3*d*-series, maxima of 3*d* orbitals $\langle r_{max} \rangle$, fall inside the maxima of the 3*s*–3*p* core and of the 4*s* orbital. With increasing Z_{eff}, the *d*-orbitals actually get buried and show gradually decreasing aptitude to undergo overlap with orbitals on other atoms. π and δ bonds are most affected.

In the fifth (and sixth) period, $\langle r_{max} \rangle$ of the valence shell *d*-orbitals are closer to $\langle r_{max} \rangle$ of relevant *s* and *p*-orbitals. π and δ- bonds are more common in these periods.

11. (a) Draw an approximate MO energy level diagram for a homonuclear diatomic species of a *d*-block element in the fifth period. Consider 4*d* and 5*s* valence shell orbitals interacting independently (i.e., no *s-d* mixing).

(b) Find the likely bond order of Mo_2 from this diagram.

Hint: See Hint to Q.10. The 4*d* orbitals on the two atoms will produce a total of ten MO-s. The 5*s* orbitals will also produce a pair of σ MOs. The separation between the 5*s* and 4*d* orbitals varies across a period and the relative positions of σ_g (5*s*) and σ_u^* (5*s*) become uncertain. Their positions also depend on the relative magnitudes of the two types of overlap. Accordingly the correct order of MOs become uncertain. A qualitative diagram (not to scale) is given (z-axis as internuclear axis).

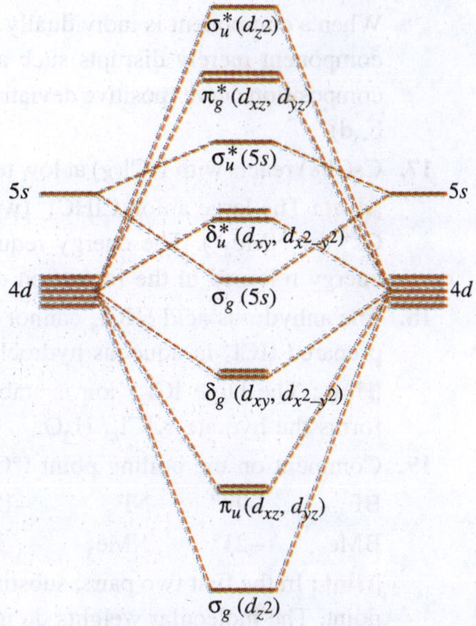

7-XIII

(b) The electron distribution for the 12 valence electrons is expected to be $\sigma_g^2\,\pi_u^4\,\delta_g^4\,\sigma_g(5s)^2$. The predicted bond order is 6. Such a *sextuple bond* has been supported by actual calculation.

12. The energy necessary to dissociate one O atom from different dioxygen species are reported as (in kJ mol^{-1}) : 623, 494, 351 and 205. Correlate these terms with the species O_2^{2-}, O_2^-, O_2 and O_2^+. Explain.

13. Do you expect the ionization energy of $C_2^{2-}(g)$ (acetylide ion) to be greater than that of C_2 (g)? Explain your answer.

14. Explain very briefly:

(a) There is no interelectronic repulsion in H_2^+.

(b) The plane perpendicular to the internuclear axis at the centre of the two nuclei in H_2^+ is a node for the first excited state.

(c) D_e is always greater than D_0 (refer Fig. 7.1).

(d) The maximum electron probability density in the ground state of H_2^+ occurs at the two nuclei.

(e) The ground electronic state of H_2 corresponds to a singlet term but that of O_2 corresponds to a triplet term.

(f) All triatomic molecules are planar.

(g) The electron affinity of C_2 is much greater (341 kJ mol^{-1}) than that of N_2 (about –212 kJ mol^{-1}) or O_2 (about 42 kJ mol^{-1}).

15. HF forms stronger H-bonds than H_2O. Still ΔH_{vap} of HF is lower than that of H_2O.—Explain.

[**Hint:** The number of H-bonds broken per mole of HF on vaporization is much less than in the case of H_2O. Each HF molecule forms two H-bonds, while each H_2O molecule forms four H-bonds. Again, the vapour of HF at its boiling point contains polymers upto $(HF)_6$, i.e. all H-bonds in liquid HF are not broken on vaporization. In case of water, vapours at boiling point contain essentially monomers, i.e., four H-bonds are broken per molecule of H_2O vaporized (the number is much less than two in case of HF).

16. Which of the following mixtures would be expected to exhibit maxima in the boiling point-composition curve? Which would show minima? Explain.

(a) methyl acetate and chloroform, (b) ethanol and n-hexane, (c) chloroform and carbon tetrachloride, (d) water and n-octane, (e) HCl and dimethyl ether, (f) chloroform and triethylamine, (g) chloroform and diethyl ether.

[**Hint:** When the components of a mixture are not individually associated through H-bonding, but may establish H-bonding in the mixture, the vapour pressure above the mixture will show a minimum at a particular composition (negative deviation from Raoult's law) and form an azeotrope with maximum boiling point (mixtures a, e, f, g.).

When a component is individually associated through H-bonding but the presence of the second component merely disrupts such association, a maximum is expected in the vapour-pressure-composition curve (positive deviation from Raoult's law) with a minimum boiling point (mixtures b, d).

17. $CsCl(s)$ reacts with $HCl(g)$ at low temperatures to form $CsHCl_2$, but $NaCl(s)$ does not. Comment.
[**Hint**: The large anion $ClHCl^-$ (weakly H-bonded) can be stabilized only by large cations like Cs^+ (or NMe_4^+). The energy required to expand the NaCl lattice is not compensated by the energy released in the formation of the $ClHCl^-$ ion.

18. The anhydrous acid $HICl_4$ cannot be isolated, but the crystalline hydrate $HICl_4$, $4H_2O$ may be prepared (ICl_3 in aqueous hydrochloric acid). Explain.
[**Hint**: The large ICl_4^- ion is stabilized by $H^+(H_2O)_4$ (i.e. $H_9O_4^+$). The corresponding K-salt forms the hydrate $KICl_4$, H_2O.

19. Comment on the boiling point (°C) trend in each of the following pairs of compound:

BF_3	−99°	NF_3	−129°	BiF_3	non-volatile	CCl_4	76.7
BMe_3	−21°	NMe_3	3.5°	$BiMe_3$	110°	$SiCl_4$	57

[**Hint**: In the first two pairs, substitution of F by CH_3 group results in a large increase in boiling point. The molecular weights do not vary appreciably for each pair of compounds, but van der Waals interaction increases due to greater polarizability of the methyl derivative. In the third pair, large electronegativity difference makes BiF_3 an ionic solid.
In the last pair, $SiCl_4$ with higher molecular weight has the lower b.p. (the trend is reverse with CH_4—SiH_4). The lower electronegativity of silicon is significant. The chlorine atoms acquire more than twice as much negative charge; this results in a substantially higher electrostatic repulson among $SiCl_4$ molecules than among CCl_4 molecules. The density of $SiCl_4$ is also lower for this reason.

20. Explain:
(a) Boiling points of noble gases increase from He—Rn.
(b) Acetic acid in the gas phase shows a lower pressure than predicted by the ideal gas law.
(c) An equimolar mixture of chloroform and acetone has a vapour pressure lower than either of the pure liquids.

8

CRYSTALLINE SOLIDS

8.1 SOLIDS

A major part of the objects in our surroundings consists of solid substances, recognized readily from their rigid shapes and incompressibility. Further observation reveals many other features possessed by one or another type of solid—their structural regularity, electrical and magnetic properties, catalytic activity and so on. In our venture to understand these macroscopic properties, we have to explore the nature of forces involved in holding the constituents of solids—the atoms, ions or molecules, in a close assembly.

8.1.1 Classification of Solids

From a structural viewpoint, we may primarily classify solids as amorphous or crystalline. An amorphous solid has no well-defined long range order in the arrangement of their constituent units like atoms, molecules or ions; only localized short range orders are present in such substances. Common glass is an amorphous solid; it has no sharp melting temperature. Rubber and many plastics such as polyethylene are also amorphous. A crystalline substance is made up of one or more crystals, each having well-defined ordered arrangement of the constituent units. Sodium chloride and sugar are crystalline substances. Such materials may occur as powder, lumps or agglomerates and appear amorphous. When examined under a microscope, the individual particles will show the characteristic interfacial angles of a crystal. Metals often occur as compact polycrystalline forms made up of individual microcrystals of atoms.

Amorphous substances are *isotropic*, that is, they have the same physical properties in all directions within their structure. Some of these physical properties are refractive index, coefficient of thermal expansion, electrical conductivity, speed of sound etc. Crystalline substances are, in general, *anisotropic*, showing difference in these physical properties along different directions. Though a finely powdered crystalline substance may show an overall isotropy owing to random orientation of the tiny crystals, a single crystal will show anisotropy in one property or another. Thus, rhombic crystals of silver nitrate show different refractive indices when measured along the three different crystal axes. The cubic crystal of NaCl has equal refractive indices along each axis and appears optically isotropic. But its response to mechanical stress is different in different directions—it undergoes cleavage only along planes containing the faces of unit cells (see later). Similarly, a piece of metal apparently appears isotropic owing to random orientation of the individual anisotropic single crystals.

The demarcation between crystalline and amorphous forms is not always sharp. There are several substances, for example, the solid synthetic polymers, which are partly crystalline and partly amorphous. **Liquid crystals** also do not belong solely to this category. They flow like liquids but unlike liquids, are anisotropic, having much of the long range order of solids.

LIQUID CRYSTAL DISPLAY (LCD)

Many liquid crystals are long organic molecules with a nonpolar chain and a polar group, for example, $H_3CO–C_6H_4–CHNC_6H_4C_4H_9$ and $C_6H_{13}C_6H_4COOH$. In a certain phase, these molecules usually have the same spatial orientation, but the intermolecular spacings are irregular. They can move about like liquids with this arrangement. In some liquid crystals, the molecules are flat (like discs) and form layers having same orientation of the molecules in a layer, but the orientation varying from one layer to another. The orientation of the molecules can be readily altered by applying a small electric field. This fact is taken advantage of in liquid crystal displays (LCD) we find in watches, calculators etc. The display consists of liquid crystals placed between glass plates whose interior surfaces are treated to orient the molecules in a particular alignment. The space between the glass plates is also provided with (i) transparent electrodes to reorient the moleculs and (ii) polarizer sheets (on both sides). Light entering through the front gets polarized and falls on the liquid crystal layer. When the voltage to a set of electrodes in certain select areas of the display is off, the liquid crystal layer in those areas can rotate the plane of the incident polarized light which falls on the second polarizer at the other side and is reflected back by a mirror; this illuminates the select display area. Where the voltage is turned on, the incident polarized light passes unchanged and does not match the second polarizer. So it is not reflected back from the mirror and the corresponding area appears black. Disc shaped liquid crystal molecules are used as temperature indicators as they show bright colours dependent on temperature.

According to the nature of the overall cohesive force, crystalline solids may be classified as *molecular, metallic, covalent network solid* and *ionic*. Table 8.1 summarizes the main features of these four classes.

Table 8.1
Types of solids

Type and constituent	Cohesive force	Binding energy/kJ mol^{-1}	Examples, mp/K	Physical nature
I. **Molecular**	(i) dispersion	0.05–40	Argon 84	soft, ins
Molecules	(ii) dipole	5–25	HCl 158	soft, ins
	(iii) H-bonding	10–40	Ice 273	hard, ins
II. **Metallic**	extensively	70–1000	Mg 923	hard but
Atoms	delocalized			malleable,
	electrons		Cu 1356	cond
III. **Covalent**	covalent	150–500	C(dia) 3773 (subl)	hard, ins
Network solid	bond		SiO_2 ~ 1873	hard, ins
Atoms	(network)		SiC infusible	hard, ins
IV. **Ionic**			NaCl 1081	hard, brittle
Ions	electrostatic	400–4000	MgO 3099	hard, brittle

ins = electrical insulator; cond = electrical conductor

I. Molecular crystals contain individual molecules held by relatively weak intermolecular forces. Accordingly, these may be subdivided into *van der Waals crystals* or *hydrogen-bonded crystals*. In the former category, the intermolecular forces involve dipole interactions and dispersion forces: Ar, CO, CO_2, HI, $C_6H_5NO_2$, $HgCl_2$, $SnCl_4$, C_{60}. Hydrogen-bonded crystals are dominated by H-bonded networks, for example, H_2O, HF, NH_3, glycine ($H_3^+NCH_2COO^-$). The intermolecular distance in such crystals is used to deduce *van der Waals radii* for atoms (section 5.3.3).

The cohesive energy or binding energy (E_c) for a molecular solid is expressed by the enthalpy of sublimation, $\Delta H°$. Some representative values in kJ mol^{-1} are (at or near respective melting points in °C given in parentheses):

Ar	Kr	CH_4	I_2	H_2O	H_2S
7·5 (–189)	10·5 (–157)	9·6 (–182)	61 (114)	51 (0)	21 (-86)

The values demonstrate the effect of size on van der Waals interaction (chapter 7). The effect of hydrogen bonding in packing is also significant.

II. Metallic crystals consist of metal atoms packed together according to their size. The interatomic distance for coordination number 12 is used to estimate the **metallic radius** (Ch. 5.3.3). The cohesive energy of a metal is the heat of sublimation of the solid to a monatomic gas at 1 bar. Some representative values in kJ mol^{-1} at 0K are given below (melting points in °C are included in parentheses).

Na	Mg	Al	Cu	W	Pt
108 (98)	146 (650)	324 (660)	337 (1083)	848 (3400)	564 (1770)

The variations are consistent with the number of valence electrons available and the effective nuclear charge (Ch. 5). The bonding in metals involves delocalization of the available valence electrons over the entire skeleton and is described by the band theory (section 8.4).

In the next section we shall use the packing of metal atoms, all uniform spheres, to illustrate the types of various crystal structures. The structures adopted by some common metals are:

Face-centered cubic (FCC, CN = 12):	**Ca, Al, Pb, Cu, Ni, Au, Pt, Pd**
Body-centered cubic (BCC, CN = 8):	**Li, Na, K, Rb, Cs, Ba, Cr, Mo, W**
Hexagonal close-packed (HCP, CN = 12):	**Be, Mg, Zn, Cd, Tl, Ti, Co**

III. Covalent network solids do not contain any individual molecule; an infinite network of atoms held together by polar or nonpolar covalent bonds extends over the entire structure which resembles a single giant molecule. Diamond, graphite, silicon, SiO_2 and SiC are common examples of such network solids. These are marked by large cohesive energies (E_c) which result from extensive covalent chemical bonds. E_c values can be obtained from $\Delta H_f°$ values of the solid and the gaseous atoms. Some representative values at 0K are (in kJ mol^{-1}):

Diamond	**Graphite**	**Si**	**SiO_2 (quartz)**	**SiC**
709	711	451	1851	1227

As expected, these substances are very hard and have high melting points.

IV. Ionic solids consist of a three-dimensional array of oppositely charged ions packed according to their relative size and charge. The cohesive energy of ionic crystals, commonly called the lattice energy or lattice enthalpy, is derived from the electrostatic interaction (both attraction and repulsion) among the ions in the bulk; some repulsion also arises from slight overlap of the electron probability densities of adjacent ions (Pauli repulsion). We shall explore the energetics and structure of ionic solids in a greater detail in subsequent sections.

Determination of crystal structure

The regular interatomic spacing in crystals may be determined by diffraction of x-ray photons. In 1912 Max von Laue realized that the interlayer separation of atoms in a crystal is of the order of 1A°; so a crystal may be used as a diffraction grating for x-rays which also have wavelengths of this order. Soon this idea was successfully applied by W. L. Bragg and W. H. Bragg (son and father) to determine the structure of sodium chloride crystals.

When a beam of monochromatic x-rays strike a layer of atoms in a plane of some crystal, most of the x-ray photons pass undeviated through this plane but a small fraction of the photons gets scattered after colliding with the electrons in the atoms present in that plane. The second layer of atoms in the next parallel plane of the crystal similarly scatters a portion of the x-rays reaching this layer and the process is repeated for successive parallel planes in the crystal. Now, the rays scattered from successive layers will reinforce each other only when they are in phase and

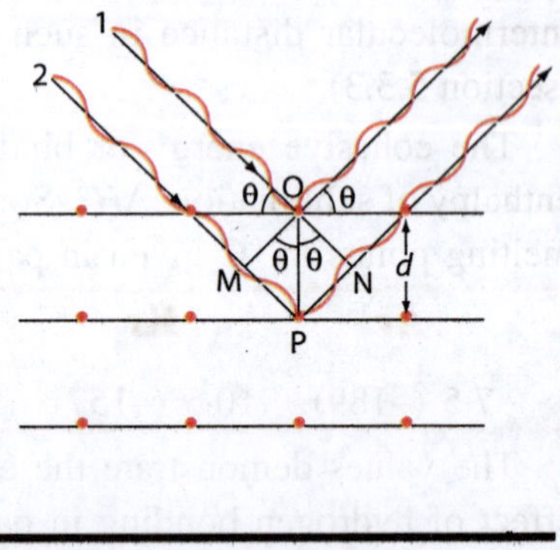

Fig.8.1: Diffraction of two waves of x-rays from two successive layers of atoms in a crystal (schematic).
Path difference for ray 2 is equal to MPN = $2d\sin\theta$.
For both of the rays to be in phase after scattering, MPN should be an integral multiple of λ, the wavelength of the x-ray. This gives $2d\sin\theta = n\lambda$, $n = 1, 2, 3, ...$

able to form constructive interference. This will result in a high intensity of the scattered beam. The condition for this situation may be understood from Fig. 8.1.

A small single crystal is mounted on a glass fiber and slowly rotated as a monochromatic x-ray beam falls on the crystal surface at various angles (θ, the glancing angle). The intensity of the beam after diffraction is followed by some suitable technique. It appears that the emerging beam shows several maxima of varying intensity caused by constructive interference for certain particular values of the glancing angle θ. The interlayer separation d is now given by the ***Bragg equation***:

$$2d \sin\theta = n\lambda \qquad n = 1, 2, 3, ...$$

where λ is the wavelength of the x-ray. n is a positive integer (1, 2, 3, ...) and is called the order of reflection. Different sets of planes with different atoms actually give rise to a complicated diffraction pattern which is resolved and skilfully analyzed to obtain lots of information on the structure of the crystal.

A set of planes will produce an intense diffracted beam when there are enough density of electrons in the plane. This requires a high density of atoms in the plane. Hydrogen atoms, which have only one electron, are not suitable for observation by this method.

Surface structure of crystals

Surface structure of crystals may now be determined with precision using several techniques developed since the 1970-s. Structures of substances adsorbed on solid surfaces may also be understood from such studies. The information received has proved to be very useful in materials science, particularly in the study of catalysts, and in biochemistry.

LEED or low energy electron diffraction study involves the scattering of low energy electrons (10–500 eV) from the surface of a crystal. A small fraction of the incident monoenergetic electrons undergoes elastic scattering (that is, their energy does not change) and the scattered electrons form a diffraction pattern through wave interference. Because the electrons have small penetrating power, their diffraction pattern is determined by the surface structure of the crystal. Metallic and covalent crystals have been mostly studied by this method.

LEED studies have revealed certain interesting facts about surfaces. Thus, it has been shown that the structure of the surface of a solid is different from the structure of corresponding planes in the bulk of the solid. The surface atoms change their locations and bonding to adopt a new structure, a phenomenon called *surface reconstruction*. LEED has also shown that in metals the distance between the surface layer and the second layer of atoms is usually 1 to 10% shorter than the distance between corresponding layers in bulk. This is called *surface relaxation*.

The scanning tunnelling microscope (STM) is another ingenious device for detailed mapping of crystal surfaces. The tip of a very sharp and precisely controlled metal needle is moved at a separation of about 5Å over the surface of an electrical conductor. A small voltage applied between the tip and the surface causes electrons to tunnel through the space between the tip and the surface, the magnitude of the tunnelling current depending on the separation between the surface and the tip. A contour map of the surface is thus obtained which can resolve even single atoms. The map shows that a crystal surface contains many terraces, steps and kinks formed during the growth of the crystal. The role of such uneven character of a surface in catalysis can be followed by STM. It has been shown that a platinum surface exposed to ethylene shows formation of deposits of carbon atoms first near the steps. Even biological specimens having sufficient conductance can be studied by STM; surfaces of viruses have been imaged by this technique.

By regulating the voltage applied, tunnelling in an STM can be restricted to specific orbitals of surface atoms when images of bonding orbitals are produced.

The atomic force microscope (AFM) scans the surface of a crystal with a sharp tip attached to a flexible cantilever arm (ever seen the arm of an old gramophone or a record player, in reality or in pictures!). A constant loading force is applied to the cantilever to keep the tip in contact with the surface. As the tip moves over the surface, the force between the tip and the surface atoms varies from place to place, causing varying deflection of the tip. This deflection is followed by a laser beam reflected from the top surface of the cantilever arm. Even individual large atoms like gold may be mapped in this manner. The surface to be scanned need not be conducting as with STM. In another variation of AFM (tapping mode), the scanning tip (the probe) taps the surface in quick succession as it scans the surface. AFM has been successfully applied to determine the shape of many biomolecules.

Electron microscopy has been successfully applied to study the structures of solid surfaces. There are different versions of electron microscopy like scanning electron microscopy (SEM), scanning tunnelling electron microscopy (STEM) and transmission electron microscopy (TEM). All of these techniques use a beam of electrons as probe.

8.2 STRUCTURES OF CRYSTALS

As early as 1665, Robert Hooke concluded (from his observations with magnifying glasses) that the various beautiful crystal forms resulted from regular packing of "globular" particles. The regular external forms of crystals are indeed related to an internal regularity in the arrangement of constituent particles. Let us briefly analyze such arrangements—again on the assumption of spherical shape for all atoms, molecules or ions.

8.2.1 Lattices and Unit Cells

A crystal consists of a basic structural unit or motif repeated in three dimensions. The basic unit may be a single atom, molecule or ion or a group of atoms or molecules which has identical environment throughout the crystal. The basic unit is also called *the basis* of a crystal. A basis has the same stoichiometric composition as the crystal itself.

Fig.8.2: The basis and the space lattice. (a) In the two dimensional arrangement, each *pair of atoms forms the basis.* (b) The centers of the small atoms (yellow) are represented by points to form the two dimensional lattice of (a). One gets the same lattice by taking the centers of the large atoms also. Actually any identical position of each basis may be used to form the lattice.

The *lattice* (or *space lattice*) is the geometrical idealization of the structure of a crystal. It is obtained by placing a point at the same location of each basic unit (or basis) in the crystal—the resulting three dimensional distribution of the points in space forms the lattice. The crystal structure may be generated by associating each point of the lattice with a basis.

When the crystal structure consists of ions or single atoms as basic units, it is conventional to associate the centers of the atoms or ions with the lattice points. In other cases any convenient point on the basic unit may be associated with the lattice, the point chosen being at the same location in each basis.

A **unit cell** of a crystal is the smallest part of the crystal which can generate the crystal structure through repetition in three dimensions by simple translational displacement (that is, without any rotation or reflection). In effect this is equivalent to constructing the whole crystal structure by stacking the unit cells together along their faces, without leaving any space inbetween.

To understand the formation of repetitive structure from a unit by translation, let us first consider a pattern in two dimensions, as shown in Fig. 8.3. (Such patterns are common in wall papers, gift wrappings and floor tiles). The repetitive pattern in the figure may be generated by translation of any of the four shaded parallelograms, two gray and two yellow, along their edges. The parallelograms in grey are preferred since they are smaller in area. Now, to generate the entire pattern, we have to translate or move the parallelogram by definite distances parallel to any of its edges (try using cardboard cutouts). Moving any distance *a* or its integral multiple towards right, we arrive at a point equivalent to the starting point. Similarly, if we move any integral multiple of *b* in the other direction (shown by arrow), we get an equivalent point. Thus, we can describe the whole pattern in two dimensions by the two vectors *a* and *b*, called the *primitive translations* of the two dimensional structure. When we represent the positions of the constituents of the two-dimensional structure by points, we get a two-dimensional lattice. The basic units of pattern are then represented by a parallelogram. Depending on the relative magnitudes of the sides *a* and *b* of the parallelogram and the angle between them, one gets five possible unit cells for a two-dimensional lattice (Fig. 8.4).

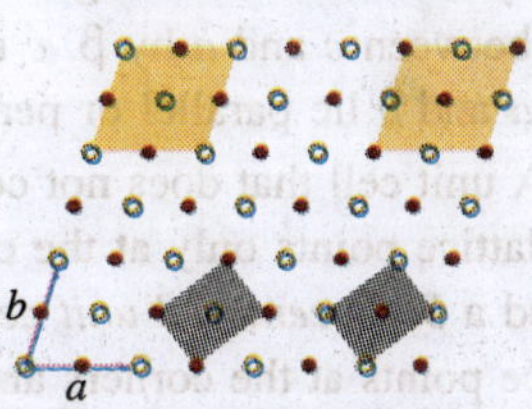

Fig.8.3: A repetitive pattern in two dimensions may be generated by translation of a parallelogram called the unit cell (gray). Two vectors *a* and *b* and the angle between them summarily describe the two-dimensional structure.

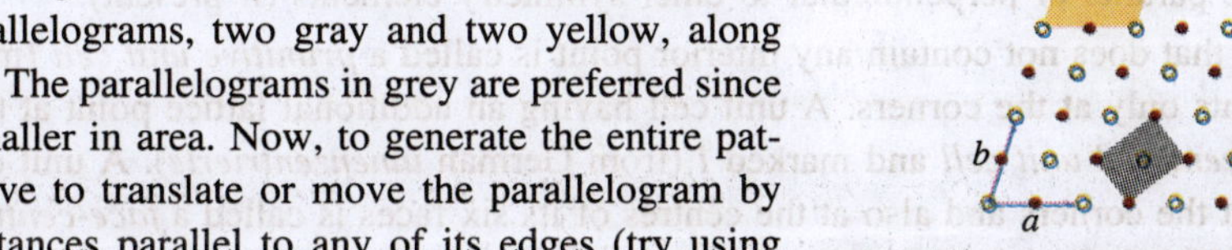

Fig.8.4: Five units of pattern (unit cells) in two dimensions.

For a pattern in three dimensions, we require a third vector out of plane of the first two vectors. These three vectors make a parallelepiped, which makes the unit cell for a repetitive pattern in three dimensions.

A unit cell of any crystal is represented by an imaginary parallelepiped, a solid whose faces are all parallelograms. Translation of the parallelepiped in three dimensions (stacking with no space left) generates the crystal structure, actually the crystal lattice (Fig. 8.5).

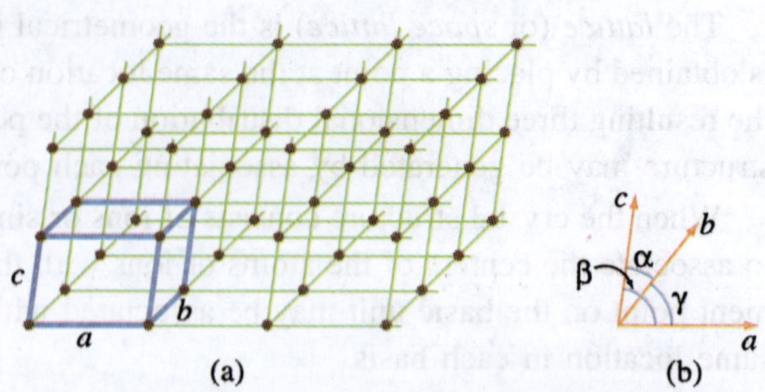

Fig.8.5: (a) A three-dimensional crystal lattice. The unit cell is shown in bold lines. (b) The three fundamental translational vectors *a*, *b*, *c* (unit cell parameters).

The selection of the unit cell is not unique; there may be alternatives. The unit cell having the maximum symmetry and the smallest volume consistent with this symmetry is normally chosen in crystallography. The measurement of the unit cell is expressed by the *unit cell parameters*: the sides of the parallelepiped *a*, *b*, *c* and the angles α, β and γ between these sides.

By convention, the angles between *a* and *b* is denoted by γ; the angle between *b* and *c* by α and that between *c* and *a* by β. *c* usually lies parallel to the highest order rotation axis in the unit cell and *a* and *b* lie parallel or perpendicular to other symmetry elements (if present).

A unit cell that does not contain any interior point is called a *primitive unit cell* (marked *P*); it has lattice points only at the corners. A unit cell having an additional lattice point at the centre is called a *body centered unit cell* and marked *I* (from German *innenzentrierte*). A unit cell that has lattice points at the corners and also at the centres of its six faces is called a *face-centred unit cell* and marked *F*. Unit lattices having lattice points at all corners and also on centres of two opposite faces are called side-centred unit cells (marked *C*; *A* or *B* are also used).

Depending on the symmetry of the unit cell, crystals may be classified into seven crystal systems represented by seven primitive unit cells.

The interrelation between the parameters *a*, *b* and *c* of the unit cells and the angles between them are shown in Table 8.2 and Fig. 8.6.

Table 8.2

The seven crystal systems (primitive unit cells)

Crystal system	Description	Symmetry criterion
1. Cubic	$a = b = c$; $\alpha = \beta = \gamma = 90°$	Four C_3 axes in a tetrahedral arrangement
2. Tetragonal	$a = b \neq c$; $\alpha = \beta = \gamma = 90°$	One C_4 axis
3. Orthorhombic	$a \neq b \neq c$; $\alpha = \beta = \gamma = 90°$	Three perpendicular C_2 axes
4. Monoclinic	$a \neq b \neq c$; $\alpha = \gamma = 90°$; $\beta \neq 90°$	One C_2 axis
5. Hexagonal	$a = b \neq c$; $\alpha = \beta = 90°$; $\gamma = 120°$	One C_6 axis
6. Rhombohedral (trigonal)	$a = b = c$; $120° > \alpha = \beta = \gamma \neq 90°$	One C_3 axis
7. Triclinic	$a \neq b \neq c$; $\alpha \neq \beta \neq \gamma \neq 90°$	None

Angle between *a* and *b* = γ; *b* and *c* = α; *c* and *a* = β.

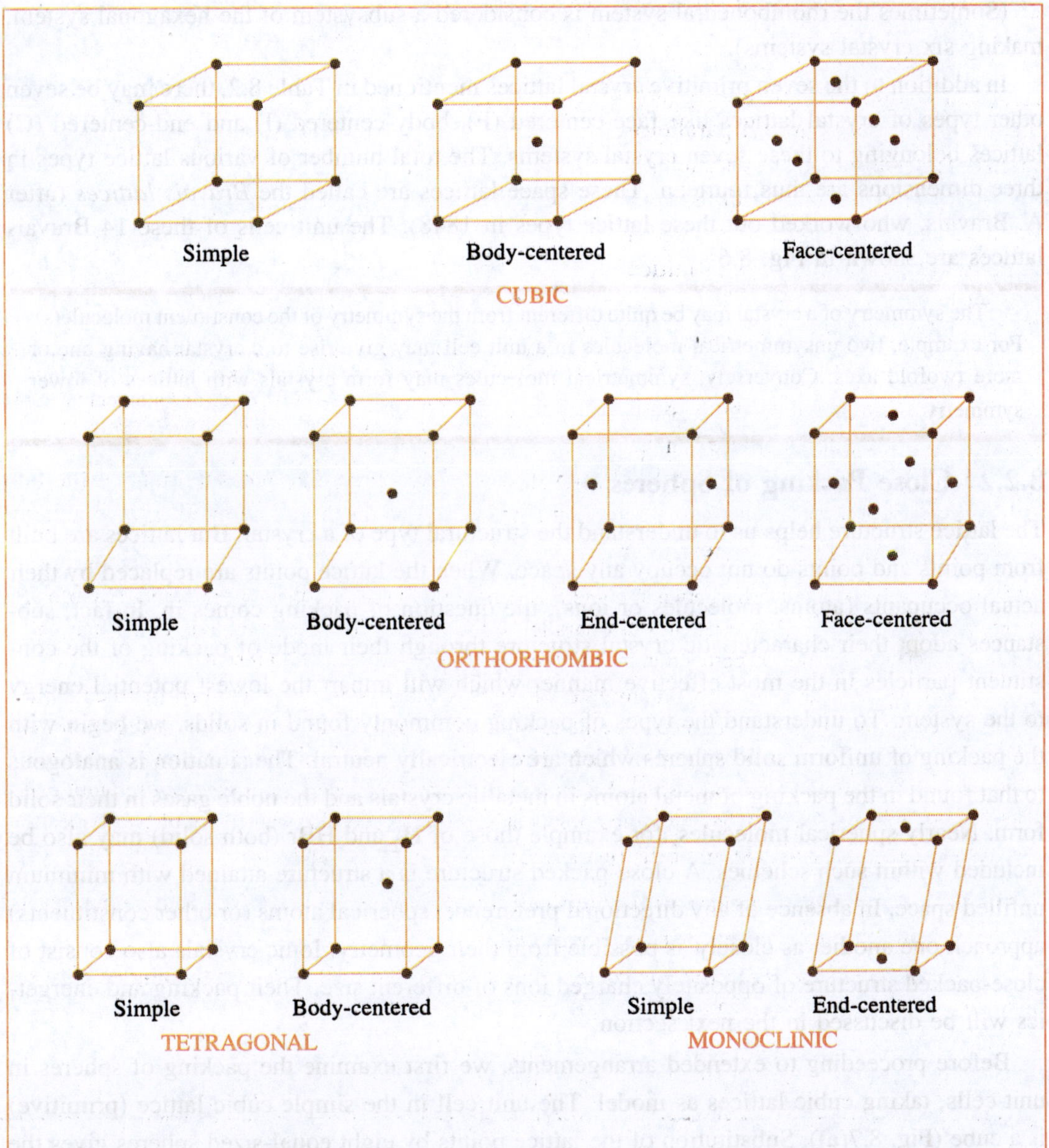

Fig.8.6: Units cells of the fourteen Bravais lattices.

(Sometimes the rhombohedral system is considered a subsystem of the hexagonal system, making six crystal systems).

In addition to the seven primitive crystal lattices mentioned in Table 8.2, there may be seven other types of crystal lattices like face-centered (F), body-centered (I) and end-centered (C) lattices belonging to these seven crystal systems. The total number of various lattice types in three dimensions are thus fourteen. These space lattices are called the *Bravais lattices* (after A. Bravais, who worked out these lattice types in 1848). The unit cells of these 14 Bravais lattices are shown in Fig. 8.6.

The symmetry of a crystal may be quite different from the symmetry of the constituent molecule(s). For example, two unsymmetrical molecules in a unit cell may give rise to a crystal having one or more twofold axes. Conversely, symmetrical molecules may form crystals with lattices of lower symmetry.

8.2.2 Close Packing of Spheres

The lattice structure helps us to understand the structural type of a crystal. But lattices are built from points and points do not occupy any space. When the lattice points are replaced by their actual occupants (atoms, molecules or ions), the question of packing comes in. In fact, substances adopt their characteristic crystal structure through their mode of packing of the constituent particles in the most effective manner which will impart the lowest potential energy to the system. To understand the types of packing commonly found in solids, we begin with the packing of uniform solid spheres which are electrically neutral. The situation is analogous to that found in the packing of metal atoms in metallic crystals and the noble gases in their solid form. Nearly spherical molecules, for example those of H_2 and HBr (both solid) may also be included within such schemes. A close-packed structure is a structure attained with minimum unfilled space. In absence of any directional preference, spherical atoms (or other constituents) approach one another as closely as possible from their geometry. Ionic crystals also consist of close-packed structure of oppositely charged ions of different size. Their packing and energetics will be discussed in the next section.

Before proceeding to extended arrangements, we first examine the packing of spheres in unit cells, taking cubic lattices as model. The unit cell in the simple cubic lattice (primitive) is a cube (Fig. 8.7(a)). Substitution of the lattice points by eight equal-sized spheres gives the arrangement shown in 8.7(b). In the extended lattice each corner sphere will be shared by eight cubes, four in the same layer and four above. Hence the effective volume occupied within the cube is equal to the volume of 8×1/8 or 1 sphere (Fig. 8.7(c)). If the radius of each sphere is r, the cube will have sides equal to $2r$ and a volume $8r^3$. The fraction (or percent) occupancy of the cube can be readily obtained from the volume of one sphere ($4/3\ \pi r^3$). The fraction comes out to be 0.524 (52.4 percent; see exercise Q.1).

The lattice of the body-centred cube (I) contains an additional point at the centre of the cube (Fig. 8.8(a)). The sphere occupying this point belongs exclusively to the cube. The space-filling model (Fig. 8.8(b)) shows that each cube effectively contains (8×1/8+1) or 2 spheres.

The size of the unit cube is now determined by the length of the body diagonal which must be equal to $4r$. If each side of the cube is a, we get $\sqrt{3}a = 4r$ (Fig. 8.10(a)).

The fraction of the volume of the cube occupied by the spheres may be calculated to be 0.680 (68.0 per cent).

Simple cube (P).
Number of atoms per unit cell
$= 8 \text{ corners} \times \frac{1}{8} = 1$

Fig.8.7: The simple cubic lattice (*Primitive, P*) and its space-filling model.

A face-centred cubic lattice (Fig. 8.9(a)) has, in addition to eight corner points, one point at each face-centre. The spheres occupying the face-centres are shared between two adjacent cubes (Fig. 8.9b and c). So the unit cell contains (8×1/8+6×1/2), a total of 4 spheres. The size correlation between the cube-edge (a) and the radius of the sphere (r) is shown in Fig. 8.10(b). The face diagonal ($= \sqrt{2}a$) now equals to $4r$.

The fraction of space occupied inside the cube is 0.740 (74.0 percent; see exercise Q.1).

Body-centred cube (I).
Number of atoms per unit cell
$= 8 \text{ corners} \times \frac{1}{8} + 1 \text{ centre} = 2$

Fig.8.8: The body-centered cubic lattice (I) and its space-filling model

Fig.8.9: The face-centred cubic lattice and its space-filling model.

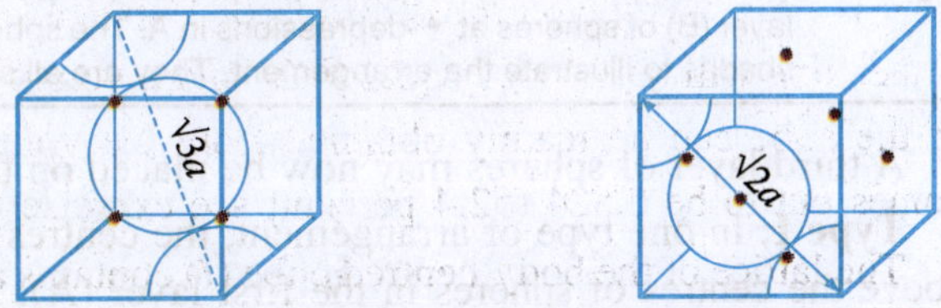

$a = (4/\sqrt{3})\,r$
(a)

$a = (2\sqrt{2})\,r$
(b)

Fig.8.10: Size-correlation in (a) body-centred (b) face-centred cubic arrangement of equal spheres. Other corner spheres have been omitted for clarity.

The characteristics of cubic lattices formed from uniform spheres are shown in Table 8.3.

Table 8.3

Characteristics of various cubic closepacking of uniform spheres.

	Simple (*P*) (Primitive)	Body-centred (*I*)	Face-centred (*F*)
Volume of unit cell	a^3*	a^3	a^3
Lattice points per cell	8	9	14
Distance to nearest neighbour (2*r**)	*a*	$\sqrt{3}a/2$	$a/\sqrt{2}$
No. of spheres per unit cell	1	2	4
Fraction of volume occupied	0·524	0·680	0·740
Coordination Number**	6	8	12

**a* = edge length of cube; *r* = radius of sphere. **Number of nearest neighbour, explained later

Next we consider the structure arising from close packing of a large number of uniform spheres layer by layer—somewhat comparable to stacking of oranges in fruit shops. In any layer, say A (Fig. 8.11a), each sphere is surrounded by six other spheres whose centres form a hexagon. This first layer has regular depressions (marked with • and ×) at each junction of three spheres. For *n* spheres in a layer, there will be 2*n* dips. We can place spheres at these depressions to form a second close-packed layer, say B. Note that we can place spheres of the second layer on half the total dips, either on all '•'-depressions or on all '×'-depressions, but not on both, for steric reasons. In Fig. 8.11(b), the second layer B has been formed by placing spheres on •-depressions of layer A. The ×-depressions of layer A are visible. The spheres in layer B also create new depressions marked o where three spheres in this layer meet.

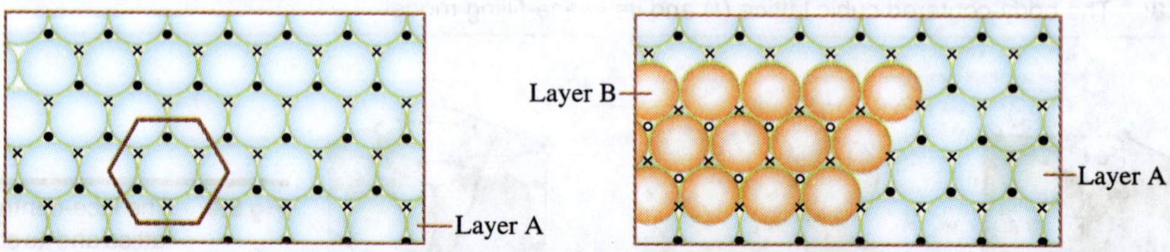

(a) (b)

Fig.8.11: (a) Close-packed layer of uniform spheres (A) (b) Two layers of spheres formed by placing a second layer (B) of spheres at '•'-depressions in A. The spheres in different layers have been shown in different shades to illustrate the arrangement. They are all same.

A third layer of spheres may now be placed on the second layer (B) in either of two ways.

Type I: In one type of arrangement, the centres of spheres in the third layer are vertically above the centres of spheres in the first layer (A). The layer B has two types of depressions marked 'o' and '×' (Fig. 8.11b). The third layer of spheres may be placed in all 'o' or all '×' depressions. If the spheres of the third layer are placed in all 'o'- depressions, we get each sphere of the third layer vertically above a sphere in layer A. Thus the third layer will be a repetition of A and the stack can be extended upward in the sequence ABABAB ... giving *hexagonal close-packing* (**HCP**) (Fig. 8.12).

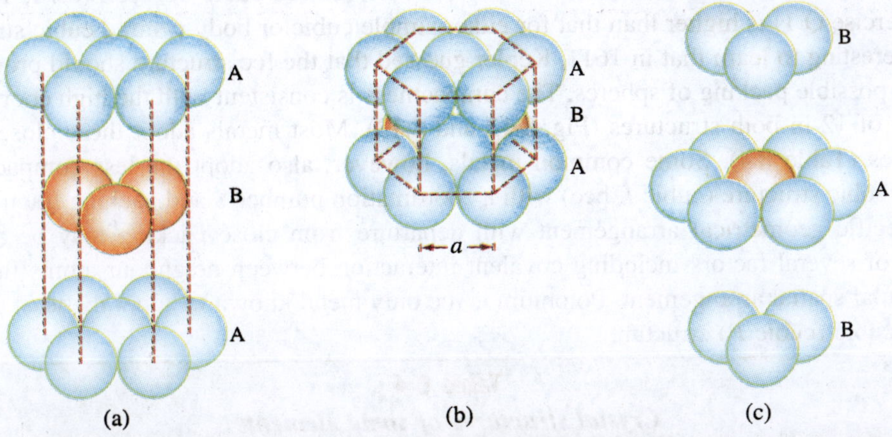

Fig.8.12: A close-packed structure of spheres in the layer sequence ABABAB. (a) Three layers of spheres shown separately (exploded view) (b) close-packed view of the layers to illustrate hexagonal close-packing. (c) Each sphere in a layer is surrounded by 6 spheres in the same plane, 3 spheres above it and 3 spheres below it—making a total of 12. This is called the coordination number.

Type II: When the spheres in the third layer are placed above the depressions which were unoccupied in the first layer (marked ×), no third layer sphere will be placed vertically above a first layer sphere. Hence the third layer will be arranged as a separate layer, say C. Hence the stack will show the sequence ABCABC ... . As shown in Fig. 8.13, such a sequence gives rise to **cubic close-packing** (CCP).

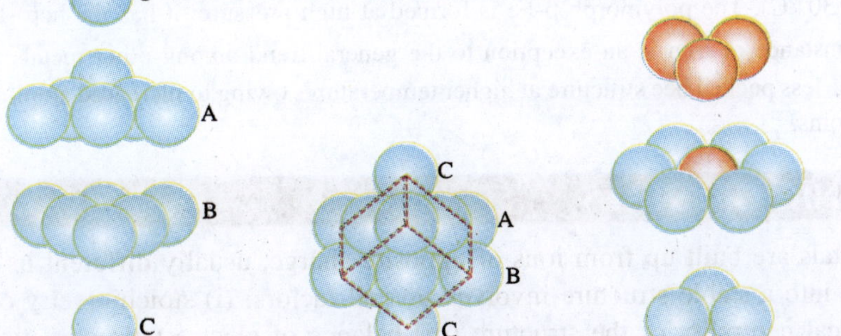

Fig.8.13: Stacking of spheres in the sequence of layers as ABCABC giving rise to cubic close-packing (ccp). (a) Separated layers. Only one sphere is shown in each C-layer for clarity. The sphere in C-layer fits the dips in both layer A and layer B.
(b) Close-packing of the layers in figure a gives rise to cubic close-packing (ccp). Each ccp unit cell has a sphere at each face, this is also a face-centred cubic (fcc) arrangement. The terms are often used interchangeably.
(c) The coordination number of each sphere (orange) is 12, 6 from the same layer, 3 above and three below.

The two types of structures (hcp and ccp) are examples of *polytypes*—structures which are same in the layers (i.e., in two dimensions) but different in three dimensions.

Both hcp and ccp structures have high packing fraction or fraction of space occupied by the spheres. For both structures, this fraction for face-centred cubic unit cell (0.74; Table 8.3 and Exercise Q.1) is higher than that for either simple cubic or body-centred cubic structures. It is interesting to learn that in 1611, Kepler guessed that the fcc structure should produce the densest possible packing of spheres. The compactness is consistent with the high coordination number of 12 in both structures (Fig. 8.12 and 8.13). Most metals adopt these close-packed structures (Table 8.4). Some common metals, however, also adopt the less compact body-centred cubic structure (cubic *I*, bcc) with a coordination number 8 and packing fraction 0.68. The specific geometrical arrangement with departure from close packing may be a consequence of several factors including covalent interaction between neighbour atoms favouring a particular spatial arrangement. Polonium is the only metal known to adopt the least compact simple cubic (cubic *P*) structure.

Table 8.4
Crystal structures of some elements

Cubic (*P*) :	Po	**hcp** :	Be, Cd, Co, Mg, Sc, Ti, Zn He
fcc (ccp, cubic *F*) :	Ag, Al, Au, Ca, Cu, Ni, Pd, Pb, Pt, Rh, Sr. Ar, Kr, Ne, Rn, Xe	**bcc (cubic *I*)** :	Ba, Cs, Cr, Fe, K, Li, Mn, Mo, Rb, Na, Ta, W, V

Metals often adopt different crystal structures under different conditions of temperature and pressure. Known as *polymorphism*, such divergent structures may be related to low directionality of the bonds formed by metals. Generally these polymorphs are labelled α, β, γ, ... as the temperature rises. Thus, iron has a bcc structure at room temperature called α-Fe which changes to a ccp structure (γ-Fe) at 906 °C. Above 1401 °C, this again reverts to α-Fe and remains in this form upto its melting point (1530 °C). The polymorph β-Fe is formed at high pressure; it has the hcp structure.

The instance of iron is an exception to the general trend among other metals which normally favour the less packed bcc structure at higher temperatures owing to increased amplitude of vibration of the atoms.

8.3 IONIC CRYSTALS

Ionic crystals are built up from ions of opposite charge, usually different in size. Packing of these ions into a stable structure involves several factors: (i) stoichiometry of the compound (ii) electrical neutrality of the structure (iii) balance of electrostatic attraction and repulsion interaction (coulombic) among the ions and (iv) Pauli repulsion (non-coulombic) between the electrons. The hard sphere assumption introduced earlier is also an oversimplification, since the ions often get deformed as a result of polarization. This will be considered separately.

Structural information on ionic crystals are obtained mainly from x-ray studies. Such studies also confirm the occurrence of ions in the crystal from electron density contours characteristic of the ions.

Our treatment of ionic crystals will be confined to structural description of some common ionic crystals on the hard sphere model and the energy factors involved in such structures.

8.3.1 Packing in Ionic Crystals

The packing of ions of different size can be understood in most cases by considering stacking of the larger ions in hexagonal or cubic close-packed structures in which the smaller ions (usually the cations) occupy tetrahedral or octahedral holes in required numbers to maintain stoichiometry and electrical neutrality. The relative sizes of the ions are thus a key factor in determining the structure of an ionic crystal. The limiting size of an octahedral or tetrahedral hole is determined by the ratio of the radii of the ions, usually $r_{\text{smaller}}/r_{\text{larger}}$ (often represented as r_+/r_- since the cations are usually smaller).

Limiting size of a tetrahedral hole

A tetrahedral hole is formed when a sphere rests on the junction of three other spheres—the centres of the four spheres now occupy the corners of a tetrahedron: three spheres form an equilateral triangle at the base and the fourth one rests on the dip formed by the three (Fig. 8.14a). The maximum size of the sphere which can be accomodated in the tetrahedral hole may be obtained from Fig. 8.14b.

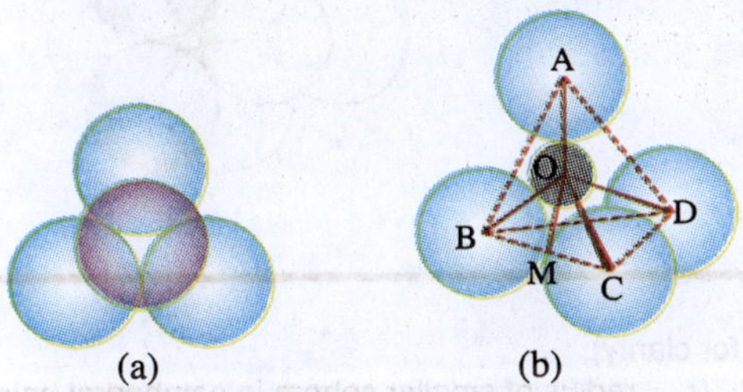

Fig.8.14: (a) A tetrahedral hole and (b) its geometry (exaggerated for clarity). The hole derives its name from the arrangement of the four surrounding nuclei (centres of spheres) which define a tetrahedron. The hole itself is formed by curved surfaces and is not tetrahedral.

If r and r_h be the radii of the larger ion and the inner sphere respectively, from the tetrahedron ABCD, OA (= OB = OC) = $r + r_h$. AB (or BC) = $2r$. BOM is a right-angled triangle where BM = MC = r. OM bisects the tetrahedral angle BOC (= 109·5°). ∠BOM = 54·75°. In the triangle BOM, sin 54·75° = BM/BO = $r/(r + r_h)$.

This readily gives r_h / r = 0·225. Thus a tetrahedral hole created by ions of radius r cannot accomodate any ion with radius greater than 0·225 r.

In any close-packed structure there will be two sets of tetrahedral holes: in one the apices of the tetrahedrons are directed up (say t) and in another the apexes are directed down (say t'). This is because each sphere in a layer is in contact with three others in the layer above it and with three others in the layer below. Close-packing of n spheres gives rise to $2n$ tetrahedral holes—n holes of each set.

Limiting size of an octahedral hole

An octahedral hole is formed between two planar triangles formed by centres of spheres in adjoining layers (Fig. 8.15). For steric reasons, neighbouring depressions in a close-packed layer cannot each accomodate a sphere of the layer above it. Each dip occupied by spheres of the second layer describes a tetrahedral hole; the dips unoccupied now form the octahedral sites as shown in Fig. 8.15. In a close-packed assembly of n spheres, there are n octahedral holes (vs. $2n$ tetrahedral holes).

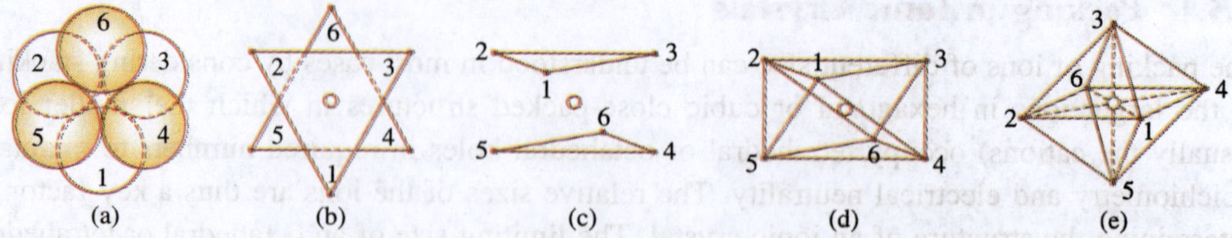

Fig.8.15: Formation of octahedral hole in close-packed structure of spheres.
We know that each face of an octahedron is an equilateral triangle. Such triangles may be formed in parallel planes of the close-packed structure by the centres of three spheres in contact (a). Two sets of three spheres, each set forming an equilateral triangle in parallel planes with vertices in opposite direction (b), may now form an "octahedral hole". Actually the centres of the spheres form an octahedron around the central cavity (c, d and e).

The geometry of the octahedral hole may be understood from Fig. 8.16. The hole cannot accomodate any ion larger than $0{\cdot}414r$ where r is the radius of the larger ion making the stack.

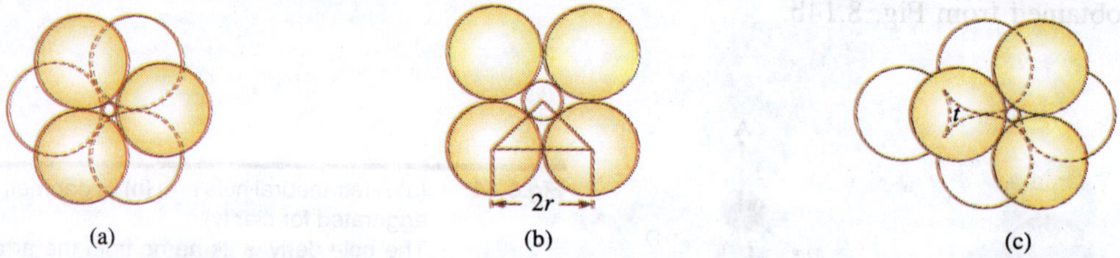

Fig.8.16: (a) An octahedral hole and
(b) its geometry (top and bottom spheres omitted for clarity).
From Pythagoras' theorem: $2(r + r_h)^2 = (2r)^2$ (r_h = radius of smaller sphere in octahedral hole
$r + r_h = \sqrt{2}r$ r = radius of large spheres forming the hole)
$r_h/r = 0.414$
(c) Neighbouring tetrahedral (t) and octahedral (o) holes

Structures derived form cubic close-packing (CCP)

Structures of many simple ionic crystals are consistent with the above scheme of close-packing. We shall cite examples mainly among compounds involving monatomic ions which may be reasonably assumed to be spherical in absence of any polarization. Thus, for compounds of the formula type AB (where A is the cation, usually smaller, and B is the anion, larger), the structure of *zinc blende* (ZnS) may be thought in terms of a close-packed fcc array of sulfide ions; the zinc ions occupy half the tetrahedral holes (Fig. 8.17a). Other compounds having similar structure are shown in Table. 8.5.

Since n spheres give rise to $2n$ tetrahedral holes, any structure involving occupancy of all tetrahedral holes should belong to the formula type AB_2 where B ions occupy all the tetrahedral holes provided by close-packing of A ions. This situation is found in the *fluorite structure* (CaF_2; Fig. 8.17b). The fluoride ions occupy the tetrahedral holes in an fcc array of calcium ions. The positions of cations and anions are interchanged in compounds of the formula type A_2B, for example, Na_2O. Here the sodium ions occupy all the tetrahedral holes in a close-packed fcc array of oxide ions. The resulting structure is called *anti-fluorite structure*, shown in Fig. 8.17c.

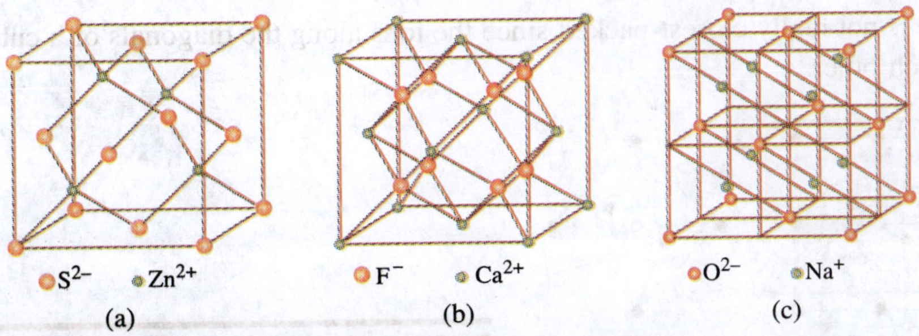

Fig.8.17: Structures derived from filling of tetrahedral holes in cubic packing.
(a) The ZnS structure (*sphalerite*: 4 Zn^{2+} ions and 4 S^{2-} ions per unit cell).
(b) The fluorite structure (4 Ca^{2+} ions and 8 F^- ions per unit cell).
(c) The antifluorite structure (4 O^{2-} ions and 8 Na^+ ions per unit cell).

Table 8.5
Structures derived from cubic close-packing

Holes occupied	Fraction filled	Name	Examples
Tetrahedral	1/2	Zinc blende	ZnS, CuCl, CuBr, CuI, AgI
Tetrahedral	1	Fluorite	CaF_2, SrF_2, BaF_2, PbF_2, UO_2
Tetrahedral	1	Antifluorite	Li_2O, Na_2O, K_2O, Rb_2O
Octahedral	1	Sodium chloride	Halides of Li, Na, K, Rb, AgF, AgCl, AgBr, Oxides and sulfides of Mg, Ca, Sr, Ba

The sodium chloride structure (rock salt structure) is interesting. The chloride ions alone form a face-centred cubic lattice; similarly the sodium ions alone form an fcc lattice. The entire crystal structure is thus formed by two interpenetrating fcc lattices of Na^+ and Cl^- ions. We can understand the structure by assuming an fcc lattice of the larger Cl^- ions in which Na^+ ions are introduced in octahedral holes (Fig. 8.18).

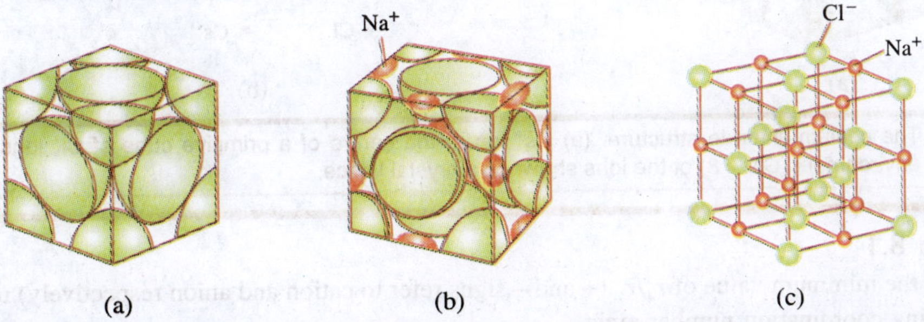

Fig.8.18: The sodium chloride structure: (a) an fcc array of Cl^- ions (b) The voids in *a* are occupied by Na^+ ions. (c) The resulting structure of NaCl: two interpenetrating fcc arrays of Na^+ and Cl^- ions (sizes reduced for clarity). The unit cell has effectively 4 Na^+ ions and 4 Cl^- ions:

Na^+ : 1 (centre) + $\frac{1}{4} \times 12$ (edge-centre) = 4; Cl^-: $\frac{1}{8} \times 8$ (corner) + $\frac{1}{2} \times 6$ (face-centre) = 4.

Each Na^+ ion in the crystal is surrounded by six Cl^- ions as nearest neighbours at the corners of an octahedron; each Cl^- ion is similarly surrounded by six Na^+ ions (6:6 coordination). The structure is not really closest-packed since the ions along the diagonals of a cubic face do not touch each other.

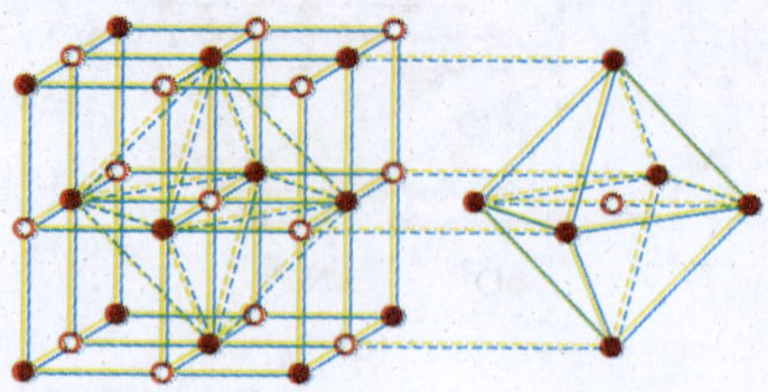

Fig.8.19: Unit lattice of sodium chloride showing octahedral coordination around Na^+ ion (o).

The cesium chloride structure (CsCl) represents an important structural type derived from cubic packing for ionic compounds of AB formula type. The structure is body centred cubic and not close-packed. The large Cs^+ ion occupies a void in a primitive cubic array of Cl^- ions. The entire structure may be supposed to be built up of two interpenetrating primitive cubic lattices of Cs^+ and Cl^- ions, both ions surrounded by eight ions of opposite charge as nearest neighbours at the corners of a cube (8 : 8 coordination). The structure is adopted by many compounds: CsBr, CsI, CaS, TlI and NH_4Cl. NH_4Cl adopts this structure in spite of the small size of NH_4^+, probably stabilized by hydrogen bonds to four chloride ions at alternate corners of the cube. Many alloys like CuZn and AlFe also have the two types of metal atoms in cesium chloride structure.

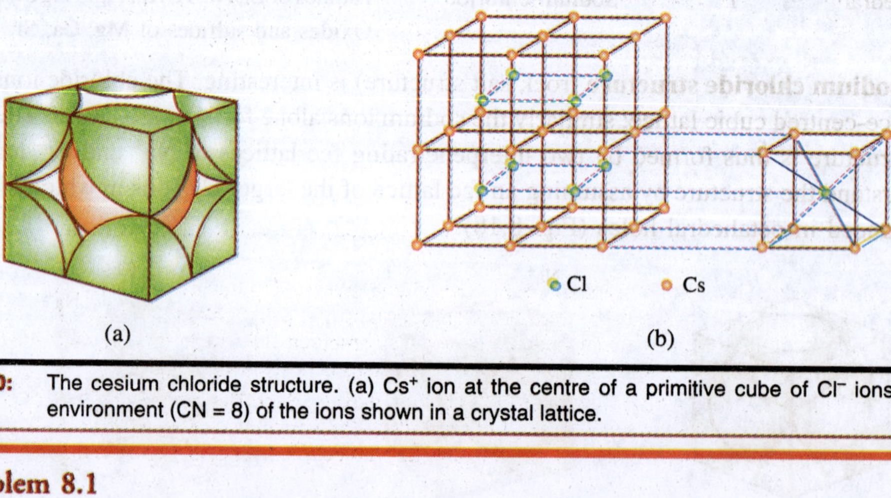

Fig.8.20: The cesium chloride structure. (a) Cs^+ ion at the centre of a primitive cube of Cl^- ions. (b) cubic environment (CN = 8) of the ions shown in a crystal lattice.

Problem 8.1

Calculate the minimum value of r_+/r_- (+ and – signs refer to cation and anion respectively) required for attaining coordination number eight.

Solution: The ratio may be calculated from the geometry of bodycentred cube (cube *I*, Fig. 8.10a). If a is the side of the cube, we have $a = 2r_-$.

Again, the body diagonal ($\sqrt{3}a$) is equal to $2r_+$ (centre) + $2r_-$ (two corners).

or, $1 + r_+/r_- = \sqrt{3} = 1{\cdot}732$; or, $r_+/r_- = 0{\cdot}732$.

Note. In CsCl the actual ratio is ~ 0·93. Many ionic compounds with r_+/r_- ratio much greater than 0·732 still adopt the NaCl structure (see later).

Structures derived from hexagonal close packing (HCP)

Structures of many common compounds are derived from hexagonal close-packing of ions. Few examples are given in Table 8.6.

Table 8.6

Structures derived from hexagonal close-packing of ions

Voids (holes) occupied	Structure type	Examples
Half tetrahedral	Wurtzite (ZnS)	ZnS, ZnO, BeO, MnS, AgI, NH_4F
All tetrahedral	No structure known; tetrahedral holes share faces	
Half octahedral	Rutile (TiO_2)*	TiO_2, MnO_2, SnO_2, MgF_2, NiF_2
All octahedral	Nickel Arsenide (NiAs)*	NiAs, NiS, FeS, CoS

* Some distortion from perfect hcp is present.

The representative structures are shown in Fig. 8.21.

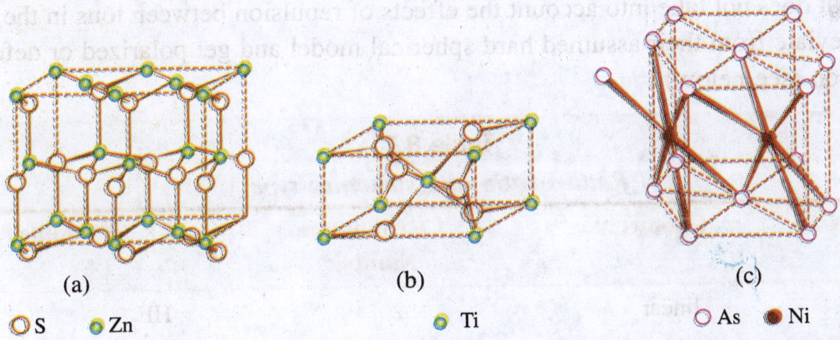

Fig.8.21: Structures derived from hexagonal close-packing.

(a) **Wurtzite:** Wurtzite is another polymorph of zinc sulfide. The sulfurs (large circles) are now in hexagonal close-packed array with zinc-atoms (green) occupying half of the tetrahedral holes (4:4 coordination). In wurtzite and sphalerite, the local symmetries of the cations and anions are identical toward their nearest neighbours but differ at the second-nearest neighbours.

(b) **Rutile:** Rutile is a mineral form of TiO_2, titanium dioxide. Each Ti is surrounded by six O ions and each O is surrounded by three Ti—6 : 3 coordination. The structure is derived from hexagonal close-packing (hcp) of anions, cations occupying only half the octahedral holes. Cassiterite, SnO_2, also has the rutile structure.

(c) **Nickel arsenide:** The structure is based on an expanded hcp array of anions (distorted). Ni-atoms occupy the octahedral holes while each As atom has a trigonal prismatic coordination of Ni atoms.

The radius ratio

Geometric requirements for close-packing of ions of different sizes (assumed spherical) can be simply expressed in terms of the radius ratio (ρ):

$$\text{radius ratio } \rho = \frac{r_{\text{smaller}}}{r_{\text{larger}}}$$

The r-terms refer to radii of the ions. Usually the cations r_+ are smaller in size compared to the anions r_- and the radius ratio is often expressed as r_+/r_-.

We have seen that the size of a cation exactly filling the tetrahedral hole formed by anions of radius r is equal to $0{\cdot}225\,r$ (Fig. 8.14). Hence the *minimum* radius ratio for coordination number four (tetrahedral) is $0{\cdot}225\,r/r$ or 0·225. A still lower value of radius ratio means a smaller cation size. In that case the anions will not be in contact with the inner cation; instead they will touch each other and destabilize the structure by electrostatic repulsion. In fact, the ions will then rearrange to attain a lower coordination number in which oppositely charged ions can make contact.

As the radius of the cation becomes larger than $0{\cdot}225\,r_-$, it will push the anions apart. When the radius ratio becomes as high as 0·414, the cation is sufficiently large to fill an octahedral hole in the close-packed array of anions (Fig. 8.16), and it attains greater stability in the increased coordination number of 6. Similar argument leads us to conclude that a radius ratio of 0·732 and higher will favour a coordination number 8 (Problem 8.1). Such correlations of structure type with radius ratio are summarized in Table 8.7. As discussed later in this section, there are certain limitations to the use of radius ratio in predicting structures. Such limitations arise mainly from two reasons: (i) radius ratio predicts the structure in the immediate neighbourhood, but does not take into account the effects of repulsion between ions in the bulk; (ii) ions often deviate from their assumed hard spherical model and get polarized or deformed by its counter-ion (see below).

Table 8.7

Radius ratio and structure type

Radius ratio	Geometry	Coordination Number	Example
< 0·155	linear	2	HF_2^-
0·155–0·225	triangular	3	BO_3^{3-}
0·225–0·414	tetrahedral	4 : 4	ZnS (Sphalerite, ccp)
			ZnS (Wurtzite, hcp)
0·414–0·732	octahedral	6 : 6	NaCl (ccp); NiAs (hcp)
		6 : 3	TiO_2
0·732–1	cubic	8 : 8	CsCl
		8 : 4	CaF_2, UO_2

There are many exceptions to these gross generalizations. Some examples are (r_+/r_- values in parentheses): BeS (0·19)—*sphalerite*; LiBr (0·38), LiI (0·34), KCl (0·77), RbCl (0·83), BaO (0·97)—all *rock salt*. Also see exercise, Question 9.

The correlation between structure and radius ratio of ionic compounds is based on the assumption that maximum stability is attained when every ion is surrounded by maximum number of ions of opposite charge. There is a practical limitation of applying this assumption to high coordination numbers. Let us consider salts of the type MX: the coordination numbers of M^{n+} and X^{n-} ions must be equal. As the coordination number increases (radius increases), more and more X^{n-} ion may be packed around each M^{n+} ion; but then each such X^{n-} should also have as many M^{n+} ions surrounding them. So the gain through electrostatic attraction will be largely cancelled by an increase in electrostatic repulsion. This is summarily reflected in the very small increase in the value of Madelung constant* for 6 and 8 coordination — 1·748 to 1·763 (less than one per cent). In fact, radius ratio suggests that twelve anions may be packed around a cation of the same radius ($r_+/r_- = 1$); but this is not practicable since each anion would then require twelve cations around them. Thus the radius-ratio provides a rough guide to crystal structure from a consideration of the *immediate neighbourhood* of a particular ion, but it does not take account of the details of a three-dimensional arrangement.

Structural predictions from radius ratio work well for complex fluorides and oxides and for the oxoanion salts which are fairly ionic in nature. But such predictions become unreliable for simple compounds such as the alkali metal halides and alkaline earth metal oxides. Such anomalies may have a number of origin like deviation from ionic model (polarization), variation of ionic radius with coordination number and anion-anion repulsion.

For LiCl, LiBr and LiI, the radius ratio lies around 0.3 suggesting fourfold coordination. Similarly for KF, KCl, RbF, RbCl, RbBr and CsF, radius ratios lie above 0.73, corresponding to coordination number 8 or higher. In practice, however, all the alkali metal haldies except CsCl, CsBr and CsI adopt 6 : 6 coordination (NaCl structure). Again, CsCl may be obtained in NaCl structure when sublimed on to an amorphous surface. On the other hand RbCl, RbBr and RbI adopt CsCl structure under high pressure.

The adoption of a particular structure naturally corresponds to maximum stability for the system under the prevailing conditions. We may reasonably assume that the $\Delta S°$ terms for the different structures would not be very different. So a comparison of $\Delta G°$ terms for the structures may be obtained from their respective $\Delta H°$ terms, the corresponding lattice enthalpies. In the CsCl structure, the higher value of the Madelung constant implies greater lattice energy; the dispersion energy would also be higher for higher coordination number. However, the interionic distance increases with increasing coordination number, thereby cancelling the effect of the higher value of the Madelung constant. In CsCl itself, the interionic distance decreases from 3.57A° in CsCl structure to 3.47A° in the NaCl structure. Similar changes in other cases show that the actual structure adopted is actually more stable, at least marginally, than the other alternative. This is illustrated below for KCl.

Problem 8.2

Why does KCl adopt the rock salt structure in spite of a radius ratio greater than 0.732? Given, r_{K^+} in C.N. 6 = 1.38 A°; in C.N. 8 = 1.51 A°. r_{Cl^-} = 1.81 A°. Madelung constant for NaCl and CsCl structures: 1.748 and 1.763 respectively (Table 8.8).

Solution: The electrostatic potential energy, the main contributor to lattice enthalpy for any structure, is directly proportional to the Madelung constant (A) for that structure and inversely proportional to the interionic distance $d(= r_+ + r_-)$. Here

$$d \text{ (NaCl structure)} = 1.38\text{A°} + 1.81\text{A°} = 3.19\text{A°}$$

$$d \text{ (CsCl structure)} = 1.51\text{A°} + 1.81\text{A°} = 3.32\text{A°}$$

* A factor that takes into account the effect of all surrounding ions in the lattice of a particular geometry (sec. 8.3.2)

The ratio of the lattice enthalpies will be roughly equal to the ratio of the respective electrostatic potential energies, i.e., the respective A/d terms. Hence,

$$\frac{U_{NaCl}}{U_{CsCl}} = \frac{A_{NaCl}}{d_{NaCl}} \times \frac{d_{CsCl}}{A_{CsCl}} = \frac{1.748 \times 3.32}{1.763 \times 3.19} = 1.03$$

This shows that the lattice for the NaCl structure would be marginally more stable and favour this structure over the CsCl structure. [The Madelung constant for CsCl structure is higher by less than 1% only whereas the interionic distance increases by nearly 4%.]

Pauling introduced a method of radius-ratio correction to account for the anion-anion repulsion in the alkali metal halides, particularly those having small cations with large anions. This helps us to understand the observed trend in many physical properties of these halides, e.g., their melting point sequence. This has been briefly discussed later (see alkali metals).

8.3.2 Energetics of Ionic Crystals

The binding energy or cohesive energy E_c of an ionic crystal is the standard molar enthalpy change (ΔH_L°) for the isothermal conversion of the crystal into its constituent ions in ideal gaseous state:

$$MX\,(c) \rightarrow M^{n+}(g) + M^{n-}(g)$$

(or its equivalent expression for other stoichiometries). The standard enthalpy change for the process (ΔH_L°) is actually the lattice enthalpy. At absolute zero, the enthalpy term equals the lattice energy; but at normal temperatures also, their difference is small (few kilojoules per mole) and may be neglected.

Lattice enthalpies are always positive since breaking of a lattice is always an energy-requiring (endothermic) process. The higher the lattice enthalpy of a substance, the more stable is the associated crystal structure. The positive signs ('+') are usually omitted from the numerical values of lattice enthalpies.

Previously lattice energies (or enthalpies) were defined for the *formation* of 1 mol of crystal from the free gaseous ions and were equivalent to the molar standard enthalpy of formation. The lattice energies were thus negative.

Lattice enthalpies are obtained from other thermodynamic quantities using a close-ended thermodynamic cyclic process, the Born-Haber cycle.

The Born-Haber cycle

It is the thermodynamic cycle correlating the heat of formation of a compound with other thermochemical quantities; among these are included the energy terms involved in various processes which may be carried on successively to form the same substance in a different path. Hess's law of constant heat summation is followed.

The construction of a Born-Haber cycle may be illustrated with reference to the formation of sodium chloride crystal from the elements in their normal states of existence. We consider the stepwise formation represented by the following processes:

Process				ΔH/kJ mol^{-1}
Na (*s*)	$\longrightarrow$	Na (*g*)	(sublimation)	+109
Na (*g*)	$\longrightarrow$	Na^+ (*g*)	(ionization)	+494

½ Cl_2 (*g*)	⟶	Cl (*g*)	(dissociation)	+121
Cl (*g*)	⟶	Cl^- (*g*)	(electron attachment)	–347
Na^+ (*g*) + Cl^- (*g*)	⟶	NaCl (*c*)	(lattice formation)	$-\Delta H_L$
NaCl (*c*)	⟶	Na (*s*) + ½ Cl_2 (*g*)	(negative of enthalpy of formation)	+414

We may construct a closed cyclic path of transformation with these processes as shown in Fig. 8.22.

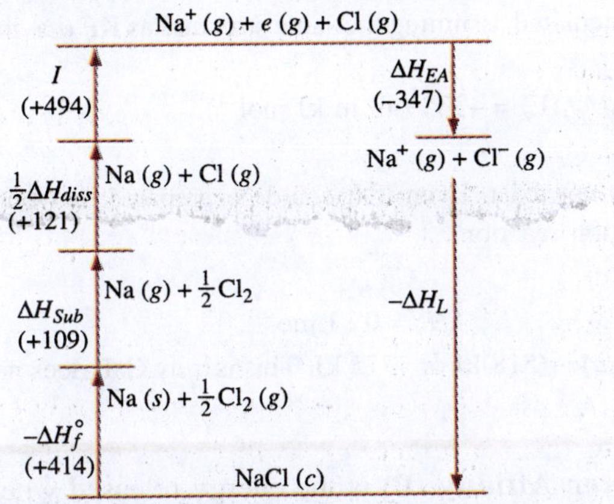

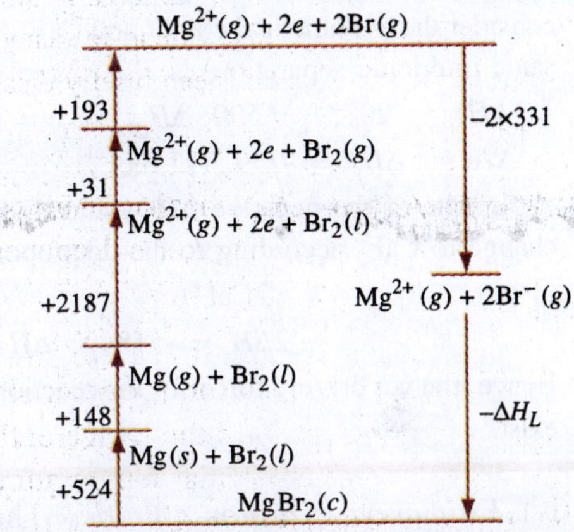

Fig.8.22: The Born-Haber cycle for NaCl (298 K).
For the closed cycle,

$$-\Delta H_f^\circ + \Delta H_{Sub} + \frac{1}{2}\Delta H_{diss} + I + \Delta H_{EA} - \Delta H_L = 0$$

or, $\Delta H_L \Rightarrow 791$ kJ mol^{-1}

Fig.8.23: Born-Haber cycle for $MgBr_2$ (*c*).

Problem 8.3

Calculate ΔH_L (lattice enthalpy) for $MgBr_2(c)$. Consult Fig. 8.23 for relevant data. All values in kJ mol^{-1}.

Ans: $\Delta H_L = 2421$ kJ mol^{-1}

Applications

(i) The Born-Haber cycle helps us in understanding the stability of many ionic compounds. Thus, formation of both Mg^{2+} and O^{2-} ions requires very large amounts of energy. Yet, MgO is a stable ionic compound because its lattice energy is very high (3800 kJ mol^{-1} at 298K). This is a consequence of the high charge on both ions.

(ii) The formula of ionic compounds as they are written may also be justified from the Born-Haber cycle.

(iii) *Estimation of electron affinities*. E.A. values for some elements are difficult to obtain directly. These can be estimated from the lattice energy of a suitable compound (calculated) with the Born-Haber cycle.

Problem 8.4

Justify the feasibility of the formation of a hypothetical compound $NaCl_2$. One theoretical value of the lattice energy is 2154 kJ mol^{-1}. Calculate the heat of formation of $NaCl_2$. The first two ionization energies (I_1 & I_2) of sodium are 494 and 4563 kJ mol^{-1}. Electron attachment enthalpy of chlorine = –347 kJ mol^{-1}; ΔH_{Sub} (Na) = 109 kJ mol^{-1}.

Hint: Construct the relevant B–H cycle similar to Fig. 8.23. This gives $\Delta H_f = +2560$ kJ mol^{-1}. ΔH_f is positive, i.e., the compound $NaCl_2$ would be highly endothermic and hence energetically unfavourable. This accounts for the non-existence of this compound.

Comment: But lattice energy alone may not always be the sole guideline for such purpose. Let us consider the hypothetical compound CaF. ΔH_L calculated assuming the same structure as KF (having same interionic separation) is 795 kJ mol^{-1}.

$\Delta H_{Sub} = 201$; $I_1 = 590$; ΔH_{diss} (F_2) = 159; ΔH_{EA}(F) = –335 (all in kJ mol^{-1}).

We get $\Delta H_f^\circ = -259.5$ kJ mol^{-1}.

Inspite of the negative ΔH_f° value, CaF does not exist. Even if it could be prepared, it would change to CaF_2 according to the disproportionation reaction

$$2CaF\,(c) \quad = \quad CaF_2\,(c) \quad + \quad Ca\,(s)$$
$$2\Delta H_f^\circ = -518 \qquad \Delta H_f^\circ = -1243 \qquad \Delta H_f^\circ = 0 \text{ kJ mol}^{-1}.$$

Hence, the net driving force for the reaction is –1243–(518)kJ or –725 kJ. This is why CaF does not exist.

(iv) *Estimation of proton affinities*. The **Proton Affinity (P)** is the energy released when a proton (*g*) is added to a species, also gaseous. Let us consider the proton affinity of NH_3 (*g*). *The proton affinity* (ΔH_{PA}°) of $NH_3(g)$ is the standard enthalpy change for the process $NH_3(g) + H^+(g) = NH_4^+(g)$. Now we may consider hypothetical thermochemical steps to start with $N_2(g)$, $H_2(g)$ and $Cl_2(g)$ to form $NH_4Cl(g)$ via $NH_3(g)$, $NH_4^+(g)$ and $Cl^-(g)$. The net enthalpy may be equated to the given enthalpy of formation of $NH_4Cl(c)$. We form a Born-Haber cycle from the following ΔH° values (kJ mol^{-1}).

Bond dissociation enthalpies, ΔH_{diss}° : H_2 436 (D_1); Cl_2 242 (D_2).

Standard enthalpy of formation, ΔH_f° : NH_3 –46; NH_4Cl (*c*) –314.

Lattice energy of NH_4Cl (*c*), ΔH°_L : 683;

Ionization enthalpy of H(*I*) : 1310; ΔH_{EA}° of Cl : –348.

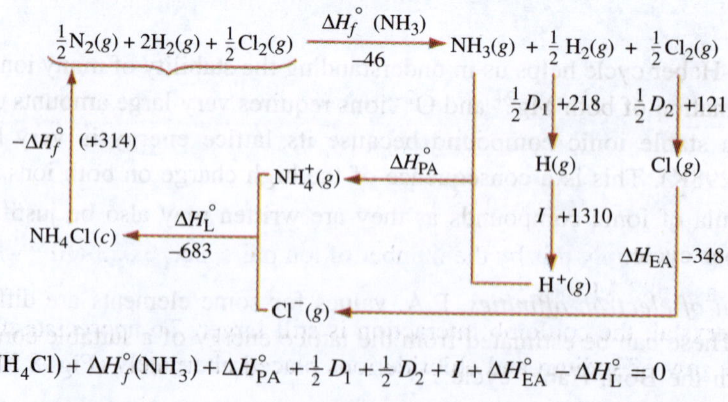

$$-\Delta H_f^\circ\,(NH_4Cl) + \Delta H_f^\circ(NH_3) + \Delta H_{PA}^\circ + \tfrac{1}{2}D_1 + \tfrac{1}{2}D_2 + I + \Delta H_{EA}^\circ - \Delta H_L^\circ = 0$$

From the given $\Delta H°$ values, $\Delta H^{\circ}_{PA} = -886$ kJ mol^{-1}.

It can be shown that the proton affinity of water is greater than that of phosphine (phosphane). That is why phosphonium compounds are readily decomposed by water.

Apart from the cases already mentioned, lattice energy data are also helpful in a number of different ways. For example:

(v) *Estimation of heat of formation*: One very interesting case is the speculation of the first true compound of xenon. The compound $O_2^+ PtF_6^-$ was known. Now ionization enthalpy of Xe (1169 kJ mol^{-1}) is nearly the same as that of O_2 (1170 kJ mol^{-1}). Also, the sizes of Xe^+ and O_2^+ are similar. This lead to the idea that the compound $Xe^+PtF_6^-$ might have a lattice energy sufficient for its formation from Xe and PtF_6. In fact, a related compound with slightly different stoichiometry was prepared by Bartlett in 1962 (see chemistry of xenon).

(vi) *Explanation of the high oxidation states of metals*: Metals form many fluorides with oxidation states higher than usual. Similarly, iodides contain the metals in a relatively lower oxidation state. This can be explained from the high lattice energies of fluorides (small size of fluoride ion) as compared to the iodides.

Theoretical calculation of lattice energy

The force of attraction F between two ions carrying unlike charges e separated at a distance r may be obtained from Coulomb's law. The corresponding potential energy U would be obtained by computing the work done in bringing the charges to this distance from infinity.

$$F = \frac{e^2}{4\pi\varepsilon_0 r^2}; \qquad U = \int_{\infty}^{r} \frac{e^2}{4\pi\varepsilon_0 r^2} = -\frac{e^2}{4\pi\varepsilon_0 r}$$

The negative sign indicates that this amount of energy is released by the system resulting in a lowering of potential energy.

However, repulsion would also develop between the electron clouds surrounding the ions. The exact expression for this cannot be framed readily as before, but it certainly raises the potential energy of the system. As the two oppositely charged gaseous ions approach each other, the net potential energy reaches a minimum at a distance r_0, the equilibrium distance between the ions. For sodium chloride, the estimated value of this distance is 238 pm. We may calculate an approximate value of the potential energy from coulomb interaction alone. For Na^+–Cl^- ion pair, the potential energy would be

$$U = -\frac{e^2}{4\pi\varepsilon_0 r_0} = -\frac{(1.602\times10^{-19}\ \text{C})^2}{4\times3.142\times8.854\times10^{-12}\ \text{J}^{-1}\text{C}^2\ \text{m}^{-1}\times238\times10^{-12}\text{m.}}$$

$$= 9.69\times10^{-15}\ \text{J.}$$

For one mole we multiply this by the number of ion pairs, i.e., 6.022×10^{23} when we get the potential energy as –583 kJ mol^{-1}.

In an actual crystal, the coulomb interaction is still larger. To appreciate this, let us first consider a single row of sodium and chloride ions placed alternately (Fig. 8.24). When such an arrangement is formed from the gaseous ions, the potential energy (coulomb) of the system is obtained from the electrostatic interaction between the ions.

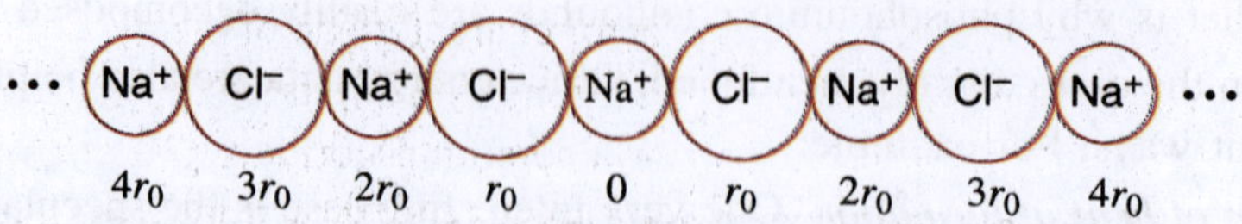

Fig.8.24: A single row of alternate sodium and chloride ions.

The central sodium ion is attracted by two chloride ions at r_0 and two more chloride ions at $3r_0$. The corresponding contributions to the potential energy would then be $-2e^2/4\pi\varepsilon_0 r_0$ and $-2e^2/4\pi\varepsilon_0 r_0 . 3r_0$ respectively. At the same time, there will be repulsion from the sodium ions, two at $2r_0$ and two at $4r_0$. This will *raise* the potential energy by $+2e^2/4\pi\varepsilon_0 . 2r_0$ and $+2e^2/4\pi\varepsilon_0 . 4r_0$ respectively. The procedure can be repeated for other ions placed similarly in the row (not shown in the diagram) and the potential energy of the central sodium ion would be obtained as the sum of an infinite number of terms:

$$U(=E_{Coul}) = -\frac{2e^2}{4\pi\varepsilon_0 r_0} + \frac{2e^2}{4\pi\varepsilon_0 2r_0} - \frac{2e^2}{4\pi\varepsilon_0 . 3r_0} + \frac{2e^2}{4\pi\varepsilon_0 4r_0} - \cdots$$

$$= -\frac{2e^2}{4\pi\varepsilon_0 r_0}\left[\left(1-\frac{1}{2}\right)+\left(\frac{1}{3}+\frac{1}{4}\right)+\cdots\right] = -\frac{2e^2}{4\pi\varepsilon_0 r_0}\ln 2.$$

The potential energy per mole will be $U \times N_A$ (Avogadro Number).

The procedure may now be extended to the three dimensional lattice of a sodium chloride crystal.

The Born-Lande equation

Consider a unit lattice of NaCl crystal (Fig. 8.25). The central Na^+ ion is surrounded by

6 Cl^- at r

12 Na^+ at $\sqrt{2}r$

8 Cl^- at $\sqrt{3}r$ etc, ...

Remembering that in an actual crystal the unit lattice is repeated in three dimensions, we shall get 6 more Na^+ ions at $2r$ (or $\sqrt{4}r$), 24 Cl^- ions at $\sqrt{5}r$, 24 Na^+ ions at $\sqrt{6}r$ and so on. The potential energy of the central sodium ion due to coulombic interaction, $E_{coul,}$ can now be written as a series of terms:

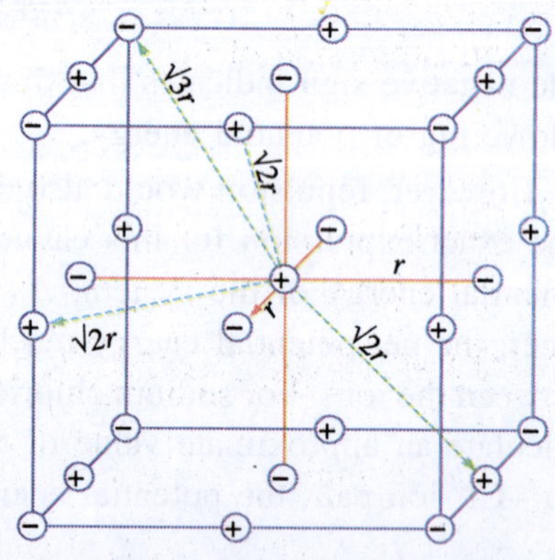

Fig. 8.25: Unit lattice of sodium chloride showing different interionic distances.

$$E_{coul} = -\frac{6e^2}{4\pi\varepsilon_0 r} + \frac{12e^2}{4\pi\varepsilon_0 \sqrt{2}.r} - \frac{8e^2}{4\pi\varepsilon_0 \sqrt{3}.r} + \frac{6e^2}{4\pi\varepsilon_0 2r} \cdots\cdots$$

$$= -\frac{e^2}{4\pi\varepsilon_0 r}\left[6 - \frac{12}{\sqrt{2}} + \frac{8}{\sqrt{3}} - \frac{6}{\sqrt{4}} + \frac{24}{\sqrt{5}} - \frac{24}{\sqrt{6}} + \ldots\right]$$

The series within the bracket converges slowly, and its limiting value is 1.74758. For a different arrangement of the ions in the lattice, the series within the bracket would be different. The limiting value of the series is thus a constant for any particular type of lattice and is called the **Madelung constant**, *A*. It is a numerical factor characteristic of a lattice type and takes into account the electrostatic interaction between all ions in a crystal of given geometry. As seen above, the value of the Madelung constant for a sodium chloride type lattice (6 : 6 coordination) is 1.748. Values of *A* for other crystal structures are shown in Table 8.8.

Table 8.8
Madelung constant (A) for some structures

Structure	A
NaCl	1.748
CsCl	1.763
Sphalerite	1.638
Wurtzite	1.641
Rutile	2.408
Fluorite	2.519

The coulomb potential energy per *mol* of NaCl may now be written using the value of the Madelung constant

$$E_{coul} = -\frac{1.748\, N_A.e^2}{4\pi\varepsilon_0 r}$$

The experimental value of *r* in the lattice is 28 pm, slightly different from the value for the gaseous ions. With this value of *r*, *E* comes out to be –867 kJ mol^{-1}. This potential energy is considerably lower than that for gaseous ion-pairs (–583 kJ mol^{-1}). The actual stabilization is, however, nearly 10% less than 867 kJ owing to repulsion. If the same crystal structure is made of ions with charges z_+e and z_-e, the potential enrgy due to coulomb interaction will have the general expression

$$E_{coul} = -\frac{A.z_+ z_- N_A e^2}{4\pi\varepsilon_0 r}$$

In addition to this coulombic interaction, there will be Pauli repulsion between the ions due to slight overlap of their electron probability densities. This repulsion energy is supposed to have an expression of the type

$$E_{rep}\ (\text{per mole}) = \frac{N_A B}{r^n} \qquad [B \text{ is a constant and } n \text{ is a high positive power}].$$

The net stabilization gained in the formation of the ionic lattice from the free gaseous ions, say, E_p, is the sum of these two energy terms for attraction and repulsion.

$$E_p = E_{coul} + E_{rep} = -\frac{A.z_+z_- N_A e^2}{4\pi\varepsilon_0 r} + \frac{N_A B}{r^n}$$

In the beginning, we defined lattice energy as the energy *required* to split a lattice into its constituent ions (gaseous). Hence the lattice energy (E_{lat}) is equal to $-E_p$ or $\left(\frac{Az_+z_-N_Ae^2}{4\pi\varepsilon_0 r} - \frac{N_AB}{r^n}\right)$.

The lattice energy (E_{lat}) is minimum for the equilibrium interionic distance in the crystal, r_0 (at 0K).

$$\left(\frac{dE_{lat}}{dr}\right)_{r=r_0} = 0 = -\frac{A.z_+z_- N.e^2}{4\pi\varepsilon_0 r_0^2} + \frac{N_A.n.B}{r_0^{n+1}};\ \text{when } B = \frac{A.z_+z_- e^2}{4\pi\varepsilon_0 n} r_0^{n-1}$$

$$\therefore\ E_{lat} = \frac{A.z_+z_- N_A.e^2}{4\pi\varepsilon_0 r}\left(1 - \frac{1}{n}\right)\ (E_{lat} = \Delta H_{lat} \text{ at } 0\text{ K}).$$

This equation is known as the **Born-Lande equation**. As mentioned before, E_{lat} is equal to lattice enthalpy ΔH_L at 0K but may be accepted to be nearly equal to ΔH_L° at ordinary temperatures.

n is called the **Born exponent**. Its value may be derived either from the measurement of the compressibility of the solid or theoretically. It depends upon the electronic configuration of the ions. Some typical values are given in Table 8.9.

Table 8.9
Born exponents for different ions

Outer electronic configuration	*n*
$1s^2$ (He)	5
$2s^2\,2p^6$ (Ne)	7
$3s^2\,3p^6$ (Ar) or $3s^2\,3p^6\,3d^{10}$ (Cu^+)	9
$4s^2\,4p^6$ (Kr) or $4s^2\,4p^6\,4d^{10}$ (Ag^+)	10
$5s^2\,5p^6$ (Xe) or $5s^2\,5p^6\,5d^{10}$ (Au^+)	12

In a crystal containing two different types of ions, the mean value of *n* is taken. In NaCl, we have Na^+ ($2s^22p^6$) and Cl^- ($3s^23p^6$) ions. So the value of *n* should be (7+9)/2 = 8.

Other refinements in the expression for lattice energy have been suggested; they allow for dipole-dipole interactions, van der Waals forces and zero-point energy of the crystal.

The Born-Mayer equation

For accurate evaluation of lattice energies, the Born-Lande equation may be further improved by replacing the repulsion term. Since orbitals decay exponentially (Ch. 4) and repulsion depends on the overlap of orbitals, the contribution of repulsion to potential energy (E_{rep}) may also be expected to be an exponential function of radius r : $E_{rep} = N_A C' e^{-r/\rho}$ where C' and ρ are constants. When ions are pressed together by an external force, repulsion should increase. Hence the parameter ρ can be obtained from the compressibility of ions. The overall potential energy of an ionic crystal is determined by the resultant of attractive and repulsive interactions. Plotted against internuclear separation (r), the total potential energy shows a minimum at an equilibrium interionic separation, say r_0 (Fig. 8.26).

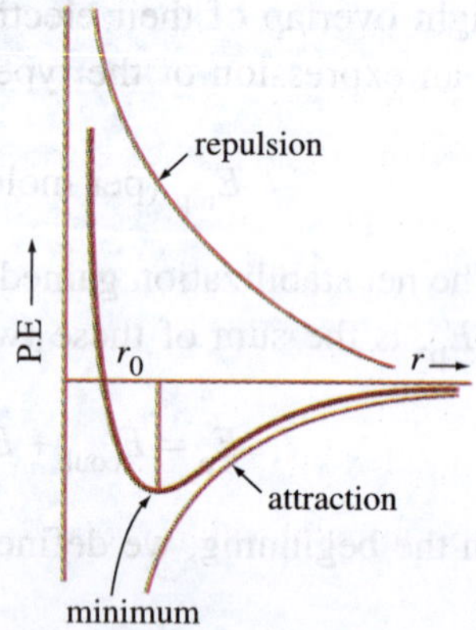

Fig.8.26: Nature of variation of potential energy of an ionic crystal with interionic separation (not to scale).

Proceeding in the same manner as in the Born-Lande equation for the coulombic interaction and including the contribution from repulsion (E_{rep}) as before, the minimum potential energy $E_{p,\,min}$ for an ionic crystal is expressed by the **Born-Mayer equation**

$$E_{p,\,min} = -\frac{N_A . A\, z_+ z_- e^2}{4\pi\varepsilon_0 r_0}\left(1 - \frac{\rho}{r_0}\right)$$

N_A is the Avogadro number, A is the Madelung constant for the structure, z_+e and z_-e are the charges on the ions at equilibrium interionic separation r_0. ε_0 is the permittivity of free space and ρ is a constant. In most cases a value of ρ equal to 34.5 pm agrees well with experimental data (r_0 in pm).

The lattice enthalpy ΔH_L is equal to $-E_{p,\,min}$ at 0K neglecting zero-point contributions. Contribution of the entropy term ($T\Delta S$) may be ignored for most practical cases and the lattice enthalpy at ordinary temperature is also taken equal to $-E_{p,\,min}$, when,

$$\Delta H_L = \frac{N_A.A.z_+z_-e^2}{4\pi\varepsilon_0 r_0}\left(1-\frac{\rho}{r_0}\right)$$

For a given lattice type, the magnitude of ΔH_L increases with the charge numbers of the ions as $|z_+ z_-|$ and decreases with interionic separation r_0.

$$\Delta H_L \propto \frac{|z_+z_-|}{r_0}$$

The quantity $|z_+ z_-|/r_0$ is called the *electrostatic parameter* ξ (xi). We conclude that ions with small radii and high charge should form crystals with high lattice enthalpy. This is consistent with our early observation on MgO where the doubly charged ions ($z_+ z_- = 4$) provide a lattice enthalpy more than four times that of NaCl, though the interionic separation is nearly equal in both. We shall discuss other consequences of lattice enthalpy in the next section.

The magnitude of lattice enthalpy also depends on the magnitude of A, the Madelung constant, which summarily represents the electrostatics effect of surrounding ions on each ion in the crystal. As we see in Table 8.8, the value of A increases with coordination number and on the charges of the ions. But the gain in lattice enthalpy from a higher Madelung constant may not be significant because a higher coordination number is always associated with slight increase in the radii of ions—the corresponding increase in r_0 will lower the lattice enthalpy. This is evident from the crystal structure of KCl which adopts the NaCl structure with 6 : 6 coordination number though the radius ratio permits the higher coordination no 8 : 8 (Problem 8.2).

As mentioned earlier, lattice energies calculated from such expressions may be further improved by inclusion of London dispersion energy and zero point energy. But their contribution is very small. The contribution of repulsive interaction is however much larger, ~11% in NaCl (from $n = 9$ in Born-Lande equation). According to a typical calculation for NaCl, the contributions are (kJ mol^{-1}): electrostatic attraction : 860; repulsion –99; London dispersion : 12; zero point energy –7; lattice energy : 766 against an experimental value of 770 (B–H cycle). A few other lattice enthalpies calculated from the ionic model (with different corrections) and the experimental values (B–H cycle) are given in Table 8.10.

Table 8.10

Few lattice enthalpy values (ΔH°_L : kJ mol^{-1})

Compound	ΔH°_L (calc)	ΔH°_L (expt)	Compound	ΔH°_L (calc)	ΔH°_L (expt)
LiF	1029	1035	CsF	748	744
LiI	730	757	CsI	601	584
NaF	915	914	CuCl	904	993
NaBr	739	728	AgF	920	953
NaI	692	681	AgCl	832	903

The agreement between calculated and experimental values is best for the alkali metal halides with small anions (F^-). Departure increases with increasing polarization of the anion, as seen for the halides of Ag(I) and LiI or CuI.

Problem 8.5

Calculate the lattice energy of AgCl using Born-Mayer equation, given r = 307 pm, A = 1.748 and ρ = 34.5 pm, Compare this with the experimental value of 903 kJ mol^{-1}. Comment.

Solution: From B–M equation

$$\text{Lattice energy} = \frac{1.748 \times 6.022 \times 10^{23}\ \text{mol}^{-1} \times (1.602 \times 10^{-19}\ \text{C})^2}{4\pi \times 8.854 \times 10^{-12}\ \text{C}^2\text{J}^{-1}\text{m}^{-1} \times 307 \times 10^{-12}\ \text{m}}\left(1 - \frac{34.5}{307}\right)$$

$$= 700\ \text{kJ mol}^{-1}.$$

The calculated lattice energy differs considerably from the experimental value, showing that AgCl has appreciable covalent character.

The Kapustinskii equation

The Madelung constants for MX and MX_2 type structures are 1.748 and 2.519 respectively, i.e., in the ratio of 2 : 3. The number of ions in the formula are also in the ratio 2 : 3. For a number of structures, the Madelung constant divided by the number of ions per formula unit (A/n) yields approximately the same value (verify from Table 8.8). This led Kapustinskii to propose (1956) that for any ionic solid there exists a hypothetical rock salt structure which would be energetically equivalent to its true structure. The lattice enthalpy for any unknown structure may then be obtained by using 0.874n as the Madelung constant where n = number of ions per formula unit (0.874 is half the Madelung constant for NaCl structure = A/n).

Kapustinskii further observed that this value of A/n increases slightly as the coordination number increases (in CsCl, for example, A/n is 1.763 ÷ 2 = 0.882; CN = 8). At the same time, the radius of an ion also increases slightly as the coordination number increases (Ch. 5). Hence the ratio A/nr_0 is expected to vary only slightly from one crystal type to another. This prompted him to use 0.874n as the value of Madelung constant for any unknown structure with an estimated value of r_0 from crystallographic radii for 6 : 6 coordination and calculate the lattice enthalpy. Proceeding in the same line of approach as described with the Born-Mayer equation, the **Kapustinskii equation** for lattice enthalpy then becomes

$$\Delta H_L = \frac{0.874 n |z_+ z_-| e^2 N_A}{4\pi\varepsilon_0 (r_+ + r_-)}\left(1 - \frac{34.5}{r_+ + r_-}\right)$$

r_+ and r_- are the crystallographic radii of the ions (Shannon) in pm. In fact, this is the lattice enthalpy of a hypothetical rock salt structure (6 : 6 coordination) which is energetically equivalent to the unknown crystal structure. Putting values of e, N_A, π and ε_0 and converting radii from pm to m, the equation may be written in the simple form

$$\Delta H_L = \frac{n|z_+ z_-|}{r_+ + r_-}\left(1 - \frac{34.5}{r_+ + r_-}\right) \times K \quad \text{where } K = 1.214 \times 10^5\ \text{kJ pm mol}^{-1}.$$

Problem 8.6

Find the value of the constant K in the Kapustinskii equation.

Hint: Compare the two expressions of ΔH_L given above. One gets

$$K = \frac{0.874\, e^2 N_A}{4\pi\varepsilon_0} = \frac{0.874 \times (1.602 \times 10^{-19}\,\text{C})^2 \times 6.022 \times 10^{23}\,\text{mol}^{-1}}{4 \times 3.142 \times 8.854 \times 10^{-12}\,\text{C}^2\text{J}^{-1}\text{m}^{-1}}$$

$$= 1.214 \times 10^{-4}\ \text{J m mol}^{-1} = 1.214 \times 10^{5}\ \text{kJ pm mol}^{-1}$$

The Kapustinskii equation provides a simple route to calculate the lattice enthalpy of a compound whose structure is not known—we need only to assume that the bonding is essentially ionic. We can also use the equation to determine the radii of ions—the best set that agrees with experimental values of lattice enthalpy (Born-Haber cycle). Such radii are called *thermochemical radii*.

8.3.3 Properties of Ionic Compounds

The properties of ionic compounds are consistent with their ionic composition. The cohesive force in an ionic solid extends throughout the entire structure of close-packed ions, making them hard. Any shearing force will disturb the equilibrium in the packing of oppositely charged ions and bring ions of similar charge close to one another. The resulting repulsion will cause rupture of the crystal structure. Ionic compounds are thus brittle. The melting points of ionic crystals are expected to rise with lattice enthalpy, which, in turn, depends upon the charges on the ions and the closeness of their approach. We have earlier mentioned the electrostatic parameter ξ (xi) which combines these effects : $\xi = |z_+ z_-|/r_0$. The hardness and melting points of some ionic solids (Table 8.11) demonstrate the role of the electrostatic parameter.

Table 8.11

Variation of hardness and melting point with charge and interionic distance

	Compound	Interionic distance / pm	Hardness (Moh's scale*)	Melting point / °C
charge increases ↓	NaF	235	3.2	990
	MgO	212	6.5	2800
	TiC	216	8.9	3140
distance increases ↓	MgO	212	6.5	2800
	CaO	240	4.5	2580
	SrO	257	3.5	2430
	BaO	277	3.3	1923

* Moh's scale: Diamond 10; Ruby, Sapphire 9; Quartz 7, Feldspar 6; Apatite 5; Fluorite 4; Calcite 3; Rock salt 2; Talc 1.

Room temperature ionic liquids. Though ionic solids generally possess high melting points, weak packing may result in low melting points of crystals containing very large cations or anions or both. $[CH_3CH_2NH_3]^+[NO_3]^-$ has a melting point of 12 °C while $[(CH_3(CH_2)_6)_4N]Cl$ shows a melting point as low as –9°C. These are liquids at room temperature. Many ionic liquids have been derived from 1, 3-dialkylimidazolium cations with counterions like $[PF_6]^-$, $[BF_4]^-$ and $[CF_3SO_3]^-$ which have melting points much less than 100°C. Ionic liquids have very low vapour pressure at room temperature and may be used as environment friendly solvents for many reactions, particularly in catalysis.

Thermal stabilities of ionic solids may be related to their lattice energies. The Gibbs free energies of decomposition of the compounds are mostly dominated by the enthalpy term—the entropy terms are either negligible or are comparable for similar compounds. Hence a comparison of thermal stability of different compounds is possible by comparing the difference in lattice enthalpies between the solid reactants and the solid products. The easy thermal decomposition of some solid metal carbonates illustrates this nicely. The smaller oxide ion provides a higher lattice enthalpy in the product oxide. Again, the lattice enthalpies of the carbonates decrease with increasing polarizing power of the metal cation—carbonates of lithium (small size) and alkaline earth metals (high charge) decompose readily on heating; carbonates of highly charged cation (Fe^{3+}, Al^{3+}) are too unstable to exist.

Problem 8.7

Which of the two compounds, $MgCO_3$ and $CaCO_3$, is expected to have a higher decomposition temperature?

Solution: The reaction concerned is : $MCO_3(s) \rightarrow MO(s) + CO_2(g)$. From the thermodynamic relation $\Delta G° = \Delta H° - T\Delta S°$, the decomposition temperature T_d corresponds to $T_d = \Delta H°/\Delta S°$. $\Delta H°$ is determined by the difference of lattice enthalpies of the oxide and the carbonates in respective cases. $\Delta S°$ may be assumed to be equal for both the decompositions.

Now, $CaCO_3$ has a higher lattice enthalpy than $MgCO_3$ owing to efficient packing of the large CO_3^{2-} ion with the larger Ca^{2+} ion. The lattice enthalpy of $MgCO_3$ is further reduced due to enhanced polarization of the large soft carbonate ion. Hence $CaCO_3$ is expected to decompose at a higher temperature than $MgCO_3$. (Actual temperature : $CaCO_3$ 840°C; $MgCO_3$ 300°C. Also, $SrCO_3$ 1100°C; $BaCO_3$ 1300°C).

Similar correlations hold for the stability of oxides and peroxides of the alkali metals (Ch.18).

The stability of high oxidation states of many elements in their fluorides and oxides may also be related to the advantage of higher lattice enthalpy. The formation of a solid metal halide MX_2 from the lower oxidation state halide MX may be expressed as

$$MX(s) + \frac{1}{2}X_2(g, l, s) \rightarrow MX_2(s) \qquad (l \text{ or } s \text{ corresponds to } Br_2/I_2)$$

The forward reaction is favoured by (i) higher lattice enthalpy, ΔH_L, of $MX_2(s)$ due to higher charge on M^{2+} (ii) electron attachment enthalpy, ΔH_{EA} of the halogen X. (Entropy changes in the process will be nearly same for all halogens and hence the ΔH terms will primarily control ΔG for the reactions). Now, the forward reaction occurs at the cost of $\frac{1}{2}\Delta H_{diss}$ of X_2 (and ΔH_{sub} or ΔH_{vap} in case of I_2 or Br_2). The dominating factor for the conversion actually lies in the gain of ΔH_L in forming $MX_2(s)$. As the radius of the halide ion increases from F^- to I^-, this gain in lattice enthalpy of $MX_2(s)$ over $MX(s)$ becomes smaller. This is consistent with the observation that there are no corresponding iodides for many fluorides:

	CuF_2	AgF_2	TlF_3	MnF_4	VF_5
but not	CuI_2	AgI_2	TlI_3^*	MnI_4	VI_5

($^*Tl^I I_3$ is known; it is $Tl^+I_3^-$)

The high oxidation state of the metals actually causes oxidation of iodide to iodine and the formation of an iodide of the metal in lower oxidation state.

Chemists have taken advantage of the low lattice enthalpy associated with large ions in stabilizing many large counter-ions against decomposition. Thus, Cs^+, N^tBu^+ etc. ions stabilize Br_3^-, ICl_4^- and many other large species; smaller cations do not favour such cases. They would favour the formation of MBr(*s*) or MCl(*s*) which have higher lattice enthalpy.

Solubility: The solubility of an ionic compound in water (or any other solvent) depends on the standard Gibbs free energy change ($\Delta G°$) of the process

$$MX(s) \rightarrow M^+(aq) + X^-(aq).$$

The standard free energy change for the dissolution of the salt, $\Delta G°_s$, is related to the solubility product K_s by the relation $\Delta G_s^\circ = -RT \ln K_s$.

The process of dissolution may be dealt with conveniently by a Born-Haber type cycle involving the following steps.

Process	*Enthalpy change*
(i) Disruption of 1 mol of solid crystal into gaseous ions: $MX(s) \rightarrow M^+(g) + X^-(g)$	Lattice enthalpy ΔH_L (energy is required).
(ii) Conversion of the gaseous cations and anions into hydrated ions in solution: $M^+(g) + X^-(g) \rightarrow M^+(aq) + X^-(aq)$	Hydration enthalpy: $\Delta H_{\text{hydration}}$ of $M^+(g) + X^-(g)$ (energy in released).

The overall enthalpy of solutions, ΔH_s° (heat of solution at infinite dilution) is clearly the sum of these enthalpy changes. For a salt of known solubility product, we can calculate ΔG_s° and obtain ΔS_s° (entropy change for dissolution) through the relation $\Delta G° = \Delta H° - T\Delta S°$.

Relative magnitudes of lattice enthalpy of the crystal and heats of hydration of the ions determine the magnitude of heat of solution—it may be positive or negative. For very soluble substances it is often (but not always) negative, corresponding to an exothermic process. Substances with a positive heat of solution may still have an overall negative free energy of dissolution by predominance of the entropy term. As the ordered arrangement of the ions in the crystal lattice is destroyed, there is a large increase in entropy. However, orientation of the randomly distributed solvent molecules around the ions decreases the entropy of the system to some extent. The overall entropy factor ($T\Delta S$) drives the process of dissolution; the work to pull the ions apart is done at the cost of the internal energy of the system (adiabatic condition), and the solution cools. If the enthaply of solution is sufficiently positive, the entropy term cannot outweigh this and the compound does not dissolve (e.g., $BaSO_4$, CaF_2 etc.).

For an ion of charge *Ze* and radius *r*, the free energy of hydration (solvation) may be expressed as

$$\Delta G^\circ_{\text{hydration}} = -\frac{Nz^2e^2}{8\pi\varepsilon_0 r}\left(1 - \frac{1}{\kappa_r}\right)$$

where ε_0 is the permittivity of free space and κ_r is the relative permittivity (dielectric constant) of water (or solvent). This shows that hydration (solvation) energy is greater in magnitude for small ions and high charge and in a solvent of high dielectric constant (Table 8.12). All these factors therefore favour dissolution of an ionic substance. Further, the attractive force between the ions in a solvent of high dielectric constant will also be smaller.

However, the equation gives only approximate results as the bulk relative permittivity of the solvent in the vicinity of an ion becomes different and the crystal radius of an ion does not represent its radius in solution. Nevertheless it has been possible to provide comparison between different ions and compilation of $\Delta G°$, $\Delta H°$ and $\Delta S°$ values for hydration of different ions (Table 8.12). The dependence of $\Delta G°_{hydration}$ on size and charge of the ions is also reflected in the values for $\Delta H°$ and $\Delta S°$. The entropy contribution ($T\Delta S°$; $\Delta S°$ values in Joules) is always much less than the enthalpy term (kJ), specially at ordinary temperature. So useful correlations between solubility and enthalpy of hydration have been made possible.

Table 8.12

$\Delta G°$, $\Delta H°$ (kJ mol^{-1}) and $\Delta S°$ (J mol^{-1}) of hydration (kJ mol^{-1}) for some ions at 298K.

	$\Delta H°$	$\Delta S°$	$\Delta G°$		$\Delta H°$	$\Delta S°$	$\Delta G°$
H^+	–1120	–121	–1084	Mg^{2+}	–1980	–293	–1895
Li^+	–544	–134	–506	Ca^{2+}	–1650	–238	–1582
Na^+	–435	–100	–406	Ba^{2+}	–1365	–188	–1310
K^+	–352	–67	–330	Al^{3+}	–4750	–506	–4600
Rb^+	–326	–54	–310	F^-	–473	–142	–432
Cs^+	–293	–50	–276	Cl^-	–339	–84	–314
				I^-	–260	–46	–247

The hydration enthalpy of a salt of the type MX may be expressed as the sum of contributions by individual ions:

$$\Delta H_{hyd} = \frac{k_1}{r_+} + \frac{k_2}{r_-} \qquad k_1, k_2\text{; constants; } r_+, r_- \text{ : radii of cation and anion.}$$

For successful dissolution of the salt, this hydration enthalpy should overcome the lattice enthalpy; the latter may also be expressed as a function of the sum of ion-radii:

$$\Delta H_L = \frac{k_3}{r_+ + r_-}. \qquad (k_3 = \text{a constant}).$$

In a series of related compounds, the variation of ΔH_{hyd} and ΔH_L would be different as the two quantities will respond differently to the variation of r_+ or r_- along the series. If for example, $r_+ << r_-$, the contribution of the anion to the hydration enthalpy will be small; the total ΔH_{hyd} of the ion-pair will be dominated by the cation alone. In a series of salts of a large anion, the hydration enthalpy will gradually decrease in magnitude with increasing cation size (e.g., $LiClO_4$, $NaClO_4$, $KClO_4$ or LiI ... CsI). But the lattice enthalpy is inversely proportional to the sum ($r_+ + r_-$). Since $r_- >> r_+$, the sum will not change significantly as r_+ increases. Consequently the lattice enthalpy will not decrease as fast as the hydration enthalpy decreases with increasing cation size. The more quickly diminishing hydration enthalpy results in a decrease in solubility along the series:

$$LiClO_4 > NaClO_4 > KClO_4 > \ldots$$

$$LiI > NaI > KI > CsI$$

$$MgSO_4 > CaSO_4 > SrSO_4 > BaSO_4$$

Some data on alkali metal halides may be found in chapter 18. Similar arguments apply to hexafluorophosphates, chloridoplatinates(IV) and hexanitrocobaltate(III). The low solubility of $K_3[Co(NO_2)_6]$ is used to detect potassium ion in solution in qualitative analysis.

With a small anion, the lattice enthalpy decreases more rapidly along a series of salts with increasing cation size and solubility increases. In general, the lattice enthalpy term determines the solubility trend when $r_+ \simeq r_-$. This is reflected in the trend of solubility among the fluorides and hydroxides : LiF < NaF < ... CsF and $Mg(OH)_2 < Ca(OH)_2$ ... $Ba(OH)_2$. Barium hydroxide (baryta) solution is used as a titrant in alkalimetry.

However, variation of solubility may have quite other different origin, too. The solubility trend among the silver halides (AgF > AgCl > AgBr > AgI) is an example (see later).

Polarization of ions

The properties of ionic compounds expected from purely ionic composition of the solid is actually an idealization. The electron probability densities ("electron clouds") of neighbouring ions do interact with one another to impart some covalent character to an ionic compound. As a consequence, ions are partly deformed or polarized from their ideal spherical shape, their electrostatic interaction is weakened and the lattice enthalpy of the compound is actually lowered. With increasing polarization, the charge clouds of neighbouring cations and anions would ultimately tend to overlap with one another as in a covalent compound. The extent of polarization is thus a measure of the tendency of the system to attain covalent character.

This trend of attaining covalent character by ionic compounds as a consequence of polarization may be generalized in terms of **Fajan's rules** (1924). According to these rules, (originally stated in terms of atomic volume), **an ion will be formed most easily** when

(i) *the electronic structure of the ion is stable,*

(ii) *the charge on the ion is small,*

and (iii) *a small atom forms the anion or a large atom forms the cation.*

Conversely, if these conditions are not fulfilled, covalent character will develop in the compounds concerned. Thus, **increased covalence is favoured by**

(i) *high charge on the ions,*

(ii) *small size of the cation and large size of the anion*

and, (iii) *cations with 18-electron structures* (i.e., 'non-inert gas' type; e.g., Cu^+).

The significance of these rules is not difficult to understand. High charge and small size of a cation imparts greater polarizing power to it. Similarly, the electron cloud of an anion is more diffuse when its charge or size is large. This makes the anion easily polarizable.

The corresponding increase of covalent character is reflected in the decreasing melting point of the compounds as illustrated in the table below. However, other factors like packing also contributes to the melting point.

Table 8.13

Variation of melting point of compounds with size and charge of ions

Compound	M.P./(°C)	Compound	M.P./(°C)	Compound	M.P./(°C)
NaF	988	$BeCl_2$	405	AlF_3	1291 (sub)
NaCl	800	$MgCl_2$	712	$AlCl_3$	180
NaBr	755	$CaCl_2$	772	$AlBr_3$	97.5
NaI	651	$SrCl_2$	872	SnF_4	705 (sub)
LiF	870	$BaCl_2$	960	$SnCl_4$	–33
LiCl	613	CaF_2	1392	$SnCl_2$	246
LiBr	547	$CaBr_2$	730	$PbCl_2$	501
LiI	446	CaI_2	575	$PbCl_4$	–15

Cartledge tried empirically to combine the effects of cation charge and cation radius into a single factor (1928). He observed that polarizing power is (i) directly proportional to the charge and (ii) inversely proportional to the radius of an ion. He then defined a term **ionic potential** (ϕ) as

$$\text{Ionic potential } (\phi) = \frac{\text{cation charge}}{\text{cation radius}}$$

Large value of ϕ for a cation indicates greater polarizing power and hence greater covalent character in a compound of that cation. With radius in A° unit, halides of a metal ion having $\sqrt{\phi} < 2.2$ are usually ionic while for metal ions with $\sqrt{\phi} > 2.2$ the halides are non-ionic. Basic character of metal oxides also decrease with increase in ϕ, as in the series $Na_2O > Li_2O > MgO > BeO$. Metal ions with $\sqrt{\phi} < 2.2$ give basic oxides; $\sqrt{\phi}$ between 2.2 to 3.2 characterizes amphoteric oxides while acidic oxides are expected with $\sqrt{\phi} > 3.2$.

Cations with 18-electron structures exert greater polarizing power than cations with 8-electron structures and same size. The additional *d*-electrons provide inadequate shielding from the nuclear charge due to relatively less penetratration of the *d*-orbitals into the inner electron core. The nuclear charge, therefore becomes more dominant and causes greater polarization.

Table 8.14

Effect of outer electronic configuration of cation on covalent character

Compound	Radius of cation (pm)*	Outer electron configuration of cation	Melting point/(°C)
KCl	133	$3s^23p^6$	776
AgCl	126	$4s^24p^64d^{10}$	455
NaCl	95	$2s^22p^6$	800
CuCl	96	$3s^23p^63d^{10}$	422
$CaCl_2$	99	$3s^23p^6$	772
$HgCl_2$	110	$5s^25p^65d^{10}$	270

* Pauling's values of crystal radii.

Besides the observed decrease of melting points, increased polarization is also responsible for a few other changes in the properties of compounds. The more important ones are:

(i) *Decrease of solubility in polar solvents.* AgI is much less soluble in water than AgF (or AgCl, or AgBr). Lattice energy considerations alone cannot explain this. Here the solubility decreases from fluoride to iodide due to increasing polarization of the anion and hence increasing covalence. The same explanation applies to the series HgF_2—$HgCl_2$—$HgBr_2$—HgI_2.

(ii) *Decrease of electrical conductance in fused state and tendency to ionize in solution, as shown by equivalent conductance* (ohm^{-1} cm^2) *of the following fused salts*:

LiCl	166	$BeCl_2$	0.086	NaCl	133	$MgCl_2$	29

(iii) *Intensification of colour.* Polarized anions seem to be responsible for colour in some compounds. As polarization increases from chloride to iodide, the colour of many metal halides are intensified, as for example:

Colourless	$HgCl_2$	AgCl	$SnCl_4$	$PbCl_2$
Coloured	HgI_2 (red)	AgI (yellow)	SnI_4(red)	PbI_4 (golden yellow)

Ferroelectric property

We have seen (Chapter 7) how the hydrogen bonded network in crystals of KH_2PO_4, $(NH_4)H_2PO_4$, Rochelle salt etc. gives rise to ferroelectric properties. Such behaviour may also arise in a crystal due to polarization caused by permanent displacenent of one sublattice with respect to another, as in barium titanate, $BaTiO_3$. It adopts the ***perovskite structure*** (mineral $CaTiO_3$) in which a small Ti^{4+} ion fits loosely in a large octahedral hole of O^{2-} ions; these O^{2-} ions again occur at the faces of a cubic close packing of Ba^{2+} ions (8-I). The Ti^{4+} can move from the centre of the cube to one side with simultaneous displacement of oxide ions, creating local dipoles due to shift of positive and negative charge centres. This gives rise to the observed ferroelectric behaviour. Above 120°C (the Curie temperature, see Ch. 27), thermal agitation cancels the effect of such polarization when the ferroelectric behaviour disappears. Layering $BaTiO_3$ with isostructural perovskite $SrTiO_3$ develops enhanced dielectric properties. Repeated nanometer thin layers of the two compounds may be built by techniques like ***chemical vapour deposition*** **(CVD)** or ***pulsed laser deposition*** **(PLD)** in a *superlattice* of total thickness in the micrometer range. Such superlattice films have many promising applications in materials science.

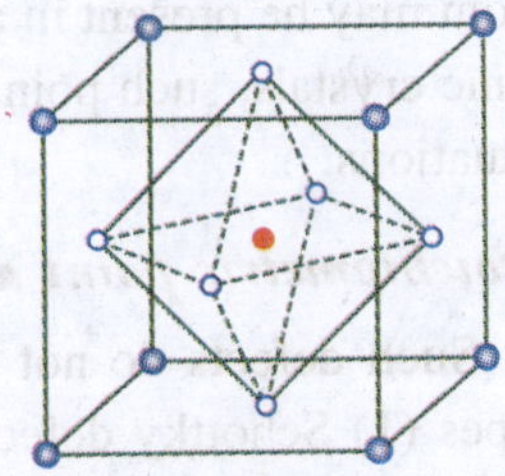

8-I

High temperature superconductors

Mixed metal oxides like $YBa_2Cu_3O_7$ and $HgBa_2Ca_2Cu_3O_{8-\delta}$ (under pressure) show superconductivity at relatively higher temperatures: Curie temperatures (T_c) 93K and 164 respectively. The phenomenon of superconductivity and the crystal structure of $YBa_2Cu_3O_7$ have been discussed in chapter 31 (sec. 31.3.6).

Variation from simple ionic structure

The simple ionic structures anticipated from radius ratio and close packing of ions are often found to be different from actual crystal structures owing to interplay of a large number of factors like polarization, van der Waals interaction, H bonding and so on. The principal types of such variations are:

1. **Layer lattices**, in which layers are often held by weak van der Waals forces. Different types of layer lattices are known, for example, $CdCl_2$, CdI_2, $CrCl_3$, BiI_3 and HgX_2. We shall mention some of these structures in relevant context.

2. **Chain lattices**, represented by compounds like $BeCl_2$, $CuCl_2$ and $PdCl_2$. Again, weak van der Waals forces may operate between chains.
3. **Silicates** which form a class by itself consisting of chains, sheets and three-dimensional networks. The structural varieties of silicates have been discussed in group 14(IV) (chapter 21).

8.3.4 Lattice Defects

A crystalline solid in bulk is built from millions and millions of the constituent particles. Hence it is not unlikely that there will be some defect or imperfection somewhere in the crystal lattice. Such lattice defects may involve only one or two lattice sites when they are called *point defects*. Alterations or displacements of a row of lattice sites are known as *line defects*. When an entire plane of lattice sites is imperfect, we call it a *plane defect*. We shall confine our discussion to point defects.

A point defect may result from a *vacancy*, that is, an atom missing from its normal lattice site. The missing atom may occupy an interstitial space in the lattice. This is an *intrinsic defect* present in pure substances. *Extrinsic defects* result from impurities or foreign atoms. A foreign atom may be present in an interstice (*interstitial*) or in a regular lattice site (*substitutional*). In ionic crystals, such point defects involve ions rather than atoms and give rise to some special situations.

Stoichiometric point defects

Such defects do not influence the overall stoichiometry of the crystal. These are of two types (1) Schottky defect and (2) Frenkel defect.

In **Schottky defect**, a vacancy arising from a missing cation is frequently accompanied by a vacancy at a nearby anion site. Thus a cation-anion pair is absent from their respective lattice points—creating a pair of "holes" or voids (Fig. 8.27). Such defects thus preserve the electrical neutrality and stoichiometry of the crystal and are formed with little change in energy. Crystals of NaCl, KCl, KBr, etc. exhibit Schottky defects.

Crystals of sodium chloride contain about 10^6 Schottky pairs per cm^3 around 130 °C. The number of ions in 1 cm^3 is about 10^{22}; so there is about one Schottky defect per 10^{16} ions. Some transition metal oxides may have much higher population of defects. An extreme example is the high temperature form of TiO which has one vacant pair per 10 formula units.

In **Frenkel defect**, an ion gets missing from its normal lattice point and occupies an interstitial void. Electrical neutrality (and stoichiometry) of the compound is not disturbed in this defect. Small cations in combination with large anions, or crystals with a rather 'open' structure exhibit this defect. Examples are AgCl, AgBr, AgI, ZnS and CaF_2.

The density of a solid remains unchanged in Frenkel defect but may be slightly lowered when large number of Schottky defects are present. Both Frenkel and Schottky defects may occur in the same compound, for example, AgBr.

In Schottky defects, the dielectric constant of the crystalline substance is not significantly changed. But in Frenkel defects, proximity of like charges (usually cations) increases the dielectric constant.

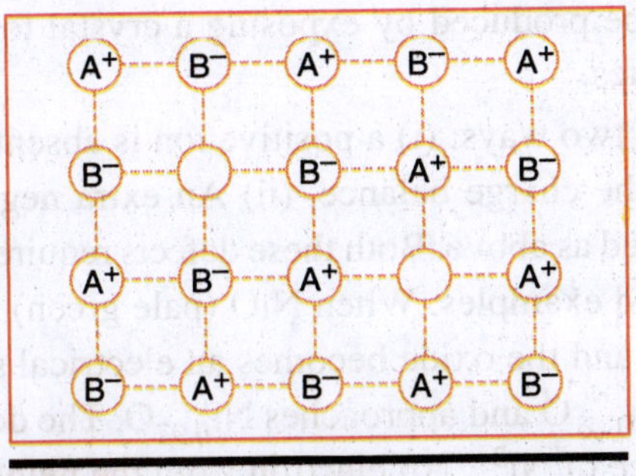

Fig.8.27: Schottky defect.

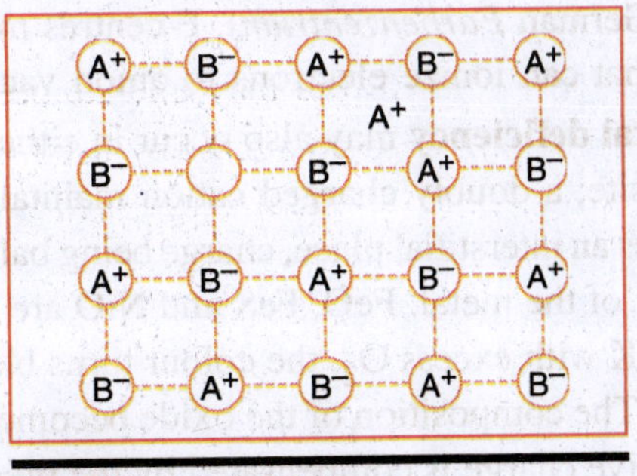

Fig.8.28: Frenkel defect.

A crystalline solid with a defective lattice may conduct electricity to a small extent. An ion moves to a "hole" from its own lattice site; a new "hole" is thus created. In this way a 'hole' migrates across the crystal—this is in effect equivalent to a charge moving in the opposite direction.

As already pointed out, Frenkel defect and Schottky defect do not alter the stoichiometry of the compound and are classed as stoichiometric defects. There are other point defects which alter the stoichiometry of the compound. These are called *non-stoichiometric defects.*

Non-stoichiometric point defects (Berthollide defects)

In compounds having such defects, the ratio of positive and negative ions differs from that indicated in their respective chemical formula. The balance of positive and negative charges is maintained by extra electrons or positive ions according to necessity. In non-stoichiometric compounds, either the metal or the nonmetal atoms may be in excess. Thus in the vanadium oxide VO_x, x may have any value between 0.6 and 1.3. Accordingly, nonstoichiometric defects may be classified as (i) *metal excess* and (ii) *metal deficiency types*. A third type of such defect may arise through substitution by impurities.

Metal excess may occur in two ways: (i) an anion may be missing from its lattice site, an electron being present there to maintain charge balance. Sodium chloride treated with sodium vapour forms such a yellow non-stoichiometric variety (see below). (ii) An extra metal atom may be present in an interstitial position. An electron in some other interstitial position balances the charge. Zinc oxide exhibits this type of metal excess defect. When heated, the lattice loses oxygen (reversibly) resulting in a nonstoichiometric phase like $Zn_{1+x}O$. The additional zinc ions cause lattice defects which trap electrons. These electrons can be subsequently excited by absorption of visible light which imparts a yellow colour to the oxide when hot. The non-stoichiometric form also shows enhanced electrical conductivity due to these electrons.

Anion vacancies are created in alkali-metal halides on heating the crystals in alkali metal vapour. The metal atoms that deposit on the surface of the crystal diffuse inward and get ionized by the crystal energy. The metal ion occupies a normal cation site while the electron released occupies an anion vacancy. The resulting composition now becomes like $Na_{1+x}Cl$ where $x << 1$. Electrons so trapped in anion vacancies give rise to interesting colours to the otherwise colourless alkali metal halides: LiCl—pink; KCl—violet; KBr—bluegreen; NaCl—orange. These electrons in halide ion vacancies are called *colour centres* or **F-centres** (from German *Farbenzentrum*). F-centres may be produced by exposing a crystal to an x-ray beam that can ionize electrons to anion vacancies.

Metal deficiency may also occur in either of two ways: (i) a positive ion is absent from its lattice site; a doubly charged cation maintains the charge balance. (ii) An extra negative ion occupies an interstitial place, charge being balanced as above. Both these defects require variable valency of the metal. FeO, FeS and NiO are good examples. When NiO (pale green) is heated at 1500K with excess O_2, the colour turns black and the oxide becomes an electrical semiconductor. The composition of the oxide becomes $Ni_{1-\delta}O$ and approaches $Ni_{0.97}O$. The deficiency of positive charge is compensated by the presence of Ni^{3+}. An electron from the neighbouring cations may jump to this Ni^{3+}, converting it to Ni^{2+} but at the same time creating a new Ni^{3+} at the other lattice point. A positive 'hole' thus migrates through the crystal, making it a *p*-type semiconductor. The second possibility to metal deficiency by gaining an extra anion is actually very unlikely since it is difficult for an anion (usually large) to occupy an interstitial site.

Nonstoichiometric defect through substitution occurs when a 'foreign' cation of comparable size replaces a cation in the lattice. A Cd^{2+}, size comparable to Ag^+, can replace two Ag^+ ions in a crystal of AgCl. When some Li_2O is dissolved in NiO, some Li^+ ions replace Ni^{2+} in the lattice. More Ni^{3+} ions are formed in the lattice to compensate the charge of univalent Li^+ ions. This largely enhances the electrical conduction in "doped" NiO, making it again a *p*-type semiconductor.

8.4 BAND THEORY OF SOLIDS

We have seen how the molecular orbital theory attempts to describe the molecular orbital energy levels of a gas-phase molecule in terms of linear combination of the atomic orbitals of constituent atoms. The idea may be extended to a crystalline solid considered as a giant molecule. Linear combination of orbitals of the constituents of the crystal—atoms, molecules or ions whatever those may be, will then approximately represent the energies of the "crystal orbitals" which extend over the entire crystal. We shall get an essence of this approach by applying it qualitatively to describe the gross bonding features of metals.

8.4.1 Bonding in Metals

Metals in general are characterized by certain distinct features. Explanation of these properties from the concepts of bonding developed so far present certain difficulties. Hardness and high melting and boiling points of most metals indicate strong cohesive force among the metal atoms. At the same time, ductility and malleability suggest that adjacent planes of metal atoms can easily glide over each other. The bonding between the metal atoms is thus strong and yet non-directional and non-rigid. Our conventional picture of covalent bonding is not in harmony with these expectations. Moreover, a metal atom possesses very few valence electrons; the number is inadequate to form electron pair bonds to the large number of neighbouring atoms which usually occur in a metallic crystal.

Again, high conductivity of the metals requires the existence of relatively free electrons within a metal. The lustre may be explained by assuming that the loosely held surface electrons immediately radiate the incident light. Photoelectric and thermionic emissions are also consistent with this idea. These aspects of a metal can be explained neither by covalency nor by an ionic model. Ionic substances are generally hard and high melting; but unlike metals they are brittle and non-malleable. They conduct electricity when fused or dissolved in a polar solvent; but metals are good conductors in the solid state. Lastly, it is difficult to understand the existence of oppositely charged ions within atoms of the same metal. With such apparent contradictions in their character, it has been necessary to suggest separate bonding schemes for the metals.

Theories of bonding in metals

(i) **The electron gas model** (Drude, 1900; Lorentz, 1923). It assumes that in a metallic crystal, the valence electrons are not bound to any particular nucleus. The positive cores of the metal atoms constitute a joint lattice: the valence electrons occupy their joint field like a gas. The free movement of electrons within a metal is thus explained, but it fails quantitatively, specially when we remember that the atomic heat of most metals is nearly constant, irrespective of the number of their valence electrons.

(ii) **The valence-bond theory** invokes elaborate resonance of electron-pair and one-electron bonds among all the nearest neighbour atoms of a metal in an aggregate. A few typical examples of such "structures" involving eight sodium atoms is shown below: the negative ion in III is formed by using *sp* hybrid orbital of the metal. Though somewhat successful, correlation of metallic

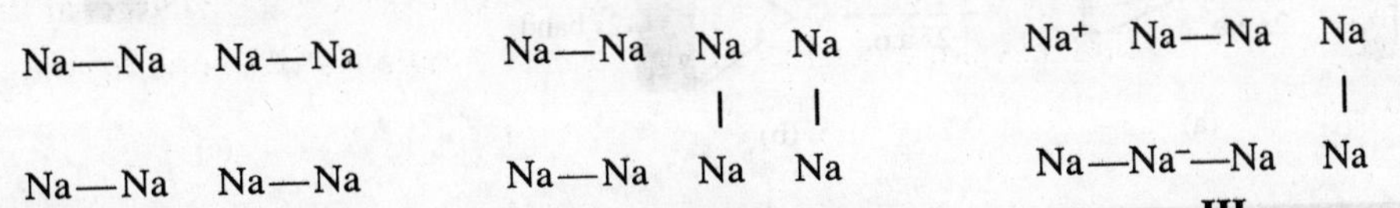

I II III

properties in this theory is not straightforward. Also metallic properties in the liquid state cannot be explained by this theory.

(iii) **The band theory** provides the most satisfactory explanation of characteristically metallic properties in a very natural way. This is in essence an extension of the molecular orbital treatment for a large number of atoms. For homonuclear diatomic species, we have seen that combination of two atomic orbitals gives rise to two molecular orbitals; these are two new energy levels for the electrons in the joint field of the two nuclei. If we extend this idea to an aggregate of a large number (say *n*) of metal atoms, each offering one orbital for combination with others, there will be a total of *n* new energy levels similar to *n* m.o.-s. These large number of energy levels will be spaced closely, one upon another, forming an energy band. The valence electrons from all the atoms will now enter these energy levels in a manner similar to the filling of molecular orbitals. A metal thus consists of energy bands formed by merging of the individual atomic orbitals. We shall illustrate two cases.

We have considered the m.o.-s formed by 2*s* atomic orbitals from two lithium atoms in Li_2. In an exactly analogous manner, three (or four) lithium atoms would give rise to three (or four) energy levels by combination of their 2*s*- a.o.-s. (Fig. 8.29(a)). Extended to *n*-atoms, all the 2*s* a.o.-s will combine to give a 2*s*-energy band with *n* energy levels (Fig. 8.29(b)). Similar combination of the *p*-a.o.-s (3 orbitals from each atom) will form a 2*p* energy band containing 3*n* energy levels.

Each lithium atom has only one valence electron. The total n electrons from n atoms will doubly occupy the lower $n/2$ energy levels in the $2s$-band (Fig. 8.29(b)). When the electrons gain thermal energy, or are placed in an electric field, the electrons move to occupy the higher unfilled energy levels; this explains the high thermal and electrical conductivity of metals. (For observed paramagnetism of alkali metals, see exercise, Q. 22).

In beryllium ($1s^2 2s^2$) each atom would contribute two valence electrons and the n energy levels in the $2s$ band will be completely filled. But the $2s$ band in beryllium merges with the empty $2p$ band just above it (Fig. 8.29(c)); the electrons can now move to these vacant energy levels under excitation (thermal, electrical). This explains the conductivity of beryllium (and similarly in magnesium by merger of $3s$ and $3p$ bands).

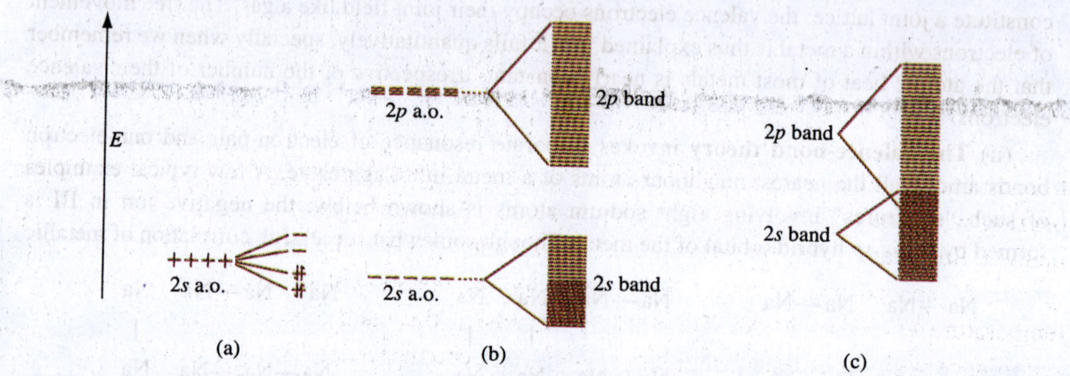

Fig.8.29: (a) Combination of four $2s$ a.o.-s. (b) Combination of n-atoms of Li. (c) Merger of the $2s$ and $2p$ bands in beryllium.

At temperatures above absolute zero, few of the electrons may be promoted to energy levels higher than the actual occupied region in the bands. The energy distribution of only these few electrons roughly follows the Maxwell-Boltzmann law. It is only these electrons, whose number is much less than the total number of valence electrons, which contribute to the heat capacity of the metal. The contribution of the electrons to the specific heat of metals is thus small. This inference is in harmony with Dulong and Petit's law.

8.4.2 The Band Theory of Solids

The band theory of metals may be extended to other nonmetallic solids as well. Conductors of electricity are in general characterized by ready availability of electrons in their structures. In metals these electrons are present in partly vacant energy bands (Li, Mg, Zn etc.). In contrast to this, let us consider a crystal of solid neon (Ne(c)) held by weak van der Waals forces. It will have a top level of filled narrow $2p$ band (V-W forces are weak); a narrow $3s$ empty band will now appear above it. Since the $2p$ and $3s$ atomic

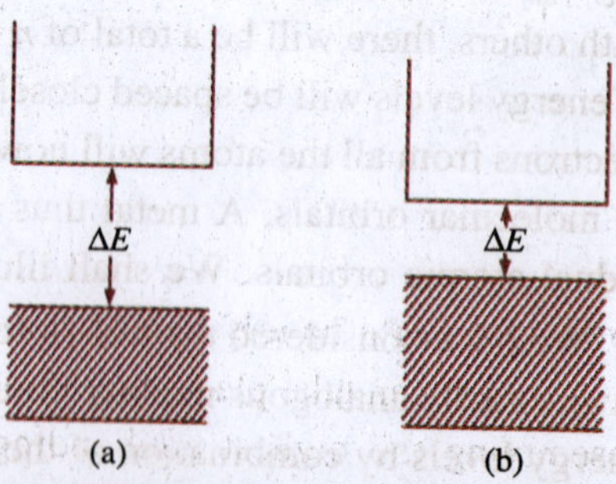

Fig.8.30: Relative positions of filled and unfilled bands in (a) insulators (b) semiconductors (arbitrary scale).

orbitals of neon are widely separated in energy, we expect the top of the 2*p*-band and bottom of the 3*s* band to be widely separated. Excitation of electrons from the 2*p* band to the 3*s* band requires a large amount of energy and we expect Ne(*c*) to be an *insulator* (Fig. 8.30). In an ionic crystal of NaCl, NaCl(*c*), we shall get bands of energy levels formed from the ions. These ions have the rare gas configuration and so we may apply the same logic as in the case of Ne(*c*). We conclude that NaCl(*c*) will be an insulator at ordinary temperature. In fact, the highest filled energy band (Cl^- 3*p* band) is separated from the next higher empty band (Na^+ 3*s* band) by about 7 eV. Molecular crystals are also expected to behave similarly, since weak van der Waals interactions among the molecules give rise to narrow bands widely separated from one another.

A covalent crystal like diamond is a bit different. In diamond itself, we would expect *n* carbon atoms to form a 2*s*-2*p* overlapping band with 4*n* energy levels. However, quantum mechanical calculation shows that the 2*s* and 2*p* atomic orbitals actually form two separate 2*s* and 2*p* bands. The top of the filled lower band and the vacant upper band is separated by about 6 eV. Hence diamond is an insulator at ordinary temperature.

Similar to the bands formed by *s* and *p*- orbitals, higher orbitals like *d* may also combine to form bands. When the bands spread over a wide range of energy, the *s*, *p* and *d* bands may overlap with one one another. Such bands are also formed by the valence orbitals in a compounds. Summarily, the energy levels in a solid material may be expressed in terms of band of closely spaced energy levels determined by the constituent nuclei. At absolute zero of temperature ($T = 0$), the individual energy levels in the band are occupied by electrons in accordance with the aufbau principle. The highest occupied energy level at $T = 0$, comparable to the HOMO in molecules, is called the **Fermi level**. Thus in a band formed from the *s* orbitals of *N* atoms providing one electron each, the *N* electrons will occupy the lower *N*/2 energy levels and the Fermi level lies close to the centre of the band. Since the band is not full, electrons close to the Fermi level can be promoted easily to the upper empty levels, thereby imparting electrical conductance in the solid. But movement of the electrons through the solid will be disrupted when the energy levels are not in order, as it really happens at higher temperatures due to atomic vibrations. The conducting electrons thus face a resistance due to "*scattering*" by atomic vibrations which increase with increase of temperature. This explains the decrease of conductance in metallic conductors with increasing temperature.

The number of energylevels in a narrow range of energy between E and $E + dE$ is called the *density of states*, $N(E)$. Such density is not uniform across a band—the greatest density occurs near the centre of the band and the lowest density occurs near its two edges. In the formation of molecular orbital energy levels in a polyatomic molecule, we found that the lowest energy MO is formed by a fully bonding combination of the atomic orbitals involved. Similarly, the highest energy MO is formed by maximum antibonding overlap of the AOs. The lower edge of a band similarly corresponds to a fully bonding combination of the concerned AOs which can take place only in one way of linear combination. The upper edge of the band, resembling the highest antibonding combination, is formed similarly. But the intermediate energy levels may be formed by several possible combinations of the contributing AOs and hence they are more densely packed. The three dimensional densities of states for *s* and *p* bands are represented by the typical parabolic regions in two-dimensional projection (Fig. 8.31).

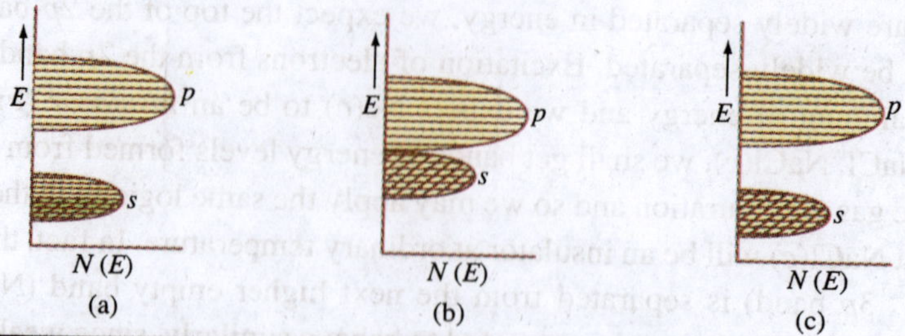

Fig.8.31: Two dimensional projections of densities of states: (a) a metal with partly filled band. (b) a full band and an empty band is adjacent to one another with a zero density of states at their junction. Solids with this band structure are called *semimetals*, e.g., graphite. (c) a filled band and an empty band separated by a large gap—a typical insulator.

The difference in energy ΔE between the highest occupied band, called the *valence band*, and the lowest vacant band (the *conduction band*) in a solid may be determined from the lowest frequency of absorption of uv or visible light by the solid. The elements of carbon family show an interesting trend.

	C (diamond)	C_{60}	Si	Ge	Sn(gray)	Pb
ΔE/eV	6	1.7	1.2	0.7	0.08	0

Lead (also Sn) is a typical metal whose electrical conductivity decreases with temperature. The band gaps in Si and Ge require special attention. The small energy gap can be overcome by thermally excited electrons. The population of the levels is approximately given by the Boltzmann distribution as proportional to $e^{-\Delta E/kT}$*. As T increases, more and more electrons gain enough excitation energy to cross the gap. Hence these substances show an increase in electrical conductivity with increasing temperature. These are called *semiconductors*. Their behaviour differs from that of metals whose electrical conductivity *decreases* with temperature. The band model provides satisfactory explanation for the properties of semiconductors.

In *intrinsic semiconductors*, the energy gap between a filled band and the next empty band is very small (Fig. 8.30b). The electrons normally cannot jump across this gap and so the substances behave as insulators at absolute zero. As the temperature is raised, the thermal energy gained by the electrons (RT = 2.5 kJ mol^{-1} at 298 K; 6.4 kJ mol^{-1} at 773 K) becomes sufficient to promote them to the next empty band. Conduction follows. Pure germanium and gray tin are intrinsic semiconductors.

Semiconductor behaviour may also be imposed in certain substances by deliberate addition of impurities. Such semiconductors, called *extrinsic semiconductors*, may be of two types:

(i) **n-type semiconductors**: addition of arsenic or antimony (with 5 valence electrons) to germanium produces an *n*-type semiconductor.

These impurities place a *filled* energy level just *below* an empty band of the metal (germanium, for example). Electrons from the impurities can be easily excited to the empty metal band. Here the conductivity results from the flow of electrons and hence it is called a negative or *n*-type semiconductor.

(ii) **p-type semiconductor**. When gallium or indium (3 valence electrons) is added to germanium as impurity, they place an *empty* band just *above* a filled metal band. Passage of

*The population (P) is actually given by the Fermi-Dirac distribution which is a version of the Boltzmann distribution with inclusion of Pauli principle.

electrons from germanium to these empty bands results in a number of vacant sites (of the electrons) on germanium and renders it positively charged. The vacant sites are called "positive holes". Adjacent electrons move to fill these positive holes, creating further positive holes behind them. In effect, there is a migration of the positive hole and we get a positive or *p*-type semiconductor.

The band gap in silicon (pure) makes it an insulator at room temperature, resistivity ~40 ohm cm at 25°C. If the Si is doped with an element of Group V(15) like P, As or Sb, an *n*-type semiconductor is obtained. Doping with an element of Group III(13) like B, Al or Ga produces a *p*-type semiconductor.

Combination of an *n*-type sample of Si with one of *p*-type gives rise to an *n-p* junction: electrons can flow from *n* to *p* and holes from *p* to *n* and the current passes more easily in one direction than the other. It can act as a diode. Two *n*-type Si separated by a weakly *p*-type Si produces an *n-p-n* junction or a transistor.

Extremely small stable single crystal transistors may be prepared by exposing selected areas on a single crystal *n*-type Si wafer to the vapour of a Group 13 element when the impurity atoms diffuse to form a layer of *p*-type Si. These areas are then covered with a protective layer of SiO_2 (heating in O_2) which is toughened by exposure to UV radiation. At other desirable places, the layer is etched with HF and exposed to the vapour of a Group 15 element. In this manner, a patterened deposit of *n-p-n* type Si is obtained which is again covered with a protective layer of SiO_2. Metals are deposited at suitable places (after reetching) to connect the *n* and *p*-regions into an integrated circuit.

Some of the most widely used semiconductors are now prepared by combining elements from Groups 13 and 15 (III-V semiconductors, e.g., GaAs) or 12 and 16 (II-VI semiconductors, e.g., CdS). The band gap decreases when heavier atoms are incorporated, e.g.,

	BP	GaP	GaAs	InAs	InSb
ΔE (kJ mol^{-1})	193	218	137	34.7	17

GaAs has an extra advantage—its carrier velocity is higher than with silicon. It is the effective average velocity with which an electron moves through the lattice. This is greatly influenced by collision with impurity atoms or with a normal atom which is away from its usual lattice site due to vibrational excitation. This superiority makes GaAs a prospective semiconductor in logic chips. Methods are being developed to obtain highly pure varieties of GaAs (and other semiconductors) by *chemical vapour deposition* (CVD) technique. Volatile compounds like $Ga(CH_3)_3$ and AsH_3 are carried in a stream of H_2 on to a heated substrate where the desired semiconductor compound is deposited, generally as a single crystal film.

The use of GaAs in light emitting diode (LED) is related to its semiconductor properties. Unlike Si, GaAs is a *direct-gap* semiconductor in which the electrons from the top of the valence band may be excited with photons to the bottom of the conduction band without any change in momentum of the electron. The probability of the reverse transition during deexcitation (i.e., from conduction band to valence band) is also correspondingly higher, with the emission of photons. In silicon, such radiative transition cannot occur without a change in the linear momentum of the electron. This involves coupling with the linear momentum of the nucleus—thermal motion of the atom is thereby stimulated, making the substance hot. The ternary system $GaAs_{1-x}P_x$ forms a more useful LED and is used widely. III-V semiconductors like InSb may be useful in converting solar energy into electricity.

ESSENTIALS

Crystalline solids may be classified as molecular, metallic, covalent (network) and ionic.

A crystal consists of a basic structural unit repeated in three dimensions, also called the ***basis of the crystal***.

A space lattice is the pattern in space formed by the points representing an identical position in each basic unit.

A unit cell of a crystal is the smallest part of the crystal which can generate the crystal structure through repetition in three dimensions by simple translational displacement. It is an imaginary parallelepiped that can generate the entire crystal structure by translational repetition.

Unit cells are classified into seven crystal systems according to their rotational symmetry as cubic, tetragonal, orthorhombic, monoclinic, hexagonal, rhombohedral and triclinic.

There may be two ways of packing equal spherical units: cubic close packing (ccp) and hexagonal close placking (hcp).

The rock salt structure (NaCl), the cesium chloride structure (CsCl) and the zinc blende structure (ZnS) are the most common representative ionic structures.

The radius ratio is the ratio r_{small}/r_{large}, usually r_{cation}/r_{anion}. It serves as a first hand indication about the coordination number of the ions and the structural type of the crystal.

A value of radius ratio between 0.414 and 0.732 suggests a coordination number six for the ions and a face-centred cubic (FCC) structure. Radius ratio above 0.732 suggests a body-centred cubic structure (BCC) having coordination number 8 for the central ion.

Formula type	Radius ratio (r_+/r_-)	C.N.	Structure in cubic system
MX	0.225–0.414	4	Sphalerite (hexagonal wurtzite)
MX	0.414–0.732	6	NaCl
MX	0.732–1.000	8	CsCl
MX_2 or	0.225–0.414	4, 2	SiO_2 and Cu_2O
M_2X	0.414–0.732	6, 3	TiO_2, CdI_2, NiS_2, FeS_2
	0.732–1.000	8, 4	CaF_2

The lattice enthalpy of an ionic crystal is the enthalpy change for complete separation (to infinity) of the constituent ions in one mole of formula unit.

The electrostatic constribution to lattice enthalpy depends directly upon the product of the ionic charges ($z_+z_-e^2$) and inversely upon the equilibrium interionic separation in the crystal (r_0).

The net electro-static effect of all surrounding ions in a given lattice type appears as a characteristic factor called the ***Madelung constant*** **(A)**.

The lattice enthalpy per mole is expressed by the Born-Mayer equation

$$\Delta H_L = \frac{N_A A z_+ z_- e^2}{4\pi\varepsilon_0 r_0}\left(1 - \frac{\rho}{r_0}\right)$$

where N_A is the Avogadro number and ρ is a constant related to compressibility of the ion; commonly ρ = 34.5 pm.

Polarization or deformation of ions leads to covalent character in ionic compounds. The extent of polarization increases when the size of the cation is small or of the anion is large and either ion has high charge.

Lattice defects arise due to imperfection in the crystal lattice. Defects involving one or two lattice points are called point defects. Point defects which do not affect the stoichiometry or composition of the crystal are of two types (1) Frenkel defect and (2) Schottky defect.

The bonding in metals is described by the band theory which is applicable to other solids also. In this theory, the valence shell atomic orbitals from all atoms in an aggregate form closely spaced energy levels or bands which are occupied by the electrons obeying the exclusion principle.

EXERCISE

1. Polonium and silver crystallizes in (a) simple cubic and (b) fcc system respectively. Calculate the percent void space in each.

   Hint: (a) See Fig. 8.7. 8 spheres at eight corners contribute $8\times\frac{1}{8}$ or 1 sphere per cube of edge length a (say). Radius of each sphere = $a/2$. Volume of cube = a^3. Volume of one sphere ; $\pi a^3/6$. Space *occupied* = $\pi/6$ fraction or $100\pi/6$ per cent.

   (b) See Fig. 8.9. 4 spheres per cube of volume a^3.
   Radius of sphere (r) related to a as $4r = \sqrt{2}a$ (Fig. 8.10b).
   Space *occupied* = $\pi/3\sqrt{2}$ fraction or 74%. [**Ans.** (a) 47.6% (b) 26%]

2. KCl crystallizes in the NaCl structure. The side a of the unit cell is 1.5 times that for CsF which crystallizes with CsCl structure. Calculate the ratio of the densities of CsF and KCl. Molar mass: CsF = 151.9; KCl = 74.6.

   Hint: $\rho = \dfrac{\sum n_i M_i}{N_0 a^3}$, where n_i is the number of i-th kind of atom of molar mass M_i. a is the side of the unit cell. KCl has $4K^+$ and $4Cl^-$ ions per unit cell. CsF has 1 Cs^+ and $1F^-$ ions per unit cell. [**Ans.** 0.58]

3. Sodium metal crystallizes in the bcc structure with a = 424 pm. Calculate the radius of Na^+.

   Hint: Body diagonal = $\sqrt{3}a = 4r$ where r = radius.

4. Given that the radii of Ag^+ and Cl^- ions are 126 pm and 181 pm respectively, what structure is expected for AgCl from radius ratio? [**Ans.** NaCl]

5. Calculate the maximum size of a sphere which can be accomodated in a tetrahedral hole formed by four spheres of radius r in close packing.

   Hint: Refer Fig. 8.14.

6. I_2 of Mg is approximately twice its I_1. Also, the second electron attachment enthalpy of oxygen is positive. Why then magnesium forms $Mg^{2+}O^{2-}$ rather than Mg^+O^- ?

   How can one show that magnesium oxide is really not Mg^+O^- ?

   Hint: Higher lattice enthalpy of the doubly charged ions more than compensates the energy requirements. Mg^+O^- would be paramagnetic.

7. Calculate the lattice enthalpy of NaCl crystal using the Born-Lande equation.

   $$r_+ + r_- = 282 \text{ pm.}$$

   Hint: $1/4\pi\varepsilon_0 = 8.99 \times 10^9$ Nm^{2}C^{-2} ; A = 1.748 and n = 8 gives ΔH_L . [**Ans.** 753 kJ mol^{-1}]

8. Calculate the lattice enthalpy of CsCl using the Born-Mayer equation. $r_+ + r_- = 356$ pm.

   Hint: Put A = 1.763. ρ = 34.5 pm and $1/4\pi\varepsilon_0$ as before. [**Ans.** 621 kJ mol^{-1}]

9. Predict the lattice types of CsF, BeO, MgO, BaO, Tl_2O and ThO_2 from the given crystal radii of their ions (in pm).

Cs^+	Be^{2+}	Mg^{2+}	Ba^{2+}	Tl^+	Th^{4+}	O^{2-}	F^-
181	59	86	149	164	108	126	119

Hint: We calculate the radius ratios r_+/r_- for each. When the ratio so obtained is > 1.00, we calculate the inverse ratio (r_-/r_+, in brackett).

CsF	BeO	MgO	BaO	Tl_2O	ThO_2
1.52	0.47	0.68	1.18	1.30	0.86
(0.66)			(0.85)	(0.77)	

CsF, BeO and MgO are expected to have 6:6 coordination and adopt the NaCl structure, while BaO should adopt the CsCl structure. In fact, CsF, MgO and BaO (also CaO, SrO) adopt the NaCl structure. BeO adopts the wurtzite structure.

Tl_2O has twice as many Tl^+ ions as the number of O^{2-} ions. We predict a 4:8 coordination, i.e., each Tl^+ ion surrounded by four O^{2-} and each O^{2-} surrounded by eight Tl^+ – the *antifluorite* structure. Actually Tl_2O adopts a structure having C.N. 6 for oxygen.

In ThO_2, we expect a coordination number 8 for Th^{4+} and 4 for O^{2-}, suggestive of a fluorite structure.

10. The estimated radius of NH_4^+ ion (148 pm) suggests a CsCl type structure for NH_4F. But NH_4F adopts the wurtzite structure. Comment.

Hint: Strong H bonding is favoured in the 4:4 tetrahedral environment of wurtzite structure. Other NH_4X crystals adopt the expected NaCl/CsCl structure.

11. Explain:

(i) Cesium iodide (CsI) is nearly ten times more soluble in water than sodium fluoride (NaF) but it is much less soluble than CsF.

(ii) Magnesium oxide has a higher melting point than sodium fluoride.

(iii) The melting point of $MgBr_2$ is 700°C while that of $AlBr_3$ is only 97°C.

(iv) The melting point of AgCl is only 455°C while that of KCl is 776°C, though the crystal radii of Ag^+ and K^+ ions are almost the same.

(v) HgI_2 is much less soluble in water than $HgCl_2$.

(vi) $LiClO_4$ forms a stable trihydrate whereas the other alkali metal perchlorates do not.

[**Hint:** relative size of Li^+ and ClO_4^- ion].

(vii) $NaClO_4$ is about 1000 times as soluble as $KClO_4$ in water.

12. Show that the potential energy of an Na^+ ion in a row of Na^+ and Cl^- ions placed alternately is lower than the potential energy in an Na^+–Cl^- ion pair.

13. Suggest reasonable crystal structures of CaF_2 and TiO_2 from the following radii (pm): Ca^{2+} : 126; F^- : 119; Ti^{4+} : 74.5; O^{2-} : 126.

[**Hint:** For CaF_2, $r_+/r_- \simeq 1.0$, suitable for 8-coordination. But charge-balance requires that the number of Ca^{2+} ions should be half the number of F^- ions. So one-half of the cation sites are unfilled in a cubic arrangement—each Ca^{2+} is surrounded cubically by eight F^- ions but each F^- ion is surrounded tetrahedrally by four Ca^{2+} ion (fluorite structure).

Similarly, in TiO_2, $r_+/r_- \simeq 0.6$, suitable for 6-coordination. In this case also, one-half of the cation sites are empty in an NaCl-type structure—the coordination is 6 : 3 (rutile structure).]

14. Calculate the lattice energy of $Mg(ClO_4)_2$ using Kapustinskii equation. Radii of Mg^{2+} and ClO_4^- ions are 86 and 226 pm respectively. $\kappa = 1.214 \times 10^5$ kJ pm mol^{-1}. [**Ans.** 2076 kJ mol^{-1}]

15. Calculate the lattice energy of ThO_2 using the Born-Lande equation. See question no. 9 for radii of Th^{4+} and O^{2-} ions. Madelung constant for fluorite structure = 2.519. Born exponent for Th^{4+} $(6s^2 6p^6)$ may be taken (by extrapolation) as 14.

Hint: $n = (14 + 7) \div 2 = 10.5$; $|z_+ z_-| = 8$. [**Ans.** 10826 kJ mol^{-1}]

Note: The factor $N_A e^2 . 10^{12}/4\pi\varepsilon_0$ appears in calculations of lattice energy by Born-Lande or Born-Mayer equation with (r_+/r_-) in pm. It has the value 138,938 kJ pm mol^{-1}.

16. Calculate the electron gain enthalpy (–*E.A.*) of chlorine from the following ΔH data (kJ mol^{-1}); $D_{Cl_2} = 242$; $I_{Na} = 494$; $\Delta H_{Sub}(\text{Na}) = 109$; ΔH_f° (NaCl) = –414. $r_{Na^+} + r_{Cl^-} = 281$ pm in NaCl (*c*).

17. (a) Calculate the lattice enthalpy of the hypothetical solid MgBr assuming $r(Mg^+) = 116$ pm and $r(Br^-) = 182$ pm in 6:6 coordination.

(b) Construct a relevant Born-Haber cycle for the formation of MgBr(*c*) using the data from Fig.8.23 and $I_1(\text{Mg}) = 738$ kJ mol^{-1}. Calculate H_f° of MgBr using the lattice energy calculated in part *a*.

18. Among $MgSO_4(c)$ and $BaSO_4(c)$, which is expected to decompose at a lower temperature according to the reaction $MSO_4(c) \rightarrow MO(c) + SO_3(g)$?

Hint: The major contribution to the enthalpy change for the reaction comes from the difference of lattice enthalpy between MSO_4 and MO. Mg^{2+}, with its smaller size, is expected to gain more in forming a lattice with O^{2-} than with the large SO_4^{2-} ion. Hence we expect $MgSO_4$ to decompose at a lower temperature.

19. What will be your choice of M among Na, K, or Cs to stabilize the product of reaction 1 which may decompose subsequently by reaction 2?

$$\textit{Reaction 1: } \text{MI}(s) + 2\text{Cl}_2(g) \rightleftharpoons \text{MICl}_4(s)$$

$$\textit{Reaction 2: } \text{MiCl}_4(s) \longrightarrow \text{MCl}(s) + \text{ICl}(g) + \text{Cl}_2(g).$$

Hint: With other factors remaining the same, decomposition of $MICl_4(s)$ to MCl(*s*) will be disfavoured when the gain in lattice enthalpy in the process is small. This suggests the use of a large cation like Cs^+ to be effective as M^+.

Note: A tetraalkylammonium ion like $^tBu_4N^+$ would be even more effective. Such large cations are also effective to stabilize anions like $[X–H–X]^-$, X = Cl, Br (see text).

20. Explain the solubility trends:

(a) $MgSO_4 > CaSO_4 > BaSO_4$

(b) $Mg(OH)_2 < Ca(OH)_2 < Ba(OH)_2$.

Hint: Large cation-large anion and small cation-small anion combinations generally favour precipitation.

21. CsF is more soluble in water than CsI while LiF is less soluble than LiI;–explain.

22. The solid alkali metals (Na, K, ...) show slight paramagnetism. Explain in the light of the band theory.

Hint: In sodium, the 2*s* band is completely filled. The 3*s* band covers a span of energy many times greater than the pairing energy of the electrons. Hence the electrons doubly occupy the lower half energy levels in the band. However, a narrow strip having width equal to about *kT* at the top of the filled levels in the band contains unpaired electrons which give rise to paramagnetism. *kT* is about 1% of the filled width of the band.

The paramagnetism decreases from top to bottom in the group.

23. Both sodium and silver dissolve readily in mercury forming amalgam but some metals like iron are essentially insoluble in mercury. Explain.

Hint: Atoms of sodium are bonded weakly in the lattice through the *s*-electrons. As *d*-orbitals are involved in forming the band, the overlap may be quite directional. This trend increases towards the

middle of each *d*-series, where strong *d-d* overlap is indicated by the increasing melting point of the metals (see text). Dissolution in liquid mercury will require rupture of the directional lattice. Hence metals in the middle of *d*-series do not readily form amalgams.

But as one proceeds towards the end of a *d*-series, the quickly increasing effective nuclear charge essentially sinks the *d*-electrons into the core. By the time Zn, Ag, Cd (and Hg) are reached, the *d*-electrons do not participate in bonding. Hence these metals are bound only by the filled *s* orbital, as reflected in their low melting points and ease of formation of amalgam.

24. Indicate the type of semiconduction (*n* or *p*) expected from the following:

(a) Ga doped Be (b) As doped Ge (c) Zn-doped GaAs (d) KCl irradiated with x-rays

(e) $In_{1+x}As_{1-x}$ ($x << 1$) (f) $WO_{2.999}$. Ans. (a) *p* (b) *n* (c) *p* (d) *n* (e) *p* (f) *n*

25. CdS has a band gap equal to 2.4 eV. How can you relate this to the colour of CdS?

Hint: ΔE = 2.4 eV ≡ 3.8 × 10^{-19} J. Maximum wavelength λ of absorption of visible light is given by $\Delta E = hc/\lambda$. λ = 523 nm falls in the green region. Transmitted light will be orange-yellow.

9

OXIDATION-REDUCTION REACTIONS

A large number of chemical reactions belong to the catagory of oxidation-reduction or redox reactions, reactions that involve change in oxidation number or oxidation state of one or more atoms of the reactants. Redox reactions take place between an electron donor (reducing agent, reductant) and an electron acceptor (oxidizing agent, oxidant). Such reactions are formally comparable to acid base interactions according to Lewis interpretation: a base is an electron donor and an acid is an electron acceptor. But in redox reactions the electron transfer results in an increase in oxidation state of some atoms(s) and an equivalent decrease in oxidation state of some other atom(s). Such electron transfers may be accompanied by atom transfer also, but electron transfer is the key feature. Redox reactions are basically enthalpy-driven. We shall appreciate later that different factors may be held responsible for making the enthalpy change favourable. Before going to that, we shall examine different aspects of redox reactions.

9.1 BASIC CONCEPTS

9.1.1 Oxidation Number

Oxidation-reduction reactions or redox reactions involve an oxidizing agent (or oxidant) and a reducing agent (or reductant). The reactions are accompanied by a decrease in oxidation number of one (or more) element in the oxidant and an equivalent increase in oxidation number of one (or more) element in the reductant.

$$\underset{\text{reductant (As, S)}}{3As_2S_3} + \underset{\text{oxidant (N)}}{28HNO_3} + 4H_2O \rightarrow 6H_3AsO_4 + 9H_2SO_4 + 28NO$$

The oxidation numbers of the elements help us to keep track of the elements (or atoms) oxidized or reduced. The oxidation number (O.C. Johnson, 1880) of an element is equal to the charge which an atom of the element *would acquire* if the compound consisted only of ions. For ionic compounds, the oxidation number of any element is thus equal to the electrical charge on its ion (monatomic) in the compound. For covalent compounds we must assume a hypothetical ionic structure and formally assign the "possession" of the electron pair in a covalent bond from electronegativity considerations. Once this is done, we may count the effective number of electrons in the valence shell of each atom. The difference between this number and the number of electrons in the valence shell of the neutral atom now gives its oxidation number. The following examples will illustrate this principle:

$$\overset{+1}{H}-\overset{-1}{F} \qquad \overset{+1}{H}-\overset{-2}{O}-\overset{+1}{H} \qquad \overset{+1}{H}-\overset{-1}{O}-\overset{-1}{O}-\overset{+1}{H}$$

$$\overset{-3}{C}\overset{+1}{H_3}\overset{-1}{C}\overset{+1}{H_2}\overset{-2}{O}\overset{+1}{H} \qquad \overset{+1}{H}\overset{0}{C}\overset{+1}{H}\overset{-2}{O} \qquad \overset{-3}{C}\overset{+1}{H_3}\overset{+1}{C}\overset{+1}{H}\overset{-2}{O}$$

The following generalizations regarding the assignment of oxidation number (O.N.) of an element may now be made:

(i) The O.N. of elements in the free state (i.e., when they are not combined to other elements) is zero. The converse, however, may not be true. O.N. may be zero in combination also, as in CH_2Cl_2 or HCHO.

(ii) The O.N. of hydrogen in combination is mostly +1. But in ionic hydrides of more electropositive metals, the O.N. of hydrogen is –1 (LiH, say).

(iii) Oxygen (combined) has the usual O.N. –2 except in peroxides, superoxides, ozonides and fluorine-compounds. In peroxides the O.N. is –1 (H_2O_2); in superoxides, –½ (KO_2). In F_2O, O.N. of oxygen is +2. In $O_2^+PtF_6^-$, O.N. of O is +½.

(iv) The O.N. of a metal in compounds is usually positive.

Exceptions may arise in typical situations. For example, metals below hydrogen in the electro-chemical series bear a negative O.N. in their hydrides: $\overset{+4}{Ge}\overset{+1}{H_4}$, $\overset{-1}{Cu}\overset{+1}{H}$. In co-ordination compounds, the O.N. of a metal depends upon the charge of the ligand.

$$\overset{0}{Ni}(CO)_4 \qquad \overset{-1}{Co}(CO)_2(NO^+) \qquad K_4[\overset{+2}{Fe}(CN)_6]$$

These assignments become clear when we remember that CO is a neutral molecule, NO^+ has one unit positive charge and CN^- one unit negative charge.

(v) The effective O.N. of an ion or radical is equal to its net charge, including sign.

(vi) The grand total of the oxidation numbers of *all* the atoms in a molecule is zero.

An element may be present in the same molecule in more than one oxidation state, e.g., $\overset{-3}{N}H_4\overset{+5}{N}O_3$ or CH_3CH_2OH (shown earlier) or $Ca(O\overset{+1}{Cl})\overset{-1}{Cl}$.

The following summary rules may also be followed:

(i) Add –1 for each bond to a less electronegative atom (or a minus charge) but –2 *if this bond is a coordinate covalent bond*. This becomes unnecessary when the coordinate covalent bond is written with charge separation and due consideration for the charge is done.

(ii) Add 0 for each bond to an identical atom or *an atom having the same electronegativity value.*

(iii) Add +1 for each bond to a more electronegative atom (or a plus charge) but +2 *if this bond is a coordinate covalent bond.*

For organic compounds, these rules may be further streamlined as follows:

(i) Add –1 for each –H; (ii) Add 0 for each –C; (iii) Add +1 for each bond to a heteroatom.

As shown earlier, the oxidation number of different carbon atoms in the same molecule may be different.

Inspection of oxidation number tells us whether any reaction is really an oxidation-reduction type, and if so, which element is getting oxidized and which one is getting reduced.

However, the concept of oxidation number is not altogether free from certain limitations. The procedure for the assignment of oxidation numbers is based on arbitrary conventions regarding charge separation. Also, the electron transfer processes actually taking place in a redox reaction may be more complicated than those suggested by mere changes of oxidation number. Thus, oxidation by the permanganate ion is very much concerned with the particular species MnO_4^-; a mere statement of the oxidation state of manganese in this species, i.e., Mn (+7) would convey very little about its chemical aptitude. Lastly, the role of the hydrogen ion (aquated) in many redox reactions in aqueous medium does not appear directly in the oxidation number procedure for balancing such reactions (compare with the *ion-electron* method). In spite of these limitations, however, the great advantage of the concept has given it wide acceptance more as a handy tool than as a theoretical guideline.

Problem 9.1

Is the following an example of a redox reaction?

$$2CCl_4 + K_2CrO_4 \rightleftharpoons 2Cl_2CO + CrO_2Cl_2 + 2KCl$$

Hint: Determine O.N. of C and Cr on both sides of the equation.

9.1.2 Balancing Redox Equations

Redox reactions involve transfer of electron from a reducing agent to an oxidizing agent. Hence, oxidation and reduction must take place simultaneously and in equivalent amounts. This provides the basis of balancing redox equations. This may be carried out in either of two ways: (i) the oxidation number method; (ii) the ion-electron method.

Oxidation number method

Step I. The mole ratio of the oxidant and reductant can be ascertained from (i) the increase of O.N. of that element in the reductant which undergoes oxidation and (ii) the decrease of O.N. of that element in the oxidant which undergoes reduction. *All* such atoms involved in each case are considered for each process and the *total* increase and decrease of oxidation number are equated by inserting appropriate factors before each species.

Step II. All the species appearing on the reactant and product side are considered and adjustment is made for atoms which do not undergo any change of oxidation number. Extra molecules of reactant and acid or alkali or water may be inserted depending on the condition and medium of the reaction.

Step III. The balancing is closed by inspection of the reactant and product side.

A few examples will illustrate the procedure.

Reaction 1. Manganese dioxide oxidizes concentrated hydrochloric acid to chlorine. Manganous chloride and water is also formed.

Step I. $\overset{+4}{MnO_2} \xrightarrow{-2\text{ unit}} \overset{+2}{MnCl_2}$; $\overset{-1}{HCl} \xrightarrow{+1\text{ unit}} \frac{1}{2}\overset{0}{Cl_2}$.

Therefore, we write: $MnO_2 + 2HCl \longrightarrow MnCl_2 + Cl_2$

Step II. The oxygen and hydrogen remaining on the left may be balanced by writing H_2O on the right. Further, the chlorine atoms in $MnCl_2$ may be justified by adding two more molecules of HCl as salt-former:

$$MnO_2 + 2HCl + 2HCl \longrightarrow MnCl_2 + Cl_2 + H_2O$$

Step III. By inspection, $\mathbf{MnO_2 + 4HCl = MnCl_2 + 2H_2O + Cl_2}$.

Reaction 2. Hydrogen peroxide oxidizes manganous sulfate in alkaline medium to manganese dioxide:

Step I. $\overset{-1}{H_2O_2} \xrightarrow{-1\text{ unit} \times 2\text{ atoms of O}} 2H_2\overset{-2}{O}$; $\overset{+2}{MnSO_4} \xrightarrow{+2\text{ units}} \overset{+4}{MnO_2} + SO_4^{2-}$

So, we write: $MnSO_4 + H_2O_2 \longrightarrow MnO_2 + 2H_2O + SO_4^{2-}$

Step II. Here the extra hydrogen atoms on the right cannot be compensated by introducing H^+ ions from acids in the left, because the medium is alkaline. So, we introduce OH^- ions on the left from say, NaOH.

$$MnSO_4 + H_2O_2 + 2NaOH \longrightarrow MnO_2 + 2H_2O + SO_4^{2-} + 2Na^+$$

Step III. By inspection, the final equation is

$$\mathbf{MnSO_4 + H_2O_2 + 2NaOH = MnO_2 + 2H_2O + Na_2SO_4}$$

Reaction 3. Oxidation of manganous sulfate solution by red lead in presence of nitric acid to pink permanganic acid.

Step I. Here the oxidant is red lead, Pb_3O_4. It is equivalent to $2PbO \cdot PbO_2$ in its reaction. It is only the PbO_2 which performs the oxidation and gets reduced to lead (II)-salts. So, we have to consider the change in O.N. of only *one atom of lead* per formula unit of Pb_3O_4.

–2 unit × 1 atom of Pb ; +5 unit

$$2PbO \cdot \overset{+4}{Pb}O_2 \longrightarrow 3\overset{+2}{Pb}{}^{2+} \quad ; \quad \overset{+2}{Mn}SO_4 \longrightarrow H\overset{+7}{Mn}O_4$$

Equating the increase and decrease of oxidation number, we have

$$5Pb_3O_4 + 2MnSO_4 \longrightarrow 15Pb^{2+} + 2HMnO_4 + 2SO_4^{2-}$$

Step II. We add requisite HNO_3 to provide necessary anions on the right:

$$5Pb_3O_4 + 2MnSO_4 + 26HNO_3 \longrightarrow 13Pb(NO_3)_2 + 2PbSO_4 + 2HMnO_4.$$

The combination of nitric acid with lead monoxide will form water also.

Step III. By inspection,

$$\mathbf{5Pb_3O_4 + 2MnSO_4 + 26HNO_3 = 13Pb(NO_3)_2 + 2PbSO_4 + 2HMnO_4 + 12H_2O.}$$

Reaction 4. Oxidation of ethyl alcohol to acetic acid by potassium dichromate in acid medium.

Step I. For organic compounds, the *sum* of the oxidation numbers of the different carbon atoms in the molecule before and after oxidation may be taken into account. Thus,

$$\overset{-3}{C}H_3\,\overset{-1}{C}H_2OH \longrightarrow \overset{-3}{C}H_3\,\overset{+3}{C}OOH. \quad \text{Total change: +4 unit (oxidation).}$$

Again, $K_2\overset{+6}{Cr_2}O_7 \longrightarrow \overset{+3}{Cr_2}(SO_4)_3$, i.e., –6 unit (reduction) per molecule of $K_2Cr_2O_7$.

Hence, $3CH_3CH_2OH + 2K_2Cr_2O_7 \longrightarrow 3CH_3COOH + 4K^+ + 4Cr^{3+}$

Sulfuric acid is generally used in the reaction.

Step II. So we write,

$$3CH_3CH_2OH + 2K_2Cr_2O_7 + 8H_2SO_4 \longrightarrow 3CH_3COOH + 2Cr_2(SO_4)_3 + 2K_2SO_4.$$

Step III. By inspection, we get the balanced equation as

$$\mathbf{3CH_3CH_2OH + 2K_2Cr_2O_7 + 8H_2SO_4 = 3CH_3COOH + 2Cr_2(SO_4)_3 + 2K_2SO_4 + 11H_2O}$$

Reaction 5. As_2S_3 dissolves in hot concentrated nitric acid to give H_3AsO_4 and H_2SO_4.

+2 × 2 atoms of As ; +8 × 3 atoms of S ; –3 unit

Step I. $\overset{+3}{As_2}\overset{-2}{S_3} + H\overset{+5}{N}O_3 \longrightarrow 2H_3\overset{+5}{As}O_4 + 3H_2\overset{+6}{S}O_4 + \overset{+2}{N}O$

Total increase in oxidation number per As_2S_3 is 28 units (As: 2 × 2; S : 8 × 3). Decrease of oxidation number per HNO_3 is 3 units.

Hence the mole ratio of reductant (As_2O_3) and oxidant (HNO_3) is 3 : 28.

We have: $3As_2S_3 + 28HNO_3 \longrightarrow 6H_3AsO_4 + 9H_2SO_4 + 28NO$

Step II. To balance H and O atoms we add $4H_2O$ on the left.

Step III. The balanced equation is

$$3As_2S_3 + 28HNO_3 + 4H_2O = 6H_3AsO_4 + 9H_2SO_4 + 28NO$$

Ion-electron method

Developed mainly by Jette and La Mer in 1927, the ion-electron method is practically limited to ionic reactions in aqueous solutions. The method is based on the electronic theory of oxidation and reduction. The procedure may be summarized as follows:

Step I. The products of the reaction are ascertained and the loss of electron by the reductant and the gain of electron by the oxidant are expressed *separately* by "partial equations".

Step II. The individual partial equations are balanced separately with respect to different atoms as well as with respect to the electrical charge on both sides. Since redox reactions encountered in aqueous solutions are chiefly concerned, only the species capable of appearing as ions in solution and actually taking part in oxidation-reduction should appear in such half-reactions. Substances which are either insoluble or slightly ionizable in water, should appear *as such* in these equations (H_2O_2, for example). H^+ and OH^- ions are inserted as required.

For electrical balance, the total charge of all the ions appearing on the two sides of the half-reactions (including those on the H^+ and OH^- ions inserted) are considered and the difference is adjusted by adding electrons (–ve charge) on the appropriate side.

Step III. The two partial equations so obtained are multiplied by suitable factors to equalize the number of electrons in them. These are then added when the electrons get cancelled. The ions are replaced by molecules making necessary adjustments.

A few illustrations are given.

Reaction 1. Cupric sulfide dissolves in hot dilute nitric acid forming cupric nitrate, nitric oxide and sulphur.

Step I. We start the partial equations by writing

(i) $NO_3^- \longrightarrow NO$ and (ii) $CuS \longrightarrow S + Cu^{2+}$

Step II. (i) For balancing this partial equation atomically, the two extra oxygen atoms on the left are adjusted by inserting two molecules of water on the right.

$$NO_3^- \longrightarrow NO + 2H_2O.$$

The four extra hydrogen atoms on the right are now balanced by placing four H^+ ions on the left. Charge balance by adding electrons gives

$$NO_3^- + 4H^+ + 3e \longrightarrow NO + 2H_2O.$$

(ii) The partial equation after charge balance is

$$CuS \longrightarrow Cu^{2+} + S + 2e$$

Step III. We combine the two partial equations so that the electrons cancel from both:

$$2NO_3^- + 8H^+ + 6e \longrightarrow 2NO + 4H_2O$$

$$3CuS \longrightarrow 3Cu^{2+} + 3S + 6e.$$

$$2NO_3^- + 8H^+ + 3CuS \longrightarrow 3Cu^{2+} + 2NO + 3S + 4H_2O.$$

To obtain the final equation, we shall have to provide 3 × 2 or 6 more NO_3^- as HNO_3:

$$\mathbf{3CuS + 8HNO_3 = 3Cu(NO_3)_2 + 2NO + 3S + 4H_2O.}$$

Reaction 2. Reduction of sodium nitrate solution by metallic aluminium in presence of sodium hydroxide.

Step I. Nitrate is reduced to ammonia and aluminium is oxidized to the aluminate ion, conventionally written AlO_2^- (though in solution, it should have been $Al(OH)_4^-$; see text):

(i) $NO_3^- \longrightarrow NH_3$ and (ii) $Al \longrightarrow AlO_2^-$ (or, $Al \longrightarrow Al(OH)_4^-$

Step II. (i) For balancing atomically, one starts as $NO_3^- \longrightarrow NH_3 + 3H_2O$.

But now the nine hydrogen atoms on the right cannot be adjusted on the left by placing $9H^+$, since the medium is *alkaline*. So we add $9H_2O$ on the left, adjust the number of H_2O and OH^- and balance the charge with electrons:

$$NO_3^- + 6H_2O + 8e \longrightarrow NH_3 + 9OH^-$$

(ii) For balancing atomically, one may write $Al + 2H_2O \longrightarrow AlO_2^- + 4H^+$

Here again, the alkalinity of the medium prompts us to add $4OH^-$ on the left and adjust the number of H_2O. Next we add electrons to balance the charge:

$$Al + 4OH^- \longrightarrow AlO_2^- + 2H_2O + 3e. \quad (Al + 4OH^- \longrightarrow Al(OH)_4^- + 3e)$$

Step III. We add the two partial equations so that the electrons cancel:

$$3NO_3^- + 18H_2O + 24e \longrightarrow 3NH_3 + 27OH^-$$

$$8Al + 32OH^- \longrightarrow 8AlO_2^- + 16H_2O + 24e$$

$$8Al + 3NO_3^- + 2H_2O + 5OH^- \longrightarrow 3NH_3 + 8AlO_2^-.$$

Adding Na^+ ions as required, the balanced equation becomes:

$$\mathbf{8Al + 3NaNO_3 + 5NaOH + 2H_2O = 3NH_3 + 8NaAlO_2.}$$

Reaction 3. Reduction of potassium ferricyanide to potassium ferrocyanide by hydrazine in alkaline solution.

Step I. Hydrazine is oxidized to nitrogen. The starting partial equations are

(i) $[Fe(CN)_6]^{3-} \longrightarrow [Fe(CN)_6]^{4-}$ (ii) $N_2H_4 \longrightarrow N_2$

Step II. (i) We have to simply balance the charge,

$$[Fe(CN)_6]^{3-} + e \longrightarrow [Fe(CN)_6]^{4-}$$

(ii) Since the medium is alkaline, it should be balanced by OH^-;

$$N_2H_4 + 4OH^- \longrightarrow N_2 + 4H_2O + 4e.$$

Step III. We combine the two partial equations as before and add K^+ ions to get

$$\mathbf{N_2H_4 + 4K_3[Fe(CN)_6] + 4KOH = N_2 + 4K_4[Fe(CN)_6] + 4H_2O}$$

Reaction 4. Oxidation of chromite ($Cr(OH)_4^-$) in basic solution by hypochlorite.

Step I. Hypochlorite is reduced to sodium chloride. Hence we start as:

(i) $ClO^- \longrightarrow Cl^-$ (ii) $Cr(OH)_4^- \longrightarrow CrO_4^{2-}$

Step II. (i) Since the medium is alkaline, we balance by OH^-. Charge balance gives

$$ClO^- + H_2O + 2e \rightleftharpoons Cl^- + 2OH^-$$

(ii) The partial equation may be written $Cr(OH)_4^- \longrightarrow CrO_4^{2-} + 4H^+ + 3e$

Since the medium is alkaline, we write $Cr(OH)_4^- + 4OH^- \longrightarrow CrO_4^{2-} + 4H_2O + 3e$

Step III. We add the two half reactions as usual:

$$3ClO^- + 2Cr(OH)_4^- + 2OH^- \rightleftharpoons 3Cl^- + 2CrO_4^{2-} + 5H_2O.$$

Taking Na^+ as the corresponding cation,

$$\mathbf{3NaClO + 2Na[Cr(OH)_4] + 2NaOH = 3NaCl + 2Na_2CrO_4 + 5H_2O}$$

Reaction 5. Oxidation of benzaldehyde to benzoic acid by potassium dichromate in acid medium.

Step I. The changes involved are

(i) $C_6H_5CHO \longrightarrow C_6H_5COOH$; (ii) $Cr_2O_7^{2-} \longrightarrow 2Cr^{3+}$

Step II. (i) The partial equation is

$$C_6H_5CHO + H_2O \rightleftharpoons C_6H_5COOH + 2H^+ + 2e$$

(ii) The partial equation is

$$Cr_2O_7^{2-} + 14H^+ + 6e \rightleftharpoons 2Cr^{3+} + 7H_2O.$$

Step III. The final balanced equation is

$$\mathbf{3C_6H_5CHO + K_2Cr_2O_7 + 8HCl \rightleftharpoons 3C_6H_5COOH + 2CrCl_3 + 2KCl + 4H_2O.}$$

Reaction 6. Oxidation of $Mn^{2+}(aq)$ to MnO_4^- by sodium bismuthate in nitric acid medium.

Step I. The starting half-reactions are

(i) $Mn^{2+} \longrightarrow MnO_4^-$ (ii) $BiO_3^- \longrightarrow Bi^{3+}$

Step II. The balanced half-reactions are

(i) $Mn^{2+} + 4H_2O \longrightarrow MnO_4^- + 8H^+ + 5e$

(ii) $BiO_3^- + 6H^+ + 2e \longrightarrow Bi^{3+} + 3H_2O$

Step III. We combine the two half-reactions as usual:

$$2Mn^{2+} + 5BiO_3^- + 14H^+ \longrightarrow 2MnO_4^- + 5Bi^{3+} + 7H_2O.$$

Inserting required number of Na^+ and NO_3^- ions, we get

$$\mathbf{2Mn(NO_3)_2 + 5NaBiO_3 + 14HNO_3 \longrightarrow 2NaMnO_4 + 5Bi(NO_3)_3 + 3NaNO_3 + 7H_2O.}$$

Merits of the ion-electron method

(i) Assignment of hypothetical oxidation numbers is not necessary; as such, no assumption regarding the distribution of valence bonds in a molecule is required.

The partial equations established for the oxidant and reductant are conceptual half-reactions but we shall soon find that they represent actual electrode processes. The oxidized and reduced species in a half-reaction are called a ***redox couple***:

$$\text{Ox (Oxidized form)} + ne \rightleftharpoons \text{Red (Reduced form)}.$$

Here, the redox couple is represented in a reduction half-reaction. They could also be written as an oxidation half-reaction. In our examples, we have established one oxidation half-reaction and one reduction half-reaction and finally added them to obtain the overall chemical equation. We could also write both half-reactions as *reduction processes* and obtain the overall equation by *subtraction*. Conventionally, this latter procedure is followed in writing electrode processes in chemical cells, as is to be found shortly.

(ii) The importance of hydrogen ion concentration of the medium is clearly understood; this is not obvious in the oxidation number method.

9.1.3 Disproportionation and Comproportionation

There are some redox reactions in which the oxidation number of an element shows both increase and decrease in the same reaction; the two processes take place to equivalent extent. Such reactions are called *disproportionation* (dismutation) reactions. Some examples are given:

(i) cuprous sulfate* in aqueous solution: $\overset{+1}{Cu_2}SO_4 \rightleftharpoons \overset{0}{Cu} + \overset{+2}{Cu}SO_4$

*prepared from cuprous oxide and dimethyl sulphate.

(ii) manganate ion in presence of acid: $3\overset{+6}{Mn}O_4^{2-} + 4H^+ \rightleftharpoons 2\overset{+7}{Mn}O_4^- + \overset{+4}{Mn}O_2 + 2H_2O$

(iii) white phosphorus on boiling with aqueous sodium hydroxide:

$$\overset{0}{P_4} + 3NaOH + 3H_2O = \overset{-3}{P}H_3 + 3NaH_2\overset{+1}{P}O_2$$

(iv) potassium chlorate on gentle heating: $4K\overset{+5}{Cl}O_3 = 3K\overset{+7}{Cl}O_4 + K\overset{-1}{Cl}$

(v) hydrogen peroxide on catalytic decomposition: $2H_2\overset{-1}{O_2} = 2H_2\overset{-2}{O} + \overset{0}{O_2}$

(vi) Mercurous ion (dimercury (2+)) in aqueous solution in presence of OH^-, CN^-, S^{2-} etc.:

$$\overset{+1}{Hg_2^{2+}} + 2CN^- \longrightarrow \overset{+2}{Hg}(CN)_2(aq) + \overset{0}{Hg}(l)$$

(vii) chlorine in contact with cold alkali: $\overset{0}{Cl_2} + 2NaOH = Na\overset{-1}{Cl} + NaO\overset{+1}{Cl} + H_2O$.

In basic solution the hypochlorite ion tends to disproportionate further as

$$3OCl^- = 2Cl^- + ClO_3^-.$$

The equilibrium constant for this reaction is very high (10^{27}), but the rate is very slow at room temperature. Around 70°C and higher, chlorate ion is the main product.

Equations for such reactions may be easily balanced by equalizing the total increase and decrease of oxidation number of the element involved.

The position of equilibrium in a disproportionation reaction may be altered by removal of reactants or products in the usual manner, e.g., neutralization, precipitation, complex formation etc. For the disproportionation of Cu(I) in aqueous solution (reaction i), the equilibrium constant is of the order of 10^6. In the presence of chloride ions, cuprous chloride is precipitated (solubility product $\simeq 10^{-7}$) and the concentration of Cu^+ ion is significantly lowered. Disproportionation is checked to cause a corresponding decrease in the concentration of Cu^{2+} ion (to maintain the equilibrium constant). Similarly, addition of cyanide ions reduces the concentration of Cu^+ ions by forming the very stable $[Cu(CN)_4]^{3-}$ ions:

$$Cu^+ + 4CN^- = [Cu(CN)_4]^{3-} \qquad K \simeq 10^{30}$$

Accordingly, disproportionation is almost checked.

Disproportionation reactions involving H^+ ions are similarly influenced by the hydrogen ion concentration of the medium. The disproportionation of manganate ion (reaction ii) is an example. The equilibrium constant is of the order of 10^{58} and disproportionation occurs in acid, neutral or even slightly basic solutions. It can be shown that manganate ions may exist only in strongly alkaline solutions (pH > 14).

The tendency of any species to disproportionate in a particular medium may be inferred from the relevant half-cell emf values. This has been exemplified later.

The reverse of disproportionation is known as *comproportionation*: an element in two different oxidation states (in the same or different species) react to form a product containing the element in an intermediate oxidation state:

$$Ag^{2+}(aq) + Ag(s) = 2Ag^+(aq)$$

$$IO_3^-(aq) + 5I^-(aq) + 6H^+(aq) \rightarrow 3I_2 + 3H_2O$$

$$NH_4NO_3 \xrightarrow{250°C} N_2O(g) + 2H_2O$$

9.1.4 Complementary and Non-complementary Redox Reactions

Redox reactions are sometimes classified as complementary and non-complementary redox reactions, particularly in connection with reactions of transition metal ions in solution. A redox reaction in which the oxidizing and reducing species change their oxidation states by an equal number is called a complementary reaction, for example, the reaction:

$$Fe(II) + Ce(IV) \longrightarrow Fe(III) + Ce(III)$$

When oxidation states of the reactants change by a different number of units, the reaction is called a non-complementary reaction:

$$2Fe(III) + Sn(II) \longrightarrow 2Fe(II) + Sn(IV)$$

Non-complementary reactions are usually slow and proceed in stages. These have been discussed in chapter 27 in connection with reactions of coordination compounds.

9.1.5 Redox Reactions and EMF

In the ion-electron method of balancing equations for redox reactions, we formed two partial equations or conceptual "half-reactions" in terms of electron loss or gain for the reductant and oxidant separately. In fact, these electron transfer processes may be made to occur at the electrodes of a suitably designed electrochemical cell (see box). The complete redox reaction now stands for the overall chemical change that takes place in the electrochemical cell. The electromotive force (EMF) developed in the cell is derived from the free energy change accompanying the cell reaction:

(decrease in free energy) $-\Delta G = nFE$ (available electrical work)

where $-\Delta G$ represents the decrease in free energy for the cell process which involves the passage of n faradays of electric charge and E is the EMF developed in the cell. A galvanic cell (also called a voltaic cell) uses a spontaneous chemical reaction to generate electricity. In this reaction, one reactant must be oxidized and another must be reduced. But the two reactants cannot be allowed to come in direct contact, for electrons would then simply flow from the reducing agent to the oxidizing agent. Instead, the oxidizing and reducing agents are physically separated and the electrons forced to flow through an external circuit.

Demonstration

Transfer of electrons accompanying any redox reaction may be demonstrated under suitable conditions. A typical experiment is described below.

A beaker containing dilute ferric chloride solution is connected with another beaker containing potassium iodide solution through a salt-bridge* (Fig. 9.1). Platinum foil electrodes dipped in each solution are then connected externally by a copper wire *via* a sensitive milliammeter. When the circuit is closed, electricity flows in the external circuit from the ferric chloride solution to the potassium iodide solution; this is *opposite* to the direction of electron-flow. After a while the yellow colour of the ferric chloride solution starts to fade out; ferrous ions can be detected in it (red colour with 0.1 per cent aqueous solution of *o*-phenanthroline in faintly acid medium). Iodine is simultaneously liberated at the other beaker, the colourless solution turning yellowish brown.

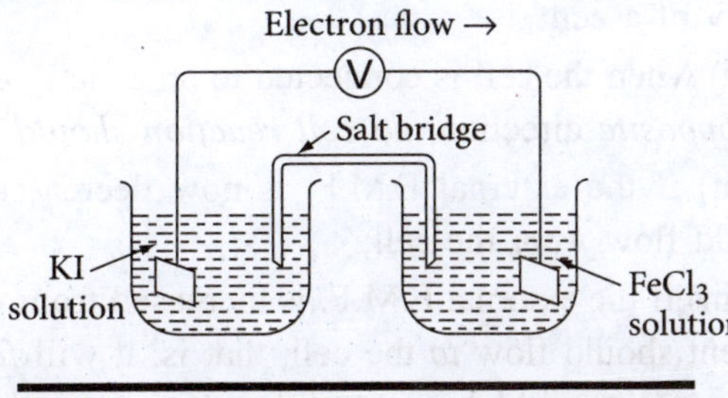

Fig.9.1: An experimental set up (schematic) to demonstrate the transfer of electrons in a redox reaction.

Similar observations may be made by taking other oxidants and reductants and using suitably devised electrodes. The flow of electricity indicates that electrons flow from the reductant to the oxidant without any physical contact between the reactants. This shows that redox reactions involve transfer of electrons and give rise to an electromotive force or e.m.f. which drives the electrons from the reductant to the oxidant. A measure of the e.m.f. developed in a particular redox reaction may then give us an idea about the mutual tendency (thermodynamic) of the oxidant and reductant to react with each other.

* The salt-bridge consists of an inverted U-tube filled with agar-agar gel containing a pinch of electrolyte like potassium chloride. It prevents mixing of the two solutions, but allows flow of ions to complete the circuit. A strip of filter-paper moistened with KCl solution will also serve the purpose.

9.2 ELECTROCHEMICAL CELLS

An electrochemical cell is established when two electrodes are placed in contact with an ionic conductor or electrolyte. Electrolytes may be a solution, a liquid or a solid. The electrodes may be of different kinds (see below) and may be placed in the same electrolyte or in different electrolytes, as seen in Fig. 9.1.

Electrochemical cells may be of two kinds. (1) In an *electrolytic cell* we use electrical energy from some external source to drive a non-spontaneous reaction (electrolysis). (2) In a *galvanic cell*, a spontaneous chemical reaction is used to produce electricity. The electromotive force may also result from a change in concentration (dilution) of some species in the cell, when it is called a concentration cell. Here we shall consider only chemical cells in which the driving force is derived from some spontaneous chemical reaction.

A chemical reaction is a consequence of the tendency of the reactants to decrease the free energy of the system and attain equilibrium. The principles of chemical equilibrium may be applied to correlate the amount of electrical energy available from a cell with the free energy change accompanying the process. For this purpose it is essential that the cell should behave reversibly. Thermodynamic reversiblity requires that (i) the driving force and the opposing force in a process should differ from each other only by an infinitesimal quantity and (ii) it should be possible to reverse any change during the process by applying a force infinitesimally greater than the driving force. Accordingly the following criterion will determine the reversibility of a cell:

(i) when the cell is connected to an exactly *equal* E.M.F. from an external source acting in the *opposite* direction, *no cell reaction should occur within it*;

(ii) if the external E.M.F. is now decreased by an *infinitesimally* small amount, current should flow *from* the cell;

(iii) if the external E.M.F. is increased from the null-point value by the same small amount, current should flow *to* the cell, that is, it will *flow through the cell in the opposite direction.* The processes which occur at the electrodes when the cell produces current should be *exactly reversed* during this operation.

Reversible and irreversible cells

A Daniell cell is an example of a reversible cell. The cell is written as

$$\text{Zn} \mid \text{ZnSO}_4 \text{ soln} \,\vdots\, \text{CuSO}_4 \text{ soln} \mid \text{Cu} \quad \xrightarrow{\hspace{2em}}$$

The arrow indicates the direction in which current ("positive current") flows *through the cell* when it produces E.M.F. by the reaction

$$Zn + CuSO_4 = ZnSO_4 + Cu.$$

Application of an external E.M.F. exactly equal to that of the cell will cause the reaction to cease. If this external E.M.F. is now slightly increased, current should flow in the opposite direction through the cell; the cell reaction will be reversed:

$$ZnSO_4 + Cu = Zn + CuSO_4.$$

Such cells behave reversibly only when infinitesimally small amounts of current are drawn from them. The functioning of the Daniell cell becomes irreversible owing to the liquid junction (shown by broken line) where the two electrolyte solutions meet one another. When the cell produces electricity, Zn^{2+} ions move into the $CuSO_4$ solution across the boundary. When the cell is forced to work in the opposite direction by an external EMF, Cu^{2+} ions will move into the $ZnSO_4$ solution. Thus the processes at the liquid junction are not reversible.

Instead of the Daniell cell considered above, let us now consider a cell composed of zinc and silver electrodes dipped in dilute sulfuric acid. When the two electrodes are connected externally by a conducting wire, the cell produces current; the reaction taking place in the cell is

$$Zn(s) + H_2SO_4 = ZnSO_4 + H_2(g).$$

But if the electrodes are connected to an external source of EMF having a higher potential difference, the cell reaction becomes

$$2\,Ag(s) + H_2SO_4 = Ag_2SO_4 + H_2(g)$$

This provides an example of an irreversible cell. Irreversibility in a cell may also be caused by several other factors.

9.2.1 Types of Reversible Electrodes

An electrode will behave reversibly only when all reactants and products of the electrode half-reaction are present at the electrode. The main types of reversible electrodes (half-cells) are mentioned below.

1. Metal-metal ion electrodes: These electrodes consist of a metal in equilibrium with its ions in solution, M^{n+} (concentration c) | M. The half-reaction (reduction) is

$$M^{n+} + ne \rightleftharpoons M$$

The zinc and copper electrodes in the Daniell cell are electrodes of this type. The potential of the electrode is determined by the activity of the ions.

Instead of the pure metals, an amalgam of a metal may be brought in equilibrium with the ions of the metal in solution, when one gets an *amalgam electrode*, for example, $Na(Hg)/Na^+(c)$. The electrode reaction would be

$$Na^+(c) + e \rightleftharpoons Na(Hg)$$

The electrode is thus reversible with respect to the sodium ion. If the amalgam is saturated with the metal, the electrode is equivalent to one containing the pure metal and the electrode potential is determined by the activity of the metal ion.

The amalgam electrodes have certain advantages over the pure metal electrodes. (i) Strains in the solid metal are absent in the amalgams. So equilibrium can be established much more rapidly with the latter and they are more readily reversible. Polarization at the electrode surface is also low. (ii) Metals like sodium, potassium etc. which react vigorously with water can be suitably moderated through amalgam formation. (iii) Small quantities of impurities which may produce anomalous behaviour in pure metals are sufficiently diluted by amalgam formation—more satisfactory and reproducible results are thereby obtained.

2. Non-metal electrodes: In such electrodes, equilibrium is established between a non-metal and its corresponding ion in solution in the presence of an unreactive metal, usually platinum or gold. Most common among such electrodes are the various *gas electrodes*, in which the gas is slowly bubbled over a platinized platinum electrode immersed in a solution containing ions to which the gas is reversible. The hydrogen electrode is a typical example of a gas electrode. It may be represented as Pt | H_2 $(g, P_{H_2})/H^+(aq)$. Hydrogen gas is slowly bubbled over a foil of platinized platinum dipping into a solution containing hydrogen ions, usually an acid. The hydrogen molecules dissociate into atoms on the catalytic platinum surface and the overall electrode reaction may be represented as

$$\frac{1}{2}H_2(g) = H^+(aq) + e$$

The electrode potential depends on the activity of hydrogen ion in solution and the pressure of the hydrogen gas.

The chlorine-chloride electrode operates similarly:

$$\frac{1}{2}Cl_2(g) + e = Cl^-(c).$$

It is reversible with respect to chloride ions.

An oxygen electrode may be similarly set up as Pt, O_2 (g, P_{O_2}) | OH^- (c). The electrode reaction is: $\frac{1}{2}O_2(g) + H_2O + 2e = 2OH^-(c)$.

The emf depends on the concentration (strictly activity) of hydroxyl ions in solution; but it does not behave reversibly since no electrode material has been found suitable which can catalyze the establishment of equilibrium between oxygen molecules and hydroxyl ions.

Other non-metal electrodes may be similarly constructed with an inert metal in contact with a liquid or solid phase at equilibrium with the corresponding ions in solution. The bromine-bromide half-cell, Pt | $Br_2(l)$ | Br^- (*aq*) or the iodine-iodide half-cell Pt | $I_2(s)$ | I^- (*aq*) are examples.

3. Metal–Insoluble salt electrodes: These electrodes consist of a metal and one of its sparingly soluble salt in contact with a solution containing the same anion as that of the salt. The calomel electrode (see later) and the silver-silver chloride electrode are typical examples. The Ag–AgCl electrode is represented as $Ag(s)/AgCl(s)/Cl^-(c)$. A silver wire coated with solid silver chloride and immersed in a solution of chloride ions (say, KCl) constitutes the electrode. The overall half-cell reaction (reduction) may be written as:

$$AgCl(s) + e \rightleftharpoons Ag(s) + Cl^-(c)$$

The electrode is reversible with respect to chloride ions. Its potential is also determined by the concentration of chloride ions in solution (equivalent to a chlorine-chloride electrode).

Similar to this type of electrodes are the *Metal-insoluble oxide electrodes* in which an oxide replaces the salt. The antimony-antimony trioxide electrode, Sb | Sb_2O_3 | $OH^-(c)$ is an example. A rod of antimony covered with a thin layer of the oxide is immersed in an alkaline solution (containing OH^- ions). The reaction in the half-cell may be written as

$$\frac{1}{2}Sb_2O_3 + \frac{3}{2}H_2O + 3e \rightleftharpoons Sb(s) + 3OH^-$$

The half-cell is reversible with respect to hydroxyl ions.

Similarly, in the mercury-mercuric oxide electrode, Hg | $HgO(s)$ | $OH^-(c)$, the reaction is

$$HgO(s) + H_2O + 2e \rightleftharpoons Hg(l) + 2OH^-(aq)$$

4. Oxidation-reduction electrodes: As mentioned in the beginning of this section, all electrode reactions involve redox processes. However, the term oxidation-reduction electrode is specifically applied to a particular type of electrodes. In this type, two ions containing the same element, but in different oxidation states are present in solution and an inert metal like platinum is immersed in it. Thus, a platinum wire dipped into a solution containing both ferrous and ferric ions forms an electrode of this type: Pt/Fe^{3+}, Fe^{2+}. The electrode reaction is:

$$Fe^{3+}(aq) + e = Fe^{2+}(aq)$$

Similarly one may have oxidation-reduction electrodes involving stannous-stannic, cerous-ceric, permanganate-manganous ions. The potential of such electrodes depends on the ratio of the concentration of the two species present (see later).

Among the different types of electrodes mentioned above, some electrodes are reversible with respect to a cation. As such they are called *cation-reversible electrodes*, the metal electrodes and the hydrogen electrode providing examples. On the other hand, there are electrodes which are *anion-reversible*. For example, the chlorine-chloride electrode, the silver-silver chloride electrode or the lead-lead sulfate electrode are anion-reversible.

Electrodes for specific functions

Besides these general types of electrodes, several electrodes have been devised to serve specific purposes in different fields, specially in biochemical, environmental and geochemical works. **Ion-selective electrodes** are most noteworthy innovations in this catagory. An ion-selective electrode is sensitive to one particular ion and consists of a membrane which responds to this particular ion. The membrane may be made from glass or a crystal or a liquid. The potential difference between the membrane and an electrolyte solution containing this ion is determined by the activity of that particular ion. Most ion-selective electrodes fall into one of the following categories.

1. **Glass electrodes:** a very thin walled (~ 0.005 cm) glass membrane permeable to H^+ and some other monovalent cations; widely used in the determination of *p*H of a solution (chapter 11).
2. **Solid state electrodes:** an inorganic salt crystal is used as the ion-sensitive membran. One common electrode of this type is the fluoride electrode that uses a crystal of LaF_3– doped with a small amount of Eu(II).
3. **Liquid-based electrodes:** a hydrophobic membrane saturated with a hydrophobic liquid ion-exchanger is used. A Ca^{2+} selective electrode consists of the ion-exchanger calcium didecylphosphate dissolved in dioctylphenylphosphonate.
4. **Compound electrodes:** these are assembled from a conventional electrode (e.g., glass electrode) with an additional membrane to isolate or generate the species which the electrode can detect. Various *gas-sensing* and *enzyme based* electrodes belong to this category. A CO_2-sensing electrode consists of an ordinary glass electrode surrounded by an electrolyte solution and enclosed by a semipermeable membrane (rubber, teflon, polythene). CO_2 diffuses through this semipermeable membrane and lowers the pH of the electrolyte solution which is followed by the glass electrode. A silver-silver chloride electrode is used as a reference. Electrodes for sensing other gases like NH_3, SO_2, H_2S and NO_x *in gaseous phase* are also known.

   Enzyme-based electrodes are another ingenious device in compound electrodes. A conventional electrode is coated with an enzyme capable of catalyzing some reaction of the analyte (substance being analysed). The product of the reaction is detected by the electrode. In this manner it is even possible to measure the concentration of a nonelectrolyte like urea. An electrode selective to NH_4^+ is coated with a gel containing the enzyme urease. Urease catalyzes the decomposition of urea to NH_4^+, whose concentration is measured.

$$NH_2CONH_2\,(aq) + 2H_2O + H^+\,(aq) \longrightarrow 2NH_4^+\,(aq) + HCO_3^-\,(aq)$$

   Blood glucose analysers used by many diabetic patients contain the enzyme glucose oxidase which specifically catalyzes the oxidation of glucose in the patient's blood (applied as a drop on a test-strip). Subsequent reactions result in a flow of electric current which can be related to the concentration of glucose in blood. The test-strip contains "printed electrodes" made from "inks" containing the electrode material. In this procedure one may have cheap, disposable electrodes of various kinds like graphite, Ag–AgCl and so on.

9.2.2 Writing Cells and Cell Reactions

As already mentioned, a chemical cell produces an emf at the expense of a chemical reaction taking place within the cell. The nature of the cell reaction can be ascertained from the half-cells forming the cell. The polarity of the half-cells in the given cell may be determined by connecting the cell to a standard cell with a galvanometer and a potentiometer. The two cells are connected to oppose each other and the electrode connected to the negative side of the standard cell is the negative electrode while the other must be the positive electrode.

Once the polarity of the half-cells is known, it is now easy to write the cell reaction. We remember that "current" flows from a cell through the external circuit from the "positive" to the "negative" while electrons flow in the opposite direction. Hence the negative electrode should act as the source of electrons. Any reaction that gives off electrons is an oxidation process, and so **oxidation always occurs at the negative electrode or anode**. Similarly, **reduction takes place at the positive electrode or cathode**. By adding the two electrode reactions so that the total number of electrons cancel each other, we obtain the overall cell reaction. As mentioned earlier, the overall cell reaction may also be obtained by writing both electrode processes as reduction and then taking their *difference*. This procedure will be helpful in following the conventions regarding electrode potentials. This has been explained later.

A shorthand notation is generally followed in representing electrochemical cells. We need two symbols: (i) a vertical line(|) indicates a phase boundary. (ii) A double vertical line(||) stands for a salt bridge between two solution.

phase boundary → | salt bridge → ||

In absence of a salt bridge, a liquid junction is represented by a broken vertical line (¦). Two species present in the same phase are separated by a comma; $Fe^{2+}(aq)$, $Fe^{3+}(aq)$.

By convention, the negative electrode or anode (the half-cell in which oxidation occurs), is written on the left.

Given a line diagram of a cell, the first task in understanding the cell reaction is to look for an element in the cell that has two oxidation states. Then we write an anode reaction for the left half-cell and a cathode reaction for the right half-cell. The following examples will be illustrative.

Example I: **Pt | $H_2(g)$ | HCl solution(*c*) | $Hg_2Cl_2(s)$ | Hg.**

Oxidation of hydrogen to hydrogen ion takes place at the negative electrode. At the positive electrode mercurous chloride is reduced to metallic mercury. Hence the cell reactions are:

$$\text{Left: } \tfrac{1}{2}H_2 \rightleftharpoons H^+(c) + e \quad ; \qquad \text{Right: } \tfrac{1}{2}Hg_2Cl_2(s) + e \rightleftharpoons Hg(l) + Cl^-(c)$$

Overall reaction: $\mathbf{\tfrac{1}{2}H_2(g) + \tfrac{1}{2}Hg_2Cl_2(s) = H^+(c) + Cl^-(c) + Hg(\mathit{l})}$ (for 1 faraday).

Example II: **Cd(*s*) | $CdCl_2(aq)$ | AgCl(*s*) | Ag(*s*).**

Anode: $Cd(s) \rightleftharpoons Cd^{2+}(aq) + 2e$

Cathode: $2AgCl(s) + 2e \rightleftharpoons 2Ag(s) + 2Cl^-(aq)$

$$\mathbf{Cd(s) + 2AgCl(s) \rightleftharpoons Cd^{2+} + 2Ag(s) + 2Cl^-(aq)}$$

One may also proceed by writing the cathode reaction as $2Ag^+(aq) + 2e \rightleftharpoons 2Ag(s)$ because some $Ag^+(aq)$ must be present from AgCl. The overall reaction now becomes:

$$\mathbf{Cd(s) + 2Ag^+(aq) \rightleftharpoons Cd^{2+}(aq) + 2Ag(s)}$$

Both reactions are equally valid descriptions of the cell and as we shall see later (Problem 9.7), both will lead to the same cell voltage. The decision to use one expression or another depends on whether the Ag^+ or Cl^- concentrations are more readily available.

Similar considerations apply to the cathode reaction in example I and elsewhere.

Example III: **Ag(*s*) | AgI(*s*) | KI soln (c_1) || $AgNO_3$ soln (c_2) | Ag(*s*).**

At the left hand electrode silver is oxidized to silver ion which forms solid silver iodide with iodide ions in the electrolyte. At the positive electrode Ag^+ is reduced to Ag.

$$Ag(s) + I^-(c_1) \rightleftharpoons AgI(s) + e \ ; \qquad Ag^+(c_2) + e \rightleftharpoons Ag(s)$$

Overall reaction: $\mathbf{Ag^+(c_2) + I^-(c_1) = AgI(s)}$ (for 1 faraday of electricity).

Example IV: **$Zn(s) \mid Zn(OH)_4^{2-}\ (a_1),\ OH^-(a_2) \mid HgO(s) \mid Hg$.**

Here zinc is oxidised to zincate ion (actually $Zn(OH)_4^{2-}$) at the negative electrode while mercuric oxide is reduced to metallic mercury at the positive electrode. Remembering that the medium is alkaline for both the reactions, we write

Left: $Zn(s) + 4OH^- \rightleftharpoons Zn(OH)_4^{2-} + 2e$; Right: $HgO(s) + H_2O + 2e \rightleftharpoons Hg + 2OH^-$

Overall reaction: $\mathbf{Zn(s) + HgO(s) + H_2O + 2OH^- = Zn(OH)_4^{2-} + Hg}$ (for 2 faradays).

Example V: **$Fe(s) \mid FeO(s) \mid KOH(aq) \mid Ni_2O_3(s) \mid NiO(s) \mid Ni(s)$.**

The reaction medium is again alkaline. The electrode reactions are

$Fe(s) + 2OH^- \rightleftharpoons FeO(s) + H_2O(l) + 2e$; $Ni_2O_3(s) + H_2O(l) + 2e \rightleftharpoons 2NiO(s) + 2OH^-$

Overall reaction: $\mathbf{Fe(s) + Ni_2O_3(s) = Fe(s) + 2NiO(s)}$ (for 2 faradays).

Example VI: **$Pt,\ H_2(g) \mid H_2O(l) \mid O_2(g),\ Pt$.**

Oxidation of hydrogen takes place at the negative electrode while at the positive electrode oxygen is reduced to hydroxyl ions. The reactions are:

Left: $\frac{1}{2}H_2(g) \rightleftharpoons H^+ + e$; Right: $\frac{1}{2}O_2(g) + H_2O + 2e \rightleftharpoons 2OH^-$

Overall reaction: $\mathbf{H_2(g) + \frac{1}{2}O_2(g) = H_2O(l)}$ (for 2 faradays).

Construction of a cell corresponding to a given reaction

This can be carried out from the same general principles as outlined above. We have to note the oxidized and reduced species in the reaction and think electrode reactions accordingly. A few examples will illustrate the procedure.

Example I: **$Cu + 2Ag^+ = Cu^{2+} + 2Ag$.**

Oxidation: $Cu - 2e = Cu^{2+}$, can be carried out in a Cu/Cu^{2+} electrode.

Reduction: $2Ag^+ + 2e = 2Ag$, can take place in an Ag/Ag^+ electrode.

Cell: **$Cu \mid Cu^{2+}(aq) \parallel Ag^+(aq) \mid Ag$**

As written, the cell reaction refers to two faradays of electricity per mol of Cu.

Example II: **$K_2Cr_2O_7 + 6FeSO_4 + 7H_2SO_4 = K_2SO_4 + Cr_2(SO_4)_3 + 3Fe_2(SO_4)_3 + 7H_2O$.**

Left: $Fe^{2+} - e \rightleftharpoons Fe^{3+}$; Right: $Cr_2O_7^{2-} + 14H^+ + 6e \rightleftharpoons 2Cr^{3+} + 7H_2O$.

Cell: **$Pt \mid Fe^{3+}(aq),\ Fe^{2+}(aq) \parallel Cr_2O_7^{2-}(aq),\ Cr^{3+}(aq),\ H^+(aq) \mid Pt$.**

Example III: **$Hg_2^{2+} + 2Cl^- = Hg_2Cl_2(s)$.**

Construction of the corresponding cell appears difficult at first sight as there appears no oxidized or reduced species in the reaction. It is understood that some common terms must have cancelled out from both sides of the equation. The formation of mercurous chloride must have been from initial oxidation of mercury to mercurous ion:

$$2Hg(l) + 2Cl^- = Hg_2Cl_2(s) + 2e.$$

But no elemental mercury appears in the final equation. On the other hand, the supplied equation bears Hg_2^{2+} on the left. So the reduction reaction may be written:

$$Hg_2^{2+} + 2e \rightleftharpoons 2Hg(l).$$

Adding the two reactions, we have $Hg_2^{2+} + 2Cl^- = Hg_2Cl_2(s)$

Cell: $\mathbf{Pt \mid Hg(\mathit{l}) \mid Hg_2Cl_2(\mathit{s}) \mid Cl^-(\mathit{aq}) \parallel Hg_2^{2+}(\mathit{aq}) \mid Hg(\mathit{l}) \mid Pt}$

Example IV: $\mathbf{2Ag + Cl_2 = 2AgCl.}$

We remember that AgCl is the product of oxidation at a silver-silver chloride electrode. So we may attempt to construct the cell by taking an Ag(*s*) | AgCl(*s*), KCl electrode as the negative one, the electrode reaction being

$$Ag(s) + Cl^- \rightleftharpoons AgCl(s) + e.$$

It now appears that a chlorine-chloride electrode may be placed as the positive electrode; in the ultimate cell reaction the chloride ion formed by the reduction of chlorine will cancel out the chloride ion appearing in the oxidation reaction.

$$\tfrac{1}{2}Cl_2(g) + e \rightleftharpoons Cl^-$$

Hence the cell may be written as

Cell: $\mathbf{Ag(\mathit{s}) \mid AgCl(\mathit{s}) \mid Cl^-(\mathit{aq}) \mid Cl_2(\mathit{g}) \mid Pt}$

Example V: $\mathbf{Pb(\mathit{s}) + PbO_2(\mathit{s}) + 2H_2SO_4 = 2PbSO_4(\mathit{s}) + 2H_2O.}$

Lead is oxidized to lead sulfate which can take place in the electrode Pb(*s*) | $PbSO_4$(*s*), H_2SO_4. In the positive electrode reduction of lead(IV) oxide takes place to lead(II) sulfate.

$$Pb(s) + H_2SO_4 \rightleftharpoons PbSO_4 + 2H^+ + 2e\ ;\ PbO_2(s) + 4H^+ + SO_4^{2-} + 2e \rightleftharpoons PbSO_4(s) + 2H_2O.$$

Cell: $\mathbf{Pb(\mathit{s}) \mid PbSO_4(\mathit{s}) \mid H_2SO_4 \mid PbSO_4(\mathit{s}) \mid PbO_2(\mathit{s}) \mid Pb(\mathit{s}).}$

Problem 9.2

An alkaline dry cell is similar to the Leclanché cell, but it has potassium hydroxide in place of ammonium chloride. The cell performs better during current drain and in cold weather. The anode is powdered zinc and the cathode is $MnO_2(s)$. The cell reactions are

Anode $Zn(s) + 2OH^-(aq) \rightleftharpoons Zn(OH)_2(s) + 2e$

Cathode $2MnO_2(s) + H_2O + 2e \rightleftharpoons Mn_2O_3(s) + 2OH^-(aq)$

Represent the cell with a line diagram.

Problem 9.3

A nickel-cadmium rechargeable battery (Nicad) has an anode of cadmium and a cathode of hydrated nickel oxide, approximately NiOOH. Write the cell reactions during discharge.

Anode $Cd(s) + 2OH^-(aq) \rightleftharpoons Cd(OH)_2(s) + 2e$

Cathode $NiOOH(s) + H_2O + e \rightleftharpoons Ni(OH)_2(s) + OH^-(aq).$

9.2.3 The Electromotive Force (EMF)

The electromotive force or EMF of a cell is expressed in terms of the potential difference (volt) established between the half-cells when no current passes through the cell.To make our study of cells meaningful, it is necessary to know the EMF of a cell and correlate it quantitatively with other fundamental properties of the system.

Measurement of EMF

The potential difference between the two terminals of a cell may be read directly with the help of a voltmeter, but the method does not give the accurate EMF of the cell. A chemical cell behaves reversibly only when an infinitesimally small current is drawn from it. A voltmeter, however sensitive, draws some finite current from the cell and the cell reaction proceeds spontaneously. The condition of reversibility is thereby disturbed. At the same time, part of the EMF will have to be utilized to overcome the internal resistance of the cell. Thus a voltmeter cannot give the actual reversible EMF of a cell.

The EMF of a cell is measured potentiometrically under a condition of balance against an equal and opposite potential difference—a condition at which the current drawn from the cell is negligible. A standard cell of precisely known EMF is used for calibration.

The standard cell used most commonly is the **Weston cell** which may be represented as

$$\underset{12.5\%}{Cd(Hg)} \,|\, CdSO_4(aq) \,|\, Hg_2SO_4(s) \,|\, Hg$$

The cadmium sulfate solution is kept saturated (4°C) with $CdSO_4, \frac{8}{3}H_2O$. The cell reaction is:

$$Cd(Hg) + Hg_2SO_4(s) + \tfrac{8}{3}H_2O = CdSO_4, \tfrac{8}{3}H_2O(s) + 2Hg.$$

Its EMF is 1·01845 V at 20°C and 1·01832 V at 25°C. The temperature coefficient of the EMF of this cell is thus very small (of the order of 10^{-5} per degree C) which is a great advantage.

Free energy and reversible EMF

When the two half-cells of a chemical cell are connected by a conducting wire, oxidation and reduction occur spontaneously at the electrodes and electrons flow through the wire from the negative to the positive electrode. The reactants in the cell thereby attain a more stable state in the products with a consequent lowering of free energy of the system. This decrease of free energy of the cell materials provides the electrical energy available from the cell. It is only for the reversible cells that the free energy change accompanying the cell process may be related quantitatively to the reversible EMF.

If E is the EMF developed by a cell reaction for n faradays (F) of electric charge, the net electrical work available would be nFE volt-coulomb or joule. When the cell operates reversibly, this work is maximum and equal to the decrease in free energy per mol of reactants. Denoting this by $-\Delta G$, we obtain the relation

$$-\Delta G = nFE \quad \text{or,} \quad \Delta G = -nFE$$

This relation is of fundamental importance in the study of cells, as we shall encounter very soon.

From thermodynamic principles we know that at constant temperature and pressure, ΔG is

(i) *negative* for any *spontaneous* process,

(ii) *positive* for any *nonspontaneous* process, and

(iii) equal to *zero* for any system *at equilibrium*.

Since $\Delta G = -nFE$, we may readily conclude that the EMF of a cell will be

(i) *positive* when the cell reaction is *spontaneous*;

(ii) *negative* when the cell reaction is *nonspontaneous*;

and (iii) equal to *zero* when the cell reaction reaches *equilibrium*.

The Nernst equation

The free energy change of a chemical reaction is dependent on the activities of the reactants and products. Accordingly, the EMF of a cell should also depend on the activities of the species taking part in the cell reaction. The Nernst equation provides this information.

Let us consider a cell with the overall reaction

$$xA + yB = pC + qD$$

ΔG of this reaction is related to the activities (a) of the reactants and products by the *reaction isotherm* as

$$\Delta G = \Delta G^\circ + RT \ln \frac{a_C^p\, a_D^q}{a_A^x\, a_B^y}$$

ΔG° is the free energy change attending the reaction when the activities of products and reactants are all unity. i.e., the substances are in their respective standard states. Now we define another quantity E° by the relation

$$\Delta G^\circ = -nFE^\circ$$

where E° is the value of the EMF corresponding to ΔG°. Substitution now gives

$$-nFE = -nFE^\circ + RT \ln \frac{a_C^p\, a_D^q}{a_A^x\, a_B^y} \qquad \text{or,} \quad E = E^\circ - \frac{RT}{nF} \ln \frac{a_C^p\, a_D^q}{a_A^x\, a_B^y}$$

This is known as the **Nernst equation**. At 25°C (298K), and converting "ln" to $\log_{10}$, putting the value of R in joules and F in coulombs, the equation may be expressed also as

$$E = E^\circ - \frac{0.05915}{n} \log_{10} \frac{a_C^p\, a_D^q}{a_A^x\, a_B^y}$$

The relation gives the dependence of the EMF of a cell on the activities of the constituents in terms of E°.E° is called the **standard EMF of the cell** and is equal to the EMF when the chemical species in the cell are in their respective standard states of unit activity.

Since $\quad \Delta G^\circ = -RT \ln K_{eq}$ *and* $\Delta G^\circ = -nFE^\circ$, $\; E^\circ = \dfrac{RT}{nF} \ln K_{eq}$

The standard EMF of a cell is thus related to the equilibrium constant of the cell reaction. This may also be obtained from the Nernst equation when we remember that at equilibrium the EMF becomes zero and the activity quotient in the equation represents the equilibrium constant.

9.3 ELECTRODE POTENTIALS

We have been able to correlate the EMF of a cell with the activities of the chemical species in it. But to use it we shall have to record data for numerous cells which may be formed by the combination of different half-cells. It would be advantageous if we could assign potential values to the different half-cells. The EMF of any cell would then be available by combining the relevant half-cell potentials in a suitable manner.

9.3.1 Single Electrode Potential: Half-cell EMF

A closer analysis of the Nernst equation shows that each cell EMF may be split into two parts, each part characterizing a particular electrode in the cell. This enables us to express single electrode potentials in terms of the components of a particular electrode. Different half-cells may now be combined with the same standard reference half-cell (taken arbitrarily) when the EMF obtained in each case would provide relative values for each half-cell potential. As a typical example, we treat the Daniell cell, in which the overall reaction is

$$Zn(s) + Cu^{2+} = Zn^{2+} + Cu(s)$$

From the Nernst equation, $E_{Cell} = E^{o}_{Cell} - \frac{RT}{2F} \ln \frac{a_{Zn^{2+}}}{a_{Cu^{2+}}}$ $\quad [a_{Zn(s)} = a_{Cu(s)} = 1]$

Let us now notionally split the E and E^{o} terms into two parts, each characterizing a particular electrode separately. That is, we assume,

$$E_{Cell} = E_{Cu} - E_{Zn} \quad \text{and} \quad E^{\circ}_{Cell} = E^{\circ}_{Cu} - E^{\circ}_{Zn}$$

[the significance of E_{Zn} or E_{Cu} (or $E^{\circ}_{Zn}, E^{\circ}_{Cu}$) will be made clear very soon.]

We may, therefore, rewrite the Nernst equation in this case as

$$E_{Cu} - E_{Zn} = \left(E^{\circ}_{Cu} - \frac{RT}{2F} \ln \frac{1}{a_{Cu^{2+}}} \right) - \left(E^{\circ}_{Zn} - \frac{RT}{2F} \ln \frac{1}{a_{Zn^{2+}}} \right)$$

The expressions in parentheses on the right may now be associated with the zinc and copper electrodes separately for respective half-cell reactions written as reduction processes:

$$Zn^{2+}(aq) + 2e \rightleftharpoons Zn(s) \quad ; \quad Cu^{2+}(aq) + 2e \rightleftharpoons Cu(s)$$

We write both electrode processes as reduction to obtain a uniform type of electrode potential ("the reduction potential"). This maintains consistency with IUPAC conventions (see below). *The actual processes* occurring at each half-cell in this cell is as usual oxidation at the zinc electrode (negative) and reduction at the copper electrode (positive). The respective reduction potentials may now be written as

$$E_{Zn} = E^{\circ}_{Zn} - \frac{RT}{2F} \ln \frac{1}{a_{Zn^{2+}}}; \quad E_{Cu} = E^{\circ}_{Cu} - \frac{RT}{2F} \ln \frac{1}{a_{Cu^{2+}}}$$

E_{Zn} and E_{Cu} are the electrode potentials of the zinc and copper electrode respectively; the difference between the two is equal to the cell EMF. E°_{Zn} and E°_{Cu} may now be looked upon as the electrode potentials of the respective electrodes when $a_{Zn^{2+}}$ or $a_{Cu^{2+}}$ is equal to unity. Consequently E°_{Zn} or E°_{Cu} has been termed the respective *standard electrode potential.*

The difficulty in measuring a single electrode potential may now be overcome in the following manner: the standard potential of the hydrogen electrode, Pt | H_2(*g*, 1 atm)/H^+(*aq*) (*a* = 1), is taken arbitrarily as zero at all temperatures. If we now form a cell with the standard hydrogen electrode and any other electrode *X*, the EMF measured would give us the potential of the electrode *X* relative to the standard hydrogen electrode. For uniformity, electrode potentials for different electrodes should be measured and compared at the same activity of the constituents of a half-cell. For this purpose, standard electrode potentials of half-cells containing their respective constituents at unit activity are always referred to.

The standard hydrogen electrode and the calomel electrode

A standard hydrogen electrode consists of a platinized platinum electrode partially immersed in a solution containing hydrogen ion at unit activity; pure hydrogen gas at 1 atm. pressure is slowly bubbled through the solution. Hydrogen gas thus surrounds the electrode and is also adsorbed by the platinized platinum. Equilibrium is thus established between hydrogen molecules and the ions in solution.

Experimentally, it is difficult to set up a standard hydrogen electrode. So, a number of subsidiary reference electrodes have been devised. These can be easily set up with accurately known potentials with respect to the standard hydrogen electrode. **The calomel electrode** is most common among them. It consists of mercury in contact with solid mercurous chloride and a solution of potassium chloride (saturated with mercurous chloride):

$$\mathrm{Hg} \mid \mathrm{Hg_2Cl_2}(s) \mid \mathrm{KCl}(a).$$

When connected with a standard hydrogen electrode, the electrode reaction is

$$\mathrm{Hg_2Cl_2}(s) + 2e \rightleftharpoons 2\mathrm{Hg}(l) + 2\mathrm{Cl^-}(a)$$

The potential of the electrode depends on the concentration of the potassium chloride solution. Three different concentrations of potassium chloride are generally used. The EMF, E_{Cal}, at 25°C for various concentrations of KCl are

(i) 1.0 N : E_{Cal} = 0.2800V; (ii) 0.1 N : E_{Cal} = 0.3338V; (iii) Saturated : E_{Cal} = 0.245V.

9.3.2 Standard Electrode Potential, E°

As per IUPAC recommendation, *the standard electrode potential of an electrode X is the EMF of a cell in which the negative electrode is a standard hydrogen electrode (SHE), arbitrarily assigned a potential of 0 V and the positive electrode is made up of the electrode X*, its components being maintained at *unit activity*. If the electrode *X* is written as X^+/X, then its standard electrode potential is the EMF of the cell

$$\mathrm{Pt} \mid \mathrm{H_2}(g, 1\ \mathrm{atm}) \mid \mathrm{H}^+_{(a=1)} \parallel \mathrm{X}^+_{(a=1)} \mid X$$

in which the cell reaction is: $\frac{1}{2}\mathrm{H_2}(g) + X^+_{(a=1)} = \mathrm{H}^+_{(a=1)} + X.$

If we place the half-cell $Ag^+(aq, a = 1) \mid Ag(s)$ in place of electrode X, the EMF observed is +0.799 V. This implies that the standard (reduction) potential for the $Ag^+ \mid Ag$ half-cell is +0.799 V. The positive sign of the cell EMF tells us that "current" flows from right to left *through the meter*, that is, electrons flow from the left half-cell (SHE) to the right half-cell through the external circuit.

Instead of the silver electrode, if one combines a $Zn^{2+}(aq) \mid Zn$ half-cell with the SHE as above, the measured voltage will be –0.763 V. The negative sign indicates that "current" now flows in the opposite direction, that is, electrons flow from the right half-cell (Zn^{++} – Zn) to the SHE through the meter. So zinc actually behaves as an anode in this set-up, undergoing oxidation during spontaneous cell reaction.

The standard (reduction) potentials ($E°$) just mentioned may be arranged in their increasing (or decreasing) order as follows (aqua ions at unit activity are implied in all cases):

Half-cell	$E°$ (V)
$Zn^{2+} + 2e \rightleftharpoons Zn(s)$	– 0.763
$H^+ + e \rightleftharpoons \frac{1}{2}H_2(g)$	0 (by definition)
$Ag^+ + e \rightleftharpoons Ag(s)$	+ 0.799

We next consider a combination of the zinc half-cell with the silver half-cell. Electrons would flow in this cell from the $Zn^{2+} \mid Zn$ half-cell to the $Ag^+ \mid Ag$ half-cell. The cell reaction is obtained by adding the half-cell reactions:

Anode	$Zn - 2e \rightleftharpoons Zn^{2+}(aq)$
Cathode	$2Ag^+(aq) + 2e \rightleftharpoons 2Ag$
Net reaction	$Zn + 2Ag^+(aq) \rightleftharpoons Zn^{2+}(aq) + 2Ag$

Thus, we may obtain a cell reaction from two half-cell reactions by turning one into an oxidation and adding it to the reduction reaction for the other half-cell. The standard EMF for the cell so written is obtained as:

$$\boldsymbol{E}^\circ_{\textbf{Cell}} = \boldsymbol{E}^\circ_{\textbf{Right}} - \boldsymbol{E}^\circ_{\textbf{Left}} = E^\circ_{Ag^+-Ag} - E^\circ_{Zn} = [0.799 - (-0.763)]\ \text{V} = 1.562\ \text{V}$$

We note that we have not multiplied the $E°$ values of the half-cells though the equations for their reactions were multiplied by integers to match the number of electrons. This is because electrical work (Joule) is the product volt (V) × coulomb (C). Hence the potential difference E between two points is the *work done per coulomb of charge* transported through that potential difference. The work *per coulomb* comes out to be the same whether we transfer 1, 2 or 100 coulombs. Of course, the total work will be different in each case.

It is a good practice to proceed via the $\Delta G°$ terms ($= -nFE°$) which are required to be multiplied by the coefficients of respective half-cell reactions and then finally added. This approach is particularly essential when we use two half-cell potentials to find the potential of another half-cell (shown later).

By this time we have certainly realized the usefulness of standard EMF data for different half-cells. A few selected data are given in Table 9.1. We shall find many half-cell emfs throughout our discussion on the chemistry of an element in its various oxidation states.

Table 9.1
E°: Standard reduction EMF (aqueous) at 25°C

Electrode	Reaction	E°(Volt)
$Li^+ \mid Li$	$Li^+ + e \rightleftharpoons Li$	−3.040
$K^+ \mid K$	$K^+ + e \rightleftharpoons K$	−2.936
$Ca^{2+} \mid Ca$	$Ca^{2+} + 2e \rightleftharpoons Ca$	−2.868
$Na^+ \mid Na$	$Na^+ + e \rightleftharpoons Na$	−2.715
$Mg^{2+} \mid Mg$	$Mg^{2+} + 2e \rightleftharpoons Mg$	−2.363
$Al^{3+} \mid Al$	$Al^{3+} + 3e \rightleftharpoons Al$	−1.662
$Zn^{2+} \mid Zn$	$Zn^{2+} + 2e \rightleftharpoons Zn$	−0.763
$Fe^{2+} \mid Fe$	$Fe^{2+} + 2e \rightleftharpoons Fe$	−0.441
$Cd^{2+} \mid Cd$	$Cd^{2+} + 2e \rightleftharpoons Cd$	−0.403
$Pt \mid Pb \mid PbSO_4(s) \mid SO_4^{2-}$	$PbSO_4(s) + 2e \rightleftharpoons Pb + SO_4^{2-}$	−0.355
$Pt \mid Cu \mid CuI(s) \mid I^-$	$CuI(s) + e \rightleftharpoons Cu + I^-$	−0.185
$Pt \mid Ag \mid AgI(s) \mid I^-$	$AgI(s) + e \rightleftharpoons Ag + I^-$	−0.152
$Pt \mid H_2(g) \mid H^+ (a = 1)$; SHE	$H^+ + e \rightleftharpoons \frac{1}{2}H_2(g)$	0.000
$Pt \mid Ag \mid AgBr(s) \mid Br^-$	$AgBr(s) + e \rightleftharpoons Ag + Br^-$	+0.071
$Pt \mid Sn^{4+} \mid Sn^{2+}$	$Sn^{4+} + 2e \rightleftharpoons Sn^{2+}$	+0.14
$Pt \mid Cu^{2+} \mid Cu^+$	$Cu^{2+} + e \rightleftharpoons Cu^+$	+0.153
$Pt \mid Ag \mid AgCl(s), Cl^-$	$AgCl(s) + e \rightleftharpoons Ag + Cl^-$	+0.222
$Hg \mid Hg_2Cl_2(s), Cl^-$	$Hg_2Cl_2(s) + 2e \rightleftharpoons 2Hg + 2Cl^-$	+0.268
$Cu^{2+} \mid Cu$	$Cu^{2+} + 2e \rightleftharpoons Cu$	0.339
$Cu^+ \mid Cu$	$Cu^+ + e \rightleftharpoons Cu$	+0.521
$Pt \mid I_3^- \mid I^-$	$I_3^- + 2e \rightleftharpoons 3I^-$	+0.536
$Pt \mid Fe^{3+}, Fe^{2+}$	$Fe^{3+} + e \rightleftharpoons Fe^{2+}$	+0.771
$Hg_2^{2+} \mid Hg(l)$	$Hg_2^{2+} + 2e \rightleftharpoons 2Hg(l)$	+0.796
$Ag^+ \mid Ag$	$Ag^+ + e \rightleftharpoons Ag$	+0.799
$Hg^{2+} \mid Hg(l)$	$Hg^{2+} + 2e \rightleftharpoons Hg(l)$	+0.852
$Pt - Ir \mid Br_2(g) \mid Br^-$	$\frac{1}{2}Br_2 + e \rightleftharpoons Br^-$	+1.065
$Pt \mid O_2(g) \mid H_2O$	$O_2 + 4H^+ + 4e \rightleftharpoons 2H_2O$	+1.229
$Pt–Ir \mid Cl_2(g) \mid Cl^-$	$\frac{1}{2}Cl_2 + e \rightleftharpoons Cl^-$	+1.359
$Pt \mid MnO_4^-, Mn^{2+}$	$MnO_4^- + 8H^+ + 5e \rightleftharpoons Mn^{2+} + 4H_2O$	+1.51
$Pt \mid Ce^{4+}, Ce^{3+}$	$Ce^{4+} + e \rightleftharpoons Ce^{3+}$	+1.72
$Pt \mid S_2O_8^{2-}, SO_4^{2-}$	$S_2O_8^{2-} + 2e \rightleftharpoons 2SO_4^{2-}$	+2.05
$Pt–Ir \mid F_2(g) \mid F^-(aq)$	$\frac{1}{2}F_2 + e = F^-$	+2.87

A good oxidizing couple is marked by a high standard reduction potential. Systems above hydrogen have low standard reduction potential and their *reduced species* are good reducing agents.

The standard electrode potential of a half-cell may be related to other thermodynamic parameters of the system. As an example, we consider a metal-metal ion electrode:

$$M^{n+}(aq) + ne \rightleftharpoons M(s)\ ;$$

we may construct a Born-Haber cycle (9-I) with the relevant enthalpy terms:

1. Sublimation of M(*s*) : ΔH°_{sub}
2. Ionization of M(*g*), (*total*) : ΔH°_I
3. Hydration of $M^{n+}(g)$: ΔH°_{hyd}

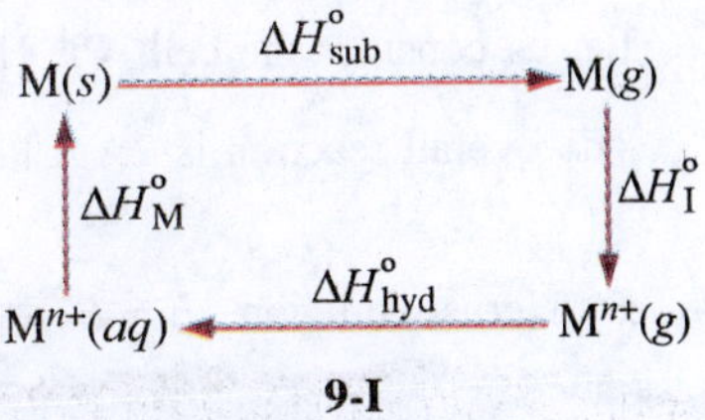

9-I

From the cycle,

$$-\Delta H^\circ_M = \Delta H^\circ_{sub} + \Delta H^\circ_I + \Delta H^\circ_{hyd}$$

For the half-cell reaction : $\Delta G^\circ = \Delta H^\circ - T\Delta S^\circ$.

In reality, the contribution of $T\Delta S^\circ$ terms for different steps is small compared to ΔH°. The standard electrode potential of the half-cell (E°_M) is actually obtained from the E°_{cell} containing an SHE as cathode and $M^{n+}(aq)$ | M as anode:

$$\left[\tfrac{1}{2}H_2(g) \rightleftharpoons H^+(aq) + e\right] \times n \qquad -\Delta H^\circ(H_2) \times n$$

$$M^{n+}(aq) + ne \rightleftharpoons M(s) \qquad \Delta H^\circ(M)$$

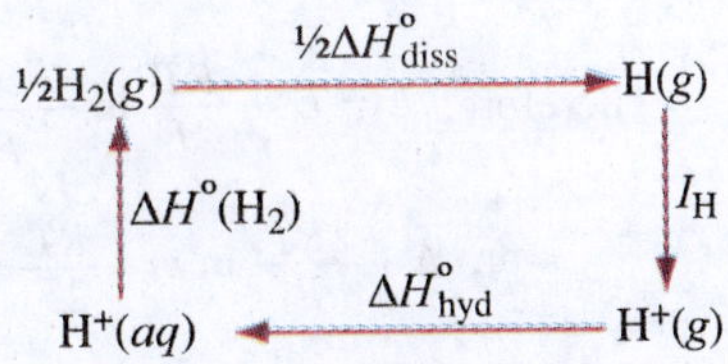

9-II

$\Delta H^\circ(H_2)$ refers to the enthalpy change for

$$H^+(aq) + e \rightleftharpoons \tfrac{1}{2}H_2(g).$$

From 9-II, $-\Delta H^\circ(H_2) = \frac{1}{2}\Delta H^\circ_{diss} + I_H + \Delta H^\circ_{hyd}(H^+)$.

Ignoring $T\Delta S$ terms, ΔG° for the cell written above is

$$\Delta G^\circ \simeq -n \times \Delta H^\circ(H_2) + \Delta H^\circ(M) = -nFE^\circ_{cell}$$

$$\text{or, } E^\circ_{cell} = (1/nF)\,[\Delta H^\circ_{sub} + \Delta H^\circ_I + \Delta H^\circ_{hyd}\,(M^{n+})] - (1/F)\,[\tfrac{1}{2}\Delta H^\circ_{diss} + I_H + \Delta H^\circ_{hyd}(H^+)]$$

Since $E^\circ_{H_2} = 0$, $E^\circ_{cell} = E^\circ_M$. One may thus calculate approximate E° of a half-cell from relevant thermodynamic data. The agreement is satisfactory.

Problem 9.4

E° for K^+ | K is –2.9 V which implies that K^+ is an extremely weak oxidizing agent. Can we expect K^+ to be a good reducing agent?

Hint: In the K^+ | K redox couple, K is the reduced form and the E° value implies that K should be a good reducing agent. To comment on the reducing power of K^+, one should think of the couple K^{2+} | K^+. E° for this is not available. But we can draw our conclusion from the high ionization energy required for converting K^+ to K^{2+}.

9.3.3 Electrode Potential Under Non-standard Conditions

We have defined the standard reduction potential as the EMF of a cell formed by connecting a standard hydrogen electrode (as anode) to the half-cell concerned with unit activities of all species involved. However, such cells are actually an idealized set-up because we have no way to adjust concentrations and ionic strengths to attain unit activities. An 1 M solution does not have unit activity. So we have to work with activities less than unity in real cells. We know that activities approach concentrations as the solution becomes infinitely dilute. Since it is not practicable to precisely measure the voltage of a cell using very dilute solutions, one takes help of the Nernst equation. The EMF of a cell is measured using different finite concentrations of the ions involved. $E°$ is then determined by extrapolation to zero concentration.

Let us consider the cell Pt | $H_2(g)$ | HCl(aq) | AgCl(s) | Ag | Pt.

The overall reaction is $2AgCl(s) + H_2(g) \rightleftharpoons 2Ag(s) + 2H^+(aq) + 2Cl^-(aq)$.

By Nernst equation, $E = E° - \frac{RT}{2F} \ln \frac{a_+^2 . a_-^2}{a_{H_2}}$;

a_+ and $a\_$ are the activities of H^+ and Cl^- ions respectively. At or near 1 bar, the activities of the pure solids are 1 and activity of H_2 may be approximately put equal to its partial pressure p (bar). a_+ and a_- are related to the stoichiometric molality of HCl through the mean ionic activity coefficient $\gamma_\pm$: $a_+a_- = \gamma_\pm^2 m^2$.

$$\text{Therefore, } E = E° - \frac{RT}{2F} \ln \frac{\gamma_\pm^4 m^4}{p} = E° - \frac{2RT}{F} \ln m + \frac{RT}{2F} \ln p - \frac{2RT}{F} \ln \gamma_\pm$$

$$\text{or, } E + \frac{2RT}{F} \ln m - \frac{RT}{2F} \ln p = E° - \frac{2RT}{F} \ln \gamma_\pm$$

The Debye-Hückel limiting law tells us that $\ln \gamma_\pm$ is proportional to $m^{1/2}$ in very dilute solutions. Hence for low molal concentrations, a plot of the left hand side of the above equation against $m^{1/2}$ gives a straight line. As $m \to 0$, $\gamma_\pm$ goes to 1 and $\ln\gamma_\pm$ tends to zero. Hence extrapolation of the straight line to $m = 0$ gives $E°$ as the intercept on the Y-axis.

Formal potential

The standard reduction potentials refer to unit activity of all reactants and products in a half-cell. If the half-cell reaction involves H^+ ions (as for example in the $MnO_4^- - Mn^{2+}$ couple), the value of $E°$ applies when $a_{H^+} (\simeq C_{H^+}) = 1$ or pH = 0. At other H^+ ion concentrations, the potential will have a term for its dependence on C_{H^+}. Similarly, effects of complexation, ionic strength etc. have their own contributions to the half-cell potential. Under a specified set of conditions of pH, ionic strength, concentration of complexing agents etc., it is possible to tag all these effects to the $E°$ term and express the Nernst equation as

$$E = E° + \text{miscellaneous contributions} - \frac{RT}{nF} \ln \frac{[\text{Reduced form}]}{[\text{Oxidized form}]}$$

$$= E°' - \frac{RT}{nF} \ln \frac{[\text{Reduced form}]}{[\text{Oxidized form}]}$$

$E^{\circ\prime}$ (E zero prime) is termed the *formal potential* of the half-cell. It is the reduction potential that applies to a specified set of conditions for a redox couple containing *equal formal concentrations* of the oxidized and reduced form. For the half-reaction

$$\text{Ox} + ne \rightleftharpoons \text{Red} + m\text{H}^+$$

where Ox = oxidized form and Red = reduced form, the Nernst equation is written

$$E = E^\circ - \frac{0 \cdot 059}{n} \log \frac{[\text{Red}]\,[\text{H}^+]^m}{[\text{Ox}]}$$

To find the formal potential, we have to rearrange it to a form in which the log term contains only the *formal concentrations* of the oxidized and reduced species:

$$E = \underbrace{E^\circ + \frac{0 \cdot 059m}{n}\,\text{pH}}_{E^{\circ\prime}} - \frac{0 \cdot 059}{n} \log \frac{[\text{Red}]}{[\text{Ox}]}$$

The formal potential may also contain other terms like those originating from ionic strength, complexation or acid/base dissociation of the species involved.

The use of formal potential is specially important to biochemists whose specification of standard states is different. Most biochemical studies are relevant for pH 7, which is the pH of most plant and animal cells. Hence *biochemists use the formal potential $E^{\circ\prime}$ appropriate for pH* 7 instead of E° (defined for pH 0). As an example, ascorbic acid is ordinarily a more powerful reducing agent than succinic acid is. However, at pH 7, this order is reversed. The change from E° to $E^{\circ\prime}$ for some redox couples involving H^+ ions is noteworthy:

Reaction	E°(V)	$E^{\circ\prime}$ (V)
$H^+ + e \rightleftharpoons \frac{1}{2} H_2(g)$	0.000	−0.414
$O_2 + 4H^+ + 4e \rightleftharpoons 2H_2O$	+1.229	+0.816
$O_2 + 2H^+ + 2e \rightleftharpoons H_2O_2$	+0.695	+0.281
Dehydroascorbate + $2H^+ + 2e \rightleftharpoons$ ascorbate + H_2O	+0.390	+0.058
Pyruvate + $2H^+ + 2e \rightleftharpoons$ lactate	+0.224	−0.190

Problem 9.5

For the half-cell

$$C_2H_2(g) + 2H^+\,(aq) + 2e \rightleftharpoons C_2H_4\,(g),\ E^\circ = 0.731\ \text{V. What will be } E^{\circ\prime} \text{ at pH} = 7?$$

Ans: By Nernst equation, at 25°C, $E = E^\circ - \frac{0 \cdot 059}{2} \log \frac{p_{C_2H_4}}{p_{C_2H_2} \cdot [H^+]^2}$

[$p_{C_2H_4}$ and $p_{C_2H_2}$ are respective partial pressures]

Rearrangement gives $E = E^\circ - 0 \cdot 059\ \text{pH} - \frac{0 \cdot 059}{2} \log \frac{p_{C_2H_4}}{p_{C_2H_2}}$

At pH = 7, $E^{\circ\prime} = [E^\circ - 0 \cdot 059 \times 7]\ \text{V} = [0 \cdot 731 - 0 \cdot 413]\ \text{V} = 0 \cdot 318\ \text{V}$.

Problem 9.6

The $Fe^{3+}(aq) - Fe^{2+}(aq)$ couple has $E° = 0.77$ V. Its formal potential ($E°'$) in different acid media are

1 M $HClO_4$: 0.73 V	1 M H_2SO_4 : 0.68 V
1 M HCl : 0.70 V	1 M H_2SO_4 + 0.5 M H_3PO_4 : 0.61 V

Comment.

Hint: ClO_4^- is the weakest complexing agent, $E°'$ is closest to $E°$.

Magnitude of $E°'$ decreases as extent of complexation increases. Thus $H_2SO_4 + H_3PO_4$ shows the least $E°'$.

Activity of electron and pE

Environmental scientists are interested to express the extent to which natural waters are oxidizing or reducing. Since redox reactions in water are often accompanied by transfer of hydrogen ions, it will be convenient to express the redox behaviour of water in a manner similar to the pH scale. This is done by expressing the electron availability in terms of the electron activity a_e similar to the pH scale:

$$pE = -\log a_e$$

In natural water, pE values may vary from −12 to 25. Water with large negative pE suggests a lot of electrons, a condition that favours reduction. Water in an anaerobic environment at the bottom of a deep ocean may have such pE values. A large positive pE, on the other hand, indicates low activity of electrons and hence a condition favouring oxidation.

The value of pE may be obtained from the $E°$ value of the concerned half-cell reaction. Thus, for the Fe^{3+}–Fe^{2+} system (*aqueous*), $E° = +0.77$ V.

$$Fe^{3+} + e \rightleftharpoons Fe^{2+} \qquad K_{eq} = [Fe^{2+}]/[Fe^{3+}] . a_e$$

$$\text{or, } \frac{1}{a_e} = K_{eq} \cdot \frac{[Fe^{3+}]}{[Fe^{2+}]} \Rightarrow pE = -\log a_e = \log K_{eq} + \log \frac{[Fe^{3+}]}{[Fe^{2+}]}$$

From $\Delta G° = -nFE°$ and $\Delta G° = -2.303\ RT \log K_{eq}$, $pE = \dfrac{nE°}{0.0591} + \log \dfrac{[Fe^{3+}]}{[Fe^{2+}]}$ at 298 K.

When $[Fe^{3+}] = [Fe^{2+}]$, $pE = pE° = 1 \times 0.77/0.0591 = 13.0$ ($n = 1$)

When $[Fe^{3+}] = 1.0 \times 10^{-5}$ M and $[Fe^{3+}] = 1.0 \times 10^{-3}$ M, pE = 15.0

For a general reaction aA + ne = bB, it is easily found that

$$pE = pE° - \frac{1}{n} \log Q \text{ (reaction quotient)} = \frac{[B]^b}{[A]^a}$$

9.3.4 Periodic Trends in $E°$

Apart from certain minor deviations, the reduction potentials of elements show some periodic trend. This is not surprising since many of the factors contributing to its magnitude, like ionization and sublimation energies show a general periodic trend. Some standard reduction potential values are given in Table 9.2 to illustrate the general trend in the periodic table. It appears that the standard reduction potential slowly falls down a particular group in the table, while it increases from left to right along a period. The same general trend becomes evident from Fig. 9.2. In both these cases, the potentials refer to a particular equilibrium, namely, those mentioned with Fig. 9.2. Many other equilibria may be established under different conditions. However, the variations observed are not very regular.

Table 9.2
Periodic variation of standard electrode potential (volt)

				F +2.87
Na –2.71	Mg –2.37		Al –1.66	Cl +1.36
K –2.94	Ca –2.87	Sc –2.03	Ga –0.53	Br +1.06
Rb –2.93	Sr –2.89	Y –2.37	In –0.34	I +0.53
Cs –2.92	Ba –2.90	La –2.37	Tl +0.72	

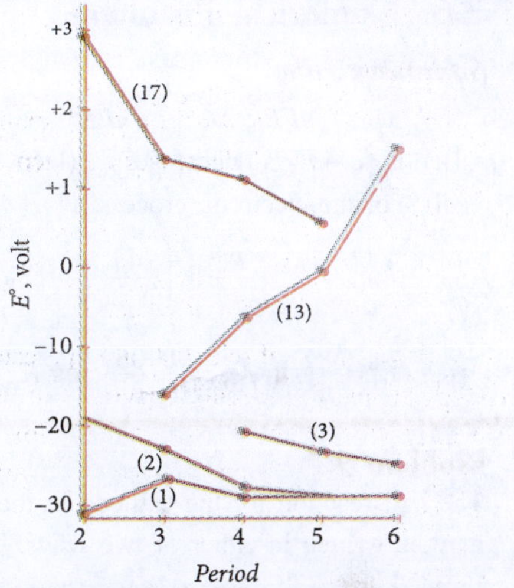

Fig. 9.2: Variation of E° among different groups of elements in the periodic table.

The values (some quoted in Table 9.2) refer to the following equilibria:

Gr 1 (Li–Cs) : $M^+(aq) + e = M(s)$;

Gr 2 (Be–Ba) : $M^{2+}(aq) + 2e = M(s)$;

Gr 3 (Sc, Y, La) : $M^{3+}(aq) + 3e = M(s)$

Gr 13 (Al, Tl) : $M^{3+}(aq) + 3e = M(s)$;

Gr 17 (F, I) : ½$\frac{1}{2}X_2(g) + e = X^-(aq)$

9.3.5 Limitations of Electrode Potential Data

Some applications of redox potential data have been discussed in section 9.5. However, inferences drawn from such electrode potential data are not altogether free from certain limitations. As already mentioned, standard electrode potential data refer to unit activities of the species involved, which is far from the actual situation existing in many practical systems. Again, replacement of activity terms by concentrations is not justifiable except when the concentrations are extremely small.

Conclusions based on electrode potential data are meaningful only from the thermodynamic point of view. But thermodynamics concerns only the initial and final states of a system. On the other hand, the practicability of a chemical process is also determined by the kinetics of the reaction involved. Thus, a redox reaction with a negative ΔG value may still not take place perceptibly if the rate of the reaction is too slow, or if there is no suitable path with a favourable energy of activation. Thus, when magnesium metal is dipped into water, no reaction seems to occur though the electrode potential for Mg^{2+}–Mg strongly suggests that magnesium should replace hydrogen ions from solution. This is due to an adherent film of oxide on the metal surface which places a large barrier of activation energy to the metal-water contact. Once the oxide layer is displaced through amalgamation, the energy of activation is lowered and reaction takes place readily.

Similarly, the rates of reactions involving simple electron transfer are kinetically more rapid than reactions in which covalent bonds have to be broken. Electrode potential data fail to take into account such factors. Thus, the $E°$ value for the $ClO_4^- - Cl^-$ couple (+1.39 V) is higher than that for the $Cr_2O_7^{2-} - Cr^{3+}$ couple (+1.33 V). So, the perchlorate ion is expected to be a stronger oxidzing agent than the dichromate ion. Contrary to this expectation, we find that ferrous ion is readily oxidized by dichromate ions in solution while it is quite stable in de-aerated perchloric acid.

9.3.6 Numerical Problems

Summary rules

(i) $\Delta G = -nFE$; $\Delta G° = -nFE°$

(ii) $E_{Cell} = E_{Red}$ (right) – E_{Red} (left)

(iii) For the electrode process $A^{n+} + ne \rightleftharpoons A$,

$$E_{Red} = E° - \frac{RT}{nF}\ln\left(\frac{a_A}{a_{A^{n+}}}\right) = E°_{Red} - \frac{0\cdot059}{n}\log\left(\frac{a_A}{a_{A^{n+}}}\right) \text{ at } 25°C$$

(iv) $E° = \dfrac{RT}{nF}\ln K_{eq} = \dfrac{0\cdot059}{n}\log K_{eq}$ at 25° C

Problem 9.7

The half-reaction for the cathode in the cell Cd(*s*) | $CdCl_2$ (0.01 M) | AgCl(*s*) | Ag(*s*) may be written in either of two ways:

1. AgCl(*s*) + *e* $\rightleftharpoons$ Ag(*s*) + Cl^- $E°$ = 0.222 V
2. $Ag^+(aq)$ + *e* $\rightleftharpoons$ Ag(*s*) $E°$ = 0.799 V

Write the complete cell reaction with both alternatives and show that they predict the same value for the cell EMF. Given $E°_{Cd} = -0.402$ V, K_{sp}(AgCl) = $1\cdot8 \times 10^{-10}$.

Solution:

1.			
Anode	Cd(*s*) $\rightleftharpoons$ $Cd^{2+}(aq)$ + 2e		$E°$(red) = – 0.402 V
Cathode	2AgCl(*s*) + 2*e* $\rightleftharpoons$ 2Ag(*s*) + $2Cl^-(aq)$		$E°$ = 0.222 V
Cell	Cd(*s*) + 2AgCl(*s*) $\rightleftharpoons$ Cd^{2+} + 2Ag(*s*) + $2Cl^-(aq)$		$E°$ = 0.624 V

By Nernst equation, $E = E° - \dfrac{0\cdot059}{2}\log[Cl^-]^2[Cd^{2+}] = 0\cdot624 - \dfrac{0\cdot059}{2}\log(0\cdot02)^2(0\cdot01) = 0\cdot783V$ at 25°C.

2.		
Anode	Cd(*s*) $\rightleftharpoons$ $Cd^{2+}(aq)$ + 2e	$E°$(red) = –0.402 V
Cathode	$2Ag^+(aq)$ + 2*e* $\rightleftharpoons$ 2Ag(*s*)	$E°$ = 0.799 V
Cell	Cd(*s*) + $2Ag^+(aq)$ $\rightleftharpoons$ $Cd^{2+}(aq)$ + 2Ag(*s*)	$E°$ = 1.201 V

By Nernst equation, $E = 1\cdot201 - \dfrac{0\cdot059}{2}\log\dfrac{[Cd^{2+}]}{[Ag^+]^2}$ at 25°C

Now, $[Ag^+] = K_{sp}/[Cl^-] = 1\cdot8\times10^{-10}/0\cdot02 = 9\times10^{-9}$ M ; $[Cd^{2+}] = 0\cdot01$ M

Substitution gives $E = 1\cdot201 - \dfrac{0\cdot059}{2}\log\dfrac{0\cdot01}{(9\times10^{-9})^2} = 0\cdot785$ V

The discrepancy with approach 1 is negligible (0.25%).

Note: The cell may be used to determine the solubility product of AgCl.

Problem 9.8

Calculate the standard EMF of the cell for the reaction $Cu + 2Ag^+ = Cu^{2+} + 2Ag$.

$E°\ (Cu^{2+} - Cu) = +0.34\ V$; $E°\ (Ag^+ - Ag) = +0.80\ V$ at 25°C.

Solution: A cell may be constructed as $Cu \mid CuSO_4(aq) \parallel AgNO_3(aq) \mid Ag$

As nothing is stated about the liquid junction, we assume it to have been eliminated by a salt bridge. As the standard EMF only is required, we write

$$E^{\circ}_{Cell} = E^{\circ}\left(Ag^{+} - Ag\right) - E^{\circ}\left(Cu^{2+} - Cu\right) = 0.80\ V - 0.34\ V = +0.46\ V.$$

Problem 9.9

Calculate the standard free energy change of the above cell-reaction.

Solution: We have the relation $\Delta G° = -nFE°$. Here the reaction takes place for two faradays of electricity per mol of copper. Hence $n = 2$.

$$\therefore \qquad \Delta G° = -2 \times 96{,}500\ C \times 0.46\ V = -88{,}780\ J.$$

Problem 9.10

The EMF of the cell Pt, $H_2(g,\ 1\ atm) \mid H^+\ (m = 1) \parallel$ mercurous ion (10^{-1} M) | Hg is 0.767 V at 25°C. Find the charge on the mercurous ion, Hg_n^{n+}. $E°$ for Hg(I) (*aq*) – Hg is +0.796 volt at 25°C.

Solution: Here, the left electrode is a SHE with $E = 0.0V$. Let E for the right half-cell be $E_{Hg(I)}$.

$$0.767 = E_{Hg(I)} - 0.0;\ \text{when}\ E_{Hg(I)} = +0.767\ V.$$

The reaction at the right electrode: $Hg_n^{n+} + ne \rightleftharpoons nHg(l)$

$$\text{So, } E_{Hg(I)} = E^{\circ}_{Hg(I)} - \frac{0.059}{n}\log\frac{1}{\left[Hg_n^{n+}\right]} \quad \text{or, } 0.767 = +0.796 - \frac{0.059}{n}\log 10.$$

which gives $n = 2$. This shows that two faradays of electricity are involved per g-ion of Hg(I). So, the charge on the mercurous ion must be 2 (represented as Hg_2^{2+}).

Problem 9.11

Construct the cell for the reaction $Co + Ni^{2+}(1M) = Co^{2+}(0.1M) + Ni$ and calculate its EMF at 298K.

$$E^{\circ}_{Co^{2+},\,Co} = -0.282\ V, \quad E^{\circ}_{Ni^{2+},\,Ni} = -0.236V.$$

Solution: A cell may be constructed as $Co \mid Co^{2+}(0.1\ M) \parallel Ni^{2+}(1\ M) \mid Ni$.

$$E_{Cell} = E_{Ni} - E_{Co} = \left[E^{\circ}_{Ni} + \frac{0.059}{2}\log\left[Ni^{2+}\right]\right] - \left[E^{\circ}_{Co} + \frac{0.059}{2}\log\left[Co^{2+}\right]\right]$$

$$= -0.236 + 0.282 + \frac{0.059}{2}\log(1/0.1) = 0.076V.$$

Problem 9.12

Calculate the EMF of the same cell as in 9.11 but with $[Co^{2+}] = 1M$ and $[Ni^{2+}] = 0.01$ M. Comment on the result obtained.

Solution: $E = E^{o}_{Ni} - E^{o}_{Co} + \frac{0.059}{2}\log\left[\frac{Ni^{2+}}{Co^{2+}}\right] = 0.046 + \frac{0.059}{2}\log\frac{0.01}{1} = -0.013$ V.

The negative value of the EMF suggests that the cell reaction as written will *not* be spontaneous. The EMF of the cell depends on the concentrations of the species in it. Since the $E°$ values of the two half-cells in this case are close to one another, the direction of the cell may be easily reversed by changing the concentration ratio of the two ions.

Problem 9.13

Calculate the equilibrium constant for the reaction

$$Fe^{2+} + Ce^{4+} \rightleftharpoons Fe^{3+} + Ce^{3+} \text{ at 298K.} \left[E^{o}_{Ce^{4+},Ce^{3+}} = +1.28 \text{ V in 1M HCl}\right]$$

Solution: At 298K, $E^{o}_{Cell} = \frac{0.059}{n}\log K_{eq}$ where K_{eq} is the equilibrium constant. Here $n = 1$.

$$E^{o}_{Cell} = E^{o}_{Ce^{4+},Ce^{3+}} - E^{o}_{Fe^{3+},Fe^{2+}} = 1.28 - 0.77 = 0.51 \text{ V.}$$

Hence, $\log K_{eq} = 0.51/0.059 = 8.64$ or, $K_{eq} = 10^{8.64} = 4.37 \times 10^{8}$.

Problem 9.14

What is the equilibrium concentration of Fe^{2+} reached when a piece of iron is dipped in a 0.1 M Zn^{2+} solution? See Table 9.1 for $E°$ values.

Solution: The reaction $Fe + Zn^{2+} = Zn + Fe^{2+}$ corresponds to the cell $Fe \mid Fe^{2+} \parallel Zn^{2+} \mid Zn$.

$$E^{o}_{Cell} = E^{o}_{Zn} - E^{o}_{Fe} = (-0.76 + 0.44) \text{ V.} = -0.32 \text{ V.}$$

(We find that the reaction will not take place spontaneously; still we may calculate K_{eq} for the reaction to find the equilibrium concentration of Fe^{2+} which must be indeed very small.)

$$E^{o}_{Cell} = \frac{0.059}{2}\log K_{eq} \quad \text{or, } \log K_{eq} = \frac{2 \times -0.32}{0.059} = -10.85 \text{ or, } K_{eq} = 10^{-10.85} = 1.41 \times 10^{-11}$$

But, from the reaction, $K_{eq} = [Fe^{2+}]/[Zn^{2+}]$, where $[Zn^{2+}] = 0.1$ M.

$$\therefore \quad [Fe^{2+}] = 1.41 \times 10^{-11} \times 10^{-1} = 1.41 \times 10^{-12} \text{ M.}$$

[**Note:** The very small concentration of Fe^{2+} ion indicates that the reaction does not proceed at all toward right.]

Problem 9.15

K_{eq} for the reaction $Fe^{3+} + Ag \rightleftharpoons Fe^{2+} + Ag^{+}$ is 0.531 (298K). Calculate $E^{o}_{Ag^{+}-Ag}$.

Solution: $E^{o}_{Cell} = \frac{0.059}{1}\log K_{eq} = 0.059 \log 0.531 = -0.016$ V.

Now, $E^{o}_{Cell} = E^{o}_{Fe} - E^{o}_{Ag}$ or, $E^{o}_{Ag} = E^{o}_{Fe} - E^{o}_{Cell} = (0.770 + 0.016)$ V $= 0.786$ V.

Calculation of standard electrode potential from free energy change

Problem 9.16

Calculate $E^o_{Mg^{2+}, Mg}$ from the following ΔG^o values (kJ mol^{-1}):

$Mg(s) + \frac{1}{2}O_2(g) = MgO(s)$: –572.9; $MgO(s) + H_2O(l) = Mg(OH)_2(s)$: –31.2;
$H_2O(l) = H_2(g) + \frac{1}{2}O_2(g)$: +241.3.

The autoprotolysis constant K_w for water at 298K = 1.00×10^{-14}. Solubility product of $Mg(OH)_2$ at 298K = 5.50×10^{-12} (Chapter 11).

Solution: This problem cannot be solved directly. We shall have to find out the ΔG^o value for a cell reaction in which E^o_{Mg} is involved. The ΔG^o value will give us E^o_{Cell} and hence E^o_{Mg}.

Now, from a study of the given set of data we find that if we add two more changes with them, the *net* change amounts to

$$Mg(s) + 2H^+(aq) = Mg^{2+}(aq) + H_2$$

Let us see how we can arrive at this:

	Reaction	ΔG^o (kJ mol^{-1})
	$Mg(s) + \frac{1}{2}O_2(g) = MgO(s)$	–572.9.
	$MgO(s) + H_2O(l) = Mg(OH)_2(s)$	–31.2
	$H_2O(l) = H_2(g) + \frac{1}{2}O_2(g)$	+241.3.
Adding,	$Mg(s) + 2H_2O(l) = Mg(OH)_2(s) + H_2(g)$	–362.8
We add further	$Mg(OH)_2(s) = Mg^{2+}(aq) + 2OH^-(aq)$	ΔG^o_I
and	$2H^+(aq) + 2OH^-(aq) = 2H_2O(l)$	ΔG^o_{II}
Overall reaction	$Mg(s) + 2H^+(aq) = Mg^{2+}(aq) + H_2$	$-362.8 + \Delta G^o_I + \Delta G^o_{II}$

This reaction corresponds to the cell Mg | $Mg^{2+}(aq)$ || $H^+(aq)$ | H_2, Pt.

The standard EMF of this cell will be: $E^o_{Cell} = E^o_{H2} - E^o_{Mg} = -E^o_{Mg}$

But we are to obtain values of ΔG^o_I and ΔG^o_{II}. These may be obtained as follows:

ΔG^o_I : The equilibrium constant for the reaction $Mg(OH)_2(s) = Mg^{2+}(aq) + 2OH^-(aq)$ is the solubility product of $Mg(OH)_2 = 5.50 \times 10^{-12}$.

$\therefore \quad \Delta G^o_I = -RT \ln K = -8.314 \text{ J. mol}^{-1} \text{ K}^{-1} \times 298\text{K} \times \ln 5.50 \times 10^{-11} = 64.24 \text{ kJ mol}^{-1}$.

ΔG^o_{II} : The reverse of this process is the autoprotolysis of water: $H_2O(l) = H^+(aq) + OH^-(aq)$

when the equilibrium constant is $K_w = a_{H^+(aq)} \times a_{OH^-(aq)}$; a_{H_2O} is unity.

For the reaction: $2H^+(aq) + 2OH^-(aq) = 2H_2O(l) \quad K = \dfrac{1}{a^2_{H^+} \times a^2_{OH^-}} = \dfrac{1}{K^2_w}$

$\therefore \quad \Delta G^o_{II} = -RT \ln K = -RT \ln K_w^{-2} = +2\,RT \ln K_w \Rightarrow -160 \text{ kJ mol}^{-1}$.

Now we calculate the ΔG^o for the overall reaction $Mg(s) + 2H^+(aq) = Mg^{2+}(aq) + H_2$

$\Delta G^o_{Total} = -362.8 + \Delta G^o_I + \Delta G^o_{II} = (-362.8 + 64.24 - 160) \text{ kJ mol}^{-1} = -458.56 \text{ kJ mol}^{-1}$

$\therefore \quad E^o_{Cell} = -\Delta G^o/2F = 2.38\text{V}$. or, $E^o_{Mg^{2+}, Mg} = -E^o_{Cell} = -2.38$ V.

Problem 9.17

Calculate $E°$ and K for the following conversion (balance as necessary) in acidic aqueous medium.

$$MnO_2(s) \rightleftharpoons Mn^{2+} + MnO_4^- \quad E°_{MnO_2-Mn^{2+}} = 1.230 \text{ V}; \quad E°_{MnO_4^- - MnO_2} = 1.692 \text{ V}.$$

Hint: Following changes in oxidation numbers, we may write two half-cell reactions as

Anode: $MnO_2(s) + 2H_2O \rightleftharpoons MnO_4^- + 4H^+ + 3e \qquad E°(\text{red}) = 1.692 \text{ V}$

Cathode: $MnO_2(s) + 4H^+ + 2e \rightleftharpoons Mn^{2+} + 2H_2O \qquad E°(\text{red}) = 1.230 \text{ V}$

$$E°_{Cell} = E°_{cathode} - E°_{anode} = 1.230 - 1.692 = -0.462 \text{ V}.$$

To find K, we write the balanced cell reaction after cancelling the electrons (6 on both side):

$$5MnO_2(s) + 4H^+ \rightleftharpoons 2MnO_4^- + 3Mn^{2+} + 2H_2O.$$

We get $n = 6$. Hence the equilibrium constant $K = 10^{6E°/0.059} = 10^{6(-0.462)/0.059} = 1.04 \times 10^{-47}$

The value of K shows that the equilibrium lies far toward left.

9.4 FACTORS AFFECTING THE EMF OF HALF-CELLS

The ultimate behaviour of a redox couple is determined by a number of factors that alter the effective concentrations of the species involved in the cell reaction. The corresponding effect on the cell potential may be calculated with the Nernst equation. Here we consider some typical situations with specific examples.

9.4.1 Effect of Concentration

For any half-cell reaction, we may write in general,

$$\text{Oxidized form} + ne \rightleftharpoons \text{Reduced form}$$

$$E = E° - \frac{0.059}{n} \log \frac{[\text{reduced form}]}{[\text{oxidized form}]}$$

This shows that a tenfold increase in the concentration of the oxidized form will *raise* the half-cell potential by $0.059/n$ volts. Thus, for the $Fe^{3+} - Fe^{2+}$ system, $E°$ is +0.77 volt; but when $[Fe^{3+}] = 1$ mol dm^{-3} and $[Fe^{2+}] = 0.0001$ mol dm^{-3}, we have

$$E = E° - \frac{0.059}{1} \log \frac{[Fe^{2+}]}{[Fe^{3+}]} = +0.77 + 0.059 \log \frac{1}{10^{-4}} = +1.006 \text{ V}.$$

A similar effect of concentration was also observed in problems 9.11 and 9.12. We shall find that such variations become important in redox titrations.

9.4.2 Effect of pH

When hydrogen or hydroxyl ions are involved in a half-cell reaction, their concentrations appear in the Nernst equation and accordingly the potential is influenced by the pH of the medium. The effect may be understood from the following typical examples.

I. The $H_3AsO_4 - H_3AsO_3$ system: The standard reduction EMF of the couple is +0.56 V. Both H_3AsO_4 and H_3AsO_3 are weak acids and their ionization in acidic and neutral medium is quite negligible. So the half-cell reaction for the couple may be fairly represented in terms of H_3AsO_4 and H_3AsO_3 which form the main species in solution:

$$H_3AsO_4 + 2H^+ + 2e \rightleftharpoons H_3AsO_3 + H_2O$$

$$\text{At } 25°\text{C}, \ E = E° - \frac{0.059}{2} \log \frac{[H_3AsO_3]}{[H_3AsO_4][H^+]^2}$$

$$= 0.56 - 0.059\,\text{pH} - \frac{0.059}{2} \log \frac{[H_3AsO_3]}{[H_3AsO_4]}$$

The influence of pH on the half-cell potential is obvious. For example, at pH = 8,

$$E = 0.09 - \frac{0.059}{2} \log \frac{[H_3AsO_3]}{[H_3AsO_4]}$$

The formal potential of the couple thus falls to 0.09 V. The actual potential will be still less if ionization of the acids in the alkaline medium is taken into consideration. In any case, the effective reduction potential of the arsenate-arsenite system now *decreases* and so also does its oxidizing power. It can now no more oxidize iodide ions to iodine. Under such condition, the $I^- - I_2$ system has a much *higher* reduction potential (+0.54 volt) and hence acts as a *stronger* oxidizing agent. Accordingly, *iodine oxidizes arsenite to arsenate at this pH.*

If a neutral or faintly alkaline solution of arsenite (sodium bicarbonate solution, pH ~ 8) is treated with a drop of blue starch-I_2 solution, the blue color is discharged as iodine is reduced by As(III) to I^-. Acidification by strong HCl restores the blue colour as the As(V) now oxidizes I^-. The test is carried out in presence of sodium bicarbonate, which maintains the pH around 8. Stronger alkalis like sodium hydroxide or sodium carbonate cannot be used since in that case (pH > 9) iodine also reacts with the alkali giving hypoiodite ion (IO^-).

II. The $MnO_4^- - Mn^{2+}$ system: The standard reduction potential of the $MnO_4^- - Mn^{2+}$ system (+1.51 volt) may at first suggest that permanganate solutions would oxidize each of chloride, bromide or iodide ions to the free halogens ($E°$ = +1.36, +1.07 and +0.54 volt respectively). But the potential of the $MnO_4^- - Mn^{2+}$ system is greatly influenced by the pH of the medium and accordingly permanganate solution may be made to oxidize the three halides selectively at different pH values:

$$MnO_4^- + 8H^+ + 5e \rightleftharpoons Mn^{2+} + 4H_2O$$

$$E = E° - \frac{0.059}{5} \log \frac{[Mn^{2+}]}{[MnO_4^-][H^+]^8}$$

$$= +1.51 - 0.094 \times pH - \frac{0.059}{5} \log \frac{[Mn^{2+}]}{[MnO_4^-]}$$

At $pH = 6$, the reduction potential falls to +0.95 volt, below the potential of either $Cl_2 - Cl^-$ or $Br_2 - Br^-$ couples. So permanganate can no more oxidize bromide or chloride ions at this pH. At pH = 3, the reduction potential shifts to about +1.23 volt, when bromide and iodide are both oxidized, but not chloride. Chloride is oxidized only below pH 1.5.

III. The $Cr_2O_7^{2-} - Cr^{3+}$ system: $E°$ for the couple is 1.33 V. The potential of the system is highly dependent on pH, as seen from the half-cell reaction:

$$Cr_2O_7^{2-} + 14H^+ + 6e \rightleftharpoons 2Cr^{3+} + 7H_2O$$

$$E = 1.33 - \frac{0.059 \times 14}{6} pH - \frac{0.059}{6} \log \frac{[Cr^{3+}]^2}{[Cr_2O_7^{2-}]} \quad \text{(at 25°C)}$$

At pH ≃ 7 (neutral solution), the effective potential falls below 0.5 V when $Cr_2O_7^{2-}$ cannot oxidize iodide.

IV. The $BrO_3^- - Br_2$ system: $E°$ is 1.52 V for the half-cell reaction

$$BrO_3^- + 6H^+ + 5e \rightleftharpoons \tfrac{1}{2}Br_2 + 3H_2O$$

$$E = E° - \frac{0.059 \times 6}{5} pH - \frac{0.059}{5} \log \frac{1}{[BrO_3^-]} \quad \text{(at 25°C)}$$

We assume $a_{Br_2} = 1$. Since $HBrO_3$ is a strong acid, its main form in solution will be BrO_3^-. The relation above shows that at pH = 7, the formal potential of the system will be around 1.0 V when BrO_3^- will not be able to oxidize Br^-.

9.4.3 Effect of Precipitation

This effect is actually due to the change of concentration of the oxidized and reduced forms as discussed earlier. The concentration change follows from precipitation of one of the reaction products due to low solubility.

The Cu^{2+}–Cu^+ system has the low reduction potential of +0.15 volt. It is not expected to oxidize iodide to iodine ($E°$ = +0.54 volt). However, the extremely low solubility of cuprous iodide reduces the concentration of cuprous ion in such a manner that the reduction potential is substantially increased:

$$Cu^{2+} + e \rightleftharpoons Cu^+$$

$$E = E° - 0.059 \log [Cu^+]/[Cu^{2+}]$$

$[Cu^+]$, the concentration of cuprous ion, is determined by the solubility product (K) of CuI: $K = [Cu^+] \times [I^-] = 1 \times 10^{-12}$. So,

$$E = E° - 0.059 \log \frac{10^{-12}}{[Cu^{2+}][I^-]}$$

$$= +0.15 + 12 \times 0.059 + 0.059 \log [Cu^{2+}][I^-]$$

$$= +0.86 + 0.059 \log [Cu^{2+}][I^-]$$

The effective potential thus rises above the reduction potential of the iodide-iodine system and iodide ion is oxidized by cupric ion in solution. In fact, the reaction,

$$2Cu^{2+} + 4I^- = 2CuI\downarrow + I_2$$

proceeds almost towards right so that copper may be estimated by titrating the liberated iodine with a standard solution of sodium thiosulfate.

The $E°$ of the ferricyanide-ferrocyanide system is +0.36 volt. The system is not expected to oxidize iodide ion. But zinc ions remove ferrocyanide ions from the system by precipitation of $K_2Zn_3[Fe(CN)_6]_2$.

$$3ZnCl_2 + 2K_4Fe(CN)_6 = K_2Zn_3[Fe(CN)_6]_2 + 6KCl$$

The reduction potential of the system, expressed as

$$E = +0.36 - 0.059 \log [Fe(CN)_6]^{4-}/[Fe(CN)_6]^{3-}$$

is thus further raised and its oxidizing power is increased. Practically, ferricyanide oxidizes iodide to iodine in presence of zinc ion and offers a method for the determination of zinc volumetrically.

Problem 9.18

25 cm^3 of 0.2M $CuSO_4$ solution is mixed with 25 cm^3 of 1.0M KI solution. Calculate the concentration of Cu^{2+} ion in the resulting solution. $E°$ ($Cu^2 - Cu^+$) = 0.15 V; $E°$ ($I_2 - I_3^-$) = 0.54 V; K_s of CuI = 1×10^{-12}.

Solution: First we attempt to calculate the equilibrium constant K for the overall reaction

$$Cu^{2+} + 2I^- \rightleftharpoons CuI(s) + \tfrac{1}{2}I_2.$$

In the cell corresponding to this reaction, Pt, I_2/I^-, $CuI(s)$ / Cu^{2+}, Pt, the electrode reactions are

$$\textbf{Left: } I^- = \tfrac{1}{2}I_2 + e\,; \qquad \textbf{Right: } Cu^{2+} + I^- + e = CuI(s)$$

$$E_{red}(\text{right}) = E°(Cu^{2+}/CuI, I^-) + 0.059 \log a_{Cu^{2+}} \times a_{I^-}\ (25°C)$$

The right hand electrode may also be considered as a Cu^{2+}/Cu^+ electrode (see Problem 9.7):

$$E_{red}(\text{right}) = E°_{(Cu^{2+}, Cu^+)} + 0.059 \log \frac{a_{Cu^{2+}}}{a_{Cu^+}} = 0.15 + 0.059 \log \frac{a_{Cu^{2+}}}{K_s / a_{I^-}}$$

Comparing the two expressions of E_{red} (right), $E°(Cu^{2+}/CuI, I^-) = 0.15 - 0.059 \log K_s$

$$= 0.86 \text{ V (from the value of } K_s).$$

$$\therefore\ E°_{Cell} = E°_{red}(\text{right}) - E°_{red}(\text{left}) = (0.86 - 0.54)\text{ V} = 0.32\text{ V}.$$

From $\Delta G° = -F E°_{cell} = -RT \ln K$ at 25°C, $\log K = 0.32/0.059 = 5.42 \Rightarrow K = 2.65 \times 10^5$

The high value of K suggests that the reaction should proceed nearly completely toward right (iodide is in excess).

Now let the concentration of Cu^{2+} ion at equilibrium be x M. Initial concentrations of both Cu^{2+} and I^- are *halved* on mixing (volume doubled). Hence the equilibrium concentration of I^- is found by subtracting the concentration of I^- reacted (mark the stoichiometry):

$$[I^-] = \{0.5 - 2(0.1 - x)\}\text{ M} = (0.3 + 2x)\text{ M} \simeq 0.3\text{ M } (x << 0.3).$$

Replacing the activities of ions by concentrations,

$$K_{eq} = 2.65 \times 10^5 = \frac{1}{[Cu^{2+}][I^-]^2} = \frac{1}{x\,(0.3)^2} \qquad \therefore\ x = 4.2 \times 10^{-5}$$

The final concentration of cupric ion in solution is thus 4.2×10^{-5} M.

9.4.4 Effect of Complex Formation

Complex formation may also bring down the effective concentration of one of the species involved in an electrode process, thereby altering the potential of the system. Ferric ion normally oxidizes iodide ion to iodine in aqueous solution. But in presence of excess fluoride ion, ferric ion is removed as the complex ion $[FeF]^{2+}$ etc.

$$Fe^{3+} + F^- \rightleftharpoons FeF^{2+}\ ;\ K_{eq} = 10^5 = \frac{[[FeF]^{2+}]}{[Fe^{3+}][F^-]}$$

The value of $[Fe^{3+}]$ from this relation may now be substituted in the expression for the reduction potential of the $Fe^{3+} - Fe^{2+}$ system:

$$E = E° - 0.059 \log \frac{[Fe^{2+}]}{[Fe^{3+}]} = +0.77 - 0.059 \log \frac{10^5 [F^-][Fe^{2+}]}{[[FeF]^{2+}]}$$

$$= +0.475 - 0.059 \log \frac{[Fe^{2+}][F^-]}{[[FeF^{2+}]]}$$

As the reduction potential decreases, the oxidizing power of Fe^{3+} decreases and it can no more oxidize iodide ion. On the other hand, copper (II) does not form any such stable complex with fluoride ion. Thus copper may be estimated selectively in an iron-copper mixture by titrating the iodine liberated from potassium iodide in presence of fluoride ion.

Problem 9.19

E = 0.321 V for the cell S.H.E || $Hg(NO_3)_2$ (0.001 M), KI (0.01 M) | $Hg(l)$

Calculate the equilibrium constant for the reaction $Hg^{2+} + 4I^- \rightleftharpoons HgI_4^{2-}$

It is assumed that no other forms of mercury are present in solution. $E°$ for $Hg^{2+} - Hg$ = +0.852 V.

Solution: The right half-cell is $Hg^{2+} - Hg$: $Hg^{2+}(aq) + 2e \rightleftharpoons Hg(l)$

$$E_{Cell} = E_{Hg^{2+}-Hg} - E_{S.H.E} \text{ or, } 0.321 = \left[E^o_{Hg^{2+}-Hg} - \frac{0.059}{2}\log\frac{1}{[Hg^{2+}]}\right] - 0$$

or, $0.321 = 0.852 + \frac{0.059}{2}\log[Hg^{2+}]$, giving $[Hg^{2+}] = 1 \times 10^{-18}$

Since $[Hg^{2+}]$ is so small, we may assume that the entire $Hg(NO_3)_2$ has been converted to $[HgI_4]^{2-}$.

Hence $[HgI_4]^{2-} = [Hg(NO_3)_2] = 0.001$ M. But each mol of Hg^{2+} has consumed 4 mol of I^-. Hence $[I^-]$ = (0.01 – 0.004) M = 0.006 M.

Now we readily get the equilibrium constant K for the complexation reaction:

$$K = \frac{[HgI_4]^{2-}}{[Hg^{2+}][I^-]^4} = \frac{0.001}{1 \times 10^{-18} \times (0.006)^4} = 7.7 \times 10^{23}$$

Stability constants of complexes

The formation of a complex from an aquated metal ion in aqueous solution may be considered to take place stepwise; with a neutral unidentate ligand L, this may be represented as (the water molecules are not shown for simplicity):

$$M^{x+} + L \rightleftharpoons ML^{x+} \qquad K_1 = [ML^{x+}]/[M^{x+}][L]$$

$$ML^{x+} + L \rightleftharpoons ML_2^{x+} \qquad K_2 = [ML_2^{x+}]/[ML^{x+}][L]$$

$$\ldots \quad \ldots \quad \ldots$$

$$ML_{n-1}^{x+} + L \rightleftharpoons ML_n^{x+} \qquad K_n = [ML_n^{x+}]/[ML_{n-1}^{x+}][L]$$

$K_1, K_2 \ldots K_n$ are the *stepwise* (or consecutive) *formation constants* expressed in terms of concentration. The product of the consecutive stability constants gives the cumulative or *overall stability constant* β_n:

$$M^{x+} + nL \rightleftharpoons ML_n^{x+} \qquad K_{eq} = \beta_n = K_1 K_2 \ldots K_n$$

The effect of stability constants of complexes on the redox potential of a system may be understood from the following examples.

Problem 9.20

The overall formation constant for $Zn(CN)_4^{2-}$ is 10^{21}. Calculate $E°$ for the process

$$Zn(CN)_4^{2-} + 2e \rightleftharpoons Zn + 4CN^-. \; (E° \text{ for } Zn^{2+}/Zn = -0.76 \text{ V})$$

Solution: We have, $Zn^{2+} + 4CN^- \rightleftharpoons Zn(CN)_4^{2-} \quad \beta_4 = 10^{21}$

and $\quad Zn^{2+} + 2e \rightleftharpoons Zn \qquad E° = -0.76 \text{ V}$

We combine these reactions to suit our purpose; at 25°C,

$$Zn(CN)_4^{2-} \rightleftharpoons Zn^{2+} + 4CN^- \qquad \Delta G° = -2.303\,RT \log 10^{-21}.$$

$$Zn^{2+} + 2e \rightleftharpoons Zn \qquad \Delta G° = -2F \times (-0.76)$$

$$Zn(CN)_4^{2-} + 2e \rightleftharpoons Zn + 4CN^- \qquad \Delta G° = 1.239F + 1.52F$$

From $\Delta G° = -nFE°$, $\quad E° = -1.38 \text{ V } (n = 2)$.

Note: In a solution initially containing C_1 molar Zn^{2+} and C_2 molar CN^- ions ($C_2 >> C_1$), the equilibrium concentration of Zn^{2+} will be very small, as seen from the high value of the formation constant. So the equilibrium concentration of the complex ion $Zn(CN)_4^{2-}$ may be taken as nearly C_1, while that of the CN^- ion will be $(C_2 - 4C_1)$. Hence the redox potential for the system at 25°C will be approximately given by

$$E = E° - \frac{0.059}{2} \log \frac{[CN^-]^4}{[Zn(CN)_4]^{2-}} = -1.38 - \frac{0.059}{2} \log \frac{(C_2 - 4C_1)^4}{C_1}$$

Problem–9.21

Predict whether $Fe(CN)_6^{3-}$ will be a stronger or weaker oxidizing agent in comparison to $Fe^{3+}(aq)$. Given $E°(Fe^{3+} - Fe^{2+}, aq) = 0.77$ V; $\log\beta_6$: $Fe(CN)_6^{3-}$, 31.0; $Fe(CN)_6^{4-}$, 24.0.

Solution: We have to compare the standard reduction potentials for the half-reactions:

$$Fe^{3+}(aq) + e = Fe^{2+}(aq) \quad E° = E_1^\circ \text{ (say)}; \qquad Fe(CN)_6^{3-} + e = Fe(CN)_6^{4-} \quad E° = E_2^\circ \text{ (say)}$$

We arrange the equilibrium processes of known $E°$ or K_{eq} to arrive at this half-reaction:

$$\begin{array}{lll} Fe^{3+}(aq) + e & = Fe^{2+}(aq) & \Delta G_I^\circ = -FE°. \\ Fe^{2+}(aq) + 6CN^- & = Fe(CN)_6^{4-} & \Delta G_{II}^\circ = -2.303RT \log \beta_6(Fe^{II}) \\ Fe(CN)_6^{3-} & = Fe^{3+}(aq) + 6CN^- & \Delta G_{III}^\circ = +2.303\,RT \log \beta_6(Fe^{III}). \\ \hline Fe(CN)_6^{3-} + e & = Fe(CN)_6^{4-} & \Delta G_I^\circ + \Delta G_{II}^\circ + \Delta G_{III}^\circ \end{array}$$

The standard reduction potential for the overall process, E_2°, is related to the overall $\Delta G°$ as $\Delta G° = -FE_2^\circ$. Expressing $\Delta G°$ in terms of ΔG_I° etc.,

$$E_2^\circ = -(-0.77F - 2.303 \times 24\,RT + 2.303 \times 31\,RT)/F \text{ V}.$$

$$= (0.77 + 0.059 \times 24 - 0.059 \times 31) \text{ V at } 25°C = 0.36 \text{ V}.$$

The standard reduction potential for Fe(III) – Fe(II) decreases significantly from the aqua species to the cyano complex (the complex of Fe^{III} has a much higher stability). Hence the oxidizing power of Fe^{III} decreases.

Note: Usually the complex of a metal in a higher oxidation state is more stable than the corresponding complex in a lower oxidation state (same coordination number, of course). Thus for hexaammine complexes of cobalt, $\log \beta_6 = 33.7$ for Co^{III} and 4.9 for Co^{II}. $E°$ drops to 0.14 from 1.84 V for the aqua ion (check yourself).

However, exceptions arise when the complex in the lower oxidation state is more stable than that in the higher oxidation state, as for example for the tris(*o*-phenanthroline) complex of iron (*o*-phenanthroline is a bidentate ligand; chapter 26). $\log \beta_3 = 21.3$ for Fe(II) and 14.1 for Fe(III). $E°$ for the couple $Fe(phen)_3^{3+}/Fe(phen)_3^{2+}$ *rises* to about 1.2 V (vs 0.77 V for the aqua ions).

Practical application: electrodeposition of brass

The standard reduction potential of Cu^{2+} ion (0.34 V) is much higher than that of Zn^{2+} ion (–0.76 V). So when an aqueous solution containing zinc and copper salts is electrolysed, copper deposits earlier. However, in presence of excess alkali cyanide, both zinc and copper form complex $Zn(CN)_4^{2-}$ and $Cu(CN)_4^{3-}$ ions. The overall stability constant of $[Cu(CN)_4]^{3-}$ is much higher (10^{30}) than that of $[Zn(CN)_4]^{2-}$ (10^{20}). Hence the concentration of free copper ions in solution is reduced to a much greater extent than for the zinc ion. Eventually the potentials of the two couples are brought close to one another when simultaneous deposition of both the metals can be achieved.

9.5 APPLICATIONS OF ELECTRODE POTENTIAL DATA

Electrode potential data provide various useful and interesting interpretation of the redox behaviour of elements not only in the laboratory but also in the geosphere and the biosphere in general. The vast field of application of redox potential and cells range from the tiny pacemaker of heartpatients to fuel cells for power generation, prevention of corrosion, extraction of elements and many other fields.

9.5.1 The Electrochemical Series

The electrochemical series is obtained by arranging the elements in the order of increasing standard reduction potential values (Table 9.1). The most electropositive elements appear at the top of the series, the less electropositive ones next, followed by the electronegative elements. Hydrogen appears at the borderline of negative and positive standard potential values ($E° = 0.0$ V). For ready reference, the series is often quoted for metals alone, only hydrogen being included within. A selected portion of the series, obtained by comparing the standard aqueous reduction EMF's at 25°C from Table 9.1 is shown in Table 9.3.

Table 9.3

The electrochemical series (part)

Elements	$E°$(V)	Elements	$E°$(V)	Elements	$E°$(V)
Li	−3.040	Mg	−2.363	Pb	−0.126
K	−2.936	Al	−1.662	H_2	0.000
Ca	−2.868	Zn	−0.763	Cu	+0.339
Na	−2.715	Fe	−0.441	Ag	+0.799

The electrochemical series helps us in understanding the general trend of reactivity among the elements. The elements appearing at the top of the series are highly reactive, reactivity decreasing gradually downward. Oxide formation, replacement of hydrogen from solutions containing hydrogen ions, and mutual replacement of one metal by another —all processes follow the same general trend in solution as well as in anhydrous conditions. The reaction employed in the thermit process:

$$Fe_2O_3 + 2Al = Al_2O_3 + 2Fe$$

may be interpreted in terms of higher electropositive character of aluminium. The preferential oxidation of iron during roasting of the copper ore copper pyrites ($CuFeS_2$) is similarly related to the higher electropositive character of iron. The principle of cathodic protection of iron against corrosion involves a metal more electropositive than iron used as a sacrificial anode.

9.5.2 Redox Stability in Water

Water may act both as an oxidizing agent and a reducing agent. In the first case it is reduced to H_2; in the second process it gets oxidized to O_2. Potentials of these two processes set a limit to the thermodynamic stability of any redox species in water.

Oxidation by water

The standard reduction potential for the hydrogen electrode is 0 V by definition. This refers to unit activity of hydrogen ion. Hence in pure water, where $[H^+] = 10^{-7}$ mol dm^{-3}, for the process $H^+(aq) + e \rightleftharpoons \frac{1}{2}H_2(g)$, at partial pressure of H_2 = 1 bar,

$$E = 0 - 0.059 \log (1/10^{-7}) \quad \text{at 298 K} = -0.414 \text{ V}.$$

Similarly in an alkaline solution having $[OH^-] = 1$ M, $[H^+] = 10^{-14}$ M, E comes out to be -0.828 V. At any pH, $E = -0.059\ pH$.

Any redox couple with $E°$ values less than these values at specified pH may be expected to get oxidized by H^+ ion with the evolution of H_2. This is what we observe in the reaction of many metals with water or acids:

$$M(s) + H^+(aq) \longrightarrow M^+(aq) + \tfrac{1}{2}H_2(g)$$

However, kinetic factors and associated overvoltage interfere and a clear margin of 0.5 – 0.6 V is often found necessary for evolution of hydrogen. Metals like zinc and iron have $E°$ values more negative then -0.414 V, but they do not liberate hydrogen from water at pH = 7; the potential difference is not sufficient to override the overvoltage for hydrogen evolution. In a more acidic medium, E_{H^+, H_2} becomes more positive, making the difference $E_{H^+, H_2} - E_{M^{2+}, M}$ large enough to exceed the overpotential when reduction of H^+ takes place.

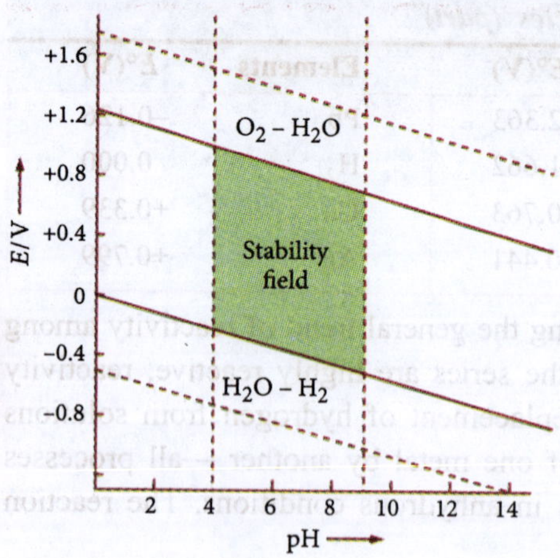

Fig.9.3: Reduction potential for $H_2O - H_2$ and $O_2 - H_2O$ couples at various pH. Redox couples with $E°$ above the upper line have the potentiality to oxidize water; those below the lower line should reduce water. Allowance for overpotential is shown by the broken lines. The region between the vertical lines at pH = 4 and pH = 9 (the normal range for natural water) is the stability field or *window of stability* for natural water.

Reduction by water

The half-cell emf for oxygen evolution from aqueous solutions is obtained from the oxygen-water half-reaction

$$\tfrac{1}{2}O_2 + 2H^+ + 2e \rightleftharpoons H_2O \qquad E° = 1.23\ \text{V}.$$

$$E = 1.23 - \frac{0.059}{2}\log\frac{1}{p_{O_2}^{1/2}[H^+]^2} \quad \text{at 298 K.}$$

$$= 1.23 - 0.059\ pH \qquad \text{at } p_{O_2} = 1\ \text{bar.}$$

At pH = 7, $E = 0.82$ V; at pH = 14, $E = 0.40$ V.

An oxidizing agent whose reduction potential lies above the reduction potential for the O_2/H_2O couple at a given pH would oxidize water and evolve oxygen. Here also, the effective stability region extends to about 0.5 V above the theoretical E values for overvoltage effects.

The strong positive value of $E° = +1.23$ V at 1M acid medium shows that acidified water is a poor reducing agent. Only a few strong oxidizing agents like $Co^{3+}(aq)$ [$E°(Co^{3+}, Co^{2+})$

= 1.92 V], $F_2(g)$ [$E°(F_2, F^-(aq)$ = 2.87 V estimated], Ag^{2+} [$E°$ (Ag^{2+} – Ag) = 1.98 V in 4N $HClO_4$] are reduced by water with the evolution of O_2:

$$4Co^{3+}(aq) + 2H_2O \rightleftharpoons 4Co^{2+}(aq) + O_2(g) + 4H^+(aq) \qquad [E° = +0.69 \text{ V}]$$

This $E°$ is above the potential necessary to counteract the notional overpotential needed for evolution of O_2. Several redox couples used in titration in aqueous solution have standard potentials greater than 1.23 V, e.g., $Cr_2O_7^{2-}$ / Cr^{3+} (1.38 V) or MnO_4^- / Mn^{2+} (+1.51 V). But they are stable in aqueous solution owing to the overpotential factor. The high overpotential is attributed to the activation barrier to the transfer of four electrons for the formation of an O_2 molecule and the establishment of an oxygen-oxygen double bond. Good catalysts for oxygen evolution are still a challenge to the inorganic chemist. Some coordination compounds of ruthenium have shown promising results. Nature has its own peculiar way of overcoming this difficulty with some metal-containing enzymes in the plant photosynthetic centre.

Problem 9.22

Refer to the following standard redox potentials: $MO_2^{2+} \xrightarrow{0.97 \text{ V}} MO_2^+ \xrightarrow{1.08 \text{ V}} M^{4+}$.

Predict whether or not an equimolar mixture of MO_2^{2+} and MO_2^+ is likely to oxidize H_2O to O_2 at p_{O_2} = 0.2 atm and (i) pH = 2.0. (ii) pH = 7.0.

Hint:

$$2MO_2^{2+} + 2e \rightleftharpoons 2MO_2^+ \qquad E° = 0.97 \text{ V}$$

$$H_2O \rightleftharpoons \tfrac{1}{2}O_2(g) + 2H^+(aq) + 2e \qquad E° = 1.23 \text{ V}$$

$$2MO_2^{2+} + H_2O \rightleftharpoons 2MO_2^+ + \tfrac{1}{2}O_2(g) + 2H^+(aq) \qquad E° = 0.97 - 1.23 = -0.26 \text{ V}$$

$$\text{At } pH = 2 \text{ (298 K)}, \; E = E° - \frac{0.059}{2} \log \frac{[MO_2^+]^2}{[MO_2^{2+}]^2} \cdot p_{O_2}^{1/2} \cdot [H^+]^2$$

$$= -0.26 - \frac{0.059}{2} \log (0.2)^{1/2} [H^+]^2 \qquad \left([MO_2^+] = [MO_2^{2+}] \text{ for equimolar mixture}\right)$$

$$= -0.26 + 0.059\, pH - \frac{0.059}{4} \log 0.2 \;= -0.13 \text{ V} \quad \text{(oxidation of water not favourable)}.$$

Similarly at pH = 7, E = +0.16 V (favourable).

9.5.3 Redox Potential Diagrams

For ready reference, it is often convenient to arrange the standard reduction potentials of different species in a diagram. We shall discuss some of these diagrams and their significance.

I. *Redox predominance diagram*

A redox predominance diagram shows the thermodynamically most stable form of an element corresponding to a given potential, conventionally $E°$(reduction). Hence the potentials refer to $a_{H^+} = 1$ or $pH = 0$. The $E°$ values are placed upward in their increasing order marking the boundary of the species which are at equilibrium at such potential. Above this boundary line (of $E°$), the predominant species is the one lying above the line; below the line, the species appearing in the lower region is the predominant one. The diagram for Fe is shown in Fig. 9.4 with corresponding half-cell reactions. Oxidation states of the element above and below each demarcation line are shown on the right. The high standard reduction potential of +2.20 V suggests that the species FeO_4^{2-} at the top of the diagram is strongly oxidizing in nature. We also observe that $E°$ values gradually decrease as the oxidation states decrease. However, construction of a redox predominance diagram may present difficulties when successive $E°$ values do not follow a regular trend. The $E°$ values for Cu may be taken as an example:

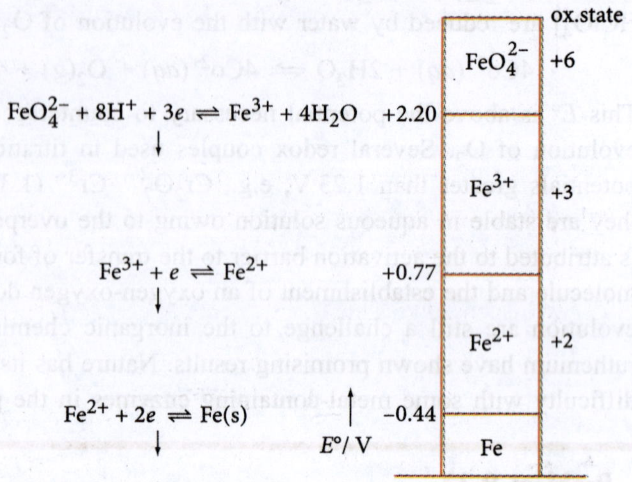

Fig.9.4: Redox predominance diagram for Fe.

$$Cu^{2+} + e \rightleftharpoons Cu^+ \qquad E_o = 0.15\ V$$
$$Cu^+ + e \rightleftharpoons Cu \qquad = 0.52\ V$$
$$Cu^{2+} + 2e \rightleftharpoons Cu \qquad = 0.34\ V$$

$E°$ for Cu^+ – Cu system is the highest among the three $E°$ values and it is not possible to construct a predominance diagram. The first $E°$ value suggests that the predominance region of Cu^+ should lie *below* 0.15 V while the second $E°$ value indicates that Cu^+ would be stable *above* 0.52 V. There is thus a gap of 0.36 V between the two "*supposed predominance*" regions of Cu^+. This gap suggests that Cu^+ is thermodynamically unstable and will undergo disproportionation:

$$Cu^+ + Cu^+ \longrightarrow Cu^{2+} + Cu.$$

E_o for this process is (0.52 – 0.15) V or 0.37 V. Similar situations are encountered with other species susceptible to disproportionation, for example, Ag^{2+}, MnO_4^{2-}, Mn^{3+} etc.

II. *Latimer diagrams*

A Latimer diagram (W. A. Latimer, 1952) summarizes the standard reduction potentials ($E°$) connecting various oxidation states of an element at a specified pH (= 0 or 14). The $E°$ values related to Fe and its aqua ions in 1M acid medium may be taken as an example:

(i) $FeO_4^{2-} + 8H^+ + 3e \rightleftharpoons Fe^{3+} + 4H_2O \qquad E° = +2.20\ V$

(ii) $Fe^{3+} + e \rightleftharpoons Fe^{2+} \qquad E° = +0.77\ V$

(iii) $Fe^{2+} + e \rightleftharpoons Fe \qquad E° = -0.44\ V$

The Latimer diagram of iron summarizes these informations without the half-reactions; the

$E°$ values(V) are written above the line joining the species. Oxidation numbers are sometimes written under (or over) the species. The Latimer diagram for Fe would be

$$\underset{+6}{FeO_4^{2-}} \xrightarrow{2.20} \underset{+3}{Fe^{3+}} \xrightarrow{0.77} \underset{+2}{Fe^{2+}} \xrightarrow{-0.44} \underset{0}{Fe}$$

$$Fe^{3+} \xrightarrow{-0.04} Fe$$

From the diagram we can write the half-reactions (i) – (iii) for adjacent species as required. We can also calculate the $E°$ value for the non-adjacent species Fe^{3+} – Fe from the given $E°$ values via $\Delta G°$ (see box).

For two successive steps 1 and 2, the overall $\Delta G^\circ_{\text{Total}}$ is the sum of individual $\Delta G°$ terms:

$$\Delta G^\circ_{\text{Total}} = \Delta G^\circ_1 + \Delta G^\circ_2$$

If steps 1 and 2 involve n_1 and n_2 faradays of electricity respectively, we have

$$\Delta G^\circ_1 = -n_1 F E^\circ_1;\quad \Delta G^\circ_2 = -n_2 F E^\circ_2;\quad \Delta G^\circ_{\text{Total}} = -F\left(n_1 E^\circ_1 + n_2 E^\circ_2\right)$$

If the potential for the combined steps 1 and 2 is written E°_{1+2}, then

$$\Delta G^\circ_{\text{Total}} = -(n_1 + n_2)\, F E^\circ_{1+2}.$$

Summarily we get $E^\circ_{1+2} = \dfrac{n_1 E^\circ_1 + n_2 E^\circ_2}{n_1 + n_2}$

For the Fe^{3+} – Fe couple, $E° = \dfrac{1 \times E^\circ_{Fe^{3+}-Fe^{2+}} + 2E^\circ_{Fe^{2+}-Fe}}{3}$

$$= \frac{1 \times (+0.77) + 2 \times (-0.44)}{3}\ \text{V} = -0.04\ \text{V}$$

The information available from the Latimer diagram of iron may now be summarized. The *reduction potential* of the couple ferrate-ferric is +2.20 V, which is appreciably high and the species FeO_4^{2-} containing Fe(VI) will be strongly oxidizing in nature.

Again, the standard reduction potential of the ferric-ferrous couple is appreciably lower, which indicates that Fe(II) will be more readily oxidized to Fe(III). So, Fe(II) will be a good reducing agent. At the same time, this reduction potential still lies above that of hydrogen; consequently ferrous ions cannot reduce hydrogen ions from aqueous solutions.

Lastly, the potential value for the Fe^{2+} – Fe couple suggests that Fe(II) is a fairly stable species in aqueous solution (in absence of oxygen or other oxidizing agents) and has no tendency to get reduced to Fe.

From this as well as other sets of electrode potential data, a few *broad generalizations* may now be made.

In aqueous solutions, systems with negative values of E°_{red} are more strongly *reducing* than hydrogen, (e.g., Fe) while systems with positive values of E°_{red} are more strongly *oxidizing* than hydrogen (e.g., Fe^{3+} in the Fe^{3+} – Fe^{2+} couple). Only the latter are therefore reduced by hydrogen (e.g., $FeCl_3$). Systems with E°_{red} values above + 0.8 volts are generally accepted as *oxidizing*; those with E°_{red} values lying around 0.8 volts (e.g., the Fe^{3+}–Fe^{2+} system: +0.77

volts) are equally oxidizing and reducing; systems with E°_{red} values below + 0.8 volts become gradually *reducing* in nature (e.g., $Sn^{4+} - Sn^{2+}$: + 0.15 V).

Similar Latimer diagrams may be drawn for basic medium (pOH = 0 or pH = 14). In this case the half-reactions have to be balanced by OH^-.

The Latimer diagram of phosphorus at pH = 14 is

$$\overset{+5}{PO_4^{3-}} \xrightarrow{-1.12} \overset{+3}{HPO_3^{2-}} \xrightarrow{-1.57} \overset{+1}{H_2PO_2^-} \xrightarrow{-2.05} \overset{0}{P} \xrightarrow{-0.89} \overset{-3}{PH_3}$$

$$HPO_3^{2-} \xrightarrow{-1.73} P$$

The half-reaction for the $PO_4^{3-} - HPO_3^{2-}$ couple should be written as

$$PO_4^{3-} + 2H_2O + 2e \rightleftharpoons HPO_3^{2-} + 3OH^- \qquad E^\circ = -1.12\ V$$

The negative value of E° suggests that HPO_3^{2-} should be a good reducing agent.

Problem 9.23

The Latimer diagram of chlorine in basic medium (pH = 14) is

$$ClO_4^- \xrightarrow{0.37} ClO_3^- \xrightarrow{0.30} ClO_2^- \xrightarrow{0.68} ClO^- \xrightarrow{0.42} Cl_2 \xrightarrow{1.36} Cl^-$$

$$ClO^- \xrightarrow{?} Cl^-$$

Calculate E° for the couple $ClO^- - Cl^-$.

Hint: The balanced equation in basic medium is

$$ClO^- + H_2O + 2e \rightleftharpoons Cl^- + 2OH^- \qquad \Delta G^\circ = -2FE^\circ$$

Calculate via ΔG° for other adjacent couples. **Ans.** +0.89 V.

Disproportionation

From the Latimer diagram it is possible to examine the thermodynamic feasibility of a species to undergo disproportionation. If any species shows a higher E° value on its right side, it is expected to disproportionate into its two neighbours. Let us consider the three species on the right end of the Latimer diagram of chlorine:

$$ClO^- \xrightarrow{0.42} Cl_2 \xrightarrow{1.36} Cl^-$$

Cl_2 shows a sharp rise in E° on its right ($Cl_2 - Cl^-$ couple). We anticipate disproportionation of Cl_2 into ClO^- and Cl^-. We represent the overall reaction of disproportionation in terms of two half-reactions

$$\frac{1}{2}Cl_2 + e \rightleftharpoons Cl^- \ \text{(reduction)} \qquad E^\circ = 1.36\ V$$

$$\frac{1}{2}Cl_2 + 2OH^- \rightleftharpoons ClO^- + H_2O + e \ \text{(oxidation)} \qquad E^\circ_{red} = 0.42\ V$$

$$Cl_2 + 2OH^- \rightleftharpoons Cl^- + ClO^- + H_2O \qquad E^\circ = 1.36 - 0.42 = +0.94\ V$$

The high positive emf indicates that the process will be highly favoured to right (ΔG° negative, $K > 1$).

The Manganese system

Oxidation states: +7, +6, +4, +3, +2, 0

$$MnO_4^- \xrightarrow{0.89} MnO_4^{2-} \xrightarrow{2.09} MnO_2 \xrightarrow{0.95} Mn^{3+} \xrightarrow{1.51} Mn^{2+} \xrightarrow{-1.18} Mn$$

($MnO_4^- \to MnO_2$: 1.69; $MnO_2 \to Mn^{2+}$: 1.23; $MnO_4^- \to Mn^{2+}$: 1.51)

The diagram shows that (i) Mn^o will be readily oxidized to Mn^{2+} ; (ii) Mn^{3+} will be readily reduced to Mn^{2+} ; (iii) MnO_4^{2-} will be readily reduced to MnO_2 ; (iv) MnO_2 will be readily reduced to Mn^{2+} ; (v) MnO_4^- will be reduced rather to MnO_2 or Mn^{2+} than to MnO_4^{2-}.

In other words, Mn^{3+}, MnO_2, MnO_4^{2-} and MnO_4^-—all will behave as strong oxidizing agents ($E^o > +0.8$ V).

Another interesting feature of the diagram is the sharp increase in E^o in the two triad systems, suggesting disproportionation of the centre species.

$$MnO_2 \overset{0.95}{\text{——}} Mn^{3+} \overset{1.51}{\text{——}} Mn^{2+}$$

and $$MnO_4^- \overset{0.89}{\text{——}} MnO_4^{2-} \overset{2.09}{\text{——}} MnO_2$$

Case I $Mn^{3+} + 2H_2O \rightleftharpoons MnO_2 + 4H^+ + e$ (ox) $E^o_{red} = +0.95$ V

$Mn^{3+} + e \rightleftharpoons Mn^{2+}$ (red) $E^o_{red} = +1.51$ V.

$2Mn^{3+} + 2H_2O \rightleftharpoons Mn^{2+} + MnO_2 + 4H^+$ $E^o = 1.51 - 0.95 = 0.56$ V.

Case II $2MnO_4^{2-} - 2e \rightleftharpoons 2MnO_4^-$ (ox) $E^o_{red} = +0.89$ V

$MnO_4^{2-} + 4H^+ + 2e \rightleftharpoons MnO_2 + 2H_2O$ (red) $E^o_{red} = +2.09$ V

$3MnO_4^{2-} + 4H^+ \rightleftharpoons MnO_2 + 2MnO_4^- + 2H_2O$ $E^o = 2.09 - 0.89 = 1.20$ V

From $\Delta G^o = -FE^o = -RT\ln K$, the equilibrium constant for the disproportionation of Mn(III) (Case I) at 298 K is found to be $10^{9.5}$ or 3.2×10^9.

This shows that the disproportionation will be almost complete. Similar calculations may be done for other cases too.

The Copper system

Oxidation states: +2 +1 0

Species: $Cu^{2+} \overset{0.15}{\text{——}} Cu^+ \overset{0.50}{\text{——}} Cu$ ($Cu^{2+} \to Cu$: 0.34)

It is found that Cu^{2+} is the most stable species in aqueous solution. On passing from Cu^{2+} to Cu^+ to Cu, the potential increases steadily, indicating possible disproportionation.

$Cu^+ - e \rightleftharpoons Cu^{2+}$ $E^o_{red} = +0.15$ V

$Cu^+ + e \rightleftharpoons Cu$ $E^o_{red} = +0.50$ V

$2Cu^+ \rightleftharpoons Cu^{2+} + Cu$ $E^o = 0.50 - 0.15 = 0.35$ V

It is suggested that Cu^+ should disproportionate in aqueous solution.

Oxidation by atmospheric oxygen

Oxidation by atmospheric oxygen results in its reduction to oxide ion or to water in presence of moisture or water. Hence the potential for the $O_2 - H_2O$ couple becomes important in this case. The Latimer diagram for oxygen in acidic solution is:

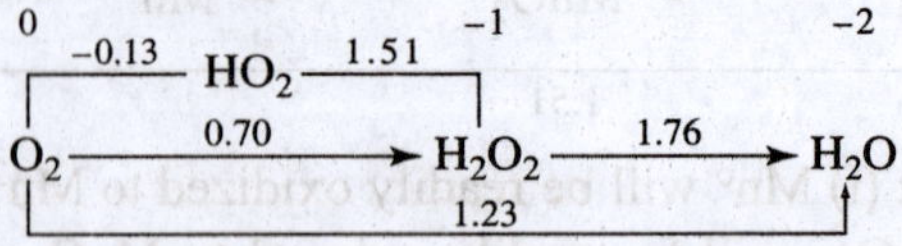

Although any reducing agent with a reduction potential lower than that for $O_2 - H_2O$ half-reaction (1.23 V at pH = 0) can theoretically be oxidized by oxygen, intermediate formation of H_2O_2 may stand in the way.

Reducing agents with reduction potentials more positive than 0.70 V produce H_2O_2 (and probably others) as intermediate which has higher energy than the reactants. This results in a high activation energy for such reactions and makes them slow. This is probably the reason why bromides ($E°$ for Br_2-Br^- = 1.05 V) are oxidized extremely slowly by air while iodides ($E°$ = 0.54 V) are rapidly oxidized.

Oxidation of Fe(II) in aqueous solution by atmospheric oxygen according to the reaction

$$4Fe^{2+}(aq) + O_2(g) + 4H^+(aq) \rightleftharpoons 4Fe^{3+}(aq) + 2H_2O(l)$$

has $E°$ = 1.23 − 0.77 = 0.46 V and is thermodynamically feasible. But this value is marginally close to the notional overpotential of 0.5 V and the process is not rapid at ordinary temperature. Hence Fe^{2+} is only slowly oxidized in aqueous solution in absence of catalyst and can be used in titrations.

Atmospheric oxidation of metallic copper according to the reaction

$$2Cu + O_2(g) + 4H^+(aq) \rightleftharpoons 2Cu^{2+}(aq) + 2H_2O$$

has $E°$ = 1.23 − 0.34 = 0.89 V and is thermodynamically favourable. However, oxidation in presence of atmospheric CO_2 and SO_2 forms an adherent protective coating of hydrated copper carbonate and sulfate on the surface and stops further oxidation.

Problem–9.24

From the Latimer diagram of oxygen given above, try to guess if H_2O_2 is likely to disproportionate into H_2O and O_2 under standard conditions.

Answer: For the overall reaction (all aquated species) $2H_2O_2 \longrightarrow 2H_2O + O_2(g)$ we add the half-reactions from the given Latimer diagram and combine the $E°_{red}$ values:

reduction:	$H_2O_2 + 2H^+ + 2e \rightleftharpoons 2H_2O$	$E° = 1.76$ V
oxidation:	$H_2O_2 \rightleftharpoons O_2 + 2H^+ + 2e$	$E° = 0.70$ V
	$2H_2O_2 \rightleftharpoons 2H_2O + O_2(g)$	$E° = 1.76 - 0.70 = 1.06$ V

The high positive value of $E°$ suggests disproportionation ($K >> 1$).

Prediction of reaction products

In addition to the cases described earlier, we consider here a reaction which yields different products in different reaction conditions. The reaction between permanganate and iodide in acid solution may be taken as an example. From the reduction potential diagrams for manganese and iodine we first eliminate the species susceptible to disproportionation, because they cannot lead to thermodynamically controlled products. If a species in a given oxidation state is a stronger oxidizing agent than it is in the next higher state, disproportionation is expected.

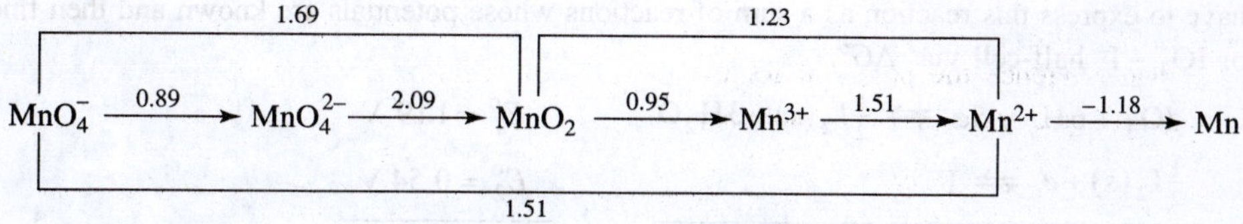

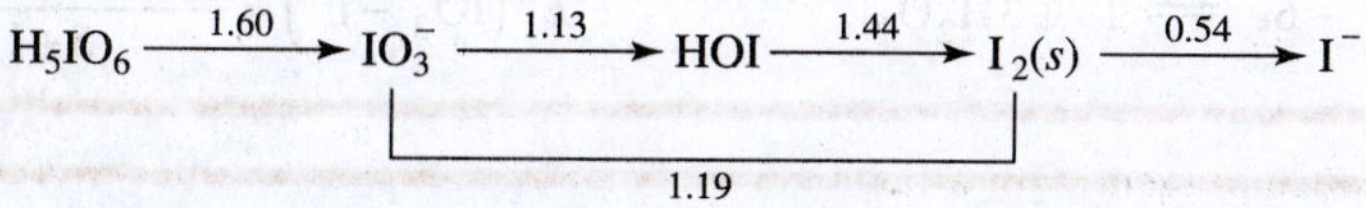

The reduction potentials on the right of MnO_4^{2-}, Mn^{3+} and HOI are higher than those on the left of each; so these are excluded to obtain the simplified diagram:

$$H_5IO_6 \xrightarrow{1.60} IO_3^- \xrightarrow{1.19} I_2 \xrightarrow{0.54} I^-$$

$$MnO_4^- \xrightarrow{1.69} MnO_2 \xrightarrow{1.23} Mn^{2+}$$

Two reaction conditions may now be adjusted in the following manner:

(i) **the permanganate solution is added dropwise to a solution containing iodide.** Since iodide ion is in excess, iodate or manganese dioxide cannot be the end product in this condition; they would be reduced by the excess iodide.

The final product should be Mn^{2+} and $I_2(I_3^-)$.

$$15I^- + 2MnO_4^- + 16H^+ \rightleftharpoons 5I_3^- + 2Mn^{2+} + 8H_2O.$$

(ii) **If an iodide solution is added dropwise to an acidic solution of permanganate,** the reaction products should correspond to the presence of excess MnO_4^-. Any Mn^{2+} formed would react with MnO_4^- to form MnO_2. Similarly, triodide formed will also be oxidized by MnO_4^-. Again, the reduction potentials of $MnO_4^- - MnO_2$ and $H_5IO_6 - IO_3^-$ couples are very close in magnitude; hence both iodate and periodic acid are expected products.

$$I^- + 2MnO_4^- + 2H^+ \longrightarrow IO_3^- + 2MnO_2 + H_2O$$

$$3I^- + 8MnO_4^- + 11H^+ + 2H_2O \longrightarrow 3H_5IO_6 + 8MnO_2$$

As already mentioned, such predictions are valid only when the equilibrium in the reaction involved is attained quickly. A slow reaction rate may not always lead to the predicted reaction product. Thus, iodine should oxidize thiosulphate to tetrathionate and also to sulphate. However, the oxidation to sulphate is extremely slow; oxidation of thiosulphate to tetrathionate may be taken as quantitative, forming the basis of titrimetric analysis.

Problem 9.25

Calculate $E°$ for the couple $IO_3^- - I^-$ from the Latimer diagram (acid medium; $E°$ in volt):

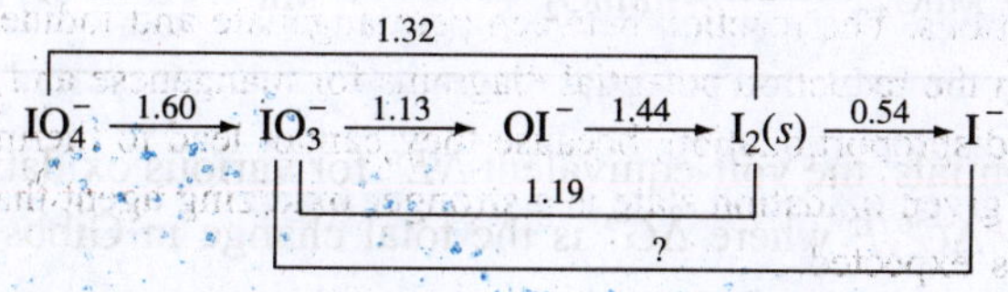

Solution: The involved reaction is $IO_3^- + 6H^+ + 6e \rightleftharpoons I^- + 3H_2O$.

We have to express this reaction as a sum of reactions whose potentials are known and then find the $E°$ for $IO_3^- - I^-$ half-cell via $\Delta G°$.

$$IO_3^- + 6H^+ + 5e \rightleftharpoons \tfrac{1}{2}I_2(s) + 3H_2O \qquad E_1^\circ = 1.19\ \text{V}$$

$$\tfrac{1}{2}I_2(s) + e \rightleftharpoons I^- \qquad E_2^\circ = 0.54\ \text{V}$$

$$IO_3^- + 6H^+ + 6e \rightleftharpoons I^- + 3H_2O \qquad E°\left(IO_3^- - I^-\right) = \frac{5 \times 1.19 + 1 \times 0.54}{5+1} = 1.08\ \text{V}$$

Problem 9.26

Calculate $E°$ for $MO_2^+ \rightarrow M^{4+}$ in 1 M acid medium from the following diagram.

$$MO_2^{2+} \xrightarrow{0.97\ \text{V}} MO_2^+ \xrightarrow{?} M^{4+} \xrightarrow{1.01\ \text{V}} M^{3+}$$

$$MO_2^{2+} \xrightarrow{1.02\ \text{V}} M^{3+}$$

Hint: Reaction: $MO_2^+ + 4H^+ + e \rightleftharpoons M^{4+} + 2H_2O$; $E° = E_1^\circ$ (say). We find E_1° from the known $E°$ values(V) via $\Delta G°$.

$$MO_2^{2+} + e \rightleftharpoons MO_2^+ \qquad E° = 0.97$$

$$MO_2^+ + 4H^+ + e \rightleftharpoons M^{4+} + 2H_2O \qquad E° = E_1^\circ$$

$$M^{4+} + e \rightleftharpoons M^{3+} \qquad E° = 1.01$$

$$MO_2^{2+} + 4H^+ + 3e \rightleftharpoons M^{3+} + 2H_2O \qquad E° = 1.02$$

From $-3F(1.02) = -F(0.97) - F\left(E_1^\circ\right) - F(1.01)$, $E_1^\circ = 1.08$ V.

III. *Frost diagrams (Ebsworth diagrams)*

A Frost diagram is a graphical plot of $NE°$ versus the oxidation number N of an element in the couple,

$$X\ (\text{O.N.} = N) + Ne \rightleftharpoons X\ (\text{O.N.} = 0) \qquad E°\ (\text{V}).$$

The actual species X containing the element in oxidation number N is the common form of the element in aqueous solution, for example Mn^{2+} (O.N. +2), MnO_4^- (O.N. +7), NH_4^+ (O.N. –3) etc.

The Frost diagram may be constructed from the $E°$ values of an element, like those available in the Latimer diagram. The Frost diagram for Mn may be taken as an illustration. We recall the Latimer diagram of manganese in acid medium:

$$MnO_4^- \xrightarrow{0.89} MnO_4^{2-} \xrightarrow{2.09} MnO_2 \xrightarrow{0.95} Mn^{3+} \xrightarrow{1.51} Mn^{2+} \xrightarrow{-1.18} Mn$$

$$MnO_4^- \xrightarrow{1.69} MnO_2 \qquad MnO_2 \xrightarrow{1.23} Mn^{2+} \qquad MnO_4^- \xrightarrow{1.51} Mn^{2+}$$

From the given potentials, the volt-equivalent $NE°$ for various oxidation states N is obtained by the relation $nE° = -\Delta G°/F$ where $\Delta G°$ is the total change in Gibbs energy for the change

X(O.N. = N) $\longrightarrow$ X(O.N. = 0) that requires n Faradays of electricity and $E°$ is the standard reduction potential for the couple. For positive oxidation states, $N = n$. For negative oxidation states, n is still positive since it is the number of Faradays ($n = |N|$). Thus the Frost diagram may also be regarded as a plot of standard reaction Gibbs energy change against oxidation number (actually $-\Delta G°/F$). For Mn, values of $NE°$ (V) are calculated below.

$N = +2$: $E°\ (Mn^{2+}, Mn) = -1.18$; $2E° = -2.36$

$N = +3$: $E°\ (Mn^{3+}, Mn)$ is not given, so we calculate it from $E°\ (Mn^{3+}, Mn^{2+})$ and $E°\ (Mn^{2+}, Mn)$.

$$E°\ (Mn^{3+}, Mn) = [1 \times 1.51 + 2 \times (-1.18)] \div 3$$

or, $3E°\ (Mn^{3+}, Mn) = -0.85$

$N = +4$: $E°\ (MnO_2, Mn^{3+}) = 0.95$

$$3E°\ (Mn^{3+}, Mn) = -0.85$$

$$4E°\ (MnO_2, Mn) = 0.10$$

$N = +6$: $E°\left(MnO_4^{2-}, MnO_2\right) = 2.09$

$$4E°\ (MnO_2, Mn) = 0.10$$

$$6E°\left(MnO_4^{2-}, Mn\right) = 2 \times 2.09 + 0.10 = 4.28$$

$N = +7$: $7E°\left(MnO_4^{-}, Mn\right) = 0.89 + 4.28 = 5.17.$

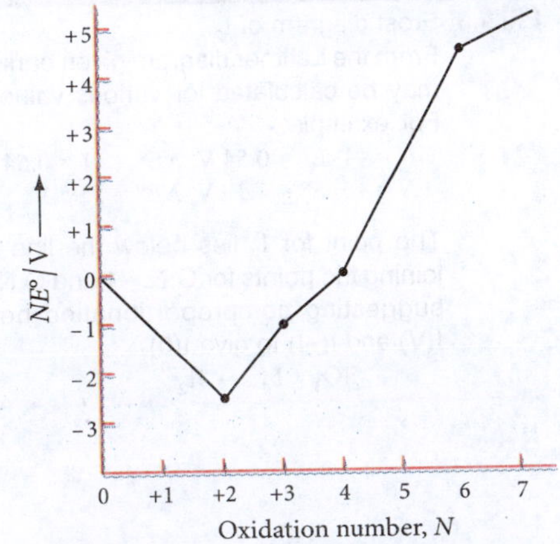

Fig.9.5: Frost diagram for Manganese (also called Ebsworth diagram).

Certain general features of the Frost diagram of an element are noteworthy.

(i) The slope of a line joining two points is equal to $E°$ for the relevant couple.

$$\text{slope} = \frac{\text{change in } NE° \text{ between two points}}{\text{corresponding change in oxidation number}}$$

The steeper the line, the higher the potential of the corresponding couple.

(ii) The oxidized form of a couple has the more positive $E°$ and hence a more positive slope; it is expected to be an oxidizing agent (e.g., MnO_4^-, MnO_2).

(iii) A species is likely to be unstable with respect to *disproportionation* when the point representing this species is *above* the line joining points immediately before and after.

Thus, (*a*) Mn^{3+} is likely to disproportionate to Mn^{2+} and MnO_2. Since both these products involve one electron transfer, the process is likely to be rapid.

(*b*) The point for Mn(VI) similarly suggests disproportionation to Mn(VII) and Mn(IV).

(iv) The most stable oxidation state of an element in aqueous solution is represented by the species lying lowest in its diagram (e.g., Mn^{2+}).

(v) If a species appears *below* the line connecting the species on its two sides, the species on the sides are likely to undergo *comproportionation* to produce the intermediate species. Examples to this effect may be found in the Frost diagram of iodine (Fig. 9.6).

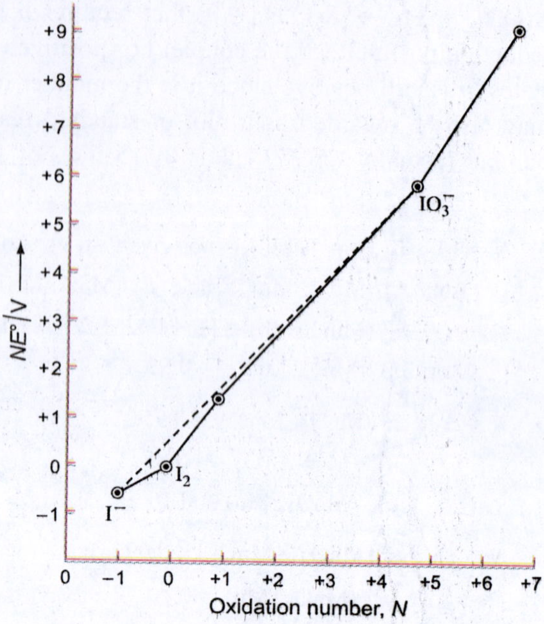

Fig.9.6: Frost diagram of I_2.
From the Latimer diagram given earlier, $NE°$ may be calculated for various values of n. For example,

$N = -1$, $E° = 0.54$ V, $NE° = -1 \times 0.54$ V
$N = +5$, $E° = 1.19$ V, $NE° = 5.95$ V.

The point for I_2 lies below the line (– – –) joining the points for O.N. –1 and O.N. = +5, suggesting comproportionation between I(V) and I(–I) to give I(0).

$$IO_3^- + 5I^- \rightarrow 3I_2$$

Problem 9.27

The Latimer diagram of silver in acid medium is

$$Ag^{2+} \xrightarrow{1.99\ V} Ag^+ \xrightarrow{0.80\ V} Ag$$

The Frost diagram (Fig. 9.7) suggests comproportionation between Ag(0) and Ag^{2+} to give Ag^+. Calculate K for this reaction.

Solution:

$Ag \rightarrow Ag^+$ $\quad E°(\text{red}) = 0.80$ V

$Ag^{2+} \rightarrow Ag^+$ $\quad E° = 1.99$ V

$Ag + Ag^{2+} \rightarrow 2Ag^+$ $\quad E° = 1.99 - 0.80 = 1.19$ V

$$\log K = \frac{1.19\ F}{2.303\ RT} = 20.1 \text{ at 298 K.}$$

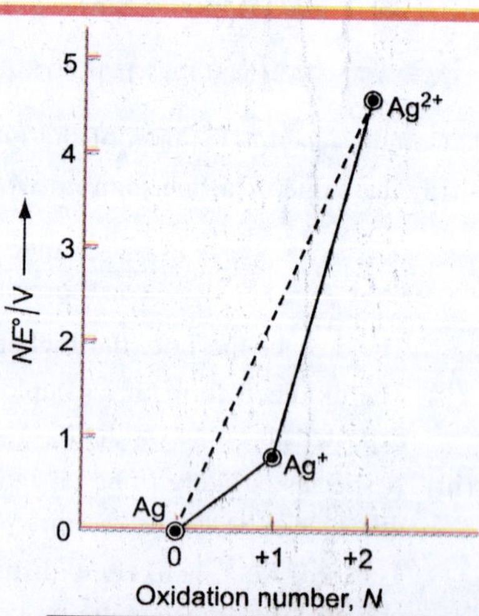

Fig.9.7: Frost diagram of silver.

The Frost diagram for nitrogen may be constructed from the Latimer diagrams. $E°$ values for acidic and basic solutions give rise to two different plots (Fig. 9.8). Both diagrams cover a wide range of oxidation number.

***Acidic solution* (pH = 0) ($E°$ in volts).**

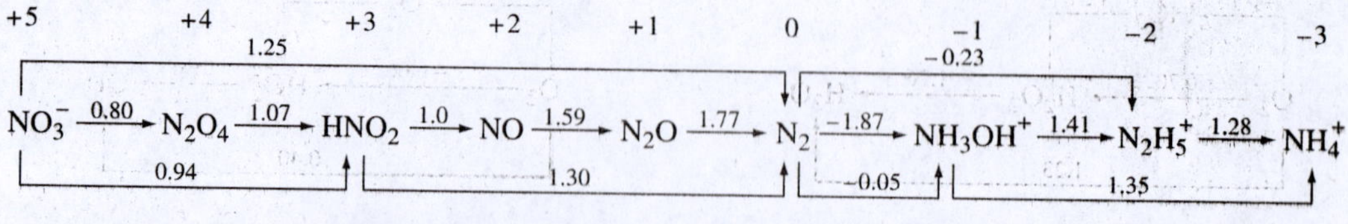

***Basic solution* (pH = 14) (*E*° in volts).**

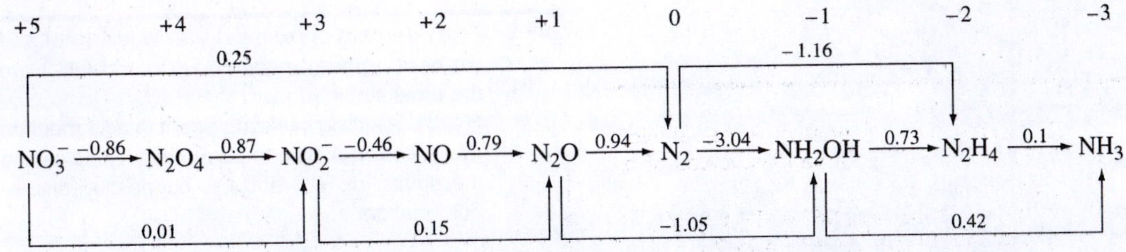

The general features of the diagram are self-explanatory. Steep slope and high E°_{red} of HNO_3, N_2O_4 or HNO_2 indicate their strong oxidizing nature. NH_4^+ is the most stable form in acid medium, followed by N_2. Again the position of N_2 is below the line joining the points for NH_4^+ and HNO_2 (or HNO_3). Comproportionation is suggested.

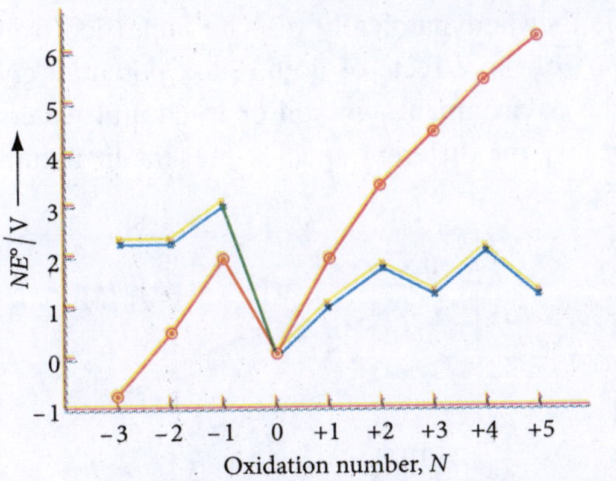

Fig.9.8: Frost diagram for N_2 in acidic (red line) and basic (green line) media.
The points in the graph are obtained from the values given in the Latimer diagram (see box).

For example, in acid medium when O.N. = −3,

$$N_2 \rightarrow N_2H_5^+ \qquad E^\circ = -0.23\text{ V}$$
$$N_2H_5^+ \rightarrow NH_4^+ \qquad E^\circ = 1.28\text{ V}$$

$$E^\circ(N_2, NH_4^+) = [2 \times (-0.23) + 1.28] \div 3 = 0.82/3\text{ V}$$

$$NE^\circ(N_2, NH_4^+) = (-3) \times \frac{0.82}{3} = -0.82\text{ V}$$

When O.N. = +2, the NE° is calculated as

$$NO \rightarrow N_2O \qquad E^\circ = 1.59\text{ V}$$
$$N_2O \rightarrow N_2 \qquad E^\circ = 1.77\text{ V}$$
$$\text{or,} \qquad NE^\circ = 3.36\text{ V}$$

The Frost diagram for oxygen (Fig. 9.9) is relatively simple since the oxidation numbers are limited. We begin with the Latimer diagrams:

***Acidic solution* (*E*° in volts)**

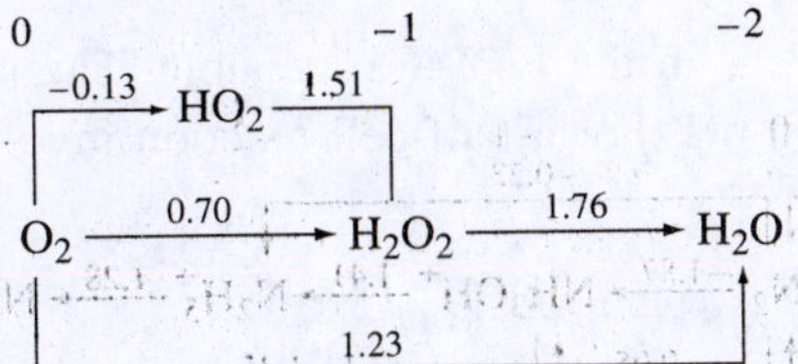

***Basic solution* (*E*° in volts)**

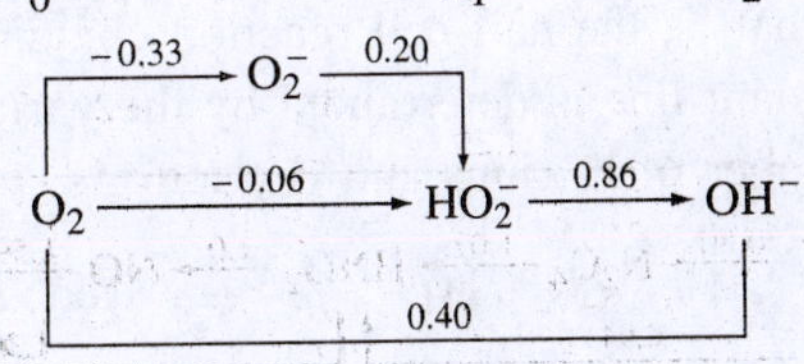

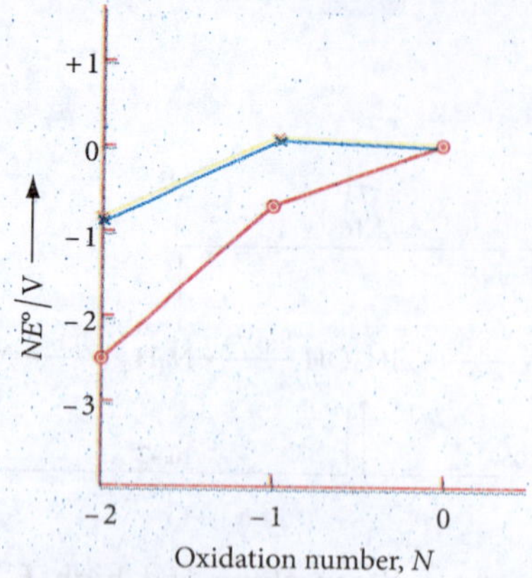

Fig.9.9: Frost diagram of oxygen in (a) acidic (red) and (b) basic (green) medium. H_2O (and OH^-) are the most stable form.
H_2O_2 is a strong oxidizing agent in acid medium (steeper slope). Its point is above the line joining the points for H_2O and O_2, suggesting disproportionation.

IV. *Pourbaix Diagrams*

We have already seen how the potentials of some couples change with pH (e.g., Fig. 9.3 for water). In the Pourbaix diagram (M. Pourbaix, 1938) of an element, we plot the potentials versus pH for various species of the element relevant to aqueous medium. Regions of pH and potential in which any species is stable thermodynamically may be identified from such plots. Thus we have a "visual representation" of the effects of both redox potential and pH on the predominant forms of an element in the environment—in soil or in natural waters. In fact, Pourbaix originally sketched these diagrams for different species present in natural water. A Pourbaix diagram, which is a merger of acid-base and redox predominance diagrams, is also known as $E°$-pH diagram or pE-pH diagram.

A sketch of the Pourbaix diagram for iron is shown is Fig. 9.10. As in the case of redox predominance diagram, potentials (E) are plotted upward from lower to higher values; the oxidation states of iron also increase upward. pH is plotted along the x-axis from 0 to 14. As already seen, the main species encountered in this pH range may be FeO_4^{2-}, Fe^{3+}, Fe^{2+}, $Fe(OH)_3$ (actually Fe_2O_3, xH_2O), $Fe(OH)_2$ and Fe.

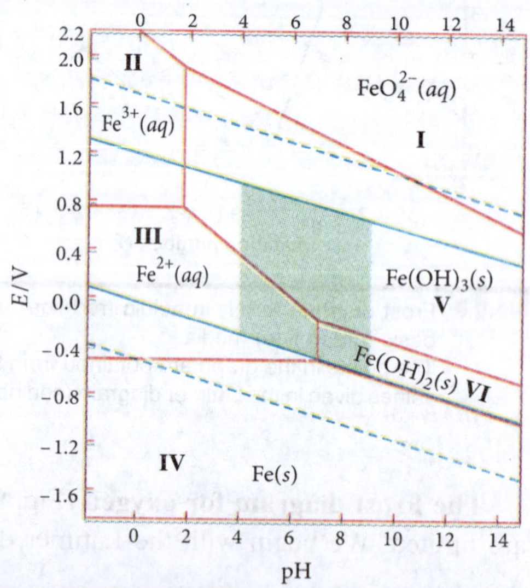

Fig.9.10: Pourbaix diagram of iron. The stability window of water has been superimposed.

The E values in the diagram refer to 1.0M concentration of all species except H^+ and OH^- ions at 25°C. The diagram consists of different regions (I, II etc.) demarcated by horizontal and sloping lines. The horizontal line implies that E for the concerned redox couple is independent of pH, that is, the half-cell reaction contains no H^+ ions, as in the Fe^{3+}–Fe^{2+} couple. The slope of a slant line is determined by the variation of E with pH. For a half-cell reaction involving x number of H^+ ions and n electrons, at 25°C,

$$\text{Ox} + xH^+ + ne \rightleftharpoons \text{Red};\ E = E° - \frac{0.059}{n}x\text{pH} - \frac{0.059}{n}\log\frac{[\text{Red}]}{[\text{Ox}]}$$

At unit activities of all other species the slope of the E–pH line can be readily understood. At all points on the line, the two species in the redox couple exist in equilibrium.

A vertical line in the diagram separates species which can coexist in equilibrium at unit activity at the indicated pH.

Now we shall have a closer look at the Pourbaix diagram of iron.

(i) The species FeO_4^{2-} with highest oxidation number (+6) gets reduced to Fe^{3+} in the pH-dependent half-reaction: $FeO_4^{2-} + 8H^+ + 3e \rightleftharpoons Fe^{3+} + 4H_2O$ ($E° = +2.20$ V).

$$E = 2.20 - \frac{0.059 \times 8}{3} pH - \frac{0.059}{3} \log \frac{[Fe^{3+}]}{[FeO_4^{2-}]} \quad \text{(at 25°C)}$$

$$= 2.20 - 0.16 \text{ pH} - 0.02 \log \frac{[Fe^{3+}]}{[FeO_4^{2-}]}.$$

Thus, E decreases with pH.

The lower boundary for FeO_4^{2-} will cut the vertical line of pH 0 at 2.20V (= $E°$). $Fe^{3+}(aq)$ will be the principal species below this line at low pH.

With increase in pH, $[Fe^{3+}]$ in solution decreases due to precipitation of the hydrated oxide. At very small concentration of Fe^{3+}, the potential decreases to much lower values than expected from the relation above. At $pH = 14$, the actual potential is near 0.7 V at $[FeO_4^{2-}] = 1$ M. The FeO_4^{2-} is stable in basic medium above the line of the E–pH plot (Region I).

The extended window of stability of water (Fig. 9.3) including overpotential has been superimposed in Fig. 9.10 (green shade). It appears that the upper boundary cuts the region of FeO_4^{2-} above pH 10. This implies that a solution containing 1 molar FeO_4^{2-} may have only marginal kinetic stability in strongly basic solutions. Eventually, the water will be oxidized:

$$4FeO_4^{2-} + 10H_2O \longrightarrow 4Fe(OH)_3 + 3O_2 + 8OH^-$$

(ii) The Fe^{3+}–Fe^{2+} couple does not involve any H^+ ions and the potential is independent of pH.

$$Fe^{3+} + e \rightleftharpoons Fe^{2+} \qquad E° = +0.77 \text{ V}$$

The E–pH plot is a horizontal straight line: above this line, Fe^{3+} is stable (Region II) and below it, Fe^{2+} is stable (Region III).

(iii) The Fe^{2+}–Fe couple is also independent of pH:

$$Fe^{2+} + 2e \rightleftharpoons Fe \qquad E° = -0.44 \text{ V}.$$

This corresponds to a horizontal line of demarcation at – 0.44 V. Above this line, Fe^{2+} is stable; below this line Fe(s) is present (Region IV).

(iv) Now we take care of the effect of pH on the various forms. The reaction

$$Fe^{3+} + 3H_2O \rightleftharpoons Fe(OH)_3(s) + 3H^+$$

is favoured to right with increasing pH. The vertical line at pH = 2 puts a boundary between Fe^{3+} and $Fe(OH)_3$. Below pH 2, Fe^{3+} is the principal species; above this, $Fe(OH)_3$ predominates (region V) (see Problem 9.28).

Problem 9.28

K_{sp} of $Fe(OH)_3 = 2.6 \times 10^{-39}$; K_{sp} of $Fe(OH)_2 = 4.9 \times 10^{-17}$. Find the pH at which $Fe(OH)_3$ and $Fe(OH)_2$ will begin to precipitate from a solution containing 1mmol L^{-1} of each Fe^{3+} and Fe^{2+}.

Solution: $[Fe^{3+}] = 10^{-3}$ mol L^{-1} ; $[Fe^{2+}] = 10^{-3}$ mol L^{-1}.

For precipitation of $Fe(OH)_3$ to begin,

$$[OH^-]^3 = \frac{K_{sp}(Fe(OH)_3)}{[Fe^{3+}]} = \frac{2.6 \times 10^{-39}}{10^{-3}} \text{ or } [OH^-] = 1.4 \times 10^{-12} \text{ or pH} = 2.1$$

For precipitation of $Fe(OH)_2$, similar calculation gives pH = 7.3.

(v) $Fe(OH)_3$ may be reduced to $Fe(OH)_2$ or Fe^{2+} according to reactions like

$$Fe(OH)_3(s) + 3H^+ + e \rightleftharpoons Fe^{2+} + 3H_2O$$

At 298 K, $E = E° - 0.059 \times 3$ pH $- 0.059 \log [Fe^{3+}]$.

Thus, the potential drops with increasing pH (boundary line between region III and V upto pH 7). Above pH 7, another equilibrium sets in:

$$Fe(OH)_3(s) + H^+ + e \rightleftharpoons Fe(OH)_2(s) + H_2O \qquad E\,(298\text{ K}) = E° - 0.059\text{ pH}.$$

The drop in potential with pH is now less steep as seen in the boundary line between regions V and region VI, that of $Fe(OH)_2(s)$.

The principal features of the Pourbaix diagram for iron may now be summarized as follows:

(i) *The strongest reducing agent* among the chemical forms of iron is the metal Fe(*s*) which has the lowest upper boundary in the diagram. In strong acid medium the region of Fe(*s*) does not overlap the stability region of water (including extension for overpotential). So metallic iron is expected to react strongly with water at low pH values.

(ii) *The most strongly oxidizing form of iron* is certainly the ferrate ion, FeO_4^{2-} whose predominance area has its lower boundary at highest potential value.

(iii) *The most abundant form of iron in neutral water* (pH 7) at a potential of 0 V may be ascertained by drawing a straight line at this potential parallel to the pH-axis. The line will intersect a vertical line at pH 7 in the area dominated by $Fe(OH)_3(s)$. So this should be the most abundant species. Since the point of intersection lies close to the regions covered by $Fe^{2+}(aq)$ and $Fe(OH)_2(s)$, these are also supposed to be present at equilibrium to a smaller extent.

Iron in natural water

Natural water in contact with the atmosphere is nearly saturated with O_2 which has strong oxidizing power ($E° = 1.23$ V). Thus iron may be present as Fe^{3+} at low pH (< 4). But pH of natural water is higher (4.5 – 8), being controlled by atmospheric CO_2 and dissolved carbonate minerals forming interrelated system of CO_2 – "H_2CO_3" – HCO_3^- – CO_3^{2-} – $MCO_3(s)$. Fe^{3+} is precipitated as $Fe(OH)_3(s)$ under such conditions. At the same time, the presence of organic matter coming through sewage water and also contact with soil rich in organic matter provides a reducing atmosphere favourable for the dissolution of iron (Fe_2O_3 or $Fe(OH)_3$) to form Fe^{2+}. As the Fe^{2+} diffuses to the surface, they may be oxidized by dissolved O_2 and again precipitated as $Fe(OH)_3$.

General features of the Pourbaix diagram

On the basis of our discussion of the Pourbaix diagram of iron, we may now generalize the main features of the Pourbaix diagram of an element.

1. *The species corresponding to any point in the Pourbaix diagram of an element represents the thermodynamically most stable and most abundant form of the element at the pH and potential of the point.* The area to which the point belongs gives the range of stability of the species within the boundaries of that area. For species in solution, the boundaries may vary with concentration as expected from the Nernst equation.

   When a species extends over a wide range of the diagram at a given pH, it is expected to show no significant redox property at that pH. Similarly, a species covering a wide range of pH at a given potential will show no acid-base properties.

2. *Strongly oxidizing forms appear at the upper region of the diagram while strongly reducing forms occupy lower regions.*

3. *Species involved in redox equilibria are separated by horizontal or slant lines.* Horizontal lines separate species whose redox equilibrium does not involve H^+ or OH^- ions in the redox equilibrium, their slope determined by the number of H^+ ions and the number of electrons involved in the half-cell reaction as already explained. (If H^+ ions appear on the right hand side of the half-cell reaction, the slope of the *E*–pH line will be naturally upward.)

4. Since strong oxidizing agents and strong reducing agents should react (thermodynamically) with one another, species separated vertically without overlap are expected to undergo redox reactions. If the gap between the predominance areas of the two species is greater than about 1 V, highly exothermic reactions are predicted.

   If the predominance area of a species in an intermediate oxidation state totally *disappears* above or below a particular pH, the species is likely to undergo *disproportionation* into the adjacent oxidation states (e.g., MnO_4^{2-} in Fig. 9.11).

5. Vertical lines separate species having the element in same oxidation state which enter acid-base equilibria. Strong acids are confined to the left of the diagram while strong bases are present in the right part of the diagram.

6. The stability of different forms of an element in water may be assessed from a superimposition of the "stability-window" or *E*–pH lines for redox behaviour of water on the Pourbaix diagram, usually shown by dotted lines. Any species above or below these lines should oxidize or reduce water, respectively. As usual, overvoltage and kinetic barriers may influence the actual reaction.

7. *It is a limitation of the Pourbaix diagram* that conclusions drawn from it are based on thermodynamic equilibrium. When the equilibrium is attained very slowly, the conclusions may not always agree with observations. At the same time, Pourbaix diagrams are concentration and temperature specific. Normally they refer to 25°C and 1M concentration or unit activity of all species except H^+ and OH^- ions. The area of predominance of a species is also susceptible to change with concentration. The vertical range changes according to change in *E* following the Nernst equation. The horizontal range also changes due to concentration effect on the solubility equilibrium of a sparingly soluble form.

We shall find further illustration of these principles in the Pourbaix diagram of manganese.

Pourbaix diagram for Manganese

The Pourbaix diagram of manganese (Fig. 9.11) is marked by the presence of several species in a variety of oxidation states and pH, namely MnO_4^-, $MnO_2(s)$, $Mn_2O_3(s)$, $Mn_3O_4(s)$, $Mn(OH)_2(s)$ and Mn^{2+}.

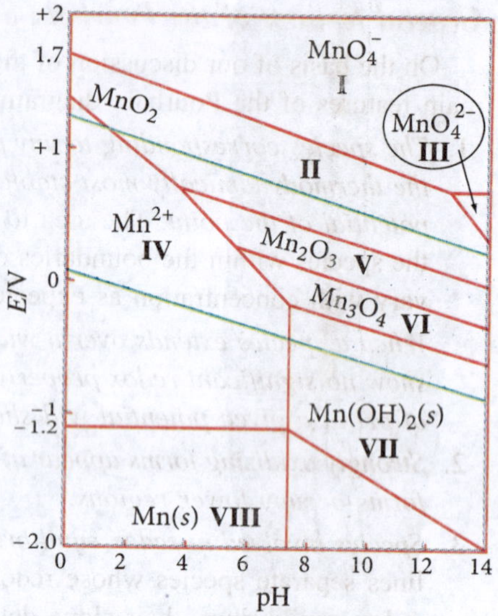

Fig.9.11: Pourbaix diagram of Mn

(i) The top region of the diagram is naturally occupied by the species with highest oxidation number (+7) and highest $E°$, MnO_4^-. MnO_4^- is reduced to Mn^{2+} only under strongly acid medium (pH near zero). The potential sharply decreases with pH (there are $8H^+$ and $5e$ in the $MnO_4^- - Mn^{2+}$ half-reaction). $E°$ for the $MnO_4^- - MnO_2$ system is also high (1.69 V) and strongly pH-dependent. At higher pH in neutral and alkaline medium the principal equilibrium rests with MnO_2 with a much lower $E°$:

$$MnO_4^- + 2H_2O + 3e \rightleftharpoons MnO_2(s) + 4OH^-$$
$$E° = 0.59 \text{ V}.$$

The upper part of the diagram is thus occupied by regions represented by MnO_4^- and $MnO_2(s)$ (I and II).

At very high pH(>13), a small region shows the presence of the manganate ion, MnO_4^{2-} (III). The region disappears below pH 13 and as mentioned under general features, MnO_4^{2-} disproportionates into the adjacent species MnO_4^- and MnO_2.

(ii) MnO_2 exists in equilibrium with Mn^{2+} in acidic medium:

$$MnO_2(s) + 4H^+ + 2e \rightleftharpoons Mn^{2+} + 2H_2O \qquad E° = 1.23 \text{ V}$$

The potential is again highly dependent on pH and marked by a slant line of demarcation between regions for MnO_2 (region II) and Mn^{2+} (region IV).

(iii) At higher pH, MnO_2 is also in equilibrium with Mn_2O_3 and Mn_3O_4 which are marked by regions V and VI respectively

$$2MnO_2(s) + H_2O + 2e \rightleftharpoons Mn_2O_3(s) + 2OH^- \qquad E° = 0.15 \text{ V}.$$

$Mn_2O_3(s)$ and $Mn_3O_4(s)$ are in equilibrium with $Mn(OH)_2(s)$ at pH > 7.5.

(iv) At pH < 7.5, $Mn(OH)_2$ enters solution as Mn^{2+}. The regions of Mn^{2+} and $Mn(OH)_2$(VII) are separated by a vertical boundary line at pH = 7.5.

(v) The system

$$Mn^{2+} + 2e \rightleftharpoons Mn(s) \qquad E° = -1.18 \text{ V}$$

lies lowest in the diagram (region VIII). Since there is no H^+ ion involved in the equilibrium, the boundary is a horizontal straight line at $E° \simeq -1.2$ V which runs upto pH 7.5. Mn^{2+} is precipitated as $Mn(OH)_2$ above this pH.

The potential now falls with pH along a slope corresponding to the half-reaction

$$Mn(OH)_2(s) + 2e \rightleftharpoons Mn(s) + 2OH^-(aq) \qquad E° = -1.56 \text{ V}.$$

9.5.4 Redox Titrations

As already shown, the EMF of a half-cell depends upon the concentrations of different chemical species present in the system. In any titration process, the relative amounts of the two reacting species are gradually varied so that the solution being titrated (titrate) acquires different potentials at different stages of the titration. This fact may be utilized to locate the equivalence point of a titration. Except when the titrant is itself coloured, one may use a suitable indicator which would get oxidized at the potential of the equivalence point (or very close to it). Such titrations are called oxidation-reduction titrations, or redox titrations and the indicators used are called redox indicators. Alternatively, one may follow the half-cell potential of the solution at different stages of the titration by connecting it to another reference half-cell of known EMF. This procedure, known as potentiometric titration, may be adapted to cover neutralization and precipitation titrations in addition to oxidation-reduction titrations.

A. *Redox titrations based on colour-change*

As a typical example, let us first consider the titration of a solution of ferrous sulfate by a solution of potassium permanganate, both of the order of (N/10). The relevant standard reduction EMF's differ widely ($E°$ for Fe^{3+}–Fe^{2+} = + 0.77 V; for $MnO_4^- - Mn^{2+}$ = + 1.51 V.) and the titration can be carried out in dilute sulfuric acid medium (~1.5 N). Phosphoric acid or NH_4HF_2 is used to mask the yellow colour of ferric ion formed by complexation. One drop of excess permanganate solution just after the equivalence point is sufficient to develop an observable pink colour in the titrated solution which marks the end point of titration.

The procedure cannot be used in presence of substances which are also oxidized by permanganate, e.g., chloride ion [$E°$ ($Cl_2 - Cl^-$) = + 1.36 V]. Estimation of iron(II) in presence of chloride may be carried out after addition of a preventive solution containing manganous sulfate. The increased concentration of Mn^{2+} seems to lower the effective reduction potential of $MnO_4^- - Mn^{2+}$ system below that of the $Cl_2 - Cl^-$.

Titrations by potassium permanganate solution can be carried out without any indicator as the titrant is able to indicate the endpoint. This is not often the case with other titrants, e.g., potassium dichromate solution. In those cases, we have to use a redox indicator like diphenylamine or its barium salt to detect the end point (sudden appearance of a blue-violet colour).

This is an example of an *internal redox indicator* which is added to the titrating system. The equivalence point is indicated by a colour change very close to the equivalence point when the indicator gets oxidized or reduced.

Alternatively, the equivalence point in a redox titration may also be located by means of an *external redox indicator*. This is not added to the titrating system, but is used externally. As an example, the titration of iron(II) with standard potassium dichromate solution may be carried out by using potassium ferricyanide as external indicator. Near the equivalence point, drops of the solution are removed from time to time and treated with freshly prepared solution of this indicator on a spot plate. When the drop first fails to produce a blue colour with the indicator, the endpoint of the titration is marked. External indicators have now been largely superseded by satisfactory internal indicators. (e.g., 1,10-phenanthroline in this case).

Redox indicators are mostly organic dyes which can undergo reversible oxidation or reduction in the titration medium imparting different colours. Usually the reduced form is colourless and the oxidized form is distinctly coloured. The redox indicator for any titration is so chosen that it gets oxidized very near the equivalence point of the titration. For example, colourless diphenylamine (1% solution in concentrated sulfuric acid) is oxidized to blue-violet diphenylbenzidine violet via diphenylbenzidine (colourless) ($E^\circ = +0.76$ V) (Fig. 9.12).

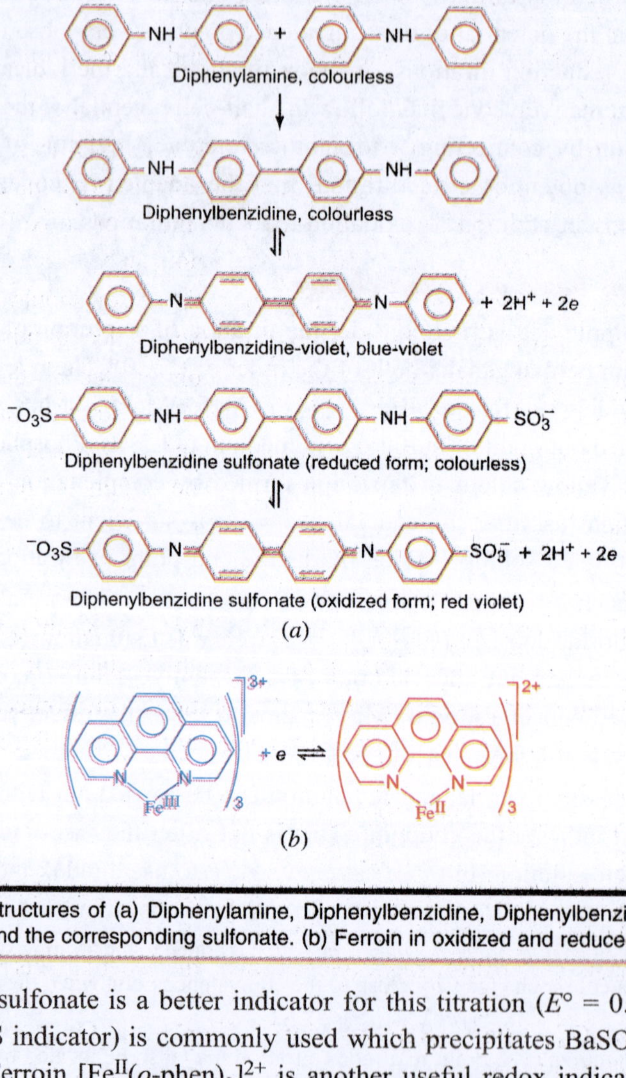

Fig.9.12: Structures of (a) Diphenylamine, Diphenylbenzidine, Diphenylbenzidine-violet, and the corresponding sulfonate. (b) Ferroin in oxidized and reduced form.

Diphenylbenzidine sulfonate is a better indicator for this titration ($E^\circ = 0.85$ V). The barium salt (known as BDS indicator) is commonly used which precipitates $BaSO_4$ and the indicator anion is released. Ferroin $[Fe^{II}(o\text{-phen})_3]^{2+}$ is another useful redox indicator.

The oxidized and reduced forms of a redox indicator establish a redox system with a characteristic electrode potential. If the two forms are expressed as In_{ox} and In_{red} respectively, we may write,

$$\underset{\text{Coloured}}{In_{ox}} + ne \rightleftharpoons \underset{\text{Colourless}}{In_{red}} \qquad E_{in} = E^\circ_{in} - \frac{0.059}{n}\log\frac{[In_{red}]}{[In_{ox}]}$$

It may be roughly assumed that the colour of the oxidized form predominates when its concentration is ten times that of the colourless form (or of the colour in the reduced state, if any). Again, the solution may be considered to attain the colour of the reduced form when this form has at least ten times greater concentration. So, for a perceptible colour of either form we have the conditions

$$E_{in} = E^{o}_{in} - \frac{0.059}{n}\log\frac{1}{10} = E^{o}_{in} + \frac{0.059}{n} \quad \text{or} \quad E_{in} = E^{o}_{in} - \frac{0.059}{n}\log\frac{10}{1} = E^{o}_{in} - \frac{0.059}{n}$$

Thus, the range of potential values across which an indicator exhibits sharp contrast of colour will be $E^{o}_{in} \pm 0.059/n$.

The value of $E°$ for diphenylamine is +0.756 V, while n is equal to 2. The range therefore lies from +0.73 V to +0.79 V. Below +0.73 V, the reduced from predominates and the solution is colourless. As the potential increases, more and more indicator is oxidized and at or above +0.79 V, the indicator will assume intense blue-violet colour, the colour of the oxidized form.

For practical purposes, the standard potential of an indicator should be intermediate between that of the solution titrated (*titrate*) and the *titrant* (see choice of indicator and Table 9.5).

The actual situation in a redox titration is further complicated by the gradual change of potential of the titration system as the titration proceeds. This is not unwarranted; as the titration proceeds, the amounts of oxidant and reductant in the system change continuously, and so also the potential. Let us illustrate the situation with the titration of ferrous ion by permanganate solution in acid medium.

Titration of Fe(II) by $KMnO_4$

Suppose we titrate 100 cm^3 of 0.1 N solution of Fe^{2+} by 0.1 N $KMnO_4$ solution. We assume that the concentration of H^+ is maintained at 1M through all stages of the titration.

Before the addition of any permanganate, the solution in the conical flask contains only ferrous ion. The potential of the system will be given by that for a ferrous-ferric couple:

$$E = E° - \frac{0.059}{1}\log\frac{[Fe^{2+}]}{[Fe^{3+}]}$$

One should not consider $[Fe^{3+}]$ to be exactly zero; traces of impurities are sufficient to make it nonzero. The potential is of course very small at the start.

When a few drops of $KMnO_4$ have been added, the reaction sets in as:

$$MnO_4^- + 5Fe^{2+} + 8H^+ \rightleftharpoons Mn^{2+} + 5Fe^{3+} + 4H_2O$$

The solution now contains Mn^{2+}, MnO_4^-, Fe^{2+} and Fe^{3+} ions in addition to the ions furnished by water and acid. *Two redox systems* are now present at equilibrium:

$$Fe^{3+} + e \rightleftharpoons Fe^{2+} \;;\; MnO_4^- + 8H^+ + 5e \rightleftharpoons Mn^{2+} + 4H_2O$$

Now, equilibrium between Fe^{2+}, Fe^{3+}, Mn^{2+} and MnO_4^- means that the two half-cell potentials for the two redox systems present are *identical*. So the potential of the solution may be expressed in terms of *either* of the two redox systems:

$$E_{Soln} = E^o_{(Fe^{3+},\ Fe^{2+})} - \frac{0.059}{1}\log\frac{[Fe^{2+}]}{[Fe^{3+}]} = E^o_{(MnO_4^-,\ Mn^{2+})} - \frac{0.059}{5}\log\frac{[Mn^{2+}]}{[MnO_4^-][H^+]^8}$$

$$= E^o_{(MnO_4^-,\ Mn^{2+})} - \frac{0.059}{5}\log\frac{[Mn^{2+}]}{[MnO_4^-]} \qquad \text{[since } [H^+] = 1 \text{ in 1 M acid.]}$$

The potential of the solution at any intermediate point of titration may be easily calculated from the expressions already given. As an example we calculate the potential at a point when 90 cm^3 of 0.1(N) $KMnO_4$ has been added to 100 cm^3 of 0.1(N) $FeSO_4$ solution in 1 M acid medium at 25°C.

$$E = E^o_{(Fe^{3+}, Fe^{2+})} - 0.059\log\frac{[Fe^{2+}]}{[Fe^{3+}]} = +0.77 - 0.059\log\frac{10}{90} = +0.83\text{ V.}$$

At the equivalence point, traces of Fe^{2+} will still exist as determined by the equilibrium and an exactly *equivalent* amount of MnO_4^- will also remain unreacted. Remembering that the mole ratio of Fe^{2+} and MnO_4^- ions in the reaction are 5 : 1, we have $[Fe^{2+}] = 5\,[MnO_4^-]$. Fe^{3+} and Mn^{2+} ions formed will also be in the same ratio, i.e., $[Fe^{3+}] = 5\,[Mn^{2+}]$.

So, we have $\dfrac{[Fe^{2+}]}{[Fe^{3+}]} = \dfrac{[MnO_4^-]}{[Mn^{2+}]}$; or, $\dfrac{[Fe^{3+}][MnO_4^-]}{[Fe^{2+}][Mn^{2+}]} = 1$

The potential at the equivalence point is given by

$$E_{eq} = E^o_{(Fe^{3+}, Fe^{2+})} - 0.059\log\frac{[Fe^{2+}]}{[Fe^{3+}]} \quad \text{or,} \quad E_{eq} = E^o_{(MnO_4^- - Mn^{2+})} - \frac{0.059}{5}\log\frac{[Mn^{2+}]}{[MnO_4^-]}$$

or $$5E_{eq} = 5E^o_{(MnO_4^- - Mn^{2+})} - 0.059\log\frac{[Mn^{2+}]}{[MnO_4^-]}$$

Hence, $$6E_{eq} = E^o_{(Fe^{3+}, Fe^{2+})} + 5E^o_{(MnO_4^-, Mn^{2+})} - 0.059\log\frac{[Fe^{2+}][Mn^{2+}]}{[Fe^{3+}][MnO_4^-]}$$

$$= E^o_{(Fe^{3+}, Fe^{2+})} + 5E^o_{(MnO_4^-, Mn^{2+})}$$

$$= 0.77 + 5\,(+1.51)\text{ V} = +8.32\text{ V. This gives } E_{eq} = +1.39\text{ V.}$$

When an excess of permanganate, say, 100.1 cm^3 has been added to the above solution, the potential may be found out by the alternative expression

$$E = E^o_{(MnO_4^-, Mn^{2+})} - \frac{0.059}{5}\log\frac{[Mn^{2+}]}{[MnO_4^-]} = +1.51 - \frac{0.059}{5}\log\frac{100}{0.1} = +1.47\text{ V.}$$

The potentials at different stages of the titration are shown in Table 9.4.

Table 9.4
Variation of potential during the titration of a 100 cm³ 0.1 N $FeSO_4$ solution by 0.1 N $KMnO_4$ solution at 25°C; $[H^+]$ = 1 M.

$KMnO_4$ added (cm^3)	$[Fe^{3+}]/[Fe^{2+}]$	$[MnO_4^-]/[Mn^{2+}]$	E/V
50.0	50 : 50		+0.77
90.0	90 : 10		+0.83
99.0	99 : 1		+0.89
99.9	99.9 : 0.1		+0.95
100.0	—	—	+1.39
100.1	—	0.1 : 100	+1.47
101.0	—	1 : 100	+1.49
110.0	—	10 : 100	+1.50

When these potentials are plotted against the volume of $KMnO_4$ solution added (in this case this is also the percentage of ferrous ion neutralized upto the equivalence point) a curve is obtained as shown in Fig. 9.13.

From the graph it is seen that there occurs a sudden rise of potential : from 0.95 V for 99.9% Fe(II) oxidized to 1.47 V for addition of only 0.1 cm^3 of excess $KMnO_4$ solution. In good redox titrations, such an abrupt change in potential around the equivalence point is desirable; the actual extent of this abruptness depends, however, on the respective potentials of the oxidant and reductant.

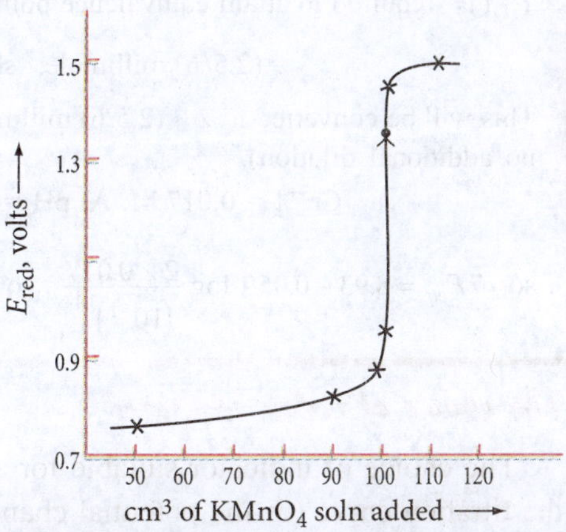

Fig.9.13: Variation of potential during the titration of a 100 cm^3 0.1 N $FeSO_4$ solution by 0.1 N $KMnO_4$ solution at 25°C

The nature of the curve is independent of dilution in the case discussed; but in other cases dilution may be a factor, as in the titration of ferrous ion by dichromate solution. One dichromate ion gives two chromic ions on reduction and the potential at 1M $[H^+]$ is

$$E = E^o_{(Cr_2O_7^{2-},\, Cr^{3+})} - \frac{0.059}{6}\log\frac{[Cr^{3+}]^2}{[Cr_2O_7^{2-}]}$$

If the volume is changed, the ratio of $[Cr^{3+}]^2/[Cr_2O_7^{2-}]$ changes due to the square term. Precipitation, complex formation, pH etc. which alter the concentration of any species maintaining equilibrium will obviously change the nature of the titration curve.

Problem 9.29

Calculate the potential of a solution at the equivalence point in the titration of 25 mL of 0.1 N $FeSO_4$ soln with 0.1 N $Cr_2O_7^{2-}$ solution. The pH of the solution may be supposed buffered to pH 1.0.

Hint: $6Fe^{2+} + Cr_2O_7^{2-} + 14H^+ \rightleftharpoons 6Fe^{3+} + 2Cr^{3+} + 7H_2O$

$$E^o_{Fe^{3+}, Fe^{2+}} = 0.77 \text{ V} \qquad E^o_{Cr_2O_7^{2-}, Cr^{3+}} = 1.36 \text{ V}$$

As explained in connection with titration of Fe^{2+} with MnO_4^-, the potential E_{eq} at the equivalence point will be same for the two redox couples.

$$E_{eq} = +0.77 - 0.059 \log \frac{[Fe^{2+}]}{[Fe^{3+}]}; \quad 6\,E_{eq} = 6 \times 1.36 - 0.059 \log \frac{[Cr^{3+}]^2}{[Cr_2O_7^{2-}][H^+]^{14}}$$

or, $$7\,E_{eq} = 8.93 - 0.059 \log \frac{[Fe^{2+}][Cr^{3+}]^2}{[Fe^{3+}][Cr_2O_7^{2-}][H^+]^{14}}$$

At the equivalence point, from the stoichiometry of the reaction, $[Fe^{3+}] = 3\,[Cr^{3+}]$ and $[Fe^{2+}] = 6\,[Cr_2O_7^{2-}]$.

Putting these relations, $7\,E_{eq} = 8.93 - 0.059 \log 2[Cr^{3+}]/[H^+]^{14}$

At the equivalence point, practically all of the $Cr_2O_7^{2-}$ added has been converted to Cr^{3+}.

$Cr_2O_7^{2-}$ required to attain equivalence point = 25 mL (of 0.1 N solution) = 25 × 0.1 milliequivalent

$\equiv$ (2.5/6) millimole, since 1 mole $Cr_2O_7^{2-} \equiv 6$ equivalent

This will be converted to 2 × (2.5/6) millmole Cr^{3+} in the final solution of volume 50 mL (assuming no additional dilution).

$\therefore$ $[Cr^{3+}] = 0.017$ M. At pH = 1, $[H^+] = 10^{-1}$ M.

so, $$7E_{eq} = 8.93 - 0.059 \log \frac{2 \times 0.017}{(10^{-1})^{14}} \quad \text{or,} \quad E_{eq} = 1.14 \text{ V.}$$

The choice of redox indicator

The choice of indicator suitable for a particular titration should be made with reference to the titration curve (or the potential change data). The indicator should be oxidized as close to the equivalence point as possible, neither too early, nor too late. So the range of potential of the indicator for detectable colour change should lie within the limits of the sharp change of potential in the titration curve.

In the titration of ferrous by permanganate, as shown above, the potential is +0.77 V when only 50% of the iron has been oxidized. Here diphenylamine would not be the right indicator as it starts changing its colour appreciably from +0.73 V. Diphenylamine would not be the right indicator in the titration of ferrous by dichromate too, as in this case also the break in the titration curve corresponds to +0.94 V to +1.3 V. However, in presence of phosphoric acid (or fluoride ion), the Fe^{3+} formed is complexed and hence the potential cannot rise above +0.71 V even when 99% of the ferrous gets oxidized. The potential crosses +0.73 V only when "all" of the ferrous is oxidized and the equivalence point is indicated without almost any error. A few redox indicators with their characteristic E^o values are given in Table 9.5.

Table 9.5

A few common redox indicators

Indicator	Colour in reduced from	Colour in oxidized from	$E°$/V at pH = 0
Methylene blue	colourless	blue	+0.52
Diphenylamine	colourless	blue-violet	+0.76
Diphenylbenzidine	colourless	violet	+0.76
Diphenylamine sulfonic acid	colourless	red-violet	+0.85
N-Phenylanthranilic acid	colourless	red-violet	+0.89
Ferroin (complex of *o*-phenanthroline with Fe^{2+}).	red	pale blue	+1.06

From the table it may be inferred that phenylanthranilic acid may be used in the titration of ferrous ion by dichromate even without phosphoric acid.

Condition for a successful titration

In order to have a quantitatively "complete" reaction (say 99.9%) at the equivalence point, the two redox systems involved in a titration must have a minimum difference between their $E°$ values. In the simplest case, we consider the half-cell reactions for two reacting systems as

System 1: $Red_1 - e \rightleftharpoons Ox_1$

System 2: $Ox_2 - e \rightleftharpoons Red_2$

$$Red_1 + Ox_2 \rightleftharpoons Ox_1 + Red_2$$

At equilibrium, the electrode potentials of the two half-cells will be equal. The reduction potentials in terms of equilibrium concentrations (25°C) are

$$E_1 = E_1^\circ - 0.059 \log [Red_1] / [Ox_1] \; ; \; E_2 = E_2^\circ - 0.059 \log [Red_2] / [Ox_2].$$

At equilibrium, $E_1 = E_2$ gives: $E_2^\circ - E_1^\circ = 0.059 \log \dfrac{[Red_2][Ox_1]}{[Red_1][Ox_2]} = 0.059 \log K_{eq}$

Now, for minimum 99.9% completion of the reaction at equilibrium,

$$[Red_1] = 0.1, [Ox_1] = 99.9; [Ox_2] = 0.1, [Red_2] = 99.9,$$

so, $[Ox_1] / [Red_1] = [Red_2] / [Ox_2] = 999$ or nearly 10^3.

$$\therefore \; K_{eq}(\text{minimum}) = \frac{[Red_2][Ox_1]}{[Red_1][Ox_2]} = 10^3 . 10^3 = 10^6.$$

$$\therefore \; E_2^\circ - E_1^\circ (\text{minimum}) = 0.059 \log 10^6 = 0.354\,\text{V}.$$

Thus, Fe(III) cannot be estimated iodometrically with high accuracy. From the $E°$ values of Fe^{3+}/Fe^{2+} and I_2/I^- systems, K(25°C) for the reaction $Fe^{3+} + I^- \rightleftharpoons Fe^{2+} + ½I_2$ is given by

$$\log K = (1/0.059)\,(0.76 - 0.53) \simeq 4 \text{ or } K = 10^4.$$

So at equilibrium, a small but significant amount of Fe^{3+} will remain unreacted.

If both the redox systems are of the type

$$Red_1 - ne \rightleftharpoons Ox_1 \text{ and } Ox_2 + ne \rightleftharpoons Red_2$$

it is easy to find that for a minimum value of 10^6 for K_{eq},

$$E_2^\circ - E_1^\circ = (0.059/n) \log K_{eq} = (0.059/n) \times 6.$$

When $n = 2$, $E_2^\circ - E_1^\circ = 0.177 \text{ V} \simeq 0.2 \text{ V}$.

For the general case involving two redox systems

$$a\,\text{Red}_1 - n_1 e \rightleftharpoons b\,\text{Ox}_1 \quad \text{and} \quad c\,\text{Ox}_2 + n_2 e \rightleftharpoons d\,\text{Red}_2$$

we have to equate the number of electrons in the overall reaction for $n_1 n_2$ electrons:

$$n_2\,a\,\text{Red}_1 + n_1\,c\,\text{Ox}_2 \rightleftharpoons n_2\,b\,\text{Ox}_1 + n_1\,d\,\text{Red}_2$$

Proceeding as before, we get, $\log K_{eq} = \dfrac{E_2^\circ - E_1^\circ}{0.059 / n_1 n_2}$

B. *Potentiometric titrations*

In this procedure, the solution to be titrated is transformed into a half-cell with suitable electrodes and is connected to another half-cell of known EMF. The EMF of the resulting cell is then measured with a potentiometer at different stages of the titration. The equivalence point is indicated by a sharp change of EMF. As already mentioned, the method can be used to follow oxidation-reduction, neutralization and precipitation titrations.

(i) Oxidation-reduction titrations. We have already discussed how the potential changes during the titration of iron (II) by potassium permanganate solution. If a platinum foil electrode is now inserted in the solution and the resulting half-cell is connected to a calomel electrode, the EMF of the total cell can be measured by a potentiometer. As the potential of the titrated solution changes at each successive addition of the titrant, so also does the EMF measured for the total cell. If we now plot the EMF(E) *vs.* the volume (V) of titrant added, we shall expect a curve similar to that in Fig. 9.13, the equivalence point being marked by a sharp change.

Instead of plotting E vs. V, one may also note ΔE, the change of EMF for the addition of the titrant in small volumes, say, ΔV. As the potential of the system changes most rapidly in the vicinity of the equivalence point, a plot of $\Delta E/\Delta V$ vs. V will show a maxima at the equivalence point (Fig. 9.14). For satisfactory results, the standard potentials of the two systems involved should differ appreciably, at least by 0.3 volt, and the solutions should be concentrated.

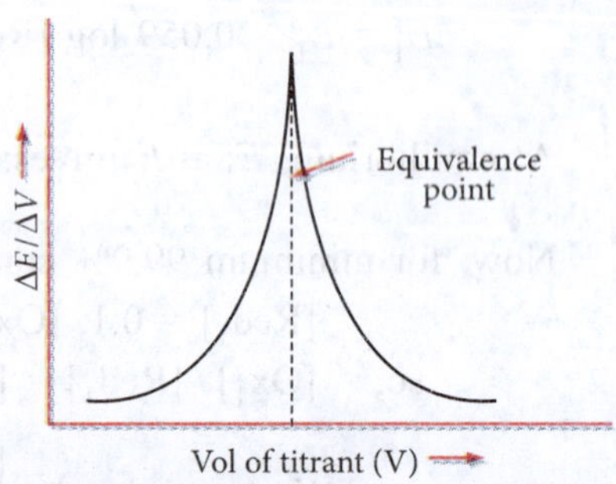

Fig.9.14: Plot of $\Delta E/\Delta V$ vs V in a potentiometric titration

(ii) Neutralization titrations. The EMF of a half-cell reversible with respect to hydrogen ion is determined by the concentration of hydrogen ion in a solution, i.e., on its pH. The EMF of such an electrode may be followed at different stages of neutralization in a titration process with reference to a half-cell of known EMF and the pH of the solution may be calculated at various stages of neutralization. The equivalence point is again marked by an abrupt change in pH. One may plot either (i) pH or E vs. the volume (V) of base added or (ii) $\Delta E/\Delta V$ vs. V as in redox titrations. The method is particularly advantageous for coloured solutions, polybasic acids and mixtures of strong and weak acids, all of which are difficult to titrate with indicators. When the ionization constants of the different stages of a polybasic acid (or of the different acids in a mixture) differ by factors of at least 10^3, the different stages of neutralization are marked distinctly in the potentiometric titration curve (Fig. 9.15).

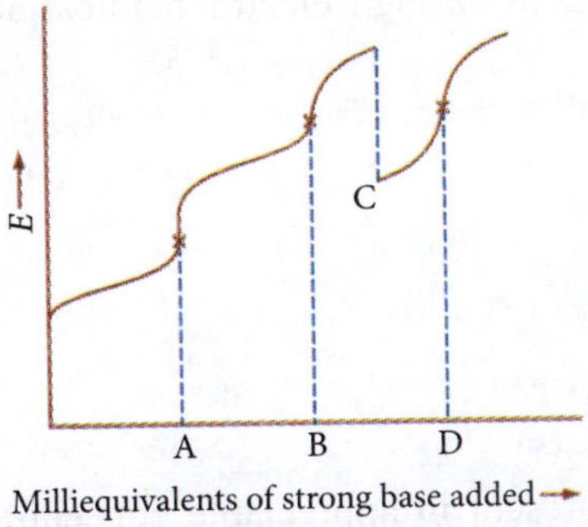

Fig.9.15: Nature of potentiometric titration curve of phosphoric acid with a strong base.
Points A and B correspond to the first and second stage of neutralization respectively.
Beyond point B, the solution becomes strongly alkaline owing to HPO_4^{2-}. Further addition of alkali fails to bring about any noticeable change in potential. Addition of salts like $CaCl_2$, $BaCl_2$ or $AgNO_3$ precipitates the metal phosphate and liberates strong acid equivalent to HPO_4^{2-}. The potential of the solution now drops (to point C, say) and further titration may be carried out to obtain the third equivalence point D.
For K_1, K_2, K_3 and pH at each stage, see chapter 11.

(iii) Precipitation titrations. When a silver nitrate solution is titrated by a sodium chloride solution, the concentration of silver ion in the titrated solution decreases continuously as titration proceeds. If the potential of this solution is measured in the usual manner from time to time, it will be observed that it changes most rapidly near the equivalence point. The same is true for any other titration leading to a sparingly soluble precipitate in which the combining ions have the same charge and the solutions are of the same concentration. Deviations for other cases are also tolerable if the concentrations are high and the precipitates are not very soluble. The method is particularly useful for the titration of halides, cyanides and their mixtures.

9.6 PRACTICAL APPLICATIONS

9.6.1 Corrosion

Corrosion involves chemical reaction of metals with the environment leading to permanent loss of the metal (that is, the metal is unrecoverable). Rusting, the commonest type of corrosion of iron, may be said to be the most important reaction of iron from an economic point of view. Thousands of crores of rupees are spent annually all over the world (the figure for India alone exceeds Rs. 2000 crores) to protect steel objects against corrosion and to replace corroded articles. Corrosion is, in one sense, the reverse process of extracton of a metal from its ore—the metal passing from its elemental state to several combined forms to be distributed in nature. There are four main types of corrosion: (i) Atmospheric corrosion (ii) Immersed corrosion (iii) Underground corrosion and (iv) Chemical corrosion.

(i) Atmospheric corrosion: As the term implies, atmospheric corrosion involves corrosion of metal surfaces exposed to the atmosphere; though the principal attacking agent is oxygen, other gases like CO_2 and SO_2 as well as dust particles and water vapour also contribute significantly to atmospheric corrosion.

Rusting of iron is the most important type of atmospheric corrosion. It is well known that a freshly cleaned surface of iron soon gets covered with a brown layer of hydrated iron(III) oxide, commonly known as rust. Both oxygen and water are necessary for rust formation, but other factors also influence the rate of rusting considerably. These include impurities or stress on the iron surface, availability of dissolved oxygen and electrolytes in contact with the iron surface.

The corrosion of iron by rusting may be summarized as formation and breakdown of an oxide film (hydrated) on the heterogeneous iron-air interface through electrochemical action by the intermediary of water:

(i) The oxygen is reduced at the cathodic region:

$$O_2 + 2H_2O + 4e \rightarrow 4OH^- \ (aq)$$

(ii) Iron is oxidized in the anodic region:

$$Fe \rightarrow Fe^{2+} \ (aq) + 2e$$

This forms iron(II) hydroxide with the hydroxyl ion

$$Fe^{2+}(aq) + 2OH^-(aq) \rightarrow Fe(OH)_2(s)$$

The electrons are taken up by hydrogen ions, present in appreciable concentration when the water/air is polluted by SO_2, NO_2 etc.:

$$2H^+(aq) + 2e \rightarrow H_2(g)$$

(ii) The $Fe(OH)_2$ is now oxidized by dissolved oxygen to hydrated Fe_2O_3 or rust:

$$2Fe(OH)_2(s) + O_2 \rightarrow Fe_2O_3 \cdot xH_2O + (H_2O)$$

The process is schematically shown in Fig. 9.16. When a drop of water covers an iron surface, cathodic regions will develop near the edges. One apparent reason for this is the excess of oxygen dissolved from the air in this region. In the anodic region near the centre of the drop where there is less oxygen, the Fe(II) ions have enough time to move away from the surface before they form $Fe(OH)_2$. This is ultimately converted to rust by oxygen, but this is not formed right on the surface. As a result, the iron surface never gets coated with the rust and the process of corrosion continues below the rusted surface. Air containing SO_2, CO_2 etc. gases and suspended particles containing, say, ammonium sulphate in towns and sodium chloride in sea shores, adds electrolyte to the system and enhances corrosion.

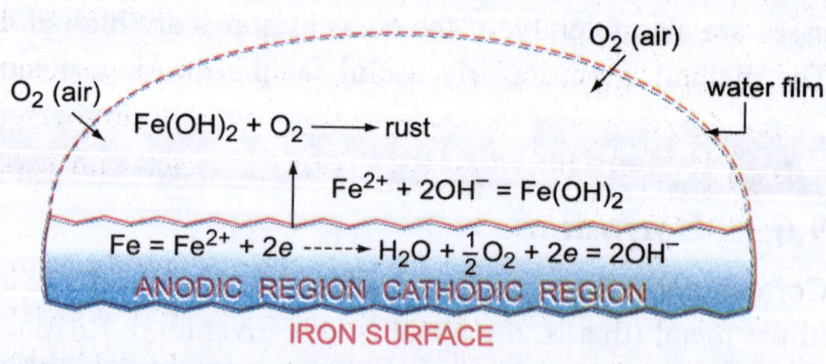

Fig.9.16: Rusting of iron (schematic).

The moisture content of air is an important factor determining the extent of atmospheric corrosion. The optimum humidity is about 70 per cent relative humidity below which such corrosion is negligible. This is often called the *critical humidity*. Iron shows a *primary critical humidity* around 50-60 per cent (relative humidity) when the primary oxide film formed on the metal surface gets ruptured and attack begins on the metal surface. Fine layers of rust now begin to deposit on the iron surface. As the humidity exceeds about 80 per cent (*secondary critical humidity*), the rate of corrosion becomes rapid with the formation of common red rust.

Certain types of atmospheric corrosion proceed even at low relative humidity, the nature of the primary film on the metal surface being different. Thus copper and silver tarnishes in air containing traces of hydrogen sulphide; copper deposits basic copper sulphate in air containing SO_2. The "fogging" of polished nickel surfaces is ascribed to the formation of a thin film of sulfuric acid on the surface.

We have just seen that the process of corrosion proceeds in a continuous manner since the product of corrosion (here the oxide layer) fails to protect the metal surface. When oxygen has ready access to the site of corrosion so that the Fe^{2+} ions are oxidized before they can leave the surface, the metal will get a protective coating of Fe(III) oxide. This is why iron does not rust in an atmosphere of oxygen at high pressure (>25 atm).

(ii) Immersed corrosion: This includes corrosion in water pipes, heating systems, boilers, river-jetties and similar structures. The situation is complicated by the involvement of metal-gas-liquid interfaces where several factors contribute to the corrosion—e.g., pH of the solution, presence of electrolytes and dissolved oxygen, the rate at which oxygen can diffuse through the liquid phase, the possibility of formation of any local galvanic cell, the overpotential of the metal and so on. The nature of the corrosion product also partly determines the rate of such corrosion. The corrosion product of immersed iron and steel consists of about 30-35% iron(II) hydroxide because oxidation is incomplete due to lack of oxygen. In highly concentrated solutions of sodium chlorides etc., low solubility of oxygen actually retards the rate of corrosion. The overall mechanism of corrosion is again electrochemical, similar to that in atmospheric corrosion. Consequently corrosion is enhanced when iron is in contact with a less electropositive metal; the iron acts as the anode and corrodes rapidly. When copper water pipes are joined to galvanized iron tanks, the iron corrodes rapidly.

(iii) Underground corrosion: This type of corrosion is mainly associated with underground pipelines and structures and is similar to immersed corrosion but the nature of the soil, stresses in the structure or the pipeline, contact with less basic metals etc. are more involved in such corrosion.

Underground pipes may corrode severely by the nearby presence of rails of electric trains and trams. When the intermediate ground is wet, a part of the current may leave the rails and travel some distance along the pipe. The point where the current leaves the pipe to return to the rails develops severe corrosion.

(iv) Chemical corrosion: Such corrosion occurs specifically by the action of chemicals like acids, acid vapours, bases, salts and their hydrolysis products etc. which are ambient in and around any chemical industry.

Differential aeration plays a great role in enhancing corrosion of iron. If a situation prevails to restrict the access of atmospheric oxygen to some part of an iron object than the remainder, the least aerated zone will become anodic to the rest and develop corrosion by the same electrolytic mechanism outlined in atmospheric corrosion.

A minute crack on an iron surface will have more oxygen near the outer surface and the interior portion (being anodic) will corrode in a cumulative manner. Underground pipelines may undergo such corrosion—the portion immersed in dry soil will have an easier access to oxygen than the part immersed in moist soil. Similarly, a leaking oil pipe will have restricted supply of oxygen near the leak owing to the oil-film on its surface and undergo rapid corrosion.

Methods used to prevent corrosion

(i) Paints are largely used to minimise corrosion. Red lead or zinc chromate primers are often used for better protection against corrosion. Engine parts and other machinery and tools are protected from corrosion by oil or grease. Another effective treatment is to treat the iron surface with phosphoric acid or acid solutions of $Mn(H_2PO_4)_2$ or $Zn(H_2PO_4)_2$.

(ii) Removal of mechanical strains in the body of a metal by annealing or heat treatment or even by simple polishing helps fight corrosion. Alloying with a suitable metal is also very helpful in preventing corrosion—stainless steel is an alloy of iron with chromium. An impervious oxide film on the surface of the metal is supposed to prevent corrosion. Corrosion of mangesium-base alloys by spray of sea-water is resisted by selenium treatment : the articles are dipped in a bath of selenious acid (or sodium selenite + acid) when a thin layer of mangesium selnide coated with elemental selenium is formed of the surface.

(iii) Application of a sacrificial metal is also adopted largely to prevent corrosion. Galvanizing, the process of protecting iron with a thin layer of zinc is very popular, though galvanized iron cannot be used for canning food-stuffs (zinc is poisonous). The more costly tin-plating is used in this latter case. However, a galvanized article lasts longer since any scratch on its surface forms a local cell with the more basic metal zinc as anode and the less basic iron as cathode. Hence it is the zinc-coating which corrodes faster. A tin-plated vessel, under similar conditions, develops a local galvanic cell with the less basic tin as cathode—the iron corrodes faster as it forms the anode.

Cathodic protection is extensively used in underground pipelines (made of steel) using sacrificial anodes made of zinc or manganese alloys placed at regular intervals along the pipe line. These form a galvanic cell with the steel pipe which becomes the cathode. The anode suffers corrosion and is replaced from time to time.

9.6.2 Electrochemical Cells

Electrochemical cells are widely used for the supply of small quantities of electrical energy as and when necessary. Electric batteries, which are series combination of such electrochemical cells, form an indispensable part of our civilization. We find their use in heart pace-makers, watches, calculators, computers and so on. These cells are of two types: (i) Primary cells and (ii) Secondary cells.

Primary cells produce electrical energy through a chemical reaction which takes place irreversibly in the cell; once exhausted, such cells cannot be utilized again by recharging with an external EMF. The common dry cells used in ordinary torchlights and radios are primary cells (1.2-1.5 V) based on the reaction of the Leclanche' cell. The overall reaction is

$$Zn + 2MnO_2 + H_2O = Zn(OH)_2 + Mn_2O_3.$$

NH_4Cl is involved in the oxidation:

$$2NH_4^+(aq) + 2MnO_2(s) + 2e \rightarrow Mn_2O_3(s) + H_2O + 2NH_3(aq)$$

An alkaline version of the cell using KOH in place of NH_4Cl performs better.

Secondary cells store electrical energy from an external supply through a reversible chemical reaction. When such cells are used to supply electrical energy, the reaction proceeds in the opposite direction and produce electrical energy. Once exhausted, these may be recharged and used again. The lead storage cell used in cars is a common example of this type. When the cell produces electricity, the reaction taking place is

$$Pb + PbO_2 + 2H_2SO_4 = 2PbSO_4(s) + 2H_2O$$

The reverse reaction takes place during recharging. Each cell produces an EMF of about 2 volts; usually six such cells are combined in series in a conventional battery.

The usefulness of a secondary cell depends mainly upon two factors. (i) The electrode reactions should be fairly fast and reversible. (ii) The cell potential should not drop significantly during continuous usage. In fact, whenever a cell is made to supply electrical energy, the potential difference between the electrodes drops somewhat from its rest emf (open circuit potential, OCP). For a good secondary cell, the potential should remain nearly steady after this drop and should not decrease rapidly during use. Besides, the materials used in a secondary cell should not be harmful to the environment after expiry of the cell.

The lead storage battery offers the advantages of fast and reversible electrode reactions and a fairly high and steady emf. Its voltage efficiency (ratio of potential difference during discharge and during charging) is 80% which is indeed very high. But there are also many disadvantages of the lead battery: (i) bulky size and heavy weight; (ii) problem of leakage of the H_2SO_4; (iii) formation of crust of $PbSO_4$ on plates during prolonged rest; (iv) disposal problem of exhausted batteries in view of lead pollution.

Some more examples of primary and secondary cells are mentioned below.

1. Ni–Cd alkaline storage cell (nicad; secondary): Cadmium serves as the negative electrode and a nickel plate with adherent layer of hydrated nickel oxide (approximately NiOOH) forms the positive electrode. The electrolyte is a solution of KOH. Cd is oxidized during discharge.

$$Cd(s) + 2NiOOH(s) + 2H_2O \underset{\text{charge}}{\overset{\text{discharge}}{\rightleftharpoons}} Cd(OH)_2(s) + 2Ni(OH)_2(s)$$

The emf obtained is nearly 1.3 V. This cell is an improvisation of the older Edison cell which had iron instead of cadmium. Ni–Cd rechargeable cells are very popular now but they are costly since the use of KOH requires stainless steel casings. Moreover, used cells pose the problem of toxic cadmium pollution if not properly dumped.

The Ni-MH batteries (common type AA) are also secondary cells with a positive electrode of NiOOH (but having a hydrogen absorbing alloy or intermetallic compound (M) as the negative electrode (instead of Cd). The electrolyte is again KOH.

$$-\text{ve:} \quad H_2O + M + e \rightleftharpoons OH^- + MH$$

$$+\text{ve:} \quad Ni(OH)_2 + OH^- \rightleftharpoons NiOOH + H_2O + e$$

M is most commonly an intermetallic compound of the type AB_5 where A is a mixture of rare earth metals (La, Ce, Nd, Pr); B may be Ni, Co, Mn and/or Al. M may have another composition like AB_2 where A is Ti and/or V and B is Zr or Ni modified with Cr, Co, Fe and/or Mn.

These cells have a capacity 2-3 times higher than an Ni-Cd cell of equivalent size. However, their energy density per volume is less compared to lithium ion batteries. The self-discharge is also high—approximately 30 per cent per month.

2. Lithium batteries: High electropositive character and low density make lithium an ideal material for use in rechargeable batteries. Several types of lithium based batteries are in wide use.

The popular rechargeable lithium polymer cell consists of a cathode of the mixed oxide $LiCoO_2$. This has a layer structure based on sheets of edge-linked CoO_6 octahedra separated by Li^+ ions. Small ionic size allows easy reversible passage of the Li^+ ion into and out of the oxide layer without major disruption of structure. During supply of power (discharging) Li^+ ions coming from the anode through the electrolyte are hosted in vacant sites in the oxide layer with concomitant reduction of Co(IV) to Co(III).

Discharge: $Li_{1-x}CoO_2 + xLi^+ + xe \longrightarrow xLiCoO_2$

(*Simply*:) $CoO_2 + Li^+ + e \longrightarrow LiCoO_2$

The electrode reaction is reversed during charging.

$LiMn_2O_4$ (spinel) may also be used as the cathode. Other cheap and nontoxic 3*d* metals are being tried to replace cobalt. Nanostructured complex oxides are also likely to be good cathode materials.

The anode in common rechargeable lithium batteries usually consists of Li-carbon (graphite) intercalation compounds. During discharge, Li is oxidized to Li^+ ions which leave the graphite layer and are carried through the solid electrolyte to the cathode.

$$C_6Li_x \longrightarrow 6C + xLi^+ + xe.$$

The electrolyte is composed of solid polyethylene oxide (PEO) with an anhydrous salt like $LiPF_6$ or $LiC(SO_2CF_3)_3$.

During charging, the sites vacated by the Li^+ ions in the graphite layer are reoccupied by Li^+ coming from the cathode. The Li^+ ions thus migrate to and fro between their sites in graphite and the mixed oxide. This is sometimes called the "rocking chair model" for the functioning of Li-batteries. The battery offers a steady high voltage of about 3.6 V during discharge and about 500 discharge/recharge cycles.

The Lithium-iodine primary cell (not rechargeable) is extensively used in batteries for cardiac pacemaker since mid 1970-s. The overall reaction is:

anode: $Li \longrightarrow Li^+ + e$; *cathode*: $\frac{1}{2}I_2 + e \longrightarrow I^-$

Solid LiI acts as the electrolyte by virtue of its Schottky defect.

Another primary cell uses thionyl chloride with $LiAlCl_4$ (or $LiGaCl_4$) as electrolyte:

anode: $4Li \longrightarrow 4Li^+ + 4e$; *cathode*: $2SOCl_2 + 4e \longrightarrow SO_2 + 4Cl^-$

The cell supplies a steady high voltage and is used in spacecrafts and military purposes. Liquid SO_2 and acetonitrile solvent are used in some other primary cells used mainly in military communication:

$$2Li(s) + 2SO_2(l) \longrightarrow Li_2S_2O_4(s)$$

Li–FeS_x battery systems use solid Li–Si alloy (–ve terminal) and FeS_x electrodes in a molten LiCl–KCl electrolyte (400°C). Such batteries may be useful in electric cars.

3. Sodium-sulfur batteries are high temperature (300-350°C) secondary cells having molten sodium (m.p. 98°C) and sulfur (m.p. 119°C) electrodes separated by a solid electrolyte sodium β-alumina (see box). These cells are thus exception to common cells in which the electrodes are solid and electrolytes are liquid. Typical sodium sulfur batteries have a central pool of liquid sodium contained in ceramic tubes made from sodium β-alumina. The wall of the tube permits the passage of Na^+ ions to the outer bath of molten sulfur in stainless steel casing. The outer compartment is packed with powdered carbon to provide electrical conduction in the sulfur. The cell reaction is

$$2Na(l) + n/8\ S_8(l) \longrightarrow Na_2S_n(l)$$

Na^+ ions produced in the central compartment pass through the electrolyte and reduce the sulfur to polysulfide. During charging, the processes are reversed.

A typical Na–S battery can store five times as much energy compared to a lead-acid battery of same weight. The open circuit voltage is also high, 2.08 V at 350°C. To keep the electrodes molten, it is necessary to maintain the necessary temperature even when the cell is not in operation. With technological improvement, sodium-sulfur cells have the potential to form the battery of choice for running pollution free electric cars.

Solid electrolytes

Electrolytes are characterized by electrolytic conduction through passage of ions. In solid electrolytes, usually ions of one kind (cations or anions) can move through a not-so-rigid crystal structure to give rise to small electrical conductance. The conductivity of solid electrolytes varies in the range 10^{-1} S m^{-1} (S = siemens or ohm^{-1}) whereas the conductivity of an 1 F aqueous solution of NaCl is 100 Sm^{-1}. The conductance of a typical ionic crystal may be of the order 10^{-16} to 10^{-2} S m^{-1}. Though considerably lower compared to that of typical electrolyte solution, the conductance of solid electrolytes have found many interesting applications, as mentioned in Li–I_2 or Na–S batteries.

Solid electrolytes resulting from mobility of cations are more common compared to anionic solid electrolytes. This is because anions are usually larger in size and can move only with difficulty through a crystal lattice. Movement of the small F^- and O^{2-} ions makes most solid electrolytes. Conductance of solid electrolytes increases rapidly with temperature owing to increasing disorder in the crystal structure.

An important solid cationic electrolyte is sodium β-alumina, prepared by heating any modification of Al_2O_3 with Na_2CO_3 to ~1500°C in a sealed vessel to avoid loss of Na_2O as Na and O_2. The compound has an idealized composition like Na_2O. $11Al_2O_3$ ($NaAl_{11}O_{17}$) but actually contains typically 20-30% excess of sodium and may be represented by a formula like $Na_{1+x}Al_{11}O_{17+x/2}$. The structure is derived from that of spinels (Ch 20, oxides of Gr 13); the Na^+ ions are accomodated between layers of O^{2-} ions which are, in turn, linked to rigid spinel frameworks on their outer sides. There are several sites in the oxygen planes which can host Na^+ ions. This permits rapid two-dimensional diffusion of Na^+ ions within the basal plane giving rise to ion conductance. The Na^+ ions may be replaced by other univalent ions like those from Li, K, Rb, Cu^I, Ag^I and even NH_4^+. The Al^{III} sites in the structure may be replaced by Fe^{III} or Ga^{III}. Inclusion of divalent cations of Mg, Ni or Zn in the structure produces an even more conducting solid electrolyte called β″-alumina. This has a greater separation between the spinel frameworks and can permit transport of divalent ions.

Sodium β-alumina and β″-alumina are examples of *framework electrolytes* in which ions can move through planes or channels in a rigid framework. A framework of ZrO_6 octahedra with SiO_4 and PO_4 tetrahedra forms the solid electrolyte NASICON (Na-Super Ionic Conductor) having the approximate composition $Na_{1+x}Zr_2P_{3-x}Si_xO_{12}$. The nonstoichiometric solid solution system has only partially filled sites of Na^+ in a three-dimensional network of channels and thus acts as a fast ionic conductor. There are many other examples of solid cationic electrolytes. Ag_2HgI_4 shows good electrical conduction above 50°C when the Ag^+ ions can move through the array of easily polarizable I^- ions. A fast cation conductor for Li^+ may be obtained from Li_4GeO_4 deped with Zn or V on some Ge sites.

Solid anionic electrolytes are found with structures containing high levels of anion vacancies —either intrinsic or deliberately created by doping other metal ions in different oxidation states. An early observation by Michael Faraday (1834) on the electrical conductance of red-hot solid PbF_2 may be interpreted in terms of migration of an F^- ion from its normal position into an interstitial position and then to a vacant F^- site. Yttrium stabilized zirconia (YSZ) is a good anionic electrolyte at high temperatures (conductivity = 1 S m^{-1} at 600°C). ZrO_2 (zirconia) has a fluorite structure at high temperature. This structure may be stabilized at room temperature when small amounts of lower valent cations like Ca^{2+} and Y^{3+} (having comparable size) are doped into it. This gives rise to vacant anion sites in the structure which provide a path for diffusion of O^{2-} ions. The composition of YSZ may be written as $Y_xZr_{1-x}O_{2-x/2}$ where x may vary between 0 and 0.15. The conductance of zirconia doped with CaO has been taken advantage of in measuring the partial pressure of oxygen in automobile exhausts. Many complex metal oxides are under study for good anionic conductance at low temperatures. Gadolinium doped cerium oxide (CGO), barium indate ($Ba_2In_2O_5$), doped lanthanum gallate (with Sr, Mg) and *d*-metal doped bismuth vanadium oxide (BIMEVOX) are a few names in this category. Some of these solid electrolytes are used to form solid oxide fuel cells (see below).

Some solid electrolytes exhibit both ionic and electronic conductivity. The mixed oxides $La_{1-x}Sr_xMO_{3-y}$ (M = Mn, Fe) conduct by migration of O^{2-} ion through the perovskite structure. They also show electronic conduction through partially filled bands arising from fractional oxidation number of the metals.

But none of these cells can supply large quantities of electricity. Side by side, the conventional procedure of generating electricity by burning a fuel (coal, diesel) can convert only about 25-35 percent of the total energy contained in the fuel to useful electrical energy. At the same time, the generating stations cause extensive atmospheric pollution. To overcome these difficulties, attempts have been made to develop fuel cells in which elcctrical energy is obtained directly by the oxidation of mainly gaseous fuels in an electrochemical cell.

Fuel Cells

A fuel cell is an electrochemical cell from which a continuous supply of electrical energy is obtained through redox reaction involving oxidation of a chemical fuel like hydrogen, methane, methanol etc. A steady supply of the fuel is an essential part of the fuel cell. The hydrogen-oxygen (H_2-O_2) fuel cell is by far the most widely studied fuel cell.

Materials used in the fuel cell differ according to type. A typical hydrogen-oxygen fuel cell operates at 200°C and 20-40 atm pressure with concentrated aqueous KOH as electrolyte and electrodes made from porous nickel sheets (compressed powder). The electrodes are coated with catalyst layers containing platinum or palladium powder or nano iron powders. Carbon nanotubes doped with nitrogen are emerging as a promising catalyst.

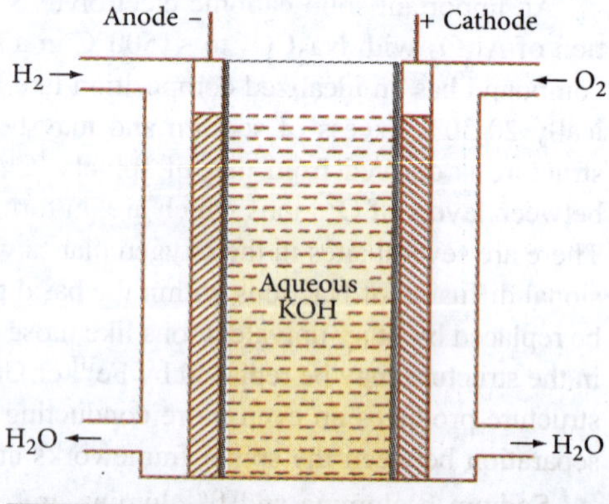

Fig.9.17: Sketch of a typical H_2-O_2 fuel cell (schematic).

The reactions taking place in the cell are:

Anode : $H_2(g) + 2OH^-(aq) \rightarrow 2H_2O(l) + 2e$ $E^\circ_{red} = -0.83$ V

Cathode : $\frac{1}{2}O_2(g) + H_2O + 2e \rightarrow 2OH^-(aq)$ $E^\circ = +0.40$ V

Overall : $H_2(g) + \frac{1}{2}O_2(g) = H_2O(l)$ $E^\circ = +1.23$ V

The exothermic reaction is less favoured thermodynamically at 200°C (vs. 25°C) but the equilibrium is largely balanced by the high pressure: $E \approx 1.2$ V is attained at 200°C and 40 atm.

Besides the alkaline fuel cells (AFC), cells with other electrolytes have also been developed. A cell with phosphoric acid as electrolyte operates with hydrogen and air at 200°C:

$$\textit{Anode} : H_2(g) \rightarrow 2H^+(aq) + 2e \quad ; \textit{Cathode} : \tfrac{1}{2}O_2(g) + 2H^+(aq) + 2e \rightarrow 2H_2O(l)$$

Such fuel cells may be useful as combined heat and power system (CHP) which utilizes the waste heat in doing useful work (including room heating).

Molten carbonates may also serve as electrolyte in fuel cells operating at around 600°C. Such cells can make direct use of natural gas.

Fuel cells based on solid electrolytes include those using proton exchange polymer membranes (PEM-FC) which can operate at about 100°C but require very pure hydrogen. On the other hand **solid oxide fuel cells (SOFC)** are based on a solid oxide ion conductor as electrolyte and can be operated at high temperatures ranging from 500°–1000°C or even higher. This provides the additional advantage of using various hydrocarbon fuels through catalytic conversion to hydrogen and carbon oxides. The efficiency of such cells is also high, making them an important environment friendly means of producing energy.

The solid oxide fuel cells operating at intermediate temperatures (500°–700°C) use an electrolyte like gadolinium-doped cerium oxide (CGO); at higher temperatures yttrium-doped zirconia (YSZ) is a better electrolyte. The anode is usually CeO_2–Ni (or YSZ–Ni at higher temperatures) while the cathode may be the mixed oxide with perovskite structure (La, Sr) (Fe, Co)O_3 (LSCF). New and better oxide ion conductors are being searched for.

In addition to pure hydrogen type, fuel cells may also be run using diesel, petrol, methanol, ethanol, methane, ammonia, hydrazine and similar hydrogen-rich compounds. With methanol, the anode reaction may be written as

$$CH_3OH(l) + 6OH^-(aq) \rightarrow 5H_2O(l) + CO_2(g) + 6e$$

IISc, Bangalore is developing one such direct methanol fuel cell (DMFC).

Biofuel cells may function partly or fully on biological systems. Enzymes may act as catalysts and available biomolecules (e.g., glucose in the bloodstream) may provide the fuel. The electricity may be extracted through molecules capable of electron transport.

The fuel cells provide a pollution-free, silent and efficient means of generating electricity; the practical efficiency may be as high as 50-60 per cent of the total energy of combustion of the fuel. Hydrogen-oxygen fuel cells have been used successfully in space-crafts. In near future, when coal and oil will get exhausted, such fuel cells may be run by hydrogen and oxygen gases obtained from sea-water by electrolysis using energy from some renewable source. Hydrogen may also be obtained from agricultural waste and other biomass by super-heating or through the action of enzymes on green algae. Storage of hydrogen is a practical problem but may be solved by use of microporous materials. Carbon nanotubes and nanofibres are excellent new materials in this context. Carbon nanofibres can store huge amounts of hydrogen—the energy released per volume of material is twice that of gasoline. The use of carbon nanotubes as catalyst has already been mentioned. Electrodes made from a network of single walled carbon nanotubes may provide a highly efficient electrode system with much higher conductivity and porosity. Automobiles run by fuel cells will share much advantage from such materials.

ESSENTIALS

The Oxidation Number of an element (O.N.) is equal to the charge which an atom of the element would acquire if the compound consisted only of ions. For covalent compounds, we must assume a hypothetical ionic structure by assigning each shared pair of electrons to the more electronegative atom.

Disproportionation reactions are those in which atoms of an element undergo simultaneous increase and decrease of oxidation number without any other redox reagent. The reverse process is known as comproportionation; two different oxidation states of an element interact mutually to an intermediate oxidation state.

A reversible cell is one in which any chemical or other process taking place when the cell produces current can be exactly reversed by applying an opposing EMF of *infinitesimally* greater magnitude.

The total cell reaction for a galvanic cell is obtained by adding the oxidation reaction at the negative electrode to the reduction reaction at the positive electrode, and cancelling the electrons on both sides. By convention, **the negative electrode is written on the left**.

The EMF of a reversible cell is related to the decrease of free energy corresponding to the cell reaction by the relation : $\Delta G = -nFE$.

The Nernst Equation: $E = E° - (RT/nF) \ln K_{eq}$, where K_{eq} is the equilibrium constant corresponding to the cell reaction. $E°$ is called the *standard* EMF *of the cell*; it is the EMF when the chemical species are at unit activity (or concentration).

The Nernst equation may also be applied to any electrode reaction to give the electrode potential (E). For a typical electrode process like $M^{n+} + ne = M$,

$$E = E° - (RT/nF) \ln [M^{n+}].$$

$E°$ is called the standard electrode potential, which refers to unit concentration of M^{n+} in solution.

The formal potential of a redox couple is the reduction potential that applies to a specific set of conditions like pH, activity coefficient etc. containing equal formal concentrations of the oxidized and reduced form. Biochemists use the formal potential for pH = 7 instead of $E°$ which refers to pH = 0.

Redox stability in water is determined by the variation of E with pH for the couples H_2O–H_2 and O_2–H_2O. In pure water, $[H^+] = 10^{-7}$ M gives E for hydrogen evolution ($H^+ \rightarrow \frac{1}{2}H_2$) equal to –0.414 V. The half-cell emf for oxygen evolution ($O_2 \rightarrow H_2O$) is given by $E = E° - 0.059$ pH = 1.23–0.059 pH. At pH = 7, E = 0.82 V.

Redox potential diagrams provide ready information about the $E°$ values relating different oxidation states of an element. In the ***Latimer diagram***, successive oxidation states are written from left to right and the $E°$ values are shown on lines connecting the states. A Frost or ***Ebsworth diagram*** is a graphical plot of $NE°$ versus the oxidation number N of an element. In the ***Pourbaix diagram***, the potentials are plotted against pH for various species of the element relevant to aqueous medium.

Redox titrations may be carried out traditionally using a redox indicator which changes color close to the equivalence point. Redox titrations may also be carried out potentiometrically by following the potential of the solution at different stages of reaction.

Electrochemical cells are widely used to supply small quantities of electricity. Primary cells derive their energy from a chemical reaction, as in disposable dry cells involving Zn, MnO_2 and NH_4Cl.

In secondary cells, electrical energy may be stored through a chemical reaction which is reversed when the cell supplies electricity. The lead acid battery using Pb, $PbO_2(s)$, $PbSO_4(s)$ and dilute H_2SO_4 is very common. The nickel cadmium battery is also used widely. Sophisticated instruments use the ***lithium battery*** consisting of $LiCoO_2$ as cathode and Li-graphite anode. Solid polyethylene oxide with anhydrous $LiPF_6$ forms the electrolyte. Li–I_2 battery is widely used in cardiac pacemakers.

Solid electrolytes provide electrical conductance through movement of ions in a non-rigid crystal structure. Sodium β-alumina having the idealized composition $Na_2O.11Al_2O_3$ contains excess Na^+ ions in a spinel structure; the ions can move through the crystal giving rise to ionic conductance. Mixed oxides of Na, Zr, P and Si also form good solid electrolytes.

A fuel cell is an electrochemical cell devised for continuous supply of electricity through redox reaction involving oxidation of a chemical fuel like hydrogen, methane, methanol etc. The hydrogen oxygen fuel cell involves reaction between H_2 and O_2 producing H_2O, a nonpolluting product. Solid oxide fuel cells provide an important enviroment-friendly means of producing energy.

EXERCISE

1. Balance the equations representing the following redox reactions by oxidation number method:
   (i) Reduction of potassium permanganate solution (to manganous salt) in acidic medium by hydrogen peroxide.
   (ii) Reaction of potassium iodide and potassium iodate in acidic solution liberating iodine.
   (iii) Reduction of nitric acid to ammonia by metallic aluminium in presence of sodium hydroxide.
   (iv) Reaction of white phosphorus with caustic soda solution forming phosphine and sodium hypophosphite.
2. Balance the equations for the following redox reactions by ion-electron method:
   (i) Oxidation of manganous sulfate solution to permanganic acid by red lead in presence of dilute nitric acid.
   (ii) Reaction of metallic tin with cold dilute nitric acid to form stannous and ammonium nitrates.
   (iii) Oxidation of ethyl alcohol to acetic acid by acidified potassium dichromate solution.
   (iv) Reaction of metallic copper with moderately concentrated nitric acid.
3. What do you mean by reversible chemical cells?
4. Give examples of (i) a gas electrode, (ii) an amalgam electrode (iii) an anion-reversible electrode and (iv) a metal-metal oxide electrode.
   Write the ionic reactions occurring when oxidation takes place at these electrodes.
5. Write the individual electrode reactions and total cell reaction for each of the following cells. Taking relevant $E°$ values (at 25°C) from tables, compute the standard EMF of the cells.
   (i) $Zn \mid Zn^{2+} \mid\mid Cu^{2+} \mid Cu$
   (ii) $Zn \mid Zn^{2+} \mid\mid Fe^{3+}, Fe^{2+} \mid Pt$
   (iii) $H_2(g) \mid H^+ \mid\mid OH^- O_2(g)$
   (iv) $Pb \mid PbSO_4(s), SO_4^{2-} \mid\mid Cu^{2+} \mid Cu$
   (v) $Ag \mid AgCl(s), Cl^- \mid\mid Cl^- \mid Cl_2(g)$
   (vi) $H_2(g) \mid H^+ \mid\mid I^-, AgI(s) \mid Ag$
6. Devise cells in which the following reactions will take place:
   (i) $Ag + ½I_2(s) = AgI(s)$
   (ii) $½Br_2(l) + Fe^{2+} = Br^- + Fe^{3+}$
   (iii) $Ag^+ + Br^- = AgBr(s)$
   (iv) $Na(s) + ½Cl_2(g) = NaCl$
7. Briefly explain:
   (i) The silver-silver chloride electrode is reversible with respect to chloride ion.
   (ii) The platinum electrode used in a hydrogen electrode is platinized.
   (iii) The magnitudes of standard electrode potentials are merely relative.
   (iv) Only a positive EMF corresponds to a spontaneous cell reaction.
   (v) Copper displaces zinc from a solution of zinc salt in presence of excess KCN.
   (vi) Solutions containing cupric ions readily oxidize potassium iodide to iodine though E^o_{red} of the $Cu^{2+} - Cu^+$ system (–0.15 V) is lower than that of the $I_2 - I^-$ system (0.54 V).
   (vii) In constructing cells, a salt-bridge is often used to eliminate liquid junction.

(viii) Electrolysis of an aqueous solution of sodium chloride using mercury cathode causes discharge of sodium ions at the cathode in preference to hydrogen ions.

(ix) Phosphoric acid is added in the titration of ferrous ion by potassium dichromate using diphenylamine as indicator.

(x) In the titration of ferrous ion by potassium permanganate in presence of hydrochloric acid, a preventive solution containing excess manganous sulfate has to be added.

8. Which of the following oxidizing agents become stronger as the concentration of hydrogen ion increases? Which are unchanged and which become weaker?

$$Cl_2,\ Cr_2O_7^{2-},\ Fe^{3+},\ MnO_4^-.$$

9. (a) Of the substances Ni, Ni^{2+}, Zn, Zn^{2+}, which one is the strongest oxidant? Which one is the strongest reductant?

(b) Will a noticeable reaction take place when (i) a nickel rod is dipped in a 1 M solution of Zn^{2+}? (ii) a zinc rod is dipped in a 1 M solution of Ni^{2+}?

(c) If the above half-cells are connected, which electrode would be negative?

10. Taking appropriate half-cell potentials from the following data, construct a cell in which the reaction would be

$$Hg^{2+} + Hg = Hg_2^{2+}$$

Also calculate the equilibrium constant for the reaction. Given E°_{ox} (volts) for

(i) $2Hg + 2Cl^- = Hg_2Cl_2 + 2e$: –0.268 V,

(ii) $2Hg + Hg_2^{2+} + 2e$: –0.788 V,

(iii) $Hg_2^{2+} = 2Hg^{2+} + 2e$: –0.92 V.

Hint: Construct cell from (ii) and (iii). **Ans.** $K = 1.7 \times 10^2$

11. Consider two redox couples as follows:

$$Ox_1 + 2e \rightleftharpoons Red_1 \quad (E^\circ = E_1^\circ) \qquad Ox_2 + 2e \rightleftharpoons Red_2 \quad (E^\circ = E_2^\circ)$$

What should be the minimum difference between E_1° and E_2° in order to have 90% complete reaction between the two systems at equilibrium at 25°C?

12. Show that for two general redox couples

$$a\,Ox_2 + n_1 e \rightleftharpoons b\,Red_2 \quad (E^\circ = E_2^\circ) \quad \text{and} \quad d\,Ox_1 + n_1 e \rightleftharpoons c\,Red_1 \quad (E^\circ = E_1^\circ)$$

the overall reaction $n_2 cRed_1 + n_1 a\,Ox_2 = n_2 dOx_1 + n_1 b\,Red_2$ has the equilibrium constant K_{eq} where $\log K_{eq} = (E_2^\circ - E_1^\circ) / (0.059/n_1 n_2)$.

13. Potassium dichromate is a common oxidizing agent in acid medium (E° = 1.38 V) for the titration of $Fe^{2+}(aq)$. Is there any harm if the solution contains large quantity of Cl^- ion (E° = 1.36 V)?

Hint: A difference of only 0.02 V suggests very slow reaction. In fact Cl^- is not oxidized.]

14. The Latimer diagram for chlorine at pH = 0 is

$$ClO_4^- \xrightarrow{1.20} ClO_3^- \xrightarrow{1.18} ClO_2 \xrightarrow{1.70} HClO \xrightarrow{1.63} Cl_2 \xrightarrow{1.36} Cl^-$$

Calculate the E° for reduction of ClO^- to Cl^- in acidic solution.

Hint: We have to proceed via ΔG° (= $-nFE^\circ$) of different steps. For oxidation numbers x, y and z differing by n_1 and n_2 electrons respectively this gives

$$E^\circ_{xz} = \frac{n_1 E^\circ_{xy} + n_2 E^\circ_{yz}}{n_1 + n_2}.$$

15. For the process $O_2(s) + 4H^+(aq) + 4e \longrightarrow 2H_2O$, E° = 1.23 V. What will be the potential at pH = 7 and P_{O_2} = 0.2 atm, the partial pressure of O_2 in air? [0.81 V]

16. In basic solution, the following potentials are given

$CrO_4^{2-}(aq) + 4H_2O + 3e \longrightarrow Cr(OH)_3(s) + 5OH^-$ $E° = -0.11$ V

$[Cu(NH_3)_2]^+ (aq) + e \longrightarrow Cu(s) + 2NH_3(aq)$ $E° = -0.10$ V.

Calculate $E°$, $\Delta G°$ and K for reduction of Cr(VI) and $[Cu(NH_3)_2]^+$ in basic solution by H_2. Why are $\Delta G°$ and K values for the two processes so different, though their $E°$ values are nearly equal?

[Cr(VI) : $K = 2.64 \times 10^{-6}$; $[Cu(NH_3)_2]^+$: $K = 2.04 \times 10^{-2}$. Note the number of electrons].

17. Consider the titration of 25 mL 0.1 N Fe^{2+} with 0.1 N $Cr_2O_7^{2-}$, the solution being buffered to pH 1.0. Calculate the potential versus a saturated Ag–AgCl electrode (0.197 V) for the following volumes of $Cr_2O_7^{2-}$ added : 10.0 mL; 20.0 mL; 24.9 mL; 25.0 mL; 25.1 mL.

Hint: $6Fe^{2+} + Cr_2O_7^{2-} + 14H^+ \rightarrow 6Fe^{3+} + 2Cr^{3+} + 7H_2O$

Fe^{2+} taken = 25 × 0.1 meq = 2.5 meq ≡ 2.5 mmol

Fe^{3+} formed at equivalence point = 2.5 mmol.

The potentials of the Fe^{3+}–Fe^{2+} couple at various stages prior to equivalence point may be calculated by the relation

$$E = 0.77 - 0.059 \log [Fe^{2+}]/[Fe^{3+}]$$

(i) For 10 mL $Cr_2O_7^{2-}$ added, Fe^{3+} formed = 10 × 0.1 meq or 1.0 meq;
Fe^{2+} remaining = (2.5 – 1.0) meq = 1.5 meq.
Hence $E = 0.77 - 0.059 \log (1.5/1.0) = 0.76$ V.
$E_{observed} = E - E_{AgCl} = 0.76 - 0.197 = 0.56$ V

(ii) E for 20.0 mL and 24.9 mL $Cr_2O_7^{2-}$ solution may be found similarly.

(iii) E_{eq} may be calculated similar to Problem 9.29.

(iv) For 25.1 mL $Cr_2O_7^{2-}$ we calculate the potential for $Cr_2O_7^{2-} - Cr^{3+}$ couple:

$$E = 1.36 - \frac{0.059}{6} \log \frac{[Cr^{3+}]^2}{[Cr_2O_7^{2-}][H^+]^{14}}$$

$[Cr^{3+}] \equiv [Fe^{3+}] \div 3$ (see stoichiometry of the reaction)
$\equiv 2.5 \div 3$ or 0.84 mmol ⇒ 16.67 mmol L^{-1}.

$[Cr_2O_7^{2-}] \equiv 0.1 \times 0.1$ meq = 0.01 ÷ 6 mmoL = 0.0017 mmol ⇒ 0.034 mmol L^{-1}.

$$E = 1.36 - \frac{0.059}{6} \log \frac{(16.67)^2}{0.034 \times (10^{-1})^{14}} = 1.18\,\text{V}$$

18. $E°$ of Ce^{4+}–Ce^{3+} couple is 1.44 V. Find E at equivalence point in the titration of 0.1 N Fe^{2+} by 0.1 N Ce^{4+} in 1 M acid medium. ($E°$ for Fe^{3+}–Fe^{2+} couple = 0.77 V).

Hint: Since both the couples involve 1e change, $E_{eq} = 1/2\ (E^o_{Ce^{4+}/Ce^{3+}} - E^o_{Fe^{3+}/Fe^{2+}})$
$= 1/2\ (1.44 + 0.77)$ or 1.10 V.

The most suitable indicator is ferroin (Table 9.5).

10

ACIDS AND BASES

10.1 INTRODUCTION

Acids (Latin *acidus* = sour) and alkalis (Arabic *al-quili* = ashes of a plant) were known as two distinct classes of substances even before chemistry was developed as a separate subject for systematic study. As early as 1663 Robert Boyle noted the general characteristics of acids as (i) sour taste, (ii) ability to change the colour of certain vegetable dyes, (iii) high dissolving ability and (iv) evolution of hydrogen with metals in most cases. Lavoisier (1787) observed that many non-metals (C, N, S, P) burn in oxygen and the product forms an acid with water. Oxygen got its name from these observations (Greek *oxus* = sour; *gennae* = I produce) and was considered to be an essential constituent of all acids. But Davy (1810) showed that hydrochloric acid contained no oxygen. He attributed acidities to the presence of hydrogen in compounds; the idea was ultimately retained.

Alkalis were known to produce a 'soapy feel' in water, affect certain vegetable dyes and "counteract" the action of acids. Certain oxides and hydroxides of metals were themselves insoluble in water but could neutralize acids forming salts. These were covered by the more general term *base*, introduced by Rouelle in mid-eighteenth century.

The span of the terms acids and bases has been extended and modified in several stages by different approaches to suit the purpose of users and to include a wide variety of chemical compounds. A very large number of chemical reactions may now be treated as essentially acid-base reactions, for example, the solvation of ions in a polar solvent and formation of complex compounds of metals. We shall first discuss the main lines of approach to develop the sense associated with the terms "acids" and "bases". These so-called "theories" of acids and bases are practically definitions, each trying to provide a wider coverage for different aspects of acid-base chemistry.

10.1.1 The Arrhenius Theory

In his theory of electrolytic dissociation (1880-1890), Arrhenius defined an acid as any hydrogen-containing compound which gives hydrogen ions in aqueous solution; a base was similarly defined as any compound which gives hydroxide in aqueous solution

$$Acid: HCl \xrightarrow{\text{water}} H^+ + Cl^- ; \qquad Base: KOH \xrightarrow{\text{water}} K^+ + OH^-$$

The theory was highly successful in explaining different aspects of acid-base behaviour in water solution. For example:

(i) the strength of an acid or base could be expressed quantitatively in terms of an ionization equilibrium of the type:

$$HA \rightleftharpoons H^+ + A^- \quad ; \; K = C_{H^+} \times C_{A^-} / C_{HA}.$$

(ii) neutralization was explained as the combination between H^+ and OH^- ions forming H_2O. This justified the observed constancy of molar heat of neutralization for any strong acid–strong base pair.

(iii) the observed catalytic property of acids in many reactions could also be correlated with the availability of hydrogen ion from an acid (i.e., its acid strength).

In spite of its wide applicability in aqueous solution, the Arrhenius theory could not be applied to nonaqueous systems. Also, the hydrogen ion, H^+, was actually shown to undergo strong hydration in aqueous solution, forming H_3O^+ or more hydrated species like $H_9O_4^+$ [i.e., $H^+(H_2O)_4$].

According to the IUPAC recommendation for nomenclature of inorganic chemistry, 2005 (see chapter 2), H_3O^+ is called the *oxonium ion* or the oxidanium ion (H_2O = oxidane), *not hydronium*. In many equations, H^+ is often used, hydration being implied. As explained in chapter 2, H^+ in its natural isotopic composition is called *hydron*; the name proton is reserved for $^1H^+$. However, for the sake of familiarity, we shall use the terms hydron and proton interchangeably.

10.1.2 The Solvent-system Definition (Franklin, 1905)

Studies on certain nonaqueous systems, particularly in liquid ammonia, revealed that reactions in these solvents may be explained by assuming auto-ionization of the solvent similar to water:

$$H_2O + H_2O \rightleftharpoons H_3O^+ + OH^- ; \qquad NH_3 + NH_3 \rightleftharpoons NH_4^+ + NH_2^-$$

Just as substances furnishing H_3O^+ ions in water act as acids in aqueous medium, substances furnishing NH_4^+ ions in liquid ammonia may also be expected to behave as an acid in that medium. Similarly, substances furnishing NH_2^- ions in liquid ammonia are formally similar to compounds giving hydroxide ions in water.

Acids in liquid ammonia : NH_4Cl, NH_4NO_3, $(NH_4)_2SO_4$ etc.

Bases in liquid ammonia : KNH_2, $NaNH_2$ etc.

In water : $\underset{\text{acid}}{HCl} + \underset{\text{base}}{NaOH} \longrightarrow \underset{\text{salt}}{NaCl} + \underset{\text{solvent}}{H_2O}$.

In liquid NH_3 : $\underset{\text{acid}}{NH_4Cl} + \underset{\text{base}}{NaNH_2} \longrightarrow \underset{\text{salt}}{NaCl} + \underset{\text{solvent}}{2NH_3}$

These observations may be generalized for other solvents as well and we may define acids and bases in general as follows:

An acid is a solute that gives the cation characteristic of the solvent (either by direct dissociation or by reaction with the solvent).

A base is a solute that furnishes the anion characteristic of the solvent (in the same manner as mentioned above).

From these definitions, acid-base systems may be considered in aqueous as well as non-aqueous media irrespective of the presence of hydrogen:

(i) **BrF_3** *Autoionization* : $BrF_3 + BrF_3 = BrF_2^+ + BrF_4^-$

Acid (SbF_5) : $SbF_5 + BrF_3 = BrF_2^+ + SbF_6^-$

Base (F^-) : $F^- + BrF_3 = BrF_4^-$

Neutralization : $BrF_2^+ + BrF_4^- = 2BrF_3$.

(ii) **$COCl_2$** *Autoionization* : $COCl_2^+ = COCl^+ + Cl^-$

Acid ($AlCl_3$) : $AlCl_3 + COCl_2 = COCl^+ + AlCl_4^-$

Base ($CaCl_2$) : $CaCl_2 = Ca^{++} + 2Cl^-$

Neutralization : $2COCl^+AlCl_4^- + CaCl_2 = Ca\,[AlCl_4]_2 + 2COCl_2$

The chief merit of the solvent system definition lies in its ability to include many nonaqueous as well as nonhydronic (nonprotonic) acid-base systems. Each solvent may be looked upon as giving rise to an acid-base system of its own.

But the definition fails to consider the inherent acid or base character of a substance independent of a solvent. At the same time, it places too much emphasis on the autoionization of the solvent.

For example, reactions in liquid sulfur dioxide may be explained in terms of the supposed autoionization

$$SO_2 + SO_2 \rightleftharpoons SO^{2+} + SO_3^{2-} \;;\quad SOCl_2 + CaSO_3 = CaCl_2 + 2SO_2$$

But $SOCl_2$ dissolved in liquid SO_2 containing ^{35}S shows no exchange of ^{35}S. Similarly, there is no exchange of ^{18}O between $SOCl_2$ and SO_2 when one of them initially contains ^{18}O. These observations do not support the ionization scheme written above.

Similarly, reactions in phosphorus oxychloride may be understood in terms of the hypothetical autoionization process:

$$OPCl_3 + OPCl_3 \rightleftharpoons OPCl_2^+ + OPCl_4^-$$

or simply, $OPCl_3 \rightleftharpoons OPCl_2^+ + Cl^-$

Accordingly, PCl_5 and $FeCl_3$ behave as acids in $OPCl_3$:

$$FeCl_3 + OPCl_3 \rightleftharpoons OPCl_2^+ + FeCl_4^-$$

Chloride ion donors like $(CH_3)_4N^+Cl^-$ will behave as a base in $OPCl_3$. In fact, one can follow the following reaction by conductometric titration in $OPCl_3$:

$$(CH_3)_4N^+Cl^- + FeCl_3 \longrightarrow [(CH_3)_4N]^+ [FeCl_4]^-$$

However, the same reaction also takes place readily in $OP(OEt)_3$ (triethyl phosphate). The chloride ion transfer may take place through stabilization of $FeCl_2^+$ by solvation

$$2FeCl_3 \xrightarrow{OP(OEt)_3} [FeCl_2\{OP(OEt)_3\}_n]^+ + FeCl_4^-$$

Similar coordination ability of — P = O group of both solvents and their dielectric constants appear to be more important than the supposed autoionization.

10.1.3 The Protonic (Hydronic) Theory

This was forwarded independently by Brönsted (Denmark) and Lowry (England) in 1923. According to this concept,

any species that tends to give up a proton (hydron) is an acid, and any substance that tends to accept a proton (hydron) is a base.

Any hydron donating species will be an acid according to this definition—molecules (HCl, HNO_3 etc.), cations (NH_4^+, H_3O^+ etc.) or anions (HSO_4^-, HCO_3^- etc.). The donation of a hydron in aqueous solution may be represented as

$$HCl + H_2O \rightleftharpoons H_3O^+ + Cl^-$$

$$NH_4^+ + H_2O \rightleftharpoons H_3O^+ + NH_3$$

$$HCO_3^- + H_2O \rightleftharpoons H_3O^+ + CO_3^{2-}$$

Similarly, any hydron-acceptor will be a base. These may be molecules (NH_3, pyridine etc.) or ions (OH^-, CO_3^{2-} etc.). The equilibrium representing transfer of a hydron to such species may be represented as

$$NH_3 + H_2O \rightleftharpoons NH_4^+ + OH^-$$

$$CO_3^{2-} + H_2O \rightleftharpoons HCO_3^- + OH^-$$

The definition can handle acid-base behaviour in hydronic nonaqueous solvents like liquid ammonia, sulfuric acid or anhydrous acetic acid (Table 10.1).

In liquid NH_3 : $CO(NH_2)_2 + NH_3 \rightleftharpoons NH_4^+ + NH_2CONH^-$

In CH_3COOH : $HCl + CH_3COOH \rightleftharpoons CH_3COOH_2^+ + Cl^-$

Reactions in gaseous phase may also be covered by this approach:

$$NH_3(g) + HCl(g) \rightleftharpoons NH_4Cl(s)$$

Table 10.1

Some hydronic acid-base systems

$Acid_1$	+	$Base_2$	$\rightleftharpoons$	$Acid_2$	+	$Base_1$	
HNO_3	+	H_2O	$\rightleftharpoons$	H_3O^+	+	NO_3^-	
H_2SO_4	+	H_2O	$\rightleftharpoons$	H_3O^+	+	HSO_4^-	.. Note 1
HSO_4^-	+	H_2O	$\rightleftharpoons$	H_3O^+	+	SO_4^{2-}	.. Note 1
$[Fe(H_2O)_6]^{3+}$	+	H_2O	$\rightleftharpoons$	H_3O^+	+	$[Fe(H_2O)_5OH]^{2+}$	
NH_4^+	+	H_2O	$\rightleftharpoons$	H_3O^+	+	NH_3	
H_2O	+	NH_3	$\rightleftharpoons$	NH_4^+	+	OH^-	
H_2O	+	CO_3^{2-}	$\rightleftharpoons$	HCO_3^-	+	OH^-	.. Note 1
H_2O	+	HS^-	$\rightleftharpoons$	H_2S	+	OH^-	.. Note 2
HCl	+	CH_3COOH	$\rightleftharpoons$	$CH_3COOH_2^+$	+	Cl^-	.. Note 3
$HClO_4$	+	H_2SO_4	$\rightleftharpoons$	$H_3SO_4^+$	+	ClO_4^-	Note 4
$HClO_4$	+	HF	$\rightleftharpoons$	H_2F^+	+	ClO_4^-	Note 4
HF	+	HNO_3	$\rightleftharpoons$	$H_2NO_3^+$	+	F^-	Note 4
$CO(NH_2)_2$	+	NH_3	$\rightleftharpoons$	NH_4^+	+	NH_2CONH^-	Note 5
HCOOH	+	$CO(NH_2)_2$	$\rightleftharpoons$	$H_2N.CO.NH_3^+$	+	$HCOO^-$	Note 5
H_2	+	OD^-	$\rightleftharpoons$	DOH	+	H^-	.. Note 6
$C_6H_5CH_3$	+	CH_3Na	$\rightleftharpoons$	CH_4	+	$C_6H_5CH_2^-Na^+$	.. Note 7

Note 1. HSO_4^- may thus behave both as an acid and a base. The same remark applies to HCO_3^- and HS^-. The conjugate base of HCO_3^- is H_2CO_3 (not shown above).

What is the conjugate base of HS^-?

Note 2. The reaction explains why an aqueous solution of NaHS will be basic and smell of H_2S.

Note 3. In pure acetic acid medium. Acetic acid acts as a *base*.

Note 4. The first reaction occurs when pure perchloric acid is dissolved in concentrated sulfuric acid. Sulfuric acid acts as a *base* !

The second and third reactions occur in anhydrous hydrogen fluoride medium. In the second reaction HF acts as a base. In the third reaction HF is the acid and HNO_3 *behaves as the base*.

Note 5. In liquid ammonia. Urea may behave both as an acid or a base.

Note 6. Occurs when hydrogen is bubbled through molten potassium deuteroxide. The hydride ion cannot, however, be detected. But the hydrogen evolving from the melt is enriched with deuterium. Note that hydrogen gas acts here as an acid !

Note 7. Occurs in hydrocarbon solvents. Note that both methane and toluene are weakly acidic, but toluene is more acidic than methane.

Conjugate acid-base pair

Donation of a hydron (proton) by a molecule of HCl in aqueous medium is represented by the equilibrium

$$\underset{\text{acid}}{HCl} + \underset{\text{base}}{H_2O} \rightleftharpoons \underset{\text{acid}}{H_3O^+} + \underset{\text{base}}{Cl^-}$$

In the forward reaction, the molecule of water functions as a base since it accepts a hydron from the HCl molecule. In the reverse reaction, the chloride ion tends to receive a hydron from the oxonium ion. Hence, the chloride ion may now be looked upon as a base and the oxonium ion as an acid. The overall reaction takes place mainly towards right, since water is a better hydron acceptor, i.e., a stronger base, than the chloride ion. The equilibrium thus involves two acid-base systems functioning simultaneously, i.e., (i) HCl (acid$_1$, say) and its corresponding base Cl^- (base$_1$; obtained by elimination of a hydron from HCl) (ii) H_2O (base$_2$, say) and its corresponding acid H_3O^+ (acid$_2$; obtained by adding a hydron to H_2O):

$$\underset{\text{acid}_1}{HCl} + \underset{\text{base}_2}{H_2O} \rightleftharpoons \underset{\text{acid}_2}{H_3O^+} + \underset{\text{base}_1}{Cl^-}$$

The acid-base pairs HCl–Cl^- or H_3O^+ – H_2O are called conjugate acid-base pairs. *Chemical species which differ from one another only to the extent of a hydron (proton) transfer are termed a conjugate acid-base pair.*

We note further that a molecule of water acts as a base towards a more strongly acidic substance and forms the weak conjugate acid H_3O^+. Similarly, it can act as an acid towards a more strongly basic substance when it gives the conjugate base OH^-. Water is thus said to be an **amphiprotic solvent**. Any solvent capable of functioning as an acid as well as a base by releasing or accepting a proton (hydron) is called an amphiprotic solvent. Liquid ammonia, anhydrous hydrofluoric acid and acetic acid are other examples of amphiprotic solvents.

The stronger the tendency of a species to donate a hydron in a given medium, the stronger will be its acid character. The corresponding conjugate base of a strong acid must be weak, in as much as its tendency towards the reverse combination is less marked. So it may be concluded that *conjugate bases of strong acids must be weak in nature*. The manifestation of the acid or base character of a substance is ultimately determined by the tendency of the medium to accept or release a hydron. Interaction between the conjugate acid-base pair of the medium and the conjugate acid-base pair of an added species leads to combination between the stronger acid and stronger base available in the system and release of the weaker acid and base.

As an example, we consider the interaction between hydrogen sulfate ion and the carbonate ion in water

$$HSO_4^- + CO_3^{2-} \rightleftharpoons SO_4^{2-} + HCO_3^-$$

Here water is both a weaker acid than HSO_4^- and a weaker base than CO_3^{2-}. So it does not take any direct role in the above equilibrium which lies mainly towards right.

Relative strengths of acids and bases can be expressed according to this theory in the same manner as in the Arrhenius definition. For the dissociation of an acid HA in water, represented as

$$HA + H_2O = H_3O^+ + A^-$$

the equilibrium constant K in terms of concentrations is given as

$$K = \frac{C_{A^-}.C_{H_3O^+}}{C_{HA}.C_{H_2O}};\ \text{or,}\ K.C_{H_2O} = \frac{C_{A^-}.C_{H_3O^+}}{C_{HA}}$$

Since C_{H_2O} is comparatively very large in an aqueous solution, it may be practically taken to be constant and we may replace $K.C_{H_2O}$ by some other constant K'. The strength of the acid HA may then be expressed in terms of K', which has the same expression as the ionization constant of an acid in the Arrhenius definition, allowing hydration of H^+. The procedure may also be applied to other media to determine relative acid-base strengths of substances. Hydrolysis of salts and catalytic activity of acids may also be explained in the light of this theory.

Kamlet and *Taft* have proposed an acid-base definition which is very similar to the Brönsted theory but does not impose the condition of hydron (proton) transfer. According to them, an acid donates a hydron (proton) within a hydrogen bond that is formed while a base donates an electron pair to a proton within a hydrogen bond that is formed. The definition gains importance with non-aqueous protic solvents (Ch 12).

10.1.4 The Lewis Theory

Lewis (1923) attributed acid-base character of a substance to its inherent electronic structure. According to him *an acid is any species which can accept a pair of electrons to form a covalent bond*; and, *a base is any species that can donate a pair of electrons to establish a covalent bond*.

A Lewis acid should therefore possess an empty orbital of low energy suitable for accepting a pair of electrons and formation of a covalent bond with the donor atom. If there were no bond formation, the donor atom would lose its share of the electrons donated and thus undergo a change in oxidation number (redox process). Boron compounds of the formula type BX_3 (X = halogen or alkyl group) are common examples of Lewis acids. There are only six bonding electrons in boron-based orbitals while the boron can accomodate eight electrons in four possible sigma bonding orbitals, attaining a coordination number of four:

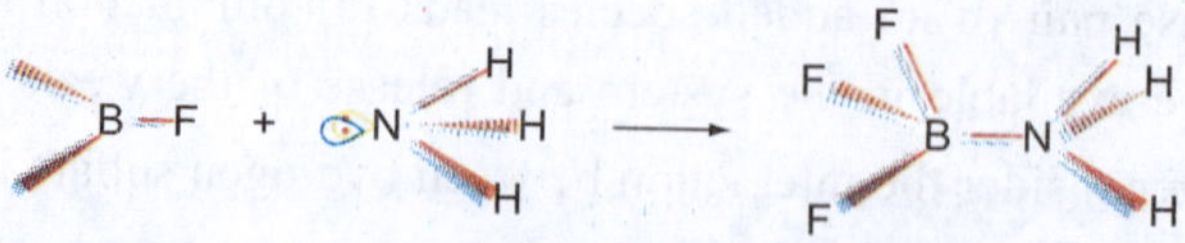

An electron pair from the highest occupied MO (HOMO) of ammonia takes part in forming a bond with boron using the lowest unoccupied MO (LUMO) on BF_3. The valence orbitals of boron are reorganized from planar to tetrahedral spatial distribution. We shall come to this point later.

The product of a Lewis acid-Lewis base combination is commonly called an *adduct*. Boron trifluoride forms a very stable adduct with the Lewis base diethyl ether by forming a bond with one of the lone pairs of electrons on oxygen:

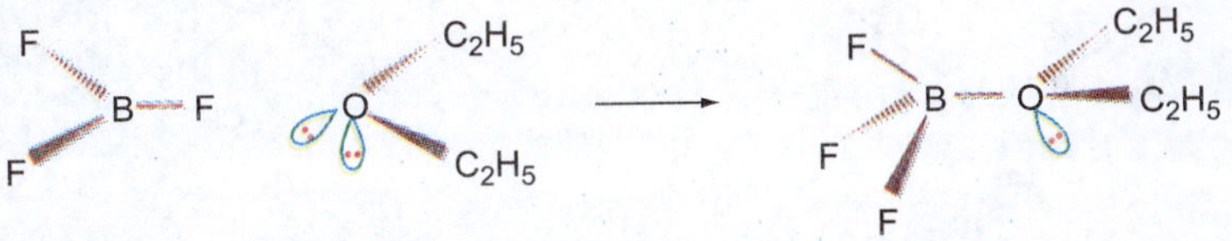

In spite of the energy required for reorganization of the geometry around boron from planar to nearly tetrahedral, the adduct is very stable — it decomposes into its components at its boiling point of 125–126°C which is significantly higher compared to the individual components: BF_3 : –99.9°C ; $(C_2H_5)_2O$: 34.5°C.

The same reasoning leads us to expect AlX_3 and SbX_5 to be good Lewis acids. The central atom in each has two electrons short of the maximum number of electrons that can be accomodated in the bonding orbitals for next higher coordination number.

Metal ions (or atoms) behave as Lewis acids when coordinated by different ligands:

$$Ag^+ + 2NH_3 \longrightarrow [Ag(NH_3)_2]^+$$

Hydration of metal ions in aqueous solution is thus a Lewis acid-base interaction.

We may summarize the situations in which a substance would behave as a Lewis acid:

(i) A molecule with an incomplete octet, e.g., $B(CH_3)_3$, $BeCl_2$ etc.

(ii) A metal atom or ion, e.g., $Ni(CO)_4$, $Na(H_2O)_6^+$ etc.

(iii) A molecule or ion capable of accepting an electron pair by rearrangement of its valence electrons. CO_2 acts as a Lewis acid by opening one of its π-bonds:

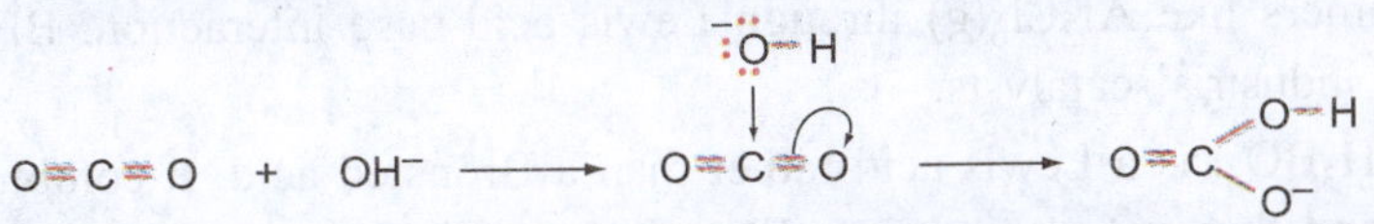

(iv) The central atom in many molecules may expand its octet to accomodate a pair of electrons, as in the formation of species like SiF_6^{2-}, PCl_6^- etc.

Lewis bases, as expected, should possess a pair of sigma-symmetry non-bonding electrons which may be donated to a Lewis acid forming a bond. Molecules or ions with one or more unshared pair(s) of electrons are thus potential Lewis bases, e.g., H^-, F^-, H_2O, NH_3 etc. However, the mere presence of unshared pair(s) of electrons is not a criterion for behaving as a Lewis base, e.g., He, Ne etc. Pi-electron donors may also function as effective Lewis bases. Ethylene, acetylene and molecules with delocalized π-electron systems like benzene and naphthalene are efficient Lewis bases. Carbon monoxide is an example of a molecule which acts as a σ-donor Lewis base and at the same time as a π-acceptor Lewis acid by accepting electron density from coordinated metal atoms through π-bonding with its vacant antibonding molecular orbital. Such compounds are known as π-acid ligands (chapter 32).

Lewis bases having several donor atoms give rise to interesting ligands. We shall find many such examples in chelates and macromolecular complexes. Chelating or macrocyclic Lewis acids are also known (Fig. 10.1) with more than one acceptor sites placed around a single Lewis base.

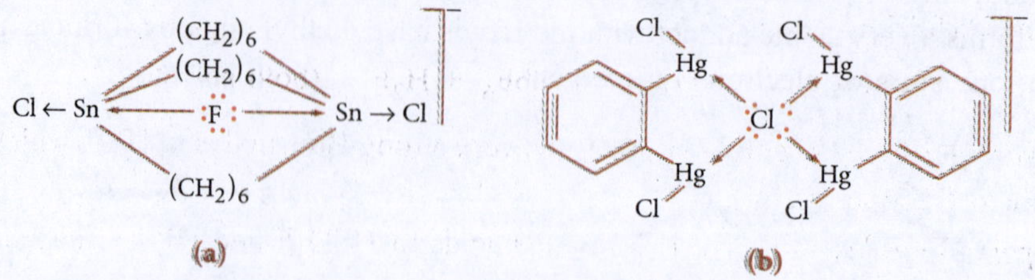

Fig.10.1: Examples of chelating or macrocyclic Lewis acids.

(a) F^- ion (base) in the adduct $[F : SnCl(C_6H_{12})_3 ClSn]^-$. The anion has been isolated as its salt with $[(C_4H_9)_4N]^+$ cation.

(b) Cl^- ion (base) in the adduct $[Cl : (C_6H_4 Hg_2Cl_2)_2]^-$, known with the cation $[Ph_4P]^+$.

Lewis acid and base properties of the main group elements and their compounds are important in understanding their chemistry:

Groups 1 and 2: The metal ions M^+ and M^{2+} may be considered Lewis acids in their aqua cations, the H_2O molecules behaving as Lewis bases. The alkali metal cations, with low charge and relatively large size, form complexes with cyclic polyethers ("crown ethers") and cryptands. The M^{2+} cations form a wide range of stable coordination compounds (see individual groups: chapters 18 and 19).

Groups 13 and 14: The boron trihalides are inherently electron deficient and good Lewis acids. Lewis acidity increases in the order

$$BF_3 < BCl_3 < BBr_3$$

as electron deficiency on B increases with increasing mismatch in energy and size of the larger halogen orbitals with the boron $2p$ orbital, making π-delocalization difficult. The aluminium halides form dimers like $Al_2Cl_6(g)$ through Lewis acid-base interaction. BF_3 and $AlCl_3$ are widely used as industrial catalysts.

Boric acid, H_3BO_3, is a Lewis acid rather than a Brönsted acid. It combines with an OH^- ion from H_2O and thus increases the acidity of the medium.

$$B(OH)_3 + 2H_2O \rightleftharpoons H_3O^+ + B(OH)_4^- \quad pK_a \simeq 9.2$$

A concentrated solution is more acidic owing to formation of a cyclic trimer:

$$3B(OH)_3 \rightleftharpoons [B_3O_3(OH)_4]^- + H_2O + H_3O^+ \quad pK_a \simeq 7.$$

Except carbon, the elements in Group 14 may behave as Lewis acids by expansion of coordination number. Among the tetrahalides, Lewis acidity follows the order consistent with electron withdrawing power of the halogen:

$$SiF_4 > SiCl_4 > SiBr_4 > SiI_4$$

Ge and Sn halides behave similarly.

$SnCl_2$ may function both as a Lewis acid and a Lewis base (see Problem 10.17):

Lewis acid: $SnCl_2 + Cl^- \longrightarrow SnCl_3^-$

The $SnCl_3^-$ ion, with an unshared pair of electrons on Sn, may behave as a Lewis base, as in $(CO)_5Mn \leftarrow SnCl_3$.

Group 15: Oxides and halides of heavier elements in the group may exhibit Lewis acidity. SbF_5 is a powerful Lewis acid:

$$SbF_5(l) + 2HF(l) \longrightarrow SbF_6^- + H_2F^+ \quad \text{(both solvated)}$$

In anhydrous HF, the reaction produces the very strong Bronsted acid H_2F^+ which is classed as a superacid (section 10.3.5).

Other pentafluorides also give rise to superacids in HSO_3F and HF. Acidity decreases as $SbF_5 > AsF_5 > TaF_5 > NbF_5 > PF_5$.

Though PCl_5 and $SbCl_5$ are quite stable, $AsCl_5$ is very unstable (see Group 15). However, many Lewis acid-base complexes of $AsCl_5$ are known:

$AsCl_4^+ AlCl_4^-$ (*preparation*: Cl_2 on $AsCl_3 + AlCl_3$)

$AsCl_5 . OPMe_3$ (*preparation*: Cl_2 on $AsCl_3 + OPMe_3$)

$Et_4N^+ AsCl_6^-$ (*preparation*: Cl_2 on $AsCl_3 + Et_4N^+Cl^-$)

$SbCl_5$ and some other MX_5 (M = Nb, Ta) molecule can serve as dibasic Lewis acids by accepting electron pairs from two donors

$$2SbCl_5 + 2MeCN \longrightarrow SbCl_4(NCMe)_2^+ + SbCl_6^-$$

Groups 16 and 17: Sulfur dioxide can function both as a Lewis acid and a Lewis base. As a Lewis acid, it accepts a pair of electrons (from, say, $:NR_3$) on the sulfur atom. While acting as a Lewis base, it may donate an electron pair either through the O atom (e.g., to SbF_5) or through the S atom (as in the complex ion $[Ru(NH_3)_4Cl(SO_2)]^+$.

Sulfur trioxide is a strong Lewis acid. It reacts with a variety of organic and inorganic molecules forming adducts, for example,

$$Ph_3P \xrightarrow{SO_3} Ph_3P.SO_3 \;;\quad Ph_3AsO \xrightarrow{SO_3} Ph_3AsO.SO_3 \;;\quad H_2O_2 \xrightarrow{SO_3} H_2SO_5$$

SO_3 can also act as a Lewis base towards strong electron-pair acceptors like AsF_3, SbF_3 etc.

The MX_4 compounds of Group 16 elements can either donate an electron pair or accept an electron pair. For example, SF_4 forms both $SF_3^+ SbF_6^-$ (Lewis base) and also SF_4.py and $Cs^+SF_5^-$ (Lewis acid). SF_4 is thus an amphoteric Lewis acid-base. Similarly $TeCl_4$ forms $TeCl_3^+AlCl_4^-$ and $PCl_4^+ TeCl_5^-$. TeF_4 is a dibasic Lewis acid:

$$2TeF_4 + 2py \longrightarrow [TeF_3(py)_2]^+ + TeF_5^-$$

The diatomic molecules Br_2 and I_2 are mild Lewis acids interacting with donor solvents like water, acetone or ethanol forming interesting charge-transfer complexes. Such charge-transfer complexes are also formed by π-donor bases like benzene or other aromatic hydrocarbons. Formation of polyhalides is also a manifestation of Lewis acid character:

$$I_2 + I^- \longrightarrow I_3^-$$

Other polyhalide ions are similarly formed (see group chemistry).

However, conventional hydronic (protonic) acids like H_2SO_4 and HCl are not covered directly by this approach as they do not apparently establish a covalent bond by accepting a pair of electrons (which a Lewis acid ought to). A compromise can be done by assuming that a hydrogen bond is first formed in the reaction of an acid HX and a base B as X–H···B; this undergoes fission to give the neutralization product.

The second major difficulty in the Lewis approach lies in ascertaining relative acid-base strengths. Here the acid or base strength of a species is dependent on the nature of reaction studied and not on the absolute power of accepting or donating a pair of electrons. Thus the fluoride complex of beryllium(II) is more stable than the fluoride complex of copper(II) ; this suggests that beryllium(II) is a stronger Lewis acid than copper(II). But the ammine complex of copper(II) is far more stable than that of beryllium(II), suggesting copper(II) to be a stronger Lewis acid.

One of the starting points of the Lewis approach was that acid-base reactions should be fast. But this is not so in practice for many reactions due to kinetic factors although they definitely involve acids and bases (e.g., the formation of $Cr(NH_3)_6^{3+}$).

Lastly, the catalytic activities of acids and bases, one of the preliminary considerations by Lewis, are not in harmony between the hydronic (protonic) acids and Lewis acids. Reactions catalyzed by Lewis acids are generally not catalyzed by the hydronic (protonic) acids.

Lewis acids and bases have been discussed further in section 10.3.

10.1.5. The Usanovich Definition

M. Usanovich (1939) proposed a very wide definition of acids and bases. According to him, *an acid is any chemical species which (i) reacts with a base, or (ii) accepts anions or electrons, or (iii) furnishes cations; a base is any chemical species that would* (i) *react with an acid,* or (ii) *combine with cations or* (iii) *furnish anions or electrons.*

The scope of the definition is thus very wide, covering all the previous definitions. Some additional examples of acid-base reactions according to this would be

Acid		Base		Salt	
SO_2	+	Na_2O	=	Na_2SO_3	SO_2 accepts the anion O^{2-}.
Cl_2	+	2Na	=	2NaCl	Chlorine accepts an electron
$Fe(CN)_2$	+	4KCN	=	$K_4Fe(CN)_6$	$Fe(CN)_2$ adds further anions (CN^-).

Usanovich considered (i) the general trend of acid-base character in the periodic table and (ii) the aspect of "coordination unsaturation". He observed that

(a) acidic character *increases* from left to right in the periodic table (Na_2O: strongly basic; Cl_2O_7: strongly acidic);

acidic character *decreases* from top to bottom in a group (As_2O_3 is less acidic than Sb_2O_3).

(b) the higher valence of an element is associated with greater acidic character (SnS_2 is more acidic than SnS.

(c) high charge of cations favours acidic property (Na_2O vs Al_2O_3), while high charge of anions promote basic property (Na_2O vs NaCl).

(ii) The term "coordination unsaturation" was used to mean the tendency of an atom to increase its covalence. Thus the sulfur atom in sulfur dioxide molecule is coordinatively unsaturated; as such, sulfur dioxide can accept an anion like O^{2-} and act as an acid.

The Usanovich definition is most general of the different acid-base definitions. It includes all acids and bases defined by the Lewis definition; at the same time it states that the donation or acceptance of electrons need not take place as shared pairs. Accordingly, oxidation-reduction reactions may be classified as acid-base reactions.

The only point raised against the Usanovich definition is its extreme generality—its scope is so large that almost all chemical reactions may be considered as acid-base reactions. The convenience of treating a particular type of compound as acids or bases is thereby lost.

10.1.6 The Lux Theory

According to this concept, an acid is an acceptor of oxide ion and a base is a donor of oxide ion:

$$\text{Base} \rightleftharpoons \text{Acid} + O^{2-}$$

This view is particularly useful in high temperature chemistry, as in the fields of ceramics and metallurgy:

$$CaO \rightleftharpoons Ca^{2+} + O^{2-}$$

$$SiO_2 + O^{2-} \rightleftharpoons SiO_3^{2-}$$

$$\underset{\textit{base}}{CaO} + \underset{\textit{acid}}{SiO_2} \rightleftharpoons \underset{\textit{salt}}{Ca\,SiO_3}.$$

Similarly, ores of Ti (as well as Ta, Nb) are dissolved in sodium pyrosulfate (an acid in the Lux sense) around 800°C:

$$TiO_2 + Na_2S_2O_7 = (TiO)SO_4 + Na_2SO_4.$$

SiO_2 displaces the more volatile bases P_4O_{10} or SO_3:

$$6\,SiO_2 + 2\,Ca_3(PO_4)_2 \longrightarrow 6CaSiO_3 + P_4O_{10}.$$

$$SiO_2 + CaSO_4 \longrightarrow CaSiO_3 + SO_3$$

Zinc oxide shows amphoterism as it can both donate or accept oxide ions:

base $ZnO + S_2O_7^{2-} = Zn^{2+} + 2SO_4^{2-}$

acid $Na_2O + ZnO = Na_2ZnO_2.$

The definition can be extended to cover acid-base reactions in terms of any negative ion transfer:

$$3NaF + AlF_3 \longrightarrow 3Na^+ + AlF_6^{3-}$$

$$Na_2S + CS_2 \longrightarrow Na_2CS_3$$

$$EtNa + Et_2Zn \longrightarrow Na^+ + ZnEt_3^-$$

$$2XeF_6 + SiO_2 \longrightarrow 2XeOF_4 + SiF_4$$

Problem 10.1

Predict how the acidity will change on adding SiO_2 to a molten mixture of Fe + FeO.

Solution: SiO_2 acts as an acceptor of oxide ion (forming $FeSiO_3$) from FeO : the acidity of the solution thus increases.

$$SiO_2 + FeO \longrightarrow FeSiO_3.$$

We have discussed several basic approaches to define, or rather describe, acids and bases. Each approach is suitable within certain domains and it would be meaningless to try to select one ultimate definition. At best we can frame a general principle of acid base reactivity that will be independent of the definition. Summarily, this principle states that the strongest acid present in a system will react with the strongest base available, without changing the formal oxidation state of any atom.

10.2 STRENGTHS OF BRÖNSTED ACIDS AND BASES

The strengths of Brönsted acids and bases may be compared by the ease with which the proton (hydron) is released by an acid or captured by a base. Such strengths may be expressed by the dissociation constant as developed by the protonic (hydronic) definition; as shown earlier, this yields the same result as the Arrhenius definition for aqueous solutions. However, such comparison of strengths actually refer to the medium or solvent. On the other hand, gas phase proton (hydron) affinities show the inherent aptitude of any species to bind a hydron in the gas phase.

10.2.1 Gas-phase Proton (hydron) Affinity

We have seen that the acidity or basicity of a Bronsted acid (or base) in solution is actually determined by the nature of the solvent. This is also true for many species covered by the solvent system definition. A comparison of acid-base strength of Bronsted acids or bases in the gas phase is provided by the gas phase proton (hydron) affinity. Analogous to electron affinity, the proton affinity (P.A.) is the energy *released* when a proton (hydron) is added to an isolated atom or molecule (or ion):

$$B(g) + H^+(g) \longrightarrow BH^+(g)$$

The thermodynamic quantity proton attachment enthalpy is equal to –P.A. and is equal to ΔH for the reverse process:

$$BH^+(g) \longrightarrow B(g) + H^+(g).$$

The proton affinity is thus a measure for both the acidity of BH^+ and basicity of B in the gas phase. A large proton affinity means that removal of the hydrogen ion from $BH^+(g)$ is difficult, that is B(g) is a strong base or $BH^+(g)$ is a weak acid.

Proton attachment enthalpies may be determined experimentally by mass spectrometry and ion-cyclotron resonance spectroscopy. A mixture of B(g) and $H_2(g)$ is subjected to an ionizing electron beam under varying voltage. From the voltage corresponding to appearance of $BH^+(g)$, one can calculate the enthalpy of formation of $BH^+(g)$. This may be combined with other enthalpy values like the enthalpy of formation of B and $H^+(g)$ to provide the desired ΔH value.

However, some species do not always yield BH^+ as a fragment. Again, the BH^+ fragments, when formed, may often contain excess energy above their normal ground states and thus introduce large uncertainties in the value of proton affinity.

We may use a Born-Haber cycle to correlate proton affinity P.A. with other familiar energy (enthalpy) quantities:

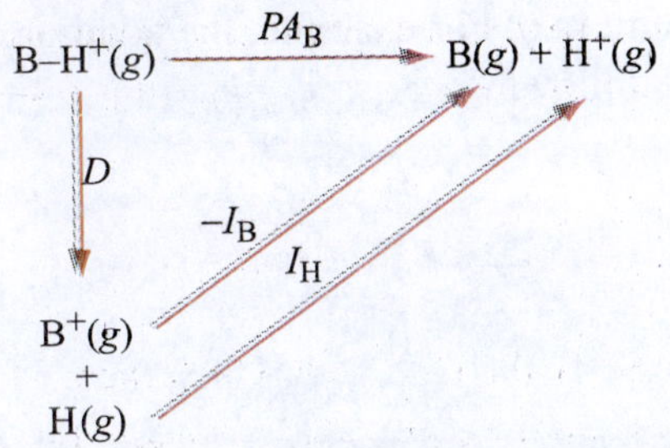

D = Bond dissociation energy of $B–H^+$

I_H = Ionization energy of hydrogen

I_B = Ionization energy of B (atom, molecule or ion)

PA_B = Proton affinity of B (gas phase)

From the cycle, $PA_B = D + I_H - I_B$

Table 10.2

Thermochemical data for estimation of gas phase proton affinity of some common hydrides (B). All values in kJ mol^{-1}.

(B =)	CH_4	CH_3^-	NH_3	NH_2^-	H_2O	OH^-	HF	F^-
D_{B-H^+}	444	435	535	456	598	497	761	569
I_B	1230	~8	979	71	1218	176	1523	331
PA_B	528	> 1695	866	1695	686	1632	548	1548
(B =)	**SiH_4**	**SiH_3^-**	**PH_3**	**PH_2^-**	**H_2S**	**HS^-**	**HCl**	**Cl^-**
D_{B-H^+}	439 ?	335	427	351	405	377	506	431
I_B	1138	100	962	121	1004	222	1230	347
PA_B	< 611	1544	774	1540	711	1464	585	1393

Problem 10.2

Sketch the appropriate thermodynamic cycle for estimating the gas phase proton affinity of (a) NH_3 (b) NH_2^-. (I_H = 1312 kJ mol^{-1}; other values are given in Table 10.2).

Solution: The relevant Born-Haber cycles are shown below:

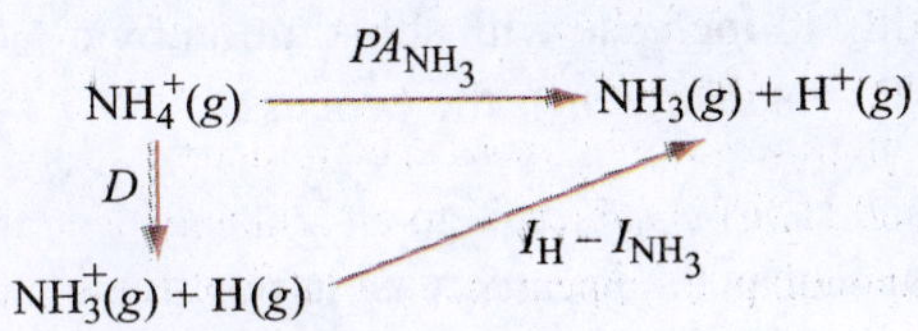

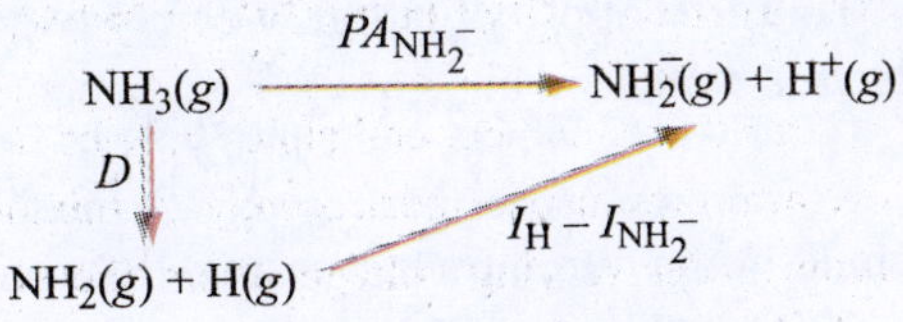

$$PA_{NH_3} = D + I_H - I_{NH_3} = 535 + 1312 - 979 = 868 \text{ kJ mol}^{-1}.$$

$$PA_{NH_2^-} = D + I_H - I_{NH_2^-} = 456 + 1312 - 71 = 1697 \text{ kJ mol}^{-1}.$$

Gas-phase proton affinities for some molecules and anions (conjugate bases) are given in Table 10.3. When the bases possess nonbonding lone-pair of electrons, they become bonding in the protonated species. When a molecule contains no lone-pair electrons, a proton may be bound by involvement of σ or π bond-pairs in a multicentre m.o. For example, the addition of a proton to methane or H_2 may involve species like the following:

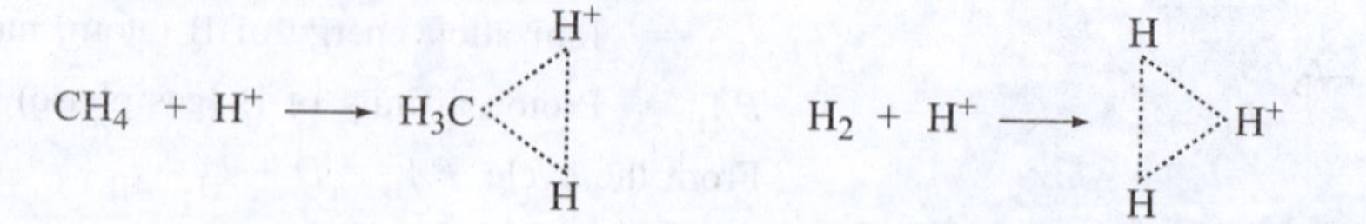

Table 10.3

Proton affinities of some molecules and anions (kJ mol^{-1})

H_2	423	NF_3	611	CH_3OH	777
H^-	1675	NCl_3	791	C_2H_5OH	797
LiOH	1008	O_2	423	CH_3O^-	1587
NaOH	1038	OH^-	1632	$C_2H_5O^-$	1470
KOH	1100	H_2O	686	CH_3SH	793
CH_4	528	H_2S	739	CF_3^-	1572
CH_3^-	1738	HS^-	1464	CCl_3	1515
SiH_4	≤ 611	HF	548	$CH_2NO_2^-$	1501
SIH_3^-	1544	F^-	1549	CH_3COO^-	1458
NH_3	866	HCl	585	CF_3COO^-	1350
NH_2^-	1695	Cl^-	1393	C_5H_5N	937
PH_3	774	HBr	590	$C_6H_5NH_2$	894
PH_2^-	1540	Br^-	1351		

Some trends of gas-phase acidic (and basic) character are interesting. The following relative acid strengths may be taken as examples.

(i) $C_2H_5OH > CH_3OH$ (iii) $CH_3SH < H_2S$ but $CH_3OH > H_2O$

(ii) $CHCl_3 > CHF_3$ (iv) $ClCH_2COOH > FCH_2COOH$

To a first approximation, we may expect the acidity to increase with substitution by a more electron withdrawing group. For a species HXR_n (R = substituent), the process

$$HXR_n(g) \longrightarrow H^+(g) + XR_n^-(g)$$

should be favoured by increasing electronegativity of R-groups, i.e., increasing positive charge on X. But this consideration alone cannot explain the acidity trends mentioned above. The energy of the ionization process, as written above, is merely the coulombic interaction energy between the proton and the anion. The actual energy term involves a second (hypothetical) factor — electronic relaxation of the anion XR_n^-; this involves redistribution of charge within the anion. The more readily the substituents can delocalize the negative charge on the anion, the greater will be its stability. This electronic relaxation, usually exothermic, will make a definite contribution to the total energy of the ionization process. Combination of these two effects now help us to understand the acidity trends given above.

(i) $C_2H_5OH > CH_3OH$. As the size of the alkyl group increases, it becomes more polarizable. This releases higher electronic relaxation energy in $C_2H_5O^-$ and makes it more stable.

(ii) $CHCl_3 > CHF_3$. There is greater delocalization of the negative formal charge on the carbon in case of $:CCl_3^-$. This increases the electronic relaxation energy and offsets the trend expected from electronegativity alone.

[Note: CH_3NO_2 is a still stronger acid as the carbon lone pair in the $:CH_2NO_2^-$ anion can be extensively delocalized into the π^* orbital of the NO_2 group.]

(iii) $CH_3OH > H_2O$ and $CH_3SH < H_2S$. In CH_3OH, we expect a greater electronic relaxation energy of CH_3O^- by way of hyperconjugation:

$$H_3C - O^- \longleftrightarrow H^- H_2C = O.$$

This effect must overcome the greater inductive effect of the methyl group, as evident from the acidity trend.

But for CH_3SH, the sizes of carbon and sulfur are greatly dissimilar, and the delocalization is not sufficient to overcome the greater electron donating ability of the methyl group. Therefore the electrostatic energy term dominates over the small electronic relaxation energy.

(iv) $ClCH_2COOH > FCH_2COOH$. The relaxation energy and the stability of the anion increase as the polarizability of the halogen atom increases from fluorine to chlorine.

A third process involving stereochemical rearrangement of the anion may also contribute to the energy of the ionization process. However, this energy is likely to be relatively small except in some unusual cases.

Gas phase base strengths also show trends different from those observed in aqueous solution. Thus, LiOH, NaOH and KOH are equally strong bases in water. But in gas phase, LiOH < NaOH < KOH; this order follows the electron releasing tendency of the cation to bind an extra proton. Similarly, proton affinity data suggest that pyridine and aniline should be stronger bases than ammonia in the gas phase, though they are much weaker base in water. A likely explanation for this behaviour lies in the strong stabilization of the ammonium ion in water through H–bonding.

Gas phase proton affinities are of limited applicability because they do not take into account the interaction with solvent, a factor that is highly significant in common reactions. In fact, gas phase proton affinity data suggest that transfer of a hydrogen ion from HCl(*g*) to $H_2O(g)$ would be highly endothermic and thus thermodynamically unfavourable.

$$HCl(g) + H_2O(g) \longrightarrow H_3O^+(g) + Cl^-(g)\ ;\quad \Delta H = 1393 - 686 = +707 \text{ kJ mol}^{-1}$$

10.2.2 Acids and Bases in Water

The acid-base behaviour of Bronsted acids in water is largely influenced by the high enthalpy of hydration of H^+ (-1130 kJ mol^{-1}). In the last section, we found that the transfer of an $H^+(g)$ from HCl(*g*) to $H_2O(g)$ is endothermic by 707 kJ mol^{-1}. We shall soon find that in aqueous solution the same transfer has a negative ΔH (-67 kJ mol^{-1}). ΔS for solution of HCl(*g*) is

–130 J mol^{-1}K^{-1} arising mainly from orientation of the water molecules around H^+ and Cl^- ions but the overall free energy change is still dominated by ΔH.

The equilibrium dissociation constant of an acid or base involving H^+ ion in solution may be taken as a measure of relative strength of the acid or base.

$$HA + H_2O \rightleftharpoons H_3O^+ + A^- ; \quad B + H_2O \rightleftharpoons BH^+ + OH^-$$

$$K_a = \frac{[H_3O^+]\,[A^-]}{[HA]} ; \qquad K_b = \frac{[BH^+]\,[OH^-]}{[B]}$$

For strong acids and bases the equilibrium lies almost completely towards right – they are said to be "*levelled*" to the acidity of H_3O^+ or basicity of OH^- (as the case may be). This is dissussed below.

Polybasic acids and polyacidic bases have more than one dissociation constant corresponding to each successive stage of dissociation. Dissociation constants for a number of common acids and bases at 298K are given in Table 10.4.

Table 10.4

Some dissociation constants in aqueous solution at 298K

	Equilibrium					K_a or K_b
Acetic acid	CH_3COOH	+	$H_2O = H_3O^+$	+	CH_3COO^-	1.8×10^{-5}
Hydrogen cyanide	HCN	+	$H_2O = H_3O^+$	+	CN^-	4.8×10^{-10}
Hydrogen fluoride	HF	+	$H_2O = H_3O^+$	+	F^-	6.8×10^{-4}
Nitrous acid	HNO_2	+	$H_2O = H_3O^+$	+	NO_2^-	4.5×10^{-4}
Monochloro acetic acid	$ClCH_2COOH$	+	$H_2O = H_3O^+$	+	$ClCH_2COO^-$	1.4×10^{-3}
Phenol	C_6H_5OH	+	$H_2O = H_3O^+$	+	$C_6H_5O^-$	1.3×10^{-10}
Hydrogen sulfide	H_2S	+	$H_2O = H_3O^+$	+	HS^-	1.1×10^{-7} (= K_1 for H_2S)
Hydrogensulfide ion	HS^-	+	$H_2O = H_3O^+$	+	$S^=$	1×10^{-14} (= K_2 for H_2S)
Carbonic acid	$H_2O + CO_2$*	+	$H_2O = H_3O^+$	+	HCO_3^-	4.2×10^{-7} (= K_1 for H_2CO_3)
Hydrogencarbonate ion	HCO_3^-	+	$H_2O = H_3O^+$	+	$CO_3^=$	4.8×10^{-11} (= K_2 for H_2CO_3)
Phosphoric acid	H_3PO_4	+	$H_2O = H_3O^+$	+	$H_2PO_4^-$	7.5×10^{-3} (= K_1 for H_3PO_4)
Dihydrogenphosphate ion	$H_2PO_4^-$	+	$H_2O = H_3O^+$	+	HPO_4^{2-}	6.2×10^{-8} (= K_2 for H_3PO_4)

Contd. to next page

* Aqueous solution of carbon dioxide does not contain as the "carbonic acid" molecule, H_2CO_3. The greater part of the dissolved CO_2 is only loosely hydrated.

Contd.

Hydrogenphosphate ion	HPO_4^{2-}	+	$H_2O = H_3O^+$	+	PO_4^{3-}	1×10^{-12} (= K_3 for H_3PO_4)
Ammonia**	NH_3	+	$H_2O = NH_4^+$	+	OH^-	1.8×10^{-5}
Methylamine	CH_3NH_2	+	$H_2O = CH_3NH_3^+$	+	OH^-	5.0×10^{-4}
Dimethylamine	$(CH_3)_2NH$	+	$H_2O = (CH_3)_2NH_2^+$	+	OH^-	7.4×10^{-4}
Trimethylamine	$(CH_3)_3N$	+	$H_2O = (CH_3)_3NH^+$	+	OH^-	7.4×10^{-3}
Aniline	$C_6H_5NH_2$	+	$H_2O = C_6H_5NH_3^+$	+	OH^-	4.0×10^{-10}

Note: 1. The successive dissociation constants of a polybasic acid become gradually smaller, as may be observed from the data for hydrogen sulfide or phosphoric acid ($K_1 > K_2 >> K_3$). It appears that the release of a positively charged proton becomes more and more difficult as the negative charge on the anion increases.

2. The values quoted are only approximate. They may differ from one book to another.

Dissociation constants may be determined from conductance measurements.

The levelling effect of water

In aqueous solutions, the extent of dissociation of an acid depends upon the tendency of the water molecules to accept the proton (hydron) i.e., on the basic nature of water. Similarly, the extent of dissociation of a base in water is dependent on the tendency of the water molecules to donate a proton (hydron) i.e., on the acidic nature of water. Thus the basic or acidic nature of water plays a significant role in determining the strengths of acids or bases.

Referring back to the equilibria

$$\text{I. } \underset{\text{acid}_1}{HA} + \underset{\text{base}_2}{H_2O} \rightleftharpoons \underset{\text{acid}_2}{H_3O^+} + \underset{\text{base}_1}{A^-} \quad \text{and} \quad \text{II. } \underset{\text{base}_1}{B} + \underset{\text{acid}_2}{H_2O} \rightleftharpoons \underset{\text{acid}_1}{BH^+} + \underset{\text{base}_2}{OH^-}$$

we may infer that all acids stronger than the oxonium ion would displace the equilibrium I far to the right while all bases stronger than the hydroxide ion would shift the equilibrium II to the right.

In fact, water acts as an appreciably good base towards all the common "strong" acids like HCl, H_2SO_4 and HNO_3. They are almost completely dissociated in aqueous solutions; as such, all of them appear equally strong in aqueous solution. Similarly, water acts as an appreciably good acid for all common "strong" bases (NaOH, KOH etc.) and they appear equally strong in water. This tendency of water to equalize the strengths of many acids to H_3O^+ or bases to OH^- is termed its **levelling effect;** *water acts here as a levelling solvent.* H_3O^+ is the strongest acid and OH^- is the strongest base that can exist in water. Comparison of acid or base strengths in water is rendered difficult by its levelling effect.

** Aqueous ammonia solutions are still commonly mentioned as "ammonium hydroxide", NH_4OH. Actually ammonia remains largely in the form of hydrated molecules.

The major strong proton (hydron) acids in water are

HBF_4	H_2SO_4	HCl	$HClO_4$
HNO_3	$H_2S_2O_7$	HBr	$HMnO_4$
HPO_2F_2	HSO_3F	HI	$H_2Cr_2O_7$
HPF_6	HSO_3Cl	$HClO_3$	
	HSO_3NH_2	$HBrO_3$	
	H_2SeO_4	HIO_3	

Except HBF_4 and HPF_6, all these strong acids contain the acidic hydrogen bound to a halogen or oxygen.

Strong proton (hydron) bases in water are furnished by anionic Lewis bases:

$$NH_2^- + H_2O \rightleftharpoons NH_3 + OH^-$$

Other examples are given below:

H^-	BO_3^{3-}	CH_3^-	N^{3-}	O^{2-}
			NH^{2-}	OCH_3^-
			NH_2^-	S^{2-}
			P^{3-}	
			PO_4^{3-}	

Proton affinity data for various species in solution provide us some insight into the levelling effect of a solvent. The proton (hydron) affinity of a species in solution is different from its gas phase proton affinity owing to solvation. For water as solvent, H-bonding contributes substantially to solvation of the H^+ ion (chapter 17), resulting in much higher proton affinity for $H_2O(l)$ compared to $H_2O(g)$:

$$H^+(g) + H_2O(g) \longrightarrow H_3O^+(g) \qquad -\Delta H = 686 \text{ kJ mol}^{-1}$$

$$H^+(g) + H_2O(l) \longrightarrow H_3O^+(aq) \qquad -\Delta H = 1130 \text{ kJ mol}^{-1}$$

Thus the effective proton affinity of $H_2O(l)$ is 1130 kJ mol^{-1}. Any hydrogen containing species having a lower effective proton affinity of its conjugate base is expected to transfer its proton (hydron) to a water molecule forming H_3O^+ in solution. The anions of common strong acids have effective proton affinity values much less than 1130 kJ mol^{-1}. Hence these acids get completely ionized in aqueous solution to give H_3O^+ and solvated to the acidity of H_3O^+ which is the strong acid that can exist in any aqueous solution. Similarly, the effective proton affinity of OH^- ion, the conjugate base of H_2O, is 1188 kJ mol^{-1}. Proton acceptors stronger than $OH^-(aq)$ would abstract an H^+ ion from water and form $OH^-(aq)$.

When the relevant species of an acid or base have effective proton affinity values lying between 1130 and 1188 kJ mol^{-1} in water, the acids or bases may be differentiated in aqueous solution according to their K_a (or K_b) values. Water now behaves as a differentiating solvent toward these acids or bases. The range or width of such discrimination may be considered as a window bound by the $H_3O^+(aq) - H_2O(l)$ and $H_2O(l) - OH^-(aq)$ couples. The pK_a values of H_3O^+ (≈ -1.74) and H_2O (≈ 15.74) suggest that the window is open to acids having approximate pK_a values between 0 and 14, covering a span equal to pK_w of water.

However, such correlations are subject to certain limitations. Proton affinities are enthalpy terms which may be used without appreciable error in gas phase reactions where the entropy terms on the two sides of a reaction nearly cancel one another. In solution, entropy contribution to ΔG° becomes appreciable owing to strong solvation effects. The acidity range available in a solvent is more reliably given by the relation $\Delta G^\circ = -RTlnK_s$ where K_s is the autoprotolysis constant of the solvent (= K_w for water).

Levelling effect of other solvents

Other protonic solvents similarly have their discrimination window and levelling boundaries set by the protonated solvent molecule or the *lyonium ion* (analogue of H_3O^+ in water) and the deprotonated solvent molecule or *lyate ion* (≡ OH^- in water). Liquid ammonia, with $pK_s = 33$ (at –50°C), has a wider range of discrimination. This permits observation of acidic behaviour for many compounds which show little or no acidity in water (chapter 12). Thus, benzoic acid is a very weak acid in aqueous solution; but in liquid ammonia it acts as a very strong acid. In general, it may be expected that acidic solvents (like acetic acid) will exert less levelling effect on acid strengths (and similarly basic solvents on base strengths). Relative strengths of the common mineral acids have been obtained by conductance measurement in anhydrous acetic acid (see sec. 12.2.3). The order found is

$$HClO_4 > HBr > H_2SO_4 > HCl > HNO_3$$

Aprotic solvents (like C_6H_6) would normally exert less levelling effect; but their dielectric constant is low. This leads to other secondary equilibria involving association of ions in such media making comparison of acid strength difficult.

On the other hand, even basic solvents like ethylenediamine have been used for comparison of acid strengths. The low dielectric constant sets the ion association equilibrium:

$$HA + xS \underset{\mathbf{I}}{\rightleftharpoons} H^+S_x + A^- \underset{\mathbf{II}}{\rightleftharpoons} H^+S_xA^- \qquad [S = \text{Solvent}]$$

Even if equilibrium I lies to right equally with diffrent acids, equilibrium II differs for each and results in varying conductance.

Acid strength of binary hydrogen compounds

In the periodic table, acidity among binary hydrogen compounds increases from left to right along a period:

$$NH_3 << H_2O << HF \;;\; PH_3 << H_2S << HCl$$

Along a group, acidity increases as one passes downward. Thus:

$$H_2O < H_2S < H_2Se < H_2Te \;;\; \text{or} \quad HF < HCl < HBr < HI$$

The variation of acidity along a period may be explained from the increasing electronegativity of the elements linked to hydrogen. Here size of the atom does not change appreciably from one element to the next; greater electronegativity of an element makes proton release easier due to greater shift of the bonding electron cloud towards it. So the acidity increases.

But consideration of electronegativity alone does not explain the increase of acidity down a group. (Rather this leads to opposite expectations). The size of the atom increases rapidly from one element to the next and upsets the electronegativity factor. As the atoms become larger, bonding with the small hydrogen atom becomes weaker (poorer overlap); the increase of acid character parallels this decrease in bond strength.

Nitrogen and chlorine have almost equal electronegativity. Yet the hydrides NH_3 and HCl are quite opposite in their acid-base behaviour in water. We may understand this difference from a comparison of the enthalpy changes (ΔH) for dissolving the gaseous hydrides in water. We consider the following Born-Haber cycle with usual notations:

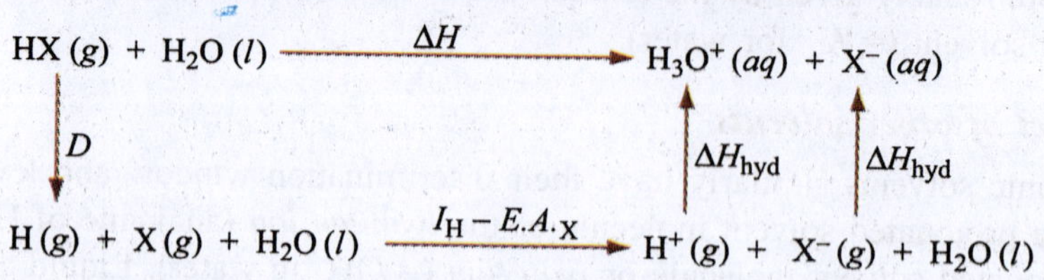

Clearly, $\Delta H = D + I_H - EA(X) + \Delta H_{hyd}(H^+) + \Delta H_{hyd}(X^-)$

$I_H = +1312 \text{ kJ mol}^{-1}$; $\Delta H_{hyd}(H^+) = -1130 \text{ kJ mol}^{-1}$

For HCl (X = Cl), $D = 431$; EA (Cl) = 349 ; $\Delta H_{hyd}(Cl^-) = -322$ (all kJ mol^{-1})

$\Delta H = 431 + 1312 - 349 - 1130 - 322 = -58 \text{ kJ mol}^{-1}$

For NH_3 (X = NH_2), $D = 389$; EA (NH_2) = 71 ; ΔH_{hyd} (NH_2^-) = – 460 (all kJ mol^{-1})

$\Delta H = 389 + 1312 - 71 - 1130 - 460 = +40 \text{ kJ mol}^{-1}$

In both cases, ΔS is negative due to higher ordering in solution but their contributions (in Joules) are expected to be similar. Hence the ΔG terms for the processes may be compared from the ΔH values calculated above. We find that dissolution of $NH_3(g)$ in water to furnish $H^+(aq)$ is favoured by the bond dissociation energy (D) and enthalpy of hydration of $NH_2^-(g)$ but the very low $E.A.$ value of $NH_2(g)$ makes ΔH positive.

Problem 10.3

Calculate ΔH for dissolution of $NH_3(g)$ in water to form $NH_4^+(aq)$ and $OH^-(aq)$ from the following ΔH values (all in kJ mol^{-1}):

Vaporization of H_2O (l) = 42; $D_{H-OH} = 498$; $I_H = 1312$; $\Delta H_{EA}(OH) = -176$; $\Delta H_{PA}(NH_3) = -866$; $\Delta H_{hyd}(NH_4^+) = -377$; $\Delta H_{hyd}(OH^-) = -460$.

Hint: The relevant Born-Haber cycle is

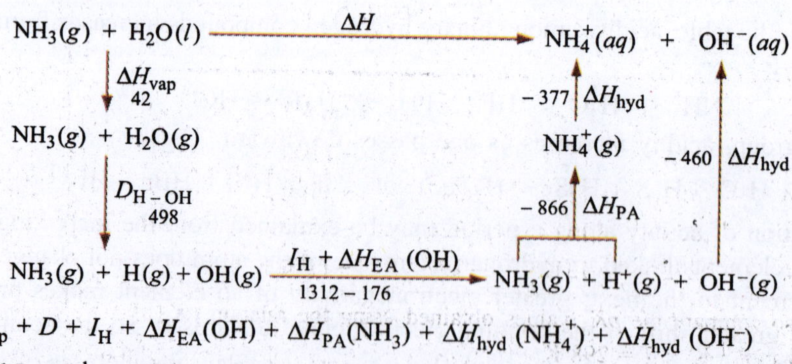

$\Delta H = \Delta H_{vap} + D + I_H + \Delta H_{EA}(OH) + \Delta H_{PA}(NH_3) + \Delta H_{hyd}(NH_4^+) + \Delta H_{hyd}(OH^-)$

$= -27 \text{ kJ mol}^{-1}$ reaction.

Note: The value of ΔH is marginally negative. The main contribution comes from the large negative enthalpy of proton attachment of $NH_3(g)$. Other terms are by and large comparable with those in the cycle for ammonia dissolving as an acid.

A comparison of the strengths of the halogen hydracids (HX) in water in terms of free energy changes (ΔG) is more meaningful since the ΔS term for HF is significantly different from others owing to strong hydrogen bonding. Also, the standard free energy change ΔG° gives the acid dissociation constant K_a by the relation $\Delta G^\circ = -RT \ln K_a$.

We proceed with a Born-Haber cycle of the type

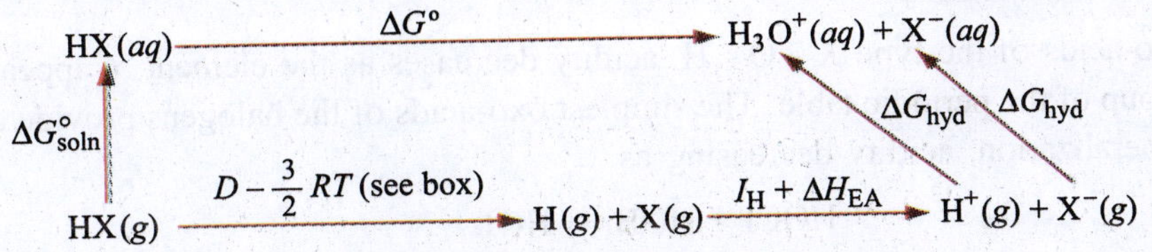

$$\Delta G^\circ = -\Delta G^\circ_{soln} + D - \tfrac{3}{2}RT + I_H + \Delta H_{EA} - \Delta G^\circ_{hyd}(H^+) - \Delta G^\circ_{hyd}(X^-).$$

For I_H and $E.A.$ the values of corresponding enthalpy terms may be used without much error since the processes are confined to the gaseous phase.

> By arguments similar to those used for ionization enthalpy (chapter 5),
>
> for the process: $HX(g) \longrightarrow H(g) + X(g), \quad \Delta H^\circ_T = \Delta H^\circ_0 + \int_0^T \Delta C_p dT$
>
> Molar heat capacity at constant pressure of H(g) and X(g) are $\frac{5}{2}R$ each. HX(g) has 3 translational, 2 rotational and 1 vibrational mode. Hence $C_p = \frac{3}{2}R + 2\times\frac{1}{2}R + R = \frac{7}{2}R$; $\Delta C_p = \frac{3}{2}R$.
>
> or, $\Delta H^\circ_T = \Delta H^\circ_0 + \frac{3}{2}RT = D_e + \frac{3}{2}RT \quad \therefore D_e = \Delta H^\circ_T - \frac{3}{2}RT$

Some data relevant to this cycle for HX (X = halogen) are shown in Table 10.5. The values show wide variation from source to source, but are nevertheless helpful in drawing a comparative account.

Table 10.5

Some thermodynamic parameters for HX (X = F, Cl, Br, I). All values in kJ mol^{-1}.

	$-\Delta G^\circ_{soln}$	$D - \frac{3}{2}RT$	I_H	$-EA_x$	$\Delta G^\circ_{hyd}(H^+)$	$\Delta G^\circ_{hyd}(X^-)$	ΔG°	K_a^*
HF	+ 25	536	1312	− 331	− 1100	− 423	+19	4.7×10^{-4}
HCl	− 4	402	1312	− 347	− 1100	− 305	− 42	2.3×10^{7}
HBr	− 4	339	1312	− 326	− 1100	− 276	− 55	4.4×10^{9}
HI	− 4	272	1312	− 297	− 1100	− 238	− 55	4.4×10^{9}

One can also compare the pK_a values obtained using the relation $pK_a = \Delta G^\circ$ (Joules)/2.303RT ; $R = 8.314$ J mol^{-1} K^{-1}; $T = 298$ K.

We observe that in aqueous solution HF behaves as a weak acid primarily because of the high H–F bond energy. The free energy change of hydration of the fluoride ion is higher than that for other halide ions but insufficient to compensate for the increased bond energy. The lower electron affinity of fluorine also contributes to its weaker acidity. At the same time the free energy of hydration of the F^- ion is considerably less favourable than its enthalpy of hydration (–464 kJ mol^{-1}) owing to extensive ordering of water molecules through strong H-bonds around the F^- ion. The strong role of H-bonding is again observed in concentrated (5 –15 M) aqueous solutions of HF which behave as strong acids owing to ionization to H_3O^+, HF_2^- and other $H_nF_{n+1}^-$ species.

Strengths of oxo-acids

For oxo-acids of the type X – O – H, acidity decreases as the element X appears lower in a given group of the periodic table. The simplest oxo-acids of the halogens provide an example of this generalization, acidity decreasing as

$$HOCl > HOBr > HOI.$$

The acidic hydrogen is linked to the halogen atom via oxygen and the acid strength is determined by the ease of breaking of the O – H bond. The electronegativity of the atom X plays the most important role in the matter, by pulling the charge cloud of the X – O bond to itself; the effect is now relayed by the oxygen atom to the O – H bond (inductive effect), facilitating release of the acidic hydrogen. This is why acidity decreases down the group with decreasing electronegativity of the element X:

Acid	Order of K_a	Acid	Order of K_a
HOCl	10^{-8}	H_2SO_3	10^{-2}
HOBr	10^{-9}	H_2SeO_3	10^{-3}
HOI	10^{-13}	H_2TeO_3	10^{-6}

When the element X has additional oxygen atoms attached to it, inductive effect of these oxygen atoms makes the acid stronger—acidity increases with the number of oxygen atoms:

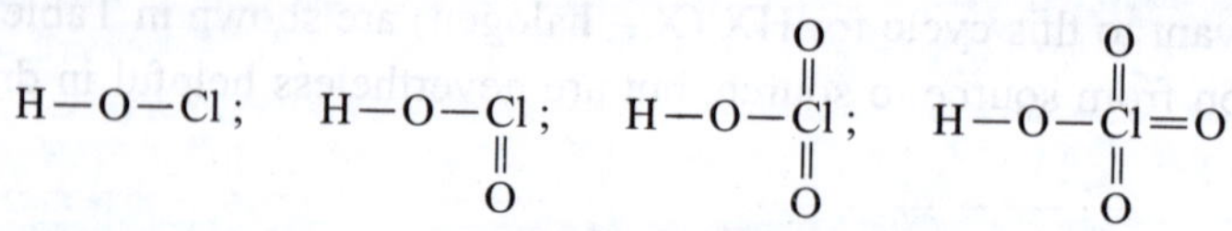

O.N. of chlorine	+1	+3	+5	+7
Order of K_a	10^{-8}	10^{-2}	Large	Very large

Similarly, sulfuric acid (H_2SO_4; $HO.SO_2.OH$) is a stronger acid than sulfurous acid (H_2SO_3; HO.SO.OH) since the sulfur atom in sulfuric acid has more coordinated oxygen atoms.

Summarily, an oxoacid will be stronger hydrogen ion (hydron) donor when the rest of the molecule exerts greater electron-withdrawing role. This is favoured by

(a) a high formal oxidation state of the central atom;

and (b) presence of a large number of terminal O-atoms or other electronegative atoms like F or Cl.

On the contrary, an oxoacid will be a weak donor of hydron (proton) when the oxoanion has more available electrons. Such situations arise due to the presence of

(a) a net negative charge;

(b) less electronegative atoms within the oxoanion;

(c) extra hydroxylic hydrogen atoms;

(d) nonbonding electron pairs and similar causes.

Among the three oxoacids of phosphorus, H_3PO_2, H_3PO_3 and H_3PO_4, acidity actually decreases as $H_3PO_2 > H_3PO_3 \geq H_3PO_4$. This is contrary to our expectation from the number of oxygen atoms alone. But a consideration of the structures of these acids explains the observed trend:

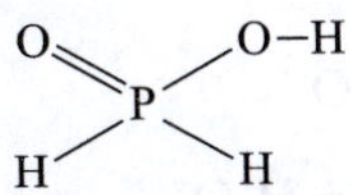
H_3PO_2

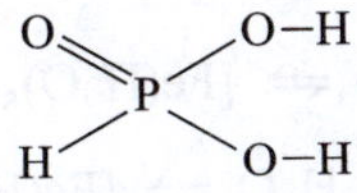
H_3PO_3

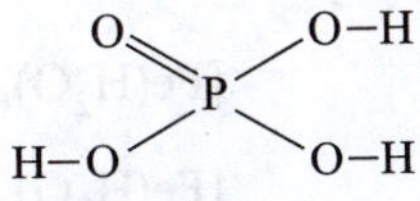
H_3PO_4

Acidity arises from the – O – H groups, one in H_3PO_2, two in H_3PO_3 and three in H_3PO_4. The inductive effect of the single oxygen atom acts on the single – O – H bond in H_3PO_2, but it is distributed over two and three – O – H bonds in H_3PO_3 and H_3PO_4 respectively. Accordingly, acidity decreases from H_3PO_2 to H_3PO_4.

Several attempts have been made to correlate the pK_a value(s) of oxoacids of formula type $O_nX(OH)_m$ with the number of hydroxylic oxygen (m) and non-hydroxylic oxygen (n). The simplest one is due to Pauling.

Pauling's rules: Pauling proposed two empirical rules to predict the strengths of mononuclear oxoacids:

(i) For an oxoacid of type $O_nX(OH)_m$, $pK_a \simeq 9 - 7n$. Subsequently it was found that the relation $pK_a = 8 - 5n$ fits better with some acids (see below).

(ii) For polyprotic acids ($m > 1$), the successive pK_a values increase by five units.

Table 10.6

Estimated pK_a values for different oxoacids (experimental values in parentheses)

	$pK_a = 9 - 7n$	$pK_a = 8 - 5n$	Examples
$n = 0$	9	8	HOCl (7.2); $Si(OH)_4$ (10)
$n = 1$	2	3	H_2CO_3 (3.6); $HClO_2$ (2.0); H_3PO_4 (2.1 ; 7.4 ; 12.7)
	$pK_1 = 2$	$pK_1 = 3$	H_3PO_3 (1.8 ; 6.6); H_3PO_2 (2.0)
	$pK_2 = 7$	$pK_2 = 8$	
	$pK_3 = 12$	$pK_3 = 13$	
$n = 2$	–5	–2	HNO_3 (–1.4) H_2SO_4 (–2.0 ; 1.9) $HClO_3$ (–1.0)
$n = 3$	–12	–7	$HClO_4$ (–10)

The importance of electronegativity is inherent in all these approaches. The highly electronegative nonhydrogenated oxygen(s) draws electrons away from the central atom and thereby increases the positive charge on the central atom. This enhanced positive charge in turn draws the electrons from the –OH bond and facilitates release of H^+. Acid strength of a molecule thus increases with the number of nonhydrogenated oxygen atoms.

Similar conclusions are also reached if we consider the conjugate base. Nonhydrogenated oxygen atoms help delocalization of the negative charge on the conjugate base and stabilize it, making it a weak base.

Acidity of aqua ions

Metal ions with high charge and small radius exhibit acidic behaviour in aqueous solution ("hydrolysis"):

$$[Fe(H_2O)_6]^{3+} + H_2O \rightleftharpoons [Fe(H_2O)_5(OH)]^{2+} + H_3O^+$$

$$[Fe(H_2O)_5(OH)]^{2+} + H_2O \rightleftharpoons [Fe(H_2O)_4(OH)_2]^+ + H_3O^+$$

In less acidic solution, hydroxide or oxide bridges may form between the metal ions and eventually the hydoxide or hydrated oxide may be precipitated. We shall come to this later.

In general the acidity of aqua ions appears to increase with increasing charge and decreasing radius of the ions—Z^2/r is a frequently used parameter for this purpose (Table 10.7).

The alkali metal ions essentially do not show any acidity, while the alkaline earth metal ions show only slight acidity. Transition metal ions, owing to their higher effective charge, are more strongly acidic, as shown by the equilibrium constants (pK_a at 298 K) for the reaction

$$[M(H_2O)_6]^{n+} + H_2O \rightleftharpoons [M(H_2O)_5(OH)]^{(n-1)+} + H_3O^+$$

A higher pK_a value indicates weaker acidity.

Table 10.7

Acidities of aqua ions

M^{n+}	Z^2/r*	pK_a	M^{n+}	Z^2/r*	pK_a
Fe^{2+}	11.2	9.5	Fe^{3+}	29.4	2.2
Cu^{2+}	11.8	8.3	Cr^{3+}	30.6	3.8
Ni^{2+}	12.4	9.3	Al^{3+}	34.2	5.0
Zn^{2+}	11.7	9.6	Sc^{3+}	26.1	5.0
Hg^{2+}	8.9	3.7	Tl^{3+}	22.5	1.2

* Values in $C^2m^{-1} \times 10^{28}$

The variation of acidity in Table 10.7 is consistent with the charge-size effect only for aqua ions of elements known to form ionic solids. For significant metal-ligand covalent interaction, the positive charge of the metal ion gets delocalized over the water molecules (ligand); the O-atom thus gains some positive charge and repels the departing H^+ ion more strongly. The greater polarizing power of the transition metal ions is shown by the stronger acidity of $Fe(H_2O)_6^{3+}$ compared to $Al(H_2O)_6^{3+}$, though the latter has a higher Z^2/r value owing to smaller radius of Al^{3+}. Disparity in the cases of Hg^{2+} and Tl^{3+} is again indicative of strong covalent interaction of the metal ions with water molecules, a behaviour quite consistent with the high effective nuclear charge (result of poor screening by inner electron cores) of the ions.

The acid-base behaviour of common cations may be summarized according to pK_a values of their hydrolytic reactions as shown below.

$pK_a \geq 14$:	*Non-acidic cations*, e.g., Na^+ (14.2), K^+ (14.5) The hydrolysis of these ions are negligible; the cation predominates at all practical pH values. No precipitation of hydroxide from 1M solution at any pH.
$pK_a \simeq 11.5 - 14$:	*Feebly acidic cations*: Li^+ (13.6), Ca^{2+} (12.8), Sr^{2+} (13.3), Ba^{2+} (13.5). pK_a values can be measured only through sensitive methods. Hydroxides may precipitate in strongly basic solutions ($p\text{H} > pK_a$).
$pK_a \simeq 6 - 11$:	*Weakly acidic cations*: Mg^{2+} (11.4), Zn^{2+} (9.6) and d-block M^{2+} ions. Solution also contains dissolved CO_2 which is nearly as strong. Biological function of the metal ions are dependent on their acidity. Insoluble metal hydroxides precipitate in neutral to slightly basic solution ($p\text{H} > pK_a$).
$pK_a \simeq 1 - 6$:	*Moderately acidic cations*: Al^{3+}, d-block M^{3+} ions (Fe^{3+}, Cr^{3+}). pK_a values comparable to those of common organic acids like acetic acid. Metal hydroxides precipitate from insufficiently acidic solutions.
$pK_a \simeq -4 - 1$:	*Strongly acidic cations*: Ti^{4+} (–4), Ce^{4+} (–1.1), Zr^{4+} (–0.3), Au^{3+} (–1.5). Form strongly acidic solutions and precipitate insoluble metal oxide or hydroxide except in very concentrated acids.

Problem 10.4

Arrange the following ions in order of increasing acidity: $[Na(H_2O)_n]^+$, $[Al(H_2O)_6]^{3+}$, $[Mn(H_2O)_6]^{2+}$, $[Ni(H_2O)_6]^{2+}$.

Solution: Considering effective nuclear charge on the metal ion, polarizing power and hence acid strength increases as $[Na(H_2O)_n]^+ < [Mn(H_2O)_6]^{2+} < [Ni(H_2O)_6]^{2+} < [Al(H_2O)_6]^{3+}$.

Anions also undergo extensive hydration by the positive ends of the water dipoles. The enthalpy of hydration of an anion thus increases with increasing charge and decreasing size of the anion (a function of Z^2/r). In many cases, the interaction is strong enough to strip a hydrogen ion from a water molecule:

$$S^{2-} + H_2O \rightleftharpoons HS^- + OH^-$$

Equilibrium constants for such proton (hydron) transfer equilibria are conventionally represented by K_b (or pK_b). Successive equilibrium constants are written as K_{b_1}, K_{b_2} ... etc.

$$S^{2-} + H_2O \rightleftharpoons SH^- + OH^- \qquad pK_{b_1} = -4$$

$$SH^- + H_2O \rightleftharpoons H_2S + OH^- \qquad pK_{b_2} = 7.11$$

A larger pK_b value again signifies weaker basicity. We may categorize the common anions according to their pK_b values.

$pK_b > 14$	Nonbasic: Cl^-, Br^-, I^-
$pK_b \approx 11.5 - 14$	Feebly basic: PH_2^- (–13)
$pK_b \simeq 6 - 11.5$	Basic: F^- (10.85); SH^- (7.11)
$pK_b \simeq 1 - 6$	Moderately basic: Te^{2-} (3.0)
$pK_b \simeq -4$ to -1	Very strongly basic: Se^{2-} (–1), S^{2-} (–4)

Anions with pK_b values less than –4 react irreversibly with water. The oxide ion is an example of such a very strongly basic anion. NH_2^- (–25) and CH_3^- (–30) are some other

$$O^{2-} + H_2O \longrightarrow 2OH^-$$

examples. These anions are levelled to the basicity of OH^- in water.

Problem 10.5

In aqueous solution, basicities of the amines show the trends

(a) $Me_2NH > MeNH_2 > Me_3N > NH_3$. (b) $Et_2NH > EtNH_2 \simeq Et_3N > NH_3$.

Comment.

Hint: In both series, the trisubstituted amines appear anomalous from consideration of +I effect of alkyl groups. The trisubstituted amines are weaker bases than expected owing to reduced hydration of the protonated cations. Hydration energies for reactions of the type

$$R_nH_{4-n}N^+(g) + H_2O \longrightarrow R_nH_{4-n}N^+(aq)$$

depend on the number of hydrogen atoms on N available for H-bonding to water. Highly substituted amines possess fewer H atoms on N and thus form fewer H-bonds. The hydration energy follows the order

$$RNH_3^+ > R_2NH_2^+ > R_3NH^+$$

Inductive effect and solvation effect thus opposes one another and results in the observed order.

Problem 10.6

Pick out and justify

(i) *Thermally most stable*:	PH_4Cl, PH_4Br, PH_4I
(ii) *Most acidic oxide*:	Ag_2O, V_2O_5, CO, N_2O_5
(iii) *Stronger base toward a proton*:	NH_2^-, PH_2^-

Hint: (i) Thermal stability refers to the general reaction type

$$PH_4X \longrightarrow PH_3 + HX$$

In a series of HX acids (X = Cl, Br, I), ease of decomposition will increase with the stability of HX, that is with decreasing acid strength. HI is the strongest acid (I^- has the lowest proton affinity) and hence PH_4I is expected to be thermally most stable. However, lattice energy would decrease with increase in anion size, *i.e.*, from PH_4Cl to PH_4I. But this does not significantly alter the primary acid strength.

(ii) The oxide with the highest positive charge on the element other than O should be most acidic. Oxidation states of V in V_2O_5 and N in N_2O_5 are both +5. But the electronegativity of N is higher, making N_2O_5 the most acidic oxide.

(iii) Bond energy (N–H > P–H) consideration suggests that NH_2^- will be the stronger base. This is also consistent with the relative strengths of the respective conjugate acid: $NH_3 < PH_3$.

Problem 10.7

Predict the nature of change, if any, on the acidity of the solution consequent to the addition of (i) HgO to an aqueous solution of potassium iodide; (ii) $CuSO_4$ to an aqueous solution of ammonium sulfate; (iii) $Al(OH)_3$ to aqueous caustic soda solution, (iv) $KHSO_4$ to sulfuric acid.

Hint: (i) Formation of the stable complex HgI_4^{2-} species; concentration of OH^- ion in the solution increases; acidity decreases.

$$HgO + 4KI + H_2O \longrightarrow HgI_4^{2-} + 2OH^- + 4K^+$$

(ii) Formation of stable $Cu(NH_3)_4{}^{2+}$ complex ion increases H^+ ion concentration; acidity increases:

$$Cu^{2+}(aq) + 4NH_4^+(aq) \longrightarrow Cu(NH_3)_4^{2+} + 4H^+(aq)$$

(iii) Formation of the complex $Al(OH)_4^-$ anion consumes OH^- ions, acidity increases.

$$Al(OH)_3(s) + OH^-(aq) \longrightarrow Al(OH)_4^-(aq)$$

(iv) $KHSO_4$ furnishes HSO_4^- ion, the conjugate base of H_2SO_4 in the first step of acid ionization. This being primarily important for the acidity of H_2SO_4, acidity decreases.

10.3 LEWIS ACIDS AND BASES

In our introductory discussion (section 10.1.4) we have seen how a wide variety of compounds and their interactions may be covered by the Lewis theory of acids and bases. Strengths of Lewis acids and bases have been correlated in terms of the enthalpy of formation of an acid-base adduct (ΔH) and acid or base parameters involving susceptibility of the species involved (i) to undergo electrostatic interaction (the *E*-parameter) and (ii) to form covalent bonds (the *C*-parameters). These will be discussed later.

The intensity of an acid-base interaction may be measured in a number of ways. We have already mentioned the remarkable change in boiling point when BF_3 forms adduct with $(C_2H_5)_2O$. Enthalpy change (ΔH) and entropy change (ΔS) accompanying an acid-base reaction may be ascertained by direct calorimetry. Infra-red spectra may provide knowledge regarding the change in bond force constant in acid-base interaction. Similarly ultraviolet or visible spectra can show changes in energy levels in the molecules after combination. Nuclear magnetic resonance coupling constants similarly indicate changes in bonding on forming an adduct.Before we proceed to compare the strengths of Lewis acids and bases, we shall have a closer look at their reactions and properties.

10.3.1 Properties of Lewis Acids and Bases

The main reaction between any acid and base is their mutual interaction; Lewis acids and bases are no exception to this fundamental criterion. We have already found several examples of Lewis acid–Lewis base interaction of the type

$$A + :B \longrightarrow A-B$$

This reaction, exclusive in the gas phase or in noncoordinating solvents, results in the *formation of an adduct or complex* (AB). The product AB is naturally different for different acids(A) and bases(B). For acids and bases in water, the combination between an acid and a base always produces a common product, H_2O (in other solvents, it would be the solvent itself). Hence we have the same enthalpy of neutralization for all strong acids and strong bases in water. But in case of Lewis acids and bases, the enthalpy change will be different for different acids and bases. Later we shall find that such enthalpy terms, which are related to the strength of the bond formed in the adduct, may be used to compare the strengths of Lewis acids and bases.

Lewis acids and bases also take part in *displacement reactions*, in which an acid (or base) displaces another from an adduct. Thus the base pyridine displaces dimethyl ether (Me_2O) from its adduct with BF_3:

$$Me_2O{\rightarrow}BF_3 + C_5H_5N \longrightarrow Me_2O + C_5H_5N{\rightarrow}BF_3$$

Similarly, the acid BF_3 displaces HCl in the reaction

$$BF_3 + NH_4Cl \longrightarrow H_3N{\rightarrow}BF_3 + HCl$$

Such displacement reactions may be identified in aqueous solutions too:

$$HSO_4^-(aq) + H_2O(l) \longrightarrow SO_4^{2-}(aq) + H_3O^+(aq)$$

H_2O, a Lewis base, displaces the base SO_4^{2-} from its "adduct" with the acid H^+.

In case of coordination compounds, displacement of one ligand (Lewis base) by another is called a substitution reaction.

The driving force for such displacement reactions may have some contribution from the relative strengths of the acids or bases, but other factors may also be involved. We shall find soon that such reactions are rationalized readily (though qualitatively) through another concept called hard and soft acids and bases.

A displacement reaction coupled with the formation of a new complex is an analog of "double decomposition" or *metathesis reaction* (from Greek 'exchange'):

$$A - B + A' - B' \longrightarrow A - B' + A' - B.$$

$$(CH_3)_3\ SiI + AgBr(s) \rightarrow (CH_3)_3\ SiBr + AgI(s)$$

Lower solubility of AgI helps to drive the equilibrium to right.

Change of bond length in acid-base interaction

Formation of a Lewis acid-base complex may have significant effect on the initial structure of the starting molecules. Usually the change is noteworthy for a Lewis acid whose central atom undergoes an expansion of coordination number. The Lewis base, in most cases, undergoes only a slight lengthening of the bond through which the donor atom operates. An interesting exception is found with S_4N_4 coordinating to $SbCl_5$ through a nitrogen atom. The structure of S_4N_4 (puckered; Gr 16) changes to a nearly planar structure except for the donor N atom (10–I).

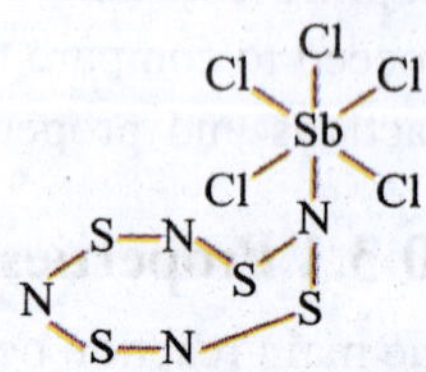

10–I

The formation of an acid-base adduct changes the adjacent bond lengths remarkably, as may be seen from the following examples:

Species	B–N distance (pm)	B–F distance (pm)
BF_3	–	130
$H_3N\,BF_3$	160	138
Me_3NBF_3	158	139

The B–F distance increases when the boron atom forms an adduct with ammonia. The distance increases further when a stronger adduct is formed by Me_3N (B–N distance decreases). This may be stated as an empirical rule (Gutmann): *the stronger and shorter the bond formed in an adduct between a Lewis acid and a Lewis base, the greater will be the lengthening of the adjacent bonds in both the donor and acceptor molecules*. The bond formed in the adduct displaces electron density from the donor atom in the base to the acceptor atom in the acid. The donor atom acquires less partial negative charge than before and the acceptor atom becomes less positive. Both these effects reduce the polarity in the bonds in the acid and base molecules–ultimately leading to lengthening of the bonds. Proceeding in the same line of approach, one can also understand that bond lengths should increase with increase in coordination number of an atom. This is supported by the following bond lengths:

Species	Bond length (pm)	Species	Bond length (pm)
SiF_4	154	SO_2	143
SiF_6^{2-}	171	SO_3^{2-}	150
$SbCl_5$	231	I_2	266
$SbCl_6^-$	247	I_3^-	283

Steric factors

The formation of an acid-base adduct usually involves rearrangement of the valence orbitals in one or both of the participating molecules. For example, the planar BF_3 molecule changes to tetrahedral BF_4^- on adding an F^- ion; the tetrahedral SiF_4 molecule changes to octahedral SiF_6^{2-} on adding two F^- ions. When the acid or base (or both) possesses bulky substituents, adduct formation may be influenced by steric effects. Thus 2-methylpyridine is a stronger base toward a hydrogen ion, as expected from the inductive effect of the methyl group. In contrast, it is a much weaker base than pyridine toward the Lewis acid $B(CH_3)_3$. The ΔH values of adduct formation are illustrative:

	pyridine	*2-methylpyridine*	*3-methylpyridine*
ΔH (kJ mol^{-1})	–71	–42	–74

We find that shifting the methyl group to the meta position largely increases the acidity. Introduction of two methyl groups in both the *ortho*-positions does not allow adduct formation to take place at all, *i.e.*, 2, 6-dimethylpyridine does not react at all with $B(CH_3)_3$.

These observations may be rationalized in terms of steric interactions or strains between substituents on the acids which appear ***"in front"*** during adduct formation. Such strain is termed **F-strain**: *in this case bulky groups interfere directly with each other on the approach of an acid or base.* F-strain is lowered when the acid and base can form adduct from a larger distance. Thus $Al(C_2H_5)_3$ forms a weak adduct with 2, 6-dimethylpyridine (ΔH = –13 kJ mol^{-1}). The longer Al–N bond reduces steric hindrance between the substituents.

Steric interference may also occur during the change in the geometry of the central atom on adduct formation. When a boron-atom changes from trigonal planar to tetrahedral disposal of its valence orbitals, the substituents bend away from the approaching base to the ***"back side"*** and come closer. If these substituents are branched, they will now introduce steric hindrance in the adduct; this is referred to as *back-strain or* **B-strain** (the crowding groups are on the "back side" of the acceptor site). Thus $B(CH_3)_3$ forms a stronger adduct with NH_3 than $B(i\text{–}C_3H_7)_3$. Trimesitylboron is inert even to such strong bases as methoxide ion because crowding of the mesityl groups prevents change of geometry of the boron atom.

Involvement of steric factors in acid-base interaction is illustrated in the following problems.

Problem 10.7

Which will be most basic toward BMe_3 in each row?

(i) Me_3N, Et_3N. (ii) 2-MePy, 4-MePy. (iii) 2-MeC_6H_4CN, C_6H_5CN.

Solution: (i) Me_3N (less F-Strain in adduct). (ii) 4-MePy (less F-strain in adduct).

(iii) 2-MeC_6H_4CN. (Inductive effect of the methyl group wins; steric effect cannot interfere since the donor nitrogen atom is situated at a larger distance (not directly in the ring as in Py).

Problem 10.8

Arrange according to increasing Lewis acid character:

(a) BF_3, BCl_3, BBr_3 (b) $B(n\text{-Bu})_3$, $B(t\text{-Bu})_3$ (c) SiF_4, $SiCl_4$, $SiBr_4$, SiI_4

Solution: (a) The compounds involve extensive π-interaction from a filled *p* orbital on the halogen and an empty *p*-orbital on B. The small F-atom forms the most stable *p–p* π-bond by efficient match of energy and size of the orbitals. Formation of an adduct with a base converts the geometry of bonds around boron from planar to tetrahedral, thus rupturing the π-bonds. The Lewis acidity therefore increases with decrease in the extent of π-conjugation, as the halogen atom increases in size giving poorer overlap, i.e., from BF_3 to BBr_3. Hence the order is

$$BF_3 < BCl_3 < BBr_3$$

(b) The highly branched tertiary butyl group involves appreciable back-strain (B-strain) when the boron atom changes to tetrahedral environment on adduct formation. This destabilizes the adduct. Hence the order is $B(t-Bu)_3 < B(n-Bu)_3$.

(c) The order in this case is the reverse of that for BX_3. π-conjugation from the halogen *p*-orbital to the Si–*d* orbital is not as extensive as in the case of BX_3. The order of acidity follows the increase in electron withdrawing power of the halogen from I to F. Hence the order is

$$SiI_4 < SiBr_4 < SiCl_4 < SiF_4.$$

Problem 10.10

Proton affinity of trimethylamine (958 kJ mol^{-1}) is significantly greater than that of triphenylamine (~ 925 kJ mol^{-1}). Comment.

Hint: In triphenylamine, the bulky substituents are bonded by nearly sp^2 hybrid orbitals (∠CNC ≃ 116°). This leaves a nearly unhybridized *p*-orbital for donor role. The overlap in the acid-base adduct is poor.

Problem 10.11

Compared to triethylamine, quinuclidine (10-II; 1-azabicyclo [2.2.2] octane) forms a more stable adduct with BMe_3, though both have nearly the same proton affinity. Explain.

Hint: F-strain is less in the adduct of BMe_3 with quinuclidine since the alkyl groups in quinuclidine are folded back. With trimethylamine, the terminal $-CH_3$ groups and the $-CH_3$ groups on B exert much V-W repulsion or F-strain. (Triethylamine involves B-strain too).

10–II

Note: For same reason, the bicyclic compound $:P(CH_2CH_2)_3CH$ is more basic than $:P(CH_2CH_3)_3$.

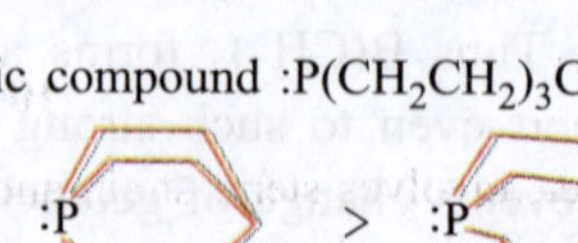

A third category of strain, called **internal strain** or **I-strain**, also exerts steric effect on the enthalpy of acid-base interaction. Such internal strains may be associated with geometric changes that occur within the acid or base when the donor acceptor bond is formed. In cyclic amines and ethers, basicity changes with the ringsize and consequent *s* or *p*-character of the hybrid orbitals involved. The bonds responsible for optimum stability of the free base molecule may be weakened when the angles between bonds change after coordination. This produces a large internal strain energy relative to the free amine. Manxine (10-III) has a nearly planar bridgehead C and N geometry. Though the molecule resembles tripropylamine at its donor centre, it has a lower proton affinity than tripropylamine. When the donor N-atom forms an N–H bond, the N atom is forced to deform the bond angles around it toward tetrahedral geometry which produces a large internal strain relative to the free amine.

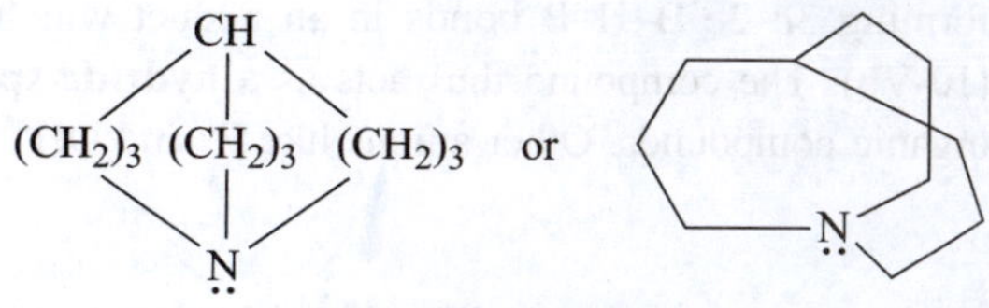

10–III (Manxine)

Compounds in which the basic centre is exocyclic, as in lactams and lactones, donor ability of the basic centre varies with the distribution of *s*- and *p*-character in the bonds in the ring. The following series of lactams may be taken as an illustration:

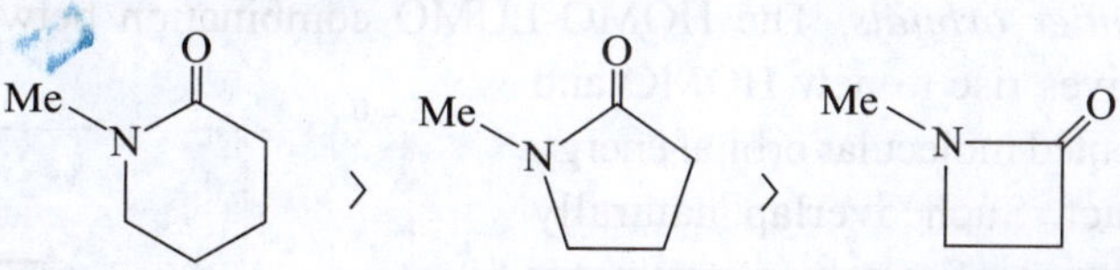

Here, basicity of the exocyclic oxygen decreases with ring size. A smaller ring size implies smaller internal bond angle and involves less *s* character in the bonds. Hence greater *s* character is allotted in the bond to the exocyclic O atom. The C atom of the ring that links the O atom exerts higher electronegativity in its bond to O and lowers the donor ability.

Sometimes it is not possible to make a clear-cut demarcation between different types of steric interferences. Triethylamine does not form an adduct with trimethylboron in spite of a favourable ΔH value (-42 kJ mol^{-1}). Though at first sight we would assign this to involve B-strain, examination of molecular models shows that the actual interference probably comes from one ethyl group which is actually twisted out to the front of the molecule.

However, in certain cases, steric strain may stabilize the adduct of an H^+ or H^- ion with a base, thereby effectively increasing the basicity. Thus, 1,8-bis(dimethylamino)naphthalene is more basic (pK_b = 1.9) compared to 1-dimethylaminonaphthalene (pK_b = 4.57) or NH_3 (pK_b = 4.74). The free base (IVa) is largely destabilized owing to repulsion between the lone pairs of electrons on N and steric hindrance of the methyl groups, The base strongly binds a proton (hydron) by symmetric three centre four electrons N···H···N hydrogen bonds (10-IVb). The compound has been given the nick-name "**proton sponge**" (also called a superbase).

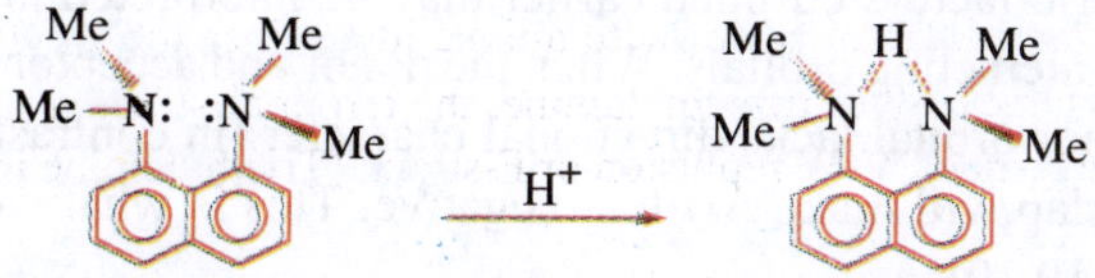

10-IVa **10-IVb**

Addition of a second proton however, involves steric strain and does not readily take place. Lewis basicity towards bulky Lewis acids is similarly reduced.

1,8-bis(dimethylboryl)naphthalene (10-Va) can bind an H^- ion through its electron-deficient B-atoms forming 3*c*–2*e* B–H–B bonds in an adduct which is both kinetically and thermodynamically stable (10-Vb). The compound thus acts as a **hydride sponge** and can be used to extract hydride ions from organic compounds. Other anions like F^- and OH^- may also be coordinated between the boron centres.

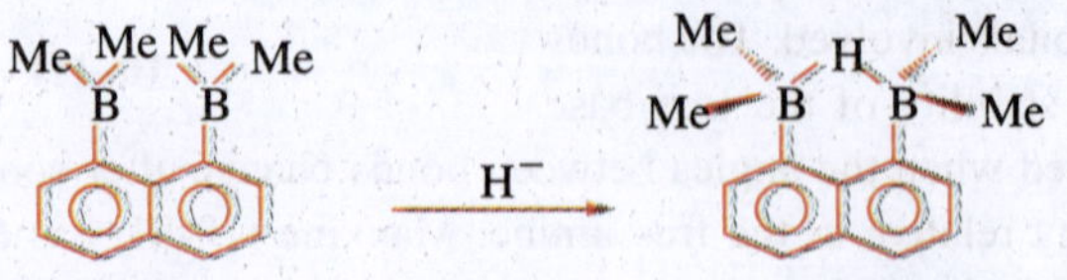

10-Va **10-Vb**

MO description of acid-base reactions

The formation of an acid-base adduct may be described in terms of the molecular orbitals of the participating species. The overall picture naturally involves the highest occupied MO (HOMO) associated with the base (donor) and the lowest unoccupied MO (LUMO) on the acid (acceptor)—the *frontier orbitals*. The HOMO-LUMO combination between the donor base and acceptor acid gives rise to new HOMO and LUMO in the re-oriented molecular orbital energy levels in the product. Such overlap naturally conforms to the criterion of matching symmetry and energy of the participating orbitals. The driving force for the combination comes from the lowering of energy of the system consequent to population of the newly formed bonding MO by the two electrons initially belonging to the HOMO of the base (Fig. 10.2).

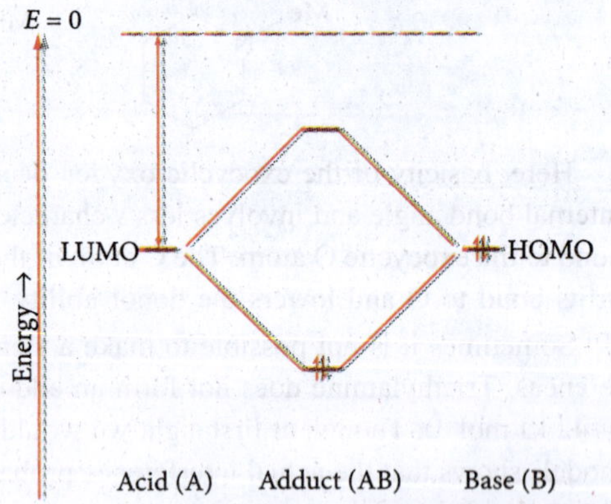

Fig. 10.2: MO representation (schematic) for the formation of an acid-base adduct between a Lewis acid (A) and Lewis base (B). The energy levels of HOMO and LUMO vary in different cases.

In terms of frontier orbitals, we may thus think of a Lewis base as one having an electron pair in a HOMO of suitable symmetry that can be stabilized by interaction with the LUMO of another species, namely, a Lewis acid. The better the energy match between the participating molecular orbitals, the stronger will be the interaction. When permitted by symmetry, it is actually the relative energies of the HOMO of one species and the LUMO of the other that decides the nature of interaction between the two.

The influence of steric factors outlined earlier may be interpreted in terms of the efficiency of overlap between the interacting orbitals. When the donor and acceptor orbitals differ considerably in size and the donor orbital lacks directional character (in contrast to a hybrid orbital, for example), reduced overlap will make ΔH less negative. This is what we termed *back strain* or B-strain (see Problem 10.10).

Van der Waals repulsion between the substituents on acid and base will also cause steric congestion and affect ΔH, a situation that we described earlier as *front strain* or F-strain.

Entropy effect in acid-base interaction

Though the overall free energy change in a Lewis acid-base interaction depends primarily on a favourable ΔH term, in some cases a positive ΔS term may make significant contribution to the overall free energy change. The following reactions may be taken as examples:

$$H_3O^+ (aq) + S^{2-} (aq) \rightleftharpoons HS^- (aq) + H_2O (l) \qquad \Delta S = +109 \text{ J mol}^{-1}\text{K}^{-1}.$$

$$Ca(H_2O)_6^{2+} (aq) + EDTA^{4-} \rightleftharpoons Ca(EDTA)^{2-} + 6H_2O (l) \qquad \Delta S = +118 \text{ J mol}^{-1}\text{K}^{-1}$$

In the first reaction (also an acid-base reaction in the Bronsted sense) lowering of net charge on the acid-base species results in poorer ordering effect on the surrounding water molecules. Hence ΔS increases. In the second reaction, chelation results in an increase in total number of species in solution, thereby increasing ΔS (see chelate effect). At 298K, the contribution of the $T\Delta S$ term (– 35 kJ per mol reaction) to ΔG is greater than that of the ΔH term (– 27 kJ).

10.3.2 Hard and Soft Acids and Bases (HSAB)

The strength of any Lewis acid or base is determined by the very nature of the reaction involved in an electron transfer process. Accordingly, assignment of any single consistent criterion for acid-base strength becomes very difficult in the Lewis definition.

A qualitative correlation between the extent of interaction between various Lewis acids and bases has been achieved by classifying the acids or bases into two classes: **hard** and **soft** (R. G. Pearson, 1963, in his theory of Hard and Soft Acids and Bases: **HSAB**). Acid-base reactions are then treated by the general principle that *hard acids prefer to combine with hard bases* and *soft acids prefer to combine with soft bases* (and vice versa).

The criterion of hardness (or softness) is ascribed to the "hardness" of the electron cloud associated with any species. A firmly held electron-cloud with low polarizability makes a species "hard", while an easily polarizable electron cloud characterizes the species as "soft". A third category with intermediate character appears in the borderline. The principal distinguishing features of hard and soft acids and bases may then be summarized as follows:

Hard acids: *Acceptor atom marked by*

(i) small size; (ii) high positive oxidation state; (iii) absence of any outer electrons which are easily excited to higher states

***Examples*:** H^+, Li^+, Na^+, K^+ ; Be^{2+}, Mg^{2+}, Ca^{2+}, Sr^{2+} ;
Fe^{3+}, Co^{3+}, Cr^{3+}, Mn^{2+} ; CO_2, SO_3 ; BF_3, BCl_3, $B(OR)_3$;
Al^{3+}, $Al(CH_3)_3$, $AlCl_3$, AlH_3 ; HX (hydrogen bonding molecules);
Ions with oxidation states of +4 or higher.

Soft acids: *Acceptor atom marked by*

(i) large size; (ii) zero or low positive oxidation state; (iii) several easily excitable valence electrons.

***Examples*:** Cu^+, Ag^+, Au^+, Hg_2^{2+}, Tl^+, $Tl(CH_3)_3$; Cd^{2+}, Pt^{2+}, Hg^{2+}, Pt^{4+} ;
I^+; BH_3, Br^+ , Cl_2, Br_2, I_2,
M^0 (metal atom) and bulk metals.
π-acceptors e.g., trinitrobenzene, tetracyanoethylene, quinones etc.

Borderline: $B(CH_3)_3$, Fe^{2+}, Co^{2+}, Ni^{2+}, Cu^{2+}, Zn^{2+}, Pb^{2+}, Bi^{3+}, SO_2, Rh^{3+}, Ir^{3+}, Ru^{3+}, Os^{2+}

Hard bases: *Donor atom marked by*

(i) high electronegativity; (ii) low polarizability; (iii) presence of filled orbitals; empty orbitals may exist at high energy level.

***Examples*:** H_2O, OH^-, F^-, Cl^-, $CH_3CO_2^-$, PO_4^{3-}, SO_4^{2-}, NO_3^-, CO_3^{2-}, ClO_4^-, ROH, R_2O, RO^-, NH_3, N_2H_4, RNH_2

Soft bases: *Donor atom marked by*

(i) low electronegativity. (ii) high polarizability. (iii) partially filled orbitals; empty orbitals are low-lying.

***Examples*:** H^-, I^-, SCN^-, CN^-, $S_2O_3^{2-}$, CO, C_2H_4, C_6H_6, R_2S, RSH, RS^-, R_3P, R_3As

Borderline: $C_6H_5NH_2$, C_5H_5N, N_3^-, N_2, NO_2^-, Br^-, SO_3^{2-}

In fact, the hard-soft idea germinated from a study of the behaviour of metals towards different nonmetal donor atoms (Ahrland, Chatt and Davies, 1958). It was observed that, certain metals prefer to form stable complexes when the first member of a given group of nonmetals in the periodic table acts as the donor atom e.g., N, O, F. These metals were marked as **class-a** (nothing to do with subgroup A of the periodic table). Alkali and alkaline earth metals, and the first row transition metals in high oxidation state (Na^+, K^+, Ca^{2+}, Fe^{3+}, Co^{3+} etc.) belong to this class.

Other metals forming their most stable complexes with heavier members of a group of nonmetals (like P, S, Cl) were marked as **class-b**. Lighter transition metals in low oxidation state and heavier transition metals belong to this class (Cu^+, Ag^+, Hg^{2+}, Pt^{2+}, Pd^{2+}, etc.).

It may be readily observed that the class-a metals are small and less polarizable; they prefer to combine with similar nonmetals. The class-b metals are larger and more easily polarizable; they combine with nonmetals having similar properties.

The hard-soft view is also compatible with Fajan's rules for development of covalent character with increasing degree of polarization in ions (Chapter 7).

A large number of chemical reactions may be understood in the hard-soft terminology. The guiding principle is that hard-hard and soft-soft combinations of acids and bases are always more stable than hard-soft combinations. A few examples will illustrate this point.

(i) Ammonia, amines, water and fluoride ions prefer to combine with Be^{2+}, Ti^{4+}, Co^{3+} etc. giving stable complexes. On the other hand, phosphines (R_3P), thioethers (R_2S) and other species with P and S as donor atoms prefer to combine with Pt^{2+}, Pd^{2+}, Hg^{2+} etc.

(ii) BF_3, a hard acid, combines readily with a further fluoride ion; but BH_3 (present as B_2H_6), a soft acid, prefer the softer base hydride ion.

$$BF_3 + F^- \rightarrow BF_4^- \ ; \ B_2H_6 + 2H^- \rightarrow 2BH_4^-$$

It is now easy to understand why the following reactions proceed as written

$$BF_3H^- + BH_3F^- \rightarrow BF_4^- + BH_4^-$$

$$CF_3H + CH_3F \rightarrow CF_4 + CH_4.$$

This tendency of fluoride or hydride ions to favour the addition of one more ion of the same kind has been termed ***symbiosis*** (Jorgensen, 1964).

(iii) Hard-hard and soft-soft combinations similarly determine the course of many other reactions; for example,

$$LiI + CsF \rightleftharpoons LiF + CsI$$

$$HgF_2 + BeI_2 \rightleftharpoons BeF_2 + HgI_2$$

These reactions illustrate the **"Pearson paradox"** of chemistry. In the Pauling concept of electronegativity, cesium (or mercury) and fluorine should form the more stable bond as their electronegativity difference is greater! In reality, however, LiF is more stable than CsF (ΔH_f°: LiF = 573; CsF: 50 kJ mol^{-1}) due to very large contribution of electrostatic interaction in the former.

In fact, the major driving force for the above reactions (and similar others) come from the stability of the hard-hard combination joining small atoms to each other by ionic and/or covalent bonding. The soft-soft combination contributes little or nothing to the driving force, except when other factors like pi-bonding are involved.

Problem 10.12

Co^{3+} is a moderately hard Lewis acid. It forms both $[Co(NH_3)_5X]^{2-}$ and $[Co(CN)_5X]^{3-}$ type complexes where X is a halide ligand. The ammine complex is more stable with $X = F^-$ than the complex with $X = I^-$. But the cyano complex with $X = F^-$ is not even known while $[Co(CN)_5I]^{3-}$ is the most stable in the series. Comment.

Hint: NH_3 is a hard base, F^- is also a hard base. CN^- is a soft base, so also is I^-.

(iv) ***Solubilities*** of many ionic salts in water may be rationalized in terms of hard-soft interaction, though with obvious limitations owing to other factors involved. Let us consider the solubilities of lithium and silver halides in water (g of salt per 100 mL water at 25°C):

	$X = F^-$	Cl^-	Br^-	I^-
LiX	0.7	64	145	165
AgX	182	10^{-4}	10^{-5}	10^{-7}

Water is a hard donor; though not as hard as F^-, it is harder than other halide ions. Consequently the solubility of fluoride salts in water is often different from those of other halides. Dissolution of the solid halides involves replacement of the halide ions by the base H_2O. The Li^+ ion, a hard acid, prefers the harder base F^- relative to H_2O, resulting in a low solubility of LiF. But Li^+ prefers H_2O over the less harder halides (Cl^- etc.) which is reflected in the higher solubility of these halides. For silver halides, the situation changes as Ag^+ is a soft acid. It leaves the harder base F^- to get hydrated by the base H_2O and we find a high solubility of AgF. Solubility of AgX decreases as X^- becomes more soft. We remember that the solubility trends discussed above may also be explained from Fajan's rules based on polarizability of ions.

Problem 10.13

TlBr is only slightly soluble in water (0.24 g L^{-1}) but the solubility of $TlBr_3$ is much higher, ~322 g L^{-1}. Comment.

Hint: Tl^+ is a soft acid; it prefers soft Br^- over hard H_2O. Tl^{3+} is much harder.

HSAB principle in qualitative analysis

The traditional scheme for separation of metal ions into different groups in qualitative analysis presents some useful correlation between HSAB theory and solubility trends, though with certain limitations. Let us have a glance at the precipitates formed in the groups by some common metal ions.

1. **Group I** — *Reagent*: HCl (dil); *precipitates*: $PbCl_2$, Hg_2Cl_2, AgCl. Cl^- is a marginally hard base while the metal ions are soft or borderline (Pb^{2+}). Favourable lattice formation over poor hydration effects appear mainly responsible for the precipitation.
2. **Group II** — *Reagent*: H_2S in dil HCl medium; low concentration of S^{2-}; *precipitates*: CuS, HgS, Bi_2S_3, As_2S_3, Sb_2S_3 etc.
   S^{2-} is a soft base; the metal ions are borderline or soft acids.
3. **Group III** — *Reagent*: H_2S in basic (ammoniacal) medium; *precipitates*: $Al(OH)_3$, $Cr(OH)_3$, ZnS, MnS, FeS, CoS, NiS.
   Al^{3+} and Cr^{3+} are hard acids and prefer to combine with hard OH^- over soft S^{2-}. Fe^{3+} is hard, but it gets reduced by H_2S; the borderline and soft metal ions are precipitated by the soft base S^{2-}.
4. **Group IV** — *Reagent*: $(NH_4)_2CO_3$ in presence of NH_3 (*aq*); *precipitates*: $CaCO_3$, $SrCO_3$, $BaCO_3$.
   Hard-hard combination favours the precipitates in this group.
5. **Group V** — *Reagent*: The cations in this group are all hard and tend to remain in solution by hydration with hard base H_2O. Mg^{2+} could be precipitated as $MgCO_3$ in Gr IV, but that is intentionally avoided using ammonium carbonate-ammonia. Na^+ and K^+ are precipitated only by specific bulky anions.

Problem 10.14

Comment on the solubility trend: $Ag_2O > Ag_2S < AgCl$.

Hint: Ag^+ is a soft acid. The S^{2-} ion (soft) is much larger than the O^{2-} ion (hard). Cl^- is also hard. The interaction between soft Ag^+ and soft S^{2-} is favoured, resulting in a low solubility of the product Ag_2S.

Problem 10.15

Predict the solubility trend among (i) $Fe(OH)_2$, $Fe(OH)_3$ and FeS.

Hint: (i) Fe^{2+} is a borderline acid; OH^- is a hard base; S^{2-} is soft. $Fe(OH)_3$ should be most insoluble (hard-hard) followed by FeS, then $Fe(OH)_2$.

(Solubility products at 25°C: $Fe(OH)_3 \sim 10^{-38}$, FeS ~ 10^{-19}, $Fe(OH)_2 \sim 10^{-16}$).

The application of hard-soft arguments to solubility correlation has its natural limitations since the role of solvent and the effects of crystal packing as well as other interactions need to be considered.

(v) ***The occurrence of different elements in nature*** in various mineral forms may also be understood in the light of the HSAB theory. In fact, the common mineral forms of the elements are closely related to the precipitates which differentiate these elements in qualitative analysis.

As early as 1796, Berzelius observed that certain metals occur in Nature as sulfides while others tend to occur as oxides, carbonates, sulfates or silicates. The hard-soft idea provides a reasonably well interpretation of this observation. Soft and borderline ions form sulfide ores while hard metal ions commonly form oxide ores. Highly basic oxides transform themselves into soluble salts in sea water or to insoluble deposits of carbonates or silicates through the action of CO_2 or SiO_2.

It's still going on

The original mineral deposits are undergoing continuous changes due to geochemical action. Oceanographic studies reveal that mineral formation is still going on deep under the seas. In mid-oceanic ridges, hot molten magma (~1200°C) rises from the mantle to form oceanic crust. During movement, the crust develops cracks through which sea water seeps. The superheated water at ~350°C, still liquid at the very high pressure underneath, brings about a number of chemical reactions. Mg^{2+} reacts with the silica in basaltic rock forming insoluble magnesium hydroxysilicate. Iron(II) silicate in the rock reduces sulfate to sulfide, which is partly transformed into iron pyrites through gross reactions like

$$2SO_4^{2-} + 4H^+ + 11Fe_2SiO_4 \rightarrow FeS_2 + 7Fe_3O_4 + 11SiO_2 + 2H_2O$$

The superheated water coming out of the vents at the top of the midoceanic region, still around 350°C, contains no Mg^{2+} or SO_4^{2-} but a good amount of *d*-block metal ions dissolved out from the basalt by hydrogen ions, together with H_2S. As this hot water cools in contact with cold ocean water, chemical reactions resembling those in qualitative analysis of cations set to work. Metal ions like Fe^{2+}, Cu^{2+}, Ni^{2+}, and Zn^{2+} are precipitated as sulfides which form a black smoker that rises like a black chimney.

MnS, however, is carried away from the smoker where it gets oxidized by dissolved O_2 and bacterial action to form nodules of MnO_2 on the ocean bed. The black smoker also remains a rich source of other minerals. Perhaps it is not far when processes will be developed for utilization of these resources.

(vi) The preference of metal ions for different ligands is also consistent with the HSAB theory. Thus the borderline acids Co^{2+}, Ni^{2+} etc. tend to bind SCN^- ion through the N-end, while the softer cations like Pd^{2+}, Pt^{2+}, Hg^+ etc. are coordinated by the same ligand preferentially through the S-end.

(vii) Lewis acid-base interactions in biological processes, particularly those involving metal ions, are to a large extent controlled by the principles of hard-soft acids and bases. These have been mentioned in the chapter on bioinorganic chemistry.

As already explained, the hard-soft classification is based on the polarizability of the electron cloud associated with a species. Accordingly, the proton or H^+ ion should be the hardest acid ("ultimate acid") as it has no more electron to ionize and has the smallest size with highest positive field density. The methylmercury cation, CH_3Hg^+, on the other hand, is a monopositive cation which acts as a typical soft acid. This makes it convenient as a reference in studying acid-base equilibria. A large size and low positive field density imparts very feeble polarizing power to the cation. The Hg–C bond is not involved in the competitive acid-base equilibrium involving a base B under study, say,

$$BH^+ + CH_3Hg^+ \rightleftharpoons CH_3HgB^+ + H^+$$

A hard base will move the equilibrium to the left while a soft base will drive it to the right.

The hardness or softness of acids and bases are quite different from their inherent acid-base strength. One may have a strong hard base in solution, such as OH^- in water, but one may also find a strong soft base like S^{2-} in water. Let us consider the reactions:

$$\text{(i) } CHCl_3.NH_3 + BMe_3.C_6H_6 \rightleftharpoons CHCl_3.C_6H_6 + BMe_3.NH_3$$

$$\text{(ii) } SO_3^{2-} + HF \rightleftharpoons HSO_3^- + F^-$$

$$\text{(iii) } OH^- + CH_3HgSO_3^- \rightleftharpoons CH_3HgOH + SO_3^{2-}$$

The equilibrium in each reaction lies largely toward right, contrary to our expectation from hard-hard and soft-soft combinations. In (i), $CHCl_3$ and NH_3 are both hard, BMe_3 is borderline, while C_6H_6 is soft. The forward reaction thus involves replacement of the hard base NH_3 from the hard acid $CHCl_3$ by the soft base C_6H_6. Similarly, reaction (ii) involves displacement of the F^- ion (hard base) from the H^+ ion (hard acid) by the softer base SO_3^{2-} (borderline). In (iii), the OH^- ion (hard base) displaces the SO_3^{2-} ion (borderline) from the soft acid CH_3Hg^+. Actually, in all the forward reactions, *the stronger base displaces the weaker one*: $NH_3 > C_6H_6$; $SO_3^{2-} > F^-$; $OH^- > SO_3^{2-}$. Thus acid-base interactions should be considered in the background of both hardness (softness) and strength. In a competitive situation involving both strength as well as hardness, the hard-soft interpretation works well:

$$CH_3HgF + HSO_3^- \rightleftharpoons CH_3HgSO_3^- + HF$$

$$CH_3HgOH + HSO_3^- \rightleftharpoons CH_3HgSO_3^- + HOH.$$

In both cases the equilibrium lies well to the right.

Problem 10.16

$(CH_3)_2N - PF_2$ has two basic sites, on the P and on the N atom. Both BH_3 and BF_3 form adduct with this. In one adduct, the B atom is bonded to the N-atom, and in the other, to the P-atom. Predict the nature of these adducts.

Hint: Hard-hard and soft-soft combination will be preferred; the N atom should donate to the B atom in BF_3 and the P atom should coordinate to the B atom in BH_3.

Problem 10.17

$SnCl_2$ can act both as a Lewis acid and a Lewis base. Explain.

Hint: $SnCl_2 \xrightarrow{Cl^-} SnCl_3^-$. This may be represented as $:SnCl_3^-$. The lone pair of electrons may be donated to a suitable acceptor.

Acid: $SnCl_3^- \xrightarrow{PtCl_4^{2-}} Pt(SnCl_3)_5^{3-}$; $SnCl_2 \xrightarrow{Me_3N} Cl_2SnNMe_3$

Base: $Cl_2SnNMe_3 \xrightarrow{BF_3} Cl_2Sn(NMe_3)(BF_3)$; $SnCl_3^- \longrightarrow (CO)_5Mn - SnCl_3^-$

[**Note:** SO_2 can similarly act both as an acid and a base. See oxides of sulfur.]

Problem 10.18

Explain: (i) the complexing ability of the halide ions show reverse trend with Al^{3+} and Hg^{2+}:

$$Al^{3+} : I^- < Br^- < Cl^- < F^-$$

$$Hg^{2+} : F^- < Cl^- < Br^- < I^-$$

(ii) The heat of reaction of I_2 with Et_2S is *greater* than that with Et_2O but the heat of reaction of C_6H_5OH with Et_2S is *less* than that with Et_2O.

(iii) The reaction $Et_3SiI + AgBr \longrightarrow Et_3SiBr + AgI$ is irreversible.

Hint: Hard-hard and soft-soft combinations are preferred.

Review

The hard-soft classification is a useful concept, but it does not lead directly to a scale for acid-base strength. Inherent acid-base strengths are also not taken into consideration. Thus the OH^- and F^- ions are both hard bases. But the former is nearly 10^{13} times as strong a base as the latter. Various approaches to correlate inherent acid-base strength with attributed hardness (and softness) factors are yet to be widely accepted.

Interpretation of many reactions by splitting the participants into acid-base fragments is also arbitrary to some extent. The very common reaction between ethanol and acetic acid may be interpreted in either of two ways:

$$CH_3COO^-H^+ + C_2H_5^+\ OH^- \quad \text{or} \quad CH_3CO^+\ OH^- + C_2H_5O^-\ H^+$$

Either interpretation is justifiable by the hard-hard combination of H^+ with OH^-. The actual involvement of the second mechanism has to be understood by other means.

Sometimes the hard-soft principle fails to keep parity with inherent acid-base strengths. The reaction:

$$CH_3^+(g) + H_2(g) \longrightarrow CH_4(g) + H^+(g)$$

should be favourable in view of the soft-soft combination of CH_3^+ with H^-. But actually the combination is endothermic by about 360 kJ per mole. (This has to be explained by the greater acidity of H^+ relative to CH_3^+).

However, we should restrict our expectations in view of the statement made by Pearson: "*It should be stressed that the HSAB principle is not a theory but is a statement about experimental facts. Accordingly an explanation of some observation in terms of hard and soft behaviour does not invalidate some other, theoretical explanation.*"

10.3.3 Strength of Lewis Acids and Bases

We have seen that hardness or softness cannot be considered intrinsic properties of a Lewis acid or base. Also we cannot readily devise a quantitative measure for the hardness or softness of a species. Pearson and Parr introduced (1983) a property called the *chemical hardness* η of a chemical species (atom, molecule or ion) as being proportional to the second derivative of the total energy (E) of the chemical system with respect to changes in the number of electrons (N) in a fixed nuclear framework (charge Z):

$$\eta = \frac{1}{2}\left(\frac{\partial^2 E}{\partial N^2}\right)_Z$$

The coefficient ½ is arbitrary. From this, an operational definition of chemical hardness was obtained in terms of the difference of ionization energy (I) and electron affinity ($E.A.$) of the atom, molecule or ion (see box).

$$\eta = \frac{1}{2}(I - E.A.)$$

By applying a three-point finite differential approximation to the second derivative, one may arrive at the operational definition as follows:

$$\eta \approx \frac{1}{2}(E_{N+1} - 2E_N + E_{N-1}) = \frac{1}{2}[(E_{N-1} - E_N) - (E_N - E_{N+1})]$$

The energy difference $(E_{N-1} - E_N)$ is equal to the ionization energy (I); the second energy difference is similarly equal to the electron affinity ($E.A.$). Therefore,

$$\eta = \frac{1}{2}(I - E.A.)$$

Ionization energies of molecules are available from photoelectron spectra and are equal to negative of the orbital energies of HOMO (we have to release the electron from the HOMO). Similarly electron affinities should be equal to the negative of the orbital energy of the LUMO; these are sometimes measurable but has to be calculated in other cases. The hardness of a molecule may then be written as

$$\eta = \frac{1}{2}(I - E.A.) = \frac{1}{2}(E_{\text{LUMO}} - E_{\text{HOMO}})$$

Thus, we expect a high value of η when the HOMO and LUMO (the frontier orbitals) are marked by a large energy gap. Such large gaps make electronic rearrangement in the molecule difficult, making it "hard". In this sense, hardness is resistance to deformation. H_2O, with an estimated hardness of 9.5 eV (HOMO : –12.5 eV ; LUMO : + 6.4 eV) is harder compared to H_2S (η = 6.2 eV). Similarly, HF (η = 11.0 eV) is much harder than HI (η = 5.3 eV).

The softness σ of a chemical species may be expressed as the inverse of hardness:

$$\sigma = 1/\eta$$

Thus, we may interpret soft acids and bases as those in which the frontier orbitals are close in energy. This implies that the valence electrons in these species can be easily promoted into the LUMO. In other words, the species are polarizable and hence 'soft'.

Values of absolute hardness for some Lewis acids and bases calculated by Pearson are given in Table 10.8.

Table 10.8

Absolute hardness of few Lewis acids and bases (gaseous)

Species	I(eV)	$E.A.$ (eV)	η	Species	I(eV)	$E.A.$ (eV)	η
Li^+	75.6	5.4	35.1	NH_3	10.7	–5.6	8.2
Na^+	47.3	5.1	21.1	PH_3	10.0	–1.9	6.0
Mg^{2+}	80.1	15.0	32.5	F^-	17.4	3.4	7.0
Ca^{2+}	50.9	11.9	19.5	Cl^-	13.0	3.6	4.7
Zn^{2+}	39.7	17.8	10.9	I^-	10.5	3.1	3.7
Al^{3+}	120.0	28.5	45.8	OH^-	13.2	1.8	5.7
H_2O	12.6	–6.4	9.5				

However, absolute hardness is calculated from ionization energies and electron-affinities, both of which refer to gaseous state of the chemical species. Hence its application to acid-base reaction is limited.

Drago and Wayland tried to correlate the enthalpy of adduct formation between Lewis acids and bases in gas phase through parameters like E and C to be explained below. For the reaction

$$A(g) + B(g) \longrightarrow A\text{–}B\ (g), \quad -\Delta H^\circ = E_A E_B + C_A C_B$$

where the A and B subscripts refer to the acid and base respectively. The E and C parameters were derived from enthalpy data in kcal mol^{-1}. To obtain ΔH° values in kJ mol^{-1}, one has to multiply the right hand side by 4.184:

$$-\Delta H^\circ \text{ (in kJ mol}^{-1}) = 4.184\ (E_A E_B + C_A C_B).$$

Alternatively, the E and C values may be adjusted by multiplying with $\sqrt{(4.184)}$ to give ΔH° in kJ mol^{-1}. The values of E and C given in Table 10.9 have been converted in this manner.

The four parameters E and C required to predict ΔH of an acid-base reaction are empirical in nature without any detailed theoretical background. The E parameters are associated with relative strength in *electrostatic bonding*, that is bonding through dipole-dipole forces. The C parameters were introduced to measure strength in *covalent bonding*. Later, two further parameters T and R have also been introduced. The T (for transmittance) parameters for the acid and base provide a measure for actual electron (charge) transfer between the acid and the base. The R term (for receptance) is similar. ΔH is then given by the **Drago equation**

$$-\Delta H = E_A E_B + C_A C_B + R_A T_B.$$

However, the last parameters are quite small for neutral acids and bases.

Values of E and C having best match for experimental ΔH values of a large number of acid-base interactions have been compiled. Some of these values, converted to kJ equivalent (by multiplying with $\sqrt{(4.184)}$) are shown in Table 10.9. Drago chose I_2 as the reference acid and set its E_A and C_A parameter to 1.00 (for ΔH in kcal). The standard for base was taken as *N,N*-dimethylacetamide assigned a value of E_B = 1.32 and diethyl sulfide assigned a value of C_B = 7.40 (ΔH in kcal). Clearly, all the values are arbitrary.

Table 10.9

E and C parameters for a few acids and bases **(kJ^{½} mol^{-1})**

	E_A	C_A		E_B	C_B
$SbCl_5$	15.1	10.5	NH_3	2.8	7.1
C_6H_5OH	8.9	0.9	C_6H_6	1.1	1.4
BF_3	20.2	3.3	$(CH_3)_2SO$	2.7	5.8
I_2	2.0	2.0	$(C_2H_5)_2O$	2.0	6.6
SO_2	1.9	1.7	CH_3NH_2	2.7	12.0
$B(CH_3)_3$	12.6	3.5	$(CH_3)_2NH$	2.2	17.8
$Al(CH_3)_3$	34.6	2.9	C_5H_5N	2.4	13.1
$CHCl_3$	6.2	0.3	$(CH_3)_2S$	0.7	15.3

Numerous ΔH values calculated using these E and C parameters are in good agreement with experimental values, barring a few exceptions (see next page).

Problem 10.19

Estimate ΔH for the reaction of I_2 with diethyl ether. E_A and C_A for $I_2 = \sqrt{4{\cdot}184}$ kJ$^{1/2}$ mol^{-1}. E_B and C_B for Et_2O = 2·0 and 6·6 respectively (all in kJ$^{1/2}$ mol^{-1}).

Solution: $-\Delta H = E_A E_B + C_A C_B = 2.04 \times 2.0 + 2.04 \times 6.6 = 17.5$ kJ mol^{-1}.

The result in problem 10.19 agrees nicely with the experimental value (17.6 kJ mol^{-1}). But all calculations may not agree to this extent. Gross discrepancies are observed in many cases from which only two are mentioned here:

1. $BMe_3 + NMe_3$: *experimental*: $-\Delta H = 73.6$ kJ mol^{-1}

calculated: $-\Delta H = E_A E_B + C_A C_B$

$= 12.6 \times 1.7 + 3.5 \times 23.6 = 104.0$ kJ mol^{-1}

The discrepancy may be largely assigned to steric hindrance (*F*-strain) in forming the adduct. Actually, the *E* and *C* parameters refer to individual acid or base; naturally they cannot take care or even predict situations which may arise on forming the adduct, as in this case.

2. $BF_3 + Et_2S$: $-\Delta H_{exp} = 12.1$ kJ mol^{-1}

$-\Delta H_{cal} = 20.2 \times 0.7 + 3.3 \times 15.3 = 64.6$ kJ mol^{-1}

Here one does not expect much *F*-strain because of the small size of the fluorine atom. The discrepancy is really large and one cannot provide any satisfactory explanation except for the reorganization energy involved in changing the geometry around B-atom from planar to tetrahedral.

The hardness or softness of an acid or base cannot be properly compared from their *E* or *C* values since these parameters are related to both hardness or softness and strength. Diethylamine has a C_B value 17.8 which is higher than that of diethylsulfide (15.3). From these figures alone, we cannot conclude that diethylamine is the softer base among the two. Even ratios of *E*/*C* do not provide a straight comparison of hardness. When both *E* and *C* of species-1 (acid or base) are higher than those of another species-2, that is $E_1 > E_2$ and $C_1 > C_2$, one may conclude that species 1 is stronger (acid or base) than species 2 in its combination with any acid or base. If $E_1 > E_2$ but $C_1 < C_2$, either acid 1 or acid 2 may appear stronger depending on the base. The same conclusion will apply while comparing bases.

Influence of solvent

The acidity of a species in a solvent is largely influenced by the solvation of the ions in solution. The Gibbs free energy of solvation of an ion of radius *r* and charge *ze* in a solvent of relative permittivity ε_r is expressed by the **Born equation**

$$\Delta G° = -\frac{Nz^2e^2}{8\pi\varepsilon_0 r}\left(1 - \frac{1}{\varepsilon_r}\right)$$

z^2/r is sometimes called the electrostatic parameter ξ(xi). We expect that small ions of high charge number will be highly stabilized in a polar solvent. Also, larger the value of ε_r, the more negative will be $\Delta G°$, favouring the formation of ions in solution. The conjugate base A^- of a neutral acid HA will be stabilized more in a polar solvent, favouring acidity. On the other hand, cationic species like NH_4^+ will be stabilized in a polar solvent with lower tendency for proton donation. H-bonding will further reduce the acidity of NH_4^+ in polar solvents like H_2O. In general polar solvents will favour proton transfer leading to charge separation. These conclusions are also applicable to Lewis acids and bases. Reactions with an increase in net charge will be favoured in a solvent of high relative permittivity.

Most common polar solvents are hard Lewis bases—e.g., water, alcohols, ethers and amines. Other important basic solvents are dmso (dimethyl sulfoxide; Me_2SO), dmf (dimethylformamide; $HCONMe_2$), and acetonitrile (MeCN). DMSO may also act as a donor through the soft S-end; it is an ambidentate solvent.

Hydrogen bonding by a solvent may influence the acidity or basicity of a substance. Thus HCl is a stronger acid in CH_3OH than in dmf. In CH_3OH the Cl^- ions are stabilized through species like Cl^- …… $H–OCH_3$. DMF has similar relative permittivity as CH_3OH but does not have a hydrogen-bond donor centre.

The basicity of solvents may be expressed quantitatively in terms of the enthalpy of adduct formation with a reference acid like the strong Lewis acid $SbCl_5$ in 1, 2-dichloroethane:

$$SbCl_5 + :B \rightarrow Cl_5Sb - B.$$

ΔH° of the above process (in kcal mol^{-1}) has been termed the **donor number** (D.N.) of the solvent. A higher donor number characterizes a stronger Lewis base.

A related parameter called the **acceptor number** (A.N.) of a solvent has also been derived with triethylphosphine oxide (Et_3 PO:) as a reference base. When the base is dissolved in the pure solvent, the nmr chemical shift of ^{31}P gives an indication of the donor-acceptor interaction. Numerical values for acceptor numbers are derived by arbitrarily setting the scale to 100 for the base in $SbCl_5$ and at 0 for the base in hexane. The higher the acceptor number, the stronger the Lewis acid.

Table 10.10

Donor number, acceptor number and relative permittivity (25°C) of some solvents

	D.N.	A.N.	ε_r
Acetone	17	12.5	20.7
Diethyl ether	19.2	3.9	4.3
Benzene	0.1	8.2	2.3
Ethanol	19.0	37.1	24.3
Pyridine	33.1	14.2	12.3
DMSO	29.8	19.3	45
Water	18	55	82

10.3.4 Superacids: Hammett Acidity Function

A superacid is more efficient proton (hydron) donor than pure H_2SO_4 (100 mass percent) and able to protonate almost every organic compound. As noted earlier, a mixture of SbF_5 and HF gives rise to the superacid H_2F^+. Superacids have high values (> 6) of $-H_0$, where H_0 is called the **Hammett acidity function**. It is defined in the same sense used for defining *p*H in dilute aqueous solution, except that the hydron (proton) donation equilibrium now refers to an indicator base (B), as explained below.

In determining the pH of solution spectrophotometrically using an indicator B, we have an equilibrium between the indicator base (B) and its protonated form (BH^+, say). Usually these two forms have different colours and the ratio $[B]/[BH^+]$ can be determined spectrophotometrically. For the system

$$B + H^+ \rightleftharpoons BH^+, \quad pH = pK_{BH^+} - \log\frac{[BH^+]}{[B]} - \log\frac{\gamma_{BH^+}}{\gamma_B}$$

Ignoring the activity coefficients, the acidity of the medium is defined by the acidity function H_0 when by definition,

$$H_0 = pK_{BH^+} - \log\frac{[BH^+]}{[B]}$$

Here B is a neutral base (zero charge) like *p*-nitroaniline; pK_a of protonated cation = 0.99.

For a system buffered with a charged species as base like B^-, the function H_- is defined in terms of a system buffered with B^- and its neutral conjugate acid HB. Similarly, one may define H_+ for an indicator system B^+/BH^{2+}.

In aqueous solution, $K_{BH^+} = [B]/[BH^+]$ gives H_0 equal to $-\log[H^+]$ or the *p*H of the solution. The acidity function actually covers a wider range of acidity and also applies to nonaqueous media. For several strong acids in aqueous solution, acidity increases with concentration and is independent of the anion. It appears that the hydrogen ion in these solutions is present as $H_9O_4^+$.

Spectrophotometric determination of *p*H using indicators offers several advantages. The pH inside microscopic vesicles of living cells may be measured by infusion of an appropriate indicator and following its spectrum. The extent of acidity of surfaces of solid catalysts may be measured when some Hammett indicators are adsorbed on the surface and their colours are studied spectrophotometrically.

Values of H_0 for several strong acids at different ranges of concentration have been determined by using suitable organic bases (like *p*-nitroaniline) and suitable indicators. Some typical values are given in Table 10.11.

Table 10.11

Values of H_0 for few acids

Acid	H_0	Acid	H_0
$HSO_3F + SbF_5$* (~14 mol%)	–21.0 to –25	HF (100%)	–11
$HF + SbF_5$ (~0.6 mol%)	–21.0 to –28	HF + NaF	–8.4
HSO_3F	–15.6	H_3PO_4	–5.0
F_3CSO_3H (Triflic acid)	–15.1	HCOOH	–2.2
H_2SO_4 (100%)	–11.9		

*Magic acid, a registered trade mark of a company. See below.

Acid media whose $-H_0$ values lie above about 6 are nearly 10^6 times more strong than 1 *M* solution of a strong acid. They are often referred to as **superacids**. The highest value of $-H_0$ is known with SbF_5 (~14 mol%) in HSO_3F, a combination of a Brönsted acid and a Lewis acid. Of the many reactions that take place in this mixture, the simplest one may be written as

$$SbF_5 + 2HSO_3F \longrightarrow [H_2SO_3F]^+ + [FSO_2O - SbF_5]^-$$

or, 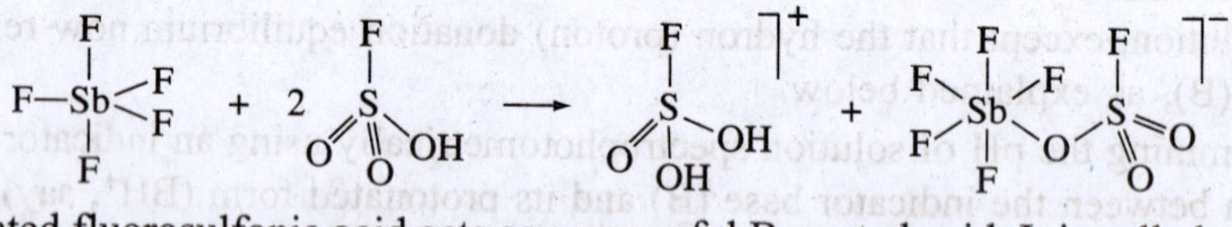

The protonated fluorosulfonic acid acts as a powerful Bronsted acid. It is called **magic acid,** which can dissolve wax (hydrocarbon).

Besides SbF_5, species which increase the concentration of $H_2SO_3F^+$ are acids:

$$HA + HSO_3F = H_2SO_3F^+ + A^-$$

Alkali metal flurophosphates and other proton (hydron) acceptors are bases:

$$KSO_3F = K^+ + SO_3F^- \; ; \; B + HSO_3F = BH^+ + SO_3F^-.$$

H_2SO_4 forms $H_3SO_4^+$ according to the second equation and thus behaves as a base.

The ClO_4^- behaves as a strong base since the conjugate acid $HClO_4$ shows no capability to donate proton:

$$ClO_4^- + HSO_3F \longrightarrow HClO_4 + SO_3F^-.$$

Organic amines, and even nitrobenzene get protonated and are bases in HSO_3F:

$$C_6H_5NO_2 + HSO_3F \longrightarrow C_6H_5NO(OH)^+ + SO_3F^-$$

Even hydrocarbons can be protonated by HSO_3F in favourable situations:

$$(CH_3)_4C + HSO_3F \longrightarrow [(CH_3)_3C\,(CH_4^+)] \longrightarrow (CH_3)_3C^+ + CH_4$$

Many unusual cations are thus formed in superacid medium:

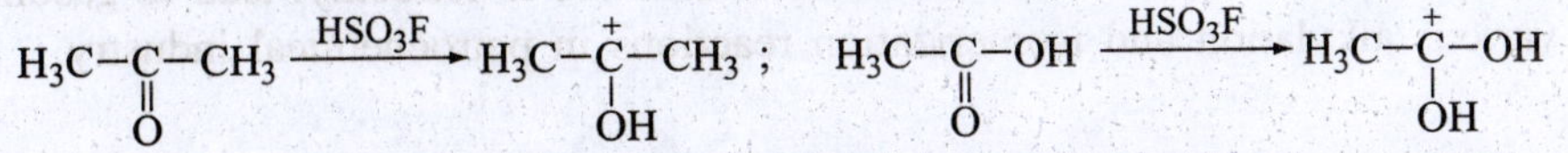

SO_2 is not protonated in SbF_5–HSO_3F mixture. It breaks the hydrogen-bonding network and reduces the viscosity of the medium but does not change the acidity.

When SO_3 and SbF_5 are dissolved in HSO_3F, SO_3 is first fluorinated to SO_3F^- which coordinates to Sb:

$$SO_3 + SbF_5 \rightleftharpoons SbF_4(SO_3F)$$

Subsequent fluorination and coordination ultimately results in $H[SbF_2(SO_3F)_4]$:

$$SbF_2(SO_3F)_3 + HSO_3F \rightleftharpoons H\,[SbF_2(SO_3F)_4]$$

It behaves as a strong acid in the SbF_5–HSO_3F superacid medium:

$$H\,[SbF_2(SO_3F)_4] + HSO_3F \longrightarrow H_2SO_3F^+ + SbF_2(SO_3F)_4^-$$

The acid is thus leveled to $H_2SO_3F^+$, the characteristic cation of the superacid medium.

The SbF_5–HSO_3F medium is also a strong fluorinating agent, particularly at high temperature. B_2O_3 and As_2O_3 are fluorinated to BF_3 and AsF_3 respectively, while P_4O_{10} gives POF_3. MnO_4^- and ClO_4^- ions are fluorinated to MnO_3F and ClO_3F respectively. These species can be distilled out of the superacid medium.

Many unusual cationic species may be prepared in superacid medium by oxidation of nonmetallic elements with oxidizing agents like peroxydisulfuryl difluoride, FSO_2OOSO_2F ($S_2O_6F_2$):

$$2I_2 + S_2O_6F_2 \longrightarrow 2I_2^+ + 2SO_3F^- \;;\; 3I_2 + S_2O_6F_2 \longrightarrow 2I_3^+ + 2SO_3F^-$$

$$Se_8 + S_2O_6F_2 \longrightarrow Se_8^{2+} + 2SO_3F^- \quad \text{etc.}$$

10.3.5 Surface Acids in Heterogeneous Catalysis

Many industrial processes take advantage of heterogeneous reactions occurring on solid sufaces acting as catalyst through Brönsted and Lewis acid-base sites. For example, the –OH groups attached to Al(III) on hydrated alumina surface (~150°C) act as Brönsted acid. Further heating to 500°C gives rise to the strong Lewis acid Al^{3+} and Lewis base O^{2-} on the surface, together with some –OH groups still remaining to function as Brönsted acid. Aluminosilicate surfaces contain bridging –OH groups between Al(III) and Si(IV). In certain clay minerals like montmorillonite which have a layered structure (see silicates, chapter 21), some Al^{3+} ions are replaced by Mg^{2+} ions, developing a negatively charged surface that can bind various exchangeable cation. Removal of these cations by washing with acids gives rise to high surface acidity, comparable to pH –5 to –8. Such clays may be used as catalysts in many organic reactions like the Friedel-Crafts reaction. Similarly, catalytic activities of three-dimensional silicates like zeolites and ZSM-5 involve both size-specific channels and exchangeable acid sites. Surface modification of the acidic sites according to requirement is also carried out to obtain desired catalytic activity: the –OH groups on silica gel may be converted to $-OSiR_3$ by treatment with R_3SiCl or R_3SiOH.

Applications of such catalysts are numerous. To mention only a few, one can cite the environment-friendly production of gasoline from methanol, isooctane from lower hydrocarbons, methyl *t*-butyl ether (MTBT) which is an alternative to tetraethyl lead as gasoline additive and various alkylation and isomerization reactions in petrochemical industry.

ESSENTIALS

Several definitions of acids and bases have been proposed, the more important ones being the following:

The Arrhenius theory: *Acid*: any hydrogen containing compound furnishing hydrogen ions in aqueous solution; *Base*–any compound which gives hydroxyl ions in aqueous solution.

The solvent system definition: *An acid* is a solute that gives the cation characteristic of the solvent. Thus NH_4Cl is an acid and $NaNH_2$ is a base in liquid ammonia which ionizes as $2NH_3 = NH_4^+ + NH_2^-$.

***The protonic theory* (Brõnsted)**: An *acid* is any species that tends to give up a proton; *a base* is any substance that tends to accept a proton. For aqueous systems, this definition is practically equivalent to the Arrhenius definition. But it covers acid-base behaviour in nonaqueous protonic solvents and gas phase.

Chemical species which differ from one another only to the extent of a proton transfer are termed **a conjugate acid-base pair**.

Any solvent capable of functioning as an acid as well as a base by releasing or accepting a proton is called an **amphiprotic solvent**. Water, ammonia (liquid) and hydrogen fluoride (anhydrous) are amphiprotic solvents.

The Lewis definition: *An acid* is any species which can accept a pair of electrons to form a covalent bond; and *a base* is any species that can donate a pair of electrons to establish a covalent bond. The theory explains acid-base interactions independent of the proton or any specific solvent:

$$NH_3(\text{base}) + BF_3(\text{acid}) = H_3N \longrightarrow BF_3(\text{adduct})$$

Strengths of protonic* (Brõnsted) *acids and bases in gas phase may be compared from gas phase proton affinity data. Relative acidity in gas phase is different from acidity in aqueous solution where hydration plays a major role.

Strengths of acids and bases in aqueous medium may be expressed in terms of respective dissociation constants.

General trends in acid-strength: *Among the hydroacids*, acid strength increases from left to right along a period: $NH_3 << H_2O << HF$. This may be explained on the basis of increasing electronegativity of the element bonded to hydrogen. Acid strength increases down any given group in the periodic table: HF < HCl < HBr < HI. The decrease in bond energy appears to be the determining factor among others.

Among oxo-acids of the types X–O–H acidity decreases as the element X appears lower in a group in the periodic table: HOCl > HOBr > HOI. The electronegativity of X acts through inductive effect.

The *net* inductive effect of additional oxygen or any other electronegative atom(s) present in a molecule also influences the release of proton from an O–H bond. Thus,

$$H_2SO_3 < H_2SO_4 \; ; \; H_3PO_3 > H_3PO_4 \; ; \; CF_3COOH > CH_3COOH.$$

According to an empirical rule proposed by Pauling, for a mononuclear oxoacid of type $O_nX\,(OH)_m$, $pK_a = 9 - 7n$ (or $8 - 5n$ in some cases).

The metal oxides become more basic as one proceeds down the periodic table: MgO < CaO < SrO.

The tendency of a cation to hydrolyze increases with increasing charge and decreasing size of the cation. Acidity of the hydrated cation increases in a parallel manner. Cations of transition metals have greater polarizing power than cations of a non-transition metal: $Fe^{3+} > Al^{3+}$

Strengths of Lewis acids and bases may be related to the strength in their adduct formed with a given base or acid, as the case may be. Steric factors often play an important role in adduct formation.

Lewis acids and bases have been classified as **hard and soft** according to the "hardness" or polarizability of the electron cloud associated with any species. In general, hard acids prefer to combine with hard bases and soft acids prefer to combine with soft bases. However, hardness (or softness) is different from inherent acid-base strength.

Drago and Wayland correlated the enthalpy of adduct formation between a Lewis acid(A) and a Lewis base(B) in gas phase through parameters representing electrostatic (E) and covalent (C) contributions to bonding:

$$-\Delta H^\circ \text{ (kJ mol}^{-1}) = 4.184\,(E_A E_B + C_A C_B)$$

The Hammett acidity function (H_0) is used to express acidity of strongly acidic media, including nonaqueous solvents. For a solution containing a neutral indicator base(B) and its protonated form (BH^+), H_0 is defined as

$$H_0 = pK_{BH^+} - \log\frac{[BH^+]}{[B]}$$

In aqueous solution, H_0 corresponds to the pH of the solution.

Superacids are stronger proton donors than even pure H_2SO_4 and have high values of $-H_0$ (> 6). Mixtures of Lewis and Brönsted acids (like SbF_5 and HSO_3F) have very high values of $-H_0$ (15 to 25). A mixture of HSO_3F and SbF_5 is called "magic acid" which can protonate even hydrocarbons. Many unusual cations like Se_8^{2+} may be formed in superacid medium.

EXERCISE

1. State the protonic definition of acids and bases.

   Show that for aqueous solutions, the protonic definition of acids is practically equivalent to the Arrhenius definition; but for bases, the protonic definition covers more compounds.

2. According to which of the various definitions of acids and bases is (a) HCl an acid in water, (b) HCl(g) an acid, (c) NH_4Cl an acid, (d) H_2SO_4 a base, (e) $NH_3(g)$ a base and (f) BF_3 an acid?

3. What are conjugate acids and bases? Write down the conjugate acids of:

   H_2O, NH_3, Cl^-, HS^-, S^{2-}, CO_3^{2-}, PO_4^{3-}, OH^-.

   Write down the conjugate bases of:

   H_2O, HCl, HS^-, NH_3, HCO_3^-, H_2SO_4, NH_4^+, HNO_3.

4. What do you mean by levelling effect of a solvent? Explain the statement: "Acetic acid exerts less levelling effect on the strengths of acids than water."

5. State, with reason, which one among each pair of the following should be a stronger acid or stronger base?
(i) $H_2O - H_3O^+$, (ii) $NH_3 - NH_4^+$, (iii) $H_2S - H_2Se$,
(iv) $HCO_3^- - CO_3^{2-}$, (v) $H_2SO_3 - H_2SO_4$, (vi) $KOH - Ba(OH)_2$,
(vii) $Be(OH)_2 - Sr(OH)_2$, (viii) $Fe(OH)_2 - Fe(OH)_3$, (ix) $HCl - HI$,
(x) $HOCl - HOI$ (xi) $H_3PO_3 - H_3PO_4$ (xii) $HClO_3 - HClO_4$.
6. Explain/Justify:
(a) F^- ion is a strong base.
(b) HNO_3, HCl and H_2SO_4 appear equally strong in water but their strengths differ in acetic acid medium.
(c) HSO_3^- may be considered both as an acid and a base.
(d) Metal cations are good Lewis acids.
7. Estimate, using Pauling's formula, the pK_a values of (a) H_3PO_4 (b) $H_2PO_4^-$ (c) HPO_4^{2-}.
8. The gas-phase proton transfer enthalpy for the following reaction is highly negative:

$$NH_4^+(g) + Cl^-(g) \rightarrow NH_3(g) + HCl(g) \quad \Delta H = -528 \text{ kJ mol}^{-1}$$

What makes the process much less favourable in aqueous solution?
9. Indicate the stronger acid among the following pairs:
(a) $[Fe(H_2O)_6]^{2+}$ and $[Fe(H_2O)_6]^{3+}$ (b) $[Al(H_2O)_6]^{3+}$ and $[Ge(H_2O)_6]^{3+}$ (c) $Si(OH)_4$ and $Ge(OH)_4$ (d) H_2CrO_4 or $HMnO_4$ (e) H_3PO_4 or H_2SO_4.
10. Arrange in order of increasing acid strength: H_3O^+, HSO_4^-, H_4SiO_4, CH_3GeH_3, NH_3, HSO_3F.
[NH_3, CH_3GeH_3, H_4SiO_4, HSO_4^-, H_3O^+, HSO_3F]
11. Arrange according to increasing basic character: Al_2O_3, B_2O_3, CO_2, SO_3, Cl_2O_7, BaO.
[Cl_2O_7, SO_3, CO_2, B_2O_3, Al_2O_3, BaO]
12. Identify the Lewis acids and bases (a) $FeCl_3 + Cl^- \rightarrow [FeCl_4]^-$ (b) $(CH_3)_2CO: + I_2 \rightarrow (CH_3)_2CO - I_2$ (c) $BrF_2 + F^- \rightarrow [BrF_4]^-$ (d) $KH + H_2O \rightarrow KOH + H_2$.
[*Acids*: $FeCl_2$, I_2, BrF_3, H^+]
13. Which of the given solvents will favour the reactions as written?
(a) $F_3B.NH_3 \rightarrow BF_3 + NH_3$ — CH_3OCH_3 or CH_3OH
(b) $CH_3Cl + I^- \rightarrow CH_3I + Cl^-$ — dmf or CH_3OH
(c) $CH_3I + NH_3 \rightarrow CH_3NH_3^+ + I^-$ — CH_3CN or CH_3OCH_3
[**Hint:** (a) Both solvents are about equally basic and does not involve any charge separation. But CH_3OH can act as an acid via H-bond donation to NH_3. Hence CH_3OH will favour the reaction more.
(b) CH_3OH acts as a stronger hard acid toward the hard base Cl^- through H-bond.
(c) Charge separation is favoured by CH_3CN, which has higher ε_r.
14. Which of the following reactions are expected to proceed to right according to the HSAB concept in gas phase or an inert solvent?
(a) $CH_3HgI + HCl \rightleftharpoons CH_3HgCl + HI$
(b) $R_3NBF_3 + R_3PBBr_3 \rightleftharpoons R_3PBF_3 + R_3NBBr_3$
(c) $[AgCl_2]^-(aq) + 2CN^-(aq) \rightleftharpoons [Ag(CN)_2]^-(aq) + 2Cl^-(aq)$
(d) $2CH_3MgF + HgF_2 \rightleftharpoons (CH_3)_2Hg + 2MgF_2$
(e) $La_2(CO_3)_3 + Tl_2S_3 \rightleftharpoons La_2S_3 + Tl_2(CO_3)_3$
(f) $(CH_3)_2O:BF_3 + (CH_3)_2S:BH_3 \rightleftharpoons (CH_3)_2O:BH_3 + (CH_3)_2S:BF_3$

15. Enthalpies of reaction of Me_3B with NH_3, $MeNH_2$, Me_2NH and Me_3N ($Me = CH_3$) are –58, –74, – 81 and –74 kJ mol^{-1} respectively. Explain the drop at Me_3N.

16. Which type of solvent would favour:

(a) displacement of Cl^- by I^- from an acid centre (b) acidity of Ag^+ over Al^{3+} (c) the reaction $2FeCl_3 + ZnCl_2 \rightarrow Zn^{2+} + 2[FeCl_4]^-$?

[(a) hard acid (e.g., H_2O (b) hard base (e.g., H_2O) (c) soft base (e.g., CH_3CN)]

17. Hg occurs in nature mostly as HgS while Zn occurs in several forms in addition to ZnS : $ZnCO_3$, ZnO etc. Comment.

18. Give examples of Lewis acid-base behaviour of (a) SO_2 (b) SO_3 (c) Xe (d) XeF_6.

Hint: (a) Lewis acid toward OH^-, F^-, R_3N:, $(CH_3)_2S = O$

Lewis base toward SbF_5, BF_3, $[Ir(PPh_3)_2(CO)(Cl)]$

(b) Lewis acid toward H_2O; Lewis base toward H_2SO_4

(c) Lewis base: $Xe + XeF^+ + SbF_5 \rightarrow Xe_2^+SbF_6^-$

(d) Lewis acid: $XeF_6 + MF \rightarrow M^+XeF_7^-$, $M_2[XeF_8]$

Lewis base: $XeF_6 + PtF_5 \rightarrow XeF_5^+PtF_6^-$

19. Explain the following order of acidity in *gas phase*:

(a) NaH > HF (b) $C_6H_5CH_3 > CH_4$ (c) $CH_3NO_2 > CH_4$.

Hint: (a) Lower bond dissociation energy of NaH dominates over higher electron affinity of F (Bond energy: NaH = 201; HF = 566; Electron affinity: Na = 53; F = 329. All in kJ mol^{-1})

(b) $C_6H_5CH_2^-$ is resonance stabilized.

20. Comment on the following acidity orders:

(a) 2,6-dimethylbenzoic acid > benzoic acid (b) *p*-fluorobenzoic acid < *p*-bromobenzoic acid

(c) $BMe_3 < AlMe_3$ (to Me_3N) (d) $BMe_3 > BPh_3$ (e) $BF_3 < B_2H_6$ (to CO and Me_2S).

Hint: (a) resonance stabilization not possible in the dimethyl compound owing to the methyl groups; the –COOH group cannot be coplanar with the benzene ring.

(b) Better match of F 2*p* orbital with C 2*p* orbital gives rise to more extensive +R effect in the fluoro compound which nearly cancels the –I effect of F.

(c) BMe_3 is largely stabilized through hyperconjugation. Adduct formation is accompanied by steric hindrance.

(d) In BPh_3, π-bonding with benzene ring dominates over hyperconjugative effect in BMe_3.

(e) BH_3 . CO may be stabilized by hyperconjugative structures like $\begin{matrix} & H^+ \\ H & \\ H & \end{matrix}\!\!>B^-{=}C{=}O$.

$Me_2S \rightarrow BH_3$ may involve hyperconjugation using vacant *d*-orbital on S.

Note: $Me_2O \rightarrow BH_3$ cannot be stabilized in this way.

21. The reaction: $HgI_2 + 2KOH \rightleftharpoons Hg(OH)_2 + 2KI$ does not show any strong tendency to proceed to one side or other and has an equilibrium constant close to 1. Comment.

Hint: Soft-soft (HgI_2) and hard-hard (KOH) combinations tend to shift the reaction toward left. But strength considerations favour the forward reaction: the stronger acid (Hg(II)) prefers to combine with the stronger base (OH^-).

11

IONIC EQUILIBRIUM IN AQUEOUS SOLUTIONS

11.1 INTRODUCTION

With a background of the acid-base behaviour of different substances, one may now be interested to find out the concentrations of various ionic species present at equilibrium in their aqueous solutions. For this purpose, one has to consider the different equilibrium processes involved in such solutions. From this, and other considerations, one may obtain as many independent equations as there are unknowns. The equations may then be solved to obtain the unknown quantities, introducing simplifications and approximations based on chemical intuition. Complex situations may also be solved through computer programmes.

11.1.1 The Ionic Product of Water

All solutions in water maintains the equilibrium involving autoionization of water, expressed in terms of the conventional "*Ionic Product of water*", K_w.

In pure water, or in any aqueous solution, the product of the concentrations (strictly activities) of hydrogen and hydroxyl ions is a constant at a given temperature. This product is known as the ionic product of water (K_w).

$$K_w = a_{H^+}.a_{OH^-} \approx C_{H^+}.C_{OH^-} \simeq C_{H_3O^+}.C_{OH^-}$$

The value of K_w at 25°C is 1×10^{-14}. It increases with temperature:

Temp (°C)	0	10	24	25	40	50
K_w	1.14×10^{-15}	2.92×10^{-15}	1.00×10^{-14}	1.01×10^{-14}	2.92×10^{-14}	5.47×10^{-14}
pK_w	14.944	14.535	14.000	13.996	13.535	13.262

Values of K_w may be calculated from the conductance of pure water. Most reliable methods are, however, obtained from EMF measurement of cells of the type

$$\text{Pt}, H_2(g) \mid \text{NaOH } (m_1), \text{NaCl } (m_2) \mid \text{AgCl } (s) \mid \text{Ag}$$

For details of procedure, any standard textbook of physical chemistry may be referred to.

The ionic product of water puts a maximum limit to the mutual concentrations of hydrogen and hydroxyl ions in any aqueous solution (at a given temperature). In pure water, the concentrations of hydrogen and hydroxyl ions are exactly equal. This is also the case for any solution which is exactly neutral.

We have, $C_{H^+} = C_{OH^-}$ and $C_{H^+}.C_{OH^-} = K_w$

$\therefore \quad C_{H^+} = K_w^{1/2}$; At 25°C, $C_{H^+} = 10^{-7}$ mol dm^{-3}.

The ionic product of water is different from the equilibrium constant (K) for the autoionization equilibrium of water (i.e., the ionization constant) at the same temperature. We have

$$K = \frac{C_{H^+}.C_{OH^-}}{C_{H_2O}^2} = \frac{K_w}{C_{H_2O}^2}$$

Assuming the density of water to be exactly unity, C_{H_2O} = (100 ÷ 18) = 55.56 mol dm^{-3}. So, at 25°C, $K = 10^{-14}/(55.56)^2 = 3.24 \times 10^{-18}$.

11.1.2 The pH Scale

In the study of acid-base equilibria in aqueous solutions we are primarily interested in the hydrogen ion concentration of a solution. Again, the solutions we deal with are usually dilute; the concentrations are accordingly small—some negative power of 10. The concentration of hydrogen ion in pure water at 25°C is 10^{-7} mol dm^{-3}; in a 0.01 M solution of hydrochloric acid, it is 10^{-2} mol dm^{-3}. Clearly, the frequent use of the negative power of ten is troublesome when we want to express the acidity (or alkalinity) of a solution readily. Sorensen suggested (1909) a technique for expressing such exponential numbers as small decimal numbers: a term called pH* was introduced as

$$\text{pH} = -\log C_{H^+} \qquad \text{(concentration in mol lit}^{-1}\text{)}.$$

* Abbreviation of *puissance d' hydrogen*, meaning pressure of hydrogen.

The pH term thus enables us to express the hydrogen ion concentration briefly and neatly. The effective range of the pH scale in water may be set as follows: for a solution containing 1 mol of H_3O^+ per dm^3, the pH is $-\log 1 = 0$. For a solution containing 1 mol of OH^- per dm^3, the concentration of H^+ will be obtained from the ionic product of water. At 25°C,

$$K_w = C_{H_3O^+}.C_{OH^-} = 10^{-14}; \quad \therefore C_{H_3O^+} = 10^{-14} / C_{OH^-} = 10^{-14} \text{ mol dm}^{-3}$$

Hence the pH $= -\log 10^{-14} = 14$. The value of the ionic product of water thus puts a practical limit to the pH scale—between 0 and 14, a neutral solution being at the midpoint of the scale at pH = 7 (at 25°C). Solutions having pH < 7.0 are acidic and those having pH > 7.0 are alkaline (at 25°C, strictly).

The concentration of OH^- ion may be similarly expressed in terms of pOH, when pOH = $-\log C_{OH^-}$. The use of pOH is however not very common to express C_{OH^-}. From the expression of K_w, we may write

$$-\log K_w = -\log C_{H^+}.C_{OH^-} \quad \text{or,} \quad pK_w = \text{pH} + \text{pOH},$$

pK_w being defined in the same way as equal to $-\log K_w$. The pH and pOH of any aqueous solution is thus inter-related via pK_w. At 25°C, $K_w = 10^{-14}$, so that $pK_w = 14$. So, pH + pOH = 14 at ordinary temperature. This may be seen in the pH scale in Fig. 11.1.

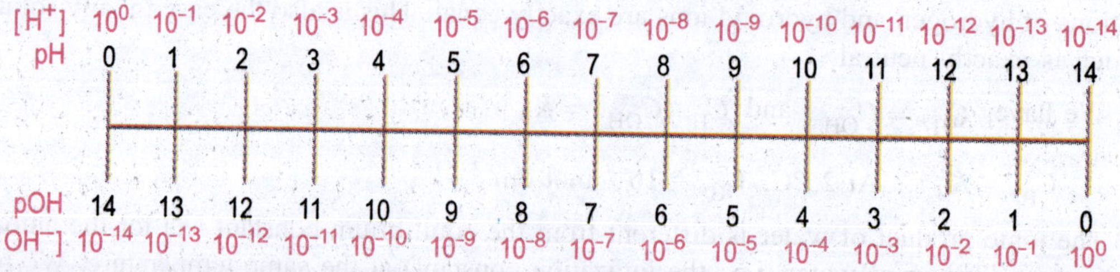

Fig.11.1: The pH scale (at 25 °C) in water.

Problem 11.1

What will be the concentration of H^+ in a solution having pH = 5.6?

Solution: pH $= -\log [H^+]$; $\therefore \log [H^+] = -$pH or, $[H^+] = 2.5 \times 10^{-6}$ mol dm^{-3}.

Problem 11.2

What is the pH of pure water at 100°C? K_w at 100°C is 5.45×10^{-13}.

Solution: In pure water, $[H^+] = [OH^-]$

$$\therefore K_w = [H^+] \times [OH^-] = [H^+]^2$$

$$\therefore \text{pH} = -\log [H^+] = -\frac{1}{2} \log K_w = 6.13$$

Note: As the pH is below 7.0, one may think that water becomes *acidic* at 100°C. But the criterion of pH = 7.0 for a neutral solution results from the value of 10^{-14} for K_w, which is correct *only at* 25°C. A neutral solution, as also *pure* water, always contains equal concentrations of H^+ and OH^- ions; $C_{H^+} = C_{OH^-} = (K_w)^{½}$. As K_w increases, C_{H^+} increases and hence pH decreases.

Problem 11.3

A solution of HCl is diluted so that its pH changes by (i) 0.3 and (ii) 1 unit. How does the concentration of H^+ ion change in each case?

Solution: As the solution is diluted, the concentration of H^+ must decrease and pH should increase. Suppose the pH changes by ΔpH when $[H^+]$ changes by a factor x.

Initially, $pH = -\log [H^+]$;

Finally, $pH + \Delta pH = -\log (x.\,[H^+]) = -\log x - \log [H^+]$ or, $\Delta pH = -\log x$

Case (i) $\Delta pH = 0.3 = -\log x$

$\therefore x = 0.5$. Thus the $[H^+]$ becomes 0.5 times the initial value.

Case (ii) $\Delta pH = 1 = -\log x$ or, $x = 10^{-1}$.

Thus, $[H^+]$ becomes 0.1 times or 1/10th the initial value.

Acid and alkali solutions in concentrations greater than 1 mol dm^{-3} do of course exist. For them, the concentrations of H^+ and OH^- ions are more conveniently expressed in molarity itself, These concentrations being greater than 10^0, their conversion to pH is not necessary; the very purpose for which the pH scale was defined does not suit the situation. Negative pH and pOH values are therefore not in usage. In concentrated acid solutions, as also in nonaqueous media, the concentration of H^+ ion is sometimes used in terms of the Hammett functions (chapter 10).

With the introduction of activities by Lewis and Randall (1923), the pH is more appropriately defined as

$$pH = -\log a_{H^+}.$$

But individual ionic activities cannot be measured; they are significant only in relation to the activity of the electrolyte. So in terms of activity, pH is better defined according to the way in which it is measured. This is called an **operational definition**. One may proceed to develop an operational definition of pH as follows:

The e.m.f. of a cell of the type

$$Pt \,|\, H_2(g,\ 1\ atm),\ H^+\ (a_{H_3O^+}) \,||\, KCl\ (1N) \,|\, Hg_2Cl_2(s),\ Hg\ (Calomel)$$

is dependent on the activity of H^+.

$$\text{Now, } E_{Cell} = E^{Red}_{Cal\,(1N)} - E^{Red}_{H2} = 0.2802 - 0.0591 \log a_{H+} \text{ (at 298 K)}$$

$$\text{or, } pH = (E_{Cell} - 0.2802)/0.0591.$$

In practice, any electrode reversible with respect to hydrogen ion can be used for this purpose. The National Bureau of Standards, USA, defines the *difference* in pH between (i) an unknown solution X (pH_x) and (ii) a standard solution S of known pH (pH_s) as

$$pH_x - pH_s = \frac{E_x - E_s}{2.3026\ RT/F}$$

where E_x and E_s are e.m.f. of the following cells:

E_x: H_2, Pt / Solution X / 3.5 M KCl / Reference electrode

E_s: H_2, Pt / Solution S / 3.5 M KCl / Reference electrode.

The standard solution S may be of

(i) potassium hydrogen tartrate (saturated): pH = 3.56 (at 25°C).

(ii) Potassium hydrogen phthalate (0.05 m): pH = 4.01 (at 25°C).

(iii) Borax (0.01 m) : pH = 9.18 (at 25°C).

(iv) 0.025 m potassium dihydrogen phosphate and 0.025 m disodium hydrogen phosphate: pH = 6.86 (at 25°C).

The pH defined in this manner slightly differs from the pH obtained by the relation pH = – log $C_{H_3O^+}$. If this latter is written by pcH, then it can be shown that pcH ≃ pH + 0.04.

Measurement of pH

The pH or hydrogen ion concentration of a solution is determined mainly by two methods:

(i) ***Indicator method***: The intensity of the colour developed by an acid-base indicator is compared to the colour developed by the same amount of indicator in some solution of known pH (buffer solutions, explained later).

(ii) ***E.M.F. method***: The experimental solution is set in equilibrium with an electrode reversible with respect to hydrogen ion. This is then connected to a reference electrode. The e.m.f. of the assembled cell is then measured by a potentiometer and the pH is calculated. The electrodes reversible with respect to hydrogen ion may be (i) a hydrogen electrode, (ii) a glass electrode or (iii) a quinhydrone electrode. The reference electrode is usually a calomel or a silver-silver chloride electrode.

The hydrogen electrode is difficult to set up. At the same time it cannot be used in the presence of (i) an oxidizing agent, (ii) ions of noble metals, (iii) unsaturated organic compounds and alkaloids.

The Glass electrode

The glass electrode is the most widely used ion-selective electrode used in measuring pH. This is made from a thin-walled bulb of low melting point glass having high electrical conductivity. The thin glass wall acts as a membrane which establishes equilibrium with hydrogen ions on both sides of it. The potential difference across the membrane is obtained by attaching reference electrodes on the two sides of the membrane. A typical setup is like the following.

$$\text{Ag}(s) \mid \text{AgCl}(s) \mid \text{Cl}^-(aq) \parallel \text{H}^+(aq) \parallel \text{H}^+(aq) \mid \text{Cl}^-(aq) \mid \text{AgCl}(s) \mid \text{Ag}(s).$$

Glass membrane

Combined electrodes are now available for use in pH meters. The glass bulb of the electrode is filled with 0.1M HCl saturated with AgCl and is fitted with two Ag–AgCl electrodes as shown in Fig. 11.2.

The monovalent cations, particularly Na^+, can move somewhat through the open silicate lattice (see Group 14) in which charged O atoms weakly coordinate to the metal ions. When immersed in an aqueous solution, the inner and outer surfaces of the glass swells to some extent and the metal cations can diffuse out of the glass membrane into the solution outside. Concomitantly, H^+ ions from the solution can diffuse through the swollen glass membrane to occupy some of the cation binding sites. Tracer studies using radioactive ^3H show that H^+ ions do not actually cross the glass membrane while the membrane still exists in equilibrium with H^+ ions on each side—charge is transported across the membrane by the Na^+ ions. In this regard, the glass electrode differs from typical ion-selective electrodes—the species adsorbed selectively (H^+) on each side of the membrane is different from the species that migrates across the membrane.

With equal concentration of chloride ions in both Ag–AgCl electrodes, the potential difference between the inner and outer Ag–AgCl electrodes depends on the potential difference across the glass membane. The activity of H^+ in the inner side of the membrane is kept constant by filling the glass bulb with 0.1M HCl (saturated with AgCl). Hence the potential of the glass electrode depends on the activity of H^+ ion in the external side, that is, the analyte solution in which the electrode is immersed. The response to change in pH fairly follows the Nernst equation and the potential E at 25°C may be expressed as

$$E = \text{Constant} - 0.059\,16 \log \frac{a_{H^+}(\text{in})}{a_{H^+}(\text{out})}$$

The constant, called ***asymmetry potential***, would be zeo if both sides of the glass were exactly identical when the electrode shows zero potential difference with equal concentrations of H^+ ions inside and outside the electrode. However, a small difference is often observed arising from asymmetry of the inner and outer surfaces of the membrane. This is adjusted by calibrating the electrode frequently (about every two hours during sustained use) using solutions of known pH. As can be seen from the expression of E, the glass electrode will develop a potential difference of 59.16 mV for each tenfold difference of H^+ ion activity, that is a change of pH by one unit.

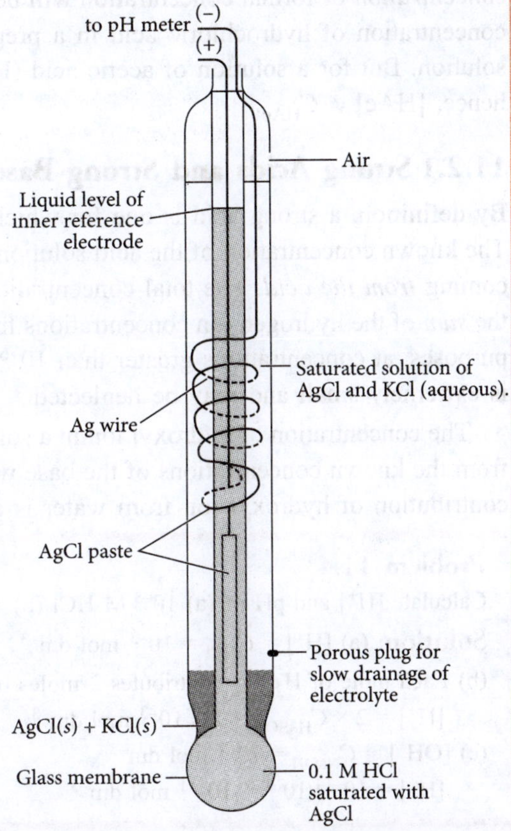

Fig.11.2: Sketch of a glass combination electrode.

11.2 EQUILIBRIUM IN SOLUTIONS OF ACIDS AND BASES

In this section, we shall first consider aqueous solutions containing a single acid-base system in addition to that provided by self-ionization of water itself. Separate treatments will be provided for strong and weak acids and bases. Equilibria involving multiple acid-base systems will be discussed subsequently.

When a Bronsted acid HA or a base B are added to water, equilibrium is established as

Acid: $HA + H_2O \rightleftharpoons H_3O^+ + A^-$ or simply, $HA \rightleftharpoons H^+ + A^-$ ... [11-1]

Base: $B + H_2O \rightleftharpoons BH^+ + OH^-$ ... [11-2]

The equilibrium set up by self-ionization of water is also present side by side. We shall now consider typical situations in which the knowledge of these equilibria may be applied to obtain equilibrium concentrations of different species in solutions, particularly of the hydrogen ion.

Henceforth, the **equilibrium concentration** of a species will be designated by a **square bracket**. Thus, **$[H_3O^+]$** will stand for the **equilibrium concentration of hydrogen ion** in a solution. The concentration at which a solute is actually taken in a solution, i.e., the *analytical concentration* or formal concentration will be designated by C (or F). Thus C_{HCl} stands for the concentration of hydrochloric acid in a prepared solution. This is completely dissociated in solution. But for a solution of acetic acid (HAc, say), only a part dissociates in solution and hence, $[HAc] \neq C_{HAc}$.

11.2.1 Strong Acids and Strong Bases

By definition, a strong acid is one for which equilibrium (11-1) lies completely to the right. The known concentration of the acid solution thus furnishes the concentration of hydrogen ion coming *from the acid.* The total concentration of hydrogen ion in the solution would now be the *sum* of the hydrogen ion concentrations from (i) the acid and (ii) water. But for all practical purposes, at concentrations greater than 10^{-6} M, the contribution of hydrogen ion from water is extremely small and may be neglected.

The concentration of hydroxyl ion in a solution of a strong base may be similarly calculated from the known concentrations of the base when the solution is not too dilute (> 10^{-6} *M*). The contribution of hydroxyl ion from water is again neglected.

Problem 11.4

Calculate $[H^+]$ and pH of (a) 10^{-3} M HCl (b) 10^{-3} M H_2SO_4 and (c) 10^{-3} *M* NaOH solutions.

Solution: (a) $[H^+] = C_{HCl} = 10^{-3}$ mol dm^{-3} $\quad \therefore$ pH = $-\log 10^{-3} = 3.0$.

(b) Each mole of H_2SO_4 contributes 2 moles of H_3O^+ ions. Hence,

$[H^+] = 2 \times C_{H_2SO_4} = 2 \times 10^{-3}$ mol dm^{-3} $\quad \therefore$ pH = $-\log(2 \times 10^{-3}) = 2.7$

(c) $[OH^-] = C_{NaOH} = 10^{-3}$ mol dm^{-3}

$[H^+] = 10^{-14}/10^{-3} = 10^{-11}$ mol dm^{-3} $\quad \therefore$ pH = $-\log 10^{-11} = 11.0$

We have mentioned that the contribution of hydrogen ion by water cannot be totally neglected for very dilute solutions ($C_{HA} < 10^{-6}$ M). To appreciate the point more directly, let us first ignore the contribution of hydrogen ion from water in such cases and see what happens.

As shown in problem 11.4, the pH of a 10^{-3} M solution of HCl is 3.0. If we neglect the contribution of hydogen ion by water at *all* concentrations, the pH value at any concentration may be obtained by the same procedure. We shall get the following results:

C_{HCl}(M)	10^{-6}	10^{-7}	10^{-8}	10^{-9}	
pH	6.0	7.0	8.0	9.0	and so on.

The results lead to the conclusion that a solution of hydrochloric acid turns *alkaline* as it is diluted beyond 10^{-7} M! Obviously there is something wrong in the approach. The situation may be handled by the general procedure outlined below.

For any aqueous solution of the acid HCl (or any acid HA), we are interested to know the equilibrium concentrations of H^+, OH^- and Cl^- ions in solution. Since there are three unknown quantities, we need three independent equations. These equations may be obtained in the following manner:

1. In any aqueous solution, equilibrium must exist between the hydrogen and hydroxyl ions as

(*Equilibrium*) $[H^+][OH^-] = K_w$ .. [11-3]

2. We must have *material balance* in the solution. The chloride ions are coming from HCl alone. Since it is fully dissociated, the concentration of chloide ions at equilibrium must be equal to the concentration of HCl taken. Suppose this is C_a. Then,

(*Material balance*) $[Cl^-] = C_a$ .. [11-4]

3. Again, there must be *charge balance* in the solution–it must be *electrically neutral.* So, the sum of the concentrations of the positive charges must be equal to the sum of the concentrations of the negative charges. The only positively charged species in the solution is H^+. There are two negatively charged species, Cl^- and OH^-. Hence,

(*Charge balance*) $[H^+] = [Cl^-] + [OH^-]$ .. [11-5]

Now we have to solve these three equations.

Putting Equation 11.3 in Equation 11.4, $[H^+] = C_a + [OH^-]$.

Putting the value of $[OH^-]$ from equation 11.3 into this relation,

$$[H^+] = C_a + K_w/[H^+] \quad \text{or,} \quad [H^+]^2 - C_a[H^+] - K_w = 0 \qquad .. [11\text{-}6]$$

Solution of this quadratic equation gives us the equilibrium concentration of hydrogen ion.

$$[H^+] = \frac{C_a \pm (C_a^2 - 4K_w)^{1/2}}{2}$$

1. When C_a^2 is appreciably larger than K_w, we may neglect K_w and write

$[H^+] = (C_a \pm C_a)/2 = C_a$ as the meaningful solution of equation 11.6. This is the situation existing in solutions having $C_a > 10^{-6}$ M. The same result is obtained if we neglect K_w in equation 11.6.

2. When C_a is around 10^{-7} M, both C_a^2 and K_w are of the same magnitude and both has to be considered in calculating $[H^+]$ (Problem 11.5).

3. As C_a falls further below 10^{-7} M, C_a^2 becomes negligible and $[H^+]$ approaches the value of $K_w^{1/2}$, i.e., 10^{-7} mol dm^{-3} (Problem 11.6). The same result is obtained from equation 11.6 when we observe that $C_a \times [H^+]$ becomes negligible in comparison to K_w and we have $[H^+]^2 = 10^{-14}$.

Aqueous solutions of a strong base may be treated similarly to obtain the equilibrium concentrations of H^+ and OH^- ions. For such solutions also, the equilibrium concentrations of H^+ and OH^- ions approach 10^{-7} mol dm^{-3} as the solution is diluted beyond 10^{-7} M (Problem 11.5c and d).

pH of aqueous solutions of strong acids and strong bases thus approach the limiting value of 7.0 with dilution, as shown in Fig. 11.3.

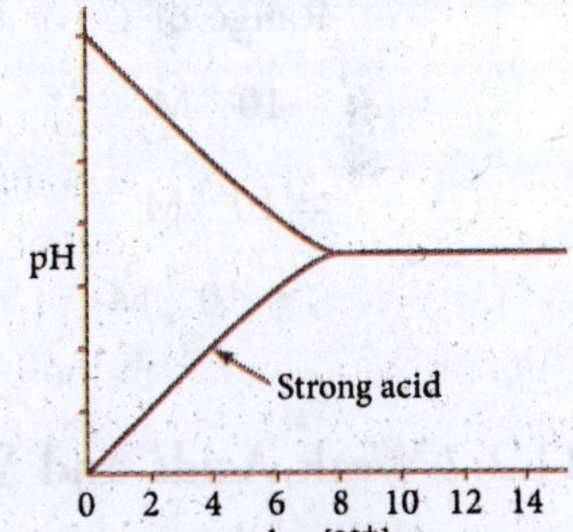

Fig.11.3: pH of aqueous solution of strong acids and bases as a function of concentration.

Problem 11.5

Calculate the pH of (a) 10^{-7} M HCl, (b) 10^{-9} M HCl (c) 10^{-7} M NaOH and (d) 10^{-9} M NaOH solutions.

Solution: (a) From equation (11.6),

$$[H^+]^2 - 10^{-7}[H^+] - 10^{-14} = 0$$

$$\text{or,}\quad [H^+] = \frac{10^{-7} \pm (10^{-14} + 4 \times 10^{-14})^{1/2}}{2} = \frac{10^{-7} \pm 2.24 \times 10^{-7}}{2} \text{ mol dm}^{-3}$$

The positive root of this equation gives $[H^+] = 1.62 \times 10^{-7}$ mol dm^{-3}, when pH = 6.79

(b) From equation (11.6), $[H^+]^2 - 10^{-9}[H^+] - 10^{-14} = 0$

$$\text{or,}\quad [H^+] = \frac{10^{-9} \pm (10^{-18} + 4 \times 10^{-14})^{1/2}}{2} = \frac{10^{-9} \pm [10^{-18} + (40{,}000 \times 10^{-18}]^{1/2}}{2}$$

The positive solution gives $[H^+] = 1.005 \times 10^{-7}$ mol dm^{-3}. or, pH = 6.9979

(c) Proceeding as in the case of a strong acid, $[H^+]$ is given in terms of the analytical concentration of the base, C_b as

$$[H^+]^2 + C_b[H^+] - K_w = 0.$$

Solution of this equation for $C_b = 10^{-7}$ M gives

$$[H^+] = \frac{-10^{-9} \pm (10^{-18} + 4 \times 10^{-14})^{1/2}}{2} = 99 \cdot 5 \times 10^{-9} \text{ mol dm}^{-3}. \quad \text{or, pH} = 7 \cdot 2457.$$

(d) From the quadratic equation given in the peceding problem, we get

$$[H^+] = \frac{-10^{-9} \pm (10^{-18} + 4 \times 10^{-14})^{1/2}}{2} = 99 \cdot 5 \times 10^{-9} \text{ mol dm}^{-3}. \quad \text{or, pH} = 7 \cdot 0022.$$

Summary Rules for solutions of Strong Acids and Bases

Our findings with solutions of strong acids and bases (present as a single system) may be summarized in the following table:

Range of C_a or C_b	$[H^+]$ for acid	$[H^+]$ for base
$> 10^{-6}$ M	C_a	$10^{-14}/C_b$
$\simeq 10^{-7}$ M	$\dfrac{C_a \pm (C_a^2 + 4K_w)^{1/2}}{2}$	$\dfrac{-C_b \pm (C_b^2 + 4K_w)^{1/2}}{2}$
$<< 10^{-7}$ M	$K_w^{1/2}$	$K_w^{1/2}$

11.2.2 Weak Acids and Weak Bases

Weak acids

A weak acid is only incompletely dissociated; at equilibrium, the solution contains a significant amount of the undissociated acid. An over-simplified method to treat such equilibrium is given at first; a closer analysis will be made subsequently.

Let us consider a solution of a weak acid HA taken at a concentration C_a M. If α is the fraction of the total amount of HA that dissociates at equilibrium, we may write:

	HA	$\rightleftharpoons$	H^+	+	A^-	
Equilibrium concentration:	$C_a(1-\alpha)$		αC_a		αC_a	mol dm^{-3}.

For the time being, we neglect the contribution of H^+ and OH^- ions by water and determine the hydrogen ion concentration of the solution from the equilibrium established by the weak acid alone:

$$K_a = \frac{[H^+][A^-]}{[HA]} = \frac{\alpha C_a \cdot \alpha C_a}{(1-\alpha)\,C_a} = \frac{\alpha^2 C_a}{(1-\alpha)}.$$

Since the acid is weak, we further assume that $\alpha << 1$, when we get

$$K_a = \alpha^2 C_a \quad \text{or,} \quad \alpha = (K_a / C_a)^{1/2}$$

$$\text{or,} \quad [H^+] = \alpha C_a = (K_a \cdot C_a)^{1/2}$$

This approximate treatment of the equilibrium of a weak acid solution works well except at some typical situations as explained later.

Problem 11.6

What is the pH of a 0.1 N solution of acetic acid? $K_a = 1.75 \times 10^{-5}$.

Solution: $[H^+] = (K_a \cdot C_a)^{1/2}$ or, pH = 2.88.

Note: When $[H^+]$ calculated in this manner is *greater than* 5% of C_a, i.e., $[H^+] > 0.05\ C_a$, this procedure *does not produce* satisfactory results. Such cases are to be treated as shown below.

Let us now consider the equilibrium existing in an aqueous solution of a weak monoprotic acid from a closer point of view. For the equilibrium

$$HA + H_2O \rightleftharpoons H_3O^+ + A^- \quad \text{or} \quad HA \rightleftharpoons H^+ + A^-$$

we have to determine the concentrations of four species at equilibrium: (a) H^+ (ii) OH^- (iii) A^- and (iv) undissociated HA. So we must have four independent equations. These basic equations may be arrived at as follows:

***Equilibrium* 1.** Equilibrium between hydrogen and hydroxyl ions:

$$[H^+]\,[OH^-] = K_w \qquad \text{.. [11-3]}$$

***Equilibrium* 2.** Equilibrium for dissociation of the weak acid HA:

$$K_a = \frac{[H^+][A^-]}{[HA]} \qquad \text{.. [11-7]}$$

3. *Material balance.* The total analytical concentration of HA must be equal to the sum of its ionized part (= $[A^-]$) and unionized part (= [HA]).

$$C_a = [HA] + [A^-] \qquad \text{.. [11-8]}$$

4. *Charge balance.*

$$[H^+] = [A^-] + [OH^-] \qquad \text{.. [11-9]}$$

The values of [HA] and $[A^-]$ from equations 11-8 and 11-9, and the value of $[OH^-]$ from equation 11-7 may be put in equation 11-8 and rearranged to give:

$$K_a = \frac{[H^+]([H^+]-[OH^-])}{C_a - ([H^+]-[OH^-])}$$

$$\text{or,} \quad [H^+]^3 + K_a\,[H^+]^2 - (K_w + C_aK_a)\,[H^+] - K_wK_a = 0$$

The cubic equation is really difficult to solve. But we may reasonably simplify it by certain assumptions:

Assumption 1. The concentration of OH^- is negligible in comparison to $[H^+]$. This is quite reasonable, since in an acid solution $[OH^-]$ is found to be much less than 10^{-7} mol dm^{-3} (provided the acid is not very very weak).

Equation 11.10 will then be modified as $[H^+] = [A^-]$; we get

$$K_a = \frac{[H^+]^2}{C_a - [H^+]} \quad \text{when } [H^+]^2 + K_a[H^+] - K_aC_a = 0 \qquad \text{.. [11-10]}$$

Assumption 2. Since the acid is weak, the concentration of H^+ produced by its dissociation will be very small in comparison to the total concentration of the acid taken, C_a. So $C_a - [H^+]$ may be fairly approximated to C_a. Then

$$[H^+]^2 = K_a \,.\, C_a \quad \text{or, } [H^+] = (K_a \,.\, C_a)^{1/2} \qquad \text{.. [11-11]}$$

This last expression is the same as that obtained earlier and applied in problem 11.6. But in a numerical problem, a first learner finds it very often difficult to decide about the "negligibility" of a term. We may arrive at a few useful guidelines from a closer analysis of the situation.

Let us first return to problem 11.6. $[H^+]$ was calculated as equation 11-11 which is based on the applicibility of both assumptions 1 and 2. Now let us check if our assumptions 1 and 2 are really justified in this case.

Since $[H^+]$ is in the range of 10^{-3}, $[OH^-]$ must be in the range of 10^{-11}. So it has been rightly neglected.

Regarding our second assumption, we put this "approximate" value of $[H^+]$ in equation 11.10.

$$K_a = \frac{[H^+]^2}{C_a - [H^+]} = \frac{[H^+]^2}{0{\cdot}1 - 1{\cdot}32 \times 10^{-3}}$$

$$\text{or, } [H^+]^2 = 0.1\, K_a - 1.32 \times 10^{-3} .\, K_a$$

$$= 0.1 \times 1.75 \times 10^{-5} - 1.32 \times 10^{-3} \times 1.75 \times 10^{-5}$$

Clearly, the second term on the right hand side has little significance, and no practical improvement in the value of $[H^+]$ is obtained.

However, there may be other cases, too, as in the next problem.

Problem 11.7

What is the pH of a (i) 0.1 and (ii) 0.01 molal solution of monochloroacetic acid? $K_a = 1.4 \times 10^{-3}$.

Solution: (i) Using the approximation formula (11-11),

$$[H^+] = (K_a \,.\, C_a)^{1/2} = (1{\cdot}4 \times 10^{-3} \times 0{\cdot}1)^{1/2}$$

$$= 1{\cdot}18 \times 10^{-2} \text{ mol dm}^{-3} \Rightarrow \text{pH} = 1{\cdot}93$$

Our first assumption is certainly valid as $[OH^-]$ will be nearly 10^{-12}.

But our second assumption, viz. $[H^+] << C_a$ has to be verified. Let us start with the approximate value of $[H^+]$ obtained above:

$$K_a = \frac{[H^+]^2}{C_a - 1.18 \times 10^{-2}} \Rightarrow [H^+] = 1.11 \times 10^{-2} \text{ mol dm}^{-3} \quad \text{or, pH} = 1{\cdot}96.$$

We find that this first order refinement gives $[H^+]$ which differs slightly from the approximate value. In fact we may try a second order refinement using this modified value of $[H^+]$; we shall be satisfied when two consecutive results do not differ too much.

Alternatively, we may obtain $[H^+]$ from equation 11-10:

$[H^+]^2 + K_a [H^+] - K_a C_a = 0$, when $[H^+] = 1.12 \times 10^{-2}$ and pH = 1.95. Note that this result differs little from our first order refinement.

Analysis: What should then be the criterion for taking $[H^+]$ into recognition in comparison to [HA] and equation 11-10 instead of equation 11-11? It is recommended that when the approximate value of $[H^+]$ obtained by using equation 11-11 is *less than* 5% *of* C_a, i.e.,

$[H^+]_{approx} < 0.05\ C_a$, equation 11-11 is alright.

But when $[H^+]_{approx} > 0.05\ C_a$, $[H^+]$ cannot be neglected in comparison to C_a and equation 11-10 should be used (or the successive refinement technique).

In problem 11.7(i), $[H^+]_{approx} = 1.18 \times 10^{-2}$; $0.05\ C_a = 0.05 \times 0.1 = 0.5 \times 10^{-2}$

So equation 11-10 should be used. We get pH = 1.95

(ii) $[H^+] = (K_a . C_a)^{1/2}$ gives $[H^+] = 3.74 \times 10^{-3}$ mol dm^{-3} or, pH = 2.43

$[OH^-]$ is clearly neglibible; assumption 1 is O.K. To check assumption 2, we calculate $0.05\ C_a$;

$0.05\ C_a = 0.05 \times 0.01 = 5 \times 10^{-4}$; $[H^+] = 3.74 \times 10^{-3} = 37 \times 10^{-4}$, i.e. $[H^+] > 0.05\ C_a$.

So, we cannot neglect $[H^+]$ with respect to C_a. We use the quadratic expression (11-10):

$[H^+]^2 + K_a [H^+] - K_a . C_a = 0.$ when, $[H^+] = 3.1 \times 10^{-3}$ mol dm^{-3}; pH = 2.51.

Alternatively, we may start with our approximate value of $[H^+]$ and achieve refinement as:

$$K_a = \frac{[H^+]^2}{C_a - 3.74 \times 10^{-3}},$$ when we get, $[H^+] = 2.96 \times 10^{-3}$; pH = 2.53.

A second refinement with this value of $[H^+]$ gives pH = 2.50.

We may summarize the procedure for problems involving solutions of weak monoprotic acids:

Note: 1. Concentration of $[H^+]$ in a solution of a weak acid HA at concentration C_a may be obtained from the simplified formula:

$[H^+] = (K_a . C_a)^{1/2}$, K_a = dissociation constant of the acid.

2. If $[H^+]$ so obtained is $< 0.05\ C_a$, things are O.K.

If $[H^+] > 0.05\ C_a$, one should either (i) use the quadratic equation to find $[H^+]$:

$[H^+]^2 + K_a [H^+] - K_a . C_a = 0$ or, (ii) refine the result by taking the approximate value of $[H^+]$. The refined value of $[H^+]$ so obtained may be used to get a second refinement and so on. When two successive calculations give nearly identical results, one may be satisfied.

3. For very very weak acids, ($K_a \simeq 10^{-9}$ or less) the $[OH^-]$ may not be negligible after all, specially in dilute solutions. In such cases, a different procedure is necessary. (See next problem).

Problem 11.8

What is the pH of a weak acid solution containing 10^{-4} mole of it per dm^3, given $K_a = 1 \times 10^{-9}$?

Analysis: First we calculate the approximate $[H^+]$: $[H^+] = (K_a . C_a)^{1/2} = 3.16 \times 10^{-7}$ mol dm^{-3}.

We note that $[H^+]$ is very small; we find $[OH^-]$: $[OH^-] = K_w/[H^+] = 3.16 \times 10^{-8}$ mol dm^{-3}.

Thus $[OH^-]$ is 10% of $[H^+]$. This should not be neglected altogether.

Should we now use the original cubic equation to get $[H^+]$? Let us go back to our fundamental equations [11-3, 11-7—11-9] at the beginning for any possible alternative.

We have, $[HA] = C_a - [A^-]$; $[A^-] = [H^+] - [OH^-]$ and $[HA] = C_a - ([H^+] - [OH^-])$

$$\therefore \text{ from Eqn. 11-7, } K_a = \frac{[H^+]([H^+]-[OH^-])}{C_a-([H^+]-[OH^-])}$$

In the present case, how much is $[H^+] - [OH^-]$? We may get the approximate value:

$[H^+] - [OH^-] = 3.16 \times 10^{-7} - 3.16 \times 10^{-8} = 2.84 \times 10^{-7}$; but $C_a = 10^{-4}$.

Obviously we can neglect $([H^+] - [OH^-])$ in the denominator. So,

$$K_a = \frac{[H^+]([H^+]-[OH^-])}{C_a} \quad \text{when } [H^+]^2 = K_a \cdot C_a + K_w$$

We should use this expression for the calculation of $[H^+]$ in dilute solutions of very weak acids.

Solution: $[H^+] = [K_a \cdot C_a + K_w)^{1/2} = (10^{-9} \times 10^{-4} + 10^{-14})^{1/2} = 3.32 \times 10^{-7}$ mol dm^{-3}.

The criterion for adopting this equation should be (i) $[H^+]$ approximately equal to $[OH^-]$

(ii) $([H^+] - [OH^-]) < 0.05\, C_a$.

In other cases, the original cubic equation will have to be solved. Fortunately, such cases are extremely rare.

Weak bases

The equilibrium in a solution of a weak base may be treated along the same line of approach as in the case of weak acids. For the equilibrium

$$\underset{C_b(1-\alpha)}{B} + H_2O \rightleftharpoons \underset{\alpha C_b}{BH^+} + \underset{\alpha C_b}{OH^-}$$

let C_b be the analytical concentration of the base B and α be the fraction dissociated at equilibrium. The dissociation constant of the base, K_b, may now be expressed after simplification as

$$K_b = \frac{[BH^+][OH^-]}{[B]} = \alpha^2 C_b \quad (\text{when } \alpha \ll 1)$$

or, $[OH^-] = \alpha C_b = (K_b \cdot C_b)^{1/2}$.

This enables one to calculate the equilibrium concentration of hydrogen ion also.

Problem 11.9

What is the pH of a 0.1 M solution of ammonia? $K_b = 1.8 \times 10^{-5}$.

Solution: $[OH^-] = (K_b \cdot C_b)^{1/2} \Rightarrow 1.34 \times 10^{-3}$ mol dm^{-3}.

$\therefore$ p OH = – log $[OH^-]$ = 2.87 or, pH = 14 – pOH = 11.13.

The applicability of this procedure is also limited to solutions in which $[OH^-]$ is very small in comparison to C_b and $[H^+]$ is negligible. A closer analysis of the equilibrium may be carried out along the same line of approach as with weak acids.

For the equilibrium : $B + H_2O \rightleftharpoons BH^+ + OH^-$
we want to know the equilibrium concentration of the species BH^+, B, OH^- and H^+.

Suppose C_b is the initial concentration of B. Proceeding exactly in the same manner as in the case of a weak acid, we have

1. ***Equilibrium consideration:*** $K_b = \dfrac{[OH^-][BH^+]}{[B]}$; $K_w = [H^+]\,[OH^-]$
2. ***Material balance:*** $C_b = [B] + [BH^+]$
3. ***Charge balance:*** $[OH^-] = [H^+] + [BH^+]$

Combining these relations, we get

$$K_b = \frac{[OH^-][BH^+]}{C_b - [BH^+]} = \frac{[OH^-]\,([OH^-]-[H^+])}{C_b - ([OH^-]-[H^+])}$$

which gives a cubic expression for $[OH^-]$. But we may simplify this considerably if we assume that (i) $[H^+]$ is now very small and (ii) $[BH^+]$ is very small in comparison to C_b. Proceeding in the same line as with weak acids, we may develop the following generalizations:

1. Concentration of OH^- in a solution of a weak base B at concentration C_b may be obtained from the simplified formula

$$[OH^-] = (K_b\,C_b)^{1/2};\quad K_b = \text{dissociation constant of B.}$$

2. The value of $[OH^-]$ obtained above should be checked for its comparability with $0.05C_b$.

If $[OH^-] < 0.05C_b$, the value of $[OH^-]$ is O.K.

If $[OH^-] > 0.05C_b$, correct value of $[OH^-]$ should be obtained from the quadratic equation

$$[OH^-]^2 + K_b\,[OH^-] - K_bC_b = 0.$$

The approximate value of $[OH^-]$ may also be refined in successive stages by insertion in the expression

$$K_b = \frac{[OH^-]^2}{C_b - [OH^-]}$$

3. For dilute solutions of very weak bases where

(i) $[OH^-] \simeq [H^+]$ (ii) $([OH^-] - [H^+]) < 0.05\,C_b$

We should use $[OH^-] = (K_b\,.\,C_b + K_w)^{1/2}$.

Problem 11.10

What is the pH of a 0.1 M solution of methylamine, $K_b = 5.0 \times 10^{-4}$?

Solution: From the simplified formula $[OH^-] = (K_b\,.\,C_b)^{1/2}$, we have

$$[OH^-] = (5 \times 10^{-4} \times 0.1)^{1/2} = 7.1 \times 10^{-3}\ \text{mol dm}^{-3}$$

when, pOH = 2.15 or pH = 14 – 2.15 = 11.85.

We check the validity of using the simplified formula:

$$[OH^-] = 7.1 \times 10^{-3};\quad 0.05\,C_b = 0.05 \times 0.1 = 5 \times 10^{-3}.$$

So, $[OH^-] > 0.05\,C_b$. So one should try the quadratic equation:

$$[OH^-]^2 + 5 \times 10^{-4}\,[OH^-] - 5 \times 10^{-4} \times 0.1 = 0.$$

$$[OH^-] = 6.82 \times 10^{-3}\ \text{mol dm}^{-3}.\quad \text{pOH} = 2.2\ ;\ \text{pH} = 11.8.$$

Note: The results obtained are close to our approximate values. This is not unusual, since $[OH^-]$ is only slightly greater than $0.05\,C_b$. But the improvement will be noteworthy as C_b decreases, as in the next problem.

Problem 11.11

What is the pH of a 0.01 M solution of methylamine? $K_b = 5.0 \times 10^{-4}$

Solution: From $[OH^-] = (K_b \cdot C_b)^{1/2}$ we have $[OH^-] = 2.24 \times 10^{-3}$; pOH = 2.65.

But here $[OH^-] > 0.05\ C_b$ ($= 0.5 \times 10^{-3}$); so we use the quadratic equation; we get

$$[OH^-] = 2 \times 10^{-3};\ \text{p OH} = 2.7;\ \text{pH} = 11.3.$$

Problem 11.12

Calculate the pH of a 0.01 N solution of ammonia, the degree of dissociation being 4.2%.

Solution: Here the degree of dissociation is given instead of K_b. For the process

$$NH_3 + H_2O \rightleftharpoons NH_4^+ + OH^-$$

1 mol of NH_4^+ and 1 mol of OH^- are formed per mole of NH_3 *dissociated*. Since the degree of dissociation is small, $[OH^-] << C_b$. Also, the solution is not too dilute. So we may neglect the contribution of OH^- from water. From the degree of dissociation, 4.2 per cent, or 0.042 fraction of the NH_3 present are dissociated. Here the solution contains 0.01 mol of NH_3 dm^{-3}. So moles of NH_3 dissociated = mol of OH^- formed $= 0.01 \times 0.042 = 4.2 \times 10^{-4}$.

$\therefore\ [OH^-] = 4.2 \times 10^{-4}$ mol dm^{-3} and pOH = 3.4 or pH = 14 – 3.4 = 10.6.

How far they react?

The extent of interaction between weak acids and weak bases in solution depends upon how "weak" the weak acids or bases are. For convenience, we may conceive of two distinct cases.

Case I: *K is large*

Fairly "strong" weak acids or bases react practically completely with large value of the equilibrium constant K. The reaction between chloroacetic acid ($pK_a = 2.85$) and methylamine ($pK_b = 3.36$) may be taken as an example:

$$\begin{array}{lcll} ClCH_2COOH & \rightleftharpoons & H^+ + ClCH_2COO^- & K = K_a \\ CH_3NH_2 + H^+ & \rightleftharpoons & CH_3NH_3^+ & K = K_b/K_w \\ \hline ClCH_2COOH + CH_3NH_2 & \rightleftharpoons & CH_3NH_3^+ + ClCH_2COO^- & K = K_aK_b/K_w = 6.2 \times 10^7 \end{array}$$

We may reasonably expect that chloroacetic acid will react almost completely with methylamine.

A solution made from 100 mL 0.02 M chloroacetic acid (≡ 2 mmol) and 60 mL 0.02 M methylamine (≡ 1.2 mmol) will finally contain 0.8 mmol of $ClCH_2COOH$ and 1.2 mmol each of $CH_3NH_3^+$ and $ClCH_2COO^-$. The solution will function as a buffer solution (see later) having pH given by

$$pH = pK_a + \log\frac{[ClCH_2COO^-]}{[ClCH_2COOH]} = 2.85 + \log\frac{1.2}{0.8} = 3.0$$

Case II: *K is small*

Such a situation may arise in a solution containing a very weak acid with a moderately weak base or vice-versa. A solution of ammonium chloride (pK_a of $NH_4^+ = 9.25$) with trimethylamine ($pK_b = 4.2$) provides an example. For the reaction

$$NH_4^+ + (CH_3)_3N \rightleftharpoons (CH_3)_3NH^+ + NH_3;\quad K = K_aK_b/K_w = 3.5$$

From the low value of K we conclude that the reaction mixture will contain appreciable amounts of the initial reactants at equilibrium.

If we mix 100.0 mL of 0.02 M NH_4Cl with 100.0 mL of 0.02 M $(CH_3)_3N$, the initial molarities of each will be halved to 0.01 M. If the final concentration of the products of reaction are x M, we write

$$\begin{array}{lcccc} & NH_4^+ & + (CH_3)_3N & \rightleftharpoons NH_3 & + (CH_3)_3NH^+ \\ \text{final } c: & 0.01-x & 0.01-x & x & x \end{array} \quad \text{or,} \quad K = 3\cdot 5 = \frac{x^2}{(0.01-x)^2} \Rightarrow x = 0.0065$$

Hence, $[NH_4^+] = [(CH_3)_3N] = 0.0035\,M$; $[NH_3] = [(CH_3)_3NH^+] = 0.0065\,M$

Note that the solution contains *two buffer systems*: (i) $NH_3 - NH_4^+$ and (ii) $(CH_3)_3N - (CH_3)_3NH^+$. We shall get the same value of pH taking any one buffer system.

$$pH = pK_a(NH_4^+) + \log\frac{[NH_3]}{[NH_4^+]} = 9.2 + \log\frac{0.0065}{0.0035} = 9.5$$

11.2.3 Polyprotic Acids and Bases

Acids which donate more than one hydrogen ion in successive stages are polyprotic. The conjugate base of the first step of dissociation acts as the acid in the second dissociation. Each stage is characterized by a dissociation constant; the successive dissociation constants decrease sharply. The case with polyprotic bases is also similar.

Exact treatment of the ionic equilibrium in a solution of polyprotic acid or base is difficult. We shall make certain generally valid approximations. Let us consider the case of a weak diprotic acid H_2A as an example. The dissociations will be

$$H_2A + H_2O \rightleftharpoons H^+ + HA^- \quad \text{or,} \quad HA \rightleftharpoons H^+ + HA^- \qquad K_1$$

$$HA^- + H_2O \rightleftharpoons H^+ + A^{2-} \quad \text{or,} \quad HA^- \rightleftharpoons H^+ + A^{2-} \qquad K_2$$

At equilibrium we shall have H_2A, HA^-, A^{2-}, H^+ and OH^-. These five unknown quantities may be evaluated by solving the five equations obtained as follows:

Equilibrium Consideration: $$K_1 = \frac{[H^+][HA^-]}{[H_2A]} \qquad \text{.. .. [11-12]}$$

$$K_2 = \frac{[H^+][A^{2-}]}{[HA^-]} \qquad \text{.. .. [11-13]}$$

$$K_w = [H^+][OH^-] \qquad \text{.. .. [11-3]}$$

Material Balance: If C_a be the initial concentration of the acid H_2A,

$$C_a = [H_2A] + [HA^-] + [A^{2-}] \qquad \text{.. .. [11-14]}$$

Charge Balance: $[H^+] = [OH^-] + [HA^-] + 2\,[A^{2-}]$ .. .. [11-15]

(Due to its double charge, each mole of A^{2-} ion will balance two moles of H^+).

We avoid the exact solution of these five equations.* We assume, of course reasonably, that $[OH^-]$ and K_2 are both negligible. At the same time the acid being a weak one, $[H^+] << C_a$. $[H^+]$ is now given by the simple expression

$$[H^+] = (K_1 . C_a)^{1/2}$$

Similarly, for a polyprotic base (first dissociation constant K_{b_1}; initial concentration C_b) We have

$$[OH^-] = (K_{b_1} . C_b)^{1/2}$$

However, calculation can always be done according to the situation without recourse to any particular formula. The following problems have been solved without putting the above formula direct, but their ultimate involvement can be easily understood.

Problem 11.13

A solution contains 0.1*f* (formality) acetic acid ($K_a \simeq 10^{-5}$). H_2SO_4 is added to make it 0.09*f* in H_2SO_4 when the pH becomes 1. The first stage of dissociation is complete. Calculate *K* for the second stage of dissociation of H_2SO_4.

Solution: At pH = 1, concentration of H^+ ion is 10^{-1} or 0.1 g-ion dm^{-3} (0.1 M).

The contribution of acetic acid ($K_a = 10^{-5}$) at this pH is clearly negligible.

Since the first stage of dissociation of H_2SO_4 is complete, 0.09 *f* H_2SO_4 will contribute 0.09 g-ion of H^+ in the *first stage* of dissociation.

The remaining (0.1 – 0.09) or 0.01 M H^+ must have been contributed by the second stage of dissociation of H_2SO_4. Hence, $[SO_4^{2-}] = 0{\cdot}01$ M.

$HSO_4^- \rightleftharpoons H^+ + SO_4^{2-}$ $\quad K = [H^+]\,[SO_4^{2-}] / [HSO_4^-]$

$[SO_4^{2-}] = 0.01\,M$; $[HSO_4^-] = 0.09 - 0.01 = 0.08$ M. $\quad \therefore K = 0.1 \times 0.01 / 0.08 = 0.0125$.

Problem 11.14

Calculate the concentration of H^+, HS^- and S^{2-} at equilibrium in a 0.1 M aqueous solution of H_2S. $K_1 = 9 \times 10^{-8}$; $K_2 = 1 \times 10^{-15}$.

Solution: We consider the dissociation stage by stage

$$H_2S \rightleftharpoons H^+ + HS^- \; ; \; K_1 = [H^+]\,[HS^-] / H_2S] = 9 \times 10^{-8}.$$

$$HS^- + H_2O \rightleftharpoons H^+ + S^{2-} \; ; \; K_2 = [H^+]\,[S^{2-}] / HS^-] = 1 \times 10^{-15}.$$

The magnitude of K_2 indicates that dissociation of HS^- is practically negligible. So the whole of HS^- formed in the first stage may be considered to be present as such. Since H^+ produced in the second stage is also negligible, we have, from the first dissociation alone, $[H^+] = [HS^-]$. The con-

* We shall have: $K_1 = \dfrac{[H^+]\left([H^+]-[OH^-]-\dfrac{2K_2([H^+]-[OH^-])}{[H^+]+2K_2}\right)}{C_a-\left([H^+]-[OH^-]-\dfrac{K_2([H^+]-[OH^-])}{[H^+]+2K_2}\right)}$

On (i) neglecting $[OH^-]$ and K_2 and (ii) assuming $[H^+] << C_a$, this reduces to $K_1 = [H^+]^2 / C_a$.

centration of H_2S will be its initial value, 0.1 mole, *less* by the amount dissociated: $[H_2S] = 0.1 - [H^+]$.

So, $$K_1 = \frac{[H^+][HS^-]}{[H_2S]} = \frac{[H^+]^2}{0.1-[H^+]} = 9 \times 10^{-8}$$

We further simplify by assuming that $[H^+] << [H_2S]$. This is justifiable in view of the small value of K_1. So, ultimately we have $K_1 \times 0.1 = [H^+]^2$.

or, $[H^+] = (9 \times 10^{-8} \times 0.1)^{1/2} = 9.5 \times 10^{-5}$ mol dm^{-3}.

This is also the value of $[HS^-]$.

As we have assumed $[H^+] = [HS^-]$, from the second dissociation equilibrium,

$$[S^{2-}] = 1 \times 10^{-15} \text{ mol dm}^{-3}.$$

Problem 11.15

What will be the pH of a solution made by dissolving 0.02 mole of CO_2 in 1000 g of water? $K_1 = 4 \times 10^{-7}$, $K_2 = 6 \times 10^{-11}$ for "carbonic acid".

Solution: CO_2 dissolves in water giving the weak "carbonic acid", represented often as H_2CO_3, though most of the CO_2 is only loosely hydrated. The dissociation of the dissolved CO_2 may however be treated in terms of the dissociation of the "carbonic acid".

$$\underbrace{H_2O + CO_2}_{\equiv H_2CO_3} \rightleftharpoons HCO_3^- + H^+; \qquad K_1 = \frac{[H^+][HCO_3^-]}{[H_2CO_3]} = 4 \times 10^{-7}$$

$$[HCO_3^-] \rightleftharpoons CO_3^{2-} + H^+; \qquad K_2 = \frac{[H^+][CO_3^{2-}]}{[HCO_3^-]} = 6 \times 10^{-11}$$

The very small value of K_2 suggests that the second dissociation may be ignored in computing the concentrations of H^+ and HCO_3^-. Also, the concentration of H_2CO_3 changes little due to dissociation. So we may put $[H^+] = [HCO_3^-]$ when,

$$K_1 = \frac{[H^+]^2}{0.02} \text{ or } [H^+] = (4 \times 10^{-7} \times 0.02)^{1/2} = 8.9 \times 10^{-5} \text{ when, pH} = 4.05$$

Note: The concentration of CO_3^{2-} is equal to the second dissociation constant, 6×10^{-11}. The contribution of H^+ from the second ionization is neglected in these two problems on the basis of $K_2 << K_1$. When this is not so, we may apply the method of successive refinement by taking the approximate value of $[H^+]$.

Amino acids

The building blocks of proteins are amino acids, which are all polyprotic. Most of them are diprotic with the general structure 11-I which spontaneously rearranges to the zwitterion form 11-II.

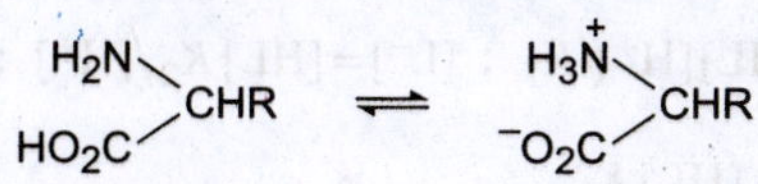

11-I **11-II**

Here the –R group has no acidic or basic functions, for example, –H (in glycine), $-CH_3$ (in α-alanine), $-CHMe_2$ (in valine) etc. The amino acids are marked by two acid dissociation constants: one for the –COOH group (say K_a^C) and one for the $-\overset{+}{N}H_3$ group (say K_a^N). For glycine, $pK_a^C = 2.35$; $pK_a^N = 9.78$.

The –R group may also bear an acidic function, for example $-CH_2COOH$ (in aspartic acid) or a basic function, for example, $-(CH_2)_3NH{-}C(=NH)NH_2$ (in arginine). They have a third pK_a value, say pK_a^R (see later).

The equilibrium in an aqueous solution of an amino acid having no acidic or basic R group may be treated as shown above for diprotic acids. As an illustration, we shall consider the case of leucine, in which $R = (CH_3)_2\,CHCH_2-$. Let us represent the zwitterion $H_3\overset{+}{N}CH(R)CO_2^-$ as HL; the following equilibria are involved:

$$H_2L^+ \rightleftharpoons HL + H^+ \qquad K_{a_1} = K_1 = 4.69 \times 10^{-3}$$

$$HL \rightleftharpoons L^- + H^+ \qquad K_{a_2} = K_2 = 1.79 \times 10^{-10}$$

One may also consider the equilibria

$$L^- + H_2O \rightleftharpoons HL + OH^- \quad K_{eq} = K_{b_1} \quad ; \quad HL + H_2O \rightleftharpoons H_2L^+ + OH^- \quad K_{eq} = K_{b_2}$$

where K_{b_1} and K_{b_2} are related to K_{a_1} (K_1) and K_{a_2} (K_2) as $K_{a_1}.K_{b_2} = K_{a_2}.K_{b_1} = K_w$

(i) In a solution having 0.05 M H_2L^+ (formal concentration), we may practically neglect the second stage of ionization from the very low value of K_2 compared to K_1. Hence the pH may be calculated by taking H_2L^+ as a monobasic acid with $K_a = K_1$.

$$\underset{0.05 - x}{H_2L^+} \rightleftharpoons \underset{x}{HL} + \underset{x}{H^+} \qquad K_1 = \frac{x^2}{0{\cdot}05 - x} = 4{\cdot}69 \times 10^{-3} \longrightarrow x = 1{\cdot}31 \times 10^{-2}$$

$$\therefore\ [HL] = [H^+] = x = 1.31 \times 10^{-2}\ M \longrightarrow pH = 1.88$$

The concentration of $[L^-]$ may be obtained from the expression of K_{a_2} :

$$K_{a_2} = \frac{[H^+][L^-]}{[HL]} = \frac{1.31 \times 10^{-2}\,[L^-]}{1.31 \times 10^{-2}} = 1.79 \times 10^{-10} \quad \text{or, } [L^-] = 1.79 \times 10^{-10}.$$

The calculations made above are quite general and do not depend on the charge types of the species. We can use the same procedure with any dibasic acid H_2A.

(ii) In a solution containing 0.5 M formal concentration of HL, we have the following equilibrium where HL may behave both as an acid and a base (amphiprotic):

$$HL \rightleftharpoons H^+ + L^- \qquad K_a = K_{a_2} \text{ of } H_2L^+ = 1.79 \times 10^{-10}$$

$$HL + H_2O \rightleftharpoons H_2L^+ + OH^- \qquad K_b = K_{b_2} \text{ of } L^- = 2.13 \times 10^{-12}$$

Since the K-values are not widely different, we have to consider both equilibria simultaneously, though equilibrium 1 will dominate slightly to make the solution acidic.

Charge balance requires: $[H^+] + [H_2L^+] = [L^-] + [OH^-]$

We can make the following substitutions:

$$[H_2L^+] = [HL][H^+]/K_1 \ ; \ [L^-] = [HL]\,K_2/[H^+] \ ; \ [OH^-] = K_w/[H^+]$$

and get $$\frac{[HL][H^+]}{K_1} - \frac{[HL]\,K_2}{[H^+]} + [H^+] - \frac{K_w}{[H^+]} = 0.$$

Multiplying by $[H^+]$, and rearranging, $$[H^+]^2 = \frac{K_1K_2\,[HL] + K_1\,K_w}{[HL] + K_1}$$

From the values of K_a and K_b, we may infer that both the reactions of HL will occur to a negligible extent. The equilibrium concentration of HL i.e., [HL] may be reasonably put equal to its analytical or formal concentration, 0.05 M. Hence,

$$[H^+] \simeq \sqrt{\frac{0.05\,K_1\,K_2 + K_1\,K_w}{K_1 + 0.05}} \longrightarrow 8.76 \times 10^{-7}\,\text{M, pH} = 6.06.$$

The procedure is general and can be applied to the intermediate form of any diprotic acid, irrespective of its charge:

$$[H^+] = \sqrt{\frac{K_1K_2F + K_1K_w}{K_1 + F}} \quad (F = \text{formal concentration of HL}) \qquad [11\text{-}16]$$

When $K_2F >> K_w$ and $K_1 << F$, one gets

$$[H^+] = \sqrt{\frac{K_1K_2F}{F}} = \sqrt{K_1K_2} \quad \text{or, } \text{pH} = \frac{1}{2}(\text{p}K_1 + \text{p}K_2) \qquad [11\text{-}17]$$

We shall find this relation useful in several occasions.

Triprotic acids

Triprotic systems (and other polyprotic systems) can be treated along the same line of approach as illustrated with the diprotic system. Thus the equilibria involved in a triprotic system are

$$H_3A \rightleftharpoons H_2A^- + H^+ \qquad K_{a_1} = K_1$$
$$H_2A^- \rightleftharpoons HA^{2-} + H^+ \qquad K_{a_2} = K_2$$
$$HA^{2-} \rightleftharpoons A^{3-} + H^+ \qquad K_{a_3} = K_3$$
$$A^{3-} + H_2O \rightleftharpoons HA^{2-} + OH^- \qquad K_{b_1} = K_w/K_{a_3}$$
$$HA^{2-} + H_2O \rightleftharpoons H_2A^- + OH^- \qquad K_{b_2} = K_w/K_{a_2}$$
$$H_2A^- + H_2O \rightleftharpoons H_3A + OH^- \qquad K_{b_3} = K_w/K_{a_1}$$

We may simplify our treatment by the following approximations:

(i) H_3A may be considered to behave as a weak monoprotic acid with $K_a = K_{a_1}$

(ii) H_2A^- behaves as the intermediate form of a diprotic acid (see calculation with leucine). pH of the solution is obtained from the relation

$$\left[H^+\right] \approx \left(\frac{K_1K_2F + K_1K_w}{K_1 + F}\right)^{1/2} \qquad \text{(11-16; see box)}$$

where F is the formal concentration of H_2A^-. As shown earlier, this may be further simplified:

$$[H^+] = (K_1K_2)^{1/2} \qquad \text{or, pH} = \frac{1}{2}(\text{p}K_1 + \text{p}K_2) \qquad \text{(11-17)}$$

(iii) $[HA^{2-}]$ may also be calculated by assuming it as an intermediate betwen H_2A^- and A^{3-}, using K_2 and K_3; $[H^+]$ is now determined by K_2 and K_3:

$$[H^+] \approx \left(\frac{K_2K_3F + K_2K_w}{K_2 + F}\right)^{1/2} \qquad \text{(11-16; } F = \text{formal concentration).}$$

Further simplification gives $[H^+] = (K_2K_3)^{1/2}$ or, $\text{pH} = \frac{1}{2}(\text{p}K_2 + \text{p}K_3)$ (11-17)

(iv) A^{3-} may be treated as a weak monoacidic base with $K_b = K_w/K_{a_3}$

Problem 11.16

Calculate the concentration of H^+, H_3PO_4, $H_2PO_4^-$, HPO_4^{2-} and PO_4^{3-} ions in a 0.1 M solution of H_3PO_4 at 25°C ($K_1 = 7.5 \times 10^{-3}$; $K_2 = 6.2 \times 10^{-8}$; $K_3 = 4.8 \times 10^{-13}$).

Hint: Write expressions for K_1, K_2 and K_3 for the three stages of ionization.

Since $K_1 >> K_2$ or K_3, $[H^+] \simeq [H_2PO_4^-]$.

If $[H^+] = x$; then $[H_3PO_4] \simeq 0.1 - x$.

From the first ionization equilibrium,

$$K_1 = 7.5 \times 10^{-3} = x^2/(0.1 - x), \text{ when } x = 0.024.$$

$$\therefore \quad [H^+] = [H_2PO_4^-] = 0.024\,M \quad \text{and} \quad [H_3PO_4] = (0.1 - 0.024)\,M = 0.076\,M$$

From the second ionization step,

$$K_2 = [H^+]\,[HPO_4^{2-}]\,/\,[H_2PO_4^-] = x\,[HPO_4^{2-}]\,/\,x$$

$$\therefore \quad [HPO_4^{2-}] = K_2 = 6.23 \times 10^{-8}\,M.$$

Again, from the third dissociation equilibrium,

$$[PO_4^{3-}] = K_3\,[HPO_4^{2-}]\,/\,[H^+] = 1.24 \times 10^{-18}\,M.$$

As already mentioned, amino acids with acidic or basic functional groups in the side chain may be treated at par with other polyprotic acids, with three acid dissociation constants, K_a^C, K_a^N and K_a^R. Usually K_a^C is many times larger compared to K_a^N or K_a^R. K_a^N may be larger or smaller than K_a^R, as shown in the following examples.

Aspartic acid (Asp: R = $-CH_2COOH$)

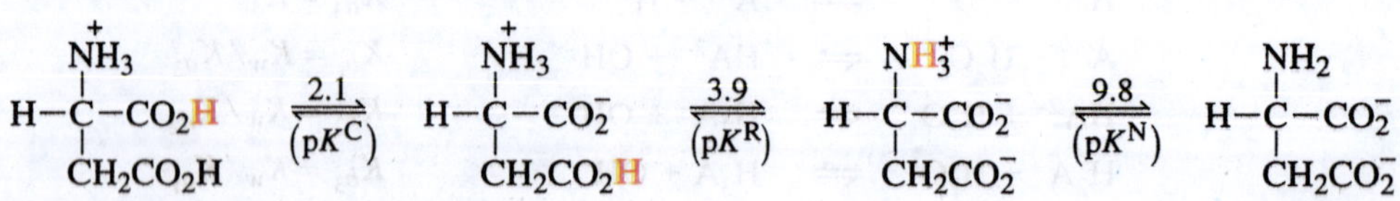

H_3Asp^+ $\quad$ H_2Asp $\quad$ $HAsp^-$ $\quad$ Asp^{2-}

Arginine $\left(\text{Arg: R} = -CH_2CH_2CH_2NH\,C{\genfrac{}{}{0pt}{}{\nearrow NH_2^+}{\searrow NH_2}}\right)$

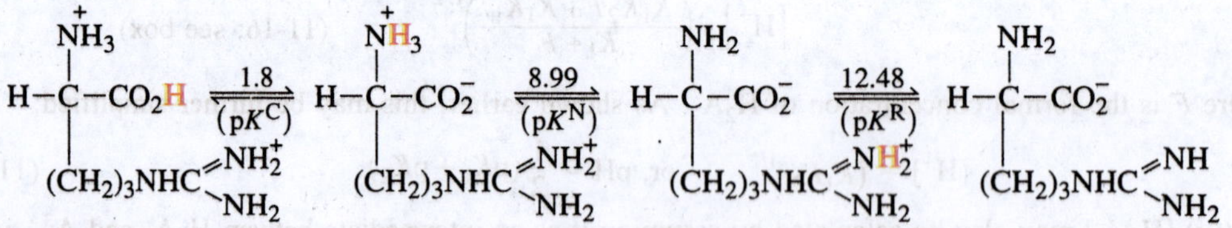

H_3Arg^{2+} $\quad$ H_3Arg^+ $\quad$ $HArg$ $\quad$ Arg^-

Our discussion on triprotic acid H_3A is valid irrespective of the charge on a species. So we may proceed in the same line and calculate the concentration of any species at a given pH. The intermediate forms (adjacent) of the amino acid constitute conjugate acid-base pairs and their relative concentrations at any pH may be obtained from the relation (see later)

$$pH = pK_a + \log\frac{[\text{conjugate base}]}{[\text{conjugate acid}]}$$

For arginine, for example, $pK_2 = pK^N = 8.99$ relates to the acidic form H_2Arg^+ and its conjugate base HArg.

At pH = 8.99, $[H_2Arg^+] = [HArg]$

At pH = 10.0, $\log [HArg] / [H_2Arg^+] \simeq 1 \longrightarrow [HArg] = 10\,[H_2Arg^+]$

Below pH 8.99, we expect H_2Arg^+ to be the principal species.

The pH of a solution containing pure H_2Arg^+ may be calculated as shown earlier:

$$pH = \frac{1}{2}(pK_1 + pK_2) = 5.40.$$

***The pH at which the average charge of a polyprotic acid is zero is called its* isoelectric pH** (or isoelectric point). Thus, a neutral polyprotic acid HA (zwitterion form) dissolves in water to give H^+, OH^-, H_2A^+ and A^- ions, in addition to HA. The average charge will be zero when $[H_2A^+] = [A^-]$. From usual equilibrium reactions:

$$[H_2A^+] = \frac{[HA][H^+]}{K_1};\quad [A^-] = \frac{K_2[HA]}{[H^+]} \quad \text{or, } [H^+] = (K_1K_2)^{1/2}$$

Hence the isoelectric $pH = \frac{1}{2}(pK_1 + pK_2)$.

The pH of a solution obtained by dissolving a pure, neutral polyprotic acid (the neutral zwitterion) in water is called the ***isoionic pH*** (or isoionic point). It was shown earlier that for a diprotic acid (leucine) $[H^+]$ is given by equation 11.16

$$[H^+] = \left(\frac{K_1K_2F + K_1K_w}{K_1 + F}\right)^{1/2}$$

When $K_1 << F$, $[H^+] \simeq (K_1K_2)^{1/2}$ or $pH = \frac{1}{2}(pK_1 + pK_2)$ which is same as the isoelectric pH.

11.3 AQUEOUS SOLUTIONS OF SALTS

The principles developed so far for aqueous solutions of various acids and bases may be applied equally to solutions of salts. Still the latter are treated separately mostly as a matter of convention, mainly to deal with acid-base behaviour known traditionally as "*hydrolysis*". The equilibrium involving sparingly soluble salts will be discussed separately in section 11.5.

11.3.1 Hydrolysis

The development of acidity or alkalinity in solutions of certain salts was at first supposed to take place as a result of "*hydrolysis*" of the salt—the term hydrolysis being defined as *the reversible decomposition of a salt by water into an acid and a base, with reversal of neutralization*. The ultimate character of the solution was then determined by the relative strengths of the acid and base formed:

Salt	+	**Water**	⇌	**Base**	+	**Acid**	
NH_4Cl	+	H_2O	⇌	NH_4OH	+	HCl	
				weak		strong	***acidic***
CH_3COONa	+	H_2O	⇌	$NaOH$	+	CH_3COOH	
				strong		weak	***alkaline***

However, the idea of "decomposition of a salt" is a bit difficult to understand when we remember that salts are completely ionized in solution. Hydrolysis is actually the reversible interaction between (i) the ions (aquated) furnished by a salt in aqueous solution and (ii) water molecules or ions present therein with the formation of weak acids or bases or both.

Pure water contains equal concentrations of hydrogen and hydroxyl ions. A salt dissolves in it furnishing its own characteristic cations and anions (aquated). If these cations and anions differ in their proton (hydron) donor or acceptor properties toward water, the solution will no more contain equal concentrations of hydrogen and hydroxyl ions. The solution will therefore react acidic or alkaline depending upon the nature of equilibrium attained.

Equilibrium in hydrolysis of salts

Hydrolysis of salts practically involves the equilibrium established by a weak acid or base provided by the cation or anion of a salt. Thus, the equilibrium existing in an aqueous solution of ammonium chloride is essentially the equilibrium set up by the ammonium ion functioning as an acid; the equilibrium established in a solution of sodium fluoride is that due to the fluoride ion functioning as a base:

acid	**base**
$NH_4^+ + H_2O \rightleftharpoons NH_3\,(aq) + H^+\,(aq)$	$F^- + H_2O \rightleftharpoons HF\,(aq) + OH^-\,(aq)$
$K = \dfrac{[NH_3][H^+]}{[NH_4^+]}$	$K = \dfrac{[HF][OH^-]}{[F^-]}$

It is customary to call these equilibrium constants the *hydrolysis constant* of the salt, K_h. We may easily correlate the hydrolysis constant with the dissociation constant of the conjugate acid or a base as follows:

The hydrolysis constant K_h of sodium fluoride, that is, of the F^- ion, may be rearranged as

$$K_h = \frac{[HF][OH^-]}{[F^-]} = \frac{[HF][OH^-][H^+]}{[F^-][H^+]} = \frac{[HF]}{[F^-][H^+]} \cdot K_w = \frac{K_w}{K_a}.$$

Similarly, the hydrolysis constant of ammonium chloride, or of the ammonium ion, may be expressed as

$$K_h = \frac{[NH_3][H^+]}{[NH_4^+]} = \frac{[NH_3][H^+][OH^-]}{[NH_4^+][OH^-]} = \frac{[NH_3]}{[NH_4^+][OH^-]} \cdot K_w = \frac{K_w}{K_b}$$

These relations are valid for other monoprotic cations and anions also. If C gram-moles of the salt are dissolved per dm^3 of solution and α is the fraction of each gram molecule of the salt hydrolysed at equilibrium (**degree of hydrolysis**), we may write:

	A^-	+	H_2O	$\rightleftharpoons$	HA	+	OH^- ;	
or,	BH^+	+	H_2O	$\rightleftharpoons$	B	+	$H^+\,(aq)$	
equilibrium concentrations	$C\,(1-\alpha)$				αC		αC	mol dm^{-3}

In either case, K_h may be correlated to α as $K_h = \dfrac{\alpha^2 C}{1-\alpha} \simeq \alpha^2 C \quad (\alpha \ll 1)$.

If we assume that the contribution of H^+ or OH^- ions from the self-ionization of water is small in comparison to that due to hydrolysis, $[H^+]$ or $[OH^-] = \alpha C$ mol dm^{-3}, as the case may be. For convenience, we shall consider three different types of hydrolyzable salts separately.

These hydrolytic equilibria may also be treated more rigorously *without neglecting* the contribution of H^+ or OH^- ions from self-ionization of water. The general procedure of considering the (i) equilibria, (ii) material balance and (iii) charge balance may be followed.

(i) Salts derived from weak acids and strong bases

As already explained, such salts produce an alkaline solution due to interactions of the type

$$A^- + H_2O \rightleftharpoons HA + OH^-$$

A^- being the anion of the weak acid HA having dissociation constant K_a. As shown above,

$$K_h = K_w/K_a$$

The degree of dissociation α is given as

$$\alpha = \left(\frac{K_h}{C}\right)^{1/2} = \left(\frac{K_w}{K_a.C}\right)^{1/2} \quad (C = \text{formal concentration of the salt})$$

$$\therefore \quad [OH^-] = \alpha C = \left(\frac{K_w.C}{K_a}\right)^{1/2} \quad \text{or, } [H^+] = \left(\frac{K_w.K_a}{C}\right)^{1/2} \quad \text{or, } \underline{pH = \frac{1}{2}pK_w + \frac{1}{2}pK_a + \frac{1}{2}\log C} \quad [11\text{-}18]$$

pH of aqueous solutions containing acetate, borate, fluoride etc. salts of strong bases may be calculated from this relation.

(ii) Salts derived from strong acids and weak bases

The conjugate acid of a weak base B constitutes the cation of the salt and results in an acidic solution due to the interaction

$$BH^+ + H_2O \rightleftharpoons B + H_3O^+ \quad \text{or} \quad BH^+ \rightleftharpoons B + H^+$$

If the dissociation constant of the weak base be K_b, we have

$$K_h = K_w/K_b.$$

The degree of dissociation of the salt, α, may be expressed as

$$\alpha = (K_h/C)^{1/2} \quad (C = \text{formal concentration of the salt in solution.})$$
$$= (K_w / K_b \,.\, C)^{1/2}$$

$$\therefore \quad [H^+] = \alpha C = \left(\frac{K_w.C}{K_b}\right)^{1/2} \quad \underline{\text{or, } pH = \tfrac{1}{2}pK_w - \tfrac{1}{2}pK_b - \tfrac{1}{2}\log C.} \quad [11\text{-}19]$$

pH of aqueous solutions of salts formed by ammonia, pyridine, aniline etc. weak bases with strong acids may be calculated from this relation.

(iii) Salts derived from weak acids and weak bases

In solutions of such salts, both the cation and the anion interact with water. Thus a salt BHA will establish the two hydrolytic reactions:

$$BH^+ + H_2O \rightleftharpoons B + H_3O^+ \quad \text{(hydrolysis by the cation)}$$
$$A^- + H_2O \rightleftharpoons HA + OH^- \quad \text{(hydrolysis by anion)}$$

$$\text{Adding,} \quad BH^+ + A^- + 2H_2O \rightleftharpoons B + HA + H_3O^+ + OH^-$$
$$\text{or} \quad BH^+ + A^- \rightleftharpoons B + HA$$

The equilibrium concentrations of B, HA, H^+ and OH^- ions are now controlled by the four equilibrium conditions.

$$K_h = \frac{[B][HA]}{[BH^+][A^-]}; \; K_w = [H^+][OH^-]; \; K_a = \frac{[H^+][A]}{[HA]}; \; K_b = \frac{[BH^+][OH^-]}{[B]}$$

These are interrelated as

$$K_h \cdot K_a \cdot K_b = K_w \quad \text{or,} \quad K_h = K_w / K_a \cdot K_b$$

B and HA are formed in solution according to the relative hydrolytic tendencies of BH^+ and A^-. If the acid BH^+ and the base A^- have different strengths, they will interact with water to different extents. Their equilibrium concentrations, as well as the amounts of B and HA formed by hydrolysis would therefore differ from one another. For simplicity, we first assume K_a and K_b to be equal to one another, at least approximately: $K_a \simeq K_b$. This gives:

$$[BH^+] = [A^-] \text{ and } [B] = [HA]$$

Now, we may write the following stoichiometric relation

$$\begin{array}{llllll} BH^+ & + A^- & \rightleftharpoons B & + HA & C = \text{concentration of salt} \\ C(1-\alpha) & C(1-\alpha) & \alpha C & \alpha C & \alpha = \text{degree of dissociation} \end{array}$$

$$\therefore \quad K_h = \frac{\alpha C . \alpha C}{(1-\alpha)^2 C^2} = \frac{\alpha^2}{(1-\alpha)^2}; \quad \text{when } \alpha << 1, \alpha = K_h^{1/2} = (K_w / K_a K_b)^{1/2}$$

If we again neglect the contribution of H^+ ion from self-ionization of water, the equilibrium concentration of H^+ may be obtained from the dissociation constant of the weak acid HA:

$$[H^+] = \frac{K_a[HA]}{[A^-]} = \frac{K_a . \alpha C}{C(1-\alpha)} = K_a . \alpha \quad (\alpha << 1).$$

From the value of α, $[H^+] = (K_w \cdot K_a / K_b)^{1/2}$

or $\quad pH = \frac{1}{2}pK_w + \frac{1}{2}pK_a - \frac{1}{2}pK_b$ [11-20]

We note that the pH is independent of C, the concentration of the salt solution.

When $K_a = K_b$, $pK_a = pK_b$; So the solution will have a pH equal to 7 ($= \frac{1}{2}pK_w$) and will be neutral.

Hydrolysis of aqua metal ions

We have already discussed (chapter 10) the acidity of aqua metal ions in solution. Taking $Fe(H_2O)_6^{3+}$ as an example, we may write the following equilibrium reaction for the first step of hydrolysis (sec. 10.2.2):

$$[Fe(H_2O)_6^{3+}] + H_2O \rightleftharpoons [Fe(H_2O)_5OH]^{2+} + H_3O^+ \text{ or simply as}$$

$$[Fe(H_2O)_6]^{3+} \rightleftharpoons [Fe(H_2O)_5OH]^{2+} + H^+; \; K_a = [H^+]\,[[Fe(H_2O)_5OH]^{2+}]/[[Fe(H_2O)_6]^{3+}]$$

The extent of hydrolysis and hence the concentration of H^+ ion in solution is determined by the acid dissociation constant K_a of the aquated metal ion. Some representative values of K_a are given in Table 11.1.

The dependence of K_a on effective charge of the metal ion is noteworthy. We also observe that some K_a values, like those of $Fe^{3+}(aq)$ and $Cr^{3+}(aq)$, are actually higher than the K_a value of acetic acid (1.8×10^{-5}).

Table 11.1

Acid dissociation constants of some aqua metal ions (25°C).

Ion	K_a	Ion	K_a
$Fe^{3+}(aq)$	6.3×10^{-3}	$Fe^{2+}(aq)$	3.2×10^{-10}
$Cr^{3+}(aq)$	1.6×10^{-4}	$Co^{2+}(aq)$	1.3×10^{-9}
$Al^{3+}(aq)$	1.1×10^{-5}	$Ni^{2+}(aq)$	5×10^{-10}
$Be^{2+}(aq)$	3.0×10^{-7}	$Zn^{2+}(aq)$	2.5×10^{-10}

The equilibrium represented above tells us that in a solution C molar with respect to $Fe^{3+}(aq)$ (analytical concentration),

$$[[Fe(H_2O)_6]^{3+}] = C - [H^+] \quad \text{and} \quad [[Fe(H_2O)_5OH]^{2+}] = [H^+]$$

Substitution in the expression of K_a gives

$$K_a = \frac{[H^+]^2}{C-[H^+]} \quad \text{or,} \quad [H^+]^2 + K_a[H^+] - K_aC = 0$$

For $C = 0.1$ M, we get $[H^+] = 2.2 \times 10^{-2}$ M or pH = 1.6.

When the aqua metal ion is only feebly acidic like $Zn^{2+}(aq)$ ($K_a \simeq 10^{-10}$), we may reasonably consider $[H^+] << C$ and get

$$K_a = \frac{[H^+]^2}{C} \quad \text{or,} \quad \text{pH} = \tfrac{1}{2}\text{p}K_a - \tfrac{1}{2}\log C$$

For a 0.1 M solution of $Zn^{2+}(aq)$, pH $= \frac{1}{2}(9.6 + 1) = 5.3$.

Numerical problems on hydrolysis of salts

We recapitulate the relations obtained in the last section: the K_h and pH of an aqueous solution of a salt may be expressed summarily according to the nature of its constituent acid and base as follows:

Weak acid–Strong base:	$K_h = K_w/K_a$;	$\text{pH} = \frac{1}{2}K_w + \frac{1}{2}\text{p}K_a + \frac{1}{2}\log C$
Strong acid–Weak base:	$K_h = K_w/K_b$;	$\text{pH} = \frac{1}{2}\text{p}K_w - \frac{1}{2}\text{p}K_b - \frac{1}{2}\log C$
Weak acid–Weak Base:	$K_h = K_w/K_a . K_b$;	$\text{pH} = \frac{1}{2}\text{p}K_w + \frac{1}{2}\text{p}K_a - \frac{1}{2}\text{p}K_b$.

But the situation may be complicated by the involvement of other equilibria involved simultaneously. Due consideration must be given for them, too, as illustrated in some problems below. Also, one may calculate the pH directly from consideration of the equilibrium without using the above formula.

Problem 11.17

Calculate the (a) hydrolysis constant, (b) the degree of hydrolysis and (c) the pH of a 0.01 molar solution of sodium acetate at 25°C. (K_a for acetic acid = 1.8×10^{-5}).

Solution: Sodium acetate is a salt of weak acid and strong base.

(a) $K_h = K_w/K_a = 1 \times 10^{-14}/1.8 \times 10^{-5} = 5.5 \times 10^{-10}$.

(b) When the degree of hydrolysis, α, is small, we have

$$\alpha = \sqrt{K_h/C} = (5.5 \times 10^{-10}/0.01)^{1/2} = 2.3 \times 10^{-4} \text{ (or, 0.023\%)}$$

(c) $\text{pH} = \frac{1}{2}\text{p}K_w + \frac{1}{2}pK_a + \frac{1}{2}\log C = 7.0 + \frac{1}{2} \times 4.74 + \frac{1}{2}(-2) = 8.4$

Problem 11.18

Calculate (i) the degree of hydrolysis and (ii) pH of a 0.02 molar solution of ammonium chloride at 25°C. [K_b for ammonia solution = 1.8×10^{-5}].

Solution: The hydrolysis may be written as: $NH_4^+ + H_2O \rightleftharpoons NH_3 + H_2O$.

(i) $K_h = K_w/K_b = 1.0 \times 10^{-14}/1.8 \times 10^{-5} = 5.6 \times 10^{-10}$, or, $\alpha = \sqrt{K_h/C} = 1.7 \times 10^{-4}$

(i) $[H^+] = \alpha C = 1.7 \times 10^{-4} \times 0.02 = 3.4 \times 10^{-6}$ mol dm^{-3}

$\therefore$ pH = – log [H$^+$] = – log $3.4 \times 10^{-6} = 5.5$.

Also, $pH = \frac{1}{2}pK_w - \frac{1}{2}pK_b - \frac{1}{2}\log C = 5.5.$

An alternative approach

The pH of a solution of NH_4Cl may also be calculated by considering NH_4^+ as an acid; the Cl^- ion does not disturb the normal ionization equlibrium of H_2O. Omitting hydration, we may simply write

$$NH_4^+ \rightleftharpoons H^+ + NH_3 \qquad K_a(NH_4^+) = \frac{[H^+][NH_3]}{[NH_4^+]} = \frac{K_w}{K_b(NH_3)}$$

From $K_b(NH_3) = 1.8 \times 10^{-5}$, we get $K_a(NH_4^+) = 5.6 \times 10^{-10}$. NH_4Cl is completely ionized in solution. In a solution of molarity C, $[NH_4^+] = C - [H^+]$; also, $[H^+] = [NH_3]$

$$\therefore\ K_a(NH_4^+) = \frac{[H^+][NH_3]}{[NH_4^+]} = \frac{[H^+]^2}{C - [H^+]}$$

If we neglect $[H^+]$ in comparison to C, we get $[H^+]^2 = K_aC$ which gives $pH = \frac{1}{2}pK_a - \frac{1}{2}\log C$. For a 0.02 M solution of NH_4Cl, pH = 5.5.

When a solution contains an anion which functions as a base, we may calculate the pH as illustrated for NH_4F (Problem 11.19).

Problem 11.19

Calculate the pH of (i) a 0.2 M NH_4F solution (ii) a 0.2 M NH_4HF_2 solution.

$$K_a\text{: HF} = 6.8 \times 10^{-4}\ ;\ NH_4^+ = 5.6 \times 10^{-10}$$

Solution: (i) The solution contains the acid NH_4^+ and the base F^-. The K_a values suggest that the conjugate base of NH_4^+, $NH_3(aq)$, is a stronger base than F^-, the conjugate base of HF. The contribution of H^+ or OH^- ions by ionization of water is expected to be negligible ($K_w = 10^{-14}$).

The salt NH_4F is completely ionized in solution. The F^- ion will abstract some H^+ ion from NH_4^+ and form the weak acid HF.

$$NH_4F \rightleftharpoons NH_4^+ + F^- \rightleftharpoons NH_3 + HF$$

For a solution containing C M NH_4F (analytical concentration), we have,

$[NH_4^+] = [F^-]$ ($\simeq C$) ; also $[HF] = [NH_3]$. Substituting these values, we get

$$K_{NH_4^+} \cdot K_{HF} = \frac{[H^+][NH_3]}{[NH_4^+]} \cdot \frac{[H^+][F^-]}{[HF]} = [H^+]^2 \quad \text{or,} \quad pH = \tfrac{1}{2}pK_{NH_4^+} + \tfrac{1}{2}pK_{HF} = 6{\cdot}2$$

The pH is independent of the concentration of the salt.

(ii) NH_4HF_2 may be considered as an equimolar mixture of the weak acid HF and its salt NH_4F. Hence the solution will behave as a buffer solution (see later) whose pH is given as

$$pH = pK_{HF} + \log\frac{[\text{salt}]}{[\text{acid}]} = pK_{HF} \text{ since } [NH_4F] = [HF].$$

From the value of K_{HF}, pH = 3.2.

Problem 11.20

Calculate (i) the degree of hydrolysis and the (ii) pH of a 0.1 molar solution of sodium sulfide at 25°C. (H_2S: $K_1 = 9.1 \times 10^{-8}$; $K_2 = 1.2 \times 10^{-15}$).

Discussion: Sodium sulfide is a salt of the weak acid H_2S and the strong base NaOH. H_2S is a dibasic acid. We first analyze the situation before we use any ready-made formula.

The hydrolysis of the sulfide ion takes place in two steps:

$$S^{2-} + H_2O \rightleftharpoons HS^- + OH^- ; \quad HS^- + H_2O \rightleftharpoons H_2S + OH^-$$

In the first step, S^{2-} is the base, HS^- being its conjugate acid. In the second step, HS^- is the base, H_2S is its conjugate acid. The extent of the first and second steps in hydrolysis is thus determined by the relative base strengths of S^{2-} and HS^-.

We know, both S^{2-} and HS^- are weak bases. From a look at the two acid dissociation constants of H_2S, we find: $K_2 << K_1$. So, HS^- is a much weaker acid than H_2S. The weaker an acid, the stronger is its conjugate base. So we may conclude that S^{2-} is a much stronger base than HS^-.

Solution: **(i)** In fact, the OH^- ions produced in the first step of hydrolysis of S^{2-} will almost completely repress the second step of hydrolysis. So the hydrolysis constant of S^{2-} is determined almost completely by the first step: $S^{2-} + H_2O \rightleftharpoons HS^- + OH^-$.

$$K_h = \frac{[HS^-][OH^-]}{[S^{2-}]}$$

Multiplying the numerator and denominator by $[H^+]$,

$$K_h = \frac{[HS^-][OH^-][H^+]}{[S^{2-}][H^+]} = \frac{1 \times 10^{-14}}{1.2 \times 10^{-15}} = 8.3 \quad (K_2 = [H^+][S^{2-}]/[HS^-])$$

The large value of K_h suggests that α, the degree of dissociation may not be negligible in comparison to unity. So we write

$$K_h = 8.3 = \frac{\alpha^2 C}{(1-\alpha)} = \frac{0.1\alpha^2}{1-\alpha} \quad \text{or, } 0.1\alpha^2 + 8.3\alpha - 8.3 = 0 \text{ when } \alpha = 0.99\ (99\%).$$

(ii) As the solution is almost completely (99%) hydrolysed, the concentration of OH^- will be almost 0.1 mol dm^{-3}.

So, pOH = – log (0.1) = 1 or, pH = 14 – 1 = 13.

Problem 11.21

Calculate the (i) pH and (ii) degree of hydrolysis of a 0.1 molar solution of Na_2CO_3. ("H_2CO_3": $K_1 = 4 \times 10^{-7}$; $K_2 = 6 \times 10^{-11}$).

Solution: The hydrolysis of the carbonate ion takes place in two steps:

$$CO_3^{2-} + H_2O \rightleftharpoons HCO_3^- + OH^-$$
$$HCO_3^- + H_2O \rightleftharpoons H_2CO_3 + OH^-$$

As explained in problem 11.20 the hydroxyl ions from the first step of hdrolysis will practically repress the second step. The OH^- ions produced by the second step may thus be ignored. The hydrolysis constant then corresponds to the first step alone:

$$K_h = \frac{[HCO_3^-][OH^-]}{[CO_3^{2-}]} = \frac{K_w}{K_2} = 1 \times 10^{-14}/6 \times 10^{-11} = 1.7 \times 10^{-4}.$$

$$K_2 = [CO_3^{2-}][H^+]/[HCO_3^-]$$

(i) $[OH^-] = \sqrt{K_h.C} = \sqrt{1.7 \times 10^{-4} \times 10^{-1}} = 4.1 \times 10^{-3}$. $\Rightarrow$ pOH = 2.4, pH = 11.6

[Also, $pH = \frac{1}{2}pK_w + \frac{1}{2}pK_a + \frac{1}{2}\log C = 11.6$. ($K_a = K_2$ of H_2CO_3)]

(ii) $K_h = \frac{\alpha^2 C}{(1-\alpha)} = \frac{\alpha^2 \times 0.1}{1-\alpha}$ or, $\alpha = 0.04$.

Problem 11.22

Calculate the pH of a 0.1 molar solution of $NaHCO_3$ [H_2CO_3 : $K_1 = 4 \times 10^{-7}$; $K_2 = 6 \times 10^{-11}$].

Discussion: We follow the same line of approach as done earlier with the diprotic acid leucine.

The bicarbonate ion hydrolyses as : $HCO_3^- + H_2O \rightleftharpoons H_2CO_3 + OH^-$

Side by side, the bicarbonate ion acts as an acid : $HCO_3^- + H_2O \rightleftharpoons CO_3^{2-} + H^+$

We may combine the two reactions to get : $2HCO_3^- \rightleftharpoons H_2CO_3 + CO_3^{2-}$, when

$$K = \frac{[H_2CO_3][CO_3^{2-}]}{[HCO_3^-]^2} = \frac{[H_2CO_3]}{[HCO_3^-][H^+]} \cdot \frac{[CO_3^{2-}].[H^+]}{[HCO_3^-]} = \frac{K_2}{K_1}$$

where K_1 and K_2 are the two acid dissociation constants of H_2CO_3.

(From the values of K_1 and K_2 given above, K_2/K_1 is nearly 10^{-4}, which means that most of the HCO_3^- remains as such.)

To know the concentration of H^+ in this solution we use the two key relations:

Charge Balance: $[Na^+] + [H^+] = [HCO_3^-] + 2\,[CO_3^{2-}] + [OH^-]$.

(The coefficient 2 before $[CO_3^{2-}]$ comes to account for the double negative charge on CO_3^{2-}).

Material Balance: $[Na^+] = [H_2CO_3] + [HCO_3^-] + [CO_3^{2-}]$

Combining these two relations, $[H^+] + [H_2CO_3] = [CO_3^{2-}] + [OH^-]$.

We replace (i) $[H_2CO_3]$ and $[CO_3^{2-}]$ by appropriate expressions from the dissociation of H_2CO_3, (ii) $[OH^-]$ by the expression obtained from self-ionization of water.

$$[H^+] + \frac{[H^+][HCO_3^-]}{K_1} = \frac{[HCO_3^-]}{[H^+]}.K_2 + \frac{K_w}{[H^+]} \quad \text{or, } [H^+]^2 = \frac{K_1(K_2[HCO_3^-] + K_w)}{K_1 + [HCO_3^-]}$$

When $[HCO_3^-]$ is not too small, this may be further simplified as follows:

(i) In the numerator, $K_w << K_2[HCO_3^-]$; (ii) In the denominator, $K_1 << [HCO_3^-]$

We get $[H^+] = (K_1K_2)^{1/2}$ or, $pH = \frac{1}{2}(pK_1 + pK_2)$.

The same expression was arrived at earlier for the diprotic acid, leucine.

Solution: Sodium bicarbonate is the acid salt of the so-called dibasic acid, H_2CO_3. The bicarbonate ion behaves both as an acid and a base in the two equilibrium processes existing simultaneously in an aqueous solution of sodium bicarbonate:

$$H_2CO_3 + H_2O \rightleftharpoons HCO_3^- + OH^- \ ; \quad HCO_3^- + H_2O \rightleftharpoons CO_3^{2-} + H^+$$

A pure solution of $NaHCO_3$ may then be regarded as an equimolar mixture of a weak acid HCO_3^- and a weak base HCO_3^-.

If the two successive dissociation constants of H_2CO_3 are K_1 and K_2, the concentration of H^+ is given as

$$[H^+] = (K_1.K_2)^{1/2}; \quad \text{or, } pH = \frac{1}{2}pK_1 + \frac{1}{2}pK_2 = 8.3$$

Problem 11.23

Calculate the hydrogen ion concentration of a 0.01 M solution of NaHS ($K_1 = 10^{-7}$; $K_2 = 10^{-15}$).

Solution: We proceed as in Problem 11.22 for $NaHCO_3$. But K_2 is less than K_w. So we write

$$[H^+]^2 = \frac{K_1(K_2[HS^-]+K_w)}{K_1+[HS^-]} \simeq \frac{K_1K_2[HS^-]+K_1K_w}{[HS^-]} \simeq \frac{K_1K_w}{[HS^-]}$$

or, $[H^+] = 3.2 \times 10^{-10}$ M.

11.3.2 Buffer Solutions

Certain solutions possess the ability to resist an appreciable change in hydrogen ion concentration (i.e., pH) when small amounts of acids or bases are added to these solutions. This resistance of a solution is known as *buffer action* and these solutions of "*reserve acidity*" or "*reserve alkalinity*" are known as *buffer solutions* or simply *buffers*. Buffer solutions usually consist of a weak acid or base together with its salt.

Thus, a solution containing acetic acid, a weak acid, and its salt, say, sodium acetate, behaves as a buffer solution. Similarly, a solution containing the weak base ammonia, and its salt, ammonium chloride constitutes a buffer mixture.

The hydrogen ion concentration of a buffer solution changes very slightly on addition of acids or alkalis. Thus when 10 cm^3 of 1 N HCl solution is added to 1 dm^3 (litre) of a buffer solution containing 0.1 M acetic acid and 0.1 M sodium acetate, the pH changes from 4.74 to 4.66 only (see Problem 11.25). *If the same amount of acid were added to* 1 dm^3 *of pure water, the pH would have changed from* 7.0 *to* 2.0.

A buffer solution maintains a steady concentration of hydrogen ion (hence pH) by adjusting equilibrium between the weak acid or base and their conjugate ions provided by the salt. Let us first consider a buffer solution consisting of a weak acid HA and its salt, MA. The salt is fully dissociated into M^+ and A^- ions while the weak acid HA is only partially ionized in solution:

$$MA = M^+ + A^- \ ; \ HA \rightleftharpoons H^+ + A^- \ ; K_a = [H^+]\,[A^-]/[HA].$$

The large concentration of A^- ions from the salt will suppress the dissociation of the weak acid by common ion effect and a very small fraction of it would ionize to give H^+ ion in solution. Now let us consider the addition of acids and alkalis to this system separately.

Addition of acid: The added H^+ ions will combine with A^- ions in solution, forming undissociated HA. The hydrogen ion concentration of the solution will thus remain steady.

$$H^+ + A^- \rightleftharpoons HA$$

Addition of alkali: The added OH^- ions will combine with H^+ ions forming water. At the same time, some HA molecules will ionize to replace the hydrogen ions so consumed and maintain its ionization equilibrium.

$$OH^- + H^+ \rightleftharpoons H_2O \ ; \ \ HA \rightleftharpoons H^+ + A^-$$

The buffer action of a solution containing a weak base B and its salt BHX (e.g., NH_3 and NH_4Cl) may be similarly explained by considering the dissociation equilibrium of the base:

$$B + H_2O \rightleftharpoons BH^+ + OH^- \ ; \quad BHX = BH^+ + X^-$$

Acid added: $H^+ + OH^- = H_2O$; $B + H_2O \rightleftharpoons BH^+ + OH^-$

Alkali added: $BH^+ + OH^- \rightleftharpoons B + H_2O$

Ammonium acetate or any other salt of a weak acid and a weak base would singly exhibit buffer action in solution. This is readily understood from the reversible equilibrium established by both the ions of the salt.

In the reactions involved in buffer action, a strong acid (or base) reacts essentially "completely" with a weak base (or acid). The equilibrium constants of such combinations are quite large. When acetic acid (HA) reacts with a strong base (OH^-), the equilibrium constant is about 10^9:

$$OH^- + HA \rightleftharpoons A^- + H_2O \qquad K = \frac{1}{K_b(A^-)} = \frac{K_a(HA)}{K_w} = 1.7 \times 10^9.$$

Similarly, when a weak base B reacts with a strong acid,

$$B + H^+ \rightleftharpoons BH^+ \qquad K = 1/K_a(BH^+)$$

When B is "tris" (see Problem 11.26), pK_a of the conjugate acid (BH^+) is 8.08. Hence,

$$K = \frac{1}{K_a(BH^+)} = 1.2 \times 10^8.$$

Reaction between a strong acid and strong base has still higher value of K. If we mix a strong acid, a strong base, a weak acid and a weak base together, the strong acid and base will react first until one is completely used up. The remaining strong acid or base will next react with the weak counterpart.

pH of a buffer solution

Let us first consider a buffer solution containing a weak acid HA at formal concentration C_{HA} and its salt MA at concentration C_S. The *equilibrium* established by the weak acid may be simply expressed as

$$HA + H_2O \rightleftharpoons H_3O^+ + A^- \text{ or } HA \rightleftharpoons H^+ + A^- \qquad K_a = \frac{[H^+][A^-]}{[HA]}$$

Material balance: $[HA] = C_{HA} - [H^+]$

(For simplicity, we neglect the contribution of hydrogen ion from the ionization of water.)

Charge balance: $[A^-] = C_S + [H^+]$

[We have again neglected the concentration of OH^- ions from the ionization of water.]

The expressions for [HA] and $[A^-]$ may now be put in the ionization constant of the acid:

$$K_a = \frac{[H^+](C_S + [H^+])}{C_{HA} - [H^+]} \quad \text{or} \quad [H^+] = K_a \times \frac{C_{HA} - [H^+]}{C_S + [H^+]}$$

This quadratic equation may be simplified for most practical purposes as follows: since the presence of the anion A^- in large concentration from the salt represses the dissociation of the acid HA, $[H^+]$ will be very small in comparison to either C_{HA} or C_S.

So, $C_{HA} - [H^+] \simeq C_{HA}$ and $C_S + [H^+] \simeq C_S$.

Hence, $[H^+] = K_a \times \frac{C_{HA}}{C_S}$ or, $pH = pK_a + \log \frac{C_S}{C_{HA}}$.

$$\text{i.e., } pH = pK_a + \log \frac{[\text{Salt}]}{[\text{Acid}]} = pK_a + \log \frac{[A^-]}{[HA]} \quad [11\text{-}21]$$

Similarly it can be shown that for a mixture of a weak base B (dissociation constant K_b) and its salt BH^+ (with a strong acid):

$$pOH = pK_b + \log \frac{[\text{Salt}]}{[\text{Base}]} \quad \text{or, } pH = pK_a(BH^+) + \log \frac{[B]}{[BH^+]} \quad [11\text{-}22]$$

These equations are known as **Henderson equations** (also Henderson-Hasselbalch equations). They give satisfactory results for solutions having pH between 4.0 and 10.0 provided they are not too dilute. In other cases the quadratic equation should be used.

These equations may give us the impression that the pH of a buffer solution depends only on the *ratio* of the concentrations of the salt and the acid, and not on their *actual* concentrations; that is, the pH of a buffer solution should not change with dilution. But this conclusion is valid only approximately. The pH changes slightly with dilution as well as with the addition of inert electrolytes to a buffer solution. This can be shown by considering activities of the ions instead of the concentrations.

The dissociation constant K_a of the acid HA is appropriately expressed using activity coefficients (f):

$$K_a = \frac{a_{H^+} . a_{A^-}}{a_{HA}} \quad \text{or, } a_{H^+} = \frac{a_{HA}}{a_{A^-}} . K_a = \frac{C_{HA} . f_{HA}}{C_{A^-} . f_{A^-}} . K_a$$

But the activity coefficient of the undissociated acid i.e., f_{HA}, is nearly unity. Hence,

$$a_{H^+} = \frac{C_{HA}}{C_{A^-} . f_{A^-}} . K_a.$$

In a buffer solution, C_{A^-} = concentration of salt (C_S). Hence,

$$pH = pK_a + \log \frac{C_S}{C_{HA}} + \log f_{A^-}$$

The activity coefficient f_{A^-} generally increases as concentration decreases. So, the pH of a buffer solution slightly *increases on dilution.*

The activity coefficient f_{A^-} is also dependent on the ionic strength of a solution. The presence of inert salts increases the ionic strength of a buffer solution and hence a decrease in f_{A^-} (Debye-Hückel limiting law). Consequently the pH will decrease.

The predominant species

From the Henderson equations, we may know the predominant species of a weak acid or weak base in solution at a given pH. For example, we want to know what fraction of ammonia is present as NH_4^+ in a solution having pH = 6. In an aqueous solution of ammonia (pK_b = 4.74), we have the equilibrium:

$$NH_3(aq) + H_2O \rightleftharpoons NH_4^+ + OH^- \quad pK_b = 4.74;\ pK_a \text{ of } NH_4^+ = 9.26.$$

$$pOH = pK_b + \log \frac{[NH_4^+]}{[NH_3]} \quad \text{or, } pH = pK_a(NH_4^+) + \log \frac{[NH_4^+]}{[NH_3]}$$

At pH = 9.26, $[NH_4^+] = [NH_3]$. At pH < 9.26, NH_4^+ will be the predominant form.

At pH = 6.0, log $[NH_4^+]/[NH_3] = 9.26 - 6.0 = 3.3$.

This gives $[NH_4^+] = 1995\ [NH_3]$ or nearly 99.9% of ammonia will be present as NH_4^+.

Next we consider a dibasic acid, say, oxalic acid (H_2Ox). We have,

$$H_2Ox \rightleftharpoons H^+ + HOx^- \qquad pK_1 = 1.25$$

$$HOx^- \rightleftharpoons H^+ + Ox^{2-} \qquad pK_2 = 4.27$$

At pH = 1.25 (= pK_1), $[H_2Ox] = [HOx^-]$; at pH = 4.27 (= pK_2), $[HOx^-] = [Ox^{2-}]$. Above pH 4.27, Ox^{2-} will predominate.

At a pH = 7.0 (say), $7.0 = 4.27 + \log\frac{[Ox^{2-}]}{[HOx^-]}$ or, $\log\frac{[Ox^{2-}]}{[HOx^-]} = 2.7$.

This gives $[Ox^{2-}] \simeq 500\,[HOx^-]$.

Examples of polyprotic acids, including amino acids, have been given earlier.

Buffer capacity

The buffer capacity of a solution is a measure of its tendency to resist change in pH on addition of acids or bases. This has been defined as the number of moles of strong acid or base which will result in unit change of pH of one dm^3 of a buffer solution. A higher buffer capacity indicates greater buffering action of a solution.

In general, the buffer capacity varies with the composition of the solution, and hence, with the pH. So it is more appropriately defined in differential terms:

$$\text{Buffer capacity } (\pi) = \frac{db}{d\text{pH}} = -\frac{da}{d\text{pH}}$$

where *db* moles of strong base or *da* moles of strong acid change the pH by an amount *d*pH. (The acid decreases the pH, hence the negative sign.)

For a buffer solution consisting of a weak acid and its salt, it can be shown that at a given total concentration of the components, buffer capacity is maximum for equal concentrations of the two components.

Many reactions in the laboaratory, specially in analytical works, have to be carried out within a narrow pH range. Buffer solutions are extensively used in such cases to maintain the desired pH range. Selective precipitation, solvent extraction or titration of a particular component in a mixture is thereby made feasible.

Buffer solutions are also important in biological systems. Various biochemical reactions take place under the catalytic action of specific enzymes. These enzymes are very sensitive to pH. Buffer systems are established within the body to maintain the critical pH range for each part. Thus the pH of human blood is maintained within the range 7.3 – 7.5 by the buffering action of carbonic acid and sodium carbonate. The pH of the gastric juice is maintained between 1.5 and 2.0.

Numerical Problems

Problem 11.24

Calculate the pH of a buffer solution containing 0.1 M acetic acid and 0.2 M sodium acetate. (K_a = 1.8×10^{-5}).

Solution: $\text{pH} = pK_a + \log\frac{[\text{salt}]}{[\text{acid}]} = 4.74 + \log 2 = 5.05$

Problem 11.25

What will be change in pH of the buffer solution mentioned in problem 11.24 when (i) 10 cm^3 of 1N hydrochloric acid and (ii) 10 cm^3 of 1N sodium hydroxide is added separately to 500 cm^3 of the buffer solution? (Neglect volume change.)

Solution: The pH of the buffer is 5.05 (calculated above).

500 cm^3 of the buffer solution contains 0.1/2, i.e., 0.05 mol acetic acid and 0.2/2, i.e., 0.1 mol sodium acetate.

(i) 10 cm^3 of 1N hydrochloric acid contains 0.01 mol of H^+ ion which combines with 0.01 mol of acetate ion (CH_3COO^-) to produce 0.01 mol of practically undissociated acetic acid.

Therefore, in 500 cm^3 solution, amount of acetic acid ***increases*** by 0.01 mol to 0.06 mol.

or, ***concentration*** of acetic acid (C_{Acid}) = 2 × 0.06 = 0.12 mol dm^{-3}.

Similarly, ***amount*** of sodium acetate = (0.1 − 0.01) = 0.09 mol ⇒ C_{Salt} = 0.18 mol dm^{-3}.

$$\therefore \quad pH = pK_a + \log\frac{[\text{salt}]}{[\text{acid}]} = 4.74 + \log\frac{0.18}{0.12} = 4.92$$

(ii) 10 cm^3 of 1N sodium hydroxide neutralizes 0.01 mol of acetic acid.

∴ ***amount*** of acetic acid = 0.05 − 0.01 = 0.04 mol. ⇒ C_{Acid} = 0.08 mol dm^{-3}.

Similarly, ***amount*** of sodium acetate = 0.01 + 0.10 = 0.11 mol. ⇒ C_{Salt} = 0.22 mol dm^{-3}.

$$\therefore \quad pH = pK_a + \log\frac{[\text{salt}]}{[\text{acid}]} = 4.74 + \log\frac{0.22}{0.08} = 5.18$$

Note (i) The amounts of salt and acid may also be put directly in the expression of pH since the volume is same for both.

(ii) Addition of same quantity of acid or alkali to 500 cm^3 of water would change the pH from 7.0 to (a) 1.70 in the first case and (b) 12.3 in the second case.

Problem 11.26

Tris(hydroxymethyl)aminomethane (11-III) (abbreviated "tris") is a widely used buffer. The formula weight of tris is 121.136, and the formula weight of tris hydrochloride is 157.597. pK_a of the conjugate acid of tris is 8.075.

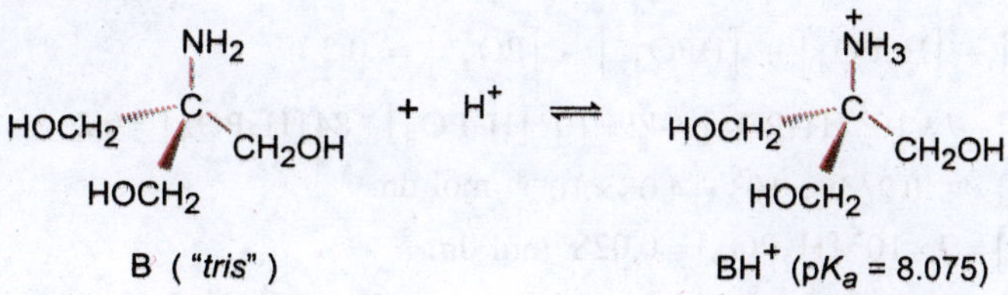

11-III **11-IV**

What will be the pH of a solution containing 12.12 g of tris plus 4.32 g of tris hydrochloride per liter?

Solution: The concentrations of tris (B) and its conjugate acid (BH^+) are

$$[B] = \frac{12.12 \text{ g L}^{-1}}{121.136 \text{ g mol}^{-1}} = 0.100 \text{ M}; \quad [BH^+] = \frac{4.32 \text{ g L}^{-1}}{157.597 \text{ g mol}^{-1}} = 0.027 \text{ M}$$

$$pH = pK_a(BH^+) + \log\frac{[B]}{[BH^+]} = 8.075 + \log\frac{0.100}{0.027} = 8.64$$

Problem 11.27

An aqueous solution contains only potassium hydrogen sulfate and potassium sulfate in the ratio 1 : 2. The pH of the solution is 2.30. Calculate pK_a for the HSO_4^- ion.

Solution: Here the HSO_4^- ion acts as the acid, its conjugate base being the SO_4^{2-} ion.

$$HSO_4^- + H_2O \rightleftharpoons H_3O^+ + SO_4^{2-} \quad \text{or,} \quad HSO_4^- \rightleftharpoons H^+ + SO_4^{2-}$$

The addition of sulfate ions to a solution containing bisulfate ions is analogous to the addition of, say, acetate ions to a solution of acetic acid.

So, $\quad pH = 2.30 = pK_a + \log\dfrac{[SO_4^{2-}]}{[HSO_4^-]}$ when $pK_a = 2.00$

Problem 11.28

Calculate the concentrations (approximate) of the various forms of orthophosphate in a buffer of total phosphate concentration equal to 0.200 mol dm^{-3} and pH = 8.00.

[H_3PO_4 : $K_1 = 7 \times 10^{-3}$; $K_2 = 6 \times 10^{-8}$; $K_3 = 2 \times 10^{-13}$.]

Solution: (i) From the pH of the solution, $[H^+] = 10^{-8}$ mol dm^{-3}

(ii) The three stages of dissociation of phosphoric acid are

$$H_3PO_4 = H^+ + H_2PO_4^- \qquad K_1 = 7 \times 10^{-3} \qquad \text{.. (I)}$$
$$H_2PO_4^- = H^+ + HPO_4^{2-} \qquad K_2 = 6 \times 10^{-8} \qquad \text{.. (II)}$$
$$HPO_4^{2-} = H^+ + PO_4^{3-} \qquad K_3 = 2 \times 10^{-13} \qquad \text{.. (III)}$$

From $K_2 = 6\times10^{-8} = \dfrac{[H^+][HPO_4^{2-}]}{[H_2PO_4^-]}$ and $[H^+] = 10^{-8}$, $\dfrac{[HPO_4^{2-}]}{[H_2PO_4^-]} = 6$ or, $[HPO_4^{2-}] = 6[H_2PO_4^-]$ .. (IV)

Again from I, $[H_2PO_4^-]/[H_3PO_4] = K_1/[H^+] = \dfrac{7\times10^{-3}}{10^{-8}} = 7\times10^5$.

or, $[H_2PO_4^-] = 7\times10^5\,[H_3PO_4]$ and $[HPO_4^{2-}] = 42\times10^5\,[H_3PO_4]$ (from IV)

From III, $[PO_4^{3-}] = K_3[HPO_4^{2-}]/[H^+]$. $\Rightarrow$ $[PO_4^{3-}] = 84\,[H_3PO_4]$

The total concentration of phosphate in the buffer (0.2 mol dm^{-3}) is equal to the sum of the concentrations of H_3PO_4, $H_2PO_4^-$, HPO_4^{2-} and PO_4^{3-}.

$\therefore \quad [H_3PO_4] + [H_2PO_4^-] + [HPO_4^{2-}] + [PO_4^{3-}] = 0.2.$

or, $\quad [H_3PO_4] + 7\times10^5\,[H_3PO_4] + 42\times10^5\,[H_3PO_4] + 84\,[H_3PO_4] = 0.2.$

or, $\quad [H_3PO_4] \simeq 0.2/49\times10^5 = 4.08\times10^{-8}$ mol dm^{-3}

$[H_2PO_4^-] = 7\times10^5\,[H_3PO_4] = 0.028$ mol dm^{-3}

$[HPO_4^{2-}] = 42\times10^5\,[H_3PO_4] = 0.171$ mol dm^{-3} and $[PO_4^{3-}] = 3.4\times10^{-6}$ mol dm^{-3}.

Note. We find that out of 0.200 mol of total orthophosphate, 0.199 mol exists in the forms $H_2PO_4^-$ and HPO_4^{2-}. The pH of a phosphate buffer is practically determined by the second dissociation constant of phosphoric acid. The $H_2PO_4^-$ ion acts as the weak acid [$K_a = K_2$ of H_3PO_4) while its conjugate base is HPO_4^{2-}.

$$pH = 7.2 + \log\frac{[HPO_4^{2-}]}{[H_2PO_4^-]}$$

The above problem may be solved approximately from this relation if we assume

$$[HPO_4^{2-}] + [H_2PO_4^-] \simeq 0.2.$$

11.4 ACID-BASE TITRATIONS

Acid-base titrations provide a very useful analytical procedure. A measured amount of an acid is added to a definite volume of base, or vice versa, and the equivalence point is located by the observed colour change* of some indicator. When the strength of one reactant is known, the strength of the other can be easily calculated. For a successful acid-base titration, it is necessary to use the right indicator which will change its colour very close to the equivalence point. To appreciate this point more clearly, we shall discuss it under two sub-headings, namely, (i) the nature of acid-base indicators and (ii) the choice of acid-base indicators on the basis of variation of hydrogen ion concentration during a titration.

***N.B.** Or the change in some other property of the system, e.g., disappearance of turbidity or appearance of luminescence using some inorganic and chemiluminescent acid-base indicators. Acid base titrations may also be carried out conductometrically.

11.4.1 Acid-Base Indicators

The common acid-base indicators are mostly weak organic acids or bases which change their colour perceptively with the change in hydrogen ion concentration of the medium. The hydrogen-ion concentration controls the relative amounts of the unionized and ionized forms of the indicator. When these two forms differ in colour, we get different colours of the indicator in acidic and alkaline media. The nature of this colour change may be understood from a consideration of the equilibrium established by an indicator in aqueous solution.

Formerly, the equilibria established by an indicator acid (HIn, say) and an indicator base (InOH, say) were treated separately. According to the Bronsted concept of acids and bases, it is unnecessary to distinguish between acid and base indicators. Instead, emphasis is placed on the charge types of the acid and alkaline forms of the indicator, which form a conjugate acid-base system.

Suppose an indicator (In) has the acidic form In_A and the basic form In_B, which differ in colour. The equilibrium between the two forms may be expressed as

$$In_A + H_2O \rightleftharpoons In_B + H_3O^+ \text{ or simply, } In_A \rightleftharpoons In_B + H^+$$

$$K_{In} = \frac{[H^+]\times[In_B]}{[In_A]} \Rightarrow pH = pK_{In} + \log\frac{[In_B]}{[In_A]}$$

pK_{In} is called the indicator constant. The relationship shows that for any indicator, the ratio of the two forms, $[In_B]/[In_A]$, is dependent on the pH. The colour of the indicator, as perceived by the eye, is dependent on this ratio. At the same time, a limitation is imposed by the capacity of the human eye to distinguish between two colours in admixture. Experience shows that it is difficult to recognize a colour when it is present below 10% of the other colour. Thus, a solution in which $[In_A]/[In_B]$ is approximately 10 or more, the solution will be characterized by the colour of In_A alone; this is popularly called the "acid colour" of the indicator. Similarly, the "basic colour" of the indicator, due mainly to In_B, appears when $[In_B]/[In_A]$ is equal to, or greater than 10. The corresponding limit of pH may now be obtained as

Acid colour: $[In_A]/[In_B] > 10;\ pH = pK_{In} + \log\frac{1}{10} = pK_{In} - 1$

Alkaline Colour: $[In_B]/[In_A] > 10;\ pH = pK_{In} + \log\frac{10}{1} = pK_{In} + 1$

Accordingly, the colour-change interval is: $pH = pK_{In} \pm 1$, over approximately two units of pH. The colour of the indicator gradually changes from one to another as the pH of the solution changes across this interval. A few common acid-base indicators with their colour-change intervals are listed in Table 11.2.

Table 11.2
Colour changes and pH range of acid-base indicators

Common Name	Chemical Name	Acid colour	Base colour	pH
Thymol blue	Thymol-sulphonephthalein	Red	Yellow	1.2–2.8
Methyl orange	Dimethylaminoazobenzene sodium sulphonate	Red	Orange	3.1–4.4
Bromocresol green	Tetrabromo-*m*-cresol sulphonephthalein	Yellow	Blue	3.8–5.4
Methyl red	*o*-Carboxybenzene-azodimethylaniline	Red	Yellow	4.8–6.0
Bromothymol blue	Dibromothymolsulphone-phthalein	Yellow	Blue	6.0–7.6
Phenolphthalein	Phenolphthalein	Colourless*	Red*	8.0–9.6

In strong acid medium (65-98% H_2SO_4) phenolphthalein turns from colourless to orange-red. Again, in strongly alkaline medium (> pH 11), the red colour of the indicator becomes colourless.

Before we use an indicator in any acid-base titration, we have to make sure that the pH change of the titration medium around the equivalence point agrees with the pH-range of the indicator. We shall return to this point in section 11.4.2 after we discuss the nature of variation of pH during any titration.

The colours of acid-base indicators are to some extent dependent on temperature. Indicators that are molecular or anionic acids (e.g. phenolphthalein, methyl red) show relatively small changes of K_{In} with temperature. Cationic indicator acids (e.g., methyl orange) exhibit larger temperature effects. Thus, at 100°C, the colour change interval of pH for phenolphthalein lies between 8.1–9.0, while for methyl orange it shifts to 2.5–3.7. The pH of the medium and K_w also change with temperature. Thus, the colour change interval of pOH for the above indicators at 100°C are 4.1–3.2 and 9.7–8.5 respectively ($pH + pOH \simeq 12$ at 100°C).

The presence of additional salts, proteins, colloids and nonelectrolytes like alcohol etc. also affect the colour of an indicator in a complicated manner. One practical way to eliminate these effects is to standardize the solutions used under identical conditions.

11.4.2 Choice of Indicator in Acid-Base Titration

Since each acid-base indicator is characterized by a definite pH-range across which its colour changes noticeably, the choice of an indicator in any acid-base titration is done with reference to the variation of pH of the titrated solution in the neighbourhood of the equivalence point. When we titrate a known volume of an acid with a base, or vice versa, hydrogen and hydroxyl ions combine to form water. The hydrogen ion concentration of the solution being titrated changes continuously in course of the titration, the concentrations of the titrant and the titrate solutions determining the pH at different stages of neutralization. Side by side, the anions and cations furnished by the acid and base, i.e., the salt formed in the reaction, may influence the pH of the solution through hydrolysis. Accordingly, the nature of variation of pH in any acid-base titration is dependent on (i) the concentration of solution and (ii) the nature of acid and base taken. This is very conveniently understood from a study of the relevant **neutralization curves** (or titration curves) obtained by plotting the pH-values at different stages of the neutralization as ordinate against the volume of alkali added as abscissae. Here we shall consider a few general cases of acid-base titrations with typical examples.

(i) Neutralization of a strong acid by a strong base (or vice versa). We assume that (i) the acids and bases are completely dissociated, (ii) the contribution of hydrogen or hydroxyl ion by ionization of water is negligible in comparison to that by the acid or base and (iii) the activity of hydrogen or hydroxyl ion is equal to their respective concentrations. As discussed earlier, the salt produced in course of the neutralization will have no effect on the hydrogen ion concentration of the solution.

As a typical example, let us consider the titration of 100 cm^3 of a 0.1 M solution of hydrochloric acid by a 0.1 M sodium hydroxide solution.

At the beginning, $[H^+] = C_{HCl} = 10^{-1}$ mol dm^{-3}; pH = 1.0.

When 50 cm^3 base have been added, the solution contains 50 cm^3 of 0.1 M HCl unneutralized, the total volume being (100 + 50) = 150 cm^3.

$\therefore$ $[H^+] = 50 \times 0.1/150 = 3.33 \times 10^{-2}$ mol dm^{-3}; pH = 1.5.

The pH at other stages of neutralization may be similarly calculated. A few typical values have been given in Table 11.3.

When 100 cm^3 base have been added, the solution will contain no unneutralized acid and its pH will be simply that of a solution of sodium chloride. In absence of dissolved carbon dioxide, it will be exactly 7.

When 100.1 cm^3 of base have been added, the solution contains 0.1 cm^3 of 0.1 M sodium hydroxide solution in excess, the total volume being 200.1 cm^3.

$\therefore$ $[OH^-] = 0.1 \times 0.1/200.1 = 5 \times 10^{-5}$ mol dm^{-3}; pOH = 4.3, pH = 9.7. The pH for other volumes of excess base added may be similarly calculated, a few typical values being given in Table 11.3.

The pH of the solution will change in a somewhat different manner if the acid and base used have some other strength. These may be calculated along the same line of approach as illustrated above for 0.1 M solutions. pH values calculated for different stages of a titration of 0.01 M hydrochloric acid with a 0.01 M sodium hydroxide solution are also shown in Table 11.3.

The data in Table 11.3 may be represented graphically to obtain the corresponding neutralization curve. In Fig. 11.4, only a portion of the curve is shown to have a closer look at the variation of pH around the equivalence point.

The results show that the pH of the solution rises very slowly in the beginning of the titration (by only 0.5 for 50 cm^3 of alkali added). But the rate of change of pH of the solution becomes very rapid as the equivalence point is approached. Between the addition of 99.9 and 100.1 cm^3 of alkali, the pH of the solution undergoes a sharp rise, e.g., from 4.3 to 9.7 for solutions of 0.1 M concentration.

Table 11.3

Variation of pH for the addition of different volume of NaOH solution to 100 cm³ of HCl solution, both in equal concentration

NaOH added, cm^3	pH 0.1 M HCl	pH 0.01 M HCl	NaOH added, cm^3	pH 0.1 M HCl	pH 0.01 M HCl
0.0	1.0	2.0	100.0	7.0	7.0
50.0	1.5	2.5	100.1	9.7	8.7
90.0	2.3	3.3	100.2	10.0	9.0
98.0	3.0	4.0	100.5	10.4	9.4
99.0	3.3	4.3	101.0	10.7	9.7
99.5	3.6	4.6	102.0	11.0	10.0
99.9	4.3	5.3	110.0	11.7	10.7

The appropriate indicator for this titration (using 0.1 M solutions) should change its colour within the pH range 4.3–9.7. Any indicator which changes colour below 4.3 would indicate the equivalence point too early; similarly, an indicator changing its colour above 9.7 would lead to over-titration. Methyl orange (pH range: 3.1–4.4) will exist chiefly in the alkaline form above pH 4.0, which corresponds to 99.8% neutralization of the solution. So, methyl orange may be used in this titration with negligible error (~ 0.2%). Similarly, phenolphthalein (8.3–10.0) will change colour when the equivalence point is just crossed and may also be used as an indicator.

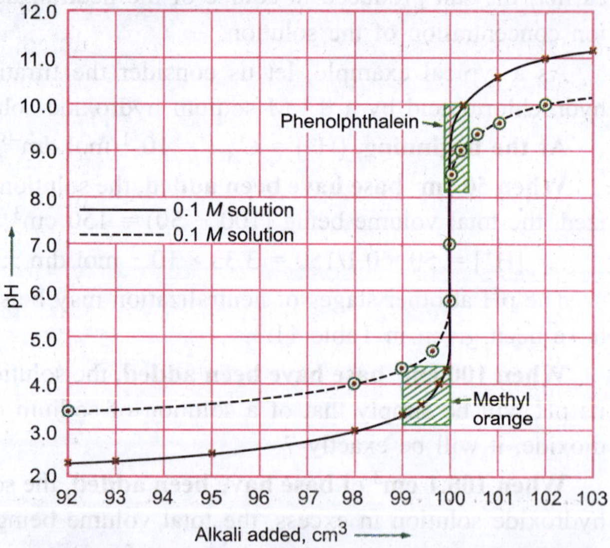

Fig.11.4: Part of the neutralization curve for the titration of 100 cm^3 of HCl by NaOH (both in equal concentration).

But with solutions of 0.01 M concentrations, pH of the solution reaches 4.0 when only 98% of the acid gets neutralized. Since methyl orange will show its alkaline colour near this pH, the titration error will be rather high, ~2%. So one should preferably choose an indicator in a slightly higher pH-range. Methyl red (4.2–6.3) or bromothymol blue (6.0–7.6) are satisfactory indicators in this case.

Carbon dioxide is usually present in such systems either from the atmosphere or from the small quantity of carbonate in sodium hydroxide. The weak "carbonic acid" introduces a small error with indicators of high pH range like phenolphthalein or thymolphthalein (~ 0.15–0.2 cm^3 of 0.1 M NaOH for the titration of 100 cm^3 solution of 0.1 M HCl).

(ii) Neutralization of a weak acid with a strong base. Let us consider the titration of 100 cm^3 of 0.1 M acetic acid with 0.1 M sodium hydroxide solution. As usual, we shall (i) neglect the contribution of hydrogen or hydroxyl ions from the ionization of water and (ii) replace activity by concentration.

At the beginning, $[H^+] = \alpha C = (K_a C)^{1/2} = 1.34 \times 10^{-3}$ mol dm^{-3}; pH = 2.9

When 50 cm^3 of the alkali have been added, the solution consists of unneutralized acetic acid and the salt formed, sodium acetate. The pH of such mixtures are given by the Henderson equation; the concentration of salt and of the acid is calculated from the volume of alkali added. In the present case,

$$\left.\begin{aligned}[\text{Acid}] &= 50 \times 0.1/150 = 3.33 \times 10^{-2} \text{ mol dm}^{-3}\\ [\text{Salt}] &= 50 \times 0.1/150 = 3.33 \times 10^{-2} \text{ mol dm}^{-3}\end{aligned}\right] \Rightarrow \text{pH} = \text{p}K_a + \log\frac{[\text{Salt}]}{[\text{Acid}]} = 4.74.$$

The pH values at different stages of the neutralization may be similarly calculated. (Below pH = 4.0 and above pH = 12.0 the quadratic form of Henderson equation should be used).

When 90 cm^3 of the alkali have been added,

$$[\text{Acid}] = 10 \times 0.1/190\ ;\quad [\text{Salt}] = 90 \times 0.1/190 \qquad (\text{mol dm}^{-3})$$

$$\therefore\ \text{pH} = \text{p}K_a + \log [\text{Salt}]/[\text{Acid}] = 4.74 + \log (9/1) = 5.7.$$

The pH at any other intermediate point of neutralization may be similarly calculated.

When 100 cm^3 of the alkali have been added, the solution contains 0.01 mole of sodium acetate in 200 cm^3 solution. Hence the concentration of sodium acetate is 0.05 M and its pH is expressed as

$$\text{pH} = \tfrac{1}{2}pK_w + \tfrac{1}{2}pK_a + \tfrac{1}{2}\log C = 7 + 2.37 + \tfrac{1}{2}\log(0.05) = 8.72.$$

When 100.1 cm^3 of alkali have been added, the solution contains excess of OH^- ions. These will repress the hydrolysis of the sodium acetate solution and the pH becomes more or less equal to that due to the excess base only. Here,

$$[OH^-] = 0.1 \times 0.1/200.1 = 5 \times 10^{-5} \text{ mol dm}^{-3} \text{ or, pOH} = 4.3; \text{pH} = 9.7.$$

pH values for different stages of the neutralization are given in Table 11.4 while a part of the neutralization curve is shown in Fig. 11.5.

Table 11.4

Neutralization of* 100 cm^3 *of* 0.1 M *acetic acid with* 0.1 M *sodium hydroxide

NaOH added cm^3	pH	NaOH added cm^3	pH
0.0	2.9	99.9	7.7
90.0	5.7	100.0	8.7
98.0	6.4	100.1	9.7
99.0	6.7	100.2	10.0
99.5	7.0	100.5	10.4
99.8	7.4	101.0	10.7

The results show that (i) the pH changes rather slowly in the vicinity of the equivalence point and (ii) the pH at the equivalence point is 8.7. So, it is necessary to use an indicator with a pH range slightly on the alkaline side, like phenolphthalein. Any indicator in the acid range, like methyl red, would change its colour even before the addition of 90% of the total alkali required; so it will be unsuitable for this titration.

An acid with still less K_a would give rise to a higher pH at the equivalence point (e.g., when $K_a = 10^{-7}$, pH = 10.0) and the rate of change of pH around the equivalence point would be more slow. Phenolphthalein would not be suitable in such cases. Thymolphthalein (9.3–10.5) may be used, though the colour change will be very much gradual. Acids with $K_a < 10^{-7}$ cannot be satisfactorily titrated with any simple indicator in 0.1 M solution.

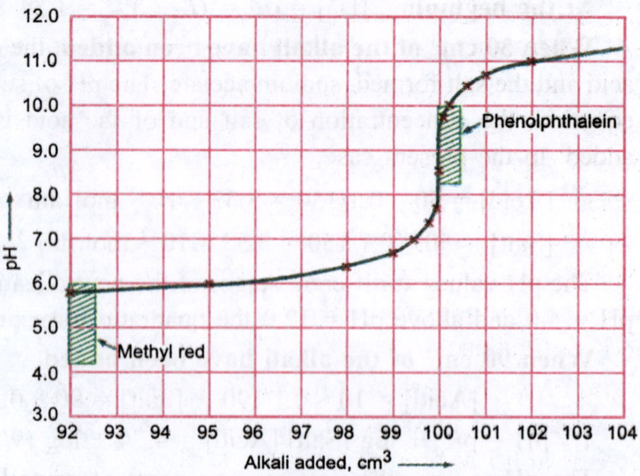

Fig.11.5: Part of the neutralization curve in the titration of acetic acid with sodium hydroxide (both 0.1 M).

(iii) Neutralization of a weak base with a strong acid. The titration of 100 cm³ of 0.1 M aqueous ammonia with 0.1 M hydrochloric acid may be considered as a typical example. With the usual assumptions, we may calculate the pH at various stages of the neutralisation as follows:

At the beginning, the solution contains only aqueous ammonia at 0.1 M concentration ($K_b = 1.8 \times 10^{-5}$).

$$[OH^-] = \alpha C = (K_b \,.\, C)^{1/2} \Rightarrow \text{pOH} = 2.9,\ \text{pH} = 11.1$$

At intermediate points of neutralization, the solution consists of the unneutralized ammonia and the ammonium chloride formed. The pH for such solutions may be obtained by the Henderson equation which gives pOH as

$$\text{pOH} = \text{p}K_b + \log\frac{[\text{Salt}]}{[\text{Base}]}$$

At the equivalence point, the solution is simply that of 0.1 mol of ammonium chloride in 200 cm³ volume, which gives a concentration of 0.05 M. Hence,

$$\text{pH} = \tfrac{1}{2}\text{p}K_w - \tfrac{1}{2}\text{p}K_b - \tfrac{1}{2}\log C = 7 - 2.37 + 0.65 = 5.28.$$

After the equivalence point, the addition of excess hydrochloric acid represses the hydrolysis of ammonium chloride and the concentration of hydrogen ion is practically that due to the excess acid.

The pH values calculated for some points of the neutralization are represented in Table 11.5 and Fig. 11.6.

Table 11.5

Neutralization of* 0.1 M *aqueous ammonia with* 0.1 M *hydrochloric acid

HCl added cm³	pH	HCl added cm³	pH
0.0	11.1	99.9	6.3
50.0	9.3	100.0	5.3
90.0	8.3	100.1	4.3
99.0	7.3	101.0	3.3
99.5	7.0		

It appears that the pH at the equivalence point is 5.3. This means that indicators like phenolphthalein cannot be used in such titrations. As the pH corresponding to the addition of 100.1 cm^3 of HCl is 4.3, indicators like methyl orange or methyl red will exist mostly in their acid form at this point. So, these may be used as indicators for this titration. (Also, bromophenol blue and bromocresol green.)

Neutralization of a weaker base ($K_b \simeq 10^{-7}$) is marked by a slower change in pH. Bromocresol green or methyl red would not produce a sharp colour-change in such cases. Bromophenol blue or methyl orange may be used for such titrations.

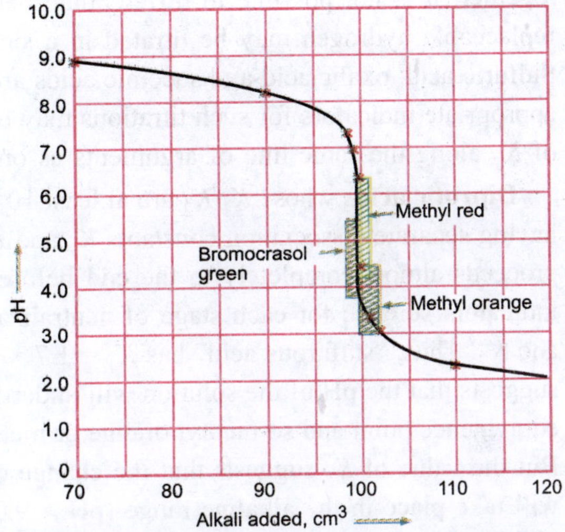

Fig.11.6: Part of the neutralization curve in the titration of aqueous ammonia by HCl (both 0.1 M).

(iv) Neutralization of a weak acid by a weak base. Let us consider the titration of 100 cm^3 of 0.1 M acetic acid by 0.1 M aqueous ammonia solution ($K_a = K_b = 1.8 \times 10^{-5}$).

Before the addition of base, the pH of the solution is 2.9, as shown earlier.

During the progress of neutralization before the equivalence point, the solution contains varying amounts of unneutralized acetic acid and the salt produced, ammonium acetate. The pH of the solution changes more or less in the same fashion as in the titration of acetic acid with sodium hydroxide solution (Fig. 11.7).

At the equivalence point, the pH is given by the relation

$$pH = \tfrac{1}{2}pK_w + \tfrac{1}{2}pK_a - \tfrac{1}{2}pK_b = 7 + 2.37 - 2.37 = 7.0.$$

Beyond the equivalence point, addition of excess alkali to the ammonium acetate solution changes the pH of the medium according to the relation

$$pOH = pK_b + \log [\text{Salt}]/[\text{Base}].$$

This is, in fact, the same expression as that obtained in the titration of ammonia solution by hydrochloric acid before the equivalence point (Table 11.5).

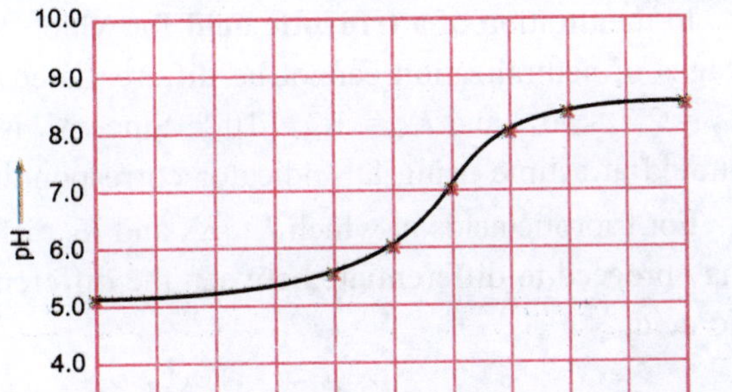

Fig. 11.7: Part of the neutralization curve for the titration of 100 cm^3 of 0.1 M acetic acid by 0.1 M ammonia solution.

Part of the neutralization curve for the titration is shown in Fig. 11.7. The change of pH is very gradual throughout the neutralization. Therefore, no single indicator is expected to produce any sharp colour change in this titration. As such the titration of a weak acid by a weak base (or vice versa) cannot be carried out using any single indicator.

However, the above titration may be carried out with a **mixed indicator**. A mixture of equal parts of neutral red and methylene blue (both as 0.1 per cent alcoholic solution) gives a sharp colour change at pH 7.0 from violet-blue to green.

(v) Neutralization of polyprotic acids with strong base.

The course of neutralization of a polyprotic acid is obviously determined by the relative magnitudes of the various dissociation constants. For diprotic acids having K_1 and K_2 close together, it is not possible to differentiate between the two stages of neutralization. All the replaceable hydrogen may be titrated in a single stage, provided K_2 is not less than $\sim 10^{-7}$. Sulfuric acid, oxalic acid and succinic acids are examples of this type of polyprotic acids. The appropriate indicators for such titrations may be selected from a knowledge of the magnitudes of K_2 along the same line of arguments as outlined earlier.

Diprotic acids whose K_1/K_2 are at least 10^4 behave like a mixture of two monoprotic acids having separate dissociation constants K_1 and K_2 respectively. The first stage of neutralization proceeds almost completely to the end before the second stage is appreciably affected. The indicators suitable for each stage of neutralization may be selected from a knowledge of K_1 and K_2. Thus, "sulfurous acid" has $K_1 = 1.7 \times 10^{-2}$ and $K_2 = 1.0 \times 10^{-7}$. The high value of K_1 suggests that the pH of the solution will undergo a sharp change in the acid range near the first equivalence point and so methyl orange or methyl red will be suitable indicators for this stage. But the value of K_2 suggests that the change of pH during the second stage of neutralization will take place in the alkaline range (pH > 9.0). Even phenolphthalein will not be a suitable indicator for this titration; thymolpthalein may be used in this case.

For "carbonic acid", $K_1 = 4.3 \times 10^{-7}$ and $K_2 = 5.6 \times 10^{-11}$, only the first stage of neutralization may be located by phenolphthalein or thymolphthalein. The second point of neutralization would change the pH of the medium extremely slowly and so no indicator will be satisfactory for this stage.

The pH at the first equivalence point of a diprotic acid is given approximately by the relation $[H^+] = (K_1K_2)^{1/2}$ or $pH = \frac{1}{2}(pK_1 + pK_2)$, provided K_1 is not very high.

In the titration of a **triprotic acid** for which K_1, K_2 and K_3 are too close together, the three stages of neutralization cannot be differentiated easily. Thus, for citric acid, $K_1 = 9.2 \times 10^{-4}$; $K_2 = 2.7 \times 10^{-5}$ and $K_3 = 1.3 \times 10^{-6}$. Since K_3 is $> 10^{-7}$, all the replaceable hydrogen may be titrated at a time using an indicator corresponding to K_3.

For triprotic acids in which K_1, K_2 and K_3 differ appreciably (by factors of 10^4 approx), one may proceed to differentiate between the different stages of neutralization. Thus, for phosphoric acid,

$$K_1 = 7 \times 10^{-3}, K_2 = 6 \times 10^{-8}; K_3 = 5 \times 10^{-13}.$$

Here, $K_1/K_2 = 1.2 \times 10^5$ and $K_2/K_3 = 1.2 \times 10^5$. So, the acid will behave as a mixture of three monoprotic acids with dissociation constants equal to K_1, K_2 and K_3 respectively. Neutralization at the first stage will be nearly complete before the second stage is appreciably affected; similarly, neutralization at the secondary stage will proceed nearly to completion before the tertiary stage sets in. Let us consider the three stages separately in a titration of 100 cm^3 of 0.1 M phosphoric acid with 0.1 M NaOH solution.

pH at the first equivalence point

The solution will contain NaH_2PO_4 and the pH will be given by pK_1 and pK_2 of phosphoric acid:

$$pH = \frac{1}{2}(pK_1 + pK_2) = 4.6.$$

Further, the magnitude of K_1 suggests that the change of pH around the first equivalence point will lie primarily in the acid range. Methyl orange, bromocresol green or congo red may be used as indicators in this titration.

pH at the second equivalence point

The solution will contain Na_2HPO_4 and the pH will be given by K_2 and K_3 of phosphoric acid:

$$pH = \frac{1}{2}(pK_2 + pK_3) = 9.7.$$

Since K_2 is less than 10^{-7}, the change of pH around the equivalence point will take place in an appreciably alkaline range and only thymolpthalein may be used as an indicator for this stage. A mixed indicator containing 2 or 3 parts phenolphthalein and 1 part 1-naphtholphthalein is also very satisfactory (from pale rose through green to violet at pH 9.6).

pH at the third equivalence point

The solution would contain Na_3PO_4. Its hydrolysis is confined mainly to the first stage:

$$PO_4^{3-} + H_2O \rightleftharpoons HPO_4^{2-} + OH^-$$

The subsequent stages of hydrolysis may be practically disregarded (compare the values of K_1, K_2 and K_3 for H_3PO_4; the hydrolysis constants for each step is given by K_w/K). So we may, at least approximately, use the formula of pH for salts of weak monoprotic acids and strong bases. The first step of the hydrolysis shown depends upon K_3 of H_3PO_4. So, we write

$$pH = \frac{1}{2}pK_w + \frac{1}{2}pK_3 + \frac{1}{2}\log C = 12.35.$$

The pH at the equivalence point as well as the very low value of K_3 ($\simeq 10^{-13}$) suggest that the change of pH around the equivalence point will be extremely gradual and in fact no indicator can be used to locate the equivalence point.

Thus, phosphoric acid cannot be titrated as a tribasic acid by usual titrimetry since the third equivalence point cannot be determined by any suitable indicator. However, all the three replaceable hydrogen in phosphoric acid may be titrated by replacing the phosphate anion as calcium phosphate according to the reaction

$$2H_3PO_4 + 3CaCl_2 = Ca_3(PO_4)_2 \downarrow + 6HCl.$$

3 mols of HCl are formed per mol of H_3PO_4 which may be titrated by a strong alkali like NaOH as usual. Barium chloride or silver nitrate may also be used in place of $CaCl_2$.

(vi) Titration of anions of weak acids (Brönsted bases) with strong acids

Anions of weak acids behave as Brönsted bases and may be titrated with strong acids. Formerly, these were considered as titrations of hydrolyzable salts, with the ultimate replacement of a weak acid by a strong acid; as such, these were called displacement titrations. For example, a solution of sodium acetate may be titrated with hydrochloric acid when the following reactions set in:

hydrolysis:	$CH_3COO^- + Na^+ + H_2O$	$\rightleftharpoons$	$CH_3COOH + OH^- + Na^+$
neutralization:	$H^+ + Cl^- + OH^-$	$\rightleftharpoons$	$H_2O + Cl^-$
Total	$CH_3COONa + HCl$	$\rightleftharpoons$	$CH_3COOH + NaCl$

In fact, we are titrating the conjugate base of a weak acid, acetate ion in this case, by a strong acid. The pH of the medium at various stages of the titration may be calculated as illustrated below for the titration of 100 cm^3 of a 0.1 M solution of sodium acetate with 0.1 M hydrochloric acid.

At the beginning, the solution in the conical flask for titration (*titrate*) contains 0.1 M sodium acetate, a salt of a weak acid and a strong base. Hence,

$$pH = \tfrac{1}{2}pK_w + \tfrac{1}{2}pK_a + \tfrac{1}{2}\log C = 7.0 + 2.37 + \tfrac{1}{2}\log(0.1) = 8.87.$$

At intermediate points of neutralization, the solution contains unreacted sodium acetate and the acetic acid formed by the reaction:

$$CH_3COONa + HCl = CH_3COOH + NaCl.$$

The pH may be calculated by the Henderson equation, viz., pH = pK_a + log [Salt]/[Acid], where [Acid] is equal to the amount of HCl added. For example, when 90 cm^3 of HCl has been added,

$$pH = 4.74 + \log\frac{[Salt]}{[Acid]} = 4.74 + \log\frac{10}{90} = 3.79.$$

But this expression of pH *ceases to be satisfactory* as the concentration of salt decreases and [H$^+$] increases. We remember that the concentration of the hydrogen ion in such a mixture was actually arrived at as (sec. 11.3.2)

$$[H^+] = K_a \times \frac{[Acid] - [H^+]}{[Salt] + [H^+]}$$

Subsequently we neglected [H$^+$] in comparison to [Acid] or [Salt], and arrived at the Henderson equation. But as [H$^+$] increases, and [Salt] decreases, this approximation is no longer valid. For 90 cm^3 of HCl added, the pH given by this approximation differs only by about 0.03 from that obtained from the quadratic expression given above. For larger amounts of HCl added, we must avoid the approximate expression. Thus for 99 cm^3 of HCl added, the approximate expression gives

$$pH = 4.74 + \log\frac{1}{99} = 2\cdot74$$

We may now calculate the pH from the quadratic expression given above. The volume of the solution after addition of 99 cm^3 of HCl becomes 199 cm^3.

$$\therefore [CH_3COOH]=\frac{99\times0.1}{199}=0.0497 \text{ mol dm}^{-3}; \ [CH_3COO\,Na]=\frac{1\times0.1}{199}=5.02\times10^{-4} \text{ mol dm}^{-3}.$$

$$[H^+]=K_a\times\frac{[CH_3COOH]-[H^+]}{[CH_3COO\,Na]+[H^+]}=1.8\times10^{-5}\times\frac{0.0497-[H^+]}{5.02\times10^{-4}+[H^+]}$$

or, $[H^+]^2 + (5.02 + 0.18) \times 10^{-4} [H^+] - 1.8 \times 0.0497 \times 10^{-5} = 0.$

$[H^+] = 7.2 \times 10^{-4}$ mol dm^{-3} or, pH = 3.14 (compare with the value of 2.74 obtained earlier).

At the equivalence point, the solution contains acetic acid (and sodium chloride) at a concentration of 0.05 M (the volume has been exactly doubled).

$$\therefore \quad [H^+]=(K_a.C)^{1/2}=(1.8\times10^{-5}\times0.05)^{1/2} \quad \text{or,} \ \text{pH}=3.02$$

Beyond the equivalence point, the excess HCl added would suppress the dissociation of acetic acid. So, one may proceed to find out the concentration of hydrogen ion by assuming the contribution of acetic acid to be negligible in comparison to that by the HCl. But this *cannot be done* until the concentration of hydrogen ion from the HCl largely supersedes that by the CH_3COOH itself.

For example, when 100.1 cm^3 of 0.1 M HCl has been added, the excess HCl alone would give $[H^+]=\frac{0.1\times0.1}{200.1}=5\times10^{-5}$ mol dm^{-3}, which gives pH = 4.3.

This indicates as if the pH has *increased* on the addition of HCl to a solution of acetic acid! This apparent anomaly is readily removed when we look at the magnitude of $[H^+]$ calculated above. Since the pH of the solution at the equivalence point was nearly 3, $[H^+] \simeq 10^{-3}$. So, the concentration of H^+ ions from the HCl is still negligible (10^{-5} vs 10^{-3}) in comparison to the hydrogen ion obtained from acetic acid itself. Let the degree of dissociation of acetic acid in presence of the added HCl be x. Remembering that the concentration of acetic acid at the equivalence point was 0.05 M, we have, at equilibrium, the following concentrations of each:

$$\underset{0.05\,(1-x)}{CH_3COOH} \rightleftharpoons \underset{0.05x}{CH_3COO^-} + \underset{0.05x}{H^+}$$

The total concentration of hydrogen ion in the solution is obtained by the sum of contributions from HCl and CH_3COOH : $[H^+] = 5 \times 10^{-5} + 0.05x$.

$$\therefore K_a=1.8\times10^{-5}=\frac{(5\times10^{-5}+0.05x)\times x}{1-x}$$

Neglecting x in the denominator, we may solve for x to get $x = 1.9 \times 10^{-2}$.

$$\therefore [H^+]=5\times10^{-5}+0.05\times1.9\times10^{-2}\Rightarrow \text{pH}=3.0.$$

Clearly, the contribution of H^+ ions from the HCl is small in comparison to that by CH_3COOH.

After the addition of 110 cm^3 of HCl, however, the concentration of hydrogen ion may be calculated by neglecting the contribution of acetic acid.

The pH values at different stages of the titration are given in Table 11.7A.

It is evident from the pH values in Table 11.7A that the change of pH is extremely gradual near the equivalence point. So it is not possible to carry out the titration satisfactorily with any indicator.

Side by side we may consider the **titration of 100 cm³ of a 0.1 M solution of sodium cyanide.** For HCN, $K_a = 7 \times 10^{-10}$ and $pK_a = 9.15$. With this small dissociation constant in view, the pH values at different stages of the titration may be calculated by the simplified Henderson equation. Beyond the equivalence point, we may totally neglect the contribution of H^+ ion from HCN and calculate the pH from the amount of excess HCl added. The pH values so obtained are recorded in Table 11.7B. It appears that the titration may be safely carried out in presence of an indicator like methyl red (pH 4.2-6.2).

Table 11.7A

***Titration of* 100 cm³ *of* 0.1 M *solution of sodium acetate with* 0.1 M HCl**

HCl added cm³	pH
0.0	8.9
50.0	4.7
90.0	3.8
99.0	3.1
99.9	3.0
100.0	3.0
100.1	3.0
101.0	2.9
110.0	2.3

Table 11.7B

***Titration of* 100 cm³ *of* 0.1 M *solution of sodium cyanide with* 0.1 M HCl**

HCl added cm³	pH
0.0	11.1
50.0	9.1
90.0	8.2
99.0	7.1
99.9	6.1
100.0	5.1
100.1	4.3
101.0	3.3
110.0	2.3

An interesting point about Table 11.7B may now be considered. The pH values correspond very closely to those given in Table 11.5 for the titration of 0.1 M aqueous ammonia solution with 0.1 M HCl. We remember that for aqueous solutions of ammonia, $pK_b = 4.74$, which is nearly equal to $14–pK_{HCN}$ (14–9.15 = 4.85). From this, and similar other calculations, one may conclude that

titration of a salt of a strong base and a weak acid of dissociation exponent pK_a *follows the same course as the titration of a weak base of dissociation constant* $14 - pK_a$.

From this generalization we may also explain why it is not possible to titrate a solution of sodium acetate. For acetic acid, $pK = 4.74$, which shows that this titration would be equivalent to the titration of a base having $pK_b = 14–4.74 \simeq 9.26$. A dissociation constant of the order of 10^{-9} is indeed very small to result in appreciable change of pH during titration. Summarily, one may satisfactorily titrate only salts of weak acids having pK equal to or greater than 7.

Exactly analogous considerations apply to the titration of **salts of weak bases and strong acids**. This is, in fact, the titration of the conjugate acid of the weak base. Parallel calculations reveal that the titration of a salt of strong acid and weak base (dissociation exponent pK_b) is equivalent to the titration of a weak acid of dissociation exponent $14 - pK_b$. Obviously, it is not possible to titrate a solution of ammonium chloride with NaOH as this is equivalent to the titration of a very weak acid having $pK_a = 14 - pK_{NH_3} = 9.25$.

Summary on choice of indicators

The choice of indicator for any titration is guided by the variation of pH of the titration medium before and after the equivalence point. Each indicator is characterized by a definite pH range ($pK_{In} \pm 1$) across which it undergoes an observable colour change. When the pH of a solution "jumps" across this characteristic pH range of an indicator just before and after the equivalence point of a titration, most satisfactory end points are obtained. This is the case in the titration of strong acids with strong bases (or reverse) and almost any indicator may be used. In other cases, the pH of the solution at and after the equivalence point tends to lie on the side of the "stronger" partner in the titration. Thus, in a strong acid-weak base titration, the pH at the equivalence point remains below 7 while in a strong base-weak acid titration, the pH lies above 7. Accordingly, indicators have to be chosen for these titrations showing colour change in the acid and alkaline range respectively. For any "weak-weak" titration, no single indicator is satisfactory. The following chart summarizes the important points regarding the choice of indicator in some typical acid base titrations using solutions at 0.1 M concentration.

Nature of Titration	pH at equivalence point	pH range	Example
(i) Strong acid–strong base	$pH = \frac{1}{2}pK_w = 7.0$	4.5 — 9.5 (5.5 — 8.5 in presence of CO_2).	Bromocresol green Methyl red Phenolphthalein
(ii) Strong acid–weak base ($K_b > 10^{-5}$)	$pH = \frac{1}{2}pK_w - \frac{1}{2}pK_b - \frac{1}{2}\log C$	3 — 7	Methyl orange Methyl red Bromocresol green
(iii) Weak acid–strong base ($K_a > 10^{-5}$)	$pH = \frac{1}{2}pK_w + \frac{1}{2}pK_a + \frac{1}{2}\log C$	7 — 10.5	Thymol blue Thymolphthalein Phenolphthalein
(iv) Polyprotic acids with strong bases	$pH = \frac{1}{2}pK_1 + \frac{1}{2}pK_2$ (first equivalence point) $pH = \frac{1}{2}pK_2 + \frac{1}{2}pK_3$ (second equivalence point)	Depends upon the relative magnitudes of K_1, K_2, K_3 etc.	
(v) Anion of a weak acid by a strong acid. ($K_a \simeq 10^{-10}$)	$pH = \frac{1}{2}pK_a - \frac{1}{2}\log C$	6.5 — 4.5	Methyl red.

Indicator error in titration

The end-point in a titration in presence of a given indicator is marked by a characteristic colour change. From the foregoing discussion, we may understand that this "end-point" lies *close* to the equivalence point, i.e., the point of 100% neutralization, but it may not exactly coincide with the equivalence point. Owing to this, a solution is somewhat over-titrated or undertitrated. This gives rise to small errors in titration, known as indicator error. The estimation of indicator error also provides an alternative way for the choice of an indicator.

The indicator error is the error caused by the difference between the pH at the equivalence point and the titration exponent of the indicator used. The latter may be defined as the pH at which titration is ended in presence of the indicator, and is often designated by pT. For methyl orange, pT = 4.0; for methyl red, pT = 5.5 and phenolphthalein, pT = 9.0 (using about 2-3 drops of the indicator in 0.1% solution).

Indicator errors are conveniently expressed in terms of the excess acid or alkali remaining at the end point. If the excess acid or alkali are strong, the errors are described as "H^+ error" or "OH^- errors". For weak acids or bases, these are called "acid error", or "alkali errors". For both the solutions at normality N, the magnitudes of these errors may be correlated to the indicator exponent pT as follows:

$$\mathbf{H^+\ error} = \frac{10^{-pT} \times 2}{N} \times 100 \text{ per cent;} \quad \mathbf{OH^-\ error} = \frac{10^{-(14-pT)}}{N} \times 2 \times 100 \text{ per cent}$$

$$\mathbf{Acid\ error} = 10^{pK - pT} \qquad \mathbf{Alkali\ error} = 10^{pK + pT - 14}.$$

Problem 11.29

Can methyl orange be used in the titration of 0.01 N HCl with 0.01 N NaOH? (pT for M.O. = 4.0).

Solution: pH at equivalence point = 7.0; pT of methyl orange = 4.0.

Hence the solution will contain some unneutralized HCl at the "end-point". The error will be an H^+ error (strong acid). The solution will be *undertitrated*, hence the error must be *negative*.

$$H^+ \text{ error} = -\frac{10^{-4} \times 200}{10^{-2}} = -2\%.$$

Clearly, the error is too large and M.O. should not be used in this titration.

Problem 11.30

What will be the error if one titrates 0.1 N aq. NH_3 solution with 0.1 N HCl in presence of phenolphthalein (pT = 9)?

Solution: pH at the equivalence point $= 7 - \frac{4\cdot8}{2} - \frac{1}{2}\log 0\cdot05 = 5\cdot25.$

pH at the "end point" = 9.

So the solution will be *undertitrated* and contain an excess of ammonia at the end point.

The alkali error $= 10^{4.8+9-14} = 10^{-0.2} = 0.63.$

$\therefore$ Out of every 1.63 parts by weight of ammonia initially taken, 0.63 part remains *unneutralized* at the so-called "end-point". An error of –0.63 (undertitration) of the 1.63 corresponds to an alkali error which is

$$-\frac{0.63}{1.63} \times 100 \approx -38\%.$$

Clearly, the indicator is unsuitable for the titration.

11.5 SOLUTIONS OF SPARINGLY SOLUBLE SALTS

Salts having solubilities less than about 0.01 mol per dm^3 (in water at ordinary temperature) may be called sparingly soluble. Silver chloride (solubility = 1.2×10^{-5} mol per dm^3) is an example of a sparingly soluble salt.

When an excess of a sparingly soluble salt is shaken up with water, part of the salt passes into solution to make it saturated at that temperature. A dynamic equilibrium is established between the solid salt and the ions furnished by it in solution. Taking the example of silver chloride, this may be represented, *assuming complete ionization of silver chloride in solution*, as

$$AgCl(s) \rightleftharpoons Ag^+ + Cl^-$$

hydration being implied. At equilibrium, the rates of the forward and the backward processes become equal. Application of the law of mass action to such systems lead to the solubility product principle which is of great significance in qualitative and quantitative analysis.

11.5.1 Solubility Product

It was found experimentally that in saturated solutions of sparingly soluble salts, the product of the total molecular concentrations of the ions is a constant at a constant temperature. This observation was incorporated by Nernst in the **solubility product principle** (1889):

In a saturated solution of a sparingly soluble electrolyte, the product of the concentrations of the constituent ions is constant at a given temperature, the ion concentrations being raised to powers equal to the respective numbers of ions of each kind furnished by one molecule of the electrolyte on complete dissociation.

Thus, for an clectrolyte, A_xB_y, ionizing as $A_xB_y \rightleftharpoons xA^+ + yB^-$, the solubility product principle states:

$$[A^+]^x \times [B^-]^y = \text{constant at a constant temperature.}$$

This product of the total molecular concentrations of the ions in a saturated solution of a sparingly soluble salt is termed the **solubility product** of the salt; it is often denoted by K_s. The solubility products of calcium fluoride and aluminium hydroxide may be taken as examples:

$$CaF_2(s) = Ca^{2+}(aq) + 2F^-(aq) \quad ; \quad K_s = [Ca^{2+}][F^-]^2$$

$$Al(OH)_3(s) = Al^{3+}(aq) + 3OH^-(aq) \; ; \quad K_s = [Al^{3+}][OH^-]^3.$$

Solubility products of a few common sparingly soluble salts are given in Table 11.8.

Table 11.8

Solubility products at 25°C

AgCl	1.8×10^{-10}	$Fe(OH)_3$	1.6×10^{-39}
AgBr	5.2×10^{-13}	$PbCl_2$	1.7×10^{-5}
$Al(OH)_3$ (α)	3.0×10^{-34}	$Mg(OH)_2$	7.1×10^{-12}
$BaSO_4$	1.1×10^{-10}	HgS (black)	2.0×10^{-53}
CdS	1.0×10^{-27}	NiS (α)	4.0×10^{-20}
CaF_2	3.9×10^{-11}	$SrSO_4$	3.2×10^{-7}
CaC_2O_4	1.3×10^{-8}	$Zn(OH)_2$	7.0×10^{-18}
CuS	8.0×10^{-37}	ZnS (amorphous)	3.0×10^{-16}

The solubility product principle follows readily from an application of the law of mass action to the equilibrium established by a sparingly soluble salt in contact with its saturated solution.

$$A_xB_y(s) \rightleftharpoons xA^+ + yB^-$$

By applying the law of mass action in terms of activities (a) of respective species,

$$K_{eq} = \frac{a^x_{A^+} \times a^y_{B^-}}{a_{A_xB_y(s)}}$$

The activity of solid A_xB_y may be taken as unity since it is the pure component. So one obtains : $K_{eq} = a^x_{A^+} \times a^y_{B^-}$

Since the solution is extremely dilute, one may reasonably replace activities by concentrations: $[A^+]^x \times [B^-]^y = \text{Constant} = K_s$.

The product $a^x_{A^+} \times a^y_{B^-}$ is called the **activity solubility product** of AB which may be represented as K^a_S. It is the actual solubility product of thermodynamic significance. We shall appreciate its importance very soon.

The tendency of A_xB_y to furnish A^+ and B^- ions depend on the surface area of the solid exposed to the solution. But the reverse tendency of the ions to crystallize from solution is also proportional to this surface area. Consequently the amount of excess solid has little effect on the equilibrium (provided, of course, a substantial amount is present in excess).

Solubility products may be evaluated from the saturation solubilities of salts in pure water. The solubilities, in turn, may be obtained by direct analysis of the saturated solutions or by conductance measurements if the ion conductances (at infinite dilution) of the constituent ions of the salt are known. If Λ_0 is the *equivalent conductance* of an electrolyte (obtained from the individual ion conductances) and κ is its measured *specific conductance* in the saturated solution containing *s g-equivalents per dm³*, then the solubility *s* is obtained in **g-equivalents per dm³** from the relation:

$$\Lambda_0 = \frac{1000\,\kappa}{s}$$

Solubility products may also be obtained from the measured EMF of suitably devised cells.

Problem 11.31

The solubility of silver chloride is 0.0015 g. dm^{-3}. Calculate its solubility product.

Solution: $AgCl\,(s) \rightleftharpoons Ag^+ + Cl^-$

1 mol of AgCl dissolves to give 1 mol each of Ag^+ and Cl^- ions. Now, solubility of AgCl in mol dm^{-3} is obtained by dividing the gram-solubility by its molecular weight 143.5.

Solubility = 0.0015/143.5 = 1.05×10^{-5} mol dm^{-3}.

$\therefore\ [Ag^+] = [Cl^-] = 1.05 \times 10^{-5}$ mol dm^{-3}.

or $K_s = [Ag^+] \times [Cl^-] = (1.05 \times 10^{-5})^2 = 1.1 \times 10^{-10}$.

Problem 11.32

The solubility product of CaF_2 is 4.9×10^{-11}. Calculate its solubility in pure water neglecting hydrolysis by the F^- ion.

Solution: $CaF_2 \rightleftharpoons Ca^{2+} + 2F^-$

For each mol of CaF_2 dissolved, we have 1 mol of Ca^{2+} ion and 2 mols of F^- ions. Let *S* be the solubility of CaF_2 in mol dm^{-3}. Then,

$[Ca^{2+}] = S$ mol dm^{-3}; $[F^-] = 2S$ mol dm^{-3} or, $K_S = [Ca^{2+}] \times [F^-]^2 = S \times (2S)^2 = 4S^3$.

So, $4S^3 = 4.9 \times 10^{-11}$ or, $S = 2.3 \times 10^{-4}$ mol dm^{-3}.

(The solubility in gram may be obtained by multiplying this value with the molar mass).

11.5.2 The Common Ion Effect

The constancy of the solubility product at a given temperature implies that if we change the concentration of an ion (in the saturated solution of the sparingly soluble salt), the concentration of the other ion will adjust itself to maintain the constancy of the solubility product. For example, let us take a saturated solution of silver chloride. We may increase the concentration of chloride ion by adding, say, sodium chloride (or any electrolyte having a common anion). If this is done, the concentration of silver ion will decrease to maintain the product $[Ag^+] \times [Cl^-]$ constant at that temperature [= K_S]. This will result in a displacement of the equilibrium towards left, i.e., some Ag^+ ions will combine with Cl^- ions to precipitate silver chloride. Similarly, when solid silver chloride is shaken with a solution of sodium chloride, a very small amount of it will pass into solution which is just sufficient to maintain the constancy of the solubility product. Thus the effect of an electrolyte giving common ions in solution as those of the sparingly soluble salt is to diminish its solubility than in pure water.

This effect of the common ion in lowering the solubility of a sparingly soluble salt is consistent with the Le Chatelier's principle. The addition of, say, sodium chloride to a saturated solution of silver chloride in water is a "stress" applied to the equilibrium maintained by the solubility of silver chloride. With the addition of chloride ions from sodium chloride, the system tends to adjust itself by way of diminishing the concentration of chloride ion. This results in the precipitation of solid silver chloride, lowering its solubility.

We shall illustrate this point further with a few typical examples.

Problem 11.33

What will be the solubility of silver chloride in a solution 0.1 M with respect to NaCl? ($K_S = 1.0 \times 10^{-10}$).

Solution: Let the solubility be equal to x mol dm^{-3} . x mol of silver chloride furnishes x mol of Ag^+ and x mol of Cl^- ions. But the solution contains additional Cl^- ions from the sodium chloride. As this is a strong electrolyte, the concentration of Cl^- ions from this is 0.1 mol dm^{-3}.

$\therefore$ The total concentration of Cl^- ion: $[Cl^-] = x + 0.1$; $[Ag^+] = x$.

$\therefore$ $K_S = 1.0 \times 10^{-10} = [Ag^+] \times [Cl^-] = x \times (x + 0.1)$.

x is very small in comparison to 0.1. $\therefore$ $K_S = x \times 0.1 = 1.0 \times 10^{-10}$ when, $x = 10^{-9}$.

Hence, the solubility of AgCl in a 0.1 M solution of NaCl would be 10^{-9} mol dm^{-3}.

Problem 11.34

Calculate the solubility of CaF_2 in a solution of 0.1 M $Ca(NO_3)_2$. $K = 4.9 \times 10^{-11}$.

Solution: Let the solubility be x mol dm^{-3}. Since each mol of CaF_2 supplies 2 mol of F^- ion in solution, the concentration of F^- ion, $[F^-] = 2x$ mol dm^{-3}.

The total concentration of Ca^{2+} ions in solution will be the sum of (i) Ca^{2+} ions from $Ca(NO_3)_2$, equal to 0.1 mol dm^{-3} and (ii) Ca^{2+} ions from CaF_2, equal to x mol dm^{-3}.

$\therefore$ $[Ca^{++}] = (0.1 + x)$ mol dm^{-3} ; $x << 0.1$ or, $[Ca^{2+}] \simeq 0.1$ mol dm^{-3}.

$\therefore$ $K_S = 4.9 \times 10^{-11} = [Ca^{2+}]\,[F^-]^2 = 0.1 \times (2x)^2 \Rightarrow x = 1.11 \times 10^{-5}$ mol dm^{-3}.

Thus, the required solubility of CaF_2 is 1.11×10^{-5} mol dm^{-3}.

11.5.3 The Salt Effect

The salt effect concerns the effect of electrolytes with no common ion upon the solubility of a sparingly soluble salt. It has been observed that addition of electrolytes having no common ion *increases* the solubility of a sparingly soluble salt. Thus, thallous chloride (TlCl) has a solubility of 0.016 mol per dm^3 in pure water at 25°C. In presence of 1.0 M KNO_3, the solubility rises to 0.031 mol per dm^3.

This can be explained when we remember that the solubility product relation in terms of concentration is applicable only in solutions of *low ionic strength*, when the activity coefficients are nearly unity. But in presence of appreciable concentration of other salts, the ionic strength of the medium becomes large and we have to introduce the activity solubility product. For a simple salt AB, we have

$$K_S^a = a_{A^+} \times a_{B^-} = [A^+] \times [B^-] \times \gamma_{A^+} \times \gamma_{B^-} = K_S \times \gamma_\pm^2$$

where the γ-terms refer to activity coefficients of the ions, $\gamma_\pm$ being the mean ionic activity coefficient.

From Debye-Hückel theory, it can be shown that for aqueous solutions at 25°C,

$$\log \gamma_\pm = -0.509\, z_+ z_- \mu^{1/2}$$

where, z_+, z_- = absolute value of the formal charge on the different types of ions

and, μ = ionic strength of the medium $= \frac{1}{2}\Sigma C_i . z_i^2$.

[C_i = concentration and z_i = formal charge of each ionic species].

Therefore, an increase of ionic strength of the medium by the added salt results in a *lowering* of the activity coefficient term ($\gamma_\pm$). In order to maintain the constancy of the activity solubility product K_S^a, K_S increases, implying greater solubility.

11.5.4 Applications of The Solubility Product Principle

The solubility product of a sparingly soluble salt sets a limit to the ultimate value of the product of the concentrations of its ions in solution at equilibrium at a given temperature. If in a solution the ionic product is different from the solubility product, the system will tend to adjust itself so as to equalize the two quantities. This has great bearing upon precipitation from solution, which is one of the principal operations of qualitative and quantitative analysis.

For any sparingly soluble salt AB, having solubility product K_S at a given temperature, the maximum concentrations of the ions, say, A^+ and B^-, capable of existing at equilibrium at this temperature is given by the relation

$$K_S = [A^+] \times [B^-].$$

If to a solution containing the ion A^+ at some arbitrary concentration (together with other ions), a large amount of the species B^- is added by some means, the solubility product principle indicates that solid AB will be precipitated until the ionic product reaches the solubility product. Using appropriate concentration of B^- ions, it is therefore possible to reduce the concentration of A^+ ions to a negligible value.

Let us consider a solution of silver nitrate at some arbitrary concentration, say, 0.01 M. To this solution we add a solution of potassium chloride, so that the concentration of chloride ion in the solution is momentarily raised to 0.1 M. The ionic product becomes $0.1 \times 0.01 = 10^{-3}$. This is greater than the solubility product of AgCl at this temperature, $\simeq 10^{-10}$. In order to maintain equilibrium, the process $Ag^+ + Cl^- \rightarrow AgCl\ (s)$ will take place, causing precipitation of solid silver chloride. The process will continue until the ionic product equals the solubility product.

One may also start with a saturated solution of silver chloride and add a solution of potassium chloride, in which case the concentration of silver ion in solution will be further reduced.

On the other hand, if we take the final solutions in the above two cases after they have attained equilibrium and add aqueous ammonia solution to them, the concentration of silver ion will be greatly reduced owing to the formation of the complex $[Ag(NH_3)_2]^+$ ion. As a result, the system will tend to restore its equilibrium by the reverse process, i.e., $AgCl(s) \rightarrow Ag^+ + Cl^-$, resulting in dissolution of the precipitated silver chloride. Using appropriate concentration of ammonia, the process may be carried out until the entire precipitate dissolves.

A few points are to be remembered before we actually apply the solubility product principle to specific cases. (i) The solubility product defines a state of equilibrium. But this does not commit anything towards the rate at which the equilibrium will be attained. In practice, some time is always necessary to start precipitation, and then a further period of variable duration must elapse before precipitation is complete. (ii) Conditions of supersaturation must be eliminated for effective precipitation. (iii) *Complete precipitation* of a sparingly soluble salt is not possible because a finite, though small, concentration of each type of ion must remain in solution to maintain the solubility product. (iv) The solubility product may somewhat vary depending upon the physical structure of the precipitate. The solubility product of freshly precipitated nickel sulfide is 3×10^{-21} while on standing the precipitates attain lower solubility product, $\sim 10^{-26}$ or even less. (v) One should not consider that a *large excess* of the precipitant (say KCl in the precipitation of AgCl from $AgNO_3$ solution) will lead to nearly complete precipitation. A certain excess of the precipitant is certainly necessary for effective precipitation, but a very large excess may actually cause some of the precipitate to dissolve owing to very high ionic strength of the medium.

A few practical applications of the solubility product principle in qualitative analysis may now be considered:

Precipitation of Sulphides: For the precipitation of a metal sulphide MS, we require that $[M^{2+}] \times [S^{=}] > K_S$ (M^{2+} is the metal ion).

In reality, there exists a wide range of solubility product K_S among the metal sulfides, For example, at ordinary temperature, some K_S values are

HgS (black)	$\simeq 10^{-53}$	CuS	$\simeq 10^{-36}$	MnS (pink)	$\simeq 10^{-11}$
PbS	$\simeq 10^{-28}$	CdS	$\simeq 10^{-27}$	ZnS	$\simeq 10^{-23}$

We observe that K_S of HgS and CuS are quite small while those of ZnS and MnS are relatively high. This suggests that if we can somehow control the concentrations of M^{2+} and S^{2-} ions in a solution containing the above metal ions, such that the ionic product exceeds the lower K_S values only, then HgS and CuS will be precipitated, while MnS and ZnS will remain in solution. The fate of lead and cadmium ions will also be determined by the relative magnitudes of their respective K_S and ionic products. We may considerably vary the concentration of sulfide ions by varying the acidity of the medium. At 0.25–0.3 M HCl medium, dissociation of H_2S is largely checked by the common ion effect of H^+ ions and $[S^{2-}]$ is approximately 10^{-22}*. So, at this acid medium, only the K_S of HgS, PbS, CuS and CdS are exceeded which are therefore precipitated. If the acid concentration be made much higher, $[S^{2-}]$ will be further reduced. CdS may not be precipitated in that situation.

If we next want to precipitate ZnS and MnS from this solution, the concentration of S^{2-} ions must be increased. In a saturated aqueous solution of H_2S, $[S^{2-}]$ is nearly 10^{-15} M**. Precipitation of ZnS will start under this condition. But the acid produced in the reaction will soon repress the dissociation of H_2S, and the precipitation will be incomplete. For complete precipitation, the concentration of hydrogen ions has to be maintained at a steady low value. This can be achieved either by (i) using ammoniacal medium or by (ii) the addition of sodium acetate. The acetate ion combines with hydrogen ions forming weak acetic acid.

Precipitation of hydroxides: The solubility products of some metal hydroxides show wide variation in range, e.g.,

$$Al(OH)_3 : \simeq 10^{-34} \; ; \quad Cr(OH)_3 : \simeq 10^{-30} \; ; \quad Mg(OH)_2 : \simeq 10^{-11}$$

* See solution to Problem 11.36. ** See solution to Problem 11.14.

It is therefore reasonable to expect that some hydroxides may be selectively precipitated by controlling the hydroxyl ion concentration in solution. Obviously sodium hydroxide cannot be used as the precipitant in such cases since it is a strong alkali and its ionization is complete and irreversible. A weak alkali like an aqueous ammonia solution is used which dissociates only incompletely and reversibly. Further, the dissociation may be considerably influenced by the addition of ammonium salts due to common ion effect of the NH_4^+ ion in suppressing the dissociation of ammonia. In fact, in presence of ammonium chloride, the hydroxyl ion concentration in dilute aqueous ammonia solution is so reduced that the $[OH^-]$ is just sufficient to exceed the solubility products of iron(III), aluminium and chromium(III), but not of cobalt, nickel, zinc, manganese or magnesium hydroxides. So only the former three metal hydroxides are precipitated, offering a means of their separation from the others in qualitative analysis.

Precipitation of carbonates. The solubility products of calcium, strontium and barium carbonates are all nearly 10^{-9} while that of magnesium carbonate is $\simeq 10^{-8}$. Selective precipitation of Ca^{2+}, Sr^{2+} and Ba^{2+} ions from solution in presence of Mg^{2+} can be done by carefully controlling the concentration of carbonate ion so that only the solubility products of the former are exceeded. Sodium carbonate, a strong alkali, cannot be used in this case as the concentration of carbonate ion from it cannot be reversibly controlled. Instead, one uses a saturated solution of ammonium carbonate in presence of ammonium chloride, when $[CO_3^{2-}]$ is just sufficient to precipitate Ca^{2+}, Sr^{2+} and Ba^{2+} but not Mg^{2+}.

Problem 11.35

A solution contains KCl and KI, both in 0.1 M concentration. Silver nitrate solution at 0.01 M concentration is added dropwise with stirring. (i) Which will be precipitated first, AgCl or AgI ? (ii) What will be the concentration of I^- in solution when AgCl begins to precipitate? K_S for AgCl $= 1.5 \times 10^{-10}$; for AgI $= 10^{-16}$.

Solution: (i) We shall have to calculate the concentration of Ag^+ ion necessary to start precipitation of AgCl and AgI.

For AgCl : $K_S = 1.5 \times 10^{-10}$; $[Cl^-] = 0.1$ mol dm^{-3}.

$\therefore\ [Ag^+] = K_S / [Cl^-] = 1.5 \times 10^{-9}$ mol dm^{-3}.

For AgI : $K_S = 1 \times 10^{-16}$; $[I^-] = 0.1$ mol dm^{-3}.

$\therefore\ [Ag^+] = K_S / [I^-] = 1 \times 10^{-15}$ mol dm^{-3}.

Since the concentration of silver ion necessary for the precipitation of silver iodide is smaller, AgI will be precipitated first.

(ii) $[Ag^+]$ necessary to start precipitation of AgCl is 1.5×10^{-9} mol dm^{-3}. At this concentration of Ag^+, the concentration of I^- at equilibrium will be

$[I^-] = K_{AgI} / [Ag^+] = 10^{-10} / 1.5 \times 10^{-9} = 6.7 \times 10^{-8}$ mol dm^{-3}.

Note: This shows that iodide ion may be separated from chloride ion almost completely by precipitation with silver nitrate.

Problem 11.36

Calculate the concentration of Cu^{++}, Cd^{++} and Mn^{++} ions remaining in solution after precipitation by excess H_2S from a 0.1 M solution of the ions in 0.3 M HCl medium. Assume K_S : CuS $= 1 \times 10^{-44}$; CdS $= 5 \times 10^{-25}$; MnS $= 1.4 \times 10^{-15}$. For H_2S, $K_1 = 9.1 \times 10^{-8}$; $K_2 = 1.2 \times 10^{-15}$.

Solution: First of all we shall calculate the $[S^{2-}]$ in 0.3 M HCl medium. $[H_2S]$ is taken as approximately the same as in aqueous solution, i.e., 0.1 M. $[H^+] = 0.3$ mol dm^{-3}.

For H_2S: $K_1 = \dfrac{[H^+][HS^-]}{[H_2S]}$; $K_2 = \dfrac{[H^+][S^{2-}]}{[HS^-]}$

$$\therefore\ [HS^-] = \frac{K_1 \times [H_2S]}{[H^+]} = \frac{9 \cdot 1 \times 10^{-8} \times 0 \cdot 1}{0.3} = 3 \cdot 03 \times 10^{-8} \text{ mol dm}^{-8}$$

$$\text{or}\quad [S^{2-}] = \frac{K_2 \times [HS^-]}{[H^+]} = \frac{1.2 \times 10^{-15} \times 3.03 \times 10^{-8}}{0.3} = 1.21 \times 10^{-23} \text{ mol dm}^{-3}.$$

For CuS : $[Cu^{2+}] \times [S^{2-}] = 1 \times 10^{-44}$ or, $[Cu^{2+}] = 8.3 \times 10^{-22}$ mol dm^{-3}.

For CdS: $[Cd^{2+}] \times [S^{2-}] = 5 \times 10^{-25}$ or, $[Cd^{2+}] = 41 \times 10^{-2}$ mol dm^{-3}.

For MnS: $[Mn^{2+}] \times [S^{2-}] = 1.4 \times 10^{-15}$ or, $[Mn^{2+}] = 1.16 \times 10^{8}$ mol dm^{-3}.

Certainly this value of $[Mn^{2+}]$ appears fictitious as the original solution contained only 0.1 mol dm^{-3}. Actually this is the *maximum* concentration of Mn^{2+} ions that can exist at equilibrium in solution under these conditions. Since the supplied concentration is less than this value, the entire amount of Mn^{2+} ions will remain in solution.

Problem 11.37

50 cm^3 of a 0.1 M solution of $MgCl_2$ is mixed with 50 cm^3 of a 0.1 M solution of ammonia. How many grams of solid NH_4Cl must be added to this solution so that the precipitate of magnesium hydroxide may just disappear?

Assume K_S of $Mg(OH)_2 = 1.0 \times 10^{-11}$; $NH_3\ (aq) : K_b = 1.8 \times 10^{-5}$.

Solution: The precipitate of $Mg(OH)_2$ will just disappear when its ionic product is just equal to its solubility product, i.e.,

$$K_S = [Mg^{2+}] \times [OH^-]^2 = 1.0 \times 10^{-11}$$

Here $[Mg^{2+}] = 0.05$ M (volume has been doubled on mixing.)

$\therefore$ $[OH^-] = (1.0 \times 10^{-11}/0.05)^{1/2} = 1.41 \times 10^{-5}$ mol dm^{-3}.

Our next task will be to ascertain the amount of amonium chloride necessary for maintaining this concentration of OH^- ions.

Suppose x moles of NH_4Cl has to be added to the 100 cm^3 solution to maintain the desired value of $[OH^-]$. Then concentration of NH_4Cl will be $10x$ mol dm^{-3} (1 dm^3 = 1000 cm^3). In an aqueous solution of ammonia,

$$K_b = 1.8 \times 10^{-5} = [NH_4^+] \times [OH^-] / [NH_3(aq)].$$

In presence of NH_4Cl, the dissociation of ammonia will be largely repressed and we may take $[NH_3(aq)]$ to be the same as the concentration of ammonia in the mixture, i.e., 0.05 M. The concentration of NH_4^+ ions from ammonia may be similarly neglected, so that

$$[NH_4^+] = C_{NH_4Cl} = 10x \text{ mol dm}^{-3}.$$

Putting $[OH^-] = 1.41 \times 10^{-5}$, we get, $1.8 \times 10^{-5} = \dfrac{10x \times 1.41 \times 10^{-5}}{0.05}$ when, $x = 6.4 \times 10^{-3}$

$\therefore$ The amount of NH_4Cl to be added is $6.4 \times 10^{-3} \times 53.5$ g $= 0.34$ g.

ESSENTIALS

The concentration of hydrogen ion in aqueous solutions containing any single acid or base at an analytical concentration C_a or C_b (i.e., the concentration at which the solution is actually *prepared*) may be obtained as follows:

(i) Strong Acids: $[H^+] = C_a$ (when $C_a > 10^{-6}$ M)

(ii) Strong Bases: $[H^+] = 10^{-14} / C_b$ (when $C_b > 10^{-6}$ M).

$[H^+]$ approaches the value of 10^{-7} mol dm^{-3} as the solutions are gradually diluted (25°C).

(iii) Weak Acids: $[H^+] \approx (K_a \cdot C_a)^{1/2}$ [K_a = dissociation constant of the acid).

If $[H^+] < 0.05\ C_a$, the calculation is satisfactory.

If $[H^+] > 0.05\ C_a$, the value of $[H^+]$ obtained is approximate and one should use the quadratic equation for precision: $[H^+]^2 + K_a \cdot [H^+] - K_a \cdot C_a = 0$.

For very very weak acids ($K_a \simeq 10^{-9}$), $[H^+] = (K_a \cdot C_a + K_w)^{1/2}$, provided (i) $[H^+] \simeq [OH^-]$ and (ii) $[H^+] - [OH^-] < 0.05\ C_a$.

(iv) Weak Bases: $[OH^-] = (K_b \cdot C_b)^{1/2}$ for $[OH^-] < 0.05\ C_b$.

The value of $[OH^-]$ obtained should be less than 0.05 C_b for satisfactory results.

If $[OH^-] > 0.05\ C_b$, one should use the quadratic equation: $[OH^-]^2 + K_b [OH^-] - K_b . C_b = 0$.

(v) Diprotic acids (H_2A)

a. *Solution of* H_2A :

$$\underset{C-x}{H_2A} \overset{K_1}{\rightleftharpoons} \underset{x}{H^+} + \underset{x}{HA^-} \quad ; \quad HA^- \overset{K_2}{\rightleftharpoons} H^+ + A^{2-}$$

H_2A may be treated as a monoprotic acid with $K_a = K_1 = x^2/(C - x)$, where C is the formal concentration of H_2A. When $K_2 << K_1$, $[H^+] = [HA^-] \simeq x$ and $[A^{2-}] = K_2$

b. *Solution of* HA^- :

$$[HA^-] \simeq C \text{ (formal concentration)} \ ; \ [H^+] = \left(\frac{K_1K_2C + K_1K_w}{K_1 + C}\right)^{1/2} \approx (K_1K_2)^{1/2} \Rightarrow pH \approx (pK_1 + pK_2)/2$$

$$[H_2A] = [HA^-][H^+]/K_1 \qquad ([HA^-] \approx C); \ [A^{2-}] = K_2[HA^-]/[H^+]$$

c. *Solution of* A^{2-} :

$$\underset{C-x}{A^{2-}} + H_2O \overset{K_{b1}}{\rightleftharpoons} \underset{x}{HA^-} + \underset{x}{OH^-}$$

A^{2-} may be treated as monobasic, with $K_b = K_{b1} = K_w / K_{a2}$.

$$[H^+] = K_w/x; [H_2A] = K_w/K_{a1} = K_{b2}.$$

(vi) Aqueous solutions of salts react acidic or alkaline when the constituent cations and anions differ in their relative acidities and basicities. Salts of strong acids and strong bases give neutral solutions. For others, the pH may be calculated from their analytical concentration C and K_a (or K_b) of the weak acid (or base).

Strong acid-weak base: $\quad pH = \frac{1}{2}K_w - \frac{1}{2}pK_b - \frac{1}{2}\log C$

Weak acid-strong base: $\quad pH = \frac{1}{2}pK_w + \frac{1}{2}pK_a + \frac{1}{2}\log C$

Weak acid-weak base: $\quad pH = \frac{1}{2}pK_w + \frac{1}{2}pK_a - \frac{1}{2}pK_b$

Solutions containing aqua metal ions of high charge react acidic owing to equilibrium like

$$[Fe(H_2O)]^{3+} \rightleftharpoons [Fe(H_2O)_5OH]^{2+} + H^+ \qquad K_a = 6.3 \times 10^{-3}$$

$K_a = [H^+]^2/(C - [H^+]$ gives the concentration of H^+, where C is the analytical concentration of Fe^{3+}.

For feebly acidic cations like $Zn^{2+}(aq)$ ($K_a \simeq 10^{-10}$), $\quad pH = \frac{1}{2}pK_a - \frac{1}{2}\log C$

Buffer solutions consist of weak acids or bases with their conjugate counterparts. The pH of a solution containing a weak acid and its conjugate base is given by the Henderson equation

$$pH = pK_a + \log [\text{conjugate base}] / [\text{acid}].$$

Acid-base indicators are mostly organic weak acids or bases. They undergo an observable colour change with the change in hydrogen ion concentration of the medium near the equivalence point. An indicator (In) exists in solution with two forms at equilibrium : an acidic form (In_A) and a basic form (In_B) which differ in colour.

$$In_A \rightleftharpoons In_B + H^+ \qquad K = K_{In}$$

The effective colour-change interval of an indicator lies around the pH range: $pH = pK_{In} \pm 1$.

The solubility product principle: In a saturated solution of a sparingly soluble electrolyte A_xB_y ionizing as $A_xB_y \rightleftharpoons xA^+ + yB^-$, the solubility product principle states:

$$[A^+]^x \times [B^-]^y = K_S \text{ (solubility product; constant at constant temperature).}$$

If a saturated solution of a sparingly soluble salt is treated with an electrolyte *with a common ion*, the ionic concentrations will adjust themselves to maintain the constancy of the solubility product at that temperature and the excess salt will be precipitated. The addition of an electrolyte *with no common ion*, however, increases the ionic strength of the medium which leads to increased solubility.

The solubility product principle has been applied in qualitative and quantitative analysis for selective precipitation.

EXERCISE

1. Calculate the pH of (i) 10^{-3} *M* HCl, (ii) 10^{-3} M NaOH and (iii) 10^{-8} M HCl. Explain the basis of your calculation in each case.

2. Explain: (i) The pH of strong acids and strong bases approach the limiting value of 7.0 on large dilution.

(ii) The pH of a dilute solution of HCl would change slightly if some sodium chloride is dissolved in it.

(iii) At ordinary temperature, aqueous solutions with pH values less than 7.0 are considered acidic.

(iv) The pH of neutral water at 100°C is 6.5 ($\frac{1}{2}pK_w$ *at this temperature*).

(v) A solution of $Al(NO_3)_3$ reacts acidic.

3. Explain the mechanism of buffer action by a solution of ammonia containing ammonium chloride. What will be the change in pH of a buffer solution, 0.1 M in NH_3 and 0.1 M in NH_4Cl if 10.0 cm^3 of 0.1 M HCl is added to 100 cm^3 of the buffer? ($K_b = 1.8 \times 10^{-5}$).

Also calculate the change in pH when the same amount of the acid is added to 100 cm^3 of water.

4. Calculate the pH of the following solutions:

(i) 50 cm^3 of 0.1 N HCl is mixed with 50 cm^3 of 0.2 N NaOH.

(ii) 50 cm^3 of 0.1 N HCl is mixed with 50 cm^3 of 0.2 N aqueous NH_3.

(iii) 50 cm^3 of 0.2 N HCl is mixed with 50 cm^3 of 0.1 N aqueous NH_3.

[**Ans.** (i) 12.7, (ii) 9.25, (iii) 1.3]

5. 0.25 mole of solid NaOH is added to 1 dm^3 of a solution containing 0.1 M NH_4Cl. What are the major ionic and molecular species at equilibrium? Calculate the equilibrium concentrations of NH_3, NH_4^+ and OH^- in the resulting solution.

[**Ans.** $[NH_3] = 0.1$ M; $[NH_4^+] = 1.2 \times 10^{-5}$ M; $[OH^-] = 0.15$ M]

6. What happens to the pH of a solution of acetic acid (say 0.1 M) when sodium acetate is added to it at constant volume? Calculate the changes in pH when the concentrations of sodium acetate becomes (i) 0.01 M, (ii) 0.05 M and (iii) 0.2 M. ($K_a = 1.8 \times 10^{-5}$).

7. A 0.1 M solution of sodium acetate is 0.008% hydrolyzed at 25°C. Calculate K_w. ($K_a = 1.8 \times 10^{-5}$).

[**Ans.** 1.15×10^{-14}]

8. Calculate the concentrations of NH_3, NH_4^+, H_3O^+ and OH^- in a 0.1 M solution of ammonia. $K_b = 1.8 \times 10^{-5}$.

[**Ans.** $[OH^-] = 4.2 \times 10^{-4}$ mol dm^{-3}]

9. Calculate the concentrations of H_3O^+, HOAc, OAc^- and OH^- in a solution containing 0.1 mole of HCl and 0.1 mole of HOAc in 1 dm^3 solution. K_b for HOAc $= 1.8 \times 10^{-5}$.

10. You are supplied with 100 cm^3 of a 0.1 M solution of acetic acid. Calculate the pH of the solution. What will be the pH of the resulting solution when (i) 10.0 cm^3 (ii) 50.0 cm^3 (iii) 99.0 cm^3 (iv) 100.0 cm^3 and (v) 101.0 cm^3 of 0.1 M solution of NaOH is added to it? How would the results change if the acid and alkali are 0.01 M?

11. Calculate the pH of the following solutions.

(a) 1.39 g 2-nitrophenol in 250 mL water. ($K_a = 6.2 \times 10^{-8}$)

(b) 0.1 M solution of elthylamine ($K_b = 4.2 \times 10^{-4}$)

(c) 0.1 M solution of ethylammonium chloride (use K_b from *b*).

Hint: (a) F.W. of 2-nitrophenol = 139.11. $C = 0.4$ M. Let $[H^+] = x$.

$$\underset{C-x}{HA} \rightleftharpoons \underset{x}{H^+} + \underset{x}{A^-} \qquad K_a = \frac{x^2}{C-x}$$

(b) $$\underset{0.1-x}{CH_3CH_2NH_2} + H_2O \rightleftharpoons \underset{x}{CH_3CH_2NH_3^+} + \underset{x}{OH^-} \qquad K_b = \frac{x^2}{0\cdot 1 - x} \Rightarrow x = 0\cdot 065 = [OH^-]$$

(c) $$CH_3CH_2NH_3^+ \rightleftharpoons CH_3CH_2NH_2 + H^+ \qquad K_a = K_w/K_b = 2.4 \times 10^{-11}$$

If $[H^+] = x$, $\dfrac{x^2}{0.1-x} = K_a \Rightarrow pH = 5.8.$

12. What will be the pH of 0.01 M $Al(NO_3)_3$? K_a of $Al(H_2O)_6^{3+} = 1.4 \times 10^{-5}$.

Hint: $pH \simeq \frac{1}{2}pK_a - \frac{1}{2}\log C = \frac{1}{2}(4.85 - \log 0.01) = 3.4.$

13. (a) Calculate the pH of a 0.02 M solution of Histidine (H His).

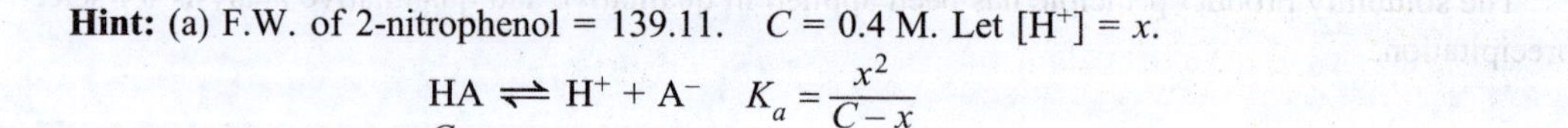

H_3His^{2+} $\quad$ H_2His^+ $\quad$ HHis $\quad$ His^-

Hint: H His is surrounded by pK_2 and pK_3 of triprotic H_3His^{2+}.

$$[H^+] = \sqrt{\frac{K_2K_3 \times 0.02 + K_2K_w}{K_2 + 0.02}} \Rightarrow 2.8 \times 10^{-8} M.$$

(b) What will be the pH when 50.0 mL of the solution in (a) is mixed with (a) 50.0 mL 0.02 M solution of a strong acid (b) 100.0 mL 0.02 M solution of a strong acid?

Hint: (a) The solution will contain H_2His^+ at formal concentration 0.01 M.

$$[H^+]=\sqrt{\frac{K_1K_2\times 0.01+K_1K_w}{K_1+0.01}}$$

(b) The solution contains H_3His^{2+} at formal concentration $\frac{50\times 0.02}{150}$ M = 0.0067 M.

$$\underset{0.0067-x}{H_3His^{2+}} \rightleftharpoons \underset{x}{H_2His^+} + \underset{x}{H^+} \quad K_1=\frac{x^2}{0.0067-x};\ [H^+]=x.$$

14. (a) How many grams of solid K_2CO_3 should be added to 1 g $NaHCO_3$ to obtain 500 mL solution at pH 10.5? H_2CO_3 : pK_1 = 6.4; pK_2 = 10.2 (see Problems 11.15 and 11.22.)

Hint: Let K_2CO_3 required = x g. F.W. of K_2CO_3 = 138.2; $NaHCO_3$ = 84.0.

$$pH=pK_2+\log\frac{[CO_3^{2-}]}{[HCO_3^-]}=10.2+\log\frac{(x/138.2)}{(1/84.0)}\Rightarrow x=3.28\,g.$$

(b) What will be the pH if 20 mL of 0.05 M HCl are added to 100 mL of the above solution?

Hint: Initial moles (in 100 mL) : $K_2CO_3=\frac{3.28}{138.2\times 5}=0.0047$; $NaHCO_3=\frac{1}{84\times 5}=0.0024$.

HCl added = 20 × 0.05 mmol = 0.0010 mol.

$$\underset{0.0047-0.0010}{CO_3^{2-}} + \underset{?}{H^+} \rightleftharpoons \underset{0.0024+0.0010}{HCO_3^-} \Rightarrow pH = 10.2+\log\frac{0.0037}{0.0034}=10.24$$

Final moles: 0.0047–0.0010 ? 0.0024 + 0.0010

15. A student titrates an unknown volume of a solution of an unknown weak acid (monoprotic) with a strong base—but the strengths of both are unknown. After adding 10.0 cm^3 of base, the solution attains a pH = 6, as shown by a pH-paper. At the equivalence point, the burette reading is 17.0 cm^3. What is the dissociation constant of the acid?

Hint: Suppose the initial solution contained x mole of the acid HA. So 17 cm^3 of base neutralizes x mole of HA at equivalence point.

∴ 10 cm^3 of base neutralized $(10x/17)$ mole of HA to form $(10x/17)$ mole of salt.

From Henderson equation, $pH = pK_a+\log\frac{[Salt]}{[Acid]}=pK_a+\log\frac{10x/17}{7x/17}$, when $pK_a = 6.8$.

16. (a) 2.05 g. of anhydrous sodium acetate is added to 100 cm^3 of 0.1 M HCl. What will be the pH of the solution? (b) If 10.0 cm^3 of 0.1 M HCl is now added to the above solution, what will be the final pH? [**Ans.** (a) 4.9]

17. Which one is greater—the solubility of silver chloride or the solubility product of silver chloride? Would AgCl precipitate if equal volumes of $AgNO_3$ and NaCl solutions, both 0.0002 M, are mixed?

18. Calculate the solubility of $Mg(OH)_2$ in pure water, given its solubility product, $K_s = 7.3\times 10^{-12}$. What will be the pH of a saturated solution of $Mg(OH)_2$ at the same temperature?

19. Calculate the solubility of CaF_2 in solutions of (i) 10^{-3} M $Ca(NO_3)_2$; (ii) 10^{-1} M NaF. $K_s = 1.7\times 10^{-10}$.

20. Excess of $Ca(OH)_2(s)$ is shaken with pure water. A 50.0 cm^3 sample of the clear saturated solution is titrated with 0.1 N HCl. Complete neutralization required 11.5 cm^3 acid. Calculate K_s for $Ca(OH)_2$ at this temperature. [**Ans.** $\simeq 6\times 10^{-6}$]

21. What is the solubility of $Al(OH)_3$ in a solution buffered at pH = 6 (ignore complex formation)? $K_s = 3 \times 10^{-34}$.

22. What concentration of NH_4^+ is needed to keep $Mn(OH)_2$ from precipitating with Fe^{3+}, Al^{3+} and Cr^{3+} from a solution containing 10 g Mn^{2+} per dm^3 when the solution is made 0.3 M in ammonia?
[K_s : $Fe(OH)_3 = 1.6 \times 10^{-39}$; $Al(OH)_3 = 3 \times 10^{-34}$; $Cr(OH)_3 = 7 \times 10^{-31}$; $Mn(OH)_2 = 1.6 \times 10^{-13}$]

23. Explain:

(i) Students often fail to obtain the precipitate of CdS in group II in qualitative analysis, though his test solution contains adequate Cd^{2+}.

(ii) NaCl and NaOH cannot be used in place of NH_4Cl and aq. NH_3 to precipitate the metals of group IIIA in qualitative analysis.

(iii) Sodium carbonate cannot be used instead of ammonium carbonate to precipitate the cations of group IV in qualitative group separation.

24. 30 cm^3 of 0.15 M $Pb(NO_3)_2$ solution is mixed with 20 cm^3 of 0.6 M KIO_3 solution. A precipitate of lead iodate, $Pb(IO_3)_2$ is obtained. What are the concentrations of Pb^{2+} and IO_3^- ions left in solution? Given $K_S = 7.9 \times 10^9$.

[**Hint:** 1 mol Pb^{2+} requires 2 mol IO_3^- for precipitation. The magniture of K_S suggest that nearly the entire Pb^{2+} will be precipitated. 4.5 mmol Pb^{2+} will bind 9 mmol of IO_3^-, leaving (12 – 9) or 3 mmol IO_3^- in 50 mL solution.

$$[IO_3^-] = 0.06 \text{ M} ; \quad [Pb^{2+}] = K_S\,[IO_3^-]^2.$$

25. How many moles of acetic acid must be added per liter to a 0.01 M solution of Na_2CO_3 to bring the pH to 6? Assume K (HAc) = 1.8×10^{-5}, K_1 and K_2 of $H_2CO_3 = 4 \times 10^{-7}$ and 6×10^{-11} respectively.

[**Hint:** At pH = 6, $[H^+] = 10^{-6}$. From K (HAc), $[Ac^-]/[HAc] = 1.8 \times 10^{-5} \div 10^{-6} = 18$. If C moles of HAc be added per liter, $\alpha C/C(1-\alpha) = 18$, or $\alpha = 0.95$. So $[Ac^-] = 0.95C$.

Similarly, from K_1 and K_2 of H_2CO_3, $[HCO_3^-]/[H_2CO_3] = K_1 / [H^+] = 0.4$ and

$$[CO_3^{2-}]/[HCO_3^-] = 6 \times 10^{-5}.$$

Evidently $[CO_3^{2-}]$ is negligible and material balance gives $[HCO_3^-] + [H_2CO_3] \simeq 0.01$ (total concentration of Na_2CO_3 taken). From $[HCO_3^-] = 0.4\,[H_2CO_3]$, $[H_2CO_3] = 0.0071$; $[HCO_3^-] = 0.0028$. Neglecting $[CO_3^{2-}]$ and $[OH^-]$, the simplified charge balance equation is $[H^+] + [Na^+] = [Ac^-] + [HCO_3^-]$ or, $10^{-6} + 2 \times 0.01 = 0.95C + 0.0028$ or $C = 0.018$.]

26. Calculate the concentration of all species present in a solution containing H_2SO_4, $NaHSO_4$, NaAc (sodium acetate) and NaOH, each at 0.1 *f* concentration. pK_2 of H_2SO_4 = 2; K (HAc) = 1.8×10^{-5}.

[**Hint:** Consider interactions among various components. 0.1 *f* H_2SO_4 + 0.1 *f* NaOH → 0.1 *f* $NaHSO_4$. But HSO_4^- is a stronger acid than HAc (from the pK values). So we write 0.1 *f* NaAc + 0.1 *f* $NaHSO_4$ → 0.1 *f* Na_2SO_4 + 0.1 *f* HAc. Final composition: 0.1 *f* $NaHSO_4$ + 0.1 *f* Na_2SO_4 + 0.1 *f* HAc. In presence of HSO_4^-, the dissociation of HAc may be neglected. Then we have a buffer containing $Na_2SO_4 - NaHSO_4$, both at equal concentration. Hence pH = pK $[HSO_4^-]$ = pK_2 (H_2SO_4). $[H^+] = 10^{-2}$.

From K(HAc), $[Ac^-]/[HAc] = 1.8 \times 10^{-3}$. Since HAc is very little dissociated, [HAc] $\simeq$ 0.1 *f*. $[Ac^-] = 1.8 \times 10^{-4}$ g ion dm^{-3}.

The various other concentrations are (gion dm^{-3}) : $[OH^-] = 10^{-12}$; $[Na^+] = 0.3$; $[SO_4^{2-}] = 0.1$; $[HSO_4^-] = 0.1$.

12

REACTIONS IN NON-AQUEOUS MEDIUM

12.1 INTRODUCTION

Most of the chemical reactions familiar to us take place in aqueous solutions. Water is unique as a solvent in view of its easy availability, long liquid range and excellent power to dissolve a wide range of substances. However, there are many reactions which cannot be carried out in aqueous solutions but can be studied in a non-aqueous medium. This is not only of laboratory interest, but is also important from an industrial point of view. So studies were initiated (since 1890's) for various non-aqueous solvents with both theoretical and practical approaches. A large number of non-aqueous solvents have been explored in the last few years; they range from very low temperature ones like liquid ammonia, hydrogen fluoride, sulfur dioxide or even liquid hydrogen cyanide and xenon to very high temperature molten salt solvents like fused potassium nitrate or sodium chloride.

The different physical and chemical processes occurring in a solvent medium are closely related to various types of interactions involving ions and/or dipoles constituting the solute and the solvent. The magnitudes of the forces involved in such interactions are largely determined by the dipole moment and dielectric constant of the solvent, which are of course interrelated.

These properties may be held responsible for the electrolytic behaviour of solvents. As such, these characteristics provide a convenient basis for the generalization of a large number of solvents.

The dipole moment of a solvent is directly related to its solvating ability. When a solvent has a large dipole moment, its molecules react readily with ionic or polar solutes and lower the force of attraction between the particles constituting the solute.

The force of attraction (F) between two opposite charges Q_1 and Q_2 placed at a distance r in a homogeneous insulating medium or *dielectric* is given by the Coulomb's law:

$$F = \frac{Q_1 \times Q_2}{4\pi \varepsilon r^2}$$

where ε is called the *permittivity* of the medium. It has the dimension of Farad per metre, $F\ m^{-1}$. The permittivity of free space, ε_0, has the value $8.854 \times 10^{-12}\ F\ m^{-1}$ (= $kg^{-1}\ m^{-3}\ s^4\ A^2$). The ratio $\varepsilon/\varepsilon_0$ is taken equal to ε_r, the *relative permittivity* or dielectric constant of a medium, which is a dimensionless quantity. It is a measure of the decrease in the intensity of electric field between two charges in presence of the dielectric compared to vacuum.

One may be interested to know how the intensity of electric field between two charges is reduced in the presence of a dielectric medium than in vacuum. Summarily, this takes place as a result of displacement of electric charges in the body of the dielectric. When the electric charges are placed in a dielectric medium composed of *polar* molecules, the molecules will arrange themselves in a preferential position in the field (*orientation polarization*). This will "absorb" a part of the electric field set up by the two charges. Hence, the greater the polarity of the molecules of a medium, the greater will be the decrease in electric field strength in that medium. Since polarity is revealed in the dipole moment of a substance while the lowering of field strength is related directly to its dielectric constant or relative permittivity, we may conclude that a large dipole moment for a solvent implies a large dielectric constant for it. This, in turn, results in significant lowering of the force operating between the constituent charged particles in a solute and hence greater solvating ability of the solvent.

Even when the molecules of a medium do not possess any permanent dipole, the intensity of the electric field will be still reduced in that medium, though slightly, due to displacement of charges in the molecules (*induced polarization*).

Dipole moments and dielectric constants for a few common solvents are given in Table 12.1 and Table 12.2. It may be observed from the tables that the relation between dipole moment and dielectric constants is not straightforward. Nevertheless, the conclusion that large dipole moments give rise to high dielectric constant of a solvent is generally valid. We reconfirm that the superiority of water as a solvent is due largely to its high dielectric constant.

Table 12.1

Dipole moments (μ) and dielectric constants (ε_r) of some typical solvents

1 Debye(D) = $3{\cdot}336 \times 10^{-30}$ Cm

Solvent	μ/D	ε_r	Solvent	μ/D	ε_r
Water	1.85	78.5 (25°C)	Ethyl alcohol	1.69	25.3 (25°C)
Hydrogen fluoride	1.91	83.6 (0°C)	Methyl alcohol	1.71	33.0 (20°C)
			Acetic acid	1.76	6 (25°C)
Ammonia	1.47	22.4 (–33°C)	Toluene	0.36	2.39 (20°C)
Sulfur dioxide	1.63	17.27 (–16.5°C)	Benzene	0	2.28 (20°C)

Classification of solvents

Solvents may be classified in general as polar or nonpolar. However, this classification is too broad and several other basis of classification have been proposed.

Protonic (hydronic) and nonprotonic (nonhydronic, aprotic) solvents: This classification is closely related to the Bronsted-Lowry concept of acids and bases. Protonic solvents yield a solvated proton on autoionization, such as water, ammonia and hydrogen fluoride. Benzene, carbon tetrachloride and liquid sulfur dioxide are examples of aprotic solvents.

Acidic and basic solvents: This is also based on the protonic theory of acids and bases. Accordingly, an acidic solvent is one having a strong tendency to proton (hydron) transfer, such as sulfuric acid and acetic acid. Basic solvents are marked by strong affinity for protons, e.g., ammonia and pyridine. Besides, we may have amphoteric solvents like water which can act both as a donor and an acceptor of protons. Solvents like benzene or carbon tetrachloride, which are inert to proton transfer may be called aprotic according to this scheme.

This is a useful scheme based on acid-base behaviour of a solvent. However, according to the protonic theory, acidity and basicity refer to specific reactions involved. So a solvent may be classified as acidic according to some reaction while it may be classified as basic in view of some other reaction.

Parent solvent Concept: This classification may be correlated to the solvent-system definition of acids and bases. Each solvent is characterized by a definite ionization scheme. The behaviour of analogous species, that is, the characteristic solvo-cations or solvo-anions, is then related to their respective parent solvents. The ionization schemes proposed for water, ammonia and sulfur dioxide are as follows:

$$H_2O + H_2O \rightleftharpoons H_3O^+ + OH^-$$

$$NH_3 + NH_3 \rightleftharpoons NH_4^+ + NH_2^-$$

$$SO_2 + SO_2 \rightleftharpoons SO^{2+} + SO_3^{2-}$$

We find that the OH^- group in the parent solvent H_2O is analogous to NH_2^- in the parent solvent NH_3 and SO_3^{2-} in the parent solvent SO_2. However, the above ionization scheme suggested for sulfur dioxide has been discarded on definite experimental grounds.

12.2 PROTONIC NON-AQUEOUS SOLVENTS

12.2.1 Liquid Ammonia

Liquid ammonia is one of the most extensively studied non-aqueous solvents.

Some of its important physical properties are:

Boiling point: –33.4°C; *Melting point*: –77.7°C.

Dielectric constant: 22.4 (at b.p.); *Specific conductance*: 5×10^{-9} ohm^{-1} m^{-1} (b.p.)

Density: 0.683 g cm^{-3} (239K) *Viscosity coefficient*: 0.254 centipoise (239.5K)

All values are considerably lower than corresponding values for water, mainly owing to weaker hydrogen bonding. Its self-ionization is also considerably lower than that of water:

$$2NH_3 \rightleftharpoons NH_4^+ + NH_2^-; \quad [NH_4^+][NH_2^-] \simeq 10^{-33} \ (\sim 50°C)$$

Solubility

The low value of the dielectric constant of ammonia (22.4) suggests a generally decreased ability to dissolve ionic compounds. Most ammonium salts are freely soluble in liquid ammonia; so also are many nitrites, nitrates, cyanides and thiocyanates. Salts containing highly charged ions, e.g., oxides, sulfides, sulfates, phosphates and carbonates are insoluble. Ionic compounds containing large, polarizable ions such as iodides and thiocyanates are quite soluble. For example, AgI, only sparingly soluble in water, is highly soluble in liquid NH_3 (207 g per 100 cm^3 at 25°C). The halides of a particular metal usually *become less soluble in going from iodide to fluoride* (for example, NaF 0.35, NaCl 3.0, NaBr 138.0, NaI 161.9 g per 100 g solvent at 25°C).

For some metal salts, solubility higher than that expected from dielectric constant alone may be observed. A stable ammine complex is formed in these cases, as with ions like Ni^{2+}, Cu^{2+} or Zn^{2+}.

Ammonia dissolves a variety of organic compounds: alcohols, amines, ethers, esters, halocarbons and other aromatics. In this sense it is a better solvent for nonionic and nonpolar substances.

Solutions of metals

The alkali metals dissolve in pure liquid ammonia without the evolution of H_2; the metals can be recovered from such solutions by evaporation. The saturation solubilities are quite high, e.g., at –33°C,

	Li	Na	K	Cs (–50°C)
Solubility (g/kg NH_3)	109	251	464	3335

Similar solutions are formed by Ca, Sr, Ba as well as the divalent lanthanoids europium and ytterbium. Solubility is favoured by low metal lattice energy, low ionization energy and high solvation energy of the cation. The electropositive metals having *aqueous* reduction potential more negative than –2.5V fall in this category.

The colour, electrical property and magnetic susceptibility of these solutions are interesting:

1. Colour: The solutions are all blue when dilute. At higher concentrations (>3M approx.) they become bronze-coloured and a metallic lustre appears.

2. Electric conductance: The solutions are extremely good conductors of electricity: they have higher conductivity than that of completely ionized salts in any known solvent. But the equivalent conductance *decreases* as the solutions become more concentrated and reaches a minimum around 0.05M. Beyond this concentration the conductivity *increases* dramatically sharply and the saturated solutions practically conduct electricity like liquid metals; specific conductivity of Hg $\simeq 10^4$ ohm^{-1}; Na in liq. NH_3 $\simeq 0.5 \times 10^4$ ohm^{-1}.

3. Density: The solutions, particularly the dilute ones, have much lower density than the pure solvent.

4. Magnetic property: The dilute solutions are paramagnetic with a susceptibility corresponding to one free electron per metal atom. But the paramagnetic susceptibility *decreases* with increasing concentration—around the region of minimum conductivity the solutions become *diamagnetic*. At still higher concentrations, the solutions again become weakly paramagnetic.

It is now generally agreed that these properties arise due to the presence of ammoniated metal cations and ammoniated electrons to various extent. At low concentrations, the equilibrium predominant is

$$M_{am} \rightleftharpoons M^+_{am} + e^-_{am} \; ; K \simeq 10^{-2}$$

The strongly polarizing electron orients ammonia molecules around it with the hydrogen atoms pointed inwards (Fig. 12.1). The formation of such cavities explains the lower density of the solution. The blue colour appears due to light absorption by the solvated electrons around 1,500 nm. The high conductivity arises from the quasi-free electrons—with mobility 280 times that of the cation. The electrons also account for the paramagnetic susceptibility.

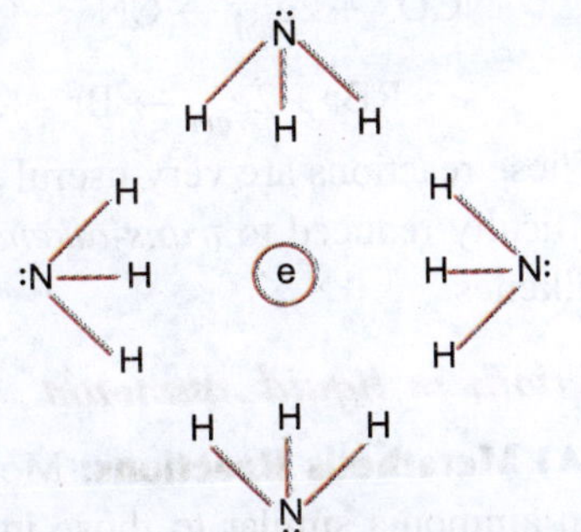

Fig.12.1: Sketch of an electron surrounded by ammonia molecules. Estimated diameter of the cavity = 300–340 pm.

As the concentration is raised, the ammoniated cations and ammoniated electrons are involved in various types of aggregations like

$$M^+_{am} + e^-_{am} \rightleftharpoons M_{am}$$

$$M_{am} + e^-_{am} \rightleftharpoons M^-_{am}$$

$$2M^+_{am} + 2e^-_{am} \rightleftharpoons (M^+_{am})_2 (e^-_{am})_2$$

These processes remove the conducting cations and electrons and account for the decrease in equivalent conductance. The formation of the dimers with appreciable spin-pairing causes the drop in paramagnetism. But in very concentrated solutions, (>1M), there are not enough ammonia to adequately solvate the electrons. The solution then conforms to the description $(M^+_{am})e$ and behaves like a molten expanded metal.

The solutions of alkali metals in liquid ammonia are not very stable; they readily form amides:

$$M + NH_3 \longrightarrow MNH_2 + \tfrac{1}{2}H_2$$

But in absence of catalytic impurities (transition metal ions) and under strictly anhydrous condition, such solutions are fairly stable. They are potential reducing agents—their reactions are virtually reactions of the ammoniated electron. Various types of reductions have been carried out, without or with bond cleavege;

Without bond cleavage:

$$O_2 + e^-_{am} \rightarrow O_2^- \; ; \qquad MnO_4^- + e^-_{am} \rightarrow MnO_4^{2-}$$

$$[Ni^{II}(CN)_4]^{2-} + 2e^-_{am} \rightarrow [\overset{0}{Ni}(CN)_4]^{4-}$$

With bond cleavage: One or two electrons may be involved initially. Subsequent ammonolysis may occur in a later step.

$$NH_3 + e^-_{am} \rightarrow NH_2^- + \tfrac{1}{2}H_2 \qquad S_8 + 2e^-_{am} \rightarrow S_8^{2-}$$

$$NH_4^+ + e^-_{am} \rightarrow NH_3 + \tfrac{1}{2}H_2 \qquad Mn_2(CO)_{10} + 2e^-_{am} \rightarrow 2[\overset{-1}{Mn}(CO)_5]^-$$

$$EtOH + e^-_{am} \rightarrow EtO^- + \tfrac{1}{2}H_2 \qquad C_6H_5NHNH_2 + 2e^-_{am} \rightarrow C_6H_5NH^- + NH_2^-$$

$$N_2O + 2e^-_{am} \rightarrow N_2 + O^{2-} \xrightarrow{NH_3} N_2 + OH^- + NH_2^- \text{ (ammonolysis)}$$

$$NCO^- + 2e^-_{am} \rightarrow CN^- + O^{2-} \xrightarrow{NH_3} CN^- + OH^- + NH_2^- \text{ (ammonolysis)}$$

$$RBr + 2e^-_{am} \rightarrow Br^- + R^- \xrightarrow{NH_3} Br^- + RH + NH_2^- \text{ (ammonolysis)}$$

These reactions are very useful in many organic synthesis. For example, alkynes are stereospecifically reduced to *trans-alkenes* while the conventional Pd/H_2 reduction would give only *cis*-alkenes.

Reactions in liquid ammonia

(A) Metathesis Reactions: Metathesis or double decomposition reactions may be carried out in ammonia similar to those in water. But the difference in solubility in the two solvents may lead to different products:

In water:	$KCl + AgNO_3 \longrightarrow AgCl\downarrow + KNO_3$
In ammonia:	$AgCl + KNO_3 \longrightarrow KCl\downarrow + AgNO_3$
In water:	$HgCl_2 + 2KOH \longrightarrow HgO\downarrow + 2KCl + H_2O$
In ammonia:	$3\,HgI_2 + 6KNH_2 \longrightarrow Hg_3N_2\downarrow + 6KI + 4NH_3$
In water:	$BaBr_2 + 2AgNO_3 \longrightarrow 2AgBr\downarrow + Ba(NO_3)_2$
In ammonia:	$Ba(NO_3)_2 + 2AgBr \longrightarrow BaBr_2\downarrow + 2AgNO_3$

The low solubility of alkali metal chlorides in liquid NH_3 may be used with advantage in preparative chemistry:

$$[Cr(NH_3)_6]\,Cl_3 + 3KBH_4 \longrightarrow [Cr(NH_3)_6]\,[BH_4]_3 + 3KCl\downarrow$$

(B) Acid-base Reactions: Such reactions in liquid ammonia may be understood on the basis of the autoionization: $2NH_3 \rightleftharpoons NH_4^+ + NH_2^-$. Accordingly, ammonium compounds behave as acids while the amides behave as bases in ammonia. Side by side, ammonia exerts its own levelling effect, in conformity with its highly basic nature. All acids which are strong acids in water, behave as strong acids in ammonia also:

$$HNO_3 + NH_3 \longrightarrow NH_4^+ + NO_3^-$$

Even the acids which are weak in water (pK_a up to about 12) react completely with ammonia and hence behave as strong acids in this medium:

$$CH_3COOH + NH_3 \longrightarrow NH_4^+ + CH_3COO^-$$

Molecules showing no acidic reaction in water may behave as weak acids in ammonia:

$$NH_2CONH_2 + NH_3 \rightleftharpoons NH_4^+ + NH_2CONH^-$$

Similarly, most species having basic character in water are either insoluble in ammonia or exhibit very weakly basic properties. Extremely strong bases may be levelled to the amide ion when they behave as strong bases:

$$H^- + NH_3 \longrightarrow NH_2^- + H_2\,;\; O^{2-} + NH_3 \longrightarrow NH_2^- + OH^-$$

As in water, the heat of neutralization between NH_4^+ and NH_2^- ions in ammonia is also constant, about 109 kJ mol^{-1}:

$$NH_4^+(am) + NH_2^-(am) \longrightarrow 2NH_3. \quad \Delta H° = -109 \text{ kJ mol}^{-1}.$$

Neutralization reactions may be run in ammonia parallel to those in water:

$$NH_4NO_3 + KNH_2 = KNO_3 + 2NH_3.$$

The course of the reaction may be followed by titration of a solution of ammonium nitrate by KNH_2 using phenolphthalein as indicator. With a slight excess of KNH_2, the solution changes from colourless to red as usual.

Amphoteric behaviour of some metals or metal compounds also parallels similar reactions in water.

Zinc dissolves in solutions of either ammonium salts or alkali metal amides in liquid ammonia, evolving hydrogen. Again, addition of potassium amide to a solution of zinc iodide results in a white precipitate of zinc amide which dissolves in excess KNH_2.

In water: $$Zn^{2+} + 2OH^- \rightarrow Zn(OH)_2 \downarrow \underset{H^+ \text{ ion}}{\overset{\text{excess } OH^-}{\rightleftharpoons}} [Zn(OH)_4]^{2-}$$

In ammonia: $$Zn^{2+} + 2NH_2^- \rightarrow Zn(NH_2)_2 \downarrow \underset{NH_4^+ \text{ ion}}{\overset{\text{excess } NH_2^-}{\rightleftharpoons}} Zn(NH_2)_4^{2-}$$

Problem 12.1

Write equations for the reactions of Zn^{2+} on gradual addition of KNH_2 in $NH_3(l)$ and on the addition of NH_4NO_3 to the resulting solution.

Ans:

$$ZnI_2 + 2KNH_2 = Zn(NH_2)_2 \downarrow + 2KI$$
$$Zn(NH_2)_2 + 2KNH_2 = K_2Zn(NH_2)_4$$
$$K_2Zn(NH_2)_4 + 2NH_4NO_3 = Zn(NH_2)_2 \downarrow + 2KNO_3 + 2NH_3.$$

Neutralization reactions in ammonia may also be followed by conductometric or potentiometric titrations. However, the speed of the NH_4^+ ion in ammonia is not as high as the speed of the H^+ (*aq*) ion in water. So the conductometric titration curve is less steep than in water.

A pH scale may be set for ammonia as follows:
pH = 0 (1M NH_4^+) ; pH = 13.5 when $[NH_4^+] = [NH_2^-]$; pH = 27 (1M NH_2^-).

***Solvolysis reactions* (*ammonolysis*)** are common in ammonia, some of which parallel those in water:

In water: $Cl_2 + 2H_2O \rightarrow HOCl + H_3O^+ + Cl^-$

In ammonia: $Cl_2 + 2NH_3 \rightarrow NH_2Cl + NH_4^+ + Cl^-$

In water: $OPCl_3 + 6H_2O \rightarrow OP(OH)_3 + 3H_3O^+ + 3Cl^-$

In ammonia: $OPCl_3 + 6NH_3 \rightarrow OP(NH_2)_3 + 3NH_4^+ + 3Cl^-$

Similarly, $SiCl_4 + 8NH_3 \rightarrow Si(NH_2)_4 + 4NH_4^+ + 4Cl^-$ (low temp.)

$$Si(NH_2)_4 \xrightarrow{0\ °C} Si(NH)(NH_2)_2 \xrightarrow{1200\ °C} Si_3N_4.$$

(C) Redox Reactions: $E°$ values in liquid ammonia (25°C) are as follows:

Acid solution (1M NH_4^+): $NH_4^+ + e = NH_3 + \frac{1}{2}H_2$ 0·0 V

$3NH_4^+ + \frac{1}{2}N_2 + 3e = 4NH_3$ −0·04 V

Basic solution (1M NH_2^-): $NH_3 + e = NH_2^- + \frac{1}{2}H_2$ 1·59 V

$2NH_3 + \frac{1}{2}N_2 + 3e = 3NH_2^-$ 1·55 V

The narrow range of 0.04 V in both acidic and basic solution is noteworthy; it implies that very few species would show any appreciable thermodynamic stability in $NH_3(l)$. Further, if all thermodynamically permissible reactions were kinetically rapid, N_2 would be the strongest oxidizing agent and H_2 would be the strongest reducing agent in this medium. However, overvoltages of about 1V extend the practical limit of reduction potential significantly: +1.0 to −1.0V in acid solution and 2.6 to 0.6 V in basic solution. This permits reactions by both strong reducing agents and strong oxidizing agents in $NH_3(l)$. Nitric acid or the nitrate ion is effectively inert towards ammonia in acid solution. In alkaline solution, nitrogen is slowly evolved:

$$3KNH_2 + 3NO_3^- \rightarrow 3KOH\downarrow + N_2 + NH_3 + 3NO_2^-$$

A few typical redox reactions in liquid ammonia are mentioned below:

Potassium permanganate oxidizes potassium amide to nitrogen:

$$6KMnO_4 + 6KNH_2 = 6K_2MnO_4 + 4NH_3 + N_2$$

Iodine, a weak oxidizing agent in liquid ammonia, oxidizes potassium ammonostannite to potassium ammonostannate.

$$K_4Sn(NH_2)_6 + I_2 = K_2Sn(NH_2)_6 + 2KI$$

The reducing action of metals in liquid ammonia solution are noteworthy. Thus,

(i) sodium reduces ammonium salts to liberate hydrogen:

$$2Na + 2NH_4Br = 2NaBr + 2NH_3 + H_2.$$

(ii) nitrous oxide is reduced to nitrogen

$$N_2O + 2K + NH_3 = KNH_2 + KOH + N_2$$

(iii) elemental sulfur is reduced to sulfide (and polysulfide): $S + 2Na = Na_2S$.

(iv) elemental oxygen is reduced to peroxide or superoxide, no cleavage of bonds between the oxygen atoms being involved. Lithium and sodium give mainly peroxides; others may be oxidized to both:

$$2K + O_2 = K_2O_2 \text{ (white)} \;;\; K_2O_2 + O_2 = 2KO_2 \text{ (yellow)}$$

(v) certain transition metal complexes may be reduced to a product containing the metal in an unusually low oxidation state:

$$K_2Ni^{II}(CN)_4 \xrightarrow[-33°C]{\text{Na-Hg or excess K}} \underset{\text{dark red; dimeric}}{K_4\left[Ni_2^{I}(CN)_6\right]} \longrightarrow \underset{\text{yellow}}{K_4[\overset{0}{Ni}(CN)_4]}$$

$[Pd(CN)_4]^{4-}$ and $[Pt(CN)_4]^{4-}$ can be similarly prepared. $[Co^{I}(CN)_4]^{3-}$ (pale yellow) and $[Co_2^{0}(CN)_8]^{8-}$ (brown-violet) have been prepared by reduction of $[Co(CN)_6]^{3-}$.

Some more examples of reduction at –33°C:

$$[Pt(NH_3)_4]Br_2 + 2K \longrightarrow [\overset{0}{Pt}(NH_3)_4] + 2KBr$$

$$Fe(CO)_5 + 2Na \longrightarrow Na_2[\overset{-2}{Fe}(CO)_4] + CO$$

Reduction of Ge, Sn, Pb and Se (M) by Na in liquid ammonia produces polyatomic anions like M_4^{2-}, M_5^{2-}, M_9^{2-} and M_9^{4-} (*Zintle anions or homopolyatomic anions*). It has been possible to isolate these anions as crystals formed with alkali crypt cations (see group 14).

12.2.2 Anhydrous Sulfuric Acid

Anhydrous sulfuric acid is a strongly acidic solvent of high dielectric constant. Some of the relevant physical characteristics are as follows:

Boiling Point: ~300°C; *Melting Point*: 10.4°C

Dielectric constant: 110 (20°C); *Specific conductance*: 1.04×10^{-2} ohm^{-1} cm^{-1} (25°C).

In addition, sulfuric acid is highly associated due to extensive hydrogen bonding and possesses very high viscosity (24.5 cP *vs* 1.01 cP for water at 20°C).

Solubility: The very high magnitude of dielectric constant suggests that sulfuric acid should be a very good solvent for ionic substances. In practice however, the very high viscosity (~ 25 times that of water) of the medium does not favour rapid diffusion of ions and dissolution or crystallization of solutes. Movement of ions through the solvent is unusually slow, except for $H_3SO_4^+$ and HSO_4^- ions which do not need to move physically through the solution. These latter ions can rapidly transfer H^+ ions through a H-bonded network (comparable to Grotthus type of conduction in water). Hence solutions containing $H_3SO_4^+$ or HSO_4^- ions have high electrical conductivity but solutions of other ions conduct electricity feebly. For the same reason, dissolution as well as crystallization processes are rather slow in pure H_2SO_4 as solvent.

Problem 12.2

In anhydrous H_2SO_4, molar conductivity of a solution of $KHSO_4$ is about half that of HNO_3. Comment.

Hint: Molar conductivities are determined almost entirely by the number of HSO_4^- ions released per mole of solute. Each HNO_3 releases two HSO_4^-.

$$KHSO_4 \rightarrow K^+ + HSO_4^- \; ; \; HNO_3 + 2H_2SO_4 \rightarrow NO_2^+ + H_3O^+ + 2HSO_4^-$$

N_2O_4 dissolves in sulfuric acid producing three HSO_4^- ions:

$$N_2O_4 + 3H_2SO_4 \rightarrow NO^+ + NO_2^+ + H_3O^+ + 3HSO_4^-$$

Raman spectrum of the solution shows peaks corresponding to the presence of NO^+ (2320 cm^{-1}), NO_2^+ (1400 cm^{-1}) and HSO_4^- (1050 cm^{-1}).

Acid-base reactions: Sulfuric acid undergoes extensive autoionization as

$$2H_2SO_4 \rightleftharpoons H_3SO_4^+ + HSO_4^- \qquad [H_3SO_4^+][HSO_4^-] = 1.7 \times 10^{-4} \text{ (10°C)}$$

A secondary reaction also occurs:

$$2H_2SO_4 \rightleftharpoons H_3O^+ + HS_2O_7^-.$$

These ions have only low mobility in the medium.

In keeping with its highly acidic nature and autoionization, H_2SO_4 offers a narrow range for discrimination of acids and bases. Any compound that can accept a hydron (proton) in any manner will bahave as a strong base. "Strong acids" in aqueous system will act as weak acids or even neutral or basic in sulfuric acid medium. Most of them are hydronated (protonated) with the formation of HSO_4^- ion, thereby behaving as base. Thus perchloric acid, one of the strongest acids known, becomes practically a nonelectrolyte and behaves as a weak acid. Nitric acid behaves as a base in sulfuric acid, as also do acetic and phosphoric acids. HCl remains undissociated.

$$HNO_3 + 2H_2SO_4 \longrightarrow NO_2^+ + H_3O^+ + 2HSO_4^-$$
$$H_3PO_4 + H_2SO_4 \longrightarrow H_4PO_4^+ + HSO_4^-$$
$$CH_3COOH + H_2SO_4 \longrightarrow CH_3C(OH)_2^+ + HSO_4^-$$

Problem 12.3

How do you expect ethanol (EtOH) to behave in H_2SO_4?

Hint: A strong base. $EtOH + 2H_2SO_4 \longrightarrow EtOSO_3H + H_3O^+ + HSO_4^-$

Water and other substances which are basic in water, are also bases in sulfuric acid:

$$H_2O + H_2SO_4 \longrightarrow H_3O^+ + HSO_4^-$$
$$OH^- + 2H_2SO_4 \longrightarrow H_3O^+ + 2HSO_4^-$$

Urea, which is a nonelectrolyte in water and an acid in ammonia, behaves as a base:

$$NH_2.CO.NH_2 + H_2SO_4 \longrightarrow HSO_4^- + NH_2CONH_3^+.$$

A solution of potassium hydrogen sulphate acts as a strong base in sulfuric acid medium. Species producing H_3O^+ or $H_3SO_4^+$ may be considered strong acids in H_2SO_4. Boric acid acts as a very strong acid in H_2SO_4.

$$B(OH)_3 + 6H_2SO_4 \longrightarrow 3H_3O^+ + [B(HSO_4)_4]^- + 2HSO_4^-$$

Disulfuric acid ("pyrosulfuric": $H_2S_2O_7$) is one of the strong acids in sulfuric acid:

$$H_2S_2O_7 + H_2SO_4 \rightleftharpoons H_3SO_4^+ + HS_2O_7^-$$

When B_2O_3 is dissolved in oleum, the ion tetrakis(hydrogensulfato)borate, $[B(HSO_4)_4]^-$ is formed extensively. Thus $HB(HSO_4)_4$ should be a strong acid in sulfuric acid:

$$B_2O_3 + 3H_2S_2O_7 + 4H_2SO_4 \longrightarrow 2H_3SO_4^+ + 2[B(HSO_4)_4]^-$$

Acid-base neutralization reactions in sulfuric acid can be followed conductimetrically. At the endpoint, conductivity virtually falls to zero owing to low ionic mobilities of the ions remaining in solution:

$$H_3SO_4^+ [B(HSO_4)_4]^- + KHSO_4 \longrightarrow K^+ [B(HSO_4)_4]^- + 2H_2SO_4.$$

Beyond the endpoint, addition of excess HSO_4^- or $H_3SO_4^+$ results in a sharp rise in conductivity.

The trend in solubilities of the common hydrogensulfates in sulfuric acid is similar to the trend shown by the solubilities of corresponding hydroxides in water:

Alkali metals > alkaline earth metals >> Mg.

All other salts undergo solvolysis in sulfuric acid in the same manner as salts of weak acids undergo hydrolysis in water. The "parent acids" of salts other than hydrogensulfates are very weak in sulfuric acid medium. Hence their anions are readily protonated and sulfuric acid is not normally suitable to study the chemistry of anions. On the other hand, sulfuric acid provides a good medium to study various cations. We have already mentioned the nitronium ion, NO_2^+ which is used as an active nitrating agent. Oxidation of iodine in H_2SO_4 produces I_3^+ and I_5^+. But the I_2^+ ion (blue) disproportionates:

$$8I_2^+ + 3HSO_4^- \longrightarrow 5I_3^+ + I(HSO_4)_3$$

Br_3^+ (brown) is also unstable in H_2SO_4. I_2^+, Br_2^+ and Br_3^+ can be produced in super-acid media like AsF_5–HSO_3F or SbF_5–SO_3–HSO_3F.

Se_8^{2+} is formed by the reaction of selenium with H_2SO_4 at 50°C; the ion can be further oxidized by SeO_2 to Se_4^{2+}. Tellurium reacts with H_2SO_4 to give red Te_4^{2+}. This is further oxidized to Te_6^{2+} in oleum.

A wide range of organic compounds dissolve in pure sulfuric acid which can be recovered unchanged by careful addition of water. In these compounds, the O, N, S etc. atoms containing lone pairs of electrons get protonated, the compounds thus acting as bases. Weak electron donors, such as nitrocompounds and nitriles also behave as weak bases in H_2SO_4 as these compounds are not completely protonated.

$$R_2C{=}O + H_2SO_4 \longrightarrow R_2C^+{-}OH + HSO_4^-$$

$$RNH_2 + H_2SO_4 \longrightarrow RNH_3^+ + HSO_4^-$$

$$\underset{\underset{OR'}{|}}{RC}{=}O + H_2SO_4 \longrightarrow \underset{\underset{OR'}{|}}{RC^+}{-}OH + HSO_4^-$$

$$\underset{\underset{O}{\|}}{RC}{-}NH_2 + H_2SO_4 \longrightarrow \underset{\underset{O}{\|}}{RC}{-}NH_3^+ + HSO_4^-$$

12.2.3 Anhydrous Acetic Acid (Ethanoic acid)

Anhydrous acetic acid provides an easily available water-like solvent which can be handled at room temperature in ordinary glass apparatus. Exclusion of moisture may be ensured in a glove-box under N_2-atmosphere. The anhydrous acid may be obtained from the 99.7% commercial acid by refluxing with acetic anhydride followed by fractional distillation from chromium(VI) oxide.

Physical properties of the solvent are given below:

Melting Point: 16.6°C; *Relative permittivity*: 6.2 (25°C);
Boiling Point: 118.1°C; *Specific conductivity*: ~$2.5 \times 10^{-9}\ \Omega^{-1}\ cm^{-1}$
Viscosity: 1.15 cP (25°C)

Low permittivity of acetic acid suggests that its molecules have weak polarity and hence provide little interaction with each other. Hence many covalent compounds are freely soluble in the acid, such as $SnCl_4$, $AsCl_3$ and H_2S. Polar molecules like water and ammonia may dissolve by H-bonding or protonation. Common anhydrous acids may dissolve through protonation of the solvent itself:

$$HClO_4 + CH_3COOH \longrightarrow CH_3C(OH)_2^+ + ClO_4^-$$

The low value of relative permittivity also suggests that metal salts should be insoluble or feebly soluble in acetic acid. However, many salts of large anions (and hence low lattice enthalpy) are fairly soluble, like the acetates, thiocyanates, cyanides and nitrates. Compounds insoluble in water are usually insoluble in acetic acid also, giving rise to precipitation reactions like the following:

$$AgNO_3 + KI \longrightarrow AgI\downarrow + KNO_3$$
$$AgNO_3 + KSCN \longrightarrow AgSCN\downarrow + KNO_3$$
$$ZnI_2 + Na_2C_2O_4 \longrightarrow ZnC_2O_4\downarrow + 2NaI$$

The very low specific conductivity of the pure solvent may be assigned to the weak self-ionization process:

$$2CH_3COOH \rightleftharpoons [CH_3C(OH)_2]^+ + CH_3COO^-$$

This is supported by rapid exchange between acetic acid and labelled (with C-14) potassium acetate. However, direct transfer of an H^+ between an acetate ion and a solvent molecule cannot be ruled out.

As expected, acetic acid reacts with bases like Na_2CO_3 or NaOH to produce salt (acetate) and water. Electropositive metals like sodium evolve hydrogen. Zinc shows amphoteric behaviour; at first it reacts slightly with the solvent to produce H_2 but the reaction soon stops owing to an insoluble coating of zinc acetate. Addition of acetate ion (solvo-base) dissolves the coating forming a soluble complex:

$$2CH_3COONa + Zn(CH_3COO)_2 \longrightarrow Na_2Zn(CH_3COO)_4$$

Neodymium triacetate also dissolves in potassium acetate forming a bluish acetato complex. In an electric field, the blue anionic complex migrates to the anode:

$$Nd(CH_3COO)_3 + nCH_3COOK \longrightarrow K_n[Nd(CH_3COO)_{3+n}]$$

As mentioned earlier, acetic acid, itself acidic in nature, exerts less levelling effect on protonic acids. In fact, it acts as a differentiating solvent for various "strong acids" (in water):

$$HClO_4 + CH_3COOH \rightleftharpoons CH_3CO_2H_2^+ + ClO_4^- \quad K = 1{\cdot}3 \times 10^{-5}$$
$$HCl + CH_3COOH \rightleftharpoons CH_3CO_2H_2^+ + Cl^- \quad K = 2{\cdot}8 \times 10^{-9}$$

The extent of protonation shows that the strengths of common inorganic acids fall in the order $HClO_4 > HBr > H_2SO_4 > HCl > HNO_3$.

The acidic nature of acetic acid results in levelling of "weak bases" (in water) to the base strength of acetate ion:

$$NH_3 + CH_3COOH \rightarrow NH_4^+ + CH_3COO^-$$

These "weak bases" can therefore be titrated in acetic acid medium against a strong acid such as perchloric acid. The progress of titration can be followed from the variation of electrical conductivity or of the potential difference between an indicator electrode and a reference electrode. The conductivity (or potential) decreases steadily upto the equivalence point:

$$CH_3COO^- + CH_3C(OH)_2^+ \longrightarrow 2CH_3COOH.$$

Beyond the equivalence point, addition of excess titrant increases the conductivity.

The constituents of a mixture of primary, secondary and tertiary aliphatic amines may be determined by potentiometric titration of three different aliquots of the mixture in acetic acid against perchloric acid.

Aliquot 1: Titrated directly to determine the total amine content.

Aliquot 2: Allowed to react with acetic anhydride for 3 hours (room temperature). The primary and secondary amines are acylated; titration gives the tertiary amine content.

Aliquot 3: Salicyladehyde is added to form a non-basic derivative with the primary amine only. The secondary and tertiary amines are now titrated.

12.3 APROTIC SOLVENTS

Aprotic solvents have either no hydrogen atoms or only noninteracting hydrogen atoms (usually nonacidic). According to their inherent character, they may be classified into four (sometimes three) categories. Starting from the lowest end of polarity, these are (i) nonpolar solvents (ii) polar, nonionizing solvents (iii) polar, ionizing solvents and (iv) molten ionic compounds.

Category I: Nonpolar solvents

This group consists of solvents which are nonpolar, essentially nonsolvating and which do not undergo autoionization. These include the hydrocarbons (aliphatic and aromatic) as well as the very feebly polar and weakly Lewis-basic chlorocarbons (CH_2Cl_2 and $CHCl_3$), carbon tetrachloride and carbon disulfide. The fats and lipids in biological systems also belong to this category. The intermolecular force in most of these solvents is essentially the feeble London dispersion forces of attraction. Accordingly, these solvents are also called *van der Waals solvents*. Both solvent-solvent interaction and solute-solvent interactions expected from these solvents are weak. Hence such solvents cannot dissolve salts and polar compounds but they can be good solvents for nonpolar hydrophobic compounds such as nonpolar organic compounds, organometallic compounds and uncharged complexes. Even charged complexes and ionic compounds can be rendered soluble in nonpolar solvents if the ions can be tagged with large hydrophobic groups. This will minimise the lattice energy which opposes dissolution. Tetraalkylammonium salts, $[R_4N]^+$ (R = methyl, ethyl and longer groups), are very important in this respect. Their usefulness may be illustrated by taking the case of potassium permanganate, an oxidizing agent frequently used in organic chemistry. $KMnO_4$ is often found insoluble in nonpolar organic solvents which might be the solvent of choice for the organic compound to be oxidized. Also, $KMnO_4$ is not highly soluble in water and even in such solutions the MnO_4^- ion remains hydrated when its oxidizing efficiency is lowered. Addition of a small, catalytic quantity of tetraalkylammonium cation will form some ion-pairs with

MnO_4^-. These ion-pairs, under the shield of the hydrophobic alkyl groups, can then dissolve in the organic solvent suitable for the organic compound. At the same time, the MnO_4^- ion is now available in an unhydrated, highly reactive form. The tetraalkylammonium salts function here as ***phase-transfer catalysts.***

One may also take the help of large hydrophobic ligands (like crown ethers and cryptates) to complex the metal cation in order to make it soluble in an organic solvent. Small concentrations of crown ethers may thus effectively function as phase-transfer catalysts.

Nonpolar solvents have prospective applications in analytical, environmental and biochemical studies. Metals present in very low concentrations in natural water may be selectively arrested by organic ligands to form an uncharged hydrophobic complex. This may be subsequently extracted with a nonpolar solvent free from other impurities in a much higher concentration suitable for estimation. Being nonpolar, such complexes have the tendency to accumulate in the fats and lipids of animal body as they are not readily excreted with watery urine. Such accumulation gets multiplied many times as one passes higher up in the food chain. This is known as *bioamplification* of the toxin.

A few examples of reactions carried out in nonpolar solvents are:

Hexane: $(C_2H_5)_6Al_2 + Sb_2O_3 \longrightarrow 2(C_2H_5)_3Sb + Al_2O_3$

CCl_4: $2HgO + 2Br_2 \longrightarrow Br_2O + HgO \cdot HgBr_2$

CS_2: $Se_8 + 4Cl_2 \longrightarrow 4Se_2Cl_2$

Category II: Polar non-ionizing solvents

These solvents do not undergo appreciable ionization but possess strong solvating ability. Liquid sulfur dioxide, dimethyl sulfoxide (DMSO, $(CH_3)_2SO$), dimethyl acetamide (DMA, $CH_3CON(CH_3)_2$) and acetonitrile (CH_3CN) are prominent examples of this category, as also are numerous organic compounds like ethers, ketones, esters, amines, amides and nitrocompounds. They contain an oxygen or nitrogen atom in a sterically prominent position and often function as Lewis-bases (hence called *Lewis-base solvents* or *coordinating solvents*). Most of them possess moderate to high dielectric constants (Table 12.2) and hence good solvent ability for ionic solutes. Their dipole moments also provide substantial ion-dipole contribution to the solvation energy. Since the smaller cationic part of the solutes (Lewis acid) provide higher solvation energy compared to the larger anions, these solvents behave primarily as Lewis bases. They may also behave as Lewis acids under suitable circumstances, depending mainly on the stable solute species. Thus, $FeCl_3$ dissolves in all such solvents in Table 12.2 when Fe^{3+} is coordinated by the solvent. The nature of the anion in such solutions, however, vary from one solvent to another. MeCN, $MeNO_2$, acetone, DMA and pyridine favour $FeCl_4^-$ as the anion. But in NMA and DMSO, the small Cl^- ion becomes the stable ion, the solvents acting as Lewis acids to this. NMA forms a hydrogen bond to the Cl^- ion from its N—H end (thus a Kamlet-Taft acid; sec. 10.1.3). DMSO can accept electrons from Cl^- in the vacant *d*-orbitals on the sulfur atom. Methanol also behaves as a Lewis acid to the Cl^- ion by forming a hydrogen bond.

Table 12.2
Some polar nonionizing solvents

Solvent	MP(°C)	BP(°C)	μ/D	ε_r
Pyridine (Py)	–42	115	2.19	12
Acetone	–95	56	2.84	20
Hexamethylphosphoramide* (HMPA)	7	232	5.37	30
Nitromethane, CH_3NO_2	–29	101	3.50	36
Dimethylformamide (DMF)	–61	153	3.82	37
Dimethylacetamide (DMA)	–20	165	3.81	38
Acetonitrile, CH_3CN	–48	82	3.92	39
Dimethylsulfoxide (DMSO)	18	189	3.96	45
Propylene carbonate (PC)	–55	240	4.98	69
N-methylacetamide (NMA)	28	206	3.73	165

*CARCINOGEN

Category III: Polar, ionizing solvents

Such solvents undergo slight autoionization. BrF_3, $POCl_3$ and $SbCl_3$ are examples of this type:

$$2BrF_3 \rightleftharpoons BrF_2^+ + BrF_4^- \qquad \kappa = 8 \times 10^{-3}\ \text{ohm}^{-1}\ \text{cm}^{-1}\ (25°C)$$
$$2POCl_3 \rightleftharpoons POCl_2^+ + POCl_4^- \qquad \kappa = 2 \times 10^{-8}\ \text{ohm}^{-1}\ \text{cm}^{-1}\ (25°C)$$
$$2SbCl_3 \rightleftharpoons SbCl_2^+ + SbCl_4^- \qquad \kappa = 8 \times 10^{-7}\ \text{ohm}^{-1}\ \text{cm}^{-1}\ (25°C)$$

The solvents have both Lewis acid and Lewis base properties (amphoteric) and are quite reactive. BrF_3, for example, reacts explosively with water and hydrocarbon grease.

As suggested by their autoionization schemes, these solvents act as chloride and fluoride ion transfer agents (*Ion-transfer solvents*) and hence they are strong chlorinating/fluorinating agents:

$$KF + BrF_3 \longrightarrow KBrF_4$$
$$SbF_5 + BrF_3 \longrightarrow [BrF_2^+][SbF_6^-]$$
$$SbCl_5 + POCl_3 \longrightarrow POCl_2^+ + SbCl_6^-$$

Some other reactions in BrF_3:

$$Au \xrightarrow{BrF_3} AuF_3\ ; \quad KCl + Ta \xrightarrow{BrF_3} KTaF_6$$

$$N_2O_4 + Sb_2O_3 \xrightarrow{BrF_3} [NO_2^+][SbF_6^-]$$

$POCl_3$ may also function simply as a coordinating solvent:

$$TiCl_4 + 2POCl_3 \longrightarrow TiCl_4(OPCl_3)_2 \rightleftharpoons [TiCl_3\ (OPCl_3)_3]^+\ Cl^-$$

Autoionization gives rise to characteristic solvo-acids and solvo-bases in such solvents. Their combination, actually the reverse of the autoionization written above, is equivalent to acid-base neutralization which can be followed by titration.

Category IV: Molten ionic salts

Molten salts are composed entirely of ions and have very high electrical conductivities. In such media, the charged ions are held in an ordered manner through coulomb interaction, though the average coordination number is less than that in a solid. In molten sodium chloride, for example, the coordination number of the ions is nearly four as against six in the solid. Though most common salts are known to melt at high temperatures, there are a few which are liquids near room temperature. Such compounds consist of very large ions so that the lattice energy is very low. *N*-butylpyridinium heptachloridoaluminate, $[C_5H_5NC_4H_9]^+ [Al_2Cl_7]^-$ is a typical example of such solvent (see room temperature ionic liquids). Both the cations and anions are weak as acids and/or bases in such solvents; this makes them useful for study of reactions between ions. Molten salts at higher temperatures also provide useful reaction media. Such studies require mostly silica cells in which the molten salt is excited by laser beam and the scattered Raman radiation is analysed. Some reactions studied in molten salts have been mentioned later.

A few typical aprotic solvents are discussed below.

12.3.1 Liquid Sulfur Dioxide

Liquid sulfur dioxide is one of the most thoroughly studied ion-transfer solvent which, however, does not undergo autoionization as expected (see below).

Some of the main physical properties of sulfur dioxide are:

Boiling point: −10.0°C *Dielectric constant*: 17.3 (−16.5°C)

Melting point: −75.5°C *Specific conductance*: 3×10^{-8} ohm^{-1} cm^{-1} (−10°C).

Solubility: Sulfur dioxide dissolves alkali metal halides, solubility decreasing from iodide to fluoride. For a particular halogen atom, solubility increases as the size of the alkali metal increases. Ionic halides of bivalent metals show low solubility. Cyanides, thiocyanates, acetates and sulphites of univalent metals are soluble to some extent. Many covalent halides like BCl_3, $AlCl_3$ and organic compounds (amines, ethers, alcohols etc.) are also soluble in sulfur dioxide.

Precipitation reactions: Many compounds may be precipitated in liquid sulfur dioxide when the requisite soluble compounds are mixed. A few representative equations are:

$$BaI_2 + Zn(CNS)_2 = Ba(CNS)_2 \downarrow + ZnI_2$$
$$AlCl_3 + 3NaI = 3NaCl \downarrow + AlI_3$$
$$2NH_4SCN + SOCl_2 = 2NH_4Cl \downarrow + SO(SCN))_2$$
$$2KI + SOCl_2 = 2KCl \downarrow + SOI_2$$

The last two reactions and other similar ones may be used to prepare various thionyl derivatives.

Solvolysis reactions. Various solvolytic reactions have been observed in sulfur dioxide. Reactions very similar to that in water are:

$$PCl_5 + SO_2 = POCl_3 + SOCl_2 \;; \quad NbCl_5 + SO_2 = NbOCl_3 + SOCl_2$$

Redox reactions: Liquid sulfur dioxide does not exhibit any pronounced oxidizing or reducing behaviour. A few redox reactions have been studied in this medium, e.g.,

$$2KI + 2FeCl_3 = 2KCl + 2FeCl_2 + I_2$$
$$6KI + 3SbCl_5 = 2K_3[SbCl_6] + SbCl_3 + 3I_2$$

Acid-base reactions: Acid-base reactions in sulfur dioxide were originally interpreted in terms of its self-ionization as

$$SO_2 + SO_2 \rightleftharpoons SO^{2+} + SO_3^{2-}$$

This scheme is supported by (i) conductance and (ii) neutralization reactions like

$$\underset{\text{acid}}{SOCl_2} + \underset{\text{base}}{CsSO_3} = \underset{\text{salt}}{2CsCl} + \underset{\text{solvent}}{2SO_2}$$

In fact, this neutralization reaction may be followed conductometrically. The fact strongly supported the ionization equilibrium written above.

However, if the above mode of self-ionization were true, exchange of ^{35}S or ^{18}O would certainly take place when $SOCl_2$ or $SOBr_2$ containing ^{35}S or ^{18}O are added to liquid SO_2. Writing $\overset{*}{S}$ for ^{35}S, one may write

$$2SO^{2+} + 2SO_3^{2-} + \overset{*}{S}O^{2+} + 2Cl^- \rightleftharpoons SO^{2+}.SO_3^{2-} + \overset{*}{S}O^{2+}.SO_3^{2-} + SO^{2+} + 2Cl^-.$$

Contrary to this expectation, no exchange is found to take place. On the other hand, rapid exchange of radioactive chlorine takes place between $SOCl_2$ and an ionic chloride in liquid sulfur dioxide medium. This suggests that the thionyl halides probably ionize as

$$SOX_2 \rightleftharpoons SOX^+ + X^-$$

Rapid exchange of radioactive sulfur between $SOCl_2$ and $SOBr_2$ also supports this view. The actual manner in which pure liquid sulfur dioxide ionizes, if at all, is yet unsettled.

In addition to the acid-base reactions shown above, one also finds amphoterism in the following series of reactions in liquid SO_2:

$$2AlCl_3 + 3SO_3^{2-} \longrightarrow Al_2(SO_3)_3\,(s)\downarrow + 6Cl^-$$

$$Al_2(SO_3)_3\,(s) + 3SO_3^{2-} \longrightarrow 2Al(SO_3)_3^{3-}$$

$$2Al(SO_3)_3^{3-} + 3SOCl_2 \longrightarrow Al_2(SO_3)_3\,(s)\downarrow + 6Cl^- + 6SO_2$$

The reactions parallel corresponding reactions of $Al^{3+}(aq)$ ions in water, though the acid-base ions SO^{2+} or SO_3^{2-} do not apparently exist in the solvent medium.

Liquid SO_2 also promotes oxide and chloride ion transfer among solutes:

$$SO_2 + (C_2H_5)_2Zn \longrightarrow ZnO + (C_2H_5)_2SO$$

$$2UCl_5 \longrightarrow UCl_4 + UCl_6$$

$$AsCl_3 + 3KF \longrightarrow AsF_3 + 3KCl$$

12.3.2 Reactions in Molten Salts

The use of molten salts as solvents dates back to ancient time when fluxes and slag-formers were used in metal extraction. At present molten salts provide a useful medium for study of chemical reactions, mostly through spectroscopic studies in silica cells. Though a few salts are liquid at or near room temperature, most salts require a high temperature to melt. Some eutectic salt-mixtures melt at lower temperatures (Table 12.3).

Table 12.3

Melting points (°C) of some salts and eutectic mixtures

LiCl	605	LiCl–KCl	350
NaCl	801	$LiNO_3$–KNO_3	125
KCl	770	$NaNO_3$–KNO_3	218
$LiNO_3$	264	$LiNO_3$–$NaNO_3$–KNO_3	120
$AlCl_3$	190 (2.5 atm)		

$HgCl_2(l)$ is also traditionally included as a molten salt though its electrical conductivity is quite low, ~10^{-3} ohm^{-1} cm^{-1}.

Metals and metalloids often dissolve in their molten halides forming two different types of solutions. Strongly electropositive metals (alkali/alkaline earth) form coloured solutions in their halides when the electrical conductivity increases many times, attributed largely to electronic rather than ionic conductance. Less electronegative metals (e.g., Cd) and metalloids dissolve in their metal halides with little or no increase in conductivity. Reduction may occur with the possibility of formation of a subhalide in some cases. Most subhalides have small clusters of metal atoms in a charged species. Thus Hg dissolves in $HgCl_2$ to form Hg_2^{2+} while Bi dissolves in $BiCl_3$ to give Bi_9^{5+}, $BiCl_5^{2-}$ and $Bi_2Cl_8^{2-}$.

$$M + MCl_3 \longrightarrow M^{2+} \quad (M = Nd, Sm, Eu)$$

$$Bi + BiCl_3 \longrightarrow Bi_9^{5+}$$

Reduction may also be carried by an electropositive metal, e.g.,

$$Cs_2UCl_6 + 2Mg \xrightarrow[550\ °C]{CsCl-MgCl_2} U\downarrow + 2MgCl_2 + 2CsCl$$

Divalent M^{II} species of many metals can be studied in molten LiCl–KCl eutectic.

$$M + LiCl - KCl\ (41:59) \xrightarrow[450\ °C]{\text{anodic oxidation}} M^{II} \quad (M = V, Cr, Fe)$$

The metals are probably formed as octahedral MCl_6^{4-} ions. V^{II} or Cr^{II} in such melts may be estimated by potentiometric titration using Fe^{III}.

$$Cr^{II} + Fe^{III} \longrightarrow Fe^{II} + Cr^{III}.$$

A variety of cations have been prepared in fused $AlCl_3$. The linear Hg_3^{2+} ion is formed by dissolving Hg in an $HgCl_2$–$AlCl_3$ melt. Yellow crystals of $Hg_3(AlCl_4)_2$ are obtained after evaporation of excess mercury and solvent. Polyatomic cations of cadmium and bismuth are formed by reducing the metal chlorides in $AlCl_3$ by the respective metals:

$$Cd + CdCl_2 \xrightarrow{AlCl_3} Cd_2\ (AlCl_4)_2$$

$$Bi + BiCl_3 \xrightarrow{AlCl_3} Bi_5\ (AlCl_4)_3 + Bi_8\ (AlCl_4)_2$$

Cations of non-metals may also be formed in molten salts. A solution of iodine in NaCl–$AlCl_3$ rich in $AlCl_3$ may be oxidized by chlorine to produce I_2^+.

Silver nitrate is very soluble in the low melting $LiNO_3$–KNO_3 eutectic. The silver is completely precipitated by addition of various anions:

$$Ag^+ + KX \longrightarrow AgX\downarrow \ (X = Cl, Br, I)$$
$$Ag^+ + KCN \longrightarrow AgCN\downarrow$$
$$Ag^+ + K_2CrO_4 \longrightarrow Ag_2CrO_4\downarrow$$

Ion-transfer reactions, mainly chlorination and fluorination are quite common in molten salts.

$HgCl_2$: $NaCl + PbCl_2 \longrightarrow Na_2PbCl_4$

KCl–$CuCl_2$ (3:1): $CuCl_2 + 3KCl + 3F_2 \longrightarrow K_3CuF_6 + Cl_2$

KF–KCl–$ZnCl_2$: $SOCl_2 + 2NaF \longrightarrow SOF_2 + 2NaCl$

$$(CH_3)_3SiCl + NaF \longrightarrow (CH_3)_3SiF + NaCl$$
$$CCl_4 + KF \longrightarrow CCl_{4-n}F_n$$

Since a molten salt consists of its constituent ions, the idea of solvoacid cation and solvobase anion loses its sense, since that would mean that the melt contains acidic and basic ions without undergoing any reaction. The Lux definition of acid-base criterion on the basis of oxide ion transfer becomes more significant in molten salts.

$$\underset{\text{base}}{CaO} + \underset{\text{acid}}{SiO_2} \longrightarrow \underset{\text{salt}}{CaSiO_3}$$

Melts involving oxoanions also show slight dissociation of oxide ion. In molten $NaNO_3$, the following reactions have been characterized

$$NO_3^- \rightleftharpoons NO_2^+ + O^{2-}$$
$$Cr_2O_7^{2-} + CO_3^{2-} \longrightarrow 2CrO_4^{2-} + CO_2$$

Molten salt media find wide applications in industry. We are already familiar with the large scale production of metals like sodium and aluminium. Preparation of fluorine and fluorocarbons or chlorofluorocarbons is also achieved in molten salts. On a large scale, methane may be blown through a melt of CuCl–$CuCl_2$–KCl at 370°– 450°C to obtain chlorinated methanes. Periodic regeneration of the melt is done by chlorine or hydrogen chloride. Under similar conditions, ethane gives chloroethene (monomer of vinyl chloride) as one of the products.

Elimination reactions involving organic compounds are also done with advantage in molten salts. Ethanol, when passed through an NaCl–$ZnCl_2$ melt, is largely (~98%) converted to ethene. A similar high yield conversion of $SiCl_4(g)$ to SiH_4 is possible in LiH dissolved in LiCl–KCl at 400°C

$$LiH + SiCl_4 \longrightarrow SiH_4 \ (100\%)$$

When the LiH is used up, the melt is electrolysed to produce lithium metal. Hydrogen is next passed into the melt to form LiH.

Room Temperature Ionic Liquid (RTIL)

Certain salts having large interionic separation (and consequent low lattice energy) remain in molten condition near room temperature (25°C). $[C_5H_5NBu]^+Al_2Cl_7^-$ has already been mentioned as an example of such salts. Addition of aluminium chloride (solid) to alkyl pyridinium chloride (solid) produces another room temperature ionic liquid (RTIL):

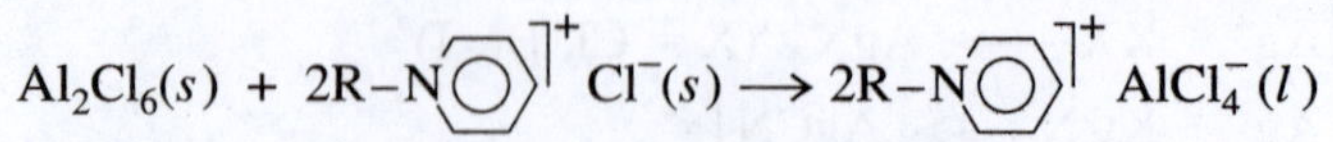

Similarly, cuprous chloride and triethylammonium chloride, both solids, may be mixed to produce an ionic liquid (RTIL):

$$Et_3NH^+Cl^-(s) + CuCl(s) \longrightarrow [Et_3NH][CuCl_2]\ (l)$$

The liquids are very sensitive to moisture which will hydrolyze the haloanion. A number of other organic cations like alkyl ammonium, phosphonium, pyrazinium and imidazolium ions (12-I, 12-II) may be combined with suitable anions like $CF_3SO_3^-$, $CH_3SO_3^-$, tosylate (*p*-toluenesulfonate), octyl sulfate or even Cl^-, Br^-, NO_3^-, BF_4^-, ClO_4^- etc. to produce suitable room temperature ionic liquids. Being nonvolatile and non-flammable, these compounds can replace volatile organic solvents which cause environmental pollution. By suitable combination of cations and anions, one can prepare a number of *tailor-made solvents* which are stable to air and moisture and can dissolve a wide range of organic and inorganic solutes. They have potential uses in various organic synthesis and separations as they are immiscible with organic nonpolar solvents and also with water. Organic ionic liquids containing halogen free anions appear to be promising *green solvents* for pollution-free chemical procedures.

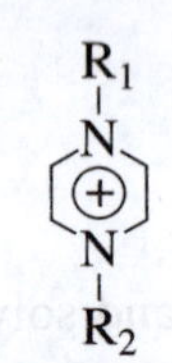

12-I
Pyrazinium cation

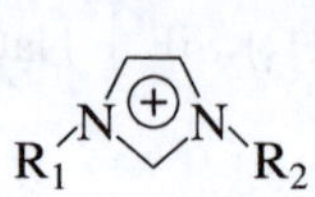

12-II
Imidazolium cation

12.3.3 Liquid Alkali Metals

The alkali metals permit a wide liquid range for study as solvent (Table 12.4). The use of liquid sodium to cool fast nuclear reactors is quite established. But study of different solute species in molten alkali metals becomes difficult since these are opaque to electromagnetic radiation. Cryoscopy, change of electrical conductance and analysis of various precipitates obtained are some of the indirect methods used to study such compositions. The molten metals are highly reactive, rapidly corroding glass at high temperature by attacking the oxide ions. At low temperature, Pyrex glass can be used for the alkali metals except lithium, but stainless steel apparatus is required at higher temperatures. Studies have to be carried out under a perfectly dry inert atmosphere, preferably argon. Precipitates may be filtered by glass-wool plugs or sintered steel discs. Excess solvent may be evaporated by vacuum distillation.

Table 12.4

Some physical properties of molten alkali metals

	Li	Na	K	Rb	Cs
M.P. (°C)	181	98	63	39	29
B.P. (°C)	1317	883	754	688	671
η (cP, 200°C)	0.57	0.45	0.30	0.35	0.35

The molten alkali metals may be simply looked as the monatomic ions in a vast sea of electrons. The electrons actually occupy numerous molecular orbitals delocalized over the entire liquid, comparable to the conduction bands of solid metals. Chemical reactions in such medium involve addition or removal of electrons of such conduction bands.

$$Ba \xrightarrow{Na(l)} Ba^{2+} + 2e \quad \text{(addition to conduction band)}$$

$$O_2 + 2e \xrightarrow{Na(l)} 2O^{2-} \quad \text{(removal from conduction band)}$$

Solvation and metathetic reactions are common in liquid alkali metals. When oxygen is passed into liquid sodium containing dissolved calcium, calcium oxide is precipitated:

$$2Ca + O_2 \xrightarrow{Na\,(l)} 2CaO \downarrow$$

The same reaction occurs between sodium oxide and calcium in solution. Only lithium forms a nitride with nitrogen.

Graphite reacts with liquid alkali metals to give a precipitate which gives acetylene (ethyne) on hydrolysis. This suggests the existence of C_2^{2-} ions in the medium. In molten lithium, these ions may be made to react with nitrogen to produce lithium cyanamide, $Li_2(NCN)$.

$$C\,(\text{graphite}) + N_2 \xrightarrow{Li\,(l)} Li_2\,(N=C=N)$$

In other reactions, polyatomic ions are rather split in liquid lithium:

$$OH^- \rightarrow O^{2-} + H^- \qquad 2CO_3^{2-} \rightarrow 6O^{2-} + C_2^{2-}$$

$$NO_3^- \rightarrow O^{2-} + N^{3-} \qquad 2CN^- \rightarrow C_2^{2-} + N^{3-}$$

Such reactions are actually dependent on the alkali metal. OH^- ions are stable in molten K, Rb or Cs. Surprisingly, when small amounts of water are added to liquid sodium, no vigorous reaction is observed (contrary to our experience of adding solid sodium metal to water). The changes taking place quietly may be represented as

$$H_2O + 2e \longrightarrow OH^- + H^-$$

$$OH^- + 2e \rightleftharpoons H^- + O^{2-}$$

ESSENTIALS

For convenience, various non-aqueous solvents may be broadly classified into: **(i) Protonic** (hydronic) and **(ii) Non-protonic or aprotic** solvents. Protonic solvents yield a solvated proton on autoionization, such as water, ammonia and hydrogen fluoride. Benzene, carbon tetrachloride and liquid sulfur dioxide are examples of aprotic solvents.

Liquid ammonia dissolves many inorganic and organic compounds. The alkali and alkaline earth metals except beryllium are highly soluble in liquid ammonia, forming blue to bronze coloured paramagnetic conducting solutions. These solutions contain ammoniated electrons.

The reactions in liquid ammonia parallel those in water. They are mostly interpreted in terms of self-ionization as $2NH_3 \rightleftharpoons NH_4^+ + NH_2^-$. Accordingly, ammonium salts act as acids in liquid NH_3 and amides behave as bases.

Anhydrous H_2SO_4 undergoes autoionization as $H_2SO_4 + H_2SO_4 \rightleftharpoons H_3SO_4^+ + HSO_4^-$.

Strong acids in aqueous medium mostly behave as moderate to strong bases in H_2SO_4. Boric acid acts as a very strong acid in H_2SO_4:

$$B(OH)_3 + 6H_2SO_4 \longrightarrow 3H_3O^+ + [B(HSO_4)_4]^- + 2HSO_4^-$$

H_2SO_4 provides a good medium for study of various cations like I_3^+, Se_8^{2+}, Te_4^{2+} etc.

Aprotic solvents may belong to several categories: (i) nonpolar, (ii) polar ionizing (iii) polar, nonionizing and (iv) molten ionic salts.

Liquid SO_2 is a widely studied aprotic solvent. It does not autoionize into SO^{2+} and SO_3^{2-}, as thought earlier. However, the sulfite ion behaves as a base and thionyl halides behave as acid in liquid SO_2.

Molten salts (and salt mixtures) also provide an interesting reaction medium. Various cations may be formed in such mixture. Ion transfer reactions, mainly chlorination and fluorination are also common. Certain large cations, mostly with organic moieties, produce room temperature ionic liquids (RTIL) which may be used with advantage in various organic synthesis. Liquid alkali metals also make good reaction medium.

EXERCISE

1. "Liquid ammonia is a water-like solvent"—justify with reference to different types of reactions studied in liquid ammonia.
2. Briefly describe the characteristics of the solutions of alkali and alkaline earth metals in liquid ammonia.
3. Explain:
   (a) Nonpolar compounds are usually soluble in strongly polar solvents.
   (b) A concentrated solution of sodium in liquid ammonia conducts electricity comparable to metals.
   (c) Both acetic acid and ammonium chloride act as strong acids in liquid ammonia.
   (d) "Nitric acid" appears as a base in anhydrous sulfuric acid.
4. Give a concise account of various reactions studied in liquid sulfur dioxide.
5. (a) Zinc amide is insoluble in liquid ammonia, but dissolves when ammonium acetate is added—Explain.
   Hint: CH_3COONH_4 is an acid in liq. NH_3. $Zn(NH_3)_4^{2+}$ is formed in solution.
   (b) Zinc acetate, insoluble in acetic acid, dissolves in presence of ammonium acetate—Explain.
   (c) Predict the effect of adding (i) Li_3N (ii) CH_3COOK to liquid ammonia.
   Hint: (i) $N^{3-} + 2NH_3 \rightarrow 3NH_2^-$: acidity decreases, (ii) no effect on acidity.
6. KOH is much more soluble in liquid ammonia containing water than in pure liquid ammonia—Explain.
   Hint: Strong H-bonding between OH^- and H_2O.
7. Suggest a method for the removal of unreacted sodium metal in liquid ammonia.
   Hint: Addition of NH_4Cl.
8. Explain:
   (a) Electrical conductivity and paramagnetism of solutions of alkali metals in liquid ammonia initially decrease with increase in concentration of the metal but again increase at higher concentration.
   (b) Addition of aqueous silver nitrate to aqueous barium bromide precipitates silver bromide but in liquid ammonia, barium nitrate reacts with silver bromide to precipitate barium bromide.
   (c) H_2SO_4 is not a good solvent for ionic solutes though it has a high dielectric constant.

(d) The acidity of pure H_2SO_4 is greatly enhanced by the additon of H_3BO_3.

(e) KF can be titrated conductimetrically by SbF_5 in BrF_3.

(f) Addition of K_2SO_3 to a solution of $AlCl_3$ in liquid SO_2 forms a white precipitate which dissolves excess K_2SO_3. The precipitate reappears on adding $SOCl_2$.

(g) Cuprous chloride and triethylammonium chloride, both solids, react to produce a liquid (at room temperature) which conducts electricity.

(h) Ammonia may be titrated with perchloric acid conductimetrically in anhydrous acetic acid medium.

9. Among $HgCl_2(aq)$ and $Hg(CH_3)_2(l)$, which one is expected to readily cross the blood-brain barrier and which one is expected to be readily excreted through urine?

Hint: $HgCl_2(aq)$ will be readily excreted through urine.
$Hg(CH_3)_2$ (l) is nonpolar. (It is also likely to undergo bioamplification).

13

RADIOACTIVITY

13.1 INTRODUCTION

Chemical changes involve rearrangement of the extranuclear part of the atom—electrons are lost, gained or delocalized in such changes while the nuclei of the atoms remain intact. Side by side, there are phenomenon like radioactivity and nuclear fission in which the nucleus undergoes a drastic change. Studies on these changes provide useful information on the nucleus, its role in the formation of different elements and the vast stock of nuclear energy.

Discovery: Radioactivity was discovered by Henri Becquerel in 1896. In 1895, Röentgen had detected x-rays by the fluorescence caused by the rays on certain materials. Becquerel was investigating if fluorescent substances could be made to emit x-rays on irradiation. He used a crystal of potassium uranyl sulfate as the fluorescent material. On exposure to sunlight, it emitted radiations which could penetrate paper, glass and aluminium sheets and ionize gases. But after a couple of cloudy days, Becquerel unexpectedly observed that the uranium salt emitted the same radiations even in darkness. It was established that uranium salts protected from all exciting radiations still emitted penetrating radiations. The nature and origin of these radiations were investigated, among others, by Becquerel, Rutherford, Pierre Curie and Marie Curie. It was found to be a property of the element uranium, not related to its physical or chemical state. The atoms of uranium changed into atoms of another element (thorium) during this change. Similar activity was also observed with thorium compounds.

The subject developed rapidly in the next few years. The Curies discovered two new elements with similar properties—polonium and radium. It is in this connection that Marie Curie proposed the term radioactivity for this special property.

Radioactivity is the phenomenon of spontaneous disintegration of certain nuclei to form new elements with the emission of "*active radiations*". These active "*radiations*" usually consist of α-particles (the nucleus of the helium atom), β-particles (electrons) and γ-rays (very short wavelength electromagnetic radiations); they are called active as they affect photographic plates and interact with electric and magnetic fields.

Radioactivity is exhibited by a large number of naturally occurring elements—uranium, thorium and radium being the most common examples. Certain isotopes of lighter elements are also radioactive, e.g., ^{40}K (abundance: 0.012%), and ^{14}C (present in the atmosphere in trace quantities due to cosmic neutron bombardment on ^{14}N). Radioactivity may also be observed in some artificially obtained isotopes.

The nature of radioactive radiations

Rutherford observed (1899-1900) that the radiations from a uranium compound consist of two components, one component being nearly 100 times more penetrating than the other. He called the less penetrating component α-rays and the other component β-rays. Subsequently their nature was established by subjecting the "*rays*" to electric and magnetic fields. A third type of radiation, called the γ-rays, was discovered and characterized later (1900).

The existence of these three types of radiations was represented by Marie Curie by means of a diagram similar to that shown in Fig. 13.1. A small amount of a radioactive substance is taken in a narrow hole drilled in a lead block. Since lead absorbs the radiations completely, only a narrow stream emerges through the opening. The lead block is placed in a large enclosure from which air is expelled as otherwise some of the radiations would be absorbed in air. A magnetic field directed *into the paper* will now split the radiations into three components, as observed by separate marks on a photographic plate.

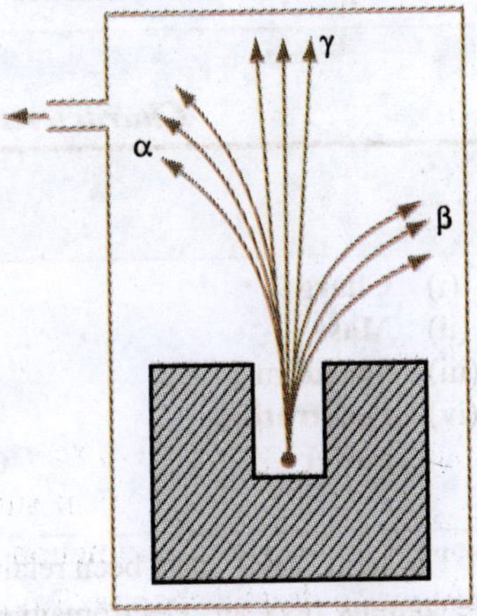

Fig.13.1: Splitting of α, β and γ rays in a magnetic field. (Magnetic field directed into the paper)

Two more types of radioactive decay, namely positron emission and K-electron capture, were recognized later.

The different types of radiations do not originate simultaneously in a single step of radioactive disintegration. As explained later, a radioactive element disintegrates to form a new element and usually emits either an α or a β-particle. The new element formed, in turn, may also be radioactive, and may give rise to another kind of radiation. γ-rays are emitted as a result of transition of the nuclei from one energy state to another. When a radioactive element is not separated from its decay products, which may also be radioactive, all three types of radiations may be emitted simultaneously from a radioactive substance.

Problem 13.1

1 g. of radium produces 0.043 cm^3 of helium gas (at NTP) in one year. The number of α-particles emitted in the same time by the same amount of radium is 11.6×10^{17}. Calculate the value of the Avogadro number.

Solution: Each α-particle gives rise to an atom of helium. So 11.6×10^{17} α-particles produce 11.6×10^{17} helium atoms or helium molecules (monatomic).

The number of helium molecules corresponding to 22,400 cm^3 or 1 mole at NTP is:

$$\frac{11.6 \times 10^{17} \times 22{,}400}{0.043} = 6.04 \times 10^{23}$$

This gives the value of the Avogadro number (vs. 6.022×10^{23}).

Though the nature of α- and β-rays were directly established from their behaviour in electric and magnetic fields, the nature of γ-rays was established over a long period of time. In 1900, Villard discovered the existence of some highly penetrating rays in radioactive radiations; these were unaffected by electric and magnetic fields. It appears that by 1903 the rays were mentioned by the term γ-rays, but controversy existed over their nature. In 1914, Rutherford and Andrade diffracted γ-rays by means of a suitable crystal. This was a convincing proof of the wave nature of γ-rays. Their wavelengths were also measured and found to be similar to those for very short wavelength x-rays. It was thus established that γ-rays are a form of electromagnetic radiation.

The important characteristics of α, β and γ-rays are tabulated in Table 13.1.

Table 13.1

Characteristics of alpha, beta and gamma rays *

	Alpha rays (α)	Beta rays (β)	Gamma rays (γ)
(i) **Charge**	$+2e$	$-e$	0
(ii) **Mass**	$4m_H$	m_e	0
(iii) **Speed,** $m\ s^{-1}$	2×10^7	2×10^8	3×10^8
(iv) **Penetrating power**	weak; 0.01 mm. of Al.	medium; 0.1 mm of Al.	high; 25 cm of steel, 5 cm of Pb.

* The term "rays" have been retained from historical usage. Actually alpha and beta rays are particles while gamms rays are electromagnetic radiations. Beta rays (β) conventionally refer to electrons of nuclear origin. The notation β is also used for positrons. In nuclear reactions electrons are written ${}^{0}_{-1}e$ or β^- and positrons as ${}^{0}_{+1}e$ or β^+. γ-rays are identical with short wave length x-rays but these are of nuclear origin while x-rays have extra-nuclear origin.

Unit of radioactivity

The SI unit of radioactivity is **becquerel**, Bq.

1 Bq = 1 distintegration per second (dis s^{-1}).

The **curie** (Ci) has been the unit of radioactivity for many years. This was defined as the quantity of any radioactive substance which has a decay rate of 3.7×10^{10} disintegrations per second (equal to that of 1g Ra; see Problem 13.2).

The corresponding subunits are:

1 millicurie (1mCi) = 3.7×10^{7} dis s^{-1}; 1 microcurie 1(μCi) = 3.7×10^{4} dis s^{-1}.

For conversion, we have: 1 Ci = 37×10^{9} Bq = 37 GBq (gigabecquerel).

The specific activity of a radionuclide is its activity per kilogram (or dm^3) of the radioactive material.

The **rutherford (rd)** is another practical unit. 1 rd = 10^{6} dis s^{-1}.

The value of curie assigned above is fixed. Increased precision in the determination of $t_{1/2}$, however, shows that 1g Ra undergoes 3.61×10^{10} dis s^{-1}.

Detection and measurement

Radioactive radiations may be detected by (i) their action on a photographic plate; (ii) production of flash of light (scintillation) when the rays strike certain materials; (ii) ionization of a gaseous medium through which the rays travel; (iv) by recording the path of the rays by some device (track detectors). The methods except under (i) may also be used in the measurement of radioactivity.

(i) Action on photographic plate: A photographic plate is affected by all the three types of radioactive radiation. The effect may be easily distinguished from ordinary light by covering the plate with some opaque screen which can be penetrated only by radioactive radiations. The method is useful in the detection of radioactivity in leaves, minerals or biological materials.

(ii) Scintillation counting: Discrete flashes of light are observed when α-particles strike a fluorescent screen of zinc sulfide. This provided an early method for counting α-rays (*spinthariscope*). A modern adaptation of this method is now widely used, specially for β and γ rays.

A variety of scintillators have now been developed, e.g., crystals of anthracene, stilbene, NaI activated with 1% TlI, etc. The last one is particularly useful for γ-ray measurements. Photomultiplier tubes are extensively used to increase the light output through secondary photoelectrons in several stages. The final counting may be done electronically.

Scintillation counters are simple and at the same time highly sensitive, having a rapid response. These facts make them useful for many practical purposes. General survey for radiation hazards is most conveniently carried out by scintillation counters as they are more efficient than Geiger Muller counters in detecting γ-rays.

(iii) Ionization methods: When an α or a β particle passes through a gaseous medium, they collide with the gas molecules, ejecting electrons and leaving the molecules as positively charged ions. γ-rays do not ionize a gas directly, but a portion of them may be absorbed by the gas molecules—resulting in the release of electrons and formation of ions. These gaseous ions may be detected from the electrical conduction of the gas, as in different ion collection methods, e.g., the ionization chamber, proportional counter or the Geiger-Muller counter.

(iv) Track detectors: The ions may serve as centres for the condensation of droplets of water from vapour and the path of the ionizing particle may be recorded photographically, as in the Wilson cloud chamber.

Detectors based on ionization

The ionization chamber: When a gas is ionized by some radiation, the electrons and positive ions formed travel toward the anode and cathode respectively. At low voltages, they may recombine before reaching the electrodes. As the applied voltage is increased, the ion current increases and reaches a point at which all the ions formed reach the electrodes. A further increase in the potential does not lead to any increase in the number of ions formed and the current reaches a constant value. This condition exists roughly between 100-400 volts; in this range the current is directly proportional to the rate of production of charged ions, and hence on the activity of the radioactive source.

The ionization chamber consists of a gas-filled metallic cylinder acting as the cathode. The anode runs axially through the cylinder and is connected through an amplifier to a counter. The gases used in the chamber are usually air or hydrogen at atmospheric pressure; higher pressures are used for detecting γ-rays.

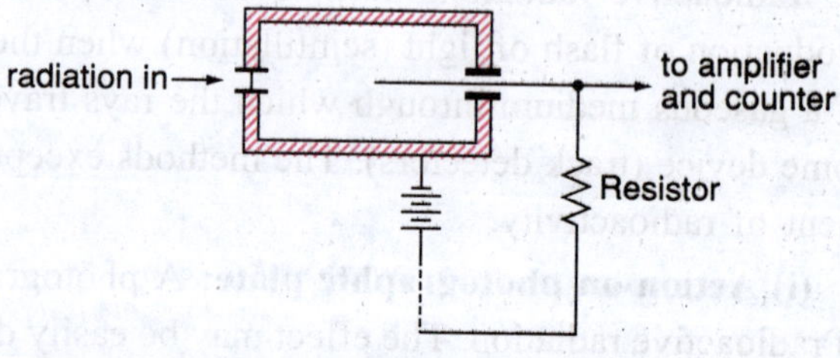

Fig.13.2: Schematic diagram of an ionization counter

A potential difference of about 200 volts is established between the electrodes. When an ionizing radiation passes through the instrument, electric pulses are created which are amplified and recorded.

The pulse size obtained is usually a few tenths of a millivolt; this corresponds to direct current in the range of 100 pA to 10 fA, or even less. (pA = picoampere, i.e., 10^{-12} A; fA = femtoampere, i.e., 10^{-15} A.)

The maximum rate of response of a counter is about 5-10 per second. When the rate of arrival of pulse is greater than this, a *scaler* is used which records 1 pulse in every 10 to 100 pulses. Multiplication by the scaling factor now readily gives the actual number of pulses.

The proportional counter: The proportional counter also consists of a cylindrical gas-filled tube with a thin wire passing through the centre as anode. The potential difference maintained between the tube and the wire is higher than that in an ionization chamber, usually in the range of 500-1000 volts. At the high voltage applied, the rapidly moving electrons undergo multiple collisions with the gas molecules and the number of ion-pairs is largely increased. The pulse size is found to depend on the applied voltage. However, for a given voltage, the size of the current pulse is proportional to the initial number of ion-pairs produced between the electrodes. The total number of ion-pairs produced by an electron from each primary ion-pair is called the gas amplification factor. It may vary from about 10 to about 10^4 in individual cases. The amount of external amplification required in a proportional counter is consequently less.

Argon or methane at 1 atmospheric pressure or a little above are commonly used in such counters. In flow-type counters, the gas flows slowly through the counter at atmospheric pressure.

Geiger-Müller counters: These are also similar in construction to a proportional counter—the anode is a thin metal wire passing axially through a metal tube acting as cathode. The tube is filled with argon at pressures below atmospheric (a little ethanol or bromine vapour is also introduced).

The Geiger-Müller counters (Geiger counter, GM counter) are operated at a high voltage range, somewhere in between 800-1500 V, depending upon the actual size and construction. The passage of a radioactive radiation causes ionization in the gas within the tube. At the very high voltage applied, each primary electron causes further ionization by collision. The resulting secondary electrons also cause ionization and ultimately there are thousands of times more ions present. Within a certain voltage range, the number of ions reaching the electrodes becomes independent of the applied voltage. This is the *characteristic voltage range* of the GM counter. Above this voltage, continuous discharge takes place within the tube.

The GM counter differs from a proportional counter in the higher voltage used; the number of ions produced are thereby largely multiplied. At the same time, ionization takes place along the entire length of the central wire (anode) while in a proportional counter it is localized at a single point.

The electrons produced by the ionization move very fast—they take less than a microsecond to reach the anode. On the other hand, the positive ions move rather slowly, a few hundred microseconds are required to reach the cathode. Consequently, the central wire (anode) remains surrounded by the positive ions for at least 100 microseconds after a pulse has been recorded due to the electrons. During this time, the field around the wire is momentarily lowered—any ionizing event occurring during this time is not recorded in the counter. Summarily, a GM counter cannot record activities greater than 10^4 per second.

Track detectors

The Wilson cloud chamber: The cloud chamber (Wilson, 1911) provides a method of recording (photographically) the path of an ionizing particle in a gaseous medium. The ions formed along the path of the particle are allowed to serve as nuclei for the condensation of water vapours—the array of the droplets may be photographed to show the track of the ionizing particle. The mass, charge and energy of the particle may be inferred from the characteristics of the track like grain density, linearity etc.

The cloud chamber basically consists of a cylinder fitted with a movable piston. The cylinder is filled with air or nitrogen saturated with water vapour (or alcohol). The piston is suddenly pulled out, when the gas expands adiabatically and gets cooled. The gas now becomes supersaturated with the water vapour at this lower temperature. Dust particles and scattered ions are carefully removed beforehand so that the supersaturation is maintained. The passage of an ionizing radiation now forms ions along its path. The ions act as nuclei for local condensation of the vapour—the droplets are arranged along the trajectory of the radiation which may be photographed. A small electrostatic gradient is used to sweep the ions between successive expansions.

The tracks of α-particles appear as straight lines of dense fog droplets, thousands of droplets being present per centimeter. The tracks of β-particles contain only a few droplets per centimeter and are curved. γ-rays produce short crooked tracks from secondary electrons.

In the **Bubble chamber** method (Glaser, 1952), a liquid is superheated above its boiling point under high pressure. The pressure is suddenly reduced when an ionizing particle enters the chamber. Boiling now takes place initially along the path of the ion—bubbles condense on the ions and the bubble-track may be suitably photographed. At low temperatures liquid hydrogen, helium or neon may be used at pressures around 5 atmosphere. Propane or freon at about 25 atmosphere pressure may also be used at room temperature. The method is particularly suitable for high energy particles having very long range in air.

13.2 THE THEORY OF RADIOACTIVE DISINTEGRATION

In 1903, Rutherford and Soddy put forward the revolutionary theory of radioactive disintegration. The theory explained many observations on radioactive decay of different elements and quantitatively correlated the rate of decay with the amount of a radioactive substance. The theory was initiated mainly by the following experimental facts.

When an excess of ammonium carbonate was added to a solution of uranyl nitrate—a precipitate formed initially but it soon redissolved almost completely. Only a small quantity of a flocculent residue was obtained. Surprisingly, this precipitate was found to be highly active. But the solution, which contained most of the uranium (as $(NH_4)_4[UO_2(CO_3)_3]$), was virtually inactive [Crookes, 1900].

It was assumed that the observed activity of the uranium salt was due to the presence of some other substance in it; this was named uranium-X. However, it was soon observed that the precipitate, supposed to contain the active uranium-X, slowly lost its activity. But the feebly active solution containing uranium gradually regained its activity in about six months.

Similar observations were made with thorium compounds. Aqueous ammonia was added to thorium nitrate solution; the thorium was precipitated as hydroxide which was inactive. On the other hand, the filtrate was found to be highly active. [Rutherford and Soddy, 1902.].

The active component of thorium nitrate solution was called thorium-X.

It was again observed that the inactive residue of thorium hydroxide slowly regained its activity in about thirty days while the active solution containing thorium-X gradually lost its activity at the same rate.

Rutherford and Soddy measured the rates of change in the activities of the different portions of the uranium and thorium compounds. They observed that in all cases the activities changed *in an exponential manner* with time, as illustrated for the thorium compounds in Fig. 13.3.

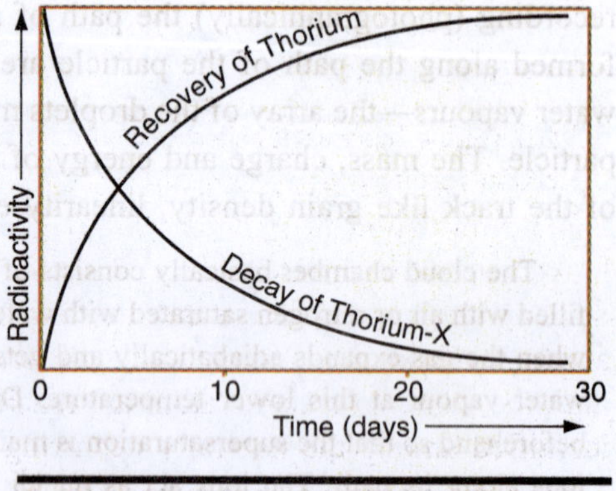

Fig.13.3: The nature of the decay-curve of Th-X and growth-curve of Th

All these observations may now be summarized as follows:

(i) Uranium is associated with uranium-X and thorium is associated with thorium-X.

(ii) Uranium-X differs in chemical and radioactive properties from uranium, so does thorium-X from thorium.

(iii) When uranium-X is removed from uranium, it is regenerated from uranium in the course of about six months. Similarly thorium isolated from thorium-X regenerates thorium-X in about a month.

(iv) These regenerated uranium-X and thorium-X may again be separated from the mother uranium or thorium respectively. But fresh quantities of uranium-X and thorium-X are regenerated in the same manner from their respective mothers and the process may be repeated almost indefinitely.

Another relevant observation in this line is the discovery of "*active emanations*". It was found that thorium salts continuously emit an active gas. The activity of this gas decays rapidly, but it leaves an "induced" activity in the containing vessel or on other surrounding matter. This active gas was called *emanation* (Rutherford, 1900). Radium and actinium were also found to emit their own characteristic emanations. It was shown that the "induced activity" was due to an active deposit left by the emanations during their decay.

The theory of radioactive disintegration

According to this theory, the atoms of a radioactive element are inherently unstable. They disintegrate spontaneously and form atoms of a new element which is physically and chemically different from its mother. The new element formed may, in turn, be again radioactive and disintegrate in a similar fashion. The process will continue until the atoms of a stable element are reached.

The theory can readily explain the observations noted earlier. Uranium is a feebly radioactive element. It disintegrates to form a new substance, uranium-X*, which is highly radioactive. It is precipitated by ammonium carbonate. In course of time, the uranium-X disintegrates to a less active product—and hence the precipitate loses activity. On the other hand, uranium retained in the solution disintegrates to form more uranium-X, and consequently the solution regains activity. The activity soon reaches an equilibrium position when the rate of formation of uranium-X is equal to the rate of its disintegration. The stages of disintegration may be summarily represented as

Uranium	$\longrightarrow$	**Uranium-X***	$\longrightarrow$	**Products**
(feebly active)		(strongly active)		(feebly active)

* It has now been established that the substance called uranium–X actually contains more than one element resulting from successive disintegration.

The formation of radioactive emanations is also explained by the theory. These may be looked upon as disintegration products of the respective elements. These emanations also disintegrate, leaving active products on the surroundings, which cause the so-called "induced activity". These products are also radioactive and they disintegrate further. The chain continues until the atoms of a stable atom are reached. Three radioactive disintegration series have been identified among the naturally occurring radioactive elements (see later).

The disintegration theory has also been successful in establishing rate expressions for radioactive decay processes, as explained below.

13.2.1 Rate of Radioactive Disintegration

Rutherford and Soddy observed that uranium-X and thorium-X decayed in an exponential manner with time (Fig. 13.3). This means that the rate of loss of activity at any instant is proportional to the amount of activity actually present at that time. Now, the rate of loss of activity must be proportional to the number of atoms which disintegrate in a unit time-interval. As the total activity at any time is also proportional to the total number of radioactive atoms present at that time, the rate of disintegration at any time should be proportional to the number of atoms present at the time.

Let n be the total number of atoms of the radioelement present in a radioactive species at any time t, *counted from any arbitrarily taken starting time* ($t = 0$). If dn atoms disintegrate between the very small time interval between t and $t + dt$, then the rate of disintegration at time t is equal to $-dn/dt$; the negative sign indicates that n decreases with time. This rate is proportional to the total number of atoms present at time t, i.e. n. Therefore,

$$-\frac{dn}{dt} \propto n \quad \text{or,} \quad -\frac{dn}{dt} = \lambda n \quad \text{or,} \quad \lambda = -\frac{1}{n}\cdot\frac{dn}{dt} \qquad (\lambda \text{ is a constant}).$$

When $dt = 1$, $\lambda = -dn/n$; this is the fraction of total atoms which disintegrate in unit time. The constant λ is called the *disintegration constant* or *decay constant* characterizing a radioelement. Its magnitude is unaffected by the physical condition or state of chemical combination of the radioelement.

The above rate equation may be slightly rearranged and integrated:

$$-dn/n = \lambda dt \quad \text{or,} \quad \ln n = -\lambda t + I \qquad (I = \text{constant of integration}).$$

Let n_0 be the number of atoms of the radioelement present at the moment from which we started to count time i.e., at $t = 0$. This gives $I = \ln n_0$.

The value of I now gives the general expression

$$\ln n = -\lambda t + \ln n_0 \;; \quad \text{or, } n = n_0 e^{-\lambda t}; \quad \text{or, } \lambda = \frac{2\cdot 303}{t}\log\frac{n_0}{n}$$

These logarithmic or exponential equations connect the number of atoms of a radioelement present across a time interval t. The equations very accurately represent the rates of disintegration of all known radioelements.

The rate laws show that a radioelement will take infinite time to decay *completely*, i.e., when n will become zero. However, the law is statistical in nature and it may not be suitable for a small number of atoms.

Half-life

The half-life is the time required for the activity of a given amount of radioelement to decay to half its initial value. The time is independent of the amount of radioactive substance present initially and is characteristic of each radioactive element. It provides a measure of the rate at which an element disintegrates.

From the decay equation deduced above, we get $t_{1/2}$ by the relation

$$\lambda = \frac{2\cdot 303}{t_{1/2}}\log\frac{n_0}{n_0/2} \quad \text{or,} \quad t_{1/2} = \frac{2\cdot 303\log 2}{\lambda} = \frac{0\cdot 693}{\lambda}.$$

This shows that the half-life is independent of n_0, i.e., the amount of radioelement taken initially. This is true for decay by *any fraction*.

It is readily observed that after time $t = t_{1/2}$, $e^{-\lambda t} = 1/2$; in general, after time $t = nt_{1/2}$, $e^{-\lambda t} = 1/2^n$ i.e., the fraction of a radioactive element becomes $1/2^n$ of the initial amount.

The average life period

The equations for radioactive decay reveal that a definite fraction of the atoms present at any instant will decay in unit time; this fraction is related to the decay constant λ. But it is not possible to foretell which particular atoms will disintegrate in a particular time interval. Thus, some of the atoms will die very early, while some others will have a very long life. In fact, the life of any radioactive atom may have all possible values from zero to infinity. Nevertheless, it is possible to estimate the average life of an aggregate of large number of atoms.

The average life t_{av} is obtained by multiplying every possible life period from zero to infinity by the corresponding number of atoms having that life, adding the terms obtained, and dividing the sum by the total number of atoms.

$$t_{av} = \frac{t_1 dn_1 + t_2 dn_2 + t_3 dn_3 + \ldots}{dn_1 + dn_2 + dn_3 + \ldots} = \frac{\sum tdn}{n_0}$$

The sum in the numerator may be obtained by integration between appropriate limits. We remember that the number of atoms may vary from zero to n_0, for a particular life time t, at least theoretically. The corresponding limits of t are also necessary. An atom may have any life-time between zero and infinity. Moreover, at $t = 0$, all the n_0 atoms are present i.e., for $t = 0$, $n = n_0$. Again, at $t = \infty$, no atoms are present, i.e., $n = 0$ for $t = \infty$. So we write,

$$t_{av} = \frac{\sum tdn}{n_0} = \frac{1}{n_0}\int_{n_0}^{0} tdn = \frac{1}{n_0}\int_{0}^{\infty} t(-\lambda ndt) = \frac{1}{n_0}\int_{0}^{\infty} -\lambda t n_0 e^{-\lambda t} dt = -\lambda\int_{0}^{\infty} te^{-\lambda t} dt = \frac{1}{\lambda}.$$

From the expression of $t_{1/2}$, we get $t_{1/2} = 0 \cdot 693 t_{av}$.

Thus, the statistical average life of a large number of atoms is equal to the inverse of the decay constant of the radioelement. This is greater than the half-life.

Decay of parent and daughter nuclides

When a radioactive nuclide A decays into another active nuclide B, the relative amounts of the parent (A) and daughter (B) at any time is determined by the relative magnitudes of the individual decay constants of A and B. At an experimental time t, the rate of decay of A is given by the relations

$$-(dN_A/dt) = \lambda_A N_A = N_A^0 e^{-\lambda_A t}$$

N and N^0 represent the amount of A at time t and at the experimental zero time respectively. B is formed at the rate at which A decays, i.e., $\lambda_A N_A$; itself, B decays at the rate $\lambda_B N_B$ (N_B = amount of B at time t). Therefore

$$dN_B/dt = \lambda_A N_A - \lambda_B N_B$$

Solution: $dN_B/dt = \lambda_A N_A - \lambda_B N_B = \lambda_A N_A^0 e^{-\lambda_A t} - \lambda_B N_B$

Multiplying by the integrating factor $e^{\lambda_B t}$ and rearranging,

$$e^{\lambda_B t} dN_B + \lambda_B N_B e^{\lambda_B t} dt = \lambda_A N_A^0 e^{(\lambda_B - \lambda_A)t} dt.$$

Integrating, $N_B e^{\lambda_B t} = [\lambda_A/(\lambda_B - \lambda_A)] N_A^0 e^{(\lambda_B - \lambda_A)t} + C$

At $t = 0$, $N_B = 0$ gives $C = -\lambda_A N_A^0/(\lambda_B - \lambda_A)$.

$$\therefore \; N_B e^{\lambda_B t} = \frac{N_A^0 \lambda_A}{\lambda_B - \lambda_A}(e^{(\lambda_B - \lambda_A)t} - 1) \quad \text{or,} \quad N_B = \frac{\lambda_A N_A^0}{\lambda_B - \lambda_A}(e^{-\lambda_A t} - e^{-\lambda_B t}) \quad \text{... [13.1]}$$

In terms of activity (say A), $A_B = \lambda_B N_B$

$$\text{From Eq 13.1, } A_B = \frac{\lambda_B}{\lambda_B - \lambda_A} \cdot A_A^0 \, (e^{-\lambda_A t} - e^{-\lambda_B t}) \quad \text{... [13.2]}$$

A_A^0 is the initial activity of A $(= \lambda_A N_A^0)$.

If the value of N_B at $t = 0$ is not equal to zero, but has a finite value N_B^0, then its contribution at time t will be $N_B^0 . e^{-\lambda_B t}$. This has to be added to the right side of eqn. 13.1.

Different situations may now be considered.

(i) ***If the parent is shorter-lived than the daughter*** $(\lambda_A > \lambda_B)$, the rate of growth of the daughter (= rate of decay of the parent) is faster than its own disintegration rate. So the activity of the daughter nuclide will gradually rise; at some time the rate of disintegration of the daughter will become equal to that of the parent. After this, the parent-activity will decrease at a rate faster than that of the daughter. Once the parent has decayed completely, the daughter will decay according to its own λ. Clearly, there cannot be any equilibrium between the parent and the daughter.

The time needed at which the daughter activity grows to a maximum is given by

$$t_{max} = \frac{2 \cdot 303}{\lambda_B - \lambda_A} \log \frac{\lambda_B}{\lambda_A}.$$

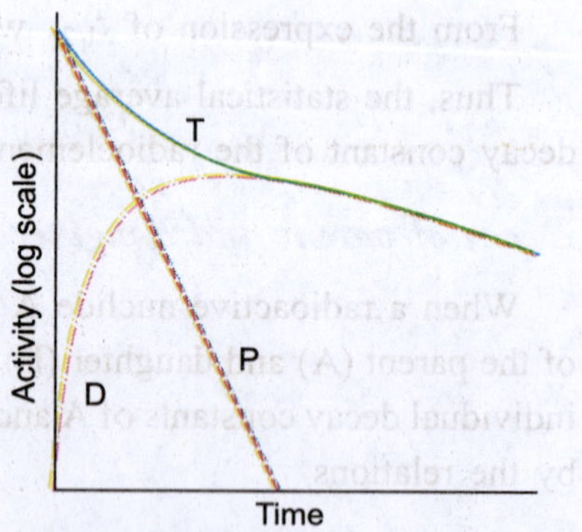

Fig.13.4: Nature of decay curves where the parent has a higher decay constant (not to scale; slopes depend on the magnitudes of λ; assumed that only pure parent exists at the experimental zero time). **P**: parent activity. **D**: daughter activity. **T**: total activity

Differentiating eqn. 13.1,

$$\frac{dN_B}{dt} = -\frac{\lambda_A^2 N_A^0}{\lambda_B - \lambda_A} e^{-\lambda_A t} + \frac{\lambda_A \lambda_B}{\lambda_B - \lambda_A} N_A^0 \, e^{-\lambda_B t}$$

At t_{max} for B, $dN_B/dt = 0$ gives the above expression for t_{max}.

Some examples of this type of decay are (with respective half-lives):

(i) $^{92}Sr \xrightarrow[2.7\text{ h}]{\beta^-} {}^{92}Y \xrightarrow[3.6\text{ h}]{\beta^-} {}^{92}Zr$ (stable) ;

(ii) $^{210}Bi \xrightarrow[5\text{ d}]{\beta^-} {}^{210}Po \xrightarrow[138\text{ d}]{\alpha} {}^{206}Pb$ (stable) ;

(iii) $^{218}Po \xrightarrow[3.05\text{ min}]{\alpha} {}^{214}Pb \xrightarrow[26.8\text{ min}]{\beta^-} {}^{214}Bi \longrightarrow \ldots$

The nuclides in example (iii) belong to the $4n + 2$ series (Table 13.3) in the conventional names Ra-A, Ra-B and Ra-C.

(ii) *The daughter has a higher decay constant than the parent*, i.e., $\lambda_A < \lambda_B$. Such radionuclides will reach a steady state after a certain time depending upon the magnitudes of λ_A and λ_B. This is often referred to as ***radioactive equlibrium***. This is actually a steady state condition among genetically related radio-nuclides where the daughter has a faster decay constant than the parent and is observed when the observation time is greater than the mean lifetime of the parent. Two situations may arise: (a) ***transient equilibrium***: the *relative amount* of the daughter, i.e., N_B/N_A remains unchanged with time but the *total activity* (of parent and daughter) slowly decreases (Fig. 13.5a). It can be shown (see box) that under this condition

$$\frac{N_B}{N_A} = \frac{\lambda_A}{\lambda_B - \lambda_A}.$$

Since $\lambda_B > \lambda_A$, $e^{-\lambda_B t}$ becomes negligible in comparison to $e^{-\lambda_A t}$ in equation 13.1 when t is sufficiently large. Then,

$$N_B = \frac{\lambda_A . N_A^0}{\lambda_B - \lambda_A} e^{-\lambda_A t} = \frac{\lambda_A N_A}{\lambda_B - \lambda_A} \quad \text{or,} \quad \frac{N_B}{N_A} = \frac{\lambda_A}{\lambda_B - \lambda_A} \quad \text{or,} \quad \frac{N_B + N_A}{N_A} = \frac{\lambda_B}{\lambda_B - \lambda_A}.$$

Since, $N_A = N_A^0 . e^{-\lambda_A t}$, $N_B + N_A = \dfrac{\lambda_B}{\lambda_B - \lambda_A} . N_A^0 e^{-\lambda_A t}$.

Thus the total activity at any time [proportional to $(N_B + N_A)$] falls according to $e^{-\lambda_A t}$, but the ratio N_B/N_A remains constant. This is referred to as **transient equilibrium.**

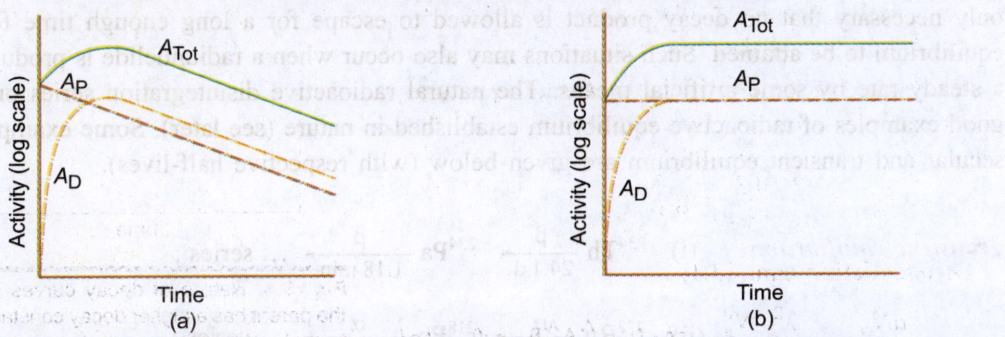

Fig.13.5: Nature of decay curves when the daughter has a higher decay constant than the parent (a)Transient equilibrium; (b) Secular equilibrium
A_P: Parent activity; **A_D**: Daughter activity; **A_T**: Total activity. (Slopes depend on relative magnitudes of λ. Assumed that only pure parent is present initially.)

(b) **secular equilibrium,** or permanent equilibrium is attained when the decay constant of the parent is negligibly small in comparison to the decay constant of the daughter, i.e. $\lambda_A << \lambda_B$.

As the parent activity does not decrease measurably during many daughter half-lives, we may write $\lambda_B - \lambda_A \simeq \lambda_B$. We get

$$\frac{N_B}{N_A} = \frac{\lambda_A}{\lambda_B} = \frac{t_{1/2}(B)}{t_{1/2}(A)}$$

Under this condition the total activity as well as the relative amount of the daughter remains unchanged with time (Fig. 13.5b).

Summary of parent-daughter relation

For the situation where both a parent radioelement A and its daughter B are radioactive

$$A \xrightarrow{\lambda_A} B \xrightarrow{\lambda_B} \ldots$$

$$A_B = \frac{\lambda_B}{\lambda_B - \lambda_A} \cdot A_A^0 \left(e^{-\lambda_A t} - e^{-\lambda_B t}\right)$$

where A_B = activity of daughter (B) at time t provided none of it was present at $t = 0$.

A_A^0 = initial activity of parent (A) ($t = 0$).

If $\lambda_B > \lambda_A$, the limiting condition of **transient equlibrium** is reached after sufficiently large t.

$$A_B = \frac{\lambda_B}{\lambda_B - \lambda_A} \cdot A_A.$$

If $\lambda_B >> \lambda_A$, the limiting condition for **secular equilibrium** is reached when

$$A_B \simeq A_A$$

The relations for radioactive equilibrium may be applied whenever several short-lived nuclides are produced in successive decays beginning with a relatively long-lived parent. It is only necessary that no decay product is allowed to escape for a long enough time for the equilibrium to be attained. Such situations may also occur when a radionuclide is produced at a steady rate by some artificial means. The natural radioactive disintegration series provide good examples of radioactive equilibrium established in nature (see later). Some examples of secular and transient equilibrium are given below (with respective half-lives).

Transient equilibrium:

(i) $^{234}Th \xrightarrow[24.1\ d]{\beta^-} {}^{234}Pa \xrightarrow[1.18\ min]{\beta^-} \ldots$ series ;

(ii) $^{222}Rn \xrightarrow[3.82\ d]{\alpha} {}^{218}Po \xrightarrow[3.05\ min]{\alpha} \ldots$ series ;

(iii) $^{140}Ba \xrightarrow[12.8\ d]{\beta^-} {}^{140}La \xrightarrow[40\ h]{\beta^-} {}^{140}Ce$ (stable).

Secular equilibrium:

(i) $$^{238}\text{U} \xrightarrow[4.5\times10^9\,\text{y}]{\alpha} {}^{234}\text{Th} \xrightarrow[24.1\,\text{d}]{\beta^-} \ldots \text{ series ;}$$

(ii) $$^{226}\text{Ra} \xrightarrow[1600\,\text{y}]{\alpha} {}^{222}\text{Rn} \xrightarrow[3.82\,\text{d}]{\alpha} \ldots \text{ series ;}$$

(iii) $$^{90}\text{Sr} \xrightarrow[28.1\,\text{y}]{\beta^-} {}^{90}\text{Y} \xrightarrow[64.1\,\text{h}]{\beta^-} {}^{90}\text{Zr (stable).}$$

If may be noted that secular equilibrium is a limiting case of transient equilibrium, while both are actually steady states. Absolute equilibrium can never occur since that would mean $dN/dt = 0$ for a radionuclide, i.e., $\lambda = 0$. Radioactive equilibrium is also different from chemical equilibrium as it is irreversible and independent of external conditions.

(iii) ***The parent and the daughter have equal decay constants***, i.e., $\lambda_A = \lambda_B$. Now the general equation 13.1 becomes indeterminate. When λ_A and λ_B are nearly equal, it can be shown that the daughter first reaches a maximum at a time equal to the mean life for B ($1/\lambda_B$) and then decays.

Such systems are rare; two examples are

(i) $$^{131}\text{Ba} \xrightarrow[11.5\text{ days}]{\text{electron capture}} {}^{131}\text{Cs} \xrightarrow[9.7\text{ days}]{\text{electron capture}} {}^{131}\text{Xe (stable) ;}$$

(ii) $$^{214}\text{Pb} \xrightarrow[26.8\text{ min}]{\beta^-} {}^{214}\text{Bi} \xrightarrow[19.7\text{ min}]{} \begin{cases} \xrightarrow{\alpha} {}^{210}\text{Tl} \longrightarrow \ldots \\ \xrightarrow{\beta^-} {}^{214}\text{Po} \longrightarrow \ldots \end{cases}$$

Branching decay

Most radionuclides decay by a single process. Some nuclides are, however, known to decay by two (or more) alternative modes. The nuclides ^{211}Bi, ^{212}Bi, ^{214}Bi decay by two modes, as shown in the natural disintegration series in Table 13.2–13.4 (under the names Actinium-C, Thorium-C and Radium-C respectively). ^{130}Cs and ^{64}Cu undergo decay by beta-emission, positron emission and K-electron capture. The half-life of the nuclide undergoing such branching decay is related to its total rate of disappearance: for the simplest case shown below, we can proceed as follows:

$$\text{A} \begin{cases} \xrightarrow{\lambda_1} \text{B} \\ \xrightarrow{\lambda_2} \text{C} \end{cases}$$

$$\frac{-dn_A}{dt} = \frac{dB}{dt} + \frac{dC}{dt}$$

$$\text{or, } \lambda_{Tot} n_A = \lambda_1 n_A + \lambda_2 n_A \quad \text{or, } \lambda_{Tot} = \lambda_1 + \lambda_2$$

$$\text{or, } t_{1/2}(\text{A}) = \frac{0\cdot693}{\lambda_{Tot}} = \frac{0\cdot693}{\lambda_1 + \lambda_2}$$

λ_1 and λ_2 may be called the partial decay constants. If the corresponding partial half-lives are $t_{1/2}(1)$ and $t_{1/2}(2)$, then $1/t_{1/2} = 1/t_{1/2}(1) + 1/t_{1/2}(2)$.

For the general case,

$$\lambda_{Tot} = \lambda_1 + \lambda_2 + \lambda_3 + \ldots$$

$$\text{or, } \quad 1/t_{Tot} = 1/t_1 + 1/t_2 + 1/t_3 + \ldots \quad (t\text{'s are respective half-lives})$$

The fraction f_i of the parent decaying by the i-th mode is (see exercise Q 15):

$$f_i = \lambda_i / \lambda_{Tot}.$$

Numerical problems

Problem 13.2

The curie was originally intended to be the activity of one gram of radium-226 ($t_{1/2}$ = 1620 y). Calculate the actual weight of radium which is equivalent to one curie.

Solution: From the value of $t_{1/2}$,

$$\lambda = \frac{0 \cdot 693}{t_{1/2}} = \frac{0 \cdot 693}{1620}\,y^{-1} = \frac{0 \cdot 693}{1620 \times 365 \times 24 \times 60 \times 60}\,s^{-1}.$$

By definition, 1 Ci *is equal to* 3.7×10^{10} *disintegrations per second*. This should be equal to $-(dn/dt)$ and hence λn, where n is the number of atoms of ^{226}Ra corresponding to this activity.

$$\text{Hence,} \quad 3 \cdot 7 \times 10^{10} = \frac{0 \cdot 693}{1620 \times 365 \times 24 \times 60 \times 60} \times n \Rightarrow n = 6 \cdot 165 \times 10^{23} \text{ atoms}$$

$$\text{Mass of radium} = \frac{6 \cdot 165 \times 10^{23}}{6 \cdot 022 \times 10^{23}} \times 226 \text{ g} = 1 \cdot 024 \text{ g}.$$

Problem 13.3

Tritium, ^3_1H, has a half-life of 3 years. What will be the specific activity of tritium gas in mCi/cm^3 (STP).

Hint: Each molecule of $^3H_2$ contains 2 atoms of ^3_1H. Hence, number of atoms (n) per cm^3 (STP)

$$= 2 \times 6 \cdot 02 \times 10^{23}/22400.$$

$$\lambda = 0 \cdot 693/3\ y^{-1} \longrightarrow 0 \cdot 693/3 \times 365 \times 24 \times 3600\ s^{-1}.$$

Activity = λn (disintegration per second; divide by $3 \cdot 7 \times 10^7$ to obtain result in mCi.

Problem 13.4

The half-life of ^{32}P is 14.3 days. Calculate the specific activity of a specimen of phosphorus containing 1.00 part per million ^{32}P.

Solution: We have to calculate the number of ^{32}P atoms in 1 kg. of phosphorus and the corresponding rate of disintegration. Atomic weight of phosphorus = 30·97.

$\therefore$ 30·97 g of phosphorus contains $6 \cdot 022 \times 10^{23}$ atoms

or, 1000 g of phosphorus contains $\frac{6 \cdot 022 \times 10^{26}}{30 \cdot 97}$ atoms = $1 \cdot 94 \times 10^{25}$ atoms.

Now, the specimen contains 1 part per million ^{32}P $\Rightarrow 1.94 \times 10^9$ ^{32}P atoms

$$\lambda = \frac{0 \cdot 693}{t_{1/2}} = \frac{0 \cdot 693}{14 \cdot 3 \times 24 \times 60 \times 60}\,s^{-1}.$$

$$\therefore \text{ Rate of disintegration} = \frac{dn}{dt} = \lambda n = \frac{0 \cdot 693}{14 \cdot 3 \times 24 \times 60 \times 60} \times 1 \cdot 94 \times 10^{19}\ s^{-1}$$

$$= 1 \cdot 088 \times 10^{13}\ s^{-1} \Rightarrow \frac{1 \cdot 088 \times 10^{13}}{3 \cdot 7 \times 10^{10}}\ \text{Ci kg}^{-1} = 294\ \text{Ci kg}^{-1}.$$

Problem 13.5

$$^{226}\text{Ra} \xrightarrow{1590\text{ y}} {}^{222}\text{Rn} \xrightarrow{4\text{ d}} {}^{218}\text{RaA}$$

A sample initially consists of pure Ra of activity = 1000 Bq. What will be the activity of Rn after (a) 12 days (b) 2 years?

Solution: The activity of Rn (daughter) is given by equation 13.2

$$A_{\text{Rn}} = \frac{\lambda_{\text{Rn}}}{\lambda_{\text{Rn}} - \lambda_{\text{Ra}}} . A^0_{\text{Ra}} \left(e^{-\lambda_{\text{Ra}} t} - e^{-\lambda_{\text{Rn}} t}\right) \quad (A = \text{activity})$$

Since both times are very short compared to the half-life of radium, $e^{-\lambda_{\text{Ra}} t}$ is essentially unity.

12 d is three half-lives of Rn, and hence $e^{-\lambda_{\text{Rn}} t} = \frac{1}{8}$. Since $\lambda_{\text{Rn}} = 0.693/4 \text{ day}^{-1}$ and $\lambda_{\text{Ra}} = 0.693/1590 \times 365 \text{ day}^{-1}$, $\lambda_{\text{Rn}}/(\lambda_{\text{Rn}} - \lambda_{\text{Ra}})$ is almost unity. So,

$$A_{\text{Rn}} = A^0_{\text{Ra}} \left(1 - \tfrac{1}{8}\right) = 1000 \times 7/8 = 875 \text{ dis s}^{-1} \text{ (Bq)}.$$

After 2 years, radon will reach secular equilibrium with the radium. So the activity of Rn will be the same as that of Ra which has decayed only slightly (2 y << 1590 y). So $A_{\text{Rn}} \simeq 1000 \text{ dis s}^{-1}$ (i.e., Bq).

Problem 13.6

$$^{95}\text{Zr} \xrightarrow{t_{1/2} = 70\text{ d}} {}^{95}\text{Nb} \xrightarrow{t_{1/2} = 35\text{ d}} \text{stable nuclide}$$

Starting with a sample of pure ^{95}Zr, what percent of the total activity will be due to ^{95}Nb after 200 d?

Hint: $\lambda_{Zr} = \dfrac{0\cdot693}{70} d^{-1} = \dfrac{2\cdot303}{200} \log \dfrac{A^0_{\text{Zr}}}{A_{\text{Zr}}} \Rightarrow A_{\text{Zr}} = A^0_{\text{Zr}} \times 0.1381$

$$A_{\text{Nb}} = \frac{\lambda_{Nb}}{\lambda_{Nb} - \lambda_{Zr}} . A^0_{\text{Zr}} \left(e^{-\lambda_{\text{Zr}} t} - e^{-\lambda_{\text{Nb}} t}\right) \Rightarrow 2 \times A^0_{\text{Zr}} \times 0.1190 = A^0_{\text{Zr}} \times 0.2380$$

$$\text{Percent contribution of Nb} = \frac{0\cdot2380 \times 100\, A^0_{\text{Zr}}}{(0\cdot2380 + 0\cdot1381)\, A^0_{\text{Zr}}} = 63\cdot3\%.$$

Problem 13.7

A and B are two genetically related radionuclides with half-lives 4d and 2d respectively. If a sample initially contains both A and B at equal activity, what will be the activity of A and B after 8 days?

Hint: Initial activities of A and B are equal, i.e., $A^0_{\text{A}} = A^0_{\text{B}}$. The activity of A after 8 days (= $2t_{1/2}$) will be 1/4th the initial activity. So $A_{\text{A}} = A^0_{\text{A}}/4$.

The initial activity of B will be reduced to $A^0_{\text{B}}/16$ after 8 days ($4t_{1/2}$). To this we have to add the activity of new B formed from A during this time (say A'_{B}).

$$A'_B = \frac{\lambda_B}{\lambda_B - \lambda_A} . A^0_{\text{A}} \left(e^{-\lambda_A t} - e^{-\lambda_B t}\right) = 2 A^0_{\text{A}} \left(\frac{1}{4} - \frac{1}{16}\right) = 3 A^0_{\text{A}}/8 = 3 A^0_{\text{B}}/8 \quad (\because A^0_{\text{A}} = A^0_{\text{B}})$$

Total activity of B after 8 days will be $A^0_{\text{B}}/16 + 3 A^0_{\text{B}}/8$.

Problem 13.8

^{152}Sm has a half-life of 10^{12} y. Calculate the number of α-particles emitted per second from 1 g of natural samarium. Abundance of ^{152}Sm is 27%; the atomic mass of natural samarium is 150.4.

Solution: We have to calculate the number of ^{152}Sm atoms in 1 g of natural samarium and find the corresponding rate of disintegration.

$$1 \text{ g of natural samarium contains } \frac{6.02 \times 10^{23}}{150.4} = 4 \times 10^{21} \text{ atoms} \Rightarrow 0.27 \times 4 \times 10^{21}$$

or, $1{\cdot}08 \times 10^{21}$ atoms of ^{152}Sm

$$\therefore \text{ Rate of disintegration} = \frac{dn}{dt} = \lambda n = \frac{0{\cdot}693 \times 1{\cdot}08 \times 10^{21}}{10^{12} \times 365 \times 24 \times 60 \times 60} \text{s}^{-1}. = 23.74 \text{ s}^{-1}.$$

The number of α-particles emitted per second must be equal to this rate of disintegration.

Problem 13.9

The half-life of radium is 1590 y and that of its daughter, radon, is 3.8 days. What volume of radon at NTP will remain in equilibrium with 1 g of radium?

Solution: From the condition of equilibrium, $n_{Ra}.\lambda_{Ra} = n_{Rn} \cdot \lambda_{Rn}$, where the n-terms represent respective number of atoms at equilibrium,

$$\frac{n_{Ra}}{n_{Rn}} = \frac{\lambda_{Rn}}{\lambda_{Ra}} = \frac{(t_{1/2})_{Ra}}{(t_{1/2})_{Rn}} = \frac{1590 \times 365}{3{\cdot}8} = 1{\cdot}527 \times 10^5.$$

Now we have to convert the ratio of atoms to mass ratio.

$$\text{From } \frac{\text{Number of gram - atoms of radium}}{\text{Number of gram - atoms of radon}} = \frac{n_{Ra}}{n_{Rn}}, \text{ and 1 g of Ra} = 1/226 \text{ g-atom of Ra,}$$

$$\text{No. of g-atoms of Rn} = \frac{n_{Rn}}{n_{Ra}} \times \frac{1}{226} = \frac{1}{1{\cdot}527 \times 10^5} \times \frac{1}{226} = 2{\cdot}9 \times 10^{-8} \Rightarrow 6{\cdot}5 \times 10^{-4} \text{ cm}^3 \text{ (NTP).}$$

Problem 13.10

^{118}Cd disintegrates as $^{118}\text{Cd} \xrightarrow{30 \text{ min}} {}^{118}\text{In} \xrightarrow{4{\cdot}5 \text{ min}} {}^{118}\text{Sn}$ (stable).

(a) Starting with pure ^{118}Cd, find the fraction of nuclei transformed into Sn after 60 min.

(b) In what proportion does the total activity diminish after this time?

Solution: Representing the successive nuclides by A, B and C respectively, we observe that C accumulates at the rate at which B (^{118}In) disintegrates: $dN_C/dt = \lambda_B N_B$.

The value of N_B may be substituted from equation 13.1:

$$\frac{dN_C}{dt} = \frac{\lambda_A \lambda_B . N_A^0}{\lambda_B - \lambda_A}\left(e^{-\lambda_A t} - e^{-\lambda_B t}\right).$$

$$\text{Integrating, } N_C = \frac{\lambda_A \lambda_B . N_A^0}{\lambda_B - \lambda_A}\left(\frac{1 - e^{-\lambda_A t}}{\lambda_A} - \frac{1 - e^{-\lambda_B t}}{\lambda_B}\right).$$

From $\lambda = 0{\cdot}693 t_{1/2}$, $\lambda_A = 0{\cdot}023\ min^{-1}$ and $\lambda_B = 0{\cdot}154\ min^{-1}$. This gives $(N_C/N_A^0) = 0{\cdot}7$ when $t = 60$ min.

This is the fraction of ^{118}Cd nuclides transformed into ^{118}Sn.

(b) Let A_A and A_B represent the activity at time t of A and B respectively. The activity of A at the beginning is A_A^0. The total activity at time t is given by

$$A_t = A_A + A_B = A_A^0\, e^{-\lambda_A t} + \lambda_B N_B.$$

Substituting the value of N_B from equation 13.1

$$A_t = A_A^0 e^{-\lambda_A t} + \frac{\lambda_A \lambda_B N_A^0}{\lambda_B - \lambda_A}\left(e^{-\lambda_A t} - e^{-\lambda_B t}\right);\ \lambda_A . N_A^0 = A_A^0 \text{ gives}$$

$$\frac{A_t}{A_A^0} = e^{-\lambda_A t} + \frac{\lambda_B}{\lambda_B - \lambda_A}\left(e^{-\lambda_A t} - e^{-\lambda_B t}\right)$$

Since only pure A was present initially, A_A^0 is also the total activity at the beginning. So (A_t/A_A^0) is the ratio of total activity at time t to the initial activity. Putting $t = 60$ minutes, $(A_t/A_A^0) = 0.55$, i.e., the total activity becomes 0.55 or (1/1.8) times the initial activity.

Problem 13.11

^{227}Ac has a half-life of 21.8 y with respect to radioactive decay. The decay follows two parallel paths: one leading to ^{227}Th (1.2%) and the other leading to ^{223}Fr (98.8%). What is the rate constant for each of the separate paths?

Hint: $\lambda_{Ac} = 0.693/21.8\ y^{-1} = 3.18 \times 10^{-2}\ y^{-1}$. Again, $\lambda_{Ac} = \lambda_{Th} + \lambda_{Fr}$.

λ_{Th} and λ_{Fr} are rate constants for the two separate paths respectively. The fractional yield of either process is obtained from the ratio of the respective rate constant to the total rate constant.

$$f_{Th} = \frac{1{\cdot}2}{100} = \frac{\lambda_{Th}}{\lambda_{Ac}} \Rightarrow \lambda_{Th} = 3{\cdot}8\times10^{-4}\ y^{-1};\quad f_{Fr} = \frac{98{\cdot}8}{100} = \frac{\lambda_{Fr}}{\lambda_{Ac}} \Rightarrow \lambda_{Fr} = 3{\cdot}1\times10^{-2}\ y^{-1}.$$

13.2.2 The Radioactive Disintegration Series

There are nearly forty naturally occurring radioactive elements. They mostly appear towards the end of the periodic table. Detailed investigation reveals that these elements form three distinct family of successive disintegration products. The elements within a family are said to form a series. The three series are:

(i) The thorium series: It begins with $^{232}_{90}Th$ and ends with $^{208}_{82}Pb$ (Table 13.2). The mass numbers of the members in the series are given by $4n$, where n is an integer. So this is also called the **4n series**.

(ii) The uranium series: It begins with $^{238}_{92}U$ and ends with $^{206}_{82}Pb$ (Table 13.3). The mass numbers of different members correspond to $4n + 2$ (n is an integer). This is also called the **4n + 2 series**.

(iii) The actinium series: It begins with $^{235}_{92}U$ (also known as actino-uranium) and ends with $^{207}_{82}Pb$ (Table 13.4). The series has been named after actinium, the first member of the series to be discovered. The mass numbers of the nuclides are given by $4n + 3$ (n is an integer). This is also called the **4n + 3 series**.

As stated earlier, emission of an α-particle in a series results in a decrease of 4 unit in the mass number and 2 unit in the atomic number of the daughter element. Emission of a β-particle results in no change in mass number but the atomic number increases by 1 unit.

Branched disintegration is observed in all three series as e.g., with Th-C in the thorium series. A definite fraction (33.7%) of it undergoes α-decay to Th-C″ while the rest (66.3%) undergoes β-decay to Th-C′. Upon subsequent disintegration, both Th-C′ and Th-C″ yield Th-D (lead).

The disintegration series provide good examples of radioactive equilibrium. The uranium series, for example, begins with ^{238}U, having a half-life of 4.5×10^9 y (4.5 billion years). Its decay constant is less than 1/1000th the λ of any other member in the series (the last stable member is ^{206}Pb, decay constant = 0). The age of the earth has been indirectly estimated to be of the order of a few billion years and so the uranium has reached equilibrium with the stable end product lead. In fact, there is no other compelling reason to accept the coexistence of lead in uranium minerals. So, one may assume that the entire lead accumulated in a uranium mineral came from the decay of uranium. Since all intermediate nuclides decay at a much faster rate, we can assume equilibrium directly between uranium and lead.

From equation 13.1, putting $\lambda_{Pb} = 0$, one gets $N_{Pb} = N_U^0 (1 - e^{-\lambda_U t})$.
Also, $N_{Pb} = N_U^0 - N_U$. From the value of λ_U and experimental lead-uranium ratio in a mineral, one can now estimate t, the age of the mineral; this is also the minimum age of the earth.

The neptunium series, also known as the **4n + 1 series**, consists of artificially prepared radionuclides. It starts with $^{214}_{94}Pu$ and ends with $^{209}_{83}Bi$ (Table 13.5). The series has been named after neptunium, the member with the longest half-life. This series does not contain any emanation.

Table 13.2
The thorium series (4n)

Historical Name	Assigned Symbol	Chemical symbol	Half-life
Thorium	$^{232}_{90}\text{Th}$	^{232}Th	1.41×10^{10} y
↓ α			
Mesothorium-1	$^{228}_{88}\text{Ms Th}_1$	^{228}Ra	5.8 y
↓ β			
Mesothorium-2	$^{228}_{89}\text{Ms Th}_2$	^{228}Ac	6.13 hour
↓ β			
Radiothorium	$^{228}_{90}\text{Ra Th}$	^{228}Th	1.91 y
↓ α			
Thorium X	$^{224}_{88}\text{Th X}$	^{224}Ra	3.66 days
↓ α			
Thoron	$^{220}_{86}\text{Tn}$	^{220}Rn	55.6 s
↓ α			
Thorium A	$^{216}_{84}\text{Th A}$	^{216}Po	0.16 s
↓ α > 95% (β → ^{216}At)			
Thorium B (^{216}At)	$^{212}_{82}\text{Th B}$	^{212}Pb (^{216}At)	10.6 hour (3×10^{-4} s)
↓ β (from ^{216}At: α)			
Thorium C	$^{212}_{83}\text{Th C}$	^{212}Bi	60.5 min.
↓ β 65%			
Thorium C′	$^{212}_{84}\text{Th C}'$	^{212}Po	3×10^{-7} s
(from Thorium C: 35% α) Thorium C″	$^{208}_{81}\text{Th C}''$	^{208}Tl	3.1 min.
↓ α (Thorium C′); β (Thorium C″)			
Thorium D (lead; stable)	$^{208}_{82}\text{Th D}$	^{208}Pb	—

Table 13.3
The uranium series (4n + 2)

Historical Name	Assigned Symbol	Chemical symbol	Half-life
Uranium-I	$^{238}_{92}$U-I	^{238}U	4.5×10^9 y
↓ α			
Uranium-X_1	$^{234}_{90}$U-X_1	^{234}Th	24.1 days
↓ β			
Uranium-X_2	$^{234}_{91}$U-X_2	^{234}Pa	6.7 h
↓ β			
Uranium-II	$^{234}_{92}$U-II	^{234}U	2.4×10^5 y
↓ α			
Ionium	$^{230}_{90}$Io	^{230}Th	7.5×10^4 y
↓ α			
Radium	$^{226}_{88}$Ra	^{226}Ra	1590 y
↓ α			
Radon	$^{222}_{86}$Rn	^{222}Rn	3.8 days
↓ α			
Radium A	$^{218}_{84}$Ra-A	^{218}Po	3.05 min.
↓ α 99.98% (to Radium B); β (to ^{218}At)			
Radium B	$^{214}_{82}$Ra-B	^{214}Pb	26.8 min
^{218}At		^{218}At	1.6 s
↓ β (Radium B); α (^{218}At)			
Radium C	$^{214}_{83}$Ra-C	^{214}Bi	19.7 min
↓ β 99.96% (to Radium C′); α 0.04% (to Radium C″)			
Radium C′	$^{214}_{84}$Ra-C′	^{214}Po	1.5×10^{-4} s
Radium C″	$^{210}_{81}$Ra-C″	^{210}Tl	1.32 min
↓ α (Radium C′); β (Radium C″)			
Radium D	$^{210}_{82}$Ra-D	^{210}Pb	22 y
↓ β			
Radium E	$^{210}_{83}$Ra-E	^{210}Bi	5 days
↓ β			
Radium F (Polonium)	$^{210}_{84}$Ra-F	^{210}Po	140 days
↓ α			
Radium G (lead; stable)	$^{210}_{82}$Ra-G	^{206}Pb	...

Table 13.4

The Actinium Series (4n + 3)

Historical Name	Assigned Symbol	Chemical symbol	Half-life
Actinouranium	$^{235}_{92}$Ac-U	^{235}U	7.04×10^8 y
↓ α			
Uranium-Y	$^{231}_{90}$U-Y	^{231}Th	25.5 hour
↓ β			
Protoactinium	$^{231}_{91}$Pa	^{231}Pa	3.3×10^4 y
↓ α			
Actinium	$^{227}_{89}$Ac	^{227}Ac	13.5 y
↓ β (98.6%); α → ^{223}Fr			
Radioactinium	$^{227}_{90}$Ra-Ac	^{227}Th	18.9 days
^{223}Fr		^{223}Fr	21.8 min
↓ α (Radioactinium); β (^{223}Fr)			
Actinium X	$^{223}_{88}$Ac-X	^{223}Ra	11.2 days
↓ α			
Actinon	$^{219}_{86}$An	^{219}Rn	3.92 s
↓ α			
Actinium A	$^{215}_{84}$Ac-A	^{215}Po	1.8×10^{-3} s
↓ α			
Actinium B	$^{211}_{82}$Ac-B	^{211}Pb	36 min.
↓ β			
Actinium C	$^{211}_{83}$Ac-C	^{211}Bi	2.16 min.
↓ α → Actinium C″; β → Actinium C′			
Actinium C″ 99.84%	$^{207}_{81}$Ac-C″	^{207}Tl	4.77 min
Actinium C′ 0.16%	$^{211}_{84}$Ac-C′	^{211}Po	0.52 s
↓ β (Actinium C″); α (Actinium C′)			
Actinium D (lead; stable)	$^{207}_{82}$Ac-D	^{207}Pb	. . .

Table 13.5

The neptunium series (4n + 1)

Name	Symbol	Half-life
Plutonium	$^{241}_{94}Pu$	14.4 y
↓ β		
Americium	$^{241}_{95}Am$	433 y
↓ α		
Neptunium	$^{237}_{93}Np$	2.1×10^6 y
↓ α		
Protoactinium	$^{233}_{91}Pa$	27 days
↓ β		
Uranium	$^{233}_{92}U$	1.6×10^5 y
↓ α		
Thorium	$^{229}_{90}Th$	7340 y
↓ α		
Radium	$^{225}_{88}Ra$	14.8 days
↓ β		
Actinium	$^{225}_{89}Ac$	10 days
↓ α		
Francium	$^{221}_{87}Fr$	4.9 min.
↓ α		
Astatine	$^{217}_{85}At$	0.032 s
↓ α		
Bismuth	$^{213}_{83}Bi$	45.6 min.
↓ β (97.8%)		
Polonium	$^{213}_{84}Po$	4.2×10^{-6} s
Bismuth ↓ α		
Thalium	$^{209}_{81}Tl$	2.2 min.
Thalium ↓ β; Polonium ↓ α		
Lead	$^{209}_{82}Pb$	3.3 hour
↓ β		
Bismuth (stable)	$^{209}_{83}Bi$	

Radioelements and the periodic table

Some forty radioactive elements have been discovered with different radioactive properties. But the mass numbers of these elements lie between 206 and 238 and this fact makes their arrangement in the periodic table difficult. It was, however, observed that some of these radioelements were chemically indifferent and all of them could be fitted into the limited space of the periodic table with some generalization. This generalization is known as the **group displacement law** (Fajans and Soddy; 1913). The law follows readily from the charge and mass of the α or β particle emitted by a radioelement. An α-particle carries 2 units (e) of positive charge and 4 units (m_H) of mass. The emission of a β-particle first involves the conversion of a neutron into a proton and an electron, together with a neutrino. The electron is expelled as the β-particle, while the additional proton results in an increase in atomic number.

(i) When an α-particle is emitted by a radioactive atom, the product nuclide has an atomic number two units less than that of its parent and it occupies a position **two places to the left of the parent element** in the periodic table.

(ii) When a β-particle is emitted by a radioactive atom, the atomic number of the product nuclide becomes one unit greater than that of its parent; the product occupies a position **one place to the right of the parent element** in the periodic table.

The law is illustrated in Fig. 13.6 for the uranium series. Successive emission of one α- and two β-particles brings an element back to its original position in the table, but with a mass number four units less.

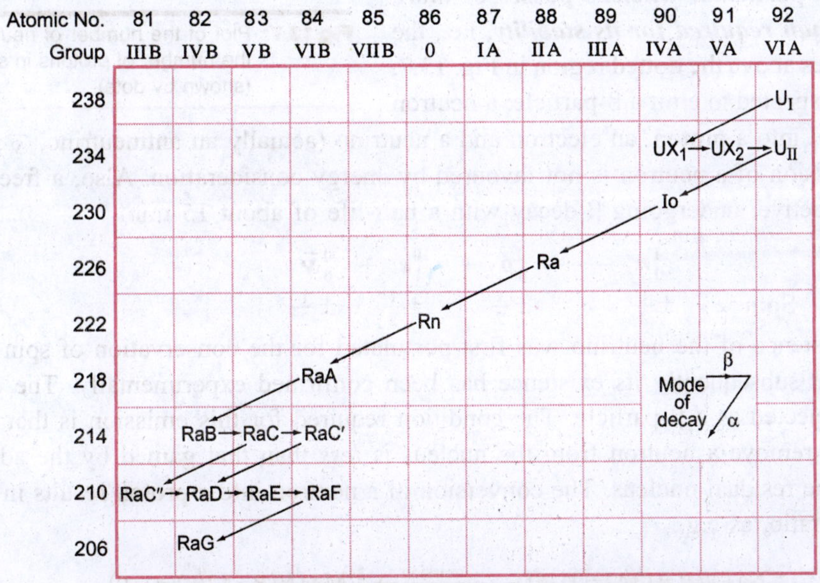

Fig.13.6: Illustration of the group displacement law for the uranium series.

13.3 THEORIES OF RADIOACTIVE DECAY

A radioactive decay process is always accompanied by a decrease in mass of the nuclide; the "lost" mass (Δm) appears as energy carried by the radioactive radiations in accordance with the mass-energy equation $E = \Delta m\, c^2$. The nature of the radiation or particle emitted in a decay process is determined by several factors, particularly the stability of the nuclides before and after decay. Certain generalizations in this respect may be made from the neutron-proton contents of stable nuclides.

13.3.1 Neutron:Proton Ratio

It is found that nuclides of stable isotopes have their neutron-to-proton ratio (*N/P*, say) within a narrow range. For stable nuclei of elements with atomic numbers less than 20, the *N/P* ratio is close to unity; for higher atomic numbers, the *N/P* ratio extends over restricted ranges depending upon the number of protons, or the atomic number *Z* (= *P*). This is revealed in the plot of the number of neutrons (*N*) in the stable nuclei for different atomic numbers (*Z*), as shown in Fig. 13.7 (dotted region). The behaviour of any unstable nucleus, lying outside this "*region of stability*", may now be explained in terms of its tendency to revert to this stable range of neutron-proton ratio by way of radioactive disintegration.

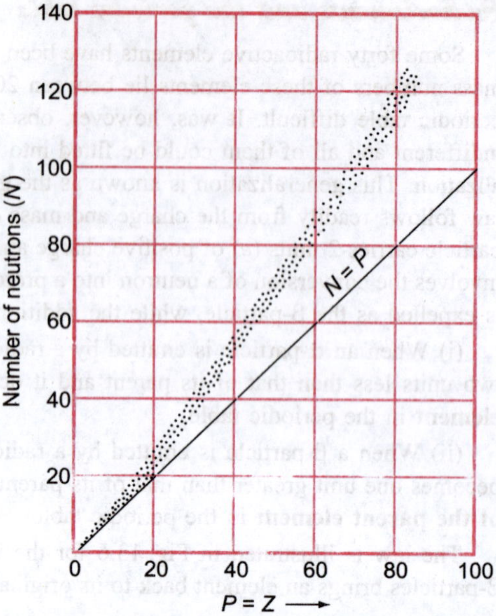

Fig.13.7: Plot of the number of neutrons against the number of protons in stable neuclei (shown by dots).

When a particular nucleus possesses more neutrons than required for its stability, i.e., the *N/P* ratio lies above the dotted region in Fig. 13.7, it may be expected to emit a β-particle; a neutron is converted into a proton, an electron and a neutrino (actually an antineutrino, $\bar{\nu}$; see later). Direct emission of a neutron is not favoured by energy consideration. Also, a free neutron is itself radioactive, undergoing β-decay with a half-life of about 15 min.

$$\begin{array}{lcccccccc} & {}^{1}_{0}n & \longrightarrow & {}^{1}_{0}p & + & {}^{0}_{-1}e & + & {}^{0}_{0}\bar{\nu} \\ \text{Spin} & \pm\frac{1}{2} & & \pm\frac{1}{2} & & \pm\frac{1}{2} & & \pm\frac{1}{2} \end{array}$$

The existence of the neutrino was first postulated for the conservation of spin during the process, but subsequently its existence has been confirmed experimentally. The electron so formed is ejected as a β-particle. The condition required for this emission is that the energy required to remove a neutron from the nucleus is *less* than that gained by the addition of a proton to the residual nucleus. The conversion of a neutron into a proton results in a decrease in the *N/P* ratio, as e.g.,

$$^{24}_{11}\text{Na}\ (N/P = 13/11 = 1\cdot18) \xrightarrow{\beta\text{-emission}} {}^{24}_{12}\text{Mg}\ (N/P = 12/12 = 1\cdot0)$$

Many artificially prepared nuclei covering the whole range of atomic numbers disintegrate by β-emission owing to high *N/P* ratio.

The energy released in the process is shared between the electron and the neutrino in a random manner. This is why the energy of a β-particle may range from almost zero to a definite maximum value equal to the total energy released in the process (see later).

A nucleus lying much above the stable *N/P* region, i.e., having a very high *N/P* ratio, may undergo a series of β-decays to attain stability:

$$^{141}_{56}\text{Ba} \xrightarrow{\beta} {}^{141}_{57}\text{La} \xrightarrow{\beta} {}^{141}_{58}\text{Ce} \xrightarrow{\beta} {}^{141}_{59}\text{Pr}$$

Alternatively, a nucleus having a high *N/P* ratio may approach stability by emitting a neutron. But this form of decay is rare, perhaps due to the very high binding energy of the neutron (~8 MeV). Only highly energetic nuclei may undergo this type of decay:

$$^{87}_{36}\text{Kr} \longrightarrow {}^{86}_{36}\text{Kr} + {}^{1}_{0}n$$

When a nucleus contains fewer number of protons than required for its stability, i.e., the *N/P* ratio lies below the stability range for this atomic number, the nucleus may tend to attain stability in either of three ways.

1. α-emission: For heavier nuclei, the common process is the emission of an α-particle, whereby 2 protons and 2 neutrons are ejected simultaneously:

$$^{238}_{92}\text{U}\ (N/P = 146/92 = 1\cdot58) \xrightarrow{\alpha\text{-decay}} {}^{234}_{90}\text{Th}\ (N/P = 144/90 = 1\cdot60)$$

It is interesting to note that no protons are emitted as such; instead, the nuclei emit α-particles which are not normal constituents of the nucleus. The energy required to remove a proton is about 8 MeV so that emission of a proton is unlikely except from highly excited nuclei. On the other hand, when two neutrons and two protons combine to form an α-particle, a large amount of energy (28.2 MeV) is released. This is the binding energy of the α-particle. When this binding energy is greater than the energy required to remove two protons and two neutrons from a particular nucleus, it is likely that the nucleus will emit an α-particle (see later). The energy required in the removal of neutrons and protons depends on the nature of the nucleus. For a large number of heavy nuclei this energy lies around 23 MeV, which is definitely less than the binding energy of the α-particle.

2. Positron emission: Sometimes the unstable nucleus may emit a positron. In this mode of decay, a proton is first converted into a neutron and a positron, together with a neutrino. This mode of decay is not found among the naturally occurring radionuclides, but has been observed among many artificially prepared radioactive species.

$$^{11}_{6}\text{C} \rightarrow {}^{11}_{5}\text{B} + {}^{0}_{+1}e + \nu; \quad {}^{19}_{10}\text{Ne} \rightarrow {}^{19}_{9}\text{F} + {}^{0}_{+1}e + \nu$$

The energy criterion for positron emission has been discussed later.

3. K-electron capture: Lastly, some artificially prepared nuclei with low *N/P* ratio have been found to attain stability by capturing an extranuclear electron.

When sufficient energy is not available for emission of a positron, the nucleus acquires stability by engulfing an extranuclear electron from the K-shell. The electron combines with a proton to form a neutron and a neutrino (conservation of half-integral spin). The vacancy in the K-shell is filled by an electron from higher shells. The K-series of characteristic x-rays of the product element is obtained as a consequence. $^{7}_{4}\text{Be}$ and $^{54}_{25}\text{Mn}$ provide examples of K-electron capture:

$$^{7}_{4}\text{Be} \xrightarrow{+{}^{0}_{-1}e} {}^{7}_{3}\text{Li} + \gamma\ (+\nu) \qquad {}^{54}_{25}\text{Mn} \xrightarrow{+{}^{0}_{-1}e\ (\text{K})} {}^{54}_{24}\text{Cr} + \gamma\ (+\nu)$$

The γ-rays are mixed with characteristic x-rays (K-series) of the product elements. $^{48}_{23}\text{V}$ decay by both positron emission (58%) and K-electron capture (42%).

13.3.2 The Theory of Alpha-decay

First of all, let us consider some of the facts related to alpha-decay. α-particles ionize the gaseous medium through which they pass. For example, in air (normal pressure) the α-particles from natural radioelements are found to cause ionization up to about 2.5 to 8.5 cm distances from the source; the ionization along the path of the particles increases to a maximum and then suddenly drops to zero. This characteristic range or sharply defined ionization path length increases with the disintegration constant of the source of the α-particles. When the logarithm of the disintegration constant (λ, in s^{-1}) is plotted against the logarithm of the range (*R*, in cm of standard air) for the α-emitting members of the same radioactive series, an approximately straight line is obtained. This is consistent with the **Geiger-Nuttall rule** which may be expressed mathematically as

$$\log \lambda = A \log R + B$$

where *A* and *B* are constants. The constant *A* gives the slope of the line which is nearly the same for the three naturally occurring series, but *B* has different values. The mean range is empirically found to be proportional to $E^{3/2}$, where *E* is the kinetic energy of the α-particle. So the rule may also be written by changing the constants:

$$\log \lambda = A' \log E + B'$$

Any theory of alpha-decay should provide an explanation for this rule.

Now we return to the question: why alpha particles are emitted rather than individual protons or other assembles of proton and neutron, *e.g.*, the $^{3}_{2}He$ nucleus? As already mentioned, the answer lies in the high binding energy of the alpha particle, a consequence of its high mass difference from the constituent particles. The total kinetic energy *Q* released in the process of an α-emission is equal to Δmc^2, where Δm is the mass difference before and after the process, i.e.,

$$\Delta m = m_i - (m_f + m_\alpha)$$

where m_i = initial mass of the nucleus; m_f = final mass of the nucleus ; m_α = mass of the alpha particle.

The kinetic energy *E* of the α-particle would be slightly *less* than Δmc^2, since the recoiling nucleus will share a part of this energy. If *A* is the mass number of the original nucleus, then

$$E \simeq (A - 4)\, Q/A.$$

Since *A* is usually large (>210), *E* is not very different from *Q*. For the α decay of $^{222}_{86}Rn$, Q = 5·59 MeV ; E = 5·49 MeV.

By similar treatment it can be shown that emission of an alpha particle is energetically more favourable than any other emission; the emission of the α-particle from $^{232}_{92}U$, for example, is accompanied by the release of 5.4 MeV energy, while the emission of a proton or $^{3}_{2}He$ nucleus would *require* 6.1 MeV or 9.6 MeV respectively. The observed disintegration energies in alpha decay are in good agreement with those predicted from the mass difference involved. The energy of the alpha particles usually lies between 4 to 9 MeV.

A paradox: At this point, we are puzzled by one more fact. $^{232}_{92}U$ emits α-particles with K.E. equal to 4.18 MeV; but when the same nucleus is bombarded with alpha-particles of much higher K.E. (e.g., 8.7 MeV, from Th C′), no disintegration takes place. If we believe (with due reasons) that the alpha particles exist momentarily within the nucleus, then the paradox is this: *an alpha particle with 4.18 MeV kinetic energy can "jump out" from a nucleus, but an alpha particle with 8.7 MeV kinetic energy cannot "jump in" to the same nucleus.*

What is the nature of interaction involved when an α-particle approaches a nucleus? Clearly, the first answer will be the coulombic repulsion; the repulsion potential energy at a distance r from a nucleus with Z protons would be

$$V(r) = \frac{2e \times Ze}{4\pi\varepsilon_0 r}$$

Using an estimated value of r for the nucleus of uranium as 8.6×10^{-15} m, $V(r)$ comes out to be about 30 MeV. Other calculations give an estimate of about 25 MeV. We may now imagine the nucleus to be surrounded with a potential energy barrier of 25 MeV. This explains why a ~9 MeV α-particle cannot disintegrate the U-nucleus. Side by side an alpha particle *inside* such a nucleus is now comparable to a particle in a box with walls of 25 MeV potential; this wall has to be surmounted by the alpha particle in the radioactive decay process. But we have already noted that such α-particles have much lower energy, 4–9 MeV and it is peculiar that they come across the barrier of ~25 MeV.

Tunnelling effect

The problem of α-emission cannot be explained from any classical analogy. Once again, quantum mechanical treatment provides a rationalization of the situation (Gamow, Gurney and Condon; 1928). It is assumed that the alpha particle exists within a heavy nucleus in a state of constant motion, undergoing frequent collisions with the wall of the nucleus (~10^{21} s^{-1}). It can be shown that there is a small, but finite, probability that the alpha particle can pass through the barrier at each collision, in spite of the height of the barrier. This "tunnelling effect" is not uncommon for waves. For example, a light wave undergoing reflection from even a perfect mirror may partially pass through it (if the surface is sufficiently thin) with an exponentially decreasing amplitude. Calculation of the transmission probability of alpha particles according to this theory was successful in explaining the variation of disintegration energy with disintegration constant. A similar correlation may also be arrived at from the empirical Geiger-Nuttall rule. The reverse process of entry of a bombarding alpha-particle into a target nucleus by tunnelling effect is also theoretically possible; but its probability would be exceedingly small.

13.3.3 Beta-rays and the Neutrino

The velocity of α-particles, and hence their momentum, can be measured by studying their deflection in a magnetic field. The study of β-particles present a different problem. It is observed that most of the β-particles have energies smaller than 4 MeV. Being much lighter than the α-particles, they travel with much greater speed: a 4 MeV β-particle would have a velocity close to 0.99 times that of light (an α-particle of same energy would have a velocity about 0.05 times). Because of such high velocities, β-particles must be treated relativistically.

When the velocities of β-particles from a radioactive nuclide are analysed by means of the magnetic spectrograph, it is found that the velocities are not constant; rather they form a continuous spectrum which passes through a maximum and then falls rapidly (Fig. 13.8). This leads to an apparent paradox. The energies of the emitted electrons may have any value between zero and the maximum energy in the continuous spectrum. Only the maximum energy electrons, which are comparatively very few in number, possess energy equal to the difference between the energy of the nucleus before and after emission, as obtained from their mass difference. But the average energy of the emitted particles is only about one-third of this maximum energy.

To verify this, a sample of beta-emitter (say $^{210}_{83}Bi$) is placed in a calorimeter and the heat evolved after a given number of decays is measured. The average energy per decay is simply the total heat evolved divided by the number of decays. In case of $^{210}_{83}Bi$, this average comes to about 0.35 MeV, close to the average value of 0.39 MeV obtained from the spectrum. This suggests that the observed continuous spectra represent the actual energy distribution of the beta-particles from the radioactive source. But the maximum energy of the beta-particles from this same source is 1.17 MeV, nearly three times the average energy.

It appears therefore that about two-thirds of the energy of disintegration have disappeared, violating the principle of conservation of mass-energy. But this principle is uniformly applicable to all known universal processes; hence, its violation in this single case appears most unlikely.

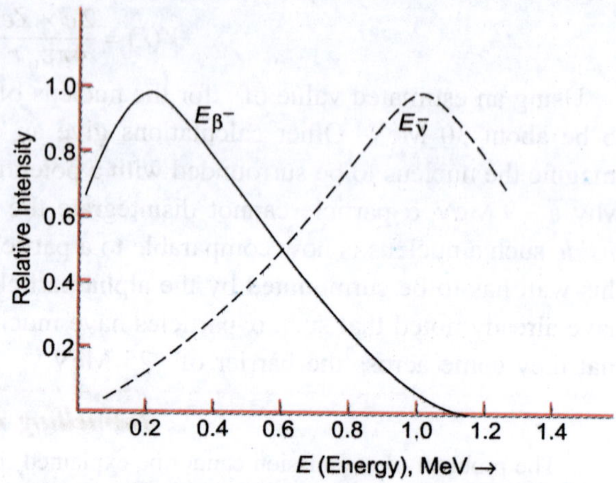

Fig.13.8: Typical energy spectrum of beta-particles from a radioactive nucleus. [Dashed curve corresponds to the energy of the antineutrino (see later).]

Beta emissions also apparently violate the principle of conservation of linear and angular momenta. The directions of the emitted beta particles and of the recoiling mother nucleus are almost never exactly opposite as required for conservation of linear momentum. Similarly, the known half-integral spins of the electron, proton and the neutron indicate that angular momentum is not conserved in the conversion of a neutron into a proton and an electron.

The neutrino hypothesis

W. Pauli, in 1931, suggested that an uncharged particle of small or zero rest mass and half-integral spin probably accompanied each beta particle during its formation from a radioactive nuclide. This idea was developed by Enrico Fermi in 1934 and the new particle was named the neutrino. According to this hypothesis, each act of beta emission releases an amount of energy equal to the maximum energy in the beta-ray spectrum. This disintegration energy is shared by the emitted beta-particle, the neutrino and the recoiling nucleus (see box). The recoil nucleus carries away negligible kinetic energy; the energy shared by the neutrino accounts well for the conservation of energy and momentum. Subsequently, two kinds of neutrino have been established—the neutrino (ν) and the antineutrino $\bar{\nu}$. The spin of the neutrino is opposite in direction to the direction of its motion. When viewed from behind, the neutrino seems to spin counterclockwise. The antineutrino spins in the same direction as its direction of motion, that is clockwise if viewed from behind. In ordinary beta-decay, an antineutrino is actually emitted:

$$n \to p + e^- + \bar{\nu}$$

Positron emission similarly involves the conversion of a nuclear proton into a neutron, a positron and a neutrino:

$$p \rightarrow n + e^+ + \nu$$

The process of K-electron capture also produces a neutrino:

$$p + e^- \rightarrow n + \nu$$

The absorption of an electron by a nucleus is equivalent to the emission of a positron from it. Hence these two nuclear processes are essentially the same.

Detection of neutrino

Since neutrinos possess no charge and mass, their interaction with matter is extremely insignificant and this is why it took nearly 25 years after Pauli's suggestion to prove the real existence of these particles. Reines and Cowan (1953 –) made an arrangement to detect antineutrinos (from the beta decays occurring in a nuclear reactor) through the *inverse beta decay*

$$p + \bar{\nu} \rightarrow n + e^+$$

The antineutrinos were allowed to interact with the protons in a tank of water containing a cadmium compound in solution. As soon as a proton absorbed a neutrino to form a neutron and a positron, the positron encountered an electron. Annihilation of these two particles gave rise to gamma radiations which were detected around the tank. The newborn neutron was also captured by the cadmium nucleus which released about 8 MeV excitation energy; but this process required several microseconds to take place. Thus the gamma ray detectors recorded the arrival of two sets of γ-ray photons at small intervals of time (~5 μs). The experiment was performed with the reactor alternately on and off and the expected variation in the frequency of neutrino-capture was confirmed.

Mass and energy change accompanying beta emission

Emission of a beta particle increases the number of protons in the nucleus by one unit while the mass number and the number of extranuclear electrons remain unchanged. The resulting cation readily captures an electron. If we express the process as

$${}^{A}_{Z}X \longrightarrow {}^{A}_{Z+1}Y^+ + {}^{0}_{-1}e + {}^{0}_{0}\bar{\nu},$$

We get, $$\Delta m = m_X - \left[m_{Y^+} + m_e\right] = m_X - m_Y.$$

We have ignored the binding energy of the electron to Y^+. The disintegration energy Q (931 × Δm MeV) is now distributed as kinetic energy of the emitted electron and antineutrino and the recoil nucleus [E_r, say]:

$$Q = 931\,\Delta m = E_e + E_{\bar{\nu}} + E_r \simeq E_e + E_{\bar{\nu}}.$$

E_r is negligible since the mass of the nucleus is many times larger compared to that of the electron.

Thus, the disintegration energy is almost completely shared by the emitted electron and the antineutrino, consistent with observation.

Beta decay cannot be accounted for by the strong interactions operating between the nucleons (see Ch. 14). Another short-range fundamental interaction seems to be responsible for such cases; this is known as *weak interaction*.

13.3.4 Gamma Decay and Internal Conversion

Gamma rays originate when a nucleus gets deexcited by releasing energy. Similar to the energy states of electrons in an atom, the nuclear energy states are also quantized. Excited nuclei return to their ground states by emitting γ-ray photons with energies equal to the difference between the various initial and final states in the transitions involved. Thus a $^{27}_{12}Mg$ nuclide may evolve beta particles ($t_{1/2}$ = 9.5 min) to leave the resulting $^{27}_{13}Al$ nucleus in either of two excited states—at 1.015 MeV or 0.834 MeV above the ground states. The excited $^{27}_{13}Al^*$ nucleus may now return to the ground state in either of two ways: (i) in one step, evolving γ-ray photons of energy equal to 1.015 MeV or 0.834 MeV; (ii) or in two steps; first with a γ-photon of energy 0.181 MeV to reach the 0.834 MeV state and subsequently emitting a γ-photon with 0.834 MeV energy.

Sometimes an excited nucleus may also return to the ground state by transferring its excitation energy to one of the ortibal electrons. This is known **as internal conversion of γ-rays (IC)**. The electron is emitted with a kinetic energy equal to the lost nuclear excitation energy *minus* the binding energy of the electron in the atom. The electrons expelled are not limited to the K-shell alone. *Discrete energies* of all the electron shells of the concerned atom is suggested by bending these electrons in a magnetic field, when a line spectrum is obtained. (β-radiations, on the other hand, produce a continuous spectrum). Many artificial radio isotopes, e.g., ^{41}Ca, ^{80}Br, ^{133}Ba exhibit internal conversion of γ-ray.

Usually the excited nuclei remain in their states of excitation for a very short time; that is, they have very short half-lives for gamma decay. But in rare cases, an excited nucleus may remain in its excited state for an appreciably long time; it is then called an *isomer* of the same nucleus in the ground state. The excited $^{87}_{38}Sr*$ nucleus is an isomer of the $^{87}_{38}Sr$ nucleus with a half-life of 2.8 hours.

13.3.5 Positron Emission and K-electron Capture

The positron was first discovered by Anderson (1932) in cosmic rays with the help of the Wilson cloud chamber described before. Positrons and electrons (also called negatrons) give the same type of track but their curvatures are opposite to one another. The mean lifetime of a positron has been estimated to be about one-tenth of a microsecond, which is much less than that of the stable proton.

When a metal foil is irradiated with high energy nuclear γ-rays, an electron-positron pair is produced simultaneously. Since two electron-masses are produced, the energy of the γ-photon must be greater than $2m_0c^2$ (m_0 = rest mass of the electron = 1/1836 u ≡ 1/1836 × 931 MeV; see chapter 14). Thus the minimum energy E of the γ-photon is

$$E = 2m_0c^2 = 2 \times 0.511 \text{ MeV} = 1.022 \text{ MeV} \Rightarrow 1{\cdot}63 \times 10^{-13} \text{ J}$$

The positron quickly recombines with the free electrons in the metal and is transformed back to γ-photons.

As already mentioned, the emission of positron by a radioactive nucleus involves the conversion of a nuclear proton into a neutron, a positron and a neutrino. The atomic number Z decreases by one unit while the mass number remains unchanged. The daughter nucleus has one more neutron than the mother nucleus but it loses one more extra-nuclear electron to maintain the balance with the lower Z. So the daughter nucleus has a mass deficit of $2m_e$ (m_e = electron mass). If we represent the process as

$$ {}^{A}_{Z}X \longrightarrow {}^{A}_{Z-1}Y^- + {}^{0}_{+1}e + {}^{0}_{0}\nu \longrightarrow {}^{A}_{Z-1}Y + {}^{0}_{-1}e + {}^{0}_{+1}e + {}^{0}_{0}\nu $$

$$ \Delta m = m_X - [m_Y + 2m_e] \quad (\text{mass of } {}^{0}_{-1}e = \text{mass of } {}^{0}_{+1}e) $$

The disintegration energy Q (= 931Δm MeV) is distributed as kinetic energy of the electron and positron; the K.E. of the recoil nucleus is negligible in view of its large mass.

As we have seen above, this amount of mass deficit corresponds to 1.022 MeV of energy. So for positron emission, the mass difference between the mother and the daughter nucleus should be greater than 1.022 MeV:

$$ m_X - m_Y \geq 1.022 \text{ MeV}. $$

In some cases, this mass-energy difference remains below 1.022 MeV when the nucleus cannot gain stability by positron emission. As seen earlier, a neutron deficient nucleus may still attain stability by increasing the number of neutrons by capturing an orbital electron (K-capture). If we represent the process as

$$ {}^{A}_{Z}X \xrightarrow{+e} {}^{A}_{Z-1}Y + {}^{0}_{0}\nu, $$

$$ \Delta m = m_X - m_Y \quad (Q \equiv 931\Delta m \text{ MeV}). $$

The small mass-difference now appears as γ rays. Thus, when the mass-energy difference in a nuclear transformation is much greater than 1.022 MeV, positrons (and neutrinos) are emitted. As the energy difference decreases but remains still greater than 1.022 MeV, the relative probability of K-capture increases; ultimately below 1.022 MeV, no positrons are emitted.

Sometimes a heavy radionuclide simultaneously undergoes all three possible modes of decay—beta emission, positron emission and K-capture:

	$\longrightarrow$	${}^{64}_{28}Ni$	K-capture, 43%
${}^{64}_{29}Cu$	$\longrightarrow$	${}^{64}_{28}Ni$	positron emission 19%
	$\longrightarrow$	${}^{64}_{30}Zn$	beta-emission 38%

The particular mode of decay by an individual nuclide is governed by chance.

Problem 13.12

The maximum kinetic energy of positrons emitted by ${}^{13}_{7}N$ is 1.20 MeV. Calculate the mass of the ${}^{13}_{7}N$ nuclide (m_N) from the masses of ${}^{13}_{6}C$ (m_C : 13·00335 u) and an electron (m_e : 0·00055 u).

Hint: $${}^{13}_{7}N \longrightarrow {}^{13}_{6}C^- + {}^{0}_{+1}e + {}^{0}_{0}\nu \longrightarrow {}^{13}_{6}C + {}^{0}_{-1}e + {}^{0}_{+1}e + {}^{0}_{0}\nu$$

$$ \Delta m = m_N - m_C - 2m_e $$

$$ \text{Also, } \Delta m = 1{\cdot}20/931\ u = 0{\cdot}00129\ u $$

$$ m_N = (0{\cdot}00129 + 13{\cdot}00335 + 2 \times 0{\cdot}00055)\ u $$

13.3.6 Nuclear Isomerism

Pairs of nuclides having the same atomic and mass numbers (isotopic and isobaric) but different radioactive properties are known as nuclear isomers and the phenomenon is called nuclear isomerism. The difference in radioactive property is marked by two modes of decay having different half-lives from the same nuclide. Though this phenomenon was first observed in U-X_2 (Table 13.3), much work has been done after the discovery of artificial radioactivity and the production of many radio-nuclides by neutron bombardment. Bombardment of ^{79}Br by slow neutrons produces ^{80}Br (natural bromine contains only ^{79}Br and ^{81}Br). The ^{80}Br undergoes β-decay with two half-lives, 4.5 hours and 18 minutes respectively. These are attributed to two isomeric states of ^{80}Br. The same nuclide produced by other means also shows this isomerism.

It has been established that the isomer of ^{80}Br with 4.5 hour half-life is in a metastable state of higher energy than the ground state. It reaches the ground state by emitting γ-rays—and the ground state isomer has the 18 minute half-life. This decays mainly by beta emission (92%) to $^{80}_{36}$Kr (stable). Positron emission (3%) and orbital electron capture (5%) to the ground state of $^{80}_{34}$Se also takes place. The two isomers of ^{80}Br are said to be genetically related.

In some cases, a nuclide may exist for a sufficiently long time in a higher energy metastable state and exhibit independent isomer decay, for example

^{52}Mn	β^+ ($t_{1/2}$ = 5.5 days and 21.3 minutes)
^{106}Ag	*K*-Capture ($t_{1/2}$ = 43 days); β^- ($t_{1/2}$ = 53 hours).

The shell model of the nucleus (chapter 14) provides some rationalization for nuclear isomerism. It is generally found that a nuclide usually exists in a metastable state if there is a large difference of spin angular momentum between the two nuclear states, but the corresponding energy difference is not great, Such situations exist near and up to the saturation limit of "shells" within the nucleus—and there occur "islands of isomerism" bounded by the magic number of nucleons: 50, 82 and 126. Fewer isomers exist immediately after these numbers. The numbers correspond to closed shells of nucleons where nuclear spins are paired and few nucleons are available to provide a large spin angular momentum.

13.4 ARTIFICIAL RADIOACTIVITY

Bombardment of stable elements with high energy α-particles, protons, neutrons, deuterons or γ-rays produce radioactive nuclides. These radionuclides do not occur naturally and may be called man-made or artificial. The radioactivity exhibited by these artificial radio-nuclides is referred to as artificial radioactivity or sometimes as induced radioactivity.

The phenomenon of artificial radioactivity was first observed by I. Curie and and F. Joliot (1934; Nobel prize 1935.). When aluminium metal was bombarded with α-particles, protons, neutrons and positrons were emitted. It was found that the emission of protons and neutrons ceased as soon as the source of α-particles was removed; but the emission of positron by the isolated aluminium target continued for a long time afterward. Chemical analysis showed that the Al-foil contained both Si and P. It was finally concluded that bombardment of the Al-foil produces in both ^{30}Si and ^{30}P; it was the latter which decayed by positron emission.

$${}^{27}_{13}\text{Al} + {}^{4}_{2}\text{He} = {}^{30}_{14}\text{Si} + {}^{1}_{1}\text{H}$$

$${}^{27}_{13}\text{Al} + {}^{4}_{2}\text{He} = {}^{30}_{15}\mathbf{P} + {}^{1}_{0}\text{n}$$

$${}^{30}_{15}\mathbf{P} \xrightarrow{t_{1/2}=3.25\text{ min}} {}^{30}_{14}\text{Si} + {}^{0}_{+1}e\,(\beta^+) + \nu$$

The radioactive phosphorus was actually isolated chemically. Dissolution of the irradiated aluminium target in HCl evolved active hydrogen gas that showed positron activity. The target aluminium was dissolved in aqua regia and treated with sodium phosphate and zirconyl nitrate—the precipitate of zirconyl phosphate carried down all the activity.

A large number of radioactive nuclides have been prepared artificially. Besides α-particles, protons, deuterons, neutrons and also γ-rays have been used as projectiles for the purpose. Irradiation by neutron is most commonly employed at present in nuclear reactors which provide a high neutron flux. The substance to be irradiated is placed in a small aluminium cylinder. It is pushed into the reactor core through a channel and irradiated for about two to three times the half-life of the product. A few examples are:

$${}^{31}_{15}\text{P} + {}^{1}_{0}n \longrightarrow \gamma + {}^{32}_{15}\mathbf{P} \xrightarrow[14.3\text{ d}]{t_{1/2}} {}^{32}_{16}\text{S} + {}^{0}_{-1}e$$

$${}^{23}_{11}\text{Na} + {}^{1}_{0}n \longrightarrow \gamma + {}^{24}_{11}\mathbf{Na} \xrightarrow[15\text{ hours}]{t_{1/2}} {}^{24}_{12}\text{Mg} + {}^{0}_{-1}e$$

Also, $${}^{27}_{13}\text{Al} + {}^{1}_{0}n \longrightarrow {}^{4}_{2}\text{He} + {}^{24}_{11}\mathbf{Na} \longrightarrow \text{(see above)}$$

$${}^{31}_{15}\text{P} + {}^{1}_{0}n \longrightarrow {}^{1}_{1}\text{H} + {}^{31}_{14}\mathbf{Si} \longrightarrow {}^{31}_{15}\text{P} + {}^{0}_{-1}e$$

$${}^{127}_{53}\text{I} + {}^{1}_{0}n \longrightarrow {}^{128}_{53}\mathbf{I} \longrightarrow {}^{128}_{54}\text{Xe} + {}^{0}_{-1}e$$

$${}^{39}_{19}\text{K} + {}^{1}_{0}n \longrightarrow 2\,{}^{1}_{0}n + {}^{38}_{19}\mathbf{K} \longrightarrow {}^{38}_{18}\text{Ar} + {}^{0}_{+1}e$$

Some more examples of artificial radionuclides are given below:

$${}^{24}_{12}\text{Mg}\,(\alpha, n) \longrightarrow {}^{27}_{14}\mathbf{Si} \xrightarrow{t_{1/2}=5\text{ s}} {}^{27}_{13}\text{Al} + {}^{0}_{+1}e + \nu$$

$${}^{10}_{5}\text{B}\,(\alpha, n) \longrightarrow {}^{13}_{7}\mathbf{N} \xrightarrow{t_{1/2}=10\text{ m}} {}^{13}_{6}\text{C} + {}^{0}_{+1}e + \nu$$

$${}^{12}_{6}\text{C}\,(d, n) \longrightarrow {}^{13}_{7}\mathbf{N} \xrightarrow{10\text{ m}} {}^{13}_{6}\text{C} + {}^{0}_{+1}e + \nu$$

$${}^{12}_{6}\text{Li}\,(d, p) \longrightarrow {}^{8}_{3}\mathbf{Li} \longrightarrow {}^{8}_{4}\text{Be} + {}^{0}_{-1}e + \bar{\nu}$$

$${}^{26}_{12}\text{Mg}\,(d, \alpha) \longrightarrow {}^{24}_{11}\mathbf{Na} \longrightarrow {}^{24}_{12}\text{Mg} + {}^{0}_{-1}e + \bar{\nu}$$

$${}^{10}_{5}\text{B}\,(p, n) \longrightarrow {}^{11}_{6}\mathbf{C} \longrightarrow {}^{11}_{5}\text{B} + {}^{0}_{+1}e + \bar{\nu}$$

$$^{12}_{6}C\,(p, \gamma) \longrightarrow {}^{13}_{7}N \longrightarrow {}^{13}_{6}C + {}^{0}_{+1}e + \nu$$

$$^{31}_{15}P\,(\gamma, n) \longrightarrow {}^{30}_{15}P \longrightarrow {}^{30}_{14}Si + {}^{0}_{+1}e + \nu$$

$$^{81}_{35}Br\,(\gamma, n) \longrightarrow {}^{80m}_{35}Br \text{ (metastable)}$$

$$\xrightarrow[\text{transition}]{\text{isomeric}} {}^{80}_{35}Br \xrightarrow{\beta} {}^{80}_{36}Kr^{*} \xrightarrow{\gamma} {}^{80}_{36}Kr$$

$$^{24}_{12}Mg\,(d, \alpha) \longrightarrow {}^{22}_{11}Na \xrightarrow[2\cdot 6\,y]{t_{1/2}} {}^{22}_{10}Ne + {}^{0}_{+1}e + \nu$$

Many of these artificial radioisotopes find widespread application in biology, medicine, agriculture and other fields.

The artificial radioelements follow the same decay rate as the natural radioelements. Their half-life periods may vary from a fraction of a second to several years, though most of them have half-lives of the order of minutes.

Majority of the artificial radioelements emit β-particles. Some of them emit positrons. Two other less frequent modes of disintegration are K-electron capture and internal conversion of γ-rays (section 13.3.4).

13.5 APPLICATIONS OF RADIOACTIVITY

Radioactivity has found widespread application in different fields of science and technology. We shall briefly mention only a few such applications, which may be broadly classified into (i) geochemical and radiocarbon dating, (ii) tracer studies.

(i) Geochemical and radiocarbon dating: These are based on the rate at which a selected radioactive nuclide undergoes decay. As an example, the age of an uranium mineral may be determined from its lead-uranium ratio (see problem 13.13). This gives the minimum age of the earth.

The age of a mineral may also be calculated from the amount of helium entrapped in its structure. Each α-particle from radioactive disintegration gives rise to one atom of helium. It has been estimated that in one year, one gram of uranium in equilibrium with its decay products forms nearly 10^{-7} cm^3 of helium while each gram of thorium present in the same mineral forms nearly 0.3 times this quantity. If each gram of a mineral contains U g. of uranium, Th g of thorium and v cm^3 of helium gas, the age of the mineral may be calculated as

$$\text{age} = \frac{v\,(\text{cm}^3 \text{ per g. mineral})}{U + 0.3\,Th} \times 10^7 \text{ years.}$$

The age obtained will be low since some helium would have invariably escaped.

The ratio of ^{40}K to ^{40}Ar or of ^{87}Rb to ^{87}Sr or of ^{235}U to ^{238}U may also be used to determine the age of a mineral and hence the earth (see problem 13.14).

Problem 13.13

An uranium mineral contains 15 g of lead for each 100 g of uranium present. What is the age of the mineral? $t_{1/2}$ of $^{238}U = 4.5 \times 10^9$ y.

Solution: It is assumed that no lead was initially present in the sample and the whole of it came from the uranium.

1 g-atom (i.e., 238 g) of U yields 1 g-atom or 206 g of Pb.

$$\therefore \frac{15}{206} \text{ g atom of Pb is obtained from } \frac{15}{206} \text{ g-atom of U} \Rightarrow 17 \text{ g of } ^{238}U.$$

Thus, during the age of the mineral, (100 + 17) or 117 g. of uranium has disintegrated to 100 g of uranium. We have to calculate the corresponding time (t).

$$\lambda_U = \frac{0 \cdot 693}{t_{1/2}} = \frac{0.693}{4.5 \times 10^9} y^{-1} = 1 \cdot 54 \times 10^{-10} \ y^{-1}.$$

Hence, $t = \frac{2 \cdot 303}{\lambda} \log \frac{117}{100} \ y \Rightarrow 1 \cdot 02 \times 10^9$ y.

Thus, the minimum age of the minerai is 1020 million years. [The *minimum age of the earth.*]

Note: Rigorous chemical analysis of different uranium and thorium minerals as well as of meteorites yield an estimated age of the earth between 2000 and 3000 million years.

Problem 13.14

The present day abundance ratio of the two main isotopes of uranium, U-235 and U-238, is nearly 1:138. Assuming that this ratio could never have been greater than unity, estimate the maximum age of the earth (Half lives: U-235:7.1 × 10^8 y; U-238: 4.5 × 10^9 y).

Hint: The maximum age of the earth corresponds to the time during which the abundance ratio has reached its present value from the maximum value of unity. Thus at t = 0, U-235/U-238 = 1.

Since $N_t = N_0 e^{-\lambda t}$ [N represents number of respective atoms]

$$N_t^{235} = N_0^{235} e^{-\lambda_{235} t} \text{ and } N_t^{238} = N_0^{238} e^{-\lambda_{238} t} \text{ or, } \frac{N_{238}}{N_{235}} \cdot \frac{N_0^{235}}{N_0^{238}} = exp \left\{ 0 \cdot 693t \left(\frac{1}{7.1 \times 10^8} - \frac{1}{4.5 \times 10^9} \right) \right\}$$

or, 138 × 1 = exp (0.693t × 10^{-9} × 1.18) ⇒ t = 6 × 10^9 y.

Problem 13.15

Find the mass of lead formed from 1 kg of ^{238}U during the age of the earth, 2.5 × 10^9 years. ($t_{1/2}$ = 4.5 × 10^9 y).

Hint: 1 atom of U finally gives 1 atom of Pb. Thus, 238 g U produces 206 g Pb.

If the amount of U–238 remaining unchanged be x kg out of 1 kg 2.5 × 10^9 years ago, then

$$\frac{0 \cdot 693}{4 \cdot 5 \times 10^9} = \frac{2 \cdot 303}{2 \cdot 5 \times 10^9} \log \frac{1}{x} \Rightarrow x = 0 \cdot 68 \text{ kg}.$$

So U disintegrated = (1 – 0.68) kg = 0.32 kg. Lead formed = (0.32/238) × 206 kg.

Radiocarbon dating

All carbon in the world's living cycle is kept uniformly radioactive through ^{14}C. This originates in the atmosphere through bombardment of nitrogen by neutrons from cosmic rays:

$$^{14}_{7}N + ^{1}_{0}n = ^{1}_{1}H + ^{14}_{6}C \xrightarrow{t_{1/2}=5730\text{ y}} ^{14}_{7}N + \beta + \overline{\nu}$$

The radio-carbon is oxidized to CO_2 and gets mixed with normal carbon dioxide of the atmosphere. The carbon dioxide enters the body of plants and animals and maintains a steady value of ^{14}C in the body. The intake of carbon dioxide ceases as soon as a plant or animal dies. The radiocarbon in the tissues of the dead body provides a method for the estimation of its 'age', i.e., the time for which the C-14 activity of the body has been undergoing decay since its death. The method is applicable to objects of wood, skin or other animal remains having a detectable ^{14}C activity which is usually the case with objects *not older than* 20,000 years.

In 1960, Libby was awarded the Nobel prize in chemistry for his work on radiocarbon dating.

The application of the C-14 dating is based on the *fundamental assumption that the intensity of cosmic rays and hence of* ^{14}C *in the atmosphere has remained constant over many thousands of years*. The validity of this assumption has been verified by comparison with other methods. The assumption implies that the activity of C-14 in a living object *today* is the same as the activity of any living object many thousand years ago. This activity is taken as the initial activity of C-14 at the time when the plant or animal died and assimilation of radioactive carbon ceased to continue. The age of a sample is now given by the expression:

$$\text{Age} = \frac{2 \cdot 303}{\lambda} \log \frac{\text{original activity}}{\text{final activity}} \text{ y.}$$

$$= \frac{2 \cdot 303 \times 5730}{0 \cdot 693} \log \frac{\text{activity of recent object}}{\text{activity of sample}} \text{ y. [for } ^{14}\text{C, } t_{1/2} = 5730 \text{ y]}$$

Problem 13.16

An archaeological specimen containing ^{14}C gives 40 counts in 5 minutes per gram of carbon. A specimen of freshly cut wood gives 20.3 counts per gram of carbon per minute. The counter used recorded a back-ground count of 5 counts per minute in absence of any ^{14}C-containing sample. What is the age of the specimen? [$t_{1/2}$ of ^{14}C = 5730 y]

Solution: The background counting rate is to be subtracted from the observed counting rate.

Hence, count rate of specimen = (40 ÷ 5) – 5 counts min^{-1}. = 3 counts min^{-1}.

Count rate of fresh wood = 20.3 – 5 counts min^{-1}. = 15.3 counts min^{-1}.

Decay constant of C-14 = 0.693/5730 y. If t be the age of the specimen, then

$$t = \frac{2 \cdot 303 \times 5730}{0 \cdot 693} \log \frac{15 \cdot 3}{3} \text{ a.} = 13474 \text{ y.}$$

(ii) Tracer studies by radioisotopes: Radioisotopes possess the same chemical properties as the stable isotopes of an element. A trace quantity of the radioisotope mixed with the stable isotope may be inserted into any system under investigation. The presence of the radioisotope in different parts of the system may be traced with a suitable counter and the mode of progress of the system may be understood with certainty. This is the basis of the tracer methods which have been widely used in chemistry, biology, agriculture, medicine and metallurgy. Only a few examples will be discussed here.

Neutron activation analysis: Many elements may be estimated by subjecting a sample containing them to (n, γ) reaction to produce a radioactive isotope. The activity of the sample is then compared to the activity of a sample containing known amounts of the element which has been subjected to the same neutron flux over the same period of time. A calibration curve is often used for the purpose. This forms the basis of this non-destructive technique. Some examples of the application of this technique in quantitative estimation are mentioned below:

Reaction	*Activity of product nuclide (half-life)*	*Estimation*
$^{50}Cr\ (n, \gamma)\ ^{51}Cr$	internal conversion (26 d)	Cr-content in ruby (for laser production)
$^{55}Mn\ (n, \gamma)\ ^{56}Mn$	β^-, γ (2·6 hr)	Mn in tea leaves
$^{75}As\ (n, \gamma)\ ^{76}As$	β^-, γ (26·5 hr)	As-content in hair

In ***isotope dilution analysis***, a known amount of a radioisotope (having moderate half-life compared to the time of study) is mixed with an unknown quantity of a sample in which the isotope may get distributed quickly. After equilibrium, the activity of the sample is measured. From the principle of conservation of total activity, one can now estimate the total amount of the sample. The method is particularly useful in biochemistry, for example, in estimating the total volume of blood in a body. Radioisotopes like ^{59}Fe ($t_{1/2}$ = 44·5 d) and ^{32}P ($t_{1/2}$ = 14·3 d) are injected into the body in small amount and the activity is measured after distribution through the body (see exercise, Q. 14).

Analytical studies: The most important applications are the study of diffusion in liquids and solids, the determination of the solubility of sparingly soluble substances, investigations on solid solution formation and adsorption on precipitates. Thus, the solubility of lead sulphate may be found by mixing a trace amount of radioactive lead sulfate with inactive lead sulfate in a known ratio and determining the activity of the mixture. The activity of the residue obtained by evaporating a solution saturated with the mixture is then measured and the solubility calculated by comparison of the activity. (See exercise, Q. 16).

Structure and mechanism

Tracer studies offer very satisfactory methods for the elucidation of structure of different compounds and verification of mechanisms of various reactions.

(a) Tracer study with ^{35}S has been used to establish the nonequivalence of the two sulfur atoms in the thiosulfate ion. Inactive sulfite ion is treated with ^{35}S (radioactive) and the resulting thiosulfate ion is decomposed by acid. It appears that the precipitated sulfur carries all the radioactivity, i.e., the ^{35}S is not distributed within the thiosulfate ion. Writing S* for ^{35}S, this may be represented as

$$SO_3^{2-} + S^* \longrightarrow S^*SO_3^{2-} \ ; \qquad S^*SO_3^{2-} + 2H^+ \longrightarrow H_2SO_3 + S^*$$

(b) The iodine atoms in the HgI_4^{2-} ion are equivalent as shown by tracer studies with radioactive iodine (I*) :

$$HgI_4 + 2NaI^* \longrightarrow Na_2HgI_4^*$$

$$Na_2HgI_4^* + 2AgNO_3 \longrightarrow 2AgI^* + HgI_2^* + 2NaNO_3$$

The radioactive iodine is distributed uniformly among the decomposition products.

(c) At one time, it was suggested that liquid sulfur dioxide undergoes self-ionization as:

$$2SO_2 \rightleftharpoons SO^{2+} + SO_3^{2-}$$

If this were really the case, $SOCl_2$ containing ^{35}S would certainly undergo exchange of radioactive sulfur with SO_2 when dissolved in liquid SO_2, since $SOCl_2$ ionizes to SO^{2+} ion: $SOCl_2 \rightleftharpoons SO^{2+} + 2Cl^-$. However, no exchange of ^{35}S is observed, indicating that the self-ionization scheme suggested for liquid sulfur dioxide is not correct (Chapter 12).

Rapid exchange of radioactive chlorine takes place between thionyl chloride and an ionic chloride in liquid SO_2, suggesting that thionyl chloride probably ionizes to $SOCl^+$ and Cl^-.

(d) ***Mechanism of reactions:*** As an example, we consider the following reaction

$$[Co(NH_3)_5Cl]^{2+} + [Cr(H_2O)_6]^{2+} + 5H_2O \longrightarrow [Co(H_2O)_6]^{2+} + [Cr(H_2O)_5Cl]^{2+} + 5NH_3$$

Cr(II) is oxidized in the reaction to Cr(III) while Co(III) is reduced to Co(II). By using radioactive chlorine in $[Co(NH_3)_5Cl]^{2+}$, it has been observed that the product Cr(III) always contains the radioactive chlorine after reaction. This has been explained by suggesting that the chloride ion of the Co(III) complex first forms a bridge between the chromium and cobalt atoms and facilitates the transfer of an electron from Co(III) to Cr(II); the intermediate will have the structure

$$[(NH_3)_5 - Co - Cl - Cr - (H_2O)_5]^{4+}$$

Once the electron is transferred, Cr(III) attracts the chlorine more strongly than Co(II) and hence the chloride ion gets attached to the Cr(III). Such reactions are often referred to as an *atom transfer reaction*.

Tracer studies have been used to throw light on the mechanism of many other reactions.

Agriculture

The uptake of phosphorus by growing plants from any soil or manure may be studied by labelling a phosphatic fertilizer with radioactive ^{32}P. Its progress through the root system to the foliage may now be followed by means of a Geiger counter. From the rate of intake of ^{32}P under varying external conditions (e.g., soil pH, temperature etc.) and at different stages of growth, one can decide about the optimum utilization of fertilizers. Growing crops, for example, have a higher rate of intake of phosphatic fertilizer (~ 60%) compared to grown up crops (~30%). Hence such fertilizers are best applied at the time of sowing.

^{60}Co has been used to study the migratory and breeding habits of insects which appear in swarms and cause havoc to crops (e.g., locusts). A few insects were dipped in a dilute solution of $CoCl_2$ containing ^{60}Co below the lethal dose. The activity of the radionuclide persisted for more than six months and provided information regarding the hibernating centre and migratory habits of the insects.

Kinetic studies with ^{14}C in the form of $^{14}CO_2$ have been used to understand the nature of photosynthesis.

Medicine and biochemistry. ^{131}I, a γ emitter, has been used in locating brain tumours and malignant thyroid tumours. The iodine is administered to the patient as iodo derivative of the dye fluorescein. The dye is preferentially absorbed on the tumour site and the position of the tumour can be located by a counter placed outside the brain. The exact position and area of the tumour is ascertained during surgery by ^{32}P as a tracer. Phosphates concentrate more on the tumour site and the short range β-particles from ^{32}P act as informers.

The γ-radiation from ^{60}Co can be used therapeutically in the treatment of deep cancerous growths.

^{51}Cr is used to label red blood corpuscles to study blood changes and blood flow.

Obstructions in blood circulation may be diagnosed by injecting sodium chloride containing ^{24}Na into a vein in the forearm of a patient. From the time lag in detecting a signal in a counter placed near the foot of the patient, it is possible to infer any impairment in blood circulation. Similarly, NaI tagged with ^{24}Na may be used to detect any abnormality in the pumping action of the heart. ^{24}Na is also used to estimate the total volume of blood in a body (see exercise, Q. 13.).

Radioisotopes emitting low energy γ-photons with moderate half-lives like ^{67}Ga, ^{99m}Tc, ^{97}Ru etc. are also widely used as diagnostic aids (see chapter 33).

Industry. Faults in industrial weldings and wear out of pistons may be detected by the high energy γ-rays emitted by ^{60}Co. The isotope may also be used in gauzing sheet thickness and also to control the height of filling during automatic packing of commercial powders. Detection of leaks in underground pipes containing liquids (e.g., petroleum) is also possible with radioisotopes. The energy of radiation from certain isotopes like ^{90}Sr may be used in charging electrical batteries in submarines and spacecrafts. The heat energy from the radiation is first utilized to create thermojunctions with high temperature gradient. Electricity is then derived from the junctions with bimetallic couples.

13.6 HAZARDS OF RADIATION

All nuclear and x-radiations as well as beams of high energy particles (e.g., neutrons) are harmful to the living system. They can severely damage the living cells and the effect may even be transmitted genetically to later generations. With increasing use of radioisotopes and the development of nuclear power stations and particle accelerators in large numbers, it is now essential to develop a general consciousness about such hazards. The recent (March 2011) disaster in the nuclear power plants in Japan following a devastative earthquake has put a red alert on this issue.

The mechanism of radiation damage to living cells is not fully understood. When penetrating radiations pass through the cells and are absorbed by the complex organic molecules present, electron(s) may be promoted to some higher level or get completely ionized. In an atomic system, an excited atom returns to its ground state as the excited electron returns to the lowest energy level. The situation in the case of complex molecules is different. During the time the molecule remains in an excited state, the constituent atoms may rearrange themselves resulting in an imbalance in the composition of the cell. The nature and extent of damage will depend on which electron(s) in the molecular structure absorbs the radiation energy and the nature and intensity of radiation. Accordingly, the effects may be short term or long term, recoverable or irrecoverable. Superficially, the radiations may affect only the skin and hair; deep within the body, they may induce blood disorders, tumours and damage the bone marrow. Some cells cannot recover from the damage and exert a cumulative effect on the tissues with a characteristic latent period. The reproductive organs are particularly sensitive to ionizing radiations at any dose. The gene arrangement in the chromosomes is specially affected and the normal gene mutation rate may be adversely and irreversibly increased—producing abnormalities in succeeding generations.

Alpha particles, because of their high charge, are not very penetrating, but they have strong ionizing power. We found that each of the three natural radioactive series contains a gaseous alpha-emitter (often called emanations)—^{219}Rn (actinon), ^{220}Rn (thoron) and ^{222}Rn (radon). So these gases are invariably inhaled by workers handling radioactive minerals. Since the gases are chemically inert, they cannot be arrested by a gas mask. Unless flushed by a very active ventilation, the gases enter the lungs of the inhaler causing severe damage. The maximum permissible safe concentration of radon in the atmosphere is only 10^{-15} percent (by volume) approximately. It has been observed that almost every miner who worked in a particular uranium mine for more than ten years has died of lung cancer.

Pierre and Marie Curie also handled large amount of uranium minerals for a very long time. But the broken shed under which they worked was probably a blessing for them—the "unwanted" wind blew away the radon.

Beta particles also possess very little penetrating and ionizing power—they are easily arrested by clothing and skin. So the beta particles are not so hazardous to the living organs.

Gamma rays (and x-rays) possess the highest penetrating power and hence are most hazardous. Neutrons (and similar accelerated particles) may give rise to (n, γ) reactions and the γ-rays may cause secondary ionization. Fast neutrons may produce knock-on protons which are capable of ionization.

Units of exposure

The units of radioactivity have already been discusseed. In addition, there are a number of units for exposure and tolerable doses.

Exposure is measured in units of *roentgen*, R. It is the quantity of x- or gamma radiation that produces 1 esu of charge [$1/(3 \times 10^9)$ C] in 1 cm^3 of dry air at STP. Rates of exposure may be expressed in roentgen per hour ($R\ h^{-1}$) or milliroentgen per hour ($mR\ h^{-1}$).

A *point source* of radioactive material having activity C curie emitting gamma ray photons of E MeV gives rise to an exposure rate I ($R\ h^{-1}$) at a distance d cm from the source where

$$I = 5300\ C E / d^2.$$

The absorbed dose of any ionizing radiation is the amount of energy imparted to unit mass of the irradiated material. **Dosage is measured in grays (Gy)** where 1 gray corresponds to energy absorbed in joules per kilogram of body tissue.

The older unit **rad** is also in current use: 1 rad (radiation absorbed dose) is equivalent to 100 ergs of absorbed energy per gram of any absorbing material. **1 rad = 10 mGy.**

(Earlier, it was observed that 1 R of x-ray exposure results in the absorption of about 97 ergs of energy per gram of soft body tissue. This led to the dose unit ***Rep***, *roentgen equivalent physical*, which is the dose of ionizing radiation resulting in the absorption of 97 ergs per gram of body tissue. Variation in the composition of the body made the unit indefinite. So it was replaced by rad).

Beams of different radiations have very different biological effects even when the body absorbs the same amount of energy from each type. So a term called the ***relative biological effectiveness*** or **RBE** has been defined for an ionizing radiation as

$$\text{RBE} = \frac{\text{absorbed dose (in rads) of x - rays (of } \sim 200 \text{ keV)}}{\text{absorbed dose (rads) of the radiation that produces the } \textit{same effect}}$$

Some representative RBE factors are

X-rays, γ-rays, β-rays	:	1	Alpha particles and		
Thermal neutrons	:	2-5	accelerated ions of O, N etc	:	10–20.
Fast neutrons	:	10			

Thus, 0.1 rad of a beam of fast neutrons will cause the same damage as 1 rad of 200 keV x-ray beam.

The product of RBE and the absorbed dose (rads) provides a convenient unit of biological dose called the **rem (roentgen equivalent man)**.

$$\textbf{Rem = RBE} \times \textbf{rad.}$$

The corresponding SI unit is ***sievert*** **(Sv). 1 Sv = RBE × grays = 100 rem.**

The health tolerance limit for occupational exposure over the whole body is set at 5 rem per year or 0.1 rem per week (maximum permissible limit or MPL). It is likely to be lowered further.

During a span of about 30 years, the average reproductive lifetime, any individual receives at least 4 R (roentgen) radiation exposure from cosmic rays and from radioactive materials found everywhere in low concentration (K–40, C–14 etc.). Exposure to x-rays adds to this *accumulated dose*. A total dose exceeding 10 R over this period may cause adverse gene mutation effects. Acute radiation exposure over the whole body gives rise to the following disorders:

50–100R	:	some blood changes.
100–250 R	:	severe illness (likely to recovery within six months).
400 R	:	fatal to 50% of the individuals affected (also called the median lethal dose, MLD or LD-50).
600 R	:	fatal to all.

Exposure to a large population in general is expressed in ***person rem*** **or** ***man-rem***. This is obtained by multiplying the total number of people exposed by the average individual dose in rems. The estimated present annual exposure to individuals from natural background is about 240 millirem per capita of world population or roughly 1100 million man-rem. Diagnostic medical exposure adds another 40–100 millirem per capita of world population per year (in individuals this may be 10–1000 millirems), amounting to 200–500 million man-rem. Each test explosion of nuclear weapons contributes about 1 millirem per capita or 500 million man-rem. Occupational exposure and nuclear power plants also add to this list, particularly after plant failure as it happened in Chernobyl (1986) or Japan (2011).

However, one need not get panicky about the hazards of radiation. Rather, the safe limits of exposure and the various precautionary measures should be kept in mind while developing methods for peaceful use of radioisotopes and nuclear energy. In fact, out of 600–800 annual incidence rate of cancer per million population, the contribution from radiation is considerably less in comparison to other factors: tobacco 30%; food 33%; viral infection 10%; alcohol 4%; industrial occupation 4%; hereditary 2% and natural background radiation: less than 0.5%. Similarly, spontaneous occurrence of inherited genetic diseases caused by mutations is about 10.7% in the population. Out of this, 7.6% is caused by environmental factors, smoking and food. About 0.1% is due to natural background radiation. Air pollution, viruses, alcohol and mutagenic chemicals also cause other genetic defects.

Problem 13.17

A patient is given a dose of 6mCi of ^{32}P (half-life = 14 days) in his blood. The blood volume is 3.6 dm^3. The phosphorus is uniformly distributed in approximately 1 hour. What would be the count rate of the patient's blood per cm^3 (a) 1 h after injection (b) 28 h after injection? The counter has an efficiency of only 20%.

Solution: Dose per cm^3 of blood = 6/3600 mCi = 1/600 mCi. This corresponds to (1/600) × 3.7 × 10^7 disintegrations per second.

(a) After 1 h, only a negligible fraction of the radiophosphorus has disintegrated. Assuming no decay, this corresponds to (1/600) × 3·7 ×10^7 dps. Since the efficiency of the counter is only 20%,

$$\text{Counts recorded} = \frac{1}{5} \times \frac{1}{600} \times 3 \cdot 7 \times 10^7 \text{ s}^{-1} = 1 \cdot 23 \times 10^4 \text{ s}^{-1}$$

(b) Since $t_{1/2}$ = 14 days, the activity is reduced to exactly 1/4 th after 28 days. Hence count rate would be 3.1 × 10^3 dis s^{-1}.

ESSENTIALS

Radioactivity is the phenomenon of spontaneous distintegration of certain nuclei to form new elements with the emission of active radiation. It is a nuclear property and is unaffected by the state of chemical combination of an element. Heavy elements like actinium, uranium, radium and certain isotopes of lighter elements like ^{40}K, ^{14}C etc. are examples of radioactive nuclides.

The unit of radioactivity is becquerel, Bq. 1 Bq = 1 disintegration per second. Previously, the unit of radioactivity was curie (Ci); 1 Ci = 3·7 × 10^{10} disintegrations per second.

The rate of radioactive disintegration follows an exponential law, i.e., the rate at any time is proportional to the number of atoms present at that time. If n_0 be the number of atoms present at any arbitrarily assumed zero time, i.e., t = 0, we have,

$$2 \cdot 303 \log \frac{n_0}{n} = \lambda \text{ or, } \lambda = \frac{2 \cdot 303}{t} \log \frac{n_0}{n}$$

where λ is a constant characterising a given radioelement, known as the **disintegration constant.**

The time required for the activity of a given amount of radioelement to decay to a definite fraction of its initial value is independent of the actual amount taken. The time for decay to half the initial activity is known as the **half-life** of a radioelement ($t_{1/2}$). $t_{1/2} = 0{\cdot}693/\lambda$.

The statistical average life of a large number of radioactive atoms, also known as its **average life** (t_{av}) is equal to $1/\lambda$.

The radioactive disintegration series are formed by the successive disintegration products in a family of radioelements. Naturally occurring radioactive elements form three disintegration series: (i) the thorium series or the **4n** series, so called because the mass numbers of the members are given by $4n$ (n is an integer). It starts with $^{232}_{90}Th$ and ends with $^{208}_{82}Pb$. (ii) the uranium series **(4n + 2)** series starting with $^{238}_{92}U$, ending in $^{206}_{82}Pb$. (iii) the actinium series, **(4n + 3)** series, beginning with $^{235}_{92}U$ and ending in $^{207}_{82}Pb$.

Artificial radioactivity is the property of radioactivity associated with artificial nuclides obtained by the bombardment of stable elements with high energy particles. When ordinary phosphorus $\left(^{31}_{15}P\right)$ is bombarded with neutrons, the product, $^{32}_{15}P$, becomes radioactive and emits β-rays.

Radioactivity has found widespread applications in (i) geochemical and radiocarbon dating and (ii) tracer studies.

Stable nuclides are characterized by a narrow range of neutron to proton (N/P) ratio. For $Z < 20$, N/P ratio is close to unity. For higher Z, N/P becomes greater than unity. When a particular nucleus possesses more neutrons than required for its stability, it may be expected to emit a β-particle by converting a neutron into a proton, an electron and a neutrino. When a nucleus contains fewer number of protons than required for its stability, it may emit an α-particle to raise its N/P ratio.

Radiation hazards have become a part of our life in the form of cosmic radiation, diagnostic procedures and nuclear power plants, **particularly after failure**.

Exposure is measured in ***roentgen***, R. 1R of x-ray or gamma radiation produces 1 esu of charge in 1 cm^3 of dry air (STP).

The amount of energy imparted to unit mass of an irradiated substance is the absorbed dose of the ionizing radiation. The older unit of dose is the ***rad*** (radiation absorbed dose): 1 rad ≡ 100 ergs of absorbed energy per gram of absorbing material. The current unit is ***grays*** (Gy): 1 Gy ≡ energy absorbed in joules per kg of body tissue.

The relative biological effectiveness (RBE) of a radiation is expressed as

$$\text{RBE} = \frac{\text{absorbed dose (in rads) of x-rays (of} \sim 200 \text{ keV)}}{\text{absorbed dose (rads) of the radiation that produces the } \textit{same effect}}$$

Rem (roentgen equivalent man) is a convenient unit of biological dose:

$$\text{Rem} = \text{RBE} \times \text{rad}.$$

The corresponding SI unit is ***sievert*** (Sv).

1 Sv = RBE × grays = 100 rem.

EXERCISE

1. Explain what do you mean by radioactive equilibrium and state the conditions under which such equilibrium is established. What do you mean by secular and transient equilibrium?
2. The half-life of a radioelement is 25 min. A specimen containing this radioelement gives 1260 counts per minute in a Geiger counter. What will be the count rate after exactly one hour?

3. 1 g of carbon made from a piece of wood from a freshly cut tree registered 12 counts per minute in a counting device (after correction for back-ground counts). 1g of carbon made from wood taken from the timbers of an old temple registered 10 counts per minute (after correction) in the same apparatus. Estimate the age of the temple. ($t_{1/2}$ of C-14 = 5730 y).
4. Give two examples of the application of radioactive isotopes in the study of structure and mechanism in Chemistry.
5. The potential energy barrier of a uranium nucleus is nearly 30 MeV. An alpha particle with about 9 MeV energy cannot disintegrate such a nucleus. But alpha particles emitted by such nuclei have energies as low as 4–9 MeV. Explain.
6. What observations in connection with beta emission lead to the neutrino hypothesis? How are neutrinos detected?
7. What is (a) nuclear isomerism (b) artificial radioactivity?
8. Calculate the mass of $^{60}_{27}Co$ having an activity of 1 mCi. $t_{1/2}$ = 5·25 y.
   Hint: $|dn/dt| = 3{\cdot}7 \times 10^7\ s^{-1} = \lambda n \Rightarrow n = 8{\cdot}85 \times 10^{15}$ atoms $\Rightarrow$ mass = $8{\cdot}8 \times 10^{-4}$ mg.
9. 1 g of ^{226}Ra emits $11{\cdot}6 \times 10^{17}$ α-particles per year. Calculate the value of the Avogadro number (N_0) given $t_{1/2}$ = 1590 y.
   Hint: From $|dn/dt| = \lambda n$, $n = 11{\cdot}6 \times 10^{17} \times 1590/0.693\ g^{-1}$.
   N_0 = No of atoms in 226 g $\Rightarrow 6{\cdot}01 \times 10^{23}$.
10. A small amount of radioactive material got inadvertently spread in an area adjoining a nuclear power plant making the level of radiation 40 times the permissible safety level. If $t_{1/2}$ of the radioactive species be 20 days, after how many days will the place be safe to live?
   Hint: If t be the time for the activity to decay to 1/40 th fraction (admissible limit),

$$t = \frac{2 \cdot 303}{\lambda} \log \frac{40}{1} \text{ days} \Rightarrow 106 \cdot 5 \text{ days.}$$

   The area will be safe for living after 107 days.
11. Calculate the decay constant, mean lifetime and half-life of a radionuclide whose activity diminishes by 10 percent during 100 days.

   [**Ans.** $\lambda = 1{\cdot}054 \times 10^{-3}\ day^{-1}$]
12. 1 g of carbon from an archaeological sample gives 9.4 counts per min. 1 g of carbon made from a recently cut tree gives 15.3 counts per min under the same conditions. How old is the sample? (Half-life of C-14 = 5730 years).
13. A small amount of solution containing ^{24}Na ($t_{1/2}$ = 15 hours) is injected into the bloodstream of a man. The initial activity is $2 \times 10^3\ s^{-1}$. The activity of 1 cm^3 of blood sample taken after 5 hours is 16 min^{-1}. Find the total volume of blood of the man.
   Hint: Note the different units of time. Initial activity, $A° = 2000\ s^{-1}$. After $t = 5 \times 60 \times 60$ s, $A = (16/60) \times V\ s^{-1}$ where V is the total volume of blood in cm^3.

$$\lambda = \frac{0 \cdot 693}{15 \times 60 \times 60}\ s^{-1} = \frac{2 \cdot 303}{5 \times 60 \times 60} \log \frac{2000}{16V/60} \Rightarrow V = 5953\ cm^3$$

14. 1 mL of blood containing ^{32}P equivalent to activity of $1{\cdot}1 \times 10^5$ counts min^{-1} is injected into a body. After about 25 minutes, equilibrium is attained when 1 mL of blood records an activity of 27 counts min^{-1}. Calculate the volume of blood in the body. $t_{1/2}$ of ^{32}P = 14·3 days.

Hint: $t_{1/2}$ of ^{32}P (14.3 days) is sufficiently large compared to the experimental time. We apply the principle of conservation of total activity. If the total volume of blood is V mL,

$$1 \times 1 \cdot 1 \times 10^5 = (V+1) \times 27 \Rightarrow V = 4074 \text{ mL} \approx 4 \cdot 1 \text{ L}.$$

15. ^{22}Na undergoes branching decay—90% through positron emission and 10% by electron capture, with an overall half-life of 2·5 years. Find the decay constants (λ_1 and λ_2) and half-lives for each mode of decay.

Hint: $^{22}Na \xrightarrow{\lambda_{Tot}}$ ($\xrightarrow{\lambda_1}$ β^+ emission (90%); $\xrightarrow{\lambda_2}$ e-capture (10%)) $\lambda_1 + \lambda_2 = \lambda_{Tot} = \dfrac{0.693}{2.5}\text{ y}^{-1} = 0.277\text{ y}^{-1}$

Now, $\dfrac{\text{decay by } \beta^+ \text{ emission}}{\text{decay by } e\text{-capture}} = \dfrac{90}{10} = \dfrac{\text{rate of } \beta^+ \text{ emission}}{\text{rate of } e\text{-capture}} = \dfrac{\lambda_1}{\lambda_2}$

Thus $\lambda_1 = 9\lambda_2 \Rightarrow \lambda_2 = 0{\cdot}0277\text{ y}^{-1}$, $\lambda_1 = 0{\cdot}249\text{ y}^{-1}$.

16. 1 mg of a precipitate of $Ba^{35}SO_4$ shows an activity of 4×10^4 counts per minute. A saturated solution of this precipitate in water was filtered and 1 mL of the filtrate was dried by evaporation. The dry residue showed an activity of 95 counts per minute. Find the solubility product of $BaSO_4$ (Ba = 137·3).

Hint: The solubility of $BaSO_4$ is estimated from the activity of the saturated solution.

1 mL solution $\equiv \dfrac{95}{4 \times 10^4}$ mg $BaSO_4$ or $2{\cdot}375 \times 10^{-3}$ mg.

Solubility of $BaSO_4$ (per litre) $= \dfrac{2 \cdot 375}{1000 \times 233 \cdot 3}$ mol $L^{-1} = 1 \cdot 02 \times 10^{-5}$

Solubility product = $1{\cdot}04 \times 10^{-10}$.

17. ^7_4Be may get converted to ^7_3Li either by positron emission or by orbital electron capture. Comment on the feasibility of these processes from the following mass data: ^7_4Be : 7·01693 u; ^7_3Li : 7·01600 u; e: 0.00055 u.

Hint: Mass difference between ^7_4Be and ^7_3Li = (7·01693 – 7·01600) u = 0·00093 u which is less than 2 × 0·00055 u (mass of 2 electrons). Hence positron emission is not feasible.

18. Predict the unstable one among the two nuclides $^{18}_9F$ and $^{18}_{10}Ne$ and its likely mode of decay. Isotopic masses are $^{18}_9F = 18 \cdot 00094\ u$, $^{18}_{10}Ne = 18 \cdot 00571\ u$, $m_e = 0 \cdot 00055\ u$.

Hint: Compare Δm and $2m_e$. $^{18}_{10}Ne$ may decay by both positron emission and K-electron capture.

19. A sample of ^{24}Na ($t_{1/2}$ = 15 h) is contaminated with some ^{31}Si ($t_{1/2}$ = 3 h). A sample of this material with an initial activity of 1200 counts per minute showed an activity of 200 counts per minute after 30 h. What percent of the initial activity was caused by ^{31}Si?

Hint: 30 hours is $2t_{1/2}$ (Na) and $10t_{1/2}$(Si). Hence the activity due to ^{31}Si is negligible after 30 h. Since the time is twice the $t_{1/2}$ of ^{24}Na, its initial activity was 4 times as much or 800 counts per minute.

The initial activity due to ^{31}Si was (1200 – 800) or 400 counts i.e., 33·3%.

20. An ampoule contains 10 cm^3 (STP) of krypton consisting of active ^{85}Kr ($t_{1/2}$ = 10 y) and stable ^{84}Kr. If the activity of the sample is 100 millicurie, calculate the percentage by weight of ^{85}Kr in the mixture.

Hint: $|dN/dt| = 100$ mCi $= 100 \times 3{\cdot}7 \times 10^7$ dis s^{-1} $= \lambda n$ (n = no. of atoms)

$\lambda = 0{\cdot}693/10 \times 365 \times 60 \times 60$ s^{-1} gives $n = 1{\cdot}68 \times 10^{18}$

$\equiv 2{\cdot}8 \times 10^{-6}$ mole ^{85}Kr $\equiv 2{\cdot}38 \times 10^{-4}$ g.

Total Krypton = 10/22400 mole $= 4{\cdot}46 \times 10^{-4}$ mole.

^{84}Kr $= 4{\cdot}46 \times 10^{-4}$ mole $- 2{\cdot}8 \times 10^{-6}$ mole $= 4{\cdot}43 \times 10^{-4}$ mole $\equiv 3{\cdot}72 \times 10^{-2}$ g.

% of ^{85}Kr = 0·64.

·**21.** In the reaction $^{23}_{11}$Na (p, n) $^{23}_{12}$Mg, $Q = -3{\cdot}5$ MeV. (a) Explain whether the product nuclide $^{23}_{12}$Mg, if radioactive, is likely to decay by β^--emission or β^+-emission. (b) Calculate the mass of $^{23}_{12}$Mg from the given mass data (u); n : 1·0087 ; p : 1·0078; ^{23}Na : 22·9920. (c) Also calculate the energy change when $^{23}_{12}$Mg reverts to $^{23}_{11}$Na by positron emission.

Hint: (a) Positron emission produces the stable $^{24}_{12}$Mg nuclide.

(b) $\Delta m = 3{\cdot}5/931\ u = 0{\cdot}0038\ u \Rightarrow m_{Mg} = 22{\cdot}9949\ u$ (Q is –ve; products have higher total mass).

(c) $\Delta m = m_{Mg} - (m_{Na} + 2m_e) = 0{\cdot}0018\ u$.

$Q = 0{\cdot}0018 \times 931$ MeV $= 1{\cdot}676$ MeV.

14

THE ATOMIC NUCLEUS

14.1 STRUCTURE AND STABILITY OF THE NUCLEUS

The extranuclear structure of the atom was explored without any detailed knowledge about the atomic nucleus. Even the composition of the nucleus was not known fully until Chadwick established the existence of neutrons in 1932. As we shall find soon, the forces that hold the nucleus are basically different from the electrostatic force operating in the extranuclear part of the atom and are, at the same time, many times greater in magnitude. The ionization of an electron or rupture of a chemical bond involve energies of few electronvolts whereas nuclear processes involve energies in the range of million electronvolts (MeV). Hence it is difficult to split a nucleus and study its interior. Our knowledge about the structure of the nucleus is thus based on other indirect evidences like study of radioactivity, analysis of particles and radiations emitted on bombarding a nucleus with high energy particles, stability of different isotopes etc.

Masses of various isotopes and particles resulting from bombardment experiments may be determined accurately (better than 1 part in 10^6) by modern mass spectrometers. Atomic masses are conventionally expressed in the mass unit (*u*) based on the mass of ^{12}C taken as exactly 12.0000 *u*:

$$1\ u = 1.6604 \times 10^{-27}\ \text{kg}.$$

Isotopes were recognized from the study of radioactive disintegration series (the term was used by Soddy in 1913), but their interpretation was delayed until the discovery of neutrons in 1932. Isotopes of an element have the same number of protons but varying number of neutrons. So they have different mass numbers but identical chemical properties. However, they may have slightly different bond energies and take part in reactions with a different rate. Each isotope of an element is a *nuclide*, a term used to refer to a particular species of nucleus.

The radioactive disintegration series contain several species with the same mass number but different atomic numbers. These nuclides contain different numbers of protons and are called **isobars**. In the uranium series, radium B, radium C and radium C′ are isobaric. The mass number of each is 214, but the number of protons in their nuclei are 82, 83 and 84 respectively. A few other examples of isobaric nuclides are ^3_1H and ^3_2He; ^6_2He and ^6_3Li; ^7_3Li and ^7_4Be. During radioactive disintegration, isobaric nuclides are formed when a nucleus emits a β-particle (β^-) or a positron (β^+):

$$^3_1H \longrightarrow {}^3_2He + {}^{\ 0}_{-1}e\,(\beta^-) + \bar{\nu}\ \text{(antineutrino)}$$

$$^{22}_{11}Na \longrightarrow {}^{22}_{10}Ne + {}^{\ 0}_{+1}e\,(\beta^+) + \nu\ \text{(neutrino)}$$

There are many nuclides having different mass numbers but the same number of neutrons. These have been termed **isotones**. ^2_1H and ^3_2He are isotones, each containing one neutron. $^{30}_{14}Si$ and $^{31}_{15}P$ are similarly isotonic, as each contains 16 neutrons. Nuclides having the same excess of neutrons over protons (*A*–2*Z*) are called **isodiapheres**, e.g., $^{37}_{17}Cl$ and $^{39}_{18}Ar$.

"Mirror nuclei" have equal mass numbers but their atomic numbers differ by 1, that is, the number of protons (*Z*) and neutrons (*N*) are interchanged: $^A_ZX_N$ and $^A_NY_Z$ are mirror nuclei ($|N - Z| = 1$). Examples are:

$$^3_1H \text{ and } ^3_2He;\ ^7_3Li \text{ and } ^7_4Be;\ ^{15}_7N \text{ and } ^{15}_8O;\ ^{31}_{15}P \text{ and } ^{31}_{16}S.$$

Though their mass numbers are equal, mirror nuclei may have slight difference in mass arising from the difference in masses of a neutron and ^1_1H. In these nuclei, one *p-p* interaction is replaced by one *n-n* interaction.

14.1.1 Composition of the Nucleus

Analysis of positive rays for different elements ultimately established that the proton is a common constituent of all nuclides. This prompted the proton-electron hypothesis which assumed that the nucleus of an atom contains enough protons to account for its mass. A nucleus whose mass (in *u*) is close to the whole number *A* should contain *A* protons. Since the actual charge of the nucleus is only $+Ze$, where Z is the atomic number, the nucleus should also contain $(Z–A)$ electrons to neutralize the excess positive charge. The number A has been called the mass number and is the integer closest to the atomic mass (in u).

The proton-electron hypothesis was apparently consistent with the emission of α- and β-particles from radioactive nuclides: β-particles are identical with electrons, while α- particles could be formed by the combination of four protons and two neutrons. One of these particles might be ejected under appropriate conditions.

However, there soon appeared a number of strong arguments against the possible existence of nuclear electrons.

A nucleus cannot contain an electron

(i) Nuclear size and the uncertainty principle: The approximate radius of a nucleus may be estimated to be about 5×10^{-15} m (see later). If we want to confine the electron to so small a region, the minimum uncertainty in its momentum should be of the order of 10^{-20} kg m s^{-1} (see problem 14.1). The corresponding kinetic energy of the electron comes out to be at least 21 MeV. But the electrons emitted during beta emission have much lower energy, only about 2–3 MeV.

Problem 14.1

Use the uncertainty principle to estimate the minimum energy an electron should possess if it has to exist within the nucleus (approximate radius 5×10^{-15} m).

Solution: First we calculate the minimum uncertainty in the momentum of the electron. The uncertainty Δx in the position may be set equal to the dimension of the nucleus, 5×10^{-15} m.

$$\Delta p \geq \frac{h}{4\pi\Delta x} \geq \frac{6\cdot626\times10^{-34}\ \text{J.s}}{4\times3\cdot142\times5\times10^{-15}\ \text{m}} \geq 1\cdot1\times10^{-20}\,\text{kg ms}^{-1}.$$

If this is the *minimum uncertainty* in the momentum, the momentum itself must be at least comparable in magnitude. In other words, the momentum (p) of the electron should be at least 1.1×10^{-20} kg ms^{-1}.

The total energy E of a particle may be expressed in terms of the momentum by the relation

$$E^2 = p^2c^2 + m_0^2c^4$$

where m_0 is the rest mass of the electron and c is the velocity of light. If $p = 1.1 \times 10^{-20}$ kg m s^{-1}, $p^2c^2 = 1.1 \times 10^{-23}$ J^2; on the other hand $m_0^2c^4 = (9\times10^{-31})^2 \times (3\times10^8)^4 = 65.6\times10^{-30}$ J^2.

Clearly, $m_0^2c^4$ is negligible in comparison to p^2c^2. Hence we may get the energy E as $E = pc = 1.1 \times 10^{-20} \times 3 \times 10^8$ J $= 3.3 \times 10^{-12}$ J $\equiv$ 21 MeV.

This limitation does not apply to a proton inside the nucleus. For a proton with a momentum equal to 1.1×10^{-20} kg, we can calculate the kinetic energy from the relation $E_k = p^2/2m$, since $E_k << m_0c^2$.

$$E_k = p^2/2m = 3.7 \times 10^{-14}\ \text{J} = 0.23\ \text{MeV}.$$

This seems to be quite reasonable.

(ii) Nuclear spin: Both protons and electrons have half-integral spin (fermions). Hence nuclei with even number of protons and neutrons should have integral spins while those having the particles in odd numbers should possess half-integral spins. Accordingly, the nucleus of deuterium, the isotope of hydrogen of mass number two, should have two protons and one electron with a resultant half-integral spin. Depending upon the orientations of the particles, the nuclear spin may be either 1/2 or 3/2. But this is not consistent with the observed spin of the deuteron nucleus, 1.

(iii) Nuclear magnetic moment. We have noted earlier that simultaneous motion of the nucleus and the electron around the common centre of gravity gives rise to hyperfine structure of the spectral lines and slight variation of the Rydberg constant with the mass of the nucleus. Detailed consideration of such hyperfine structure leads to the assumption (Pauli, 1924) that the nucleus has a spin angular momentum. Now, this angular momentum has an associated magnetic moment as in the case of the electron. As discussed in section 14.1.2, magnetic moments of nuclei would be considerably large if they contained electrons.

(iv) Electron-nuclear interaction. Though the exact nature of nuclear forces is not known, it has been estimated that such forces give rise to binding energies (see later) of the order of 8 MeV per particle. If the nucleus contains electrons which are capable of such strong interaction with the protons, it is not clear why the orbital electrons should interact only electrically with the protons of the nucleus. Moreover, scattering of fast electrons by nuclei indicate the involvement of electric forces only; fast protons on similar scattering show that something other than mere electric interaction is involved.

Discovery of the neutron

Bombardment of light elements with alpha particles produced certain highly penetrating "radiations" (Bothe and Becker; 1930). This "radiation" was initially thought to be gamma rays. But Chadwick, a former student of Rutherford, showed that the "radiations" ejected protons from many substances like hydrogen, carbon, helium, etc. These substances, rather the atoms in them, were thus disintegrated by these so-called radiations. Actually this was something too much to be expected from mere radiations. From the speeds of the particles involved, Chadwick showed (1932) that the so-called radiation consisted of a stream of particles having no charge but mass equal to that of a proton.

Actually, Rutherford had already predicted such a chargeless particle nearly ten years earlier. Because of their electrical neutrality, the particles were named **neutrons**. Their high penetrating power is a consequence of this neutrality; they undergo very little electromagnetic interaction when passing through matter. A proton with the same mass is easily stopped by matter due to the interaction of its positive charge with the electrons present in the substance.

Neutrons, present in the nucleus of an atom, play an important role in determining the stability of nuclides. Neutrons and protons (commonly called nucleons) constitute the nucleus of an atom consistent with the charge and mass of the nuclide. Isotopes of an element contain the same number of protons but different numbers of neutrons. The mass (1.0086654 *u*) and half-integral spin of the neutron fit all observed properties of nuclei. The role of the neutron in nuclear structure will be discussed separately. It may be mentioned here that the neutron is not a stable particle outside the nucleus. The free neutron has a mean life of 15.5 min. for radioactive decay into a proton, an electron and an antineutrino.

In contrast to free neutrons, aggregates of neutrons in a highly condensed state do not undergo radioactive decay. Such aggregates are believed to be present in **neutron stars**—the remnants of supernova explosion in big stars. The atoms in such neutron stars have been so compressed that the protons and electrons have fused into neutrons. The density of matter in these stars is comparable to the density of typical nuclei (see below)—million million tons per cm^3; a neutron star packs the mass of one or two sun in a sphere of only 10 km radius. Rotating neutron stars emit electromagnetic radiations at regular intervals—known as **pulsars**.

14.1.2 Shape and Size of the Nucleus

For all practical purposes, the nucleus is supposed to be spherical. How do we know it? If the charge distribution within a nucleus is not spherically symmetrical, the nucleus will have an electric quadrupole moment. Interaction of this moment with the orbital electrons will give rise to slight changes in the atomic energy levels, resulting in hyperfine splitting of spectral lines. By careful observation it is possible to distinguish such hyperfine splittings from those caused by nuclear magnetic moment. It has been found that nuclei whose spin quantum numbers are 1 or more do deviate slightly from sphericity.

Nuclear size

An approximate idea about the size of the nucleus was obtained from the scattering of α-particles. Suppose an α-particle of mass M_α and velocity v approaches a nucleus of charge Ze. The charge on the α-particle is $2e$. The closest distance of approach (d_o) in a head-on collision is determined at the point where the α-particle reverses its direction. At this point, the kinetic energy of the α-particle is exactly balanced by the potential energy due to repulsion.

$$\text{Force of repulsion} = \frac{2e\,Ze}{4\pi\varepsilon_0 d_0^2}\text{; corresponding potential energy} = \frac{Ze^2}{2\pi\varepsilon_0 d_0}$$

$$\text{At the point of closest approach,}\quad \frac{Ze^2}{\pi\,2\varepsilon_0 d_0} = \tfrac{1}{2}M_\alpha v^2 \quad\text{or,}\quad d_0 = \frac{Ze^2}{\pi\,\varepsilon_0 M_\alpha v^2}$$

$$\left[\text{In CGS units, } \frac{2Ze^2}{d_0} = \tfrac{1}{2}M_\alpha v^2;\; d_0 = \frac{4Ze^2}{M_\alpha v^2}\right]$$

α-particles from natural radioactive sources have velocities in the range $1.3–1.9 \times 10^7$ m s^{-1}. This gives the value of d_0 around 5×10^{-15} m, i.e., 5 fm (femtometer, 10^{-15} m*) for aluminium. Scattering of fast neutrons gives more dependable information on the radius R of a nucleus as the distance from the nuclear centre within which nuclear forces act. This is given by the empirical relation

$$R = R_0 \cdot A^{1/3}$$

where R_0 is a constant equal to 1.4 fm and A is the atomic mass number of the nucleus. The radius of the aluminium nucleus is thus estimated to be 4.2 fm. Similar results are obtained

* This is equal to the unit **"fermi"** (abbreviated as F). $1F = 10^{-15}$ m. Nuclear dimensions are often expressed in this unit, in honour of Enrico Fermi, the famous nuclear physicist.

from scattering of high-energy electrons. The figures suggest that the nucleus is many times smaller than the atom as a whole.

Density of the nucleus

As noted above, the radius R of a nucleus is proportional to $A^{1/3}$, where A is its mass number. Clearly, the volume of the nucleus, given by $4/3\ \pi R^3$ on the assumption of a spherical shape, is proportional to A, the mass number. This leads to the important conclusion that nuclear volumes are proportional to nuclear masses and hence *all nuclei should have approximately the same density*. Since almost the entire mass of the atom is located inside the small nucleus, the density of the nucleus should be very high—a rough estimate gives a value as high as 10^{11} kg. per cubic centimeter (see problem).

Problem 14.2

Calculate the density of the aluminium nucleus, given its radius = 5 fm.

Solution: Volume of the nucleus = $\frac{4}{3}\pi R^3 \Rightarrow 5 \cdot 24 \times 10^{-43}\ \text{m}^3$.

The mass of the nucleus = $27 \div 6{\cdot}02 \times 10^{23}$ g. $\Rightarrow 4 \cdot 48 \times 10^{-26}$ kg.

(Alternatively, mass = $27u = 27 \times 1{\cdot}66 \times 10^{-27}$ kg = $4{\cdot}48 \times 10^{-26}$ kg).

$\therefore$ Density = mass/volume $\Rightarrow 8{\cdot}5 \times 10^{16}\ \text{kg m}^{-3}$
$= 8.5 \times 10^{13}\ \text{g cm}^{-3}$.

[Thus, even if it were possible to isolate and collect 1 cm³ of nuclei, there would be no container or no crane which can withstand this mass.]

Magnetic properties of the nucleus

To start with, let us consider the magnetic moments of the constituents of the nucleus, namely, the protons and neutrons, each having half-integral spin($\frac{1}{2}$). Though the overall angular momentum of the nucleus arises from the orbital and spin contributions of the nucleons, the total angular momentum of the nucleus is often associated with the spin alone. Analogous to s and m_s for electrons (chapter 4), a nucleus may be assigned two spin quantum numbers, I and M_I. The magnitude of the nuclear spin angular momentum **I** is related to I as

$$\mathbf{I} = \sqrt{I(I+1)}\ \frac{h}{2\pi}$$

where I is the nuclear spin quantum number. The value of I may be zero, half-integral or integral depending on the composition of the nucleus:

Z even, N even	A (= Z + N) odd	Z odd, N odd
$I = 0$	$I = \frac{1}{2}, \frac{3}{2}, \frac{5}{2}, \ldots$	$I = 1, 2, 3, \ldots$

The angular momentum vector **I** is subject to space quantization (compare magnetic quantum number) through the quantam number M_I when the permitted values of the components of **I** along a specific direction (say z) are given by $M_I h/2\pi$ where M_I may vary from $+I$ to $-I$.

$$\mathbf{I}_z = M_I h/2\pi\,;\ M_I = -I, (-I+1), \ldots (I-1), I$$

By analogy with the magnetic moment of the electron, one might expect the magnetic moment of a proton to be equal to $eh/4\pi M_p$ (M_p = mass of proton); the magnetic moment of the neutron should then be zero since it has no charge. However, the results obtained quantum mechanically are different, owing mainly to the composite structure of the nucleus.

The magnetic (dipole) moment of a nucleus is expressed as

$$\mu_I = g_N \sqrt{I(I+1)} \frac{e\hbar}{2M_p} \equiv \gamma \sqrt{I(I+1)} \frac{h}{2\pi}$$

where g_N is called the nuclear *g*-factor (= 5.58569 for a proton) and γ ($= g_N e/2M_p$) is the *gyromagnetic* (or *magnetogyric*) *ratio* of the nucleus, $\gamma = g_N \times 4{\cdot}78942 \times 10^7$ HzT^{-1}. Values of g_N are determined experimentally—these cannot be predicted from present theories of nuclear structure. The quantity $e\hbar/2M_p$ is called the *nuclear magneton* (μ_N; also called *Rabi magneton*). It may be considered the unit of nuclear magnetic moment (analogous to the Bohr magneton).

$$\mu_N = \frac{e\hbar}{2M_p} = \frac{eh}{4\pi M_p} = 5{\cdot}0508 \times 10^{-27}\,\mathrm{JT^{-1}} \quad \text{(Joule per Tesla)}$$

Considering the mass of the proton, we find that μ_N is only 1/1836 times μ_B (Bohr magneton). The proton has a magnetic moment of about 2.8 μ_N, which is only about 0.15 percent that of the electron. Values of nuclear magnetic moment vary from about 0 to 5 μ_N.

The magnetic property of neutrons is difficult to understand from any classical analogy. Its g_N value of $-3{\cdot}826$ and γ equal to $-18{\cdot}32 \times 10^7$ T^{-1} s^{-1} imply that the magnetic moment of the neutron is opposite in sign to that of the proton. Superficially one may trace this difference to the mesons emitted and absorbed by neutrons (π^0 and π^-; see later). Protons emit and absorb π^0 and π^+ mesons.

When a molecule containing nuclei with nonzero spin is subjected to an external magnetic field of strength *B* (along the *z*-axis, say), the energy of nuclear magnetic dipole is given by

$$E = -\gamma\,|\,\mathbf{I}\,|\,B\cos\theta = -\gamma \mathbf{I}_z B$$

where θ is the angle subtended by the angular momentum vector **I** with the *z*-axis. Since $\mathbf{I}_z$ is quantized with the possible value $M_I\hbar$, the nuclear magnetic moment now gives rise to a set of quantized energy levels as

$$E = -\gamma\hbar B M_I, \quad M_I = -I,\ -(I-1)\cdots+I$$

Transitions between these nuclear-spin energy levels are subject to the selection rule $\Delta M_I = \pm 1$ and may be observed by subjecting the sample to electromagnetic radiation of appropriate frequency, usually in the radiofrequency region of the electromagnetic spectrum (~500 MHz) in the magnetic fields normally used (~12T). This forms the basis of NMR (*nuclear magnetic resonance*) spectroscopy. NMR spectroscopy differs from electron paramagnetic resonance (EPR) spectroscopy where magnetic fields of about 0·3 T are used with electromagnetic waves in the microwave region (a wavelength of a few cm; frequency ~10 GHz).

NMR spectroscopy has been widely applied in studying the structure of molecules and the nature of chemical bonds. The chemical environment (rather the electronic environment) of a nucleus has a minute but precisely observable influence on the transitions in NMR spectra which can be interpreted by various techniques to get a detailed knowledge about the surrounding bonds. Proton (^1H) NMR spectroscopy is used in MRI (*Magnetic Resonance Imaging*) for diagnosis of disorders in our body. Other nuclides may also be used in NMR spectroscopy for structural investigation, e.g., ^{13}C, ^{11}B, ^{15}N, ^{19}F and ^{31}P.

14.1.3 Mass Defect and Packing Fraction

With the discovery of isotopes and the development of Aston's mass spectrograph, it was found that the great majority of the elements consisted of isotopic nuclides whose masses were nearly integers. From this, Aston suggested his "whole number rule": the true atomic weights of all elements are whole numbers. Fractional atomic weights were explained by the existence of isotopes in varying abundance. This was practically a revival of Prout's hypothesis, proposed a hundred years earlier, which stated that all elements are built up of hydrogen. However, it was soon observed that though isotopic masses were close to whole numbers, they *always differed slightly from whole numbers*. Aston generalized these deviations in terms of the so-called "mass defect" and "packing fraction", as outlined below.

Mass defect and packing fraction

The nearest integer to the mass of a nuclide (in the atomic mass scale) was called the **mass number** of that isotope. The difference between the actual isotopic mass and the mass number was termed the **mass defect** of the nuclide (see section 14.1.4 for true mass defect).

At present, the mass number of any isotope may be looked upon as the *total number of neutrons and protons* in it. (Neutrons were not discovered before 1932, while Aston suggested these terms around 1927). Accordingly, the deviation of isotopic mass from whole numbers, the so called 'mass defect', may be understood in terms of mass-energy interconversion (sec. 14.1.4). The term mass defect is now preferably replaced by the term **mass excess,** ΔM, which may be either positive or negative. For a nuclide of mass M (in atomic mass scale) and mass number A, we may write

$$M = A + \Delta M \quad \text{or,} \quad \Delta M = M - A$$

i.e., **Mass Excess** ('Mass Defect') = **Isotopic Mass – Mass Number.**

Aston introduced a term "packing fraction" for each nuclide to compare their mass excess. It was defined as

$$\textbf{Packing Fraction (f)} = \frac{\textbf{Isotopic Mass } (M) - \textbf{Mass Number } (A)}{\textbf{Mass Number } (A)} \times 10^4 = \frac{\Delta M}{A} \times 10^4.$$

The factor of 10^4 was used to obtain figures which are easy to plot.

The packing fraction does not have any precise theoretical significance, but it gives an indication about the stability of a nucleus. A negative packing fraction indicates that the isotopic mass is actually less than the nearest whole number (ΔM negative). This suggests that a small fraction of the mass has been converted to the energy of formation of that nucleus. If we now want to split the nucleons from that nucleus, the same amount of energy would have to be supplied. So, the more negative the packing fraction of an element,

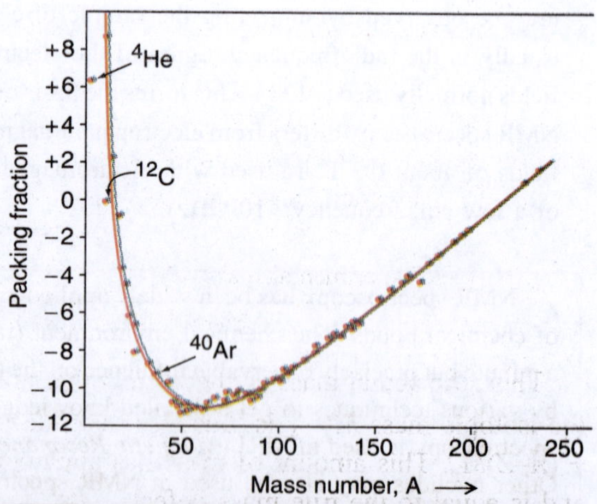

Fig.14.1: Plot of packing fraction against mass number

the more stable should be the nucleus. On the other hand, a positive packing fraction indicates that the concerned nucleus should be less stable. These generalizations are mostly revealed by the plot of packing fractions against mass numbers for different stable nuclides (Fig. 14.1). The observations on this curve may be summarized as follows:

(i) Packing fractions of the stable nuclides lie more or less on a smooth curve.

(ii) The curve passes through a minima in the mass number range 50-60, implying maximum stability.

(iii) After the minima, packing fraction steadily increases and ultimately becomes positive for elements of high mass number. This is consistent with the observed radioactivity, i.e., inherent instability, of these elements.

However, the idea of the packing fraction is based primarily on the observed deviations of isotopic masses from integral values. It does not throw any light regarding the nature of binding or packing of the nucleons. After the discovery of the neutron, practically the same generalizations were made in terms of nuclear binding energies.

14.1.4 Nuclear Binding Energy

After the discovery of neutrons, the mass of any atom could be estimated in terms of the masses of the constituent electrons, protons and neutrons. (The latter quantities were obtained from mass spectrometry with precisions of the order of a few parts per million). However, this estimated mass of a nuclide was always found to be greater than the precisely determined experimental mass of that isotope. This difference between (i) the mass of the constituents of an atom and (ii) the experimental atomic mass may be called *the true mass defect* of an isotope. This may be illustrated by the following examples:

Example I: The helium atom contains 2 protons, 2 neutrons and 2 electrons. The mass of 2 protons and 2 electrons should equal the mass of 2 hydrogen atoms $\left({}^{1}_{1}\mathrm{H}\right)$. Hence,

$$M_{\mathrm{He}} = 2M_{\mathrm{H}} + 2M_{\mathrm{n}}. \text{ (Subscripts refer to respective species).}$$

In atomic mass units we have $M_{\mathrm{H}} = 1.007825\, u$; $M_{\mathrm{n}} = 1.008665\, u$.

$\therefore M_{\mathrm{He}} = 2 \times (1.007825 + 1.008665) = 4.032980\ u$.

But the mass spectrometric mass of the helium atom $= 4.002603\ u$.

Mass defect $= 0.030377\ u$

Example II: Let us consider an atom with a negative packing fraction, say ${}^{35}_{17}\mathrm{Cl}$. Its mass should be equal to the sum of the masses of 17 protons, 17 electrons and 18 neutrons.

In atomic mass units, $M_{\mathrm{Cl}} = 17M_{\mathrm{H}} + 18\ M_{\mathrm{n}} = 17 \times 1.007825\ u + 18 \times 1.008665\ u$.

$= 35.288995\ u$

Experimental mass of ${}^{35}_{17}\mathrm{Cl} = 34.968851\ u$

Mass defect $= 0.320144\ u$

Thus, the actual mass is always *less* than the estimated mass. For any isotopic species ${}^{A}_{Z}X$, the isotopic mass $M_{Z,\,A}$ is thus less than the total mass of the constituents, obtained as $ZM_{\mathrm{H}} + (A-Z)M_{\mathrm{n}}$. This amount of mass has apparently disappeared in the formation of the nuclide and is equal to the true mass defect.

True mass defect $= [Z \cdot M_{\mathrm{H}} + (A-Z)\, M_{\mathrm{n}}] - M_{Z,\,A}$.

This "lost" mass may be considered as the mass which has been converted to the energy released in the hypothetical formation of the nucleus from individual neutrons and protons. Since the true mass of an atomic species has been compared with its estimated mass, the mass of the electrons had to be considered. This is why we used the mass of the hydrogen atom, not that of the proton alone, in calculating true mass defect. However, we have neglected the very small change in mass accompanying the formation of a hydrogen atom from a proton and an electron. The binding energy of the electrons to the nucleus is also neglected.

The magnitude of the binding energy may now be estimated from the true mass defect of a nuclide using the mass-energy equivalence: $E = \Delta m \,.\, c_0^2$ where Δm = mass lost, E = energy released and c_0 = velocity of light in vacuum.

Let us calculate the binding energy of the chlorine-35 nucleus.

$$\Delta m = \text{True mass defect of } {}^{35}\text{Cl} = 0.320144\ u.$$
$$= 0.320144 \times 1.660 \times 10^{-27}\ \text{kg}$$
$$\therefore\quad E = 0.320144 \times 1.660 \times 10^{-27}\ \text{kg} \times (2.9979 \times 10^{8}\ \text{m s}^{-1})^2 = 4.776 \times 10^{-11}\ \text{J}.$$
$$\Rightarrow 4.776 \times 10^{-11}\ \text{J} \div 1.602 \times 10^{-19}\ \text{J per eV.} = 298\ \text{MeV.}$$

The binding energy corresponding to 1 atomic mass unit, i.e. u, is equal to

$$1.660 \times 10^{-27}\ \text{kg} \times (2.9979 \times 10^{8}\ \text{m s}^{-1})^2$$
$$= 14.9191 \times 10^{-11}\ \text{J.} \equiv 931.28\ \text{MeV}$$

For most calculations, 931 MeV is a good approximation. So, the binding energy of any nucleus in MeV is equal to the true mass defect in atomic mass unit, u, multiplied by 931.

Binding energy $(B) = 931 \times [Z \,.\, M_H + (A\text{–}Z)\, M_n - M_{z,\,A}]$ MeV.

The average binding energy per nucleon, B, is obtained by dividing the binding energy by the number of nucleons. That is

Binding energy per nucleon, $(\overline{B}) = \dfrac{\text{Total binding energy of the nucleus}}{\text{Total number of nucleons}} = \dfrac{B}{A}.$

Problem 14.3

Calculate the binding energy of the helium nucleus, given that its true mass defect is 0.030377 u. Also calculate the average binding energy per nucleon.

Solution: Binding energy = 931 × 0.030377 MeV.

$$= 28.281\ \text{MeV.}\ \ (= 45.31 \times 10^{-13}\ \text{J.})$$

Since there are two protons and two neutrons in the nucleus, $\overline{B} = 11.327 \times 10^{-13}$ J.

Problem 14.4

Calculate the binding energy per nucleon of the ${}^{40}_{18}$Ar nucleus. The experimental mass of ${}^{40}_{18}$Ar is 39.962384 u. M_H = 1.007825 u ; M_n = 1.008665 u.

Solution: There are 18 protons and 22 neutrons in the nucleus and 18 extranuclear electrons.

∴ Estimated mass of ${}^{40}_{18}$Ar is equal to the mass of 18 hydrogen atoms and 22 neutrons.

$$= 18 \times 1{\cdot}007825\ u + 22 \times 1{\cdot}008665\ u = 40{\cdot}331480\ u$$

Isotopic mass of $^{40}_{18}Ar$ = 39·962384 *u*; difference = 0·369096 *u*

B.E. = 0·369096 × 931 MeV ⇒ 5·50 × 10^{-11} J.

∴ Binding energy per nucleon = 5·50 × 10^{-11} J ÷ 40 = 1·38 × 10^{-12} J.

A plot of average binding energy per nucleon against mass number gives a nearly smooth curve (Fig. 14.2). The curve is practically an inversion of the curve obtained by plotting packing fractions (Fig. 14.1). Maximum binding energy per nucleon is indicated in the mass number range 55-60. The average binding energies for ^4He, ^{12}C and ^{16}O lie above the curve, showing that they are exceptionally stable. It may be observed from the curve that the mean binding energy per nucleon is about 8 MeV over a considerable range of mass numbers.

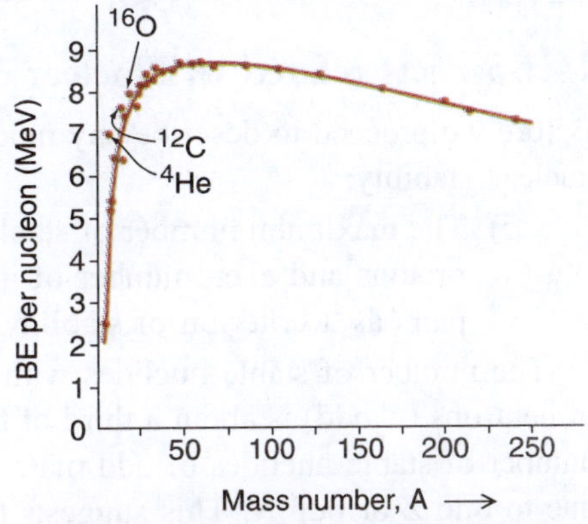

Fig.14.2: The nature of plot of average binding energy per nucleon against mass number

The overall shape of figure 14.2 represents another interesting feature. With nuclei of small mass numbers, B.E. first increases with increase in mass number. This suggests that energy will be *released* if one could fuse lighter nuclei into the more stable ones. This forms the basis of *nuclear fusion* processes. In the high mass number range (say above 100), however, B.E. decreases sharply with mass number, showing that the heavier nuclei are less stable. Energy would also be *released* if one could split such heavier nuclei into some lighter, stable fragments. This is, in fact, the basis of *nuclear fission* reactions. We shall return to these topics later.

Magic numbers

It has been observed that nuclei containing some specific number of neutrons and protons are exceptionally stable. These numbers are 2, 8, 20, 28, 50, 82 and 126. They have been called *magic numbers*.* The stability of nuclides containing either neutrons or protons equal to any magic number may be understood from the following examples:

(i) Elements having the largest number of stable isotopes contain 20 and 50 protons or 20, 50 and 82 neutrons. Tin (50 protons) has the maximum number of isotopes—ten. Calcium, with 20 protons, has six stable isotopes.

(ii) Nuclides containing magic number of either neutrons or protons or both are more abundant in the universe than others in the same mass number region, e.g., $^{40}_{20}$**Ca** (20*p* and 20*n*); $^{88}_{38}$**Sr**, $^{89}_{39}$**Y**, $^{90}_{40}$**Zr** (all containing 50 *n*); $^{118}_{50}$**Sn** (50 *p*); $^{138}_{56}$**Ba**, $^{139}_{57}$La, $^{140}_{58}$Ce (all containing 82 *n*); $^{208}_{82}$**Pb** (82 *p*).

(iii) The nuclides $^{17}_{8}O$, $^{87}_{36}Kr$ and $^{137}_{54}Xe$ have one neutron in excess of a magic number. They exhibit the unusual property of neutron emission to attain the magic number.

(iv) Nuclides containing magic numbers of neutrons are very reluctant to increase their neutron content, as shown by their low neutron capture cross section.

* By Haxel, Jensen and others (1949). Later, the same authors used the term "shell number".

(v) The three main radioactive series (ch. 13) end in lead, which contains 82 protons. $^{208}_{82}Pb$ is the most stable isotope of lead, with 82 protons and 126 neutrons.

(vi) The energy of an emitted alpha particle in a radioactive change is exceptionally large when the decay product belongs to the magic number family.

These and other similar observations have been explained on the basis of a shell model of the nucleus (sec. 14.1.6). The magic numbers correspond to closed shells of nucleons and accordingly nuclei containing these numbers of nucleons are stable, just like elements possessing closed shell of extranuclear electrons.

14.1.5 Facts related to Nuclear Stability: a Summary

Before we proceed to describe any model for the nucleus, we may summarize the facts about nuclear stability:

(i) The maximum number of stable, naturally occurring nuclides contain even number of protons and even number of neutrons. This suggests a tendency to form *p–p* or *n–n* pairs as a criterion of stability.

The number of stable nuclides with either odd number of protons (*Z* odd) or odd number of neutrons (*N* odd) is about a third of those having both *Z* or *N* even (sec. 1.3.1). Further, the number of stable nuclides of odd mass number ($A = Z + N$) is about the same whether this is due to odd *Z* or odd *N*. This suggests that protons and neutrons behave independent of their charge in determining nuclear stability.

The stability of nuclides with "magic numbers" of neutrons or protons or both is also a point of concern in connection with nuclear stability.

(ii) Except $^{1}_{1}H$, the neutron-proton ratio (*N/Z*) is nearly equal to 1 for all stable nuclides upto $^{40}_{20}Ca$; thereafter the ratio rises slowly and widens to a zone for stable nuclides. This indicates that nuclear stability is associated with an optimum range of *N/Z* ratio.

(iii) The binding energy per nucleon is essentially similar in magnitude for all nuclides, independent of mass number. This shows that a nucleon does not interact with *all* other nucleons present in the nucleus. (In that case binding energy would increase gradually with mass number).

The independence of nuclear density on mass number also leads to a similar conclusion. One may now try to analyze other systems with analogous characteristics, for example:

(a) a drop of a liquid, like liquid argon, has a density and a binding energy per atom which are independent of the size of the drop. Attraction from van der Waals force between the atoms is significant only for the nearest neighbours. At very close approach, repulsion predominates and hence limits the number of nearest neighbours. Similar short range attractive and repulsive forces are also likely to be present in the nucleus.

(b) In a solid like diamond, the density and binding energy per atom are also independent of the mass. However, in this case the binding force is limited by the number of valence electrons of carbon. The coulombic potential energy of a pair of protons at a distance *r* is $e^2/4\pi\varepsilon_0 r$. When *r* lies in the range of nuclear size, say 3 fm, the energy is about 0.48 MeV, much less than the average binding energy per nucleon (~ 8 MeV). Gravitational force between the nucleons will also be negligible in comparison to the observed binding energy. This leads us to believe that short range attractive forces operate between all the nucleons within the nucleus independent of their charge. Just outside the nucleus, the force abruptly disappears and coulombic interaction dominates. This gives us the idea of a nuclear potential similar to a square well (we recall the particle in a box case).

Summarily, any model for the nucleus should aim to answer the following questions:

(i) What is the force that holds the protons and neutrons together? This force must be stronger than the coulomb repulsive force between the protons. At the same time, there is no central field of force within the nucleus similar to that exerted on the extra-nuclear electrons.

(ii) Why are the densities and binding energies per nucleon independent of mass numbers?

(iii) Why are the nuclides corresponding to the magic numbers exceptionally stable?

(iv) Why do nuclei emit α- and β-particles although they contain only protons and neutrons?

With these, and other similar points in view, scientists have forwarded different models of the nucleus. Of these, the liquid drop model and the shell model will be mentioned here in brief.

14.1.6 Nuclear Models

The Liquid-drop model (Bohr, 1936)

The model is based on the observed similarities in the properties of a drop of liquid and the nucleus. Some of these are repeated below:

(i) The density of a liquid is constant throughout.

(ii) Each liquid possesses a definite latent heat of vaporization. This resembles the binding energy of the nucleon.

(iii) Radioactive properties of certain nuclei may be compared to the evaporation of a drop of liquid.

(iv) In a drop of a liquid, each molecule attracts other molecules in the immediate vicinity, but has little influence on the distant molecules. In the nucleus also, each nucleon attracts only those which are present in its immediate neighbourhood. If all nucleons were to interact with all other nucleons, the binding energy per nucleon would increase with increasing number of nucleons. This is contrary to our observation.

According to the liquid-drop model, each nucleon in a nucleus behaves in the same manner as the molecules in a liquid-drop. Each nucleon interacts strongly only with its adjacent nucleons. Hence any excess energy associated with an individual nucleon would be readily shared among its neighbours. The energy levels of the nucleus are thus assumed to be quantum states of the nucleus as a whole, not of any single nucleon.

Some other similarities suggested by the model between a liquid drop and the nucleus are also interesting:

(v) The nucleus is homogeneous and incompressible like a liquid drop. Density and other properties of a nucleus are same throughout except at the surface. The volume of the nucleus is thus proportional to its mass and hence, the mass number *A*.

(vi) Similar to an ideal solution, in which the solvent and solute particles interact without any distinction, the force between nucleons is the same irrespective of their charge and spin.

(vii) Since the nucleons interact strongly only with its neighbours, the nuclear force should be short range and tend to saturate as the number of nucleons increases. The interaction energy is simply proportional to mass number A.

(viii) Similar to a drop of liquid, the nucleus also shows surface tension property, the force being proportional to the surface area of the nucleus, hence to $A^{2/3}$.

(ix) If a nucleus is bombarded with a high energy particle, the latter may be captured to form a compound nucleus in an excited state. Deexcitation of this compound nucleus may now occur in several ways similar to a liquid drop (*given in parenthesis*):

(a) emission of radiation (*cooling*); (b) emission of one or more nucleons (*evaporation of a part*); (c) fission into two nuclei (*rupture into two droplets*).

Interpretation of nuclear fission by the liquid drop model has been discussed in sec. 14.2.1.

(x) Fusion of smaller drops of liquid into a large one is similar to the fusion of smaller nuclei into a large one.

The liquid drop model has been successfully developed to calculate the average binding energy per nucleon (see box). The results are in fair agreement with observed values. The model has the merit of explaining nuclear fissions and the properties of many nuclides. However, the model is not consistent with the observed stability of nuclides with *p–p* and *n–n* pairing (even Z, even N). Also it does not explain the stability of magic number nuclides, which strongly suggest some closed shell arrangement within the nucleus. Independent motions of the nucleons, as well as their individual spin, parity and magnetic moment effects are also ignored.

Calculation of average binding energy per nucleon.

Suppose the energy of each nucleon-nucleon interaction = U. Though inherently negative, U is written conventionally as positive for the sake of convenience. Since U operates between two nucleons, the binding energy per nucleon = $\frac{1}{2}U$.

As seen before (chapter 8), in a close-packed structure of spheres of same size, a central sphere has 12 other spheres in contact. Hence the binding energy for a nucleon in a close-packed arrangement would be $\frac{1}{2}U \times 12 = 6U$.

For a nuclide of mass number A, there are A nucleons and hence the total binding energy would be equal to $6AU$. This is often called the volume energy E_v : $E_v = 6AU$. As a matter of convenience, E_v is written as $E_v = a_1 A$ (a_1 = constant).

But a nucleon on the surface of a closepacked spherical aggregate has fewer than 12 neighbours. The number of such "surface nucleons" is clearly proportional to the surface area.

From section 14.1.2, surface area = $4\pi R^2 = 4\pi R_0^2 A^{2/3}$.

So, we may consider a quantity called the surface energy E_s which would *reduce* the binding energy and hence be negative:

$$E_s = -a_2 A^{2/3}$$

The binding energy between nucleons is further decreased by the coulomb repulsion between protons. For a nuclide having Z protons, the coulomb energy E_c for $Z(Z-1)/2$ pairs of protons would be

$$E_c = \frac{Z(Z-1)}{2}\left(-\frac{e^2}{4\pi\varepsilon_0 r_{av}}\right)$$

where r_{av} is the distance averaged over all proton pairs. If the protons are supposed to be uniformly distributed in a nucleus of radius R, r_{av} would be proportional to R and hence to $A^{1/3}$. One may thus express the coulomb energy E_c as

$$E_c = -a_3 \frac{Z(Z-1)}{A^{1/3}}$$

The total binding energy for A nucleons is obtained by the net sum of E_v, E_s and E_c:

$$E_{tot} = a_1 A - a_2 A^{2/3} - a_3 \frac{Z(Z-1)}{A^{1/3}}$$

The binding energy per nucleon is obtained by dividing E_{tot} by A.

The expression agrees well with experimental figures. A still better agreement may be achieved by including some corrections which are beyond the perview of the liquid drop model. One correction involves the so-called asymmetry energy E_a which arises when there are more neutrons than protons ($N > Z$):

$$E_a = -a_4 \frac{(A-2Z)^2}{A}.$$

Another correction involves the pairing energy E_p arising from the tendency of protons and neutrons to form pairs within each category. We found that even-even nuclei are most stable, suggestive of higher binding energies. Odd-odd nuclei are least stable, presumably having low binding energies. E_p is positive for even-even nuclei and negative for odd-odd nuclei and appears to vary with $A^{-3/4}$. Odd-even and even-odd nuclei have zero pairing energy.

$$E_p = (\pm \text{ or } 0) . \frac{a_5}{A^{3/4}}$$

The overall binding energy E_{tot} may now be obtained by including these corrections to our previous expression:

$$E_{tot} = a_1 A - a_2 A^{2/3} - a_3 \frac{Z(Z-1)}{A^{1/3}} - a_4 \frac{(A-2Z)^2}{A} \pm (1, 0) \frac{a_5}{A^{3/4}}$$

This is known as the ***Weizsäcker equation*** (1935) which gives a satisfactory estimate of nuclear binding energy with the following suitable values of the coefficients a_1, a_2, ... (all in MeV):

a_1	a_2	a_3	a_4	a_5	
14	13	0.60	19	34	MeV

The Shell model

The extranuclear structure of the atom was successfully described in terms of electron shells. A similar arrangement of the nuclear constituents in shells was suggested even before the discovery of the neutron. The idea was developed after the discovery of the neutron in 1932 by several scientists. However, the picture gained little importance in comparison to the liquid drop model. In 1948 Maria Goeppert-Mayer (USA) observed that many properties of the nucleus may be rightly interpreted in terms of the shell model. Some of the evidences favouring the shell model may be summarized as follows.

1. Unusual stability of nuclides containing "magic numbers" of neutrons or protons. This suggests that these nuclides probably correspond to closed shells of neutrons and protons.
2. Studies on nuclear spins and magnetic moments.

Many nuclei apparently spin about their axes like a top. Since the nucleus carries a charge, this corresponds to an electric current and gives rise to a magnetic moment. The spins and the magnetic moments of the nucleons and of the nucleus can be measured. It appears that the spin of nuclei is also quantized like that of the electrons: when measured in the same units ($h/2\pi$) it is always a whole number or half a whole number. The spin of the neutron and proton has been determined to be 1/2 unit. Their magnetic properties have been discussed earlier.

It has been observed that (i) the spins and magnetic moments of nuclei with an even number of protons and neutrons are zero. (ii) The spins of nuclei with an odd number of particles are all half-integral. (iii) Two isotopes with the same odd number of protons but different even numbers of neutrons have the same spin and magnetic moments. For example,both $^{113}_{49}In$ (49 protons, 64 neutrons) and $^{115}_{49}In$ (49 protons, 66 neutrons) have the same spin and magnetic moment. The additional pair of neutrons makes no contribution to the spin and magnetic moment.

Again, the isotope $^{87}_{38}Sr$ has 38 protons (even) and 49 neutrons (odd). It has the same spin, i.e. 9/2, as In-113 or In-115, which contain 49 protons. The magnetic moments are however, not same. Regarding spin alone, we find that 49 neutrons behave just as 49 protons.

These observations suggest that the spin and magnetic moment of "odd nuclei" (which contain odd number of either neutrons or protons) are entirely due to the last odd particle, be it a proton or a neutron. This signifies that all the other nucleons play no role except that they provide a potential energy field in which the odd-particle moves. Thus the protons and neutrons appear to behave with considerable independence of each other.

We recall that the arrangement of extranuclear electrons may be described in terms of the quantum numbers n, l and m derived from wave mechanics. There we have to assume that each electron is essentially independent of others; it simply moves in an electrostatic potential field due to the nucleus and all the other electrons.

In an analogous manner, one may try to obtain the single-particle orbitals for the nucleons by solving the appropriate Schrödinger equation. The number of neutrons and protons filling each level may then be determined by the Pauli exclusion principle.

But solution of the wave equation requires a knowledge of the potential field in the nucleus. The simplest guess we can make about the potential is to compare it to a well. No force acts on a particle outside the well and none inside. A nucleon also experiences no force inside the spherical nucleus, or outside it. But at the surface of the spherical shell, a nucleon experiences a strong attraction, corresponding to the barrier at the edge of a well. The potential will change abruptly at the edge if we consider a "square well": a nucleon experiences no interaction until it reaches a certain distance from the centre of the nucleus; then the full effect of the potential field is experienced. But the potential will change gradually if the well has sloping edges. Good results are obtained by considering (i) a potential energy distribution intermediate between a square well and a well with sloping edges and (ii) coupling of spin and orbital angular momentum for the nucleons.

With either form of potential, it is found that the first three energy levels (starting from the lowest) can accomodate 2, 6 and 12 nucleons respectively. This suggests that closed shells will occur with 2, (2 + 6) and (2 + 6 + 12) nucleons; the first three magic numbers 2, 8 and 20 are thus readily explained. The higher magic numbers may be explained from consideration of spin orbit coupling.

Similar to coupling of spin and orbital angular momentum for extranuclear electrons, the shell theory assumes that the *LS* coupling scheme holds for the lighter nuclei having small values of *l*. The total spin angular momenta S (calculated separately for the neutrons and protons) and the total orbital angular momenta L (again calculated separately for the two groups of particles) finally combine to give the total angular momenta J. As in the case with extranuclear electrons, the *jj* coupling holds for the heaviest nuclei, with an intermediate coupling scheme for elements inbetween.

A nuclear particle can align its own 1/2-unit spin parallel or anti-parallel with the angular momentum (*l*) of its orbit. This splits the orbit into two sublevels having total angular momentum quantum number j equal to $l + \frac{1}{2}$ and $l - \frac{1}{2}$. Each sublevel can accommodate $2j + 1$ neutrons and $2j + 1$ protons. As the levels are gradually filled, the $l + \frac{1}{2}$ levels are filled before the $l - \frac{1}{2}$ levels. The separation between these levels increases with l and may become quite large. Consequently, the nucleons occupying the $l - \frac{1}{2}$ levels may be placed at a different shell from those which enter the $l + \frac{1}{2}$ level. The pattern of filling is shown in Table 14.1. It may be observed that the permitted values of the quantum numbers are different from those obtained earlier for extranuclear electrons. This is natural, since the potential energy field is different. Each horizontal line indicates a sharp change in binding energy—and therefore implies the beginning of a new shell. The total number of neutrons or protons required to complete the shells (last column) corresponds to the magic numbers.

Problem 14.5

With the help of Table 14.1, write the distribution of the nucleons in various energy levels for the following nuclides

$${}^{7}_{3}\text{Li},\ {}^{16}_{8}\text{O},\ {}^{17}_{8}\text{O},\ {}^{41}_{20}\text{Ca}$$

Solution: Neutrons and protons fill the energy levels in the sequence shown in Table 14.1. Occupancy of neutrons and protons are shown separately. Closed shells are indicated by third brackett.

	protons	*neutrons*
${}^{7}_{3}\text{Li}$	$\left[(1s_{1/2})^2\right](2p_{3/2})^1$	$\left[(1s_{1/2})^2\right](2p_{3/2})^2$
${}^{16}_{8}\text{O}$	$\left[(1s_{1/2})^2(2p_{3/2})^4(2p_{1/2})^2\right] = [8]$	$\left[(1s_{1/2})^2(2p_{3/2})^4(2p_{1/2})^2\right] = [8]$
${}^{17}_{8}\text{O}$	[8]	$[8](3d_{5/2})^1$
${}^{41}_{20}\text{Ca}$	$[8]\left[(3d_{5/2})^6(2s_{1/2})^2(3d_{3/2})^4\right] = [20]$	$[20](4f_{7/2})^1$

The total angular momentum quantum number I of a nuclide may be readily ascertained from Table 14.1 when we remember that paired neutrons and paired protons do not contribute to the spin of the nucleus. Three different cases may arise:

i. ***Nuclides with even number of neutrons and protons***: Such nuclides, including the magic number nuclides, have all nucleons paired and hence have $I = 0$. ${}^{4}_{2}\text{He}$, ${}^{40}_{20}\text{Ca}$, ${}^{6}_{2}\text{He}$ are such examples.

ii. ***Nuclides with odd mass number (A)***: Only one unpaired nucleon (neutron or proton) contributes to the spin. The value of I is equal to the j value of the sublevel occupied by the unpaired nucleon (Table 14.1). Thus in ${}^{37}_{17}\text{Cl}$, the odd proton occupies the $3d_{3/2}$ level when $I = 3/2$.

Table 14.1

Sequence of energy levels for neutrons and protons (nucleons) according to the Shell model.

Energy (not to scale) ↑

States *Without spin-orbit coupling*	States *With spin-orbit coupling*	Nucleons *in the level* $2j + 1$		Total nucleous *upto and including a level*
	$8j_{15/2}$	16		
	$5d_{3/2}$	4		
4s	$4s_{1/2}$	2		
5d	$6g_{7/2}$	8		
6g	$7i_{11/2}$	12		
	$5d_{5/2}$	6		
7i	$6g_{9/2}$	10		
	$7i_{13/2}$	14		
	$4p_{1/2}$	2		
4p	$4p_{3/2}$	4	44	126
5f	$5f_{5/2}$	6		
	$5f_{7/2}$	8		
6h	$6h_{9/2}$	10		
	$6h_{11/2}$	12		
3s	$3s_{1/2}$	2		
4d	$4d_{3/2}$	4	32	82
	$4d_{5/2}$	6		
5g	$5g_{7/2}$	8		
	$5g_{9/2}$	10		
3p	$3p_{1/2}$	2	22	50
	$4f_{5/2}$	6		
4f	$3p_{3/2}$	4		
	$4f_{7/2}$	8	8	28
	$3d_{3/2}$	4		20
2s	$2s_{1/2}$	2	12	
3d	$3d_{5/2}$	6		
	$2p_{1/2}$	2		
2p	$2p_{3/2}$	4	6	8
1s	$1s_{1/2}$	2	2	2

iii. Nuclides with odd number of protons and odd number of neutrons: Few nuclides belong to this category (five examples in sec. 1.3.1). They contain one unpaired proton and one unpaired neutron. If l_p, l_n and j_p, j_n are the corresponding quantum numbers of the unpaired proton and neutron respectively for the orbital angular momenta and total angular momenta, I is determined as follows.

$$l_p + l_n + j_p + j_n = \text{even}, \quad I = |j_p - j_n| \ ; \quad l_p + l_n + j_p + j_n = \text{odd}, \quad I = |j_p + j_n|$$

Thus $^{76}_{33}As$ with 33 protons and 43 neutrons has $l_p + l_n + j_p + j_n = 3 + 4 + 5/2 + 9/2 = 14$ (even) (see Table 14.1). Hence $I = 9/2 - 5/2 = 2$.

The ***parity*** of a nuclide is often a useful designation. The term refers to the behaviour of a wave function under an inversion in space (sec. 4.6):

$$\psi(x, y, z) = \psi(-x, -y, -z): \ \textit{even}\ (+)\ \textit{parity}$$

Orbitals with $l = 0, 2, 4, 6$, i.e., *s*, *d*, *g* etc. have '+' parity.

$$\psi(x, y, z) = -\psi(-x, -y, -z): \ \textit{odd}\ (-)\ \textit{parity}$$

Orbitals with $l = 1, 3, 5$ (odd) i.e., *p*, *f*, *h*, .. etc. have '–' parity.

In $^{41}_{20}Ca$, for example, the last unpaired neutron occupies a $4f_{7/2}$ level (see Table 14.1 and problem 14.5). For *f*-level, l is odd, indicating '–' parity.

Besides explaining the magic numbers, the shell model provides fairly satisfactory interpretation of nuclear spin and magnetic moment. However, the model appears somewhat oversimplified when it assumes a perfectly spherical distribution of charge within the nucleus. In fact, nuclei undergo deformation as a result of nucleon-nucleon interactions. One of the models which considers this interaction is the collective model.

The collective model was developed in 1951 by A. Bohr (son of Niels Bohr) and Mottelson (Nobel Prize in Physics, 1975). It postulates mutual interaction or collective action among the nucleons. As a consequence, the nucleus is visualized again as a liquid drop which may suffer deformations. This ultimately gives rise to oscillation and rotation of the nucleus as a whole.

Deviation of the charge distribution in a nucleus from the spherical form may be measured in terms of nuclear quadrupole moments. The shell model could explain such moments with limited success. The collective model can account for such moments with better quantitative agreement. At the same time, it enables accurate calculation of the excited states of the nucleus.

The unified model (S. G. Nilsson, 1955) combines the desirable features of both the shell model and the collective model and provides a more satisfactory interpretation of nuclear properties.

14.1.7 Forces Within the Nucleus

When we survey the different nuclear models, the question that automatically comes to our mind is: "what holds the nucleus together?" To answer this question, let us recapitulate the characteristics of nuclear forces.

1. **Nuclear forces cannot be electric or gravitational in nature**. In the first place, the electric force between two protons is repulsive. Secondly, the binding of the neutral particle, neutron, cannot be covered by electric forces. Side by side, both electric and gravitational forces are many times weaker than the estimated magnitude of nuclear binding energies.

2. **Nuclear forces must be very short range**, operating over very short distances of the order of 1 fermi (10^{-15} m). As already discussed, the nearly constant binding energy per nucleon indicates that each nucleon is attracted only by its immediate neighbours. With long-range forces like gravitational and electrostatic forces, the binding energy would increase with increasing number of nucleons.

3. **Nuclear forces are charge-independent**. The force between any two nucleons is the same, irrespective of their nature. The approximate constancy of the nuclear binding energy per nucleon and nuclear density supports this conclusion.

The existence of proton-proton (*p–p*), proton-neutron (*p–n*) and neutron-neutron (*n–n*) attractions may be readily understood by considering the following common nuclei:

Nuclide	Composition	Probable interaction	Binding energy of the nucleus
$^{2}_{1}H$	1*p*, 1*n*	(*p–n*)	2.22 MeV
$^{3}_{1}H$	1*p*, 2*n*	2 (*p–n*) 1 (*n–n*)	8.48 MeV
$^{3}_{2}He$	2*p*, 1*n*	2 (*p–n*) 1 (*p–p*)	7.72 MeV

Considering the *p–p* repulsion in $^{3}_{2}He$, it comes out that the *n–n* and *p–p* attractive forces in the tritium and helium nuclei are essentially identical. Similar conclusions may be reached from the scattering of neutrons or protons of various energies by neutrons or protons in a target. In one such experiment, a beam of neutrons is directed at a piece of matter containing hydrogen. Neutrons colliding with the protons in the hydrogen are scattered in various directions. By observing the number of neutrons scattered in each direction and by using neutrons of different speed, it has been possible to calculate the magnitude of the *n–p* force. Comparison of the binding energies of certain isobaric nuclides. e.g., ^{10}Be, ^{10}B and ^{10}C also confirms that *n–n*, *p–p* and *n–p* forces are of similar magnitude.

4. The magnitude of the force between two nucleons also depends on the spins of the nucleons. Some relevant evidences are:

(i) The deuterium nucleus with 1 proton and 1 neutron is stable, having a spin of one unit. That is, the spins of the neutron and the proton are parallel. Scattering shows that an *n–p* combination with antiparallel spin is not stable.

Contrary to this, the species dineutron and diproton are not known. The exclusion principle requires that the spins of the two nucleons in these species should be antiparallel.

(ii) When neutrons are scattered by protons, two distinct types of scattering are observed: neutrons with spins parallel to those of the protons and neutrons with spins opposite to those of the protons. Similarly, *ortho* and *para* hydrogen give different types of neutron scattering.

5. Interaction between spin and orbital angular momenta of nucleons also contributes to the force between two nucleons. Evidence for this comes from scattering of protons by protons: the scattered beam contains two streams of protons—in one case the proton spins are parallel to the direction of the orbital angular momentum vector and in the other case they are opposite.

6. Apart from the electrostatic repulsion between the protons, other repulsive forces are also present between nucleons like that between the molecules of a liquid. In absence of this repulsion, the volume of nuclei would increase less rapidly. The repulsive force between two nucleons becomes dominant at distances less than 0.5 fermi.

The nature of nuclear force thus appears to be very complex.

The meson theory of nuclear forces

It is now generally accepted that the force between nucleons arises from the interchange of a new type of particle, the *meson*, between the nucleons. The suggestion was first made by Yukawa in 1935 from an analysis of the nature of electromagnetic forces.

The interaction between electrically charged bodies may be interpreted in terms of quantum mechanics. It appears that a charged body emits light quanta or photons which are absorbed by another. In the process, the light quanta transmit energy and momentum from the first object to the second. This is equivalent to the exertion of an electric force via the photons, though the photons are themselves chargeless. According to Yukawa, the role of photons in electric forces is played by some other particle in the nuclear field. At the time of his suggestion, such particles were hypothetical. Subsequently they have been identified as mesons.

According to the meson theory, a meson field surrounds the nucleons, just as a photon-field surrounds electrically charged bodies. The nucleons interact through interchange of mesons— which have masses nearly 200 times the mass of an electron and may have single unit of either positive or negative charge. The mass and charge of the particle is in harmony with short range exchange forces. When a proton and a neutron interact, the proton emits a positive meson which is absorbed by the neutron. Thereby the proton is converted into a neutron while the neutron is transformed into a proton. The same effect is obtained if the neutron emits a negative meson which is absorbed by the proton. Representing the positive and negative mesons by π^+ and π^- respectively, one may represent the exchange as

$$p + n \longrightarrow (n + \pi^+) + n \longrightarrow n + (\pi^+ + n) \longrightarrow n + p$$

Transfer of positive meson

$$n + p \longrightarrow (p + \pi^-) + p \longrightarrow p + (\pi^- + p) \longrightarrow p + n$$

Transfer of negative meson

In 1947, such mesons with mass 273 times the mass of an electron and carrying single positive or negative charge were discovered in cosmic rays. They are known as π-mesons or pions. Subsequently, another π-meson with no charge, designated by π°, was predicted and discovered. These neutral mesons have a mass 264 times the mass of an electron; they are supposed to be exchanged between two protons (or two neutrons).

$$p + p \longrightarrow (p + \pi^\circ) + p \longrightarrow p + (\pi^\circ + p) \longrightarrow p + p$$

Transfer of neutral meson

$$n + n \longrightarrow (n + \pi^\circ) + n \longrightarrow n + (\pi^\circ + n) \longrightarrow n + n$$

Transfer of neutral meson

The meson family

Before the discovery of the π-meson, during 1936 and 1937, another kind of meson was detected in cosmic rays. Known as μ-mesons or muons, these particles have masses nearly 206 times the mass of an electron and carries single positive or negative charge. It was however shown that these particles did not interact strongly with nucleons and therefore could not be accepted as the media for transfer of nuclear forces.

K-mesons or kaons with masses 966 times that of an electron have also been discovered. They may be neutral or charged. But they have no significant role in nuclear forces.

Outside the nucleus, $\pi^{\pm}$ mesons have a mean life of $2{\cdot}6 \times 10^{-8}$ s and are readily transformed into μ-mesons. The latter ultimately decay to yield a positron (or electron), a neutrino, an antineutrino and energy.

$$\pi^+ \rightarrow \mu^+ + \text{neutrino} + \text{energy} \quad ; \quad \pi^- \rightarrow \mu^- + \text{antineutrino} + \text{energy}$$

π^0 mesons have a mean life of $8{\cdot}4 \times 10^{-7}$ s.

$$\mu^+ \text{ (or } \mu^-) \rightarrow e^+ \text{ (or } e^-) + \text{neutrino} + \text{antineutrino} + \text{energy.}$$

At present, mesons may be produced artificially in cyclotrons and studied through scattering experiments. The mass and charge of mesons are sufficient to explain the short range, exchange nature and charge independence of nuclear forces. Other aspects of the meson theory can explain the dependence of nuclear forces on spin, the repulsion between nucleons at very short distances and the simultaneous interaction between more than two nucleons. Calculations based on the observed mass of the meson give correct figures for the range of nuclear forces.

Mass of pi-mesons

The emission and absorption of pions by nucleons take place so rapidly that the masses of neutrons and protons remain practically unchanged. The uncertainty principle in the form $\Delta E \, \Delta t \geq h/4\pi$ limits the time Δt (= $h/4\pi\Delta E$) of an event during which the energy conservation may be relaxed by an amount ΔE. This enables us to estimate the mass of pions.

We assume that a pion of mass m_π travels from one nucleon to another at a speed v which is nearly equal to c (actually $v < c$). The temporary discrepancy in energy for the emission of the mass m_π would be $\Delta E \simeq m_\pi c^2$, neglecting the kinetic energy of the pion. The time Δt required by the pion to move a distance r is r/v. Since we have assumed $v \simeq c$, $\Delta t \simeq r/c$. Nuclear forces have a maximum range of about 1.7 fm. This may be approximately taken as the value of r. Now we write the uncertainty relation in its approximate form

$$\Delta E \; \Delta t \simeq h/2\pi$$

$$\text{or,} \quad (m_\pi c^2)\,\frac{r}{c} \simeq \frac{h}{2\pi}$$

$$\text{when,} \quad m_\pi \simeq \frac{h}{2\pi rc} = \frac{6.626 \times 10^{-34}\ \text{J s}}{2 \times 3.142 \times 1.7 \times 10^{-15}\ \text{m} \times 3 \times 10^{8}\ \text{m s}^{-1}}$$

$$= 2.1 \times 10^{-28}\ \text{kg.}$$

Thus the mass of the pion is about 230 times the mass of an electron.

Contribution from still heavier mesons will be limited to still shorter ranges.

14.2 ARTIFICIAL NUCLEAR REACTIONS

Towards the end of the First World War, Rutherford was investigating the scattering of α-particles by gases. In 1929, he observed that nitrogen gas always produced fast protons when bombarded with α-particles of kinetic energy around 7.7 MeV (from RaC′). He suggested that these fast protons could appear only from the disintegration of the nitrogen nucleus.

$$^{4}_{2}He + ^{14}_{7}N \longrightarrow ^{17}_{8}O + ^{1}_{1}H$$

Rutherford's guess was subsequently confirmed. With this was opened a new horizon in the field of nuclear science: one element could be transformed into another. Since then, a large number of artificial nuclear reactions have been studied.

The reaction written above may also be represented as $^{14}_{7}N\ (\alpha, p)\ ^{17}_{8}O$. Here $^{14}_{7}N$ stands for the **target nucleus**, α- is the bombarding particle or **projectile**, *p*- represents the emitted particle or *ejectile* and $^{17}_{8}O$ symbolizes the product or **recoil nucleus**.

However, we should not think that all nitrogen atoms bombarded with α-particles undergo this transformation. It only takes place when an α-particle makes an almost *head-on collision* with a nitrogen nucleus. This takes place very rarely—one in about 50,000 α-particles cause such disintegration.

In the next few years, several other (α, *p*) reactions were studied, using light element targets. It was realized that α-particles from natural sources did not possess sufficient energy to disintegrate heavier nuclei. Attempts were made to devise accelerating machines for other particles which could be used as projectiles. In 1930-32, Cockroft and Walton developed a proton accelerator. Many new nuclear reactions were studied using accelerated protons as projectiles. Other accelerating machines were soon developed. Deuterons, α-particles and other heavy particles could be accelerated and used as projectiles. This opened a vast field of nuclear reactions.

14.2.1 Different Types of Nuclear Reactions

A necessary condition for two nuclei to interact is that they must be brought very close to one another—so close that the short range nuclear forces may overcome the electrostatic force of repulsion between the positively charged nuclei. When a projectile possesses sufficient energy to interact with a suitable target nucleus, the two particles combine to form a compound nucleus. The compound nucleus disintegrates within a very short time (10^{-12} s or less) to produce more stable nuclei. The process is usually accompanied by the expulsion of one or more particles, e.g., a proton (*p*), deuteron (*d*), an α-particle or a neutron (*n*). Sometimes the compound nucleus may also attain stability by the loss of its excess energy in the form of γ-rays. The particular mode of disintegration of a compound nucleus depends on its state of excitation and on the energy required to expel different particles. Accordingly nuclear reactions have been classified in a variety of ways.

To begin with, let us first classify nuclear reactions according to the projectiles as (i) **charged particle reactions** and (ii) **neutral particle reactions**. In either category, the projectile may be captured (*projectile-capture reaction*) with the emission of γ-rays or, one or more particles may be emitted (*projectile-capture particle emission reaction*).

Charged particle reactions

Such reactions may be carried out by accelerated protons, deuterons or alpha particles. Depending upon the nature of the emitted particle, we may have different types of reactions:

Projectile	***Reaction Type***
Proton	(p, n), (p, α), (p, γ), (p, d)
Deuteron	(d, p), (d, n), (d, α), (d, γ)
Alpha particles	(α, n), (α, p).

Example:

p, n: $^{23}_{11}Na + ^{1}_{1}H = ^{23}_{12}Mg + ^{1}_{0}n$

p, α: $^{7}_{3}Li + ^{1}_{1}H = ^{4}_{2}He + ^{4}_{2}He$ ... (1)

p, γ: $^{27}_{13}Al + ^{1}_{1}H = ^{28}_{14}Si + \gamma$

p, d: $^{9}_{4}Be + ^{1}_{1}H = ^{8}_{4}Be + ^{2}_{1}H$

d, p: $^{27}_{13}Al + ^{2}_{1}H = ^{28}_{13}Al + ^{1}_{1}H$

d, n: $^{27}_{13}Al + ^{2}_{1}H = ^{28}_{14}Si + ^{1}_{0}n$

d, α: $^{6}_{3}Li + ^{2}_{1}H = ^{4}_{2}He + ^{4}_{2}He$ ... (2)

d, γ: $^{63}_{29}Cu + ^{2}_{1}H = ^{65}_{30}Zn + \gamma$

α, n: $^{9}_{4}Be + ^{4}_{2}He = ^{12}_{6}C + ^{1}_{0}n$ ... (3)

α, p: $^{14}_{7}N + ^{4}_{2}He = ^{17}_{8}O + ^{1}_{1}H$ ... (4)

Note: (1) The first artificial nuclear reaction carried out with **artificially** accelerated projectiles.

(2) This reaction gives one of the highest non-fission energies (22.2 MeV) recorded for light atoms.

(3) The neutron was discovered through this reaction.

(4) The first artificial nuclear reaction studied.

Neutral particle reactions

Neutrons are generally used as projectiles in such reactions. When a neutron approaches a nucleus, no repulsion develops and the potential energy remains unaltered. Only when the neutron is very close to the nucleus, the nuclear attractive forces become effective and the potential energy decreases sharply. So it is not necessary to use accelerated neutrons. In fact slow neutrons with energies less than an electronvolt (~0.026 eV) are found to be more effective than fast neutrons. However, when the ejection of a particle requires a high excitation energy of the compound nucleus, high energy neutrons become important. Neutrons of high energy are obtained from different artificial nuclear reactions. These are slowed down through **moderators** like graphite, heavy water, paraffin etc. Much of the energy of the neutrons is transferred to these substances through elastic scattering.

Neutron-induced reactions may be of (n, γ), (n, p) or (n, α) type. The first one is of the projectile capture type (also known as radiative capture). The (n, p) reactions require somewhat faster neutrons. Neutrons of high energy may produce (n, α) reactions; however, slow neutrons may also give such reaction with targets of low atomic number.

$$n, \gamma: \quad {}^{238}_{92}U + {}^{1}_{0}n = {}^{239}_{92}U + \gamma \qquad \text{... (1)}$$

$$n, p: \quad {}^{14}_{7}N + {}^{1}_{0}n = {}^{14}_{6}C + {}^{1}_{1}H$$

$$n, \alpha: \quad {}^{6}_{3}Li + {}^{1}_{0}n = {}^{4}_{2}He + {}^{3}_{1}H \qquad \text{... (2)}$$

Note: (1) Important for the production of plutonium.

(2) Important for the production of **tritium** for various experimental work. It is also obtained as

$$ {}^{2}_{1}H + {}^{2}_{1}H = {}^{3}_{1}H + {}^{1}_{1}H; \quad {}^{9}_{4}Be + {}^{2}_{1}H = {}^{8}_{4}Be + {}^{3}_{1}H$$

The most important type of reaction brought about by neutrons is nuclear fission. This has been discussed separately.

Photonuclear reactions may also be considered as neutral particle reactions. These are initiated by photons. The energy of the photon must be in the range of million electronvolts (in the range of nuclear binding energy). γ-rays from various radioactive materials can give (γ, *n*) type reactions with light nuclei like ${}^{2}_{1}H$ or ${}^{9}_{4}Be$, where the binding energy for the neutron is low (~ 2 MeV).

$$ {}^{2}_{1}H + \gamma = {}^{1}_{1}H + {}^{1}_{0}n; \quad {}^{9}_{4}Be + \gamma = {}^{8}_{4}Be + {}^{1}_{0}n$$

The second reaction serves as a source of photoneutrons.

For most other nuclei, the binding energy per nucleon is around 8 MeV. γ-rays of such high energy are not commonly available from natural sources. Such high energy photons are obtained as x-rays by striking suitable anticathodes with high energy electrons. (γ, *p*), (γ, 2*n*) or (γ, *np*) type reactions are also possible with such photons.

Nuclear reactions may also be classified according to the overall transformation. We have already found many examples of this type. Photonuclear reactions form a category in such classifications.

1. *Projectile capture (Radiative capture) reactions*

The projectile is absorbed by the target nucleus and no massive particle other than the product nucleus is emitted. γ-radiation may or may not be emitted.

$$ {}^{85}_{37}Rb + {}^{1}_{0}n = {}^{86}_{37}Rb + \gamma; \quad {}^{59}_{27}Co + {}^{1}_{0}n = {}^{60}_{27}Co; \quad {}^{12}_{6}C + {}^{1}_{1}H = {}^{13}_{7}C + \gamma$$

2. *Projectile capture particle emission reactions*

A massive particle is commonly ejected in addition to the product nucleus. With projectiles of high energy, more than one particle may also be produced.

$$ {}^{14}_{7}N + {}^{1}_{0}n = {}^{14}_{6}C + {}^{1}_{1}H; \quad {}^{11}_{5}B + {}^{1}_{1}H = {}^{11}_{6}C + {}^{1}_{0}n$$

Adequately accelerated heavy ions may be used as projectiles to synthesize a variety of nuclides:

$$ {}^{27}_{13}Al + {}^{14}_{7}N \xrightarrow{\text{nucleon transfer}} {}^{25}_{13}Al + {}^{16}_{7}N \text{ or, } {}^{27}_{12}Mg + {}^{16}_{8}O$$

$$ {}^{51}_{23}V + {}^{16}_{8}O \xrightarrow{\text{particle emission}} {}^{65}_{31}Ga + 2n$$

3. *Nuclear Fission*

In 1939, Otto Hahn and F. Strassmann observed that when uranium is bombarded with neutrons, an isotope of barium is formed. It was suggested (Meitner and Frisch) that the uranium nucleus captures the neutron and divides itself into two nuclei of roughly equal size. The idea was confirmed by several workers in different countries. The process of the division of a heavy nucleus into two roughly equal parts on bombardment has been termed nuclear fission*. Such fission reactions are accompanied by the release of a large amount of energy. It has been observed that uranium-235 absorbs neutrons of *all energies* and undergoes fission. It is thus said to be "*fissionable*" and is the only naturally occurring fissionable nuclide with a long half-life. Plutonium-239 and uranium-233, both produced artificially, are also fissionable. Uranium-238 may also undergo fission with neutrons having energy in excess of 1 MeV (see later). Fission of lighter nuclei like bismuth, lead, thallium, platinum and even tantalum has been achieved with accelerated α-particles (~ 400 MeV) and deuterons (~ 200 MeV). However, fission reactions by neutrons have been studied and applied most widely.

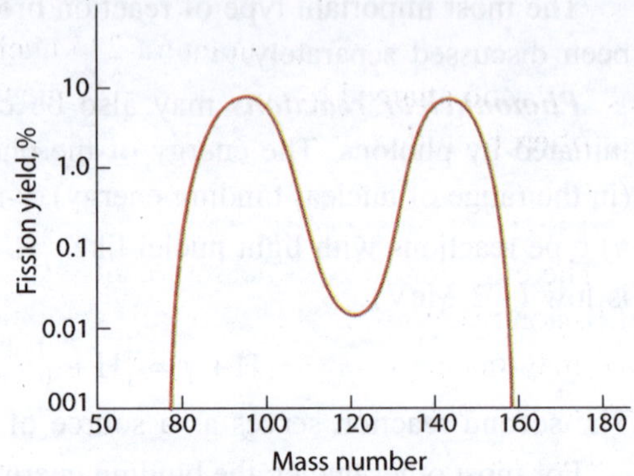

Fig.14.3: Nature of distribution of mass number among the fission fragments of U-235. The **fission yield** of a pro-duct is the percentage of the fission that produces that particular nuclide or its isobar.

The products obtained from the fission of a heavy nucleus are not fixed—they may vary considerably in mass and nuclear charge. With uranium-235, nearly 6.4 percent of the nuclei undergoing fission yield products with mass numbers 95 and 139 (Fig. 14.3). In only 0.01 percent cases does symmetrical fission occur, giving products of mass numbers 117. In fact, a given species may undergo almost forty different kinds of fission, giving eighty diffe-rent primary fission products, often termed *fission fragments*. These fission fragments are also not stable. Each of them undergoes three stages of β-decay on an average—ultimately giving nearly 250 *fission products*.

Each act of fission is accompanied by the release of 2 to 3 neutrons and about 200 MeV energy on an average. A typical fission process may be represented as

$$^{235}_{92}\text{U} + {}^{1}_{0}n \rightarrow {}^{92}_{36}\text{Kr} + {}^{141}_{56}\text{Ba} + 3\,{}^{1}_{0}n + 200\ \text{MeV}.$$

The energy released and the number of neutrons emitted in any fission reaction depend on the mode of fission. The average energy per nucleus is 200 MeV. The weighted average of the number of neutrons released is 2.5 for ^{233}U and ^{235}U. It is 2.9 for plutonium.

A rough estimate of the fission energy may be obtained as follows:

Average binding energy per nucleon around mass number of 235 = 7.6 MeV .. (Fig. 14.2)

∴ **Energy released in the formation of a U-235 nucleus = 7.6 × 235 MeV.**

Again, average binding energy per nucleon in the central mass number region (corresponding to the fission fragments) = 8.5 MeV.

* By analogy with the division of cells in living organisms. The term was suggested by W. Arnold, an American biologist.

∴ **Energy released in the formation of the nuclei of the fission fragments = 8.5 × 235 MeV.**

The excess energy in the latter case should be the *approximate* energy released when a U-235 nucleus undergoes fission to the lighter fragments (ignoring the emission of neutrons).

∴ **Approximate energy released in the fission of a U-235 nucleus = (8.5 – 7.6) × 235 = 211.5 MeV.**

Thus, the fission of a *single heavy nucleus* releases nearly 200 MeV of energy. This corresponds to about **8.2 × 10^7 kJ energy per gram of U-235.**

The emission of neutrons in fission may be similarly understood from the relevant neutron: proton ratios (*N*/*P*). For uranium-235, *N*/*P* is 1.55. For barium, *N*/*P* is 1.44. There is thus an excess of neutrons when a uranium-235 nucleus undergoes fission to the lighter nuclei. Again, the *N*/*P* ratio attained by the fission fragment is still high for stability. So, these product nuclei are often β-emitters.

Fission chain

The emission of two or more neutrons in each act of fission has opened a new possibility of fission reactions. If at least one of the emitted neutrons can be made to cause further fission, one may obtain a self-sustaining chain process. The actual condition required for this is that the number of neutrons produced in each act of fission should be at least equal to (i) the number of neutrons that escape from the material *plus* (ii) the number of neutrons absorbed in various fission and nonfission reactions. Since neutrons escape only through the surface of a fissionable material but is produced throughout its body, the size of the fissionable material may be increased to decrease the ratio of area to volume. The fraction of neutrons lost by escape is thereby reduced. This condition implies that there should be a **critical size** of the fissionable material to maintain a fission chain. This, in turn, requires a minimum **critical mass** of the fissionable material.

An atomic bomb is produced with the objective of liberating a tremendous amount of fission energy within a very short time. This is carried out by bringing together very quickly (~ 10^{-6} s) two separate masses of uranium-235 each of which are below its critical mass. The combined mass suddenly exceeds the critical mass and uncontrolled fission sets in—resulting in an explosion. Temperature suddenly rises to several million degrees and the pressure to several million atmospheres—causing widespread destruction. **A hydrogen bomb**, about one thousand times as powerful as an atom bomb, is obtained by surrounding an atom bomb with a mixture of deuterium and tritium. Fission raises the temperature of the gas mixture and fusion sets in—liberating tremendous amount of energy (see later).

The critical mass of a fissionable nuclide depends upon several factors like (i) the nature and abundance of the particular nuclide; (ii) presence of other materials which absorb neutrons; (iii) presence of moderators which slow down the fast neutrons formed; (iv) shape and arrangement of the material and so on. All these things are top secret for any particular plant. Ordinarily, it may be estimated that with graphite moderators, more than 200 kg of uranium-235 should be present in a total mass of about 30 tons of natural uranium to maintain a fission chain. It has been possible to design systems in which a self-sustaining fission chain may be maintained under control to supply atomic power (see Nuclear Reactors).

Natural uranium does not undergo spontaneous fission in spite of the presence of uranium-235 and the availability of neutrons. This is because adequate number of neutrons are not available to maintain a fission chain. A large proportion is absorbed for nonfission reactions (radiative capture) by U-238 as well as U-235, together with other nuclides present. At the same time, a major fraction of neutrons escapes from the system.

Mechanism of fission

The mechanism of nuclear fission induced by neutrons may be understood by the liquid drop model of the nucleus. The nuclear force may be compared with the force of surface tension maintaining the shape of a liquid drop. Splitting of the drop requires additional energy which would overcome the surface tension. The division of a liquid drop into two fragments may then be supposed to take place through a series of successive deformations.

The fission of a nucleus may also be considered to take place in an analogous manner. When a neutron is absorbed by a target nucleus, the binding energy of the additional neutron (plus kinetic energy, if any) raises the compound nucleus to an excited state. The excess energy may be emitted as γ-rays or lead to the expulsion of particles. But it may also set up strong oscillations within the nucleus. These oscillations tend to deform the nucleus. If the nuclear forces are stronger, the deformation cannot be large (just as surface tension restores the shape of a liquid drop). But when the deformation is large, it will ultimately lead to splitting of the nucleus into two fragments. The course of fission of the compound nucleus may be understood by imagining successive stages of deformation of the nucleus as shown in Fig. 14.4.

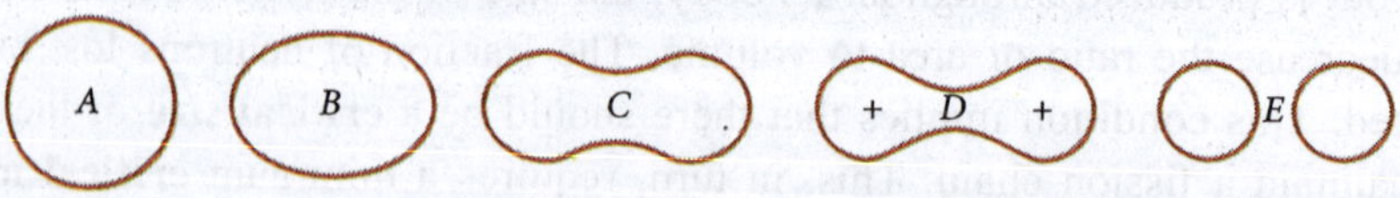

Fig.14.4: Interpretation of nuclear fission by the liquid drop model. ***A***: Spherical compound nucleus; ***B***: Initial deformation to an ellipsoidal; ***C***: Narrowing; ***D***: Critical State; ***E***: Separation.

The analogy with a liquid drop is obvious. When the oscillation is so violent that the nucleus is deformed to the extent shown in *D*, it is difficult for the nucleus to return to its original shape. The positive charges at the two ends repel each other and the nucleus ultimately splits as shown in *E*. Thus, a nucleus undergoing fission must pass through a critical state as shown in *D*. Clearly, this corresponds to a critical energy of the compound nucleus.

The critical energy required for the final splitting of a compound nucleus may be obtained from the liquid drop model as

$$E_{crit} = 0.89\, A^{2/3} - 0.02\, \frac{Z(Z-1)}{A^{1/3}}\ \text{MeV},$$

where *A* is the mass number of the compound nucleus and *Z* its atomic number. When a neutron enters the nucleus of U-235, the compound nucleus has a mass number 236; in case of a U-238 atom, the mass number of the compound nucleus becomes 239. The corresponding critical energies are:

U-235 $\quad E_{crit} = 0.89 \times (236)^{2/3} - \dfrac{0.02 \times 92 \times 91}{236^{1/3}}\ \text{MeV} = 6.9\ \text{MeV}.$

$$\textbf{U-238} \qquad E_{crit} = 0.89 \times (239)^{2/3} - \frac{0.02 \times 92 \times 91}{239^{1/3}} \text{ MeV} = 7.3 \text{ MeV}.$$

This critical energy has to be supplied by the kinetic energy and binding energy of the added neutron. We can easily calculate the gain in binding energy in the two cases:

For $^{235}U + {}^{1}n \rightarrow {}^{236}U$,

mass difference = 236.0440 – (235.0423 + 1.0092) u = 0.0075 u.

additional binding energy = 0.0075 × 931 or 6.98 MeV.

For $^{238}U + {}^{1}n \rightarrow {}^{239}U$,

mass difference = 239.0530 – (238.0493 + 1.0092) u = 0.0055 u.

additional binding energy = 0.0055 × 931 or 5.12 MeV.

Thus, in case of U-235, the additional binding energy of the added neutron supplies 6.98 MeV, which is more than the critical energy required for fission (6.9 MeV; the figures are approximate). Therefore, the compound nucleus can undergo fission with slow neutrons. The situation for U-238 is clearly unfavourable for fission—with a deficit of about 2.2 MeV from the critical energy. It will require high energy neutrons to cause fission. (Also see sec. 14.2.3).

U-235 has 92 protons and 143 neutrons (even-odd nuclide). U-238 is an even-even nuclide (92 protons + 146 neutrons). For a given value of Z, liberation of fission energy is easier for an even-odd nucleus forming a compound nucleus in even-even combination with higher energy. Other nuclides with even number of protons and odd number of neutrons are similarly fissionable by slow neutrons—e.g., $^{233}_{92}U$, $^{239}_{94}Pu$. Even-even nuclides like $^{238}_{92}U$ and $^{232}_{90}Th$ are fissionable with fast neutrons.

4. Spallation

It was observed by Seaborg and his coworkers (1947) that bombardment with high energy charged particles (e.g., 400 MeV protons or 200 MeV deuterons) can break some target nuclei into several smaller nuclei with the emission of a large number of nucleons (10 to 20 or even more). Such reactions have been termed **spallation*** reactions. Spallation may occur with light as well as heavy nuclei. In heavier nuclei, fission and spallation occur at the same time. The mass distribution between the fragments obtained is different for the two types of reactions. In fission, the masses are very much less than that of the target. In spallation, the masses differ by up to 10 or 20 units from that of the target. Typical examples of spallation are:

$$^{75}_{33}As + {}^{2}_{1}H = {}^{56}_{25}Mn + 9\,{}^{1}_{1}H + 12\,{}^{1}_{0}n\,; \qquad {}^{238}_{92}U + {}^{4}_{2}He = {}^{187}_{74}W + 20\,{}^{1}_{1}H + 35\,{}^{1}_{0}n.$$

5. Nuclear fusion

In this type of nuclear reaction, certain light nuclei may fuse together with the liberation of tremendous amount of energy. To achieve this, the colliding nuclei must possess enough kinetic energy to overcome the initial repulsion between the positively charged cores. This

* To spall = to break up by chipping off small fragments.

energy may be made available by raising the temperature of the reacting system to several million degrees. Such reactions are therefore also known as **thermonuclear reactions**. A typical example is

$$^{2}_{1}H + ^{3}_{1}H = ^{4}_{2}He + ^{1}_{0}n + 17.6 \text{ MeV}.$$

The energy accompanying a fusion process is derived from the difference of binding energies between the product and the reactant nuclei. The high temperature required to initiate such reactions may be attained initially through a fission process. Once started, the energy released maintains the temperature and the reaction may continue in a self-sustaining process, much like the combustion of a fuel. The genesis of different elements in the universe is believed to be due to series of fusion reactions as described earlier (section 1.2).

The energy source of the stars, including the sun, is believed to be continuous fusion processes at the very high temperature present. Thus, the energy radiated by the sun (nearly 10^6 J s^{-1}) is believed to occur by the fusion of protons to helium (proton-proton cycle: H. A. Bethe, 1939) at the core temperature of about $1{\cdot}5 \times 10^7$ K.

$$^{1}_{1}H + ^{1}_{1}H \rightarrow ^{2}_{1}D + ^{0}_{+1}e(\beta^+) + \nu$$
$$^{1}_{1}H + ^{2}_{1}D \rightarrow ^{3}_{2}He$$
$$^{1}_{1}H + ^{3}_{2}He \rightarrow ^{4}_{2}He + ^{0}_{+1}e(\beta^+) + \nu$$
$$4^{1}_{1}H \rightarrow ^{4}_{2}He + 2^{0}_{+1}e(\beta^+) + 2\nu + E\ (\equiv 26{\cdot}7 \text{ MeV})$$

The energy liberated in the process, ~27 MeV per four protons, on conversion, leads to the high value of $6{\cdot}5 \times 10^8$ kJ per gram of hydrogen. It is estimated that all the protons in the sun will be converted to helium in about 3×10^{10} years.

At higher stellar temperatures an alternative carbon-nitrogen cycle has been proposed. The cycle involves a higher coulombic energy barrier than the proton-proton cycle and requires a higher temperature (~$1{\cdot}6 \times 10^7$ K).

$$^{12}_{6}C + ^{1}_{1}H \rightarrow ^{13}_{7}N + \gamma$$
$$^{13}_{7}N \rightarrow ^{13}_{6}C + ^{0}_{+1}e(\beta^+) + \nu \quad (t_{1/2} = 10{\cdot}1 \text{ s})$$
$$^{13}_{6}C + ^{1}_{1}H \rightarrow ^{14}_{7}N + \gamma$$
$$^{14}_{7}N + ^{1}_{1}H \rightarrow ^{15}_{8}O$$
$$^{15}_{8}O \rightarrow ^{15}_{7}N + ^{0}_{+1}e(\beta^+) + \nu \quad (t_{1/2} = 2{\cdot}05 \text{ min})$$
$$^{1}_{1}H + ^{15}_{7}N \rightarrow ^{12}_{6}C + ^{4}_{+2}He$$
$$4^{1}_{1}H \rightarrow ^{4}_{2}He + 2^{0}_{+1}e(\beta^+) + 2\nu \quad + E\ (\equiv 24{\cdot}7 \text{ MeV})$$

The C-12 acts virtually as a catalyst in the process as it is regenerated at the end of the cycle. The energy liberated in this fusion process is nearly the same as in the previous cycle.

The fusion reactions which take place so vigorously in the stars would be extremely slow on the earth due to small quantity of the fusion material available here. Thermonuclear reactions which are likely to proceed rapidly on the earth are of the type:

$$^{2}_{1}H + ^{2}_{1}H = ^{3}_{2}He + ^{1}_{0}n + 3.2 \text{ MeV} \qquad ^{2}_{1}H + ^{2}_{1}H = ^{3}_{1}H + ^{1}_{1}H + 4.0 \text{ MeV}$$
$$^{3}_{1}H + ^{2}_{1}H = ^{4}_{2}He + ^{1}_{0}n + 17.6 \text{ MeV} \qquad ^{2}_{1}H + ^{3}_{2}He = ^{4}_{2}He + ^{1}_{1}H + 18.3 \text{ MeV}$$

The amount of heavy water present in natural water is sufficient to provide a lasting stock of deuterium for these reactions. The problem then rests on the success of attaining the initial temperature required for such reactions, calculated to be roughly 4.5×10^7 K or higher. Attempts are being made to set up fusion reactors in which controlled thermonuclear reactions may be carried out. This requires sufficient knowledge about plasma—a state of coexistence of positive nuclei and electrons at the very high temperature involved.

The hydrogen bomb

In the hydrogen bomb, fusion reaction between various hydrogen nuclides is initiated by a fission reaction (^{235}U or ^{239}Pu) in the core. Probable reactions are

$$^{3}H + {}^{2}H \longrightarrow {}^{4}He + n + 17\text{ MeV}; \quad {}^{3}H + {}^{3}H \longrightarrow {}^{4}He + 2n + 11\text{ MeV};$$

$$^{3}H + {}^{1}H \longrightarrow {}^{4}He + 20\text{ MeV}$$

There is no critical size of the H-bomb. In the *cobalt bomb*, the H-bomb is covered with a thick sheath of cobalt which absorbs the emitted neutrons and produces radioactive ^{60}Co by (n, γ) reaction.

Some other classification of nuclear reactions appear useful in different contexts.

Elastic scattering: Kinetic energy is conserved between the projectile and ejectile, both same nuclides, e.g.,

$$^{9}Be\ (n, n)\ {}^{9}Be\ ; \quad {}^{12}C\ (n, n)\ {}^{12}C.$$

Inelastic scattering: The projectile and the ejectile are again the same nuclide, but a portion of the kinetic energy of the projectile is spent to excite the target nucleus. The ejectile is formed in an excited state. Deexcitation follows by γ-radiation or *isomeric transition* (IT).

$$^{238}U + n \longrightarrow n + {}^{238}U^* \longrightarrow {}^{238}U + \gamma.$$

Evaporation: This refers to ejection of a large number of nucleons, usually neutrons, in a lot:

$$^{209}_{83}Bi\ (d, 12n)\ {}^{199}_{84}Po^* \longrightarrow {}^{100}_{42}Mo + {}^{99}_{42}Mo \xrightarrow{2\beta} {}^{99}_{44}Ru$$

Stripping: Sometimes the joint field of the target nuclide and the projectile splits the approaching projectile; a split part now takes part in the nuclear reaction. On approaching a target ^{63}Cu nuclide, a low energy deuteron splits into a proton and a neutron. The neutron penetrates the target nucleus

$$^{63}Cu + {}^{2}_{1}H \xrightarrow{2\beta} {}^{1}_{1}H + ({}^{63}Cu + {}^{1}_{0}n) \longrightarrow {}^{64}Cu$$

14.2.2 Energy and Cross Section of Nuclear Reactions

Nuclear reactions involve large amounts of energy—nearly million times greater than the energy of ordinary chemical reactions. The difference of mass between the reactants and products is significantly large in nuclear reactions and can be known in most cases from the isotopic masses of the nuclei involved. On the other hand, chemical reactions are accompanied by very small mass changes which cannot be detected by any known experimental technique. The energy of a nuclear reaction (E) may be conveniently estimated from the mass-energy equivalence, i.e.,

$$E = \Delta m\ .\ c^2$$

where Δm represents the difference in mass between the reactant and product nuclei. As already seen, the amount of energy equivalent to one atomic mass unit (u) is nearly 931 MeV.

Problem 14.6

Calculate the energy released in the reaction $^{35}_{17}Cl + {}^{1}_{0}n \rightarrow {}^{35}_{16}S + {}^{1}_{1}H$. Masses ($u$): Cl-35 = 34.9688; $^{1}_{0}n$ = 1.0087; S-35 = 34.9690; ^{1}H = 1.0078.

Hint: Total mass of reacting particles = Mass of Cl-35 + Mass of ${}_0^1n$ = 35.9775 *u*

Total mass of product particles = Mass of S-35 + Mass of ${}^1H$ = 35.9768 *u*

∴ Mass decrease (Δm) = 0.0007 *u* ⇒ Energy released = 0.6517 MeV.

Problem 14.7

Calculate the maximum energy of a β-particle emitted from a ${}^{14}C$ nucleus which does not emit any γ-radiations. The masses of C-14 and N-14 isotopes are 14.003242 *u* and 14.003074 *u* respectively.

Solution: The maximum energy of the β-particle may be calculated from the supplied mass difference assuming that no energy is shared by the accompanying neutrino.

$$ {}^{14}C \xrightarrow{-{}_{-1}^{0}e} {}^{14}N \qquad \Delta m = 0{\cdot}000168\ u $$

∴ Energy released in the process = 0.000168 × 931 or 0.3554 MeV.

This is the maximum energy of the β-particle.

Note. In solving the above problem, **the mass of the emitted β-particle has not been taken into account**. This is because the atom of N-14 produced contains *one more extranuclear electron* than the C-14 atom.

But when **a positron is emitted, two electron masses have to be taken into account**: (i) one for the positron emitted and (ii) one for the extra electron that leaves the extranuclear shell to maintain electrical neutrality (see section 13.3.4).

However, this estimate of the energy released in a nuclear reaction is only approximate. Except for reactions carried by slow neutrons, the initial kinetic energy of the projectile has to be sufficient to overcome the coulombic potential barrier between the projectile and the target, both positively charged. In such cases, due consideration is necessary for the original kinetic energies of the reacting particles. Sometimes the product of a nuclear reaction exists in an excited state and subsequently reaches the ground state by emitting γ-radiations. In such cases, the kinetic energy of the product particles is certainly less than the energy released in the process by the amount of energy radiated in the form of γ-rays.

Nuclear reaction energies may also be estimated directly from the energies of the interacting particles and the product particles. These may be determined from the ranges of these particles or from their deflections in electric or magnetic fields of known strengths.

Studies on the mass-energy interconversion in nuclear reactions have led to the revision of many isotopic masses.

Cross section of nuclear reactions

The probability for the reaction between a projectile particle and a target nucleus may be supposed to be proportional to the cross-sectional target area presented by the nucleus. The probability of a nuclear reaction is thus conventionally expressed in terms of a cross sectional area σ.

Suppose a given nucleus offers a cross-sectional area of σ fm² for a particular reaction under consideration. When a projectile traverses 1 fm length of the target material, it sweeps a volume equal to σ fm³. If the material contains *N* target nuclei per fm³, there will be *N*σ nuclei in σ fm³. Hence, the probability that the projectile will react with a nucleus in traversing 1 fm of the target is *N*σ.

For a beam of projectiles striking a *thin* target (through which the intensity of the projectile particles does not change appreciably with the thickness traversed) the rate of the particular nuclear process is given by the expression

$$ \textbf{Rate}\ (s^{-1}) = IN\sigma x $$

where *I* = number of incident projectiles per second and *x* = thickness of the target in fm.

The unit of nuclear cross-section is the **barn**; 1 barn (b) = 100 $fm^2 = 10^{-28}$ m^2. Since nuclear diameters are of the order of 10^{-14} m, the geometrical cross-sections are about 10^{-28} m^2.

If instead of a thin target, we have a thick target in which the intensity of the projectile particle decreases by dI in traversing a thickness dx, we have

$$-dI = IN\sigma dx.$$

Assuming σ to remain constant, $I = I_o . e^{-N\sigma x}$

where I_o = Intensity of incident beam

I = Intensity of the beam after traversing a target thickness x.

$\therefore$ Number of reactions per second $= I_o - I = I_o (1 - e^{-N\sigma x})$.

The idea of a nuclear cross-section holds fairly well for reactions carried out by fast neutrons. But charged particles used as projectiles have to overcome a coulomb barrier. For these and for slow neutrons, the concept of the nuclear reaction cross section does not work well. Nevertheless, it is a useful measure of the probability for any nuclear reaction.

Nuclear reaction cross sections generally vary with the energy of the projectile. For charged particles, it generally increases with increasing energy of the projectile. But with neutrons used as projectiles, the cross section usually decreases as the energy of the projectile increases. Thus, slow neutrons are more effective in causing fission than fast neutrons.

For many nuclides, the neutron capture cross section exhibits an abrupt and marked increase for certain values of the energy of the neutron. This is known as ***resonance***.

14.2.3 Nuclear Reactors

Nuclear reactors are designed to obtain energy from artificial nuclear reactions in a controlled manner. Both nuclear fission and fusion reactions appear to be suitable for the purpose. But reactors based on fusion reactions are still at an experimental stage for certain practical difficulties. On the other hand, reactors based on a self-sustaining fission chain have been successfully installed and operated in different countries, including India. Such reactors may be used for different purposes—(i) research, (ii) production of fissile materials from non-fissionable nuclides or (iii) conversion of nuclear energy into useful power, i.e., to generate electricity.

Fissionable materials suitable for use in a nuclear reactor are called nuclear fuels. Uranium-235 is the most commonly used nuclear fuel—it occurs to the extent of about 0.7 per cent by weight in natural uranium. It undergoes fission by both slow and fast neutrons, though the cross section for fission by fast neutrons is much smaller. Other important nuclear fuels are uranium-233 and plutonium-239. Both these isotopes are obtained through (n, γ) reactions on natural ^{238}U and ^{232}Th respectively (see breeder reactors).

$$^{238}_{92}U + ^{1}_{0}n \longrightarrow ^{239}_{92}U \xrightarrow[t_{1/2} = 23 \text{ min}]{\beta^-} ^{239}_{93}Np \xrightarrow[t_{1/2} = 2.3 \text{ days}]{\beta^-} ^{239}_{94}Pu$$

$$^{232}_{90}Th + ^{1}_{0}n \longrightarrow ^{233}_{90}Th \xrightarrow[t_{1/2} = 23.5 \text{ min}]{\beta^-} ^{233}_{91}Pa \xrightarrow[t_{1/2} = 27.4 \text{ days}]{\beta^-} ^{233}_{92}U.$$

U-238 or Th-232 undergo fission by fast neutrons with energy greater than 1 MeV. But these cannot be used as fuels as it is not possible to maintain a self-sustaining chain process. There are mainly two reasons for this: (i) neutrons are largely captured for non-fission reactions (n, γ type). (ii) As a result of inelastic collisions with uranium or thorium nuclei, the energy of the fission neutrons is very rapidly reduced to less than 1 MeV, when they cannot cause fresh fission of ^{238}U or ^{232}Th nuclei.

Again, U-238 has only a small cross-section for the capture of slow neutrons, while the cross-section of U-235 for slow-neutron induced fission is very high ~ 582 barns. In reactor fuel, the content of U-235 is enriched through gaseous diffusion or gas centrifuge of UF_6.

When a large amount of uranium-235 is brought together (> critical mass), fission may be initiated by a single stray neutron—for example one from cosmic radiation or from a spontaneous fission. This generates ~2.5 neutrons on an average per fission. Loss of neutrons through the reactor surface or their absorption without causing fission is minimized by (i) increasing the size of the reactor (a larger object has smaller surface area with respect to its volume) and (ii) by surrounding the fuel with materials which will scatter the neutrons back to the fuel without absorbing them.

The design of a nuclear reactor is very complicated. Essentially, it consists of an active core where the fission reaction proceeds and generates heat. If the fission is intended to be carried out by slow neutrons, moderators are included in the core to retard the fast neutrons produced in fission. The moderators commonly used are water, heavy water and graphite. In **thermal reactors**, sufficient amount of moderators are used to convert most of the energies of the fission neutrons into utilizable heat energy. On the other hand in **fast reactors**, no moderators are used; the majority of the fissions are carried out by neutrons with energies of 0.5 MeV or more. Though the cross section for such reactions is much smaller, yet they are useful as **breeder reactors** in which fertile materials like thorium-232 and uranium-238 are converted to fissile uranium-233 and plutonium-239 respectively. Loss of neutrons from the core is reduced by using a neutron reflector.

The production of energy in a reactor is controlled in either of two ways: (i) surplus neutrons are removed by means of a neutron absorber like boron or cadmium in the form of rods. When the rods are pushed into the reactor, the rate of fission, and hence the energy liberated, comes to a minimum. The rods are pulled out as and when necessary to maintain a steady output of energy. (ii) Alternatively, parts of the core or neutron reflector may be moved to allow neutrons to escape from the reactor.

The heat energy produced in a reactor has to be continuously removed by circulating a suitable coolant. Water, heavy water, liquid sodium or organic compounds of high boiling point are suitable for the purpose. Gases like helium, air or carbon dioxide are also used as coolants. The coolant, in turn, is allowed to transmit its heat to a power generating system.

In a *pressurized water reactor* (**PWR**) water at about 150 atm. pressure (so that it cannot boil) is circulated around the core and the heat abstracted is utilized through a heat exchanger to produce steam. In a *boiling water reactor* (**BWR**; e.g., Tarapur 1 and 2 in India) water at about 68 atm. pressure directly forms steam which is used to drive a turbine.

Because deuterons are less likely to capture neutrons than protons, heavy water is a better moderator. With heavy water as moderator, natural uranium may be used as the fuel instead of enriched uranium. India has now selected *pressurized heavy water reactor* (**PHWR**; from Kota onward) with natural uranium as fuel and heavy water as moderator and coolant. *Advanced Heavy Water Reactor* (**AHWR**) are also under construction. Together with electricity, these reactors produce plutonium fuel with 96% unused U-238. In the second stage, *Prototype* fast breeder reactors (PFBR) will use plutonium as fuel. These will produce more plutonium from U-238, kept as a blanket. It will also produce U-233 from thorium, used as blanket. In the third stage, there are plans to use U-233 as fuel and thorium as blanket, producing more U-233. India's nuclear power programme upto 2010 AD is shown in Table 14.3.

Operation of nuclear power plants has to be maintained strictly within safety limits for public life and the environment. The chance of radiation contamination is particularly high during decommissioning of a plant after its usual life time. Chances of accident are present in every industry. But this is of a slightly different type for nuclear power plants. Once such an accident takes place, the aftereffect may continue for years to follow with severe radiation damage to coming generations over the entire world through radioactive ashes. However, the world record of accidents over the last three decades in operating about 430 civilian nuclear power plants has been very small before the recent disaster in Japan (March 2011). Just two major accidents are worth mention. The Three Mile Island (TMI–2) in USA in March 1979 and the Chernobyl (Unit–4) in USSR in April 1986. Both were due to operating error. In TMI–2, there were no casualties nor radioactive fallout in the public domain. At Chernobyl, un-controlled power excursion killed at least thirtyone persons inside the plant with an estimated death toll of 9000(WHO). The reactor had exploded sending a plume of deadly radioactive particles 30,000 feet high into the sky. The graphite rods used as moderators burnt for four days, sending radioactivity streaming up and directly affecting 150,000 sq km area in Belarus, Russia and Ukraine; another 45,000 sq km area was also endangered. The amount of radioactivity spread was estimated at 5.2 million TBq (tarabecquerel = 10^{12} Bq). The disaster was graded 7, the highest level of nuclear hazard.

The recent disaster in Japan (March 11, 2011) has also been assigned level 7 in view of the widespread and lasting effect on public life and environment. The Fukushima-Daiichi power station, lying close to the epicentre of a 9.0 magnitude earthquake was badly damaged by high tsunami tide. The power supply to the plant collapsed, causing the reactors and storage pools to overheat and nuclear fuel rods to partially melt. Radioactive water (> 1000 millisieverts per hour) leaking through cracks seeped into sea and radioactivity was transmitted to other countries. Radioactivity was also detected in tap water, milk and other agricultural products. Divergent estimates of the amount of radioactivity released in one month from the affected plant give figures ranging between 30,000 to 630,000 TBq, far less than the Chernobyl figure. However, the extent of damage is yet to be ascertained over years but during this time consolations and commitments from the heads of different countries will be little fruitful in wiping the panic and doubt out of public mind. While all countries will now thoroughly review the safety aspects of their nuclear power plants, it is high time that we also actively consider other possible alternatives to nuclear power. Among others, geothermal energy seems to be a likely substitute in the future.

The Indian scenario

India has 20 nuclear power plants in operation (April 2011), with plant capacity of 4780 MW electricity. Six others under construction are expected to start commercial operation between June 2011 and December 2016, with a plant capacity of 4800 MW. Further 15 projects are under active consideration.

Table 14.3
***Nuclear power plants of India* (April 2011)**

Narora (Uttar Pradesh)	PHWR	2 × 220 MW
Rawatbhata (Rajasthan)	PHWR	1 × 100 MW 1 × 200 MW 4 × 220 MW
Kakrapar (Gujrat)	PHWR	2 × 220 MW
Tarapur (Maharastra)	BWR, PHWR	2 × 160 MW 2 × 540 MW
Kaiga (Karnataka)	PHWR	4 × 220 MW
Kalpakkam (Tamilnadu)	PHWR	2 × 220 MW

As of 2010, India stands 6th in the world in terms of number of operational nuclear power reactors with an average of about 75% output of their installed capacity. The cost of production is also competitive with that from thermal power plants. In 2008-09, the average tariff of nuclear power has been estimated at Rs 2.34 per kWh. India has a vast domestic reserve of about 80,000 to 112,000 tons of uranium (~ 1% of global reserve). This is large enough to supply all of India's commercial and military reactors. Currently, India's nuclear power reactors consume, at most, 478 metric tons of U per year. Even if India were to quadruple its power output to 20 GW by 2020, power generation would consume about 2000 metric tons of U per year. The reserve is thus sufficient to run the power plants for 40 to 50 years. With reprocessing and use of breeder technique, this can be further stretched. Uranium required for India's nuclear arsenal (~ 32 tons) are only about 1/15th of that required for power generation. Presently, India produces 568 billion units of electricity. Of this, 64 percent is thermal power, 23 percent is hydro-power and 10 percent is produced by non-conventional methods; nuclear power contributes only three percent of the total production.

Fission and fusion as possible sources of energy

The fission of a single heavy nucleus of ^{235}U releases about 200 MeV or $200 \times 1.6 \times 10^{-13}$ J of energy. Multiplying by the Avogadro number and etc., the energy released in the fission of all the nuclei in 1 kg of U-235 would be 8.2×10^{10} kJ, which is roughly equal to the energy released in the explosion of 20,000 tons of TNT.

We may readily express this fission energy in conventional power units. Since 1 joule = 1 watt-sec, the fission energy from 1 kg of U-235 is equivalent to 2.3×10^{7} kilo-watt hour (kWh) or approximately 1000 megawatt-day. If this amount of heat energy can be converted to electrical energy with an efficiency of 30%, we would have an energy supply of about 300 megawatt per day; this is equivalent to the output of a large thermal power plant consuming 2500 tons of coal per day.

Technology for controlled release of fission power and its conversion to electrical energy at low cost is now available. Side by side, rapid development of such technologies has its own associated problems like waste-handling, radiation hazards, misuse of nuclear power in war etc.

Nuclear fusion reactions which may occur at an appreciable rate on the earth are based on the fusion of deuterium nuclei (see before). Calculations show that the energy output per deuteron burned is approximately 7.2 MeV. Though this energy is much less than that obtained from the fission of a single heavy nucleus of ^{235}U, the energy released per kilogram of deuterium is higher than that from one kilogram of U-235.

1 kg $D_2 \equiv 1000/4$ or 250 mole.

Fusion energy output per kg of $D_2 = 250 \times 6.02 \times 10^{23} \times 7.2$ MeV $\equiv 1.7 \times 10^{14}$ J.

This is nearly double the energy obtained from fission of 1 kg of U-235. The large amount of deuterium present in the earth (0.15 percent in sea-water—a total of about 10^5 tons) may be utilised as an alternative source of energy in the future. But suitable technologies are yet to be developed to carry out thermonuclear reactions in a controlled manner. At the high temperatures required for such processes (10^8K), the reactants get fully ionized and remain in the plasma state; it is difficult to achieve a sufficiently high particle density n for a long enough time τ which will give a net energy yield ($n\tau \simeq 10^{20}$ m^{-3} s). Such conditions are attained only in the interiors of stars and in hydrogen bombs.

The hot reactive plasma can be contained in a strong magnetic field (magnetic bottle) in the form of a torus (doughnut). Such arrangements are under trial in the setups *Tokomak* (Russia) and *Stellarater* (USA). Scientists are also exploring the possibility of simultaneously heating and compressing tiny deuterium-tritium pellets in a series of hydrogen-bomb-like explosions. Laser beams as well as electron and proton beams are being tried for the purpose. Once the technology of fusion-reactors is developed, fusion will be a superior energy source than fission: the fuel is readily available, possibility of use in war is very little and the production of radioactive waste is negligible.

14.3 APPLICATION OF ISOTOPES IN ANALYSIS

In the previous chapter we have discussed many applications of radioactive isotopes in analysis. Similar to radioactive isotopes, nonradioactive isotopes have also been extensively used to trace the course of many chemical processes. In this case, presence of the nonradioactive isotope in different stages of the chemical process is ascertained by its mass difference from the common isotopes, detected by mass spectrometric studies. A few examples are given below.

(i) The mechanism of photosynthesis by plants has been investigated by ^{18}O. The essential step in this process involves the combination between water and carbon dioxide to form sugars and oxygen:

$$x\,CO_2 + x\,H_2O \xrightarrow{h\nu} x\,C_6H_{12}O_6 + O_2$$

Formerly, it was believed that the oxygen comes from the carbon dioxide molecule. But some aquatic plants were grown in water enriched with ^{18}O in presence of carbon dioxide containing ^{16}O. It was found that the evolved oxygen contained only ^{18}O, showing that it originated entirely from the water. But when the same plants were grown in presence of carbon dioxide containing ^{18}O and water containing only ^{16}O, the ^{18}O isotope was retained in the plant body. The overall reactions are thus

$$x\,C^{16}O_2 + x\,H_2{}^{18}O \rightarrow x\,C_6H_{12}{}^{16}O_6 + {}^{18}O_2$$

and $$x\,C^{18}O_2 + x\,H_2{}^{16}O \rightarrow x\,C_6H_{12}{}^{18}O_6 + {}^{16}O_2$$

(ii) Hydrolysis of esters carried out with water enriched in the isotope ^{18}O indicates that after hydrolysis, excess ^{18}O is associated with the acid formed. So one should expect that the reaction proceeds by rupturing the C—OR′ bond in the ester:

$$R-\underset{\underset{O}{\|}}{C}-OR' + H_2\,{}^{18}O \longrightarrow R-\underset{\underset{O}{\|}}{C}-{}^{18}OH + R'OH.$$

(iii) The reaction of hydrogen peroxide with acidified permanganate has been studied by exchange methods. When hydrogen peroxide containing ^{18}O is oxidized by potassium permanganate (which does not contain any ^{18}O), it is found that the evolved oxygen contains the entire ^{18}O; no ^{18}O is left in the solution after reaction. This shows that the permanganate oxidizes the hydrogen peroxide molecule without breaking the $-O-O-$ bond:

$$2KMn^{16}O_4 + 5H_2{}^{18}O_2 + 3H_2SO_4 \rightarrow 2MnSO_4 + K_2SO_4 + 8H_2{}^{16}O + 5{}^{18}O_2.$$

(iv) The absence of any acidic hydrogen in NaH_2PO_2 has been established by crystallizing the salt from heavy water. No exchange of hydrogen by deuterium is observed. On the other hand, compounds containing hydrogen bonded to oxygen (or nitrogen) always exchange their hydrogen for deuterium under similar treatment. The lack of exchange in the case of sodium hypophosphite (NaH_2PO_2) indicates that no more $-O-H$ bonds are present in the $H_2PO_2^-$ ion. So, the hydrogen atoms must be linked to the phosphorus atom, as confirmed by the $-P-H$ stretching frequency in the IR spectrum of the salt.

(v) Isotopes differ in their zero-point energies and as a consequence, the activation energy for a chemical reaction is slightly higher with the heavier isotope. Accordingly, a species containing the heavier isotope reacts more slowly. The overall rate of a chemical reaction is influenced by the substitution of one isotope by another in the reacting species provided the bond to the isotopic element is involved in the rate-determining step. This effect, also known as **the kinetic isotope effect**, has been widely investigated to understand the mechanism of many chemical reactions. The effect is very prominent and measurable when ^3H, ^2H, ^{14}C, ^{15}N and ^{18}O substitute their respective lighter isotopes. With heavier elements, however, the effect becomes gradually less significant.

As an example, we may consider the rates of nitration of benzene by nitric acid and sulfuric acid mixtures. When benzene is prepared using ^3_1H (tritium) and the tritiated benzene is nitrated, no difference in the rate of nitration is observed from ordinary benzene. This suggests that the C—H bond is not broken in the rate determining step. The mechanism suggested is

$$\text{(i) } C_6H_6 + NO_2^+ \xrightarrow{\text{slow}} C_6H_6NO_2^+ \quad \text{(ii) } C_6H_6NO_2^+ \xrightarrow{\text{fast}} C_6H_5NO_2 + H^+$$

(vi) Isotopic substitution has been largely applied to spectroscopy and other methods of structure determination. The difference in the moments of inertia of isotopically substituted molecules has been utilized in the determination of interatomic distances and bond angles in molecules by microwave spectroscopy. Again, in the different vibrational spectra (Raman and infra-red), the vibration frequency depends on the square root of the reduced masses of molecules. Hence isotopically substituted molecules present different vibrational spectra and help in the elucidation of many structure problems. Nuclear magnetic resonance spectroscopy (NMR) and neutron differaction studies have also been developed by the use of suitably chosen isotopes.

ESSENTIALS

Isotopes are atoms whose nuclei contain different numbers of neutrons but the same number of protons.

Isobars are atoms having different numbers of protons but the same mass number.

Isotones: Nuclides having different mass number but the same number of neutrons are known as isotones. ${}_{1}^{2}H$ and ${}_{2}^{3}He$ or ${}_{14}^{30}Si$ and ${}_{15}^{31}P$ are isotonic.

The size of the nucleus is very small, its radius being of the order of 5×10^{-15} m. The density of a nucleus would be correspondingly very high—of the order of 10^{11} kg per cm^3. Nuclear volumes are proportional to nuclear masses and hence all nuclei have approximately the same density.

Isotopic masses (in atomic mass scale) differ from the total mass number (no. of protons, no. of neutrons) of a nuclide. Accordingly, a quantity called "mass defect" or "mass excess" (ΔM) is defined:

Mass Excess = Isotopic Mass – Mass Number.

The mass excess of different nuclides may be compared from their packing fractions, defined as

$$\textbf{Packing fraction} = \frac{\text{Isotopic Mass} - \text{Mass Number}}{\text{Mass Number}} \times 10^4.$$

The true mass defect of a nuclide may be obtained from the difference between (i) the sum of the masses of the constituent nucleons and (ii) the actual mass of a nuclide. The "lost" mass (Δm) may be considered as the mass which has been converted to the energy released in the hypothetical formation of the nucleus from the individual neutrons and protons. This is the **binding energy of a nucleus** and is equal to the energy required to split the nucleus into its component nucleons. The binding energy (E) of a nucleus may be estimated from the relation $E = \Delta m \cdot c^2$. The magnitude of E per atomic mass unit (u) is approximately 931 MeV.

Magic Numbers: Nuclei containing neutrons and protons in numbers equal to 2, 8, 20, 28, 50, 82 and 126 are exceptionally stable. These numbers are called magic numbers. They correspond to closed shells of nucleons.

According to **the liquid-drop model of the nucleus**, each nucleon in a nucleus behaves in the same manner as the molecules in a liquid-drop. Each nucleon interacts strongly only with its adjacent nucleons. The model can successfully interpret many observed properties of the nucleus.

The shell model assumes that nucleons occur in definite shells or quantum levels within a nucleus which may be obtained from the solution of appropriate form of the Schrödinger equation. For this purpose, the potential field within the nucleus is compared to a well, either with a square edge, or with a rounded edge. This model has the merit of explaining the magic numbers.

Nuclear forces must be short range and charge independent. It is now generally accepted that the force between the nucleons arises from the interchange of a new type of particle, the meson, between the nucleons. A meson field is assumed to surround the nucleons, just as a photon-field surrounds electrically charged bodies. The nucleons interact through interchange of mesons which have masses nearly 200 times the mass of an electron and may have single unit of either positive or negative charge. Neutral mesons have also been discovered subsequently. The exchange of meson (π^+ or π^-) may be represented as

transfer of π^+: $\quad p + \pi \rightarrow (n + \pi^+) + n \rightarrow n + (\pi^+ + n) \rightarrow n + p.$

This is equivalent to the transfer of a positive meson (π^+) from a proton to a neutron. Similarly, a negative meson may be transferred from a neutron to a proton:

transfer of π^-: $\quad n + p \rightarrow (p + \pi^-) + p \rightarrow p + (\pi^- + p) \rightarrow p + n.$

Proton-proton and neutron-neutron interactions may be similarly explained by the exchange of neutral mesons (π^0).

Nuclear reactions may be classified in terms of the overall transformation taking place, e.g.,

1. **Projectile capture:** $^{12}_{6}C + ^{1}_{1}H = ^{13}_{7}N + \gamma$

2. **Projectile capture particle emission:** $^{14}_{7}N + ^{1}_{0}n = ^{14}_{6}C + ^{1}_{1}H$

3. **Nuclear fission:** A heavy nucleus is divided into two roughly equal parts on bombardment with a suitable projectile. The process releases a large amount of energy and several neutrons.

$$^{235}_{92}U + ^{1}_{0}n \rightarrow ^{92}_{36}Kr + ^{141}_{56}Ba + 3^{1}_{0}n + 200 \text{ MeV}.$$

4. **Spallation:** Bombardment with high energy charged particles may sometimes break a target nucleus into several smaller nuclei with the emission of a large number of nucleons.

$$^{238}_{92}U + ^{4}_{2}He = ^{187}_{74}W + 20^{1}_{1}H + 35^{1}_{0}n.$$

5. **Nuclear fusion:** Certain light nuclei may fuse together at very high temperatures (~ 10^6 K) with the liberation of tremendous amount of energy.

$$^{2}_{1}H + ^{3}_{1}H = ^{4}_{2}He + ^{1}_{0}n + 17 \cdot 6 \text{ MeV}.$$

The energy accompanying a fusion process is derived from the difference of binding energies between the product and the reactant nuclei. The energy source of the stars, including the sun, is believed to be continuous fusion processes taking place at the high stellar temperature.

Nuclear Reactors have been designed to obtain energy from controlled nuclear fission reactions in a self-sustaining chain process. The fissionable material is usually uranium-235. A portion of the neutrons liberated in each act of fission is slowed down by passage through moderators like water. These slow neutrons are allowed to cause further fission. Control is achieved by absorbing the surplus neutrons in substances like boron or cadmium. The heat energy produced in a reactor is continuously removed by suitable coolants like liquid sodium, water or high boiling organic compounds.

EXERCISE

1. Explain with examples: Isotopes, Isotones, Isobars, Isodiapheres and Mirror nuclei.
2. ^{13}N decays by positron emission. The maximum kinetic energy of the positron is 1.20 MeV. What is the nuclidic mass of ^{13}N?
   [Given m (^{13}C) = 13.00335 u, m_e = 0.00055 u.]
3. (a) Explain the terms packing fraction and mass defect. How is mass defect related to nuclear binding energy?
   (b) Why packing fraction may be positive or negative whereas mass defect cannot?
   [**Hint:** P. F. is defined on the basis of mass excess which is related to the mass number. Mass defect arises from conversion of a part of the constituent mass to nuclear binding energy.]
4. Calculate the binding energies of the following isobars and their average binding energies per nucleon.
   $^{64}_{28}Ni = 63 \cdot 9280\, u;\ ^{64}_{29}Cu = 63 \cdot 9298\, u.$ [577 · 2 and 9 · 02 MeV for Ni]
5. Calculate the average binding energy per nucleon in $^{3}_{1}H$ (mass = 3·016050 u) and $^{3}_{2}He$ (mass = 3·016030 u).

6. Fission of a nucleus of mass number 240 gives four nuclei, each having mass number 60. Approximately how much energy is released? Average BE per nucleon in the mass number range 240 = 7.6 MeV; in the mass number range 60 = 8.6 MeV.

7. Explain

(i) The true mass of an atom (in u) is always less than the sum of the masses of its constituents.

(ii) The density of the nucleus is many times greater than that of iron.

(iii) Nuclear forces cannot be electrical in nature.

(iv) Neutrons are most effective as projectiles.

(v) U-238 cannot be commonly used as a nuclear fuel.

(vi) A nucleus does not contain electrons.

(vii) A nucleus resembles a drop of liquid.

(viii) Fusion produces more energy per gram of fuel consumed but fusion-based reactors have not yet been developed.

(ix) Fission reactions are different from nuclear spallation.

(x) Nuclei containing 2, 20, 50 etc. number of neutrons and/or protons are exceptionally stable.

(xi) Average binding energy per nucleon is almost constant (~ 8.4 MeV) for all but the lightest nuclei.

(xii) Fission of a heavy nucleus by a single neutron gives rise to more than one fission neutrons.

8. The electrostatic potential energy E operating between two equal nuclei formed in the fission of a heavy nucleus at the moment of their separation may be calculated by Coulomb's law. Calculate this energy in the case of U-235.

Hint: For two charges Ze at a distance d, $E = Z^2e^2/4\pi\varepsilon_0 d$. Here $Z = 92/2 = 46$; $d = 2R$ where R is the radius of the product nucleus. R may be evaluated from the relation $R = R_0A^{1/3}$ where $R_0 = 1.37$ fm and A is the mass number of the nucleus (here $A = 118$).

[**Ans.** 3.642×10^{-11} J ≡ 228 MeV.]

Note. If the products are not equal, say of mass numbers A_1 and A_2 and atomic numbers Z_1 and Z_2, we may separately find R for each and put $d = R_1 + R_2$. Z^2e^2 will be replaced by $Z_1Z_2e^2$.

9. How many fissions per second are required in a reactor producing 600 MW power per day with an efficiency of 30%? Average energy release per fission = 200 MeV. [**Ans.** $7{\cdot}23 \times 10^{14}$]

10. How much petrol would provide the same energy as that obtained from 1 dm^3 of water used in a thermonuclear reactor? Energy value of petrol = 10 kWh kg^{-1}.

11. What is the probable energy source of the sun?

12. Briefly discuss the advantages and limitations of nuclear fission and fusion processes as probable alternative to fossil fuels as energy source.

13. Compare the density of the nucleus of $^2_1$H with that of the atom.

Hint: Mass of the nucleus ≃ Mass of atom (ignoring the electrons) = $2u = 2 \times 1{\cdot}66 \times 10^{-27}$ kg.

Volume of the nucleus = $\frac{4}{3}\pi R^3$ $(R = 1\cdot4 \times A^{1/3}$ fm) (A = mass number = 2). Volume of the atom = $\frac{4}{3}\pi a_0^3$ where a_0 = first Bohr radius = $0{\cdot}53 \times 10^{-10}$ m.

$\rho_{\text{nucleus}}/\rho_{\text{atom}} = 1\cdot45 \times 10^{17}/5\cdot35 \times 10^{3}$.

14. For $^{12}_{6}$C, nuclear binding energy per nucleon = 7·683 MeV. Find the mass defect.

Hint: Total binding energy = 7·683 × 12 = 92·196 MeV ≡ 0·099 u

Since the isotopic mass of ^{12}C is 12·000 u, mass defect = 0·099 u.

15. Calculate the energy liberated in the fission of 1 kg of ^{235}U. Average binding energy per nucleon (MeV) : $^{235}U = 7{\cdot}6$; fission products = 8·5.

Hint: Energy liberated per nucleon = 8·5 – 7·6 = 0·9 MeV.

Energy liberated by 1 g-atom or 235 g $^{235}U = 235 \times 0{\cdot}9 \times 6{\cdot}02 \times 10^{23}$ MeV

Energy liberated by 1 kg of $^{235}U \Rightarrow 5{\cdot}42 \times 10^{26}$ MeV $\equiv 8{\cdot}67 \times 10^{13}$ J.

16. Calculate the binding energy of the last proton and the last neutron in ^{12}C from the following mass data (u)

$^{11}B = 11{\cdot}0093$; $^{11}C = 11{\cdot}0014$; $^{1}H = 1{\cdot}0078$; $^{1}n = 1{\cdot}0087$.

Hint: *Addition of last proton*: $^{11}B + {}^{1}H \rightarrow {}^{12}C$ $\Delta m = 11{\cdot}0093 + 1{\cdot}0078 - 12{\cdot}0000 = 0{\cdot}0171\ u$

Binding energy of last proton = 0·0171 × 931 MeV

Addition of last neutron: $^{11}C + {}^{1}n \rightarrow {}^{12}C$ $\Delta m = 0{\cdot}0101\ u$

Binding energy of last neutron = 0·0101 × 931 MeV.

17. The mass of $^{55}_{25}Mn$ nuclide is 54·988 u. Calculate the total nuclear binding energy of the nuclide and the average binding energy per nucleon. (Given: mass (u): e = 0·000 5486 ; ^{1}H = 1·007825; ^{1}n = 1·008665).

18. The binding energy of the last neutron in ^{16}O is 15·64 MeV. Find the mass of the ^{15}O nuclide from the following mass data (u) : ^{16}O = 15·9949; n = 1·0087). [15·0030 u]

APPENDIX 1

The elements with their year of discovery and main isotopes. Relative abundances of the nuclides and their radioactive nature (if present) with half-lives are also included. Many other isotopes are known for several elements.

Z	Element (year)	Symbol	*A*	Nuclide mass, *u*	Relative abundance, %	Half-life
1	Hydrogen (1766)	**H**	1	1.007 825	9.985	
			2	2.014 102	0.015	
			3	3.016 050	0	12.3 y (β^-)
2	Helium (1895)	**He**	3	3.016 029	0.0001	
			4	4.002 603	99.9999	
			6	6.018 891	0	0.81 s (β^-)
3	Lithium (1817)	**Li**	6	6.015 123	7.5	
			7	7.016 004	92.5	
			8	8.022 487	0	0.84 s (β^-)
4	Beryllium (1797)	**Be**	9	9.012 182	100	
5	Boron (1808)	**B**	10	10.012 938	20	
			11	11.009 305	80	
6	Carbon (*pre-history*)	**C**	10	10.016 858		19.3 s (β^+)
			11	11.011 433		20.3 min (β^+)
			12	12.000 000	98.89	
			13	13.003 355	1.11	
			14	14.003 242		5730 y (β^-)

Z	Element (year)	Symbol	*A*	Nuclide mass, *u*	Relative abundance, %	Half-life
7	Nitrogen (1772)	N	14	14.003 074	99.63	
			15	15.000 109	0.37	
			16	16.006 099		7.10 s (β^-)
8	Oxygen (1774)	O	16	15.994 915	99.758	
			17	16.999 131	0.038	
			18	17.999 159	0.204	
9	Fluorine (1886)	F	19	18.998 403	100	
10	Neon, Ne (1898)		20	19.992 439	90.51	
			21	20.993 845	0.57	
			22	21.991 384	9.22	
11	Sodium (1807)	Na	22	21.994 435		2.60 y (β^+)
			23	22.989 770	100	
			24	23.990 963		15.0 h (β^-)
12	Magnesium (1755)	Mg	24	23.985 045	78.99	
			25	24.985 839	10.00	
			26	25.982 595	11.01	
13	Aluminium (1825)	Al	27	26.981 541	100	
14	Silicon (1824)	Si	28	27.976 928	92.23	
			29	28.976 496	4.67	
			30	29.973 772	3.10	
15	Phosphorus (1669)	P	30	29.978 310		2.50 min (β^+)
			31	30.973 763	100	
16	Sulfur (*pre-history*)	S	32	31.972 072	95.02	
			33	32.971 459	0.75	
			34	33.967 868	4.21	
			36	35.967 079	0.017	
17	Chlorine (1774)	Cl	35	34.968 853	75.77	
			36	35.968 307		3.01×10^5 y (β^+, β^-, K)
			37	36.965 903	24.23	
18	Argon (1894)	Ar	36	35.967 546	0.337	
			37	36.966 776		34.8 d (K)
			38	37.962 732	0.063	
			39	38.964 315		269 y (β^-)
			40	39.962 383	99.60	
19	Potassium, (1807)	K	39	38.963 708	93.26	
			40	38.963 999	0.01	1.28×10^9 y (β^-)
			41	40.961 825	6.73	

Z	Element (year)	Symbol	A	Nuclide mass, u	Relative abundance, %	Half-life
20	Calcium (1808)	Ca	40	39.962 591	96.94	
			41	40.962 278		1.3×10^5 y (K)
			42	41.958 622	0.647	
			43	42.958 770	0.135	
			44	43.955 485	2.09	
			45	44.956 189		163 d (β^-)
			46	45.953 689	0.0035	
			47	46.954 543		4.5 d (β^-)
			48	47.952 532	0.187	
21	Scandium (1879)	Sc	45	44.955 914	100	
22	Titanium (1791)	Ti	46	45.952 633	8.25	
			47	46.951 765	7.45	
			48	47.947 947	73.7	
			49	48.947 871	5.4	
			50	49.944 786	5.2	
23	Vanadium (1801)	V	48	47.952 257	0	16 d (β^+, γ, K)
			50	49.947 161	0.25	
			51	50.943 962	99.75	
24	Chromium (1780)	Cr	50	49.946 046	4.35	
			52	51.940 510	83.79	
			53	52.940 651	9.50	
			54	53.938 882	2.36	
25	Manganese (1774)	Mn	54	53.940 360	0	312.5 d (K)
			55	54.938 046	100	
26	Iron (> 2500 B.C.)	Fe	54	53.939 612	5.8	
			56	55.934 939	91.8	
			57	56.935 396	2.1	
			58	57.933 278	0.3	
27	Cobalt (1735)	Co	58	57.935 755		70.8 d (β^+, γ)
			59	58.933 198	100	
			60	59.933 820		5.3 y (β^-)
28	Nickel (1751)	Ni	58	57.935 347	68.3	
			60	59.930 789	26.1	
			61	60.931 059	1.1	
			62	61.928 346	3.6	
			64	63.927 968	0.9	
29	Copper (5000 B.C.)	Cu	63	62.929 599	69.2	
			64	63.929 766		12.7 h (β^-, β^+, K)
			65	64.927 792	30.8	

Z	Element (year)	Symbol	*A*	Nuclide mass, *u*	Relative abundance, %	Half-life
30	Zinc	Zn	64	63.929 145	48.6	
	(> 1500 B.C.)		65	64.929 244		244 d (β^+, K, γ)
			66	65.926 035	27.9	
			67	66.927 129	4.1	
			68	67.924 846	18.8	
			70	69.925 325	0.6	
31	Gallium	Ga	69	68.925 581	60.1	
	(1875)		71	70.924 701	39.9	
32	Germanium	Ge	70	69.924 250	20.5	
	(1896)		72	71.922 080	27.4	
			73	72.923 464	7.8	
			74	73.921 179	36.5	
			76	75.921 403	7.8	
33	Arsenic	As	74	73.923 930		17.8 d (β^-, β^+, γ)
	(~ 1250)		75	74.921 596	100	
34	Selenium	Se	74	73.922 477	0.9	
	(1817)		76	75.919 207	9.0	
			77	76.919 908	7.6	
			78	77.917 304	23.5	
			80	79.916 520	49.8	
			82	81.916 709	9.2	
35	Bromine	Br	79	78.918 336	50.7	
	(1826)		80	79.918 528		17.7 min (β^-, β^+)
			81	80.916 290	49.3	
36	Krypton	Kr	78	77.920 397	0.35	
	(1898)		80	79.916 375	2.25	
			81	80.916 578		2.1×10^5 y (K)
			82	81.913 483	11.6	
			83	82.914 134	11.5	
			84	83.911 506	57.0	
			86	85.910 614	17.3	
37	Rubidium	Rb	85	84.911 800	72.2	
	(1861)		87	86.909 184	27.8	49×10^{10} y (β^-)
38	Strontium	Sr	84	83.913 428	0.6	
	(1808)		86	85.909 273	9.8	
			87	86.908 890	7.0	2.7 h (γ)
			88	87.905 625	82.6	
39	Yttrium		89	88.905 856	100	
	(1794)					
40	Zirconium	Zr	90	89.904 708	51.5	
	(1789)		91	90.905 644	11.2	
			92	91.905 039	17.1	

Z	Element (year)	Symbol	A	Nuclide mass, u	Relative abundance, %	Half-life
			94	93.906 319	17.4	
			96	95.908 272	2.8	
41	Niobium (1801)	Nb	93	92.906 378	100	
42	Molybdenum (1781)	Mo	92	91.906 809	14.8	
			94	93.905 086	9.3	
			95	94.905 838	15.9	
			96	95.904 675	16.7	
			97	96.906 018	9.6	
			98	97.905 405	24.1	
			100	99.907 473	9.6	
43	Technetium (1937)	Tc	99	98.906 252		2.1×10^5 y (β^-)
44	Ruthenium (1808)	Ru	96	95.907 596	5.5	
			98	97.905 287	1.9	
			99	98.905 937	12.7	
			100	99.904 217	12.6	
			101	100.905 581	17.0	
			102	101.904 347	31.6	
			104	103.905 422	18.7	
45	Rhodium (1803)	Rh	103	102.905 503	100	
46	Palladium (1803)	Pd	102	101.905 609	1.0	
			104	103.904 026	11.0	
			105	104.905 075	22.2	
			106	105.903 475	27.3	
			108	107.903 894	26.7	
			110	109.905 169	11.8	
47	Silver (~ 3000 B.C.)	Ag	107	106.905 095	51.8	
			108	107.905 956		2.41 min (β^-)
			109	108.904 754	48.2	
48	Cadmium (1817)	Cd	106	105.906 461	1.3	
			108	107.904 186	0.9	
			110	109.903 007	12.5	
			111	110.904 182	12.8	
			112	111.902 761	24.1	
			113	112.904 401	12.2	9×10^{15} y (β^-)
			114	113.903 361	28.7	
			116	115.904 758	7.5	
49	Indium (1863)	In	113	112.904 056	4.3	
			115	114.903 875	95.7	5×10^{14} y (γ)

Z	Element (year)	Symbol	A	Nuclide mass, u	Relative abundance, %	Half-life
50	Tin (~ 2100 B.C.)	Sn	112	111.904 823	1.0	
			114	113.902 781	0.7	
			115	114.903 344	0.4	
			116	115.901 743	14.7	
			117	116.902 954	7.7	14.5 d (γ)
			118	117.901 607	24.3	
			119	118.903 310	8.6	245 d (γ)
			120	119.902 199	32.4	
			122	121.903 440	4.6	
			124	123.905 271	5.6	
51	Antimony (~ 1600 B.C.)	Sb	121	120.903 824	57.3	
			123	122.904 222	42.7	
52	Tellurium (1783)	Te	120	119.904 021	0.1	
			122	121.903 055	2.5	
			123	122.904 278	0.9	~1.2 × 10^{13} y (K)
			124	123.902 825	4.6	
			125	124.904 435	7.0	
			126	125.903 310	18.7	
			127	126.905 222		9.4 h (β^-)
			128	127.904 464	31.7	
			130	129.906 229	34.5	
53	Iodine (1811)	I	127	126.904 477	100	
			131	130.906 119		8.0 d (β^-)
54	Xenon (1898)	Xe	124	123.906 12	0.1	
			126	125.904 281	0.1	
			128	127.903 531	1.9	
			129	128.904 780	26.4	
			130	129.903 509	4.1	
			131	130.905 076	21.2	
			132	131.904 148	26.9	
			134	133.905 395	10.4	
			136	135.907 219	8.9	
55	Cesium (1860)	Cs	133	132.905 433	100	
56	Barium (1808)	Ba	130	129.906 277	0.1	
			132	131.905 042	0.1	
			134	133.904 490	2.4	
			135	134.905 668	6.6	
			136	135.904 556	7.9	
			137	136.905 816	11.2	
			138	137.905 236	71.7	

Z	Element (year)	Symbol	A	Nuclide mass, u	Relative abundance, %	Half-life
57	Lanthanum (1839)	La	138	137.907 114	0.1	
			139	138.906 355	99.9	
58	Cerium (1803)	Ce	136	135.907 14	0.2	
			138	137.905 996	0.2	
			140	139.905 442	88.5	
			142	141.909 249	11.1	
59	Praseodymium (1885)	Pr	141	140.907 657	100	
60	Neodymium (1885)	Nd	142	141.907 731	27.2	
			143	142.909 823	12.2	
			144	143.910 096	23.8	2.1×10^{15} y (α)
			145	144.912 582	8.3	
			146	145.913 126	17.2	
			148	147.916 901	5.7	
			150	149.920 900	5.6	
61	Promethium (1945)	Pm	147	146.915 148		2.6 y (β^-)
62	Samarium (1879)	Sm	144	143.912 009	3.1	
			147	146.914 907	15.1	1.1×10^{11} y (α)
			148	147.914 832	11.3	$> 1.2 \times 10^{15}$ y (α)
			149	148.917 193	13.9	$> 10^{15}$ y (α)
			150	149.917 285	7.4	
			152	151.919 741	26.7	
			154	153.922 218	22.6	
63	Europium (1901)	Eu	151	150.919 860	47.9	
			153	152.921 243	52.1	
64	Gadolinium (1880)	Gd	152	151.919 803	0.2	
			154	153.920 876	2.1	
			155	154.922 629	14.8	
			156	155.922 130	20.6	
			157	156.923 967	15.7	
			158	157.924 111	24.8	
			160	159.927 061	21.8	
65	Terbium (1843)	Tb	159	158.925 350	100	
66	Dysprosium (1886)	Dy	156	155.924 287	0.1	
			158	157.924 412	0.1	
			160	159.925 203	2.3	
			161	160.926 939	19.0	
			162	161.926 805	25.5	
			163	162.928 737	24.9	
			164	163.929 183	28.1	

Z	Element (year)	Symbol	A	Nuclide mass, u	Relative abundance, %	Half-life
67	Holmium (1878)	Ho	165	164.930 332	100	
68	Erbium (1842)	Er	162	161.928 787	0.1	
			164	163.929 211	1.6	
			166	165.930 305	33.4	
			167	166.932 061	22.9	
			168	167.932 383	27.1	
			170	169.935 476	14.9	
69	Thulium (1879)	Tm	169	168.934 225	100	
70	Ytterbium (1878)	Yb	168	167.933 908	0.1	
			170	169.934 774	3.2	
			171	170.936 338	14.4	
			172	171.936 393	21.9	
			173	172.938 222	16.2	
			174	173.938 873	31.6	
			176	175.942 576	12.6	
71	Lutetium (1907)	Lu	175	174.940 785	97.4	
			176	176.942 694	2.6	2.6×10^{10} y (β^-)
72	Hafnium (1923)	Hf	174	173.940 065	0.2	2.0×10^{15} y (α)
			176	175.941 420	5.2	
			177	176.943 233	18.6	
			178	177.943 710	27.1	
			179	178.945 827	13.7	
			180	179.946 561	35.2	
73	Tantalum (1802)	Ta	180	179.947 489	0.01	
			181	180.948 014	99.99	
74	Tangsten (1783)	W	180	179.946 727	0.1	
			182	182.948 225	26.3	
			183	182.950 245	14.3	
			184	183.950 953	30.7	
			186	185.954 377	28.6	
75	Rhenium (1925)	Re	185	184.952 977	37.4	
			187	186.955 765	62.6	5×10^{10} y (β^-)
76	Osmium (1803)	Os	184	183.952 514	0.02	
			186	185.953 852	1.6	
			187	186.955 762	1.6	
			188	187.955 850	13.3	
			189	188.958 156	16.1	
			190	189.958 455	26.4	
			192	191.961 487	41.0	

Z	Element (year)	Symbol	A	Nuclide mass, u	Relative abundance, %	Half-life
77	Iridium (1803)	**Ir**	191	190.960 603	37.3	
			193	192.962 942	62.7	
78	Platinum (> 1700)	**Pt**	190	189.959 937	0.01	6.9×10^{11} y (α)
			192	191.961 049	0.79	~ 10^{15} y (α)
			194	193.962 679	32.9	
			195	194.964 785	33.8	
			196	195.964 947	25.3	
			198	197.967 879	7.2	
79	Gold (~ 3000 B.C.)	**Au**	197	196.966 560	100	
80	Mercury (~ 1500 B.C.)	**Hg**	196	195.965 812	0.2	
			198	197.966 760	10.0	
			199	198.968 269	16.8	
			200	199.968 316	23.1	
			201	200.970 293	13.2	
			202	201.970 632	29.8	
			204	203.973 481	6.9	
81	Thallium (1861)	**Tl**	203	202.972 336	29.5	
			205	204.974 410	70.5	
82	Lead (~ 1000 B.C.)	**Pb**	204	203.973 037	1.4	
			206	205.974 455	24.1	
			207	206.975 885	22.1	
			208	207.976 641	52.4	
			210	209.984 178		22.3 y (β^-)
			214	213.999 764		26.8 min (β^-)
83	Bismuth (~ 1500)	**Bi**	209	208.980 388	100	
			212	211.991 267		60.6 min (α, β^-)
84	Polonium (1898)	**Po**	210	209.982 876		138 d (α, γ)
			214	213.995 191		0.16 ms (α)
			216	216.001 790		0.15 s (α)
			218	218.008 930		3.05 min (α)
85	Astatine (1940)	**At**	218	218.008 607		1.8 s (α)
86	Radon (1900)	**Rn**	220	220.011 401		56 s (α)
			222	222.017 574		3.824 d (α)
87	Francium (1939)	**Fr**	223	223.019 73		22 min (β^-)
88	Radium (1898)	**Ra**	226	226.025 406		1.60×10^3 y (α)
89	Actinium (1899)	**Ac**	227	227.027 751		21.8 y (α, β^-)

Z	Element (year)	Symbol	*A*	Nuclide mass, *u*	Relative abundance, %	Half-life
90	Thorium (1815)	**Th**	228	228.028 750		1.9 y (α)
			230	230.033 131		7.7×10^4 y (α)
			232	232.038 054	100	1.4×10^{10} y (α)
			233	233.041 580		22.2 min (β^-)
91	Protactinium (1917)	**Pa**	233	233.040 244		27 d (β^-)
92	Uranium (1789)	**U**	232	232.037 168		72 y (α)
			233	233.039 629		1.6×10^5 (α)
			234	234.040 947		2.4×10^5 y (α)
			235	235.043 925	0.72	7.04×10^8 y (α)
			238	238.050 786	99.28	4.47×10^9 y (α)
93	Neptunium (1940)	**Np**	237	237.048 169		2.14×10^6 y (α)
			239	239.052 932		2.4 d (β^-)
94	Plutonium (1940)	**Pu**	239	239.052 158		2.4×10^4 y (α)
			240	240.053 809		6.6×10^3 y (α)
95	Americium (1944)	**Am**	243	243.061 374		7.7×10^3 y (α)
96	Curium (1944)	**Cm**	247	247.070 349		1.6×10^7 y (α)
97	Berkelium (1949)	**Bk**	247	247.070 300		1.4×10^3 y (α)
98	Californium (1950)	**Cf**	249	249.0748		360 y (α)
			251	251.079 581		900 y (α)
99	Einsteinium (1952)	**Es**	252	252.082 82		472 d
			253			20.7 d (α)
100	Fermium (1952)	**Fm**	257	257.095 103		100.5 d (α)
101	Mendelevium (1955)	**Md**	258	258.098 57		56 d (K)
102	Nobelium (1958)	**No**	259	259.100 941		58 m (α)
103	Lawrencium (1961)	**Lr**	260	260.105 36		~ 3.0 m (α)
104	Rutherfordium	**Rf**	261	261.108 69		1.1 m
105	Hahnium	**Ha**	262	262.113 84		0.7 m

Adapted from *The Elements*, Emsley, John, OUP, 2nd edition and *Concepts of Modern Physics*, Beiser, A, 4th edition, McGraw-Hill International.

APPENDIX 2

Separation of variables in the hydrogen-atom wave function

In chapter 4 (sec. 4.3.3) we simply stated that the Schrödinger equation in polar form for an electron in the H-atom can be separated into three functions according to the variables r, θ and ϕ and written as equation 4.18. This is repeated in eqn. A-1.

$$\psi(r, \theta, \phi) = \psi(r)\,\psi(\theta)\,\psi(\phi) = R(r)\,\Theta(\theta)\,\Phi(\phi) \qquad \text{[A-1]}$$

Here $R(r)$ describes the dependence of ψ on the radial distance from the nucleus at constant θ and ϕ. The function $\Theta(\theta)$ describes how ψ varies with θ (zenith angle, see Fig. A-I) along a meridian on a sphere having the nucleus at its centre (r and ϕ constant). The function $\Phi(\phi)$ similarly expresses the dependence of ψ on the azimuthal angle ϕ along a circle on the sphere (centered at the nucleus) at constant r and θ.

Now we proceed to separate the variables as assumed in equation A-1.

Equation A-1 may be written in the simpler form

$$\psi = R\Theta\Phi$$

Since each of the functions R, Θ and Φ now depends only on the respective variables r, θ and ϕ, we may change from partial derivatives ($\partial/\partial r$ etc.) to ordinary derivatives and write

$$\frac{\partial \psi}{\partial r} = \Theta\Phi\frac{\partial R}{\partial r} = \Theta\Phi\frac{dR}{dr}$$

$$\frac{\partial \psi}{\partial \theta} = R\Phi\frac{\partial \Theta}{\partial \theta} = R\Phi\frac{d\Theta}{d\theta}$$

$$\frac{\partial^2 \psi}{\partial \phi^2} = R\Theta\frac{\partial^2 \Phi}{\partial \phi^2} = R\Theta\frac{d^2\Phi}{d\phi^2}$$

The Schrödinger equation for the H-atom (eqn. 4.16) may be multiplied by $r^2\sin^2\theta$ to give*

$$\sin^2\theta\frac{\partial}{\partial r}\left(r^2\frac{\partial\psi}{\partial r}\right)+\sin\theta\frac{\partial}{\partial\theta}\left(\sin\theta\frac{\partial\psi}{\partial\theta}\right)+\frac{\partial^2\psi}{\partial\phi^2}+\frac{8\pi^2mr^2\sin^2\theta}{h^2}\left(E+\frac{e^2}{4\pi\varepsilon_0 r}\right)\psi=0. \quad [A\text{-}2]$$

We substitute ψ by $R\Theta\Phi$, put the expression of $(\partial\psi/\partial r)$ etc. as written above and finally divide the entire equation by $R\Theta\Phi$ to get

$$\frac{\sin^2\theta}{R}\frac{d}{dr}\left(r^2\frac{dR}{dr}\right)+\frac{\sin\theta}{\Theta}\frac{d}{d\theta}\left(\sin\theta\frac{d\Theta}{d\theta}\right)+\frac{1}{\Phi}\frac{d^2\Phi}{d\phi^2}+\frac{8\pi^2mr^2\sin^2\theta}{h^2}\left(\frac{e^2}{4\pi\varepsilon_0 r}+E\right)=0 \quad [A\text{-}3]$$

The third term of Eqn. A-3 is a function of azimuth angle ϕ only; other terms are functions of r and θ only. Eqn. A-3 may be rearranged as

$$\frac{\sin^2\theta}{R}\frac{d}{dr}\left(r^2\frac{dR}{dr}\right)+\frac{\sin\theta}{\Theta}\frac{d}{d\theta}\left(\sin\theta\frac{d\Theta}{d\theta}\right)+\frac{8\pi^2mr^2\sin^2\theta}{h^2}\left(\frac{e^2}{4\pi\varepsilon_0 r}+E\right)=-\frac{1}{\Phi}\frac{d^2\Phi}{d\phi^2} \quad [A\text{-}4]$$

Since the two sides of the equation are functions of different variables, the equation can be correct only if both sides of it are equal to the same constant. We shall see that it is convenient to set this constant as m_l^2. The differential equation for the function ϕ now gets the form

$$-\frac{1}{\Phi}\frac{d^2\Phi}{d\phi^2}=m_l^2 \quad [A\text{-}5]$$

Next we substitute m_l^2 for the right-hand side of Eqn. [A-4], divide the entire equation by $\sin^2\theta$, and rearrange the various terms. We get

$$\frac{1}{R}\frac{d}{dr}\left(r^2\frac{dR}{dr}\right)+\frac{8\pi^2mr^2}{h^2}\left(\frac{e^2}{4\pi\varepsilon_0 r}+E\right)=\frac{m_l^2}{\sin^2\theta}-\frac{1}{\Theta\sin\theta}\frac{d}{d\theta}\left(\sin\theta\frac{d\Theta}{d\theta}\right) \quad [A\text{-}6]$$

Eqn. A-6 also constains functions of different variables on each side. Again we have to equalize both sides of the equation to some other new constant. Let us put this constant equal to L (the reason for such choice will be understood later). We are now able to write two separate equations for the Θ and Φ functions:

$$\frac{m_l^2}{\sin^2\theta}-\frac{1}{\Theta\sin\theta}\frac{d}{d\theta}\left(\sin\theta\frac{d\Theta}{d\theta}\right)=L \quad [A\text{-}7]$$

$$\frac{1}{R}\frac{d}{dr}\left(r^2\frac{dR}{dr}\right)+\frac{8\pi^2mr^2}{h^2}\left(\frac{e^2}{4\pi\varepsilon_0 r}+E\right)=L \quad [A\text{-}8]$$

Each of the equations A-5, A-7 and A-8 are an ordinary differential equation for a single function of a single variable. The equ^tions are conventionally written in the following forms

Equation for Φ
$$\frac{d^2\Phi}{d\phi^2}+m_l^2\Phi=0 \quad [A\text{-}9]$$

Equation for Θ
$$\frac{1}{\sin\theta}\frac{d}{d\theta}\left(\sin\theta\frac{d\Theta}{d\theta}\right)+\left[L-\frac{m_l^2}{\sin^2\theta}\right]\Theta=0 \quad [A\text{-}10]$$

Equation for R
$$\frac{1}{r^2}\frac{d}{dr}\left(r^2\frac{dR}{dr}\right)+\left[\frac{8\pi^2m}{h^2}\left(\frac{e^2}{4\pi\varepsilon_0 r}+E\right)-\frac{L}{r^2}\right]R=0 \quad [A\text{-}11]$$

* In eqn. 4.17 (p. 104), ψ should be outside the paranthesis in the last term, as in eqn. A-2.

We have thus separated the variables in the Schrödinger's equation for an electron in the hydrogen atom, beginning with a partial differential equation for a function ψ of three variables. Now let us see how the separation helps us to arrive at the permissible values of the quantum numbers n, l and m_l.

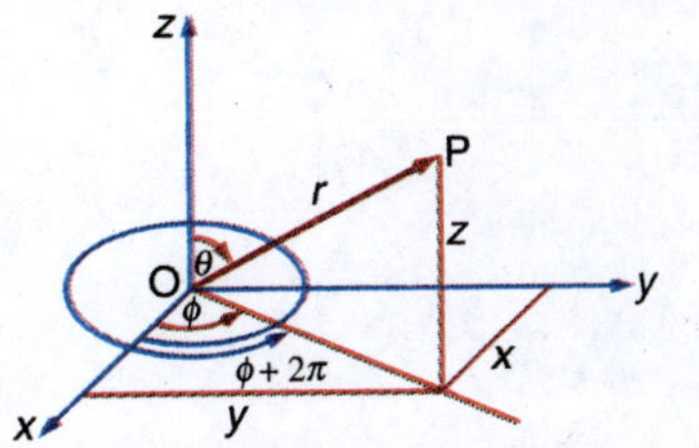

Fig.A-I: The angles ϕ and $\phi + 2\pi$ both identify the same meridian plane.

Quantum numbers

The magnetic quantum number

Eqn. A-9 is readily solved. The result is

$$\Phi(\phi) = Ae^{im_l\phi} \qquad (A \text{ is the constant of integration.}) \qquad \text{[A-12]}$$

The function Φ, which is a component of the complete well-behaved wave function ψ, must have a single value at a given point in space. From Fig. A-I it is evident that ϕ and $\phi + 2\pi$ both identify the same meridian plane. $\Phi(\phi)$ should therefore be equal to $\Phi(\phi + 2\pi)$. In other words,

$$Ae^{im_l\phi} = Ae^{im_l(\phi+2\pi)}$$

This requires that m_l should be 0 or a positive or negative integer, ±1, ±2, ⋯.

$$m_l = 0, \pm1, \pm2, \ldots.$$

The constant m_l is thus identified as the magnetic quantum number.

Orbital quantum number

Using the acceptable values of m_l, the differential equation for $\Theta(\theta)$ [Eqn A-10] can be solved to obtain well-behaved solutions for the function $\Theta(\theta)$. It is found that L must be of the form $l(l + 1)$ where l is an integer equal to or greater than $|m_l|$, the absolute value of m_l. This requirement leads to the condition on the values of l and m_l in the form

$$l = |m_l|, |m_l| + 1, |m_l| + 2 \ \ldots \quad \text{or, } m_l = 0, \pm1, \pm2, \cdots, \pm l$$

The constant l is known as the *orbital quantum number*.

Principal quantum number

We next use the values of L in Eqn. A-11 for the radial part $R(r)$ of the hydrogen-atom wave function ψ. Though all positive values of E are found to be associated with well-behaved R functions, we are interested in negative values of E which signifies that the electron is bound to the atom. It is found that well-behaved solutions for $R(r)$ are obtained only for negative values of E_n specified by

$$E_n = -\frac{me^4}{8\varepsilon_0^2 h^2}\left(\frac{1}{n^2}\right) \qquad n = l+1, l+2, l+3, \ldots$$

This is precisely the same formula for the energy levels of the Bohr orbits of hydrogen. Here n is called the principal quantum number which sets a limit to the permissible values of l for any given n.

$$l = 0, 1, 2, \ldots (n-1).$$

The three quantum numbers n, l and m_l thus appear as a natural consequence of quantum mechanical treatment of the hydrogen atom.

APPENDIX 3

Separation of Isotopes

All procedures for the separation of isotopes of an element are based on the difference in the mass of the various isotopes. The effect of mass difference on different physical properties of the isotopes is readily understandable—as in their rates of diffusion or behaviour in electric and magnetic fields. These differences give rise to the **physical methods** of separation of isotopes. On the other hand, the difference of masses of isotopes also influences the fundamental vibration frequencies of isotopic molecules and hence their zero-point energy. As a consequence, the activation energy for a chemical reaction is slightly higher with the heavier isotope. Accordingly, a species containing the heavier isotope reacts more slowly. The overall rate of a chemical reaction is influenced by the substitution of one isotope by another in the reacting species provided the bond to the isotopic nuclide is involved in the rate-determining step. This gives rise to the **chemical methods** of separation of isotopes.

Here we shall first mention the physical and chemical methods for the separation of *stable isotopes*. A few methods applicable for *unstable* isotopes will be mentioned at the end.

The separation factor

The separation factor, s, signifies the extent of enrichment in any separation procedure. Suppose a mixture contains n_1 atoms of the light isotope and n_2 atoms of the heavier isotope (per gram of mixture, say). Let the corresponding numbers be n'_1 and n'_2 after a particular separation process. The separation factor for the process is then given as

$$s = \frac{\text{ratio of atoms after separation}}{\text{ratio of atoms before separation}} = \frac{n'_1/n'_2}{n_1/n_2}$$

The separation factor may apply to a single stage or an overall procedure. For x successive single-stage separations having a factor s, the ultimate enrichment achieved is expressed by the overall separation factor S, where $S = s^x$.

Physical methods

A. Gaseous diffusion

Two isotopic species diffuse through a porous membrane* at rates which are inversely proportional to the square root of their respective molecular weights, say M_1 and M_2:

$$\frac{\text{Rate}_1}{\text{Rate}_2} = \left(\frac{M_2}{M_1}\right)^{1/2}$$

Hence the molecules containing the lighter isotope will diffuse more rapidly than the others, resulting in some separation. The separation factor s for the process would be $s = (M_1/M_2)^{1/2}$, where M_1 is the molecular weight of the heavier gas and M_2 is the molecular weight of the lighter gas. Thus, s will be theoretically equal to $2^{1/2}$ or 1·4 for a mixture containing D_2 and H_2; but s will fall gradually with higher molecular weights. For $^{235}UF_6$ and $^{238}UF_6$, s is only $(352/349)^{1/2} = 1{\cdot}0043$. The actual value is still less for several practical difficulties. To obtain practical separation, a cascade system involving many successive diffusion stages at reduced pressure is used and the enriched material is recycled many times. ^{20}Na and ^{22}Ne; ^{12}C and ^{13}C; and $^{235}UF_6$ and $^{238}UF_6$ have been separated by this method. The method is used for the large scale enrichment of ^{235}U for reactors. About 4000 diffusion stages are required to obtain 99 per cent $^{235}UF_6$.

Thermal diffusion has also been applied successfully for the separation of isotopes. From the kinetic theory of gases, it can be shown that if a mixture of light and heavy molecules is subjected to a temperature gradient, the heavy molecules tend to diffuse towards the region of lower temperature while the lighter molecules move to the region of higher temperature.The process continues until the effects of ordinary diffusion counterbalance it.In actual practice,the effect of thermal diffusion is enhanced by convection current.Long (1—36 m), vertical tubes with axially fitted nichrome wire for heating to about 500°C are used in succession. The hotter gas in the centre flows downward. This ultimately leads to a concentration of the lighter isotope at the top of the tube and of the heavier one at the bottom. $H^{37}Cl$, $^{13}CH_4$, $^{15}N_2$ and $^{18}O_2$ have been enriched by this method.

The separation factor in thermal diffusion is a function of the masses of the molecules (m_1 and m_2, say $m_1 > m_2$) and the temperature gradient. It T_1 and T_2 be the temperatures in the wall of the column and the hot wire respectively, then

$$\ln s = k\frac{m_1 - m_2}{m_1 + m_2}\ln\frac{T_2}{T_1} \quad (k = \text{constant}).$$

B. The electromagnetic method

The principle of this method is primarily based on the studies initiated by J. J. Thomson one hundred year ago (1910) on the deflection of positive ions (from a discharge tube, the so-called positive rays or anode rays) in electric and magnetic fields. Ions having different charge-to-

* The diameter of the holes should be much smaller than the mean free paths of the molecules.

mass ratio were made to strike different points on a photographic plate, producing mass spectrographs. The method was subsequently developed by Aston and others to a high degree of precision. In modern mass spectrometry, ions produced in a variety of ways are accelerated in an electric field and then deflected by a magnetic field. Ions with a lower mass-to-charge ratio are deflected more than the heavier ions and collected in suitably designed ion-collectors. The method is also used to study fragmentation and ionization of various molecules, including biomolecules. Though the method offers complete separation in a single stage, it is not always economic to use this method for large-scale separations. Its application is mainly restricted to structural and analytical studies. Minute amounts of the isotopes of lithium, potassium and rubidium were separated by this method. ^{235}U for experimental purposes was first obtained by this method.

C. *Other physical methods*

Distillation, centrifugal separation, ion migration and gas-liquid chromatography have been adopted for the separation of isotopes.

Distillation under reduced pressure has been used to separate the isotopes of mercury, zinc and potassium. Mercury was taken *between the two walls* of a Dewar flask while liquid air was kept in the inner flask. The inner flask was fixed at about 1-2 cm. above the mercury surface, which is near the mean path of gaseous mercury atoms. Mercury was allowed to evaporate under reduced pressure at about 40°-60°C. The vapours condensed to solid mercury in contact with the outer wall of the inner flask. The frozen distillate was removed and the process repeated several times–when the residual mercury got enriched in the heavier isotope.

Fractional distillation has been used for liquid hydrogen, oxygen and neon.

Chemical methods

As already mentioned, zero-point energies of isotopic molecular species are different. This leads to small differences in the free energies of reactions involving isotopes of different masses. This fact has been utilised in the chemical exchange method for the separation of isotopes. The following reaction will be illustrative:

$$^{15}NH_3(g) + {}^{14}NH_4^+(aq) \rightleftharpoons {}^{14}NH_3(g) + {}^{15}NH_4^+(aq). \quad K = 1.031$$

Since the equilibrium constant is slightly greater than unity, the aqueous phase will have a slightly greater concentration of ^{15}N. In actual practice, ammonia gas is blown up a packed column; ammonium sulfate solution is fed at the top of this column. The solution reaching the bottom of the column is treated with alkali and the regenerated ammonia is blown up the column again. The process is repeated several times, using more than one column for successive enrichment.

Similar exchange between $^{13}CO_2(g)$ and $H^{12}CO_3^-(aq)$ leads to an enrichment of ^{13}C in the bicarbonate solution:

$$^{13}CO_2(g) + H^{12}CO_3^-(aq) \rightleftharpoons {}^{12}CO_2(g) + H^{13}CO_3^-(aq) \qquad K = 1.012$$

The exchange is carried out under pressure in presence of fibre glass or alumina catalyst. Exchange reactions in purely gaseous systems are also known:

$$^{15}NO + {}^{14}NO_2 \rightleftharpoons {}^{14}NO + {}^{15}NO_2 \ ; \ {}^{12}CO_2 + {}^{13}CO \rightleftharpoons {}^{13}CO_2 + {}^{12}CO.$$

Separation of unstable isotopes

At present, a large number of isotopes are prepared artificially through specially designed nuclear reactions. These reactions usually involve bombardment of a suitable target nucleus with neutrons or other accelerated particles. Separation of the product isotope after its formation in such processes usually involves chemical separation together with volatilization, solvent extraction, electrodeposition, ion-exchange or precipitation on a suitable carrier.

For example, ^{32}P is prepared by the action of neutrons on ^{32}S. The target is dissolved in aqua regia when phosphorus is oxidised to phosphoric acid. Addition of ferric ions followed by partial neutralization precipitates ferric hydroxide. This carries down the phosphate with it. The precipitate is redissolved in acid and the iron is removed by an ion-exchange column—the remaining solution contains the radiophosphorus as phosphoric acid.

The fissile plutonium ^{239}Pu is produced in nuclear reactors by the action of slow neutrons on ^{238}U. After irradiation, the rods of uranium are dissolved in nitric acid. Both plutonium and uranium goes into solution, forming plutonyl (PuO_2^{2+}) and uranyl (UO_2^{2+}) nitrates. These are separated from the rest of the solution by solvent extraction—usually with tributyl phosphate. A mild reducing agent is then applied which preferentially reduces the plutonium (VI), thereby separating it from the uranium.

With Grateful Acknowledgement

Books Consulted

1. Alberty, Robert, A. ... Physical Chemistry
2. Alexeyev, V. ... Quantitative Analysis
3. Arnikar, H. J. ... Essentials of Nuclear Chemistry
4. Barnard, A. K. ... Theoretical Basis of Inorganic Chemistry
5. Beiser, A. ... Concepts of Modern Physics
6. Brown, G. I. ... A New Guide to Modern Valency Theory
7. Chandra, M. ... Atomic Structure and Chemical Bonding
8. Connelly, Neil, G., Damhus, Ture.; Hartshorn, Richard, M.; Hutton, Alan, T ... Nomenclature of Inorganic Chemistry IUPAC Recommendations, 2005
9. Cotton, F. A. & Wilkinson, G. ... Advanced Inorganic Chemistry
10. Cotton, F. A., Wilkinson, G. & Gaus, P. L. ... Basic Inorganic Chemistry
11. Das, Asim, K. ... Fundamental Concepts of Inorganic Chemistry
12. Day, M. C. & Selbin, J. ... Theoretical Inorganic Chemistry
13. Douglas, B. E., McDaniel, D. H. & Alexander, J. J. ... Concepts and Models of Inorganic Chemistry
14. Drago, R. S. ... Physical Methods in Inorganic Chemistry
15. Emeleus, H. J. & Sharpe, A. G. ... Modern Aspects of Inorganic Chemistry
16. Emsley, John. ... The Elements
17. Friendlander, G., Kennedy, J. W. & Miller, J. M. ... Nuclear and Radio Chemistry
18. Ghosh, P. K. & Shukla, P. K. ... Atomic Eletronic Structure-Atomic Orbitals
19. Glasstone, S. ... Text-book of Physical Chemistry
20. Glasstone, S. ... Sourcebook on Atomic Energy
21. Gould, E. S. ... Inorganic Reactions and Structure
22. Greenwood, N. N. & Earnshaw, A. ... Chemistry of the Elements
23. Harvey, K. B. & Porter, G. B. ... Introduction to Physical Inorganic Chemistry
24. Heslop, R. B. ... Numerical Aspects of Inorganic Chemistry
25. Huheey, J. E. ... Inorganic Chemistry
26. Jenkins, F. A. & White, H. E. ... Fundamentals of Optics
27. Jolly, W. L. ... Principles of Inorganic Chemistry
28. Levine, Ira, N. ... Physical Chemistry
29. Littlefield, T. A. & Thorley, N. ... Atomic & Nuclear Physics—An Introduction
30. Mackay, K. M. & Mackay, R. A. ... Modern Inorganic Chemistry
31. Mahan, B. H. ... University Chemistry
32. Maron, S. H. & Prutton, C. F. ... Principles of Physical Chemistry
33. Miessler, Gary, L. & Tarr, Donald, A ... Inorganic Chemistry
34. Mingos, D. M. P. ... Essential Trends in Inorganic Chemistry

35. Moeller, T. ... Inorganic Chemistry
36. Moore, W. J. ... Physical Chemistry
37. Mukherjee, G. N. ... Semimicro Qualitative Inorganic Analysis
38. Mukherjee, Nirad Gopal ... Elementary Quantum Chemistry
39. Porterfield, William, W. ... Inorganic Chemistry—A Unified Approach
40. Rao, C. N. R. (edited) ... University General Chemistry
41. Robbins, Omer ... Ionic Reactions and Equilibrium
42. Sharpe, A. G. ... Inorganic Chemistry
43. Shriver, D. F., Atkins, P. W. & Langford, C. H. ... Inorganic Chemistry
44. Wehr, M. R., Richards, J. A. & Adair T. W. ... Physics of the Atom
45. Wulfsberg, Gary ... Inorganic Chemistry

Multipliers for SI Units

a	atto-	10^{-18}	da	deka-	10^{1}
f	femto-	10^{-15}	h	hecto-	10^{2}
p	pico-	10^{-12}	k	kilo-	10^{3}
n	nano-	10^{-9}	M	mega-	10^{6}
μ	micro-	10^{-6}	G	giga-	10^{9}
m	milli	10^{-3}	T	tera-	10^{12}
c	centi-	10^{-2}	P	peta-	10^{15}
d	deci	10^{-1}	E	exa-	10^{18}

Greek Alphabets in Common Use

α	alpha	ξ	xi
β	beta	ν	nu
γ	gamma	Π, π	pi
Δ, δ	delta	ρ	rho
ε	epsilon	Σ, σ	sigma
η	eta	τ	tau
Θ, θ	theta	Φ, ϕ	phi
κ	kappa	χ	chi
Λ, λ	lambda	Ψ, ψ	psi
μ	mu	Ω, ω	omega

INDEX